D0138941

Math in a Real World Context

Binge Watching TV

With the availability of streaming video, an increasing number of people are choosing to binge watch several episodes of a single TV series in one sitting. Fractions can be used to make predictions about TV viewers who binge watch. See Exercises 115 and 116 on page 24.

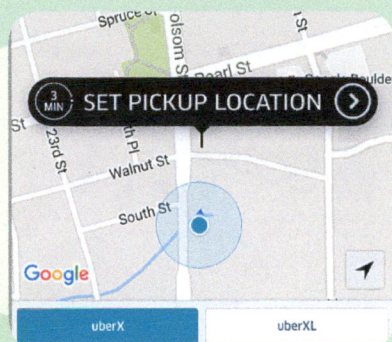

Analyzing Unicorn Companies

Companies that have private venture backing and are worth $1 billion or more (such as Uber) are known as "unicorn" companies. Recently, the total number of unicorn companies in existence during specified months could be modeled by a linear equation. Solving this equation helps us learn more about such companies. See Example 4 on page 100.

Monitoring Heart Rate

When an athlete stops a strenuous exercise routine, his or her heart rate decreases rapidly at first and then more gradually after a few minutes. This situation can be modeled by a polynomial function. Graphs of polynomial functions can be used when straight lines do not apply. See Example 6 on page 570.

Waiting in Traffic

Mathematics shows that in certain conditions, if the number of cars on the road increases even slightly, then the movement of traffic can slow dramatically. To understand this phenomenon more fully, we use a rational equation to model the situation. See Example 6 on page 438.

Taking the Ice Bucket Challenge

During a one-month period when the Ice Bucket Challenge for ALS became a viral fundraising phenomenon, donations increased by more than 20% per *day*. An exponential function can be used to model total donations during the month. See Example 10 on page 831.

Eating Pizza

Eating your first piece of pizza usually provides a lot of satisfaction. A second piece may be almost as good. After a few more pieces, there is often a point where satisfaction starts to level off, and can even go down. A parabola models this situation. See Example 4 on page 733.

Posting on Social Networks

When a link is posted on a social network such as Facebook, the majority of those who click on the link do so in the first few hours. After that, the number of people clicking the link decreases by a factor of one-half every few hours. This is an example of exponential decay. See the Chapter 12 Opener on page 805.

Using the Internet

Of American adults who *don't* use the Internet, about 33% are either "just not interested" or they "don't have a computer." With a little more information, we can use a system of linear equations to learn more about the reasons given in this survey. See Example 7 on page 263.

Owning a Pet

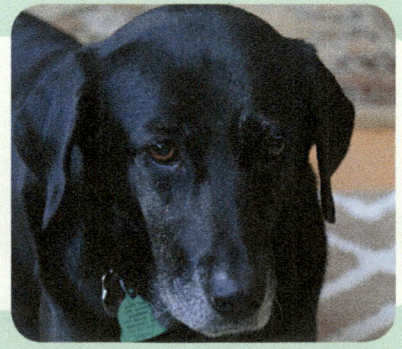

There are both one-time and recurring costs associated with owning a pet. From a table listing such costs for cats and dogs, we can write linear equations that can be used to determine the life-time cost of ownership for each animal. See the Group Activity on page 225.

Pricing Coffee Drinks

If the cost of a coffee drink is too high, demand will decrease because consumers want to save money. If the price of the drink is too low, supply will decrease because suppliers want to make money. To find a good price for the coffee drink, we can use a system of two linear equations. See Example 6 on page 262.

KV 04.10.2019 0817

EDITION 4

Beginning and Intermediate Algebra

with Applications and Visualization

Gary K. Rockswold
Minnesota State University, Mankato

Terry A. Krieger
Rochester Community and Technical College

with

Jessica C. Rockswold

Editorial Director *Chris Hoag*
Editor in Chief *Michael Hirsch*
Executive Editor *Cathy Cantin*
Editorial Assistant *Shannon Bushee*
Content Producer *Lauren Morse*
Managing Producer *Scott Disanno*
Product Marketing Manager *Jennifer Edwards*
Field Marketing Managers *Lauren Schur and Jenny Crum*
Product Marketing Assistant *Hanna Lafferty*
Media Producer *Erin Carreiro*
Executive Content Manager, MathXL *Rebecca Williams*
MathXL QA Supervisor, TestGen *Mary Durnwald*
Senior Author Support/Technology Specialist *Joe Vetere*
Rights and Permissions Project Manager *Gina Cheselka*
Manufacturing Buyer *Carol Melville, LSC Communications*
Associate Director of Design *Blair Brown*
Text and Cover Design, *Tamara Newnam*
Production Coordination and Composition, *Cenveo® Publisher Services*

Copyright © 2018, 2013, 2009 by Pearson Education, Inc. All Rights Reserved. Printed in the United
States of America. This publication is protected by copyright, and permission should be obtained
from the publisher prior to any prohibited reproduction, storage in a retrieval system, or transmission
in any form or by any means, electronic, mechanical, photocopying, recording, or otherwise. For
information regarding permissions, request forms and the appropriate contacts within the Pearson
Education Global Rights & Permissions department, please visit www.pearsoned.com/permissions/.

Acknowledgments of third party content appear on page **G-8**, which constitutes an extension of this
copyright page.

PEARSON, ALWAYS LEARNING, and MYMATHLAB are exclusive trademarks owned by
Pearson Education, Inc. or its affiliates in the U.S. and/or other countries.

Unless otherwise indicated herein, any third-party trademarks that may appear in this work are the
property of their respective owners and any references to third-party trademarks, logos or other trade
dress are for demonstrative or descriptive purposes only. Such references are not intended to imply
any sponsorship, endorsement, authorization, or promotion of Pearson's products by the owners
of such marks, or any relationship between the owner and Pearson Education, Inc. or its affiliates,
authors, licensees or distributors.

Library of Congress Cataloging-in-Publication Data
Names: Rockswold, Gary K. | Krieger, Terry A.
Title: Beginning and intermediate algebra with applications and visualization
  / Gary K. Rockswold, Minnesota State University, Mankato, Terry A.
  Krieger, Rochester Community and Technical College.
Description: Fourth edition. | Boston : Pearson Education, [2018] | Includes index.
Identifiers: LCCN 2016036042 | ISBN 9780134474304 (se hardcover : alk. paper)
Subjects: LCSH: Algebra–Textbooks.
Classification: LCC QA152.3 .R595 2018 | DDC 512.9–dc23 LC record available at
https://lccn.loc.gov/2016036042

Pearson

ISBN 13: 978-0-13-447430-4
ISBN 10: 0-13-447430-9

About the Authors

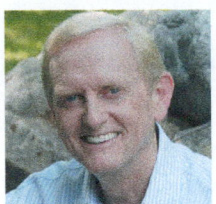

Gary Rockswold has been a professor and teacher of mathematics, computer science, astronomy, and physical science for over 35 years. Not only has he taught at the undergraduate and graduate college levels, but he has also taught middle school, high school, vocational school, and adult education. He received his BA degree with majors in mathematics and physics from St. Olaf College and his PhD in applied mathematics from Iowa State University. He has been a principal investigator at the Minnesota Supercomputer Institute, publishing research articles in numerical analysis and parallel processing. He is currently an emeritus professor of mathematics at Minnesota State University–Mankato. He is an author for Pearson Education and has numerous textbooks at the developmental and precalculus levels. Making mathematics accessible to students and professing the power of mathematics are special passions for Gary. He frequently gives keynote and invited addresses at regional, national, and international math conferences. In his spare time he enjoys sailing, doing yoga, hiking, and spending time with his family.

Terry Krieger has taught mathematics for over 20 years at the middle school, high school, vocational, community college, and university levels. His undergraduate degree in secondary education is from Bemidji State University in Minnesota, where he graduated summa cum laude. He received his MA in mathematics from Minnesota State University–Mankato. In addition to his teaching experience in the United States, Terry has taught mathematics in Tasmania, Australia, and in a rural school in Swaziland, Africa, where he served as a Peace Corps volunteer. Terry has been involved with various aspects of mathematics textbook publications throughout his career. In his free time, Terry enjoys spending time with his wife and two boys, physical fitness, wilderness camping, and trout fishing.

For two amazing exchange students from Colombia,

Santiago Posada and Maria Camila Calderon.

Siempre serán parte de la familia.

– TK

Contents

1   Introduction to Algebra   1

v

3 Graphing Equations 169

4 Systems of Linear Equations in Two Variables 256

5   Polynomials and Exponents   310

8   Introduction to Functions   516

11   Quadratic Functions and Equations   727

14   Sequences and Series   917

Preface

*B*eginning and Intermediate Algebra with Applications and Visualization, Fourth Edition, gives meaning to the numbers that students encounter by *developing concepts in context* through the use of applications, multiple representations, and visualization. By seeing the concept in context *before* being given the mathematical abstraction, students make math part of their own experiences instead of just memorizing techniques. Research shows that this method is essential to empowering students. Seamlessly integrated real-life connections, graphs, tables, charts, and meaningful data help students deepen understanding and prepare for future math courses—and life—by teaching them critical thinking and problem-solving skills.

By adding new design features and reworking existing content in this text, we increased the amount of content that is delivered in a real-world context. We made the text more accessible by replacing paragraphs of text with visual information. Recognizing that many users want a customizable text, we put new emphasis on clearly identifying objectives and content in both the exposition and the exercise sets.

It is essential that all students achieve competency in mathematics so that they can realize their vocational dreams and contribute to society. A purely traditional/abstract approach to teaching math does not work for most of today's students, and it has led to exclusivity. Our approach promotes inclusivity and diversity within the discipline and beyond. Our goal in *Beginning and Intermediate Algebra with Applications and Visualization* is to help all students succeed.

This textbook is one part of our comprehensive program:

- *Prealgebra*
- *Beginning Algebra with Applications and Visualization,* Third Edition
- *Beginning & Intermediate Algebra with Applications and Visualization,* Fourth Edition
- *Intermediate Algebra with Applications and Visualization,* Fifth Edition
- *Developmental Math with Applications and Visualization: Prealgebra, Beginning Algebra, and Intermediate Algebra,* Second Edition
- *MyMathLab for Interactive Developmental Mathematics*

New To This Edition

We enriched our program—the text, the supplements, and MyMathLab—in the following areas to enhance our support for conceptual understanding. See also Resources for Success: MyMathLab online course (pp. xx–xxii) for additional information.

- **Section Introduction Videos** Every section in the text is now coupled with a short video (in MyMathLab) that introduces the section's concepts in a contextual setting.
- **See the Concept** Throughout the text, we replaced paragraphs of text with clearly labeled infographics that visually—and concisely—walk students through concepts. Every See the Concept in the text has a companion video in MyMathLab to help bring the concept to life.
- **Assignable Content** To assess student understanding when they encounter new material, many text features are assignable in MyMathLab, including every Section Introduction Video and See the Concept Video and the majority of the Reading Checks, Making Connections, and Critical Thinking features.

- **Math in Context** Students are more engaged when mathematics is tied to current and relevant topics. This new feature helps students recognize when and how math is used in the real world.
- **Connecting Concepts with Your Life** Students are often already equipped with the knowledge needed to understand a new concept. Our new Connecting Concepts with Your Life feature gives meaning to mathematics by relating common life experiences that students already understand.
- **Objective List** At the beginning of each section, we list objectives to give a clear outline of the section contents and to make it easier to customize the course.
- **Comment Balloons** When appropriate, comments were inserted next to steps and procedures to make them more (immediately) understandable. Now students don't always have to read a paragraph of text to understand the concept.
- **Modeling Data** The data in hundreds of exercises and examples were updated to keep the applications relevant and fresh. Each chapter opens with a real-data application, and many new examples were added, including discussions about the Internet, social networks, tablet computers, and other contemporary topics.
- *Guided Workbook* Keyed to the text by section and objective, the *Workbook* leads students through the course, giving them the opportunity to record key information, work practice problems, and show and keep their work for reference—as well as taking conceptual understanding one step further by asking students to explain "Why?" after select questions.

Refinement of Content

CHAPTER 1: New exercises were added throughout the chapter, with an emphasis on expanding the coverage of translating words to expressions and identifying parts of a fraction.

CHAPTER 2: A completely new section on unit analysis was added. At the request of reviewers, inequality answers now include interval notation. Over 50 new exercises were also added throughout the chapter, with an emphasis on linear equations and their solutions.

CHAPTER 3: New examples and exercises were added to cover graphing nonlinear equations by hand. Also, at the request of reviewers, intercepts are now identified as ordered pairs rather than a real number.

CHAPTER 4: Supply and demand examples and exercises involving systems of linear equations were added to this chapter.

CHAPTER 5: The exercises pertaining to rules for exponents were expanded and more clearly categorized in the exercise sets.

CHAPTER 6: New examples and exercises that require students to rearrange terms before factoring by grouping were added to this chapter.

CHAPTER 7: The description of least common multiple now includes a more visual, contextual explanation involving the listing method.

CHAPTER 8: We put more emphasis on solving absolute value inequalities symbolically. A new objective that more fully discusses the concept of a relation was inserted before the definition of a function, and exercises were added that give students practice differentiating between relations and functions. Also, a new objective on piecewise-defined functions was added in Section 8.3, with corresponding exercises in both Sections 8.3 and 8.4.

CHAPTER 9: New exercises that ask students to represent simple social networks with matrices were added to this chapter.

CHAPTER 10: The coverage of using zeros to write the formula for a function was expanded in Sections 10.5 and 10.7.

CHAPTER 11: A new graphical description for deriving the formula to find the vertex of a parabola is now included. Also, a real-world example was added to highlight a common situation in which a quadratic equation might have zero, one, or two solutions. This chapter additionally includes increased coverage of complex solutions to quadratic equations.

CHAPTER 12: A discussion connecting social network posts with half-life and exponential decay is now included to give students a relevant and accessible example of this important concept. At the request of reviewers, we expanded the coverage of changing between exponential and logarithmic forms of equations.

CHAPTER 13: New exercises involving the equation of a circle were added.

CHAPTER 14: In this final chapter, the coverage of writing a formula for a sequence from given terms of the sequence was increased.

Hallmark Features

Experiencing Math in the Real World

In our program, we emphasize teaching algebra in context. Students typically understand the math best when the concepts and skills are tied to the real world and presented visually. We believe that meaningful applications and visualization of important concepts are pathways to success in mathematics.

- **Chapter Openers** Each chapter opens with a contemporary application that motivates students by offering insights into the relevance of that chapter's mathematical concepts.
- **NEW! Math in Context** Where appropriate, we expand on specific math topics and their connections to everyday life.
- **Applications** We integrate applications (many with actual data) into both the discussions and exercises to help students become more effective problem solvers.
- **Modeling Data** We provide opportunities for students to model real and relevant data with their own functions. **Online Exploration** exercises invite students to find their own data on the Internet and use mathematics to analyze them.
- **NEW! Section Introduction Videos** Every section in the text is now coupled with a short video (in MyMathLab) that introduces the section's concepts in a contextual setting.
- **NEW! Connecting Concepts with Your Life** This new feature helps students understand mathematics by relating it to common life experiences.

Understanding the Concepts

Conceptual understanding is critical to success in mathematics.

- **Learning the Math from Multiple Perspectives** Throughout the text, we present concepts by means of **verbal, graphical, numerical,** and **symbolic representations** to support multiple learning styles and problem-solving methods.
- **New Vocabulary** At the beginning of each section, we direct students' attention to important terms before they are discussed in context.
- **Reading Checks** These questions appear in the text along with important concepts, ensuring that students understand the material they have just read before moving on. Selected Reading Checks (RC) are also available to be assigned in MyMathLab.
- **Making Connections** Throughout the text, we help students understand the relationships between previously learned concepts and new ones. Selected Making Connections (MC) are also available to be assigned in MyMathLab.
- **Critical Thinking** One or more of these exercises appear in most sections, posing questions that can be used for class discussion, group work, or homework assignments. Selected Critical Thinking (CT) features are also available to be assigned in MyMathLab.
- **Putting It All Together** At the end of each section, we summarize the techniques just covered and reinforce the mathematical concepts presented in the section.
- **NEW! See the Concept** Throughout the text, we replaced paragraphs of text with clearly labeled infographics that visually—and concisely—walk students through concepts. Every See the Concept in the text has a companion video in MyMathLab to help bring the concept to life.
- **NEW! Comment Balloons** Comments are included next to steps and procedures to make them more (immediately) understandable.

Practicing the Concepts

A variety of exercise types support the application-based and conceptual nature of this text. Designed to reinforce the skills students need to be able to move on to the next concept, the comprehensive section and chapter exercise sets cover basic concepts, skill-building, writing, applications, and conceptual mastery. These exercise sets are further enhanced by several special types of exercises integrated throughout the text:

- **Now Try** Suggested exercises follow every example for immediate reinforcement of skills and concepts.
- **Concepts and Vocabulary exercises** appear in every section.
- **Checking Basic Concepts** These mixed-review exercise sets appear after every other section and can be used for individual or group review.
- **Thinking Generally** Appearing in most section exercise sets, these open-ended conceptual questions encourage students to synthesize what they have just learned.
- **Writing about Mathematics** At the end of most section exercise sets, students are asked to explain the concepts behind the mathematics procedures they have just learned.
- **NEW! Assignable Content** To assess student understanding when they encounter new material, many text features are assignable in MyMathLab, including every Section Introduction and See the Concept Video and the majority of the Reading Checks, Making Connections, and Critical Thinking features.
- **Group Activities:** This feature occurs once or twice per chapter, and provides an opportunity for students to work collaboratively on a problem. Most activities can be completed with limited use of class time.
- **Using a Graphing Calculator**
  - **Graphing Calculator Exercises** The ▦ icon denotes an optional exercise that requires students to have access to a graphing calculator.
  - **Technology Notes** Throughout the text, these optional notes offer students guidance, suggestions, and cautions on the use of the graphing calculator.

Mastering the Concepts

By reviewing the material and putting their abilities to the test, students will be able to assess their level of mastery as they complete each chapter.

- **Chapter Summary** In a quick but thorough review, we combine key terms, topics, and procedures with illuminating examples to assist students as they prepare for a test.
- **Chapter Review Exercises** Students can work these exercises to gain confidence that they have mastered the material.
- **Chapter Test** Students can reduce math anxiety by using these tests as a rehearsal for the real thing.
- **Chapter Test Prep Videos** These videos offer further help by showing an instructor's step-by-step solutions to every exercise in each Chapter Test.
- **Cumulative Review Exercises** Starting with Chapter 2 and appearing in all subsequent chapters, Cumulative Review Exercises help students see the big picture of math by reviewing topics and skills they have already learned.

Resources for Success

MyMathLab® Online Course (access code required)

Pearson

The course for *Beginning and Intermediate Algebra with Applications and Visualization*, 4e, includes all of MyMathLab's robust features and functionality, plus these additional new features:

Expanded Video Program

Each section in the text is now supported by a short *new* **Section Introduction Video** that presents the section concepts in context. For key concepts throughout the text, infographics visually—and concisely—walk students through the concepts. Companion **See the Concept Videos** help bring these concepts to life.

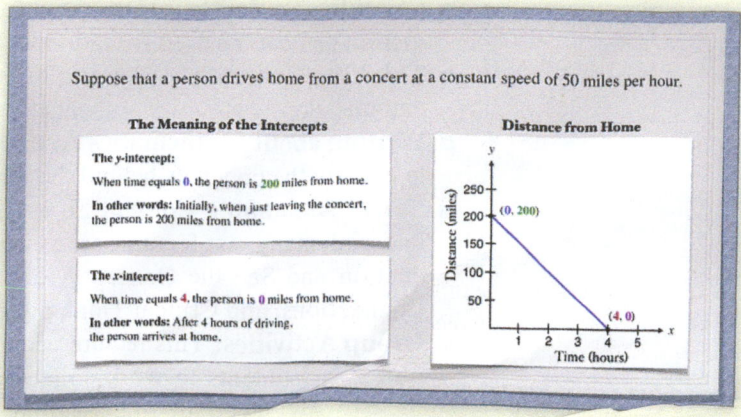

Suppose that a person drives home from a concert at a constant speed of 50 miles per hour.

The Meaning of the Intercepts

The y-intercept:

When time equals **0**, the person is **200** miles from home.

In other words: Initially, when just leaving the concert, the person is 200 miles from home.

The x-intercept:

When time equals **4**, the person is **0** miles from home.

In other words: After 4 hours of driving, the person arrives at home.

Distance from Home

New Workspace

Workspace Assignments allow students to work through an exercise step-by-step, showing their mathematical reasoning. Students receive immediate feedback after they complete each step, and helpful hints and videos are available for guidance as needed. When students access workspace using a mobile device, handwriting recognition software allows them to write out answers naturally using their fingertip or a stylus.

New Learning Catalytics

Learning Catalytics uses students' mobile devices for an engagement, assessment, and classroom intelligence system that gives instructors real-time feedback on student learning.

www.mymathlab.com

Resources for Success

MyMathLab® Online Course (access code required)

New Skill Builder Adaptive Practice

When a student is struggling with the assigned homework, Skill Builder exercises offer just-in-time additional adaptive practice. The adaptive engine tracks student performance and delivers questions to each individual that adapt to his or her level of understanding. When the system has determined that the student has a high probability of successfully completing the assigned exercise, it suggests that the student return to the assignment. This new feature allows instructors to assign fewer questions for homework, allowing students to complete as many or as few questions as needed.

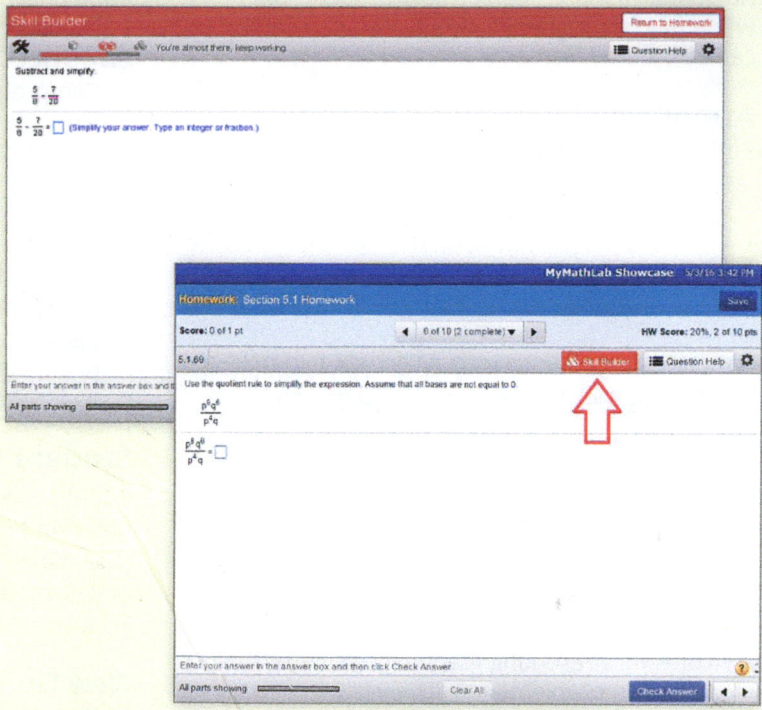

Interactive Exercises

MyMathLab's hallmark interactive exercises help build problem-solving skills and foster conceptual understanding. For this Fourth Edition, Guided Solutions (GS) exercises were added to reinforce the step-by-step problem-solving process, while the *new* Drag & Drop (DD) functionality was applied to matching exercises throughout the course to better assess a student's understanding of the concepts.

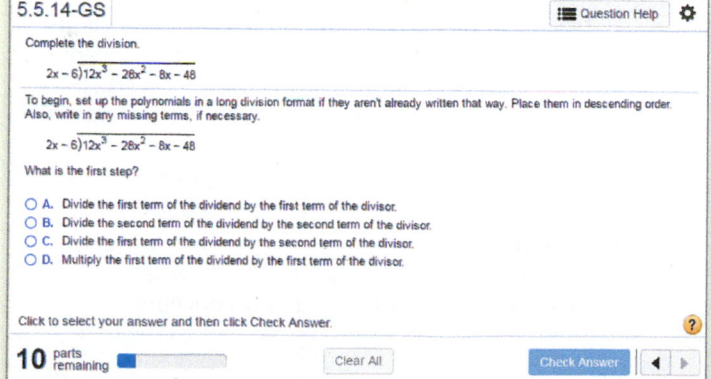

www.mymathlab.com

Resources for Success

Pearson

Instructor Resources

Additional resources can be downloaded from **www.pearsonhighered.com** or hardcopy resources can be ordered from your sales representative.

Annotated Instructor's Edition

- Contains Teaching Tips, Teaching Examples, and answers to every exercise in the textbook (excluding Writing about Mathematics).
- Answers that do not fit on the same page as the exercises are supplied in the Instructor's Answer Appendix in the back of the textbook.

Instructor's Resource Manual with Tests and Mini-Lectures (Download only)

Includes resources designed to help both new and adjunct faculty with course preparation and classroom management.

- *New!* **Activities** Multi-step Activities for group work, classroom instruction, and homework are available for key concepts.
- Mini-lectures for each text section.
- Notes for presenting graphing calculator topics as well as supplemental activities.
- Teaching Tips and additional exercises for selected content.
- Three free-response alternative test forms and one multiple-choice test form per chapter; one free-response and one multiple-choice final exam.
- Five sets of Cumulative Review Exercises (Chapters 1–3, 1–6, 1–9, 1–12, and 1–14).

Instructor's Solutions Manual (Download only)

Solutions to section-level exercises (excluding Writing about Mathematics), Checking Basic Concepts, Chapter Review, Chapter Test, and Cumulative Review.

PowerPoint® Lecture Slides

Fully editable slides that correlate to the textbook.

TestGen®

TestGen® (www.pearsoned.com/testgen) enables instructors to build, edit, print, and administer tests using a computerized bank of questions developed to cover all the objectives of the text.

Student Resources

Additional resources to help student success.

Enhanced Video Resources

- Comprehensive video coverage of the course available at the section and objective levels, with new Section Introduction and See the Concept Videos.
- Ideal for variety of course formats, including lecture-based, lab-based, online, self-paced, hybrid, and flipped classroom.
- Chapter Test Prep Videos let students watch an instructor work through step-by-step solutions to all Chapter Test exercises from the text. Available in MyMathLab and on YouTube.

Student's Solutions Manual

Contains solutions to the following text exercises: odd-numbered Section and Chapter Review (excluding Writing about Mathematics), Concepts and Vocabulary, Checking Basic Concepts, Chapter Test, and Cumulative Review.

New Guided Workbook

- Keyed to the text by section and objective.
- Helps the students stay on track and reinforces conceptual understanding throughout the course.
- Promotes active learning with space for students to answer questions about the See the Concept videos, write about key terms, solve exercises and show and keep their work, and reflect about their understanding of the math.
- Binder-ready (3-hole-punched and unbound), it can form the basis of a course notebook.
- **Activities** Multi-step Activities for group work, classroom instruction, and homework are available in MyMathLab for key concepts.

Acknowledgments

We thank the many individuals who contributed to the development of our program over its four editions by participating in focus groups, completing surveys, completing reviews, and contacting us with suggestions. We thank the following instructors whose comments and suggestions informed our preparation of *Beginning and Intermediate Algebra with Applications and Visualization*, Fourth Edition:

Gurdial Arora, *Xavier University of Louisiana*

Susan Barbitta, *Guilford Technical Community College*

Patricia Blus, *National Louis University*

Susan Cooper-Nguyen, *Harrisburg Area Community College*

Laurie DeHerrera, *Pikes Peak Community College*

Sarah Endel, *Rochester Community and Technical College*

Thomas Fitzkee, *Francis Marion University*

Kim Ghiselin, *State College of Florida, Manatee-Sarasota*

Ryan Girard, *Kauai Community College*

Melissa Hedlund, *Christopher Newport University*

Marsha Hodge Cardena, *University of Texas at El Paso*

Roland "Bud" Jenkins, *Hawaii Pacific University*

Nena Kabranski, *Tarrant County College*

Chris Knight, *Walters State Community College*

Doug Mace, *Kirtland Community College*

Marvin F. Mascarenas, *New Mexico Highlands University*

Cindy McCallum, *Tarrant County College*

Alejandro Mena, *University of Texas at El Paso*

Linda Myers, *Harrisburg Area Community College*

David Price, *Tarrant County College, SE*

Elise Price, *Tarrant County College, SE*

Jorge Sarmiento, *County College of Morris*

Salvatore Sciandra, *Niagara County Community College*

Janet E. Teeguarden, *Ivy Tech Community College*

Marissa Wolfe, *Yavapai College*

We would like to welcome David Atwood to our team. He has been instrumental in this revision, providing assistance with answers, proofing, and video creation.

Jennifer Blue, Perian Herring, Hal Whipple, and Mark Rockswold deserve special credit for their help with proofing and accuracy checking. Without the excellent cooperation from the professional staff at Pearson Education, this project would not have been possible. Thanks goes to Michael Hirsch for his support. Particular recognition is due Cathy Cantin, who gave essential advice and assistance. The outstanding contributions of Lauren Morse, Joe Vetere, Jennifer Edwards, Erin Carreiro, Rebecca Williams, and Alison Oehmen are greatly appreciated. Special thanks go to Chere Bemelmans at Cenveo who was instrumental in the success of this project.

Thanks go to Wendy Rockswold, Carrie Krieger, and Burgess Johns, whose unwavering encouragement and support made this project possible. We also thank the many students and instructors who used the previous editions of this textbook. Their suggestions were insightful and helpful.

Please feel free to send us your comments and questions. You can find our contact information at www.garyrockswold.net. Your opinion is important to us.

Gary Rockswold
Terry Krieger
Jessica Rockswold

Index of Applications

Environment

General Interest

Geometry

Health

1

Introduction to Algebra

Numbers are all around us, and we need mathematics to understand them. For example, binge watching TV shows is rapidly becoming mainstream and the new normal. Over 60% of TV viewers binge watch regularly, which is defined as watching 2–3 episodes in one setting. For $\frac{3}{4}$ of binge watchers, watching several episodes at once is a welcome change from their busy lives. And $\frac{4}{5}$ of viewers say that binge watching makes the show better. (See Exercises 115 and 116 in the second section of this chapter.)

Whether or not you binge watch, interpreting numbers is a central part of everyone's lives. Even though we have calculators and computers, the need to learn mathematics has never been greater. In the next decade, there will be over 1 million unfilled jobs that require some type of mathematical understanding. No matter what career path you choose, a solid background in mathematics will provide you with the tools to reach your full potential.

1.1 Numbers, Variables, and Expressions

NEW VOCABULARY

☐ Natural numbers
☐ Whole numbers
☐ Product
☐ Factors
☐ Prime number
☐ Composite number
☐ Prime factorization
☐ Variable
☐ Algebraic expression
☐ Equation
☐ Formula

CRITICAL THINKING 1

Give an example from everyday life of natural number or whole number use.

STUDY TIP

Bring your book, notebook, and a pen or pencil to every class. Write down major concepts presented by your instructor. Be sure that you understand the new vocabulary words.

1 Identifying Natural Numbers and Whole Numbers

One important set of numbers is the set of **natural numbers**. These numbers are the *counting numbers* and may be expressed as follows.

$$1, 2, 3, 4, 5, 6, \ldots$$

Because there are infinitely many natural numbers, three dots are used to show that the list continues without end. A second set of numbers is called the **whole numbers**, and may be expressed as follows.

$$0, 1, 2, 3, 4, 5, \ldots$$

Whole numbers include all of the natural numbers and the number 0.

Connecting Concepts with Your Life Natural numbers and whole numbers can be used when data are not broken into fractional parts. For example, the bar graph in **FIGURE 1.1** shows the number of apps on a student's iPad for the first 5 months after buying the device. Note that both natural numbers and whole numbers are appropriate to describe these data because a fraction of an app is not possible.

FIGURE 1.1

2 Classifying Numbers as Prime or Composite

When two natural numbers are multiplied, the result is another natural number. For example, multiplying the natural numbers 3 and 4 results in 12, a natural number. The result 12 is called the **product** and the numbers 3 and 4 are **factors** of 12.

$$3 \cdot 4 = 12$$
factor   factor   product

NOTE: Products can be expressed in several ways. For example, the product $3 \cdot 4 = 12$ can also be written as $3 \times 4 = 12$ and $3(4) = 12$. ∎

A natural number greater than 1 that has *only* itself and 1 as natural number factors is a **prime number**. For example, the number 7 is prime because the only natural number factors of 7 are 1 and 7. The following is a partial list of prime numbers. There are *infinitely many* prime numbers.

$$2, \quad 3, \quad 5, \quad 7, \quad 11, \quad 13, \quad 17, \quad 19, \quad 23, \quad 29$$

READING CHECK 1

- How do prime numbers differ from composite numbers?

A natural number greater than 1 that is not prime is a **composite number**. For example, the natural number 15 is a composite number because $3 \cdot 5 = 15$. In other words, 15 has factors other than 1 and itself.

Every composite number can be written as a product of prime numbers. For example, in the following See the Concept, we use a *factor tree* to find the prime factors of the composite number 120.

See the Concept 1   **PRIME FACTORIZATION OF 120**

The prime factorization is $120 = 2 \cdot 2 \cdot 2 \cdot 3 \cdot 5$.

Ⓐ Each composite number branches into a product that includes the smallest prime factor of that number.

Ⓑ Bring each prime number down one level in the tree using a single branch.

Ⓒ Stop when the numbers at the bottom of the tree are all prime.

WATCH VIDEO IN MML.

CRITICAL THINKING 2

Suppose that you draw a tree diagram for the prime factorization of a prime number. Describe what the tree will look like.

NOTE: Every composite number has a unique prime factorization, and it is customary to write the prime factors in order from smallest to largest. ∎

MAKING CONNECTIONS 1

Factor Trees and Prime Factorization

The prime factors of 120 can also be found using the tree shown. Even though this tree is different than the one used in See the Concept 1 above, the prime factors it reveals are the same. No matter what tree is used, a prime factorization is *always* unique.

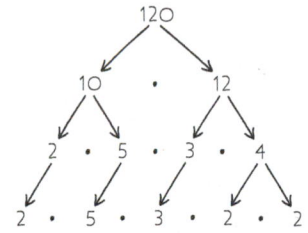

EXAMPLE 1   **Classifying numbers as prime or composite**

Classify each number as prime or composite, if possible. If a number is composite, write it as a product of prime numbers.
(a) 31   **(b)** 1   **(c)** 35   **(d)** 200

Solution
(a) The only factors of 31 are itself and 1, so the number 31 is prime.
(b) The number 1 is neither prime nor composite because prime and composite numbers must be greater than 1.

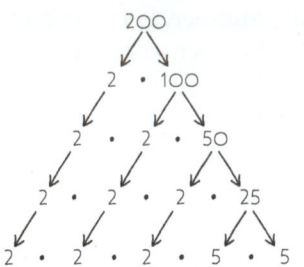

FIGURE 1.2 Prime Factorization of 200

(c) The number 35 is composite because 5 and 7 are factors. It can be written as a product of prime numbers as $35 = 5 \cdot 7$.

(d) The number 200 is composite because 10 and 20 are factors. A factor tree can be used to write 200 as a product of prime numbers, as shown in **FIGURE 1.2**. The factor tree reveals that 200 can be factored as $200 = 2 \cdot 2 \cdot 2 \cdot 5 \cdot 5$.

Now Try Exercises 21, 23, 25, 29

3 Evaluating Algebraic Expressions and Formulas

Connecting Concepts with Your Life There are 12 inches in 1 foot, so 5 feet equal $5 \cdot 12 = 60$ inches. Similarly, in 3 feet there are $3 \cdot 12 = 36$ inches. To convert feet to inches frequently, **TABLE 1.1** might help.

Converting Feet to Inches

Feet	1	2	3	4	5	6	7
Inches	12	24	36	48	60	72	84

TABLE 1.1

However, this table is not helpful in converting 8 or more feet to inches. Expanding the table to accommodate every possible value for feet would be impossible. *Variables* are often used in mathematics when tables of numbers are inadequate.

A **variable** is a symbol, typically an italic letter such as x, y, z, or F, used to represent an unknown quantity. In the preceding example, the number of feet could be represented by the variable F, and the corresponding number of inches could be represented by the variable I. The number of inches in F feet is given by the *algebraic expression* $12 \cdot F$. That is, to calculate the number of inches in F feet, multiply F by 12. The relationship between feet F and inches I is shown using the *equation* or *formula*

$$I = 12 \cdot F.$$

A dot ($\cdot$) is used to indicate multiplication because a multiplication sign ($\times$) can be confused with the variable x. Many times the multiplication sign is omitted altogether. Thus all three formulas

$$I = 12 \cdot F, \quad I = 12F, \quad \text{and} \quad I = 12(F)$$

represent the same relationship between feet and inches. If we wish to find the number of inches in 10 feet, for example, we can replace F in one of these formulas with the number 10. If we let $F = 10$ in the first formula,

$$I = 12 \cdot 10 = 120.$$

That is, there are 120 inches in 10 feet.

READING CHECK 2

- In mathematics, variables are used to represent what type of quantity?

MAKING CONNECTIONS 2

Algebraic Expressions, Equations, and Formulas

An **algebraic expression** consists of numbers, variables, operation symbols such as $+$, $-$, $\cdot$, and $\div$, and grouping symbols such as parentheses. An **equation** is a mathematical statement that two algebraic expressions are equal. Equations *always* contain an equals sign. A **formula** is a special type of equation that expresses a relationship between two or more quantities. For example, the formula $I = 12F$ states that to calculate the number of inches in F feet, multiply F by 12.

1 Introduction to Algebra

Numbers are all around us, and we need mathematics to understand them. For example, binge watching TV shows is rapidly becoming mainstream and the new normal. Over 60% of TV viewers binge watch regularly, which is defined as watching 2–3 episodes in one setting. For $\frac{3}{4}$ of binge watchers, watching several episodes at once is a welcome change from their busy lives. And $\frac{4}{5}$ of viewers say that binge watching makes the show better. (See Exercises 115 and 116 in the second section of this chapter.)

Whether or not you binge watch, interpreting numbers is a central part of everyone's lives. Even though we have calculators and computers, the need to learn mathematics has never been greater. In the next decade, there will be over 1 million unfilled jobs that require some type of mathematical understanding. No matter what career path you choose, a solid background in mathematics will provide you with the tools to reach your full potential.

1.1   Numbers, Variables, and Expressions

NEW VOCABULARY

☐ Natural numbers
☐ Whole numbers
☐ Product
☐ Factors
☐ Prime number
☐ Composite number
☐ Prime factorization
☐ Variable
☐ Algebraic expression
☐ Equation
☐ Formula

CRITICAL THINKING 1

Give an example from everyday life of natural number or whole number use.

STUDY TIP

Bring your book, notebook, and a pen or pencil to every class. Write down major concepts presented by your instructor. Be sure that you understand the new vocabulary words.

1 Identifying Natural Numbers and Whole Numbers

One important set of numbers is the set of **natural numbers**. These numbers are the *counting numbers* and may be expressed as follows.

$$1, 2, 3, 4, 5, 6, \ldots$$

Because there are infinitely many natural numbers, three dots are used to show that the list continues without end. A second set of numbers is called the **whole numbers**, and may be expressed as follows.

$$0, 1, 2, 3, 4, 5, \ldots$$

Whole numbers include all of the natural numbers and the number 0.

Connecting Concepts with Your Life Natural numbers and whole numbers can be used when data are not broken into fractional parts. For example, the bar graph in **FIGURE 1.1** shows the number of apps on a student's iPad for the first 5 months after buying the device. Note that both natural numbers and whole numbers are appropriate to describe these data because a fraction of an app is not possible.

iPad Apps

FIGURE 1.1

2 Classifying Numbers as Prime or Composite

When two natural numbers are multiplied, the result is another natural number. For example, multiplying the natural numbers 3 and 4 results in 12, a natural number. The result 12 is called the **product** and the numbers 3 and 4 are **factors** of 12.

$$\underset{\text{factor}}{3} \quad \cdot \quad \underset{\text{factor}}{4} \quad = \quad \underset{\text{product}}{12}$$

NOTE: Products can be expressed in several ways. For example, the product $3 \cdot 4 = 12$ can also be written as $3 \times 4 = 12$ and $3(4) = 12$. ∎

A natural number greater than 1 that has *only* itself and 1 as natural number factors is a **prime number**. For example, the number 7 is prime because the only natural number factors of 7 are 1 and 7. The following is a partial list of prime numbers. There are *infinitely many* prime numbers.

$$2, \quad 3, \quad 5, \quad 7, \quad 11, \quad 13, \quad 17, \quad 19, \quad 23, \quad 29$$

READING CHECK 1

• How do prime numbers differ from composite numbers?

A natural number greater than 1 that is not prime is a **composite number**. For example, the natural number 15 is a composite number because $3 \cdot 5 = 15$. In other words, 15 has factors other than 1 and itself.

Every composite number can be written as a product of prime numbers. For example, in the following See the Concept, we use a *factor tree* to find the prime factors of the composite number 120.

See the Concept 1   **PRIME FACTORIZATION OF 120**

B Bring down the prime factors.

A Write each composite number as a product.

C The prime factorization is in the bottom row.

The prime factorization is $120 = 2 \cdot 2 \cdot 2 \cdot 3 \cdot 5$.

A Each composite number branches into a product that includes the smallest prime factor of that number.

B Bring each prime number down one level in the tree using a single branch.

C Stop when the numbers at the bottom of the tree are all prime.

WATCH VIDEO IN MML .

CRITICAL THINKING 2

Suppose that you draw a tree diagram for the prime factorization of a prime number. Describe what the tree will look like.

NOTE: Every composite number has a unique prime factorization, and it is customary to write the prime factors in order from smallest to largest. ■

MAKING CONNECTIONS 1

Factor Trees and Prime Factorization

The prime factors of 120 can also be found using the tree shown. Even though this tree is different than the one used in See the Concept 1 above, the prime factors it reveals are the same. No matter what tree is used, a prime factorization is *always* unique.

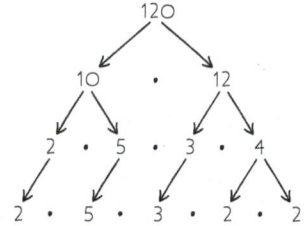

EXAMPLE 1   **Classifying numbers as prime or composite**

Classify each number as prime or composite, if possible. If a number is composite, write it as a product of prime numbers.
(a) 31   (b) 1   (c) 35   (d) 200

Solution
(a) The only factors of 31 are itself and 1, so the number 31 is prime.
(b) The number 1 is neither prime nor composite because prime and composite numbers must be greater than 1.

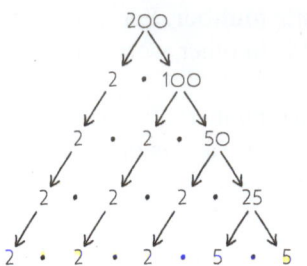

FIGURE 1.2 Prime Factorization of 200

(c) The number 35 is composite because 5 and 7 are factors. It can be written as a product of prime numbers as $35 = 5 \cdot 7$.

(d) The number 200 is composite because 10 and 20 are factors. A factor tree can be used to write 200 as a product of prime numbers, as shown in **FIGURE 1.2**. The factor tree reveals that 200 can be factored as $200 = 2 \cdot 2 \cdot 2 \cdot 5 \cdot 5$.

Now Try Exercises 21, 23, 25, 29

3 Evaluating Algebraic Expressions and Formulas

Connecting Concepts with Your Life There are 12 inches in 1 foot, so 5 feet equal $5 \cdot 12 = 60$ inches. Similarly, in 3 feet there are $3 \cdot 12 = 36$ inches. To convert feet to inches frequently, **TABLE 1.1** might help.

Converting Feet to Inches

Feet	1	2	3	4	5	6	7
Inches	12	24	36	48	60	72	84

TABLE 1.1

READING CHECK 2

• In mathematics, variables are used to represent what type of quantity?

However, this table is not helpful in converting 8 or more feet to inches. Expanding the table to accommodate every possible value for feet would be impossible. *Variables* are often used in mathematics when tables of numbers are inadequate.

A **variable** is a symbol, typically an italic letter such as x, y, z, or F, used to represent an unknown quantity. In the preceding example, the number of feet could be represented by the variable F, and the corresponding number of inches could be represented by the variable I. The number of inches in F feet is given by the *algebraic expression* $12 \cdot F$. That is, to calculate the number of inches in F feet, multiply F by 12. The relationship between feet F and inches I is shown using the *equation* or *formula*

$$I = 12 \cdot F.$$

A dot ($\cdot$) is used to indicate multiplication because a multiplication sign ($\times$) can be confused with the variable x. Many times the multiplication sign is omitted altogether. Thus all three formulas

$$I = 12 \cdot F, \quad I = 12F, \quad \text{and} \quad I = 12(F)$$

represent the same relationship between feet and inches. If we wish to find the number of inches in 10 feet, for example, we can replace F in one of these formulas with the number 10. If we let $F = 10$ in the first formula,

$$I = 12 \cdot 10 = 120.$$

That is, there are 120 inches in 10 feet.

MAKING CONNECTIONS 2

Algebraic Expressions, Equations, and Formulas

An **algebraic expression** consists of numbers, variables, operation symbols such as $+$, $-$, $\cdot$, and $\div$, and grouping symbols such as parentheses. An **equation** is a mathematical statement that two algebraic expressions are equal. Equations *always* contain an equals sign. A **formula** is a special type of equation that expresses a relationship between two or more quantities. For example, the formula $I = 12F$ states that to calculate the number of inches in F feet, multiply F by 12.

EXAMPLE 2 **Evaluating algebraic expressions that have one variable**

Evaluate each algebraic expression for $x = 4$.

(a) $x + 5$ **(b)** $5x$ **(c)** $15 - x$ **(d)** $\dfrac{x}{(x - 2)}$

Solution
(a) Replace x with 4 in the expression $x + 5$ to obtain $4 + 5 = 9$.
(b) The expression $5x$ indicates multiplication of 5 and x. Thus $5x = 5 \cdot 4 = 20$.
(c) $15 - x = 15 - 4 = 11$
(d) Perform all arithmetic operations inside parentheses first.

$$\frac{x}{(x - 2)} = \frac{4}{(4 - 2)} = \frac{4}{2} = 2$$

Now Try Exercises 45, 47, 51, 53

🌐 **Math in Context** *Driving* Some algebraic expressions contain more than one variable. For example, if a car travels 120 miles on 6 gallons of gasoline, then the car's *mileage* is $\frac{120}{6} = 20$ miles per gallon. In general, if a car travels M miles on G gallons of gasoline, then its mileage is given by the expression $\frac{M}{G}$. Note that $\frac{M}{G}$ contains two variables, M and G, whereas the expression $12F$ contains only one variable, F. In the next example, we evaluate expressions that contain two variables.

EXAMPLE 3 **Evaluating algebraic expressions that have two variables**

Evaluate each algebraic expression for $y = 2$ and $z = 8$.

(a) $3yz$ **(b)** $z - y$ **(c)** $\dfrac{z}{y}$

Solution
(a) Replace y with 2 and z with 8 to obtain $3yz = 3 \cdot 2 \cdot 8 = 48$.
(b) $z - y = 8 - 2 = 6$

(c) $\dfrac{z}{y} = \dfrac{8}{2} = 4$

Now Try Exercises 55, 57, 59

EXAMPLE 4 **Evaluating formulas**

Find the value of y for $x = 15$ and $z = 10$.

(a) $y = x - 3$ **(b)** $y = \dfrac{x}{5}$ **(c)** $y = 8xz$

Solution
(a) Substitute 15 for x and then evaluate the right side of the formula to find y.

$$\begin{aligned} y &= x - 3 &&\text{Given formula} \\ &= 15 - 3 &&\text{Replace } x \text{ with 15.} \\ &= 12 &&\text{Subtract.} \end{aligned}$$

(b) $y = \dfrac{x}{5} = \dfrac{15}{5} = 3$

(c) $y = 8xz = 8 \cdot 15 \cdot 10 = 1200$

Now Try Exercises 61, 67, 69

4 Translating Words to Expressions

Many times in mathematics, algebraic expressions are not given; rather, we must write our own expressions. To accomplish this task, we often translate words to symbols. The symbols $+$, $-$, $\cdot$, and $\div$ have special mathematical words associated with them. When two numbers are added, the result is called the *sum*. When one number is subtracted from another number, the result is called the *difference*. Similarly, multiplying two numbers results in a *product* and dividing two numbers results in a *quotient*. TABLE 1.2 lists many of the words commonly associated with these operations.

READING CHECK 3

- Which of the following words is *not* associated with subtraction: minus, twice, difference, or fewer?

Words Associated with Arithmetic Symbols

Symbol	Associated Words
$+$	add, plus, more, sum, total
$-$	subtract, minus, less, difference, fewer
$\cdot$	multiply, times, twice, double, triple, product
$\div$	divide, divided by, quotient

TABLE 1.2

EXAMPLE 5   **Translating words to expressions**

Translate each phrase to an algebraic expression. Specify what each variable represents.
(a) Four more than a number
(b) Ten less than the president's age
(c) A number plus 10, all divided by a different number
(d) The product of 6 and a number

Solution
(a) If the number were 20, then four more than the number would be $20 + 4 = 24$. If we let n represent the number, then four more than the number would be $n + 4$.
(b) If the president's age were 55, then ten less than the president's age would be $55 - 10 = 45$. If we let A represent the president's age, then ten less than the president's age would be $A - 10$.
(c) Let x be the first number and y be the other number. Then the expression can be written as $(x + 10) \div y$. Note that parentheses are used around $x + 10$ because it is "all" divided by y. This expression could also be written as $\frac{x + 10}{y}$.
(d) If n represents the number, then the product of 6 and n is $6 \cdot n$ or $6n$.

Now Try Exercises 73, 79, 83, 85

EXAMPLE 6   **Translating words to expressions**

Translate each phrase to an algebraic expression. Use x as the variable.
(a) Nine times the sum of a number and 3
(b) Four less than the product of 5 and her score

Solution
(a) The sum of a number and 3 can be written $x + 3$. To find 9 times this sum, we write the sum in parentheses and multiply by 9. The expression is $9(x + 3)$.
(b) The product of 5 and a score can be written $5x$. To find 4 less than this product, we subtract 4 from it. The expression is $5x - 4$.

Now Try Exercises 87, 89

🌐 *Math in ᴄˡⁱᵐᵃᵗᵗ Context* Cities are made up of large amounts of concrete and asphalt that heat up in the daytime from sunlight but do not cool off completely at night. As a result, urban areas tend to be warmer than surrounding rural areas. This effect is called the *urban heat island* and has been documented in cities throughout the world. The next example discusses the impact of this effect in Phoenix, Arizona.

EXAMPLE 7   **Translating words to an equation**

For each year after 1990, the average nighttime temperature in Phoenix has increased by about $0.15\,°C$.
(a) What was the increase in the nighttime temperature after 20 years, or in 2010?
(b) Write a formula (or equation) that gives the increase T in average nighttime temperature x years after 1990.
(c) Use your formula to estimate the increase in nighttime temperature in 2015.

Solution
(a) The average nighttime temperature has increased by $0.15\,°C$ per year, so after **20** years the temperature increase would be $0.15 \cdot \textbf{20} = 3.0\,°C$.
(b) To calculate the nighttime increase in temperature, multiply the number of years past 1990 by 0.15. Thus $T = 0.15x$, where x represents the number of years *after* 1990.
(c) Because $2015 - 1990 = 25$, let $x = \textbf{25}$ in the formula $T = 0.15x$ to get

$$T = 0.15(\textbf{25}) = 3.75.$$

The average nighttime temperature increased by $3.75\,°C$ from 1990 to 2015.

Now Try Exercise 95

Another use of translating words to formulas is in finding the areas of various shapes.

EXAMPLE 8   **Finding the area of a rectangle**

The area A of a rectangle equals its length L times its width W, as illustrated in the accompanying figure.
(a) Write a formula that shows the relationship between these three quantities.
(b) Find the area of a standard sheet of paper that is 8.5 inches wide and 11 inches long.

Solution
(a) The word *times* indicates that the length and width should be multiplied. The formula is given by $A = LW$.
(b) $A = 11 \cdot 8.5 = 93.5$ square inches

Now Try Exercise 101

1.1   Putting It All Together

STUDY TIP

Putting It All Together gives a summary of important concepts in each section. Be sure that you have a good understanding of these concepts.

CONCEPT	COMMENTS	EXAMPLES
Natural Numbers	Sometimes referred to as the *counting numbers*	$1, 2, 3, 4, 5, \ldots$
Whole Numbers	Includes the natural numbers and 0	$0, 1, 2, 3, 4, \ldots$
Products and Factors	When two numbers are multiplied, the result is called the product. The numbers being multiplied are called factors.	$6 \cdot 7 = 42$ **factor** **factor** **product**
Prime Number	A natural number greater than 1 whose only factors are itself and 1; there are infinitely many prime numbers.	2, 3, 5, 7, 11, 13, 17, and 19 are the prime numbers less than 20.
Composite Number	A natural number greater than 1 that is *not* a prime number; there are infinitely many composite numbers.	4, 9, 25, 39, 62, 76, 87, 91, 100
Prime Factorization	Every composite number can be written as a product of prime numbers. This unique product is called the prime factorization.	$60 = 2 \cdot 2 \cdot 3 \cdot 5$, $84 = 2 \cdot 2 \cdot 3 \cdot 7$
Variable	Represents an unknown quantity	x, y, z, A, F, and T
Algebraic Expression	May consist of variables, numbers, operation symbols such as $+, -, \cdot$, and $\div$, and grouping symbols such as parentheses	$x + 3$, $\frac{x}{y}$, $2z + 5$, $12F$, $x + y + z$, $x(y + 5)$
Equation	An equation is a statement that two algebraic expressions are equal. An equation always includes an equals sign.	$2 + 3 = 5$, $x + 5 = 7$, $I = 12F$, $y = 0.1x$
Formula	A formula is a special type of equation that expresses a relationship between two or more quantities.	$I = 12F$, $y = 0.1x$, $A = LW$, $F = 3Y$

1.1 Exercises

MyMathLab®

CONCEPTS AND VOCABULARY

1. The _____ numbers comprise the counting numbers.

2. A whole number is either a natural number or the number _____.

3. The factors of a prime number are itself and _____.

4. A natural number greater than 1 that is not a prime number is called a(n) _____ number.

5. Because $3 \cdot 6 = 18$, the numbers 3 and 6 are _____ of 18.

6. A(n) _____ is a special type of equation that expresses a relationship between two or more quantities.

7. A symbol or letter used to represent an unknown quantity is called a(n) _____.

8. Equations always contain a(n) _____.

9. When one number is added to another number, the result is called the _____.

10. When one number is multiplied by another number, the result is called the _____.

11. The result of dividing one number by another is called the _____.

12. The result of subtracting one number from another is called the _____.

IDENTIFYING QUANTITIES DESCRIBED BY WHOLE NUMBERS

Exercises 13–20: State whether the given quantity could accurately be described by the whole numbers.

13. The population of a country

14. The cost of a gallon of gasoline in dollars

15. A student's grade point average

16. The Fahrenheit temperature in Antarctica

17. The number of apps on an iPad

18. The number of students in a class

19. The winning time in a 100-meter sprint

20. The number of bald eagles in the United States

CLASSIFYING NUMBERS AS PRIME OR COMPOSITE

Exercises 21–32: Classify the number as prime, composite, or neither. If the number is composite, write it as a product of prime numbers.

21. 4   22. 36

23. 1   24. 0

25. 29   26. 13

27. 92   28. 69

29. 225   30. 900

31. 149   32. 101

Exercises 33–44: Write the composite number as a product of prime numbers.

33. 6   34. 8

35. 12   36. 20

37. 32   38. 100

39. 39   40. 51

41. 294   42. 175

43. 300   44. 455

EVALUATING ALGEBRAIC EXPRESSIONS AND FORMULAS

Exercises 45–54: Evaluate the expression for the given value of x.

45. $3x$ $x = 5$ 46. $x + 10$ $x = 8$

47. $9 - x$ $x = 4$ 48. $13x$ $x = 0$

49. $\dfrac{x}{8}$ $x = 32$ 50. $\dfrac{5}{(x - 3)}$ $x = 8$

51. $3(x + 1)$ $x = 5$ 52. $7(6 - x)$ $x = 3$

53. $\left(\dfrac{x}{2}\right) + 1$ $x = 6$ 54. $3 - \left(\dfrac{6}{x}\right)$ $x = 2$

Exercises 55–60: Evaluate the expression for the given values of x and y.

55. $x + y$ $x = 8, \ y = 14$

56. $5xy$ $x = 2, \ y = 3$

57. $6 \cdot \dfrac{x}{y}$ $x = 8, \ y = 4$

58. $y - x$ $\qquad$ $x = 8, \quad y = 11$

59. $y(x - 2)$ $\qquad$ $x = 5, \quad y = 3$

60. $(x + y) - 5$ $\qquad$ $x = 6, \quad y = 3$

Exercises 61–64: Find the value of y for the given value of x.

61. $y = x + 5$ $\qquad$ $x = 0$

62. $y = x \cdot x$ $\qquad$ $x = 7$

63. $y = 4x$ $\qquad$ $x = 7$

64. $y = 2(x - 3)$ $\qquad$ $x = 3$

Exercises 65–68: Find the value of F for the given value of z.

65. $F = z - 5$ $\qquad$ $z = 12$

66. $F = \dfrac{z}{4}$ $\qquad$ $z = 40$

67. $F = \dfrac{30}{z}$ $\qquad$ $z = 6$

68. $F = z \cdot z \cdot z$ $\qquad$ $z = 5$

Exercises 69–72: Find the value of y for the given values of x and z.

69. $y = 3xz$ $\qquad$ $x = 2, \quad z = 0$

70. $y = x + z$ $\qquad$ $x = 3, \quad z = 15$

71. $y = \dfrac{x}{z}$ $\qquad$ $x = 9, \quad z = 3$

72. $y = x - z$ $\qquad$ $x = 9, \quad z = 1$

TRANSLATING WORDS TO EXPRESSIONS

Exercises 73–90: Translate the phrase to an algebraic expression. State what each variable represents.

73. Five more than a number

74. Four less than a number

75. Three times the cost of a soda

76. Twice the cost of a gallon of gasoline

77. The sum of a number and 5

78. The quotient of two numbers

79. Two hundred less than the population of a town

80. The total number of dogs and cats in a city

81. A number divided by six

82. A number divided by another number

83. The product of a car's speed and traveling time

84. The difference between 220 and a person's heart rate

85. A number plus seven, all divided by a different number

86. One-fourth of a number increased by one-tenth of a different number

87. Five times the sum of a number and 4

88. Eight times the difference of a number and 2

89. Three less than the product of 7 and her age

90. Six more than the product of his age and 4

APPLICATIONS INVOLVING FORMULAS

91. **Dollars to Pennies** Write a formula that converts D dollars to P pennies.

92. **Quarters to Nickels** Write a formula that converts Q quarters to N nickels.

93. **Yards to Feet** Make a table of values that converts y yards to F feet. Let $y = 1, 2, 3, \ldots, 7$. Write a formula that converts y yards to F feet.

94. **Gallons to Quarts** Make a table of values that converts g gallons to Q quarts. Let $g = 1, 2, 3, \ldots, 6$. Write a formula that converts g gallons to Q quarts.

95. **NASCAR Speeds** On the fastest speedways, some NASCAR drivers reach average speeds of 3 miles per minute. Write a formula that gives the number of miles M that such a driver would travel in x minutes. How far would this driver travel in 36 minutes?

96. **NASCAR Speeds** On slower speedways, some NASCAR drivers reach average speeds of 2 miles per minute. Write a formula that gives the number of miles M that such a driver would travel in x minutes. How far would this driver travel in 42 minutes?

97. **Thinking Generally** If there are 6 blims in every drog, is the formula that relates B blims and D drogs $D = 6B$ or $B = 6D$?

98. **Heart Rate** The resting heart rate of a person is 70 beats per minute. Write a formula that gives the number of beats B that occur in x minutes. How many beats are there in an hour?

99. **Cost of Album Downloads** The table lists the cost C of downloading x albums. Write an equation that relates C and x.

Albums (x)	1	2	3	4
Cost (C)	\$12	\$24	\$36	\$48

100. Gallons of Water The table lists the gallons G of water coming from a garden hose after m minutes. Write an equation that relates G and m.

Minutes (m)	1	2	3	4
Gallons (G)	4	8	12	16

101. Area of a Rectangle The area of a rectangle equals its length times its width. Find the area of the rectangle shown in the figure.

9 ft

22 ft

102. Area of a Square A square is a rectangle whose length and width have equal measures. Find the area of a square with length 14 inches.

WRITING ABOUT MATHEMATICS

103. Give an example in which the whole numbers are not sufficient to describe a quantity in real life. Explain your reasoning.

104. Explain what a prime number is. How can you determine whether a number is prime?

105. Explain what a composite number is. How can you determine whether a number is composite?

106. When are variables used? Give an example.

1.2 Fractions

Objectives

1 Identifying Fractions, Numerators, and Denominators

2 Simplifying Fractions to Lowest Terms

3 Multiplying and Dividing Fractions

4 Adding and Subtracting Fractions (Like Denominators)

5 Adding and Subtracting Fractions (Unlike Denominators)

6 Solving Applications Involving Fractions

1 Identifying Fractions, Numerators, and Denominators

 Connecting Concepts with Your Life If we divide a circular pie into 6 equal slices, as shown in **FIGURE 1.3**, then each piece represents one-sixth of the pie and can be represented by the fraction $\frac{1}{6}$. Five slices of the pie would represent five-sixths of the pie and can be represented by the fraction $\frac{5}{6}$.

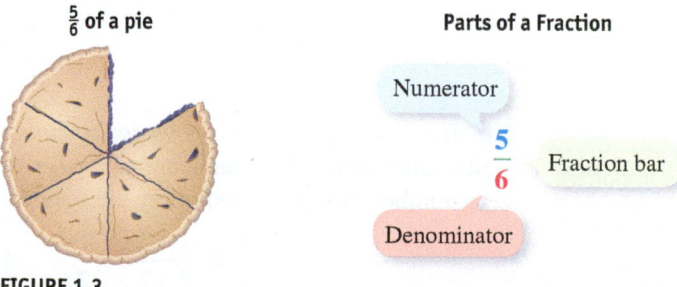

$\frac{5}{6}$ of a pie

Parts of a Fraction

Numerator

$\frac{5}{6}$ Fraction bar

Denominator

FIGURE 1.3

NEW VOCABULARY

☐ Equivalent fractions
☐ Lowest terms
☐ Greatest common factor (GCF)
☐ Basic principle of fractions
☐ Multiplicative inverse or reciprocal
☐ Least common denominator (LCD)

Sometimes we can represent a general fraction by using variables. The fraction $\frac{a}{b}$ can represent *any* fraction with **numerator** a and **denominator** b. However, the value of b cannot equal 0, which is denoted $b \neq 0$. (The symbol $\neq$ means "not equal to.")

NOTE: The fraction bar represents division. For example, the fraction $\frac{1}{2}$ represents the result when 1 is divided by 2, which is 0.5. We discuss this concept further in the fourth section of this chapter. ∎

EXAMPLE 1   **Identifying numerators and denominators**

Give the numerator and denominator of each fraction.

(a) $\dfrac{6}{13}$ (b) $\dfrac{ac}{b}$ (c) $\dfrac{x-5}{y+z}$

Solution

(a) The numerator is 6, and the denominator is 13.
(b) The numerator is ac, and the denominator is b.
(c) The numerator is $x-5$, and the denominator is $y+z$.

Now Try Exercises 15, 17, 19

2   Simplifying Fractions to Lowest Terms

 Connecting Concepts with Your Life Consider the amount of pizza shown in each of the three pies in **FIGURE 1.4**. The first pie is cut into sixths with three pieces remaining, the second pie is cut into fourths with two pieces remaining, and the third pie is cut into only two pieces with one piece remaining. In all three cases, half a pizza remains.

Three Ways to Represent One-Half

       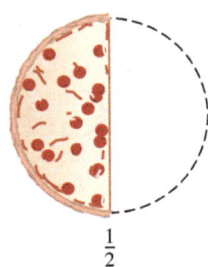

$\dfrac{3}{6}$ $\dfrac{2}{4}$ $\dfrac{1}{2}$

FIGURE 1.4 Equivalent Fractions

READING CHECK 1

• How can you tell if a fraction is written in lowest terms?

FIGURE 1.4 illustrates that the fractions $\frac{3}{6}$, $\frac{2}{4}$, and $\frac{1}{2}$ are equal, or **equivalent fractions**. The fraction $\frac{1}{2}$ is in **lowest terms** because its numerator and denominator have no factors in common (other than 1), whereas the fractions $\frac{3}{6}$ and $\frac{2}{4}$ are not in lowest terms. In the fraction $\frac{3}{6}$, the numerator and denominator have a common factor of 3.

$$\frac{3}{6} = \frac{1 \cdot 3}{2 \cdot 3} \qquad \text{Common factor: 3}$$

$$= \frac{1}{2}$$

STUDY TIP

A positive attitude is important. The first step to success is believing in yourself.

To simplify $\frac{3}{6}$, we used the *basic principle of fractions*: The value of a fraction is unchanged if the numerator and denominator of the fraction are multiplied (or divided) by the same nonzero number. We can also simplify the fraction $\frac{2}{4}$ to $\frac{1}{2}$ by using the basic principle of fractions.

$$\frac{2}{4} = \frac{1 \cdot 2}{2 \cdot 2} = \frac{1}{2} \qquad \text{Basic principle of fractions}$$

When simplifying fractions, we usually factor out the *greatest common factor* of the numerator and the denominator. The **greatest common factor (GCF)** of two or more numbers is the largest factor that is common to those numbers. For example, to simplify $\frac{27}{36}$, we first find the greatest common factor of 27 and 36. The numbers **3** and **9** are *common* factors of 27 and 36 because

READING CHECK 2

• What is the greatest common factor of 27 and 36?

$$3 \cdot 9 = 27 \text{ and } 3 \cdot 12 = 36 \quad \text{and} \quad 9 \cdot 3 = 27 \text{ and } 9 \cdot 4 = 36.$$

However, the *greatest* common factor is **9** because it is the largest number that divides evenly into both 27 and 36. The fraction $\frac{27}{36}$ simplifies to lowest terms as

$$\frac{27}{36} = \frac{3 \cdot 9}{4 \cdot 9} = \frac{3}{4}. \qquad \text{Basic principle of fractions}$$

SIMPLIFYING FRACTIONS

To simplify a fraction to lowest terms, factor out the greatest common factor c in the numerator and in the denominator. Then apply the **basic principle of fractions**:

$$\frac{a \cdot c}{b \cdot c} = \frac{a}{b}.$$

NOTE: This principle is true because multiplying a fraction by $\frac{c}{c}$ or 1 does not change the value of the fraction. (Fractions are multiplied in the next subsection.) ■

The greatest common factor for two numbers is not always obvious. The next example demonstrates two different methods that can be used to find the GCF.

EXAMPLE 2 **Finding the greatest common factor**

Find the greatest common factor (GCF) for each pair of numbers.
(a) 24, 60   **(b)** 36, 54

Solution

(a) One way to determine the greatest common factor is to find the prime factorization of each number using factor trees, as shown in **FIGURE 1.5**.

Prime Factorizations of 24 and 60

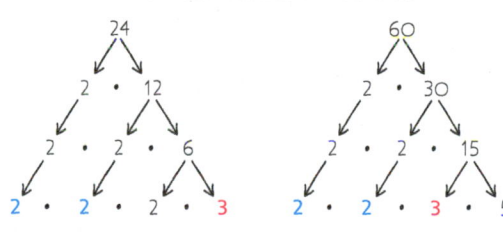

FIGURE 1.5

The prime factorizations have two **2**s and one **3** *in common*.

$$24 = \mathbf{2 \cdot 2} \cdot 2 \cdot \mathbf{3}$$
$$60 = \mathbf{2 \cdot 2 \cdot 3} \cdot 5$$

Thus the GCF of 24 and 60 is $\mathbf{2 \cdot 2 \cdot 3} = 12$.

(b) Another way to find the greatest common factor is to create a factor step diagram. Working downward from the top, the numbers in each step are found by dividing the two numbers in the previous step by their *smallest common prime factor*. The process continues until no common prime factor can be found. A factor step diagram for 36 and 54 is shown in **FIGURE 1.6**.

Factor Step Diagram for 36 and 54

2	36	54	← Start: Divide 36 and 54 by 2.
3	18	27	← Divide 18 and 27 by 3.
3	6	9	← Divide 6 and 9 by 3.
	2	3	← Stop: No common prime factors.

Smallest common prime factor.

FIGURE 1.6

The greatest common factor is the product of the prime numbers along the left side of the diagram. So the GCF of 36 and 54 is $\mathbf{2 \cdot 3 \cdot 3} = 18$.

Now Try Exercises 21, 23

EXAMPLE 3   **Simplifying fractions to lowest terms**

Simplify each fraction to lowest terms.

(a) $\dfrac{24}{60}$ (b) $\dfrac{42}{105}$

Solution

(a) From Example 2(a), the GCF of 24 and 60 is 12. Thus

$$\frac{24}{60} = \frac{2 \cdot \mathbf{12}}{5 \cdot \mathbf{12}} = \frac{2}{5}.$$

> The GCF is 12.

CRITICAL THINKING 1

Describe a situation from everyday life in which fractions would be needed.

(b) The prime factorizations of 42 and 105 are

$$42 = 2 \cdot \mathbf{3} \cdot \mathbf{7} \quad \text{and} \quad 105 = 5 \cdot \mathbf{3} \cdot \mathbf{7}.$$

The GCF of 42 and 105 is $\mathbf{3} \cdot \mathbf{7} = 21$. Thus

> The GCF is 21.

$$\frac{42}{105} = \frac{2 \cdot \mathbf{21}}{5 \cdot \mathbf{21}} = \frac{2}{5}.$$

Now Try Exercises 35, 39

MAKING CONNECTIONS 1

Simplifying Fractions in Steps

Sometimes a fraction can be simplified to lowest terms in multiple steps. By using *any* common factor, a new fraction in *lower* terms will result.

> 3 is a common factor of 42 and 105.   > 7 is a common factor of 14 and 35.

$$\frac{42}{105} = \frac{14 \cdot \mathbf{3}}{35 \cdot \mathbf{3}} = \frac{14}{35} = \frac{2 \cdot \mathbf{7}}{5 \cdot \mathbf{7}} = \frac{2}{5}$$

---

3 Multiplying and Dividing Fractions

Connecting Concepts with Your Life Suppose we cut *half* an apple into *thirds*, as shown in **FIGURE 1.7**. Then each piece represents one-sixth of the original apple. One-third of one-half is given by the following product.

Half an Apple Cut into Thirds

FIGURE 1.7

> $\frac{1}{3}$ of $\frac{1}{2}$

$$\frac{1}{3} \cdot \frac{1}{2} = \frac{1}{6}$$

This example demonstrates the following results.

- The numerator of the product of two fractions is found by multiplying the numerators of the two fractions.
- The denominator of the product of two fractions is found by multiplying the denominators of the two fractions.

For example, the product of $\frac{2}{3}$ and $\frac{5}{7}$ is found as follows.

> Multiply numerators.

$$\frac{2}{3} \cdot \frac{5}{7} = \frac{\mathbf{2 \cdot 5}}{\mathbf{3 \cdot 7}} = \frac{10}{21}$$

> Multiply denominators.

NOTE: The word "of" in mathematics often indicates multiplication. For example, the phrases "one-fifth **of** the cookies," "twenty percent **of** the price," and "half **of** the money" all suggest **multiplication**. ∎

MULTIPLICATION OF FRACTIONS

The product of $\frac{a}{b}$ and $\frac{c}{d}$ is given by

$$\frac{a}{b} \cdot \frac{c}{d} = \frac{ac}{bd},$$

where b and d are not 0.

EXAMPLE 4

Multiplying fractions

Multiply. Simplify the result when appropriate.

(a) $\frac{4}{5} \cdot \frac{6}{7}$ (b) $\frac{8}{9} \cdot \frac{3}{4}$ (c) $3 \cdot \frac{5}{9}$ (d) $\frac{x}{y} \cdot \frac{z}{3}$

Solution

(a) $\frac{4}{5} \cdot \frac{6}{7} = \frac{4 \cdot 6}{5 \cdot 7} = \frac{24}{35}$ The product is in lowest terms.

(b) $\frac{8}{9} \cdot \frac{3}{4} = \frac{8 \cdot 3}{9 \cdot 4} = \frac{24}{36}$; the GCF of 24 and 36 is 12, so

Simplify for lowest terms.

$$\frac{24}{36} = \frac{2 \cdot 12}{3 \cdot 12} = \frac{2}{3}.$$

(c) Start by writing 3 as $\frac{3}{1}$.

Simplify to lowest terms.

$$3 \cdot \frac{5}{9} = \frac{3}{1} \cdot \frac{5}{9} = \frac{3 \cdot 5}{1 \cdot 9} = \frac{15}{9}$$

The GCF of 15 and 9 is 3, so

$$\frac{15}{9} = \frac{5 \cdot 3}{3 \cdot 3} = \frac{5}{3}.$$

(d) $\frac{x}{y} \cdot \frac{z}{3} = \frac{x \cdot z}{y \cdot 3} = \frac{xz}{3y}$

NOTE: When we write the product of a variable and a number, such as $y \cdot 3$, we typically write the number first, followed by the variable. That is, $y \cdot 3 = 3y$. ∎

Now Try Exercises 41, 45, 49, 53

MAKING CONNECTIONS 2

Multiplying and Simplifying Fractions

When multiplying fractions, sometimes it is possible to change the order of the factors to rewrite the product so that it is easier to simplify. In Example 4(b) the product could be written as

$$\frac{8}{9} \cdot \frac{3}{4} = \frac{3 \cdot 8}{9 \cdot 4} = \frac{3}{9} \cdot \frac{8}{4} = \frac{1}{3} \cdot \frac{2}{1} = \frac{2}{3}.$$

Instead of simplifying $\frac{24}{36}$, which contains larger numbers, the fractions $\frac{3}{9}$ and $\frac{8}{4}$ were simplified first and then multiplied.

EXAMPLE 5 **Finding fractional parts**

Find each fractional part.

(a) One-fifth of two-thirds **(b)** Four-fifths of three-sevenths **(c)** Three-fifths of ten

Solution

(a) The phrase "one-fifth of" indicates multiplication by one-fifth. The fractional part is

$$\frac{1}{5} \cdot \frac{2}{3} = \frac{1 \cdot 2}{5 \cdot 3} = \frac{2}{15}.$$

(b) $\frac{4}{5} \cdot \frac{3}{7} = \frac{4 \cdot 3}{5 \cdot 7} = \frac{12}{35}$

(c) $\frac{3}{5} \cdot 10 = \frac{3}{5} \cdot \frac{10}{1} = \frac{30}{5} = 6$

Now Try Exercises 55, 57, 59

🌐 *Math in* ₍Education₎ *Context* Fractions can be used to describe particular parts of the U.S. population. In the next application, we use fractions to find the portion of the population that has completed a bachelor's degree.

EXAMPLE 6 **Estimating college completion rates**

About $\frac{22}{25}$ of the U.S. population over the age of 25 has a high school diploma. About $\frac{15}{44}$ of those people have gone on to complete a bachelor's degree. What fraction of the U.S. population over the age of 25 has completed a bachelor's degree? (*Source:* U.S. Census Bureau.)

Solution

We need to find $\frac{15}{44}$ of $\frac{22}{25}$.

$$\frac{15}{44} \cdot \frac{22}{25} = \frac{15 \cdot 22}{44 \cdot 25} = \frac{330}{1100} = \frac{3 \cdot 110}{10 \cdot 110} = \frac{3}{10}.$$

About $\frac{3}{10}$ of the U.S. population over the age of 25 has completed a bachelor's degree.

Now Try Exercise 125

Numbers and Their Reciprocals

Number	Reciprocal
3	$\frac{1}{3}$
$\frac{1}{4}$	4
$\frac{3}{2}$	$\frac{2}{3}$
$\frac{21}{37}$	$\frac{37}{21}$

TABLE 1.3

The **multiplicative inverse**, or **reciprocal**, of a nonzero number a is $\frac{1}{a}$. **TABLE 1.3**, lists several numbers and their reciprocals. Note that the product of a number and its reciprocal is always 1. For example, the reciprocal of 2 is $\frac{1}{2}$, and their product is $2 \cdot \frac{1}{2} = 1$.

⚛ **Connecting Concepts with Your Life** Suppose that a person wants to buy raffle tickets that cost a half dollar each. If she has 4 dollars, then the number of raffle tickets that can be purchased equals the number of half dollars that there are in 4 dollars. Thus 8 raffle tickets can be purchased.

This calculation is given by $4 \div \frac{1}{2}$. To divide a number by a fraction, multiply the number by the reciprocal of the fraction.

$$4 \div \frac{1}{2} = 4 \cdot \frac{2}{1} \qquad \text{Multiply by the reciprocal of } \tfrac{1}{2}.$$

$$= \frac{4}{1} \cdot \frac{2}{1} \qquad \text{Write 4 as } \tfrac{4}{1}.$$

$$= \frac{8}{1} \qquad \text{Multiply the fractions.}$$

$$= 8 \qquad \tfrac{a}{1} = a \text{ for all values of } a.$$

TECHNOLOGY NOTE

When entering a mixed number into a calculator, it is usually easiest first to convert the mixed number to an improper fraction. For example, enter $2\frac{2}{3}$ as $\frac{8}{3}$. Otherwise, enter $2\frac{2}{3}$ as $2 + \frac{2}{3}$. See the accompanying figure.

```
8/3
         2.666666667
2+2/3
         2.666666667
```

Justification for multiplying by the reciprocal when dividing two fractions is shown here.

$$\frac{a}{b} \div \frac{c}{d} = \frac{\dfrac{a}{b}}{\dfrac{c}{d}} = \frac{\dfrac{a}{b} \cdot \dfrac{d}{c}}{\dfrac{c}{d} \cdot \dfrac{d}{c}} = \frac{\dfrac{a}{b} \cdot \dfrac{d}{c}}{1} = \frac{a}{b} \cdot \frac{d}{c}$$

These results are summarized as follows.   Basic principle of fractions

DIVISION OF FRACTIONS

For real numbers a, b, c, and d, with b, c, and d not equal to 0,

$$\frac{a}{b} \div \frac{c}{d} = \frac{a}{b} \cdot \frac{d}{c}.$$

NOTE: The step where dividing by $\frac{c}{d}$ changes to multiplying by $\frac{d}{c}$ is sometimes referred to as "invert and multiply." ∎

EXAMPLE 7   **Dividing fractions**

Divide. Simplify the result when appropriate.

(a) $\frac{1}{3} \div \frac{3}{5}$ **(b)** $\frac{4}{5} \div \frac{4}{5}$ **(c)** $5 \div \frac{10}{3}$ **(d)** $\frac{x}{2} \div \frac{y}{z}$

Solution

(a) To divide $\frac{1}{3}$ by $\frac{3}{5}$, multiply $\frac{1}{3}$ by $\frac{5}{3}$, which is the reciprocal of $\frac{3}{5}$.

"Invert and multiply."

$$\frac{1}{3} \div \frac{3}{5} = \frac{1}{3} \cdot \frac{5}{3} = \frac{1 \cdot 5}{3 \cdot 3} = \frac{5}{9}$$

(b) $\frac{4}{5} \div \frac{4}{5} = \frac{4}{5} \cdot \frac{5}{4} = \frac{4 \cdot 5}{5 \cdot 4} = \frac{20}{20} = 1$. Note that when we divide any nonzero number by itself, the result is 1.

(c) $5 \div \frac{10}{3} = \frac{5}{1} \cdot \frac{3}{10} = \frac{15}{10} = \frac{3}{2}$ **(d)** $\frac{x}{2} \div \frac{y}{z} = \frac{x}{2} \cdot \frac{z}{y} = \frac{xz}{2y}$

Now Try Exercises 65, 69, 73, 77

NOTE: In Example 7(c), the answer $\frac{3}{2}$ is an *improper fraction*, which could be written as the mixed number $1\frac{1}{2}$. However, in algebra, fractions are often left as improper fractions. ∎

EXAMPLE 8   **Writing a problem**

Describe a problem for which the solution could be found by dividing 5 by $\frac{1}{6}$.

Solution

One possible problem could be stated as follows. If five pies are each cut into sixths, how many pieces of pie are there? See **FIGURE 1.8**. (There are thirty $\frac{1}{6}$'s in 5.)

Five Pies Cut into Sixths

FIGURE 1.8

Now Try Exercise 129

$\frac{5}{8}$ of a Sheet of Paper

FIGURE 1.9

4 Adding and Subtracting Fractions (Like Denominators)

Connecting Concepts with Your Life Suppose that a person cuts a sheet of paper into eighths. If that person picks up two pieces and another person picks up three pieces, then together they have

$$\frac{2}{8} + \frac{3}{8} = \frac{5}{8}$$

of a sheet of paper, as illustrated in **FIGURE 1.9**. When the denominator of one fraction is the same as the denominator of a second fraction, the sum of the two fractions can be found by adding their numerators and keeping the common denominator.

Similarly, if someone picks up 5 pieces of paper and gives 2 away, then that person has

$$\frac{5}{8} - \frac{2}{8} = \frac{3}{8}$$

of a sheet of paper. To subtract two fractions with common denominators, subtract their numerators and keep the common denominator.

ADDITION AND SUBTRACTION OF FRACTIONS

To add or subtract fractions with a common denominator d, use the equations

$$\frac{a}{d} + \frac{b}{d} = \frac{a + b}{d} \quad \text{and} \quad \frac{a}{d} - \frac{b}{d} = \frac{a - b}{d},$$

where d is not 0.

EXAMPLE 9 **Adding and subtracting fractions with common denominators**

Add or subtract as indicated. Simplify your answer to lowest terms when appropriate.

(a) $\frac{5}{13} + \frac{12}{13}$ **(b)** $\frac{11}{8} - \frac{5}{8}$

Solution
(a) Because the fractions have a common denominator, add the numerators and keep the common denominator.

Add the numerators.

$$\frac{5}{13} + \frac{12}{13} = \frac{5 + 12}{13} = \frac{17}{13}$$

Keep the denominator.

(b) Because the fractions have a common denominator, subtract the numerators and keep the common denominator.

Subtract the numerators.

$$\frac{11}{8} - \frac{5}{8} = \frac{11 - 5}{8} = \frac{6}{8}$$

Keep the denominator.

The fraction $\frac{6}{8}$ can be simplified to $\frac{3}{4}$.

Now Try Exercise 81

5 Adding and Subtracting Fractions (Unlike Denominators)

 Connecting Concepts with Your Life Suppose that one person mows half a large lawn while another person mows a fourth of the lawn. To determine how much they mowed together,

Lawn Mowing

$\frac{1}{2} + \frac{1}{4}$

(a)

$\frac{2}{4} + \frac{1}{4} = \frac{3}{4}$

(b)

FIGURE 1.10

we need to find the sum $\frac{1}{2} + \frac{1}{4}$. See **FIGURE 1.10(a)**. Before we can add fractions with unlike denominators, we must write each fraction with a common denominator. The *least common denominator* of 2 and 4 is 4. Thus we need to write $\frac{1}{2}$ as $\frac{?}{4}$ by multiplying the numerator and denominator by the *same nonzero number*.

$$\frac{1}{2} = \frac{1}{2} \cdot \frac{2}{2} \qquad \text{Multiply by 1.}$$

$$= \frac{2}{4} \qquad \text{Multiply fractions.}$$

Now we can find the needed sum.

$$\frac{1}{2} + \frac{1}{4} = \frac{2}{4} + \frac{1}{4} = \frac{3}{4}$$

Together, the two people mow three-fourths of the lawn, as illustrated in **FIGURE 1.10(b)**.

To add or subtract fractions with unlike denominators, we first find the least common denominator for the fractions. The **least common denominator (LCD)** for two or more fractions is the smallest number that is divisible by every denominator.

FINDING THE LEAST COMMON DENOMINATOR (LCD)

STEP 1: Find the prime factorization for each denominator.

STEP 2: List each factor that appears in one or more of the factorizations. If a factor is repeated in any of the factorizations, list this factor the maximum number of times that it is repeated.

STEP 3: The product of this list of factors is the LCD.

READING CHECK 3

• What is the LCD of 4 and 6?

The next example demonstrates two different methods that can be used to find the LCD. In part (a), the LCD is found using the three-step method shown above, and in part (b), a factor step diagram is used instead.

EXAMPLE 10 **Finding the least common denominator**

Find the LCD for each set of fractions.

(a) $\frac{5}{6}, \frac{3}{4}$ (b) $\frac{7}{36}, \frac{5}{54}$

Solution
(a) **STEP 1:** For the fractions $\frac{5}{6}$ and $\frac{3}{4}$, the prime factorizations of the denominators are

$$6 = 2 \cdot 3 \qquad \text{and} \qquad 4 = 2 \cdot 2.$$

STEP 2: List the factors: **2, 2, 3**. Note that, because the factor 2 appears a maximum of two times, it is listed twice.
STEP 3: The LCD is the product of this list, or $2 \cdot 2 \cdot 3 = 12$.

NOTE: Finding an LCD is equivalent to finding the smallest number that each denominator divides into evenly. Both 6 and 4 divide into 12 evenly, and 12 is the smallest such number. Thus 12 is the LCD for $\frac{5}{6}$ and $\frac{3}{4}$. ■

(b) The *same* factor step diagram that was used in Example 2(b) to find the GCF of 36 and 54 can be used to find the LCD for $\frac{7}{36}$ and $\frac{5}{54}$; however, the final step differs slightly. As in Example 2(b), we find the numbers in each step by dividing the two numbers in the previous step by their *smallest common prime factor*. The process continues until no common prime factor can be found, as shown in **FIGURE 1.11** on the next page.

Factor Step Diagram for Finding the LCD

FIGURE 1.11

The process for finding the LCD differs from that used to find the GCF in that we find the LCD by multiplying not only the numbers along the side of the diagram but also the numbers at the bottom of the diagram. The LCD for $\frac{7}{36}$ and $\frac{5}{54}$ is $2 \cdot 3 \cdot 3 \cdot 2 \cdot 3 = 108$.

Now Try Exercises 85, 89

Once the LCD has been found, the next step in the process for adding or subtracting fractions with unlike denominators is to rewrite each fraction with the LCD.

EXAMPLE 11 **Rewriting fractions with the LCD**

Rewrite each set of fractions using the LCD.

(a) $\frac{5}{6}, \frac{3}{4}$ **(b)** $\frac{7}{12}, \frac{5}{18}$

Solution

(a) From Example 10(a), the LCD is 12. To write $\frac{5}{6}$ with a denominator of 12, we multiply the fraction by 1 in the form $\frac{2}{2}$.

$$\frac{5}{6} \cdot \frac{2}{2} = \frac{5 \cdot 2}{6 \cdot 2} = \frac{10}{12}$$

To write $\frac{3}{4}$ with a denominator of 12, we multiply the fraction by 1 in the form $\frac{3}{3}$.

$$\frac{3}{4} \cdot \frac{3}{3} = \frac{3 \cdot 3}{4 \cdot 3} = \frac{9}{12}$$

Thus $\frac{5}{6}$ can be rewritten as $\frac{10}{12}$ and $\frac{3}{4}$ can be rewritten as $\frac{9}{12}$.

(b) The LCD for $\frac{7}{12}$ and $\frac{5}{18}$ is 36. To write $\frac{7}{12}$ with a denominator of 36, multiply the fraction by $\frac{3}{3}$. To write $\frac{5}{18}$ with a denominator of 36, multiply the fraction by $\frac{2}{2}$.

$$\frac{7}{12} \cdot \frac{3}{3} = \frac{7 \cdot 3}{12 \cdot 3} = \frac{21}{36} \quad \text{and} \quad \frac{5}{18} \cdot \frac{2}{2} = \frac{5 \cdot 2}{18 \cdot 2} = \frac{10}{36}$$

Thus $\frac{7}{12}$ can be rewritten as $\frac{21}{36}$ and $\frac{5}{18}$ can be rewritten as $\frac{10}{36}$.

Now Try Exercises 95, 99

The next example demonstrates how the concepts shown in the last two examples can be used to add or subtract fractions with unlike denominators.

EXAMPLE 12 **Adding and subtracting fractions with unlike denominators**

Add or subtract as indicated. Simplify your answer to lowest terms when appropriate.

(a) $\frac{5}{6} + \frac{3}{4}$ **(b)** $\frac{7}{12} - \frac{5}{18}$ **(c)** $\frac{1}{4} + \frac{2}{5} + \frac{7}{10}$

Solution

(a) From Example 10(a), the LCD is 12. Begin by writing each fraction with a denominator of 12, as demonstrated in Example 11(a).

$$\frac{5}{6} + \frac{3}{4} = \frac{5}{6} \cdot \frac{2}{2} + \frac{3}{4} \cdot \frac{3}{3} \qquad \text{Change to LCD of 12.}$$

$$= \frac{10}{12} + \frac{9}{12} \qquad \text{Multiply the fractions.}$$

$$= \frac{10 + 9}{12} \qquad \text{Add the numerators.}$$

$$= \frac{19}{12} \qquad \text{Simplify.}$$

(b) Using the results from Example 11(b), we perform the following steps.

$$\frac{7}{12} - \frac{5}{18} = \frac{7}{12} \cdot \frac{3}{3} - \frac{5}{18} \cdot \frac{2}{2} \qquad \text{Change to LCD of 36.}$$

$$= \frac{21}{36} - \frac{10}{36} \qquad \text{Multiply the fractions.}$$

$$= \frac{21 - 10}{36} \qquad \text{Subtract the numerators.}$$

$$= \frac{11}{36} \qquad \text{Simplify.}$$

(c) The LCD for 4, 5, and 10 is 20.

$$\frac{1}{4} + \frac{2}{5} + \frac{7}{10} = \frac{1}{4} \cdot \frac{5}{5} + \frac{2}{5} \cdot \frac{4}{4} + \frac{7}{10} \cdot \frac{2}{2} \qquad \text{Change to LCD of 20.}$$

$$= \frac{5}{20} + \frac{8}{20} + \frac{14}{20} \qquad \text{Multiply the fractions.}$$

$$= \frac{5 + 8 + 14}{20} \qquad \text{Add the numerators.}$$

$$= \frac{27}{20} \qquad \text{Simplify.}$$

Now Try Exercises 103, 109, 113

6 Solving Applications Involving Fractions

🌐 *Math in Context* ^Construction In building projects, boards are often cut into equal lengths. The next example illustrates a situation where fractions occur in carpentry.

EXAMPLE 13 **Applying fractions to carpentry**

A board measuring $35\frac{3}{4}$ inches is cut into four equal pieces, as depicted in **FIGURE 1.12**. Find the length of each piece.

Board in Four Sections

$35\frac{3}{4}$ in.

FIGURE 1.12

Solution

Begin by writing $35\frac{3}{4}$ as the improper fraction $\frac{143}{4}$ (because $4 \cdot 35 + 3 = 143$). Because the board is to be cut into four equal parts, the length of each piece should be

$$\frac{143}{4} \div 4 = \frac{143}{4} \cdot \frac{1}{4} = \frac{143}{16}, \quad \text{or} \quad 8\frac{15}{16} \text{ inches.}$$

> Multiply by the reciprocal of 4.

Now Try Exercise 119

1.2   Putting It All Together

CONCEPT	COMMENTS	EXAMPLES
Fraction	The fraction $\frac{a}{b}$ has numerator a and denominator b.	The fraction $\frac{xy}{2}$ has numerator xy and denominator 2.
Lowest Terms	A fraction is in lowest terms if the numerator and denominator have no factors in common.	The fraction $\frac{3}{8}$ is in lowest terms because 3 and 8 have no factors in common.
Greatest Common Factor (GCF)	The GCF of two numbers equals the largest number that divides into both evenly.	The GCF of 12 and 18 is 6 because 6 is the largest number that divides into 12 and 18 evenly.
Simplifying Fractions	Use the principle $$\frac{a \cdot c}{b \cdot c} = \frac{a}{b}$$ to simplify fractions, where c is the GCF of the numerator and denominator.	The GCF of 24 and 32 is 8, so $$\frac{24}{32} = \frac{3 \cdot 8}{4 \cdot 8} = \frac{3}{4}.$$ The GCF of 20 and 8 is 4, so $$\frac{20}{8} = \frac{5 \cdot 4}{2 \cdot 4} = \frac{5}{2}.$$
Multiplicative Inverse or Reciprocal	The reciprocal of $\frac{a}{b}$ is $\frac{b}{a}$, where a and b are not zero. The product of a number and its reciprocal is 1.	The reciprocals of 5 and $\frac{3}{4}$ are $\frac{1}{5}$ and $\frac{4}{3}$, respectively, because $5 \cdot \frac{1}{5} = 1$ and $\frac{3}{4} \cdot \frac{4}{3} = 1$.
Multiplication and Division of Fractions	$$\frac{a}{b} \cdot \frac{c}{d} = \frac{ac}{bd}$$ $$\frac{a}{b} \div \frac{c}{d} = \frac{a}{b} \cdot \frac{d}{c}$$	$$\frac{3}{5} \cdot \frac{4}{9} = \frac{12}{45} = \frac{4}{15}$$ $$\frac{3}{2} \div \frac{6}{5} = \frac{3}{2} \cdot \frac{5}{6} = \frac{15}{12} = \frac{5}{4}$$
Addition and Subtraction of Fractions with Like Denominators	$$\frac{a}{d} + \frac{c}{d} = \frac{a+c}{d}$$ $$\frac{a}{d} - \frac{c}{d} = \frac{a-c}{d}$$	$$\frac{3}{5} + \frac{4}{5} = \frac{3+4}{5} = \frac{7}{5}$$ $$\frac{17}{12} - \frac{11}{12} = \frac{17-11}{12} = \frac{6}{12} = \frac{1}{2}$$
Least Common Denominator (LCD)	The LCD of two fractions equals the smallest number that both denominators divide into evenly.	The LCD of $\frac{5}{12}$ and $\frac{7}{18}$ is 36 because 36 is the smallest number that both 12 and 18 divide into evenly.
Addition and Subtraction of Fractions with Unlike Denominators	First write each fraction with the least common denominator. Then add or subtract the numerators.	The LCD of $\frac{3}{4}$ and $\frac{7}{10}$ is 20. $$\frac{3}{4} + \frac{7}{10} = \frac{3}{4} \cdot \frac{5}{5} + \frac{7}{10} \cdot \frac{2}{2}$$ $$= \frac{15}{20} + \frac{14}{20}$$ $$= \frac{29}{20}$$

1.2   Exercises

MyMathLab®

CONCEPTS AND VOCABULARY

1. A small pie is cut into 4 equal pieces. If someone eats 3 of the pieces, what fraction of the pie does the person eat? What fraction of the pie remains?

2. In the fraction $\frac{11}{21}$ the numerator is _____ and the denominator is _____.

3. In the fraction $\frac{a}{b}$, the variable b cannot equal _____.

4. The fraction $\frac{7}{12}$ is in _____ terms because 7 and 12 have no factors in common.

5. (True or False?) The numerator of the product of two fractions is found by multiplying the numerators of the two fractions.

6. (True or False?) The denominator of the sum of two fractions is found by adding the denominators of the two fractions.

7. $\frac{ac}{bc} =$ _____, by the basic principle of fractions.

8. In the phrase "two-fifths of one-third," the word *of* indicates that we _____ the fractions $\frac{2}{5}$ and $\frac{1}{3}$.

9. What is the reciprocal of a, provided $a \neq 0$?

10. To divide $\frac{3}{4}$ by 5, multiply $\frac{3}{4}$ by _____.

11. $\frac{a}{b} \cdot \frac{c}{d} =$ _____

12. $\frac{a}{b} \div \frac{c}{d} =$ _____

13. $\frac{a}{b} + \frac{c}{b} =$ _____

14. $\frac{a}{b} - \frac{c}{b} =$ _____

IDENTIFYING NUMERATORS AND DENOMINATORS

Exercises 15–20: Give the numerator and the denominator of the fraction.

15. $\frac{19}{21}$

16. $\frac{9}{7}$

17. $\frac{a}{bd}$

18. $\frac{8a}{11c}$

19. $\frac{a+2}{a-5}$

20. $\frac{x-7}{x+7}$

SIMPLIFYING FRACTIONS TO LOWEST TERMS

Exercises 21–26: Find the greatest common factor.

21. 4, 12

22. 3, 27

23. 50, 75

24. 45, 105

25. 100, 60, 70

26. 36, 48, 72

Exercises 27–30: Use the basic principle of fractions to simplify the expression.

27. $\frac{3 \cdot 4}{5 \cdot 4}$

28. $\frac{2 \cdot 7}{9 \cdot 7}$

29. $\frac{3 \cdot 8}{8 \cdot 5}$

30. $\frac{7 \cdot 16}{16 \cdot 3}$

Exercises 31–40: Simplify the fraction to lowest terms.

31. $\frac{4}{8}$

32. $\frac{4}{12}$

33. $\frac{10}{25}$

34. $\frac{5}{20}$

35. $\frac{12}{36}$

36. $\frac{16}{24}$

37. $\frac{12}{30}$

38. $\frac{60}{105}$

39. $\frac{19}{76}$

40. $\frac{17}{51}$

MULTIPLYING AND DIVIDING FRACTIONS

Exercises 41–54: Multiply and simplify to lowest terms when appropriate.

41. $\frac{3}{4} \cdot \frac{1}{5}$

42. $\frac{3}{2} \cdot \frac{5}{8}$

43. $\frac{5}{3} \cdot \frac{3}{5}$

44. $\frac{21}{32} \cdot \frac{32}{21}$

45. $\frac{5}{6} \cdot \frac{18}{25}$

46. $\frac{7}{9} \cdot \frac{3}{14}$

47. $4 \cdot \frac{3}{5}$

48. $5 \cdot \frac{7}{9}$

49. $2 \cdot \frac{3}{8}$

50. $10 \cdot \frac{1}{100}$

51. $\frac{x}{y} \cdot \frac{y}{x}$

52. $\frac{x}{y} \cdot \frac{y}{z}$

53. $\frac{a}{b} \cdot \frac{3}{2}$

54. $\frac{5}{8} \cdot \frac{4x}{5y}$

Exercises 55–60: Find the fractional part.

55. One-fourth of three-fourths

56. Three-sevenths of nine-sixteenths

57. Two-thirds of six

58. Three-fourths of seven

59. One-half of two-thirds

60. Four-elevenths of nine-eighths

Exercises 61–64: Give the reciprocal of each number.

61. (a) 5   (b) 7   (c) $\frac{4}{7}$ (d) $\frac{9}{8}$

62. (a) 3   (b) 2   (c) $\frac{6}{5}$ (d) $\frac{3}{8}$

63. (a) $\frac{1}{2}$ (b) $\frac{1}{9}$ (c) $\frac{12}{101}$ (d) $\frac{31}{17}$

64. (a) $\frac{1}{5}$ (b) $\frac{7}{3}$ (c) $\frac{23}{64}$ (d) $\frac{63}{29}$

Exercises 65–80: Divide and simplify to lowest terms when appropriate.

65. $\frac{1}{2} \div \frac{1}{3}$ **66.** $\frac{3}{4} \div \frac{1}{5}$

67. $\frac{3}{4} \div \frac{1}{8}$ **68.** $\frac{6}{7} \div \frac{3}{14}$

69. $\frac{4}{3} \div \frac{4}{3}$ **70.** $\frac{12}{21} \div \frac{4}{7}$

71. $\frac{32}{27} \div \frac{8}{9}$ **72.** $\frac{8}{15} \div \frac{2}{25}$

73. $10 \div \frac{5}{6}$ **74.** $8 \div \frac{4}{3}$

75. $\frac{9}{10} \div 3$ **76.** $\frac{32}{27} \div 16$

77. $\frac{a}{b} \div \frac{2}{b}$ **78.** $\frac{3a}{b} \div \frac{3}{c}$

79. $\frac{x}{y} \div \frac{x}{y}$ **80.** $\frac{x}{3y} \div \frac{x}{3}$

ADDING AND SUBTRACTING FRACTIONS (LIKE DENOMINATORS)

Exercises 81–84: Add or subtract. Write each answer in lowest terms.

81. (a) $\frac{5}{12} + \frac{1}{12}$ (b) $\frac{5}{12} - \frac{1}{12}$

82. (a) $\frac{3}{2} + \frac{1}{2}$ (b) $\frac{3}{2} - \frac{1}{2}$

83. (a) $\frac{18}{29} + \frac{7}{29}$ (b) $\frac{18}{29} - \frac{7}{29}$

84. (a) $\frac{5}{33} + \frac{2}{33}$ (b) $\frac{5}{33} - \frac{2}{33}$

ADDING AND SUBTRACTING FRACTIONS (UNLIKE DENOMINATORS)

Exercises 85–94: Find the least common denominator.

85. $\frac{4}{9}, \frac{2}{15}$ **86.** $\frac{1}{11}, \frac{1}{2}$

87. $\frac{2}{5}, \frac{3}{15}$ **88.** $\frac{8}{21}, \frac{3}{7}$

89. $\frac{1}{6}, \frac{5}{8}$ **90.** $\frac{1}{9}, \frac{5}{12}$

91. $\frac{1}{2}, \frac{1}{3}, \frac{1}{4}$ **92.** $\frac{2}{5}, \frac{2}{3}, \frac{1}{6}$

93. $\frac{1}{4}, \frac{3}{8}, \frac{1}{12}$ **94.** $\frac{2}{15}, \frac{7}{20}, \frac{1}{30}$

Exercises 95–102: Rewrite each set of fractions with the least common denominator.

95. $\frac{1}{2}, \frac{2}{3}$ **96.** $\frac{3}{4}, \frac{1}{5}$

97. $\frac{7}{9}, \frac{5}{12}$ **98.** $\frac{5}{13}, \frac{1}{2}$

99. $\frac{1}{16}, \frac{7}{12}$ **100.** $\frac{5}{18}, \frac{1}{24}$

101. $\frac{1}{3}, \frac{3}{4}, \frac{5}{6}$ **102.** $\frac{4}{15}, \frac{2}{9}, \frac{3}{5}$

Exercises 103–114: Add or subtract. Write your answer in lowest terms.

103. $\frac{5}{8} + \frac{3}{16}$ **104.** $\frac{1}{9} + \frac{2}{15}$

105. $\frac{25}{24} - \frac{7}{8}$ **106.** $\frac{4}{5} - \frac{1}{4}$

107. $\frac{11}{14} + \frac{2}{35}$ **108.** $\frac{7}{8} + \frac{4}{15}$

109. $\frac{5}{12} - \frac{1}{18}$ **110.** $\frac{9}{20} - \frac{7}{30}$

111. $\frac{3}{100} + \frac{1}{300} - \frac{1}{200}$ **112.** $\frac{43}{36} + \frac{4}{9} + \frac{1}{4}$

113. $\frac{7}{8} - \frac{1}{6} + \frac{5}{12}$ **114.** $\frac{9}{40} - \frac{3}{50} - \frac{1}{100}$

APPLICATIONS INVOLVING FRACTIONS

Exercises 115–116: **Binge Watching TV** *It is no longer a trend, but rather the mainstream, to watch several episodes of a single TV series in one sitting. About $\frac{3}{5}$ of people binge watch regularly, watching an average of 2.3 episodes per sitting, and $\frac{4}{5}$ of binge viewers say binge watching makes the show better. Use the above information to answer the following questions.*

115. In a sample of 1000 viewers, how many viewers binge watch?

116. In a sample of 10,000 binge viewers, how many viewers say that their binge watching makes their TV shows better?

117. **American Flag** According to Executive Order 10834, the length of an official American flag should be $1\frac{9}{10}$ times the width. If an official flag has a width that measures $2\frac{1}{2}$ feet, find its length.

118. **American Flag** The blue rectangle containing the stars on an American flag is called the *union*. On an official American flag, the width of the union should be $\frac{7}{13}$ of the width of the flag. If an official flag has a width of $32\frac{1}{2}$ inches, what is the width of the union?

119. **Carpentry** A board measuring $64\frac{5}{8}$ inches is cut in half. Find the length of each half.

120. **Cutting Rope** A rope measures $15\frac{1}{2}$ feet and needs to be cut into four equal parts. Find the length of each piece.

121. **Geometry** Find the area of the triangle shown with base $1\frac{2}{3}$ yards and height $\frac{3}{4}$ yard. (*Hint:* The area of a triangle equals half the product of its base and height.)

122. **Geometry** Find the area of the rectangle shown.

123. **Distance** Use the map to find the distance between Smalltown and Bigtown by traveling through Middletown.

124. **Distance** An athlete jogs $1\frac{3}{8}$ miles, $5\frac{3}{4}$ miles, and $3\frac{5}{8}$ miles. In all, how far does the athlete jog?

125. **Vegetarian Diets** About $\frac{4}{125}$ of U.S. adults follow a vegetarian diet. Within this part of the population, $\frac{1}{200}$ is vegan. What fraction of the adult U.S. population is vegan? (*Source: Vegetarian Times—2010.*)

126. **Streaming Music** Recently, about $\frac{6}{29}$ of all music industry revenues were from streaming. About $\frac{11}{72}$ of streaming revenues were from on-demand streaming. What fraction of music industry revenues were from on-demand streaming?

127. **Accidental Deaths** For the age group 15 to 24, motor vehicle accidents account for $\frac{31}{42}$ of all accidental deaths, and firearms account for $\frac{31}{1260}$ of all accidental deaths. What fraction of all accidental deaths do vehicle accidents *and* firearms account for? (*Source: National Safety Council.*)

128. **Illicit Drug Use** For the age group 18 to 25, the fraction of people who used illicit drugs during their lifetime is $\frac{3}{5}$, whereas the fraction who used illicit drugs during the past year is $\frac{7}{20}$. What fraction of this population has used illicit drugs but not during the past year? (*Source: Department of Health and Human Services.*)

WRITING ABOUT MATHEMATICS

129. Describe a problem in real life for which the solution could be found by multiplying 30 by $\frac{1}{4}$.

130. Describe a problem in real life for which the solution could be found by dividing $2\frac{1}{2}$ by $\frac{1}{3}$.

SECTIONS 1.1 and 1.2   **Checking Basic Concepts**

1. Classify each number as prime, composite, or neither. If a number is composite, write it as a product of primes.
   (a) 19 (b) 28 (c) 1 (d) 180

2. Evaluate $\frac{10}{(x + 2)}$ for $x = 3$.

3. Find y for $x = 5$ if $y = 6x$.

4. Translate the phrase "a number x plus five" into an algebraic expression.

5. Write a formula that converts F feet to I inches.

6. Find the greatest common factor.
   (a) 3, 18   (b) 40, 72

7. Simplify each fraction to lowest terms.
   (a) $\frac{25}{35}$ (b) $\frac{26}{39}$

8. Give the reciprocal of $\frac{4}{3}$.

9. Evaluate each expression. Write each answer in lowest terms.
   (a) $\frac{2}{3} \cdot \frac{3}{4}$ (b) $\frac{5}{6} \div \frac{10}{3}$
   (c) $\frac{3}{10} + \frac{1}{10}$ (d) $\frac{3}{4} - \frac{1}{6}$

10. A recipe needs $1\frac{2}{3}$ cups of flour. How much flour should be used if the recipe is doubled?

1.3   Exponents and Order of Operations

Objectives

1 **Evaluating Natural Number Exponents**

2 **Applying Order of Operations**

3 **Translating Words to Expressions**

NEW VOCABULARY

☐ Exponential expression
☐ Base
☐ Exponent

Area Equals 25 Square Inches

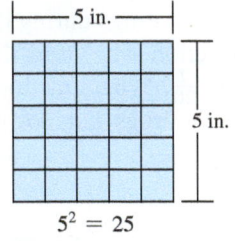

$5^2 = 25$

FIGURE 1.13

READING CHECK 1

• What type of expression represents a fast way to multiply?

1   Evaluating Natural Number Exponents

In elementary school, you learned addition. Later, you learned that multiplication is a fast way to add. For example, rather than adding

$$3 + 3 + 3 + 3, \quad \longleftarrow \text{4 terms}$$

we can multiply $3 \cdot 4$. Similarly, exponents represent a fast way to multiply. Rather than multiplying

$$3 \cdot 3 \cdot 3 \cdot 3, \quad \longleftarrow \text{4 factors}$$

we can evaluate the *exponential expression* 3^4. We begin by discussing natural numbers as exponents.

The area of a square equals the length of one of its sides times itself. If the square is 5 inches on a side, then its area is

$$5 \cdot 5 = 5^2 = 25 \text{ square inches.}$$

with Exponent pointing to the 2 and Base pointing to the 5.

See **FIGURE 1.13**. The expression 5^2 is an **exponential expression** with *base* 5 and *exponent* 2. Exponential expressions occur in a variety of applications. For example, suppose an investment doubles 3 times. Then, the calculation

$$2 \cdot 2 \cdot 2 = 2^3 = 8$$

with Exponent 3 pointing to the exponent and **3** Factors of **2** bracketing $2 \cdot 2 \cdot 2$.

shows that the final value is 8 times as large as the original investment. For example, if $10 doubles 3 times, it becomes $20, $40, and finally $80, which is 8 times the original investment of $10. **TABLE 1.4** contains examples of exponential expressions.

Exponential Expressions

Repeated Multiplication	Exponential Expression	Base	Exponent
$2 \cdot 2 \cdot 2 \cdot 2$	2^4	2	4
$4 \cdot 4 \cdot 4$	4^3	4	3
$9 \cdot 9$	9^2	9	2
$\frac{1}{2}$	$\left(\frac{1}{2}\right)^1$	$\frac{1}{2}$	1
$b \cdot b \cdot b \cdot b \cdot b$	b^5	b	5

TABLE 1.4

Read 9^2 as "9 squared," 4^3 as "4 cubed," and 2^4 as "2 to the fourth power." The terms *squared* and *cubed* come from geometry as illustrated in the following See the Concept.

See the Concept 1   **SQUARING AND CUBING**

3 SQUARED

The area is 9 *square* units.

$$3 \cdot 3 = 3^2 = 9$$

Three *squared*

4 CUBED

The volume is 64 *cubic* units.

$$4 \cdot 4 \cdot 4 = 4^3 = 64$$

Four *cubed*

WATCH VIDEO IN MML .

NOTE: The expressions 3^2 and 4^3 can also be read as "3 to the second power" and "4 to the third power," respectively. In general, the expression x^n is read as "x to the nth power" or "the nth power of x." ■

EXPONENTIAL NOTATION

The expression b^n, where n is a natural number, means

$$b^n = \underbrace{b \cdot b \cdot b \cdot \cdots \cdot b}_{n \text{ factors}}.$$

The **base** is b and the **exponent** is n.

EXAMPLE 1   **Writing products in exponential notation**

Write each product as an exponential expression.

(a) $7 \cdot 7 \cdot 7 \cdot 7$ **(b)** $\dfrac{1}{4} \cdot \dfrac{1}{4} \cdot \dfrac{1}{4}$ **(c)** $x \cdot x \cdot x \cdot x \cdot x$

Solution
(a) Because there are **four** factors of **7**, the exponent is **4** and the base is **7**.
Thus $7 \cdot 7 \cdot 7 \cdot 7 = 7^4$.
(b) Because there are three factors of $\frac{1}{4}$, the exponent is 3 and the base is $\frac{1}{4}$.
Thus $\frac{1}{4} \cdot \frac{1}{4} \cdot \frac{1}{4} = \left(\frac{1}{4}\right)^3$.
(c) Because there are five factors of x, the exponent is 5 and the base is x.
Thus $x \cdot x \cdot x \cdot x \cdot x = x^5$.

Now Try Exercises 13, 15, 17

EXAMPLE 2   **Evaluating exponential notation**

Evaluate each expression.

(a) 3^4 **(b)** 10^3 **(c)** $\left(\dfrac{3}{4}\right)^2$

Solution
(a) The exponential expression 3^4 indicates that 3 is to be multiplied times itself 4 times.

$$3^4 = \underbrace{3 \cdot 3 \cdot 3 \cdot 3}_{4 \text{ factors}} = 81$$

CALCULATOR HELP

To evaluate an exponential expression with a calculator, see Appendix A (page AP-1).

(b) $10^3 = 10 \cdot 10 \cdot 10 = 1000$ **(c)** $\left(\frac{3}{4}\right)^2 = \frac{3}{4} \cdot \frac{3}{4} = \frac{9}{16}$

Now Try Exercises 25, 27

EXAMPLE 3

Writing numbers in exponential notation

Use the given base to write each number as an exponential expression. Check your results with a calculator, if one is available.

(a) 100 (base 10)   **(b)** 16 (base 2)   **(c)** 27 (base 3)

Solution

(a) $100 = 10 \cdot 10 = 10^2$

(b) $16 = 4 \cdot 4 = 2 \cdot 2 \cdot 2 \cdot 2 = 2^4$

(c) $27 = 3 \cdot 9 = 3 \cdot 3 \cdot 3 = 3^3$

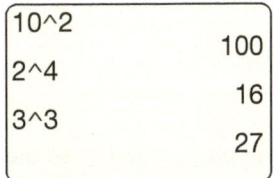

FIGURE 1.14

These values are supported in **FIGURE 1.14**, where exponential expressions are evaluated with a calculator by using the "^" key. (Note that some calculators may have a different key for evaluating exponential expressions.)

Now Try Exercises 29, 33, 35

Math in Context (Computer) Computer memory is often measured in bytes, with each *byte* capable of storing one letter of the alphabet. For example, it takes four bytes to store the word "math" in a computer. Bytes of computer memory are often manufactured in amounts equal to powers of 2, as illustrated in the next example.

EXAMPLE 4

Analyzing computer memory

In computer technology, 1 K (kilobyte) of memory equals 2^{10} bytes, and 1 MB (megabyte) of memory equals 2^{20} bytes. Determine whether 1 K of memory equals exactly one thousand bytes and whether 1 MB equals exactly one million bytes.

Solution

FIGURE 1.15 shows that $2^{10} = 1024$ and $2^{20} = 1{,}048{,}576$. Thus 1 K represents slightly more than one thousand bytes, and 1 MB represents more than one million bytes.

Now Try Exercise 75

FIGURE 1.15

CRITICAL THINKING 1

One gigabyte of memory equals 2^{30} bytes and is often referred to as 1 billion bytes. If you have a calculator available, determine whether 1 gigabyte is exactly 1 billion bytes.

2 Applying Order of Operations

Connecting Concepts with Your Life If there are 15 energy drinks on a store shelf and 3 boxes of 24 drinks each in the storage room, then the total number of energy drinks is given by the expression $15 + 3 \cdot 24$. How would you find the total number of energy drinks? Is the total

$$15 + 3 \cdot 24 \stackrel{?}{=} 18 \cdot 24 \stackrel{?}{=} 432 \quad \text{or} \quad 15 + 3 \cdot 24 \stackrel{?}{=} 15 + 72 \stackrel{?}{=} 87?$$

(add first, then multiply)   (multiply first, then add)

FIGURE 1.16

In this section, we discuss the *order of operations agreement*, which can be used to show that the expression $15 + 3 \cdot 24$ evaluates to 87. See **FIGURE 1.16**.

Because arithmetic expressions may contain parentheses, exponents, absolute values, and several operations, it is important to evaluate these expressions consistently. (Absolute value will be discussed in the fourth section of this chapter.) To ensure that we all obtain the same result when evaluating an arithmetic expression, the following rules are used.

STUDY TIP

Ideas and procedures written in boxes like the one to the right are major concepts. Be sure that you understand them.

ORDER OF OPERATIONS

Use the following order of operations. First perform all calculations within parentheses and absolute values, or above and below the fraction bar.

1. Evaluate all exponential expressions.
2. Do all multiplication and division from *left to right*.
3. Do all addition and subtraction from *left to right*.

EXAMPLE 5 **Evaluating arithmetic expressions**

Evaluate each expression by hand.

(a) $10 - 4 - 3$ (b) $10 - (4 - 3)$ (c) $5 + \dfrac{12}{3}$ (d) $\dfrac{4+1}{2+8}$

Solution

(a) There are no parentheses, so we evaluate subtraction from *left to right*.

$$10 - 4 - 3 = 6 - 3 = 3$$

(b) Note the similarity between this part and part (a). The difference is the parentheses, so subtraction inside the parentheses must be performed first.

$$10 - (4 - 3) = 10 - 1 = 9$$

(c) We perform division before addition.

$$5 + \frac{12}{3} = 5 + 4 = 9$$

(d) Evaluate the expression as though both the numerator and the denominator have parentheses around them.

$$\frac{4+1}{2+8} = \frac{(4+1)}{(2+8)} = \frac{5}{10} = \frac{1}{2}$$

Now Try Exercises 43, 45, 47, 57

EXAMPLE 6 **Evaluating arithmetic expressions**

Evaluate each expression by hand.

(a) $25 - 4 \cdot 6$ (b) $6 + 7 \cdot 2 - (4 - 1)$ (c) $\dfrac{3 + 3^2}{14 - 2}$ (d) $5 \cdot 2^3 - (3 + 2)$

Solution

(a) Multiplication is performed before subtraction, so evaluate the expression as follows.

$$25 - 4 \cdot 6 = 25 - 24 \qquad \text{Multiply.}$$
$$= 1 \qquad \text{Subtract.}$$

(b) Start by performing the subtraction within the parentheses first and then perform the multiplication. Finally, perform the addition and subtraction from left to right.

$$6 + 7 \cdot 2 - (4 - 1) = 6 + 7 \cdot 2 - 3 \quad \text{Subtract within parentheses.}$$
$$= 6 + 14 - 3 \quad \text{Multiply.}$$
$$= 20 - 3 \quad \text{Add.}$$
$$= 17 \quad \text{Subtract.}$$

(c) First note that parentheses are implied around the numerator and denominator.

$$\frac{3 + 3^2}{14 - 2} = \frac{(3 + 3^2)}{(14 - 2)} \qquad \text{Insert parentheses.}$$

$$= \frac{(3 + 9)}{(14 - 2)} \qquad \text{Evaluate the exponent first.}$$

$$= \frac{12}{12} \qquad \text{Add and subtract.}$$

$$= 1 \qquad \text{Simplify.}$$

(d) Begin by evaluating the expression inside parentheses.

$$5 \cdot 2^3 - (3 + 2) = 5 \cdot 2^3 - 5 \qquad \text{Add within parentheses.}$$

$$= 5 \cdot 8 - 5 \qquad \text{Evaluate the exponent.}$$

$$= 40 - 5 \qquad \text{Multiply.}$$

$$= 35 \qquad \text{Subtract.}$$

Now Try Exercises 41, 59, 61, 63

3 Translating Words to Expressions

Sometimes before we can solve a problem we must translate words into mathematical expressions. For example, if a prepaid cell phone plan allows for 500 minutes of call time each month and 376 minutes have already been used, then "five hundred minus three hundred seventy-six," or $500 - 376 = 124$, is the number of minutes remaining for the month.

EXAMPLE 7 **Writing and evaluating expressions**

Translate each phrase into a mathematical expression and then evaluate it.
(a) Two to the fourth power plus ten
(b) Twenty decreased by five times three
(c) Ten cubed divided by five squared
(d) Sixty divided by the quantity ten minus six

Solution
(a) $2^4 + 10 = 2 \cdot 2 \cdot 2 \cdot 2 + 10 = 16 + 10 = 26$
(b) $20 - 5 \cdot 3 = 20 - 15 = 5$
(c) $\dfrac{10^3}{5^2} = \dfrac{1000}{25} = 40$
(d) Here, the word "quantity" indicates that parentheses should be used.

$$60 \div (10 - 6) = 60 \div 4 = 15$$

Now Try Exercises 65, 67, 69, 73

1.3 Putting It All Together

CONCEPT	COMMENTS	EXAMPLES
Exponential Expression	If n is a natural number, then b^n equals $\underbrace{b \cdot b \cdot b \cdot \cdots \cdot b}_{n \text{ factors}}$ and is read "b to the nth power."	$5^1 = 5$, $7^2 = 7 \cdot 7 = 49$, $4^3 = 4 \cdot 4 \cdot 4 = 64$, and $k^4 = k \cdot k \cdot k \cdot k$

CONCEPT	COMMENTS	EXAMPLES
Base and Exponent	The base in b^n is b and the exponent is n.	7^4 has base 7 and exponent 4, and x^3 has base x and exponent 3.
Order of Operations	First, perform all calculations within parentheses and absolute values, or above and below a fraction bar. **1.** Evaluate all exponential expressions. **2.** Do all multiplication and division from *left to right*. **3.** Do all addition and subtraction from *left to right*.	$10 \div 5 + 3 \cdot 7 = 2 + 21$ $= 23$ $27 \div (4 - 1)^2 = 27 \div 3^2$ $= 27 \div 9$ $= 3$

1.3   Exercises

MyMathLab®

CONCEPTS AND VOCABULARY

1. Exponents represent a fast way to _____.

2. In the expression 2^5, there are five factors of _____ being multiplied.

3. In the expression 5^3, the number 5 is called the _____ and the number 3 is called the _____.

4. Use symbols to write "6 squared."

5. Use symbols to write "8 cubed."

6. When evaluating the expression $5 + 6 \cdot 2$, the result is _____ because _____ is performed before _____.

7. When evaluating the expression $10 - 2^3$, the result is _____ because _____ are evaluated before _____ is performed.

8. The expression $10 - 4 - 2$ equals _____ because subtraction is performed from _____ to _____.

9. (True or False?) The expressions 2^3 and 3^2 are equal.

10. (True or False?) The expression 5^2 equals $5 \cdot 2$.

EVALUATING NATURAL NUMBER EXPONENTS

Exercises 11–20: Write the product as an exponential expression.

11. $3 \cdot 3 \cdot 3 \cdot 3$

12. $10 \cdot 10$

13. $2 \cdot 2 \cdot 2 \cdot 2 \cdot 2$

14. $4 \cdot 4 \cdot 4$

15. $\frac{1}{2} \cdot \frac{1}{2} \cdot \frac{1}{2} \cdot \frac{1}{2}$

16. $\frac{5}{7} \cdot \frac{5}{7} \cdot \frac{5}{7} \cdot \frac{5}{7} \cdot \frac{5}{7}$

17. $a \cdot a \cdot a \cdot a \cdot a$

18. $b \cdot b \cdot b \cdot b$

19. $(x + 3) \cdot (x + 3)$

20. $(x - 4) \cdot (x - 4) \cdot (x - 4)$

Exercises 21–28: Evaluate each expression.

21. (a) 2^4 (b) 4^2

22. (a) 3^2 (b) 5^3

23. (a) 6^1 (b) 1^6

24. (a) 17^1 (b) 1^{17}

25. (a) 2^5 (b) 10^3

26. (a) 10^5 (b) 3^4

27. (a) $\left(\frac{2}{3}\right)^2$ (b) $\left(\frac{1}{2}\right)^5$

28. (a) $\left(\frac{1}{10}\right)^3$ (b) $\left(\frac{4}{3}\right)^1$

Exercises 29–40: (Refer to Example 3.) Use the given base to write the number as an exponential expression. Check your result if you have a calculator available.

29. 8 (base 2)

30. 9 (base 3)

31. 25 (base 5)

32. 32 (base 2)

33. 49 (base 7)

34. 81 (base 3)

35. 1000 (base 10)

36. 256 (base 4)

37. $\frac{1}{16}$ $\left(\text{base } \frac{1}{2}\right)$

38. $\frac{9}{25}$ $\left(\text{base } \frac{3}{5}\right)$

39. $\frac{32}{243}$ $\left(\text{base } \frac{2}{3}\right)$

40. $\frac{216}{343}$ $\left(\text{base } \frac{6}{7}\right)$

APPLYING ORDER OF OPERATIONS

Exercises 41–64: Evaluate the expression by hand.

41. $5 + 4 \cdot 6$

42. $6 \cdot 7 - 8$

43. $6 \div 3 + 2$

44. $20 - 10 \div 5$

45. $100 - \frac{50}{5}$

46. $\frac{200}{100} + 6$

47. $10 - 6 - 1$

48. $30 - 9 - 5$

49. $20 \div 5 \div 2$

50. $500 \div 100 \div 5$

51. $3 + 2^4$

52. $10 - 3^2 + 1$

53. $4 \cdot 2^3$

54. $100 - 2 \cdot 3^3$

55. $(3 + 2)^3$

56. $5 \cdot (3 - 2)^8 - 5$

57. $\frac{4 + 8}{1 + 3}$

58. $5 - \frac{3 + 1}{3 - 1}$

59. $\frac{2^3}{4 - 2}$

60. $\frac{10 - 3^2}{2 \cdot 4^2}$

61. $10^2 - (30 - 2 \cdot 5)$

62. $5^2 + 3 \cdot 5 \div 3 - 1$

63. $\left(\frac{1}{2}\right)^4 + \frac{5 + 4}{3}$

64. $\left(\frac{7}{9}\right)^2 - \frac{6 - 5}{3}$

TRANSLATING WORDS TO EXPRESSIONS

Exercises 65–74: Translate the phrase into a mathematical expression and then evaluate it.

65. Two cubed minus eight

66. Five squared plus nine

67. Thirty decreased by four times three

68. One hundred plus five times six

69. Four squared divided by two cubed

70. Three cubed times two squared

71. Forty divided by ten, plus two

72. Thirty times ten, minus three

73. One hundred times the quantity two plus three

74. Fifty divided by the quantity eight plus two

APPLICATIONS INVOLVING EXPONENTS

75. Email Attachments (Refer to Example 4.) In Gmail, users can send attachments up to 25MB. Determine the number of bytes in an attachment of this size.

76. iPhone Memory Determine the number of bytes in a 64GB iPhone 6s option. (*Hint:* One gigabyte equals 2^{30} bytes.)

77. Population by Gender One way to measure the gender balance in a given population is to find the number of males for every 100 females in the population. In 1900, the western region of the United States was significantly out of gender balance. In this region, there were 128 males for every 100 females. (*Source:* U.S. Census Bureau.)

(a) Find an exponent k so that $2^k = 128$.

(b) During this time, how many males were there for every 25 females?

78. Solar Eclipse In early December 2048, there will be a total solar eclipse visible in parts of Botswana. Find an exponent k so that $2^k = 2048$. (*Source:* NASA.)

79. Rule of 72 Investors sometimes use the *rule of 72* to determine the time required to double an investment. If 72 is divided by the annual interest rate earned on an investment, the result approximates the number of years needed to double the investment. For example, an investment earning 6% annual interest will double in value approximately every $72 \div 6 = 12$ years.

(a) Approximate the number of years required to double an investment earning 9% annual interest.

(b) If an investment of $10,000 earns 12% annual interest, approximate the value of the investment after 18 years.

80. Doubling Effect Suppose that a savings account containing $1000 doubles its value every 7 years. How much money will be in the account after 28 years?

WRITING ABOUT MATHEMATICS

81. Explain how exponential expressions are related to multiplication. Give an example.

82. Explain why agreement on the order of operations is necessary.

Group Activity

1. Arithmetic Operations Insert one of the symbols $+$, $-$, $\cdot$, or $\div$ in each blank to obtain the given answer. Do not use any parentheses.

$$2 \_\_ 2 \_\_ 2 \_\_ 2 = 0$$
$$3 \_\_ 3 \_\_ 3 \_\_ 3 = 10$$
$$4 \_\_ 4 \_\_ 4 \_\_ 4 = 1$$
$$6 \_\_ 6 \_\_ 6 \_\_ 6 = 36$$
$$7 \_\_ 7 \_\_ 7 \_\_ 7 = 63$$

2. Magic Squares The following square is called a "magic square" because it contains each of the numbers from 1 to 9 and the numbers in each row, column, and diagonal sum to 15.

8	3	4
1	5	9
6	7	2

Complete the following magic square having 4 rows and 4 columns by arranging the numbers 1 through 16 so that each row, column, and diagonal sums to 34. The four corners will also sum to 34.

	2		13
5			
		6	
4			1

1.4 Real Numbers and the Number Line

Objectives

1 Identifying Signed Numbers and Opposites
2 Classifying Integers and Rational Numbers
3 Calculating Square Roots
4 Classifying Real Numbers and Irrational Numbers
5 Graphing on the Number Line
6 Finding Absolute Values
7 Determining Inequality

NEW VOCABULARY

☐ Opposite (additive inverse)
☐ Negative sign
☐ Integers
☐ Rational number
☐ Square root
☐ Principal square root
☐ Real number
☐ Irrational numbers
☐ Approximately equal
☐ Average
☐ Origin
☐ Absolute value
☐ Less than/greater than
☐ Less than or equal to
☐ Greater than or equal to

1 Identifying Signed Numbers and Opposites

Connecting Concepts with Your Life The idea that numbers could be negative was a difficult concept for many mathematicians. As late as the eighteenth century, negative numbers were not readily accepted by everyone. After all, how could a person have -5 oranges?

However, negative numbers make more sense to someone working with money. If you owe someone 100 dollars (a debt), this amount can be thought of as -100, whereas if you have a balance of 100 dollars in your checking account (an asset), this amount can be thought of as $+100$. (The positive sign is usually omitted.)

The **opposite**, or **additive inverse**, of a number a is $-a$. That is, a **negative sign** is used to form the opposite of a. For example, the opposite of 25 is -25, the opposite of -5 is $-(-5)$, or 5, and the opposite of 0 is 0 because 0 is neither positive nor negative. The following *double negative rule* is helpful in simplifying expressions containing negative signs.

DOUBLE NEGATIVE RULE

Let a be any number. Then $-(-a) = a$.

Thus $-(-8) = 8$ and $-\left(-(-10)\right) = -(10) = -10$.

READING CHECK 1

• What is the opposite of x?

EXAMPLE 1 **Finding opposites (or additive inverses)**

Find the opposite of each expression.

(a) 13   **(b)** $-\dfrac{4}{7}$ **(c)** $-(-7)$

Solution

(a) The opposite of 13 is -13.
(b) The opposite of $-\frac{4}{7}$ is $\frac{4}{7}$.
(c) $-(-7) = 7$, so the opposite of $-(-7)$ is -7.

Now Try Exercises 17, 19, 21

NOTE: To find the opposite of an exponential expression, *evaluate the exponent first*. For example, the opposite of 2^4 is

Evaluate the exponent first.

$$-2^4 = -(2 \cdot 2 \cdot 2 \cdot 2) = -16. \ \blacksquare$$

EXAMPLE 2 **Finding an additive inverse (or opposite)**

Find the additive inverse of $-t$, if $t = -\frac{2}{3}$.

Solution

Because $-(-t) = t$ by the double negative rule, the additive inverse of $-t$ is $t = -\frac{2}{3}$.

Now Try Exercise 25

2 Classifying Integers and Rational Numbers

In the opening section of this chapter, we discussed natural numbers and whole numbers. Because these sets of numbers do not include negative numbers, fractions, or decimals, other sets of numbers are needed. The **integers** include the natural numbers, zero, and the opposites of the natural numbers. The integers are given by the following.

$$\ldots, -3, -2, -1, 0, 1, 2, 3, \ldots \quad \boxed{\text{Integers}}$$

A **rational number** is any number that can be expressed as the ratio of two integers, $\frac{p}{q}$, where $q \neq 0$. Rational numbers can be written as fractions, and they include all integers. Rational numbers may be positive, negative, or zero. Some examples of rational numbers are

$$\frac{2}{3}, \quad -\frac{3}{5}, \quad \frac{-7}{2}, \quad 1.2, \quad \text{and} \quad 3. \quad \boxed{\text{Rational numbers}}$$

STUDY TIP

The word "NOTE" is used to draw attention to important concepts that may otherwise be overlooked.

The numbers 1.2 and 3 are both rational numbers because they can be written as $\frac{12}{10}$ and $\frac{3}{1}$.

NOTE: The fraction $\frac{-7}{2}$ can also be written as $\frac{7}{-2}$ and $-\frac{7}{2}$. The position of the negative sign does not affect the value of the fraction. $\blacksquare$

The fraction bar can be thought of as a division symbol. As a result, rational numbers have decimal equivalents. For example, $\frac{1}{2}$ is equivalent to $1 \div 2$. The division

$$\begin{array}{r} 0.5 \\ 2\overline{)1.0} \end{array}$$

READING CHECK 2

• When a number is written in decimal form, how do we know if the number is a rational number?

shows that $\frac{1}{2} = 0.5$. In general, a rational number may be expressed in a decimal form that either *repeats* or *terminates*. The fraction $\frac{1}{3}$ may be expressed as $0.\overline{3}$, a repeating decimal, and the fraction $\frac{1}{4}$ may be expressed as 0.25, a terminating decimal. The overbar indicates that $0.\overline{3} = 0.3333333\ldots$.

🌐 *Math in Context* (Population) Integers and rational numbers are used to describe quantities such as change in population. **FIGURE 1.17** shows the change in population from 2000 to 2012 for selected U.S. cities. Note that both positive and negative numbers are used to describe these population changes. (*Source:* U.S. Census Bureau.)

Population Change 2000–2012

FIGURE 1.17

| EXAMPLE 3 | **Classifying numbers** |

Classify each number as one or more of the following: natural number, whole number, integer, or rational number.

(a) $\dfrac{12}{4}$ (b) -3 (c) 0 (d) $-\dfrac{9}{5}$

Solution
(a) Because $\frac{12}{4} = 3$, the number $\frac{12}{4}$ is a natural number, whole number, integer, and rational number.
(b) The number -3 is an integer and a rational number, but it is not a natural number or a whole number.
(c) The number 0 is a whole number, integer, and rational number but not a natural number.
(d) The fraction $-\frac{9}{5}$ is a rational number because it is the ratio of two integers. It is not a natural number, a whole number, or an integer.

Now Try Exercises 47, 49, 51, 53

3 Calculating Square Roots

The number b is a **square root** of a number a if $b \cdot b = a$. Every *positive* number has one positive square root and one negative square root. For example, the positive square root of 9 is 3 because $3 \cdot 3 = 9$. The negative square root of 9 is -3. (We will show that $-3 \cdot (-3) = 9$ in Section 6 of this Chapter.) If a is a positive number, then the **principal square root** of a, denoted $\sqrt{a}$, is the positive square root of a. For example, $\sqrt{25} = 5$ because $5 \cdot 5 = 25$ and the number 5 is positive. Note that the principal square root of 0 is 0. That is, $\sqrt{0} = 0$.

| EXAMPLE 4 | **Calculating principal square roots** |

Evaluate each square root. Approximate your answer to three decimal places as needed.
(a) $\sqrt{36}$ (b) $\sqrt{100}$ (c) $\sqrt{5}$

CALCULATOR HELP

To evaluate square roots with a calculator, see Appendix A (page AP-1).

Solution
(a) $\sqrt{36} = 6$ because $6 \cdot 6 = 36$ and 6 is positive.
(b) $\sqrt{100} = 10$ because $10 \cdot 10 = 100$ and 10 is positive.
(c) We can estimate the value of $\sqrt{5}$ with a calculator. **FIGURE 1.18** reveals that $\sqrt{5}$ is *approximately* equal to 2.236. However, 2.236 does not exactly equal $\sqrt{5}$ because $2.236 \cdot 2.236 = 4.999696$, which does *not* equal 5.

Now Try Exercises 41, 43, 45

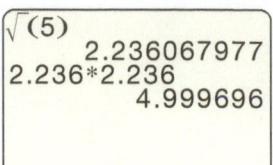

```
√(5)
              2.236067977
2.236*2.236
              4.999696
```

FIGURE 1.18

4 Classifying Real Numbers and Irrational Numbers

If a number can be represented by a decimal number, then it is a **real number**. Every fraction has a decimal form, so real numbers include all rational numbers. However, some real numbers cannot be expressed by fractions. They are called **irrational numbers**. The numbers $\sqrt{2}$, $\sqrt{15}$, and π are examples of irrational numbers. Every irrational number has a decimal representation that does not terminate or repeat.

NOTE: For any positive integer a, if $\sqrt{a}$ is not an integer, then $\sqrt{a}$ is an irrational number. So $\sqrt{3}$, $\sqrt{12}$, and $\sqrt{21}$ are all irrational numbers, whereas $\sqrt{4}$, $\sqrt{25}$, and $\sqrt{100}$ are not. ∎

Examples of real numbers include

Exact Real Numbers

$$-17, \quad \frac{4}{5}, \quad -\sqrt{3}, \quad 21\frac{1}{2}, \quad 57.63, \quad \text{and} \quad \sqrt{7}.$$

Any real number may be *approximated* by a terminating decimal. The symbol $\approx$ is used to represent **approximately equal**. Each of the following real numbers has been approximated to two *decimal places*.

Approximate Real Numbers

$$\frac{1}{7} \approx 0.14, \quad 2\pi \approx 6.28, \quad \text{and} \quad \sqrt{60} \approx 7.75$$

FIGURE 1.19 shows the relationships among the different sets of numbers. Note that each real number is either a rational number or an irrational number but not both. All natural numbers, whole numbers, and integers are rational numbers.

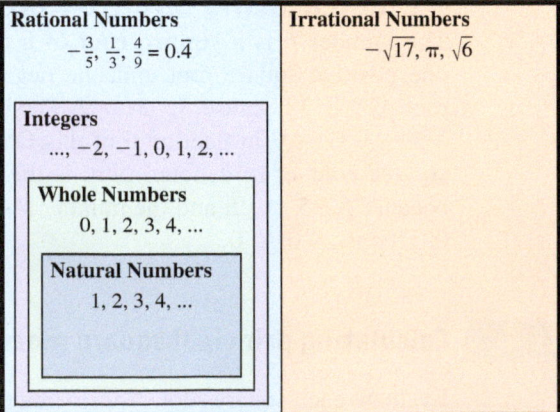

Real Numbers

Rational Numbers	**Irrational Numbers**
$-\frac{3}{5}, \frac{1}{3}, \frac{4}{9} = 0.\overline{4}$	$-\sqrt{17}, \pi, \sqrt{6}$

Integers
..., −2, −1, 0, 1, 2, ...

Whole Numbers
0, 1, 2, 3, 4, ...

Natural Numbers
1, 2, 3, 4, ...

FIGURE 1.19

> **MAKING CONNECTIONS 1**
>
> **Rational and Irrational Numbers**
>
> Both rational and irrational numbers can be written as decimals. However, rational numbers can be represented by either terminating or repeating decimals. For example, $\frac{1}{2} = 0.5$ is a terminating decimal and $\frac{1}{3} = 0.333\ldots$ is a repeating decimal. Irrational numbers are represented by decimals that *neither* terminate *nor* repeat.

EXAMPLE 5   **Classifying numbers**

Identify the natural numbers, whole numbers, integers, rational numbers, and irrational numbers in the following list.

$$-\sqrt{5}, \quad 9, \quad -3.8, \quad \sqrt{49}, \quad \frac{11}{4}, \quad -41$$

Solution

Natural numbers: 9 and $\sqrt{49} = 7$
Whole numbers: 9 and $\sqrt{49} = 7$
Integers: 9, $\sqrt{49} = 7$, and -41
Rational numbers: 9, -3.8, $\sqrt{49} = 7$, $\frac{11}{4}$, and -41
Irrational number: $-\sqrt{5}$

Now Try Exercises 55, 59, 63

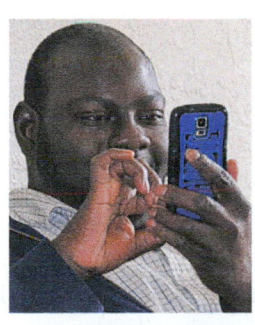

🌐 *Math in Context* (Cell Phone) Even though a data set may contain only integers, we often need decimals to describe it. For example, positive integers are used to represent the number of mobile subscribers in the United States for various years. However, to describe the *average* number of subscribers over a longer time period, decimals may be helpful. Recall that the **average** of a set of numbers is found by adding the numbers and then dividing by how many numbers there are in the set.

EXAMPLE 6   **Analyzing cell phone subscriber data**

TABLE 1.5 lists the number of mobile subscribers, in millions, in the United States, for various years. Find the average number of subscribers for these years. Is the result (in millions) a natural number, a rational number, or an irrational number?

U.S. Mobile Subscribers (millions)

Year	2000	2005	2010	2015
Subscribers	97	194	292	468

TABLE 1.5 *Source: CTIA.*

Solution

The average number of subscribers is

$$\frac{97 + 194 + 292 + 468}{4} = \frac{1051}{4} = 262.75 \text{ million.}$$

The average of these four natural numbers is an integer divided by an integer, which is a rational number. However, 262.75 is neither a natural number nor an irrational number.

Now Try Exercise 105

CRITICAL THINKING 1

Give an example in which the sum of two irrational numbers is a rational number.

5 Graphing on the Number Line

The real numbers can be represented *visually* by using a number line, as shown in **FIGURE 1.20**. Each real number corresponds to a unique point on the number line. The point associated with the real number 0 is called the **origin**. The positive integers are equally spaced to the right of the origin, and the negative integers are equally spaced to the left of the origin. The number line extends indefinitely both to the left and to the right.

A Number Line

FIGURE 1.20

Other real numbers can also be located on the number line. For example, the number $\frac{1}{2}$ can be identified by placing a dot halfway between the integers 0 and 1. The numbers $-\sqrt{2} \approx -1.41$ and $\frac{5}{4} = 1.25$ can also be placed (approximately) on this number line. (See **FIGURE 1.20**.)

EXAMPLE 7   **Plotting numbers on a number line**

Plot each real number on a number line.

(a) $-\dfrac{3}{2}$ **(b)** $\sqrt{3}$ **(c)** π

Plotting Numbers

FIGURE 1.21

Solution

(a) $-\frac{3}{2} = -1.5$. Place a dot halfway between -2 and -1, as shown in **FIGURE 1.21**.

(b) A calculator gives $\sqrt{3} \approx 1.73$. Place a dot between 1 and 2 so that it is about three-fourths of the way toward 2, as shown in **FIGURE 1.21**.

(c) $\pi \approx 3.14$. Place a dot just past the integer 3, as shown in **FIGURE 1.21**.

Now Try Exercises 69, 73

The following See the Concept shows how to graph the decimal 3.68 on a number line with increasing accuracy.

See the Concept 1   **GRAPHING DECIMALS**

Graph 3.68 on a number line.

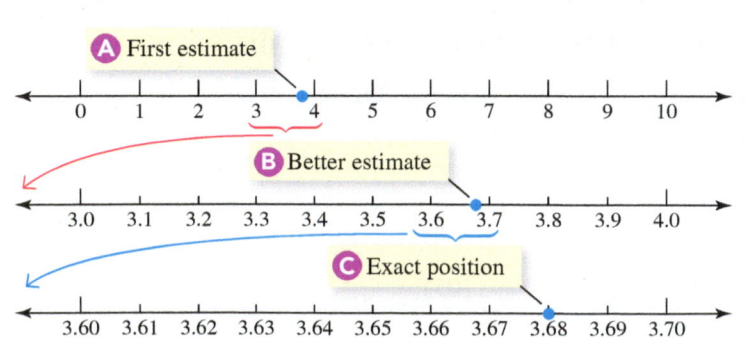

Ⓐ Estimate the position of the decimal between consecutive integers and place a dot on the number line.

Ⓑ Make a new number line with the space between the consecutive integers from Ⓐ divided into 10 equal parts, making tenths.

Ⓒ Make a new number line with the space between the consecutive tenths from Ⓑ divided into 10 equal parts, making hundredths.

Continue this process until the desired accuracy is obtained.

WATCH VIDEO IN MML .

$|-3| \quad = \quad |3|$

FIGURE 1.22

6 Finding Absolute Values

The **absolute value** of a real number equals its distance on the number line from the origin. Because distance is never negative, the absolute value of a real number is *never negative*. The absolute value of a real number a is denoted $|a|$ and is read "the absolute value of a." **FIGURE 1.22** shows that the absolute values of -3 and 3 are both equal to 3 because both have distance **3** from the origin. That is, $|-3| = $ **3** and $|3| = $ **3**.

EXAMPLE 8 **Finding the absolute value of a real number**

Evaluate each expression.
(a) $|3.1|$ **(b)** $|-7|$ **(c)** $|0|$

Solution
(a) $|3.1| = 3.1$ because the distance between the origin and 3.1 is 3.1.
(b) $|-7| = 7$ because the distance between the origin and -7 is 7.
(c) $|0| = 0$ because the distance is 0 between the origin and 0.

Now Try Exercises 75, 77, 79

Our results about absolute value can be summarized as follows.

$$|a| = a, \qquad \text{if } a \text{ is positive or 0.}$$

$$|a| = -a, \quad \text{if } a \text{ is negative.}$$

7 Determining Inequality

If a real number a is located to the left of a real number b on the number line, we say that a is **less than** b and write $a < b$. Similarly, if a real number b is located to the right of a real number a, we say that b is **greater than** a and write $b > a$. See **FIGURE 1.23**.

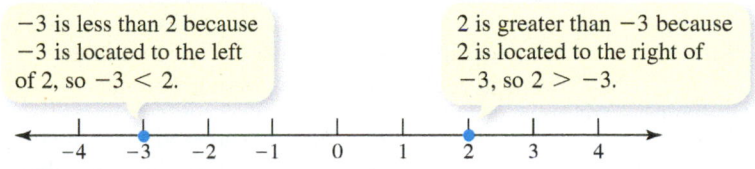

−3 is less than 2 because −3 is located to the left of 2, so −3 < 2.

2 is greater than −3 because 2 is located to the right of −3, so 2 > −3.

FIGURE 1.23

NOTE: Any negative number is always less than any positive number, and any positive number is always greater than any negative number. ■

We say that a is **less than or equal to** b, denoted $a \le b$, if either $a < b$ or $a = b$ is true. Similarly, a is **greater than or equal to** b, denoted $a \ge b$, if either $a > b$ or $a = b$ is true.

Inequalities are often used to compare the relative sizes of two quantities. This can be illustrated visually as shown in the figure below.

Visualizing Inequality by Relative Size

EXAMPLE 9   **Ordering real numbers**

List the given numbers from least to greatest. Then plot the numbers on a number line.

$$-2, \quad -\pi, \quad \sqrt{2}, \quad 0, \quad 2.5$$

Solution

First note that $-\pi \approx -3.14 < -2$. The two negative numbers are less than 0, and the two positive numbers are greater than 0. Also, $\sqrt{2} \approx 1.41$, so $\sqrt{2} < 2.5$. Listing the numbers from least to greatest results in

$$-\pi, \quad -2, \quad 0, \quad \sqrt{2}, \quad \text{and} \quad 2.5.$$

These numbers are plotted on the number line shown in **FIGURE 1.24**. Note that these numbers *increase from left to right* on the number line.

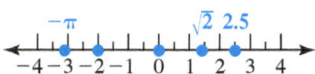

FIGURE 1.24

Now Try Exercise 103

1.4   Putting It All Together

CONCEPT	COMMENTS	EXAMPLES
Integers	Include the natural numbers, their opposites, and 0	$\ldots, -2, -1, 0, 1, 2, \ldots$
Rational Numbers	Include integers, all fractions $\frac{p}{q}$, where p and $q \neq 0$ are integers, and all repeating and terminating decimals	$\frac{1}{2}, -3, \frac{128}{6}, -0.335, 0, 0.25 = \frac{1}{4}$, and $0.\overline{3} = \frac{1}{3}$
Irrational Numbers	Any decimal numbers that neither terminate nor repeat; real numbers that are not rational	$\pi, \sqrt{3},$ and $\sqrt{15}$
Real Numbers	Any numbers that can be expressed in decimal form; include the rational and irrational numbers	$\pi, \sqrt{3}, -\frac{4}{7}, 0, -10, 0.\overline{6} = \frac{2}{3}, 1000$, and $\sqrt{15}$
Average	To find the average of a set of numbers, find the sum of the numbers and then divide the sum by how many numbers there are in the set.	The average of 2, 6, and 10 is $$\frac{2 + 6 + 10}{3} = \frac{18}{3} = 6.$$
Number Line	A number line can be used to visualize the real number system. The point associated with the number 0 is called the origin.	
Absolute Value	The absolute value of a number a equals its distance on the number line from the origin. If $a \geq 0$, then $|a| = a$. If $a < 0$, then $|a| = -a$. $|a|$ is *never* negative.	$|17| = 17$ $|-12| = 12$ $|0| = 0$
Inequality	If a number a is located to the left of a number b on a number line, then a is less than b (written $a < b$). If a is located to the right of b, then a is greater than b (written $a > b$). Symbols of inequality include: $<, >, \leq, \geq,$ and $\neq$.	$-3 < -2$ less than $6 > 4$ greater than $-5 \leq -5$ less than or equal $18 \geq 0$ greater than or equal $7 \neq 8$ not equal

1.4 Exercises

MyMathLab®

CONCEPTS AND VOCABULARY

1. The opposite of the number b is _____.

2. The integers include the natural numbers, zero, and the opposites of the _____ numbers.

3. A number that can be written as $\frac{p}{q}$, where p and q are integers with $q \neq 0$, is a(n) _____ number.

4. If a number can be written in decimal form, then it is a(n) _____ number.

5. If a real number is not a rational number, then it is a(n) _____ number.

6. (True or False?) A rational number can be written as a repeating or terminating decimal.

7. (True or False?) An irrational number cannot be written as a repeating or terminating decimal.

8. Write 0.272727... using an overbar.

9. The decimal equivalent for $\frac{1}{4}$ can be found by dividing _____ by _____.

10. The equation $4 \cdot 4 = 16$ indicates that $\sqrt{16} =$ _____.

11. The positive square root of a positive number is called the _____ square root.

12. The symbol $\neq$ is used to indicate that two numbers are _____ .

13. The symbol $\approx$ is used to indicate that two numbers are _____ .

14. The origin on the number line corresponds to the number _____.

15. Negative numbers are located to the (left/right) of the origin on the number line.

16. The absolute value of a number a gives its distance on the number line from the _____.

IDENTIFYING SIGNED NUMBERS AND OPPOSITES

Exercises 17–24: Find the opposite of each expression.

17. (a) 9   (b) -9

18. (a) -6 (b) 6

19. (a) $\frac{2}{3}$ (b) $-\frac{2}{3}$

20. (a) $-\left(\frac{-4}{5}\right)$ (b) $-\left(\frac{-4}{-5}\right)$

21. (a) $-(-8)$ (b) $-\left(-(-8)\right)$

22. (a) $-\left(-(-2)\right)$ (b) $-(-2)$

23. (a) a (b) $-a$

24. (a) $-b$ (b) $-(-b)$

25. Find the additive inverse of t, if $-t = 6$.

26. Find the additive inverse of $-t$, if $t = -\frac{4}{5}$.

27. Find the additive inverse of $-b$, if $b = \frac{1}{2}$.

28. Find the additive inverse of b, if $-b = \frac{5}{-6}$.

WRITING RATIONAL NUMBERS IN DECIMAL FORM

Exercises 29–40: Find the decimal equivalent for the rational number.

29. $\frac{1}{4}$ 30. $\frac{3}{5}$

31. $\frac{7}{8}$ 32. $\frac{3}{10}$

33. $\frac{3}{2}$ 34. $\frac{3}{50}$

35. $\frac{1}{20}$ 36. $\frac{3}{16}$

37. $\frac{2}{3}$ 38. $\frac{2}{9}$

39. $\frac{7}{9}$ 40. $\frac{5}{11}$

CALCULATING SQUARE ROOTS

Exercises 41–46: Evaluate the square root. Approximate your answer to three decimal places when appropriate.

41. $\sqrt{25}$ 42. $\sqrt{81}$

43. $\sqrt{49}$ 44. $\sqrt{64}$

45. $\sqrt{7}$ 46. $\sqrt{11}$

CLASSIFYING REAL NUMBERS

Exercises 47–54: Classify the number as one or more of the following: natural number, whole number, integer, or rational number.

47. 8   48. -8

49. $\frac{16}{4}$ 50. $\frac{5}{7}$

51. 0   52. $-\frac{15}{31}$

53. $-\frac{7}{6}$ 54. $-\frac{10}{5}$

Exercises 55–66: Classify the number as one or more of the following: natural number, integer, rational number, or irrational number.

55. -4.5

56. π

57. $\frac{3}{7}$

58. $\sqrt{25}$

59. $\sqrt{11}$

60. $-\sqrt{3}$

61. $\frac{8}{4}$

62. -5

63. $\sqrt{49}$

64. $3.\overline{3}$

65. $1.\overline{8}$

66. $\frac{9}{3}$

GRAPHING ON THE NUMBER LINE

Exercises 67–74: Plot each number on a number line.

67. (a) 0 (b) -2 (c) 3

68. (a) $-\frac{3}{2}$ (b) $\frac{3}{2}$ (c) 0

69. (a) $\frac{1}{2}$ (b) $-\frac{1}{2}$ (c) 2

70. (a) 1.3 (b) -2.5 (c) 0.7

71. (a) -10 (b) -20 (c) 30

72. (a) 5 (b) 10 (c) -10

73. (a) π (b) $\sqrt{2}$ (c) $-\sqrt{3}$

74. (a) $\sqrt{11}$ (b) $-\sqrt{5}$ (c) $\sqrt{4}$

FINDING ABSOLUTE VALUES

Exercises 75–82: Evaluate the expression.

75. $|5.23|$

76. $|\pi|$

77. $|-8|$

78. $|-\sqrt{2}|$

79. $|2 - 2|$

80. $|\frac{2}{3} - \frac{1}{3}|$

81. $|\pi - 3|$

82. $|3 - \pi|$

83. Thinking Generally Find $|b|$, if b is negative.

84. Thinking Generally Find $|-b|$, if b is positive.

DETERMINING INEQUALITY

Exercises 85–96: Insert the symbol $>$ or $<$ to make the statement true.

85. 5 _____ 7

86. -5 _____ 7

87. -5 _____ -7

88. $\frac{3}{5}$ _____ $\frac{2}{5}$

89. $-\frac{1}{3}$ _____ $-\frac{2}{3}$

90. $-\frac{1}{10}$ _____ 0

91. -1.9 _____ -1.3

92. 5.1 _____ -6.2

93. $|-8|$ _____ 3

94. 4 _____ $|-1|$

95. $|-2|$ _____ $|-7|$

96. $|-15|$ _____ $|32|$

Exercises 97–100: **Thinking Generally** *Insert the symbol $<$, $=$, or $>$ to make the statement true.*

97. If a number a is located to the right of a number b on the number line, then a _____ b.

98. If $b > 0$ and $a < 0$, then b _____ a.

99. If $a \geq b$, then either $a > b$ or a _____ b.

100. If $a \geq b$ and $b \geq a$, then a _____ b.

Exercises 101–104: List the given numbers from least to greatest.

101. $-3, 0, 1, -9, -2^3$

102. $4, -2^3, \frac{1}{2}, -\frac{3}{2}, \frac{3}{2}$

103. $-2, \pi, \frac{1}{3}, -\frac{3}{2}, \sqrt{5}$

104. $9, 14, -\frac{1}{12}, -\frac{3}{16}, \sqrt{7}$

APPLICATIONS

105. Music Sales   The table lists the digital share of the total music industry.

Year	2009	2010	2011	2012	2013
Digital Share (%)	40	46	51	59	64

Source: RIAA.

(a) What was this share in 2011?

(b) Calculate the average digital share for these selected years. Is this result any of the following: natural number, rational number, or irrational number?

106. Pet Adoption   The table lists the number of pet adoptions facilitated by the organization PAWS Chicago during selected years.

Year	2010	2011	2012	2013
Adoptions	4042	4268	5606	5872

Source: PAWS.

(a) How many adoptions were there in 2012?

(b) Calculate the average number of adoptions for these selected years. Is this result any of the following: natural number, rational number, or irrational number?

107. Numbers in Real Life Categorize the following real-life values as one or more of the following: natural number, whole number, integer, rational number, or irrational number
  (a) Shoe sizes in the U.S.
  (b) Freezing point of water, 0 °C
  (c) One third of a pizza

108. Numbers in Real Life Categorize the following real-life values as one or more of the following: natural number, whole number, integer, or rational number

(a) Coldest recorded U.S. temperature, $-80°F$
(b) Population of the world
(c) Distance from Chicago to Minneapolis, 408.9 miles

WRITING ABOUT MATHEMATICS

109. What is a rational number? Is every integer a rational number? Why or why not?

110. Explain why $\frac{3}{7} > \frac{1}{3}$. Now explain, in general, how to determine whether $\frac{a}{b} > \frac{c}{d}$. Assume that a, b, c, and d are natural numbers.

SECTIONS 1.3 and 1.4	Checking Basic Concepts

1. Write each product as an exponential expression.
  (a) $5 \cdot 5 \cdot 5 \cdot 5$ **(b)** $7 \cdot 7 \cdot 7 \cdot 7 \cdot 7$

2. Evaluate each expression.
  (a) 2^3 **(b)** 10^4 **(c)** $\left(\frac{2}{3}\right)^3$

3. Use the given base to write the number as an exponential expression.
  (a) 64 (base 4)   **(b)** 64 (base 2)

4. Evaluate each expression without a calculator.
  (a) $6 + 5 \cdot 4$ **(b)** $6 + 6 \div 2$
  (c) $5 - 2 - 1$ **(d)** $\frac{6 - 3}{2 + 4}$
  (e) $12 \div (6 \div 2)$ **(f)** $2^3 - 2\left(2 + \frac{4}{2}\right)$

5. Translate the phrase "five cubed divided by three" to an algebraic expression.

6. Find the opposite of each expression.
  (a) -17 **(b)** a

7. Find the decimal equivalent for the given rational number.
  (a) $\frac{3}{20}$ **(b)** $\frac{5}{8}$

8. Classify each number as one or more of the following: natural number, integer, rational number, or irrational number.
  (a) $\frac{10}{2}$ **(b)** -5
  (c) $\sqrt{5}$ **(d)** $-\frac{5}{6}$

9. Plot each number on the same number line.
  (a) 0   **(b)** -3 **(c)** 2
  (d) $\frac{3}{4}$ **(e)** $-\sqrt{2}$

10. Evaluate each expression.
  (a) $|-12|$ **(b)** $|6 - 6|$

11. Insert the symbol $>$ or $<$ to make the statement true.
  (a) 4 ___ 9 **(b)** -1.3 ___ -0.5
  (c) $|-3|$ ___ $|-5|$

12. List the following numbers from least to greatest.
$$\sqrt{3}, \quad -7, \quad 0, \quad \frac{1}{3}, \quad -1.6, \quad 3^2$$

1.5 Addition and Subtraction of Real Numbers

Objectives

1 Adding Real Numbers
 • Adding Integers Visually
2 Subtracting Real Numbers
3 Solving Applications Involving Addition and Subtraction
 • Words Associated with Addition and Subtraction

1 Adding Real Numbers

In an addition problem, the two numbers added are called **addends**, and the answer is called the **sum**. For example, in the following addition problem the numbers **3** and **5** are the **addends** and the number **8** is the **sum**.

$$3 \;+\; 5 \;=\; 8$$

Addend   Addend   Sum

The *opposite* (or *additive inverse*) of a real number a is $-a$. When we add opposites, the result is 0. For example, $4 + (-4) = 0$. In general, the equation $a + (-a) = 0$ is true for every real number a.

READING CHECK 1

• What is the answer to an addition problem called?

EXAMPLE 1

NEW VOCABULARY

☐ Addends
☐ Sum
☐ Difference

Adding opposites

Find the opposite of each number and calculate the sum of the number and its opposite.

(a) 45 **(b)** $\sqrt{2}$ **(c)** $-\dfrac{1}{2}$

Solution

(a) The opposite of 45 is -45. Their sum is $45 + (-45) = 0$.
(b) The opposite of $\sqrt{2}$ is $-\sqrt{2}$. Their sum is $\sqrt{2} + (-\sqrt{2}) = 0$.
(c) The opposite of $-\frac{1}{2}$ is $\frac{1}{2}$. Their sum is $-\frac{1}{2} + \frac{1}{2} = 0$.

Now Try Exercises 11, 13, 15

 Connecting Concepts with Your Life When adding real numbers, it may be helpful to think of money. A positive number represents income, and a negative number indicates debt. The sum $9 + (-5) = 4$ would represent being paid \$9 and owing \$5. In this case, \$4 would be left over. Similarly, the sum $-3 + (-6) = -9$ would represent debts of \$3 and \$6, resulting in a total debt of \$9. To add two real numbers, we can use the following rules.

STUDY TIP

Do you know your instructor's name? Do you know how to contact or visit your instructor? Make sure that you can get help when needed.

ADDITION OF REAL NUMBERS

To add two real numbers with *like* signs, do the following.

1. Find the sum of the absolute values of the numbers.
2. The common sign of the two numbers is the sign of the sum.

To add two real numbers with *unlike* signs, do the following.

1. Find the absolute values of the numbers.
2. Subtract the smaller absolute value from the larger absolute value.
3. The sign of the number with the larger absolute value is the sign of the sum.

EXAMPLE 2   **Adding real numbers**

Find each sum by hand.

(a) $-2 + (-4)$ (b) $-\dfrac{2}{5} + \dfrac{7}{10}$ (c) $6.2 + (-8.5)$

Solution

(a) The numbers are both negative, so we add the absolute values $|-2|$ and $|-4|$ to obtain 6. The signs of the addends are both negative, so the answer is -6. That is, $-2 + (-4) = -6$. If we owe \$2 and then owe \$4, the total amount owed is \$6.

(b) The numbers have opposite signs, so we subtract their absolute values to obtain

$$\frac{7}{10} - \frac{2}{5} = \frac{7}{10} - \frac{4}{10} = \frac{3}{10}.$$

The sum is positive because $\left|\frac{7}{10}\right|$ is greater than $\left|-\frac{2}{5}\right|$ and $\frac{7}{10}$ is positive. That is, $-\frac{2}{5} + \frac{7}{10} = \frac{3}{10}$. If we spend \$0.40 $\left(-\frac{2}{5} = -0.4\right)$ and receive \$0.70 $\left(\frac{7}{10} = 0.7\right)$, we have \$0.30 $\left(\frac{3}{10} = 0.3\right)$ left.

(c) $6.2 + (-8.5) = -2.3$ because $|-8.5|$ is 2.3 more than $|6.2|$. If we have \$6.20 and we owe \$8.50, we are short \$2.30.

Now Try Exercises 17, 23, 25

ADDING INTEGERS VISUALLY One way to add integers visually is to use the symbol ⌒ to represent a positive unit and to use the symbol ⌣ to represent a negative unit. Now adding opposites visually results in "zero," as shown.

    $1 + (-1) = 0$

For example, to add $-3 + 5$, we draw three negative units and five positive units.

   $-3 + 5 = 2$

Because the "zeros" add no value, the sum is two positive units, or **2**.

EXAMPLE 3   **Adding integers visually**

Add visually, using the symbols ⌒ and ⌣.

(a) $3 + 2$ (b) $-6 + 4$ (c) $2 + (-3)$ (d) $-5 + (-2)$

Solution

(a) Draw three positive units and then draw two more positive units.

   $3 + 2 = 5$

Because no zeros were formed, the sum is five positive units, or 5.

(b) Draw six negative units and then draw four positive units.

   $-6 + 4 = -2$

Ignoring the zeros that were formed, the sum is two negative units, or -2.

(c) Draw two positive and three negative units.

   $2 + (-3) = -1$

The sum is -1.

(c) $-5 + (-2) = -7$

The sum is -7.

Now Try Exercises 55, 57, 59, 61

Another way to add integers visually is to use a number line as illustrated in the following See the Concept.

See the Concept 1   ADDING ON A NUMBER LINE

(A) Starting at the origin, complete step B for the first addend.

(B) Draw an arrow with a length equal to the absolute value of this addend. If the addend is **positive**, draw the arrow to the **right**; if it is **negative**, draw the arrow to the **left**.

(C) Starting at the end of the first arrow, repeat step B for the second addend.

(D) The **sum** is located at the end of the second arrow.

WATCH VIDEO IN $\boxed{\text{MML}}$.

READING CHECK 2

• What is the answer to a subtraction problem called?

2 Subtracting Real Numbers

The answer to a subtraction problem is the **difference**. When subtracting two real numbers, changing the problem to an addition problem may be helpful.

SUBTRACTION OF REAL NUMBERS

For any real numbers a and b,

$$a - b = a + (-b).$$

To subtract b from a, add a and the opposite of b.

EXAMPLE 4   Subtracting real numbers

Find each difference by hand.
(a) $10 - 20$ (b) $-5 - 2$ (c) $-2.1 - (-3.2)$ (d) $\frac{1}{2} - \left(-\frac{3}{4}\right)$

Solution
(a) $10 - 20 = 10 + (-20) = -10$
(b) $-5 - 2 = -5 + (-2) = -7$
(c) $-2.1 - (-3.2) = -2.1 + 3.2 = 1.1$
(d) $\frac{1}{2} - \left(-\frac{3}{4}\right) = \frac{1}{2} + \frac{3}{4} = \frac{2}{4} + \frac{3}{4} = \frac{5}{4}$

Change subtraction to addition.

Now Try Exercises 29, 31, 35, 37

In the next example, we show how to add and subtract groups of numbers.

EXAMPLE 5   **Adding and subtracting real numbers**

Evaluate each expression by hand.
(a) $5 - 4 - (-6) + 1$ **(b)** $\frac{1}{2} - \frac{3}{4} + \frac{1}{3}$ **(c)** $-6.1 + 5.6 - 10.1$

Solution
(a) Rewrite the expression in terms of addition only, and then find the sum.

$$5 - 4 - (-6) + 1 = 5 + (-4) + 6 + 1$$

> Change subtraction to addition.

$$= 1 + 6 + 1$$
$$= 8$$

(b) Begin by rewriting the fractions with the LCD of 12.

$$\frac{1}{2} - \frac{3}{4} + \frac{1}{3} = \frac{6}{12} - \frac{9}{12} + \frac{4}{12}$$

> Change subtraction to addition.

$$= \frac{6}{12} + \left(-\frac{9}{12}\right) + \frac{4}{12}$$
$$= -\frac{3}{12} + \frac{4}{12}$$
$$= \frac{1}{12}$$

CRITICAL THINKING 1

Explain how subtraction of real numbers can be performed on a number line.

(c) The expression can be evaluated by changing subtraction to addition.

$$-6.1 + 5.6 - 10.1 = -6.1 + 5.6 + (-10.1)$$
$$= -0.5 + (-10.1)$$

> Change subtraction to addition.

$$= -10.6$$

Now Try Exercises 43, 49, 51

3 Solving Applications Involving Addition and Subtraction

WORDS ASSOCIATED WITH ADDITION AND SUBTRACTION In application problems, some words indicate that we should add, while other words indicate that we should subtract. **TABLE 1.6** and **1.7** show examples of such words along with sample phrases.

Words Associated with Addition

Words	Sample Phrase
add	add the two temperatures
plus	her age plus his age
more than	5 cents more than the cost
sum	the sum of two measures
total	the total of the four prices
increased by	height increased by 3 inches

TABLE 1.6

Words Associated with Subtraction

Words	Sample Phrase
subtract	subtract dues from the price
minus	his income minus his taxes
fewer than	18 fewer flowers than shrubs
difference	the difference in their heights
less than	his age is 4 less than yours
decreased by	the weight decreased by 7

TABLE 1.7

READING CHECK 3

- Give three words that are associated with addition.
- Give three words that are associated with subtraction.

🌐 *Math in Context* ^Real world Sometimes temperature differences are found by subtracting positive and negative real numbers. At other times, addition of positive and negative real numbers occurs at banks if positive numbers represent deposits and negative numbers represent withdrawals. The next two examples illustrate these situations.

EXAMPLE 6 **Calculating temperature differences**

The hottest outdoor temperature ever recorded in the shade was $134\,°F$ in Death Valley, and the coldest outside temperature ever recorded was $-136\,°F$ in Antarctica. Find the difference between these two temperatures. (*Source: Guinness Book of Records.*)

Solution
The word *difference* indicates subtraction. We must subtract the two temperatures.

$$134 - (-136) = 134 + 136 = 270\,°F.$$

Now Try Exercise 85

EXAMPLE 7 **Balancing a checking account**

The initial balance in a checking account is $285. Find the final balance if the following represents a list of withdrawals and deposits: $-$15$, $-$20$, 500, and $-$100$.

Solution
Find the sum of the five numbers.

$$285 + (-15) + (-20) + 500 + (-100) = 270 + (-20) + 500 + (-100)$$
$$= 250 + 500 + (-100)$$
$$= 750 + (-100)$$
$$= 650$$

The final balance is $650. This result may be supported with a calculator, as illustrated in **FIGURE 1.25**. Note that the expression has been evaluated two different ways.

Now Try Exercise 81

```
285+(-15)+(-20)+
500+(-100)
                 650
285-15-20+500-10
0
                 650
```

FIGURE 1.25

TECHNOLOGY NOTE

Subtraction and Negation
Calculators typically have *different* keys to represent subtraction and negation. Be sure to use the correct key. Many graphing calculators have the following keys for subtraction and negation.

$$\boxed{-} \qquad \boxed{(-)}$$
Subtraction *Negation*

1.5   Putting It All Together

CONCEPT	COMMENTS	EXAMPLES
Addition	To add real numbers, use a number line or follow the rules presented in this section.	$-2 + 8 = 6$ $0.8 + (-0.3) = 0.5$ $-\frac{1}{7} + \left(-\frac{3}{7}\right) = \frac{-1 + (-3)}{7} = -\frac{4}{7}$ $-4 + 4 = 0$ $-3 + 4 + (-2) = -1$
Subtraction	To subtract real numbers, transform the problem to an addition problem by adding the opposite. $a - b = a + (-b)$	$6 - 8 = 6 + (-8) = -2$ $-3 - 4 = -3 + (-4) = -7$ $-\frac{1}{2} - \left(-\frac{3}{2}\right) = -\frac{1}{2} + \frac{3}{2} = \frac{2}{2} = 1$ $-5 - (-5) = -5 + 5 = 0$ $9.4 - (-1.2) = 9.4 + 1.2 = 10.6$

1.5   Exercises

MyMathLab®

CONCEPTS AND VOCABULARY

1. The solution to an addition problem is the _____.

2. When you add opposites, the sum is always _____.

3. (True or False?) If two positive numbers are added, the sum is always a positive number.

4. (True or False?) If two negative numbers are added, the sum is always a negative number.

5. If two numbers with opposite signs are added, the sum has the same sign as the number with the larger _____.

6. The answer to a subtraction problem is the _____.

7. When subtracting two real numbers, it may be helpful to change the problem to a(n) _____ problem.

8. To subtract b from a, add the _____ of b to a. That is, $a - b = a +$ _____.

9. The words *sum*, *more*, and *plus* indicate that _____ should be performed.

10. The words *difference*, *less than*, and *minus* indicate that _____ should be performed.

ADDING AND SUBTRACTING REAL NUMBERS

Exercises 11–16: Find the opposite of the number and then calculate the sum of the number and its opposite.

11. 25

12. $-\frac{1}{2}$

13. $-\sqrt{21}$

14. $-\pi$

15. 5.63

16. -6^2

Exercises 17–28: Find the sum.

17. $5 + (-4)$

18. $-9 + 7$

19. $-1 + (-6)$

20. $-10 + (-23)$

21. $\frac{3}{4} + \left(-\frac{1}{2}\right)$

22. $-\frac{5}{12} + \left(-\frac{1}{6}\right)$

23. $-\frac{6}{7} + \frac{3}{14}$

24. $-\frac{2}{9} + \left(-\frac{1}{12}\right)$

25. $0.6 + (-1.7)$

26. $4.3 + (-2.4)$

27. $-52 + 86$

28. $-103 + (-134)$

Exercises 29–40: Find the difference.

29. $5 - 8$

30. $3 - 5$

31. $-2 - (-9)$

32. $-10 - (-19)$

33. $\frac{6}{7} - \frac{13}{14}$

34. $-\frac{5}{6} - \frac{1}{6}$

35. $-\frac{1}{10} - \left(-\frac{3}{5}\right)$

36. $-\frac{2}{11} - \left(-\frac{5}{11}\right)$

37. $0.8 - (-2.1)$

38. $-9.6 - (-5.7)$

39. $-73 - 91$

40. $201 - 502$

Exercises 41–54: Evaluate the expression.

41. $10 - 19$

42. $5 + (-9)$

43. $19 - (-22) + 1$

44. $53 + (-43) - 10$

45. $-3 + 4 - 6$

46. $-11 + 8 - 10$

47. $100 - 200 + 100 - (-50)$

48. $-50 - (-40) + (-60) + 80$

49. $1.5 - 2.3 + 9.6$

50. $10.5 - (-5.5) + (-1.5)$

51. $-\frac{1}{2} + \frac{1}{4} - \left(-\frac{3}{4}\right)$

52. $\frac{1}{4} - \left(-\frac{2}{5}\right) + \left(-\frac{3}{20}\right)$

53. $|4 - 9| - |1 - 7|$

54. $|-5 - (-3)| - |-6 + 8|$

ADDING NUMBERS VISUALLY

Exercises 55–62: Refer to Example 3. Find the sum visually.

55. $1 + 3$

56. $3 + 1$

57. $4 + (-2)$

58. $-4 + 6$

59. $-1 + (-2)$

60. $-2 + (-2)$

61. $-3 + 7$

62. $5 + (-6)$

Exercises 63–70: Use a number line to find the sum.

63. $-1 + 3$

64. $3 + (-1)$

65. $4 + (-5)$

66. $2 + 6$

67. $-10 + 20$

68. $15 + (-5)$

69. $-50 + (-100)$

70. $-100 + 100$

TRANSLATING WORDS TO EXPRESSIONS

Exercises 71–80: Write an arithmetic expression for the given phrase and then simplify it.

71. The sum of two and negative five

72. Subtract ten from negative six

73. Negative five increased by seven

74. Negative twenty decreased by eight

75. The additive inverse of the quantity two cubed

76. Five minus the quantity two cubed

77. The difference between negative six and seven (*Hint:* Write the numbers for the subtraction problem in the order given.)

78. The difference between one-half and three-fourths

79. Six plus negative ten minus five

80. Ten minus seven plus negative twenty

SOLVING APPLICATIONS INVOLVING ADDITION AND SUBTRACTION

81. Checking Account The initial balance in a checking account is $358. Find the final balance resulting from the following withdrawals and deposits: −$45, $37, $120, and −$240.

82. Savings Account A savings account has $1245 in it. Find the final balance resulting from the following withdrawals and deposits: −$189, $975, −$226, and −$876.

83. Football Stats A running back carries the ball five times. Find his total yardage if the carries were 9, −2, −1, 14, and 5 yards.

84. Tracking Weight A person weighs himself every Monday for four weeks. He gains one pound the first week, loses three pounds the second week, gains two pounds the third week, and loses one pound the last week.
  (a) Using positive numbers to represent weight gains and negative numbers to represent weight losses, write a sum that gives the total gain or loss over this four-week period.
  (b) If he weighed 170 pounds at the beginning of this process, what was his weight at the end?

85. Deepest and Highest The deepest point in the ocean is the Mariana Trench, which is 35,839 feet below sea level. The highest point on Earth is Mount Everest, which is 29,029 feet above sea level. What is the difference in height between Mount Everest and the Mariana Trench? (*Source:* The Guinness Book of Records.)

86. Greatest Temperature Ranges The greatest temperature range on Earth occurs in Siberia, where the temperature can vary between 98 °F in the summer and −90 °F in the winter. Find the difference between these two temperatures. (*Source:* The Guinness Book of Records.)

87. PC Growth In the first quarter of 2010, Global PC shipments *increased* 36% from the same quarter the previous year. In the first quarter of 2013, shipments *decreased* by 13% over the previous year. Find the difference between the 2010 and 2013 growth percentages? (*Source:* Business Insider.)

88. Samsung Profits In the first quarter of 2013, Samsung's operating profits *increased* 54% from the same quarter the previous year. In the first quarter of 2014, profits *decreased* by 4% over the previous year. Find the difference between the 2013 and 2014 growth percentages? (*Source:* Samsung Electronics.)

89. Online Exploration Use the Internet to find the highest and lowest points in the continental United States, and then find the difference in their heights.

90. Online Exploration In 1972, residents of Loma, Montana, experienced the largest 24-hour temperature swing ever recorded in the United States. Use the Internet to find the starting and ending temperatures, and then write a subtraction equation that shows the difference.

WRITING ABOUT MATHEMATICS

91. Explain how to add two negative numbers. Give an example.

92. Explain how to subtract a negative number from a positive number. Give an example.

1.6 Multiplication and Division of Real Numbers

Objectives

1 Multiplying Real Numbers
2 Dividing Real Numbers
  • Converting Between Fractions and Decimals
  • Fractions and Calculators
3 Solving Applications Involving Multiplication and Division

1 Multiplying Real Numbers

In a multiplication problem, the numbers multiplied are called the *factors*, and the answer is called the *product*.

$$7 \quad \cdot \quad 4 \quad = \quad 28$$
Factor   Factor   Product

Multiplication is a fast way to perform addition. For example, $5 \cdot 2 = 10$ is equivalent to finding the sum of five 2s, or

$$2 + 2 + 2 + 2 + 2 = 10.$$

NEW VOCABULARY

☐ Dividend
☐ Divisor
☐ Quotient
☐ Reciprocal (multiplicative inverse)

READING CHECK 1

• What is the answer to a multiplication problem called?

Similarly, the product $5 \cdot (-2)$ is equivalent to finding the sum of five -2s, or

$$(-2) + (-2) + (-2) + (-2) + (-2) = -10.$$

Thus $5 \cdot (-2) = -10$. In general, the product of a positive number and a negative number is a negative number.

What sign should the product of two negative numbers have? To answer this question, consider multiplying -4 times consecutive integers. What values should replace the question marks to continue the pattern?

$$-4 \cdot 3 = -12$$
Decrease by 1 ——— Increase by 4
$$-4 \cdot 2 = -8$$
Decrease by 1 ——— Increase by 4
$$-4 \cdot 1 = -4$$
Decrease by 1 ——— Increase by 4
$$-4 \cdot 0 = 0$$
Decrease by 1
$$-4 \cdot (-1) = \; ?$$
Decrease by 1
$$-4 \cdot (-2) = \; ?$$

Continuing the pattern results in $-4 \cdot (-1) = 4$ and $-4 \cdot (-2) = 8$. This pattern suggests that if we multiply two negative real numbers, the product is positive.

When multiplying two real numbers, the sign of the product is determined by the signs of the factors.

STUDY TIP

Studying with other students can improve your chances of success. Consider exchanging phone numbers or email addresses with some of your classmates. If possible, set up a regular meeting time and invite other classmates to join you.

SIGNS OF PRODUCTS

The product of two numbers with *like* signs is positive. The product of two numbers with *unlike* signs is negative.

NOTE: Multiplying a number by -1 results in the additive inverse (opposite) of the number. For example, $-1 \cdot 4 = -4$, and -4 is the additive inverse of 4. Similarly, $-1 \cdot (-3) = 3$, and 3 is the additive inverse of -3. In general, $-1 \cdot a = -a$. ∎

EXAMPLE 1 **Multiplying real numbers**

Find each product by hand.
(a) $-9 \cdot 7$ **(b)** $\frac{2}{3} \cdot \frac{5}{9}$ **(c)** $-2.1(-40)$ **(d)** $(-2.5)(4)(-9)(-2)$

Solution
(a) The resulting product is negative because the factors -9 and 7 have unlike signs. Thus $-9 \cdot 7 = -63$.
(b) The product is positive because both factors are positive.

$$\frac{2}{3} \cdot \frac{5}{9} = \frac{2 \cdot 5}{3 \cdot 9} = \frac{10}{27}$$

(c) As both factors are negative, the product is positive. Thus $-2.1(-40) = 84$.
(d) $(-2.5)(4)(-9)(-2) = (-10)(-9)(-2) = (90)(-2) = -180$

Now Try Exercises 13, 19, 27, 29

The following See the Concept shows how to determine the sign of a product containing several integers.

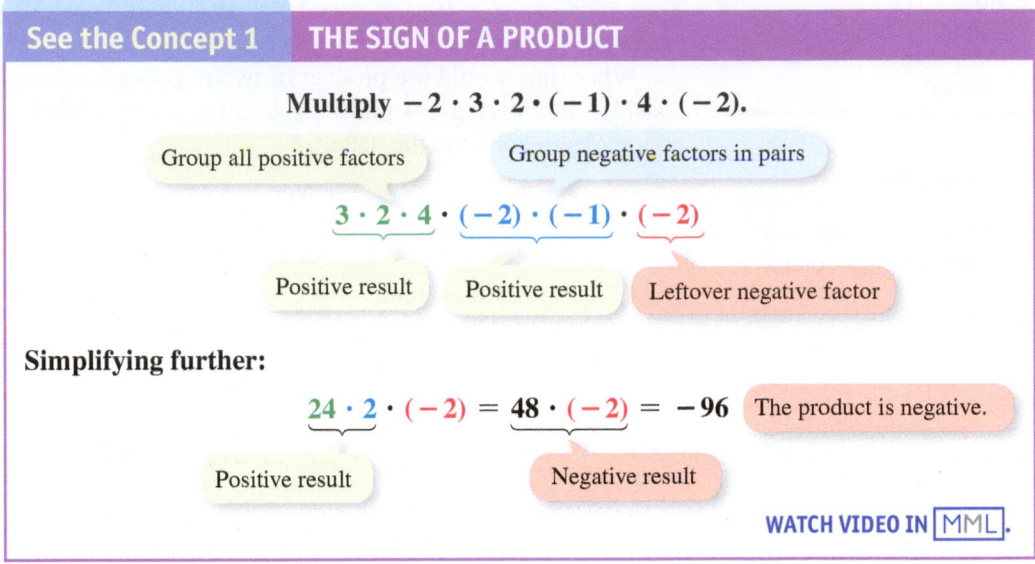

See the Concept 1   **THE SIGN OF A PRODUCT**

Multiply $-2 \cdot 3 \cdot 2 \cdot (-1) \cdot 4 \cdot (-2)$.

Group all positive factors   Group negative factors in pairs

$3 \cdot 2 \cdot 4 \cdot (-2) \cdot (-1) \cdot (-2)$

Positive result   Positive result   Leftover negative factor

Simplifying further:

$24 \cdot 2 \cdot (-2) = 48 \cdot (-2) = -96$ The product is negative.

Positive result   Negative result

WATCH VIDEO IN MML.

NOTE: If a product contains an *even* number of negative factors, the answer will be positive. If a product contains an *odd* number of negative factors, the answer will be negative. ∎

When evaluating expressions such as

-5^2 (**Square** and then **negate**)   and $(-5)^2$ (**Negate** and then **square**)

it is important to note that the first represents the opposite of five squared, while the second indicates that negative five should be squared. The next example illustrates the difference between the opposite of an exponential expression and a power of a negative number.

EXAMPLE 2   **Evaluating real numbers with exponents**

Evaluate each expression by hand.
(a) $(-4)^2$ **(b)** -4^2 **(c)** $(-2)^3$ **(d)** -2^3

Solution
(a) Because the exponent is outside of parentheses, the base of the exponential expression is -4. The expression is evaluated as

$$(-4)^2 = (-4)(-4) = 16.$$

(b) This is the negation of an exponential expression with base 4. Evaluating the exponent before negating results in $-4^2 = -(4)(4) = -16$.
(c) $(-2)^3 = (-2)(-2)(-2) = -8$ **(d)** $-2^3 = -(2)(2)(2) = -8$

Now Try Exercises 35, 37, 39, 41

MAKING CONNECTIONS 1

Negative Square Roots

Because the product of two negative numbers is positive, $(-3) \cdot (-3) = 9$. That is, -3 is a square root of 9. As discussed earlier in this chapter, every positive number has one positive square root and one negative square root. For a positive number a, the positive square root is called the *principal square root* and is denoted $\sqrt{a}$. The negative square root is denoted $-\sqrt{a}$. For example, $\sqrt{4} = 2$ and $-\sqrt{4} = -2$.

READING CHECK 2

• What is the answer to a division problem called?

2 Dividing Real Numbers

In a division problem, the **dividend** is divided by the **divisor**, and the result is the **quotient**.

$$20 \div 4 = 5$$

Dividend   **Divisor**   **Quotient**

This division problem can also be written in fraction form as $\frac{20}{4} = 5$. Division of real numbers can be defined in terms of multiplication and reciprocals. The **reciprocal**, or **multiplicative inverse**, of a nonzero real number a is $\frac{1}{a}$. The number 0 has *no reciprocal*.

TECHNOLOGY NOTE

Try dividing 5 by 0 with a calculator. On most calculators, dividing a number by 0 results in an error message.

DIVIDING REAL NUMBERS

For real numbers a and b with $b \neq 0$,

$$\frac{a}{b} = a \cdot \frac{1}{b}.$$

That is, to divide a by b, multiply a by the reciprocal of b.

NOTE: Division by 0 is undefined because 0 has no reciprocal. ∎

EXAMPLE 3 **Dividing real numbers**

Evaluate each expression by hand.

(a) $-16 \div \frac{1}{4}$ **(b)** $\dfrac{\frac{3}{5}}{-8}$ **(c)** $\dfrac{-8}{-36}$ **(d)** $9 \div 0$

Solution

(a) $-16 \div \frac{1}{4} = \frac{-16}{1} \cdot \frac{4}{1} = \frac{-64}{1} = -64$ The reciprocal of $\frac{1}{4}$ is $\frac{4}{1}$.

(b) $\dfrac{\frac{3}{5}}{-8} = \frac{3}{5} \div (-8) = \frac{3}{5} \cdot \left(-\frac{1}{8}\right) = -\frac{3}{40}$ The reciprocal of -8 is $-\frac{1}{8}$.

(c) $\dfrac{-8}{-36} = -8 \cdot \left(-\frac{1}{36}\right) = \frac{8}{36} = \frac{2}{9}$ The reciprocal of -36 is $-\frac{1}{36}$.

(d) $9 \div 0$ is undefined.   The number 0 has no reciprocal.

Now Try Exercises 49, 51, 59, 63

When determining the sign of a quotient, the following rules may be helpful. Note that these rules are similar to those for signs of products.

SIGNS OF QUOTIENTS

The quotient of two numbers with *like* signs is positive. The quotient of two numbers with *unlike* signs is negative.

NOTE: To see why a negative number divided by a negative number is a positive number, remember that *division is a fast way to perform subtraction*. For example, because

$$20 - 4 - 4 - 4 - 4 - 4 = 0,$$

a total of **five** 4s can be subtracted from **20**, so $\frac{20}{4} = 5$. Similarly, because

$$-20 - (-4) - (-4) - (-4) - (-4) - (-4) = 0,$$

a total of **five** -4s can be subtracted from -20, so $\frac{-20}{-4} = 5$. ∎

CONVERTING BETWEEN FRACTIONS AND DECIMALS In a business such as an auto parts store, employees may need to convert fractions and mixed numbers to decimal numbers. The next example illustrates this process.

EXAMPLE 4 Converting fractions to decimals

Convert each measurement to a decimal number.
(a) $2\frac{3}{8}$-inch bolt   (b) $\frac{15}{16}$-inch diameter   (c) $1\frac{1}{3}$-cup flour

Solution

(a) First divide 3 by 8.

$$
\begin{array}{r}
0.375 \\
8\overline{)3.000} \\
\underline{24} \\
60 \\
\underline{56} \\
40 \\
\underline{40} \\
0
\end{array}
$$

So, $2\frac{3}{8} = 2.375$.

(b) Divide 15 by 16.

$$
\begin{array}{r}
0.9375 \\
16\overline{)15.0000} \\
\underline{144} \\
60 \\
\underline{48} \\
120 \\
\underline{112} \\
80 \\
\underline{80} \\
0
\end{array}
$$

So, $\frac{15}{16} = 0.9375$.

(c) First divide 1 by 3.

$$
\begin{array}{r}
0.333\ldots \\
3\overline{)1.000} \\
\underline{9} \\
10 \\
\underline{9} \\
10 \\
\underline{9} \\
1
\end{array}
$$

So, $1\frac{1}{3} = 1.\overline{3}$.

Now Try Exercises 69, 71, 75

In the next example, numbers that are expressed as terminating decimals are converted to fractions.

EXAMPLE 5 Converting decimals to fractions

Convert each decimal number to a fraction in lowest terms.
(a) 0.06   (b) 0.375   (c) 0.0025

Solution

(a) The decimal 0.06 equals six hundredths, or $\frac{6}{100}$. Simplifying this fraction gives

$$\frac{6}{100} = \frac{3 \cdot 2}{50 \cdot 2} = \frac{3}{50}.$$

(b) The decimal 0.375 equals three hundred seventy-five thousandths, or $\frac{375}{1000}$. Simplifying this fraction gives

$$\frac{375}{1000} = \frac{3 \cdot 125}{8 \cdot 125} = \frac{3}{8}.$$

(c) The decimal 0.0025 equals twenty-five ten thousandths, or $\frac{25}{10,000}$. Simplifying this fraction gives

$$\frac{25}{10,000} = \frac{1 \cdot 25}{400 \cdot 25} = \frac{1}{400}.$$

Now Try Exercises 81, 85, 87

FRACTIONS AND CALCULATORS Many calculators have the capability to perform arithmetic on fractions and express the answer as either a decimal or a fraction. The next example illustrates this capability.

EXAMPLE 6 **Performing arithmetic operations with technology**

Use a calculator to evaluate each expression. Express your answer as a decimal and as a fraction.

(a) $\frac{1}{3} + \frac{2}{5} - \frac{4}{9}$ (b) $\left(\frac{4}{9} \cdot \frac{3}{8}\right) \div \frac{2}{3}$

Solution

(a) **FIGURE 1.26(a)** shows that

$$\frac{1}{3} + \frac{2}{5} - \frac{4}{9} = 0.2\overline{8}, \text{ or } \frac{13}{45}.$$

(b) **FIGURE 1.26(b)** shows that $\left(\frac{4}{9} \cdot \frac{3}{8}\right) \div \frac{2}{3} = 0.25$, or $\frac{1}{4}$.

CALCULATOR HELP

To find fraction results with a calculator, see Appendix A (pages AP-1 and AP-2).

```
(1/3)+(2/5)-(4/9
)
          .2888888889
(1/3)+(2/5)-(4/9
)▶Frac
             13/45
```

```
((4/9)*(3/8))/(2
/3)
              .25
((4/9)*(3/8))/(2
/3)▶Frac
              1/4
```

(a)   (b)

FIGURE 1.26

Now Try Exercises 89, 91

NOTE: Generally, it is a good idea to put parentheses around fractions when you are using a calculator. ∎

3 Solving Applications Involving Multiplication and Division

🌐 *Math in Context* Entertainment There are many instances when we may need to multiply or divide real numbers. In the next example, we multiply a whole number and a fraction to compare top-grossing movies.

EXAMPLE 7 **Comparing top-grossing movies**

Even though *The Hunger Games: Catching Fire* (2013) was an extremely successful film, it ranks only 93rd among top-grossing movies of all time when calculated by using estimated total admissions. The total admissions for *The Hunger Games: Catching Fire* are $\frac{9}{25}$ of the total admissions for *The Sound of Music* (1965), which ranks third. (*Source:* Box Office Mojo.)

(a) If *The Sound of Music* had estimated total admissions of 142 million people, find the estimated total admissions for *The Hunger Games: Catching Fire*.

(b) The top-grossing movie of all time is *Gone With the Wind* (1939), which had total admissions of 202 million. How many more people saw *Gone With the Wind* than *The Hunger Games: Catching Fire*?

Solution

(a) To find the total admissions for *The Hunger Games: Catching Fire*, we multiply the real numbers 142 and $\frac{9}{25}$ to obtain $142 \cdot \frac{9}{25} = \frac{1278}{25} \approx 51$, or about 51 million people.

(b) The difference is $202 - 51 = 151$ million people.

Now Try Exercise 97

🌐 *Math in ⌃Context* (Business) When fractions describe data in business, it is sometimes helpful to convert the fractions to decimals. In the next example, we convert a fraction that describes online advertising revenue to a decimal.

EXAMPLE 8   **Analyzing online ad revenue**

It is estimated that $\frac{31}{200}$ of online ad revenue is generated by mobile ads. Write this fraction as a decimal. (*Source:* Business Insider)

Solution

One method for writing the fraction as a decimal is to divide 31 by 200, using long division. An alternative method is to multiply the fraction by $\frac{5}{5}$ so that the denominator becomes 1000. Then write the numerator in the thousandths place in the decimal.

$$\frac{31}{200} \cdot \frac{5}{5} = \frac{155}{1000} = 0.155$$

Now Try Exercise 99

MAKING CONNECTIONS 2

The Four Arithmetic Operations

If you know how to add real numbers, then you also know how to subtract real numbers because subtraction is defined in terms of addition. That is,

$$a - b = a + (-b).$$

If you know how to multiply real numbers, then you also know how to divide real numbers because division is defined in terms of multiplication. That is,

$$\frac{a}{b} = a \cdot \frac{1}{b}.$$

1.6   Putting It All Together

CONCEPT	COMMENTS	EXAMPLES
Multiplication	The product of two numbers with like signs is positive, and the product of two numbers with unlike signs is negative.	$(-6) \cdot (-7) = 42$ $6 \cdot (-7) = -42$
Division	For real numbers a and b, with $b \neq 0$, $$\frac{a}{b} = a \cdot \frac{1}{b}.$$ The quotient of two numbers with like signs is positive, and the quotient of two numbers with unlike signs is negative.	$\dfrac{-42}{-6} = 7$ $\dfrac{-42}{6} = -7$
Converting Fractions to Decimals	The fraction $\frac{a}{b}$ is equivalent to $a \div b$, where $b \neq 0$.	The fraction $\frac{2}{9}$ is equivalent to $0.\overline{2}$, because $2 \div 9 = 0.222\ldots$.
Converting Terminating Decimals to Fractions	Write the decimal as a fraction with a denominator equal to a power of 10 and then simplify this fraction.	$0.55 = \dfrac{55}{100} = \dfrac{11 \cdot 5}{20 \cdot 5} = \dfrac{11}{20}$

1.6 Exercises

MyMathLab®

CONCEPTS AND VOCABULARY

1. The solution to a multiplication problem is the _____.

2. The product of a positive number and a negative number is a(n) _____ number.

3. The product of two negative numbers is _____.

4. The solution to a division problem is the _____.

5. The reciprocal of a nonzero number a is _____.

6. Division by zero is undefined because zero has no _____.

7. To divide a by b, multiply a by the _____ of b.

8. In general, $-1 \cdot a =$ _____.

9. $\dfrac{a}{b} = a \cdot$ _____.

10. A negative number divided by a negative number is a(n) _____ number.

11. A negative number divided by a positive number is a(n) _____ number.

12. To convert $\frac{5}{8}$ to a decimal, divide _____ by _____.

MULTIPLYING REAL NUMBERS

Exercises 13–32: Multiply.

13. $-3 \cdot 4$
14. $-5 \cdot 7$
15. $6 \cdot (-3)$
16. $2 \cdot (-1)$
17. $0 \cdot (-2.13)$
18. $-2 \cdot (-7)$
19. $-6 \cdot (-10)$
20. $-3 \cdot (-1.7) \cdot 0$
21. $-\frac{1}{2} \cdot \left(-\frac{2}{4}\right)$
22. $-\frac{3}{4} \cdot \left(-\frac{5}{12}\right)$
23. $-\frac{3}{7} \cdot \frac{7}{3}$
24. $\frac{5}{8} \cdot \left(-\frac{4}{15}\right)$
25. $-10 \cdot (-20)$
26. $1000 \cdot (-70)$
27. $-0.5 \cdot 100$
28. $-0.5 \cdot (-0.3)$
29. $-2 \cdot 3 \cdot (-4) \cdot 5$
30. $-3 \cdot (-5) \cdot (-2) \cdot 10$
31. $-6 \cdot \frac{1}{6} \cdot \frac{7}{9} \cdot \left(-\frac{9}{7}\right) \cdot \left(-\frac{3}{2}\right)$
32. $-\frac{8}{5} \cdot \frac{1}{8} \cdot \left(-\frac{5}{7}\right) \cdot -7$

33. **Thinking Generally** Is the product given by the expression $a \cdot (-a) \cdot (-a) \cdot a \cdot (-a)$ positive or negative if $a > 0$?

34. **Thinking Generally** Is the product given by the expression $a \cdot (-a) \cdot (-a) \cdot a \cdot (-a)$ positive or negative if $a < 0$?

Exercises 35–44: Evaluate the expression.

35. $(-5)^2$
36. -5^2
37. $(-1)^3$
38. $(-6)^2$
39. -2^4
40. $-(-4)^2$
41. $-(-2)^3$
42. $3 \cdot (-3)^2$
43. $5 \cdot (-2)^3$
44. -1^4

DIVIDING REAL NUMBERS

Exercises 45–68: Divide.

45. $-10 \div 5$
46. $-8 \div 4$
47. $-20 \div (-2)$
48. $-15 \div (-3)$
49. $\frac{-12}{3}$
50. $\frac{-25}{-5}$
51. $-16 \div \frac{1}{2}$
52. $10 \div \left(-\frac{1}{3}\right)$
53. $0 \div 3$
54. $\frac{0}{-5}$
55. $\frac{-1}{0}$
56. $\frac{0}{-2}$
57. $\frac{1}{2} \div (-11)$
58. $-\frac{3}{4} \div (-6)$
59. $\dfrac{-\frac{4}{5}}{-3}$
60. $\dfrac{\frac{7}{8}}{-7}$
61. $\frac{5}{6} \div \left(-\frac{8}{9}\right)$
62. $-\frac{11}{12} \div \left(-\frac{11}{4}\right)$
63. $-\frac{1}{2} \div 0$
64. $-9 \div 0$
65. $-0.5 \div \frac{1}{2}$
66. $-0.25 \div \left(-\frac{3}{4}\right)$
67. $-\frac{2}{3} \div 0.5$
68. $\frac{1}{6} \div 1.5$

CONVERTING FRACTIONS TO DECIMALS

Exercises 69–80: Write the number as a decimal.

69. $\frac{1}{2}$
70. $\frac{3}{4}$
71. $\frac{3}{16}$
72. $\frac{1}{9}$
73. $3\frac{1}{2}$
74. $2\frac{1}{4}$
75. $5\frac{2}{3}$
76. $6\frac{7}{9}$
77. $1\frac{7}{16}$
78. $6\frac{1}{12}$
79. $\frac{7}{8}$
80. $\frac{11}{16}$

CONVERTING DECIMALS TO FRACTIONS

Exercises 81–88: Write the decimal number as a fraction in lowest terms.

81. 0.25

82. 0.8

83. 0.16

84. 0.35

85. 0.625

86. 0.0125

87. 0.6875

88. 0.21875

USING A CALCULATOR

Exercises 89–96: Use a calculator to evaluate each expression. Express your answer as a decimal and as a fraction.

89. $\left(\frac{1}{3} + \frac{5}{6}\right) \div \frac{1}{2}$

90. $\frac{4}{9} - \frac{1}{6} + \frac{2}{3}$

91. $\frac{4}{5} \div \frac{2}{3} \cdot \frac{7}{4}$

92. $4 - \frac{7}{4} \cdot 2$

93. $\frac{15}{2} - 4 \cdot \frac{7}{3}$

94. $\frac{1}{6} - \frac{3}{5} + \frac{7}{8}$

95. $\frac{17}{40} + 3 \div 8$

96. $\frac{3}{4} \cdot \left(6 + \frac{1}{2}\right)$

APPLICATIONS INVOLVING MULTIPLICATION AND DIVISION

97. Windows Devices In 2015, Windows was the second most popular operating system of total global connected-device shipments. Find the total number of Windows shipments in millions if they are $\frac{7}{25}$ of the number of Android shipments, which had a total of 1,358 million. (*Source:* Statista.)

98. Planet Climate Saturn has a cold average surface temperature of $-220\,°\text{F}$. Neptune has a colder average surface temperature that is $\frac{3}{2}$ times that of Saturn. Find the average surface temperature on Neptune. (*Source:* NASA.)

99. Teleportation About $\frac{14}{25}$ of U.S. adults do not believe that scientists will be able to perform teleportation in the next 50 years. Write this fraction as a decimal.

100. Smartphone Size In 2012, smartphones with a screen size greater than 4 inches accounted for 40% of the total market growth in global smartphone sales. Write this percentage as a fraction.

WRITING ABOUT MATHEMATICS

101. Division is a fast way to subtract. Consider the division problem $\frac{-6}{-2}$, whose quotient represents the number of -2s in -6. Using this idea, explain why the answer is a positive number.

102. Explain how to determine whether the product of three integers is positive or negative.

SECTIONS 1.5 and 1.6 — Checking Basic Concepts

1. Find each sum.
(a) $-4 + 4$
(b) $-10 + (-12) + 3$

2. Evaluate each expression.
(a) $\frac{2}{3} - \left(-\frac{2}{9}\right)$
(b) $-1.2 - 5.1 + 3.1$

3. Write an arithmetic expression for the given phrase and then simplify it.
(a) The sum of negative one and five
(b) The difference between four and negative three

4. The hottest temperature ever recorded at International Falls, Minnesota, was $99\,°\text{F}$, and the coldest temperature ever recorded was $-46\,°\text{F}$. What is the difference between these two temperatures?

5. Find each product.
(a) $-5 \cdot (-7)$
(b) $-\frac{1}{2} \cdot \frac{2}{3} \cdot \left(-\frac{4}{5}\right)$

6. Evaluate each expression.
(a) -3^2
(b) $4 \cdot (-2)^3$
(c) $(-5)^2$

7. Evaluate each expression.
(a) $-5 \div \frac{2}{3}$
(b) $-\frac{5}{8} \div \left(-\frac{4}{3}\right)$

8. What is the reciprocal of $-\frac{7}{6}$?

9. Simplify each expression.
(a) $\frac{-10}{2}$
(b) $\frac{10}{-2}$
(c) $-\frac{10}{2}$
(d) $\frac{-10}{-2}$

10. Convert each fraction or mixed number to a decimal number.
(a) $\frac{3}{5}$
(b) $3\frac{7}{8}$

1.7   Properties of Real Numbers

Objectives

1 Applying the Commutative Properties
2 Applying the Associative Properties
3 Applying the Distributive Properties
4 Recognizing Identity and Inverse Properties
5 Performing Mental Calculations

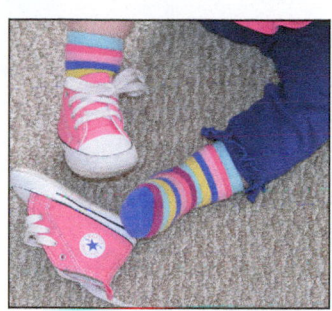

NEW VOCABULARY

☐ Commutative property for addition
☐ Commutative property for multiplication
☐ Associative property for addition
☐ Associative property for multiplication
☐ Distributive properties
☐ Identity property of 0
☐ Additive identity
☐ Identity property of 1
☐ Multiplicative identity

1 Applying the Commutative Properties

Connecting Concepts with Your Life The order in which you perform actions is often important. For example, putting on your socks and then your shoes is not the same as putting on your shoes and then your socks. In mathematical terms, the action of putting on footwear is not *commutative* with respect to socks and shoes. However, the order in which you tie your shoes and put on a sweatshirt does not matter. So, these two actions are *commutative*. In mathematics some operations are commutative and others are not.

The **commutative property for addition** states that two numbers, a and b, can be added in any order and the result will be the same. That is, $a + b = b + a$. For example, if a person buys 4 DVDs and then buys 2 DVDs or first buys 2 DVDs and then buys 4 DVDs, as shown in **FIGURE 1.27**, the result is the same. Either way the person buys a total of $4 + 2 = 2 + 4 = 6$ DVDs.

Visualizing the Commutative Property for Addition

FIGURE 1.27 $4 + 2 = 2 + 4$

There is also a **commutative property for multiplication**. It states that two numbers, a and b, can be multiplied in any order and the result will be the same. That is, $a \cdot b = b \cdot a$. For example, if one person rolls 3 dice, each resulting in 6, and another person rolls 6 dice, each resulting in 3, as shown in **FIGURE 1.28**, then each person has rolled a total of $3 \cdot 6 = 6 \cdot 3 = 18$.

Visualizing the Commutative Property for Multiplication

FIGURE 1.28 $3 \cdot 6 = 6 \cdot 3$

We can summarize these results as follows.

COMMUTATIVE PROPERTIES

For any real numbers a and b,

$$a + b = b + a \quad \text{and} \quad a \cdot b = b \cdot a.$$

Addition   Multiplication

EXAMPLE 1   Applying the commutative properties

Use a commutative property to rewrite each expression.
(a) $15 + 100$ **(b)** $a \cdot 8$

Solution
(a) By the commutative property for addition, $15 + 100$ can be written as $100 + 15$.
(b) By the commutative property for multiplication, $a \cdot 8$ can be written as $8 \cdot a$ or $8a$.

Now Try Exercises **13, 19**

While there are commutative properties for addition and multiplication, the operations of subtraction and division are *not* commutative. **TABLE 1.8** shows each of the four arithmetic operations along with examples illustrating whether or not each operation is commutative.

READING CHECK 1

• Which of the four arithmetic operations are not commutative?

Commutativity of Operations

Operation	Commutative?	Example
$+$	Yes	$4 + 9 = 9 + 4$
$-$	No	$5 - 3 \neq 3 - 5$
$\cdot$	Yes	$8 \cdot 5 = 5 \cdot 8$
$\div$	No	$4 \div 2 \neq 2 \div 4$

TABLE 1.8

2 Applying the Associative Properties

The *associative properties* allow us to change how numbers are grouped. For example, if a person buys 1, 2, and 4 energy drinks, as shown in **FIGURE 1.29**, then the total number of drinks can be calculated either as

$$(1 + 2) + 4 = 3 + 4 = 7 \text{ or as } 1 + (2 + 4) = 1 + 6 = 7.$$

Visualizing the Associative Property for Addition

FIGURE 1.29 $(1 + 2) + 4 = 1 + (2 + 4)$

In either case, we obtain the same answer, 7 drinks, which is the result of the **associative property for addition**. We did not change the order of the numbers; we only changed how the numbers were grouped. There is also an **associative property for multiplication**, which can be illustrated by considering the total number of flowers shown in **FIGURE 1.30**, where 2 shelves hold 3 pots each, and each pot contains 4 flowers. The total number of flowers can be calculated either as

> Calculate the number of pots first (6) and then multiply by the number of flowers in each pot (4).

> Calculate the number of flowers on one shelf first (12) and then multiply by the number of shelves (2).

$$(2 \cdot 3) \cdot 4 = 6 \cdot 4 = 24 \quad \text{or as} \quad 2 \cdot (3 \cdot 4) = 2 \cdot 12 = 24.$$

Visualizing the Associative Property for Multiplication

FIGURE 1.30

We can summarize these results as follows.

ASSOCIATIVE PROPERTIES

For any real numbers a, b, and c,

$$(a + b) + c = a + (b + c) \quad \text{and} \quad (a \cdot b) \cdot c = a \cdot (b \cdot c).$$

Addition   Multiplication

NOTE: Sometimes we omit the multiplication dot: $a \cdot b = ab$ and $5 \cdot x \cdot y = 5xy$. ∎

EXAMPLE 2 Applying the associative properties

Use an associative property to rewrite each expression.
(a) $(5 + 6) + 7$ **(b)** $x(yz)$

Solution
(a) The given expression is equivalent to $5 + (6 + 7)$.
(b) The given expression is equivalent to $(xy)z$.

Now Try Exercises 21, 23

EXAMPLE 3 Identifying properties of real numbers

State the property that each equation illustrates.
(a) $5 \cdot (8y) = (5 \cdot 8)y$ **(b)** $3 \cdot 7 = 7 \cdot 3$ **(c)** $x + yz = yz + x$

Solution
(a) This equation illustrates the associative property for multiplication because the grouping of the numbers has been changed.
(b) This equation illustrates the commutative property for multiplication because the order of the numbers 3 and 7 has been changed.
(c) This equation illustrates the commutative property for addition because the order of the terms x and yz has been changed.

Now Try Exercises 53, 55, 63

While there are associative properties for addition and multiplication, the operations of subtraction and division are *not* associative. **TABLE 1.9** shows each of the four arithmetic operations along with examples illustrating whether or not each operation is associative.

Associativity of Operations

Operation	Associative?	Example
$+$	Yes	$(3 + 6) + 7 = 3 + (6 + 7)$
$-$	No	$(10 - 3) - 1 \neq 10 - (3 - 1)$
$\cdot$	Yes	$(4 \cdot 5) \cdot 3 = 4 \cdot (5 \cdot 3)$
$\div$	No	$(16 \div 8) \div 2 \neq 16 \div (8 \div 2)$

TABLE 1.9

READING CHECK 2

• Which of the four arithmetic operations are associative?

STUDY TIP

The information in Making Connections ties the current concepts to those studied earlier. By reviewing your notes often, you can gain a better understanding of mathematics.

MAKING CONNECTIONS 1

Commutative and Associative Properties

Both the commutative and associative properties work for addition and multiplication. However, neither property works for subtraction or division.

---

3 Applying the Distributive Properties

The **distributive properties** are used frequently in algebra to simplify expressions. Arrows are often used to indicate that a distributive property is being applied.

$$4(3 + 2) = 4 \cdot 3 + 4 \cdot 2$$

The 4 must be multiplied by *both* the 3 and the 2—not just the 3.

The distributive property remains valid when addition is replaced with subtraction.

$$4(3 - 2) = 4 \cdot 3 - 4 \cdot 2$$

The following See the Concept shows how the distributive property can be visualized geometrically.

See the Concept 1   **THE DISTRIBUTIVE PROPERTY**

There are 20 circles on the left side of the equals sign (12 blue and 8 red) and there are also 20 circles on the right side of the equals sign (again, 12 blue and 8 red).

When multiplying $4 \cdot (3 + 2)$, we can distribute the 4 over the sum in parentheses.

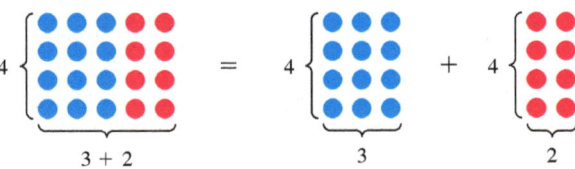

First Add 3 + 2, and then Multiply

$$4(3 + 2) = 4(5) = 20$$

First Distribute the 4, and then Add

$$4(3 + 2) = 4 \cdot 3 + 4 \cdot 2 = 12 + 8 = 20$$

The results are the same.

WATCH VIDEO IN $\boxed{\text{MML}}$.

DISTRIBUTIVE PROPERTIES

For any real numbers a, b, and c,

$$a(b + c) = ab + ac \quad \text{and} \quad a(b - c) = ab - ac.$$

NOTE: Because multiplication is commutative, the distributive properties can also be written as

$$(b + c)a = ba + ca \quad \text{and} \quad (b - c)a = ba - ca. \quad \blacksquare$$

EXAMPLE 4   **Applying the distributive properties**

Apply a distributive property to each expression.
(a) $3(x + 2)$ **(b)** $-6(a - 2)$ **(c)** $-(x + 7)$ **(d)** $15 - (b + 4)$ **(e)** $(x - 2)2$

Solution
(a) Both the x and the 2 must be multiplied by 3.

$$3(x + 2) = 3 \cdot x + 3 \cdot 2 = 3x + 6$$

(b) $-6(a - 2) = -6 \cdot a - (-6) \cdot 2 = -6a + 12$

(c) $-(x + 7) = (-1)(x + 7)$ In general, $-a = -1 \cdot a$.

   $= (-1) \cdot x + (-1) \cdot 7$ Distributive property

   $= -x - 7$ Multiply.

(d) $15 - (b + 4) = 15 + (-1)(b + 4)$ Change subtraction to addition.

   $= 15 + (-1) \cdot b + (-1) \cdot 4$ Distributive property

   $= 15 - b - 4$ Multiply.

   $= 11 - b$ Simplify.

NOTE: To simplify the expression $15 - (b + 4)$, we subtract *both* the b and the 4. Thus we can quickly simplify the given expression to

$$15 - (b + 4) = 15 - b - 4. \blacksquare$$

(e) $(x - 2)2 = 2x - 4$

Now Try Exercises 31, 35, 37, 39, 41

EXAMPLE 5   **Inserting parentheses using the distributive property**

Use a distributive property to insert parentheses in the expression and then simplify the result.
(a) $5a + 2a$ **(b)** $3x - 7x$ **(c)** $-4y + 5y$

Solution
(a) Because a is a factor in both $5a$ and $2a$, we use the distributive property to write the a outside of parentheses.

$$5a + 2a = (5 + 2)a \quad \text{Distributive property}$$

$$= 7a \quad \text{Simplify.}$$

(b) $3x - 7x = (3 - 7)x = -4x$
(c) $-4y + 5y = (-4 + 5)y = 1y = y$

Now Try Exercises 45, 47, 51

EXAMPLE 6   **Identifying properties of real numbers**

State the property or properties illustrated by each equation.
(a) $4(5 - x) = 20 - 4x$ **(b)** $(4 + x) + 5 = x + 9$
(c) $5z + 7z = 12z$ **(d)** $x(y + z) = zx + yx$

Solution

(a) The equation $4(5 - x) = 20 - 4x$ illustrates the distributive property (subtraction).

$$4(5 - x) = 4 \cdot 5 - 4 \cdot x = 20 - 4x$$

(b) The equation $(4 + x) + 5 = x + 9$ illustrates the commutative and associative properties for addition.

$$
\begin{aligned}
(4 + x) + 5 &= (x + 4) + 5 && \text{Commutative property for addition} \\
&= x + (4 + 5) && \text{Associative property for addition} \\
&= x + 9 && \text{Simplify.}
\end{aligned}
$$

(c) The equation $5z + 7z = 12z$ illustrates the distributive property (addition).

$$5z + 7z = (5 + 7)z = 12z$$

(d) The equation $x(y + z) = zx + yx$ illustrates the distributive property (addition) and commutative properties for addition and multiplication.

$$
\begin{aligned}
x(y + z) &= xy + xz && \text{Distributive property} \\
&= xz + xy && \text{Commutative property for addition} \\
&= zx + yx && \text{Commutative property for multiplication}
\end{aligned}
$$

Now Try Exercises 57, 59, 61, 65

4 Recognizing Identity and Inverse Properties

READING CHECK 3

• What number is the additive identity?

• What number is the multiplicative identity?

The **identity property of 0** states that if 0 is added to any real number a, the result is a. The number 0 is called the **additive identity**. Examples include

$$-4 + 0 = -4 \quad \text{and} \quad 0 + 11 = 11.$$

The **identity property of 1** states that if any number a is multiplied by 1, the result is a. The number 1 is called the **multiplicative identity**. Examples include

$$-3 \cdot 1 = -3 \quad \text{and} \quad 1 \cdot 8 = 8.$$

IDENTITY PROPERTIES

For any real number a,

$$a + 0 = 0 + a = a \quad \text{and} \quad a \cdot 1 = 1 \cdot a = a.$$

Additive identity: 0   Multiplicative identity: 1

The *additive inverse*, or *opposite*, of a number a is $-a$. The number 0 is its own opposite. The sum of a number a and its additive inverse equals the additive identity 0. Thus $-5 + 5 = 0$ and $x + (-x) = 0$.

The *multiplicative inverse*, or *reciprocal*, of a nonzero number a is $\frac{1}{a}$. The number 0 has no multiplicative inverse. The product of a number and its multiplicative inverse equals the multiplicative identity 1. For example, $-\frac{5}{4} \cdot \left(-\frac{4}{5}\right) = 1$.

INVERSE PROPERTIES

For any real number a,

$$a + (-a) = 0 \quad \text{and} \quad -a + a = 0. \quad \text{Additive inverse: } -a$$

For any *nonzero* real number a,

$$a \cdot \frac{1}{a} = 1 \quad \text{and} \quad \frac{1}{a} \cdot a = 1. \quad \text{Multiplicative inverse: } \frac{1}{a}$$

EXAMPLE 7 **Recognizing identity and inverse properties**

State the property or properties illustrated by each equation.

(a) $0 + xy = xy$ 　　　　　　　　(b) $\frac{36}{30} = \frac{6}{5} \cdot \frac{6}{6} = \frac{6}{5}$

(c) $x + (-x) + 5 = 0 + 5 = 5$ 　　(d) $\frac{1}{9} \cdot 9y = 1 \cdot y = y$

Solution

(a) This equation illustrates use of the identity property for 0.

(b) Because $\frac{6}{6} = 1$, these equations illustrate how a fraction can be simplified by using the identity property for 1.

(c) These equations illustrate use of the additive inverse property and the identity property for 0.

(d) These equations illustrate use of the multiplicative inverse property and the identity property for 1.

Now Try Exercises 67, 69, 71, 73

5 Performing Mental Calculations

Properties of numbers can be used to simplify calculations. For example, to find the sum

$$4 + 7 + 6 + 3,$$

we might apply the commutative and associative properties for addition to obtain

$$(4 + 6) + (7 + 3) = 10 + 10 = 20.$$

Suppose that we are to add $128 + 19$ mentally. One way is to add 20 to 128 and then subtract 1.

$$
\begin{aligned}
128 + 19 &= 128 + (20 - 1) &&19 = 20 - 1 \\
&= (128 + 20) - 1 &&\text{Associative property} \\
&= 148 - 1 &&\text{Add.} \\
&= 147 &&\text{Subtract.}
\end{aligned}
$$

The distributive property can be helpful when we multiply mentally. For example, to determine the number of people in a marching band with 7 columns and 23 rows, we need to find the product $7 \cdot 23$. To evaluate the product mentally, think of 23 as $20 + 3$.

$$
\begin{aligned}
7 \cdot 23 &= 7(20 + 3) &&23 = 20 + 3. \\
&= 7 \cdot 20 + 7 \cdot 3 &&\text{Distributive property} \\
&= 140 + 21 &&\text{Multiply.} \\
&= 161 &&\text{Add.}
\end{aligned}
$$

EXAMPLE 8 **Performing calculations mentally**

Use properties of real numbers to calculate each expression mentally.

(a) $21 + 15 + 9 + 5$ 　(b) $\frac{1}{2} \cdot \frac{2}{3} \cdot 2 \cdot \frac{3}{2}$ 　(c) $523 + 199$ 　(d) $6 \cdot 55$

Solution

(a) Use the commutative and associative properties to group numbers into pairs that sum to a multiple of 10.

$$21 + 15 + 9 + 5 = (21 + 9) + (15 + 5) = 30 + 20 = 50$$

(b) Use the commutative and associative properties to group numbers with their reciprocals.

$$\frac{1}{2} \cdot \frac{2}{3} \cdot 2 \cdot \frac{3}{2} = \left(\frac{1}{2} \cdot 2\right) \cdot \left(\frac{2}{3} \cdot \frac{3}{2}\right) = 1 \cdot 1 = 1$$

(c) Instead of adding 199, add 200 and then subtract 1.

$$523 + \mathbf{199} = 523 + \mathbf{200} - \mathbf{1} = 723 - 1 = 722$$

(d) Think of 55 as $50 + 5$ and then apply the distributive property.

$$6 \cdot \mathbf{55} = 6 \cdot (\mathbf{50 + 5}) = 300 + 30 = 330$$

Now Try Exercises 77, 81, 85, 89

CRITICAL THINKING 1

How could you quickly calculate $5283 - 198$ without a calculator?

🌐 *Math in Context* (Olympics) The next example illustrates how the commutative and associative properties for multiplication can be used together to simplify a product that gives the volume of a swimming pool.

EXAMPLE 9 **Finding the volume of a swimming pool**

An Olympic swimming pool is 50 meters long, 25 meters wide, and 2 meters deep. The volume V of the pool is found by multiplying 50, 25, and 2. Use the commutative and associative properties for multiplication to calculate the volume of the pool mentally.

Solution
Because $50 \cdot 2 = 100$ and multiplication by 100 is relatively easy, it may be convenient to order and group the multiplication as $(50 \cdot 2) \cdot 25 = 100 \cdot 25 = 2500$. Thus the pool contains 2500 cubic meters of water.

Now Try Exercise 103

1.7   Putting It All Together

PROPERTY	DEFINITION	EXAMPLES
Commutative	For any real numbers a and b, $$a + b = b + a \quad \text{and}$$ $$a \cdot b = b \cdot a.$$	$4 + 6 = 6 + 4$ $4 \cdot 6 = 6 \cdot 4$
Associative	For any real numbers a, b, and c, $$(a + b) + c = a + (b + c) \quad \text{and}$$ $$(a \cdot b) \cdot c = a \cdot (b \cdot c).$$	$(3 + 4) + 5 = 3 + (4 + 5)$ $(3 \cdot 4) \cdot 5 = 3 \cdot (4 \cdot 5)$
Distributive	For any real numbers a, b, and c, $$a(b + c) = ab + ac \quad \text{and}$$ $$a(b - c) = ab - ac.$$	$5(x + 2) = 5x + 10$ $5(x - 2) = 5x - 10$
Identity (0 and 1)	The identity for addition is 0, and the identity for multiplication is 1. For any real number a, $a + 0 = a$ and $a \cdot 1 = a$.	$5 + 0 = 5 \quad \text{and} \quad 5 \cdot 1 = 5$
Inverse	The additive inverse of a is $-a$, and $a + (-a) = 0$. The multiplicative inverse of a nonzero number a is $\frac{1}{a}$, and $a \cdot \frac{1}{a} = 1$.	$8 + (-8) = 0 \quad \text{and} \quad \frac{2}{3} \cdot \frac{3}{2} = 1$ Note that if $a = \frac{2}{3}$, then $\frac{1}{a} = \frac{3}{2}$.

1.7 Exercises

MyMathLab®

CONCEPTS AND VOCABULARY

1. $a + b = b + a$ illustrates the _____ property for _____.

2. $a \cdot b = b \cdot a$ illustrates the _____ property for _____.

3. $(a + b) + c = a + (b + c)$ illustrates the _____ property for _____.

4. $(a \cdot b) \cdot c = a \cdot (b \cdot c)$ illustrates the _____ property for _____.

5. (True or False?) Addition and multiplication are commutative.

6. (True or False?) Subtraction and division are associative.

7. $a(b + c) = ab + ac$ illustrates the _____ property.

8. $a(b - c) = ab - ac$ illustrates the _____ property.

9. The equations $a + 0 = a$ and $0 + a = a$ each illustrate the _____ property for _____.

10. The equations $a \cdot 1 = a$ and $1 \cdot a = a$ each illustrate the _____ property for _____.

11. The additive inverse, or opposite, of a is _____.

12. The multiplicative inverse, or reciprocal, of a nonzero number a is _____.

APPLYING THE COMMUTATIVE PROPERTIES

Exercises 13–20: Use a commutative property to rewrite the expression. Do not simplify.

13. $-6 + 10$

14. $23 + 7$

15. $-5 \cdot 6$

16. $25 \cdot (-46)$

17. $a + 10$

18. $b + c$

19. $b \cdot 7$

20. $a \cdot 23$

APPLYING THE ASSOCIATIVE PROPERTIES

Exercises 21–28: Use an associative property to rewrite the expression. Do not simplify.

21. $(1 + 2) + 3$

22. $-7 + (5 + 15)$

23. $2 \cdot (3 \cdot 4)$

24. $(9 \cdot (-4)) \cdot 5$

25. $(a + 5) + c$

26. $(10 + b) + a$

27. $(x \cdot 3) \cdot 4$

28. $5 \cdot (x \cdot y)$

29. **Thinking Generally** Use the commutative and associative properties to show that $a + b + c = c + b + a$.

30. **Thinking Generally** Use the commutative and associative properties to show that $a \cdot b \cdot c = c \cdot b \cdot a$.

APPLYING THE DISTRIBUTIVE PROPERTIES

Exercises 31–42: Use a distributive property to rewrite the expression. Then simplify the expression.

31. $4(3 + 2)$

32. $5(6 - 9)$

33. $a(b - 8)$

34. $3(x + y)$

35. $-4(t - z)$

36. $-1(a + 6)$

37. $-(5 - a)$

38. $12 - (4u - b)$

39. $(a + 5)3$

40. $(x + y)7$

41. $12 - (a - 5)$

42. $4x - 2(3y - 5)$

43. **Thinking Generally** Use properties of real numbers to show that the distributive property can be extended as follows.

$$a \cdot (b + c + d) = ab + ac + ad.$$

44. **Thinking Generally** Use properties of real numbers to show that the distributive property can be extended as follows.

$$a \cdot (b - c - d) = ab - ac - ad.$$

Exercises 45–52: Use the distributive property to insert parentheses in the expression and then simplify the result.

45. $6x + 5x$

46. $4y - y$

47. $-4b + 3b$

48. $2b + 8b$

49. $3a - a$

50. $-2x + 5x$

51. $13w - 27w$

52. $25a - 21a$

RECOGNIZING PROPERTIES OF REAL NUMBERS

Exercises 53–66: State the property or properties that the equation illustrates.

53. $x \cdot 5 = 5x$

54. $7 + a = a + 7$

55. $(a + 5) + 7 = a + 12$

56. $(9 + a) + 8 = a + 17$

57. $4(5 + x) = 20 + 4x$

58. $3(5 + x) = 3x + 15$

59. $x(3 - y) = 3x - xy$

60. $-(u - v) = -u + v$

61. $6x + 9x = 15x$ **62.** $9x - 11x = -2x$

63. $3 \cdot (4 \cdot a) = 12a$ **64.** $(x \cdot 3) \cdot 5 = 15x$

65. $-(t - 7) = -t + 7$

66. $z \cdot 5 - y \cdot 6 = 5z - 6y$

RECOGNIZING IDENTITY AND INVERSE PROPERTIES

Exercises 67–76: State the property or properties that are illustrated.

67. $0 + x = x$ **68.** $5x + 0 = 5x$

69. $1 \cdot a = a$ **70.** $\frac{1}{7} \cdot (7 \cdot a) = a$

71. $\frac{25}{15} = \frac{5}{3} \cdot \frac{5}{5} = \frac{5}{3}$ **72.** $\frac{50}{40} = \frac{5}{4} \cdot \frac{10}{10} = \frac{5}{4}$

73. $\frac{1}{xy} \cdot xy = 1$ **74.** $\frac{1}{a+b} \cdot (a + b) = 1$

75. $-xyz + xyz = 0$ **76.** $\frac{1}{y} + \left(-\frac{1}{y}\right) = 0$

PERFORMING MENTAL CALCULATIONS

Exercises 77–92: Use properties of real numbers to calculate the expression mentally.

77. $4 + 2 + 9 + 8 + 1 + 6$

78. $21 + 32 + 19 + 8$

79. $45 + 43 + 5 + 7$

80. $5 + 7 + 12 + 13 + 8$

81. $129 + 49$ **82.** $87 + 99$

83. $178 - 99$ **84.** $500 - 101$

85. $6 \cdot 15$ **86.** $4 \cdot 56$

87. $8 \cdot 102$ **88.** $5 \cdot 999$

89. $\frac{1}{2} \cdot \frac{1}{2} \cdot \frac{1}{2} \cdot 2 \cdot 2 \cdot 2$ **90.** $\frac{1}{2} \cdot \frac{4}{5} \cdot \frac{7}{3} \cdot 2 \cdot \frac{5}{4}$

91. $\frac{7}{6} \cdot \frac{1}{2} \cdot \frac{1}{2} \cdot \frac{1}{2} \cdot \frac{8}{7}$ **92.** $\frac{4}{11} \cdot \frac{11}{6} \cdot \frac{6}{7} \cdot \frac{7}{4}$

MULTIPLYING AND DIVIDING BY POWERS OF 10 MENTALLY

93. Multiplying by 10 To multiply an integer by 10, attach one 0 to the number. For example, $10 \cdot 23 = 230$. Simplify each expression mentally.
  (a) $10 \cdot 41$ **(b)** $10 \cdot 997$
  (c) $-630 \cdot 10$ **(d)** $-14,000 \cdot 10$

94. Multiplying by 10 To multiply a decimal number by 10, move the decimal point one place to the right. For example, $10 \cdot 23.45 = 234.5$. Simplify each expression at the top of the next column mentally.

  (a) $10 \cdot 101.68$ **(b)** $10 \cdot (-1.235)$
  (c) $-113.4 \cdot 10$ **(d)** $0.567 \cdot 10$
  (e) $10 \cdot 0.0045$ **(f)** $-0.05 \cdot 10$

95. Multiplying by Powers of 10 To multiply an integer by a power of 10 in the form

$$10^k = \underbrace{100\ldots0},$$

attach k zeros to the number. Some examples of this are $100 \cdot 45 = 4500$, $1000 \cdot 235 = 235{,}000$, and $10{,}000 \cdot 12 = 120{,}000$. Simplify each expression mentally.
  (a) $1000 \cdot 19$ **(b)** $100 \cdot (-451)$
  (c) $10{,}000 \cdot 6$ **(d)** $-79 \cdot 100{,}000$

96. Multiplying by Powers of 10 To multiply a decimal number by a power of 10 in the form

$$10^k = \underbrace{100\ldots0},$$

move the decimal point k places to the right. For example, $100 \cdot 1.234 = 123.4$. Simplify each given expression mentally.
  (a) $1000 \cdot 1.2345$ **(b)** $100 \cdot (-5.1)$
  (c) $45.67 \cdot 1000$ **(d)** $0.567 \cdot 10{,}000$
  (e) $100 \cdot 0.0005$ **(f)** $-0.05 \cdot 100{,}000$

97. Dividing by 10 To divide a number by 10, move the decimal point one place to the left. For example, $78.9 \div 10 = 7.89$. Simplify each expression mentally.
  (a) $12.56 \div 10$ **(b)** $9.6 \div 10$
  (c) $0.987 \div 10$ **(d)** $-0.056 \div 10$
  (e) $1200 \div 10$ **(f)** $4578 \div 10$

98. Dividing by Powers of 10 To divide a decimal number by a power of 10 in the form

$$10^k = \underbrace{100\ldots0},$$

move the decimal point k places to the left. For example, $123.4 \div 100 = 1.234$. Simplify each expression mentally.
  (a) $78.89 \div 100$ **(b)** $0.05 \div 1000$
  (c) $5678 \div 10{,}000$ **(d)** $-9.8 \div 1000$
  (e) $-101 \div 100{,}000$ **(f)** $7.8 \div 100$

APPLICATIONS INVOLVING PROPERTIES OF REAL NUMBERS

99. Earnings Earning $100 one day and $75 the next day is equivalent to earning $75 the first day and $100 the second day. What property of real numbers does this example illustrate?

100. Leasing a Car An advertisement for a lease on a new car states that it costs $2480 down and $201 per month for 20 months. Mentally calculate the cost of the lease. Explain your reasoning.

101. Gasoline Mileage A car travels 198 miles on 10 gallons of gasoline. Mentally calculate the number of miles that the car travels on 1 gallon of gasoline.

102. Gallons of Water A wading pool is 50 feet long, 20 feet wide, and 1 foot deep.
(a) Mentally determine the number of cubic feet in the pool. (*Hint:* Volume equals length times width times height.)
(b) One cubic foot equals about 7.5 gallons. Mentally calculate the number of gallons of water in the pool.

103. Dimensions of a Pool A small pool of water is 13 feet long, 5 feet wide, and 2 feet deep. The volume *V* of the pool in cubic feet is found by multiplying 13 by 5 by 2.
(a) To do this calculation mentally, would you multiply $(13 \cdot 5) \cdot 2$ or $13 \cdot (5 \cdot 2)$? Why?
(b) What property allows you to do either calculation and still obtain the correct answer?

104. Digital Images of Io The accompanying picture of Jupiter's moon Io is a digital picture, created by using a rectangular pattern of small pixels. This image is 400 pixels wide and 500 pixels high, so the total number of pixels in it is $400 \cdot 500 = 200,000$ pixels. (*Source:* NASA.)

(a) Find the total number of pixels in an image 500 pixels wide and 400 pixels high.
(b) Suppose that a picture is *x* pixels wide and *y* pixels high. What property states that it has the same number of pixels as a picture *y* pixels wide and *x* pixels high?

WRITING ABOUT MATHEMATICS

105. To determine the cost of tuition, a student tries to compute $16 \cdot \$96$ with a calculator and gets $\$15,360$. How do you know that this computation is not correct?

106. A student performs the following computation by hand.

$$20 - 6 - 2 + 8 \div 4 \div 2 \stackrel{?}{=} 20 - 4 + 8 \div 2$$
$$\stackrel{?}{=} 20 - 4 + 4$$
$$\stackrel{?}{=} 20$$

Find any incorrect steps and explain what is wrong. What is the correct answer?

107. The computation $3 + 10^{20} - 10^{20}$ performed on a calculator gives a result of 0. (Try it.) What is the correct answer? Why is it important to know properties of real numbers even though you have a calculator?

108. Does the "distributive property"

$$a + (b \cdot c) \stackrel{?}{=} (a + b) \cdot (a + c)$$

hold for all numbers *a*, *b*, and *c*? Explain your reasoning.

Group Activity

Winning the Lottery In the multistate lottery game *Powerball*, there are 175,223,510 possible number combinations, only one of which is the grand prize winner. The cost of a single ticket (one number combination) is $2. (*Source:* Powerball.com.)

Suppose that a *very* wealthy person decides to buy tickets for every possible number combination to be assured of winning a $400 million grand prize.

(a) If this individual could purchase one ticket every second, how many hours would it take to buy all of the tickets? How many years is this?
(b) If there were a way for this individual to buy all possible number combinations quickly, discuss reasons why this strategy might lose money.

1.8   Simplifying and Writing Algebraic Expressions

Objectives

1 **Identifying Terms**
2 **Combining Like Terms**
3 **Simplifying Expressions**
4 **Writing Expressions**

NEW VOCABULARY

☐ Term
☐ Coefficient
☐ Like terms

READING CHECK 1

• Identify the coefficient for the term $-5x^2y$.

1 Identifying Terms

One way to simplify expressions is to combine like terms. A **term** is a number, a variable, or a *product* of numbers and variables raised to natural number powers. Examples include

$$4, \quad z, \quad 5x, \quad \frac{2}{5}z, \quad -4xy, \quad -x^2, \quad \text{and} \quad 6x^3y^4. \quad \boxed{\text{Terms}}$$

Terms do not contain addition or subtraction signs, but they can contain negative signs. The **coefficient** of a term is the number that appears in the term. If no number appears, then the coefficient is understood to be either -1 or 1. **TABLE 1.10** shows examples of terms and their corresponding coefficients.

Terms and Coefficients

Term	Coefficient
$\frac{1}{2}xy$	$\frac{1}{2}$
$-z^2$	-1
12	12
$-5w^4$	-5

TABLE 1.10

EXAMPLE 1   **Identifying terms**

Determine whether each expression is a term. If it is a term, identify its coefficient.
(a) 51   **(b)** $5a$ **(c)** $2x + 3y$ **(d)** $-3x^2$

Solution
(a) A number is a term. This term's coefficient is 51.
(b) The product of a number and a variable is a term. This term's coefficient is 5.
(c) The sum (or difference) of two terms is not a term.
(d) The product of a number and a variable with an exponent is a term. This term's coefficient is -3.

Now Try Exercises 7, 9, 11, 13

STUDY TIP

One of the best ways to prepare for class is to read a section before it is covered by your instructor. Reading ahead gives you the chance to formulate any questions you might have about the concepts in the section.

MAKING CONNECTIONS 1

Factors and Terms

When variables and numbers are multiplied, they are called *factors*. When they are added or subtracted, they are called *terms*.

Expression	Expression
$4xy$	$a + 5ab + 1$
Three factors: 4, x, and y	Three terms: a, $5ab$, and 1

NOTE: The term $5ab$ in the second expression has three factors: 5, a, and b. ∎

READING CHECK 2

• Which property of real numbers is used to combine like terms?

2 Combining Like Terms

 Connecting Concepts with Your Life Suppose that we have two boards with lengths $2x$ and $3x$, where the value of x could be any length, such as a foot. The following See the Concept illustrates how we can determine that the combined length is $5x$ by applying the distributive property. The lengths $2x$ and $3x$ are **like terms** because they contain the same variables raised to the same powers. However, if their lengths were $2x$ and $3v$, where x might be a yard and v might be a meter, then the terms $2x$ and $3v$ are **unlike terms** and cannot be combined.

See the Concept 1　LIKE AND UNLIKE TERMS

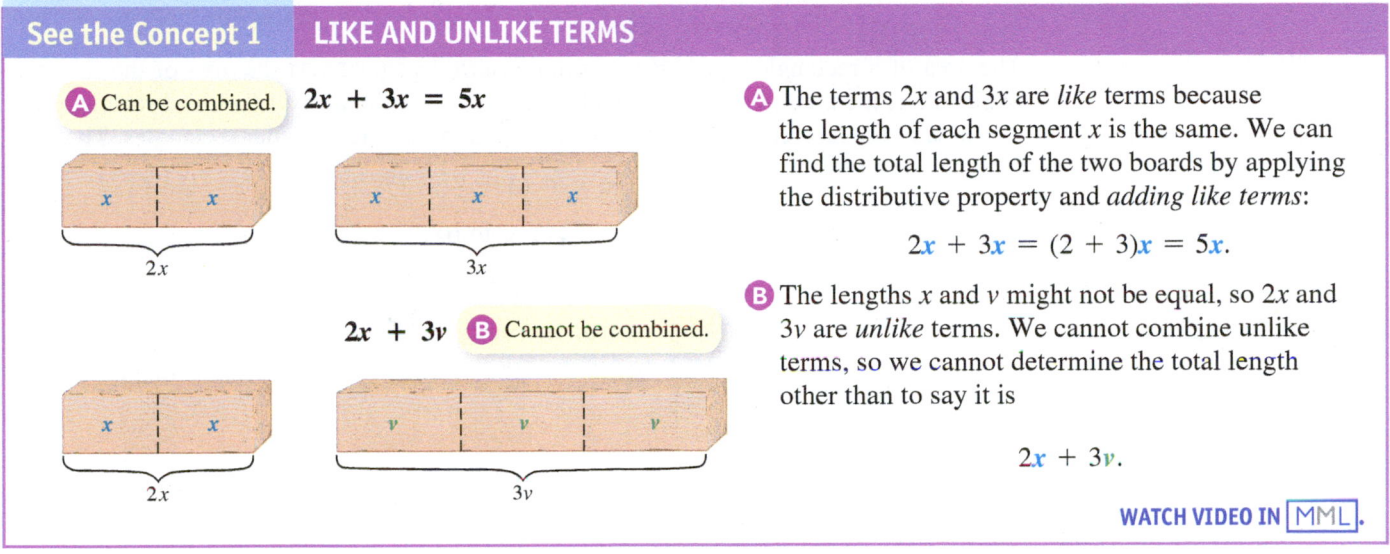

Ⓐ Can be combined.　$2x + 3x = 5x$

Ⓐ The terms $2x$ and $3x$ are *like* terms because the length of each segment x is the same. We can find the total length of the two boards by applying the distributive property and *adding like terms*:

$$2x + 3x = (2 + 3)x = 5x.$$

$2x + 3v$　Ⓑ Cannot be combined.

Ⓑ The lengths x and v might not be equal, so $2x$ and $3v$ are *unlike* terms. We cannot combine unlike terms, so we cannot determine the total length other than to say it is

$$2x + 3v.$$

WATCH VIDEO IN MML .

READING CHECK 3

Which of the following pairs of expressions are like terms?

$$-5x^2, x^2; \quad xy, -xy; \quad 4a^2b, 4ab^2; \quad 5y, -9y$$

EXAMPLE 2　**Identifying like terms**

Determine whether the terms are like or unlike.

(a) $-4m, 7m$　**(b)** $8x^2, 8y^2$　**(c)** $\frac{1}{2}z, -3z^2$　**(d)** $5, -4n$

Solution

(a) The variable in both terms is m (with power 1), so they are like terms.

(b) The variables are different, so they are unlike terms.

(c) The term $-3z^2$ contains $z^2 = z \cdot z$, whereas the term $\frac{1}{2}z$ contains only z. Thus they are unlike terms.

(d) The term 5 has no variable, whereas the term $-4n$ contains the variable n. They are unlike terms.

Now Try Exercises 19, 21, 23, 27

EXAMPLE 3　**Combining like terms**

Combine terms in each expression, if possible.

(a) $-3x + 5x$　**(b)** $-x^2 + 5x^2$　**(c)** $\frac{1}{2}y - 3y^3$

Solution

(a) Combine terms in $-3x + 5x$ by applying a distributive property.

$$-3x + 5x = (-3 + 5)x = 2x$$

(b) Note that $-x^2$ can be written as $-1x^2$. Combine like terms in $-x^2 + 5x^2$ as follows.

$$-x^2 + 5x^2 = (-1 + 5)x^2 = 4x^2$$

(c) The terms in $\frac{1}{2}y - 3y^3$ are unlike terms, so they cannot be combined.

Now Try Exercises 31, 39, 43

3 Simplifying Expressions

The area of a rectangle equals length times width. In **FIGURE 1.31** the area of the first rectangle is $3x$, the area of the second rectangle is $2x$, and the area of the third rectangle is x. The area of the last rectangle equals the total area of the three smaller rectangles. That is,

$$3x + 2x + x = (3 + 2 + 1)x = 6x.$$

The expression $3x + 2x + x$ can be *simplified* to $6x$.

CRITICAL THINKING 1

Use rectangles to explain how to add $xy + 2xy$.

Simplifying an Expression Visually: $3x + 2x + x = 6x$

FIGURE 1.31

EXAMPLE 4 **Simplifying expressions**

Simplify each expression.
(a) $2 + x - 6 + 5x$ **(b)** $2y - (y + 3)$ **(c)** $\dfrac{-1.5x}{-1.5}$

Solution
(a) Combine like terms by applying the properties of real numbers.

$$
\begin{aligned}
2 + x - 6 + 5x &= 2 + (-6) + x + 5x && \text{Commutative property} \\
&= 2 + (-6) + (1 + 5)x && \text{Distributive property} \\
&= -4 + 6x && \text{Add.}
\end{aligned}
$$

(b)
$$
\begin{aligned}
2y + (-1)(y + 3) &= 2y + (-1)y + (-1) \cdot 3 && \text{Distributive property} \\
&= 2y - 1y - 3 && \text{Definition of subtraction} \\
&= (2 - 1)y - 3 && \text{Distributive property} \\
&= y - 3 && \text{Subtract.}
\end{aligned}
$$

Change subtraction to adding the opposite.

(c)
$$
\begin{aligned}
\frac{-1.5x}{-1.5} &= \frac{-1.5}{-1.5} \cdot \frac{x}{1} && \text{Multiplication of fractions} \\
&= 1 \cdot x && \text{Simplify the fractions.} \\
&= x && \text{Multiplicative identity}
\end{aligned}
$$

Now Try Exercises 47, 59, 75

NOTE: The expression in Example 4(c) can be simplified directly by using the basic principle of fractions: $\frac{ac}{bc} = \frac{a}{b}$. ■

EXAMPLE 5 | **Simplifying expressions**

Simplify each expression.
(a) $7 - 2(5x - 3)$ (b) $3x^3 + 2x^3 - x^3$

(c) $3y^2 - z + 4y^2 - 2z$ (d) $\dfrac{12x - 8}{4}$

Solution

(a) $\quad 7 - 2(5x - 3) = 7 + (-2)(5x + (-3))$ Change subtraction to addition.

$\qquad\qquad\qquad\quad = 7 + (-2)(5x) + (-2)(-3)$ Distributive property

$\qquad\qquad\qquad\quad = 7 - 10x + 6$ Multiply.

$\qquad\qquad\qquad\quad = 13 - 10x$ Combine like terms.

(b) $\quad 3x^3 + 2x^3 - x^3 = (3 + 2 - 1)x^3$ Distributive property

$\qquad\qquad\qquad\quad = 4x^3$ Add and subtract.

(c) $\quad 3y^2 - z + 4y^2 - 2z = 3y^2 + 4y^2 + (-1z) + (-2z)$ Commutative property

$\qquad\qquad\qquad\quad = (3 + 4)y^2 + (-1 + (-2))z$ Distributive property

$\qquad\qquad\qquad\quad = 7y^2 - 3z$ Add.

(d) $\qquad \dfrac{12x - 8}{4} = \dfrac{12x}{4} - \dfrac{8}{4}$ Subtraction of fractions

$\qquad\qquad\qquad\quad = 3x - 2$ Simplify fractions.

Now Try Exercises 57, 65, 71, 79

READING CHECK 4

What two properties make it possible to simplify an expression directly by allowing us to rearrange a sum in any order?

Recall that the commutative and associative properties of addition allow us to rearrange a sum in any order. For example, if we write the expression

$$2x - 3 - 4x + 10 \quad \text{as the sum} \quad 2x + (-3) + (-4x) + 10,$$

the terms can be arranged and grouped as

$$(2x + (-4x)) + (-3 + 10).$$

Applying the distributive property and adding within each grouping results in

$$(2 + (-4))x + (-3 + 10) = -2x + 7.$$

Note that the terms in the result can be found directly by combining like (same color) terms in the given expression, where addition indicates that a term is positive and subtraction indicates that a term is negative. In the next example, we simplify an expression in this way.

EXAMPLE 6 | **Simplifying an expression directly**

Simplify the expression $6x + 5 - 2x - 8$.

Solution
The like terms and their indicated signs are shown.

$$6x + 5 - 2x - 8$$

By combining like terms, $6x - 2x$ and $5 - 8$, the expression can be simplified as

$$4x - 3.$$

Now Try Exercise 53

4 Writing Expressions

 Connecting Concepts with Your Life In real-life situations, we often have to translate words to symbols. For example, to calculate federal and state income tax, we might have to multiply taxable income by 0.15 and by 0.05 and then find the sum. If we let x represent taxable income, then the total federal and state income tax is $0.15x + 0.05x$. This expression can be simplified with a distributive property.

$$0.15x + 0.05x = (0.15 + 0.05)x$$
$$= 0.20x$$

Thus the total income tax on a taxable income of $x = \$20,000$ would be

$$0.20(20,000) = \$4000.$$

EXAMPLE 7 **Writing and simplifying an expression**

A sidewalk has a constant width w and several short sections with lengths 12, 6, and 5 feet, as illustrated in **FIGURE 1.32**.
(a) Write and simplify an expression that gives the number of square feet of sidewalk.
(b) Find the area of the sidewalk if its width is 3 feet.

Finding the Area of a Sidewalk

6 ft

5 ft

w

12 ft

FIGURE 1.32

Solution
(a) The area of each sidewalk section equals its length times its width w. The total area of the sidewalk is

$$12w + 6w + 5w = (12 + 6 + 5)w = 23w.$$

(b) When $w = 3$, the area is $23w = 23 \cdot 3 = 69$ square feet.

Now Try Exercise 91

1.8 Putting It All Together

CONCEPT	COMMENTS	EXAMPLES	
Term	A term is a number, variable, or product of numbers and variables raised to natural number powers.	$12, -10, x, -y$ $-3x, 5z, xy, 6x^2$ $10y^3, \frac{3}{4}x, 25xyz^2$	
Coefficient of a Term	The coefficient of a term is the number that appears in the term. If no number appears, then the coefficient is either 1 or -1.	12	Coefficient is 12
		x	Coefficient is 1
		$-xy$	Coefficient is -1
		$-4x$	Coefficient is -4

CONCEPT	COMMENTS	EXAMPLES
Like Terms	Like terms have the same variables raised to the same powers.	$5m$ and $-6m$ x^2 and $-74x^2$ $2xy$ and $-xy$
Combining Like Terms	Like terms can be combined by using a distributive property.	$5x + 2x = (5 + 2)x = 7x$ $y - 3y = (1 - 3)y = -2y$ $8x^2 + x^2 = (8 + 1)x^2 = 9x^2$

1.8 Exercises

MyMathLab®

CONCEPTS AND VOCABULARY

1. A(n) _____ is a number, a variable, or a product of numbers and variables raised to natural number powers.

2. The number 7 in the term $7x^2y$ is called the _____ of the term.

3. When variables and numbers are multiplied, they are called _____. When they are added or subtracted, they are called _____.

4. If two terms contain the same variables raised to the same powers, they are (like/unlike) terms.

5. We can add or subtract (like/unlike) terms.

6. We can combine like terms in an expression by applying a(n) _____ property.

IDENTIFYING TERMS

Exercises 7–16: Determine whether the expression is a term. If the expression is a term, identify its coefficient.

7. 91

8. -12

9. $-6b$

10. $9z$

11. $x + 10$

12. $20 - 2y$

13. x^2

14. $4x^3$

15. $4x - 5$

16. $5z + 6x$

COMBINING LIKE TERMS

Exercises 17–28: Determine whether the terms are like or unlike.

17. $6, -8$

18. $2x, 19$

19. $5x, -22x$

20. $19y, -y$

21. $14, 14a$

22. $-33b, -3b$

23. $18x, 18y$

24. $-6a, -6b$

25. $x^2, -15x^2$

26. $y, 19y$

27. $3x^2, \frac{1}{5}x$

28. $12y^2, -y^2$

29. **Thinking Generally** Are the terms $4ab$ and $-3ba$ like or unlike?

30. **Thinking Generally** Are the terms $-xyz^2$ and $3yz^2x$ like or unlike?

Exercises 31–46: Combine terms, if possible.

31. $-4x + 7x$

32. $6x - 8x$

33. $19y - 5y$

34. $22z + z$

35. $28a + 13a$

36. $41b - 17b$

37. $11z - 11z$

38. $4y + 4y$

39. $5x - 7y$

40. $3y + 3z$

41. $5 + 5y$

42. $x + x^2$

43. $5x^2 - 2x^2$

44. $25z^3 - 10z^3$

45. $8y - 10y + y$

46. $4x^2 + x^2 - 5x^2$

SIMPLIFYING EXPRESSIONS

Exercises 47–82: Simplify the expression.

47. $5 + x - 3 + 2x$

48. $x - 5 - 5x + 7$

49. $-\frac{3}{4} + z - 3z + \frac{5}{4}$

50. $\frac{4}{3}z - 100 + 200 - \frac{1}{3}z$

51. $4y - y + 8y$

52. $14z - 15z - z$

53. $-3 + 6z + 2 - 2z$

54. $19a - 12a + 5 - 6$

55. $-2(3z - 6y) - z$

56. $6\left(\frac{1}{2}a - \frac{1}{6}b\right) - 3b$

57. $2 - \frac{3}{4}(4x + 8)$

58. $-5 - (5x - 6)$

59. $-x - (5x + 1)$

60. $2x - 4(x + 2)$

61. $1 - \frac{1}{3}(x + 1)$

62. $-3 - 3(4 - x)$

63. $\frac{3}{5}(x + y) - \frac{1}{5}(x - 1)$ **64.** $-5(a + b) - (a + b)$

65. $0.2x^2 + 0.3x^2 - 0.1x^2$ **66.** $32z^3 - 52z^3 + 20z^3$

67. $2x^2 - 3x + 5x^2 - 4x$ **68.** $\frac{5}{6}y^2 - 4 + \frac{1}{12}y^2 + 3$

69. $a + 3b - a - b$ **70.** $2z^2 - z - z^2 + 3z$

71. $8x^3 + 7y - x^3 - 5y$ **72.** $4y - 6z + 2y - 3z$

73. $\frac{8x}{8}$ **74.** $\frac{-0.1y}{-0.1}$

75. $\frac{-3y}{-y}$ **76.** $\frac{2x}{7x}$

77. $\frac{-108z}{-108}$ **78.** $\frac{3xy}{-6xy}$

79. $\frac{9x - 6}{3}$ **80.** $\frac{18y + 9}{9}$

81. $\frac{14z + 21}{7}$ **82.** $\frac{15x - 20}{5}$

WRITING EXPRESSIONS

Exercises 83–88: Translate the phrase into a mathematical expression and then simplify. Use the variable x.

83. The sum of five times a number and six times the same number

84. The sum of a number and three times the same number

85. The sum of a number squared and twice the same number squared

86. One-half of a number minus three-fourths of the same number

87. Six times a number minus four times the same number

88. Two cubed times a number minus three squared times the same number

APPLICATIONS INVOLVING EXPRESSIONS

89. Snowblowers Two snowblowers are being used to clear a driveway. The first blower can remove 20 cubic feet per minute, and the second blower can remove 30 cubic feet per minute.
   (a) Write a simplified expression that gives the number of cubic feet of snow that can be removed in x minutes.
   (b) Use your expression to find the total number of cubic feet of snow that can be removed in 48 minutes.

   (c) How many minutes would it take to remove the snow from a driveway that measures 30 feet by 20 feet, if the snow is 2 feet deep?

90. Winding Rope Two motors are winding up rope. The first motor can wind up rope at 2 feet per second, and the second motor can wind up rope at 5 feet per second.
   (a) Write a simplified expression that gives the length of rope wound by both motors in x seconds.
   (b) Use your expression to find the total length of rope that can be wound up by the two motors in 3 minutes.
   (c) How many minutes would it take to wind up 2100 feet of rope by using both motors?

91. Street Dimensions (Refer to Example 7.) A street has a constant width w and several straight sections with lengths 600, 400, 350, and 220 feet.

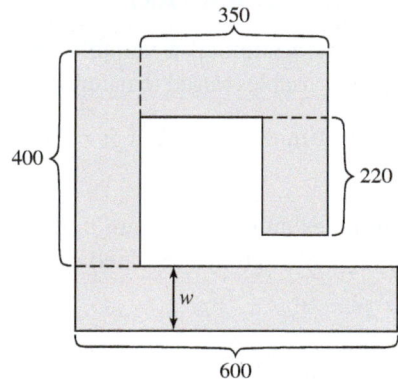

   (a) Write a simplified expression that gives the square footage of the street.
   (b) Find the area of the street if its width is 42 feet.

92. Sidewalk Dimensions A sidewalk has a constant width w and several short sections with lengths 12, 14, and 10 feet.
   (a) Write a simplified expression that gives the number of square feet of sidewalk.
   (b) Use your expression to find the area of the sidewalk if its width is 5 feet.

WRITING ABOUT MATHEMATICS

93. The following expression was simplified *incorrectly*.

$$3(x - 5) + 5x \overset{?}{=} 3x - 5 + 5x$$
$$\overset{?}{=} 3x + 5x - 5$$
$$\overset{?}{=} (3 + 5)x - 5$$
$$\overset{?}{=} 8x - 5$$

Find the error and explain what went wrong. What should the final answer be?

94. Explain how to add like terms. What property of real numbers is used?

Checking Basic Concepts

1. Use a commutative property to rewrite each expression.
   (a) $y \cdot 18$ **(b)** $10 + x$

2. Use an associative and a commutative property to simplify $5 \cdot (y \cdot 4)$.

3. Simplify each expression.
   (a) $10 - (5 + x)$ **(b)** $5(x - 7)$

4. State the property that the equation $5x + 3x = 8x$ illustrates.

5. Simplify $-4xy + 4xy$.

6. Mentally evaluate each expression.
   (a) $32 + 17 + 8 + 3$ **(b)** $\frac{5}{6} \cdot \frac{7}{8} \cdot \frac{6}{5} \cdot 8$
   (c) $567 - 199$

7. Determine whether the terms are like or unlike.
   (a) $-3x, -3z$ **(b)** $4x^2, -2x^2$

8. Combine like terms in each expression.
   (a) $5z + 9z$
   (b) $5y - 4 - 8y + 7$

9. Simplify each expression.
   (a) $2y - (5y + 3)$
   (b) $-4(x + 3y) + 2(2x - y)$
   (c) $\dfrac{20x}{20}$ **(d)** $\dfrac{35x^2}{x^2}$

10. Write "the sum of three times a number and five times the same number" as a mathematical expression with the variable x and then simplify the expression.

CHAPTER 1 Summary

SECTION 1.1 ■ NUMBERS, VARIABLES, AND EXPRESSIONS

Sets of Numbers

Natural Numbers $1, 2, 3, 4, \ldots$

Whole Numbers $0, 1, 2, 3, \ldots$

Products and Factors   Two numbers that are multiplied are called factors and the result is called the product.

    Example: $3 \cdot 5 = 15$
               Factor Factor Product

Prime and Composite Numbers

Prime Number   A natural number greater than 1 whose only natural number factors are itself and 1

    Examples: 2, 3, 5, 7, 11, 13, and 17

Composite Number and Prime Factorization   A natural number greater than 1 that is not prime; a composite number can be written as a product of two or more prime numbers.

    Examples: $24 = 2 \cdot 2 \cdot 2 \cdot 3$ and $18 = 2 \cdot 3 \cdot 3$

Important Terms

Variable   A symbol or letter that represents an unknown quantity

    Examples: $a, b, F, x,$ and y

Algebraic Expression   Consists of numbers, variables, arithmetic symbols, and grouping symbols

    Examples: $3x + 1$ and $5(x + 2) - y$

Equation   A statement that two algebraic expressions are equal; an equation always contains an equals sign.

    Examples: $1 + 2 = 3$ and $z - 7 = 8$

Formula A special type of equation that expresses a relationship between two or more quantities

Examples: $P = 2l + 2w$ (Perimeter of a rectangle)
$A = lw$ (Area of a rectangle)

SECTION 1.2 ■ FRACTIONS

Fractions and Lowest Terms

Fraction A fraction has a numerator, a denominator, and a fraction bar.

Example: Numerator $\searrow$ $\dfrac{4}{9}$ $\leftarrow$ Fraction bar Denominator $\nearrow$

Lowest Terms A fraction is in lowest terms if the numerator and denominator have no factors in common.

Example: The fraction $\frac{3}{5}$ is in lowest terms because 3 and 5 have no factors in common.

Simplifying Fractions
$$\frac{a \cdot c}{b \cdot c} = \frac{a}{b} \quad \text{(Basic principle of fractions)}$$

Example: $\frac{20}{35} = \frac{4 \cdot 5}{7 \cdot 5} = \frac{4}{7}$

Multiplicative Inverse, or Reciprocal The reciprocal of a nonzero number a is $\frac{1}{a}$.

Examples: The reciprocal of -5 is $-\frac{1}{5}$, and the reciprocal of $\frac{2x}{y}$ is $\frac{y}{2x}$, provided $x \neq 0$ and $y \neq 0$.

Multiplication and Division

$$\frac{a}{b} \cdot \frac{c}{d} = \frac{ac}{bd} \quad \text{and} \quad \frac{a}{b} \div \frac{c}{d} = \frac{a}{b} \cdot \frac{d}{c}$$

b and d are nonzero. b, c, and d are nonzero.

Examples: $\frac{3}{4} \cdot \frac{7}{5} = \frac{21}{20}$ and $\frac{3}{4} \div \frac{7}{5} = \frac{3}{4} \cdot \frac{5}{7} = \frac{3 \cdot 5}{4 \cdot 7} = \frac{15}{28}$

Addition and Subtraction with Like Denominators

$$\frac{a}{b} + \frac{c}{b} = \frac{a + c}{b} \quad \text{and} \quad \frac{a}{b} - \frac{c}{b} = \frac{a - c}{b}$$

Examples: $\frac{3}{5} + \frac{1}{5} = \frac{3 + 1}{5} = \frac{4}{5}$ and $\frac{3}{5} - \frac{1}{5} = \frac{3 - 1}{5} = \frac{2}{5}$

Addition and Subtraction with Unlike Denominators Write the expressions with the LCD. Then add or subtract the numerators, keeping the denominator unchanged.

Examples: $\frac{2}{9} + \frac{1}{6} = \frac{2}{9} \cdot \frac{2}{2} + \frac{1}{6} \cdot \frac{3}{3} = \frac{4}{18} + \frac{3}{18} = \frac{7}{18}$
$\frac{2}{9} - \frac{1}{6} = \frac{2}{9} \cdot \frac{2}{2} - \frac{1}{6} \cdot \frac{3}{3} = \frac{4}{18} - \frac{3}{18} = \frac{1}{18}$ $\Big\rangle$ LCD is 18.

SECTION 1.3 ■ EXPONENTS AND ORDER OF OPERATIONS

Exponential Expression

Base $\rightarrow 5^3 \leftarrow$ Exponent

Example: $3^4 = 3 \cdot 3 \cdot 3 \cdot 3 = 81$

Order of Operations

Use the following order of operations. First perform all calculations within parentheses and absolute values, or above and below the fraction bar.

1. Evaluate all exponential expressions.
2. Do all multiplication and division from *left to right*.
3. Do all addition and subtraction from *left to right*.

NOTE: Negative signs are evaluated after exponents, so $-2^4 = -16$. ∎

Examples: $100 - 5^2 \cdot 2 = 100 - 25 \cdot 2$ and $\dfrac{4+2}{5-3} \cdot 4 = \dfrac{6}{2} \cdot 4$

$$= 100 - 50 \qquad\qquad = 3 \cdot 4$$

$$= 50 \qquad\qquad\quad = 12$$

SECTION 1.4 ■ REAL NUMBERS AND THE NUMBER LINE

Opposite or Additive Inverse The opposite of the number a is $-a$.

Examples: The opposite of 5 is -5, and the opposite of -8 is $-(-8) = 8$.

Sets of Numbers

Integers
$$\ldots, -3, -2, -1, 0, 1, 2, 3, \ldots$$

Rational Numbers
$\dfrac{p}{q}$, where p and $q \neq 0$ are integers; rational numbers can be written as decimal numbers that either repeat or terminate.

Examples: $-\dfrac{3}{4}, 5, -7.6, \dfrac{6}{3}$, and $\dfrac{1}{3}$

Real Numbers
Numbers that can be written as decimal numbers

Examples: $-\dfrac{3}{4}, 5, -7.6, \dfrac{6}{3}, \pi, \sqrt{3}$, and $-\sqrt{7}$

Irrational Numbers
Real numbers that are not rational

Examples: $\pi, \sqrt{3}$, and $-\sqrt{7}$

Average To calculate the average of a set of numbers, find the sum of the numbers and then divide the sum by how many numbers there are in the set.

Example: The average of 4, 5, 20, and 11 is

$$\frac{4 + 5 + 20 + 11}{4} = \frac{40}{4} = 10.$$

We divide by 4 because we are finding the average of 4 numbers.

The Number Line

The origin corresponds to the number 0.

Absolute Value If a is positive or 0, then $|a| = a$, and if a is negative, then $|a| = -a$. The absolute value of a number is *never* negative.

Examples: $|5| = 5, \quad |-5| = 5, \quad$ and $\quad |0| = 0$

Inequality If a is located to the left of b on the number line, then $a < b$. If a is located to the right of b on the number line, then $a > b$.

Examples: $-3 < 6$ and $-1 > -2$

Addition of Opposites

$$a + (-a) = 0$$

Examples: $3 + (-3) = 0$ and $-\frac{5}{7} + \frac{5}{7} = 0$

Addition of Real Numbers
To add two real numbers with *like* signs, do the following.

1. Find the sum of the absolute values of the numbers.
2. The common sign of the two numbers is the sign of the sum.

To add two real numbers with *unlike* signs, do the following.

1. Find the absolute values of the numbers.
2. Subtract the smaller absolute value from the larger absolute value.
3. The sign of the number with the larger absolute value is the sign of the sum.

Examples: $-4 + 5 = 1$, $3 + (-7) = -4$, $-4 + (-2) = -6$, and $8 + 2 = 10$

Subtraction of Real Numbers For any real numbers a and b, $a - b = a + (-b)$.

Examples: $5 - 9 = 5 + (-9) = -4$ and $-4 - (-3) = -4 + 3 = -1$

Important Terms
Factors — Numbers multiplied in a multiplication problem

Example: 5 and 4 are factors of 20 because $5 \cdot 4 = 20$.

Product — The answer to a multiplication problem

Example: The product of 5 and 4 is 20.

Dividend, Divisor, and Quotient — If $\frac{a}{b} = c$, then a is the dividend, b is the divisor, and c is the quotient.

Example: In the division problems $30 \div 5 = 6$ and $\frac{30}{5} = 6$, 30 is the dividend, 5 is the divisor, and 6 is the quotient.

Dividing Real Numbers For real numbers a and b with $b \neq 0$,

$$\frac{a}{b} = a \cdot \frac{1}{b}.$$

Examples: $\dfrac{8}{\frac{1}{2}} = \dfrac{8}{1} \cdot \dfrac{2}{1} = 16$, $14 \div \dfrac{2}{3} = \dfrac{14}{1} \cdot \dfrac{3}{2} = \dfrac{42}{2} = 21$, and $5 \div 0$ is undefined.

Signs of Products or Quotients The product or quotient of two numbers with like signs is positive. The product or quotient of two numbers with unlike signs is negative.

Examples: $-4 \cdot 6 = -24$, $-2 \cdot (-5) = 10$, $\dfrac{-18}{6} = -3$, and $-4 \div (-2) = 2$

Writing Fractions as Decimals To write the fraction $\frac{a}{b}$ as a decimal, divide b into a.

Example: $\frac{4}{9} = 0.\overline{4}$ because dividing 4 by 9 gives the repeating decimal $0.4444\ldots$.

Writing Decimals as Fractions Write the decimal as a fraction with a denominator equal to a power of 10 and then simplify this fraction.

Example: $0.45 = \frac{45}{100} = \frac{9 \cdot 5}{20 \cdot 5} = \frac{9}{20}$

SECTION 1.7 ■ PROPERTIES OF REAL NUMBERS

Important Properties

Commutative

$a + b = b + a$ and $a \cdot b = b \cdot a$

Examples: $3 + 4 = 4 + 3$ and $-6 \cdot 3 = 3 \cdot (-6)$

Associative

$(a + b) + c = a + (b + c)$ and $(a \cdot b) \cdot c = a \cdot (b \cdot c)$

Examples: $(2 + 3) + 4 = 2 + (3 + 4)$
$(2 \cdot 3) \cdot 4 = 2 \cdot (3 \cdot 4)$

Distributive

$a(b + c) = ab + ac$ and $a(b - c) = ab - ac$

Examples: $3(x + 5) = 3 \cdot x + 3 \cdot 5$
$4(5 - 2) = 4 \cdot 5 - 4 \cdot 2$

Identity

$a + 0 = 0 + a = a$ Additive identity is 0.
$a \cdot 1 = 1 \cdot a = a$ Multiplicative identity is 1.

Examples: $5 + 0 = 0 + 5 = 5$ and $1 \cdot (-4) = -4 \cdot 1 = -4$

Inverse

$a + (-a) = 0$ and $-a + a = 0$
$a \cdot \frac{1}{a} = 1$ and $\frac{1}{a} \cdot a = 1$

Examples: $5 + (-5) = 0$ and $\frac{1}{2} \cdot 2 = 1$

NOTE: The commutative and associative properties apply to addition and multiplication but *not* to subtraction and division. ■

SECTION 1.8 ■ SIMPLIFYING AND WRITING ALGEBRAIC EXPRESSIONS

Important Terms

Term

A term is a number, a variable, or a product of numbers and variables raised to natural number powers.

Examples: 5, $-10x$, $3xy$,   and x^2

Coefficient

The number portion of a term

Examples: The coefficients for the terms $3xy$, $-x^2$, and -7 are 3, -1, and -7, respectively.

Like Terms

Terms containing the same variables raised to the same powers; their coefficients may be different.

Examples: The following pairs are like terms:
$5x$ and $-x$; $6x^2$ and $-2x^2$; $3y$ and $-\dfrac{1}{2}y$.

Combining Like Terms To add or subtract like terms, apply a distributive property.

Examples: $4x + 5x = (4 + 5)x = 9x$ and $5y - 7y = (5 - 7)y = -2y$

CHAPTER 1 Review Exercises

SECTION 1.1

Exercises 1–4: Classify the number as prime, composite, or neither. If the number is composite, write it as a product of prime numbers.

1. 29

2. 27

3. 108

4. 1

Exercises 5 and 6: Evaluate the expression for the given values of x and y.

5. $2x - 5$ $x = 4$

6. $\dfrac{2x}{x - y}$ $x = 6, \quad y = 4$

Exercises 7 and 8: Find the value of y for the given values of x and z.

7. $y = x - 5$ $x = 12$

8. $y = xz + 1$ $x = 2, \quad z = 3$

Exercises 9–12: Translate the phrase into an algebraic expression. State precisely what each variable represents when appropriate.

9. Three squared increased by five

10. Two cubed divided by the quantity three plus one

11. The product of three and a number

12. The difference between a number and four

SECTION 1.2

Exercises 13 and 14: Find the greatest common factor.

13. 15, 35

14. 12, 30, 42

15. Use the basic principle of fractions to simplify each expression.
(a) $\dfrac{5 \cdot 7}{8 \cdot 7}$ (b) $\dfrac{3a}{4a}$

16. Simplify each fraction to lowest terms.
(a) $\dfrac{9}{12}$ (b) $\dfrac{36}{60}$

Exercises 17–20: Multiply and then simplify the result to lowest terms when appropriate.

17. $\dfrac{3}{4} \cdot \dfrac{5}{6}$

18. $\dfrac{12}{11} \cdot \dfrac{22}{23}$

19. $4 \cdot \dfrac{5}{8}$

20. $\dfrac{2}{3} \cdot \dfrac{9x}{4y}$

21. Find the fractional part: one-fifth of three-sevenths.

22. Find the reciprocal of each number.
(a) 8   (b) 1   (c) $\dfrac{5}{19}$ (d) $\dfrac{3}{2}$

Exercises 23–26: Divide and then simplify to lowest terms when appropriate.

23. $\dfrac{3}{2} \div \dfrac{1}{6}$

24. $\dfrac{9}{10} \div \dfrac{7}{5}$

25. $8 \div \dfrac{2}{3}$

26. $\dfrac{4x}{3y} \div \dfrac{9x}{5}$

Exercises 27 and 28: Find the least common denominator for the fractions.

27. $\dfrac{1}{8}, \dfrac{5}{12}$

28. $\dfrac{3}{14}, \dfrac{1}{21}$

Exercises 29–32: Add or subtract and then simplify to lowest terms when appropriate.

29. $\dfrac{2}{15} + \dfrac{3}{15}$

30. $\dfrac{5}{4} - \dfrac{3}{4}$

31. $\dfrac{11}{12} - \dfrac{1}{8}$

32. $\dfrac{1}{6} + \dfrac{2}{3} - \dfrac{1}{9}$

SECTION 1.3

Exercises 33–36: Write the expression as an exponential expression.

33. $5 \cdot 5 \cdot 5 \cdot 5 \cdot 5 \cdot 5$

34. $\dfrac{7}{6} \cdot \dfrac{7}{6} \cdot \dfrac{7}{6}$

35. $x \cdot x \cdot x \cdot x \cdot x$

36. $(a - 5) \cdot (a - 5) \cdot (a - 5)$

37. Use multiplication to rewrite each expression, and then evaluate the result.
(a) 4^3 (b) 7^2 (c) 8^1

38. Find a natural number n such that $2^n = 32$.

Exercises 39–48: Evaluate the expression by hand.

39. $7 + 3 \cdot 6$

40. $15 - 5 - 3$

41. $24 \div 4 \div 2$

42. $30 - 15 \div 3$

43. $18 \div 6 - 2$

44. $\dfrac{18}{4 + 5}$

45. $9 - 3^2$

46. $3^2 - 4(5 - 3)$

47. $7 - \dfrac{4 + 6}{2 + 3}$

48. $3^3 - 2^3$

SECTION 1.4

Exercises 49 and 50: Find the opposite of each expression.

49. (a) -8 (b) $-(-(-3))$

50. (a) $-\left(\dfrac{-3}{7}\right)$ (b) $\dfrac{-2}{-5}$

Exercises 51 and 52: Find the decimal equivalent for the rational number.

51. (a) $\frac{4}{5}$ (b) $\frac{3}{20}$

52. (a) $\frac{5}{9}$ (b) $\frac{7}{11}$

Exercises 53–58: Classify the number as one or more of the following: natural number, whole number, integer, rational number, or irrational number.

53. 0   **54.** $-\frac{5}{6}$

55. -7 **56.** $\sqrt{17}$

57. π **58.** 3.4

59. Plot each number on the same number line.
    (a) 0   (b) -2 (c) $\frac{5}{4}$

60. Evaluate each expression.
    (a) $|-5|$ (b) $|-\pi|$ (c) $|4 - 4|$

61. Insert the symbol $>$ or $<$ to make each statement true.
    (a) -5 ____ 4   (b) $-\frac{1}{2}$ ____ $-\frac{5}{2}$
    (c) -3 ____ $|-9|$ (d) $|-8|$ ____ $|-1|$

62. List the numbers $\sqrt{3}, -3, 3, -\frac{2}{3},$ and $\pi - 1$ from least to greatest.

SECTIONS 1.5 AND 1.6

Exercises 63 and 64: Use a number line to find the sum.

63. $-1 + 2$ **64.** $-2 + (-3)$

Exercises 65–74: Evaluate the expression.

65. $5 + (-4)$ **66.** $-9 - (-7)$

67. $11 \cdot (-4)$ **68.** $-8 \cdot (-5)$

69. $11 \div (-4)$ **70.** $-4 \div \frac{4}{7}$

71. $-\frac{5}{9} - \left(-\frac{1}{3}\right)$ **72.** $-\frac{1}{2} + \left(-\frac{3}{4}\right)$

73. $-\frac{1}{3} \cdot \left(-\frac{6}{7}\right)$ **74.** $\frac{3}{8} \div (-0.5)$

Exercises 75 and 76: Write an arithmetic expression for the given phrase and then simplify.

75. Three plus negative five

76. Subtract negative four from two

Exercises 77 and 78: Write the fraction or mixed number as a decimal.

77. $\frac{7}{9}$ **78.** $2\frac{1}{5}$

Exercises 79 and 80: Write the decimal number as a fraction in lowest terms.

79. 0.6   **80.** 0.375

SECTION 1.7

Exercises 81 and 82: Use a commutative property to rewrite the expression. Do not simplify.

81. $3 + 16$ **82.** $14 \cdot (-x)$

Exercises 83 and 84: Use an associative property to rewrite the expression by changing the parentheses. Do not simplify.

83. $-4 + (1 + 3)$ **84.** $(x \cdot y) \cdot 5$

Exercises 85 and 86: Use a distributive property to rewrite the expression. Then simplify the expression.

85. $5(x + 12)$ **86.** $-(a - 3)$

Exercises 87–90: State the property or properties that are illustrated by the given equation.

87. $y + 0 = y$ **88.** $b \cdot 1 = b$

89. $\frac{1}{4} \cdot 4 = 1$ **90.** $-3a + 3a = 0$

Exercises 91–98: State the property that is illustrated by the given equation.

91. $z \cdot 3 = 3z$

92. $6 + (7 + 5x) = (6 + 7) + 5x$

93. $1 \cdot a = a$ **94.** $3 \cdot (5x) = (3 \cdot 5)x$

95. $12 - (x + 7) = 12 - x - 7$

96. $a + 0 = a$ **97.** $-5x + 5x = 0$

98. $-5 \cdot \left(-\frac{1}{5}\right) = 1$

Exercises 99–102: Use properties of real numbers to evaluate the expression mentally.

99. $7 + 9 + 12 + 8 + 1 + 3$

100. $500 - 199$

101. $25 \cdot 99$ **102.** $4356 \div 100$

SECTION 1.8

Exercises 103–104: Determine whether the expression is a term. If the expression is a term, identify its coefficient.

103. $55x$ **104.** $x - 7$

Exercises 105–112: Simplify the expression.

105. $-10x + 4x$ **106.** $19z - 4z$

107. $3x^2 + x^2$ **108.** $7 + 2x - 6 + x$

109. $4x^2 - 3 + 5x^2 - 3$ **110.** $3x^2 + 4x^2 - 7x^2$

111. $\dfrac{35a}{7a}$ **112.** $\dfrac{24x - 60}{12}$

APPLICATIONS

113. Painting a Wall Two people are painting a large wall. The first person paints 3 square feet per minute, while the second person paints 4 square feet per minute.
  (a) Write a simplified expression that gives the total number of square feet the two people can paint in x minutes.
  (b) Find the number of square feet the two people paint in 1 hour.
  (c) How many minutes would it take for them to paint a wall 8 feet tall and 21 feet wide?

114. Area of a Triangle Find the area of the triangle shown.

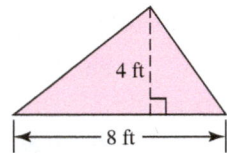

115. Gallons to Pints There are 8 pints in 1 gallon. Let $G = 1, 2, 3, \ldots, 6$, and make a table of values that converts G gallons to P pints. Write a formula that converts G gallons to P pints.

116. Google Search Each day, Google search has an average of 43 million active users. The table shows the cumulative number of active users, U, in millions after x days.

x	1	2	3	4
U	43	86	129	172

Source: Business Insider.

  (a) Write an equation that relates U and x.
  (b) How many active users, in millions, are there after 30 days?

117. Aging in the United States In 2050, about $\frac{1}{5}$ of the population will be age 65 or over and about $\frac{1}{20}$ of the population will be age 85 or over. Estimate the fraction of the population that will be between the ages of 65 and 85 in 2050. (*Source:* U.S. Census Bureau.)

118. Rule of 72 (Refer to Exercise 79 in Section 1.3.) If an investment of $25,000 earns 9% annual interest, approximate the value of the investment after 24 years.

119. Carpentry A board measures $5\frac{3}{4}$ feet and needs to be cut into five equal pieces. Find the length of each of the pieces.

120. Distance Over four days, an athlete jogs $3\frac{1}{8}$ miles, $4\frac{3}{8}$ miles, $6\frac{1}{4}$ miles, and $1\frac{5}{8}$ miles. How far does the athlete jog in all?

121. Checking Account The initial balance in a checking account is $1652. Find the final balance resulting from the following sequence of withdrawals and deposits: $-\$78$, $-\$91$, $\$256$, and $-\$638$.

122. Temperature Range The highest temperature ever recorded in Amarillo, Texas, is $108\,°\text{F}$ and the lowest is $-16\,°\text{F}$. Find the difference between these two temperatures. (*Source:* The Weather Almanac.)

CHAPTER 1   Test   MyMathLab® YouTube

1. Classify the number as prime or composite. If the number is composite, write it as a product of prime numbers.
   (a) 29   (b) 56

2. Evaluate the expression $\dfrac{5x}{2x-1}$ for $x = -3$.

3. Translate the phrase "four squared decreased by three" to an algebraic expression. Then find the value of the expression.

4. Simplify $\frac{24}{32}$ to lowest terms.

5. Find the fractional part: two-tenths of one-sixth.

6. Find the least common denominator for $\frac{3}{8}$ and $\frac{5}{12}$.

7. Evaluate each expression. Write your answer in lowest terms.
   (a) $\frac{5}{8} + \frac{1}{8}$ (b) $\frac{5}{9} - \frac{3}{15}$ (c) $\frac{3}{5} \cdot \frac{10}{21}$
   (d) $6 \div \frac{8}{5}$ (e) $\frac{5}{12} + \frac{4}{9}$ (f) $\frac{10}{13} \div 5$

8. Write $y \cdot y \cdot y \cdot y$ as an exponential expression.

9. Evaluate each expression.
   (a) $6 + 10 \div 5$ (b) $4^3 - (3 - 5 \cdot 2)$
   (c) $-6^2 - 6 + \frac{4}{2}$ (d) $11 - \frac{1+3}{6-4}$

10. Write $\frac{7}{20}$ as a decimal.

11. Classify the number as one or more of the following: natural number, whole number, integer, rational number, or irrational number.

    (a) -1 (b) $\sqrt{5}$

12. Plot each number on the same number line.

    (a) -2 (b) $\frac{1}{3}$ (c) $\sqrt{7}$

13. Insert the symbol $>$ or $<$ to make each statement true.

    (a) 2 _____ $|-5|$ (b) $|-1|$ _____ $|0|$

14. Evaluate the expression.

    (a) $-5 \div \frac{5}{6}$ (b) $-7 \cdot (-3)$

15. Write 0.75 as a fraction in lowest terms.

16. State the property or properties that each equation illustrates.

    (a) $6x - 2x = 4x$ (b) $12 \cdot (3x) = 36x$
    (c) $4 + x + 8 = 12 + x$

17. Use properties of real numbers to evaluate $17 \cdot 102$ mentally.

18. Simplify each expression.

    (a) $5 - 5z + 7 + z$ (b) $12x - (6 - 3x)$

    (c) $5 - 4(x + 6) + \dfrac{15x}{3}$

19. **Mowing a Lawn** Two people are mowing a lawn. The first person has a riding mower and can mow $\frac{4}{3}$ acres per hour; the second person has a push mower and can mow $\frac{1}{4}$ acre per hour.

    (a) Write a simplified expression that gives the total number of acres that the two people mow in x hours.

    (b) Find the total acreage that they can mow in an 8-hour work day.

20. **Cutting Wire** A wire $7\frac{4}{5}$ feet long is to be cut in 3 equal parts. How long should each part be?

21. **Cost Equation** The table lists the cost C of buying x tickets to a hockey game.

Tickets (x)	3	4	5	6
Cost (C)	$39	$52	$65	$78

    (a) Find an equation that relates C and x.
    (b) What is the cost of 17 tickets?

22. **Banking** The initial balance for a savings account is $892. Find the final balance resulting from withdrawals and deposits of $-\$57$, $\$150$, and $-\$345$.

2 Linear Equations and Inequalities

Mathematics lets us use numbers to describe the intensity of ultraviolet light from the sun. The table shows the maximum ultraviolet intensity measured in milliwatts per square meter for various latitudes and dates.

The Sun's Ultraviolet Intensity (mW/m^2)

Latitude	Mar. 21	June 21	Sept. 21	Dec. 21
0°	325	254	325	272
10°	311	275	280	220
20°	249	292	256	143
30°	179	248	182	80
40°	99	199	127	34
50°	57	143	75	13

If a student from Chicago, located at a latitude of 42°, spends spring break in Hawaii with a latitude of 20°, the sun's ultraviolet rays in Hawaii will be approximately $\frac{249}{99} \approx 2.5$ times as intense as they are in Chicago. See Exercises 75 and 76 in the first section of this chapter. Equations can be used to describe, or model, the intensity of the sun at various latitudes. In this chapter, we will focus on *linear equations* and the related concept of *linear inequalities*. **Source:** J. Williams, *The USA Today Weather Almanac.*

2.1 Introduction to Equations

Objectives

1. Solving Equations and Finding Solutions
2. Applying the Addition Property of Equality
   • Checking a Solution
3. Applying the Multiplication Property of Equality

NEW VOCABULARY

☐ Solution
☐ Solution Set
☐ Equivalent equations

1 ▶ Solving Equations and Finding Solutions

Connecting Concepts with Your Life Suppose that during a storm it rains 2 inches before noon and 1 inch per hour thereafter until 5 P.M. **TABLE 2.1** lists the total rainfall R after various elapsed times x, where $x = 0$ corresponds to noon.

Rainfall x Hours Past Noon

Elapsed Time: x (hours)	0	1	2	3	4	5
Total Rainfall: R (inches)	2	3	4	5	6	7

TABLE 2.1

The data suggest that the total rainfall R in inches is **2** more than the elapsed time x. An equation that *models*, or *describes*, the rainfall x hours past noon is given by

$$R = x + 2. \qquad \boxed{\text{Total rain } x \text{ hours past noon}}$$

For example, **3** hours past noon, or at 3 P.M.,

$$R = 3 + 2 = 5$$

inches of rain have fallen. Even though $x = 4.5$ does not appear in the table, we can calculate the amount of rainfall at 4:30 P.M. with the equation as

$$R = 4.5 + 2 = 6.5 \text{ inches.}$$

The advantage that an equation has over a table of values is that an equation can be used to calculate the rainfall at *any* time x, not just at the times listed in the table.

In the rainfall example above, how can we determine when 6 inches of rain have fallen? From **TABLE 2.1** the *solution* is 4, or 4 P.M. To find this solution without the table, we can *solve* the equation

$$\boxed{\text{What plus 2 equals 6? The answer is the solution.}} \quad x + 2 = 6.$$

READING CHECK 1

• What is a solution to an equation?

An equation can be either true or false. For example, the equation $1 + 2 = 3$ is true, whereas the equation $1 + 2 = 4$ is false. When an equation contains a variable, the equation may be true for some values of the variable and false for other values of the variable.

The following See the Concept shows how to recognize solutions and solution sets.

See the Concept 1 | **SOLUTIONS AND SOLUTION SETS**

Equation	*Solution(s)*	*Not a Solution*	*Solution Set*
$9 - n = 1$	$9 - 8 = 1$	$9 - 6 \overset{?}{=} 1$	$\{8\}$
	A Solution	**B** Not a Solution	**C** Solution Set
$x^2 = 4$	$(-2)^2 = 4$ and $2^2 = 4$	$3^2 \overset{?}{=} 4$	$\{-2, 2\}$
	A Solutions	**B** Not a Solution	**C** Solution Set

A Any value of the variable that makes an equation **true** is a **solution** to the equation.

B Any value of the variable that makes an equation **false** is *not* a solution to the equation.

C The *set of all solutions* is called the **solution set**. (Braces { } are used to denote a set.)

NOTE: *Solving an equation* means finding its solution set. ■   **WATCH VIDEO IN** MML .

EXAMPLE 1 **Determining if a value is a solution**

Determine if 3 is a solution to each equation.
(a) $7x - 5 = 16$ (b) $1 + 2x = 9$

Solution

Replace the variable with **3** and determine if the resulting equation is true.

(a) $7x - 5 = 16$ Given equation
$7(3) - 5 \stackrel{?}{=} 16$ Replace x with 3.
$21 - 5 \stackrel{?}{=} 16$ Multiply.
$16 = 16$ ✓   Simplify.
          (True equation)

(b) $1 + 2x = 9$ Given equation
$1 + 2(3) \stackrel{?}{=} 9$ Replace x with 3.
$1 + 6 \stackrel{?}{=} 9$ Multiply.
$7 \stackrel{?}{=} 9$ ✗   Simplify.
          (False equation)

The value 3 is a solution.   The value 3 is not a solution.

Now Try Exercises 13, 15

Many times we cannot solve an equation simply by looking at it. In these situations, we must use a step-by-step procedure. During each step, an equation is transformed into a different but equivalent equation. **Equivalent equations** are equations that have the same solution set. For example, the equations

$$x + 2 = 5 \quad \text{and} \quad x = 3$$

are equivalent equations because the solution set for both equations is {3}.

EXAMPLE 2 **Determining if two equations are equivalent**

Determine if the two equations are equivalent.
(a) $x - 5 = 3$ and $x = 9$ (b) $2x + 1 = 7$ and $2x = 6$

Solution

(a) Since $8 - 5 = 3$, the solution to the first equation is **8**. However, the solution to the equation $x = 9$ is clearly **9**. The two equations are *not* equivalent because they have different solution sets.

(b) Since $2(3) + 1 = 7$ and $2(3) = 6$, the two equations have the same solution, **3**. The two equations are equivalent because they have the same solution set.

Now Try Exercises 19, 21

READING CHECK 2

• What is the difference between an equation and an expression?

MAKING CONNECTIONS 1

Equations and Expressions

Although the words "equation" and "expression" occur frequently in mathematics, they are *not* interchangeable. An equation *always* contains an equals sign, but an expression *never* contains an equals sign. We often want to *solve* an equation, whereas an expression can sometimes be *simplified*. Furthermore, the equals sign in an equation separates two expressions. For example, $3x - 5 = x + 1$ is an equation where $3x - 5$ and $x + 1$ are each expressions.

When solving equations, it is often helpful to transform a more complicated equation into an equivalent equation that has an obvious solution, such as $x = 3$. The *addition property of equality* and the *multiplication property of equality* can be used to transform an equation into an equivalent equation that is easier to solve.

2 Applying the Addition Property of Equality

The following See the Concept shows how the addition property of equality is used to make equivalent equations. We can sometimes solve an equation by using this property to change a given equation into an equivalent equation.

See the Concept 2   **THE ADDITION PROPERTY OF EQUALITY**

We can think of an equation as a scale, with the equals sign at the balancing point. Just as a scale remains *balanced* when the same amount of weight is added to (or removed from) each side, an equation remains *true* (and an equivalent equation results) when the same number is added to (or subtracted from) each side.

True Equation	*Equivalent Equation*	*Not Balanced*
$2 = 2$	$2 + 1 = 2 + 1$	$2 + 1 \neq 2 + 0$
$=$	$=$	$\neq$

WATCH VIDEO IN MML.

STUDY TIP

The addition property of equality is used to solve equations throughout the remainder of the text. Be sure that you understand this important property.

When solving an equation, we apply the same operation to each side of the equation so that the equation stays "balanced." For example, one way to solve the equation

$$x + 2 = 5$$

is to add -2 to each side. This step results in isolating the x on one side of the equation.

Equivalent equations
$$
\begin{cases}
x + 2 = 5 & \text{Given equation} \\
x + 2 + (-2) = 5 + (-2) & \text{Add } -2 \text{ to each side.} \\
x + 0 = 3 & \text{Simplify each side.} \\
x = 3 & \text{Additive identity}
\end{cases}
$$

These four equations are equivalent, but the solution is easiest to see in the last equation. When -2 is added to each side of the given equation, the addition property of equality is used.

READING CHECK 3

- Why do we use the addition property of equality?

ADDITION PROPERTY OF EQUALITY

If a, b, and c are real numbers, then

$$a = b \quad \text{is equivalent to} \quad a + c = b + c.$$

That is, adding the same number to each side of an equation results in an equivalent equation.

NOTE: Because any subtraction problem can be changed to an addition problem, *the addition property of equality also works for subtraction.* For example, adding -2 to each side or subtracting 2 from each side are equivalent steps. That is, if the same number is subtracted from each side of an equation, the result is an equivalent equation. ■

| EXAMPLE 3 | **Solving an equation using the addition property of equality** |

Solve each equation.
(a) $x + 10 = 7$ (b) $t - 4 = 3$ (c) $\frac{1}{2} = -\frac{3}{4} + y$

Solution

(a) When solving an equation, we try to isolate the variable on one side of the equation. If we add -10 to (or **subtract 10** from) each side of the given equation, we get an equivalent equation that gives the value of x.

$$x + 10 = 7 \qquad\qquad \text{Given equation}$$
$$x + 10 - \mathbf{10} = 7 - \mathbf{10} \qquad \text{Subtract 10 from each side.}$$
$$x + 0 = -3 \qquad\qquad \text{Simplify each side.}$$
$$x = -3 \qquad\qquad\quad \text{Additive identity}$$

The solution is -3.

(b) To isolate the variable t, **add 4** to each side.

$$t - 4 = 3 \qquad\qquad \text{Given equation}$$
$$t - 4 + \mathbf{4} = 3 + \mathbf{4} \qquad \text{Add 4 to each side.}$$
$$t + 0 = 7 \qquad\qquad \text{Simplify each side.}$$
$$t = 7 \qquad\qquad\quad \text{Additive identity}$$

The solution is 7.

(c) To isolate the variable y, **add $\frac{3}{4}$** to each side.

$$\frac{1}{2} = -\frac{3}{4} + y \qquad\qquad \text{Given equation}$$
$$\frac{1}{2} + \mathbf{\frac{3}{4}} = -\frac{3}{4} + \mathbf{\frac{3}{4}} + y \qquad \text{Add } \tfrac{3}{4} \text{ to each side.}$$
$$\frac{5}{4} = 0 + y \qquad\qquad \text{Simplify each side.}$$
$$\frac{5}{4} = y \qquad\qquad\quad \text{Additive identity}$$

The solution is $\frac{5}{4}$.

Now Try Exercises 31, 33, 37

CHECKING A SOLUTION To check a solution, substitute it in the *given* equation to find out if a true statement results. To check the solution for Example 3(c), substitute $\frac{5}{4}$ for y in the given equation. Note that a question mark is placed over the equals sign when a solution is being checked.

CRITICAL THINKING 1

When you are checking a solution, why do you substitute your answer in the *given* equation?

$$\frac{1}{2} = -\frac{3}{4} + y \qquad \text{Given equation}$$
$$\frac{1}{2} \overset{?}{=} -\frac{3}{4} + \frac{5}{4} \qquad \text{Replace } y \text{ with } \tfrac{5}{4}.$$
$$\frac{1}{2} \overset{?}{=} \frac{2}{4} \qquad\qquad \text{Add fractions.}$$
$$\frac{1}{2} = \frac{1}{2} \; \checkmark \qquad\qquad \text{Simplify } \tfrac{2}{4}; \text{ the answer checks.}$$

The answer of $\frac{5}{4}$ checks because the resulting equation is true.

EXAMPLE 4   **Solving an equation and checking a solution**

Solve the equation $-5 + y = 3$ and then check the solution.

Solution

Isolate y by **adding 5** to each side.

$$-5 + y = 3 \qquad \text{Given equation}$$
$$-5 + 5 + y = 3 + 5 \qquad \text{Add 5 to each side.}$$
$$0 + y = 8 \qquad \text{Simplify each side.}$$
$$y = 8 \qquad \text{Additive identity}$$

The solution is 8. To check this solution, substitute **8** for **y** in the given equation.

$$-5 + y = 3 \qquad \text{Given equation}$$
$$-5 + 8 \stackrel{?}{=} 3 \qquad \text{Replace } y \text{ with 8.}$$
$$3 = 3 \checkmark \qquad \text{Add; the answer checks.}$$

Now Try Exercise 35

3 Applying the Multiplication Property of Equality

STUDY TIP

The multiplication property of equality is used to solve equations throughout the remainder of the text. Be sure that you understand this important property.

We can illustrate the multiplication property of equality by considering a formula that converts yards to feet. Because there are 3 feet in 1 yard, the formula $F = 3Y$ computes F, the number of feet in Y yards. For example, if $Y = 5$ yards, then $F = 3 \cdot 5 = 15$ feet.

Now consider a reverse computation such as converting 27 feet to yards. The answer to this conversion corresponds to the solution to the equation

$$27 = 3Y.$$

To find the solution, **multiply** each side of the equation by the reciprocal of 3, or $\frac{1}{3}$.

Equivalent equations
$$\begin{cases} 27 = 3Y & \text{Given equation} \\ \frac{1}{3} \cdot 27 = \frac{1}{3} \cdot 3 \cdot Y & \text{Multiply each side by } \frac{1}{3}. \\ 9 = 1 \cdot Y & \text{Simplify each side.} \\ 9 = Y & \text{Multiplicative identity} \end{cases}$$

Thus **27** feet are equivalent to **9** yards.

> **MULTIPLICATION PROPERTY OF EQUALITY**
>
> If a, b, and c are real numbers with $c \neq 0$, then
>
> $$a = b \quad \text{is equivalent to} \quad ac = bc.$$
>
> That is, multiplying each side of an equation by the same nonzero number results in an equivalent equation.

READING CHECK 4

• Why do we use the multiplication property of equality?

NOTE: Because any division problem can be changed to a multiplication problem, *the multiplication property of equality also works for division*. For example, multiplying each side by $\frac{1}{2}$ or dividing each side by 2 are equivalent steps. That is, if each side of an equation is divided by the same nonzero number, the result is an equivalent equation. ∎

EXAMPLE 5   **Solving an equation using the multiplication property of equality**

Solve each equation.

(a) $\frac{1}{3}x = 4$ (b) $-4y = 8$ (c) $5 = \frac{3}{4}z$

Solution
(a) We start by **multiplying** each side of the equation by **3**, the reciprocal of $\frac{1}{3}$.

$$\frac{1}{3}x = 4 \qquad \text{Given equation}$$

$$3 \cdot \frac{1}{3}x = 3 \cdot 4 \qquad \text{Multiply each side by 3.}$$

$$1 \cdot x = 12 \qquad \text{Simplify each side.}$$

$$x = 12 \qquad \text{Multiplicative identity}$$

The solution is 12.

(b) The coefficient of the y-term is -4, so we can either multiply each side of the equation by $-\frac{1}{4}$ or **divide** each side by -4. This step will make the coefficient of y equal to 1.

$$-4y = 8 \qquad \text{Given equation}$$

$$\frac{-4y}{-4} = \frac{8}{-4} \qquad \text{Divide each side by } -4.$$

$$y = -2 \qquad \text{Simplify fractions.}$$

The solution is -2.

(c) To change the coefficient of z from $\frac{3}{4}$ to 1, **multiply** each side of the equation by $\frac{4}{3}$, the reciprocal of $\frac{3}{4}$.

$$5 = \frac{3}{4}z \qquad \text{Given equation}$$

$$\frac{4}{3} \cdot 5 = \frac{4}{3} \cdot \frac{3}{4}z \qquad \text{Multiply each side by } \frac{4}{3}.$$

$$\frac{20}{3} = 1 \cdot z \qquad \text{Simplify each side.}$$

$$\frac{20}{3} = z \qquad \text{Multiplicative identity}$$

The solution is $\frac{20}{3}$.

Now Try Exercises 51, 57, 59

EXAMPLE 6   **Solving an equation and checking a solution**

Solve the equation $\frac{3}{4} = -\frac{3}{7}t$ and then check the solution.

Solution
Multiply each side of the equation by $-\frac{7}{3}$, the reciprocal of $-\frac{3}{7}$.

$$\frac{3}{4} = -\frac{3}{7}t \qquad \text{Given equation}$$

$$-\frac{7}{3} \cdot \frac{3}{4} = -\frac{7}{3} \cdot \left(-\frac{3}{7}\right)t \qquad \text{Multiply each side by } -\frac{7}{3}.$$

$$-\frac{7}{4} = 1 \cdot t \qquad \text{Simplify each side.}$$

$$-\frac{7}{4} = t \qquad \text{Multiplicative identity}$$

The solution is $-\frac{7}{4}$. To check this answer, substitute $-\frac{7}{4}$ for t in the given equation.

$$\frac{3}{4} = -\frac{3}{7}t \qquad \text{Given equation}$$

$$\frac{3}{4} \overset{?}{=} -\frac{3}{7} \cdot \left(-\frac{7}{4}\right) \qquad \text{Replace } t \text{ with } -\frac{7}{4}.$$

$$\frac{3}{4} = \frac{3}{4} \checkmark \qquad \text{Multiply; the answer checks.}$$

Now Try Exercise 61

🌐 Math in Context *Social Networking* In its early years, Twitter's popularity increased dramatically and new accounts were added at an amazing rate, as shown in the next example. (*Source:* Twitter.)

EXAMPLE 7   **Analyzing Twitter account data**

For several months during one of its early growth periods, Twitter added an average of 0.3 million new accounts every day.
(a) Write a formula that gives the number of new Twitter accounts T added in x days.
(b) At this rate, how many days would be needed to add 18 million new accounts?

Solution
(a) In **1** day $0.3 \cdot 1 = 0.3$ million new accounts were added, in **2** days $0.3 \cdot 2 = 0.6$ million new accounts were added, and in x days $0.3 \cdot x = 0.3x$ new accounts were added. So the formula is $T = 0.3x$, where x is in days and T is in millions.
(b) To find the number of days needed for Twitter to add 18 million new accounts, replace the variable T in the formula with **18** and solve the resulting equation.

$$T = 0.3x \qquad \text{Formula from part (a)}$$

$$18 = 0.3x \qquad \text{Replace } T \text{ with 18.}$$

$$\frac{18}{0.3} = \frac{0.3x}{0.3} \qquad \text{Divide each side by 0.3.}$$

$$60 = x \qquad \text{Simplify.}$$

At this rate, it takes 60 days to add 18 million accounts.

Now Try Exercise 73

2.1   Putting It All Together

CONCEPT	COMMENTS	EXAMPLES
Equation	An equation is a mathematical statement that two expressions are equal. An equation can be either true or false.	The equation $2 + 3 = 5$ is true. The equation $1 + 3 = 7$ is false.
Solution	A value for a variable that makes an equation a true statement	The solution to $x + 5 = 20$ is 15, and the solutions to $x^2 = 9$ are -3 and 3.
Solution Set	The set of all solutions to an equation	The solution set to $x + 5 = 20$ is $\{15\}$, and the solution set to $x^2 = 9$ is $\{-3, 3\}$.

continued on next page

continued from previous page

CONCEPT	COMMENTS	EXAMPLES
Equivalent Equations	Two equations are equivalent if they have the same solution set.	The equations $$2x = 14 \quad \text{and} \quad x = 7$$ are equivalent because the solution set to both equations is $\{7\}$.
Addition Property of Equality	The equations $$a = b \quad \text{and} \quad a + c = b + c$$ are equivalent. This property is used to solve equations.	To solve $x - 3 = 8$, add 3 to each side of the equation. $$x - 3 + 3 = 8 + 3$$ $$x = 11$$ The solution is 11.
Multiplication Property of Equality	When $c \neq 0$, the equations $$a = b \quad \text{and} \quad a \cdot c = b \cdot c$$ are equivalent. This property is used to solve equations.	To solve $\frac{1}{5}x = 10$, multiply each side of the equation by 5. $$5 \cdot \frac{1}{5}x = 5 \cdot 10$$ $$x = 50$$ The solution is 50.
Checking a Solution	Substitute the solution for the variable in the given equation and then simplify each side to see if a true statement results.	To show that 8 is a solution to $$x + 12 = 20,$$ substitute 8 for x. $$8 + 12 \stackrel{?}{=} 20$$ $$20 = 20 \checkmark \quad \text{True}$$

2.1   Exercises

MyMathLab®

CONCEPTS AND VOCABULARY

1. Each value of a variable that makes an equation true is called a(n) _____.

2. The equation $1 + 3 = 4$ is (true/false).

3. The equation $2 + 3 = 6$ is (true/false).

4. The _____ is the set of all solutions to an equation.

5. To solve an equation, find all _____.

6. Equations with the same solution set are called _____ equations.

7. If $a = b$, then $a + c =$ _____ .

8. Because any subtraction problem can be changed to an addition problem, the addition property of equality also works for _____.

9. If $a = b$ and $c \neq 0$, then $ac =$ _____.

10. Because any division problem can be changed to a multiplication problem, the multiplication property of equality also works for _____.

11. To solve an equation, transform the equation into a(n) _____ equation that is easier to solve.

12. To check a solution, substitute it for the variable in the _____ equation.

IDENTIFYING EQUATIONS AND SOLUTIONS

Exercises 13–18: Determine if the given value for x is a solution to the equation.

13. $x = 4$; $-x - 5 = -1$

14. $x = 10$; $8 - x = -2$

15. $x = -5$; $8 + 3x = -7$

16. $x = -8$; $15 + 2x = 1$

17. $x = 3$; $9x - 5 = 22$

18. $x = 4$; $-6x + 5 = 29$

Exercises 19–22: Determine if the equations are equivalent.

19. $x - 8 = 7$ and $x = 15$

20. $x + 9 = -7$ and $x = -2$

21. $3x - 1 = 5$ and $3x = 4$

22. $2x + 5 = -7$ and $2x = -12$

Exercises 23–26: Classify each as an equation or an expression. If it is an equation find the solution. If it is an expression, simplify it.

23. $x - 4 = 5$ **24.** $5 + x = 9$

25. $2x + 4x$ **26.** $5x - 3x$

APPLYING THE ADDITION PROPERTY OF EQUALITY

27. To solve $x - 22 = 4$, add _____ to each side.

28. To solve $\frac{5}{6} = \frac{1}{6} + x$, add _____ to each side.

29. To solve $x + 3 = 13$, subtract _____ from each side.

30. To solve $\frac{3}{4} = \frac{1}{4} + x$, subtract _____ from each side.

Exercises 31–44: Solve the equation. Check your answer.

31. $x + 5 = 0$ **32.** $x + 3 = 7$

33. $a - 12 = -3$ **34.** $a - 19 = -11$

35. $9 = y - 8$ **36.** $97 = -23 + y$

37. $\frac{1}{5} = z - \frac{3}{2}$ **38.** $\frac{3}{4} + z = -\frac{1}{2}$

39. $t - 0.8 = 4.3$ **40.** $y - 1.23 = -0.02$

41. $4 + x = 1$ **42.** $16 + x = -2$

43. $1 = \frac{1}{3} + y$ **44.** $\frac{7}{2} = -2 + y$

45. **Thinking Generally** To solve $x - a = b$ for x, add _____ to each side.

46. **Thinking Generally** To solve $x + a = b$ for x, subtract _____ from each side.

APPLYING THE MULTIPLICATION PROPERTY OF EQUALITY

47. To solve $5x = 4$, multiply each side by _____.

48. To solve $\frac{4}{3}y = 8$, multiply each side by _____.

49. To solve $6x = 11$, divide each side by _____.

50. To solve $0.2x = 4$, divide each side by _____.

Exercises 51–66: Solve the equation. Check your answer.

51. $5x = 15$ **52.** $25x = 0$

53. $-7x = 0$ **54.** $-2x = 8$

55. $-35 = -5a$ **56.** $-32 = -4a$

57. $-18 = 3a$ **58.** $-70 = 10a$

59. $\frac{1}{2}x = \frac{3}{2}$ **60.** $\frac{3}{4}x = \frac{5}{8}$

61. $\frac{1}{2} = \frac{2}{5}z$ **62.** $-\frac{3}{4} = -\frac{1}{8}z$

63. $0.5t = 3.5$ **64.** $2.2t = -9.9$

65. $-1.7 = 0.2x$ **66.** $6.4 = 1.6x$

67. **Thinking Generally** To solve $\frac{1}{a} \cdot x = b$ for x, multiply each side by _____.

68. **Thinking Generally** To solve $ax = b$ for x, where $a \neq 0$, divide each side by _____.

APPLICATIONS INVOLVING EQUATIONS

69. **Rainfall** On a stormy day, it rains 3 inches before noon and 1 inch per hour thereafter until 6 P.M.
   (a) Make a table that shows the total rainfall R in inches, x hours past noon, ending at 6 P.M.
   (b) Write an equation that calculates R.
   (c) Use your equation to calculate the total rainfall at 3 P.M. Does the answer agree with the value in your table from part (a)?
   (d) How much rain has fallen by 2:15 P.M.?

70. **Cold Weather** A furnace is turned on at midnight when the temperature inside a cabin is 0 °F. The cabin warms at a rate of 10 °F per hour until 7 A.M.
   (a) Make a table that shows the cabin temperature T in degrees Fahrenheit, x hours past midnight, ending at 7 A.M.
   (b) Write an equation that calculates T.
   (c) Use your equation to calculate the temperature at 5 A.M. Does the answer agree with the value in your table from part (a)?
   (d) Find the cabin temperature at 2:45 A.M.

71. Football Field A football field is 300 feet long.
 (a) Write an equation that gives the length L in feet of x football fields.
 (b) Write an equation whose solution gives the number of football fields in 870 feet.
 (c) Solve your equation from part (b).

72. Acreage An acre equals 43,560 square feet.
 (a) Write an equation that converts A acres to S square feet.
 (b) Write an equation whose solution gives the number of acres in 871,200 square feet.
 (c) Solve your equation from part (b).

73. Twitter Accounts (Refer to Example 7.) Suppose Twitter added 0.3 million new accounts every day. At this rate, how many days would be needed to add 15 million new Twitter accounts? *(Source: Twitter.)*

74. Facebook "Likes" Every day in 2014, there were about 4.5 billion "likes" on Facebook. At this rate, about how many days would it take for there to be 1 trillion "likes" on Facebook?

75. Online Exploration The city of Winnipeg is located in the province of Manitoba in Canada.
 (a) Use the Internet to find the latitude of Winnipeg to the nearest degree.
 (b) Use the table from the chapter opener to determine how many times as intense the sun's ultraviolet rays are at the equator (latitude 0°) on March 21 compared to the sun's intensity in Winnipeg. Round your answer to 1 decimal place.

76. Online Exploration Columbus is a city located in the center of Ohio. It is the state's capital city.
 (a) Use the Internet to find the latitude of Columbus to the nearest degree.
 (b) Use the table from the chapter opener to determine how many times as intense the sun's ultraviolet rays are in Limon, Costa Rica, (latitude 10°N) on June 21 compared to the sun's intensity in Columbus. Round your answer to 1 decimal place.

77. Cost of a Car When the cost of a car is multiplied by 0.07 the result is $1750. Find the cost of the car.

78. Raise in Salary If an employee's salary is multiplied by 1.06, which corresponds to a 6% raise, the result is $58,300. Find the employee's current salary.

WRITING ABOUT MATHEMATICS

79. A student solves an equation as follows.

$$x + 30 = 64$$
$$x \stackrel{?}{=} 64 + 30$$
$$x \stackrel{?}{=} 94$$

Identify the student's mistake. What is the solution?

80. What is a good first step for solving the equation $\frac{a}{b}x = 1$, where a and b are natural numbers? What is the solution? Explain your answers.

2.2 Linear Equations

Objectives

1 Identifying Linear Equations in One Variable
2 Solving Linear Equations
   • Solving with Tables
   • Solving Symbolically
3 Solving Linear Equations by Applying the Distributive Property
4 Solving Linear Equations by Clearing Fractions and Decimals
5 Identifying Equations with No Solution, One Solution, or Infinitely Many Solutions

1 Identifying Linear Equations in One Variable

Connecting Concepts with Your Life Suppose that a bicyclist is 5 miles from home, riding *away* from home at 10 miles per hour, as shown in **FIGURE 2.1**. The distance between the bicyclist and home for various elapsed times is shown in **TABLE 2.2**.

Riding Away from Home

←—5 mi—→

FIGURE 2.1

Distance from Home

Elapsed Time (hours)	0	1	2	3
Distance (miles)	5	15	25	35

TABLE 2.2   10 mi   10 mi   10 mi

The bicyclist is moving at a constant speed, so the distance increases by 10 miles every hour. The distance D from home after x hours can be calculated by the equation

$$D = 10x + 5.$$

NEW VOCABULARY

☐ Linear equation
☐ Identity
☐ Contradiction

For example, after **2** hours the distance is

$$D = 10(2) + 5 = 25 \text{ miles.}$$

TABLE 2.2 verifies that the bicyclist is 25 miles from home after 2 hours. However, the table is less helpful if we want to find the elapsed time when the bicyclist is 18 miles from home. To answer this question, we could begin by substituting **18** for D in the distance equation to obtain

$$18 = 10x + 5.$$

The equation $18 = 10x + 5$ can be written in a different form by applying the addition property of equality. Subtracting 18 from each side gives an equivalent equation.

Equivalent equations
$$\begin{cases} 18 - 18 = 10x + 5 - 18 & \text{Subtract 18 from each side.} \\ 0 = 10x - 13 & \text{Simplify.} \\ 10x - 13 = 0 & \text{Rewrite the equation.} \end{cases}$$

Even though these steps did not result in a solution to the equation $18 = 10x + 5$, applying the addition property of equality allowed us to rewrite the equation as $10x - 13 = 0$, which is an example of a *linear equation*. (See Example 3(a) for a solution to the equation $10x - 13 = 0$.)

LINEAR EQUATION IN ONE VARIABLE

A **linear equation** in one variable is an equation that can be written in the form

$$ax + b = 0,$$

where a and b are constants with $a \neq 0$.

NOTE: Linear equations can model applications in which things move or change at a *constant rate*. ∎

If an equation is linear, writing it in the form $ax + b = 0$ should not require any properties or processes other than the following:

- using the distributive property to clear any parentheses
- combining like terms
- applying the addition property of equality

TABLE 2.3 gives examples of linear equations and values for a and b.

Examples of Linear Equations

Equation	$ax + b = 0$ Form	a	b
$x = 1$	$x - 1 = 0$	1	-1
$-5x + 4 = 3$	$-5x + 1 = 0$	-5	1
$2.5x = 0$	$2.5x + 0 = 0$	2.5	0

TABLE 2.3

MAKING CONNECTIONS 1

READING CHECK 1

- Name three things that tell you that an equation is not a linear equation.

Equations That Are Not Linear

An equation *cannot* be written in the form $ax + b = 0$ (and is not a *linear* equation) if after clearing parentheses and combining like terms, any of the following statements are true.

1. The variable has an exponent other than 1.
2. The variable appears in a denominator of a fraction.
3. The variable appears under the symbol $\sqrt{\ }$ or within an absolute value.

EXAMPLE 1   **Determining whether an equation is linear**

Determine whether the equation is linear. If the equation is linear, give values for a and b that result when the equation is written in the form $ax + b = 0$.
(a) $4x + 5 = 0$ (b) $5 = -\frac{3}{4}x$ (c) $4x^2 + 6 = 0$ (d) $\frac{3}{x} + 5 = 0$

Solution
(a) The equation is linear because it is in the form $ax + b = 0$ with $a = 4$ and $b = 5$.
(b) The equation can be rewritten as follows.

$$5 = -\frac{3}{4}x \qquad \text{Given equation}$$

$$\frac{3}{4}x + 5 = \frac{3}{4}x + \left(-\frac{3}{4}x\right) \qquad \text{Add } \tfrac{3}{4}x \text{ to each side.}$$

$$\frac{3}{4}x + 5 = 0 \qquad \text{Additive inverse}$$

The given equation is linear because it can be written in the form $ax + b = 0$ with $a = \frac{3}{4}$ and $b = 5$.

NOTE: If 5 had been subtracted from each side, the result would be $0 = -\frac{3}{4}x - 5$, which is an equivalent linear equation with $a = -\frac{3}{4}$ and $b = -5$. ∎

(c) The equation is *not* linear because it cannot be written in the form $ax + b = 0$. The variable has exponent 2.
(d) The equation is *not* linear because it cannot be written in the form $ax + b = 0$. The variable appears in the denominator of a fraction.

Now Try Exercises 9, 11, 13, 15

2 Solving Linear Equations

Every linear equation $(ax + b = 0, a \neq 0)$ has *exactly one* solution. Showing that this is true is left as an exercise (see Exercise 59). Solving a linear equation means finding the (one) value of the variable that makes the equation true.

SOLVING WITH TABLES One way to solve a linear equation is to make a table of values. A table provides an organized way of checking possible values of the variable to see if there is a value that makes the equation true. For example, if we want to solve the equation

$$2x - 5 = -7,$$

we substitute various values for x in the left side of the equation. If one of these values results in -7, then the value makes the equation true and is the solution. In the next example, a table of values is used to solve this equation.

EXAMPLE 2   **Using a table to solve an equation**

Complete **TABLE 2.4** for the given values of x. Then solve the equation $2x - 5 = -7$.

Listing Possible Values for $2x - 5$

x	-3	-2	-1	0	1	2	3
$2x - 5$	-11						

TABLE 2.4

Solution
To complete the table, substitute $x = -2, -1, 0, 1, 2,$ and 3 into the expression $2x - 5$. For example, if $x = -2$, then $2x - 5 = 2(-2) - 5 = -9$. The other values shown in **TABLE 2.5** can be found similarly.

TECHNOLOGY NOTE

Graphing Calculators and Tables

Many graphing calculators have the capability to make tables. Table 2.5 is shown in the accompanying figure.

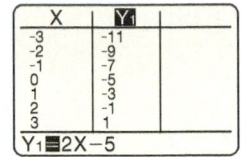

CALCULATOR HELP

To make a table on a calculator, see Appendix A (pages AP-2 and AP-3).

STUDY TIP

We will be solving equations throughout the remainder of the text. Spend a little extra time practicing these steps so that they are easy to recall when needed later.

Solving $2x - 5 = 7$ Numerically

x	-3	-2	-1	0	1	2	3
$2x - 5$	-11	-9	-7	-5	-3	-1	1

TABLE 2.5

We see in **TABLE 2.5** that $2x - 5$ equals -7 when $x = -1$. The solution to $2x - 5 = -7$ is -1.

Now Try Exercise 25

SOLVING SYMBOLICALLY Although tables can be used to solve some linear equations, the process of creating a table that contains the solution can take a significant amount of time. For example, the solution to the equation $9x - 4 = 0$ is $\frac{4}{9}$. However, creating a table that reveals this solution would be quite challenging.

The following strategy, which involves the addition and multiplication properties of equality, is a method for solving linear equations symbolically.

SOLVING A LINEAR EQUATION SYMBOLICALLY

STEP 1: Use the distributive property to clear any parentheses on each side of the equation. Combine any like terms on each side.

STEP 2: Use the addition property of equality to get all of the terms containing the variable on one side of the equation and all other terms on the other side of the equation. Combine any like terms on each side.

STEP 3: Use the multiplication property of equality to isolate the variable by multiplying each side of the equation by the reciprocal of the number in front of the variable (or divide each side by that number).

STEP 4: Check the solution by substituting it in the given equation.

When a linear equation does not contain parentheses, we can start with STEP 2 in the strategy shown above. This is the case for the three equations in the next example.

EXAMPLE 3   **Solving linear equations**

Solve each linear equation. Check the answer for part (b).
(a) $10x - 13 = 0$ **(b)** $\frac{1}{2}x + 3 = 6$ **(c)** $5x + 7 = 2x + 3$

Solution
(a) First, isolate the x-term on the left side of the equation by adding 13 to each side.

$$10x - 13 = 0 \qquad \text{Given equation}$$

$$10x - 13 + 13 = 0 + 13 \qquad \text{Add 13 to each side. (STEP 2)}$$

$$10x = 13 \qquad \text{Add the real numbers.}$$

To obtain a coefficient of 1 on the x-term, divide each side by 10.

$$\frac{10x}{10} = \frac{13}{10} \qquad \text{Divide each side by 10. (STEP 3)}$$

$$x = \frac{13}{10} \qquad \text{Simplify.}$$

The solution is $\frac{13}{10}$.

(b) Start by subtracting 3 from each side.

$$\frac{1}{2}x + 3 = 6 \qquad \text{Given equation}$$

$$\frac{1}{2}x + 3 - 3 = 6 - 3 \qquad \text{Subtract 3 from each side. (STEP 2)}$$

$$\frac{1}{2}x = 3 \qquad \text{Subtract the real numbers.}$$

$$2 \cdot \frac{1}{2}x = 2 \cdot 3 \qquad \text{Multiply each side by 2. (STEP 3)}$$

$$x = 6 \qquad \text{Multiply the real numbers.}$$

The solution is **6**. To check it, substitute **6** for x in the equation, $\frac{1}{2}x + 3 = 6$.

$$\frac{1}{2} \cdot \mathbf{6} + 3 \stackrel{?}{=} 6 \qquad \text{Replace } x \text{ with 6. (STEP 4)}$$

$$3 + 3 \stackrel{?}{=} 6 \qquad \text{Multiply.}$$

$$6 = 6 \checkmark \qquad \text{Add; the answer checks.}$$

(c) Since this equation has two x-terms, we need to get all x-terms on one side of the equation and all real numbers on the other side. To do this, begin by subtracting $2x$ from each side.

$$5x + 7 = 2x + 3 \qquad \text{Given equation}$$

$$5x - 2x + 7 = 2x - 2x + 3 \qquad \text{Subtract } 2x \text{ from each side. (STEP 2)}$$

$$3x + 7 = 3 \qquad \text{Combine like terms.}$$

$$3x + 7 - 7 = 3 - 7 \qquad \text{Subtract 7 from each side. (STEP 2)}$$

$$3x = -4 \qquad \text{Simplify.}$$

$$\frac{3x}{3} = \frac{-4}{3} \qquad \text{Divide each side by 3. (STEP 3)}$$

$$x = -\frac{4}{3} \qquad \text{Simplify the fractions.}$$

The solution is $-\frac{4}{3}$.

Now Try Exercises 31, 33, 37

 Math in Context *Financial* The number of private venture-backed companies worth $1 billion or more is increasing. The rise of these "unicorns" has been dramatic. One such example is Uber, a popular ride-sharing service worth billions. (*Source:* Statista.)

EXAMPLE 4 **Solving an application involving a linear equation**

The number N of "unicorn" companies worth $1 billion or more can be modeled by the equation $N = 3.625x + 42$, where $0 \le x \le 20$ and x is the number of months after January 2014. Use this equation to determine the x-value when N was equal to 100. Interpret the meaning of this solution.

Solution
Begin by solving for x when $N = 100$. That is, solve $3.625x + 42 = 100$.

$$3.625x + 42 = 100 \qquad \text{Equation to solve}$$

$$3.625x = 58 \qquad \text{Subtract 42 from each side.}$$

$$x = \frac{58}{3.625} \qquad \text{Divide each side by 3.625.}$$

$$x = 16 \qquad \text{Simplify}$$

The value $x = 16$ represents 16 months after January 2014, or May 2015. Thus there were 100 venture-backed private companies worth \$1 billion or more in May 2015.

Now Try Exercise 73

READING CHECK 2

• In the strategy for solving linear equations symbolically, which step involves the use of the distributive property?

3 Solving Linear Equations by Applying the Distributive Property

Sometimes the distributive property is helpful in solving linear equations. The next example demonstrates how to apply the distributive property in such situations. Use of the distributive property appeared in STEP 1 of the strategy for solving linear equations discussed earlier.

EXAMPLE 5 Applying the distributive property

Solve each linear equation. Check the answer for part (a).
(a) $4(x - 3) + x = 0$ (b) $2(3z - 4) + 1 = 3(z + 1)$

Solution
(a) Begin by applying the distributive property.

$$4(x - 3) + x = 0 \qquad \text{Given equation}$$

$4 \cdot x = 4x$
$4 \cdot 3 = 12$

$$4x - 12 + x = 0 \qquad \text{Distributive property (STEP 1)}$$

$$5x - 12 = 0 \qquad \text{Combine like terms.}$$

$$5x - 12 + 12 = 0 + 12 \qquad \text{Add 12 to each side. (STEP 2)}$$

$$5x = 12 \qquad \text{Add the real numbers.}$$

$$\frac{5x}{5} = \frac{12}{5} \qquad \text{Divide each side by 5. (STEP 3)}$$

$$x = \frac{12}{5} \qquad \text{Simplify.}$$

To see if $\frac{12}{5}$ is the solution, substitute $\frac{12}{5}$ for x in the equation $4(x - 3) + x = 0$.

$$4\left(\frac{12}{5} - 3\right) + \frac{12}{5} \stackrel{?}{=} 0 \qquad \text{Replace } x \text{ with } \frac{12}{5}. \text{ (STEP 4)}$$

$$4\left(\frac{12}{5} - \frac{15}{5}\right) + \frac{12}{5} \stackrel{?}{=} 0 \qquad \text{Common denominator}$$

$$4\left(-\frac{3}{5}\right) + \frac{12}{5} \stackrel{?}{=} 0 \qquad \text{Subtract within parentheses.}$$

$$-\frac{12}{5} + \frac{12}{5} \stackrel{?}{=} 0 \qquad \text{Multiply.}$$

$$0 = 0 \checkmark \qquad \text{Add; the answer checks.}$$

(b) Begin by applying the distributive property to each side of the equation. Then get all z-terms on the left side and terms containing only real numbers on the right side.

$$2(3z - 4) + 1 = 3(z + 1) \quad \text{Given equation}$$

$$6z - 8 + 1 = 3z + 3 \quad \text{Distributive property (STEP 1)}$$

$$6z - 7 = 3z + 3 \quad \text{Add the real numbers.}$$

$$3z - 7 = 3 \quad \text{Subtract } 3z \text{ from each side. (STEP 2)}$$

$$3z = 10 \quad \text{Add 7 to each side. (STEP 2)}$$

$$\frac{3z}{3} = \frac{10}{3} \quad \text{Divide each side by 3. (STEP 3)}$$

$$z = \frac{10}{3} \quad \text{Simplify.}$$

The solution is $\frac{10}{3}$.

Now Try Exercises 43, 45

4 Solving Linear Equations by Clearing Fractions and Decimals

Some people prefer to do calculations without fractions or decimals. For this reason, clearing an equation of fractions or decimals before solving it can be helpful. To clear fractions or decimals, multiply each side of the equation by the least common denominator (LCD).

EXAMPLE 6 **Solving a linear equation by clearing fractions**

Solve each linear equation.

(a) $\frac{1}{7}x - \frac{5}{7} = \frac{3}{7}$ (b) $\frac{2}{3}x - \frac{1}{6} = x$

Solution

(a) Multiply each side of the equation by the LCD **7** to clear (remove) fractions.

$$\frac{1}{7}x - \frac{5}{7} = \frac{3}{7} \quad \text{Given equation}$$

The LCD is 7.
$$7\left(\frac{1}{7}x - \frac{5}{7}\right) = 7 \cdot \frac{3}{7} \quad \text{Multiply each side by 7.}$$

$$\frac{7}{1} \cdot \frac{1}{7}x - \frac{7}{1} \cdot \frac{5}{7} = \frac{7}{1} \cdot \frac{3}{7} \quad \text{Distributive property}$$

$$x - 5 = 3 \quad \text{Simplify.}$$

$$x = 8 \quad \text{Add 5 to each side.}$$

The solution is 8.

(b) The LCD for 3 and 6 is 6. Multiply each side of the equation by **6**.

$$\frac{2}{3}x - \frac{1}{6} = x \quad \text{Given equation}$$

The LCD is 6.
$$6\left(\frac{2}{3}x - \frac{1}{6}\right) = 6 \cdot x \quad \text{Multiply each side by 6.}$$

$$\frac{6}{1} \cdot \frac{2}{3}x - \frac{6}{1} \cdot \frac{1}{6} = 6 \cdot x \quad \text{Distributive property}$$

$$4x - 1 = 6x \quad \text{Simplify.}$$

$$-1 = 2x \quad \text{Subtract } 4x \text{ from each side.}$$

$$-\frac{1}{2} = x \quad \text{Divide each side by 2.}$$

The solution is $-\frac{1}{2}$.

Now Try Exercises 55, 57

EXAMPLE 7   Solving a linear equation by clearing decimals

Solve each linear equation.
(a) $0.2x - 0.7 = 0.4$
(b) $0.01x - 0.42 = -0.2x$

Solution
(a) The least common denominator for 0.2, 0.7, and 0.4 $\left(\text{or } \frac{2}{10}, \frac{7}{10}, \text{ and } \frac{4}{10}\right)$ is **10**. Multiply each side by 10. When multiplying by 10, move the decimal point 1 place to the right.

$$0.2x - 0.7 = 0.4 \qquad \text{Given equation}$$

The LCD is 10.
$$\mathbf{10}(0.2x - 0.7) = \mathbf{10}(0.4) \qquad \text{Multiply each side by 10.}$$

$$\mathbf{10}(0.2x) - \mathbf{10}(0.7) = \mathbf{10}(0.4) \qquad \text{Distributive property}$$

$$2x - 7 = 4 \qquad \text{Simplify.}$$

$$2x = 11 \qquad \text{Add 7 to each side.}$$

$$x = \frac{11}{2} \qquad \text{Divide each side by 2.}$$

The solution is $\frac{11}{2}$, or 5.5.

(b) The least common denominator for 0.01, 0.42, and 0.2 $\left(\text{or } \frac{1}{100}, \frac{42}{100}, \text{ and } \frac{2}{10}\right)$ is **100**. Multiply each side by 100. To do this, move the decimal point 2 places to the right.

$$0.01x - 0.42 = -0.2x \qquad \text{Given equation}$$

The LCD is 100.
$$\mathbf{100}(0.01x - 0.42) = \mathbf{100}(-0.2x) \qquad \text{Multiply each side by 100.}$$

$$\mathbf{100}(0.01x) - \mathbf{100}(0.42) = \mathbf{100}(-0.2x) \qquad \text{Distributive property}$$

$$x - 42 = -20x \qquad \text{Simplify.}$$

$$x - 42 \,{\color{red}+\, 20x + 42} = -20x \,{\color{red}+\, 20x + 42} \qquad \text{Add } 20x \text{ and } 42.$$

$$21x = 42 \qquad \text{Combine like terms.}$$

$$x = 2 \qquad \text{Divide each side by 21.}$$

The solution is 2.

Now Try Exercises 51, 53

MAKING CONNECTIONS 2

Clearing Decimals and the LCD

When clearing decimals from a linear equation, the LCD is always a power of **10** determined by the maximum number of decimal places in the coefficients of the terms in the equation. For example, for the equation in Example 7(b), the maximum number of such decimal places is **2**.

2 decimal places ——— $0.\mathbf{01}x - 0.\mathbf{42} = -0.2x$ ——— 1 decimal place

The LCD is $\mathbf{10^2}$, or **100**. That is, decimals are cleared from the equation by multiplying each term by **100**. As a result, the decimal point in each term is moved **2** places to the right.

---

5 Identifying Equations with No Solutions, One Solution, or Infinitely Many Solutions

Some equations that appear to be linear are not because when they are written in the form $ax + b = 0$ the value of a is 0 and no x-term appears. This type of equation can have *no solutions* or *infinitely many solutions*. The following See the Concept explains how we can tell if an equation has no solutions, one solution, or infinitely many solutions.

See the Concept 1   THE NUMBER OF SOLUTIONS

	No Solutions	*One Solution*	*Infinitely Many Solutions*
Equation	$x + 1 = x$	$4x - 1 = 7$	$5x = 2x + 3x$
$ax + b = 0$ *form*	$0x + 1 = 0$ Ⓐ	$4x - 8 = 0$ Ⓐ	$0x + 0 = 0$ Ⓐ
Simplified form	$1 = 0$ Ⓑ	$x = 2$ Ⓑ	$0 = 0$ Ⓑ

Equations with No Solutions:

Ⓐ When the equation is written in $ax + b = 0$ form, the value of a is 0.
Ⓑ The simplified equation is *always false* for any value of the variable. An equation that is always false is called a **contradiction**.

Equations with One Solution:

Ⓐ When the equation is written in $ax + b = 0$ form, the value of a is *not* 0.
Ⓑ The simplified equation is true for *only one* value of the variable. This is called a **conditional equation**.

Equations with Infinitely Many Solutions:

Ⓐ When the equation is written in $ax + b = 0$ form, the value of a is 0.
Ⓑ The simplified equation is *always true* for any value of the variable. An equation that is always true is called an **identity**.

WATCH VIDEO IN MML .

READING CHECK 3

• How can you tell when an equation will have infinitely many solutions?

EXAMPLE 8 **Determining numbers of solutions**

Determine whether the equation has no solutions, one solution, or infinitely many solutions.
(a) $3x = 2(x + 1) + x$ **(b)** $2x - (x + 1) = x - 1$ **(c)** $5x = 2(x - 4)$

Solution
(a) Start by applying the distributive property.

$$3x = 2(x + 1) + x \qquad \text{Given equation}$$
$$3x = 2x + 2 + x \qquad \text{Distributive property}$$
$$3x = 3x + 2 \qquad \text{Combine like terms.}$$
$$0 = 2 \qquad \text{Subtract } 3x \text{ from each side.}$$

The equation $0 = 2$ is always false, so the given equation is a contradiction and it has no solutions.

(b) Start by applying the distributive property.

$$2x - (x + 1) = x - 1 \qquad \text{Given equation}$$
$$\underbrace{-1(x + 1)}\quad 2x - x - 1 = x - 1 \qquad \text{Distributive property}$$
$$x - 1 = x - 1 \qquad \text{Combine like terms.}$$
$$x = x \qquad \text{Add 1 to each side.}$$
$$0 = 0 \qquad \text{Subtract } x \text{ from each side.}$$

The equation $0 = 0$ is always true, so the given equation is an identity that has infinitely many solutions. Note that the solution set contains all real numbers.

(c) Start by applying the distributive property.

$$5x = 2(x - 4) \qquad \text{Given equation}$$
$$5x = 2x - 8 \qquad \text{Distributive property}$$
$$3x = -8 \qquad \text{Subtract } 2x \text{ from each side.}$$
$$x = -\frac{8}{3} \qquad \text{Divide each side by 3.}$$

Thus there is one solution, $-\frac{8}{3}$.

Now Try Exercises 61, 63, 65

CRITICAL THINKING 1

What must be true about b and d for the equation

$$bx - 2 = dx + 7$$

to have no solutions? What must be true about b and d for this equation to have exactly one solution?

2.2   Putting It All Together

CONCEPT	COMMENTS	EXAMPLES
Linear Equation	Can be written as $$ax + b = 0,$$ where $a \neq 0$; has one solution	The equation $5x - 8 = 0$ is linear, with $a = 5$ and $b = -8$. The equation $2x^2 + 4 = 0$ is not linear.
Solving Linear Equations Numerically	To solve a linear equation numerically, complete a table for various values of the variable and then select the solution from the table, if possible.	The solution to $2x - 4 = -2$ is **1**. 	x	-1	0	1	 	---	---	---	---	 	$2x - 4$	-6	-4	-2	
Solving Linear Equations Symbolically	Use the addition and multiplication properties of equality to isolate the variable. See the four-step approach to solving a linear equation in this section.	$5x - 8 = 0$ Given equation $5x = 8$ Add 8 to each side. $x = \dfrac{8}{5}$ Divide each side by 5.
Equations with No Solutions	Some equations that appear to be linear have no solutions. Solving will result in an equivalent equation that is always false.	The equation $$x = x + 5$$ has no solutions because it is equivalent to the equation $0 = 5$, which is always false.
Equations with Infinitely Many Solutions	Some equations that appear to be linear have infinitely many solutions. Solving will result in an equivalent equation that is always true.	The equation $$2x = x + x$$ has infinitely many solutions because it is equivalent to the equation $0 = 0$, which is always true.

2.2 Exercises

MyMathLab®

CONCEPTS AND VOCABULARY

1. Linear equations can model applications in which things move or change at a(n) _____ rate.

2. A linear equation can be written in the form _____ with $a \neq 0$.

3. How many solutions does a linear equation in one variable have?

4. When a table of values is used to solve a linear equation, the equation is being solved _____.

5. What two properties of equality are frequently used to solve linear equations?

6. To clear fractions or decimals from an equation, multiply each side by the _____.

7. If solving an equation results in $0 = 4$, how many solutions does it have?

8. If solving an equation results in $0 = 0$, how many solutions does it have?

IDENTIFYING LINEAR EQUATIONS IN ONE VARIABLE

Exercises 9–22: (Refer to Example 1.) Is the equation linear? If it is linear, give values for a and b that result when the equation is written in the form $ax + b = 0$.

9. $3x - 7 = 0$ 10. $-2x + 1 = 4$

11. $\frac{1}{2}x = 0$ 12. $-\frac{3}{4}x = 0$

13. $4x^2 - 6 = 11$ 14. $-2x^2 + x = 4$

15. $\frac{6}{x} - 4 = 2$ 16. $2\sqrt{x} - 1 = 0$

17. $1.1x + 0.9 = 1.8$ 18. $-5.7x - 3.4 = -6.8$

19. $2(x - 3) = 0$ 20. $\frac{1}{2}(x + 4) = 0$

21. $|3x| + 2 = 1$ 22. $3x = 4x^3$

SOLVING LINEAR EQUATIONS NUMERICALLY

Exercises 23–28: Complete the table for the given values of x. Then solve the given equation numerically.

23. $x - 3 = -1$

x	−1	0	1	2	3
$x - 3$	−4				

24. $-2x = 0$

x	−2	−1	0	1	2
$-2x$	4				

25. $-3x + 7 = 1$

x	0	1	2	3	4
$-3x + 7$	7				

26. $5x - 2 = 3$

x	−1	0	1	2	3
$5x - 2$	−7				

27. $4 - 2x = 6$

x	−2	−1	0	1	2
$4 - 2x$	8				

28. $9 - (x + 3) = 4$

x	−2	−1	0	1	2
$9 - (x + 3)$	8				

SOLVING LINEAR EQUATIONS SYMBOLICALLY

Exercises 29–40: Solve and check the solution.

29. $11x = 3$ 30. $-5x = 15$

31. $x - 18 = 5$ 32. $8 = 5 + x$

33. $\frac{1}{2}x - 1 = 13$ 34. $\frac{1}{4}x + 3 = 9$

35. $-6 = 5x + 5$

36. $31 = -7x - 4$

37. $3z + 2 = z - 5$

38. $z - 5 = 5z - 3$

39. $12y - 6 = 33 - y$

40. $-13y + 2 = 22 - 3y$

SOLVING LINEAR EQUATIONS BY APPLYING THE DISTRIBUTIVE PROPERTY

Exercises 41–50: Solve and check the solution.

41. $4(x - 1) = 5$

42. $-2(2x + 7) = 1$

43. $1 - (3x + 1) = 5 - x$

44. $6 + 2(x - 7) = 10 - 3(x - 3)$

45. $5t - 6 = 2(t + 1) + 2$

46. $-2(t - 7) - (t + 5) = 5$

47. $3(4z - 1) - 2(z + 2) = 2(z + 1)$

48. $-(z + 4) + (3z + 1) = -2(z + 1)$

49. $4y - 2(y + 1) = 0$

50. $(15y + 20) - 5y = 5 - 10y$

SOLVING LINEAR EQUATIONS BY CLEARING DECIMALS

Exercises 51–54: Solve and check the solution.

51. $7.3x - 1.7 = 5.6$ **52.** $5.5x + 3x = 51$

53. $-9.5x - 0.05 = 10.5x + 1.05$

54. $0.04x + 0.03 = 0.02x - 0.1$

SOLVING LINEAR EQUATIONS BY CLEARING FRACTIONS

Exercises 55–58: Solve and check the solution.

55. $\frac{1}{2}x - \frac{3}{2} = \frac{5}{2}$

56. $-\frac{1}{4}x + \frac{5}{4} = \frac{3}{4}$

57. $-\frac{3}{8}x + \frac{1}{4} = \frac{1}{8}$

58. $\frac{1}{3}x + \frac{1}{4} = \frac{1}{6} - x$

59. Thinking Generally A linear equation has exactly one solution. Find the solution to the equation $ax + b = 0$, where $a \neq 0$, by solving for x.

60. Thinking Generally Solve the linear equation given by $\frac{1}{a}x - b = 0$ for x.

IDENTIFYING EQUATIONS WITH NO SOLUTIONS, ONE SOLUTION, OR INFINITELY MANY SOLUTIONS

Exercises 61–70: Determine whether the equation has no solutions, one solution, or infinitely many solutions.

61. $5x = 5x + 1$ **62.** $4x = 5(x + 3) - x$

63. $8x = 0$ **64.** $9x = x + 1$

65. $2(x - 3) = 2x - 6$

66. $5x = 15 - 2(x + 7)$

67. $x - (3x + 2) = 15 - 2x$

68. $4(x + 2) - 2(2x + 3) = 10$

69. $5(2x + 7) - (10x + 5) = 30$

70. $2x - (x + 5) = x - 5$

APPLICATIONS INVOLVING LINEAR EQUATIONS

71. Distance Traveled A bicyclist is 4 miles from home, riding away from home at 8 miles per hour.
   (a) Make a table that shows the bicyclist's distance D from home after 0, 1, 2, 3, and 4 hours.
   (b) Write an equation that calculates the distance D after x hours.
   (c) Determine D when $x = 3$ hours. Does your answer agree with the value found in your table?
   (d) Find x when $D = 22$ miles. Interpret the result.

72. Distance Traveled An athlete is 16 miles from home, running *toward* home at 6 miles per hour.
   (a) Write an equation that calculates the distance D the athlete is from home after x hours.
   (b) Determine D when $x = 1.5$ hours.
   (c) Find x when $D = 5.5$ miles. Interpret the result.

73. Internet Users The number of Internet users I in millions during year x, where $x \geq 2007$, can be approximated by the equation

$$I = 241x - 482{,}440.$$

Approximate the year in which there were 1730 million (1.73 billion) Internet users.

74. HIV Infections The cumulative number of HIV infections N in thousands for the United States in year x can be approximated by the equation

$$N = 42x - 83{,}197,$$

where $x \geq 2000$. Approximate the year when this number reached 970 thousand. (*Source:* CDC)

75. Working-Age Population The U.S. working age population P in millions during year x can be approximated by $P = 1.2x - 2205$, where x is any year between 1970 and 2050. Find the year when the U.S. working age population was 207 million. (*Source:* UN)

76. Working-Age Population The sub-Saharan African working age population P in millions during year x can be approximated by $P = 19.44x - 38{,}661$, where x is any year between 2014 and 2050. Estimate the year when the sub-Saharan African working age population will be 705 million. (*Source:* UN)

77. Internet Ad Spending The amount A spent worldwide on Internet advertising in billions of dollars during year x can be approximated by $A = 1.885x - 3774$, where x is any year between 2006 and 2015. Determine the year when Internet advertising spending was $20.5 billion.

78. Global Oil Demand The global demand D for oil in millions of barrels per day during year x can be approximated by $D = 0.75x - 1420$, where x is any year from 2012 to 2020. Estimate the year when global demand for oil was 92 million barrels per day.

WRITING ABOUT MATHEMATICS

79. A student says that the equation $4x - 1 = 1 - x$ is not a linear equation because it is not in the form $ax + b = 0$. Is the student correct? Explain.

80. A student solves a linear equation as follows.

$$4(x + 3) = 5 - (x + 3)$$
$$4x + 3 \stackrel{?}{=} 5 - x + 3$$
$$4x + 3 \stackrel{?}{=} 8 - x$$
$$5x \stackrel{?}{=} 5$$
$$x \stackrel{?}{=} 1$$

Identify and explain the errors that the student made. What is the correct answer?

SECTIONS 2.1 and 2.2   ## Checking Basic Concepts

1. Determine whether the equation is linear.
   (a) $4x^3 - 2 = 0$ **(b)** $2(x + 1) = 4$

2. Complete the table for each value of x. Then use the table to solve $4x - 3 = 13$.

x	3	3.5	4	4.5	5
$4x - 3$	9				17

3. Solve each equation and check your answer.
   (a) $x - 12 = 6$
   (b) $\frac{3}{4}z = \frac{1}{8}$
   (c) $0.6t + 0.4 = 2$
   (d) $5 - 2(x - 2) = 3(4 - x)$

4. Determine whether each equation has no solutions, one solution, or infinitely many solutions.
   (a) $x - 5 = 6x$
   (b) $-2(x - 5) = 10 - 2x$
   (c) $-(x - 1) = -x - 1$

5. Distance Traveled A driver is 300 miles from home and is traveling toward home on a freeway at a constant speed of 75 miles per hour.
   (a) Write an equation to calculate the distance D that the driver is from home after x hours.
   (b) Write an equation whose solution gives the hours needed for the driver to reach home.
   (c) Solve the equation from part (b).

2.3   Introduction to Problem Solving

Objectives

1   Identifying Steps for Solving a Problem

2   Translating Sentences into Equations

3   Solving Number Problems and Applications

4   Solving Percent Problems
   • Percent, Fraction, and Decimal Notation

5   Calculating Percent Change

6   Solving Distance Problems

7   Solving Mixture Problems

1   Identifying Steps for Solving a Problem

Word problems can be challenging because formulas and equations are not usually given. To solve such problems, we need a strategy. The following steps are based on George Polya's (1888–1985) four-step process for problem solving.

STEPS FOR SOLVING A PROBLEM

STEP 1: Read the problem carefully and be sure that you understand it. (You may need to read the problem more than once.) Assign a variable to what you are being asked to find. If necessary, write other quantities in terms of this variable.

STEP 2: Write an equation that relates the quantities described in the problem. You may need to sketch a diagram or refer to known formulas.

STEP 3: Solve the equation. Use the solution to determine the solution(s) to the original problem. Include any necessary units.

STEP 4: Check your solution in the original problem. Does it seem reasonable?

NEW VOCABULARY

□ Percent Change

READING CHECK 1

• Why doesn't the problem-solving strategy stop after the equation is solved in STEP 3?

STUDY TIP

STEP 4 in the problem-solving strategy provides good advice for working any math problem. Checking your answer, especially when taking an exam, can lead to fewer errors and better scores.

2 Translating Sentences into Equations

Even if we understand the problem that we are trying to solve (STEP 1), we may not be able to find a solution if we cannot write an appropriate equation (STEP 2). **TABLE 2.6** lists common words that are associated with the math symbols needed to write equations.

Math Symbols and Associated Words

Symbol	Associated Words
$+$	add, plus, more, sum, total, increase
$-$	subtract, minus, less, difference, fewer, decrease
$\cdot$	multiply, times, twice, double, triple, product
$\div$	divide, divided by, quotient, per
$=$	equals, is, gives, results in, is the same as

TABLE 2.6

EXAMPLE 1   **Translating sentences into equations**

Translate the sentence into an equation using the variable x. Then solve the resulting equation.
- (a) Three times a number minus 6 is equal to 18.
- (b) The sum of half a number and 5 is zero.
- (c) Sixteen is 4 less than twice a number.

Solution

(a) The phrase "Three times a number" indicates that we multiply x by 3 to get $3x$. The word "minus" indicates that we then subtract 6 from $3x$ to get $3x - 6$. This expression "equals" 18, so the equation is $3x - 6 = 18$. The solution is 8 as shown.

> Three times a number minus 6 is equal to 18.

$$3x - 6 = 18 \qquad \text{Equation to be solved}$$
$$3x = 24 \qquad \text{Add 6 to each side.}$$
$$\frac{3x}{3} = \frac{24}{3} \qquad \text{Divide each side by 3.}$$
$$x = 8 \qquad \text{Simplify the fractions.}$$

(b) The word "sum" indicates that we add "half a number" and 5 to get $\frac{1}{2}x + 5$. The word "is" implies equality, so the equation is $\frac{1}{2}x + 5 = 0$. The solution is -10 as shown.

> The sum of half a number and 5 is zero.

$$\frac{1}{2}x + 5 = 0 \qquad \text{Equation to be solved}$$
$$\frac{1}{2}x = -5 \qquad \text{Subtract 5 from each side.}$$
$$x = -10 \qquad \text{Multiply each side by 2.}$$

(c) To translate "4 less than twice a number" into a mathematical expression, we write $2x - 4$. If this seems backwards, consider how you would calculate "4 less than your age." The equation is $16 = 2x - 4$ and the solution is 10 as shown.

> Sixteen is 4 less than twice a number.

$$16 = 2x - 4 \qquad \text{Equation to be solved}$$
$$20 = 2x \qquad \text{Add 4 to each side.}$$
$$\frac{20}{2} = \frac{2x}{2} \qquad \text{Divide each side by 2.}$$
$$x = 10 \qquad \text{Simplify and rewrite.}$$

Now Try Exercises 13, 15, 19

3 Solving Number Problems and Applications

In the next example, we apply the four-step process to a word problem that involves three consecutive numbers. Two integers (or natural numbers) are *consecutive* if they differ by 1. For example, 21 and 22 are consecutive, as are -137 and -136.

EXAMPLE 2   **Solving a number problem**

The sum of three consecutive natural numbers is 81. Find the three numbers.

Solution

STEP 1: Start by assigning a variable n to an unknown quantity.

$$n: \text{smallest of the three natural numbers}$$

Next, write the other two natural numbers in terms of n.

$$n + 1: \text{next consecutive natural number}$$

$$n + 2: \text{largest of the three consecutive natural numbers}$$

STEP 2: Write an equation that relates these unknown quantities. The sum of the three consecutive natural numbers is **81**, so the needed equation is

$$n + (n + 1) + (n + 2) = 81.$$

STEP 3: Solve the equation in STEP 2.

$$n + (n + 1) + (n + 2) = 81 \quad \text{Equation to be solved}$$

$$(n + n + n) + (1 + 2) = 81 \quad \text{Commutative and associative properties}$$

$$3n + 3 = 81 \quad \text{Combine like terms.}$$

$$3n = 78 \quad \text{Subtract 3 from each side.}$$

$$n = 26 \quad \text{Divide each side by 3.}$$

The smallest of the three numbers is 26, so the three numbers are 26, 27, and 28.
STEP 4: To check this solution, we can add the three numbers to see if their sum is 81.

$$26 + 27 + 28 = 81 \checkmark$$

Now Try Exercise 21

🌐 *Math in Context* (World Health) The infant mortality rate for a country measures the number of deaths of infants under one year of age per 1000 live births in a given year. A high infant mortality rate may indicate a lack of good quality health care for infants. In the next example, a number problem is solved to find the infant mortality rate in Iceland, which has one of the lowest rates in the world.

EXAMPLE 3   **Finding an infant mortality rate**

Sierra Leone, in West Africa, has one of the highest infant mortality rates in the world at 75. This rate is 15 more than 20 times the rate in Iceland. Find the infant mortality rate in Iceland. (*Source:* World Population Prospectus.)

Solution

STEP 1: Let r represent the unknown infant mortality rate in Iceland.
STEP 2: Since the rate 75 in Sierra Leone is 15 more than 20 times the rate r in Iceland, the equation to solve is $75 = 15 + 20r$.

STEP 3: Rewrite the equation in STEP 2 and then solve it.

$$20r + 15 = 75 \quad \text{Equation to be solved (rewritten)}$$
$$20r = 60 \quad \text{Subtract 15 from each side.}$$
$$r = 3 \quad \text{Divide each side by 20.}$$

The infant mortality rate in Iceland is 3 deaths per 1000 live births.

STEP 4: To check this solution, we verify that 15 more than 20 times the Iceland infant mortality rate of 3 gives the Sierra Leone infant mortality rate of 75.

$$20(3) + 15 = 60 + 15 = 75 \checkmark$$

Now Try Exercise 33

4 Solving Percent Problems

Problems involving percentages occur in everyday life. Wage increases, sales tax, and government data all make use of percentages. To solve percent problems, it is important that we know how to convert between percent, fraction, and decimal notation.

PERCENT, FRACTION, AND DECIMAL NOTATION   Percent notation can be changed to either fraction or decimal notation as described below.

PERCENT NOTATION

The expression $x\%$ represents the fraction $\frac{x}{100}$ or the decimal number $x \cdot 0.01$.

NOTE: To write $x\%$ as a decimal number, move the decimal point in the number x two places to the *left* and then remove the % symbol. For example, 13.6% can be written as a decimal number as follows.

$$13.6\% \rightarrow 13.6\% \rightarrow 0.136 \ \blacksquare$$

READING CHECK 2

• How do we write $x\%$ as a decimal?

EXAMPLE 4   **Converting percent notation**

Convert each percentage to fraction and decimal notation.
(a) 23%   (b) 5.2%   (c) 0.3%

Solution
(a) *Fraction Notation*: To convert to fraction notation, divide 23 by 100. Thus $23\% = \frac{23}{100}$.
Decimal Notation: Move the decimal point two places to the left and then remove the % symbol: $23\% = 0.23$.
(b) *Fraction Notation*: $5.2\% = \frac{5.2}{100} = \frac{52}{1000} = \frac{13 \cdot 4}{250 \cdot 4} = \frac{13}{250}$
Decimal Notation: Move the decimal point two places to the left and then remove the % symbol: $5.2\% = 0.052$.
(c) *Fraction Notation*: $0.3\% = \frac{0.3}{100} = \frac{3}{1000}$
Decimal Notation: Move the decimal point two places to the left and then remove the % symbol: $0.3\% = 0.003$.

Now Try Exercises 49, 53, 55

EXAMPLE 5   Converting to percent notation

Convert each real number to a percentage.
(a) 0.234   **(b)** $\frac{1}{4}$ **(c)** 2.7

Solution
(a) Move the decimal point two places to the *right* and then insert the % symbol to obtain
$0.234 = 23.4\%$.
(b) $\frac{1}{4} = 0.25$, so $\frac{1}{4} = 25\%$.
(c) Move the decimal point two places to the right and then insert the % symbol to obtain
$2.7 = 270\%$. Note that percentages can be greater than 100%.

Now Try Exercises 57, 59, 63

As we previously indicated, percentages frequently occur in applications.

EXAMPLE 6   Solving an application involving a percentage

In 2008, the global revenue for recorded music in physical form was $22 billion. At that time, this number was forecast to be only 31.8% of its 2008 value by 2017, largely because of digital download sales. Find the forecasted revenue for music in physical form in 2017.

Solution
STEP 1: Let R represent the revenue in 2017.
STEP 2: R equals 31.8% of $22 billion. Because $31.8\% = 0.318$ and the word "of" in percent problems often indicates multiplication, we can find R using the equation $R = 0.318(22)$.
STEP 3: We calculate R as follows.

$$R = 0.318(22) = 6.996 \approx 7$$

The revenue R in 2017 was forecast to be about $7 billion.
STEP 4: To check this answer we can write the fraction $\frac{7}{22}$ as a percent.

$$\frac{7}{22} \times 100 \approx 31.8\% \checkmark$$

CRITICAL THINKING 1

If your salary increased 200%, by what factor did it increase?

Now Try Exercise 73

EXAMPLE 7   Solving an application involving a percent

In 2025, about 10% of the world's population, or 800 million people, will be older than 65. Find the estimated population of the world in 2025. (*Source:* World Population Prospectus.)

Solution
STEP 1: Let P represent the world's population in millions in 2025.
STEP 2: Since 10% of P equals 800 million, this information is described by the equation.

$$0.10P = 800.$$

STEP 3: To solve the equation in STEP 2, divide each side by 0.10.

$$0.10P = 800 \qquad \text{Equation to be solved}$$

$$P = 8000 \qquad \text{Divide each side by 0.10.}$$

In 2025, the estimated world population is 8000 million, or 8 billion.
STEP 4: To check the answer, determine whether 10% of 8 billion is 800 million.

$$(0.10)8,000,000,000 = 800,000,000 \checkmark$$

Now Try Exercise 75

5 Calculating Percent Change

 Connecting Concepts with Your Life When prices increase (or decrease), the actual amount of the increase (or decrease) is often not as significant as the *percent change* in the price. For example, consider the $1000 increases shown in **FIGURE 2.2**.

Home Price:
$1000 Increase

Tuition:
$1000 Increase

$250,000 → $251,000

$4500 → $5500

FIGURE 2.2

Even though both prices increased by $1000, the percent increase in the price of the home is four-tenths of 1% and the percent increase in the cost of tuition is more than 22%. The increase in tuition is much more dramatic than the increase in the price of the home.

If a quantity changes from an **old amount** to a **new amount**, the **percent change** is given by

$$\text{Percent Change} = \frac{\text{new amount} - \text{old amount}}{\text{old amount}} \cdot 100.$$

The reason for multiplying by 100 is to change the decimal representation to a percentage.

NOTE: A positive percent change corresponds to an increase and a negative percent change corresponds to a decrease. ■

READING CHECK 3

• What does it mean when a percent change is negative?

EXAMPLE 8   **Calculating percent increase in Skype subscriptions**

Over a three-year period, the number of users paying for a Skype subscription increased from 4.6 million to 8.1 million. Calculate the percent increase in paying Skype subscribers over this period of time. (*Source:* Skype SEC filing.)

Solution
The **old amount** is **4.6** and the **new amount** is **8.1**. The percent increase is

$$\frac{\text{new amount} - \text{old amount}}{\text{old amount}} \cdot 100 = \frac{8.1 - 4.6}{4.6} \cdot 100 \approx 76\%.$$

Now Try Exercise 79

6 Solving Distance Problems

 Connecting Concepts with Your Life If a person drives on an interstate highway at 70 miles per hour for 3 hours, then the total distance traveled is $70 \cdot 3 = 210$ miles. In general,

$$d = rt,$$

where *d* is the distance traveled, *r* is the rate (or speed), and *t* is time. In this example, the distance is in miles, the time is in hours, and the rate is expressed in miles per hour. In general, the rate in a distance problem is expressed in units of distance per unit of time.

EXAMPLE 9 **Solving a distance problem**

A person drives for 2 hours and 30 minutes at a constant speed and travels 180 miles. See **FIGURE 2.3**. Find the speed of the car in miles per hour.

Traveling 180 Miles in 2 Hours and 30 Minutes

180 mi
2 hr 30 min

FIGURE 2.3

Solution

STEP 1: Let r represent the car's rate, or speed, in miles per hour.

STEP 2: The rate is to be given in miles per hour, so change 2 hours and 30 minutes to 2.5 or $\frac{5}{2}$ hours. Because $d = 180$ and $t = \frac{5}{2}$, the equation $d = rt$ becomes

$$180 = r \cdot \frac{5}{2}.$$

STEP 3: Solve the equation in STEP 2 for r by multiplying each side of the equation by $\frac{2}{5}$, which is the reciprocal of $\frac{5}{2}$.

$$180 = \frac{5}{2} \cdot r \qquad \text{Equation to be solved}$$

Reciprocal of $\frac{5}{2}$ $\qquad \frac{2}{5} \cdot 180 = r \cdot \frac{5}{2} \cdot \frac{2}{5} \qquad \text{Multiply each side by } \frac{2}{5}.$

$$72 = r \qquad \text{Simplify.}$$

The speed of the car is 72 miles per hour.

STEP 4: Because 2 hours and 30 minutes is equivalent to $\frac{5}{2}$ hours, traveling for 2 hours and 30 minutes at a constant rate of **72** miles per hour results in a distance of

$$d = rt = 72 \cdot \frac{5}{2} = 180 \text{ miles.} \checkmark$$

Now Try Exercise 93

EXAMPLE 10 **Solving a distance problem**

An athlete jogs at two speeds, covering a distance of 7 miles in $\frac{3}{4}$ hour. If the athlete runs $\frac{1}{4}$ hour at 8 miles per hour, find the second speed.

Solution

STEP 1: Let r represent the second speed of the jogger in miles per hour.

STEP 2: The total time spent jogging is $\frac{3}{4}$ hour, so the time spent jogging at the second speed must be $\frac{3}{4} - \frac{1}{4} = \frac{1}{2}$ hour. See **FIGURE 2.4**.

Jogging 7 Miles in $\frac{3}{4}$ Hour

$\frac{1}{4}$ hr at
8 mph

$\frac{1}{2}$ hr at
r mph

7 mi

FIGURE 2.4

The **distance** traveled at 8 miles per hour for $\frac{1}{4}$ hour is given by $8 \cdot \frac{1}{4}$ and the **distance** traveled at r miles per hour for $\frac{1}{2}$ hour is given by $r \cdot \frac{1}{2}$. The sum of these distances must equal **7** miles. Thus

$$8 \cdot \frac{1}{4} + r \cdot \frac{1}{2} = 7.$$

STEP 3: Solve the equation in STEP 2 for r.

$$8 \cdot \frac{1}{4} + r \cdot \frac{1}{2} = 7 \qquad \text{Equation to be solved}$$

$$2 + \frac{r}{2} = 7 \qquad \text{Simplify.}$$

$$\frac{r}{2} = 5 \qquad \text{Subtract 2 from each side.}$$

$$r = 10 \qquad \text{Multiply each side by 2.}$$

The athlete's second speed is 10 miles per hour.

STEP 4: Jogging at 8 miles per hour for $\frac{1}{4}$ hour results in a distance of $8 \cdot \frac{1}{4} = 2$ miles. Jogging at 10 miles per hour for $\frac{1}{2}$ hour results in a distance of $10 \cdot \frac{1}{2} = 5$ miles. The total distance is $2 + 5 = 7$ miles and the total time is $\frac{1}{4} + \frac{1}{2} = \frac{3}{4}$ hour. ✓

Now Try Exercise 97

7 Solving Mixture Problems

🌐 *Math in Context* (*Health Science*) Many applied problems involve linear equations. For example, right after people have their wisdom teeth pulled, they may need to rinse their mouth with salt water. In the next example, we use linear equations to determine how much water must be added to dilute a concentrated saline solution.

EXAMPLE 11   **Diluting a saline solution**

A solution contains 4% salt. How much pure water should be added to 30 ounces of the solution to dilute it to a 1.5% solution?

Solution

STEP 1: FIGURE 2.5 illustrates this situation. Assign a variable x as follows.

x: ounces of pure water (0% salt solution)   **30**: ounces of 4% salt solution

$x + 30$: ounces of 1.5% salt solution

Mixing a Saline Solution

| x ounces | 30 ounces | $x + 30$ ounces |

FIGURE 2.5

STEP 2: Note that the amount of salt in the first two beakers must equal the amount of salt in the third beaker. TABLE 2.7 on the next page can be used to organize our calculations. The amount of salt in a solution equals the concentration times the solution amount, as shown in the last column of the table.

Mixing a Saline Solution

Solution Type	Concentration (as a decimal)	Solution Amount (ounces)	Salt (ounces)
Pure Water	0% = 0.00	x	$0.00x$
Initial Solution	4% = 0.04	30	$0.04(30)$
Diluted Solution	1.5% = 0.015	$x + 30$	$0.015(x + 30)$

TABLE 2.7

The amount of salt in the first two beakers is

$$0.00x + 0.04(30) = 0 + 1.2 = 1.2 \text{ ounces.}$$

The amount of salt in the final beaker is

$$\mathbf{0.015(x + 30)} \text{ ounces.}$$

Because the amounts of salt in the solutions before and after mixing must be equal, the following equation must hold.

$$\mathbf{0.015(x + 30) = 1.2}$$

STEP 3: Solve the equation in STEP 2.

$$
\begin{aligned}
0.015(x + 30) &= 1.2 && \text{Equation to be solved} \\
0.015x + 0.45 &= 1.2 && \text{Distributive property} \\
0.015x &= 0.75 && \text{Subtract 0.45 from each side.} \\
\frac{0.015x}{0.015} &= \frac{0.75}{0.015} && \text{Divide each side by 0.015.} \\
x &= 50 && \text{Simplify fractions.}
\end{aligned}
$$

The amount of water that should be added is 50 ounces.

STEP 4: Adding 50 ounces of water will yield $50 + 30 = 80$ ounces of water containing $0.04(30) = 1.2$ ounces of salt. The concentration is $\frac{1.2}{80} = 0.015$ or 1.5%. ✓

Now Try Exercise 105

🌐 *Math in Context* *Financial* Many times interest rates for student loans vary. In the next example, we present a situation in which a student has to borrow money at two different interest rates. This problem can be thought of as a mixture problem where the mixture is made up of two loans with different rates.

EXAMPLE 12 **Calculating interest on college loans**

A student takes out a loan for a limited amount of money at 5% interest and then must pay 7% for any additional money. If the student borrows $2000 more at 7% than at 5%, then the total interest for one year is $440. How much does the student borrow at each rate?

Solution

STEP 1: Assign a variable x as follows.

$$x: \text{loan amount at 5\% interest}$$

$$x + 2000: \text{loan amount at 7\% interest}$$

STEP 2: The amount of interest paid for the 5% loan is 5% of x, or $0.05x$. The amount of interest paid for the 7% loan is 7% of $x + 2000$, or $0.07(x + 2000)$. The total interest equals $440, so we solve the equation

$$0.05x + 0.07(x + 2000) = 440.$$

STEP 3: Solve the equation in STEP 2 for x.

$$0.05x + 0.07(x + 2000) = 440 \qquad \text{Equation to be solved}$$
$$0.05x + 0.07x + 140 = 440 \qquad \text{Distributive property}$$
$$0.12x + 140 = 440 \qquad \text{Combine like terms.}$$
$$0.12x = 300 \qquad \text{Subtract 140 from each side.}$$
$$x = 2500 \qquad \text{Divide each side by 0.12.}$$

The student borrows $2500 at 5% and $2500 + 2000 = \$4500$ at 7%.

STEP 4: The amount of interest on $2500 borrowed at 5% is $0.05(2500) = \$125$ and the amount of interest on $4500 borrowed at 7% is $0.07(4500) = \$315$. Thus the total interest is $125 + 315 = \$440$. Furthermore, the amount borrowed at 7% is $2000 more than the amount borrowed at 5%. ✓

Now Try Exercise 107

2.3   Putting It All Together

In this section, we presented a four-step approach to problem solving. However, because no approach works in every situation, solving mathematical problems takes time, effort, and creativity.

CONCEPT	COMMENTS	EXAMPLES
Percent Notation	$x\%$ represents either the fraction $\frac{x}{100}$ or the decimal $x \cdot 0.01$.	$17\% = \frac{17}{100} = 0.17$ $1.5\% = \frac{1.5}{100} = \frac{15}{1000} = 0.015$ $234\% = \frac{234}{100} = 2.34$
Converting Fractions to Percent Notation	Divide the numerator by the denominator, multiply by 100, and insert the % symbol.	$\frac{1}{2} = 0.5 = 50\%$ $\frac{2}{3} = 0.\overline{6} \approx 66.7\%$ $\frac{8}{5} = 1.6 = 160\%$
Percent Change	If a quantity changes from an old amount to a new amount, then the percent change is $$\frac{\text{new amount} - \text{old amount}}{\text{old amount}} \cdot 100.$$	If a price changes from $1.50 to $1.35, the percent change is $$\frac{1.35 - 1.50}{1.50} \times 100 = -10\%.$$ The price decreases by 10%.
Distance Problems	Distance d, rate r, and time t are related as expressed in the equation $$d = rt.$$	If a car travels 60 mph for 3 hours, then the distance d is $$d = rt = 60 \cdot 3 = 180 \text{ miles.}$$

2.3 Exercises

MyMathLab®

CONCEPTS AND VOCABULARY

1. When you are solving a word problem, what is the last step?

2. The words *more*, *sum*, and *increase* are associated with the symbol _____.

3. The words *is*, *gives*, and *results in* are associated with the symbol _____.

4. Given an integer n, what are the next two consecutive integers?

5. The expression $x\%$ equals the fraction _____.

6. The expression $x\%$ equals the decimal $x \cdot$ ____.

7. To write 63.2% as a decimal, move the decimal point 2 places (left/right) and then remove the % symbol.

8. To write 0.349 as a percentage, move the decimal point 2 places (left/right) and then insert the % symbol.

9. If a price changes from A to B, then the percent change equals _____.

10. A positive percent change corresponds to a(n) _____ and a negative percent change corresponds to a(n) _____ .

11. If a car travels at speed r for time t, then distance d is given by $d =$ _____.

12. In general, the rate in a distance problem is expressed in units of _____ per unit of _____ .

TRANSLATING SENTENCES INTO EQUATIONS

Exercises 13–20: Using the variable x, translate the sentence into an equation. Solve the resulting equation.

13. The sum of 2 and a number is 12.

14. A number divided by 5 equals the number decreased by 24.

15. Twice a number plus 7 equals 9.

16. 25 times a number is 125.

17. A number subtracted from 7 is −2.

18. A number subtracted from 8 is 5.

19. The quotient of a number and 2 is 17.

20. The product of 5 and a number equals 95.

SOLVING NUMBER PROBLEMS AND APPLICATIONS

Exercises 21–28: Find the number or numbers.

21. The sum of three consecutive natural numbers is 96.

22. The sum of three consecutive integers is −123.

23. Three times a number equals 102.

24. A number plus 18 equals twice the number.

25. Five times a number is 24 more than twice the number.

26. Three times a number is 18 less than the number.

27. Six times a number divided by 7 equals 18.

28. Two less than twice a number, divided by 5, equals 4.

29. **Finding Age** In 10 years, a child will be 3 years older than twice her current age. What is the current age of the child?

30. **Finding Age** A mother is 15 years older than twice her daughter's age. If the mother is 49 years old, how old is the daughter?

31. **Weight Loss** After losing 30 pounds, an individual weighs 110 pounds more than one-third his previous weight. Find the previous weight of the individual.

32. **Weight Gain** After gaining 25 pounds, a person is 115 pounds lighter than double his previous weight. How much did the person weigh before gaining 25 pounds?

33. **World Billionaires** In 2012, New York City was home to the most billionaires. This is 10 more than 5 times the number of billionaires in Tokyo. How many billionaires were in Tokyo if there were 70 billionaires in New York City? (*Source: The Economist.*)

34. **World Millionaires** In 2012, Tokyo was home to the most millionaires. This is 317 thousand less than twice the number of millionaires in New York City. How many millionaires were in New York City if there were 461 thousand in Tokyo? (*Source: The Economist.*)

35. **Endangered Species** There were 80 birds on the endangered species list in 2014. This is 10 more than 5 times the number of reptiles on the list. How many reptiles were on the endangered species list in 2014? (*Source: U.S. Fish & Wildlife Service.*)

36. Immigrant Religion Of those who immigrated to the United States between 1992 and 2012, 12.7% self-identified as Christian. This is 0.8% more than 7 times the percentage that self-identified as Muslim. What percentage of immigrates self-identified as Muslim? (*Source:* Pew Research Center.)

37. Musical Revenues From their opening dates, *Phantom of the Opera* and *Les Miserables* have together made $8.2 billion worldwide. *Phantom of the Opera* has made $0.4 billion more than twice what *Les Miserables* has made. Find the revenue of each musical (*Source:* The Economist.)

38. Museums The number of museums in Canada and the United States combined is 20 thousand. The number of museums in the United States is 2.5 thousand less than 8 times the number in Canada. Find the number of museums in each country. (*Source:* The Economist.)

39. Highest Mountains By 2013, there were 5656 successful ascents of Mount Everest. This is 148 more than 18 times the number of successful ascents of K2. How many successful ascents of K2 were there? (*Source:* The Economist.)

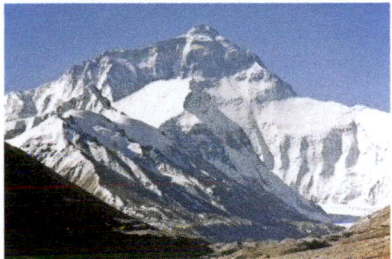

40. Highest Mountains By 2013, there were 29 deaths on K2 for every 100 safe returns. This is 1 more than 7 times the number of deaths for every 100 safe returns on Mount Everest. How many deaths for every 100 safe returns were there on Mount Everest? (*Source:* The Economist.)

41. Geometry If the perimeter of the rectangle shown is 106 inches, find the value of *x*.

x

$x + 5$

42. Geometry If the perimeter of the following triangle is 24 inches, find the value of *x*.

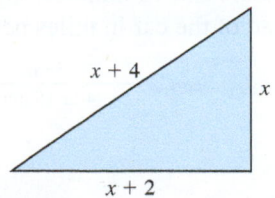

$x + 4$

x

$x + 2$

43. Rectangle Dimensions The length of a rectangle is 7 inches longer than the width. If the perimeter of the rectangle is 62 inches, find the measures of the length and width.

44. Triangle Dimensions The shortest side of a triangle measures 15 feet less than the longest side. If the third side is 6 feet shorter than the longest side and the perimeter is 102 feet, find the measures of the three sides of the triangle.

45. Facebook vs. Twitter In June 2013, the number of monthly active Facebook users in the U.S. was 2 million more than 4 times the number of monthly active Twitter users. If the two sites had a total of 247 million active monthly users, how many people used each site? (*Source:* Company Reports.)

46. Blackberry Shares In June 2008, Blackberry share prices (in Canadian dollars) peaked at $149.90. This is $4.46 less than 17 times what the share price was in September 2013. What was the share price in September 2013? (*Source:* Reuters.)

47. Screen Time In 2013, the monthly time spent watching traditional TV was 10 hours less than 6 times the monthly time spent using the Internet on a computer. If the time spent watching TV was 146 hours, what was the time spent using the Internet on a computer? (*Source:* Nielsen.)

48. iPhone Prices At launch, the price in Brazil of the 16GB iPhone 5S was $1200. This is $214 less than twice the price of the same phone in the United States. Find the price in the United States? (*Source:* Mobile Unlocked.)

CONVERTING BETWEEN PERCENTAGE, FRACTION, AND DECIMAL NOTATION

Exercises 49–56: Convert the percentage to fraction and decimal notation.

49. 37%   **50.** 52%

51. 148%   **52.** 252%

53. 6.9%   **54.** 8.1%

55. 0.05%   **56.** 0.12%

Exercises 57–68: Convert the number to a percentage.

57. 0.45

58. 0.08

59. 1.8

60. 2.97

61. 0.006

62. 0.0001

63. $\frac{2}{5}$

64. $\frac{1}{3}$

65. $\frac{3}{4}$

66. $\frac{7}{20}$

67. $\frac{5}{6}$

68. $\frac{53}{50}$

SOLVING PERCENT PROBLEMS

69. Meal Plus Tip A meal cost $20.50. If a person includes an 18% tip, find the total cost of the meal.

70. Tree Trimming It cost a homeowner $320 to have a tree trimmed, excluding sales tax. If the sales tax is 9%, how much is the final bill?

71. Wages A part-time instructor is receiving $950 per credit taught. If the instructor receives a 4% increase, how much will the new per credit compensation be?

72. Tuition Increase Tuition is currently $125 per credit. There are plans to raise tuition by 8% for next year. What will the new tuition be per credit?

73. Digital Entertainment In 2008, worldwide spending on digital media and entertainment was $300 billion, but was expected to increase 200% by 2017. Find the spending in 2017. (*Source:* PricewaterhouseCoopers.)

74. Organ Transplants In 1989, there were 20 thousand people awaiting organ transplants. This is about 16.7% of the number of people awaiting an organ transplant in 2014. How many people were awaiting a transplant in 2014? (*Source:* Organ Procurement and Transplantation Network.)

75. Weight Loss A person's weight decreases by 9% to 136.5 pounds. Find the person's initial weight.

76. Weight Gain A person's weight increases by 4% to 182 pounds. Find the person's initial weight.

77. Sales Tax A suit costs $160.50, including 7% sales tax. Find the price without any sales tax.

78. Sales Tax A pre-owned car costs $5724, including 8% sales tax. Find the price without any sales tax.

CALCULATING PERCENT CHANGE

79. Gasoline Prices The price of gasoline increases from $3.49 to $3.85. Find the percent change in the price.

80. Cable TV After 6 months, the price of one cable television service plan increases from $79.95 to $89.95. Find the percent change in the price.

81. Stock Value The value of a mutual fund was initially $55 per share. During the first week, the fund decreased by 20%, and during the next week, it increased by 25%. Find the final value of a share of stock. What was the percent change over the two-week period?

82. Stock Value The value of a mutual fund was initially $125 per share. During the first week, the fund decreased by 20%, and during the next week, it increased by 20%. Find the final value of a share of stock. What was the percent change over the two-week period?

83. Newspaper Ads In 2004, newspaper ad revenue was $47 billion, but $20 billion by 2012. Find the percent change in revenue. (*Source:* Business Insider.)

84. Google Ads In 2004, Google ad revenue was $3 billion, but $21 billion by 2012. Find the percent change in revenue. (*Source:* Business Insider.)

85. Thinking Generally Calculate the percent change if the price of an item increases from $1.20 to $1.50. Now calculate the percent change if the price of the item decreases from $1.50 to $1.20.

86. Thinking Generally Refer to the previous exercise. Suppose an amount increases from A_1 to A_2 and the percent change is calculated to be 30%. If that amount now decreases from A_2 back to A_1, will the percent change be -30%?

SOLVING DISTANCE PROBLEMS

Exercises 87–92: Use the formula $d = rt$ to find the value of the missing variable.

87. $r = 4$ mph, $t = 2$ hours

88. $r = 70$ mph, $t = 2.5$ hours

89. $d = 1000$ feet, $t = 50$ seconds

90. $d = 1250$ miles, $t = 5$ days

91. $d = 200$ miles, $r = 40$ mph

92. $d = 1700$ feet, $r = 10$ feet per second

93. Driving a Car A person drives a car at a constant speed for 4 hours and 15 minutes, traveling 255 miles. Find the speed of the car in miles per hour.

255 mi
4 hr 15 min

94. Flying an Airplane A pilot flies a plane at a constant speed for 5 hours and 30 minutes, traveling 715 miles. Find the speed of the plane in miles per hour.

95. Jogging Speeds One runner passes another runner traveling in the same direction on a hiking trail. The faster runner is jogging 2 miles per hour faster than the slower runner. Determine how long it will be before the faster runner is $\frac{3}{4}$ mile ahead of the slower runner.

96. Distance Running An athlete runs 8 miles, first at a slower speed and then at a faster speed. The total time spent running is 1 hour. If the athlete runs $\frac{1}{3}$ hour at 6 miles per hour, find the second speed.

97. Jogging Speeds At first an athlete jogs at 5 miles per hour and then jogs at 8 miles per hour, traveling 7 miles in 1.1 hours. How long does the athlete jog at each speed? (*Hint:* Let t represent the amount of time the athlete jogs at 5 mph. Then $1.1 - t$ represents the amount of time the athlete jogs at 8 mph.)

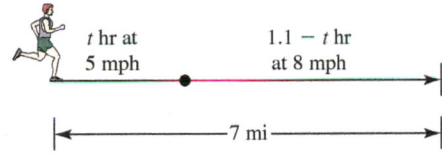

98. Distance and Time A bus is 160 miles east of the North Dakota–Montana border and is traveling west at 70 miles per hour. How long will it take for the bus to be 295 miles west of the border?

99. Finding Speeds Two cars pass on a straight highway while traveling in opposite directions. One car is traveling 6 miles per hour faster than the other car. After 1.5 hours, the two cars are 171 miles apart. Find the speed of each car.

100. Distance and Time A plane is 300 miles west of Chicago, Illinois, and is flying west at 500 miles per hour. How long will it take for the plane to be 2175 miles west of Chicago?

101. Average Speed A driver travels at 50 mph for the first hour and then travels at 70 mph for the second hour. What is the average speed of the car?

102. Average Speed A bicyclist rides 1 mile uphill at 5 mph and then rides 1 mile downhill at 10 mph. Find the average speed of the bicyclist. Does your answer agree with what you expected?

103. Average Speed At a 3-mile cross-country race, an athlete runs 2 miles at 8 mph and 1 mile at 10 mph. What is the athlete's average speed?

104. Average Speed A pilot flies an airplane between two cities and travels half the distance at 200 mph and the other half at 100 mph. Find the average speed of the airplane.

SOLVING MIXTURE PROBLEMS

105. Saline Solution (Refer to Example 11.) A solution contains 3% salt. How much water should be added to 20 ounces of this solution to make a 1.2% solution?

106. Acid Solution A solution contains 15% hydrochloric acid. How much water should be added to 50 milliliters of this solution to dilute it to a 2% solution?

107. College Loans (Refer to Example 12.) A student takes out two loans, one at 5% interest and the other at 6% interest. The 5% loan is $1000 more than the 6% loan, and the total interest for 1 year is $215. How much is each loan?

108. Bank Loans Two bank loans, one for $5000 and the other for $3000, cost a total of $550 in interest for one year. The $5000 loan has an interest rate 3% lower than the interest rate for the $3000 loan. Find the interest rate for each loan.

109. Coffee Mixture A more expensive brand of organic coffee costs $12 per pound and a cheaper brand is $7 per pound. How many pounds of each coffee would it take to make an 8-pound bag of coffee that costs $83.50?

110. Coin Mixture A sample of 23 dimes and nickels has a value of $1.45. How many of each type of coin are there?

111. Running Speeds A jogger runs a total distance of 9.9 miles at 8 miles per hour and at 9 miles per hour. If the jogger runs for 1 hour and 12 minutes, how many minutes did the jogger run at each speed.

112. Traveling Speeds A truck driver on an interstate highway travels a total distance of 285 miles at 65 miles per hour and at 75 miles per hour. If it takes 4 hours, how many hours did the truck driver travel at each speed.

113. Ticket Sales At a show, student tickets cost $7 and adult tickets cost $12. If 151 tickets were sold for a total of $1532, how many tickets of each type were sold?

114. Ticket Sales At a ball game, student tickets cost $15 and adult tickets cost $20. If 260 tickets were sold for a total of $4575, how many tickets of each type were sold?

115. Mixing Antifreeze How many gallons of 70% antifreeze should be mixed with 10 gallons of 30% antifreeze to obtain a 45% antifreeze mixture?

116. Mixing Antifreeze How many gallons of 65% antifreeze and how many gallons of 20% antifreeze should be mixed to obtain 50 gallons of a 56% mixture of antifreeze? (*Hint:* Let x represent the number of gallons of 65% antifreeze. Then $50 - x$ represents the amount of 20% antifreeze.)

117. Hydrocortisone Cream A pharmacist needs to make a 1% hydrocortisone cream. How many grams of 2.5% hydrocortisone cream should be added to 15 grams of cream base (0% hydrocortisone) to make the 1% cream?

118. Credit Card Debt A person carries a balance on two credit cards, one with a monthly interest rate of 1.5% and the other with a monthly rate of 1.75%. The balance on the 1.5% card is $600 less than the balance on the 1.75% card. If the total interest for the month is $49.50, what is the balance on each card?

WRITING ABOUT MATHEMATICS

119. State the four steps for solving a word problem.

120. The cost of living has increased about 600% during the past 50 years. Does this percent change correspond to a cost of living increase of 6 times? Explain.

Group Activity

Exercises 1–5: **Climate Change** If the global climate were to warm significantly, the Arctic ice cap in Greenland would start to melt. This ice cap contains the equivalent of about 680,000 cubic miles of water. More than 200 million people live on land that is less than 3 feet above sea level. In the United States, several large cities have low average elevations. Three examples are Boston (14 feet), New Orleans (4 feet), and San Diego (13 feet). In the following exercises you are to estimate the rise in the sea level if the Arctic ice cap were to melt and to determine whether this event would have a significant impact on people living in coastal areas.

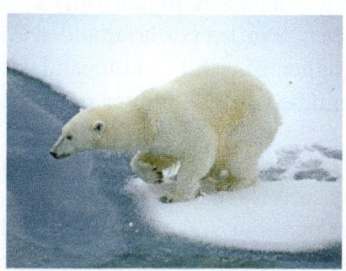

1. The surface area of a sphere is given by the formula $4\pi r^2$, where r is its radius. Although the shape of Earth is not exactly spherical, it has an average radius of 3960 miles. Use the formula to estimate the surface area of Earth.

2. Oceans cover approximately 71% of the total surface area of Earth. How many square miles of Earth's surface area are covered by oceans?

3. Approximate the potential rise in sea level by dividing the total volume of the water from the ice cap by the surface area of the oceans. Convert your answer from miles to feet.

4. Discuss the implications of your calculation. How would cities such as Boston, New Orleans, and San Diego be affected?

5. The Antarctic ice cap contains about 6,300,000 cubic miles of water. Estimate how much the sea level would rise if this ice cap melted. (*Source:* Department of the Interior Geological Survey.)

2.4 Formulas

NEW VOCABULARY

☐ Degree
☐ Isosceles triangle
☐ Circumference

Rectangle

7 cm

12 cm

FIGURE 2.6

1 Recognizing a Formula

A **formula** is an equation relating two or more variables, making it possible to calculate a quantity by using known values of other quantities. Many formulas are written with a single variable on the left side of the equals sign and a mathematical expression containing one or more variables on the right side. Examples of formulas include

$$A = \frac{1}{2}bh, \qquad I = Prt, \qquad \text{and} \qquad F = 3Y.$$

Area of triangle   Simple interest   Convert yards to feet

2 Using Formulas from Geometry

Formulas from geometry are frequently used in various fields, including surveying and construction. In this subsection, we discuss several important formulas from geometry.

RECTANGLES If a rectangle has length l and width w, then its perimeter P and its area A are given by the following formulas.

$$P = 2l + 2w \quad \text{and} \quad A = lw$$

For example, the rectangle in **FIGURE 2.6** has length **12** centimeters and width **7** centimeters. Its perimeter is

$$P = 2(12) + 2(7) = 24 + 14 = 38 \text{ centimeters},$$

and its area is

$$A = 12 \cdot 7 = 84 \text{ square centimeters}.$$

NOTE: All measurements used in a geometry formula must have the same units. ■

READING CHECK 1

• What do l and w stand for in the formulas for rectangles?

TRIANGLES If a triangle has base b and height h, as shown in **FIGURE 2.7(a)**, then its area A is given by the following formula.

$$A = \frac{1}{2}bh$$

For example, to find the area of a triangle with base **4** feet and height **18** inches as illustrated in **FIGURE 2.7(b)**, we begin by making sure that all measurements have the same units. By converting 18 inches to **1.5** feet, the area of the triangle is

$$A = \frac{1}{2}bh = \frac{1}{2}(4)(1.5) = 3 \text{ square feet}.$$

If the base had been converted to **48** inches, the area is

$$A = \frac{1}{2}bh = \frac{1}{2}(48)(18) = 432 \text{ square inches}.$$

READING CHECK 2

• What do b and h stand for in the triangle area formula?

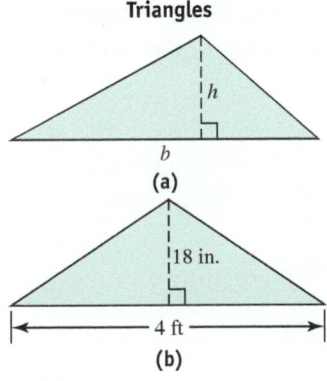

Triangles

h

b

(a)

18 in.

4 ft

(b)

FIGURE 2.7

EXAMPLE 1 **Calculating area of a region**

A residential lot is shown in **FIGURE 2.8**. It comprises a rectangular region and an adjacent triangular region.

(a) Find the area of this lot.

(b) An acre contains 43,560 square feet. How many acres are there in this lot?

235 ft

405 ft   124 ft

FIGURE 2.8

Solution

(a) The rectangular portion of the lot has length **405** feet and width **235** feet. Its area A_R is

$$A_R = lw = \mathbf{405} \cdot \mathbf{235} = \mathbf{95{,}175} \text{ square feet.}$$

The triangular region has base **124** feet and height **235** feet. Its area A_T is

$$A_T = \frac{1}{2}bh = \frac{1}{2} \cdot \mathbf{124} \cdot \mathbf{235} = \mathbf{14{,}570} \text{ square feet.}$$

The total area A of the lot equals the sum of A_R and A_T.

$$A = \mathbf{95{,}175} + \mathbf{14{,}570} = 109{,}745 \text{ square feet}$$

(b) Each acre equals 43,560 square feet, so divide 109,745 by 43,560 to calculate the number of acres.

$$\frac{109{,}745}{43{,}560} \approx 2.5 \text{ acres}$$

Now Try Exercise 27

Angle sum is 180°

FIGURE 2.9

ANGLES Angles are often measured in degrees. A **degree** (°) is $\frac{1}{360}$ of a revolution, so there are 360° in one complete revolution. In any triangle, the sum of the measures of the angles equals 180°. In **FIGURE 2.9**, triangle ABC has angles with measures x, y, and z.

$$x + y + z = 180°$$

EXAMPLE 2 **Finding angles in a triangle**

A triangle with at least two angles of the same measure is called an **isosceles triangle**. Suppose that in an isosceles triangle, the two smaller angles are equal in measure and are each half the measure of the largest angle. Find the measure of each angle.

Solution

Let x represent the measure of each of the two smaller angles, as illustrated in **FIGURE 2.10**. Then the measure of the largest angle is $2x$, and the sum of the measures of the three angles is given by

$$x + x + 2x = 180°.$$

This equation can be solved as follows.

Isosceles Triangle

FIGURE 2.10

$$x + x + 2x = 180° \quad \text{Equation to be solved}$$

$$4x = 180° \quad \text{Combine like terms.}$$

$$\frac{4x}{4} = \frac{180°}{4} \quad \text{Divide each side by 4.}$$

$$x = 45° \quad \text{Divide the real numbers.}$$

The measure of the largest angle is $2x = 2 \cdot 45° = 90°$. Thus the measures of the three angles are 45°, 45°, and 90°.

Now Try Exercise 33

CIRCLES The *radius* of a circle is the distance from its center to a point on the circle. See **FIGURE 2.11**. The perimeter of a circle is called its **circumference**. If r is the radius of the circle, then the circumference C and area A are given by the following formulas.

$$C = 2\pi r \quad \text{and} \quad A = \pi r^2$$

The distance across a circle, through its center, is called the *diameter*. Note that a circle's radius is half of its diameter. (Recall that $\pi \approx 3.14$.)

Circle

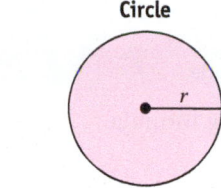

FIGURE 2.11

> **EXAMPLE 3** **Finding the circumference and area of a circle**
>
> A circle has a diameter of 25 inches. Find its circumference and area.
>
> **Solution**
> The radius is half the diameter, or **12.5** inches.
>
> $$\text{Circumference: } C = 2\pi r = 2\pi(12.5) = 25\pi \approx 78.5 \text{ inches.}$$
>
> $$\text{Area: } A = \pi r^2 = \pi(12.5)^2 = 156.25\pi \approx 491 \text{ square inches.}$$
>
> **Now Try Exercise 37**

READING CHECK 3

• What is the approximate value of π when rounded to the nearest hundredth?

CALCULATOR HELP

To evaluate π on a calculator, see Appendix A (page AP-1).

CRITICAL THINKING 1

Write formulas for the circumference and area of a circle having diameter d.

Trapezoid

FIGURE 2.12

TRAPEZOIDS A trapezoid such as the one shown in **FIGURE 2.12** has bases a and b, and height h. The area A of a trapezoid is given by the following formula.

$$A = \frac{1}{2}(a + b)h$$

> **EXAMPLE 4** **Finding the area of a trapezoid**
>
> Find the area of the trapezoid shown in **FIGURE 2.13**.
>
> **Solution**
> The bases of the trapezoid shown in **FIGURE 2.13** are **126** millimeters and **182** millimeters, and its height is **95** millimeters. Substituting these values in the area formula gives an area of

FIGURE 2.13

$$A = \frac{1}{2}(a + b)h = \frac{1}{2}(126 + 182) \cdot 95 = \frac{1}{2}(308) \cdot 95 = 14{,}630 \text{ square millimeters.}$$

Now Try Exercise 19

Rectangular Box

FIGURE 2.14

BOXES The box in **FIGURE 2.14** has length l, width w, and height h. Its volume V is given by the following formula.

$$V = lwh$$

The surface of the box comprises six rectangular regions: top and bottom, front and back, and left and right sides. The total surface area S of the box is given by

$$S = \underset{\text{top}}{lw} + \underset{\text{bottom}}{lw} + \underset{\text{front}}{wh} + \underset{\text{back}}{wh} + \underset{\text{left side}}{lh} + \underset{\text{right side}}{lh}.$$

When we combine like terms, this expression simplifies to the following formula.

$$S = 2lw + 2wh + 2lh$$

EXAMPLE 5   **Finding the volume and surface area of a box**

Find the volume and surface area of the box shown in **FIGURE 2.15**.

Solution
FIGURE 2.15 shows that the box has length $l = 10$ inches, width $w = 8$ inches, and height $h = 5$ inches. The volume of the box is

$$V = lwh = 10 \cdot 8 \cdot 5 = 400 \text{ cubic inches.}$$

Since $l = 10$, $w = 8$, and $h = 5$, the surface area of the box is

$$\begin{aligned}
S &= 2lw + 2wh + 2lh \\
&= 2(10)(8) + 2(8)(5) + 2(10)(5) \\
&= 160 + 80 + 100 \\
&= 340 \text{ square inches.}
\end{aligned}$$

Now Try Exercise 41

FIGURE 2.15

CYLINDERS A soup can is usually made in the shape of a cylinder. See **FIGURE 2.16**. The volume of a cylinder that has radius r and height h is given by the following formula.

$$V = \pi r^2 h$$

Cylinder

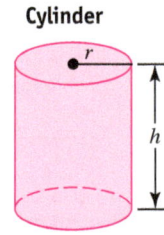

FIGURE 2.16

EXAMPLE 6   **Calculating the volume of a soda can**

A cylindrical soda can has a radius of $1\frac{1}{4}$ inches and a height of $4\frac{3}{8}$ inches.
(a) Find the volume of the can.
(b) If 1 cubic inch equals 0.554 fluid ounce, find the number of fluid ounces in the can.

Solution
(a) Changing mixed numbers to improper fractions gives the radius as $r = \frac{5}{4}$ inches and the height as $h = \frac{35}{8}$ inches.

$$\begin{aligned}
V &= \pi r^2 h & &\text{Volume of the soda can} \\
&= \pi \left(\frac{5}{4}\right)^2 \left(\frac{35}{8}\right) & &\text{Substitute.} \\
&= \pi \left(\frac{875}{128}\right) & &\text{Multiply the fractions.} \\
&\approx 21.48 \text{ cubic inches} & &\text{Approximate.}
\end{aligned}$$

(b) To calculate the number of fluid ounces in 21.48 cubic inches, multiply by 0.554.

$$21.48(0.554) \approx 11.9 \text{ fluid ounces}$$

Note that a typical aluminum soda can holds 12 fluid ounces.

Now Try Exercise 49

3 Solving a Formula for a Variable

Sometimes a formula must be rewritten to solve for the needed variable. For example, if the area A and the width w of a rectangular region are given, then its length l can be found by solving the formula $A = lw$ for l.

$$A = lw \qquad \text{Area formula}$$

$$\frac{A}{w} = \frac{lw}{w} \qquad \text{Divide each side by } w.$$

$$\frac{A}{w} = l \qquad \text{Simplify the fraction.}$$

$$l = \frac{A}{w} \qquad \text{Rewrite the equation.}$$

If the area A of a rectangle is 400 square inches and its width w is 16 inches, then the rectangle's length l is

$$l = \frac{A}{w} = \frac{400}{16} = 25 \text{ inches.}$$

Once the area formula has been solved for l, the resulting formula can be used to find the length of *any* rectangle whose area and width are known.

EXAMPLE 7

Finding the base of a trapezoid

The area of a trapezoid is given by

$$A = \frac{1}{2}(a + b)h,$$

where a and b are the bases of the trapezoid and h is the height.
(a) Solve the formula for b.
(b) A trapezoid has area $A = 36$ square inches, height $h = 4$ inches, and base $a = 8$ inches. Find b.

STUDY TIP

If you are studying with classmates, make sure that they do not "do the work for you." A classmate with the best intentions may give too many verbal hints while helping you work through a problem. Remember that members of your study group will not be giving hints during an exam.

Solution
(a) To clear the equation of the fraction, multiply each side by 2.

$$A = \frac{1}{2}(a + b)h \qquad \text{Trapezoid area formula}$$

$$2A = (a + b)h \qquad \text{Multiply each side by 2.}$$

$$\frac{2A}{h} = a + b \qquad \text{Divide each side by } h.$$

$$\frac{2A}{h} - a = b \qquad \text{Subtract } a \text{ from each side.}$$

$$b = \frac{2A}{h} - a \qquad \text{Rewrite the formula.}$$

(b) Let $A = 36$, $h = 4$, and $a = 8$ in $b = \frac{2A}{h} - a$. Then

$$b = \frac{2(36)}{4} - 8 = 18 - 8 = 10 \text{ inches.}$$

Now Try Exercise 59

EXAMPLE 8   **Solving for a variable**

Solve each equation for the indicated variable.

(a) $c = \frac{a+b}{2}$ for b **(b)** $ab - bc = ac$ for c

Solution

(a) To clear the equation of the fraction, multiply each side by 2.

$$c = \frac{a+b}{2} \qquad \text{Given equation}$$

$$2c = a + b \qquad \text{Multiply each side by 2.}$$

$$2c - a = b \qquad \text{Subtract } a.$$

The equation solved for b is $b = 2c - a$.

(b) In $ab - bc = ac$, the variable c appears in two terms. We will combine the terms containing c by using the distributive property. Begin by moving the term on the left side containing c to the right side of the equation.

$$ab - bc = ac \qquad \text{Given equation}$$

$$ab - bc + \mathbf{bc} = ac + \mathbf{bc} \qquad \text{Add } bc \text{ to each side.}$$

$$ab = (a+b)c \qquad \text{Combine terms; distributive property.}$$

$$\frac{ab}{a+b} = \frac{(a+b)c}{(a+b)} \qquad \text{Divide each side by } (a+b).$$

$$\frac{ab}{a+b} = c \qquad \text{Simplify the fraction.}$$

The equation solved for c is $c = \frac{ab}{a+b}$.

Now Try Exercises 63, 67

CRITICAL THINKING 2

Are the equations $c = \frac{1}{a-b}$ and $c = \frac{-1}{b-a}$ equivalent? Why?

4 ▶ Finding Grade Point Average and Other Formulas

To calculate a student's grade point average, or GPA, the number of credits earned with a grade of A, B, C, D, and F must be known. If A, B, C, D, and F represent these credit counts respectively, then the following formula calculates GPA.

$$\text{GPA} = \frac{4A + 3B + 2C + D}{A + B + C + D + F}$$

This formula is based on the assumption that a 4.0 GPA is an A, a 3.0 GPA is a B, and so on.

EXAMPLE 9   **Calculating a student's GPA**

A student has earned 16 credits of A, 32 credits of B, 12 credits of C, 2 credits of D, and 5 credits of F. Calculate the student's GPA to the nearest hundredth.

Solution

Let $A = \mathbf{16}$, $B = \mathbf{32}$, $C = \mathbf{12}$, $D = \mathbf{2}$, and $F = \mathbf{5}$. Then

$$\text{GPA} = \frac{4 \cdot \mathbf{16} + 3 \cdot \mathbf{32} + 2 \cdot \mathbf{12} + \mathbf{2}}{\mathbf{16} + \mathbf{32} + \mathbf{12} + \mathbf{2} + \mathbf{5}} = \frac{186}{67} \approx 2.78.$$

The student's GPA is 2.78.

Now Try Exercise 71

EXAMPLE 10   **Converting temperature scales**

In the United States, temperature is measured with either the Fahrenheit or the Celsius temperature scale. To convert Fahrenheit degrees F to Celsius degrees C, the formula $C = \frac{5}{9}(F - 32)$ can be used.

(a) Solve the formula for F to find a formula that converts Celsius degrees to Fahrenheit degrees.
(b) If the outside temperature is 20 °C, find the equivalent Fahrenheit temperature.

Solution
(a) The reciprocal of $\frac{5}{9}$ is $\frac{9}{5}$, so multiply each side by $\frac{9}{5}$.

$$C = \frac{5}{9}(F - 32) \qquad \text{Given equation}$$

$$\frac{9}{5}C = \frac{9}{5} \cdot \frac{5}{9}(F - 32) \qquad \text{Multiply each side by } \tfrac{9}{5}.$$

$$\frac{9}{5}C = F - 32 \qquad \text{Multiplicative inverses}$$

$$\frac{9}{5}C + 32 = F \qquad \text{Add 32 to each side.}$$

The required formula is $F = \frac{9}{5}C + 32$.

(b) If $C = \mathbf{20}$ °C, then $F = \frac{9}{5}(\mathbf{20}) + 32 = 36 + 32 = 68$ °F.

Now Try Exercise 75

2.4   Putting It All Together

CONCEPT	FORMULA	EXAMPLES
Area and Perimeter of a Rectangle	$A = lw$ and $P = 2l + 2w$, where l is the length and w is the width.	If $l = 10$ feet and $w = 5$ feet, then $A = 10 \cdot 5 = 50$ square feet and $P = 2(10) + 2(5) = 30$ feet. 5 ft 10 ft
Area of a Triangle	$A = \frac{1}{2}bh$, where b is the base and h is the height.	If $b = 5$ inches and $h = 6$ inches, then the area is $A = \frac{1}{2}(5)(6) = 15$ square inches. 6 in. 5 in.
Angle Measure in a Triangle	$x + y + z = 180°$, where x, y, and z are the angle measures.	If $x = 40°$ and $y = 60°$, then $z = 80°$ because $40° + 60° + 80° = 180°$ z 40° 60°

continued on next page

continued from previous page

CONCEPT	FORMULA	EXAMPLES
Circumference and Area of a Circle	If a circle has radius r, then its circumference is $$C = 2\pi r$$ and its area is $$A = \pi r^2.$$	If $r = 6$ inches, $C = 2\pi(6) = 12\pi \approx 37.7$ inches and $A = \pi(6)^2 = 36\pi \approx 113.1$ square inches. 6 in.
Area of a Trapezoid	$$A = \tfrac{1}{2}(a + b)h,$$ where a and b are the bases and h is the height.	If $a = 4$, $b = 6$, and $h = 3$, then the area is $$A = \frac{1}{2}(4 + 6)(3) = 15 \text{ square units.}$$ 4 3 6
Volume and Surface Area of a Box	If a box has length l, width w, and height h, then its volume is $$V = lwh$$ and its surface area is $$S = 2lw + 2wh + 2lh.$$	If a box has dimensions $l = 4$ feet, $w = 3$ feet, and $h = 2$ feet, then $$V = 4(3)(2) = 24 \text{ cubic feet}$$ and $$S = 2(4)(3) + 2(3)(2) + 2(4)(2)$$ $$= 52 \text{ square feet.}$$ 2 ft 4 ft 3 ft
Volume of a Cylinder	$$V = \pi r^2 h,$$ where r is the radius and h is the height.	If $r = 5$ inches and $h = 20$ inches, then the volume is $$V = \pi(5^2)(20) = 500\pi \text{ cubic inches.}$$ 5 in. 20 in.
Calculating Grade Point Average (GPA)	GPA is calculated by $$\frac{4A + 3B + 2C + D}{A + B + C + D + F},$$ where A, B, C, D, and F represent the number of credits earned with grades of A, B, C, D, and F, respectively.	10 credits of A, 8 credits of B, 6 credits of C, 12 credits of D, and 8 credits of F results in a GPA of $$\frac{4(10) + 3(8) + 2(6) + 12}{10 + 8 + 6 + 12 + 8} = 2.0.$$
Converting Between Fahrenheit and Celsius Degrees	$$C = \frac{5}{9}(F - 32)$$ $$F = \frac{9}{5}C + 32$$	212 °F is equivalent to $$C = \frac{5}{9}(212 - 32) = 100\text{ °C.}$$ 100 °C is equivalent to $$F = \frac{9}{5}(100) + 32 = 212\text{ °F.}$$

2.4   Exercises

MyMathLab®

CONCEPTS AND VOCABULARY

1. A(n) _____ can be used to calculate one quantity by using known values of other quantities.

2. The area A of a rectangle with length l and width w is
$A =$ _____.

3. The area A of a triangle with base b and height h is
$A =$ _____.

4. One degree equals _____ of a revolution.

5. There are _____ degrees in one revolution.

6. The sum of the measures of the angles in a triangle equals _____ degrees.

7. The volume V of a box with length l, width w, and height h is $V =$ _____.

8. The surface area S of a box with length l, width w, and height h is $S =$ _____.

9. The circumference C of a circle with radius r is
$C =$ _____.

10. The area A of a circle with radius r is $A =$ _____.

11. The volume V of a cylinder with radius r and height h is $V =$ _____.

12. The area A of a trapezoid with height h and bases a and b is $A =$ _____.

USING FORMULAS FROM GEOMETRY

Exercises 13–20: Find the area of the region shown.

13.

3 ft
6 ft

14.

2.5 yd
4 yd

15.

3 in.
6 in.

16.

1 mi
3 mi

17.

4 cm

18.

6 km

19.

6 mm
2 mm
5 mm

20.

3 ft
2 ft
4 ft

21. Find the area of a rectangle that has a 7-inch width and a 13-inch length.

22. Find the area of a triangle that has a 9-centimeter base and a 72-centimeter height.

23. Find the area of a rectangle that has a 5-foot width and a 7-yard length.

24. Find the area of a triangle that has a 12-millimeter base and a 6-millimeter height.

25. Find the circumference of a circle that has an 8-inch diameter.

26. Find the area of a circle that has a 9-foot radius.

27. Area of a Lot Find the area of the lot shown, which consists of a square and a triangle.

73 ft
52 ft
52 ft

28. Area of a Lot Find the area of the lot shown, which consists of a rectangle and two triangles.

60 yd
30 yd
20 yd
10 yd

Exercises 29 and 30: **Angle Measure** *Find the measure of the third angle in the triangle.*

29.   **30.**

31. A triangle contains two angles with measures of 23° and 76°. Find the measure of the third angle.

32. The measures of the angles in an *equilateral triangle* are equal. Find their measure.

33. The measures of the angles in a triangle are x, $2x$, and $2x$. Find the value of x.

34. The measures of the angles in a triangle are $3x$, $4x$, and $11x$. Find the value of x.

35. In a triangle, the two smaller angles are equal in measure and are each one third of the measure of the larger angle. Find the measure of each angle.

36. In a triangle, the two larger angles differ by 10°. The smaller angle is 50° less than the largest angle. Find the measure of each angle.

37. The diameter of a circle is 12 inches. Find its circumference and area.

38. The radius of a circle is $\frac{5}{4}$ feet. Find its circumference and area.

39. The circumference of a circle is 2π inches. Find its radius and area.

40. The circumference of a circle is 13π feet. Find its radius and area.

Exercises 41–44: A box with a top has length l, width w, and height h. Find the volume and surface area of the box.

41. $l = 22$ inches, $w = 12$ inches, $h = 10$ inches

42. $l = 5$ feet, $w = 3$ feet, $h = 6$ feet

43. $l = \frac{2}{3}$ yard, $w = \frac{2}{3}$ foot, $h = \frac{3}{2}$ feet

44. $l = 1.2$ meters, $w = 0.8$ meter, $h = 0.6$ meter

Exercises 45–48: Use the formula $V = \pi r^2 h$ to find the volume of a cylindrical container for the given r and h. Leave your answer in terms of π.

45. $r = 2$ inches, $h = 5$ inches

46. $r = \frac{1}{2}$ inch, $h = \frac{3}{2}$ inches

47. $r = 5$ inches, $h = 2$ feet

48. $r = 2.5$ feet, $h = 1.5$ yards

49. Volume of a Can (Refer to Example 6.)
(a) Find the volume of a can with a radius of $\frac{3}{4}$ inch and a height of $2\frac{1}{2}$ inches.
(b) Find the number of fluid ounces in the can if one cubic inch equals 0.554 fluid ounces.

50. Volume of a Barrel (Refer to Example 6.) A cylindrical barrel has a diameter of $1\frac{3}{4}$ feet and a height of 3 feet. Find the volume of the barrel.

SOLVING A FORMULA FOR A VARIABLE

Exercises 51–68: Solve the formula for the given variable.

51. $9x + 3y = 6$ for y **52.** $-2x - 2y = 10$ for y

53. $4x + 3y = 12$ for y **54.** $5x - 2y = 22$ for y

55. $A = lw$ for w **56.** $A = \frac{1}{2}bh$ for b

57. $V = \pi r^2 h$ for h **58.** $V = \frac{1}{3}\pi r^2 h$ for h

59. $A = \frac{1}{2}(a + b)h$ for a **60.** $C = 2\pi r$ for r

61. $V = lwh$ for w **62.** $P = 2l + 2w$ for w

63. $s = \dfrac{a + b + c}{2}$ for b **64.** $t = \dfrac{x - y}{3}$ for x

65. $\dfrac{a}{b} - \dfrac{c}{b} = 1$ for b **66.** $\dfrac{x}{y} + \dfrac{z}{y} = 5$ for z

67. $ab = cd + ad$ for a

68. $S = 2lw + 2lh + 2wh$ for w

69. Perimeter of a Rectangle If the width of a rectangle is 5 inches and its perimeter is 40 inches, find the length of the rectangle.

70. Perimeter of a Triangle Two sides of a triangle have lengths of 5 feet and 7 feet. If the triangle's perimeter is 21 feet, what is the length of the third side?

FINDING GRADE POINT AVERAGE AND OTHER FORMULAS

Exercises 71–74: (Refer to Example 9.) Let the variable A represent the number of credits with a grade of A, B the number of credits with a grade of B, and so on. Calculate the corresponding grade point average (GPA). Round your answer to the nearest hundredth.

71. $A = 30, B = 45, C = 12, D = 4, F = 4$

72. $A = 70, B = 35, C = 5, D = 0, F = 0$

73. $A = 0, B = 60, C = 80, D = 10, F = 6$

74. $A = 3, B = 5, C = 8, D = 0, F = 22$

Exercises 75–78: (Refer to Example 10.) Convert the Celsius temperature to an equivalent Fahrenheit temperature.

75. 25 °C   **76.** 100 °C   **77.** −40 °C   **78.** 0 °C

Exercises 79–82: (Refer to Example 10.) Convert the Fahrenheit temperature to an equivalent Celsius temperature.

79. 23 °F   **80.** 98.6 °F   **81.** −4 °F   **82.** −31 °F

83. Gas Mileage The formula $M = \frac{D}{G}$ can be used to calculate a car's gas mileage M after it has traveled D miles on G gallons of gasoline. Suppose that a truck driver leaves a gas station with a full tank of gas and the odometer showing 87,625 miles. At the next gas stop, it takes 38 gallons to fill the tank and the odometer reads 88,043 miles. Find the gas mileage for the truck.

84. Gas Mileage (Refer to Exercise 83.) A car that gets 34 miles per gallon is driven 578 miles. How many gallons of gasoline are used on this trip?

85. Lightning If there is an x-second delay between seeing a flash of lightning and hearing the thunder, then the lightning is $D = \frac{x}{5}$ miles away. Suppose that the time delay between a flash of lightning and the sound of thunder is 12 seconds. How far away is the lightning?

86. Lightning (Refer to Exercise 85.) Doppler radar shows an electrical storm 2.5 miles away. If you see the lightning from this storm, how long will it be before you hear the thunder?

WRITING ABOUT MATHEMATICS

87. A student solves the formula $A = \frac{1}{2}bh$ for h and obtains the formula $h = \frac{1}{2}bA$. Explain the error that the student is making. What is the correct answer?

88. Give an example of a formula that you have used and explain how you used it.

SECTIONS 2.3 and 2.4	Checking Basic Concepts

1. Translate the sentence into an equation containing the variable x. Then solve the resulting equation.
  (a) The product of 3 and a number is 36.
  (b) A number subtracted from 35 is 43.

2. When three consecutive integers are added, the sum is −93. Find the three integers.

3. Convert 9.5% to a decimal.

4. Convert $\frac{5}{4}$ to a percentage.

5. Violent Crime In 2012, New York City experienced 53,037 violent crimes. This figure represented a 71% decrease from the number in 1993. Find the number of violent crimes in 1993. (*Source:* New York State Crime Update.)

6. Driving a Car How many hours does it take the driver of a car to travel 390 miles at 60 miles per hour?

7. College Loans A student takes out two loans, one at 6% and the other at 7%. The 6% loan is $2000 more than the 7% loan, and the total interest for one year is $510. Find the amount of each loan.

8. Gas Mileage A car that gets 28 miles per gallon is driven 504 miles. How many gallons of gasoline are used on this trip?

9. Height of a Triangle The area of a triangle that has a base of 6 inches is 36 square inches. Find the height of the triangle.

10. Area and Circumference Find the area and circumference of the circle shown.

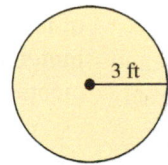

3 ft

11. Angles Find the value of x in the triangle shown.

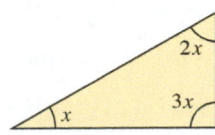

$2x$

$3x$

x

12. Solving a Formula Solve $A = \pi r^2 + \pi r l$ for l.

2.5 Unit Analysis

NEW VOCABULARY

☐ Inch, foot, yard, mile
☐ Millimeter, centimeter, meter, kilometer
☐ Unit Fraction
☐ Capacity/volume
☐ Ounce, cup, pint, quart, gallon
☐ Milliliter, liter
☐ Mass/weight
☐ Ounce, pound, ton
☐ Milligram, gram, kilogram
☐ Second, minute, hour, day, week, month, year
☐ Speed

STUDY TIP

Unit analysis can be valuable in other classes, especially the sciences. Take some extra time to fully learn this technique.

READING CHECK 1

- When converting 3 miles to feet, what unit fraction should be used?

1 Applying Unit Analysis and Unit Fractions

There are many situations that require the conversion of one unit of measure to another. To do this, we use a method called **unit analysis** (also called *dimensional analysis*). This method is based on the fact that multiplying a measurement by 1 does not change the value of the measurement. Since 1 can be written in many forms, it is important to write it in the form of a *unit fraction*. A **unit fraction** is a fraction that is equivalent to 1.

Unit Fractions

$$\frac{12 \text{ in.}}{1 \text{ ft}}, \quad \frac{1 \text{ c}}{8 \text{ oz.}}, \quad \frac{100 \text{ cm}}{1 \text{ m}}, \quad \frac{16 \text{ oz}}{1 \text{ lb}}, \quad \text{and} \quad \frac{1 \text{ km}}{1000 \text{ m}}$$

Note that the numerator value and the denominator value represent the same measurement, just with different units. Thus a unit fraction is always equal to 1.

When a measurement is multiplied by a unit fraction, the units in the denominator of the unit fraction are *divided out* and the units in the numerator of the unit fraction are *multiplied in*.

CONVERTING FROM ONE UNIT OF MEASURE TO ANOTHER

To convert from one unit of measure to another, do the following.

STEP 1: Write the given measure (including its units) over 1.

STEP 2: Multiply by a unit fraction whose denominator contains the given (unwanted) units and whose numerator contains the new (desired) units.

It may be necessary to multiply by more than one unit fraction to get the desired units in the final result.

2 Converting Units of Length

While the **inch (in.)**, **foot (ft)**, **yard (yd)**, and **mile (mi)** are the most commonly used units of length in the American system of measurement, the **millimeter (mm)**, **centimeter (cm)**, **meter (m)**, and **kilometer (km)** are most common in the metric system. These measurements are related in the following ways. Recall that $\approx$ means *approximately equal*.

UNITS OF LENGTH

American Units

1 ft = 12 in.   1 yd = 36 in.   1 yd = 3 ft   1 mi = 1760 yd   1 mi = 5280 ft

Metric Units

1 cm = 10 mm   1 m = 100 cm   1 m = 1000 mm   1 km = 1000 m

American-Metric Conversions

1 in. = 2.54 cm   1 m ≈ 1.09 yd   1 km ≈ 0.62 mi

EXAMPLE 1

Converting from one unit of length to another

Convert as indicated. Round to 2 decimal places as needed.
(a) 75 inches to feet   **(b)** 3.6 kilometers to meters   **(c)** 38 centimeters to inches

Solution
(a) The given (unwanted) units are inches, and the new (desired) units are feet. To make this conversion, write 75 inches over 1 and then multiply by a unit fraction with inches in the denominator and feet in the numerator.

$$75 \text{ in.} = \frac{75 \text{ in.}}{1} \cdot \frac{1 \text{ ft}}{12 \text{ in.}} = \frac{75}{12} \text{ ft} = \frac{25 \cdot 3}{4 \cdot 3} \text{ ft} = \frac{25}{4} \text{ ft} = 6\frac{1}{4} \text{ ft, or } 6.25 \text{ ft}$$

(b) Since 1 km = 1000 m, we use the unit fraction $\frac{1000 \text{ m}}{1 \text{ km}}$, which has the unwanted units (km) in the denominator and the desired units (m) in the numerator.

$$3.6 \text{ km} = \frac{3.6 \text{ km}}{1} \cdot \frac{1000 \text{ m}}{1 \text{ km}} = 3.6(1000)\text{m} = 3600 \text{ m}$$

(c) $38 \text{ cm} = \frac{38 \text{ cm}}{1} \cdot \frac{1 \text{ in.}}{2.54 \text{ cm}} = \frac{38}{2.54} \text{ in.} \approx 14.96 \text{ in.}$

Now Try Exercises 9, 17, 19

If you do not know the direct relationship between two units of measure, then it may be necessary to use more than one unit fraction during a conversion. The next example illustrates this process.

EXAMPLE 2

Converting by using more than one unit fraction

Convert 1.4 miles to inches by using more than one unit fraction.

Solution
Since most people do not know how many inches are in a mile, we will first convert miles to feet and then convert feet to inches. To convert from miles to feet, we use a unit fraction with feet in the numerator and miles in the denominator. Then, to convert from feet to inches, we use a unit fraction with inches in the numerator and feet in the denominator.

$$1.4 \text{ mi} = \frac{1.4 \text{ mi}}{1} \cdot \frac{5280 \text{ ft}}{1 \text{ mi}} \cdot \frac{12 \text{ in.}}{1 \text{ ft}} = \frac{1.4 \cdot 5280 \cdot 12}{1} \text{ in.} = 88,704 \text{ in.}$$

Miles "out," feet "in"   Feet "out," inches "in"

Now Try Exercise 13

When we do not have a unit fraction to convert directly from one unit to another, it may be necessary to complete some conversions *within* each system first. For example, a measurement in miles could be converted to meters by first converting miles to kilometers and then converting kilometers to meters.

$$\text{miles} \rightarrow \text{kilometers} \rightarrow \text{meters}$$

Another option would be to convert miles to yards and then convert yards to meters.

$$\text{miles} \rightarrow \text{yards} \rightarrow \text{meters}$$

In either case, the red arrow represents a unit fraction that converts American system units to metric system units. In the next example, we convert miles to meters in this manner.

EXAMPLE 3   **Completing American–metric conversions of length**

Convert 2.5 miles to meters using the given sequence. Round answers to 2 decimal places.
(a) Miles → kilometers → meters   **(b)** Miles → yards → meters

Solution
(a) Use the unit fraction $\frac{1 \text{ km}}{0.62 \text{ mi}}$ to convert miles to kilometers, and then use $\frac{1000 \text{ m}}{1 \text{ km}}$ to convert kilometers to meters.

$$2.5 \text{ mi} \approx \frac{2.5 \text{ mi}}{1} \cdot \frac{1 \text{ km}}{0.62 \text{ mi}} \cdot \frac{1000 \text{ m}}{1 \text{ km}} = \frac{2.5(1000)}{0.62} \text{ m} \approx 4032.26 \text{ m}$$

(b) Use $\frac{1760 \text{ yd}}{1 \text{ mi}}$ to convert miles to yards, and then use $\frac{1 \text{ m}}{1.09 \text{ yd}}$ to convert yards to meters.

$$2.5 \text{ mi} \approx \frac{2.5 \text{ mi}}{1} \cdot \frac{1760 \text{ yd}}{1 \text{ mi}} \cdot \frac{1 \text{ m}}{1.09 \text{ yd}} = \frac{2.5(1760)}{1.09} \text{ m} \approx 4036.70 \text{ m}$$

Now Try Exercise 23

NOTE: The two results for Example 3 are not exactly the same because some of the unit fractions used contain approximations that do not give exact results. ∎

3 Converting Units of Area

READING CHECK 2

• Why do the results from parts (a) and (b) of Example 3 differ slightly?

While there are 3 feet in 1 yard, this does *not* mean that there are 3 *square* feet in 1 *square* yard. The square yard shown in **FIGURE 2.17** can be used to visualize the correct relationship between square feet and square yards.

As **FIGURE 2.17** shows, there are 9 square feet in one square yard. When converting from one square unit to another, we can use the relationships for length that we already know. For example, we can write square yards as either

$$\textbf{yd}^2 \quad \text{or} \quad \textbf{yd} \cdot \textbf{yd}.$$

Nine Square Feet in One Square Yard

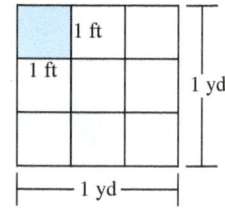

FIGURE 2.17

READING CHECK 3

• When converting from one square unit of measure to another, how many times is a unit fraction used?

To see that there are 9 square feet in 1 square yard, we find a known unit fraction that involves feet and yards and then use it twice.

$$1 \text{ yd}^2 = \frac{1 \text{ yd} \cdot \text{yd}}{1} \cdot \frac{3 \text{ ft}}{1 \text{ yd}} \cdot \frac{3 \text{ ft}}{1 \text{ yd}} = 3 \text{ ft} \cdot 3 \text{ ft} = 9 \text{ ft}^2$$

Unit fraction used twice

EXAMPLE 4   **Converting from one unit of area to another**

Convert as indicated.
(a) 6 square feet to square inches   **(b)** 1,200,000 square centimeters to square meters

Solution
(a) Use a unit fraction relating feet and inches, twice.

$$6 \text{ ft}^2 = \frac{6 \text{ ft} \cdot \text{ft}}{1} \cdot \frac{12 \text{ in.}}{1 \text{ ft}} \cdot \frac{12 \text{ in.}}{1 \text{ ft}} = 6 \cdot 12 \text{ in.} \cdot 12 \text{ in.} = 864 \text{ in}^2$$

(b) Use a unit fraction relating centimeters and meters, twice.

$$1,200,000 \text{ cm}^2 = \frac{1,200,000 \text{ cm} \cdot \text{cm}}{1} \cdot \frac{1 \text{ m}}{100 \text{ cm}} \cdot \frac{1 \text{ m}}{100 \text{ cm}} = \frac{1,200,000}{100 \cdot 100} \text{ m}^2 = 120 \text{ m}^2$$

Now Try Exercises 33, 37

READING CHECK 4

• Explain how capacity and volume differ.

4 Converting Units of Capacity and Volume

Even though the words *capacity* and *volume* are often used interchangeably, they do not have exactly the same meaning. **Capacity** is a measure of the amount of substance that a container can hold, while **volume** is a measure of the actual amount of the substance present. For example, if a 16-gallon gas tank is half full, the capacity of the tank is 16 gallons, and the volume of gas in the tank is 8 gallons.

In the American system of measurement, the units used to measure *both* capacity and volume of liquids are **fluid ounce (oz)**, **cup (c)**, **pint (pt)**, **quart (qt)**, and **gallon (gal)**. The units most commonly used to measure capacity and volume in the metric system are **liter (L)** and **milliliter (mL)**. These measurements are related in the following ways.

UNITS OF CAPACITY AND VOLUME

American Units
$$1 \text{ c} = 8 \text{ oz} \qquad 1 \text{ pt} = 2 \text{ c} \qquad 1 \text{ qt} = 2 \text{ pt} \qquad 1 \text{ gal} = 4 \text{ qt}$$

Metric Units
$$1 \text{ L} = 1000 \text{ mL}$$

American-Metric Conversions
$$1 \text{ L} \approx 1.06 \text{ qt} \qquad 1 \text{ oz} \approx 29.57 \text{ mL}$$

EXAMPLE 5 **Converting from one unit of capacity to another**

Convert as indicated. Round answers to 2 decimal places as needed.
(a) 2.5 gallons to cups   **(b)** 25,000 milliliters to liters   **(c)** 6 gallons to liters

Solution
(a) Convert from gallons to quarts, then quarts to pints, and finally from pints to cups.

$$2.5 \text{ gal} = \frac{2.5 \text{ gal}}{1} \cdot \frac{4 \text{ qt}}{1 \text{ gal}} \cdot \frac{2 \text{ pt}}{1 \text{ qt}} \cdot \frac{2 \text{ c}}{1 \text{ pt}} = 2.5 \cdot 4 \cdot 2 \cdot 2 \text{ c} = 40 \text{ c}$$

(b) $$25,000 \text{ mL} = \frac{25,000 \text{ mL}}{1} \cdot \frac{1 \text{ L}}{1000 \text{ mL}} = \frac{25,000}{1000} \text{ L} = 25 \text{ L}$$

(c) gallon $\rightarrow$ quarts $\rightarrow$ liters

$$6 \text{ gal} \approx \frac{6 \text{ gal}}{1} \cdot \frac{4 \text{ qt}}{1 \text{ gal}} \cdot \frac{1 \text{ L}}{1.06 \text{ qt}} = \frac{6(4)}{1.06} \text{ L} \approx 22.64 \text{ L}$$

Now Try Exercises 45, 49, 57

---

5 Converting Units of Mass (Weight)

Although the terms *weight* and *mass* are sometimes used interchangeably, they do not always have *exactly* the same meaning. While **mass** is a measure of the amount of matter in an object, **weight** is a measure of the force on an object due to gravity. The mass of an object does not change, but its weight may change depending on its distance from the center of Earth.

The **ounce (oz)**, **pound (lb)**, and **ton (T)** are the most common units used in the American system to measure weight or mass. In the metric system, the **milligram (mg)**, **gram (g)**, and **kilogram (kg)** are commonly used. These measurements are related in the following ways.

UNITS OF MASS (WEIGHT)

American Units
1 lb = 16 oz   1 T = 2000 lb

Metric Units
1 g = 1000 mg   1 kg = 1000 g

American-Metric Conversions
1 kg ≈ 2.2 lb   1 oz ≈ 28.35 g

EXAMPLE 6 **Converting from one unit of weight to another**

Convert as indicated. Round answers to 2 decimal places as needed.
(a) 0.04 ton to ounces   **(b)** 10,000 milligrams to grams   **(c)** 135 pounds to kilograms

Solution
(a) Convert from tons to pounds and then from pounds to ounces.

$$0.04 \text{ T} = \frac{0.04 \text{ T}}{1} \cdot \frac{2000 \text{ lb}}{1 \text{ T}} \cdot \frac{16 \text{ oz}}{1 \text{ lb}} = 0.04 \cdot 2000 \cdot 16 \text{ oz} = 1280 \text{ oz}$$

(b) $10,000 \text{ mg} = \dfrac{10,000 \text{ mg}}{1} \cdot \dfrac{1 \text{ g}}{1000 \text{ mg}} = \dfrac{10,000}{1000} \text{ g} = 10 \text{ g}$

(c) $135 \text{ lb} \approx \dfrac{135 \text{ lb}}{1} \cdot \dfrac{1 \text{ kg}}{2.2 \text{ lb}} = \dfrac{135}{2.2} \text{ kg} \approx 61.36 \text{ kg}$

Now Try Exercises 63, 69, 75

MAKING CONNECTIONS 1

Ounces of Capacity and Ounces of Weight

The term *ounce* is used to name two different units of measure: a unit of capacity and a unit of weight. For this reason, we must clearly understand whether a problem involves capacity or weight. To avoid confusion, some people use the term *fluid ounces* (fl oz) for capacity and the term *ounces* (oz) for weight.

---

6 Converting Units of Time

The **second (sec)**, **minute (min)**, **hour (hr)**, **day (d)**, **week (wk)**, **month (mo)**, and **year (yr)** are the units most commonly used for measuring time, and these units are related to each other as follows.

UNITS OF TIME

1 min = 60 sec   1 hr = 60 min   1 d = 24 hr

1 wk = 7 d   1 yr = $365\dfrac{1}{4}$ d (or 365.25 d)

NOTE: While we often hear that there are 52 weeks in 1 year, the relationship is not exactly accurate because a 52-week period is actually 7 · 52 = 364 days long, which is not equal to 1 year, or 365.25 days. Also, months should not be used in relationships because the number of days in a month can vary. ∎

EXAMPLE 7 **Converting from one unit of time to another**

Convert as indicated. Express answers in decimal form when needed.
(a) 3.5 minutes to seconds   **(b)** 102 hours to days   **(c)** 5 days to minutes

Solution

(a) $3.5 \text{ min} = \dfrac{3.5 \text{ min}}{1} \cdot \dfrac{60 \text{ sec}}{1 \text{ min}} = 3.5(60) \text{ sec} = 210 \text{ sec}$

(b) $102 \text{ hr} = \dfrac{102 \text{ hr}}{1} \cdot \dfrac{1 \text{ d}}{24 \text{ hr}} = \dfrac{102}{24} \text{ d} = 4.25 \text{ d}$

(c) $5 \text{ d} = \dfrac{5 \text{ d}}{1} \cdot \dfrac{24 \text{ hr}}{1 \text{ d}} \cdot \dfrac{60 \text{ min}}{1 \text{ hr}} = 5 \cdot 24 \cdot 60 \text{ min} = 7200 \text{ min}$

Now Try Exercises 79, 81, 85

7 ▸ Converting Units of Speed

A **rate** is a ratio used to compare two different kinds of quantities. If one quantity is distance and the other is time, then the rate represents *speed*. **Speed** is a rate that gives a distance traveled in an amount of time. A unit of speed is expressed as a ratio with a unit of distance in the numerator and a unit of time in the denominator.

🌐 *Math in Context* (Driving) One recognizable example of speed in everyday life can be found posted along roads and highways on speed limit signs. In the United States, we measure the speed of a vehicle in miles per hour, or mi/hr (also abbreviated as mph). In countries using the metric system, the speed of a vehicle is measured in kilometers per hour, or km/hr. Other examples of speed include

ft/sec,   m/min,   mi/sec,   and   cm/sec.

It is possible to convert from one unit of speed to another by converting the distance unit, the time unit, or both units. The next example illustrates how to convert from one unit of speed to another using unit analysis.

EXAMPLE 8 **Converting units of speed**

Convert as indicated.
(a) 120 miles per hour to miles per minute   **(b)** 66 feet per second to miles per hour

Solution
(a) The distance unit does not change, so we only need to convert the time unit from hours to minutes. Since the hours unit is in the denominator of the given speed, we need a unit fraction that relates hours and minutes and has hours in the numerator.

$$120 \text{ mi/hr} = \dfrac{120 \text{ mi}}{1 \text{ hr}} \cdot \dfrac{1 \text{ hr}}{60 \text{ min}} = \dfrac{120 \text{ mi}}{60 \text{ min}} = \dfrac{2 \text{ mi}}{1 \text{ min}} = 2 \text{ mi/min}$$

(b) The distance unit must be converted from **feet to miles**, and the time unit must be converted from **seconds to hours**.

$$66 \text{ ft/sec} = \dfrac{66 \text{ ft}}{1 \text{ sec}} \cdot \dfrac{1 \text{ mi}}{5280 \text{ ft}} \cdot \dfrac{60 \text{ sec}}{1 \text{ min}} \cdot \dfrac{60 \text{ min}}{1 \text{ hr}} = \dfrac{66 \cdot 60 \cdot 60}{5280} \text{ mi/hr} = 45 \text{ mi/hr}$$

Now Try Exercises 89, 93

READING CHECK 5

• Which of the following speeds are equal to 60 miles per hour?
$88 \frac{\text{ft}}{\text{sec}}, 180 \frac{\text{yd}}{\text{hr}}, 1 \frac{\text{mi}}{\text{hr}}, 800 \frac{\text{in.}}{\text{sec}}$

8 ▸ Solving an Application Involving Unit Analysis

Sometimes, speed is more easily interpreted when the units change. The next example illustrates how to convert from an American unit of speed to a metric unit of speed.

EXAMPLE 9   **Analyzing speed limit**

After crossing the border into Canada, an American driver uses cruise control set at 60 miles per hour. If the posted speed limit on the Canadian road is 90 kilometers per hour, is the American driving over the speed limit?

Solution

To determine if the driver is speeding, convert 60 miles per hour to kilometers per hour.

$$60 \text{ mi/hr} \approx \frac{60 \text{ mi}}{1 \text{ hr}} \cdot \frac{1 \text{ km}}{0.62 \text{ mi}} = \frac{60}{0.62} \text{ km/hr} \approx 96.8 \text{ km/hr}$$

The American is driving nearly 97 kilometers per hour, which is over the speed limit.

Now Try Exercise 123

2.5   Putting It All Together

CONCEPT	COMMENTS	EXAMPLES
Unit Fraction	A unit fraction is a fraction that is equivalent to 1. Unit fractions can be used to convert from one unit of measure to another.	$\dfrac{3 \text{ ft}}{1 \text{ yd}}$, $\dfrac{1 \text{ gal}}{4 \text{ qt}}$, and $\dfrac{2000 \text{ lb}}{1 \text{ T}}$
Units of Length	Inch, foot, yard, mile Millimeter, centimeter, meter, kilometer ***Relationships:*** 1 ft = 12 in. 1 yd = 36 in. 1 yd = 3 ft 1 mi = 1760 yd 1 mi = 5280 ft 1 cm = 10 mm 1 m = 100 cm 1 m = 1000 mm 1 km = 1000 m 1 in. = 2.54 cm 1 m ≈ 1.09 yd 1 km ≈ 0.62 mi	Convert 24 inches to feet. $\dfrac{24 \text{ in.}}{1} \cdot \dfrac{1 \text{ ft}}{12 \text{ in.}} = \dfrac{24}{12} \text{ ft} = 2 \text{ ft}$
Units of Area	Units of area are squares of the units of length. To convert from one unit of area to another, use the corresponding unit fraction relating units of length, twice.	Convert 2 square yards to square feet. $\dfrac{2 \text{ yd} \cdot \text{yd}}{1} \cdot \dfrac{3 \text{ ft}}{1 \text{ yd}} \cdot \dfrac{3 \text{ ft}}{1 \text{ yd}} = 18 \text{ ft}^2$
Units of Capacity and Volume	Ounce, cup, pint, quart, gallon Milliliter, liter ***Relationships:*** 1 c = 8 oz 1 qt = 2 pt 1 pt = 2 c 1 gal = 4 qt 1 L = 1000 mL 1 L ≈ 1.06 qt 1 oz ≈ 29.57 mL	Convert 4 gallons to quarts. $\dfrac{4 \text{ gal}}{1} \cdot \dfrac{4 \text{ qt}}{1 \text{ gal}} = 16 \text{ qt}$
Units of Mass (Weight)	Ounce, pound, ton Milligram, gram, kilogram ***Relationships:*** 1 lb = 16 oz 1 T = 2000 lb 1 g = 1000 mg 1 kg = 1000 g 1 kg ≈ 2.2 lb 1 oz ≈ 28.35 g	Convert 64 ounces to pounds. $\dfrac{64 \text{ oz}}{1} \cdot \dfrac{1 \text{ lb}}{16 \text{ oz}} = \dfrac{64}{16} \text{ lb} = 4 \text{ lb}$

CONCEPT	COMMENTS	EXAMPLES
Units of Time	Second, minute, hour, day, week, month, and year ***Relationships:*** 1 min = 60 sec 1 hr = 60 min 1 d = 24 hr 1 wk = 7 d 1 yr = $365\frac{1}{4}$ d (or 365.25 d)	Convert 3.5 days to hours. $$\frac{3.5 \text{ d}}{1} \cdot \frac{24 \text{ hr}}{1 \text{ d}} = 84 \text{ hr}$$
Units of Speed	A unit of speed is expressed as a ratio with a unit of distance in the numerator and a unit of time in the denominator.	Convert 10,560 feet per hour to miles per hour. $$\frac{10{,}560 \text{ ft}}{1 \text{ hr}} \cdot \frac{1 \text{ mi}}{5280 \text{ ft}} = 2 \text{ mi/hr}$$

2.5 Exercises

MyMathLab®

CONCEPTS AND VOCABULARY

1. A(n) _____ fraction is a fraction that is equivalent to 1.

2. When using a unit fraction to convert from one unit of measure to another, write the given (unwanted) unit in the _____ of the unit fraction.

3. When using a unit fraction to convert from one unit of measure to another, write the new (desired) unit in the _____ of the unit fraction.

4. When converting from one unit of _____ to another, use the corresponding unit fraction relating units of length, twice.

5. The amount of a substance that a container can hold is called the container's _____.

6. The actual amount of a substance that is present in a container is called the _____ of the substance.

7. A rate that gives a distance traveled in an amount of time is called _____.

8. A unit of speed has a unit of _____ in its numerator and a unit of _____ in its denominator.

CONVERTING UNITS OF LENGTH

Exercises 9–18: Convert the length as indicated.

9. 48 inches to feet

10. 880 yards to miles

11. $\frac{2}{3}$ yard to inches

12. 87 feet to yards

13. 0.025 mile to inches

14. 47,520 inches to miles

15. 53.6 centimeters to millimeters

16. 2007 millimeters to meters

17. 14.59 meters to kilometers

18. 0.6 meter to centimeters

Exercises 19–22: Convert the length as indicated. Round answers to 2 decimal places.

19. 17 inches to centimeters

20. 98 centimeters to inches

21. 240 yards to meters

22. 84 miles to kilometers

Exercises 23–26: Convert the given length as indicated. Round answers to 2 decimal places.

23. 5800 centimeters to yards
    (a) Centimeters → inches → yards
    (b) Centimeters → meters → yards

24. 8000 yards to kilometers
    (a) Yards → meters → kilometers
    (b) Yards → miles → kilometers

25. 4500 meters to miles
    (a) Meters → kilometers → miles
    (b) Meters → yards → miles

26. 0.5 yard to centimeters
    (a) Yards → inches → centimeters
    (b) Yards → meters → centimeters

Exercises 27–32: Convert the length as indicated. Round answers to 2 decimal places. Answers may vary slightly.

27. 27 feet to meters

28. 3400 centimeters to feet

29. 0.25 meter to inches

30. 15,840 feet to kilometers

31. 0.003 kilometer to feet

32. 11 feet to centimeters

CONVERTING UNITS OF AREA

Exercises 33–40: Convert the area as indicated.

33. 10 square feet to square inches

34. 0.01 square mile to square yards

35. 324 square feet to square yards

36. 278,784 square feet to square miles

37. 0.004 square meter to square centimeters

38. 30 square millimeters to square centimeters

39. 0.0025 square kilometer to square meters

40. 0.006 square meter to square millimeters

CONVERTING UNITS OF CAPACITY AND VOLUME

Exercises 41–52: Convert the capacity as indicated.

41. 48 ounces to cups

42. $\frac{1}{2}$ quart to pints

43. 10.25 gallons to quarts

44. 26 cups to pints

45. $\frac{3}{4}$ gallon to pints

46. 0.5 gallon to ounces

47. 5 gallons to cups

48. 800 ounces to quarts

49. 90,000 milliliters to liters

50. 9.63 milliliters to liters

51. 1.08 liters to milliliters

52. 0.09 liter to milliliters

Exercises 53–56: Convert the capacity as indicated. Round answers to 2 decimal places.

53. 16 ounces to milliliters

54. 7.5 liters to quarts

55. 56 quarts to liters

56. 1500 milliliters to ounces

Exercises 57–60: Convert the capacity as indicated. Round answers to 2 decimal places. Answers may vary slightly.

57. 75 liters to gallons

58. 16 pints to liters

59. 956 milliliters to cups

60. 3.6 gallons to liters

CONVERTING UNITS OF MASS (WEIGHT)

Exercises 61–70: Convert the mass as indicated.

61. 440 ounces to pounds

62. $7\frac{5}{8}$ pounds to ounces

63. 0.00025 ton to ounces

64. $\frac{1}{100}$ ton to pounds

65. 1,760,000 ounces to tons

66. 22,500 pounds to tons

67. 6500 grams to kilograms

68. 0.045 kilogram to grams

69. 550 milligrams to grams

70. 5 grams to milligrams

Exercises 71–74: Convert the mass as indicted. Round answers to 2 decimal places.

71. 420 grams to ounces

72. 98 pounds to kilograms

73. 61.2 kilograms to pounds

74. 3 ounces to grams

Exercises 75–78: Convert the mass as indicated. Round answers to 2 decimal places. Answers may vary slightly.

75. 0.135 ounce to milligrams

76. 2.6 kilograms to ounces

77. 1.45 pounds to grams

78. 0.0064 pound to milligrams

CONVERTING UNITS OF TIME

Exercises 79–88: Convert as indicated. Express answers in decimal form when needed.

79. 5.25 hours to minutes

80. 3 days to hours

81. 90 minutes to hours

82. 63 days to weeks

83. 4.1 hours to seconds

84. 0.25 day to minutes

85. 99,360 seconds to days

86. 19,800 seconds to hours

87. 4383 hours to years

88. 252 hours to weeks

CONVERTING UNITS OF SPEED

Exercises 89–96: Convert as indicated. Round answers to one decimal place, as needed.

89. 34 yards per second to feet per second

90. 4 kilometers per hour to meters per hour

91. 12 meters per min to meters per second

92. 144 miles per hour to miles per min

93. 30 miles per hour to feet per second

94. 50 feet per day to inches per hour

95. 110 feet per second to miles per hour

96. 3 meters per second to kilometers per hour

Exercises 97–100: Convert as indicated. Round answers to the nearest whole number.

97. 11 meters per second to feet per second

98. 70 miles per hour to kilometers per hour

99. 92 kilometers per hour to miles per hour

100. 190 feet per second to meters per second

SOLVING APPLICATION INVOLVING UNIT ANALYSIS

101. Weight in Stones The *stone* is a unit of weight in the Imperial system of measurement. One stone is equivalent to 14 pounds. If a person weighs 154 pounds, find his weight in stones.

102. Length in Rods The *rod* is a unit of length often used on wilderness maps to describe the length of a portage (trail between bodies of water). A rod is equivalent to 16.5 feet. Find the length in rods of a portage that is 1 mile long.

103. Cost of Carpet For a square room that measures 15 feet on a side, do the following.
(a) Find the area of the room in square feet.
(b) Convert your answer to square yards.
(c) Determine the cost of carpet for the room if the carpet sells for $32 per square yard.

104. Cost of Carpet For a rectangular room that measures 15 feet by 18 feet, do the following.
(a) Find the area of the room in square feet.
(b) Convert your answer to square yards.
(c) Determine the cost of carpet for the room if the carpet sells for $46 per square yard.

105. Serving Size The label on a 1-gallon jug of fruit juice states that the serving size is 8 oz. How many servings are in the jug?

106. Serving Size The label on a 2-pound bag of chips states that the serving size is 4 oz. How many servings are in the bag?

107. Soda Bottle A person buys a 500-milliliter bottle of soda from a vending machine. How many liters of soda did the person buy?

108. Servings in a Bottle A large bottle contains 3 liters of fruit juice. How many 200-milliliter servings are in this bottle?

109. Metric Tons The mass of very heavy objects can be measured in metric tons. A metric ton (t) is equal to 1000 kilograms. If an African elephant weighs 5500 kilograms, what is its weight in metric tons?

110. Metric Tons Refer to the previous exercise. Find the number of grams in one metric ton.

111. Serving Size If the label on a 1-kilogram can of mixed nuts states that the serving size is 40 grams, how many servings are in the can?

112. Serving Size If the label on a 2-liter bottle of fruit drink states that the serving size is 250 milliliters, how many servings are in the bottle?

113. Triple Jump The world record in the triple jump was set in 1995 by Jonathan Edwards, with a jump of 18.29 meters. Using the sequence

Meters $\rightarrow$ centimeters $\rightarrow$ inches $\rightarrow$ feet,

is this jump over 60 feet? (*Source:* International Association of Athletics Federations.)

114. Blue Whale An adult blue whale can weigh more than 172,700 kilograms. Find this weight to the nearest ton. (*Source: Smithsonian National Zoological Park.*)

115. Super Soaker An Aquashock Hydroblitz Super Soaker has a 101-ounce water reservoir. To the nearest whole number, how many liters of water does this water gun hold? (*Source: Hasbro.*)

116. Long Hair The person with the world's longest hair is Xie Qiuping of China. Her hair is 5.627 meters long. Is her hair over 20 feet long? (*Source: Guinness Book of World Records.*)

117. Obesity An adult male who is 5 feet 10 inches tall is considered obese if his weight is above 95.5 kilograms. What is this weight to the nearest pound? (*Source: Centers for Disease Control.*)

118. iPod Nano The fourth-generation iPod Nano is 90.7 millimeters long and 38.7 millimeters wide. What are these two measurements to the nearest tenth of an inch? (*Source: Apple.*)

119. NASCAR Speeds Some NASCAR drivers reach average speeds of 3 miles per minute. Convert this speed to miles per hour.

120. NASCAR Speeds On slower speedways, NASCAR drivers reach average speeds of 120 miles per hour. Convert this speed to miles per minute.

121. Animal Speeds With top speeds of about 103 feet per second, the cheetah is the fastest land animal. To the nearest whole number, what is the top speed of a cheetah in miles per hour?

122. Human Speeds Usain Bolt set a world record in the 100-meter sprint during the 2008 Olympics by running an average speed of 34 feet per second. To the nearest whole number, what was Bolt's average speed in miles per hour?

123. Speed Limits An American drives 30 miles per hour after crossing the border into a Canadian town. If the posted speed limit is 50 kilometers per hour, is the driver exceeding the speed limit?

124. Speed Limits A Canadian drives 75 kilometers per hour after crossing the border into an American town. If the posted speed limit is 45 miles per hour, is the driver exceeding the speed limit?

WRITING ABOUT MATHEMATICS

125. A student converts 3 pounds to cups as follows.

$$3 \text{ lb} = \frac{3 \text{ lb}}{1} \cdot \frac{16 \text{ oz}}{1 \text{ lb}} \cdot \frac{1 \text{ c}}{8 \text{ oz}} = 6 \text{ c}$$

Explain the error in this computation.

126. A classmate tells you that there are 36 square inches in 1 square yard because there are 36 inches in 1 yard. How do you convince your classmate that this is incorrect?

2.6 Linear Inequalities

Objectives

1. Graphing Solution Sets on Number Lines
   • Checking Solutions

2. Solving Linear Inequalities with Tables

3. Writing Solution Sets Using Interval Notation

4. Applying the Addition Property of Inequalities

5. Applying the Multiplication Property of Inequalities

6. Writing Solution Sets Using Set-Builder Notation

7. Translating Words to Inequalities

8. Solving Applications Involving Inequalities

1 Graphing Solution Sets on Number Lines

At an amusement park, a particular ride might be restricted to people at least 48 inches tall. A child who is x inches tall may go on the ride if $x \geq 48$ but may not go on the ride if $x < 48$. A height of 48 inches represents the *boundary* between being allowed on the ride and being denied access to the ride.

Solving linear inequalities is closely related to solving linear equations because equality is the boundary between *greater than* and *less than*. In this section, we discuss techniques used to solve linear inequalities.

A **linear inequality** results whenever the equals sign in a linear equation is replaced with any one of the symbols $<$, $\leq$, $>$, or $\geq$. Examples of linear equations include

$$x = 5, \quad 2x + 1 = 0, \quad 1 - x = 6, \quad \text{and} \quad 5x + 1 = 3 - 2x.$$

Therefore examples of linear inequalities include

$$x > 5, \quad 2x + 1 < 0, \quad 1 - x \geq 6, \quad \text{and} \quad 5x + 1 \leq 3 - 2x.$$

TABLE 2.8 shows how each of the inequality symbols is read.

NEW VOCABULARY

☐ Linear inequality
☐ Solution
☐ Solution set
☐ Interval notation
☐ Set-builder notation

Inequality Symbols

Symbol	How the Symbol Is Read
$>$	greater than
$<$	less than
$\geq$	greater than or equal to
$\leq$	less than or equal to

TABLE 2.8

Inequalities often have infinitely many solutions. A **solution** to an inequality is a value of the variable that makes the inequality statement true. The set of all solutions is called the **solution set**. Two inequalities are *equivalent* if they have the same solution sets.

The following See the Concept shows how to graph inequalities on number lines.

See the Concept 1   GRAPHING THE SOLUTION SET TO AN INEQUALITY

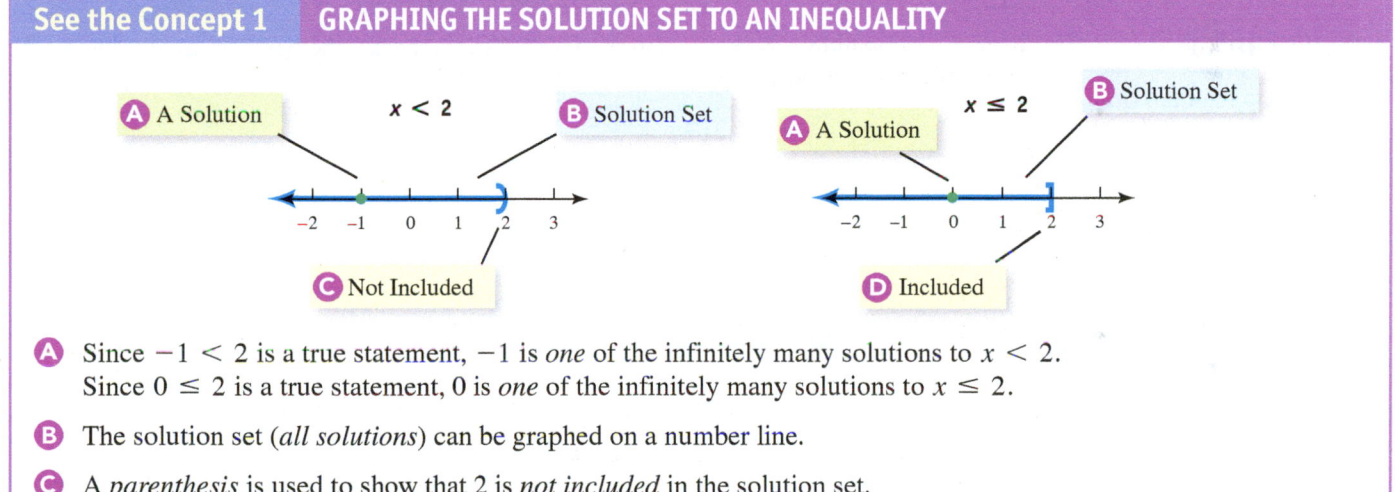

A Since $-1 < 2$ is a true statement, -1 is *one* of the infinitely many solutions to $x < 2$.
Since $0 \leq 2$ is a true statement, 0 is *one* of the infinitely many solutions to $x \leq 2$.

B The solution set (*all solutions*) can be graphed on a number line.

C A *parenthesis* is used to show that 2 is *not included* in the solution set.

D A *bracket* is used to show that 2 is *included* in the solution set.

WATCH VIDEO IN MML.

EXAMPLE 1   Graphing inequalities on a number line

Use a number line to graph the solution set to each inequality.
(a) $x > 0$ (b) $x \geq 0$ (c) $x \leq -1$ (d) $x < 3$

Solution
(a) First locate $x = 0$ (or the origin) on a number line. Numbers greater than 0 are located to the right of the origin, so shade the number line to the right of the origin. Because the inequality is $x > 0$, the number 0 is not included, so place a parenthesis "(" at 0, as shown in **FIGURE 2.18(a)**.
(b) **FIGURE 2.18(b)** is similar to the graph in part (a) except that a bracket "[" is placed at the origin because the inequality symbol is $\geq$ and 0 is included in the solution set.

(a) $x > 0$ (b) $x \geq 0$

FIGURE 2.18

(c) First locate $x = -1$ on the number line. Numbers less than -1 are located to the left of -1. Because -1 is included, a bracket "]" is placed at -1, as shown in **FIGURE 2.18(c)**.

(d) Real numbers less than 3 are graphed in **FIGURE 2.18(d)**.

(c) $x \le -1$ (d) $x < 3$

FIGURE 2.18

Now Try Exercises 19, 21, 23, 25

CHECKING SOLUTIONS We can check possible solutions to an inequality in the same way that we checked possible solutions to an equation. For example, to check whether 5 is a solution to the equation $2x + 3 = 13$, we substitute **5** for x in the equation.

$$2(\mathbf{5}) + 3 \overset{?}{=} 13 \qquad \text{Replace } x \text{ with 5.}$$
$$13 = 13 \checkmark \qquad \text{A true statement}$$

READING CHECK 1

• How many solutions do inequalities often have?

Thus 5 is a solution to this equation. Similarly, to check whether 7 is a solution to the inequality $2x + 3 > 13$, we substitute **7** for x in the inequality.

$$2(\mathbf{7}) + 3 \overset{?}{>} 13 \qquad \text{Replace } x \text{ with 7.}$$
$$17 > 13 \checkmark \qquad \text{A true statement}$$

Thus 7 is a solution to the inequality.

EXAMPLE 2   **Checking possible solutions**

Determine whether the given value of x is a solution to the inequality.
(a) $3x - 4 < 10$, $x = 6$ (b) $4 - 2x \le 8$, $x = -2$

Solution
(a) Substitute **6** for x and simplify.

$$3(\mathbf{6}) - 4 \overset{?}{<} 10 \qquad \text{Replace } x \text{ with 6.}$$
$$14 < 10 \ \textcolor{red}{\times} \qquad \text{A false statement}$$

Thus 6 is *not* a solution to the inequality.

(b) Substitute **−2** for x and simplify.

$$4 - 2(\mathbf{-2}) \overset{?}{\le} 8 \qquad \text{Replace } x \text{ with } -2.$$
$$8 \le 8 \checkmark \qquad \text{A true statement}$$

Thus -2 is a solution to the inequality.

Now Try Exercises 13, 15

2 ▸ Solving Linear Inequalities with Tables

Just as solving a linear equation means finding the value of the variable that makes the equation true, solving a linear inequality means finding the *values* of the variable that make the inequality true.

Making a table is an organized way of checking possible values of the variable to see if there are values that make an inequality true. In the next example, we use a table to find solutions to an equation and related inequalities.

| EXAMPLE 3 | **Finding solutions to equations and inequalities** |

In **TABLE 2.9**, the expression $2x - 6$ has been evaluated for several values of x. Use the table to determine any solutions to each equation or inequality.

(a) $2x - 6 = 0$ (b) $2x - 6 > 0$ (c) $2x - 6 \geq 0$ (d) $2x - 6 < 0$

Solving an Inequality with a Table

x	0	1	2	3	4	5	6
$2x - 6$	-6	-4	-2	0	2	4	6

TABLE 2.9

Solution
(a) From **TABLE 2.9**, $2x - 6$ equals **0** when $x = 3$.
(b) The values of x in the table that make the expression $2x - 6$ greater than 0 are 4, 5, and 6. These values are all greater than 3, which is the solution found in part (a). It follows that $2x - 6 > 0$ when $x > 3$.
(c) The values of x in the table that make the expression $2x - 6$ greater than or equal to 0 are 3, 4, 5, and 6. It follows that $2x - 6 \geq 0$ when $x \geq 3$.
(d) The expression $2x - 6$ is less than 0 when $x < 3$.

Now Try Exercise 33

3 ▸ Writing Solution Sets Using Interval Notation

Rather than always graphing a solution set on a number line, a convenient notation called **interval notation** can be used to represent the solutions to an inequality. The following See the Concept shows how interval notation is written.

| See the Concept 2 | **INTERVAL NOTATION** |

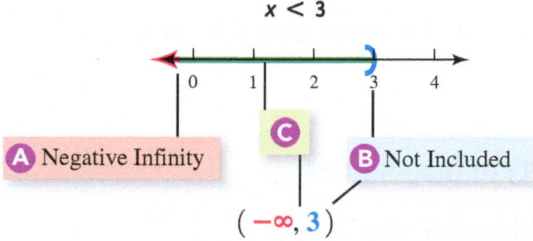

$x < 3$

Interval Notation

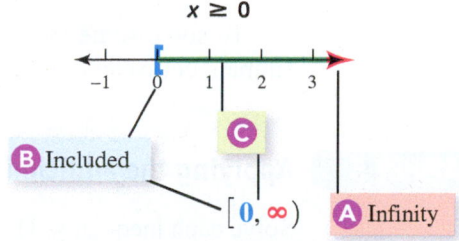

$x \geq 0$

Interval Notation

Ⓐ The left arrow corresponds to negative infinity, denoted $-\infty$ in interval notation. The right arrow corresponds to infinity, denoted ∞ in interval notation.

Ⓑ The *parenthesis* used in both the graph and interval notation shows that 3 is *not included*. The *bracket* used in both the graph and interval notation shows that 0 is *included*.

Ⓒ A comma is used in interval notation to separate the two ends of the interval.

WATCH VIDEO IN MML

NOTE: In interval notation, a *parenthesis* is always used with ∞ or $-\infty$. ∎

EXAMPLE 4   **Writing solution sets in interval notation**

Write the solution set to each inequality in interval notation.
(a) $x > 4$ **(b)** $y \le -3$ **(c)** $z \ge -1$

Solution
(a) Real numbers greater than 4 are represented by the interval $(4, \infty)$.
(b) Real numbers less than or equal to -3 are represented by the interval $(-\infty, -3]$.
(c) The solution set is represented by the interval $[-1, \infty)$.

Now Try Exercises 41, 43, 45

4 Applying the Addition Property of Inequalities

Connecting Concepts with Your Life Suppose that the speed limit on a country road is 55 miles per hour, and this is 25 miles per hour faster than the speed limit in town. If x represents lawful speeds in town, then x satisfies the inequality

$$x + 25 \le 55.$$

To solve this inequality, we **add** -25 to (or subtract 25 from) each side of the inequality.

$$x + 25 + (-25) \le 55 + (-25) \qquad \text{Add } -25 \text{ to each side.}$$
$$x \le 30 \qquad\qquad\qquad \text{Add the real numbers.}$$

Thus drivers are obeying the speed limit in town when they travel at 30 miles per hour or less. To solve this inequality, the addition property of inequalities was used.

> ### ADDITION PROPERTY OF INEQUALITIES
>
> Let a, b, and c be expressions that represent real numbers. The inequalities
>
> $$a < b \quad \text{and} \quad a + c < b + c$$
>
> are equivalent. That is, the same number may be added to (or subtracted from) each side of an inequality. Similar properties exist for $>$, $\le$, and $\ge$.

To solve some inequalities, we apply the addition property of inequalities to obtain a simpler, equivalent inequality.

EXAMPLE 5   **Applying the addition property of inequalities**

Solve each inequality. Then graph the solution set.
(a) $x - 1 > 4$ **(b)** $3 + 2x \le 5 + x$

Solution
(a) Begin by adding 1 to each side of the inequality.

$$x - 1 > 4 \qquad\qquad \text{Given inequality}$$
$$x - 1 + 1 > 4 + 1 \qquad \text{Add 1 to each side.}$$
$$x > 5 \qquad\qquad\quad \text{Add the real numbers.}$$

The solution set is given by $x > 5$ and is graphed as follows.

(b) Begin by subtracting x from (or adding $-x$ to) each side of the inequality.

$$3 + 2x \leq 5 + x \qquad \text{Given inequality}$$

$$3 + 2x - x \leq 5 + x - x \qquad \text{Subtract } x \text{ from each side.}$$

$$3 + x \leq 5 \qquad \text{Combine like terms.}$$

$$3 + x - 3 \leq 5 - 3 \qquad \text{Subtract 3 from each side.}$$

$$x \leq 2 \qquad \text{Subtract the real numbers.}$$

The solution set is given by $x \leq 2$ and is graphed as follows.

Now Try Exercises 47, 51

The following See the Concept illustrates the addition property of inequalities by using a scale.

See the Concept 3   THE ADDITION PROPERTY OF INEQUALITIES

To understand the addition property of inequalities, we can think of an inequality as a scale that is not balanced. Just as a scale remains *in the same position* when the same amount of weight is added to (or subtracted from) each side, an inequality remains *true* (and an equivalent inequality results) when the same number is added to (or subtracted from) each side.

True Inequality	*Equivalent Inequality*	*Different Position*
$1 < 2$	$1 + 1 < 2 + 1$	$1 + 2 > 2 + 0$
$<$	$<$	$>$

WATCH VIDEO IN $\boxed{\text{MML}}$.

EXAMPLE 6   Applying the addition property of inequalities

Solve $5 + \frac{1}{2}x \leq 3 - \frac{1}{2}x$. Then graph the solution set.

Solution
Begin by subtracting 5 from each side of the inequality.

$$5 + \frac{1}{2}x \leq 3 - \frac{1}{2}x \qquad \text{Given inequality}$$

$$5 + \frac{1}{2}x - 5 \leq 3 - \frac{1}{2}x - 5 \qquad \text{Subtract 5 from each side.}$$

$$\frac{1}{2}x \leq -\frac{1}{2}x - 2 \qquad \text{Subtract real numbers.}$$

$$\frac{1}{2}x + \frac{1}{2}x \leq -\frac{1}{2}x + \frac{1}{2}x - 2 \qquad \text{Add } \tfrac{1}{2}x \text{ to each side.}$$

$$x \leq -2 \qquad \text{Combine like terms.}$$

The solution set is given by $x \leq -2$ and is graphed as follows.

Now Try Exercise 53

STUDY TIP

If you miss something in class, section video lectures provide a short lecture for each section in this text. These videos are available as streaming videos and video podcasts within MyMathLab.

5 ▸ Applying the Multiplication Property of Inequalities

The multiplication property of inequalities is not exactly the same as the multiplication property of *equality*. When we multiply each side of an inequality by the same nonzero number, we may need to reverse the inequality symbol to make sure that the resulting inequality remains true. **TABLE 2.10** shows various results that occur when each side of a true inequality is multiplied by the same nonzero number.

Determining When the Inequality Symbol Should Be Reversed

True Statement	Multiply Each Side By	Resulting Inequality	Is the Result True or False?	Reverse the Inequality Symbol
$-3 < 5$	4	$-12 \overset{?}{<} 20$	True	Not needed
$7 > -1$	-2	$-14 \overset{?}{>} 2$	**False**	$-14 < 2$
$-2 > -5$	3	$-6 \overset{?}{>} -15$	True	Not needed
$4 < 9$	-11	$-44 \overset{?}{<} -99$	**False**	$-44 > -99$

TABLE 2.10

TABLE 2.10 indicates that the inequality symbol must be reversed when each side of the given inequality is *multiplied by a negative number*. This result is summarized below.

READING CHECK 2

• When solving an inequality, when does it become necessary to reverse the inequality symbol?

> **MULTIPLICATION PROPERTY OF INEQUALITIES**
>
> Let a, b, and c be expressions that represent real numbers with $c \neq 0$.
>
> 1. If $c > 0$, then the inequalities $a < b$ and $ac < bc$ are equivalent. That is, each side of an inequality may be multiplied (or divided) by the same positive number.
> 2. If $c < 0$, then the inequalities $a < b$ and $ac > bc$ are equivalent. That is, each side of an inequality may be multiplied (or divided) by the same negative number, provided the inequality symbol is reversed.
>
> Note that similar properties exist for $\leq$ and $\geq$.

NOTE: Remember to reverse the inequality symbol when either multiplying or *dividing* by a negative number. ■

EXAMPLE 7 **Applying the multiplication property of inequalities**

Solve each inequality. Then graph the solution set.

(a) $3x < 18$ **(b)** $-7 \leq -\frac{1}{2}x$

Solution
(a) To solve for x, divide each side by 3.

$$3x < 18 \quad \text{Given inequality}$$
$$\frac{3x}{3} < \frac{18}{3} \quad \text{Divide each side by 3.}$$
$$x < 6 \quad \text{Simplify fractions.}$$

The solution set is given by $x < 6$ and is graphed as follows.

(b) To isolate x in $-7 \leq -\frac{1}{2}x$, multiply each side by -2 and reverse the inequality symbol.

$$-7 \leq -\frac{1}{2}x \qquad \text{Given inequality}$$

$$-2(-7) \geq -2\left(-\frac{1}{2}\right)x \qquad \text{Multiply by } -2 \text{; reverse the inequality.}$$

$$14 \geq 1 \cdot x \qquad \text{Multiply the real numbers.}$$

$$x \leq 14 \qquad \text{Rewrite the inequality.}$$

The solution set is given by $x \leq 14$ and is graphed as follows.

Now Try Exercises 55, 57

6 Writing Solution Sets Using Set-Builder Notation

Because $x \leq 14$ is an inequality with infinitely many solutions and is not itself a set of solutions, a notation called **set-builder notation** has been devised for writing the solutions to an inequality as a set. For example, the solution set consisting of "all real numbers x such that x is less than or equal to 14" can be written as $\{x \mid x \leq 14\}$. The vertical line segment in this notation "$\mid$" is read "such that."

In the next example, the solution sets are expressed in set-builder notation. However, this notation is not widely used throughout this text.

EXAMPLE 8 **Applying both properties of inequalities**

Solve each inequality. Write the solution set in set-builder notation.
(a) $4x - 7 \geq -6$ **(b)** $-8 + 4x \leq 5x + 3$ **(c)** $0.4(2x - 5) < 1.1x + 2$

Solution
(a) Start by adding 7 to each side.

$$4x - 7 \geq -6 \qquad \text{Given inequality}$$

$$4x - 7 + 7 \geq -6 + 7 \qquad \text{Add 7 to each side.}$$

$$4x \geq 1 \qquad \text{Add real numbers.}$$

$$\frac{4x}{4} \geq \frac{1}{4} \qquad \text{Divide each side by 4.}$$

$$x \geq \frac{1}{4} \qquad \text{Simplify.}$$

In set-builder notation, the solution set is $\left\{x \mid x \geq \frac{1}{4}\right\}$.
(b) Begin by adding 8 to each side.

$$-8 + 4x \leq 5x + 3 \qquad \text{Given inequality}$$

$$-8 + 4x + 8 \leq 5x + 3 + 8 \qquad \text{Add 8 to each side.}$$

$$4x \leq 5x + 11 \qquad \text{Add real numbers.}$$

$$4x - 5x \leq 5x + 11 - 5x \qquad \text{Subtract } 5x \text{ from each side.}$$

$$-x \leq 11 \qquad \text{Combine like terms.}$$

$$-1 \cdot (-x) \geq -1 \cdot 11 \qquad \text{Multiply by } -1 \text{; reverse the inequality.}$$

$$x \geq -11 \qquad \text{Simplify.}$$

The solution set is $\{x \mid x \geq -11\}$.

(c) Begin by applying the distributive property.

$$0.4(2x - 5) < 1.1x + 2 \quad \text{Given inequality}$$
$$0.8x - 2 < 1.1x + 2 \quad \text{Distributive property}$$
$$0.8x < 1.1x + 4 \quad \text{Add 2 to each side.}$$
$$-0.3x < 4 \quad \text{Subtract } 1.1x \text{ from each side.}$$
$$x > -\frac{4}{0.3} \quad \text{Divide by } -0.3; \text{ reverse the inequality.}$$

Since $-\frac{4}{0.3} = -\frac{40}{3}$, the solution set is $\left\{ x \mid x > -\frac{40}{3} \right\}$.

Now Try Exercises 89, 91, 93

CRITICAL THINKING 1

Solve

$$-5 - 3x > -2x + 7$$

without having to reverse the inequality symbol.

7 Translating Words to Inequalities

To solve applications involving inequalities, we often have to translate words or phrases to mathematical statements. **TABLE 2.11** lists words and phrases that are associated with each inequality symbol.

Words and Phrases Associated with Inequality Symbols

Symbol	Associated Words and Phrases
$>$	greater than, more than, exceeds, above, over
$<$	less than, fewer than, below, under
$\geq$	greater than or equal to, at least, is not less than
$\leq$	less than or equal to, at most, does not exceed

TABLE 2.11

EXAMPLE 9 **Translating words to inequalities**

Translate each phrase to an inequality. Let the variable be x.
(a) A number that is more than 30
(b) An age that is at least 18
(c) A grade point average that is at most 3.25

Solution
(a) The inequality $x > 30$ represents a number x that is more than 30.
(b) The inequality $x \geq 18$ represents an age x that is at least 18.
(c) The inequality $x \leq 3.25$ represents a grade point average x that is at most 3.25.

Now Try Exercises 95, 97, 101

8 Solving Applications Involving Inequalities

🌐 *Math in Weather Context* In the atmosphere, the air temperature generally becomes colder as the altitude increases. One mile above Earth's surface the temperature is about 19 °F colder than the ground-level temperature. As the air cools, there is an increased chance of clouds forming. In the next example we estimate the altitudes where clouds may form. (*Source:* A. Miller and R. Anthes, *Meteorology.*)

EXAMPLE 10   **Finding the altitude of clouds**

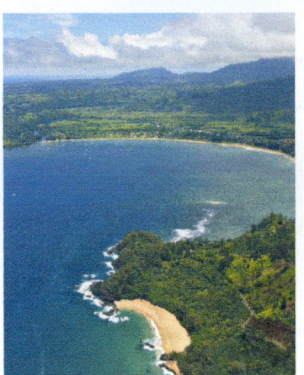

If the ground temperature is 79 °F, then the temperature T above Earth's surface is given by the formula $T = 79 - 19x$, where x is the altitude in miles. Suppose that clouds form only where the temperature is 3 °F or colder. Solve an inequality to determine the heights at which clouds may form.

Solution
Clouds may form when $T \leq 3$ degrees. Since $T = 79 - 19x$, we can substitute the expression $79 - 19x$ for T to obtain the inequality $79 - 19x \leq 3$.

$$79 - 19x \leq 3 \qquad \text{Inequality to be solved}$$
$$-19x \leq -76 \qquad \text{Subtract 79 from each side.}$$
$$\frac{-19x}{-19} \geq \frac{-76}{-19} \qquad \text{Divide by } -19; \text{ reverse the inequality.}$$
$$x \geq 4 \qquad \text{Simplify the fractions.}$$

Clouds may form at 4 miles or higher.

Now Try Exercise 115

EXAMPLE 11   **Calculating revenue, cost, and profit**

For a computer company, the cost to produce one high-end laptop computer (variable cost) is $1320 plus a one-time cost (fixed cost) of $200,000 for research and development. The revenue received from selling one laptop computer is $1850.
(a) Write a formula that gives the cost C of producing x laptop computers.
(b) Write a formula that gives the revenue R from selling x laptop computers.
(c) Profit equals revenue minus cost. Write a formula that calculates the profit P from selling x laptop computers.
(d) How many computers need to be sold to yield a positive profit?

Solution
(a) The cost of producing the first laptop is

$$1320 \times 1 + 200{,}000 = \$201{,}320.$$

The cost of producing two laptops is

$$1320 \times 2 + 200{,}000 = \$202{,}640.$$

And, in general, the cost of producing x laptops is

$$1320 \times x + 200{,}000 = 1320x + 200{,}000.$$

Thus $C = 1320x + 200{,}000$.
(b) Because the company receives $1850 for each laptop, the revenue for x laptops is given by $R = 1850x$.
(c) Profit equals revenue minus cost, so

$$P = R - C$$
$$= 1850x - (1320x + 200{,}000)$$
$$= 530x - 200{,}000.$$

Thus $P = 530x - 200{,}000$.

(d) To determine how many laptops need to be sold to yield a positive profit, we must solve the inequality $P > 0$.

$$530x - 200{,}000 > 0 \qquad \text{Inequality to be solved}$$
$$530x > 200{,}000 \qquad \text{Add 200,000 to each side.}$$
$$x > \frac{200{,}000}{530} \qquad \text{Divide each side by 530.}$$

Because $\frac{200{,}000}{530} \approx 377.4$, the company must sell at least 378 laptops. Note that the company cannot sell a fraction of a laptop.

Now Try Exercise 111

2.6   Putting It All Together

CONCEPT	COMMENTS	EXAMPLES
Linear Inequality	If the equals sign in a linear equation is replaced with $<$, $>$, $\leq$, or $\geq$, a linear inequality results.	*Linear Equation* *Linear Inequality* $\quad 4x - 1 = 0 \qquad 4x - 1 > 0$ $\quad 2 - x = 3x \qquad 2 - x \leq 3x$ $4(x+3) = 1 - x \quad 4(x+3) < 1 - x$
Solution to an Inequality	A value for the variable that makes the inequality a true statement	6 is a solution to $2x > 5$ because $2(6) > 5$ is a true statement.
Solution Set to an Inequality	The set of all solutions to an inequality	The solution set to $x + 1 > 5$ is given by $x > 4$ and can be written in set-builder notation as $\{x \mid x > 4\}$.
Number Line Graphs	The solutions to an inequality can be graphed on a number line.	$x < 2$ is graphed as follows. $x \geq -1$ is graphed as follows.
Addition Property of Inequalities	$a < b$ is equivalent to $a + c < b + c$, where a, b, and c represent real number expressions.	$x - 5 \geq 6 \qquad$ Given inequality $x \geq 11 \qquad$ Add 5. $3x > 5 + 2x \quad$ Given inequality $x > 5 \qquad$ Subtract $2x$.
Multiplication Property of Inequalities	$a < b$ is equivalent to $ac < bc$ when $c > 0$, and is equivalent to $ac > bc$ when $c < 0$.	$\dfrac{1}{2}x \geq 6 \qquad$ Given inequality $x \geq 12 \qquad$ Multiply by 2. $-3x > 5 \qquad$ Given inequality $x < -\dfrac{5}{3} \qquad$ Divide by -3; reverse the inequality symbol.

CONCEPT	COMMENTS	EXAMPLES
Interval Notation	A notation that can be used to identify the solution set to an inequality as an interval of real numbers	The inequality $x < 5$ is written in interval notation as $(-\infty, 5)$. The inequality $x \geq -1$ is written in interval notation as $[-1, \infty)$.
Set-Builder Notation	A notation that can be used to identify the solution set to an inequality as a set of real numbers	The solution set for $x - 2 < 5$ can be written as $\{x \mid x < 7\}$ and is read "the set of real numbers x such that x is less than 7."

2.6   Exercises

MyMathLab®

CONCEPTS AND VOCABULARY

1. A linear inequality results whenever the equals sign in a linear equation is replaced by any one of the symbols _____, _____, _____ or _____.

2. Equality is the boundary between _____ and _____.

3. A(n) _____ is a value of the variable that makes an inequality statement true.

4. Two linear inequalities are _____ if they have the same solution set.

5. (True or False?) When a linear equation is solved, the solution set contains one solution.

6. (True or False?) When a linear inequality is solved, the solution set contains infinitely many solutions.

7. The solution set to a linear inequality can be graphed by using a(n) _____.

8. The addition property of inequalities states that if $a > b$, then $a + c$ _____ $b + c$.

9. The multiplication property of inequalities states that if $a < b$ and $c > 0$, then ac _____ bc.

10. The multiplication property of inequalities states that if $a < b$ and $c < 0$, then ac _____ bc.

GRAPHING SOLUTION SETS ON NUMBER LINES

Exercises 11–18: Determine whether the given value of the variable is a solution to the inequality.

11. $x + 5 > 5$, $x = 4$

12. $-3x \leq -8$, $x = -2$

13. $5x \geq 25$, $x = 3$

14. $4y - 3 \leq 5$, $y = -3$

15. $3y + 5 \geq -8$, $y = -3$

16. $-(z + 7) > 3(6 - z)$, $z = 2$

17. $5(z + 1) < 3z - 7$, $z = -4$

18. $\frac{3}{2}t - \frac{1}{2} \geq 1 - t$, $t = \frac{3}{5}$

Exercises 19–26: Use a number line to graph the solution set to the inequality.

19. $x < 0$ 20. $x > -2$

21. $x > 1$ 22. $x < -\frac{5}{2}$

23. $x \leq 1.5$ 24. $x \geq -3$

25. $z \geq -2$ 26. $z \leq -\pi$

Exercises 27–32: Express the set of real numbers graphed on the number line as an inequality.

27.   28.

29.   30.

31.   32.

SOLVING LINEAR INEQUALITIES WITH TABLES

Exercises 33–36: Use the table to solve the inequality.

33. $3x + 6 > 0$

x	-4	-3	-2	-1	0
$3x + 6$	-6	-3	0	3	6

34. $6 - 3x \leq 0$

x	1	2	3	4	5
$6 - 3x$	3	0	−3	−6	−9

35. $-2x + 7 > 5$

x	−1	0	1	2	3
$-2x + 7$	9	7	5	3	1

36. $5(x - 3) \leq 4$

x	3.2	3.4	3.6	3.8	4
$5(x - 3)$	1	2	3	4	5

Exercises 37–40: Complete the table. Then use the table to solve the inequality.

37. $-2x + 6 \leq 0$

x	1	2	3	4	5
$-2x + 6$	4				

38. $3x - 1 < 8$

x	0	1	2	3	4
$3x - 1$	−1				

39. $5 - x > x + 7$

x	−3	−2	−1	0	1
$5 - x$	8				4
$x + 7$	4				8

40. $2(3 - x) \geq -3(x - 2)$

x	−2	−1	0	1	2
$2(3 - x)$					
$-3(x - 2)$					

WRITING INTERVAL NOTATION

Exercises 41–46: Write the solution set to the inequality in interval notation.

41. $x \geq 6$

42. $x < 3$

43. $y > -2$

44. $y \geq 1$

45. $z \leq 7$

46. $z < -5$

APPLYING THE ADDITION PROPERTY OF INEQUALITIES

Exercises 47–54: Use the addition property of inequalities to solve the inequality. Graph the solution set.

47. $x - 3 > 0$

48. $x + 6 < 3$

49. $3 - y \leq 5$

50. $8 - y \geq 10$

51. $12 + z < 4 + 2z$

52. $2z \leq z + 17$

53. $5 - 2t \geq 10 - t$

54. $-2t > -3t + 1$

APPLYING THE MULTIPLICATION PROPERTY OF INEQUALITIES

Exercises 55–62: Solve the inequality using the multiplication property of inequalities. Graph the solution set.

55. $2x < 10$

56. $3x > 9$

57. $-\frac{1}{2}t \geq 1$

58. $-5t \leq -6$

59. $\frac{3}{4} > -5y$

60. $10 \geq -\frac{1}{7}y$

61. $-\frac{2}{3} \leq \frac{1}{7}z$

62. $-\frac{3}{10}z < 11$

SOLVING LINEAR INEQUALITIES SYMBOLICALLY

Exercises 63–88: Solve the linear inequality. First write all solutions for x as an inequality. Then use interval notation to write to the solution set.

63. $3x + 1 < 22$

64. $4 + 5x \leq 9$

65. $5 - \frac{3}{4}x \geq 6$

66. $10 - \frac{2}{5}x > 0$

67. $45 > 6 - 2x$

68. $69 \geq 3 - 11x$

69. $5x - 2 \leq 3x + 1$

70. $12x + 1 < 25 - 3x$

71. $-x + 24 < x + 23$

72. $6 - 4x \leq x + 1$

73. $-(x + 1) \geq 3(x - 2)$

74. $5(x + 2) > -2(x - 3)$

75. $3(2x + 1) > -(5 - 3x)$

76. $4x \geq -3(7 - 2x) + 1$

77. $1.6x + 0.4 \leq 0.4x$

78. $-5.1x + 1.1 < 0.1 - 0.1x$

79. $0.8x - 0.5 < x + 1 - 0.5x$

80. $0.1(x + 1) - 0.1 \leq 0.2x - 0.5$

81. $-\frac{1}{2}\left(\frac{2}{3}x + 4\right) \geq x$

82. $-5x > \frac{4}{5}\left(\frac{10}{3}x + 10\right)$

83. $\frac{3}{7}x + \frac{2}{7} > -\frac{1}{7}x - \frac{5}{14}$

84. $\frac{5}{6} - \frac{1}{3}x \geq -\frac{1}{3}\left(\frac{5}{6}x - 1\right)$

85. $\frac{x}{3} + \frac{5x}{6} \leq \frac{2}{3}$

86. $\frac{3x}{4} - \frac{x}{2} < 1$

87. $\frac{6x}{7} < \frac{1}{3}x + 1$

88. $\frac{5x}{8} - \frac{3x}{4} \leq 8$

WRITING SOLUTION SETS USING SET-BUILDER NOTATION

Exercises 89–94: Solve the linear inequality and write the solution in set-builder notation.

89. $x + 6 > 7$

90. $x + 4 < 1$

91. $-3x \le 21$

92. $4x \ge -20$

93. $2x - 3 < 9$

94. $-5x + 4 < 44$

TRANSLATING WORDS TO INEQUALITIES

Exercises 95–102: Translate each phrase to an inequality. Let x be the variable.

95. A speed that is greater than 60 miles per hour

96. A speed that is at most 60 miles per hour

97. An age that is at least 21 years old

98. An age that is less than 21 years old

99. A salary that is more than $40,000

100. A salary that is less than or equal to $40,000

101. A speed that does not exceed 70 miles per hour

102. A speed that is not less than 70 miles per hour

APPLICATIONS INVOLVING INEQUALITIES

103. **Geometry** Find all values for x so that the perimeter of the rectangle is less than 50 feet.

104. **Geometry** A triangle with height 12 inches is to have area less than 120 square inches. What must be true about the base of the triangle?

105. **Grade Average** A student scores 74 out of 100 on a test. If the maximum score on the next test is also 100 points, what score does the student need to maintain at least an average of 80?

106. **Grade Average** A student scores 65 and 82 on two different 100-point tests. If the maximum score on the next test is also 100 points, what score does the student need to maintain at least an average of 70?

107. **Parking Rates** Parking in a student lot costs $2 for the first half hour and $1.25 for each hour thereafter. A partial hour is charged the same as a full hour. What is the longest time that a student can park in this lot for $8?

108. **Parking Rates** Parking in a student lot costs $2.50 for the first hour and $1 for each hour thereafter. A nearby lot costs $1.25 for each hour. In both lots a partial hour is charged as a full hour. In which lot can a student park the longest for $5? For $11?

109. **Car Rental** A rental car costs $25 per day plus $0.20 per mile. If someone has $200 to spend and needs to drive the car 90 miles each day, for how many days can that person rent the car? Assume that the car cannot be rented for part of a day.

110. **Car Rental** One car rental agency charges $20 per day plus $0.25 per mile. A different agency charges $37 per day with unlimited mileage. For what mileages is the second rental agency a better deal?

111. **Revenue and Cost** (Refer to Example 11.) The cost to produce one compact disc is $1.50 plus a one-time fixed cost of $2000. The revenue received from selling one compact disc is $12.
    (a) Write a formula that gives the cost C of producing x compact discs. Be sure to include the fixed cost.
    (b) Write a formula that gives the revenue R from selling x compact discs.
    (c) Profit equals revenue minus cost. Write a formula that calculates the profit P from selling x compact discs.
    (d) How many compact discs need to be sold to yield a positive profit?

112. **Revenue and Cost** The cost to produce one laptop computer is $890 plus a one-time fixed cost of $100,000 for research and development. The revenue received from selling one laptop computer is $1520.
    (a) Write a formula that gives the cost C of producing x laptop computers.
    (b) Write a formula that gives the revenue R from selling x laptop computers.
    (c) Profit equals revenue minus cost. Write a formula that calculates the profit P from selling x laptop computers.
    (d) How many computers need to be sold to yield a positive profit?

113. **Distance and Time** Two athletes are jogging in the same direction along an exercise path. After x minutes, the first athlete's distance in miles from a parking lot is given by $\frac{1}{6}x$ and the second athlete's distance is given by $\frac{1}{8}x + 2$.
    (a) When are the athletes the same distance from the parking lot?
    (b) When is the first athlete farther from the parking lot than the second?

114. Altitude and Dew Point If the dew point on the ground is $65\,°F$, then the dew point x miles high is given by $D = 65 - 5.8x$. Determine the altitudes at which the dew point is greater than $36\,°F$. (*Source*: A. Miller.)

115. Altitude and Temperature (Refer to Example 10.) If the temperature on the ground is $90\,°F$, then the air temperature x miles high is given by $T = 90 - 19x$. Determine the altitudes at which the air temperature is less than $4.5\,°F$. (*Source*: A. Miller.)

116. Size and Weight of a Fish If the length of a bass is between 20 and 25 inches, its weight W in pounds can be estimated by the formula $W = 0.96x - 14.4$, where x is the length of the fish. (*Source*: Minnesota Department of Natural Resources.)

(a) What length of bass is likely to weigh 7.2 pounds?

(b) What lengths of bass are likely to weigh less than 7.2 pounds?

WRITING ABOUT MATHEMATICS

117. Explain each of the terms and give an example.
(a) Linear equation
(b) Linear inequality

118. Suppose that a student says that a linear equation and a linear inequality can be solved the same way. How would you respond?

| SECTIONS 2.5 and 2.6 | Checking Basic Concepts |

1. Convert each length as indicated.
(a) 4048 yards to miles
(b) 65.7 centimeters to millimeters
(c) 0.75 mile to inches
(d) 13,716 centimeters to feet

2. Convert each area as indicated.
(a) 477 square feet to square yards
(b) 5.2 square meters to square centimeters

3. Convert each capacity as indicated.
(a) 5.25 gallons to pints
(b) 3.4 liters to milliliters

4. Convert each mass as indicated.
(a) 6.7 pounds to ounces
(b) 820 milligrams to grams

5. Convert as indicated.
(a) 5 days to minutes
(b) 5256 hours to years

6. Convert as indicated.
(a) 5 miles per hour to feet per minute
(b) 85 inches per second to centimeters per second

7. Use a number line to graph the solution set to the inequality $x + 1 \geq -1$.

8. Express the set of real numbers graphed on the number line by using an inequality.

9. Determine whether -3 is a solution to the inequality $4x - 5 \leq -15$.

10. Complete the table. Then use the table to solve the inequality $5 - 2x \leq 7$.

x	-2	-1	0	1	2
$5 - 2x$					1

11. Solve each inequality.
(a) $x + 5 > 8$
(b) $-\frac{5}{7}x \leq 25$
(c) $3x \geq -2(1 - 2x) + 3$

12. Translate the phrase "a price that is not more than \$12" to an inequality using the variable x.

13. Geometry The length of a rectangle is 5 inches longer than twice its width. If the perimeter of the rectangle is more than 88 inches, find all possible widths for the rectangle.

CHAPTER 2   Summary

SECTION 2.1 ■ INTRODUCTION TO EQUATIONS

Equations An equation is a mathematical statement that two expressions are equal. Every equation contains an equals sign. An equation can be either true or false.

Important Terms

Solution	A value for a variable that makes the equation true
Solution Set	The set of all solutions
Equivalent Equations	Equations that have the same solution set
Checking a Solution	Substituting the solution in the given equation to verify that the solution makes the equation true

Example: 3 is the solution to $4x - 2 = 10$ because $4(3) - 2 = 10$ is a true statement.

Properties of Equality

Addition Property	$a = b$ is equivalent to $a + c = b + c$.

Example: $x - 3 = 0$ and $x - 3 + 3 = 0 + 3$ are equivalent equations.

Multiplication Property	$a = b$ is equivalent to $ac = bc$, provided $c \neq 0$.

Example: $2x = 5$ and $2x \cdot \frac{1}{2} = 5 \cdot \frac{1}{2}$ are equivalent equations.

SECTION 2.2 ■ LINEAR EQUATIONS

Linear Equation Can be written in the form $ax + b = 0$, where $a \neq 0$

Examples: $3x - 5 = 0$ is linear, whereas $5x^2 + 2x = 0$ is *not* linear.

Solving Linear Equations with a Table To solve a linear equation numerically, complete a table for various values of the variable and then select the solution from the table.

Example: The solution to $3x - 4 = -1$ is 1.

x	-1	0	1
$3x - 4$	-7	-4	-1

Solving Linear Equations Symbolically The following steps can be used as a guide for solving linear equations symbolically.

STEP 1: Use the distributive property to clear any parentheses on each side of the equation. Combine any like terms on each side.

STEP 2: Use the addition property of equality to get all of the terms containing the variable on one side of the equation and all other terms on the other side of the equation. Combine any like terms on each side.

STEP 3: Use the multiplication property of equality to isolate the variable by multiplying each side of the equation by the reciprocal of the number in front of the variable (or divide each side by that number).

STEP 4: Check the solution by substituting it in the given equation.

Distributive Properties $a(b + c) = ab + ac$ or $a(b - c) = ab - ac$

Examples: $5(2x + 3) = 10x + 15$ and $5(2x - 3) = 10x - 15$

Clearing Fractions and Decimals When fractions or decimals appear in an equation, multiplying each side by the least common denominator can be helpful.

Examples: Multiply each side of $\frac{1}{3}x - \frac{1}{6} = \frac{2}{3}$ by 6 to obtain $2x - 1 = 4$.

Multiply each side of $0.04x + 0.1 = 0.07$ by 100 to obtain $4x + 10 = 7$.

Number of Solutions Equations that can be written in the form $ax + b = 0$, where a and b are *any* real number, can have no solutions, one solution, or infinitely many solutions.

Examples: $x + 3 = x$ is equivalent to $3 = 0$.   (No solutions)

$2y + 1 = 9$ is equivalent to $y = 4$.   (One solution)

$z + z = 2z$ is equivalent to $0 = 0$.   (Infinitely many solutions)

SECTION 2.3 ■ INTRODUCTION TO PROBLEM SOLVING

Steps for Solving a Problem The following steps can be used for solving word problems.

STEP 1: Read the problem carefully and be sure that you understand it. (You may need to read the problem more than once.) Assign a variable to what you are being asked to find. If necessary, write other quantities in terms of this variable.

STEP 2: Write an equation that relates the quantities described in the problem. You may need to sketch a diagram or refer to known formulas.

STEP 3: Solve the equation. Use the solution to determine the solution(s) to the original problem. Include any necessary units.

STEP 4: Check your solution in the original problem. Does it seem reasonable?

Percent Problems

The expression x%   Represents the fraction $\frac{x}{100}$ or the decimal number given by $x \cdot 0.01$

Examples: $45\% = \dfrac{45}{100} = 0.45$

$7.1\% = \dfrac{7.1}{100} = \dfrac{71}{1000} = 0.071$

Changing Fractions to Percents   Divide the numerator by the denominator, multiply by 100, and insert the % symbol.

Example: Since $3 \div 4 = 0.75$, the fraction $\frac{3}{4}$ is written in percent notation as $0.75 \cdot 100 = 75\%$.

Percent Change   If a quantity changes from an old amount to a new amount, then the percent change equals $\dfrac{(\text{new amount}) - (\text{old amount})}{(\text{old amount})} \cdot 100$.

Example: If a price increases from \$2 to \$3, then the percent change equals $\frac{3 - 2}{2} \cdot 100 = 50\%$.

Distance Problems

If an object travels at speed (rate) r for time t, then the distance d traveled is calculated by $d = rt$.

Example: A car moving at 65 mph for 2 hours travels

$$d = rt = 65 \cdot 2 = 130 \text{ miles.}$$

SECTION 2.4 ■ FORMULAS

Formula A formula is an equation that can be used to calculate a quantity by using known values of other quantities.

Formulas from Geometry

Area of a Rectangle	$A = lw$, where l is the length and w is the width.
Perimeter of a Rectangle	$P = 2l + 2w$, where l is the length and w is the width.
Area of a Triangle	$A = \frac{1}{2}bh$, where b is the base and h is the height.
Degree Measure	There are $360°$ in one complete revolution.
Angle Measure	The sum of the measures of the angles in a triangle equals $180°$.
Circumference	$C = 2\pi r$, where r is the radius.
Area of a Circle	$A = \pi r^2$, where r is the radius.
Area of a Trapezoid	$A = \frac{1}{2}(a + b)h$, where h is the height and a and b are bases of the trapezoid.
Volume of a Box	$V = lwh$, where l is the length, w is the width, and h is the height.
Surface Area of a Box	$S = 2lw + 2wh + 2lh$, where l is the length, w is the width, and h is the height.
Volume of a Cylinder	$V = \pi r^2 h$, where r is the radius and h is the height.

Other Formulas

Grade Point Average (GPA) $\quad$ $\text{GPA} = \frac{4A + 3B + 2C + D}{A + B + C + D + F}$, where A represents the number of A credits earned, B the number of B credits earned, and so on.

Temperature Scales $\quad$ $F = \frac{9}{5}C + 32$ and $C = \frac{5}{9}(F - 32)$, where F is the Fahrenheit temperature and C is the Celsius temperature.

See Putting It All Together in Section 2.4 for examples.

SECTION 2.5 ■ UNIT ANALYSIS

Unit Fraction

A unit fraction is a fraction that is equivalent to 1.

Examples: $\dfrac{12 \text{ in.}}{1 \text{ ft}}$, $\dfrac{1 \text{ lb}}{16 \text{ oz}}$, and $\dfrac{2 \text{ c}}{1 \text{ pt}}$

Units of Length

Inch, foot, yard, mile, millimeter, centimeter, meter, kilometer

Relationships: 1 ft $=$ 12 in., 1 yd $=$ 36 in., 1 yd $=$ 3 ft,

1 mi $=$ 1760 yd, 1 mi $=$ 5280 ft

1 cm $=$ 10 mm, 1 m $=$ 100 cm, 1 m $=$ 1000 mm, 1 km $=$ 1000 m

1 in. $=$ 2.54 cm, 1 m $\approx$ 1.09 yd, 1 km $\approx$ 0.62 mi

Example: Convert 7 meters to yards.

$$7 \text{ m} \approx \frac{7 \ \cancel{\text{m}}}{1} \cdot \frac{1.09 \text{ yd}}{1 \ \cancel{\text{m}}} = 7.63 \text{ yd}$$

Units of Area

Units of area are the squares of units of length. To convert one unit of area to another, use the corresponding unit fraction relating units of length, twice.

Example: Convert 3 square feet to square inches.

$$3 \text{ ft}^2 = \frac{3 \cancel{\text{ft}} \cdot \cancel{\text{ft}}}{1} \cdot \frac{12 \text{ in}}{1 \cancel{\text{ft}}} \cdot \frac{12 \text{ in}}{1 \cancel{\text{ft}}} = 432 \text{ in}^2$$

Units of Capacity and Volume

Ounce, cup, pint, quart, gallon, milliliter, liter

Relationships: 1 c = 8 oz, 1 pt = 2 c, 1 qt = 2 pt, 1 gal = 4 qt

1 L = 1000 mL, 1 L ≈ 1.06 qt, 1 oz ≈ 29.57 mL

Examples: Convert 6 cups to pints.

$$6 \text{ c} = \frac{6 \cancel{\text{c}}}{1} \cdot \frac{1 \text{ pt}}{2 \cancel{\text{c}}} = \frac{6}{2} \text{ pt} = 3 \text{ pt}$$

Convert 5 liters to quarts.

$$5 \text{ L} \approx \frac{5 \cancel{\text{L}}}{1} \cdot \frac{1.06 \text{ qt}}{1 \cancel{\text{L}}} = 5.3 \text{ qt}$$

Units of Mass (Weight)

Ounce, pound, ton, milligram, gram, kilogram

Relationships: 1 lb = 16 oz, 1 T = 2000 lb, 1 g = 1000 mg,

1 kg = 1000 g, 1 kg ≈ 2.2 lb, 1 oz ≈ 28.35 g

Examples: Convert 3.5 tons to pounds.

$$3.5 \text{ T} = \frac{3.5 \cancel{\text{T}}}{1} \cdot \frac{2000 \text{ lb}}{1 \cancel{\text{T}}} = 7000 \text{ lb}$$

Convert 8.8 pounds to kilograms.

$$8.8 \text{ lb} \approx \frac{8.8 \cancel{\text{lb}}}{1} \cdot \frac{1 \text{ kg}}{2.2 \cancel{\text{lb}}} = 4 \text{ kg}$$

Units of Time

Second, minute, hour, day, week, month, year

Relationships: 1 min = 60 sec, 1 hr = 60 min, 1 d = 24 hr, 1 wk = 7 d, 1 yr = $365\frac{1}{4}$ d (or 365.25 d)

Example: Convert 105 minutes to hours.

$$105 \text{ min} = \frac{105 \cancel{\text{min}}}{1} \cdot \frac{1 \text{ hr}}{60 \cancel{\text{min}}} = 1.75 \text{ hr}$$

Units of Speed

A unit of speed is expressed as a ratio with a distance in the numerator and time in the denominator.

Example: Convert 450 feet per minute to yards per minute.

$$450 \text{ ft/min} = \frac{450 \cancel{\text{ft}}}{1 \text{ min}} \cdot \frac{1 \text{ yd}}{3 \cancel{\text{ft}}} = 150 \text{ yd/min}$$

SECTION 2.6 ■ LINEAR INEQUALITIES

Linear Inequality When the equals sign in a linear equation is replaced with any one of the symbols $<$, $\leq$, $>$, or $\geq$, a linear inequality results.

Examples: $x > 0$, $6 - \frac{2}{3}x \leq 7$, and $4(x - 1) < 3x - 1$

Solution to an Inequality Any value for the variable that makes the inequality a true statement

Example: 3 is a solution to $2x < 9$ because $2(3) < 9$ is a true statement.

Number Line Graphs A number line can be used to graph the solution set to a linear inequality.

Example: The graph of $x \leq 1$ is shown in the figure.

Properties of Inequalities

Addition Property

$a < b$ is equivalent to $a + c < b + c$.

Example: $x - 3 < 0$ and $x - 3 + 3 < 0 + 3$ are equivalent inequalities.

Multiplication Property

When $c > 0$, $a < b$ is equivalent to $ac < bc$.

When $c < 0$, $a < b$ is equivalent to $ac > bc$.

Examples: $2x < 6$ is equivalent to $2x\left(\frac{1}{2}\right) < 6\left(\frac{1}{2}\right)$ or $x < 3$.

$-2x < 6$ is equivalent to $-2x\left(-\frac{1}{2}\right) > 6\left(-\frac{1}{2}\right)$ or $x > -3$.

CHAPTER 2   Review Exercises

SECTION 2.1

Exercises 1–8: Solve the equation. Check your solution.

1. $x + 9 = 3$

2. $x - 4 = -2$

3. $x - \frac{3}{4} = \frac{3}{2}$

4. $x + 0.5 = 0$

5. $4x = 12$

6. $3x = -7$

7. $-0.5x = 1.25$

8. $-\frac{1}{3}x = \frac{7}{6}$

SECTION 2.2

Exercises 9 and 10: Decide whether the equation is linear. If the equation is linear, give values for a and b so that it can be written in the form $ax + b = 0$.

9. $-4x + 3 = 2$

10. $\frac{3}{8}x^2 - x = \frac{1}{4}$

Exercises 11–20: Solve the equation. Check the solution.

11. $4x - 5 = 3$

12. $7 - \frac{1}{2}x = -4$

13. $5(x - 3) = 12$

14. $3 + x = 2x - 4$

15. $2(x - 1) = 4(x + 3)$

16. $1 - (x - 3) = 6 + 2x$

17. $3.4x - 4 = 5 - 0.6x$

18. $2.1x - 0.3(2 - x) = 0.5 + x$

19. $\frac{2}{3}x - \frac{1}{6} = \frac{5}{12}$

20. $-\frac{1}{3}(3 - 6x) = -(x + 2) + 1$

Exercises 21–24: Determine whether the equation has no solutions, one solution, or infinitely many solutions.

21. $4(3x - 2) = 2(6x + 5)$

22. $5(3x - 1) = 15x - 5$

23. $8x = 5x + 3x$

24. $9x - 2 = 8x - 2$

Exercises 25 and 26: Complete the table. Then use the table to solve the given equation.

25. $-2x + 3 = 0$

x	0.5	1.0	1.5	2.0	2.5
$-2x + 3$	2				

26. $-(x + 1) + 3 = 2$

x	-2	-1	0	1	2
$-(x + 1) + 3$					

SECTION 2.3

Exercises 27–30: Using the variable x, translate the sentence into an equation. Solve the resulting equation.

27. The product of a number and 6 is 72.

28. The sum of a number and 18 is -23.

29. Twice a number minus 5 equals the number plus 4.

30. The sum of a number and 4 equals the product of the number and 3.

Exercises 31 and 32: Find the number or numbers.

31. Five times a number divided by 3 equals 15.

32. The sum of three consecutive integers is -153.

Exercises 33–36: Convert the percentage to fraction and decimal notation.

33. 85%   **34.** 5.6%

35. 0.03%   **36.** 342%

Exercises 37–40: Convert the number to a percentage.

37. 0.89   **38.** 0.005

39. 2.3   **40.** 1

Exercises 41–44: Use the formula d = rt to find the value of the missing variable.

41. $r = 8$ miles per hour, $t = 3$ hours

42. $r = 70$ feet per second, $t = 55$ seconds

43. $d = 500$ yards, $t = 20$ seconds

44. $d = 125$ miles, $r = 15$ miles per hour

SECTION 2.4

Exercises 45 and 46: Find the area of the region shown.

45.

46.

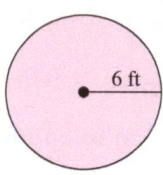

47. Find the area of a circle that has a 5-inch radius.

48. Find the perimeter of a rectangle that has a 13-inch length and a 7-inch width.

49. Find the circumference of a circle that has an 18-foot diameter.

50. Find the area of a rectangle that has a 24-inch width and a 3-foot length.

51. Find the measure of the third angle in the triangle.

52. The angles in a triangle have measures x, $3x$, and $4x$. Find the value of x.

53. If a cylinder has radius 5 inches and height 25 inches, find its volume. (*Hint:* $V = \pi r^2 h$.)

54. Find the area of a trapezoid with height 5 feet and bases 3 feet and 18 inches. (*Hint:* $A = \frac{1}{2}(a + b)h$.)

Exercises 55 and 56: Find the area of the figure shown.

55.

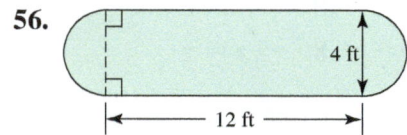

56.

Exercises 57–62: Solve the given formula for the specified variable.

57. $3x = 5 + y$ for y **58.** $16 = 2x + 2y$ for y

59. $z = 2xy$ for y **60.** $S = \dfrac{a + b + c}{3}$ for b

61. $T = \dfrac{a}{3} + \dfrac{b}{4}$ for b **62.** $cd = ab + bc$ for c

Exercises 63 and 64: Let the variable A represent the number of credits with a grade of A, B the number of credits with a grade of B, and so on. Calculate the grade point average (GPA). Round your answer to the nearest hundredth.

63. $A = 20, B = 25, C = 12, D = 4, F = 4$

64. $A = 64, B = 32, C = 20, D = 10, F = 3$

65. Convert 15 °C to an equivalent Fahrenheit temperature.

66. Convert 113 °F to an equivalent Celsius temperature.

SECTION 2.5

67. Convert each length as indicated.
   (a) 125 centimeters to meters
   (b) 1.2 miles to inches

68. Convert each area as indicated.
   (a) 333 square feet to square yards
   (b) 1.2 square miles to square feet

69. Convert each capacity as indicated.
   (a) 8.5 gallons to pints
   (b) 1200 milliliters to liters

70. Convert each mass as indicated.
   (a) 120,000 ounces to tons
   (b) 510 milligrams to grams

71. Convert as indicated.
   (a) 9 days to minutes
   (b) 315,360 seconds to days

72. Convert as indicated.
   (a) 60 miles per hour to feet per second
   (b) 5 inches per second to centimeters per second

SECTION 2.6

Exercises 73–76: Use a number line to graph the solution set to the inequality.

73. $x < 2$ **74.** $x > -1$

75. $y \geq -\frac{3}{2}$ **76.** $y \leq 2.5$

Exercises 77 and 78: Express the set of real numbers graphed on the number line with an inequality.

77.   **78.**

Exercises 79 and 80: Determine whether the given value of x is a solution to the inequality.

79. $1 - (x + 3) \geq x$ $x = -2$

80. $4(x + 1) < -(5 - x)$ $x = -1$

Exercises 81 and 82: Complete the table and then use the table to solve the inequality.

81. $5 - x > 3$

x	0	1	2	3	4
$5 - x$	5				

82. $2x - 5 \leq 0$

x	1	1.5	2	2.5	3
$2x - 5$					

Exercises 83–88: Solve the inequality.

83. $x - 3 > 0$

84. $-2x \leq 10$

85. $5 - 2x \geq 7$

86. $3(x - 1) < 20$

87. $5x \leq 3 - (4x + 2)$

88. $3x - 2(4 - x) \geq x + 1$

Exercises 89–92: Translate the phrase to an inequality. Let x be the variable.

89. A speed that is less than 50 miles per hour

90. A salary that is at most $45,000

91. An age that is at least 16 years old

92. A year before 1995

APPLICATIONS

93. Rainfall On a stormy day, 2 inches of rain fall before noon and $\frac{3}{4}$ inch per hour fall thereafter until 5 P.M.
   (a) Make a table that shows the total rainfall at each hour starting at noon and ending at 5 P.M.
   (b) Write a formula that calculates the rainfall R in inches, x hours past noon.
   (c) Use your formula to calculate the total rainfall at 5 P.M. Does your answer agree with the value in your table from part (a)?
   (d) How much rain had fallen at 3:45 P.M.?

94. Cost of a Laptop A 5% sales tax on a laptop computer amounted to $106.25. Find the cost of the laptop.

95. Tax Rules In 1969, there were 16,500 pages of federal tax rules; in 2011, there were about 72,536. Find the percent change in the number of pages of federal tax rules between 1969 and 2011. (*Source:* The Economist.)

96. Distance Traveled At noon, a bicyclist is 50 miles from home, riding toward home at 10 miles per hour.
(a) Make a table that shows the bicyclist's distance D from home after 1, 2, 3, 4, and 5 hours.
(b) Write a formula that calculates the distance D from home after x hours.
(c) Use your formula to determine D when $x = 3$ hours. Does your answer agree with the value shown in your table?
(d) For what times was the bicyclist at least 20 miles from home? Assume that $0 \le x \le 5$.

97. Car Speeds One car passes another car on a freeway. The faster car is traveling 12 miles per hour faster than the slower car. Determine how long it will be before the faster car is 2 miles ahead of the slower car.

98. Dimensions of a Rectangle The width of a rectangle is 10 inches less than its length. If the perimeter is 112 inches, find the dimensions of the rectangle.

99. Saline Solution A saline solution contains 3% salt. How much water should be added to 100 milliliters of this solution to dilute it to a 2% solution?

100. Investment Money A student invests two sums of money, $500 and $800, at different interest rates, receiving a total of $55 in interest after one year. The $500 investment receives an interest rate 2% lower than the interest rate for the $800 investment. Find the interest rate for each investment.

101. Geometry A triangle with height 8 inches is to have an area that is not more than 100 square inches. What lengths are possible for the base of the triangle?

102. Grade Average A student scores 75 and 91 on two different tests of 100 points. If the maximum score on the next test is also 100 points, what score does the student need to maintain an average of at least 80?

103. Parking Rates Parking in a lot costs $2.25 for the first hour and $1.25 for each hour thereafter. A partial hour is charged the same as a full hour. What is the longest time that someone can park for $9?

104. Profit The cost to produce one DVD player is $85 plus a one-time fixed cost of $150,000. The revenue received from selling one DVD player is $225.
(a) Write a formula that gives the cost C of producing x DVD players.
(b) Write a formula that gives the revenue R from selling x DVD players.
(c) Profit equals revenue minus cost. Write a formula that calculates the profit P from selling x DVD players.
(d) What numbers of DVD players sold will result in a loss? (*Hint:* A loss corresponds to a negative profit.)

CHAPTER 2   Test   MyMathLab® You Tube

Exercises 1–4: Solve the equation. Check your solution.

1. $9 = 3 - x$

2. $4x - 3 = 7$

3. $4x - (2 - x) = -3(2x + 6)$

4. $\frac{1}{12}x - \frac{2}{3} = \frac{1}{2}\left(\frac{3}{4} - \frac{1}{3}x\right)$

Exercises 5 and 6: Determine the number of solutions to the given equation.

5. $6(2x - 1) = -4(3 - 3x)$

6. $4(2x - 1) = 8x - 4$

7. Complete the table. Then use the table to solve the equation $6 - 2x = 0$.

x	0	1	2	3	4
$6 - 2x$	6				

Exercises 8 and 9: Translate the sentence into an equation, using the variable x. Then solve the resulting equation.

8. The sum of a number and -7 is 6.

9. Twice a number plus 6 equals the number minus 7.

10. The sum of three consecutive natural numbers is 336. Find the three numbers.

11. Convert 3.2% to fraction and decimal notation.

12. Convert 0.345 to a percentage.

13. Use the formula $d = rt$ to find r when $d = 200$ feet and $t = 4$ seconds.

14. Area Find the area of the triangle shown.

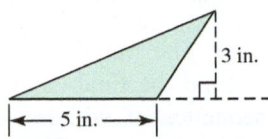

15. Find the circumference and area of a circle with a 30-inch diameter.

16. The measures of the angles in a triangle are x, $2x$, and $3x$. Find the value of x.

Exercises 17 and 18: Solve the formula for x.

17. $z = y - 3xy$

18. $R = \dfrac{x}{4} + \dfrac{y}{5}$

19. Convert 45 yards to inches.

20. Convert 10,512 hours to years.

21. Use a number line to graph the solution set to $x \leq 3$.

22. Express the set of real numbers graphed on the number line with an inequality.

Exercises 23 and 24: Solve the inequality.

23. $-3x + 9 \geq x - 15$

24. $3(6 - 5x) < 20 - x$

25. Snowfall Suppose 5 inches of snow fall before noon and 2 inches per hour fall thereafter until 10 P.M.
  (a) Write a formula that calculates the snowfall S in inches, x hours past noon.
  (b) Use your formula to calculate the total snowfall at 8 P.M.
  (c) How much snow had fallen at 6:15 P.M.?

26. Mixing Acid A solution is 45% hydrochloric acid. How much water should be added to 1000 milliliters of this solution to dilute it to a 15% solution?

27. Energy Demand In 2011, the primary energy demand for non-OECD countries was 7.4 billion tonnes of oil equivalent. By 2035, this is expected to rise to 11.4 billion tonnes. Find the percent increase in primary energy demand for non-OECD countries over this time period. (*Source: International Energy Agency.*)

CHAPTERS 1–2 Cumulative Review Exercises

Exercises 1 and 2: Classify the number as prime or composite. If the number is composite, write it as a product of prime numbers.

1. 45

2. 37

Exercises 3 and 4: Multiply or divide and then simplify to lowest terms when appropriate.

3. $\dfrac{4}{3} \cdot \dfrac{3}{8}$

4. $\dfrac{2}{3} \div 6$

Exercises 5 and 6: Add or subtract and then simplify to lowest terms when appropriate.

5. $\dfrac{11}{12} - \dfrac{3}{8}$

6. $\dfrac{2}{3} + \dfrac{1}{5}$

Exercises 7 and 8: Classify the number as one or more of the following: natural number, whole number, integer, rational number, or irrational number.

7. -1

8. $\sqrt{3}$

Exercises 9–12: Evaluate the expression by hand.

9. $15 - 4 \cdot 3$

10. $30 \div 6 \cdot 2$

11. $23 - 4^2 \div 2$

12. $11 - \dfrac{3 + 1}{6 - 4}$

Exercises 13 and 14: Simplify the expression.

13. $5x^3 - x^3$

14. $4 + 2x - 1 + 3x$

Exercises 15–18: Solve the equation.

15. $x - 3 = 11$

16. $4x - 6 = -22$

17. $5(6y + 2) = 25$

18. $11 - (y + 2) = 3y + 5$

Exercises 19 and 20: Determine whether the equation has no solutions, one solution, or infinitely many solutions.

19. $6x + 2 = 2(3x + 1)$

20. $2(3x - 4) = 6(x - 1)$

21. Find three consecutive integers whose sum is 90.

22. Convert 4.7% to decimal notation.

23. Convert 0.17 to a percentage.

24. Find the average speed of a car that travels 325 miles in 5 hours.

25. The angles in a triangle have measures $2x$, $3x$, and $4x$. Find the value of x.

26. Find the area of a circle that has a 10-inch diameter.

Exercises 27 and 28: Solve the formula for x.

27. $a = 3xy - 4$

28. $A = \dfrac{x + y + z}{3}$

Exercises 29 and 30: Solve the inequality and graph the solution set.

29. $7 - 3x > 4$

30. $6x \leq 5 - (x - 9)$

31. **Yards to Inches** Write a formula that converts Y yards to I inches.

32. **Checking Account** The initial balance in a checking account is $468. Find the final balance resulting from the following sequence of withdrawals and deposits: $-\$14$, $\$200$, $-\$73$, $-\$21$, and $\$58$.

33. **Acid Solution** How much of a 4% acid solution should be added to 150 milliliters of a 10% acid solution to dilute it to a 6% acid solution?

34. **Grocery Spending** In 2011, in the United States, about 1.25% of online spending was for groceries. This is 0.1% more than half the percent of online spending for groceries in 2016. Find the percent of online spending for groceries in 2016. (*Source:* The Economist.)

3 Graphing Equations

Since entering the 21st century, the U.S. economy has experienced both incredible growth and substantial decline. One way that economists monitor changes in the economy is by tracking median home prices. The *line graph* in the accompanying figure shows median new home prices displayed in two-year intervals. Graphs such as this one provide an excellent way to visualize data trends. In this chapter, we discuss line graphs and other types of graphs that can be used to represent mathematical information visually. *Source:* U.S. Census Bureau.

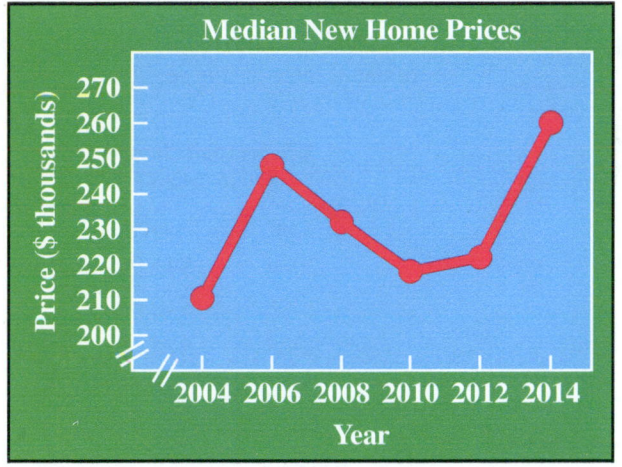

169

3.1   Introduction to Graphing

Objectives

NEW VOCABULARY

☐ Rectangular coordinate
   system (xy-plane)
☐ x-axis
☐ y-axis
☐ Origin
☐ Quadrants
☐ Ordered pair
☐ Scatterplot
☐ x-coordinate
☐ y-coordinate
☐ Line graph

STUDY TIP

Have you been completing all of the assigned homework on time? Regular and timely practice is one of the keys to having a successful experience in any math class.

1 Introducing Tables and Graphs

When data are displayed in a table, it is often difficult to recognize any trends. In order to visualize information provided by a set of data, it can be helpful to create a graph. In math, we visualize data by plotting points to make *scatterplots* and *line graphs*.

🌐 Math in Context *(Financial)* **TABLE 3.1** lists the per capita (per person) income for the United States for selected years. In **TABLE 3.1**, we can easily see that income first increased, and then remained relatively flat before it increased again. However, if a table contained 1000 data values, determining trends would be extremely difficult.

Rather than always using tables to display data, presenting data on a graph is often more useful. For example, the data in **TABLE 3.1** are graphed in **FIGURE 3.1**. This line graph is more visual than the table and shows the trend at a glance. Line graphs will be discussed in more detail later in this section.

U.S. per Capita Income (2014 Dollars)

Year	Amount
2004	$41,929
2006	$46,444
2008	$48,407
2010	$48,358
2012	$51,749

TABLE 3.1 *Source:* World Bank

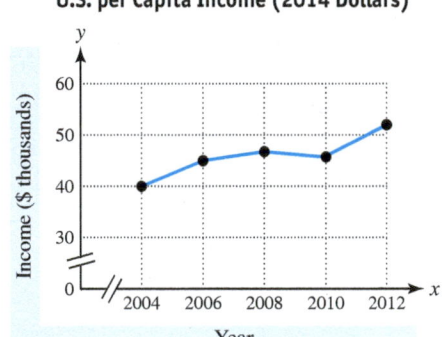

FIGURE 3.1

NOTE: In **FIGURE 3.1**, the double hash marks // on the graph indicate a break in data values. The years before 2004 and incomes below $30,000 have been skipped. ■

2 Understanding the Rectangular Coordinate System

One common way to graph data is to use the **rectangular coordinate system**, or **xy-plane**. The following See the Concept shows important information about the *xy*-plane.

See the Concept 1   THE xy-PLANE

Two axes divide the *xy*-plane into four regions called **quadrants**, which are numbered **I**, **II**, **III**, and **IV** counterclockwise, as shown in the figure.

Ⓐ The horizontal axis is called the *x*-axis.

Ⓑ The vertical axis is called the *y*-axis.

Ⓒ The axes intersect at the origin, which is the point associated with zero on each axis.

Negative values are located to the left of the origin along the *x*-axis and below the origin along the *y*-axis.

Positive values are located to the right of the origin along the *x*-axis and above the origin along the *y*-axis.

WATCH VIDEO IN MML.

PLOTTING AND READING ORDERED PAIRS Before we can plot data, we must understand the concept of an **ordered pair** (x, y). In **TABLE 3.1**, we can let x-values correspond to the year and y-values correspond to the per capita income. Then the fact that the per capita income in **2004** was $\$41{,}929$ can be summarized by the ordered pair $(2004, 41929)$. Similarly, the ordered pair $(2008, 48407)$ indicates that the per capita income was $\$48{,}407$ in 2008.

Order is important in an ordered pair. The ordered pairs given by $(1950, 2025)$ and $(2025, 1950)$ are different. The first ordered pair indicates that the per capita income in 1950 was $\$2025$, whereas the second ordered pair indicates that the per capita income in 2025 will be $\$1950$ (which is probably not correct).

To plot an ordered pair such as $(-2, 3)$ in the xy-plane, begin at the origin and move left to locate $x = -2$ on the x-axis. Then move upward until a height of $y = 3$ is reached. Thus the point $(-2, 3)$ is located 2 units left of the origin and 3 units above the origin. In **FIGURE 3.2**, the point $(-2, 3)$ is plotted in quadrant II.

NOTE: A point that lies on an axis is not located in a quadrant. ∎

Plotting a Point

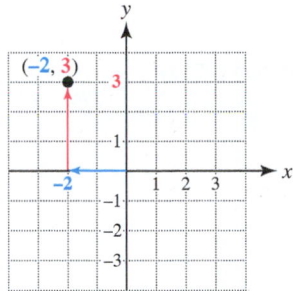

FIGURE 3.2

READING CHECK 1

- Is the order of the numbers in an ordered pair important? Explain.

EXAMPLE 1

Plotting points

Plot the following ordered pairs on the same xy-plane. State the quadrant in which each point is located, if possible.

(a) $(3, 2)$ **(b)** $(-2, -3)$ **(c)** $(-3, 0)$

Solution

(a) The point $(3, 2)$ is located in quadrant I, 3 units to the right of the origin and 2 units above the origin. See **FIGURE 3.3**.

(b) The point $(-2, -3)$ is located 2 units to the left of the origin and 3 units below the origin. **FIGURE 3.3** shows the point $(-2, -3)$ in quadrant III.

(c) The point $(-3, 0)$ is not in any quadrant because it is located 3 units left of the origin on the x-axis, as shown in **FIGURE 3.3**.

Now Try Exercise 13

FIGURE 3.3

 Math in Business Context In the next example, a scatterplot is used to analyze historical data concerning the valuation of the social network Twitter.

EXAMPLE 2

Reading a graph containing Twitter data

FIGURE 3.4 shows the value of Twitter in billions of dollars during its first 6 years. Use the graph to estimate the value of Twitter in 2009 and 2013. (*Source:* Business Insider.)

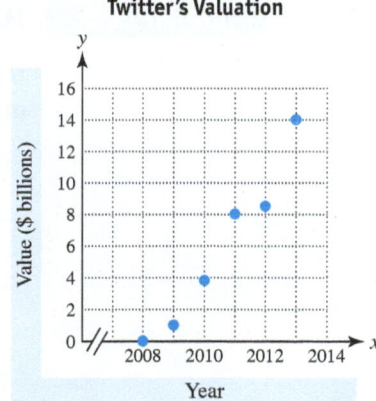

FIGURE 3.4

Solution

To find the value of Twitter in 2009, start by locating 2009 on the *x*-axis and move **vertically** upward to the data point. From the data point, move **horizontally** to the *y*-axis. **FIGURE 3.5(a)** shows that Twitter's value was about $1 billion in 2009. Similarly, the value was about $14 billion by 2013, as shown in **FIGURE 3.5(b)**.

Twitter's Valuation (2009)

(a)

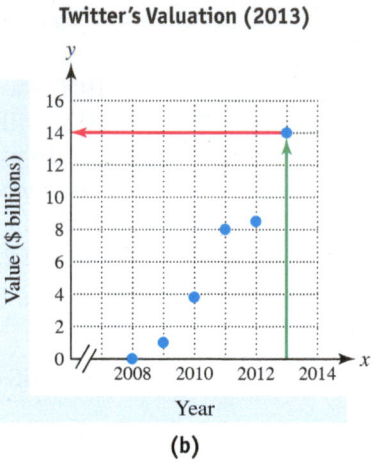

Twitter's Valuation (2013)

(b)

FIGURE 3.5

Now Try Exercise 35

3 Making and Reading Scatterplots

If distinct points are plotted in the *xy*-plane, then the resulting graph is called a **scatterplot**. **FIGURE 3.5(a)** is an example of a scatterplot that displays information about Twitter. A different scatterplot is shown in **FIGURE 3.6**, in which the points $(1, 3)$, $(2, 2)$, $(3, 1)$, $(4, 4)$, and $(5, 1)$ are plotted.

A Scatterplot

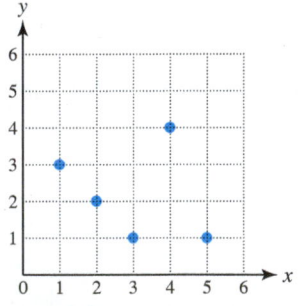

FIGURE 3.6

CHOOSING A SCALE FOR A GRAPH Choosing appropriate scales for the axes is important when plotting points and making graphs. This can be accomplished by looking at the *coordinates* of the ordered pairs to be plotted. When plotting in the *xy*-plane, the first value in an ordered pair is called the **x-coordinate** and the second value is called the **y-coordinate**. In **FIGURE 3.6**, the *x*-coordinates of the points are 1, 2, 3, 4, and 5 and the *y*-coordinates are 1, 2, 3, and 4. Because no coordinate is more than 6 units from the origin, the scale shown in **FIGURE 3.6** is appropriate.

READING CHECK 2

• Which value in the ordered pair $(1, -2)$ is the *x*-coordinate, and which is the *y*-coordinate?

🌐 *Math in Context* (Economics) While gasoline prices can vary from day to day, historical data show a general upward trend in gas prices. The next example shows how a scatterplot can be used to visualize this trend.

EXAMPLE 3 **Making a scatterplot of gasoline prices**

TABLE 3.2 lists the average price of a gallon of gasoline for selected years. Make a scatterplot of the data. These prices have *not* been adjusted for inflation.

Average Price of Gasoline

Year	1960	1970	1980	1990	2000	2010
Cost (per gal)	31¢	36¢	119¢	115¢	156¢	273¢

TABLE 3.2 *Source:* Department of Energy.

Solution

The data point $(1960, 31)$ can be used to indicate that the average cost of a gallon of gasoline in 1960 was 31¢. Plot the six data points $(1960, 31)$, $(1970, 36)$, $(1980, 119)$,

$(1990, 115)$, $(2000, 156)$, and $(2010, 273)$ in the *xy*-plane. The *x*-values vary from 1960 to 2010, so label the *x*-axis from 1960 to 2020 every 10 years. The *y*-values vary from 31 to 273, so label the *y*-axis from 0 to 350 every 50¢. Note that the *x*- and *y*-scales must be large enough to accommodate every data point. **FIGURE 3.7** shows the scatterplot.

In **FIGURE 3.7**, the *increment* on the *x*-axis is 10 because each step from one vertical grid line to the next represents a change of 10 years. Similarly, the increment on the *y*-axis is 50 because each step from one horizontal grid line to the next represents a 50-cent change in price. This example demonstrates that the scale and increment on one axis are not always the same as those on the other axis.

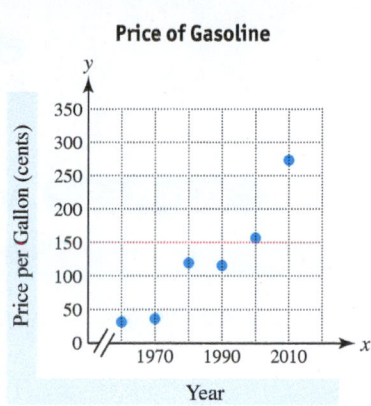

Price of Gasoline

FIGURE 3.7

Now Try Exercise 25

4 Making and Reading Line Graphs

Sometimes it is helpful to connect consecutive data points in a scatterplot with line segments. This type of graph visually emphasizes changes in the data and is called a **line graph**. When making a line graph, be sure to plot all of the given data points *before* connecting the points with line segments. The points should be connected consecutively from left to right on the scatterplot, even if the data are given "out of order" in a table.

EXAMPLE 4   **Making a line graph**

Use the data in **TABLE 3.3** to make a line graph.

Data Points for a Line Graph

x	-2	-1	0	1	2
y	1	2	-2	-1	1

TABLE 3.3

Solution

The data in **TABLE 3.3** are represented by the five ordered pairs $(-2, 1)$, $(-1, 2)$, $(0, -2)$, $(1, -1)$, and $(2, 1)$. Plot these points and then connect consecutive points with line segments, as shown in **FIGURE 3.8**.

Line Graph

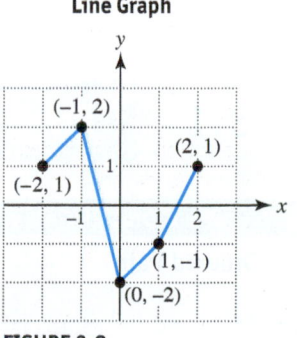

FIGURE 3.8

Now Try Exercise 37

CALCULATOR HELP

To make a scatterplot or a line graph, see Appendix A (pages AP-3 and AP-4).

TECHNOLOGY NOTE

Scatterplots and Line Graphs

Graphing calculators are capable of creating line graphs and scatterplots. The line graph in **FIGURE 3.8** is shown (below) to the left, and the corresponding scatterplot is shown to the right.

🌐 *Math in Context* The next example illustrates how a line graph can be used to analyze U.S. energy use over a 50-year period.

EXAMPLE 5 | **Analyzing a line graph**

U.S. Energy Consumption

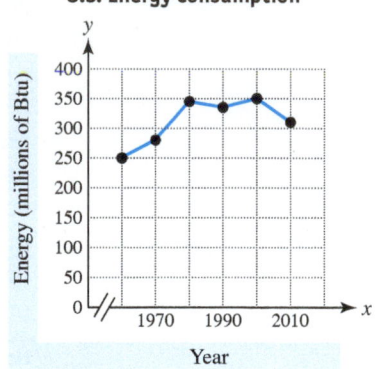

FIGURE 3.9

The line graph in **FIGURE 3.9** shows the per capita energy consumption in the United States from 1960 to 2010. Units are in millions of Btu, where 1 Btu equals the amount of energy necessary to heat 1 pound of water 1 °F. (*Source:* Department of Energy.)

(a) Did energy consumption ever decrease during this time period? If so, when?

(b) Estimate the per capita energy consumption in 1960 and in 2000.

(c) Estimate the percent change in energy consumption from 1960 to 2000.

READING CHECK 3

• What is the main difference between a scatterplot and a line graph?

CRITICAL THINKING 1

When analyzing data, do you prefer a table of values, a scatterplot, or a line graph? Explain your answer.

Solution

(a) Yes, energy consumption decreased slightly between 1980 and 1990, and again between 2000 and 2010.

(b) From the graph, per capita energy consumption in 1960 was about 250 million Btu. In 2000 it was about 350 million Btu.

NOTE: When different people read a graph, the values they obtain may vary slightly. ∎

(c) The percent change from 1960 to 2000 was

$$\frac{350 - 250}{250} \cdot 100 = 40,$$

so the increase was 40%.

Now Try Exercise 49

3.1 Putting It All Together

CONCEPT	EXPLANATION	EXAMPLES
Ordered Pair	Has the form (x, y), where the order of x and y is important	$(1, 2)$, $(-2, 3)$, $(2, 1)$ and $(-4, -2)$ are distinct ordered pairs.
Rectangular Coordinate System, or xy-plane	Consists of a horizontal x-axis and a vertical y-axis that intersect at the origin Has four quadrants, which are numbered counterclockwise as I, II, III, and IV	*(coordinate plane showing Quadrant II, Quadrant I, Origin, Quadrant III, Quadrant IV)*

CONCEPT	EXPLANATION	EXAMPLES
Scatterplot	Individual points that are plotted in the xy-plane	The points $(1, 1)$, $(2, 3)$, $(3, 2)$, $(4, 5)$, and $(5, 4)$ are plotted in the graph.
Line Graph	Similar to a scatterplot except that line segments are drawn between consecutive data points	

3.1 Exercises

MyMathLab®

CONCEPTS AND VOCABULARY

1. Another name for the rectangular coordinate system is the _____.

2. The point where the x-axis and y-axis intersect is called the _____.

3. In the xy-plane, the origin corresponds to the ordered pair _____.

4. How many quadrants are there in the xy-plane?

5. (True or False?) Every point in the xy-plane is located in one of the quadrants.

6. In the xy-plane, the first value in an ordered pair is called the _____-coordinate and the second value is called the _____-coordinate.

7. If distinct points are plotted in the xy-plane, the resulting graph is called a(n) _____.

8. If the consecutive points in a scatterplot are connected with line segments, the resulting graph is called a(n) _____ graph.

UNDERSTANDING THE RECTANGULAR COORDINATE SYSTEM (xy-PLANE)

Exercises 9–12: Identify the coordinates of each point in the graph.

9.

10.

11.

12.
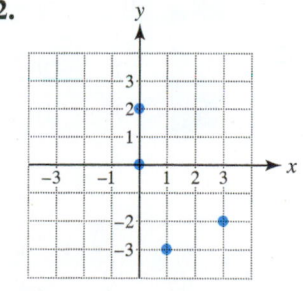

Exercises 13–16: Plot the given ordered pairs in the same xy-plane. If possible, state the quadrant in which each of the points is located.

13. $(1, 3)$, $(0, -3)$, and $(-2, 2)$

14. $(4, 0)$, $(-3, -4)$, and $(2, -3)$

15. $(0, 6)$, $(8, -4)$, and $(-6, -6)$

16. $(-4, 8)$, $(6, 8)$, and $(-8, 0)$

Exercises 17–22: If possible, identify the quadrant in which each point is located.

17. (a) $(1, 4)$ **(b)** $(-1, -4)$

18. (a) $(-2, -3)$ **(b)** $(2, -3)$

19. (a) $(7, 0)$ **(b)** $(0.1, 7)$

20. (a) $(100, -3)$ **(b)** $(-100, 3)$

21. (a) $\left(-\frac{1}{2}, \frac{3}{4}\right)$ **(b)** $\left(\frac{3}{4}, -\frac{1}{2}\right)$

22. (a) $(1.2, 0)$ **(b)** $(0, -1.2)$

23. Thinking Generally Which of the four quadrants contain points whose *x*- and *y*-coordinates have the same sign?

24. Thinking Generally Which of the four quadrants contain points whose *x*- and *y*-coordinates have different signs?

MAKING AND READING SCATTERPLOTS

Exercises 25–30: Make a scatterplot by plotting the given points. Be sure to label each axis.

25. $(0, 0)$, $(1, 2)$, $(-3, 2)$, $(-1, -2)$

26. $(0, -3)$, $(-2, 1)$, $(2, 2)$, $(-4, -4)$

27. $(-1, 0)$, $(4, -3)$, $(0, -1)$, $(3, 4)$

28. $(1, 1)$, $(-2, 2)$, $(-3, -3)$, $(4, -4)$

29. $(2, 4)$, $(-4, 4)$, $(0, -4)$, $(-6, 2)$

30. $(4, 8)$, $(8, 4)$, $(-8, -4)$, $(-4, 0)$

Exercises 31–34: Use the table to make a scatterplot.

31.

x	5	5	-10	10
y	0	-5	-20	-10

32.

x	10	-20	40	-30
y	30	10	0	-10

33.

x	0	0.2	-0.1
y	0.1	-0.3	0.4

34.

x	1.5	-1	-2.5
y	2.5	-1.5	0

Exercises 35 and 36: **Real Data Graphs** *Identify the coordinates of each point in the graph. Then explain what the coordinates of the first point indicate.*

35. Cigarette consumption in the United States (*Source:* U.S. Department of Health and Human Services.)

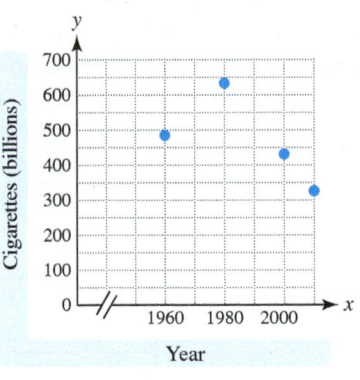

U.S. Cigarette Consumption

36. Number of vinyl albums sold (*Source:* Recording Industry Association of America.)

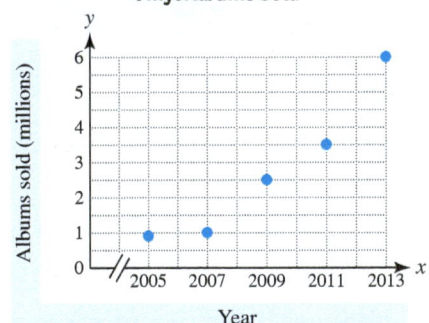

Vinyl Albums Sold

MAKING AND READING LINE GRAPHS

Exercises 37–42: Use the table to make a line graph.

37.

x	-2	-1	0	1	2
y	2	1	0	-1	-2

38.

x	-4	-2	0	2	4
y	4	-2	3	-1	2

39.

x	-10	-5	0	5	10
y	20	-10	10	0	-20

40.

x	1	2	3	4	5
y	2	3	1	5	4

41.

x	−5	5	−10	10	0
y	10	20	30	20	40

42.

x	3	−2	2	1	−3
y	4	3	3	−2	−3

GRAPHING REAL DATA

Exercises 43–48: **Graphing Real Data** *The table contains real data.*

(a) *Make a line graph of the data. Be sure to label the axes.*

(b) *Comment on any trends in the data.*

43. Average annual cost C of a four-year residential college degree in thousands of dollars during year t

t	1980	1990	2000	2010	2015
C	10	12	16	21	24

Source: National Center for Education Statistics.

44. Federal income tax receipts I in billions during year t

t	1970	1980	1990	2000	2010
I	90	244	467	1003	2165

Source: Office of Management and Budget.

45. Welfare beneficiaries B in millions during year t

t	1970	1980	1990	2000	2010
B	7	11	12	6	4

Source: Administration for Children and Families.

46. Detroit city population P in millions during year t

t	1900	1920	1940	1960	1980	2000
P	0.3	1.0	1.6	1.7	1.2	0.9

Source: Census Bureau.

47. U.S. Internet users y in millions during year x

x	2010	2011	2012	2013	2014
y	221	229	237	244	251

Source: eMarketer.

48. Number of farms in Iowa F in thousands during year x

x	2003	2005	2007	2009	2011
F	90	89	92	92	92

Source: U.S. Department of Agriculture.

49. Infant Mortality The line graph shows the U.S. infant mortality rate for selected years.

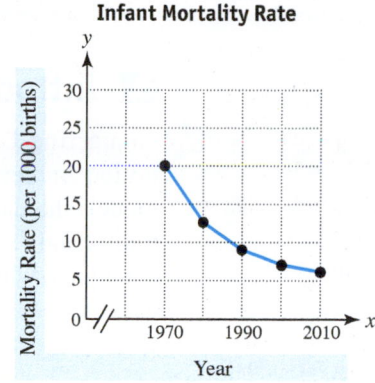

Infant Mortality Rate

(a) Comment on any trends in the data.

(b) Estimate the infant mortality rate in 1990.

(c) Estimate the percent change in the infant mortality rate from 1970 to 2010.

50. Population The line graph shows the population of the midwestern states, in millions, for selected years.

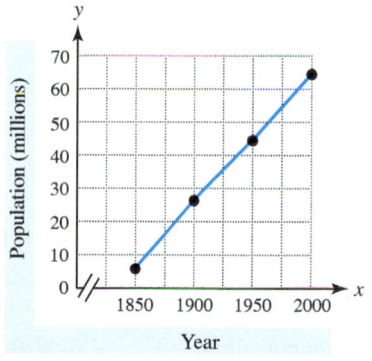

Population of Midwestern States

(a) Comment on any trends in the data.

(b) Estimate this population in 1900.

(c) Estimate the percent change in population from 1850 to 2000.

WRITING ABOUT MATHEMATICS

51. Explain how to identify the quadrant that a point lies in if it has coordinates (x, y).

52. Explain the difference between a scatterplot and a line graph. Give an example of each.

3.2 Equations in Two Variables

NEW VOCABULARY

☐ Standard form (of a linear equation in two variables)
☐ Linear equation in two variables

STUDY TIP

If you have tried to solve a homework problem but need help, ask a question in class. Other students will likely have the same question.

1 Introducing Equations in Two Variables

Connecting Concepts with Your Life **FIGURE 3.10** shows a scatterplot of average college tuition and fees at public colleges and universities (in 2010 dollars), together with a line that models the data. If we could find an equation for this line, then we could use it to estimate tuition and fees for years without data points, such as 2015. In this section, we discuss linear equations, whose graphs are lines. Linear equations and lines are often used to approximate data. (*Source:* The College Board.)

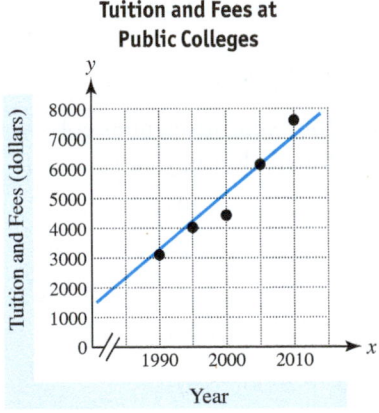

FIGURE 3.10

NOTE: The graph shown in **FIGURE 3.10** is *not* a line graph that has consecutive points connected with line segments. Rather, it is a scatterplot together with a single straight line that closely models the data. ∎

Equations can have any number of variables. Previously, we solved equations with *one* variable. The following are examples of equations with *two* variables.

$$y = 2x, \quad 3x + 2y = 4, \quad z = t^2, \quad \text{and} \quad a - b = 1$$

Equations with two variables

2 Recognizing Ordered Pairs as Solutions

A solution to an equation with two variables consists of two numbers, one for each variable, which can be expressed as an ordered pair. For example, one solution to the equation $y = 5x$ is given by $x = 1$ and $y = 5$ because $5 = 5(1)$ is a true statement. This solution can be expressed as the ordered pair $(1, 5)$. **TABLE 3.4** lists several ordered pairs and shows whether each ordered pair is a solution to the equation $y = 5x$.

READING CHECK 1

• How is a solution to an equation in two variables expressed?

Checking Ordered-Pair Solutions to $y = 5x$

Ordered Pair	Check: $y = 5x$	Is it a solution?
$(1, 5)$	$5 \overset{?}{=} 5(1)$	Yes, because $5 = 5$
$(2, 12)$	$12 \overset{?}{=} 5(2)$	No, because $12 \neq 10$
$(-2, -10)$	$-10 \overset{?}{=} 5(-2)$	Yes, because $-10 = -10$
$(4, 20)$	$20 \overset{?}{=} 5(4)$	Yes, because $20 = 20$

TABLE 3.4

EXAMPLE 1   **Testing solutions to equations**

Determine whether the given ordered pair is a solution to the given equation.

(a) $y = x + 3$, $(1, 4)$ **(b)** $2x - y = 5$, $\left(\frac{1}{2}, -4\right)$ **(c)** $-4x + 5y = 20$, $(-5, 1)$

Solution

(a) Let $x = 1$ and $y = 4$ in the given equation.

$$y = x + 3 \qquad \text{Given equation}$$
$$4 \stackrel{?}{=} 1 + 3 \qquad \text{Substitute.}$$
$$4 = 4 \checkmark \qquad \text{The solution checks.}$$

The ordered pair $(1, 4)$ is a solution.

(b) Let $x = \frac{1}{2}$ and $y = -4$ in the given equation.

$$2x - y = 5 \qquad \text{Given equation}$$
$$2\left(\frac{1}{2}\right) - (-4) \stackrel{?}{=} 5 \qquad \text{Substitute.}$$
$$1 + 4 \stackrel{?}{=} 5 \qquad \text{Simplify the left side.}$$
$$5 = 5 \checkmark \qquad \text{The solution checks.}$$

The ordered pair $\left(\frac{1}{2}, -4\right)$ is a solution.

(c) Let $x = -5$ and $y = 1$ in the given equation.

$$-4x + 5y = 20 \qquad \text{Given equation}$$
$$-4(-5) + 5(1) \stackrel{?}{=} 20 \qquad \text{Substitute.}$$
$$20 + 5 \stackrel{?}{=} 20 \qquad \text{Simplify the left side.}$$
$$25 \neq 20 \; \text{✗} \qquad \text{The solution does } not \text{ check.}$$

The ordered pair $(-5, 1)$ is *not* a solution.

Now Try Exercises 11, 15, 17

3   Making a Table of Solutions

A table can be used to list solutions to an equation. For example, **TABLE 3.5** lists solutions to $x + y = 5$, where the sum of each xy-pair equals 5.

x + y = 5

x	-2	-1	0	1	2
y	7	6	5	4	3

Sum equals 5

TABLE 3.5

READING CHECK 2

• When is it helpful to have a table that lists a few solutions to an equation?

Most equations in two variables have infinitely many solutions, so it is impossible to list all solutions in a table. However, when you are graphing an equation, having a table that lists a few solutions to the equation is often helpful. The next two examples demonstrate how to complete a table for a given equation.

EXAMPLE 2

Completing a table of solutions

Complete the table for the equation $y = 2x - 3$.

x	-4	-2	0	2
y				

Solution
Start by determining the corresponding y-value for each x-value in the table. For example, when $x = -2$, the equation $y = 2x - 3$ implies that $y = 2(-2) - 3 = -4 - 3 = -7$. Filling in the y-values results in **TABLE 3.6**.

$$y = 2x - 3$$

x	-4	-2	0	2
y	-11	-7	-3	1

TABLE 3.6

Now Try Exercise 21

EXAMPLE 3

Making a table of solutions

Use $y = 0, 5, 10,$ and 15 to make a table of solutions to $5x + 2y = 10$.

Solution
Begin by listing the required y-values in the table. Next determine the corresponding x-values for each y-value by using the equation $5x + 2y = 10$.

When $y = 0$,
$5x + 2(0) = 10$
$5x + 0 = 10$
$5x = 10$
$x = 2$

When $y = 5$,
$5x + 2(5) = 10$
$5x + 10 = 10$
$5x = 0$
$x = 0$

When $y = 10$,
$5x + 2(10) = 10$
$5x + 20 = 10$
$5x = -10$
$x = -2$

When $y = 15$,
$5x + 2(15) = 10$
$5x + 30 = 10$
$5x = -20$
$x = -4$

Filling in the x-values results in **TABLE 3.7**.

$$5x + 2y = 10$$

x	2	0	-2	-4
y	0	5	10	15

TABLE 3.7

Now Try Exercise 23

 Math in ~Context~ [Health Care] Newspapers, magazines, and books often list numbers in a table rather than presenting a formula for the reader to use. The next example illustrates a situation in health care in which a table might be preferable to a formula.

EXAMPLE 4

Calculating appropriate lengths of crutches

People with leg injuries often need crutches. An appropriate crutch length L in inches for an injured person who is t inches tall is estimated by $L = 0.72t + 2$. (*Source: Journal of the American Physical Therapy Association.*)

(a) Complete the table. Round values to the nearest inch.

t	60	65	70	75	80
L					

(b) Use the table to determine the appropriate crutch length for a person 5 feet 10 inches tall.

Solution

(a) In the formula $L = 0.72t + 2$, if $t = 60$, then $L = 0.72(60) + 2 = 43.2 + 2 = 45.2$, or about **45** inches. If $t = 65$, then $L = 0.72(65) + 2 = 46.8 + 2 = 48.8 \approx 49$. Other values in **TABLE 3.8** are found similarly.

Crutch Lengths

t	60	65	70	75	80
L	45	49	52	56	60

TABLE 3.8

(b) A person who is 5 feet 10 inches tall is $5 \cdot 12 + 10 = 70$ inches tall. **TABLE 3.8** shows that a person 70 inches tall needs crutches that are about 52 inches long.

Now Try Exercise 79

4 ▸ Graphing Linear Equations in Two Variables

Many times graphs are used in mathematics to make concepts easier to visualize and understand. Graphs can be either curved or straight; however, graphs of *linear* equations in two variables are always *straight* lines.

The following See the Concept shows how to graph the equation $y = 2x$.

See the Concept 1   **GRAPHING $y = 2x$**

Start by making a table of values. A table can be either vertical or horizontal. We will use a vertical table.

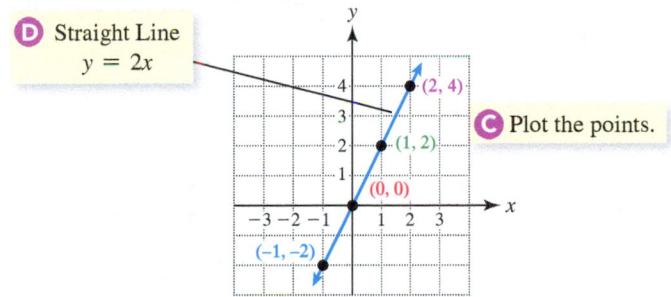

D Straight Line
$y = 2x$

C Plot the points.

A Select a few convenient values for x, such as $x = -1, 0, 1$ and 2. *Any* numbers may be chosen for x-values. In this example, we choose values that are easy to evaluate in the equation $y = 2x$.

B Calculate the y-values for each x-value. Since $y = 2x$, each y-value is found by doubling the x-value.

C The table indicates that the points $(-1, -2)$, $(0, 0)$, $(1, 2)$, and $(2, 4)$ lie on the line. Plot these points. Since $y = 2x$ is a *linear equation*, the points should lie on a *straight line*.

D Draw a straight line that passes through the plotted points. This line is the graph of $y = 2x$.

WATCH VIDEO IN $\boxed{\text{MML}}$.

NOTE: Because the graph of a linear equation in two variables is a straight line, you may want to use a ruler or other straight edge when graphing this kind of equation. ∎

EXAMPLE 5 **Graphing linear equations**

Graph each linear equation.
(a) $y = \frac{1}{2}x - 1$ (b) $x + y = 4$

Solution

(a) The equation $y = \frac{1}{2}x - 1$ is a linear equation in two variables and its graph is a line. Two points determine a line. However, it is a good idea to plot three points to be sure that the line is graphed correctly. Start by choosing three values for x and then calculate the corresponding y-values, as shown in **TABLE 3.9**. In **FIGURE 3.11**, the points $(-2, -2)$, $(0, -1)$, and $(2, 0)$ are plotted and the line passing through these points is drawn.

$y = \frac{1}{2}x - 1$

x	y
-2	-2
0	-1
2	0

TABLE 3.9

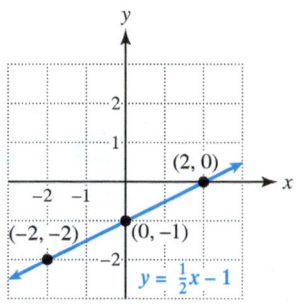

FIGURE 3.11

(b) The equation $x + y = 4$ is a linear equation in two variables, so its graph is a line. If an ordered pair (x, y) is a solution to the given equation, then the sum of x and y is 4. **TABLE 3.10** shows three examples. In **FIGURE 3.12**, the points $(0, 4)$, $(2, 2)$, and $(4, 0)$ are plotted with the line passing through each one.

$x + y = 4$

x	y
0	4
2	2
4	0

TABLE 3.10

The sum is 4.

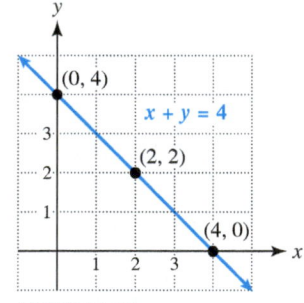

FIGURE 3.12

Now Try Exercises 43, 57

TECHNOLOGY NOTE

CALCULATOR HELP

To graph an equation, see Appendix A (page AP-5).

Graphing Equations
Graphing calculators can be used to graph equations. Before graphing the equation $x + y = 4$, solve the equation for y to obtain $y = 4 - x$. A calculator graph of **FIGURE 3.12** is shown, except that the three points have not been plotted.

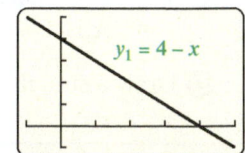

MAKING CONNECTIONS 1

Line Graphs and Graphs of Linear Equations

Do not confuse a line graph with the graph of a linear equation. A *line graph* results when consecutive points from a table of values are connected with line segments. However, the *graph of a linear equation* is a single, straight line that passes through points from a table of values and continues indefinitely in two directions.

READING CHECK 3

- Is a line graph the same as a graph of a linear equation?

STANDARD FORM FOR A LINEAR EQUATION Not every equation with two variables has a graph that is a straight line. Those equations whose graphs are straight lines are called *linear* equations in two variables. Every linear equation in two variables can be written in the following **standard form**.

LINEAR EQUATION IN TWO VARIABLES

A **linear equation in two variables** can be written as

$$Ax + By = C,$$

where A, B, and C are fixed numbers (constants) and A and B are not both equal to 0. The graph of a linear equation in two variables is a line.

The equation $y = 2x$ was presented earlier in this section in See the Concept 1. This equation is a linear equation in two variables because it can be written in standard form by adding $-2x$ to each side.

$$
\begin{aligned}
y &= 2x & &\text{Given equation} \\
-2x + y &= -2x + 2x & &\text{Add } -2x \text{ to each side.} \\
-2x + y &= 0 & &\text{Simplify.}
\end{aligned}
$$

The equation $-2x + y = 0$ is a linear equation in two variables because it is in the form $Ax + By = C$ with $A = -2$, $B = 1$, and $C = 0$.

SOLVING STANDARD FORM FOR A VARIABLE When a linear equation in two variables is given in standard form, it is sometimes difficult to create a table of solutions. Solving an equation for y often makes it easier to select x-values for the table that will make the y-values simpler to calculate. This is demonstrated in the next example.

EXAMPLE 6   **Solving for y and then graphing**

Graph the linear equation $4x - 3y = 12$ by solving for y first.

Solution
First solve the given equation for y.

$$
\begin{aligned}
4x - 3y &= 12 & &\text{Given equation (standard form)} \\
-3y &= -4x + 12 & &\text{Subtract } 4x \text{ from each side.} \\
\frac{-3y}{-3} &= \frac{-4x + 12}{-3} & &\text{Divide each side by } -3. \\
\frac{-3y}{-3} &= \frac{-4x}{-3} + \frac{12}{-3} & &\text{Property of fractions, } \frac{a+b}{c} = \frac{a}{c} + \frac{b}{c} \\
y &= \frac{4}{3}x - 4 & &\text{Simplify fractions.}
\end{aligned}
$$

Divide each term by -3.

Select *multiples of 3* (the denominator of $\frac{4}{3}$) as x-values for the table of solutions. For example, if $x = 6$ is chosen, $y = \frac{4}{3}(6) - 4 = \frac{24}{3} - 4 = 8 - 4 = 4$. **TABLE 3.11** lists the solutions $(0, -4)$, $(3, 0)$, and $(6, 4)$, which are plotted in **FIGURE 3.13** on the next page with the line passing through each one.

$y = \frac{4}{3}x - 4$

x	y
0	−4
3	0
6	4

TABLE 3.11

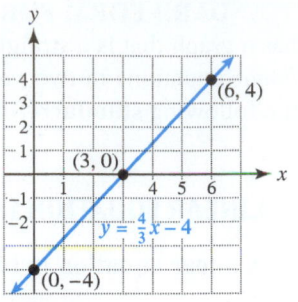

FIGURE 3.13

Now Try Exercise 59

MAKING CONNECTIONS 2

Graphs and Solution Sets

A graph visually depicts the set of solutions to an equation. Each point (x, y) on the graph represents one solution to the equation. In a linear equation, each x-value determines a unique y-value. Because there are infinitely many x-values, there are infinitely many points located on the graph of a linear equation. Thus the graph of a linear equation is a *continuous* line with no breaks.

---

5   Graphing Nonlinear Equations in Two Variables

Some equations in two variables have graphs that are *not* straight lines. Such equations are called **nonlinear equations**. It is possible to graph some nonlinear equations by completing a table of values and plotting points. However, because these graphs can take on many different shapes, it is important to plot enough points so that an accurate graph can be obtained.

EXAMPLE 7   **Graphing a nonlinear equation**

Graph the nonlinear equation $y = x^2$.

Solution

Complete a table of values as shown in **TABLE 3.12**. Be sure to include both positive and negative x-values to get an accurate graph. Plot the points from the table and draw a smooth curve that passes through each point, as shown in **FIGURE 3.14**.

$y = x^2$

x	y
-2	4
-1	1
0	0
1	1
2	4

TABLE 3.12

Graph of $y = x^2$

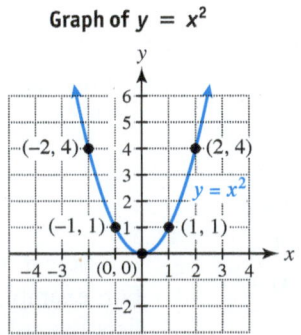

FIGURE 3.14

Now Try Exercise 71

3.2   Putting It All Together

CONCEPT	EXPLANATION	EXAMPLES
Equation in Two Variables	An equation that has two variables	$y = 4x + 5$, $4x - 5y = 20$, and $u - v = 100$
Solution to an Equation in Two Variables	An ordered pair (x, y) whose x- and y-values satisfy the equation Equations in two variables often have infinitely many solutions.	$(1, 3)$ is a solution to $3x + y = 6$ because $3(1) + 3 = 6$ is a true statement. The equation $y = 2x$ has infinitely many solutions, such as $(1, 2)$, $(2, 4)$, $(3, 6)$, and so on.
Table of Solutions	A table can be used to list solutions to an equation in two variables.	The following table lists solutions to the equation $y = 3x - 1$. <table><tr><td>x</td><td>−1</td><td>0</td><td>1</td><td>2</td></tr><tr><td>y</td><td>−4</td><td>−1</td><td>2</td><td>5</td></tr></table>
Linear Equation in Two Variables (Standard Form)	Can be written as $$Ax + By = C,$$ where A, B, and C are fixed numbers and A and B are not both equal to 0	$3x + 4y = 5$, $y = 4 - 3x$, $x = 2y + 1$ (or $3x + y = 4$) (or $x - 2y = 1$) The graph of each equation is a line.

3.2   Exercises

MyMathLab®

CONCEPTS AND VOCABULARY

1. (True or False?) The equation $4x + 6y = 24$ is an equation in two variables.

2. (True or False?) The equation $4x - 3x = 10$ is a linear equation in two variables.

3. A solution to an equation with two variables consists of two numbers expressed as a(n) _____.

4. A(n) _____ equation in two variables can be written in the form $Ax + By = C$.

5. (True or False?) A table of solutions lists all of the solutions to an equation in two variables.

6. (True or False?) A linear equation in two variables has infinitely many solutions.

7. An equation's _____ visually depicts its solution set.

8. The graph of a linear equation in two variables is a(n) _____

9. (True or False?) Every equation in two variables has a graph that is a straight line.

10. (True or False?) A line graph and the graph of a linear equation in two variables are the same thing.

RECOGNIZING ORDERED PAIRS AS SOLUTIONS

Exercises 11–20: Determine whether the ordered pair is a solution to the given equation.

11. $y = x + 1$, $(5, 6)$

12. $y = 4 - x$, $(6, 2)$

13. $y = 4x + 7$, $(2, 13)$

14. $y = -3x + 2$, $(-2, 8)$

15. $4x - y = -13$, $(-2, 3)$

16. $3y + 2x = 0$, $(-2, 3)$

17. $y - 6x = -1$, $\left(\frac{1}{2}, 2\right)$

18. $\frac{1}{2}x + \frac{3}{2}y = 0$, $\left(-\frac{3}{2}, \frac{1}{2}\right)$

19. $0.31x - 0.42y = -9$, $(100, 100)$

20. $0.5x - 0.6y = 4$, $(20, 10)$

MAKING A TABLE OF SOLUTIONS

Exercises 21–28: Complete the table for the given equation.

21. $y = 4x$

x	-2	-1	0	1	2
y	-8				

22. $y = \frac{1}{2}x - 1$

x	0	1	2	3	4
y	-1				

23. $3y + 2x = 6$

x					
y	-2	0	2	4	8

24. $3x - 5y = 30$

x					
y	-9	-6	-3	0	3

25. $y = x + 4$

x	y
-8	-4
0	
4	
8	
12	

26. $2x - y = 1$

x	y
-1	-3
	-1
	0
	1
	3

27. $2x - 6y = 12$

x	y
-6	
0	
6	
12	
18	

28. $4x + 3y = 12$

x	y
9	
6	
3	
0	
-3	

Exercises 29–32: Use the given values of the variable to make a horizontal table of solutions for the equation.

29. $y = 3x$ $x = -3, 0, 3, 6$

30. $y = 1 - 2x$ $x = 0, 1, 2, 3$

31. $x + y = 6$ $y = -2, 0, 2, 4$

32. $2x - 3y = 9$ $y = -3, 0, 1, 2$

Exercises 33–36: Use the given values of the variable to make a vertical table of solutions for the equation.

33. $y = \dfrac{x + 4}{2}$ $x = -8, -4, 0, 4$

34. $y = \dfrac{x}{3} - 1$ $x = 0, 2, 4, 6$

35. $y - 4x = 0$ $y = -2, -1, 0, 1$

36. $-4x = 6y - 4$ $y = -1, 0, 1, 2$

37. Thinking Generally If a student wishes to avoid fractional y-values when making a table of solutions for $y = \frac{3}{5}x - 7$, what must be true about any selected integer x-values?

38. Thinking Generally If a student wishes to avoid fractional y-values when making a table of solutions for $y = \frac{a}{b}x$, where a and b are natural numbers, what must be true about any selected integer x-values?

GRAPHING LINEAR EQUATIONS IN TWO VARIABLES

Exercises 39–44: Make a table of solutions for the equation, and then use the table to graph the equation.

39. $y = -2x$ **40.** $y = 2x - 1$

41. $x = 3 - y$ **42.** $x = y + 1$

43. $x + 2y = 4$ **44.** $2x - y = 1$

Exercises 45–58: Graph the equation.

45. $y = x$ **46.** $y = \frac{1}{2}x$

47. $y = \frac{1}{3}x$ **48.** $y = -2x$

49. $y = x + 3$ **50.** $y = x - 2$

51. $y = 2x + 1$ **52.** $y = \frac{1}{2}x - 1$

53. $y = 4 - 2x$ **54.** $y = 2 - 3x$

55. $y = 7 + x$ **56.** $y = 2 + 2x$

57. $y = -\frac{1}{2}x + \frac{1}{2}$ **58.** $y = -\frac{3}{4}x + 2$

SOLVING STANDARD FORM FOR A VARIABLE

Exercises 59–70: Graph the linear equation by solving for y first.

59. $2x + 3y = 6$ **60.** $3x + 2y = 6$

61. $x + 4y = 4$ **62.** $4x + y = -4$

63. $-x + 2y = 8$ **64.** $-2x + 6y = 12$

65. $y - 2x = 7$

66. $3y - x = 2$

67. $5x - 4y = 20$

68. $4x - 5y = -20$

69. $3x + 5y = -9$

70. $5x - 3y = 10$

GRAPHING NONLINEAR EQUATIONS IN TWO VARIABLES

Exercises 71–78: (See Example 7.) Make a convenient table of values for the given equation. Then plot the points and graph the equation.

71. $y = x^2 - 1$

72. $y = x^2 + 2$

73. $y = 2x^2$

74. $y = \frac{1}{2}x^2$

75. $y = |x| - 1$

76. $y = |x - 1|$

77. $y = \sqrt{x}$

78. $y = 2\sqrt{x}$

APPLICATIONS INVOLVING EQUATIONS IN TWO VARIABLES

79. U.S. Population For the years 2010 to 2050, the projected percentage P of the U.S. population that will be over the age of 65 during year t is estimated by $P = 0.178t - 344.6$. (*Source:* U.S. Census Bureau.)

  (a) Complete the table. Round each resulting value to the nearest tenth.

t	2010	2020	2030	2040	2050
P					

  (b) Use the table to find the year when the percentage of the population over the age of 65 is expected to reach 16.7%.

80. U.S. Population For the years 2010 to 2050, the projected percentage P of the U.S. population that will be 18 to 24 years old during year t is estimated by $P = -0.025t + 60.35$. (*Source:* U.S. Census Bureau.)

  (a) Complete the table. Round each resulting value to the nearest tenth.

t	2010	2020	2030	2040	2050
P					

  (b) Use the table to find the year when the percentage of the population that is 18 to 24 years old is expected to be 9.4%.

81. Solid Waste in the Past In 1960, the amount A of garbage in pounds produced after t days by the average American is given by $A = 2.7t$. (*Source:* Environmental Protection Agency.)

  (a) Graph the equation for $t \geq 0$.

  (b) How many days did it take for the average American in 1960 to produce 100 pounds of garbage?

82. Solid Waste Today (See Exercise 81.) Today the amount A of garbage in pounds produced after t days by the average American is given by $A = 4.5t$. (*Source:* Environmental Protection Agency.)

  (a) Graph the equation for $t \geq 0$.

  (b) How many days does it take for the average American to produce 100 pounds of garbage today?

83. Web Browsers From 2008 to 2014, the percent P of the global desktop web-browser market held by Internet Explorer can be modeled by $P = -7.333t + 14,793$, where t is the year. (*Source:* Business Insider.)

  (a) Evaluate P for $t = 2008$ and $t = 2014$. Interpret your answers.

  (b) Use your results from part (a) to graph the equation for P from 2008 to 2014.

  (c) In what year was $P = 50\%$?

84. U.S. HIV Infections The cumulative number of HIV infections I in thousands during year t is modeled by the equation $I = 42t - 83,197$, where $t \geq 2000$. (*Source:* Department of Health and Human Services.)

  (a) Evaluate I for $t = 2000$ and for $t = 2005$.

  (b) Use your results from part (a) to graph the equation from 2000 to 2005.

  (c) In what year was $I = 971$?

WRITING ABOUT MATHEMATICS

85. The number of welfare beneficiaries B in millions during year t is shown in the table. Discuss whether a linear equation might work to approximate these data during this period of time.

t	1970	1980	1990	2000	2010
B	7	11	12	6	4

Source: Administration for Children and Families.

86. The Asian-American population P in millions during year t is shown in the table. Discuss whether a linear equation might model these data.

t	2002	2004	2006	2008	2010
P	12.0	12.8	13.6	14.4	15.2

Source: U.S. Census Bureau.

Checking Basic Concepts

1. Identify the coordinates of the four points in the graph. State the quadrant, if any, in which each point lies.

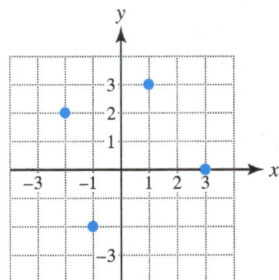

2. Make a scatterplot of the five points $(-2, -2)$, $(-1, -3)$, $(0, 0)$, $(1, 2)$, and $(2, 3)$.

3. **U.S. Population** The table gives the percentage P of the U.S. population that was over the age of 85 during year t. Make a line graph of the data in the table, and then comment on any trends. (*Source:* U.S. Census Bureau.)

t	1970	1980	1990	2000	2010
P	0.7	1.0	1.2	1.5	2.1

4. Determine whether $(-2, -3)$ is a solution to the equation $-2x - y = 7$.

5. Complete the table for the equation $y = -2x + 1$.

x	-2	-1	0	1	2
y	5				

6. Graph the equation.
   (a) $y = \frac{1}{2}x$ (b) $4x + 6y = 12$

7. **Windows 8 Performance** The percent P of the global market share of Windows 8 in the first 12 months after its release can be approximated by $P = 0.73t + 0.27$, where t is months after the release date.
   (a) Find P when $t = 1$ and $t = 12$. Interpret each result.
   (b) Use your results from part (a) to graph the equation for P from month 1 to month 12. Be sure to label the axes.
   (c) How many months after its launch did Windows 8 represent 4% of the global market share?

3.3   Intercepts; Horizontal and Vertical Lines

Objectives

1 Finding Intercepts
2 Using Intercepts to Graph Lines
3 Recognizing Horizontal Lines and Their Equations
4 Recognizing Vertical Lines and Their Equations

NEW VOCABULARY

☐ x-intercept
☐ y-intercept

1 Finding Intercepts

When things move or change at a *constant* rate, linear equations can often be used to describe or model the situation. For example, when a car moves at a constant speed, a graph of a linear equation can be used to visualize its driver's distance from a particular location. In real-world situations such as this, the *x*- and *y-intercepts* of the graph often provide important information.

Connecting Concepts with Your Life Suppose that someone leaves a family gathering at a state park and drives home at a constant speed of 50 miles per hour. The graph in **FIGURE 3.15** shows the distance of the driver from home at various times. The graph intersects the *y*-axis at $(0, 200)$, which is called the *y-intercept*. In this situation, the *y*-intercept tells us that the initial distance (when $x = 0$) between the driver and home is 200 miles. The graph also intersects the *x*-axis at $(4, 0)$, which is called the *x-intercept*. This intercept tells us that the elapsed time is 4 hours when the distance of the driver from home is 0 miles (when $y = 0$.) In other words, the driver arrived at home after 4 hours of driving.

FIGURE 3.15

STUDY TIP

Before getting help, be sure that you have tried a problem. Organize your questions so that you can be specific about the part of the problem that is giving you difficulty.

READING CHECK 1

- Where are the intercepts located on a graph?

FINDING x- AND y-INTERCEPTS

A point where a graph intersects the x-axis is an **x-intercept**. To find an x-intercept, let $y = 0$ in the equation and solve for x.

A point where a graph intersects the y-axis is a **y-intercept**. To find a y-intercept, let $x = 0$ in the equation and solve for y.

EXAMPLE 1   **Using a table to find intercepts**

Complete the table for the equation $x - y = -1$. Then determine the x-intercept and y-intercept for the graph of the equation $x - y = -1$.

x	-2	-1	0	1	2
y					

Solution

Substitute -2 for x in $x - y = -1$ to find the corresponding y-value.

$$-2 - y = -1 \quad \text{Let } x = -2.$$
$$-y = 1 \qquad \text{Add 2 to each side.}$$
$$y = -1 \qquad \text{Multiply each side by } -1.$$

The other y-values can be found similarly. See **TABLE 3.13**.

The x-intercept corresponds to a point on the graph whose y-coordinate is 0. **TABLE 3.13** shows that the y-coordinate is 0 when $x = -1$. So the x-intercept is $(-1, 0)$. Similarly, the y-intercept corresponds to a point on the graph whose x-coordinate is 0. **TABLE 3.13** shows that the x-coordinate is 0 when $y = 1$. So the y-intercept is $(0, 1)$.

READING CHECK 2

- How can the intercepts of a graph be found using a table of values?

$$x - y = -1$$

x	-2	-1	0	1	2
y	-1	0	1	2	3

TABLE 3.13

Now Try Exercise 17

CRITICAL THINKING 1

If a line has no x-intercept, what can you say about the line?
If a line has no y-intercept, what can you say about the line?

2 Using Intercepts to Graph Lines

It is often convenient to use intercepts to graph lines. The following See the Concept shows how to graph a linear equation using its intercepts.

See the Concept 1 **USING INTERCEPTS TO GRAPH $3x + 2y = 6$**

To graph the linear equation $3x + 2y = 6$, start by finding the intercepts.

A Find the x-intercept.

$$3x + 2(\mathbf{0}) = 6 \quad \text{Let } y = 0.$$
$$3x = 6 \quad \text{Simplify.}$$
$$x = \mathbf{2} \quad \text{Divide each side by 3.}$$

B Find the y-intercept.

$$3(\mathbf{0}) + 2y = 6 \quad \text{Let } x = 0.$$
$$2y = 6 \quad \text{Simplify.}$$
$$y = \mathbf{3} \quad \text{Divide each side by 2.}$$

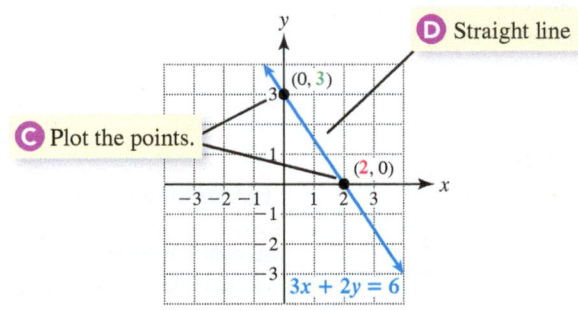

C Plot the points.

D Straight line

A Find the x-intercept: Let $y = \mathbf{0}$ in the given equation and solve for x. The line passes through $(\mathbf{2}, 0)$.

B Find the y-intercept: Let $x = \mathbf{0}$ in the given equation and solve for y. The line passes through $(0, \mathbf{3})$.

C Plot the points: Plot the points $(\mathbf{2}, 0)$ and $(0, \mathbf{3})$ on the xy-plane.

D Draw the line: Draw a straight line passing through the intercepts.

WATCH VIDEO IN $\boxed{\text{MML}}$.

EXAMPLE 2 **Using intercepts to graph a line**

Use intercepts to graph $2x - 6y = 12$.

Solution

The x-intercept is found by letting $y = \mathbf{0}$.

$$2x - 6(\mathbf{0}) = 12 \quad \text{Let } y = 0.$$
$$x = \mathbf{6} \quad \text{Solve for } x.$$

The y-intercept is found by letting $x = \mathbf{0}$.

$$2(\mathbf{0}) - 6y = 12 \quad \text{Let } x = 0.$$
$$y = \mathbf{-2} \quad \text{Solve for } y.$$

Therefore the graph passes through the points $(\mathbf{6}, 0)$ and $(0, \mathbf{-2})$, as shown in **FIGURE 3.16**.

Now Try Exercise 21

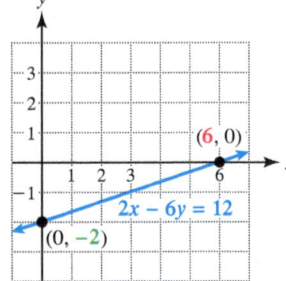

FIGURE 3.16

EXAMPLE 3 **Modeling the velocity of a toy rocket**

A toy rocket is shot vertically into the air. Its velocity v in feet per second after t seconds is given by $v = 160 - 32t$. Assume that $t \geq 0$ and $t \leq 5$.
(a) Graph the equation by finding the intercepts. Let t correspond to the horizontal axis (x-axis) and v correspond to the vertical axis (y-axis).
(b) Interpret each intercept.

Solution

(a) To find the *t*-intercept, let $v = 0$.

$$0 = 160 - 32t \qquad \text{Let } v = 0.$$
$$32t = 160 \qquad \text{Add } 32t \text{ to each side.}$$
$$t = 5 \qquad \text{Divide each side by 32.}$$

To find the *v*-intercept, let $t = 0$.

$$v = 160 - 32(0) \qquad \text{Let } t = 0.$$
$$v = 160 \qquad \text{Simplify.}$$

Therefore the graph passes through the intercepts $(5, 0)$ and $(0, 160)$, as shown in **FIGURE 3.17**.

Velocity of a Toy Rocket

$v = 160 - 32t$

FIGURE 3.17

(b) The *t*-intercept indicates that the rocket had a velocity of 0 feet per second after **5** seconds. The *v*-intercept indicates that the rocket's initial velocity was **160** feet per second.

Now Try Exercise 77

3 Recognizing Horizontal Lines and Their Equations

 Connecting Concepts with Your Life Suppose that someone drives a car on a freeway at a constant speed of 70 miles per hour. **TABLE 3.14** shows the speed *y* after *x* hours.

Speed of a Car

x	1	2	3	4	5
y	70	70	70	70	70

TABLE 3.14

We can make a scatterplot of the data by plotting the points $(1, 70)$, $(2, 70)$, $(3, 70)$, $(4, 70)$, and $(5, 70)$, as shown in **FIGURE 3.18(a)**. The speed is always 70 miles per hour and the graph of the car's speed is a horizontal line, as shown in **FIGURE 3.18(b)**. The equation of this line is $y = 70$ with *y*-intercept $(0, 70)$. There are no *x*-intercepts.

Speed of a Car

(a)

Speed of a Car

$y = 70$

(b)

FIGURE 3.18

In general, the equation of a horizontal line is $y = b$, where *b* is a constant that corresponds to the *y*-coordinate of the *y*-intercept. Examples of horizontal lines are shown in **FIGURE 3.19** on the next page. Note that every point on the graph of $y = 3$ in **FIGURE 3.19(a)** has a *y*-coordinate of **3**, and that every point on the graph of $y = -2$ in **FIGURE 3.19(b)** has a *y*-coordinate of -2.

Horizontal Lines: $y = b$

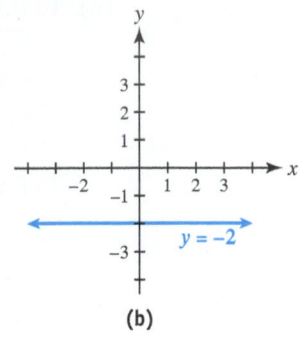

(a)　　　　　　　　　　　　　(b)

FIGURE 3.19

READING CHECK 3

• Which variable, x or y, is *not* present in the equation for a horizontal line?

HORIZONTAL LINE

The equation of a horizontal line with y-intercept $(0, b)$ is $y = b$.

EXAMPLE 4 **Graphing a horizontal line**

Graph the equation $y = -1$ and identify its y-intercept.

Solution

The graph of $y = -1$ is a horizontal line passing through the point $(0, -1)$, as shown in **FIGURE 3.20**. Its y-intercept is $(0, -1)$, and it has no x-intercept.

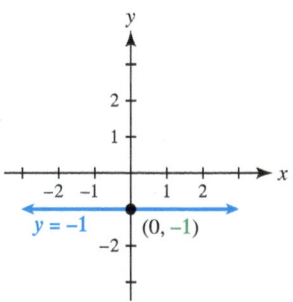

FIGURE 3.20

Now Try Exercise 45(a)

4 Recognizing Vertical Lines and Their Equations

Text Messages

x (months)	y (texts)
1	1631
1	12
1	2314
1	359

TABLE 3.15

Connecting Concepts with Your Life A parent looks at the number of text messages sent in one month by each of the four people on the family's data plan. **TABLE 3.15** shows the results, where x represents the number of months and y represents the number of text messages.

We can make a scatterplot of the data by plotting the points $(1, 1631)$, $(1, 12)$, $(1, 2314)$, and $(1, 359)$, as shown in **FIGURE 3.21(a)**. In each case the time is always 1 month and each point lies on the graph of a vertical line, as shown in **FIGURE 3.21(b)**. This vertical line has the equation $x = 1$ because each point on the line has an x-coordinate of **1** and there are no restrictions on the y-coordinate. This line has x-intercept $(1, 0)$ but no y-intercept.

FIGURE 3.21

In general, the graph of a vertical line is $x = k$, where k is a constant that corresponds to the x-coordinate of the x-intercept. Examples of vertical lines are shown in **FIGURE 3.22**. Note that every point on the graph of $x = 3$ in **FIGURE 3.22(a)** has an x-coordinate of **3** and that every point on the graph of $x = -2$ shown in **FIGURE 3.22(b)** has an x-coordinate of **-2**.

Vertical Lines: $x = k$

CALCULATOR HELP

To graph a vertical line, see Appendix A (page AP-5).

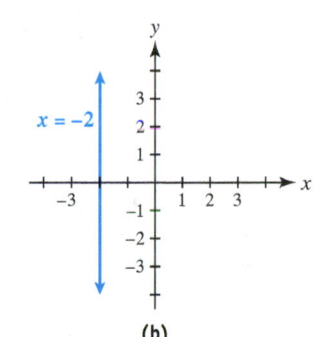

FIGURE 3.22

READING CHECK 4

- Which variable, x or y, is *not* present in the equation for a vertical line?

VERTICAL LINE

The equation of a vertical line with x-intercept $(k, 0)$ is $x = k$.

EXAMPLE 5 **Graphing a vertical line**

Graph the equation $x = -3$, and identify its x-intercept.

Solution

The graph of $x = -3$ is a vertical line passing through the point $(-3, 0)$, as shown in **FIGURE 3.23**. Its x-intercept is $(-3, 0)$ and it has no y-intercept.

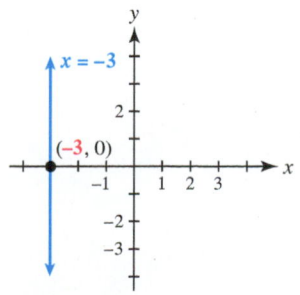

FIGURE 3.23

Now Try Exercise 45(b)

EXAMPLE 6 **Writing equations of horizontal and vertical lines**

Write the equation of the line shown in each graph.

(a)

(b)

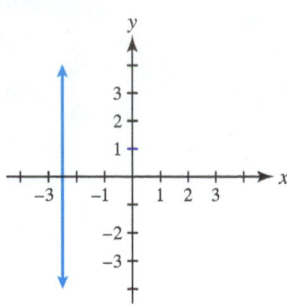

Solution

(a) The graph is a horizontal line with y-intercept $(0, 2)$. Its equation is $y = 2$.

(b) The graph is a vertical line with x-intercept $(-2.5, 0)$. Its equation is $x = -2.5$.

Now Try Exercises 51, 53

EXAMPLE 7 **Writing equations of horizontal and vertical lines**

Find an equation for a line satisfying the given conditions.

(a) Vertical, passing through $(2, -3)$

(b) Horizontal, passing through $(3, 1)$

(c) Perpendicular to $x = 3$, passing through $(-1, 2)$

Solution

(a) A vertical line passing through $(2, -3)$ has x-intercept $(2, 0)$ as shown in **FIGURE 3.24(a)**. The equation of a vertical line with x-intercept $(2, 0)$ is $x = 2$.

(b) A horizontal line passing through $(3, 1)$ has y-intercept $(0, 1)$, as shown in **FIGURE 3.24(b)**. The equation of a horizontal line with y-intercept $(0, 1)$ is $y = 1$.

(c) Because the line $x = 3$ is vertical, a line that is perpendicular to this line is horizontal, as shown in **FIGURE 3.24(c)**. The equation of a horizontal line passing through the point $(-1, 2)$ is $y = 2$.

Horizontal and Vertical Lines

(a)

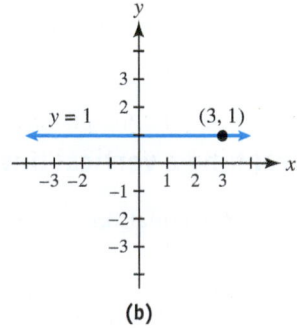

(b)

(c)

FIGURE 3.24

Now Try Exercises 65, 67, 69

The following See the Concept shows how any line can be written in the standard form $Ax + By = C$.

See the Concept 2 | **STANDARD FORM:** $Ax + By = C$

Every straight line in the xy-plane must be exactly one of the following.

Horizontal line $(A = 0, B \neq 0)$	**Vertical line** $(A \neq 0, B = 0)$	**General (Slanted) Line** $(A \neq 0, B \neq 0)$

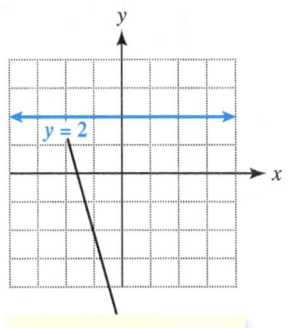

$A = 0, B = 1, C = 2$

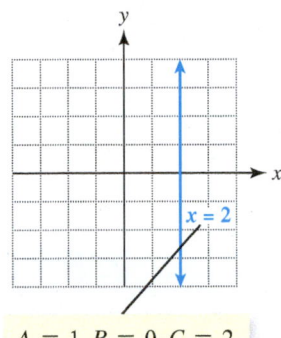

$A = 1, B = 0, C = 2$

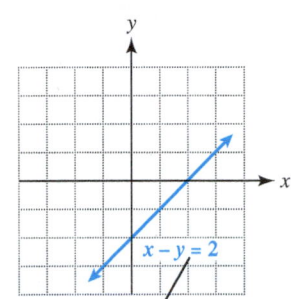

$A = 1, B = -1, C = 2$

The equation of *any* line can be written in the standard form $Ax + By = C$.　　**WATCH VIDEO IN** MML.

3.3　Putting It All Together

CONCEPT	EXPLANATION	EXAMPLES
x- and y-Intercepts	A point at which a graph intersects the x-axis is called an x-intercept. A point at which a graph intersects the y-axis is called a y-intercept.	![graph showing y-intercept (0, 2) and x-intercept (-3, 0)]
Finding Intercepts	To find x-intercepts, let $y = 0$ in the equation and solve for x. To find y-intercepts, let $x = 0$ in the equation and solve for y.	Let $4x - 5y = 20$. x-intercept:　$4x - 5(0) = 20$ 　　　　　　　　　　$x = 5$ y-intercept:　$4(0) - 5y = 20$ 　　　　　　　　　　$y = -4$ The x-intercept is $(5, 0)$ and the y-intercept is $(0, -4)$.
Horizontal Line	A horizontal line has equation $y = b$, where b is a constant. It also has y-intercept $(0, b)$ and no x-intercept when $b \neq 0$.	**y-intercept $(0, b)$** [graph showing horizontal line $y = b$]

continued on next page

continued from previous page

CONCEPT	EXPLANATION	EXAMPLES
Vertical Line	A vertical line has equation $x = k$, where k is a constant. It also has x-intercept $(k, 0)$ and no y-intercept when $k \neq 0$.	**x-intercept $(k, 0)$** 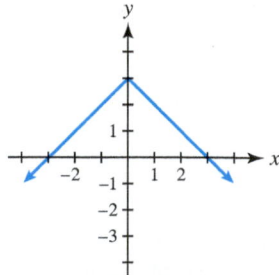

3.3 Exercises

MyMathLab®

CONCEPTS AND VOCABULARY

1. A point where a graph intersects the x-axis is called a(n)_____.

2. To find an x-intercept, let $y =$ _____ and solve for x.

3. A point where a graph intersects the y-axis is called a(n)_____.

4. To find a y-intercept, let $x =$ _____ and solve for y.

5. The graph of the linear equation $Ax + By = C$ with $A = 0$ and $B = 1$ is a(n) _____ line.

6. A horizontal line with y-intercept $(0, b)$ has equation _____.

7. The graph of the linear equation $Ax + By = C$ with $A = 1$ and $B = 0$ is a(n) _____ line.

8. A vertical line with x-intercept $(k, 0)$ has equation _____.

FINDING INTERCEPTS

Exercises 9–16: Identify any x-intercepts and y-intercepts in the graph.

9.

10.

11.

12.

13.

14.

15.

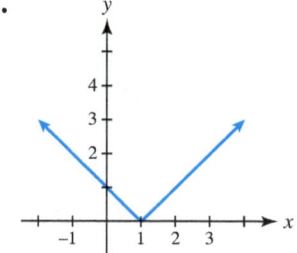

16.

Exercises 17–20: Complete the table. Then determine the x-intercept and the y-intercept for the graph of the equation.

17. $y = x + 2$

x	-2	-1	0	1	2
y					

18. $y = 2x - 4$

x	−2	−1	0	1	2
y					

19. $-x + y = -2$

x	−4	−2	0	2	4
y					

20. $x + y = 1$

x	−2	−1	0	1	2
y					

USING INTERCEPTS TO GRAPH LINES

Exercises 21–38: Find any intercepts for the graph of the linear equation and then graph the equation.

21. $-2x + 3y = -6$ **22.** $4x + 3y = 12$

23. $x - 3y = 6$ **24.** $5x + y = -5$

25. $6x - y = -6$ **26.** $5x + 7y = -35$

27. $3x + 7y = 21$ **28.** $-3x + 8y = 24$

29. $40y - 30x = -120$ **30.** $10y - 20x = 40$

31. $\frac{1}{2}x - y = 2$ **32.** $x - \frac{1}{2}y = 4$

33. $-\dfrac{x}{4} + \dfrac{y}{3} = 1$ **34.** $\dfrac{x}{3} - \dfrac{y}{4} = 1$

35. $\dfrac{x}{3} + \dfrac{y}{2} = 1$ **36.** $\dfrac{x}{5} - \dfrac{y}{4} = 1$

37. $0.6y - 1.5x = 3$ **38.** $0.5y - 0.4x = 2$

39. Thinking Generally Find any intercepts for the graph of $Ax + By = C$.

40. Thinking Generally Find any intercepts for the graph of $\frac{x}{A} + \frac{y}{B} = 1$.

RECOGNIZING HORIZONTAL AND VERTICAL LINES

Exercises 41–44: Write an equation for the line that passes through the points shown in the table.

41.

x	−2	−1	0	1	2
y	1	1	1	1	1

42.

x	0	1	2	3	4
y	−10	−10	−10	−10	−10

43.

x	−6	−6	−6	−6	−6
y	5	4	3	2	1

44.

x	20	20	20	20	20
y	−2	−1	0	1	2

Exercises 45–50: Graph each equation.

45. (a) $y = 2$ **(b)** $x = 2$

46. (a) $y = -4$ **(b)** $x = -4$

47. (a) $y = 0$ **(b)** $x = 0$

48. (a) $y = -\frac{1}{2}$ **(b)** $x = -\frac{1}{2}$

49. (a) $y = \frac{3}{2}$ **(b)** $x = \frac{3}{2}$

50. (a) $y = -1.5$ **(b)** $x = -1.5$

Exercises 51–58: Write an equation for the line shown in the graph.

51.

52.

53.

54.

55.

56.

57.

58.
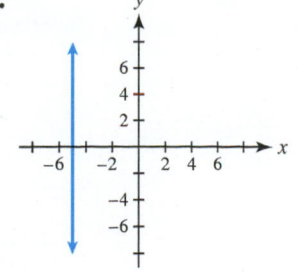

Exercises 59–64: Write the equations of a horizontal line and a vertical line that pass through the given point. (Hint: Make a sketch.)

59. $(1, 2)$

60. $(-3, 4)$

61. $(20, -45)$

62. $(-5, 12)$

63. $(0, 5)$

64. $(-3, 0)$

Exercises 65–72: Find an equation for a line satisfying the following conditions.

65. Vertical, passing through $(-1, 6)$

66. Vertical, passing through $(2, -7)$

67. Horizontal, passing through $\left(\frac{3}{4}, -\frac{5}{6}\right)$

68. Horizontal, passing through $(5.1, 6.2)$

69. Perpendicular to $y = \frac{1}{2}$, passing through $(4, -9)$

70. Perpendicular to $x = 2$, passing through $(3, 4)$

71. Parallel to $x = 4$, passing through $\left(-\frac{2}{3}, \frac{1}{2}\right)$

72. Parallel to $y = -2.1$, passing through $(7.6, 3.5)$

73. Thinking Generally Write the equation of the *x*-axis. (*Hint:* The *x*-axis is a horizontal line.)

74. Thinking Generally Write the equation of the *y*-axis. (*Hint:* The *y*-axis is a vertical line.)

APPLICATIONS INVOLVING LINES

*Exercises 75 and 76: **Distance** The distance of a driver from home is illustrated in the graph.*

  (a) Find the intercepts.
  (b) Interpret each intercept.

75. **Distance from Home**

76. **Distance from Home**

*Exercises 77 and 78: **Modeling a Toy Rocket** (Refer to Example 3.) The velocity v of a toy rocket in feet per second after t seconds of flight is given. Assume that $t \geq 0$ and t does not take on values greater than the t-intercept.*

  (a) Find the intercepts and then graph the equation.
  (b) Interpret each intercept.

77. $v = 128 - 32t$

78. $v = 96 - 32t$

*Exercises 79 and 80: **Water in a Pool** The amount of water in a swimming pool is depicted in the graph.*

  (a) Find the intercepts.
  (b) Interpret each intercept.

79. **Water in a Pool**

80. **Water in a Pool**

WRITING ABOUT MATHEMATICS

81. Given an equation, explain how to find an *x*-intercept and a *y*-intercept.

82. The form $\frac{x}{a} + \frac{y}{b} = 1$ is called the **intercept form** of a linear equation. Explain how you can use this equation to find the intercepts. (*Hint:* Graph $\frac{x}{2} + \frac{y}{3} = 1$ and find its intercepts.)

Group Activity

1. Radio Stations The approximate number of radio stations on the air for each decade from 1960 to 2010 is shown in the table.

x (year)	1960	1970	1980
y (stations)	4100	6800	8600

x (year)	1990	2000	2010
y (stations)	10,800	12,600	14,500

Source: M. Street Corporation.

Make a line graph of the data. Be sure to label both axes.

2. Estimation Discuss ways to estimate the number of radio stations on the air in 1975. Compare your estimates with the actual value of 7700 stations. Repeat this estimate for 1985 and compare it to the actual value of 10,400. Discuss your results.

3. Modeling Equation Substitute each *x*-value from the table into the equation $y = 220x - 427{,}100$ and determine the corresponding *y*-value. Do these *y*-values give reasonable approximations to the *y*-values in the table? Explain your answer.

4. Making Estimates Use $y = 220x - 427{,}100$ to estimate the number of radio stations on the air in 1975 and 1985. Compare the results to your answer in Exercise 2.

3.4 Slope and Rates of Change

Objectives

1 **Introducing Slope**
2 **Recognizing Positive, Negative, Zero, and Undefined Slope**
3 **Finding Slopes of Lines**
 • **Using Slope to Sketch a Line**
 • **Using Slope to Complete a Table**
4 **Understanding Slope as a Rate of Change**

NEW VOCABULARY

☐ Rise
☐ Run
☐ Slope
☐ Positive slope
☐ Negative slope
☐ Zero slope
☐ Undefined slope
☐ Rate of change

1 Introducing Slope

Take a moment to look at the graphs in **FIGURE 3.25**, where the horizontal axis represents time.

(a)

(b)

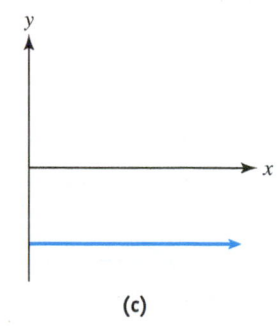
(c)

FIGURE 3.25

Which graph might represent:

- the distance traveled by you if you are walking at a constant rate?
- the temperature in your freezer?
- the value of a laptop computer after it is purchased?

To be able to answer these questions, you probably used the concept of slope. In mathematics, slope is a real number that measures the "tilt" of a line.

Connecting Concepts with Your Life Suppose you are using a stairway where each tread is **12** inches wide and the rise between steps is **6** inches high, as illustrated in **FIGURE 3.26(a)**. We say that this stairway has a *rise* of 6 inches for each 12 inches of *run*. Next, if we draw a blue line along this stairway and include an *x*- and *y*-axis, as shown in **FIGURE 3.26(b)**, then this blue line has a *slope* of $\frac{6}{12}$, or $\frac{1}{2}$. That is, the slope of a line along this stairway equals *rise over run*.

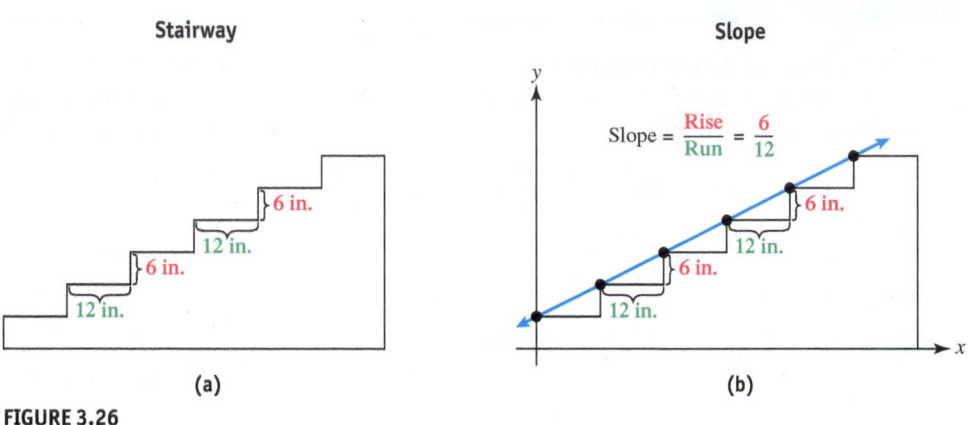

FIGURE 3.26

NOTE: For a line, the ratio $\frac{\text{rise}}{\text{run}}$ equals the slope and is always the same regardless of where one is on the line. ■

2 Recognizing Positive, Negative, Zero, and Undefined Slope

By visually scanning across the graph of a line from *left to right*, we can determine if the graph has a positive slope, a negative slope, a slope of zero, or an undefined slope. The following See the Concept shows how this is done.

See the Concept 1   **POSITIVE, NEGATIVE, ZERO, AND UNDEFINED SLOPE**

In mathematics, it is common to use the letter *m* to represent slope. Visually scan across a graph from left to right to determine the "character" of its slope, *m*.

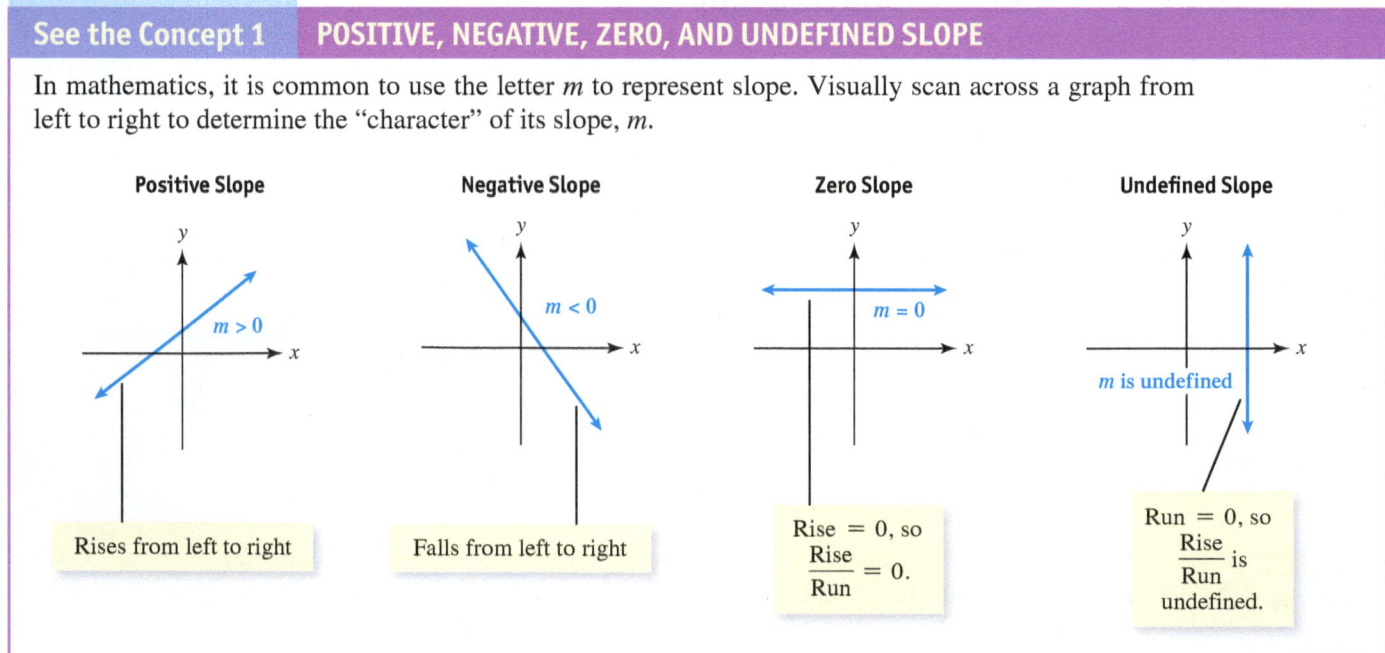

WATCH VIDEO IN MML.

> **SLOPE OF A LINE**
>
> 1. A line that rises from left to right has **positive slope**.
> 2. A line that falls from left to right has **negative slope**.
> 3. A horizontal line has **zero slope**.
> 4. A vertical line has **undefined slope**.

3 Finding Slopes of Lines

 Connecting Concepts with Your Life The graph shown in **FIGURE 3.27** illustrates the cost of parking for x hours. The graph tilts upward from left to right, which indicates that the cost increases as the number of hours increases.

Cost of Parking

FIGURE 3.27

The graph *rises* 2 units vertically for every horizontal unit of *run*, and the ratio $\frac{\text{rise}}{\text{run}}$ equals the *slope* of the line. The slope m of this line is **2**, which indicates that the cost of parking is $2 per hour. In applications, slope indicates a *rate of change*.

If we know the coordinates of any two points on a (non-vertical) line, we can calculate the slope of the line. A general case is shown in the following See the Concept, where a line passes through the points (x_1, y_1) and (x_2, y_2).

See the Concept 2   **CALCULATING SLOPE USING TWO POINTS**

Any two *distinct* points on a *non-vertical* line can be used to find the line's slope.

Equivalent Slope Formulas

$$m = \frac{\textbf{Rise}}{\textbf{Run}}$$

$$m = \frac{\textbf{Change in } y}{\textbf{Change in } x}$$

$$m = \frac{y_2 - y_1}{x_2 - x_1} \quad (\text{where } x_1 \neq x_2)$$

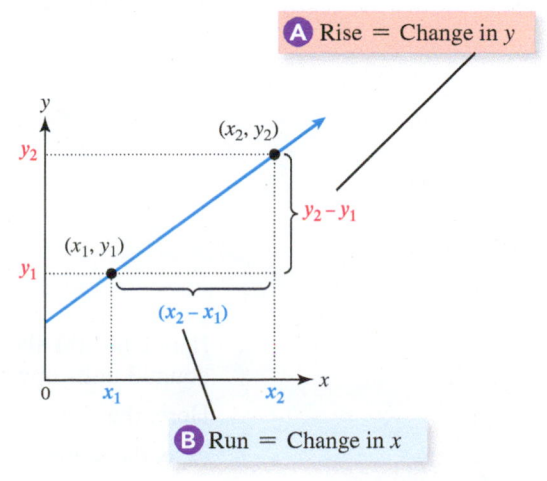

Ⓐ The **rise** is found by calculating the **change in y**. To do this, subtract $y_2 - y_1$.

Ⓑ The **run** is found by calculating the **change in x**. To do this, subtract $x_2 - x_1$.

Ⓐ Rise = Change in y

Ⓑ Run = Change in x

WATCH VIDEO IN MML.

NOTE: The symbol x_1 has a *subscript* of 1 and is read "*x* sub one" or "*x* one". Thus x_1 and x_2 are used to denote two different *x*-values. Similar comments apply to y_1 and y_2. ■

READING CHECK 1

• By looking at the *x*-coordinates of two points on a line, how can you tell if the line has undefined slope?

SLOPE

The **slope** m of the line passing through the points (x_1, y_1) and (x_2, y_2) is

$$m = \frac{\text{Rise}}{\text{Run}} = \frac{y_2 - y_1}{x_2 - x_1},$$

where $x_1 \neq x_2$. That is, slope equals *rise over run*.

NOTE: If $x_1 = x_2$, the line is vertical and the slope is undefined. ■

EXAMPLE 1 **Calculating the slope of a line**

Use the two points labeled in **FIGURE 3.28** to find the slope of the line. What are the rise and run between these two points? Interpret the slope in terms of rise and run.

FIGURE 3.28

Solution
The line passes through the points $(-3, 1)$ and $(3, -2)$, so let $(x_1, y_1) = (-3, 1)$ and $(x_2, y_2) = (3, -2)$. The slope is

$$m = \frac{y_2 - y_1}{x_2 - x_1} = \frac{-2 - 1}{3 - (-3)} = \frac{-3}{6} = -\frac{1}{2}.$$

Starting at the point $(-3, 1)$, count 3 units downward and then 6 units to the right to return to the graph at the point $(3, -2)$. Thus the "rise" is -3 units and the run is 6 units. See **FIGURE 3.29(a)**. The ratio $\frac{\text{rise}}{\text{run}}$ is $\frac{-3}{6}$, or $-\frac{1}{2}$.

**Finding Slope Moving
Left to Right**

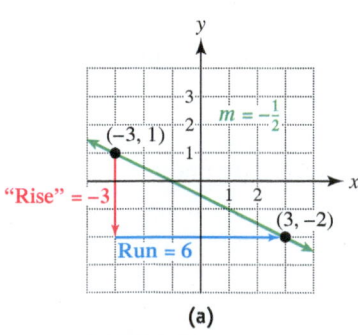

(a)

**Finding Slope Moving
Right to Left**

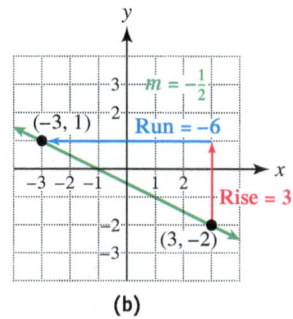

(b)

FIGURE 3.29

FIGURE 3.29 (b) shows an alternate way of finding this slope. Starting at the point $(3, -2)$, count 3 units upward and then 6 units to the left to return to the graph at the point $(-3, 1)$. Here, the rise is 3 units and the run is -6 units so that the ratio $\frac{\text{rise}}{\text{run}}$ is $\frac{3}{-6}$, or $-\frac{1}{2}$. In either case, the slope is $-\frac{1}{2}$.

Now Try Exercise 27

NOTE: In Example 1, the same slope would result if we let $(x_1, y_1) = (3, -2)$ and $(x_2, y_2) = (-3, 1)$. In this case, the calculation would be

$$m = \frac{y_2 - y_1}{x_2 - x_1} = \frac{1 - (-2)}{-3 - 3} = \frac{3}{-6} = -\frac{1}{2}. \quad \blacksquare$$

A graph is not needed to find slope. Any two points on a line can be used to find the line's slope, as demonstrated in the next example.

EXAMPLE 2 **Calculating the slope of a line**

Calculate the slope of the line passing through each pair of points. Graph the line.
(a) $(-2, 3), (2, 1)$ (b) $(-1, 3), (2, 3)$ (c) $(-3, 3), (-3, -2)$

Solution

(a) $m = \frac{y_2 - y_1}{x_2 - x_1} = \frac{1 - 3}{2 - (-2)} = \frac{-2}{4} = -\frac{1}{2}$. This slope indicates that the line falls 1 unit for every 2 units of horizontal run, as shown in **FIGURE 3.30 (a)**.

(b) $m = \frac{y_2 - y_1}{x_2 - x_1} = \frac{3 - 3}{2 - (-1)} = \frac{0}{3} = 0$. The line is horizontal, as shown in **FIGURE 3.30 (b)**.

(c) Because $x_1 = x_2 = -3$, the slope formula does not apply. If we try to use it, we obtain
$m = \frac{y_2 - y_1}{x_2 - x_1} = \frac{-2 - 3}{-3 - (-3)} = \frac{-5}{0}$, which is an **undefined** expression. The line has undefined slope and is vertical, as shown in **FIGURE 3.30 (c)**.

(a)

(b)

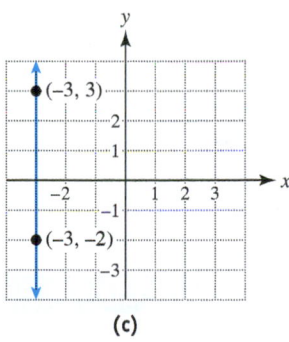
(c)

FIGURE 3.30

Now Try Exercises 41, 43, 45

EXAMPLE 3 **Finding slope from a graph**

Find the slope of each line.
(a)

(b)

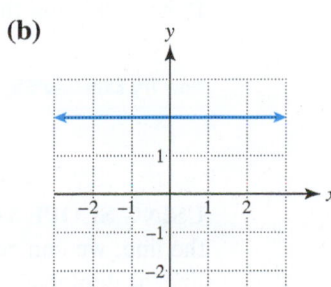

Solution

(a) The graph rises 30 units for every 10 units of run. For example, the graph passes through $(10, 10)$ and $(20, 40)$, so the line rises $40 - 10 = 30$ units while the run is $20 - 10 = 10$ units, as shown in **FIGURE 3.31**.

$$m = \frac{\text{Rise}}{\text{Run}} = \frac{30}{10} = 3$$

(b) The line is horizontal, so the slope is 0.

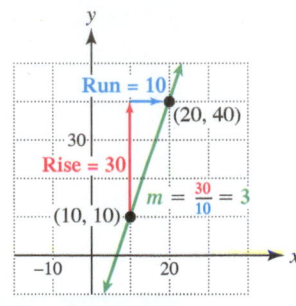

FIGURE 3.31

Now Try Exercises 19, 29

USING SLOPE TO SKETCH A LINE A point and a slope also determine a line, as illustrated in the next example.

EXAMPLE 4 Sketching a line with a given slope

Sketch a line that passes through the point $(1, 4)$ and has slope $-\frac{2}{3}$.

Slope: $-\frac{2}{3}$; Through: $(1, 4)$

Solution

Start by plotting the point $(1, 4)$. A slope of $-\frac{2}{3}$ can be written as $\frac{-2}{3}$, which indicates that the line *falls* 2 units for every 3-unit increase in the run. Because the line passes through $(1, 4)$, a 2-unit decrease in y and a 3-unit increase in x result in the line passing through the point $(1 + 3, 4 - 2)$ or $(4, 2)$. See **FIGURE 3.32**.

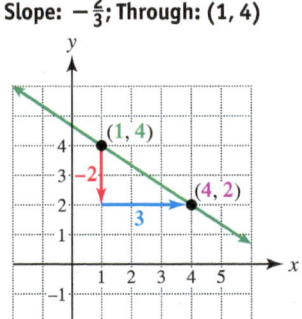

FIGURE 3.32

Now Try Exercise 59

EXAMPLE 5 Sketching a line with a given *y*-intercept

Sketch a line with slope -2 and *y*-intercept $(0, 3)$.

Slope: -2; *y*-intercept: $(0, 3)$

Solution

First plot the *y*-intercept, $(0, 3)$. The slope of -2 can be written as $\frac{-2}{1}$, so the *y*-values decrease 2 units for each unit increase in x. Increasing the *x*-value in the point $(0, 3)$ by 1 and decreasing the *y*-value by 2 result in the point $(0 + 1, 3 - 2)$ or $(1, 1)$. Plot $(1, 1)$ and then sketch the line, as shown in **FIGURE 3.33**.

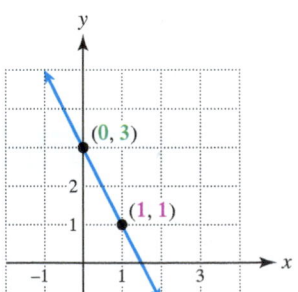

FIGURE 3.33

Now Try Exercise 55

USING SLOPE TO COMPLETE A TABLE If we know the slope of a line and a point on the line, we can complete a table of values that gives the coordinates of other points on the line, as demonstrated in the next example.

EXAMPLE 6 **Completing a table of values**

A line has slope 2 and passes through the first point listed in the table. Complete the table so that each point lies on the line.

x	-2	-1	0	1
y	1			

> Complete the table so that the slope between consecutive points is 2.

Solution

Slope 2 indicates that $\frac{\text{Rise}}{\text{Run}} = 2$. Because consecutive x-values in the table increase by 1 unit, the run from one point in the table to the next is 1 unit. Substituting **1** for the **run** in the slope equation results in $\frac{\text{Rise}}{1} = 2$. Thus the **rise** is **2** and consecutive y-values shown in **TABLE 3.16** increase by 2 units.

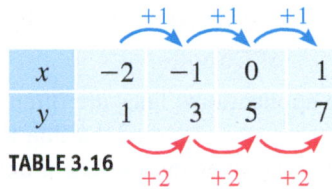

x	-2	-1	0	1
y	1	3	5	7

TABLE 3.16

> The slope between each pair of points is 2.

Now Try Exercise 63

4 ▸ Understanding Slope as a Rate of Change

When lines are used to model physical quantities in applications, their slopes provide important information. Slope measures the **rate of change** in a quantity.

EXAMPLE 7 **Interpreting slope**

The distance y in miles that an athlete training for a marathon is from home after x hours is shown in **FIGURE 3.34**.
(a) Find the y-intercept. What does it represent?
(b) The graph passes through the point $(1, 10)$. Discuss the meaning of this point.
(c) Find the slope of this line. Interpret the slope as a rate of change.

Distance from Home

FIGURE 3.34

Solution
(a) The y-intercept is $(0, 20)$, so the athlete is initially 20 miles from home.
(b) The point $(1, 10)$ means that after 1 hour the athlete is 10 miles from home.

(c) The line passes through the points $(0, 20)$ and $(1, 10)$. Its slope is

$$m = \frac{10 - 20}{1 - 0} = -10.$$

Slope -10 means that the athlete is running *toward* home at 10 miles per hour. A *negative* slope indicates that the distance between the runner and home is decreasing.

Now Try Exercise 87

NOTE: The units for a rate of change are determined by putting the y-axis units over the x-axis units. In Example 7, the y-units are **miles** and the x-units are **hours**, so the units for the rate of change are $\frac{\text{miles}}{\text{hour}}$, or miles per hour. ∎

READING CHECK 2

• When a line models a real-world situation, how are the units of the slope found?

MAKING CONNECTIONS 1

Average Rates of Change

In real-world applications, lines are often used to model data that may not be perfectly linear. In such cases, the slope of the line that models the situation gives an *average* rate of change for the quantities involved. For example, **FIGURE 3.35** illustrates the actual number of words typed as a student writes a term paper. Because the student does not type at a *constant* rate, a line does not fit the situation perfectly. However, the line shown in **FIGURE 3.36** can be used to *model* the student's typing and its slope gives the *average* number of words typed per minute.

FIGURE 3.35

The student types at an *average* rate of **50** words per minute.

FIGURE 3.36

EXAMPLE 8   **Interpreting slope**

When a company manufactures 2000 MP3 players, its profit is \$10,000, and when it manufactures 4500 MP3 players, its profit is \$35,000.
(a) Find the slope of the line passing through $(2000, 10000)$ and $(4500, 35000)$.
(b) Interpret the slope as a rate of change.

Solution

(a) $m = \dfrac{35,000 - 10,000}{4500 - 2000} = \dfrac{25,000}{2500} = 10$

(b) When 2000 to 4500 MP3 players are manufactured, profit is, *on average*, \$10 per MP3 player made and sold.

Now Try Exercise 91

🌐 Math in Context Slope can be used to estimate *average rates of change* in the occurrences of a disease. In the next example, this concept is applied to tetanus.

EXAMPLE 9 Analyzing tetanus cases

TABLE 3.17 lists numbers of reported cases of tetanus in the United States for selected years.

Cases of Tetanus in the United States

Year	1960	1970	1980	1990	2000	2010
Cases of Tetanus	368	148	95	64	45	18

TABLE 3.17 *Source:* Department of Health and Human Services.

(a) Make a line graph of the data.
(b) Find the slope of each line segment.
(c) Interpret each slope as a rate of change.

Solution

(a) A line graph connecting the points $(1960, 368)$, $(1970, 148)$, $(1980, 95)$, $(1990, 64)$, $(2000, 45)$, and $(2010, 18)$ is shown in FIGURE 3.37.

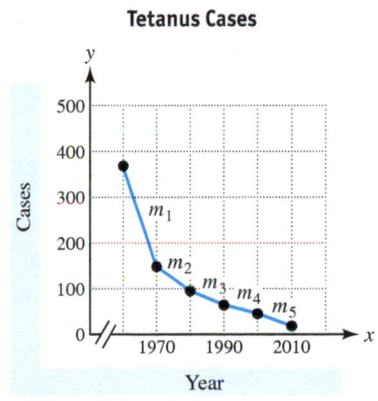

Tetanus Cases

FIGURE 3.37

(b) The slope of each line segment may be calculated as follows.

$$m_1 = \frac{148 - 368}{1970 - 1960} = -22.0, \qquad m_2 = \frac{95 - 148}{1980 - 1970} = -5.3,$$

$$m_3 = \frac{64 - 95}{1990 - 1980} = -3.1, \qquad m_4 = \frac{45 - 64}{2000 - 1990} = -1.9$$

$$m_5 = \frac{18 - 45}{2010 - 2000} = -2.7$$

(c) Slope $m_1 = -22.0$ indicates that, *on average*, the number of tetanus cases *decreased* by 22.0 cases per year from 1960 to 1970. The other four slopes can be interpreted similarly.

NOTE: The number of tetanus cases did not decrease by *exactly* 22.0 cases each year between 1960 and 1970. However, the yearly *average* decrease was 22.0 cases. ∎

Now Try Exercise 89

EXAMPLE 10 Sketching a model

During a storm, rain falls at the average rates of 2 inches per hour from 1 A.M. to 3 A.M., 1 inch per hour from 3 A.M. to 4 A.M., and $\frac{1}{2}$ inch per hour from 4 A.M. to 6 A.M.

(a) Sketch a graph that shows the total accumulation of rainfall from 1 A.M. to 6 A.M.
(b) What does the slope of each line segment represent?

Solution

(a) At 1 A.M. the accumulated rainfall is 0, so place a point at $(1, 0)$. Rain falls at an average rate of 2 inches per hour for the next 2 hours, so at 3 A.M. the total rainfall is 4 inches. Place a point at $(3, 4)$. Sketch a line segment from $(1, 0)$ to $(3, 4)$, as shown in **FIGURE 3.38**. Similarly, during the next hour 1 inch of rain falls, so draw a line segment from $(3, 4)$ to $(4, 5)$. Finally, 1 inch of rain falls from 4 A.M. to 6 A.M., so draw a line segment from $(4, 5)$ to $(6, 6)$.

(b) The slope of each line segment represents the average rate at which rain is falling. For example, the first segment has slope 2 because rain falls at an average rate of 2 inches per hour during this period of time.

Total Rainfall

FIGURE 3.38

Now Try Exercise 81

3.4   Putting It All Together

CONCEPT	COMMENTS	EXAMPLE
Rise, Run, and Slope	Rise is a vertical change in a line, and run is a horizontal change in a line. The ratio $\frac{\text{rise}}{\text{run}}$ is the slope m when run is nonzero. 1. A line that rises from left to right has positive slope. 2. A line that falls from left to right has negative slope. 3. A horizontal line has zero slope. 4. A vertical line has undefined slope.	$m = \dfrac{\text{Rise}}{\text{Run}} = \dfrac{1}{2}$ Run = 2 Rise = 1
Calculating Slope	For any two points (x_1, y_1) and (x_2, y_2), slope m is $$m = \frac{y_2 - y_1}{x_2 - x_1},$$ where $x_1 \neq x_2$.	The slope of the line passing through $(-2, 3)$ and $(1, 5)$ is $$m = \frac{5 - 3}{1 - (-2)} = \frac{2}{3}.$$ The line rises vertically 2 units for every 3 horizontal units of run.
Slope as a Rate of Change	Slope measures the "tilt" of a line. In applications, slope measures the rate of change in a quantity.	Slope $m = -66\frac{2}{3}$ indicates that water is *leaving* the pool at the rate of $66\frac{2}{3}$ gallons per hour. **Water in a Pool** $m = -66\frac{2}{3}$

3.4 Exercises

MyMathLab®

CONCEPTS AND VOCABULARY

1. (True or False?) The change in the horizontal distance along a line is called run.

2. (True or False?) The change in the vertical distance along a line is called rise.

3. Slope m of a line equals _____ over _____.

4. Slope 0 indicates that a line is _____ .

5. Undefined slope indicates that a line is _____ .

6. If a line passes through (x_1, y_1) and (x_2, y_2) where $x_1 \neq x_2$, then $m =$ _____ .

7. A line that rises from left to right has _____ slope.

8. A line that falls from left to right has _____ slope.

9. When a line models a physical quantity in an application, its slope measures the _____ of change in the quantity.

10. If a line that models the distance between a hiker and camp has a negative slope, is the hiker moving away from or toward camp?

RECOGNIZING SLOPE

Exercises 11–18: State whether the slope of the line is positive, negative, zero, or undefined.

11.

12.

13.
14.

15.
16.

17.
18.
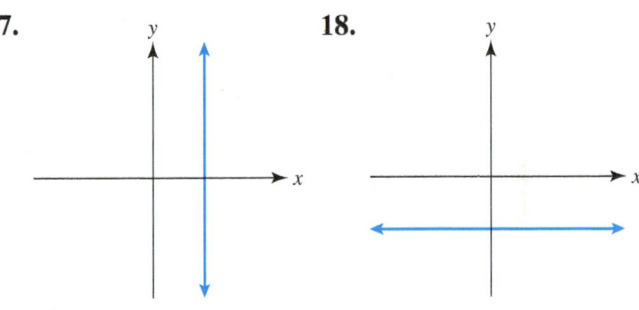

FINDING SLOPES OF LINES

Exercises 19–30: If possible, find the slope of the line. Interpret the slope in terms of rise and run.

19.
20.

21.
22.

23.

24.

25.

26.

27.

28.

29.

30.

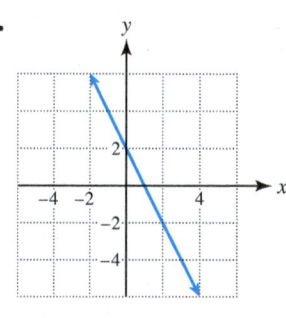

Exercises 31–38: If possible, find the slope of the line passing through the two points. Graph the line.

31. $(1, 2), (2, 4)$ **32.** $(-4, 7), (7, -4)$

33. $(2, 1), (2, 4)$ **34.** $(-1, 3), (-1, -1)$

35. $(1, 3), (-2, 5)$ **36.** $(0, -4), (-4, 6)$

37. $(2, -1), (-2, -1)$ **38.** $(-2, 3), (1, 3)$

Exercises 39–54: If possible, find the slope of the line passing through the two points.

39. $(4, -2), (-3, -9)$ **40.** $(15, -3), (20, 9)$

41. $(-3, 4), (4, -2)$ **42.** $(1, -3), (3, -5)$

43. $(-3, 5), (2, 5)$ **44.** $(-3, 3), (-5, 3)$

45. $(-1, 6), (-1, -4)$ **46.** $\left(\frac{1}{2}, -\frac{2}{7}\right), \left(\frac{1}{2}, \frac{13}{17}\right)$

47. $(1980, 5), (2000, 18)$

48. $(1989, 10), (1999, 16)$

49. $(1950, 6.1), (2000, 10.6)$

50. $(1900, 10), (1950, 35)$

51. $\left(\frac{1}{3}, -\frac{2}{7}\right), \left(-\frac{2}{3}, \frac{3}{7}\right)$

52. $(-1.3, 5.6), (-2.6, -2.5)$

53. $(12, -34), (14, 64)$

54. $(-25, 105), (60, 55)$

SKETCHING LINES

Exercises 55–62: (Refer to Example 4.) Sketch a line that passes through the point and has slope m.

55. $(0, 2), m = -1$ **56.** $(0, -1), m = 2$

57. $(1, 1), m = 3$ **58.** $(1, -1), m = -2$

59. $(-2, 3), m = -\frac{1}{2}$ **60.** $(-1, -2), m = \frac{3}{4}$

61. $(-3, 1), m = \frac{1}{2}$ **62.** $(-2, 2), m = -3$

COMPLETING TABLES

Exercises 63–68: (Refer to Example 6.) A line has the given slope m and passes through the first point listed in the table. Complete the table so that each point in the table lies on the line.

63. $m = 2$

x	0	1	2	3
y	-4			

64. $m = -\frac{1}{2}$

x	0	1	2	3
y	2			

65. $m = -3$

x	1	2	3	4
y	4			

66. $m = -1$

x	-1	0	1	2
y	10			

67. $m = \frac{3}{2}$

x	-4	-2	0	2
y	0			

68. $m = 3$

x	-2	0	2	4
y	-4			

UNDERSTANDING SLOPE AS A RATE OF CHANGE

Exercises 69–72: **Modeling** *Choose the graph (a.–d.) that models the situation best.*

69. Cost of buying x gum balls at a price of 25¢ each

70. Total number of computers with CD/DVD drives purchased during the past 5 years

71. Average cost of a new car over the past 30 years

72. Height of the Empire State Building after x people have entered it

a.

b.

c.

d.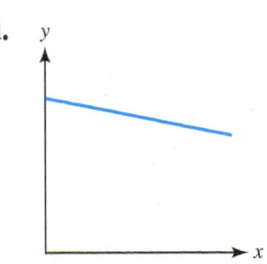

Exercises 73 and 74: **Modeling** *The line graph represents the gallons of water in a small swimming pool after x hours. Assume that there is a pump that can either add water to or remove water from the pool at a constant rate.*

> *(a) Estimate the slope of each line segment.*
> *(b) Interpret each slope as a rate of change.*
> *(c) Describe what happened to the amount of water in the pool.*

73.

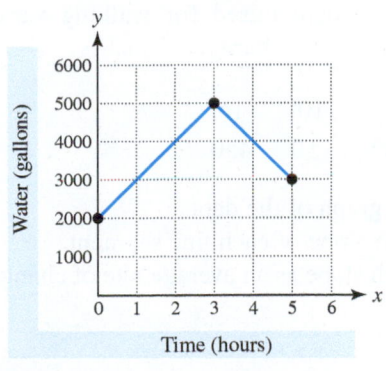

Water in a Pool

74.

Water in a Pool

Exercises 75 and 76: **Modeling** *An individual is driving a car at a constant speed along a straight road. The graph shows the distance that the driver is from home after x hours.*

> *(a) Find the slope of each line segment in the graph.*
> *(b) Interpret each slope as a rate of change.*
> *(c) Describe both the motion of the car and its distance from home.*

75.

Distance from Home

76.

Distance from Home

77. Thinking Generally If a line is used to model physical data where the x-axis is labeled "Time (minutes)" and the y-axis is labeled "Volume (cubic meters)," what are the units for the rate of change represented by the slope of the line?

78. Thinking Generally If a line is used to model physical data where the x-axis is labeled "Cookies" and the y-axis is labeled "Chocolate Chips," what are the units for the rate of change represented by the slope of the line?

Exercises 79–82: **Sketching a Model** *Sketch a graph that models the given situation.*

79. The distance that a boat is from a harbor if the boat is initially 6 miles from the harbor and arrives at the harbor after sailing at a constant speed for 3 hours

80. The distance that a person is from home if the person starts at home, walks away from home at 4 miles per hour for 90 minutes, and then walks back home at 3 miles per hour

81. The distance that an athlete is from home if the athlete jogs for 1 hour to a park that is 7 miles away, stays for 30 minutes, and then jogs home at the same pace

82. The amount of oil in a 55-gallon drum that is initially full, then is drained at a rate of 5 gallons per minute for 4 minutes, is left for 6 minutes, and then is emptied at a rate of 7 gallons per minute

83. Online Exploration Search the "Fast Facts" pages on the U.S. Census Bureau's Web site to complete the following.
(a) Rounding to the nearest million, what was the U.S. population in 1900?
(b) Rounding to the nearest million, what was the U.S. population in 2000?
(c) Using the concept of slope, find the *average* rate of change in the U.S. population over this time.

84. Online Exploration Search the "Fast Facts" pages on the U.S. Census Bureau's Web site to complete the following.
(a) Find the population of New York City in 1900. Round to the nearest hundred-thousand.
(b) Find the population of New York City in 2000. Round to the nearest hundred-thousand.
(c) Using the concept of slope, find the *average* rate of change in the New York City population over this time.

APPLICATIONS INVOLVING SLOPE

85. Twitter Followers One year after a celebrity first began posting comments on Twitter, she had 25,600 followers. Four years after this celebrity first began posting comments, she had 148,000 followers.
(a) Find the slope of the line that passes through the points $(1, 25600)$ and $(4, 148000)$.
(b) Interpret the slope as an average rate of change.

86. Profit from Tablet Computers When a company manufactures 500 tablet computers, its profit is $100,000, and when it manufactures 1500 tablet computers, its profit is $400,000.
(a) Find the slope of the line passing through the points $(500, 100000)$ and $(1500, 400000)$.
(b) Interpret the slope as an average rate of change.

87. Revenue The graph shows revenue received from selling x flash drives.

Flash Drive Revenue

(a) Find the slope of the line shown.
(b) Interpret the slope as a rate of change.

88. Electricity The graph shows how voltage is related to amperage in an electrical circuit. The slope corresponds to the resistance in ohms. Find the resistance in this electrical circuit.

Voltage

89. Walking for Charities The table lists the amount of money M in dollars raised for walking various distances x in miles for a charity.

x	0	5	10	15
M	0	100	250	450

(a) Make a line graph of the data.
(b) Calculate the slope of each line segment.
(c) Interpret each slope as an average rate of change.

90. Insect Population The table lists the number N of black flies in thousands per acre after x weeks.

x	0	2	4	6
N	3	4	10	18

(a) Make a line graph of the data.
(b) Calculate the slope of each line segment.
(c) Interpret each slope as an average rate of change.

91. Median Family Income In 2000, median family income was about \$42,000, and in 2012, it was about \$51,000. (*Source:* Department of the Treasury.)
(a) Find the slope of the line passing through the points $(2000, 42000)$ and $(2012, 51000)$.
(b) Interpret the slope as an average rate of change.
(c) If this trend continues, estimate the median family income in 2018.

92. Social Media Ads In 2012, revenue from social media ads was \$4.3 billion, and by 2017, it is projected to be \$12 billion. (*Source:* BIA/Kelsey.)
(a) Find the slope of the line that passes through the points $(2012, 4.3)$ and $(2017, 12)$.
(b) Interpret the slope as an average rate of change.
(c) If this trend continues, estimate the revenue from ads in 2020.

93. Rate of Change Suppose that $y = -2x + 10$ is graphed in the first quadrant of the xy-plane where the x-axis is labeled "Time (minutes)" and the y-axis is labeled "Distance (feet)." If this graph represents the distance y that an ant is from a stone after x minutes, answer each of the following.
(a) Is the ant moving toward or away from the stone?
(b) Initially, how far from the stone is the ant?
(c) At what rate is the ant moving?
(d) What is the value of x (time) when the ant reaches the stone?

94. Rate of Change Suppose that we graph $y = 15x + 8$ in the first quadrant of the xy-plane where the x-axis is labeled "Time (minutes)" and the y-axis is labeled "Distance (feet)." If this graph represents the distance y that a frog is from a tree after x minutes, answer each of the following.
(a) Is the frog moving toward or away from the tree?
(b) Initially, how far from the tree is the frog?
(c) At what rate is the frog moving?
(d) What is the value of x when the frog is 53 feet from the tree?

WRITING ABOUT MATHEMATICS

95. If you are given two points and the slope formula $m = \frac{y_2 - y_1}{x_2 - x_1}$, does it matter which point is (x_1, y_1) and which point is (x_2, y_2)? Explain.

96. Suppose that a line approximates the distance y in miles that a person drives in x hours. What does the slope of the line represent? Give an example.

97. Describe the information that the slope m gives about a line. Be as complete as possible.

98. Could one line have two slopes? Explain.

SECTIONS 3.3 and 3.4 **Checking Basic Concepts**

1. Identify the x- and y-intercepts in the graph.

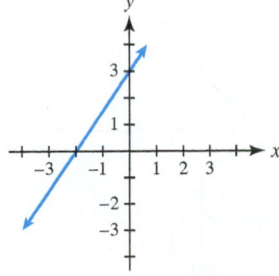

2. Complete the table for the equation $2x - y = 2$. Then determine the x- and y-intercepts.

x	-2	-1	0	1	2
y	-6				

3. Find any intercepts for the graphs of the equations and then graph each linear equation.
(a) $x - 2y = 6$ (b) $y = 2$ (c) $x = -1$

4. Write the equations of a horizontal line and a vertical line that pass through the point $(-2, 4)$.

5. If possible, find the slope of the line passing through each pair of points.
(a) $(-2, 3), (2, 6)$
(b) $(-5, 3), (0, 3)$
(c) $(1, 5), (1, 8)$

6. Find the slope of the line shown.

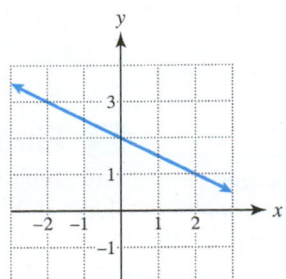

continued on next page

continued from previous page

7. Sketch a line that passes through the point $(-3, 1)$ and has slope 2.

8. **Modeling** The line graph to the right shows the depth of water in a small pond before and after a rain storm.
   (a) Estimate the slope of each line segment.
   (b) Interpret each slope as a rate of change.
   (c) Describe what happened to the amount of water in the pond.

Water in a Pond

3.5   Slope–Intercept Form

Objectives

1 **Determining a Line**
2 **Finding Slope–Intercept Form**
3 **Working with Parallel and Perpendicular Lines**
   • **Parallel Lines**
   • **Perpendicular Lines**

NEW VOCABULARY

☐ Slope–intercept form
☐ Negative reciprocals

STUDY TIP

A new concept is often easier to learn when we can find a relationship between the concept and our personal experience. Try to list two real-life situations that have an initial value and a constant rate of change.

READING CHECK 1

• How many points does it take to determine a line?

1 Determining a Line

Connecting Concepts with Your Life When a line models a real-world situation, slope represents a rate of change and the y-intercept often represents an initial value. For example, if there are initially **30** gallons of water in a small wading pool, and the pool is being filled by a garden hose at a constant rate of **2** gallons per minute, a line representing this situation has a slope of **2** and the y-coordinate of its y-intercept is **30**.

For any two points in the xy-plane, we can draw a unique line passing through them, as illustrated in **FIGURE 3.39(a)**. Another way we can determine a unique line is to know the y-intercept and the slope. For example, if a line has y-intercept $(0, 2)$ and slope $m = 1$, then the resulting line is shown in **FIGURE 3.39(b)**.

Two Points Determine a Line

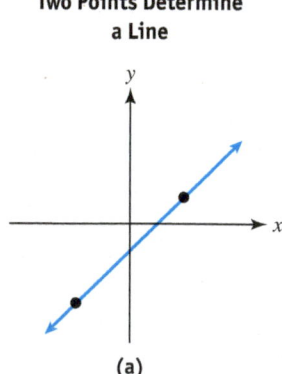

(a)

One Point and a Slope Determine a Line

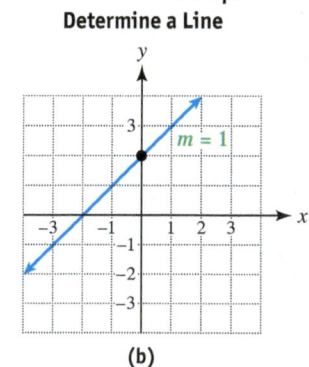

(b)

FIGURE 3.39

2 Finding Slope–Intercept Form

The graph of $y = 2x + 3$ passes through $(0, 3)$ and $(1, 5)$, as shown in **FIGURE 3.40**. The slope of this line is

$$m = \frac{5 - 3}{1 - 0} = 2.$$

Slope 2, y-intercept (0, 3)

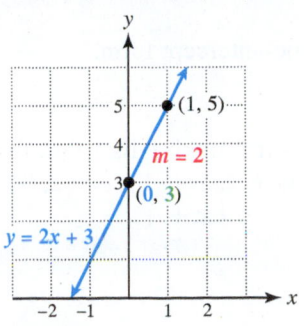

FIGURE 3.40

If $x = 0$ in $y = 2x + 3$, then $y = 2(0) + 3 = 3$. Thus the graph of $y = 2x + 3$ has slope **2** and y-intercept $(0, 3)$. In general, the graph of $y = mx + b$ has slope m and y-intercept $(0, b)$. The form $y = mx + b$ is called the *slope–intercept form*.

SLOPE–INTERCEPT FORM

The line with slope m and y-intercept $(0, b)$ is given by

$$y = mx + b,$$

the **slope–intercept form** of a line.

TABLE 3.18 shows several equations in the form $y = mx + b$ and lists the corresponding slope and y-intercept for the graph associated with each.

READING CHECK 2

- In $y = mx + b$, which value is the slope and what is the y-intercept?

$y = mx + b$ **Form**

Equation	Slope	y-intercept
$y = 4x - 3$	4	$(0, -3)$
$y = 12$	0	$(0, 12)$
$y = -x - \frac{5}{8}$	-1	$\left(0, -\frac{5}{8}\right)$
$y = -\frac{2}{3}x$	$-\frac{2}{3}$	$(0, 0)$

TABLE 3.18

EXAMPLE 1   **Using a graph to write the slope–intercept form**

For each graph, write the slope–intercept form of the line.

(a)

(b)

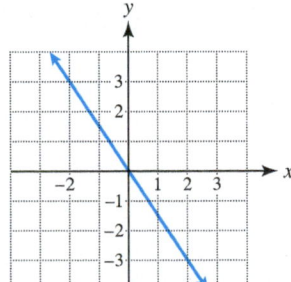

Solution

(a) The graph intersects the y-axis at the y-intercept, $(0, -2)$. Because the graph rises **1** unit for each **2**-unit increase in x, the slope is $\frac{1}{2}$. The slope–intercept form of the line is $y = \frac{1}{2}x - 2$.

(b) The graph intersects the y-axis at the y-intercept, $(0, 0)$. Because the graph falls 3 units for each 2-unit increase in x, the slope is $-\frac{3}{2}$. The slope–intercept form of the line is $y = -\frac{3}{2}x + 0$, or $y = -\frac{3}{2}x$.

Now Try Exercises 15, 19

EXAMPLE 2   ## Sketching a line

Sketch a line with slope $-\frac{1}{2}$ and y-intercept $(0, 1)$. Write its slope–intercept form.

Solution

For the y-intercept, plot the point $(\mathbf{0}, \mathbf{1})$. Slope $-\frac{1}{2}$ indicates that the graph falls **1** unit for each **2**-unit increase in x. Thus the line passes through the point $(\mathbf{0} + \mathbf{2}, \mathbf{1} - \mathbf{1})$, or $(2, 0)$, as shown in **FIGURE 3.41**. The slope–intercept form of this line is $y = -\frac{1}{2}x + 1$.

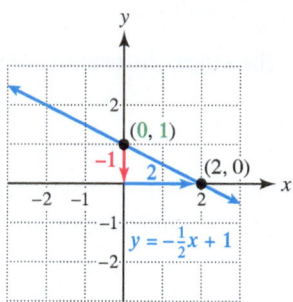

FIGURE 3.41

Now Try Exercise 23

When a linear equation is not given in slope-intercept form, the coefficient of the x-term may not represent the slope and the constant term may not represent the y-intercept. For example, the graph of $2x + 3y = 12$ does *not* have slope 2 and it does *not* have y-intercept $(0, 12)$. To find the correct slope and y-intercept, the equation can first be written in slope–intercept form. This is demonstrated in the next example.

EXAMPLE 3   ## Writing an equation in slope–intercept form

Write each equation in slope–intercept form. Then give the slope and y-intercept of the line.
(a) $2x + 3y = 12$ **(b)** $x = 2y + 4$

Solution

(a) To write the equation in slope–intercept form, solve for y.

$$2x + 3y = 12 \qquad \text{Given equation (standard form)}$$

$$3y = -2x + 12 \qquad \text{Subtract } 2x \text{ from each side.}$$

$$y = -\frac{2}{3}x + 4 \qquad \text{Divide each side by 3.}$$

The slope of the line is $-\frac{2}{3}$, and the y-intercept is $(0, 4)$.

(b) This equation is *not* in slope–intercept form because it is solved for x, not y.

$$x = 2y + 4 \qquad \text{Given equation}$$

$$x - 4 = 2y \qquad \text{Subtract 4 from each side.}$$

$$\frac{1}{2}x - 2 = y \qquad \text{Divide each side by 2.}$$

$$y = \frac{1}{2}x - 2 \qquad \text{Rewrite the equation.}$$

The slope of the line is $\frac{1}{2}$, and the y-intercept is $(0, -2)$.

Now Try Exercises 39, 41

EXAMPLE 4 **Writing an equation in slope–intercept form and graphing it**

Write the equation $2x + y = 3$ in slope–intercept form and then graph it.

Solution
First write the given equation in slope–intercept form.

$$2x + y = 3 \qquad \text{Given equation (standard form)}$$

$$y = -2x + 3 \qquad \text{Subtract } 2x \text{ from each side.}$$

The slope–intercept form is $y = -2x + 3$, with slope -2 and y-intercept $(0, 3)$. To graph this equation, plot the y-intercept as the point $(0, 3)$. The line falls **2** units for each **1**-unit increase in x, so plot the point $(0 + 1, 3 - 2) = (1, 1)$. Sketch a line passing through $(0, 3)$ and $(1, 1)$, as shown in **FIGURE 3.42**.

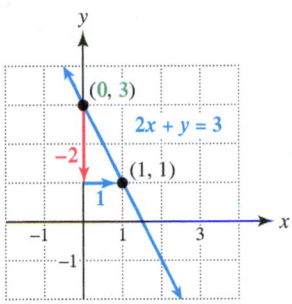

FIGURE 3.42

Now Try Exercise 51

MAKING CONNECTIONS 1

Two Ways to Graph the Same Line

In Example 4, the line $2x + y = 3$ was graphed by finding its slope–intercept form. A second way to graph this line is to find the x- and y-intercepts of the line. If $y = 0$, then $x = \frac{3}{2}$ makes this equation true, and if $x = 0$, then $y = 3$ makes this equation true. Thus the x-intercept is $\left(\frac{3}{2}, 0\right)$ and the y-intercept is $(0, 3)$. Note that the line in **FIGURE 3.42** passes through these points, which could be used to graph the line.

EXAMPLE 5 **Modeling cell phone costs**

Suppose that international calls with a cell phone cost $5 for the initial connection and $0.50 per minute.
(a) If someone talks for 23 minutes, what is the charge?
(b) Write the slope–intercept form that gives the cost of talking for x minutes.
(c) If the charge is $8.50, how long did the person talk?

Solution
(a) The charge for 23 minutes at $**0.50** per minute plus $**5** would be

$$\textbf{0.50} \cdot 23 + \textbf{5} = \$16.50.$$

(b) The rate of increase is $0.50 per minute with an initial cost of $5. Let $y = \textbf{0.5}x + \textbf{5}$, where the slope or rate of change is **0.5** and the y-intercept is $(0, \textbf{5})$.

(c) To determine how long a person can talk for $8.50, we solve the following equation.

$$0.5x + 5 = 8.5 \quad \text{Equation to solve}$$

$$0.5x = 3.5 \quad \text{Subtract 5 from each side.}$$

$$x = \frac{3.5}{0.5} \quad \text{Divide each side by 0.5.}$$

$$x = 7 \quad \text{Simplify.}$$

The person talked for 7 minutes. Note that this solution is based on the assumption that the phone company did not round up a fraction of a minute.

Now Try Exercise 79

Parallel Lines Have the Same Slope

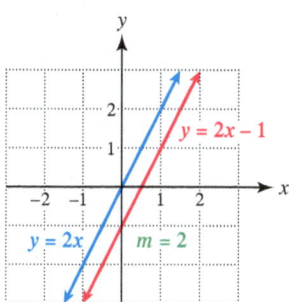

FIGURE 3.43

3 Working with Parallel and Perpendicular Lines

Slope is important for determining whether two lines are parallel or perpendicular.

PARALLEL LINES If two lines have the same slope, they are parallel. For example, the lines $y = 2x$ and $y = 2x - 1$ are parallel because they both have slope 2, as shown in **FIGURE 3.43**.

> **PARALLEL LINES**
>
> Two lines with the same slope are parallel.
> Two nonvertical parallel lines have the same slope.

NOTE: Two vertical lines are parallel and the slope of each is undefined. ■

EXAMPLE 6   **Finding parallel lines**

Find the slope–intercept form of a line parallel to $y = -2x + 3$ and passing through the point $(-2, 3)$. Sketch each line in the same xy-plane.

Solution
Because the line $y = -2x + 3$ has slope -2, any parallel line also has slope -2 with slope–intercept form $y = -2x + b$ for some value of b. The value of b can be found by substituting the point $(-2, 3)$ in the slope–intercept form.

$$y = -2x + b \quad \text{Slope–intercept form}$$

$$3 = -2(-2) + b \quad \text{Let } x = -2 \text{ and } y = 3.$$

$$3 = 4 + b \quad \text{Multiply}$$

$$-1 = b \quad \text{Subtract 4 from each side.}$$

The y-intercept is $(0, -1)$, and so the slope–intercept form is $y = -2x - 1$. The graphs of the equations $y = -2x + 3$ and $y = -2x - 1$ are shown in **FIGURE 3.44**. Note that they are parallel lines, both with slope -2 but with different y-intercepts, $(0, 3)$ and $(0, -1)$.

Now Try Exercise 69

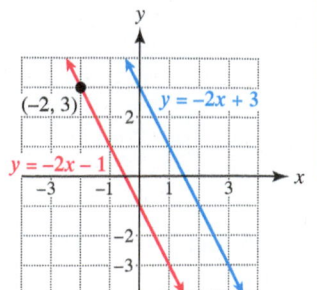

FIGURE 3.44

PERPENDICULAR LINES The lines shown in **FIGURE 3.45** are perpendicular because they intersect at a 90° angle. Rather than measure the angle between two intersecting lines, we can determine whether two lines are perpendicular from their slopes. The slopes of perpendicular lines satisfy the properties given in the box below **FIGURE 3.45**.

Perpendicular Lines

FIGURE 3.45

PERPENDICULAR LINES

If two perpendicular lines have nonzero slopes m_1 and m_2, then $m_1 \cdot m_2 = -1$.

If two lines have slopes m_1 and m_2 such that $m_1 \cdot m_2 = -1$, then they are perpendicular lines.

NOTE: A vertical line and a horizontal line are perpendicular. ∎

The following See the Concept shows how to determine if two lines are perpendicular by looking at their slopes.

See the Concept 1   SLOPES OF PERPENDICULAR LINES

A See the graph.

Negative Reciprocals

m_1	m_2	$m_1 \cdot m_2$
1	-1	-1
2	$-\frac{1}{2}$	**-1**
-4	$\frac{1}{4}$	-1
$\frac{2}{3}$	$-\frac{3}{2}$	-1
$-\frac{3}{4}$	$\frac{4}{3}$	-1

B $m_1 = -\frac{1}{m_2}$ **B** $m_2 = -\frac{1}{m_1}$

C Product is always -1.

Perpendicular Lines

$y_2 = -\frac{1}{2}x + 1$

$y_1 = 2x - 3$

A $2 \cdot \left(-\frac{1}{2}\right) = -1$

In the table, m_1 and m_2 represent the slopes of two *nonvertical* perpendicular lines. Since $m_1 \cdot m_2 = -1$ for each pair of slopes, we say that the slopes are **negative reciprocals** of each other.

A For perpendicular lines, $m_1 \cdot m_2 = -1$.

B Solving $m_1 \cdot m_2 = -1$ for one of the slopes gives $m_1 = -\frac{1}{m_2}$ or $m_2 = -\frac{1}{m_1}$.

C When the slopes of perpendicular lines are multiplied, the result is always -1.

WATCH VIDEO IN MML.

READING CHECK 3

• What does it mean for one number to be a negative reciprocal of another?

EXAMPLE 7 **Finding negative reciprocals**

Find the negative reciprocal of each number.
(a) $\frac{2}{3}$ (b) -6 (c) $-\frac{1}{4}$

Solution

(a) The negative reciprocal is found by inverting the fraction $\frac{2}{3}$ and finding its opposite, which results in $-\frac{3}{2}$. We can verify this result by showing that $\frac{2}{3} \cdot \left(-\frac{3}{2}\right) = -1$.

(b) After first writing -6 in fraction form as $-\frac{6}{1}$, invert the fraction and find its opposite. The reciprocal of $-\frac{6}{1}$ is $\frac{1}{6}$. Note that $-\frac{6}{1} \cdot \frac{1}{6} = -1$.

(c) Since $-\frac{1}{4} \cdot \frac{4}{1} = -1$, the negative reciprocal of $-\frac{1}{4}$ is $\frac{4}{1}$ or **4**.

Now Try Exercises 61, 63

We can use the concept of negative reciprocals to find equations of perpendicular lines, as illustrated in the next two examples.

EXAMPLE 8 **Finding equations of perpendicular lines**

For each of the given lines, find the slope–intercept form of a line passing through the origin that is perpendicular to the given line.
(a) $y = 3x$ (b) $y = -\frac{2}{5}x + 5$ (c) $-3x + 4y = 24$

Solution

(a) If a line passes through the origin, then its y-intercept is $(0, 0)$ and its slope–intercept form is $y = mx$. The given line $y = 3x$ has slope $m_1 = 3$, so a line perpendicular to it has a slope that is the negative reciprocal of 3.

$$m_2 = -\frac{1}{m_1} = -\frac{1}{3} = -\frac{1}{3}$$

The required slope–intercept form is $y = -\frac{1}{3}x$.

(b) The given line $y = -\frac{2}{5}x + 5$ has slope $m_1 = -\frac{2}{5}$, so a line perpendicular to it has slope $m_2 = \frac{5}{2}$ because $-\frac{2}{5} \cdot \frac{5}{2} = -1$. The required slope–intercept form is $y = \frac{5}{2}x$.

(c) To find the slope of the given line, write $-3x + 4y = 24$ in slope–intercept form.

$$-3x + 4y = 24 \qquad \text{Given equation}$$
$$4y = 3x + 24 \qquad \text{Add } 3x \text{ to each side.}$$
$$y = \frac{3}{4}x + 6 \qquad \text{Divide each side by 4.}$$

The slope of the given line is $m_1 = \frac{3}{4}$, so a line perpendicular to it has slope $m_2 = -\frac{4}{3}$. The required slope–intercept form is $y = -\frac{4}{3}x$.

Now Try Exercise 73

EXAMPLE 9 **Finding a perpendicular line equation**

Find the slope–intercept form of the line perpendicular to $y = -\frac{1}{2}x + 1$ and passing through the point $(1, -1)$. Sketch each line in the same xy-plane.

Solution

The line $y = -\frac{1}{2}x + 1$ has slope $m_1 = -\frac{1}{2}$. Any line perpendicular to it has slope $m_2 = 2$ $\left(\text{because } -\frac{1}{2} \cdot 2 = -1\right)$ with slope–intercept form $y = 2x + b$ for some value of b. The value of b can be found by substituting the point $(1, -1)$ in the slope–intercept form.

$$y = 2x + b \qquad \text{Slope–intercept form}$$
$$-1 = 2(1) + b \qquad \text{Let } x = 1 \text{ and } y = -1.$$
$$-1 = 2 + b \qquad \text{Multiply.}$$
$$-3 = b \qquad \text{Subtract 2 from each side.}$$

The slope–intercept form is $y = 2x - 3$. The graphs of $y = -\frac{1}{2}x + 1$ and $y = 2x - 3$ are shown in **FIGURE 3.46**. Note that the point $(1, -1)$ lies on the graph of $y = 2x - 3$.

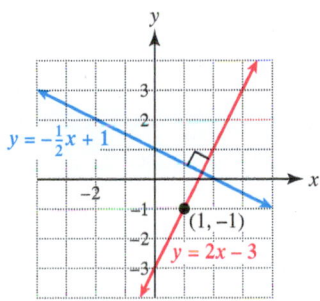

FIGURE 3.46

Now Try Exercise 75

3.5 Putting It All Together

CONCEPT	COMMENTS	EXAMPLE
Slope–Intercept Form $y = mx + b$	A unique equation for a line, determined by the slope m and the y-intercept $(0, b)$	The equation of the line with slope $m = 3$ and y-intercept $(0, -5)$ is $y = 3x - 5$.
Parallel Lines	$y = m_1x + b_1$ and $y = m_2x + b_2$, where $m_1 = m_2$ Nonvertical parallel lines have the same slope. Two vertical lines are parallel.	The lines $y = 2x - 1$ and $y = 2x + 2$ are parallel because they both have slope 2.

continued on next page

continued from previous page

CONCEPT	COMMENTS	EXAMPLE
Perpendicular Lines	$y = m_1x + b_1$ and $y = m_2x + b_2$, where $m_1m_2 = -1$ Perpendicular lines that are neither vertical nor horizontal have slopes whose product equals -1. A vertical line and a horizontal line are perpendicular.	The lines $y = 3x - 1$ and $y = -\frac{1}{3}x + 2$ are perpendicular because $$m_1m_2 = 3\left(-\frac{1}{3}\right) = -1.$$

3.5 Exercises

MyMathLab®

CONCEPTS AND VOCABULARY

1. The slope–intercept form of a line is _____ .

2. In the slope–intercept form of a line, m represents the _____ of the line.

3. In the slope–intercept form of a line, b represents the _____ of the line.

4. If $b = 0$ in the slope–intercept form of a line, then its graph passes through the _____ .

5. Two lines with the same slope are _____ .

6. Two nonvertical parallel lines have the same _____ .

7. If m_1 and m_2 are the slopes of two lines where $m_1 \cdot m_2 = -1$, the lines are _____ .

8. If two perpendicular lines have nonzero slopes, the slopes are negative _____ of each other.

DETERMINING A LINE

Exercises 9–14: Match the description of the line with its graph (a.–f.).

9. A line with positive slope and negative y-coordinate of the y-intercept

10. A line with positive slope and positive y-coordinate of the y-intercept

11. A line with negative slope and y-coordinate of the y-intercept 0

12. A line with negative slope and nonzero y-coordinate of the y-intercept

13. A line with no x-intercept

14. A line with no y-intercept

a.

b.

c.

d.

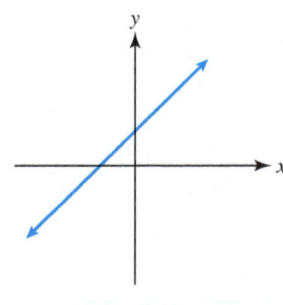

e.

f.

FINDING SLOPE–INTERCEPT FORM

Exercises 15–22: Write the slope–intercept form for the line shown in the graph.

15.

16.

17.

18.

19.

20.

21.

22.

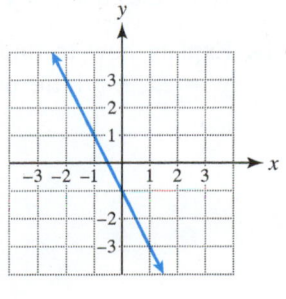

Exercises 23–32: Sketch a line with the given slope m and y-coordinate of the y-intercept b. Write the slope–intercept form of the line.

23. $m = 1, b = 2$ **24.** $m = -1, b = 3$

25. $m = 2, b = -1$ **26.** $m = -3, b = 2$

27. $m = -\frac{1}{2}, b = -2$ **28.** $m = -\frac{2}{3}, b = 0$

29. $m = \frac{1}{3}, b = 0$ **30.** $m = 3, b = -3$

31. $m = 0, b = 3$ **32.** $m = 0, b = -3$

Exercises 33–44: Do the following.

  (a) Write the equation in slope–intercept form.
  (b) Give the slope and y-intercept of the line.

33. $x + y = 4$ **34.** $x - y = 6$

35. $2x + y = 4$ **36.** $-4x + y = 8$

37. $x - 2y = -4$ **38.** $x + 3y = -9$

39. $2x - 3y = 6$ **40.** $4x + 5y = 20$

41. $x = 4y - 6$ **42.** $x = -3y + 2$

43. $\frac{1}{2}x + \frac{3}{2}y = 1$ **44.** $-\frac{3}{4}x + \frac{1}{2}y = \frac{1}{2}$

Exercises 45–56: Graph the equation.

45. $y = -3x + 2$ **46.** $y = \frac{1}{2}x - 1$

47. $y = \frac{1}{3}x$ **48.** $y = -2x$

49. $y = 2$ **50.** $y = -3$

51. $x + y = 3$ **52.** $-\frac{2}{3}x + y = -2$

53. $x + 2y = 2$ **54.** $-2x - y = -2$

55. $x = 2 - y$ **56.** $x = -\frac{1}{3}y + \frac{2}{3}$

Exercises 57–60: The table shows points that all lie on the same line. Find the slope–intercept form for the line.

57.

x	0	1	2
y	2	4	6

58.

x	-1	0	1
y	4	8	12

59.

x	-2	0	2
y	-4	-2	0

60.

x	0	2	4
y	6	3	0

WORKING WITH PARALLEL AND PERPENDICULAR LINES

Exercises 61–66: Find the negative reciprocal of the number.

61. 5   **62.** -4

63. $-\frac{4}{5}$ **64.** $\frac{7}{5}$

65. -0.25 **66.** 0.4

Exercises 67–76: Find the slope–intercept form of the line satisfying the given conditions.

67. Slope $\frac{4}{7}$, y-intercept $(0, 3)$

68. Slope $-\frac{1}{2}$, y-intercept $(0, -7)$

69. Parallel to $y = 3x + 1$, passing through $(0, 0)$

70. Parallel to $y = -2x$, passing through $(0, 1)$

71. Parallel to $2x + 4y = 5$, passing through $(1, 2)$

72. Parallel to $-x - 3y = 9$, passing through $(-3, 1)$

73. Perpendicular to $y = -\frac{1}{2}x - 3$, passing through $(0, 0)$

74. Perpendicular to $y = \frac{3}{4}x - \frac{1}{2}$, passing through $(3, -2)$

75. Perpendicular to $x = -\frac{1}{3}y$, passing through $(-1, 0)$

76. Perpendicular to $6x - 3y = 18$, passing through $(4, -3)$

APPLICATIONS INVOLVING SLOPE–INTERCEPT FORM

77. Rental Cars Driving a rental car x miles costs $y = 0.25x + 25$ dollars.
  (a) How much would it cost to rent the car but not drive it?
  (b) How much does the cost increase for each additional mile driven?
  (c) What is the y-intercept of $y = 0.25x + 25$? What does it represent?
  (d) What is the slope of $y = 0.25x + 25$? What does it represent?

78. Calculating Rainfall The total rainfall y in inches that fell x hours past noon is given by $y = \frac{1}{2}x + 3$.
  (a) How much rainfall was there at noon?
  (b) At what rate was rain falling in the afternoon?
  (c) What is the y-intercept of $y = \frac{1}{2}x + 3$? What does it represent?
  (d) What is the slope of $y = \frac{1}{2}x + 3$? What does it represent?

79. Land Line Phone Plan A phone plan in a rural area costs $3.95 per month plus $0.07 per minute. (Assume that a partial minute is not rounded up.)
  (a) During July, a person talks a total of 50 minutes. What is the charge?
  (b) Write an equation in slope–intercept form that gives the monthly cost C of talking on this plan for x minutes.

(c) If the charge for one month amounts to $8.64, how much time did the person spend talking on the phone?

80. Electrical Rates Electrical service costs $8 per month plus $0.10 per kilowatt-hour of electricity used. (Assume that a partial kilowatt-hour is not rounded up.)
  (a) If the resident of an apartment uses 650 kilowatt-hours in 1 month, what is the charge?
  (b) Write an equation in slope–intercept form that gives the cost C of using x kilowatt-hours in 1 month.
  (c) If the monthly electrical bill for the apartment's resident is $43, how many kilowatt-hours of electricity were used?

81. Cost of Driving The cost of driving a car includes both fixed costs and mileage costs. Assume that it costs $164.30 per month for insurance and car payments and $0.35 per mile for gasoline, oil, and routine maintenance.
  (a) Find values for m and b so that the equation $y = mx + b$ models the monthly cost of driving the car x miles.
  (b) What does the value of b represent?

82. Antarctic Ozone Layer The ozone layer occurs in Earth's atmosphere between altitudes of 12 and 18 miles and is an important filter of ultraviolet light from the sun. The thickness of the ozone layer is frequently measured in Dobson units. An average value is 300 Dobson units. In 2013, the reported minimum in the antarctic *ozone hole* was about 133 Dobson units. (*Source:* NASA.)
  (a) The equation $T = 0.01D$ describes the thickness T in millimeters of an ozone layer that is D Dobson units. How many millimeters thick was the ozone layer over the antarctic in 2013?
  (b) What is the average thickness of the ozone layer in millimeters?

WRITING ABOUT MATHEMATICS

83. Explain how the values of m and b can be used to graph the equation $y = mx + b$.

84. Explain how to find the value of b in the equation $y = 2x + b$ if the point $(3, 4)$ lies on the line.

Group Activity

Exercises 1–4: **Pet Ownership** *In following table, you will see different costs associated with having a dog or a cat. Use the table to complete each of the following.*

1. Find the annual costs and the fixed costs for each type of pet.

2. Write a linear equation for each type of pet that models the total cost of owning the pet over a period of *x* years.

3. How much will the first year of owning a dog cost? How much will the first year of owning a cat cost?

4. If the average dog is expected to live 13 years, and the average cat is expected to live 15 years, what is the total expense of owning each type of animal over its entire lifetime?

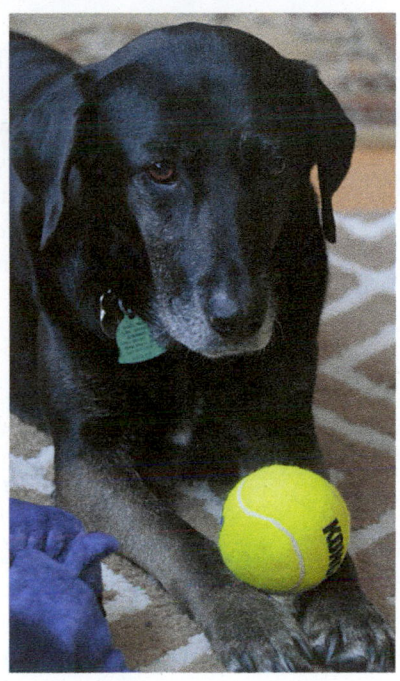

Expenses	Dog Expenses ($)	Cat Expenses ($)
Food	120	115
Recurring medical	235	160
Spay/neuter	200	145
Other initial medical	70	130
Litter	0	165
Toys/treats	55	25
Yearly health insurance	225	175
Scratching post	0	15
Yearly licence	15	0
Initial training class	110	0
Carrier bag/crate	95	40
Collar/leash	30	10
Litter box	0	25
Other recurring costs	45	30

Source: ASPCA.

3.6 Point–Slope Form

Slope of a Line

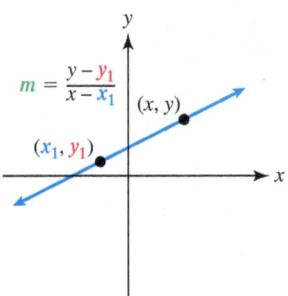

FIGURE 3.47

NEW VOCABULARY

☐ Equation of a line
☐ Point–slope form

1 Deriving the Point–Slope Form

If we know the slope and y-intercept of a line, we can write its slope–intercept form, $y = mx + b$, which is an example of an **equation of a line**. The point–slope form is a different type of equation of a line.

Suppose that a (nonvertical) line with slope m passes through the point (x_1, y_1). If (x, y) is a different point on this line, then $m = \frac{y - y_1}{x - x_1}$. See **FIGURE 3.47**. Using this slope formula, we can find the point–slope form of the equation of a line.

$$m = \frac{y - y_1}{x - x_1} \qquad \text{Slope formula}$$

$$m \cdot (x - x_1) = \frac{y - y_1}{x - x_1} \cdot (x - x_1) \qquad \text{Multiply each side by } (x - x_1).$$

$$m(x - x_1) = y - y_1 \qquad \text{Simplify.}$$

$$y - y_1 = m(x - x_1) \qquad \text{Rewrite the equation.}$$

POINT–SLOPE FORM

The line with slope m passing through the point (x_1, y_1) is given by

$$y - y_1 = m(x - x_1),$$

the **point–slope form** of a line.

NOTE: The equation $y - y_1 = m(x - x_1)$ is traditionally called the *point–slope form*. By adding y_1 to each side of this equation, we get $y = m(x - x_1) + y_1$, which is an equivalent form that is helpful when graphing. ∎

2 Finding Point–Slope Form

In the next example, we find a point–slope form for a line. Note that *any* point that lies on the line can be used in its point–slope form.

EXAMPLE 1 **Finding a point–slope form**

Use the labeled point in each figure to write a point–slope form for the line and then simplify it to the slope–intercept form.

(a)

(b)

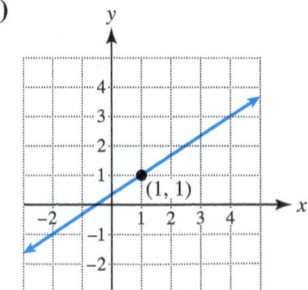

Solution

(a) The graph rises 3 units for each unit of horizontal run, so the slope is $\frac{3}{1}$, or 3. Let $m = 3$ and $(x_1, y_1) = (1, 2)$ in the point–slope form.

$$y - y_1 = m(x - x_1) \qquad \text{Point–slope form}$$

$$\boxed{\text{Point–slope form}} \quad y - 2 = 3(x - 1) \qquad \text{Let } m = 3, x_1 = 1 \text{ and } y_1 = 2.$$

To obtain slope–intercept form, apply the distributive property to $y - 2 = 3(x - 1)$.

$$y - 2 = 3x - 3 \qquad \text{Distributive property}$$

$$\boxed{\text{Slope–intercept form}} \quad y = 3x - 1 \qquad \text{Add 2 to each side.}$$

(b) The graph rises 2 units for each 3 units of horizontal run, so the slope is $\frac{2}{3}$. Let $m = \frac{2}{3}$ and $(x_1, y_1) = (1, 1)$ in the point–slope form.

$$y - y_1 = m(x - x_1) \qquad \text{Point–slope form}$$

$$\boxed{\text{Point–slope form}} \quad y - 1 = \frac{2}{3}(x - 1) \qquad \text{Let } m = \frac{2}{3}, x_1 = 1, \text{ and } y_1 = 1.$$

To obtain slope–intercept form, apply the distributive property to $y - 1 = \frac{2}{3}(x - 1)$.

$$y - 1 = \frac{2}{3}x - \frac{2}{3} \qquad \text{Distributive property}$$

$$\boxed{\text{Slope–intercept form}} \quad y = \frac{2}{3}x + \frac{1}{3} \qquad \text{Add 1, or } \frac{3}{3}, \text{ to each side.}$$

Now Try Exercises 13, 15

EXAMPLE 2 Finding a point–slope form

Find a point–slope form for a line passing through the point $(-2, 3)$ with slope $-\frac{1}{2}$. Does the point $(2, 1)$ lie on this line?

Solution

Let $m = -\frac{1}{2}$ and $(x_1, y_1) = (-2, 3)$ in the point–slope form.

$$y - y_1 = m(x - x_1) \qquad \text{Point–slope form}$$

$$y - 3 = -\frac{1}{2}(x - (-2)) \qquad x_1 = -2, y_1 = 3, \text{ and } m = -\frac{1}{2}$$

$$\boxed{\text{Point–slope form}} \quad y - 3 = -\frac{1}{2}(x + 2) \qquad \text{Simplify.}$$

To determine whether $(2, 1)$ lies on the line, substitute $x = 2$ and $y = 1$ in the equation.

$$1 - 3 \stackrel{?}{=} -\frac{1}{2}(2 + 2) \qquad \text{Let } x = 2 \text{ and } y = 1.$$

$$-2 \stackrel{?}{=} -\frac{1}{2}(4) \qquad \text{Simplify.}$$

$$\boxed{\text{Yes, } (2, 1) \text{ lies on the line.}} \quad -2 = -2 \checkmark \qquad \text{A true statement}$$

Now Try Exercises 9, 19

In the next example, we use the point–slope form to find an equation of a line passing through two points.

EXAMPLE 3   **Finding an equation of a line passing through two points**

Use the point–slope form to find an equation of the line passing through the points $(1, -4)$ and $(-2, 5)$.

Solution
Before we can apply the point–slope form, we must find the slope.

$$m = \frac{y_2 - y_1}{x_2 - x_1} \qquad \text{Slope formula}$$

$$= \frac{5 - (-4)}{-2 - 1} \qquad x_1 = 1, y_1 = -4, x_2 = -2, \text{ and } y_2 = 5$$

$$= -3 \qquad \text{Simplify.}$$

We can let either the point $(1, -4)$ or the point $(-2, 5)$ be (x_1, y_1) in the point–slope form. If $(x_1, y_1) = (\mathbf{1}, \mathbf{-4})$, then the equation of the line becomes the following.

$$y - y_1 = m(x - x_1) \qquad \text{Point–slope form}$$

$$y - (\mathbf{-4}) = \mathbf{-3}(x - \mathbf{1}) \qquad x_1 = 1, y_1 = -4, \text{ and } m = -3$$

$$y + 4 = -3(x - 1) \qquad \text{Simplify.}$$

If we let $(x_1, y_1) = (-2, 5)$, the point–slope form becomes

$$y - 5 = -3(x + 2).$$

Now Try Exercise 25

NOTE: Although the two point–slope forms in Example 3 might appear to be different, they actually are equivalent because they simplify to the *same* slope–intercept form.

$$\begin{array}{lll}
y + 4 = -3(x - 1) & y - 5 = -3(x + 2) & \text{Point–slope forms} \\
y = -3(x - 1) - 4 & y = -3(x + 2) + 5 & \text{Addition property} \\
y = -3x + 3 - 4 & y = -3x - 6 + 5 & \text{Distributive property} \\
\underbrace{y = -3x - 1} & \underbrace{y = -3x - 1} & \text{Same slope–intercept form} \blacksquare
\end{array}$$

A line can have only one slope–intercept form.

MAKING CONNECTIONS 1

Slope–Intercept and Point–Slope Forms

READING CHECK 1

• Is the point–slope form of the equation of a line unique?

The slope–intercept form, $y = mx + b$, is *unique* because any nonvertical line has one slope m and one y-intercept $(0, b)$. The point–slope form, $y - y_1 = m(x - x_1)$, is *not* *unique* because (x_1, y_1) can be any point that lies on the line. However, any point–slope form can be simplified to a unique slope–intercept form.

EXAMPLE 4   **Finding equations of lines**

Find the slope–intercept form for the line that satisfies the conditions.
(a) Slope $\frac{1}{2}$, passing through $(-2, 4)$
(b) x-intercept $(-3, 0)$, y-intercept $(0, 2)$
(c) Perpendicular to $y = -\frac{2}{3}x$, passing through $\left(\frac{2}{3}, 3\right)$

Solution

(a) Substitute $m = \frac{1}{2}$, $x_1 = -2$, and $y_1 = 4$ in the point–slope form.

$$y - y_1 = m(x - x_1) \qquad \text{Point–slope form}$$

$$y - 4 = \frac{1}{2}(x + 2) \qquad \text{Substitute and simplify.}$$

$$y - 4 = \frac{1}{2}x + 1 \qquad \text{Distributive property}$$

$$y = \frac{1}{2}x + 5 \qquad \text{Add 4 to each side.}$$

(b) The line passes through the points $(-3, 0)$ and $(0, 2)$. The slope of the line is

$$m = \frac{2 - 0}{0 - (-3)} = \frac{2}{3}.$$

Because the line has slope $\frac{2}{3}$ and y-intercept $(0, 2)$, the slope–intercept form is

$$y = \frac{2}{3}x + 2.$$

(c) The slope of $y = -\frac{2}{3}x$ is $m_1 = -\frac{2}{3}$, so the slope of a line perpendicular to it is the negative reciprocal of $-\frac{2}{3}$, or $\frac{3}{2}$. Let $m = \frac{3}{2}$, $x_1 = \frac{2}{3}$, and $y_1 = 3$ in the point–slope form.

$$y - y_1 = m(x - x_1) \qquad \text{Point–slope form}$$

$$y - 3 = \frac{3}{2}\left(x - \frac{2}{3}\right) \qquad \text{Substitute.}$$

$$y - 3 = \frac{3}{2}x - 1 \qquad \text{Distributive property}$$

$$y = \frac{3}{2}x + 2 \qquad \text{Add 3 to each side.}$$

Now Try Exercises 43, 47, 51

In the next example, the point–slope form is used to find the slope–intercept form of a line that passes through several points given in a table.

EXAMPLE 5   **Using a table to find slope–intercept form**

The points in **TABLE 3.19** lie on a line. Find the slope–intercept form of the line.

x	2	4	6	8
y	2	1	0	−1

TABLE 3.19

Solution

The y-values in the table decrease 1 unit for every 2-unit increase in the x-values, so the line has a "rise" of -1 when the run is 2. The slope is $m = \frac{\text{Rise}}{\text{Run}} = \frac{-1}{2} = -\frac{1}{2}$. *Any* point from the table can be used to obtain a point–slope form of the line, which can then be simplified to slope–intercept form. Letting $(x_1, y_1) = (2, 2)$ and $m = -\frac{1}{2}$ in the point–slope form yields the following result.

$$y - y_1 = m(x - x_1) \qquad \text{Point–slope form}$$

$$y - 2 = -\frac{1}{2}(x - 2) \qquad \text{Substitute.}$$

$$y - 2 = -\frac{1}{2}x + 1 \qquad \text{Distributive property}$$

$$y = -\frac{1}{2}x + 3 \qquad \text{Add 2 to each side.}$$

This result can be checked by substituting each x-value from the table in the equation. For example, when $x = 4$, the corresponding y-value is $y = -\frac{1}{2}(4) + 3 = -2 + 3 = 1$. This agrees with the table.

Now Try Exercise 57

MAKING CONNECTIONS 2

Finding Slope–Intercept Form Without Using Point–Slope Form

The equation in Example 5 can be obtained without using the point–slope form. The value of b can be found by letting $(x, y) = (2, 2)$ and $m = -\frac{1}{2}$ in the slope–intercept form.

$$y = mx + b \qquad \text{Slope–intercept form}$$

$$2 = -\frac{1}{2}(2) + b \qquad \text{Substitute.}$$

$$2 = -1 + b \qquad \text{Multiply.}$$

$$3 = b \qquad \text{Add 1 to each side.}$$

Because $m = -\frac{1}{2}$ and $b = 3$, the slope–intercept form is $y = -\frac{1}{2}x + 3$.

In the next example, we use the process shown in the above Making Connections to find the slope–intercept form of a line equation without using the point–slope form.

EXAMPLE 6 **Finding slope–intercept form directly using $y = mx + b$**

If a line has slope 4 and passes through the point $(2, 6)$, find its slope–intercept form *without* using the point–slope equation.

Solution
Find the value of b in the slope–intercept form by letting $(x, y) = (2, 6)$ and $m = 4$ in the equation $y = mx + b$.

$$y = mx + b \qquad \text{Slope–intercept form}$$

$$6 = 4(2) + b \qquad x = 2, y = 6 \text{ and } m = 4.$$

$$6 = 8 + b \qquad \text{Multiply.}$$

$$-2 = b \qquad \text{Subtract 8 from each side.}$$

Since $m = 4$ and $b = -2$, the slope–intercept form is $y = 4x - 2$.

Now Try Exercise 61

3 Solving Applications Involving Equations of Lines

 Math in Gender Context   In 1995, there were 690 female officers in the Marine Corps, and by 2010, this number had increased to about 1110. This growth is illustrated in **FIGURE 3.48**, where the line passes through the points $(1995, 690)$ and $(2010, 1110)$. Because two points determine a unique line, we can find the equation of this line and use it to *estimate* the number of female officers in other years. (*Source:* Department of Defense.)

Female Officers in the Marine Corps

A **linear approximation** for the growth in the number of female Marine Corps officers

FIGURE 3.48

EXAMPLE 7

Modeling numbers of female officers

Refer to **FIGURE 3.48** above.
(a) Use the point $(1995, 690)$ to find a point–slope form of the line.
(b) Interpret the slope as a rate of change.
(c) Use **FIGURE 3.48** to estimate the number of female officers in 2006. Then use your equation from part (a) to approximate this number. How do your answers compare?

Solution
(a) The slope of the line passing through $(1995, 690)$ and $(2010, 1110)$ is

$$m = \frac{1110 - 690}{2010 - 1995} = 28.$$

If we let $x_1 = 1995$ and $y_1 = 690$, then the point–slope form becomes

$$y - 690 = 28(x - 1995).$$

(b) Slope $m = 28$ indicates that the number of female officers increased, *on average,* by about 28 officers per year.
(c) From **FIGURE 3.48**, it appears that the number of female officers in 2006 was about 1000. To estimate this value, let $x = 2006$ in the equation found in part (a).

$$y - 690 = 28(2006 - 1995) \quad \text{or} \quad y = 998$$

Although the graphical estimate and calculated answers are not exactly equal, they are approximately equal. Estimations made from a graph usually are not exact.

Now Try Exercise 73

In the next example, we review several concepts related to lines.

EXAMPLE 8 **Modeling water in a pool**

A small swimming pool is being emptied by a pump that removes water at a constant rate. After 1 hour, the pool contains 5000 gallons, and after 3 hours, it contains 3000 gallons.
(a) How fast is the pump removing water?
(b) Find the slope–intercept form of a line that models the amount of water in the pool. Interpret the slope.
(c) Find the y-intercept and the x-intercept. Interpret each.
(d) Sketch a graph of the amount of water in the pool during the first 6 hours.
(e) The point $(2, 4000)$ lies on the graph. Explain its meaning.

Solution
(a) The pump removes $5000 - 3000 = 2000$ gallons of water in 2 hours, or 1000 gallons per hour.
(b) The line passes through the points $(1, 5000)$ and $(3, 3000)$, so the slope is

$$m = \frac{3000 - 5000}{3 - 1} = -1000.$$

One way to find the slope–intercept form is to use the point–slope form.

$$y - y_1 = m(x - x_1) \qquad \text{Point–slope form}$$
$$y - 5000 = -1000(x - 1) \qquad m = -1000, x_1 = 1, \text{ and } y_1 = 5000$$
$$y - 5000 = -1000x + 1000 \qquad \text{Distributive property}$$
$$y = -1000x + 6000 \qquad \text{Add 5000 to each side.}$$

Slope -1000 means that the pump is *removing* 1000 gallons of water per hour.
(c) The y-intercept is $(0, 6000)$ and indicates that the pool initially contained 6000 gallons of water. To find the x-intercept, let $y = 0$ in the slope–intercept form.

$$0 = -1000x + 6000 \qquad \text{Let } y = 0.$$
$$1000x = 6000 \qquad \text{Add } 1000x \text{ to each side.}$$
$$x = \frac{6000}{1000} \qquad \text{Divide by 1000.}$$
$$x = 6 \qquad \text{Simplify.}$$

An x-intercept of $(6, 0)$ indicates that the pool is empty after 6 hours.
(d) Sketch a line from the x-intercept $(6, 0)$ to the y-intercept $(0, 6000)$, as shown in **FIGURE 3.49**.

CRITICAL THINKING 1

Suppose that a line models the amount of water in a swimming pool. What does a negative slope indicate?

Water in a Pool

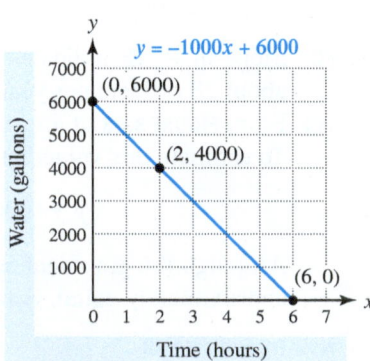

FIGURE 3.49

(e) The point $(2, 4000)$ indicates that after 2 hours the pool contains 4000 gallons of water.

Now Try Exercise 67

3.6   Putting It All Together

CONCEPT	COMMENTS	EXAMPLE
Point–Slope Form $$y - y_1 = m(x - x_1)$$	Used to find an equation of a line, given two points or one point and the slope Can always be simplified to slope–intercept form Does *not* provide a unique equation for a line because *any* point on the line can be used An equivalent form that is useful for graphing is $$y = m(x - x_1) + y_1.$$	For two points $(1, 2)$ and $(3, 5)$, first compute $m = \frac{5 - 2}{3 - 1} = \frac{3}{2}$. An equation of this line is $$y - 2 = \frac{3}{2}(x - 1).$$

3.6   Exercises

MyMathLab®

CONCEPTS AND VOCABULARY

1. (True or False?) One line is determined by two distinct points.

2. (True or False?) One line is determined by a point and a slope.

3. Give the slope–intercept form of a line.

4. Give the point–slope form of a line.

5. If the point–slope form is written for a line passing through $(1, 3)$, then $x_1 = $ _____ and $y_1 = $ _____ .

6. To write a point–slope equation in slope–intercept form, use the _____ property to clear the parentheses.

7. Is the slope–intercept form of a line unique? Explain.

8. Is the point–slope form of a line unique? Explain.

WORKING WITH POINT–SLOPE FORM

Exercises 9–12: Determine whether the given point lies on the line.

9. $(3, -11)$ $y + 1 = -2(x + 3)$

10. $(1, 4)$ $y - 3 = -(x - 1)$

11. $(0, 4)$ $y = \frac{1}{2}(x + 4) + 2$

12. $(2, -5)$ $y = -\frac{1}{3}(x - 5) - 6$

Exercises 13–18: Use the labeled point to write a point–slope form for the line.

13.

14.

15.

16.

17.

18.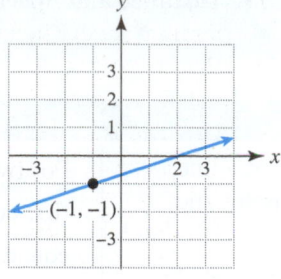

Exercises 19–30: Find a point–slope form for the line that satisfies the stated conditions. When two points are given, use the first point in the point–slope form.

19. Slope 4, passing through $(-3, 1)$

20. Slope -3, passing through $(1, -2)$

21. Slope $\frac{1}{2}$, passing through $(-5, -3)$

22. Slope $-\frac{2}{3}$, passing through $(-3, 6)$

23. Slope 1.5, passing through $(2010, 30)$

24. Slope -10, passing through $(2014, 100)$

25. Passing through $(2, 4)$ and $(-1, -3)$

26. Passing through $(3, -1)$ and $(1, -4)$

27. Passing through $(5, 0)$ and $(0, -3)$

28. Passing through $(-2, 0)$ and $(0, -1)$

29. Passing through $(2003, 15)$ and $(2013, 65)$

30. Passing through $(2009, 5)$ and $(2014, 30)$

Exercises 31–42: Write the point–slope form in slope–intercept form.

31. $y - 4 = 3(x - 2)$ **32.** $y - 3 = -2(x + 1)$

33. $y + 2 = \frac{1}{3}(x + 6)$ **34.** $y - 1 = \frac{2}{5}(x + 10)$

35. $y - \frac{3}{4} = \frac{2}{3}(x - 1)$ **36.** $y + \frac{2}{3} = -\frac{1}{6}(x - 2)$

37. $y = -2(x - 2) + 5$ **38.** $y = 4(x + 3) - 7$

39. $y = \frac{3}{5}(x - 5) + 1$ **40.** $y = -\frac{1}{2}(x + 4) - 6$

41. $y = -16(x + 1.5) + 5$

42. $y = -15(x - 1) + 100$

Exercises 43–56: Find the slope–intercept form for the line satisfying the conditions.

43. Slope -2, passing through $(4, -3)$

44. Slope $\frac{1}{5}$, passing through $(-2, 5)$

45. Passing through $(3, -2)$ and $(2, -1)$

46. Passing through $(8, 3)$ and $(-7, 3)$

47. x-intercept $(3, 0)$, y-intercept $(0, \frac{1}{3})$

48. x-intercept $(2, 0)$, y-intercept $(0, -3)$

49. Parallel to $y = 2x - 1$, passing through $(2, -3)$

50. Parallel to $y = -\frac{3}{2}x$, passing through $(0, 20)$

51. Perpendicular to $y = -\frac{1}{2}x + 3$, passing through the point $(6, -3)$

52. Perpendicular to $y = \frac{3}{5}(x + 1) + 3$, passing through the point $(1, -2)$

53. Passing through $(2, 6)$ and parallel to the line passing through $(4, 1)$ and $(-2, -2)$

54. Passing through $(1, 3)$ and parallel to the line passing through $(2, 1)$ and $(4, 5)$

55. Passing through $(-1, -5)$ and perpendicular to the line passing through $(2, -3)$ and $(-2, -1)$

56. Passing through $(-3, 2)$ and perpendicular to the line passing through $(0, -1)$ and $(3, 8)$

Exercises 57–60: The points in the table lie on a line. Find the slope–intercept form of the line.

57.

x	1	2	3	4
y	-3	-5	-7	-9

58.

x	2	3	4	5
y	5	8	11	14

59.

x	-1	1	3	5
y	-3	-2	-1	0

60.

x	-1	5	11	17
y	1	-3	-7	-11

Exercises 61–64: (Refer to Example 6.) Without using point–slope form, find the slope–intercept form of the line that has slope m and passes through the given point.

61. $m = 2$, $(1, -3)$ **62.** $m = -3$, $(-1, 1)$

63. $m = -\frac{1}{2}$, $(-1, -1)$ **64.** $m = \frac{2}{3}$, $(4, 3)$

65. Thinking Generally Find the y-intercept of the line given by $y - y_1 = m(x - x_1)$.

66. Thinking Generally Find the x-intercept of the line given by $y - y_1 = m(x - x_1)$.

INTERPRETING GRAPHS OF LINES

67. Change in Temperature The outside temperature was 40 °F at 1 A.M. and 15 °F at 6 A.M. Assume that the temperature changed at a constant rate.
(a) At what rate did the temperature change?
(b) Find the slope–intercept form of a line that models the temperature T at x A.M. Interpret the slope as a rate of change.
(c) Assuming that your equation is valid for times after 6 A.M., find and interpret the x-intercept.
(d) Sketch a graph that shows the temperature from 1 A.M. to 9 A.M.
(e) The point $(4, 25)$ lies on the graph. Explain its meaning.

68. Cost of Fuel The cost of buying 5 gallons of fuel oil is $12 and the cost of buying 15 gallons of fuel oil is $36.
(a) What is the cost of a gallon of fuel oil?
(b) Find the slope–intercept form of a line that models the cost of buying x gallons of fuel oil. Interpret the slope as a rate of change.
(c) Find and interpret the x-intercept.
(d) Sketch a graph that shows the cost of buying 20 gallons or less of fuel oil.
(e) The point $(11, 26.40)$ lies on the graph. Explain its meaning.

69. Water and Flow The graph shows the amount of water y in a 500-gallon tank after x minutes have elapsed.
(a) Is water entering or leaving the tank? How much water is in the tank after 4 minutes?
(b) Find the y-intercept. Explain its meaning.
(c) Find the slope–intercept form of the line. Interpret the slope as a rate of change.
(d) After how many minutes will the tank be full?

70. Water and Flow A hose is used to fill a 100-gallon barrel. If the hose delivers 5 gallons of water per minute, sketch a graph of the amount A of water in the barrel during the first 20 minutes.

71. Distance and Speed A person is driving a car along a straight road. The graph shows the distance y in miles that the driver is from home after x hours.
(a) Is the person traveling toward or away from home?
(b) The graph passes through $(1, 250)$ and $(4, 100)$. Discuss the meaning of these points.
(c) How fast is the driver traveling?
(d) Find the slope–intercept form of the line. Interpret the slope as a rate of change.

72. Distance and Speed A person rides a bicycle at 10 miles per hour, first away from home for 1 hour and then toward home for 1 hour. Sketch a graph that shows the distance d between the bicyclist and home after x hours.

SOLVING APPLICATIONS INVOLVING EQUATIONS OF LINES

73. Mobile Internet Users The graph models the number of mobile Internet users worldwide in millions from 2009 to 2015. (*Source:* ComScore.)
(a) The line passes through the points $(2009, 700)$ and $(2015, 1900)$. Explain the meaning of the first point.
(b) Use the first point to write a point–slope form for the equation of this line.
(c) Use the graph to estimate the number of mobile Internet users in 2013. Then use your equation from part (b) to estimate the number of mobile Internet users in 2013.
(d) Interpret the slope as a rate of change.

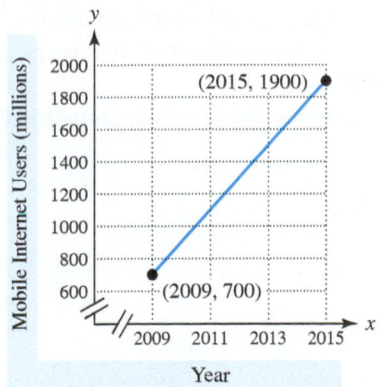

74. Wyoming Tuition The graph models average tuition at Wyoming public 2-year colleges after the year 2000, where 1 corresponds to 2001, 2 corresponds to 2002, and so on. (*Source:* SSTI *Weekly Digest.*)

(a) The line passes through the points $(3, 1557)$ and $(13, 2607)$. What is the meaning of the first of these two points?

(b) Use the first point to write a point–slope form for the equation of this line.

(c) Use the graph to estimate average tuition in 2010. Then use your equation from part (b) to estimate tuition in 2010.

(d) Interpret the slope as a rate of change.

Year (0 corresponds to 2000)

75. International Adoption In 2012, there were 1567 children adopted into the United States from Ethiopia, and this number was decreasing by 165 children per year. (*Source:* U.S. Department of State.)

(a) Determine a point–slope equation of a line that approximates the number of children who were adopted into the United States from Ethiopia during year x, where $x \geq 2012$.

(b) Estimate the number of children adopted into the United States from Ethiopia in 2014.

76. Median Household Income In 2012, median income for U.S. households was $50,099, and this number was increasing at a rate of $941 per year. (*Source:* Department of the Treasury.)

(a) Determine a point–slope equation of a line that approximates median family income during year x, where $x \geq 2012$.

(b) Estimate median family income in 2022.

77. Cigarette Consumption For any year x from 1975 to 2015, the number of cigarettes y consumed in the United States is modeled by $y = -10.33x + 21,087$, where y is in billions. Interpret the slope as a rate of change. (*Source:* The Tobacco Outlook Report.)

78. Worldwide Smartphones If x represents a year from 2010 to 2016, then the number y of smartphones sold worldwide in millions is approximated by $y = 216.67x - 435,200$. (*Source:* Business Insider.)

(a) Interpret the slope as a rate of change.

(b) Estimate the number of smartphones sold worldwide in 2015.

79. Rise in Sea Level The highest projection for the rise in sea level by year 2100 is that it will be 1.5 meters higher than the level in 1940. (*Source:* The Economist.)

(a) Find the slope–intercept equation of a line that approximates the rise in sea level above the 1940 level for year x, where x is between 1940 and 2100.

(b) Interpret the slope.

(c) Estimate the rise in sea level by 2020.

80. Hospitals In 2002, there were 3039 U.S. hospitals with more than 100 beds, and this number was decreasing at a rate of 33 hospitals per year. (*Source:* AHA Hospital Statistics.)

(a) Determine a slope–intercept equation of a line that approximates the number of U.S. hospitals with more than 100 beds during year x, where $x \geq 2002$.

(b) Estimate the number of hospitals with more than 100 beds in 2012.

WRITING ABOUT MATHEMATICS

81. Explain how to find the equation of a line passing through two points with coordinates (x_1, y_1) and (x_2, y_2). Give an example.

82. Explain how slope is related to rate of change. Give an example.

Checking Basic Concepts

1. Write the slope–intercept form for the line shown in the graph.

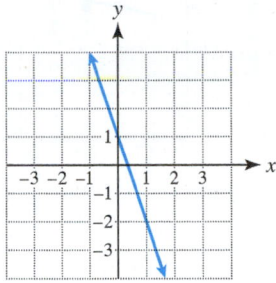

2. Write $4x - 5y = 20$ in slope–intercept form. Give the slope and y-intercept.

3. Graph $y = \frac{1}{2}x - 3$

4. Write the slope–intercept form of a line that satisfies the given conditions.
   (a) Slope 3, passing through $(0, -2)$
   (b) Perpendicular to $y = \frac{2}{3}x$, passing through the point $(-2, 3)$
   (c) Passing through $(1, -4)$ and $(-2, 3)$

5. Write a point–slope form for a line with slope -2, passing through $(-1, 3)$.

6. Write the equation $y + 3 = -2(x - 2)$ in slope–intercept form.

7. Find the slope–intercept form of the line passing through the points in the table.

x	-3	-1	1	3
y	-3	1	5	9

8. **Distance and Speed** A bicyclist is riding at a constant speed and is 36 miles from home at 1 P.M. Two hours later, the bicyclist is 12 miles from home.
   (a) Find the slope–intercept form of a line passing through $(1, 36)$ and $(3, 12)$.
   (b) How fast is the bicyclist traveling?
   (c) When will the bicyclist arrive home?
   (d) How many miles was the bicyclist from home at noon?

9. **Snowfall** The total amount of snowfall S in inches t hours past noon is given by $S = 2t + 5$.
   (a) How many inches of snow fell by noon?
   (b) At what rate did snow fall in the afternoon?
   (c) What is the S-intercept for the graph of this equation? What does it represent?
   (d) What is the slope for the graph of this equation? What does it represent?

3.7 Introduction to Modeling

Objectives

1 Introducing Linear Models

2 Modeling Linear Data

1 Introducing Linear Models

For centuries people have tried to understand the world around them by creating models. A model is an *abstraction* of something that people observed. Not only should a good model describe *known* data, but it should also be able to predict *future* data.

🌐 *Math in Context* Real-World **FIGURE 3.50(a)** on the next page shows a scatterplot of the number of inmates in the federal prison system from 2007 to 2013. The four points in the graph appear almost to lie on the same line. An example of such a line is shown in **FIGURE 3.50(b)** on the next page. Using mathematical modeling, we can find an equation for such a line. Once we have found this equation, we can use it to make estimates about the federal inmate population. (See Exercise 49 at the end of this section.)

STUDY TIP

There is often more than one way to accurately model data. Don't be afraid to try different methods. Part of the modeling process is trying to justify the reasoning behind a particular modeling choice.

READING CHECK 1

• What are some uses for a
  mathematical model?

Federal Prison Population

(a)

A Linear Model

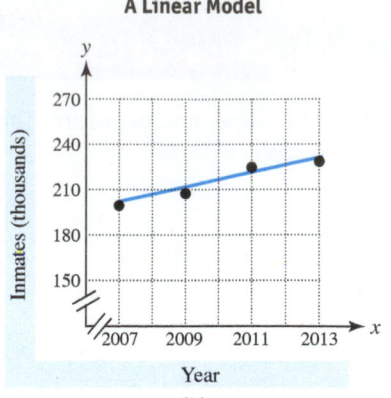

(b)

FIGURE 3.50 *Source:* U.S. Department of Justice.

Generally, mathematical models are not *exact* representations of data. Although the line in **FIGURE 3.50(b)** appears to touch every point, it does not pass through all four points exactly. **FIGURE 3.51(a)** shows data modeled *exactly* by a line, whereas **FIGURE 3.51(b)** shows data modeled *approximately* by a line. In applications, a model is more likely to be approximate than exact.

Exact Linear Model   **Approximate Linear Model**

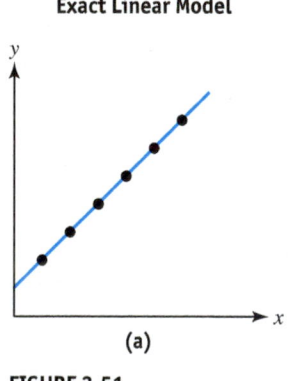

(a)   (b)

READING CHECK 2

• Does a mathematical
  model always touch every
  point in a scatterplot?

FIGURE 3.51

EXAMPLE 1 **Determining whether a model is exact**

TABLE 3.20 shows the worldwide spending on Internet advertising S in billions dollars for selected years x. Does the equation $S = 2x - 4006$ model the data exactly? Explain.

Spending on Internet Advertising

x	2009	2011	2013	2015
S	13	16	20	24

TABLE 3.20

Solution
To determine whether the equation models the data exactly, let $x = $ 2009, 2011, 2013, and 2015 in the given equation.

$$x = \mathbf{2009}: \qquad S = 2(\mathbf{2009}) - 4006 = 12$$
$$x = \mathbf{2011}: \qquad S = 2(\mathbf{2011}) - 4006 = 16$$
$$x = \mathbf{2013}: \qquad S = 2(\mathbf{2013}) - 4006 = 20$$
$$x = \mathbf{2015}: \qquad S = 2(\mathbf{2015}) - 4006 = 24$$

The model is *not exact*. It does not predict a spending of $13 billion in 2009.

Now Try Exercise 13

2 Modeling Linear Data

A line can model linear data. In the next example, we use a line to model gas mileage.

EXAMPLE 2 **Determining gas mileage**

TABLE 3.21 shows the number of miles y traveled by an SUV on x gallons of gasoline.

x	2	4	6	8
y	30	60	90	120

TABLE 3.21

(a) Plot the data in the xy-plane. Be sure to label each axis.
(b) Sketch a line that models the data. (You may want to use a ruler.)
(c) Find the equation of the line and interpret the slope of the line.
(d) How far could this SUV travel on 11 gallons of gasoline?

Solution
(a) Plot the points $(2, 30)$, $(4, 60)$, $(6, 90)$, and $(8, 120)$, as shown in **FIGURE 3.52(a)**.

(a)   (b)

FIGURE 3.52

(b) Sketch a line similar to the one shown in **FIGURE 3.52(b)**. This particular line passes through each data point.
(c) First find the slope m of the line by choosing two points that the line passes through, such as $(2, 30)$ and $(8, 120)$.

$$m = \frac{120 - 30}{8 - 2} = \frac{90}{6} = 15$$

Now find the equation of the line passing through $(2, 30)$ with slope 15.

$$y - y_1 = m(x - x_1) \quad \text{Point–slope form}$$

$$y - 30 = 15(x - 2) \quad x_1 = 2, y_1 = 30, \text{ and } m = 15$$

$$y - 30 = 15x - 30 \quad \text{Distributive property}$$

Linear Model $\quad y = 15x \quad\quad$ Add 30 to each side.

The data are modeled by the equation $y = 15x$. Slope **15** indicates that the mileage of this SUV is **15** miles per gallon.
(d) On **11** gallons of gasoline the SUV could go $y = 15(11) = 165$ miles.

Now Try Exercise 51

EXAMPLE 3   **Modeling linear data**

TABLE 3.22 contains ordered pairs that can be modeled *approximately* by a line.
(a) Plot the data. Could a line pass through all five points?
(b) Sketch a line that models the data and then determine its equation.

x	1	2	3	4	5
y	3	5	6	10	11

TABLE 3.22

Solution

(a) Plot the ordered pairs $(1, 3)$, $(2, 5)$, $(3, 6)$, $(4, 10)$, and $(5, 11)$, as shown in **FIGURE 3.53(a)**. The points are not collinear, so it is impossible to sketch a single line that passes through *all* five points.

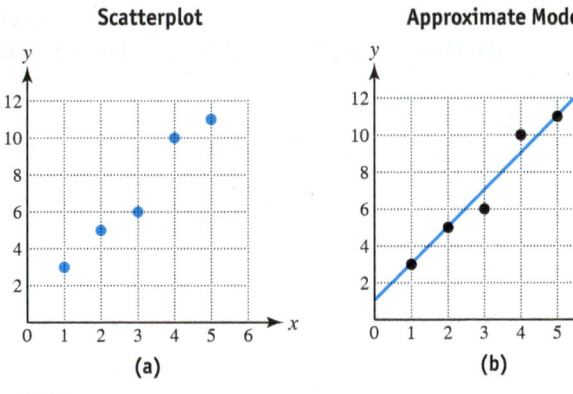

FIGURE 3.53

(b) One possibility for a line is shown in **FIGURE 3.53(b)**. This line passes through three of the five points, and is above one point and below another point. To determine the equation of this line, pick two points that the line passes through. For example, the points $(1, 3)$ and $(5, 11)$ lie on the line. The slope of this line is

$$m = \frac{11 - 3}{5 - 1} = \frac{8}{4} = 2.$$

The equation of the line passing through $(\mathbf{1}, \mathbf{3})$ with slope $\mathbf{2}$ can be found as follows.

$$y - y_1 = m(x - x_1) \quad \text{Point–slope form}$$
$$y - \mathbf{3} = \mathbf{2}(x - \mathbf{1}) \quad x_1 = 1, y_1 = 3, \text{ and } m = 2$$
$$y - 3 = 2x - 2 \quad \text{Distributive property}$$
$$\boxed{\text{Linear Model}} \quad y = 2x + 1 \quad \text{Add 3 to each side.}$$

The equation of this line is $y = 2x + 1$.

Now Try Exercise 33

NOTE: The equation found in Example 3 represents *one* possible linear model. Other linear models are possible. ■

🌐 *Math in Context*   *World Health*   When a quantity increases at a constant rate, it can be modeled with the linear equation $y = mx + b$. This concept is illustrated in the next example.

EXAMPLE 4 **Modeling worldwide HIV/AIDS cases in children**

In 2012, a total of 3.3 million children (under age 15) were living with HIV/AIDS. The rate of new infections was 0.26 million per year.
(a) Write a linear equation $C = mx + b$ that models the total number of children C in millions that were living with HIV/AIDS, x years after 2012.
(b) Estimate C in 2015.

Solution
(a) In the equation $C = mx + b$, the rate of change in HIV infections (**0.26** million per year) corresponds to the slope m, and the initial number of cases (**3.3** million) in 2012 corresponds to b. Therefore the equation $C = 0.26x + 3.3$ models the data.
(b) Since 2015 is 3 years after 2012, let $x = 3$.

$$C = 0.26(3) + 3.3 = 4.08 \text{ million}$$

Now Try Exercise 47

The following box offers a strategy for modeling data that have a constant rate of change.

MODELING WITH A LINEAR EQUATION

To model a quantity y that has a constant rate of change, use the equation

$$y = mx + b,$$

where $m = $ (constant rate of change) and $b = $ (initial amount).

EXAMPLE 5 **Modeling with linear equations**

Find a linear equation in the form $y = mx + b$ that models the quantity y after x days.
(a) A quantity y is initially 500 and increases at a rate of 6 per day.
(b) A quantity y is initially 1800 and decreases at a rate of 25 per day.
(c) A quantity y is initially 10,000 and remains constant.

Solution
(a) In the equation $y = mx + b$, the y-intercept $(0, b)$ represents the initial amount and the slope m represents the rate of change. Therefore $y = 6x + 500$.
(b) The quantity y is decreasing at the rate of 25 per day with an initial amount of 1800, so $y = -25x + 1800$.
(c) The quantity is constant, so $m = 0$. The equation is $y = 10,000$.

Now Try Exercises 25, 27, 29

3.7 Putting It All Together

CONCEPT	COMMENTS	EXAMPLE
Linear Model	Used to model a quantity that has a constant rate of change	If a total of 2 inches of rain falls before noon, and if rain falls at the rate of $\frac{1}{2}$ inch per hour, then $y = \frac{1}{2}x + 2$ models the total rainfall x hours past noon.

continued on next page

continued from previous page

CONCEPT	COMMENTS	EXAMPLE
Modeling Linear Data with a Line	1. Plot the data. 2. Sketch a line that passes either through or nearly through the points. 3. Pick two points on the line and find the equation of the line.	To model $(0, 4)$, $(1, 3)$, and $(2, 2)$, plot the points and sketch a line as shown in the accompanying figure. Many times one line cannot pass through all the points. The equation of the line is $y = -x + 4$.

3.7   Exercises

MyMathLab®

CONCEPTS AND VOCABULARY

1. Linear data are modeled by a(n) _____ equation.

2. If a line passes through all the data points, it is a(n) _____ model.

3. If a line passes near but not through each data point, it is a(n) _____ model.

4. Linear models are used to describe data that have a(n) _____ rate of change.

5. If a quantity is modeled by the equation $y = mx + b$, then m represents the _____.

6. If a quantity is modeled by the equation $y = mx + b$, then b represents the _____.

UNDERSTANDING LINEAR MODELS

Exercises 7–12: **Modeling** *Match the situation to the graph (a.–f.) that models it best.*

7. College tuition from 2000 to 2015

8. Average temperature in a freezer in degrees Celsius.

9. Profit from selling boxes of candy if it costs $200 to make the candy

10. Height of Mount Hood in Oregon over time

11. Total amount of water delivered by a garden hose if it flows at a constant rate

12. Sales of music on CDs from 2000 to 2010

a.

b.

c.

d.

e.

f.
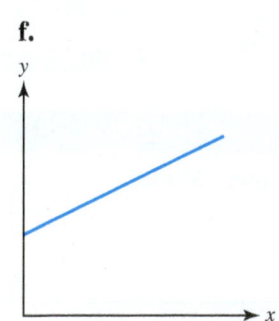

Exercises 13–18: State whether the ordered pairs in the table are modeled exactly by the linear equation.

13. $y = 2x + 2$

x	0	1	2
y	2	4	6

14. $y = -2x + 5$

x	0	1	2
y	5	3	0

15. $y = -4x$

x	−1	0	1
y	4	0	−8

16. $y = 5 - x$

x	−2	1	4
y	7	4	1

17. $y = 1.4x - 4$

x	0	5	10
y	−4	3	9

18. $y = -\frac{4}{3}x - \frac{13}{3}$

x	−7	−4	−1
y	5	1	−3

Exercises 19–24: State whether the linear model in the graph is exact or approximate. Then find the equation of the line.

19.

20.

21.

22.

23.

24.

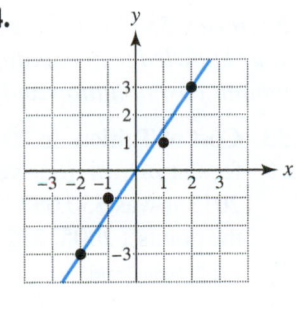

MODELING LINEAR DATA

Exercises 25–30: Find an equation $y = mx + b$ that models the quantity y after x units of time.

25. A quantity y is initially 40 and increases at a rate of 5 per minute.

26. A quantity y is initially -60 and increases at a rate of 1.7 per minute.

27. A quantity y is initially -5 and decreases at a rate of 20 per day.

28. A quantity y is initially 5000 and decreases at a rate of 35 per day.

29. A quantity y is initially 8 and remains constant.

30. A quantity y is initially -45 and remains constant.

Exercises 31–38: **Modeling Data** *For the ordered pairs in the table, do the following.*

(a) *Plot the data. Could a line pass through all five points?*

(b) *Sketch a line that models the data.*

(c) *Determine an equation of the line. For data that are not exactly linear, answers may vary.*

31.

x	0	1	2	3	4
y	4	2	0	−2	−4

32.

x	0	1	2	2	3
y	7	6	5	4	3

33.

x	−2	−1	0	1	2
y	4	1	0	−1	−4

34.

x	−2	−1	0	1	2
y	7	5	1	−3	−5

35.

x	−6	−4	−2	0	2
y	1	0	−1	−2	−3

36.

x	−4	−2	0	2	4
y	1	2	3	4	5

37.

x	−6	−3	0	3	6
y	−3	−2	−0.5	0	0.5

38.

x	−3	−2	−1	0	1
y	1	2	3	4	5

APPLICATIONS INVOLVING LINEAR MODELS

Exercises 39–46: Write an equation that models the described quantity. Specify what each variable represents.

39. A barrel contains 200 gallons of water and is being filled at a rate of 5 gallons per minute.

40. A barrel contains 40 gallons of gasoline and is being drained at a rate of 3 gallons per minute.

41. An athlete has run 5 miles and is jogging at 6 miles per hour.

42. A new car has 141 miles on its odometer and is traveling at 70 miles per hour.

43. A worker has already earned $200 and is being paid $8 per hour.

44. A gambler has lost $500 and is losing money at a rate of $150 per hour.

45. A carpenter has already shingled 5 garage roofs and is shingling roofs at a rate of 1 per day.

46. A hard drive has been spinning for 2 minutes and is spinning at 7200 revolutions per minute.

47. Kilimanjaro Glacier Mount Kilimanjaro is located in Tanzania, Africa, and has an elevation of 19,340 feet. In 1912, the glacier on its peak covered 5 acres. By 2002, this glacier had melted to only 1 acre. (*Source:* NBC News.)
  (a) Assume that this glacier melted at a constant rate each year. Find this *yearly* rate.
  (b) Use your answer in part (a) to write a linear equation that gives the acreage A of this glacier t years past 1912.

48. World Population In 1987, the world's population reached 5 billion people, and by 2012, the world's population reached 7 billion people. (*Source:* U.S. Census Bureau.)
  (a) Find the average yearly increase in the world's population from 1987 to 2012.
  (b) Write a linear equation that estimates the world's population P in billions x years after 1987.

49. Prison Population The points shown in **FIGURE 3.50** include $(2007, 200)$, $(2009, 208)$, $(2011, 218)$, and $(2013, 219)$, where the y-coordinates are in thousands.
  (a) Use the first and last data points to determine a line that models the data. Write the equation in slope–intercept form.
  (b) Use the line to estimate the population in 2017.

50. Niagara Falls The average flow of water over Niagara Falls is 212,000 cubic feet of water per second.
  (a) Write an equation that gives the number of cubic feet of water F that flow over Niagara Falls in x seconds.
  (b) How many cubic feet of water flow over Niagara Falls in 1 minute?

51. Gas Mileage The table shows the number of miles y traveled by a car on x gallons of gasoline.

x (gallons)	3	6	9	12
y (miles)	60	120	180	240

  (a) Plot the data in the xy-plane. Label each axis.
  (b) Sketch a line that models these data. (You may want to use a ruler.)
  (c) Calculate the slope of the line. Interpret the slope.
  (d) Find an equation of the line.
  (e) How far could this car travel on 7 gallons of gasoline?

52. Air Temperature Generally, the air temperature becomes colder as the altitude above the ground increases. The table lists typical air temperatures x miles high when the ground temperature is 80 °F.

x (miles)	0	1	2	3
y (°F)	80	61	42	43

  (a) Plot the data in the xy-plane. Label each axis.
  (b) Sketch a line that models these data. (You may want to use a ruler.)
  (c) Calculate the slope of the line. Interpret the slope.
  (d) Find the slope–intercept form of the line.
  (e) Estimate the air temperature 5 miles high.

Exercises 53–56: In this set of exercises, you are to use your knowledge of equations of lines to model the average annual cost of tuition and fees.

53. Cost of Tuition In 2005, the average cost of tuition and fees at *private* four-year colleges was $21,235, and in 2013 it was $30,083. Sketch a line that passes through the points $(2005, 21235)$ and $(2013, 30083)$. (*Source:* The College Board.)

54. Rate of Change in Tuition Calculate the slope of the line in your graph. Interpret this slope as a rate of change.

55. Modeling Tuition Find the slope–intercept form of the line in your sketch. What is the *y*-intercept and does it have meaning in this situation?

56. Predicting Tuition Use your equation to estimate tuition and fees in 2010 and compare it to the known value of $27,290. Estimate tuition and fees in 2015.

WRITING ABOUT MATHEMATICS

57. In Example 2, the gas mileage of an SUV is modeled with a linear equation. Explain why it is reasonable to use a linear equation to model this situation.

58. Explain the steps for finding the equation of a line that models data points in a table.

SECTION 3.7   Checking Basic Concepts

1. State whether the ordered pairs shown in the table are modeled exactly by $y = -5x + 10$.

x	-2	-1	0	1
y	20	15	10	5

2. State whether the linear model in the graph is exact or approximate. Find the equation of the line.

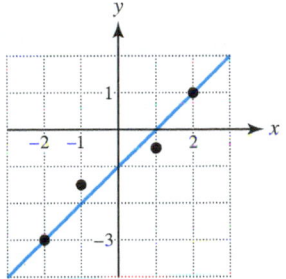

3. Find an equation, $y = mx + b$, that models the quantity *y* after *x* units of time.
   (a) A quantity *y* is initially 50 pounds and increases at a rate of 10 pounds per day.
   (b) A quantity *y* is initially 200 °F and decreases at a rate of 2 °F per minute.

4. The table contains ordered pairs.
   (a) Plot the data.
   (b) Sketch a line that models the data.
   (c) Determine the equation of the line.

x	-2	0	2	4
y	2	1	0	-1

5. Climate Change Since 1945, the average annual recorded temperature on the Antarctic Peninsula has increased by 0.075 °F per year.
   (a) Write an equation that models the average temperature *increase T*, *x* years after 1945.
   (b) Use your equation to calculate the temperature increase between 1945 and 2017.

CHAPTER 3   Summary

SECTION 3.1 ■ INTRODUCTION TO GRAPHING

The Rectangular Coordinate System (xy-plane)

Points   Plotted as (x, y) ordered pairs (see the *xy*-plane below).

Four Quadrants The *x*- and *y*-axes divide the *xy*-plane into quadrants I, II, III, and IV.

NOTE: A point on an axis, such as $(1, 0)$, does not lie in a quadrant. ■

SECTION 3.2 ▪ EQUATIONS IN TWO VARIABLES

Equations in Two Variables An equation with two variables and possibly some constants

Examples: $y = 3x + 7$ and $x + y = 100$

Solution to an Equation in Two Variables The solution to an equation in two variables is an ordered pair that makes the equation a true statement.

Example: $(1, 2)$ is a solution to $2x + y = 4$ because $2(1) + 2 = 4$ is true.

Graphing a Linear Equation in Two Variables

Linear Equation $y = mx + b$ or $Ax + By = C$ (standard form)

Graphing   Plot at least three points and sketch a line passing through each one.

Example: $y = 2x - 1$

x	y
0	−1
1	1
2	3

SECTION 3.3 ▪ INTERCEPTS; HORIZONTAL AND VERTICAL LINES

Intercepts

x-Intercept   A point at which a graph intersects the x-axis; to find an x-intercept, let $y = 0$ in the equation and solve for x.

y-Intercept   A point at which a graph intersects the y-axis; to find a y-intercept, let $x = 0$ in the equation and solve for y.

Example: $x + 3y = 3$

x-intercept: Solve $x + 3(0) = 3$ to find the x-intercept of $(3, 0)$.
y-intercept: Solve $0 + 3y = 3$ to find the y-intercept of $(0, 1)$.

Horizontal and Vertical Lines

The equation of a horizontal line with y-intercept $(0, b)$ is $y = b$.

The equation of a vertical line with x-intercept $(k, 0)$ is $x = k$.

Example: The horizontal line $y = -1$ has y-intercept $(0, -1)$ and no x-intercepts.
The vertical line $x = 2$ has x-intercept $(2, 0)$ and no y-intercepts.

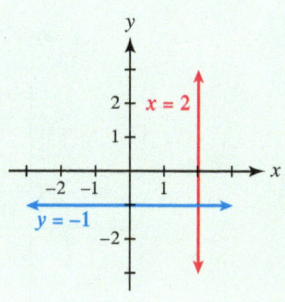

SECTION 3.4 ▪ SLOPE AND RATES OF CHANGE

Slope The ratio $\frac{\text{rise}}{\text{run}}$, or $\frac{\text{change in } y}{\text{change in } x}$, is the slope m of a line when run (change in x) is nonzero. A positive slope indicates that a line rises from left to right, and a negative slope indicates that a line falls from left to right.

Example: Slope $\frac{2}{3}$ indicates that a line rises 2 units for every 3 units of horizontal run.

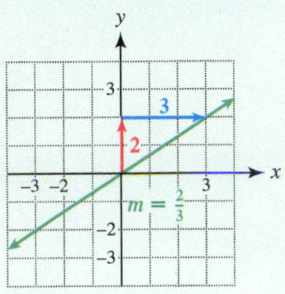

Calculating Slope   A line passing through (x_1, y_1) and (x_2, y_2) has slope

$$m = \frac{y_2 - y_1}{x_2 - x_1}, \text{ where } x_1 \neq x_2.$$

Example: The line through $(-2, 2)$ and $(3, 4)$ has slope

$$m = \frac{4 - 2}{3 - (-2)} = \frac{2}{5}.$$

Horizontal Line   Has zero slope

Vertical Line   Has undefined slope

Slope as a Rate of Change Slope measures the "tilt" of a line. In applications, slope measures the rate of change in a quantity.

Example: The line shown in the graph has slope -2 and depicts an initial outside temperature of $6\,°F$. Slope -2 indicates that the temperature is *decreasing* at a rate of $2\,°F$ per hour.

SECTION 3.5 ▪ SLOPE–INTERCEPT FORM

Slope–Intercept Form

For the line given by $y = mx + b$, the slope is m and the y-intercept is $(0, b)$.

Example: $y = -\frac{1}{2}x + 2$ has slope $-\frac{1}{2}$ and y-intercept $(0, 2)$, as shown in the graph on the following page.

Parallel Lines Lines with the same slope are parallel; nonvertical parallel lines have the same slope. Two vertical lines are parallel.

Example: The equations $y = -2x + 1$ and $y = -2x$ determine two parallel lines because $m_1 = m_2 = -2$. See the graph on the following page.

Perpendicular Lines

If two perpendicular lines have nonzero slopes m_1 and m_2, then $m_1 \cdot m_2 = -1$.

If two lines have nonzero slopes satisfying $m_1 \cdot m_2 = -1$, then they are perpendicular.

A vertical line and a horizontal line are perpendicular.

Example: The equations $y = -\frac{1}{2}x$ and $y = 2x - 2$ determine perpendicular lines because $m_1 \cdot m_2 = -\frac{1}{2} \cdot 2 = -1$. See the graph shown on the following page.

Slope–intercept Form	Parallel Lines	Perpendicular Lines
		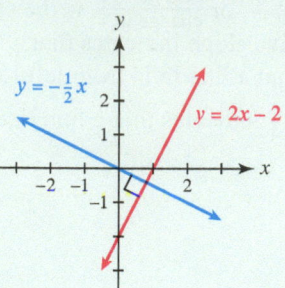

SECTION 3.6 ■ POINT–SLOPE FORM

Point–Slope Form An equation of the line passing through (x_1, y_1) with slope m is $y - y_1 = m(x - x_1)$.

Example: If $m = -2$ and $(x_1, y_1) = (-2, 3)$, then the point–slope form is

$$y - 3 = -2(x + 2).$$

Example: To find an equation of a line passing through $(-2, 5)$ and $(4, 2)$, first find the slope.

$$m = \frac{2 - 5}{4 - (-2)} = \frac{-3}{6} = -\frac{1}{2}$$

Either $(-2, 5)$ or $(4, 2)$ may be used in the point–slope form. The point $(4, 2)$ results in the equation $y - 2 = -\frac{1}{2}(x - 4)$.

SECTION 3.7 ■ INTRODUCTION TO MODELING

Mathematical Modeling Mathematics can be used to describe or approximate the behavior of real-world phenomena.

Exact Model The equation describes the data precisely without error.

Example: The equation $y = 3x$ models the data in the table exactly.

x	0	1	2	3
y	0	3	6	9

Approximate Model The modeling equation describes the data approximately. An approximate model occurs most often in applications.

Example: The line in the graph models the data approximately.

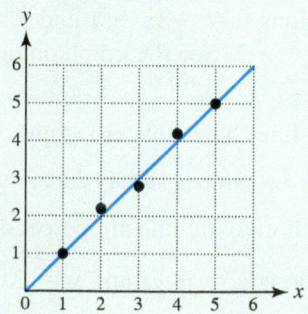

Modeling with a Linear Equation To model a quantity y that has a constant rate of change, use the equation $y = mx + b$, where

$$m = (\text{constant rate of change}) \quad \text{and} \quad b = (\text{initial amount}).$$

Example: If the temperature is initially $100\,^\circ F$ and cools at $5\,^\circ F$ per hour, then

$$T = -5x + 100$$

models the temperature T after x hours.

CHAPTER 3   Review Exercises

SECTION 3.1

1. Identify the coordinates of each point in the graph. Identify the quadrant, if any, in which each point lies.

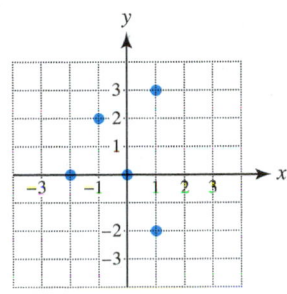

2. Make a scatterplot by plotting the following four points: $(-2, 3)$, $(-1, -1)$, $(0, 3)$, and $(2, -1)$.

Exercises 3 and 4: If possible, identify the quadrant in which each point is located.

3. (a) $(-4, 3)$ (b) $\left(\frac{1}{3}, -\frac{1}{2}\right)$

4. (a) $(0, 3.2)$ (b) $(-5, -1.7)$

Exercises 5 and 6: Use the table of xy-values to make a line graph.

5.
x	-2	-1	0	1	2
y	-3	2	-1	-2	3

6.
x	-10	-5	0	5	10
y	5	-10	10	-5	0

SECTION 3.2

Exercises 7 and 8: Determine whether the ordered pair is a solution for the given equation.

7. $y = x - 3$ $(6, 3)$

8. $3x - y = 3$ $(-1, 6)$

Exercises 9 and 10: Complete the table for the given equation.

9. $y = -3x$

x	-2	-1	0	1	2
y					

10. $2x + y = 5$

x					
y	-3	-1	0	1	3

Exercises 11 and 12: Use the given values of the variable to make a horizontal table of solutions for the equation.

11. $y = 3x + 2$ $x = -2, 0, 2, 4$

12. $2y + x = 1$ $y = 1, 2, 3, 4$

Exercises 13–16: Graph the equation.

13. $y = 2x$ 14. $y = -3x + 2$

15. $x + y = 2$ 16. $3x - 2y = 6$

SECTION 3.3

Exercises 17 and 18: Identify the x- and y-intercepts.

17.

18.

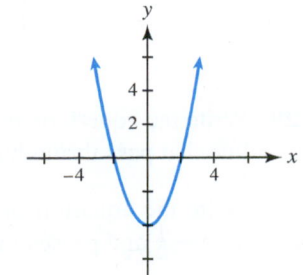

Exercises 19 and 20: Complete the table. Then determine the x- and y-intercepts for the graph of the equation.

19. $y = 2 - x$

x	-2	-1	0	1	2
y					

20. $x - 2y = 4$

x	-4	-2	0	2	4
y					

Exercises 21–24: Find any intercepts for the graph of the equation and then graph the linear equation.

21. $2x - 3y = 6$

22. $5x - y = 5$

23. $0.1x - 0.2y = 0.4$

24. $\dfrac{x}{2} + \dfrac{y}{3} = 1$

Exercises 25 and 26: Write an equation for the line that passes through the points shown in the table.

25.

x	-2	-1	0	1	2
y	1	1	1	1	1

26.

x	3	3	3	3	3
y	-2	-1	0	1	2

27. Graph each equation.
   (a) $y = 1$ **(b)** $x = -3$

28. Write an equation for each line shown in the graph.

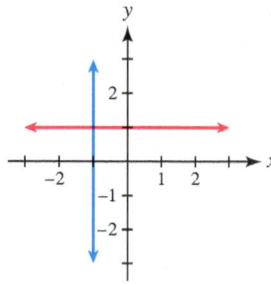

29. Write the equations of a horizontal line and a vertical line that pass through the point $(-2, 3)$.

30. Write the equation of a line that is perpendicular to $y = -\frac{1}{2}$ and passes through $(4, 1)$.

31. Write the equation of a line that is parallel to $y = 3$ and passes through $(-6, -5)$.

32. Distance The distance a driver is from home is illustrated in the graph.
   (a) Find the intercepts.
   (b) Interpret each intercept.

SECTION 3.4

Exercises 33–36: Find the slope, if possible, of the line passing through the two points.

33. $(2, 3), (4, 7)$ **34.** $(-3, 1), (2, -1)$

35. $(2, 1), (5, 1)$ **36.** $(-5, 6), (-5, 10)$

Exercises 37 and 38: Find the slope of the line shown in the graph.

37.

38.

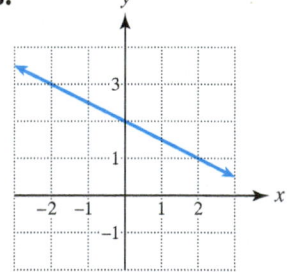

Exercises 39 and 40: Do the following.

   (a) *Graph the linear equation.*
   (b) *What are the slope and y-intercept of the line?*

39. $y = -2x$ **40.** $2x - 3y = -6$

Exercises 41 and 42: Sketch a line that passes through the given point and has slope m.

41. $(0, -3), m = 2$ **42.** $(2, 2), m = 1$

43. A line with slope $\frac{1}{2}$ passes through the first point shown in the table. Complete the table so that each point in the table lies on the line.

x	0	1	2	3
y	1			

44. If a line models the cost of buying x coffee drinks at $3.49 each, does the line have positive or negative slope?

SECTION 3.5

Exercises 45 and 46: Write the slope–intercept form for the line shown in the graph.

45.

46.

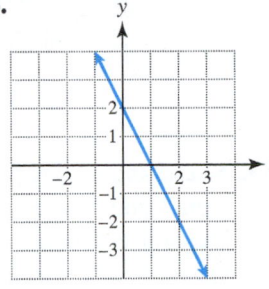

Exercises 47 and 48: Sketch a line with slope m and y-coordinate of the y-intercept b. Write the slope–intercept form of the equation of the line.

47. $m = 2, b = -2$

48. $m = -\frac{3}{4}, b = 3$

Exercises 49 and 50: Do the following.

   (a) *Write the equation in slope–intercept form.*
   (b) *Give the slope and y-intercept of the line.*

49. $x + y = 3$

50. $5x - 6y = 30$

Exercises 51–54: Graph the equation.

51. $y = \frac{1}{2}x + 1$

52. $y = 3x - 2$

53. $y = 4 - x$

54. $y = 2 - \frac{2}{3}x$

Exercises 55 and 56: All the points shown in the table lie on the same line. Find the slope–intercept form for the line.

55.

x	0	1	2
y	-5	0	5

56.

x	-1	0	1
y	2	0	-2

Exercises 57–60: Find the slope–intercept form for the line satisfying the given conditions.

57. Slope $-\frac{5}{6}$, y-intercept $(0, 2)$

58. Parallel to $y = -2x + 1$, passing through $(1, -5)$

59. Perpendicular to $y = -\frac{3}{2}x$, passing through $(3, 0)$

60. Perpendicular to $y = 5x - 3$, passing through the point $(0, -2)$

SECTION 3.6

Exercises 61 and 62: Determine whether the given point lies on the line.

61. $(-3, 1)$ $y - 1 = 2(x + 3)$

62. $(3, -8)$ $y = -3(x - 1) + 2$

Exercises 63–68: Find a point–slope form for the line that satisfies the conditions given. When two points are given, use the first point in the point–slope form.

63. Slope 5, passing through $(1, 2)$

64. Passing through $(20, -30)$ and $(40, 30)$

65. x-intercept $(3, 0)$, y-intercept $(0, -4)$

66. x-intercept $(\frac{1}{2}, 0)$, y-intercept $(0, -1)$

67. Parallel to $y = 2x$, passing through $(5, 7)$

68. Perpendicular to $y - 4 = \frac{3}{2}(x + 1)$, and passing through the point $(-1, 0)$

Exercises 69 and 70: Write the given point–slope form in slope–intercept form.

69. $y - 2 = 3(x + 1)$ **70.** $y = -\frac{1}{4}(x - 8) + 1$

SECTION 3.7

71. State whether the ordered pairs shown in the table are modeled exactly by $y = -x + 4$.

x	0	1	2	2
y	4	3	2	1

72. State whether the linear model shown in the graph is exact or approximate. Find the equation of the line.

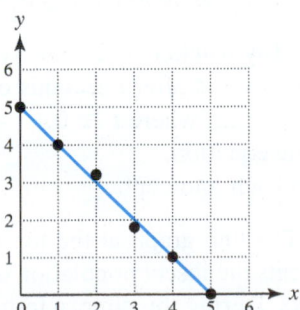

Exercises 73–76: Find an equation $y = mx + b$ that models y after x units of time.

73. y is initially 40 pounds and decreases at a rate of 2 pounds per minute.

74. y is initially 200 gallons and increases at 20 gallons per hour.

75. y is initially 50 and remains constant.

76. y is initially 20 feet *below* sea level and rises at 5 feet per second.

Exercises 77 and 78: For the ordered pairs in the table, do the following.

(a) *Plot the data. Could a line pass through all five points?*

(b) *Sketch a line that models the data.*

(c) *Determine an equation of the line. For data that are not exactly linear, answers may vary.*

77.

x	0	1	2	3	4
y	10	6	2	-2	-6

78.

x	-4	-2	0	2	4
y	1	2.1	3	3.9	5

APPLICATIONS

79. Graphing Real Data The table contains real data on the number of divorces D in millions during year t.

(a) Make a line graph of the data. Label the axes.

(b) Comment on any trends in the data.

t	1970	1980	1990	2000	2010
D	0.7	1.2	1.2	1.2	1.0

Source: National Center for Health Statistics.

80. Water Usage The average American uses 100 gallons of water each day.

(a) Write an equation that gives the gallons G of water that a person uses in t days.

(b) Graph the equation for $t \geq 0$.

(c) How many days does it take for the average American to use 5000 gallons of water?

81. Modeling a Toy Rocket The velocity v of a toy rocket in feet per second after t seconds of flight is given by $v = 160 - 32t$, where $t \geq 0$.

(a) Graph the equation.

(b) Interpret each intercept.

82. Modeling The line graph at the top of the next column represents the insect population on 1 acre of land after x weeks. During this time, a farmer sprayed pesticides on the land.

(a) Estimate the slope of each line segment.

(b) Interpret each slope as a rate of change.

(c) Describe what happened to the insect population.

83. Sketching a Graph An athlete jogs 4 miles away from home at a constant rate of 8 miles per hour and then turns around and jogs back home at the same speed. Sketch a graph that shows the distance that the athlete is from home. Be sure to label each axis.

84. Music Downloads In 2008, global recorded-music revenue from downloads was $3 billion, and it was projected to be $9 billion by 2017. (*Source:* The Economist.)

(a) Calculate the slope of the line passing through $(2008, 3)$ and $(2017, 9)$.

(b) Interpret the slope as a rate of change.

85. Rental Cars The cost C in dollars for driving a rental car x miles is $C = 0.2x + 35$.

(a) How much would it cost to rent the car but not drive it?

(b) How much does the cost increase for each additional mile driven?

(c) Determine the C-intercept for the graph of $C = 0.2x + 35$. What does this C-intercept represent?

(d) What is the slope of the graph of $C = 0.2x + 35$? What does it represent?

86. Distance and Speed A person is driving a car along a straight road. The graph at the top of the next page shows the distance y in miles that the driver is from home after x hours.

(a) Is the person traveling toward or away from home? Why?

(b) The graph passes through $(1, 200)$ and $(3, 100)$. Discuss the meaning of these points.

(c) Find the slope–intercept form of the line. Interpret the slope as a rate of change.

(d) Use the graph to estimate the distance from home after 2 hours. Then check your answer by using your equation from part (c).

Distance (miles) / Time (hours)

(1, 200)
(3, 100)

(b) Find the slope–intercept form of a line that models the revenue R in year t. Interpret the slope of this line.

(c) Is the line you found in part (b) an exact model for the data in the table?

(d) If this trend continues, what will the revenue be in 2018?

88. Gas Mileage The table shows the number of miles y traveled by a car on x gallons of gasoline.

x	2	4	8	10
y	40	79	161	200

(a) Plot the data in the xy-plane. Be sure to label each axis.

(b) Sketch a line that models these data. (You may want to use a ruler.)

(c) Calculate the slope of the line. Interpret the slope of this line.

(d) Find the equation of the line. Is your line an *exact* model?

(e) How far could this car travel on 9 gallons of gasoline?

87. Social Ad Revenue Social display advertising is advertising seen next to content on social media pages, such as Facebook. The table gives the U.S. ad revenue R in billions of dollars for social display ads during year t.

t	2013	2014	2015	2016
R	4.3	4.9	5.5	6.1

(a) Make a scatterplot of the data.

CHAPTER 3   Test   MyMathLab® YouTube™

1. Identify the coordinates of each point in the graph. State the quadrant, if any, in which each point lies.

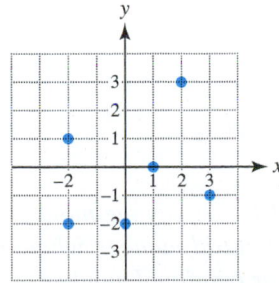

2. Make a scatterplot by plotting the four points $(0, 0)$, $(-2, -2)$, $(3, 0)$, and $(3, -2)$.

3. Complete the table for the equation $y = 2x - 4$. Then determine the x- and y-intercepts for the graph of the equation.

x	-2	-1	0	1	2
y					

4. Determine whether the ordered pair $(1, -3)$ is a solution for the equation $2x - y = 5$.

5. Sketch a line passing through the point $(2, 1)$ and having slope $-\frac{1}{2}$.

6. Find the x- and y-intercepts for the graph of the equation $5x - 3y = 15$.

Exercises 7–10: Graph the equation.

7. $y = 2$

8. $x = -3$

9. $y = -3x + 3$

10. $4x - 3y = 12$

11. Write the slope–intercept form for the line shown in the graph.

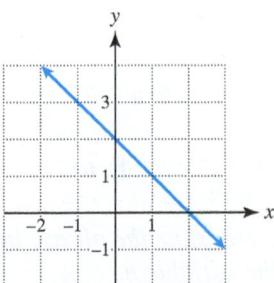

12. Write the equation $-4x + 2y = 1$ in slope–intercept form. Give the slope and the y-intercept.

13. Write the equation of a horizontal line and a vertical line passing through the point $(1, -5)$.

14. Find the slope of a line passing through the points $(-4, 3)$ and $(5, 1)$.

Exercises 15–18: Find the slope–intercept form for the line that satisfies the given conditions.

15. Slope $-\frac{4}{3}$, y-intercept $(0, -5)$

16. Parallel to $y = 3x - 1$, passing through $(2, -5)$

17. Perpendicular to $y = \frac{1}{3}x$, passing through $(1, 2)$

18. Passing through $(-4, 2)$ and $(2, -1)$

19. Write $y - 3 = \frac{1}{2}(x + 4)$ in slope–intercept form.

20. Do all of the points shown in the table lie on the same line. If so, find the slope–intercept form for the line.

x	-2	-1	0	1	2
y	-8	-5	-2	1	4

21. Write a point–slope form for a line that is parallel to $y = -3x$ and passes through $(-2, 7)$.

22. State whether the linear model shown is exact or approximate. Then find the equation of the line.

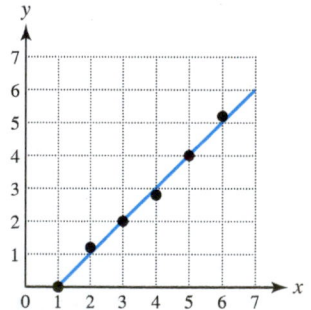

23. Sketching a Graph A cyclist rides a bicycle at 10 miles per hour for 2 hours and then at 8 miles per hour for 1 hour. Sketch a graph that shows the total distance d traveled after x hours. Be sure to label each axis.

24. Modeling The line graph represents the total fish population P in a small lake after x years. One winter the lake almost froze solid.
  (a) Estimate the slope of each line segment.
  (b) Interpret each slope as a rate of change.
  (c) Describe what happened to the fish population.

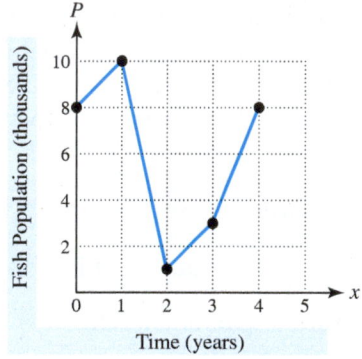

25. Modeling Insect Population Write an equation in slope–intercept form that models the number of insects N after x days if there are initially 2000 insects and the population increases at a rate of 100 insects per day.

CHAPTERS 1–3	**Cumulative Review Exercises**

Exercises 1 and 2: Classify the number as prime or composite. If a number is composite, write it as a product of prime numbers.

1. 40   **2.** 61

Exercises 3 and 4: Translate the phrase into an algebraic expression using the variable n.

3. Ten more than a number

4. A number squared decreased by 2

Exercises 5 and 6: Evaluate by hand and then simplify to lowest terms.

5. $\frac{3}{4} \div \frac{9}{8}$ **6.** $\frac{7}{10} - \frac{2}{15}$

Exercises 7 and 8: Evaluate the expression by hand.

7. $20 - 2 \cdot 3$ **8.** -3^2

Exercises 9 and 10: Classify the number as one or more of the following: natural number, whole number, integer, rational number, or irrational number.

9. $-\frac{4}{5}$ **10.** $\sqrt{3}$

Exercises 11 and 12: Simplify the expression.

11. $3 + 4x - 2 + 3x$ **12.** $2(x - 1) - (x + 2)$

Exercises 13 and 14: Solve the equation. Check your solution.

13. $3t - 5 = 1$ **14.** $2(x - 3) = -6 - x$

Exercises 15 and 16: Find the area of the figure shown.

15.   **16.**

17. Complete the table. Then use the table to solve the equation $6 - 2x = 4$.

x	-2	-1	0	1	2
$6 - 2x$					

18. Translate the sentence "Twice a number increased by 2 equals the number decreased by 5" to an equation, using the variable n. Then solve the equation.

19. If $r = 10$ mph and $d = 80$ miles, use the formula $d = rt$ to find t.

20. Use an inequality to express the set of real numbers graphed.

Exercises 21 and 22: Solve the inequality. Write the solution set in set-builder notation.

21. $3 - 6x < 3$

22. $2x \le 1 - (2x - 1)$

23. Identify the coordinates of each point in the graph. State the quadrant, if any, in which each point lies.

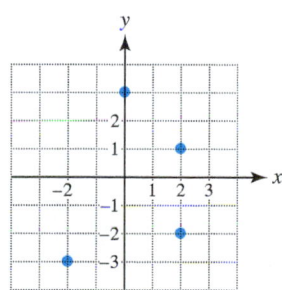

24. Complete the table for the equation $x + 2y = 4$.

x	-2	-1	0	1	2
y					

Exercises 25 and 26: Graph the equation.

25. $2x + 3y = -6$ **26.** $y = -\frac{3}{2}$

27. Determine the intercepts for the graph of the equation $-4x + 5y = 40$.

28. Write an equation for each line shown in the graph.

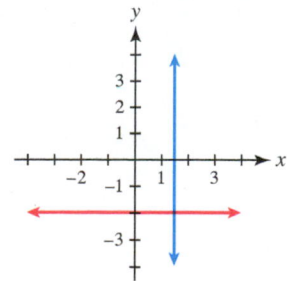

29. The table lists points located on a line. Write the slope–intercept form of the line.

x	-1	0	1	2
y	-6	-3	0	3

30. Write the equation $3x - 5y = 15$ in slope–intercept form. Graph the equation.

Exercises 31 and 32: Find the slope–intercept form for the line satisfying the given conditions.

31. Perpendicular to $3x - 2y = 6$, passing through the point $(0, -3)$

32. Passing through $(-1, 3)$ and $(2, -3)$

33. Let the initial value of y be 100 pounds, increasing at 5 pounds per hour. Find an equation in slope–intercept form that models y after x hours.

34. An insect population is initially 20,000 and is increasing at 5000 per day. Write an equation that gives the number of insects I after x days.

APPLICATIONS

35. Pizzas The table lists the cost C of buying x large pepperoni pizzas. Write an equation that relates C and x.

x	2	3	4
C	\$16	\$24	\$32

36. Photo-Sharing Sites In 2013, among the top four photo-sharing sites, Snapchat and Instagram made up about $\frac{14}{25}$ of the daily photo uploads. For every Instagram upload, there were 7 Snapchat uploads. Estimate the fraction of the uploads that were on Snapchat.

37. Investment Money A student invests two sums of money at 3% and 4% interest, receiving a total of \$110 in interest after 1 year. Twice as much money is invested at 4% as at 3%. Find the amount invested at each interest rate.

38. Rental Cars The cost C in dollars for driving a rental car x miles is $C = 0.3x + 25$.
(a) How much does it cost to rent the car and drive it 200 miles?
(b) How much does it cost to rent the car but not drive it?
(c) How much does the cost increase for each additional mile driven?

4 Systems of Linear Equations in Two Variables

With a land-line telephone, information is typically sent along a single wire, or path. However, if a person is talking on a cell phone in a car, any *one* path could fail at any time because the location of the person's cell phone is always changing. As a result, information must be sent simultaneously along several paths so that this information can be reconstructed even when one or more paths fail during transmission. Mathematics, and *systems of equations* in particular, play an important role in determining how information is sent. (*Source:* Effros, M., and A. Goldsmith, *Scientific American.*)

In this chapter we discuss graphical, numerical, and symbolic methods for solving systems of linear equations.

4.1 Solving Systems of Linear Equations Graphically and Numerically

1 Solving Equations Graphically and Numerically

SOLVING GRAPHICALLY In the previous chapter, we showed that the graph of $y = mx + b$ is a line with slope m and y-intercept $(0, b)$, as illustrated in **FIGURE 4.1**. Each point on this line represents a solution to the equation $y = mx + b$. Because there are infinitely many points on a line, there are infinitely many solutions to this equation. However, many applications require that we find one particular solution to a linear equation. One way to find such a solution is to graph a second line in the same xy-plane and determine the point of intersection (if one exists).

Connecting Concepts with Your Life Consider the following application of a line. If renting a moving truck for one day costs $25 plus $0.50 per mile driven, then the equation $C = 0.5x + 25$ represents the cost C in dollars of driving the rental truck x miles. The graph of this line is shown in **FIGURE 4.2(a)** for $x \geq 0$.

Truck Rental Cost

Rental Cost of $75

Driving **100** miles costs **$75**.

(a)   (b)

FIGURE 4.2

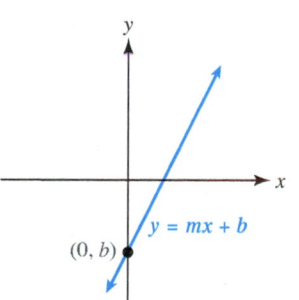

FIGURE 4.1

NEW VOCABULARY

□ Intersection-of-graphs method
□ System of linear equations in two variables
□ Solution to a system
□ Inconsistent system
□ Consistent system with independent equations
□ Consistent system with dependent equations

Suppose that we want to determine the number of miles that the truck is driven when the rental cost is $75. One way to solve this problem *graphically* is to graph both $C = 0.5x + 25$ and $C = 75$ in the same coordinate plane, as shown in **FIGURE 4.2(b)**. The lines intersect at the point $(\mathbf{100}, \mathbf{75})$, which is a solution to $C = 0.5x + 25$ and to $C = 75$. That is, if the rental cost is **$75**, then the mileage must be **100** miles. This graphical technique for solving two equations is sometimes called the **intersection-of-graphs method**. To find a solution with this method, we locate a point where two graphs intersect.

STUDY TIP

Try to find a consistent time and place to study your notes and do your homework. When the time comes to study for an exam, do so at your usual study time in your usual place rather than "pulling an all-nighter" in unfamiliar surroundings.

EXAMPLE 1   **Solving an equation graphically**

Use a graph to find the x-value when $y = 3$.
(a) $y = 2x - 1$ (b) $-3x + 2y = 12$

Solution

(a) Begin by graphing the equations $y = 2x - 1$ and $y = 3$. The graph of $y = 2x - 1$ is a line with slope 2 and y-intercept $(0, -1)$. The graph of $y = 3$ is a horizontal line with y-intercept $(0, 3)$. In **FIGURE 4.3(a)**, the graphs intersect at the point $(2, 3)$. Therefore an x-value of **2** corresponds to a y-value of **3**.

    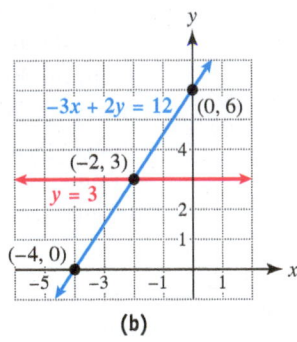

(a)   (b)

FIGURE 4.3

(b) One way to graph $-3x + 2y = 12$ is to write this equation in slope–intercept form.

$$-3x + 2y = 12 \qquad \text{Given equation}$$
$$2y = 3x + 12 \qquad \text{Add } 3x \text{ to each side.}$$
$$y = \frac{3}{2}x + 6 \qquad \text{Divide each side by 2.}$$

The line has slope $\frac{3}{2}$ and y-intercept $(0, 6)$. Its graph and the graph of $y = 3$ are shown in **FIGURE 4.3(b)**. The graphs intersect at $(-2, 3)$. Therefore an x-value of **−2** corresponds to a y-value of **3**.

Now Try Exercises 15, 17

EXAMPLE 2 **Solving an equation graphically**

The equation $P = 10x$ calculates an employee's pay for working x hours at $10 per hour. Use the intersection-of-graphs method to find the number of hours that the employee worked if the amount paid is $40.

Employee Wages

Solution

Begin by graphing the equations $P = 10x$ and $P = 40$, as illustrated at right.

The graphs intersect at the point $(4, 40)$. Since the x-coordinate represents the number of hours worked and the P-coordinate represents pay, the point $(4, 40)$ indicates that the employee must work **4** hours to earn **$40**.

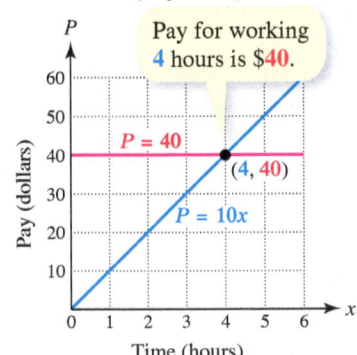

READING CHECK 1

• How is the intersection-of-graphs method used to find a solution to two equations?

Now Try Exercise 73 (a), (b)

SOLVING NUMERICALLY WITH A TABLE OF VALUES Equations can be solved in more than one way. In Example 2, we determined graphically that $P = 10x$ is equal to 40 when $x = 4$. We could also solve this problem by making a table of values, as illustrated by **TABLE 4.1**. Note that, when $x = 4$, $P = 40. A table of values provides a *numerical* solution.

Wages Earned at $10 per Hour

x (hours)	0	1	2	3	4	5	6
P (pay)	$0	$10	$20	$30	**$40**	$50	$60

TABLE 4.1

Find the x-value where $P = 40$.

NOTE: Although graphical and numerical methods are different, both methods should give the same solution. However, slight variations may occur because reading a graph precisely may be difficult, or a needed value may not appear in a table. ■

TECHNOLOGY NOTE

Intersection of Graphs and Table of Values
A graphing calculator can be used to find the intersection of the two graphs shown in Example 2. It can also be used to create **TABLE 4.1**. The accompanying figures illustrate how a calculator can be used to determine that $y_1 = 10x$ equals $y_2 = 40$ when $x = 4$.

CALCULATOR HELP

To find a point of intersection, see Appendix A (page AP-6). To make a table, see Appendix A (pages AP-2 and AP-3).

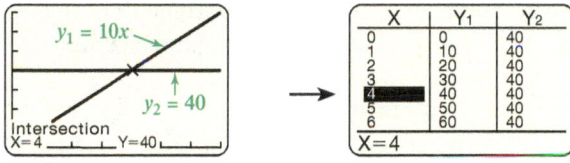

2 Introducing Systems of Linear Equations

In Example 1(b), we determined the x-value when $y = 3$ in the equation $-3x + 2y = 12$. This problem can be thought of as solving the following *system of two equations in two variables*.

$$-3x + 2y = 12$$
$$y = 3$$

READING CHECK 2

• How do we express the solution to a system of equations in two variables?

The solution is the ordered pair $(-2, 3)$, which indicates that when $x = -2$ and $y = 3$, each equation is a true statement.

$$-3(-2) + 2(3) = 12 \checkmark \quad \text{A true statement}$$
$$3 = 3 \checkmark \quad \text{A true statement}$$

Suppose that the sum of two numbers is 10 and that their difference is 4. If we let x and y represent the two numbers, then the equations

$$x + y = 10 \quad \text{Sum is 10.}$$
$$x - y = 4 \quad \text{Difference is 4.}$$

describe this situation. Each equation is a linear equation in two variables, so we call these equations a **system of linear equations in two variables**. Its graph typically consists of two lines. A **solution to a system** of two equations is an ordered pair (x, y) that makes *both* equations true. If a single solution exists, the ordered pair gives the coordinates of a point where the two lines intersect.

NOTE: For two distinct lines, there can be no more than one intersection point. If such an intersection point exists, the ordered pair corresponding to it represents the only solution to the system of linear equations. In this case, we say the ordered pair is *the* solution to the system of equations. ■

TYPES OF EQUATIONS AND NUMBER OF SOLUTIONS When a system of two linear equations in two variables is graphed, exactly one of the following situations will result.

See the Concept 1 **THREE TYPES OF LINEAR SYSTEMS**

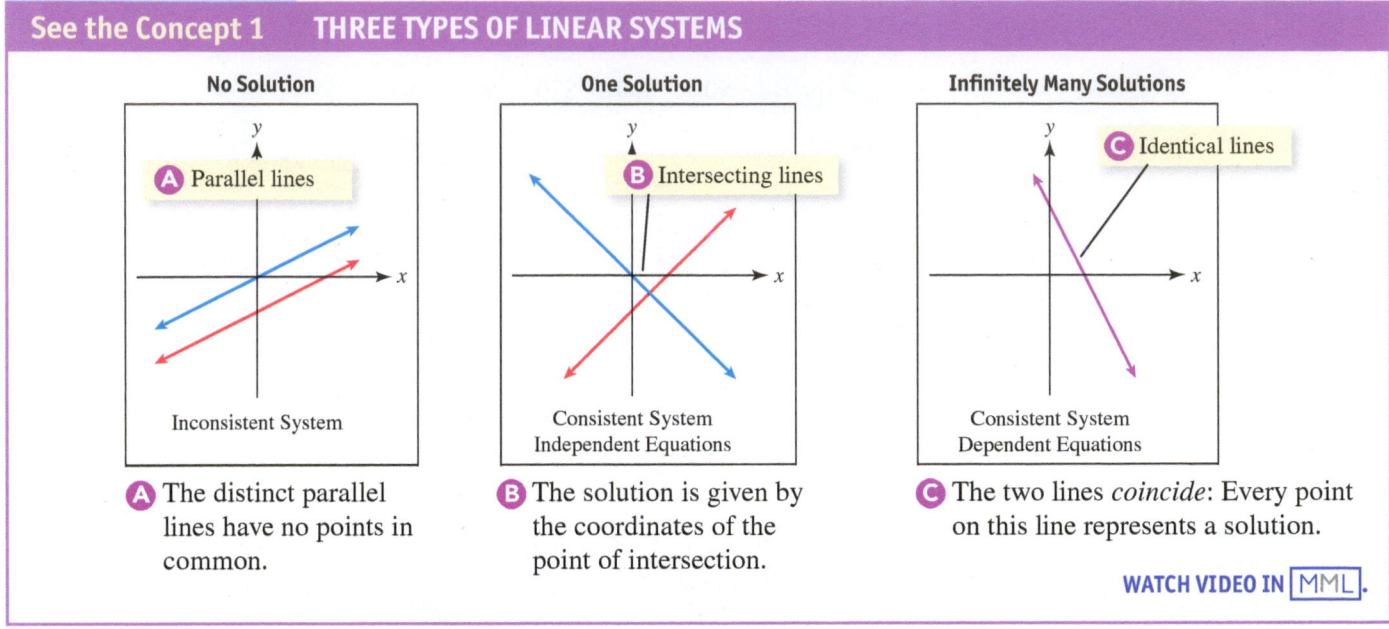

A The distinct parallel lines have no points in common.

B The solution is given by the coordinates of the point of intersection.

C The two lines *coincide*: Every point on this line represents a solution.

WATCH VIDEO IN MML.

In the first situation, there are no solutions, and it is an **inconsistent system**. In the second situation, there is exactly one solution, and it is a **consistent system** with **independent equations**. In the third situation, there are infinitely many solutions, and it is a **consistent system** with **dependent equations**. This information is summarized in **TABLE 4.2**.

READING CHECK 3

• How many solutions are possible for a system of linear equations in two variables?

Types of Systems of Equations

Type of Graph	Number of Solutions	Type of System	Type of Equations
Parallel Lines	0	Inconsistent	—
Intersecting Lines	1	Consistent	Independent
Identical Lines	Infinitely many	Consistent	Dependent

TABLE 4.2

EXAMPLE 3 **Identifying types of equations**

Graphs of two equations are shown. State the number of solutions to each system of equations. Then state whether the system is consistent or inconsistent. If it is consistent, state whether the equations are dependent or independent.

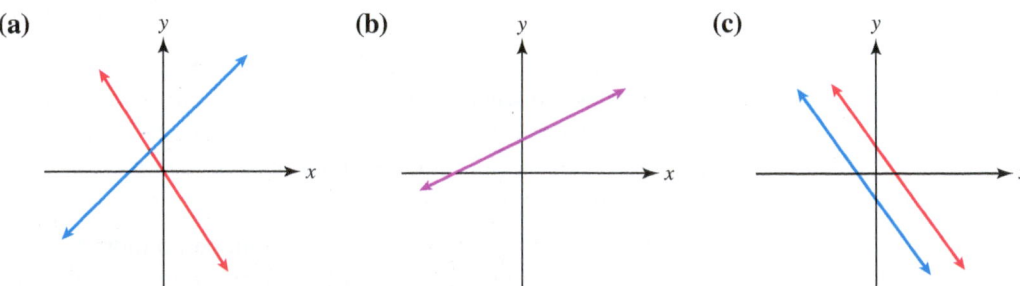

Solution
(a) The lines intersect at one point, so there is one solution. The system is consistent, and the equations are independent.
(b) There is only one line, which indicates that the graphs are identical, or coincide, so there are infinitely many solutions. The system is consistent and the equations are dependent.
(c) The lines are parallel, so there are no solutions. The system is inconsistent.

Now Try Exercises 25, 27, 29

CHECKING A SOLUTION TO A SYSTEM OF EQUATIONS For the remainder of this section, we focus on systems of linear equations in two variables that have exactly one solution. We will solve the systems by graphing and by making a table of values. Methods for solving two equations symbolically will be discussed later in the chapter.

Recall that an ordered pair is a solution to a system of two equations if its coordinates make *both* equations true. In the next example, we test possible solutions.

EXAMPLE 4 | **Checking possible solutions**

Determine whether $(4, 6)$ or $(7, 3)$ is the solution to the system of equations

$$x + y = 10$$
$$x - y = 4.$$

Solution
To determine whether $(4, 6)$ is the solution, substitute $x = 4$ and $y = 6$ in each equation. It must make *both* equations true.

$x + y = 10$	$x - y = 4$	Given equations
$4 + 6 \overset{?}{=} 10$	$4 - 6 \overset{?}{=} 4$	Let $x = 4, y = 6$.
$10 = 10$ (True) ✓	$-2 = 4$ (False) ✗	Second equation is false.

Because $(4, 6)$ does not satisfy *both* equations, it is not the solution for the system of equations. Next let $x = 7$ and $y = 3$ to determine whether $(7, 3)$ is the solution.

$x + y = 10$	$x - y = 4$	Given equations
$7 + 3 \overset{?}{=} 10$	$7 - 3 \overset{?}{=} 4$	Let $x = 7, y = 3$.
$10 = 10$ (True) ✓	$4 = 4$ (True) ✓	Both are true.

Because $(7, 3)$ makes *both* equations true, it is the solution to the system of equations.

Now Try Exercise 31

SOLVING SYSTEMS GRAPHICALLY AND NUMERICALLY In the next example, we find the solution to a system of linear equations graphically and numerically.

EXAMPLE 5 | **Solving a system graphically and numerically**

Solve the system of linear equations

$$x + 2y = 4$$
$$2x - y = 3$$

with a graph and with a table of values.

Solution
Graphically Begin by writing each equation in slope–intercept form.

$x + 2y = 4$	First equation	$2x - y = 3$	Second equation
$2y = -x + 4$	Subtract x.	$-y = -2x + 3$	Subtract $2x$.
$y = -\dfrac{1}{2}x + 2$	Divide by 2.	$y = 2x - 3$	Multiply by -1.

The graphs of $y = -\frac{1}{2}x + 2$ and $y = 2x - 3$ are shown in **FIGURE 4.4**. The graphs intersect at the point $(2, 1)$; thus, $(2, 1)$ is the solution.

Numerically **TABLE 4.3** on the next page shows the equations $y = -\frac{1}{2}x + 2$ and $y = 2x - 3$ evaluated for various values of x. Note that when $x = 2$, both equations have a y-value of **1**. Thus $(2, 1)$ is the solution.

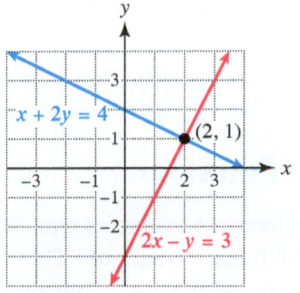

FIGURE 4.4

A Numerical Solution

x	−1	0	1	**2**	3
$y = -\frac{1}{2}x + 2$	2.5	2	1.5	**1**	0.5
$y = 2x - 3$	−5	−3	−1	**1**	3

TABLE 4.3

Find an x-value where the y-values are equal.

Now Try Exercise 51

 Math in ~~Context~~ Business In business, linear equations are sometimes used to model supply and demand for a product. For example, if the price of a coffee drink is too high, the demand for the drink will decrease because consumers are interested in saving money. Similarly, if the price of the coffee drink is too low, supply will decrease because suppliers are interested in making money. To find an appropriate price for the coffee drink, a system of linear equations can be solved.

EXAMPLE 6 | **Solving a supply and demand problem graphically**

The quantity of coffee drinks a company will supply at price P is $Q_s = 3P$ and the quantity of coffee drinks customers will demand at price P is $Q_d = -2P + 10$, where quantity is in hundreds of cups of coffee and price is in dollars. Graph each equation and determine the number of cups of coffee the company should sell and the corresponding price per cup.

Solution

The equation $Q_s = 3P$ is a line passing through the origin with slope 3. The equation $Q_d = -2P + 10$ is a line passing through the point $(0, 10)$ with slope -2. As shown in **FIGURE 4.5**, their intersection point is $(2, 6)$, meaning that the company should sell **6**00 cups of coffee at a price of $**2** per cup.

Cups of Coffee

Intersection at $(2, 6)$

FIGURE 4.5

Now Try Exercise 77

3 Solving Applications of Systems of Linear Equations

A FOUR-STEP PROCESS FOR SOLVING APPLICATIONS In the next example, we use a four-step process to solve an application involving a system of linear equations. These steps are based on the four-step process discussed earlier in this text, with Step 3 split into two parts to emphasize the importance of using the solution to the system of equations to determine the solution to the given problem.

| EXAMPLE 7 | **Reasons for not using the Internet** |

Of American adults who don't use the Internet, about 33% are either "Just not interested" or "Don't have a computer." There are 7% more adults who are "Just not interested" than those who "Don't have a computer." Find the percentage of adults that are "Just not interested" and the percentage that "Don't have a computer." (*Source:* Pew Center for the Internet.)

Solution

STEP 1:   *Identify each variable.*

x: percentage of adults who are "Just not interested"

y: percentage of adults who "Don't have a computer"

STEP 2:   *Write a system of equations.* Of those who don't use the Internet, the total percentage of adults who are either "Just not interested" or "Don't have a computer" is 33%, so $x + y = 33$. Because 7% more adults were "Just not interested" than those who "Don't have a computer," we also know that $x - y = 7$. Thus a system of equations representing this situation is

$$x + y = 33$$
$$x - y = 7.$$

STEP 3A: *Solve the system of equations.* To solve this system graphically, write each equation in slope–intercept form.

$$y = -x + 33$$
$$y = x - 7$$

Their graphs intersect at the point $(20, 13)$, as shown in **FIGURE 4.6**.

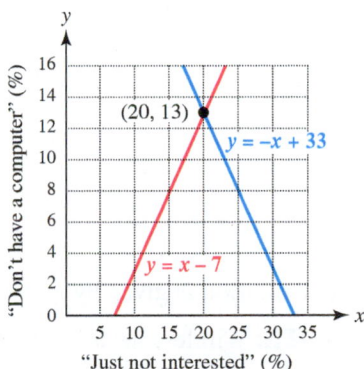

Non-Internet Users

FIGURE 4.6

STEP 3B: *Determine the solution to the problem.* The point $(20, 13)$ corresponds to $x = 20$ and $y = 13$. Thus about 20% of adults who don't use the Internet are "Just not interested" and 13% "Don't have a computer."

STEP 4:   *Check your solution.* Note that $20 + 13 = 33\%$ of adults who don't use the Internet are either "Just not interested" or "Don't have a computer," and $20 - 13 = 7\%$ more stated that they are "Just not interested" compared to those who "Don't have a computer."

Now Try Exercise 75

CALCULATOR HELP

To find a point of intersection, see Appendix A (page AP-6).

TECHNOLOGY NOTE

Checking Solutions

The solution to a system can be checked with a graphing calculator, as shown in the accompanying figure where the graphs of $y_1 = -x + 50$ and $y_2 = x - 10$ intersect at the point $(30, 20)$. However, a graphing calculator cannot read an application problem and write down the system of equations. A human mind is needed for these tasks.

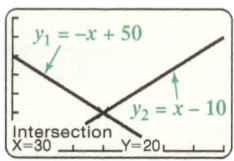

🌐 *Math in Context* In the opening discussion for this chapter, we considered how systems of equations are used to transmit information reliably over wireless networks. To better understand how this is done, consider the following simple example.

Suppose we want to send the number 35 from one cell phone to another. Begin by letting $x = 3$ and $y = 5$. Instead of sending both 3 and 5 over one path, which could fail, we will use four paths. Over the first path we send $x = 3$, over the second $y = 5$, over the third $x + y = 8$, and over the fourth $2x + y = 11$. See the figure below.

Systems of Equations in Wireless Networks

Sending the Number 35

In the figure, any two paths can fail and the receiving phone can still determine the two numbers, 3 and 5. For example, suppose that the second and fourth paths shown fail. Then the receiving phone gets the information only from the first and third paths, $x = 3$ and $x + y = 8$. However, it can still determine the values x and y by solving the following system of linear equations.

$$x = 3 \qquad \text{First path information}$$
$$x + y = 8 \qquad \text{Third path information}$$

The solution is given by $x = 3$ and $y = 5$, so the number is 35.

NOTE: Wireless network systems are often designed so that any pair of equations is consistent and independent, and thus the solution (information) is unique. ∎

EXAMPLE 8 | ### Sending cell phone messages

A person's age is being transmitted over a wireless network in a manner similar to that illustrated in the figure above. Let x be the tens digit and y be the ones digit. Suppose that the first and second paths fail, resulting in the following system that must be solved.

$$x + y = 3 \qquad \text{Third path information}$$
$$2x + y = 5 \qquad \text{Fourth path information}$$

Solve this system graphically and find the person's age.

Decoding Cell Phone Message

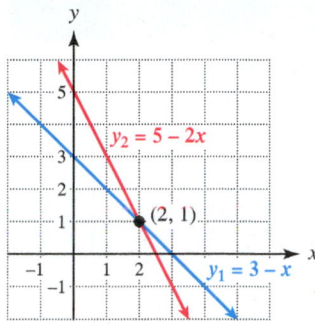

FIGURE 4.7 Person's Age: 21

Solution

Begin by solving each equation for y.

$$y = 3 - x \qquad \text{Subtract } x \text{ in third path equation.}$$

$$y = 5 - 2x \qquad \text{Subtract } 2x \text{ in fourth path equation.}$$

Graph $y_1 = 3 - x$ and $y_2 = 5 - 2x$, as in **FIGURE 4.7**. Their graphs intersect at $(2, 1)$. It follows that $x = 2$, $y = 1$, and the person's age is 21.

Now Try Exercise 81

4.1 Putting It All Together

CONCEPT	EXPLANATION	EXAMPLE
System of Linear Equations in Two Variables	Can be written as $$Ax + By = C$$ $$Dx + Ey = F$$	$$x + y = 8$$ $$x - y = 2$$
Solution to a System of Equations	An ordered pair (x, y) that satisfies *both* equations	The solution to the preceding system is $(5, 3)$ because, when $x = 5$ and $y = 3$ are substituted, both equations are true. $$5 + 3 \overset{?}{=} 8 \ \checkmark \quad \text{True}$$ $$5 - 3 \overset{?}{=} 2 \ \checkmark \quad \text{True}$$
Graphical Solution to a System of Equations	Graph each equation. A point of intersection represents a solution.	The graphs of $y = -x + 8$ and $y = x - 2$ intersect at $(5, 3)$.
Numerical Solution to a System of Equations	Make a table for each equation. A solution occurs when one x-value gives the same y-values in both equations.	Make a table for $y = -x + 8$ and $y = x - 2$. When $x = 5$, $y = 3$ in both equations, so $(\mathbf{5}, \mathbf{3})$ is the solution.

x	4	5	6
$y = -x + 8$	4	3	2
$y = x - 2$	2	3	4

continued on next page

continued from previous page

THREE TYPES OF LINEAR SYSTEMS

No Solution	One Solution	Infinitely Many Solutions
		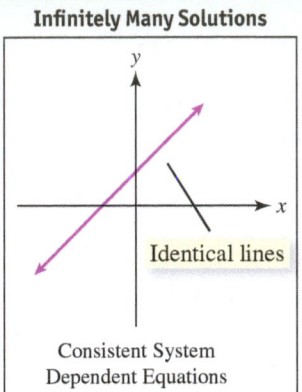
Inconsistent System	Consistent System Independent Equations	Consistent System Dependent Equations
The distinct parallel lines have no points in common.	The solution is given by the coordinates of the point of intersection.	The two lines *coincide*: Every point on this line represents a solution.

4.1   Exercises   MyMathLab®

CONCEPTS AND VOCABULARY

1. A solution to a system of two equations in two variables is a(n) _____ pair.

2. A graphical technique for solving a system of two equations in two variables is the _____ method.

3. A system of linear equations can have _____, _____, or _____ solutions.

4. If a system of linear equations has at least one solution, then it is a(n) (consistent/inconsistent) system.

5. If a system of linear equations has no solutions, then it is a(n) (consistent/inconsistent) system.

6. If a system of linear equations has exactly one solution, then the equations are (dependent/independent).

7. If a system of linear equations has infinitely many solutions, then the equations are (dependent/independent).

8. To find a numerical solution to a system, start by creating a(n) _____ of values for the equations.

9. If a graphical method and a numerical method (table of values) are used to solve the same system of equations, then the two solutions should be (the same/different).

10. One way to graph a line is to write its equation in slope–intercept form. A second method is to find the x- and y- _____.

SOLVING AN EQUATION GRAPHICALLY

Exercises 11–18: Determine graphically the x-value when $y = 2$ in the given equation.

11. $y = 2x$

12. $y = \frac{1}{3}x$

13. $y = 4 - x$

14. $y = -2 - x$

15. $y = -\frac{1}{2}x + 1$

16. $y = 3x - 1$

17. $-3x + 4y = 11$

18. $2x + y = 6$

SOLVING AN EQUATION NUMERICALLY WITH A TABLE OF VALUES

Exercises 19–24: Complete the table. Then use it to solve the given equation.

19. $y = 4$

x	0	1	2	3
$y = x + 3$				

20. $y = -1$

x	−1	0	1	2
$y = 2x - 1$				

21. $y = -3$

x	−4	−2	2	4
$y = 3 - 3x$				

22. $y = -42$

x	-10	-5	5	10
$y = 8 - 5x$				

23. $y = 0$

x	-3	0	3	6
$2x + 3y = 6$				

24. $y = -4$

x	-3	0	3	6
$4x - 3y = 12$				

DETERMINING TYPES OF EQUATIONS AND NUMBER OF SOLUTIONS

Exercises 25–30: The graphs of two equations are shown. State the number of solutions to each system of equations. Then state whether the system is consistent or inconsistent. If it is consistent, state whether the equations are dependent or independent.

25.

26.

27.

28.

29.

30.

CHECKING A SOLUTION TO A SYSTEM OF EQUATIONS

Exercises 31–36: Determine which ordered pair is a solution to the system of equations.

31. $(0, 0), (1, 1)$
$x + y = 2$
$x - y = 0$

32. $(-1, 2), (1, -2)$
$2x + y = 0$
$x - 2y = -5$

33. $(-1, -1), (2, -3)$
$2x + 3y = -5$
$4x - 5y = 23$

34. $(2, -1), (-2, -2)$
$-x + 4y = -6$
$6x - 7y = 19$

35. $\left(\frac{1}{5}, \frac{4}{5}\right), \left(\frac{1}{4}, \frac{1}{3}\right)$
$-5x + 5y = 3$
$4x + 9y = 8$

36. $\left(\frac{1}{2}, \frac{3}{2}\right), \left(\frac{3}{4}, \frac{5}{4}\right)$
$x + y = 2$
$3x - y = 0$

SOLVING SYSTEMS GRAPHICALLY AND NUMERICALLY

Exercises 37–42: The graphs of two equations are shown. Use the intersection-of-graphs method to identify the solution to both equations. Then check your answer.

37.

38.

39.

40.

41.

42.
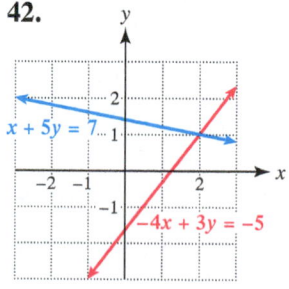

Exercises 43–46: A table for two equations is given. Identify the solution to both equations.

43.

x	1	2	3	4
$y = 2x$	2	4	6	8
$y = 4$	4	4	4	4

44.

x	2	3	4	5
$y = 6 - x$	4	3	2	1
$y = x - 2$	0	1	2	3

45.

x	1	2	3	4
$y = 4 - x$	3	2	1	0
$y = x - 2$	−1	0	1	2

46.

x	1	2	3	4
$y = 6 - 3x$	3	0	−3	−6
$y = 2 - x$	1	0	−1	−2

Exercises 47 and 48: Complete the table for each equation. Then identify the solution to both equations.

47.

x	0	1	2	3
$y = x + 2$				
$y = 4 - x$				

48.

x	−5	−4	−3	−2
$y = 2x + 1$				
$y = x - 1$				

Exercises 49–54: Use the specified method to solve the system of equations.

(a) *Graphically*
(b) *Numerically (table of values)*

49. $y = 2x + 3$
$y = 1$

50. $y = 2 - x$
$y = 0$

51. $y = 4 - x$
$y = x - 2$

52. $y = 2x$
$y = -\frac{1}{2}x$

53. $y = 3x$
$y = x + 2$

54. $y = 2x - 3$
$y = -x + 3$

Exercises 55–66: Solve the system of equations graphically.

55. $x + y = -3$
$x - y = 1$

56. $x - y = 3$
$2x + y = 3$

57. $2x - y = 3$
$3x + y = 2$

58. $x + 2y = 6$
$-x + 3y = 4$

59. $-4x + 2y = 0$
$x - y = -1$

60. $4x - y = 2$
$y = 2x$

61. $2x - y = 4$
$x + 2y = 7$

62. $x - y = 2$
$\frac{1}{2}x + y = 4$

63. $x = -y + 4$
$x = 3y$

64. $x = 2y$
$y = -\frac{1}{2}x$

65. $x + y = 3$
$x = \frac{1}{2}y$

66. $2x - 4y = 8$
$\frac{1}{2}x + y = -4$

WRITING AND SOLVING SYSTEMS

Exercises 67–72: **Number Problems** *For each problem, complete each of the following.*

(a) *Write a system of equations for the problem.*
(b) *Find the unknown numbers by solving the system of equations graphically.*

67. The sum of two numbers is 4 and their difference is 0.

68. The sum of two numbers is −5 and their difference is 1.

69. The sum of twice a number and another number is 7. Their difference is 2.

70. Three times a number subtracted from another number results in 1. Their sum is 5.

71. One number is triple another number. Their difference is 4.

72. Half of a number added to another number equals 5. Their difference is 1.

APPLICATIONS INVOLVING SYSTEMS OF EQUATIONS

73. **Renting a Truck** A rental truck costs $50 plus $0.50 per mile.
   (a) Write an equation that gives the cost C of driving the truck x miles.
   (b) Use the intersection-of-graphs method to determine the number of miles that the truck is driven if the rental cost is $80.
   (c) Solve part (b) numerically with a table of values.

74. **Renting a Car** A rental car costs $25 plus $0.25 per mile.
   (a) Write an equation that gives the cost C of driving the car x miles.
   (b) Use the intersection-of-graphs method to determine the number of miles that the car is driven if the rental cost is $100.
   (c) Solve part (b) numerically with a table of values.

75. **Recorded Music** In 2013, digital downloads and streaming subscriptions accounted for 60% of all music revenues. Digital downloads revenues were double the streaming subscriptions revenues. (*Source:* Recording Industry Association of America.)

(a) Let *x* be the percentage of revenues due to digital downloads and *y* be the percentage of revenues due to streaming subscriptions. Write a system of two equations that describes the given information.

(b) Solve your system graphically and interpret your solution.

76. iPhone Colors About 70% of iPhone 5S users have a phone that is either gold or space gray. About 16% more users have a phone that is space gray than a phone that is gold.

(a) Let *x* be the percentage of users who have a space gray phone and *y* be the percentage of users who have a gold phone. Write a system of two equations that describes the given information.

(b) Solve your system graphically and interpret your solution.

77. Supply and Demand The quantity of e-readers a company supplies per week at price *P* is $Q_s = 1.5P$ and the quantity of e-readers their customers demand at price *P* is $Q_d = 300 - P$, where *P* is in dollars. Determine graphically the number of e-readers the company should sell and the corresponding price per e-reader.

78. Supply and Demand The quantity of donuts a bakery supplies per week at price *P* is $Q_s = P$ and the quantity of donuts their customers demand at price *P* is $Q_d = 9 - 2P$, where price is in dollars and quantity is in hundreds of donuts. Determine graphically the number of donuts the bakery should sell and the corresponding price per donut.

79. Dimensions of a Rectangle A rectangle is 4 inches longer than it is wide. Its perimeter is 28 inches.

(a) Write a system of two equations in two variables that describes this information. Be sure to specify what each variable means.

(b) Solve your system graphically. Interpret your results.

80. Dimensions of a Triangle An isosceles triangle has a perimeter of 17 inches, with its two shorter sides equal in length. The longest side measures 2 inches more than either of the shorter sides.

(a) Write a system of two equations in two variables that describes this information. Be sure to specify what each variable means.

(b) Solve your system graphically. Explain what your results mean.

Exercises 81–84: **Wireless Networks** *(Refer to Example 8.) The system of four equations represents a wireless transmission of a person's age.*

$$x = 2 \quad (1)$$
$$y = 3 \quad (2)$$
$$x + y = 5 \quad (3)$$
$$3x + y = 9 \quad (4)$$

Solve the given pair of equations and find the age of the person. Do you get the same answer in each case?

81. Equations (1) and (3)   **82.** Equations (2) and (3)

83. Equations (2) and (4)   **84.** Equations (3) and (4)

WRITING ABOUT MATHEMATICS

85. Use the intersection-of-graphs method to help explain why you typically expect a linear system in two variables to have one solution.

86. Could a system of two linear equations in two variables have exactly two solutions? Explain your reasoning.

87. Give one disadvantage of using a table to solve a system of equations. Explain your answer.

88. Do the equations $y = 2x + 1$ and $y = 2x - 1$ have a common solution? Explain your answer.

4.2   Solving Systems of Linear Equations by Substitution

Objectives

1 **Using the Substitution Method**
   • Solving for a Variable Before Using Substitution

2 **Recognizing Systems with No Solutions or Infinitely Many Solutions**

3 **Solving Applications Using the Substitution Method**

NEW VOCABULARY

☐ Method of substitution

READING CHECK 1

• What is an advantage of using the method of substitution rather than using a graph or table to solve a system of linear equations?

STUDY TIP

Questions on exams do not always come in the order that they are presented in the text. When studying for an exam, choose review exercises randomly so that the topics are studied in the same random way they may appear on an exam.

1 Using the Substitution Method

In the previous section, we solved systems of linear equations by using graphs and tables. A disadvantage of a graph is that reading the graph precisely can be difficult. A disadvantage of using a table is that locating the solution can be difficult when it is either a fraction or a large number. In this subsection, we introduce the *method of substitution*, in which we solve systems of equations symbolically. The advantage of this method is that the *exact* solution can always be found (provided it exists).

Connecting Concepts with Your Life   Suppose that you and a friend earned $120 together. If x represents how much your friend earned and y represents how much you earned, then the equation $x + y = 120$ describes this situation. Now, if we also know that you earned twice as much as your friend, then we can include a second equation, $y = 2x$. The amount that each of you earned can now be determined by *substituting* $2x$ for y in the first equation.

$$x + y = 120 \qquad \text{First equation}$$
$$x + 2x = 120 \qquad \text{Substitute } 2x \text{ for } y.$$
$$3x = 120 \qquad \text{Combine like terms.}$$
$$x = 40 \qquad \text{Divide each side by 3.}$$

So your friend earned $40, and you earned twice as much, or $80.

This technique of substituting an expression for a variable and solving the resulting equation is called the **method of substitution**. The following four-step process can be used when using the method of substitution.

STEPS FOR USING THE METHOD OF SUBSTITUTION

STEP 1: Solve one of the equations for one variable.

STEP 2: Substitute the expression found in Step 1 in the other equation and solve the resulting equation to find the value of one variable.

STEP 3: Use the solution from Step 2 to find the value of the other variable. Write the solution to the system of equations as an ordered pair.

STEP 4: Check the solution.

EXAMPLE 1   Using the method of substitution

Solve each system of equations.

(a) $2x + y = 10$
     $y = 3x$

(b) $-2x + 3y = -8$
     $x = 3y + 1$

Solution

(a) **STEP 1:** The second equation is already solved for y.

STEP 2: From the second equation, substitute $3x$ for y in the first equation.

$$2x + y = 10 \qquad \text{First equation}$$
$$2x + 3x = 10 \qquad \text{Substitute } 3x \text{ for } y.$$
$$5x = 10 \qquad \text{Combine like terms.}$$
$$x = 2 \qquad \text{Divide each side by 5.}$$

STEP 3: The solution to this system is an *ordered pair*, so we must also find y. Because $y = 3x$ and $x = 2$, it follows that $y = 3(2) = 6$. The solution is $(2, 6)$.

STEP 4: To check the solution, substitute 2 for x and 6 for y in each equation.

$$2x + y = 10 \qquad y = 3x \qquad \text{Given equations}$$
$$2(2) + 6 \stackrel{?}{=} 10 \qquad 6 \stackrel{?}{=} 3(2) \qquad \text{Let } x = 2 \text{ and } y = 6.$$
$$10 = 10 \checkmark \qquad 6 = 6 \checkmark \qquad \text{Both are true.}$$

(b) STEP 1: The second equation, $x = 3y + 1$, is solved for x.

STEP 2: Substitute $(3y + 1)$ for x in the first equation. Include parentheses around the expression $3y + 1$ since this entire expression is to be multiplied by -2.

$$-2x + 3y = -8 \qquad \text{First equation}$$
$$-2(3y + 1) + 3y = -8 \qquad \text{Substitute } (3y + 1) \text{ for } x.$$
$$-6y - 2 + 3y = -8 \qquad \text{Distributive property}$$

> Be sure to include parentheses.

$$-3y - 2 = -8 \qquad \text{Combine like terms.}$$
$$-3y = -6 \qquad \text{Add 2 to each side.}$$
$$y = 2 \qquad \text{Divide each side by } -3.$$

STEP 3: To find x, substitute 2 for y in $x = 3y + 1$ to obtain $x = 3(2) + 1 = 7$. The solution is $(7, 2)$.

STEP 4: Substitute 7 for x and 2 for y in each equation.

$$-2x + 3y = -8 \qquad x = 3y + 1 \qquad \text{Given equations}$$
$$-2(7) + 3(2) \stackrel{?}{=} -8 \qquad 7 \stackrel{?}{=} 3(2) + 1 \qquad \text{Let } x = 7 \text{ and } y = 2.$$
$$-8 = -8 \checkmark \qquad 7 = 7 \checkmark \qquad \text{Both are true.}$$

Now Try Exercises 5, 11

READING CHECK 2

- When substituting an expression with two or more terms for a single variable, why is it important to use parentheses?

NOTE: When an expression contains two or more terms, it is usually best to place parentheses around it when substituting it for a single variable in an equation. In Example 1(b), the distributive property would not have been applied correctly without the parentheses. ∎

SOLVING FOR A VARIABLE BEFORE USING SUBSTITUTION Sometimes it is necessary to solve one of the equations for a variable before substitution can be used, as demonstrated in the next example.

EXAMPLE 2 **Using the method of substitution**

Solve each system of equations.

(a) $\quad x + y = 8$ **(b)** $\quad 3a - 2b = 2$
$\qquad 2x - 3y = 6$ $\qquad\qquad a + 4b = 3$

Solution

(a) STEP 1: Neither equation is solved for a variable, but we can easily solve the first equation for y.

$$x + y = 8 \qquad \text{First equation}$$
$$y = 8 - x \qquad \text{Subtract } x \text{ from each side.}$$

STEP 2: Now we can substitute $(8 - x)$ for y in the second equation.

Include parentheses.	$2x - 3y = 6$	Second equation
	$2x - 3(\mathbf{8 - x}) = 6$	Substitute $(8 - x)$ for y.
	$2x - 24 + 3x = 6$	Distributive property
	$5x = 30$	Combine like terms; add 24.
	$x = 6$	Divide each side by 5.

STEP 3: Because $y = 8 - x$ and $x = \mathbf{6}$, we know that $y = 8 - \mathbf{6} = \mathbf{2}$. The solution is $(\mathbf{6, 2})$.

STEP 4: To check the solution, substitute **6** for x and **2** for y in each equation.

$x + y = 8$	$2x - 3y = 6$	Given equations
$\mathbf{6} + \mathbf{2} \stackrel{?}{=} 8$	$2(\mathbf{6}) - 3(\mathbf{2}) \stackrel{?}{=} 6$	Let $x = 6$ and $y = 2$.
$8 = 8 \checkmark$	$6 = 6 \checkmark$	Both are true.

(b) STEP 1: Although we could solve either equation for either variable, solving the second equation for a is easiest because the coefficient of a is 1 and we can avoid fractions.

$a + 4b = 3$	Second equation
$a = 3 - 4b$	Subtract $4b$ from each side.

STEP 2: Now substitute $(3 - 4b)$ for a in the first equation.

$3a - 2b = 2$	First equation
$3(\mathbf{3 - 4b}) - 2b = 2$	Substitute $(3 - 4b)$ for a.
$9 - 12b - 2b = 2$	Distributive property
$-14b = -7$	Combine like terms; subtract 9.
$b = \dfrac{1}{2}$	Divide each side by -14.

STEP 3: Substitute $b = \frac{1}{2}$ in $a = 3 - 4b$ to obtain $a = \mathbf{1}$. The solution is $\left(\mathbf{1, \frac{1}{2}}\right)$.

STEP 4: Substitute **1** for a and $\frac{1}{2}$ for b in each equation.

$3a - 2b = 2$	$a + 4b = 3$	Given equations
$3(\mathbf{1}) - 2\left(\dfrac{\mathbf{1}}{\mathbf{2}}\right) \stackrel{?}{=} 2$	$\mathbf{1} + 4\left(\dfrac{\mathbf{1}}{\mathbf{2}}\right) \stackrel{?}{=} 3$	Let $a = 1$ and $b = \frac{1}{2}$.
$2 = 2 \checkmark$	$3 = 3 \checkmark$	Both are true.

Now Try Exercises 17, 23

NOTE: When a system of equations contains variables other than x and y, we will list them alphabetically in an ordered pair. ∎

2 Recognizing Systems with No Solutions or Infinitely Many Solutions

A system of linear equations typically has exactly one solution. However, in the next example we see how the method of substitution can be used on systems that have no solutions or infinitely many solutions.

EXAMPLE 3   **Solving other types of systems**

If possible, use substitution to solve the system of equations. Then use graphing to help explain the result.

(a) $\begin{array}{r} 3x + y = 4 \\ 6x + 2y = 2 \end{array}$ **(b)** $\begin{array}{r} x + y = 2 \\ 2x + 2y = 4 \end{array}$

Solution

(a) Solve the first equation for y to obtain $y = 4 - 3x$. Next substitute $(4 - 3x)$ for y in the second equation.

$$6x + 2y = 2 \qquad \text{Second equation}$$
$$6x + 2(4 - 3x) = 2 \qquad \text{Substitute } (4 - 3x) \text{ for } y.$$
$$6x + 8 - 6x = 2 \qquad \text{Distributive property}$$
$$8 = 2 \text{ (False)} \qquad \text{Combine like terms.}$$

The equation $8 = 2$ is *always false*, which indicates that there are *no solutions*. One way to graph each equation is to write the equations in slope–intercept form.

$$3x + y = 4 \qquad \text{First equation} \qquad \bigg| \qquad 6x + 2y = 2 \qquad \text{Second equation}$$
$$y = -3x + 4 \qquad \text{Subtract } 3x. \qquad \bigg| \qquad y = -3x + 1 \qquad \text{Subtract } 6x; \text{ divide by } 2.$$

The graphs of these equations are parallel lines with slope -3, as shown in **FIGURE 4.8(a)**. Because the lines *do not intersect*, there are *no solutions* to the system of equations.

(b) Solve the first equation for y to obtain $y = 2 - x$. Now substitute $(2 - x)$ for y in the second equation.

$$2x + 2y = 4 \qquad \text{Second equation}$$
$$2x + 2(2 - x) = 4 \qquad \text{Substitute } (2 - x) \text{ for } y.$$
$$2x + 4 - 2x = 4 \qquad \text{Distributive property}$$
$$4 = 4 \text{ (True)} \qquad \text{Combine like terms.}$$

The equation $4 = 4$ is *always true*, which means that there are *infinitely many solutions*. One way to graph these equations is to write them in slope–intercept form.

$$x + y = 2 \qquad \text{First equation} \qquad \bigg| \qquad 2x + 2y = 4 \qquad \text{Second equation}$$
$$y = -x + 2 \qquad \text{Subtract } x. \qquad \bigg| \qquad y = -x + 2 \qquad \text{Subtract } 2x; \text{ divide by } 2.$$

Because the equations have the same slope–intercept form, their graphs are identical, resulting in a single line, as shown in **FIGURE 4.8(b)**. *Every point* on this line *represents a solution* to the system of equations, so there are infinitely many solutions.

Now Try Exercises 33, 35

(a) No Solutions

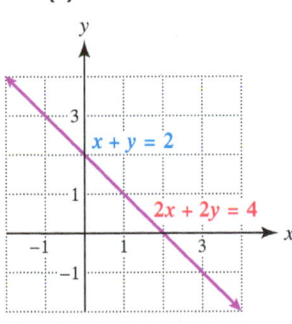

(b) Infinitely Many Solutions

FIGURE 4.8

3 ▸ Solving Applications Using the Substitution Method

🌐 *Math in Context* (Messaging) Suppose that a boy and a girl sent a total of 760 text messages in one month. With only this information, it is impossible to know how many messages were sent by each person. The boy may have sent 432 messages while the girl sent 328. However, 11 and 749 also total 760, as do many other possibilities. If we are told that the boy sent three times as many messages as the girl, a system of linear equations can be written and solved to find the answer.

EXAMPLE 4   **Finding numbers of text messages**

A boy and a girl sent a total of 760 text messages in one month. If the boy sent three times as many messages as the girl, how many text messages did each person send?

Solution

If x represents the number of text messages sent by the girl and y represents the number of text messages sent by the boy, then the equation $x + y = 760$ can be written because the total number of messages is 760. Also, the equation $y = 3x$ can be included because the boy sent three times as many messages as the girl. The system is

$$x + y = 760$$
$$y = 3x.$$

Substitute $3x$ for y in the first equation.

$$x + y = 760 \quad \text{First equation}$$
$$x + 3x = 760 \quad \text{Substitute } 3x \text{ for } y.$$
$$4x = 760 \quad \text{Combine like terms.}$$
$$x = 190 \quad \text{Divide each side by 4.}$$

The girl sent 190 text messages and the boy sent 3 times as many, or $3(190) = 570$. This answer checks because $190 + 570 = 760$ and 570 is three times 190.

Now Try Exercise 71

The next two examples illustrate how the method of substitution can be used to solve applications. In these examples, we apply the same four-step process that was used to solve application problems in the previous section.

EXAMPLE 5   **Determining time spent online**

In December 2013, the total time Americans spent using tablets or smartphones to access the Internet was 566 billion minutes. The amount of online time for smartphones was about 3.565 times the amount of online time for tablets. Find the amount of time spent accessing the Internet on each type of device. (*Source: ComScore.*)

Solution

STEP 1:   *Identify each variable.* Clearly identify what each variable represents.

x: tablet time online, in billions of minutes

y: smartphone time online, in billions of minutes

STEP 2:   *Write a system of equations.*

$$x + y = 566 \quad \text{Total time is 566 billion minutes.}$$
$$y = 3.565x \quad \text{Time on smartphones is 3.565 times the amount of time on tablets.}$$

STEP 3A: *Solve the system of linear equations.* Substitute $3.565x$ for y in the first equation.

$$x + y = 566 \quad \text{First equation}$$
$$x + 3.565x = 566 \quad \text{Substitute } 3.565x \text{ for } y.$$
$$4.565x = 566 \quad \text{Combine like terms.}$$
$$x = \frac{566}{4.565} \quad \text{Divide each side by 4.565.}$$
$$x \approx 124 \quad \text{Simplify.}$$

Because $y = 3.565x$, it follows that $y = 3.565(124) \approx 442$.

STEP 3B: *Determine the solution to the problem.* The solution $x = 124$ and $y = 422$ indicates that online time for tablets was 124 billion minutes and online time for smartphones was 442 billion minutes in December 2013.

STEP 4: *Check the solution.* The sum of these times is $124 + 442 = 566$ billion minutes, and the amount of time spent on smartphones is $\frac{442}{124} \approx 3.565$ times the amount of time spent on tablets. The answer checks.

Now Try Exercise 57

EXAMPLE 6   **Determining airplane speed and wind speed**

An airplane flies 2400 miles into (or against) the wind in 8 hours. The return trip takes 6 hours. Find the speed of the airplane with no wind and the speed of the wind.

Solution

STEP 1: *Identify each variable.*

x: the speed of the airplane without wind

y: the speed of the wind

STEP 2: *Write a system of equations.* If we solve the distance formula $d = rt$ for the variable r, the result is $r = \frac{d}{t}$. Using this formula, the speed (rate) of the airplane against the wind is $\frac{2400}{8} = 300$ miles per hour, because it traveled 2400 miles in 8 hours. The wind slowed the plane, so $x - y = 300$. Similarly, the airplane flew $\frac{2400}{6} = 400$ miles per hour with the wind because it traveled 2400 miles in 6 hours. The wind made the plane fly faster, so $x + y = 400$.

$$x - y = 300 \quad \text{Speed against the wind}$$

$$x + y = 400 \quad \text{Speed with the wind}$$

STEP 3A: *Solve the system of linear equations.* Solve the first equation for x to obtain $x = y + 300$. Substitute $(y + 300)$ for x in the second equation.

$$x + y = 400 \quad \text{Second equation}$$

$$(\boldsymbol{y + 300}) + y = 400 \quad \text{Substitute } (y + 300) \text{ for } x.$$

$$2y = 100 \quad \text{Combine like terms; subtract 300.}$$

$$y = 50 \quad \text{Divide each side by 2.}$$

Because $x = y + 300$, it follows that $x = 50 + 300 = 350$.

STEP 3B: *Determine the solution to the problem.* The solution $x = 350$ and $y = 50$ indicates that the airplane can fly 350 miles per hour with no wind, and the wind speed is 50 miles per hour.

STEP 4: *Check the solution.* The plane flies $350 - 50 = 300$ miles per hour into the wind, taking $\frac{2400}{300} = 8$ hours. The plane flies $350 + 50 = 400$ miles per hour with the wind, taking $\frac{2400}{400} = 6$ hours. The answers check.

Now Try Exercise 63

CRITICAL THINKING 1

A boat travels 10 miles per hour upstream and 16 miles per hour downstream. How fast is the current?

4.2   Putting It All Together

CONCEPT	EXPLANATION	EXAMPLE
Method of Substitution	Can be used to solve a system of equations Gives the exact solution, provided one exists **STEP 1:** Solve one equation for one variable. **STEP 2:** Substitute the result in the other equation and solve. **STEP 3:** Use the solution for the first variable to find the other variable. **STEP 4:** Check the solution.	$x + y = 5$ Sum is 5. $x - y = 1$ Difference is 1. **STEP 1:** Solve for x in the second equation. $$x = y + 1$$ **STEP 2:** Substitute $(y + 1)$ in the first equation for x. $$(y + 1) + y = 5$$ $$2y = 4$$ $$y = 2$$ **STEP 3:** $x = 2 + 1 = 3$ **STEP 4:** $3 + 2 = 5$ ✓ $3 - 2 = 1$ ✓ $(3, 2)$ checks.

4.2   Exercises

MyMathLab®

CONCEPTS AND VOCABULARY

1. One advantage of solving a linear system using the method of substitution rather than graphical or numerical methods is that the _____ solution can always be found (provided it exists).

2. When substituting an expression that contains two or more terms for a single variable in an equation, it is usually best to place _____ around it.

3. Suppose that the method of substitution results in the equation $1 = 1$. What does this indicate about the number of solutions to the system of equations?

4. Suppose that the method of substitution results in the equation $0 = 1$. What does this indicate about the number of solutions to the system of equations?

USING THE METHOD OF SUBSTITUTION

Exercises 5–32: Use the method of substitution to solve the system of linear equations.

5. $x + y = 9$
   $y = 2x$

6. $x + y = -12$
   $y = -3x$

7. $x + 2y = 4$
   $x = 2y$

8. $-x + 3y = -12$
   $x = 5y$

9. $2x + y = -2$
   $y = x + 1$

10. $-3x + y = -10$
    $y = x - 2$

11. $x + 3y = 3$
    $x = y + 3$

12. $x - 2y = -5$
    $x = 4 - y$

13. $3x + 2y = \frac{3}{2}$
    $y = 2x - 1$

14. $-3x + 5y = 4$
    $y = 2 - 3x$

15. $2x - 3y = -12$
    $x = 2 - \frac{1}{2}y$

16. $\frac{3}{4}x + \frac{1}{4}y = -\frac{7}{4}$
    $x = 1 - 2y$

17. $2x - 3y = -4$
    $3x - y = 1$

18. $\frac{1}{2}x - y = -1$
    $2x - \frac{1}{2}y = \frac{13}{2}$

19. $x - 5y = 26$
    $2x + 6y = -12$

20. $4x - 3y = -4$
    $x + 7y = -63$

21. $\frac{1}{2}y - z = 5$
    $y - 3z = 13$

22. $3y - 7z = -2$
    $5y - z = 2$

23. $10r - 20t = 20$
$r + 60t = -29$

24. $-r + 10t = 22$
$-10r + 5t = 30$

25. $3x + 2y = 9$
$2x - 3y = -7$

26. $5x - 2y = -5$
$2x - 5y = 19$

27. $2a - 3b = 6$
$-5a + 4b = -8$

28. $-5a + 7b = -1$
$3a + 2b = 13$

29. $-\frac{1}{2}x + 3y = 5$
$2x - \frac{1}{2}y = 3$

30. $3x - \frac{1}{2}y = 2$
$-\frac{1}{2}x + 5y = \frac{19}{2}$

31. $3a + 5b = 16$
$-8a + 2b = 34$

32. $5a - 10b = 20$
$10a + 5b = 15$

RECOGNIZING OTHER TYPES OF SYSTEMS

Exercises 33–50: Use the method of substitution to solve the system of linear equations. These systems may have no solutions, one solution, or infinitely many solutions.

33. $x + y = 9$
$x + y = 7$

34. $x - y = 8$
$x - y = 4$

35. $x - y = 4$
$2x - 2y = 8$

36. $2x + y = 5$
$4x + 2y = 10$

37. $x - y = 3$
$2x - y = 7$

38. $x + y = 4$
$x - y = 2$

39. $x - y = 7$
$-x + y = -7$

40. $2u - v = 6$
$-4u + 2v = -12$

41. $u - 2v = 5$
$2u - 4v = -2$

42. $3r + 3t = 9$
$2r + 2t = 4$

43. $5x - y = -1$
$2x - 7y = 7$

44. $2r + 3t = 1$
$r - 3t = -5$

45. $y = 5x$
$y = -3x$

46. $a = b + 1$
$a = b - 1$

47. $5a = 4 - b$
$5a = 3 - b$

48. $3y = x$
$3y = 2x$

49. $2x + 4y = 0$
$3x + 6y = 5$

50. $-5x + 10y = 3$
$\frac{1}{2}x - y = 1$

51. Thinking Generally If the method of substitution results in an equation that is always true, what can be said about the graphs of the two equations?

52. Thinking Generally If the method of substitution results in an equation that is always false, what can be said about the graphs of the two equations?

SOLVING APPLICATIONS USING THE METHOD OF SUBSTITUTION

53. Rectangle A rectangular garden is 10 feet longer than it is wide. Its perimeter is 72 feet.
   (a) Let W be the width of the garden and L be the length. Write a system of linear equations whose solution gives the width and length of the garden.
   (b) Use the method of substitution to solve the system. Check your answer.

54. Angles The measures of the two smaller angles in a triangle are equal and their sum equals the largest angle.
   (a) Let x be the measure of each of the two smaller angles and y be the measure of the largest angle. Write a system of linear equations whose solution gives the measures of these angles.
   (b) Use the method of substitution to solve the system. Check your answer.

55. Complementary Angles The smaller of two angles that sum to $90°$ is half the measure of the larger angle.
   (a) Let x be the measure of the smaller angle and y be the measure of the larger angle. Write a system of linear equations whose solution gives the measures of these angles.
   (b) Use the method of substitution to solve your system.
   (c) Use graphing to solve the system.

56. Supplementary Angles The smaller of two angles that sum to $180°$ is one-fourth the measure of the larger angle.
   (a) Let x be the measure of the smaller angle and y be the measure of the larger angle. Write a system of linear equations whose solution gives the measures of these angles.
   (b) Use the method of substitution to solve the system.

57. Average Room Prices In 2013, the average room price for a hotel in Cancun, Mexico, was $30 more than it was in 2012. The 2012 room price was 86.4% of the 2013 room price. (*Source:* Hotel Price Index.)
   (a) Let x be the average room price in 2013 and y be this price in 2012. Write a system of equations whose solution gives the average room prices for each year.
   (b) Use the method of substitution to solve the system to the nearest cent.

58. Ticket Prices Two hundred tickets were sold for a baseball game, which amounted to $840. Student tickets cost $3, and adult tickets cost $5.
   (a) Let x be the number of student tickets and y be the number of adult tickets. Write a system of linear equations whose solution gives the number of each type of ticket.
   (b) Use the method of substitution to solve the system of equations.

59. NBA Basketball Court An official NBA basketball court is 44 feet longer than it is wide. If its perimeter is 288 feet, find its dimensions.

60. Football Field A U.S. football field is 139.5 feet longer than it is wide. If its perimeter is 921 feet, find its dimensions.

61. Number Problem The sum of two numbers is 70. The larger number is two more than three times the smaller number. Find the two numbers.

62. Number Problem The difference of two numbers is 12. The larger number is one less than twice the smaller number. Find the two numbers.

63. Speed Problem A tugboat goes 120 miles upstream in 15 hours. The return trip downstream takes 10 hours. Find the speed of the tugboat without a current and the speed of the current.

64. Speed Problem An airplane flies 1200 miles into the wind in 3 hours. The return trip takes 2 hours. Find the speed of the airplane without a wind and the speed of the wind.

65. Mixture Problem A chemist has 20% and 50% solutions of acid available. How many liters of each solution should be mixed to obtain 10 liters of a 40% acid solution?

66. Mixture Problem A mechanic needs a radiator to have a 40% antifreeze solution. The radiator currently is filled with 4 gallons of a 25% antifreeze solution. How much of the antifreeze mixture should be drained from the car if the mechanic replaces it with pure antifreeze?

67. Mixture Problem A 15-pound mixture of coffee that contains both decaffeinated and regular coffee costs $53. The decaffeinated coffee costs $3 per pound and the regular coffee costs $4 per pound.
   (a) Let x be the weight of the decaffeinated coffee and y be the weight of the regular coffee. Write a system of equations whose solution gives the amount of each type of coffee in the mixture.
   (b) Use the method of substitution to solve the system of equations.

68. Real Estate Sites In 2014, the top two real estate websites were Zillow and Trulia. Combined, they made up 25.4% of the U.S. market share of visits. Zillow had 7.6% more market share than Trulia. (*Source:* Experian Marketing Services.)
   (a) Let x be the Zillow market share of visits and y be the Trulia market share of visits. Write a system of equations whose solution gives the market share of visits for each site.
   (b) Use the method of substitution to solve the system of equations.

69. Exercise and Calories In a 48-minute workout, a student burns a total of 450 calories on a tread climber and a stationary bicycle. The student burns 12 calories per minute on the tread climber and 5 calories per minute on the stationary bicycle.
   (a) Let x be the time spent on the tread climber and y be the time spent on the stationary bicycle. Write a system of equations whose solution gives the number of minutes spent on each machine.
   (b) Use the method of substitution to solve the system of equations.
   (c) How many calories were burned on each machine?

70. Exercise and Calories In 1.2 hours, a student travels 10.2 miles by running at 8 miles per hour and at 10 miles per hour.
   (a) Let x be the time spent running at 8 miles per hour and y be the time spent running at 10 miles per hour. Write a system of equations whose solution gives the time spent running at each speed.
   (b) Use the method of substitution to solve the system of equations.

71. Great Lakes Together, Lake Superior and Lake Michigan cover 54 thousand square miles. Lake Superior is approximately 10 thousand square miles larger than Lake Michigan. Find the size of each lake. (*Source:* National Oceanic and Atmospheric Administration.)

72. Longest Rivers The two longest rivers in the world are the Nile and the Amazon. Together, they are 8145 miles, with the Amazon being 145 miles shorter than the Nile. Find the length of each river. (*Source:* National Oceanic and Atmospheric Administration.)

WRITING ABOUT MATHEMATICS

73. State one advantage that the method of substitution has over the intersection-of-graphs method. Explain your answer.

74. When applying the method of substitution, how do you know that there are no solutions?

75. When applying the method of substitution, how do you know that there are infinitely many solutions?

76. When applying the intersection-of-graphs method, how do you know that there are no solutions?

SECTIONS 4.1 and 4.2

Checking Basic Concepts

1. Determine graphically the x-value in each equation when $y = 2$.
  (a) $y = 1 - \frac{1}{2}x$ **(b)** $2x - 3y = 6$

2. Determine whether $(-1, 0)$ or $(4, 2)$ is a solution to the system

$$2x - 5y = -2$$
$$3x + 2y = 16.$$

3. Solve the system of equations graphically. Check your answer.

$$x - y = 1$$
$$2x + y = 5$$

4. If possible, use the method of substitution to solve each system of equations.

(a) $x + y = -1$
   $y = 2 - x$

(b) $4x - y = 5$
   $-x + y = -2$

(c) $x + 2y = 3$
   $-x - 2y = -3$

5. Room Prices A hotel rents single and double rooms for $150 and $200, respectively. The hotel receives $55,000 for renting 300 rooms.
  (a) Let x be the number of single rooms rented and let y be the number of double rooms rented. Write a system of linear equations whose solution gives the values of x and y.
  (b) Use the method of substitution to solve the system. Check your answer.

4.3   Solving Systems of Linear Equations by Elimination

NEW VOCABULARY

☐ Elimination Method

STUDY TIP

Much of mathematics builds on previous knowledge. We learned about the addition property of equality in Chapter 2. In this section, we see how we can use this property to solve a system of linear equations.

READING CHECK 1

• What is eliminated from both of the equations when the elimination method is used to solve a system of linear equations?

1   Using the Elimination Method

The **elimination method** is based on the addition property of equality. If

$$a = b \quad \text{and} \quad c = d,$$

then

$$a + c = b + d.$$

For example, if the sum of two numbers is 20 and their difference is 4, then the system of equations

$$x + y = 20$$
$$x - y = 4$$

describes these two numbers. By the addition property of equality, the sum of the left sides of these equations equals the sum of their right sides.

$$(x + y) + (x - y) = 20 + 4$$

Add the left sides.   Add the right sides.

$$2x = 24 \quad \text{Combine terms.}$$

Note that the y-variable is eliminated when the left sides are added. The resulting equation, $2x = 24$, simplifies to $x = 12$. Thus the value of x in the solution to the system of equations is 12. The value of y can be found by substituting 12 for x in either of the given equations. Substituting 12 for x in the first equation, $x + y = 20$, results in $12 + y = 20$ or $y = 8$. The solution to the system of equations is $x = 12, y = 8$, which can be written as the ordered pair $(12, 8)$.

To organize the elimination method better, we can carry out the addition vertically.

$$
\begin{array}{rll}
x + y = 20 & & \\
x - y = 4 & & \text{Add the right sides.}\\
\hline
2x + 0y = 24 & & \text{Add the equations.}\\
2x = 24 & & \text{Simplify.}\\
x = 12 & & \text{Divide each side by 2.}
\end{array}
$$

Add the left sides.

Once the value of one variable is known, in this case $x = 12$, don't forget to find the value of the other variable by substituting this known value in either of the *given* equations. By substituting 12 for x in the second equation, we obtain $12 - y = 4$ or $y = 8$. The solution is the ordered pair $(12, 8)$.

EXAMPLE 1   Applying the elimination method

Solve each system of equations. Check each solution.

(a) $2x + y = 1$
     $3x - y = 9$

(b) $-2a + b = -3$
     $2a + 3b = 7$

Solution

(a) Adding these two equations eliminates the y-variable.

$$
\begin{array}{rll}
2x + y = 1 & & \text{First equation}\\
3x - y = 9 & & \text{Second equation}\\
\hline
5x = 10 \quad \text{or} \quad x = 2 & & \text{Add and solve for } x.
\end{array}
$$

To find y, substitute **2** for x in either of the *given* equations. We will use $2x + y = 1$.

$$2(\mathbf{2}) + y = \quad 1 \qquad \text{Let } x = 2 \text{ in first equation.}$$
$$y = \ -\mathbf{3} \qquad \text{Subtract 4 from each side.}$$

The solution is the *ordered pair* $(\mathbf{2}, -\mathbf{3})$, which can be checked by substituting 2 for x and -3 for y in each of the given equations.

$2x + y = 1$	$3x - y = 9$	Given equations
$2(\mathbf{2}) + (-\mathbf{3}) \overset{?}{=} 1$	$3(\mathbf{2}) - (-\mathbf{3}) \overset{?}{=} 9$	Let $x = 2$ and $y = -3$.
$4 - 3 \overset{?}{=} 1$	$6 + 3 \overset{?}{=} 9$	Simplify.
$1 = 1$ ✓	$9 = 9$ ✓	The solution checks.

(b) Adding these two equations eliminates the a-variable.

$$-2a + \ b = -3 \qquad\qquad\qquad \text{First equation}$$
$$\underline{2a + 3b = \quad 7} \qquad\qquad\qquad \text{Second equation}$$
$$4b = \quad 4 \quad \text{or} \quad b = 1 \qquad \text{Add and solve for } b.$$

To find a, substitute **1** for b in either of the *given* equations. We will use $2a + 3b = 7$.

$$2a + 3(\mathbf{1}) = 7 \qquad \text{Let } b = 1 \text{ in second equation.}$$
$$2a = 4 \qquad \text{Subtract 3 from each side.}$$
$$a = \mathbf{2} \qquad \text{Divide each side by 2.}$$

The solution is the *ordered pair* $(\mathbf{2}, \mathbf{1})$, which can be checked by substituting 2 for a and 1 for b in each of the given equations.

$-2a + b = -3$	$2a + 3b = 7$	Given equations
$-2(\mathbf{2}) + \mathbf{1} \overset{?}{=} -3$	$2(\mathbf{2}) + 3(\mathbf{1}) \overset{?}{=} 7$	Let $a = 2$ and $b = 1$.
$-4 + 1 \overset{?}{=} -3$	$4 + 3 \overset{?}{=} 7$	Simplify.
$-3 = -3$ ✓	$7 = 7$ ✓	The solution checks.

Now Try Exercises 15, 17

MULTIPLYING BEFORE APPLYING ELIMINATION Adding two equations does not always eliminate a variable. For example, adding the following equations eliminates neither variable.

$$3x - 2y = 11 \qquad \text{First equation}$$
$$\underline{4x + \ y = 11} \qquad \text{Second equation}$$
$$7x - \ y = 22 \qquad \text{Add the equations.}$$

READING CHECK 2

• In using the elimination method, when is it necessary to apply the multiplication property of equality?

However, by the multiplication property of equality, we can multiply the second equation by 2. Then adding the equations eliminates the y-variable.

$$3x - 2y = 11 \qquad\qquad \text{First equation}$$
$$\underline{8x + 2y = 22} \qquad\qquad \text{Multiply the second equation by 2.}$$
$$11x \qquad = 33 \quad \text{or} \quad x = 3 \qquad \text{Add and solve for } x.$$

Opposites

EXAMPLE 2 **Multiplying before applying elimination**

Solve each system of equations.

(a) $5x - \ y = -11$ **(b)** $3x + 2y = 1$
      $2x + 3y = \ -1$ $2x - 3y = 5$

Solution

(a) We multiply $5x - y = -11$ by 3 and then add to eliminate the y-variable.

$$
\begin{aligned}
15x - 3y &= -33 & &\text{Multiply the first equation by 3.} \\
2x + 3y &= -1 & &\text{Second equation} \\
\hline
17x \quad\; &= -34 \quad \text{or} \quad x = -2 & &\text{Add and solve for } x.
\end{aligned}
$$

Opposites

We can find y by substituting -2 for x in the second equation, $2x + 3y = -1$.

$$
\begin{aligned}
2(-2) + 3y &= -1 & &\text{Let } x = -2 \text{ in second equation.} \\
3y &= 3 & &\text{Add 4 to each side.} \\
y &= 1 & &\text{Divide each side by 3.}
\end{aligned}
$$

The solution is $(-2, 1)$.

(b) We must apply the multiplication property to both equations. If we multiply the first equation, $3x + 2y = 1$, by 3, and the second equation, $2x - 3y = 5$, by 2, then the coefficients of the y-variables will be opposites. Adding eliminates the y-variable.

$$
\begin{aligned}
9x + 6y &= 3 & &\text{Multiply the first equation by 3.} \\
4x - 6y &= 10 & &\text{Multiply the second equation by 2.} \\
\hline
13x \quad\; &= 13 \quad \text{or} \quad x = 1 & &\text{Add and solve for } x.
\end{aligned}
$$

Opposites

To find y, substitute **1** for x in the first *given* equation, $3x + 2y = 1$.

$$
\begin{aligned}
3(1) + 2y &= 1 & &\text{Let } x = 1 \text{ in first equation.} \\
2y &= -2 & &\text{Subtract 3 from each side.} \\
y &= -1 & &\text{Divide each side by 2.}
\end{aligned}
$$

The solution is $(1, -1)$.

Now Try Exercises 29, 33

In practice, it is possible to eliminate *either* variable from a system of linear equations. It is often best to choose the variable that requires the least amount of computation to complete the elimination. In the next example, we solve a system of equations twice—first by using the multiplication property of equality to eliminate the x-variable and then by using it to eliminate the y-variable.

EXAMPLE 3 **Multiplying before applying elimination**

Solve the system of equations two times, first by eliminating x and then by eliminating y.

$$
\begin{aligned}
2y &= -6 - 5x \\
2x &= -5y + 6
\end{aligned}
$$

Solution

It is best to write each equation in the standard form: $Ax + By = C$.

$$
\begin{aligned}
5x + 2y &= -6 & &\text{First equation in standard form} \\
2x + 5y &= 6 & &\text{Second equation in standard form}
\end{aligned}
$$

Eliminate x If we multiply the first equation in standard form by -2 and the second equation in standard form by 5, then we can eliminate the x-variable by adding.

$$
\begin{aligned}
-10x - 4y &= 12 & &\text{Multiply the first equation by } -2. \\
10x + 25y &= 30 & &\text{Multiply the second equation by 5.} \\
\hline
21y &= 42 \quad \text{or} \quad y = 2 & &\text{Add and solve for } y.
\end{aligned}
$$

To find x, substitute **2** for **y** in the first *given* equation, $2y = -6 - 5x$.

$$2(2) = -6 - 5x \qquad \text{Let } y = 2 \text{ in first equation.}$$
$$10 = -5x \qquad \text{Add 6 to each side.}$$
$$-2 = x \qquad \text{Divide each side by } -5.$$

The solution is $(-2, 2)$.

Eliminate y If we multiply the first equation in standard form by -5 and the second equation in standard form by 2, then we can eliminate the y-variable by adding.

$$-25x - 10y = 30 \qquad \text{Multiply the first equation by } -5.$$
$$\underline{4x + 10y = 12} \qquad \text{Multiply the second equation by 2.}$$
$$-21x \quad\quad = 42 \quad \text{or} \quad x = -2 \qquad \text{Add and solve for } x.$$

To find y, substitute -2 for x in the second given equation, $2x = -5y + 6$.

$$2(-2) = -5y + 6 \qquad \text{Let } x = -2 \text{ in second equation.}$$
$$-10 = -5y \qquad \text{Subtract 6 from each side.}$$
$$2 = y \qquad \text{Divide each side by } -5.$$

The solution is $(-2, 2)$.

Now Try Exercise 31

SOLVING A SYSTEM SYMBOLICALLY, GRAPHICALLY, AND NUMERICALLY In the next example, we use three different methods to solve a system of equations.

EXAMPLE 4 **Solving a system with different methods**

Solve the system of equations symbolically, graphically, and numerically.

$$x + y = 2$$
$$x - 3y = 6$$

Solution

Symbolic Solution Both the method of substitution and the elimination method are symbolic methods. The elimination method is used here. We can solve the system by multiplying the second equation by -1 and adding to eliminate the x-variable.

$$x + y = 2 \qquad \text{First equation}$$
$$\underline{-x + 3y = -6} \qquad \text{Multiply the second equation by } -1.$$
$$4y = -4 \quad \text{or} \quad y = -1 \qquad \text{Add and solve for } y.$$

We can find x by substituting -1 for y in the first equation, $x + y = 2$.

$$x + (-1) = 2 \qquad \text{Let } y = -1 \text{ in first equation.}$$
$$x = 3 \qquad \text{Add 1 to each side.}$$

The solution is $(3, -1)$.

Graphical Solution For a graphical solution, we solve each equation for y to obtain the slope–intercept form.

$x + y = 2$ First equation	$x - 3y = 6$ Second equation
$y = -x + 2$ Subtract x.	$-3y = -x + 6$ Subtract x.
	$y = \dfrac{1}{3}x - 2$ Divide by -3.

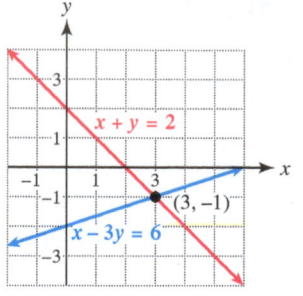

FIGURE 4.9

The graphs of $y = -x + 2$ and $y = \frac{1}{3}x - 2$ are shown in **FIGURE 4.9**. They intersect at $(3, -1)$. (These graphs could also be obtained by finding the x- and y-intercepts for each equation.)

Numerical Solution A numerical solution consists of a table of values, as shown in **TABLE 4.4**. Note that when $x = 3$, both y-values equal -1. Therefore the solution is $(3, -1)$.

x	0	1	2	**3**	4
$y = -x + 2$	2	1	0	-1	-2
$y = \frac{1}{3}x - 2$	-2	$-\frac{5}{3}$	$-\frac{4}{3}$	-1	$-\frac{2}{3}$

TABLE 4.4

Find the x-value where the two y-values are equal.

Now Try Exercise 41

MAKING CONNECTIONS 1

Solving Systems Using Different Methods

Systems of linear equations can be solved symbolically, graphically, and numerically. Symbolic solutions are *always exact*, whereas graphical and numerical solutions are *sometimes approximate*. The following example shows how to use each method to solve the system

$$-x + 2y = 3$$
$$x + 6y = 5.$$

For the graphical and numerical solutions, solve each equation for y to obtain

$$y = \frac{1}{2}x + \frac{3}{2}$$

$$y = -\frac{1}{6}x + \frac{5}{6}.$$

Symbolic Solution (elimination)

$$-x + 2y = 3$$
$$\underline{x + 6y = 5}$$
$$8y = 8 \quad \text{or} \quad y = \mathbf{1}$$
$$-x + 2(\mathbf{1}) = 3$$
$$-x = 1$$
$$x = \mathbf{-1}$$

The solution is $(-\mathbf{1}, \mathbf{1})$.

Graphical Solution

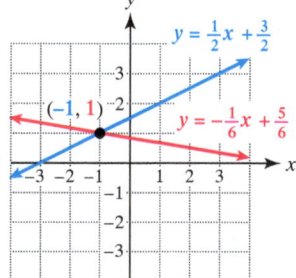

The solution is $(-\mathbf{1}, \mathbf{1})$.

Numerical Solution

x	-2	-1	0	1
$y = \frac{1}{2}x + \frac{3}{2}$	$\frac{1}{2}$	1	$\frac{3}{2}$	2
$y = -\frac{1}{6}x + \frac{5}{6}$	$\frac{7}{6}$	1	$\frac{5}{6}$	$\frac{2}{3}$

The solution is $(-\mathbf{1}, \mathbf{1})$.

---

2 Recognizing Systems with No Solutions or Infinitely Many Solutions

In the first section of this chapter, we discussed how a system of linear equations can have no solutions, one solution, or infinitely many solutions. Elimination can also be used on systems that have no solutions or infinitely many solutions.

EXAMPLE 5 **Solving other types of systems**

Solve each system of equations by using the elimination method. Then graph the system.

(a) $\quad x - 2y = 4$
$\quad -2x + 4y = -8$

(b) $3x + 3y = 6$
$\quad x + y = 1$

Solution

(a) We multiply the first equation by 2 and then add, which eliminates both variables.

$$2x - 4y = 8 \qquad \text{Multiply the first equation by 2.}$$
$$\underline{-2x + 4y = -8} \qquad \text{Second equation}$$
$$0 = 0 \quad \textbf{(True)} \qquad \text{Add.}$$

The equation $0 = 0$ is *always true*, which indicates that the system has *infinitely many solutions*. A graph of the two equations is shown in **FIGURE 4.10**. The two lines are identical so there actually is only one line, and *every point on this line represents a solution*. For example, $(0, -2)$ and $(4, 0)$ lie on the line and are both solutions.

(b) We multiply the second equation by -3 and add, eliminating both variables.

$$3x + 3y = 6 \qquad \text{First equation}$$
$$\underline{-3x - 3y = -3} \qquad \text{Multiply the second equation by } -3.$$
$$0 = 3 \quad \textbf{(False)} \qquad \text{Add.}$$

The equation $0 = 3$ is *always false*, which indicates that the system has *no solutions*. A graph of the two equations is shown in **FIGURE 4.11**. Note that the two lines are parallel and thus do not intersect.

Now Try Exercises 51, 55

FIGURE 4.10

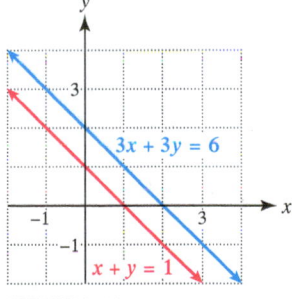

FIGURE 4.11

3 Solving Applications Using the Elimination Method

 🌐 Math in Context In the next two examples, we use elimination to solve applications relating to connected devices and to burning calories during exercise.

EXAMPLE 6 **Determining numbers of Android and Windows devices**

In 2015, there were 1737 million (1.737 billion) connected devices shipped globally that ran either a Windows or Android operating system. There were 979 million more Android devices shipped than Windows devices. How many connected devices were shipped with each operating system?

Solution

STEP 1: *Identify each variable.*

$\quad x$: Android devices, in millions

$\quad y$: Windows devices, in millions

STEP 2: *Write a system of equations.*

$$x + y = 1737 \quad \text{The total number of devices is 1737 million.}$$

$$x - y = 979 \quad \text{Android devices outnumber Windows devices by 979 million.}$$

STEP 3A: *Solve the system of linear equations.* Add the two equations to eliminate the y-variable.

$$
\begin{array}{r}
x + y = 1737 \\
\underline{x - y = 979} \\
2x = 2716 \quad \text{or} \quad x = 1358
\end{array}
$$

Substituting **1358** for x in the first equation results in **1358** $+ y = 1737$ or $y = $ **379**. The solution is $x = 1358$ and $y = 379$.

STEP 3B: *Determine the solution to the problem.* A total of **1358** million connected devices shipped running Android and **379** million shipped running Windows.

STEP 4: *Check the solution.* The total number of devices using these operating systems is $1358 + 379 = 1737$ million. The number of Android devices exceeds the number of Windows devices by $1358 - 379 = 979$ million. The answer checks.

Now Try Exercise 57

EXAMPLE 7 ## Burning calories during exercise

During strenuous exercise, an athlete can burn 10 calories per minute on a rowing machine and 11.5 calories per minute on a stair climber. If an athlete uses both machines and burns 433 calories in a 40-minute workout, how many minutes does the athlete spend on each machine? (*Source: Runner's World.*)

Solution

STEP 1: *Identify each variable.*

x: number of minutes on a rowing machine

y: number of minutes on a stair climber

STEP 2: *Write a system of equations.* The total workout takes 40 minutes, so $x + y = 40$. The athlete burns $10x$ calories on the rowing machine and $11.5y$ calories on the stair climber. Because the total number of calories equals 433, it follows that $10x + 11.5y = 433$.

$$x + y = 40 \qquad \text{Workout is 40 minutes.}$$

$$10x + 11.5y = 433 \qquad \text{Total calories is 433.}$$

STEP 3A: *Solve the system of linear equations.* Multiply the first equation by -10 and add the two equations.

$$
\begin{array}{ll}
-10x - 10y = -400 & \text{Multiply first equation by } -10. \\
\underline{10x + 11.5y = 433} & \text{Second equation} \\
 1.5y = 33 \quad \text{or} \quad y = \dfrac{33}{1.5} = 22 & \text{Add and solve for } y.
\end{array}
$$

Because $x + y = 40$ and $y = 22$, it follows that $x = 18$.

STEP 3B: *Determine the solution to the problem.* The athlete spends 18 minutes on the rowing machine and 22 minutes on the stair climber.

STEP 4: *Check your answer.* Because $18 + 22 = 40$, the athlete works out for 40 minutes. Also, the athlete burns $10(18) + 11.5(22) = 433$ calories. The answer checks.

Now Try Exercise 59

4.3   Putting It All Together

CONCEPT	EXPLANATION	EXAMPLE
Elimination Method	Is based on the addition property of equality If $a = b$ and $c = d$, then $$a + c = b + d.$$ May be used to solve systems of equations	$$\begin{array}{l} x + y = 5 \\ \underline{x - y = -1} \\ 2x = 4 \quad \text{or} \quad x = 2 \quad \text{Add.} \end{array}$$ Because $x + y = 5$ and $x = 2$, it follows that $y = 3$. The solution is $(2, 3)$.
Systems with No Solutions	Elimination can be used to recognize systems that have no solutions.	$$\begin{array}{l} -x - y = -4 \\ \underline{x + y = 2} \\ 0 = -2 \quad \text{Add.} \end{array}$$ Because $0 = -2$ is always false, there are no solutions.
Systems with Infinitely Many Solutions	Elimination can be used to recognize systems that have infinitely many solutions.	$$\begin{array}{l} x + y = 4 \\ 2x + 2y = 8 \end{array}$$ Multiply the first equation by -2. $$\begin{array}{l} -2x - 2y = -8 \\ \underline{2x + 2y = 8} \\ 0 = 0 \quad \text{Add.} \end{array}$$ Because $0 = 0$ is always true, there are infinitely many solutions.

4.3   Exercises

CONCEPTS AND VOCABULARY

1. Name two symbolic methods for solving a system of linear equations.

2. The elimination method is based on the _____ property of equality.

3. The addition property of equality states that if $a = b$ and $c = d$, then $a + c$ _____ $b + d$.

4. The multiplication property of equality states that if $a = b$, then ca _____ cb.

5. Suppose that the elimination method results in the equation $0 = 0$. What does this indicate about the number of solutions to the system of equations?

6. Suppose that the elimination method results in the equation $0 = 1$. What does this indicate about the number of solutions to the system of equations?

USING THE ELIMINATION METHOD

Exercises 7–14: If possible, use the given graph to solve the system of equations. Then use the elimination method to verify your answer.

7. $\begin{aligned} x - y &= 0 \\ x + y &= 2 \end{aligned}$

8. $\begin{aligned} x + y &= 6 \\ 2x - y &= 3 \end{aligned}$

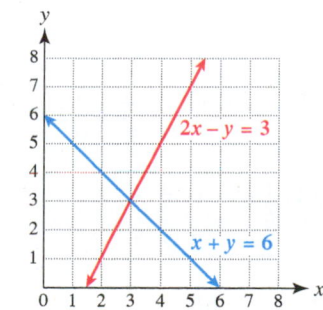

9. $2x + 3y = -1$
$2x - 3y = -7$

10. $-2x + y = -3$
$4x - 3y = 7$

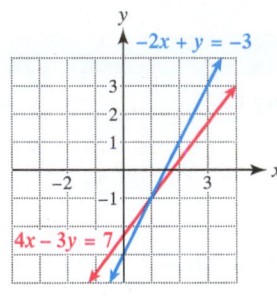

11. $x + y = 3$
$x + y = -1$

12. $2x - y = 4$
$-2x + y = -4$

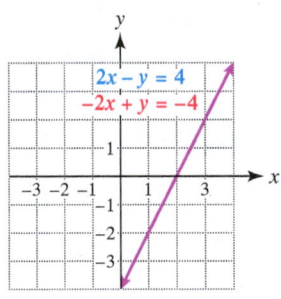

13. $2x + 2y = 6$
$x + y = 3$

14. $-x + 3y = 4$
$x - 3y = 3$

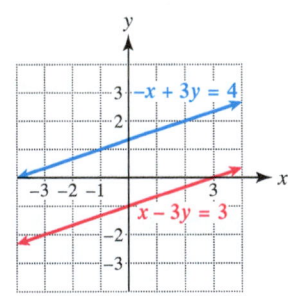

Exercises 15–36: Use the elimination method to solve the system of equations.

15. $x + y = 7$
$x - y = 5$

16. $x - y = 8$
$x + y = 4$

17. $-x + y = 5$
$x + y = 3$

18. $x - y = 10$
$-x - y = 20$

19. $2x + y = 8$
$3x - y = 2$

20. $-x + 2y = 3$
$x + 6y = 5$

21. $-2x + y = -3$
$2x - 4y = 0$

22. $2x + 6y = -5$
$7x - 6y = -4$

23. $\frac{1}{2}x - y = 3$
$\frac{3}{2}x + y = 5$

24. $x - \frac{1}{4}y = 4$
$-4x + \frac{1}{4}y = -9$

25. $a + 6b = 2$
$a + 3b = -1$

26. $3r - t = 7$
$2r - t = 2$

27. $5a - 6b = -2$
$5a + 5b = 9$

28. $-r + 2t = 0$
$3r + 2t = 8$

29. $3u + 2v = -16$
$2u + v = -9$

30. $5u - v = 0$
$3u + 3v = -18$

31. $5x - 7y = 5$
$-2x + 2y = -2$

32. $2x + 7y = 6$
$4x - 3y = -22$

33. $5x - 3y = 4$
$3x + 2y = 10$

34. $-3x - 8y = 1$
$2x + 5y = 0$

35. $-5x - 10y = -22$
$10x + 15y = 35$

36. $-15x + 4y = -20$
$5x + 7y = 90$

SOLVING A SYSTEM USING A TABLE OF VALUES

Exercises 37–40: A table of values is given for two linear equations. Use the table to solve this system.

37.

x	0	1	2	3	4
$y = -x + 5$	5	4	3	2	1
$y = 2x - 4$	-4	-2	0	2	4

38.

x	-3	-2	-1	0	1
$y = x + 1$	-2	-1	0	1	2
$y = -x - 3$	0	-1	-2	-3	-4

39.

x	-2	-1	0	1	2
$y = 3x + 1$	-5	-2	1	4	7
$y = -x + 1$	3	2	1	0	-1

40.

x	-2	-1	0	1	2
$y = 2x$	-4	-2	0	2	4
$y = -x$	2	1	0	-1	-2

SOLVING A SYSTEM SYMBOLICALLY, GRAPHICALLY, AND NUMERICALLY

Exercises 41–46: Solve the system of equations

   (a) symbolically,
   (b) graphically, and
   (c) numerically.

41. $2x + y = 5$
$x - y = 1$

42. $-x + y = 2$
$3x + y = -2$

43. $2x + y = 5$
$x + y = 1$

44. $-x + y = 2$
$3x - y = -2$

45. $6x + 3y = 6$
$-2x + 2y = -2$

46. $-x + 2y = 5$
$2x + 2y = 8$

RECOGNIZING OTHER TYPES OF SYSTEMS

Exercises 47–56: Use elimination to determine whether the system of equations has no solutions, one solution, or infinitely many solutions. Then graph the system.

47. $2x - 2y = 4$
$-x + y = -2$

48. $-2x + y = 4$
$4x - 2y = -8$

49. $x - y = 0$
$x + y = 0$

50. $x - y = 2$
$x + y = 2$

51. $x - y = 4$
$x - y = 1$

52. $-2x + 3y = 5$
$4x - 6y = 10$

53. $x - 3y = 2$
$-x + 3y = 4$

54. $6x + 9y = 18$
$4x + 6y = 12$

55. $4x - 8y = 24$
$6x - 12y = 36$

56. $x - y = 5$
$2x - y = 4$

SOLVING APPLICATIONS USING THE ELIMINATION METHOD

57. Unique Site Visitors In March 2014, the websites healthcare.gov and ncaa.com saw large percentage increases in unique visitors over the previous month. There were a combined 20.4 million unique visitors to ncaa.com and healthcare.gov. There were 3 million more visitors to healthcare.gov than ncaa.com. How many unique visitors were there to each site in March 2014? (*Source:* Compete.com.)

58. Millionaires In 2012, there were 210,000 millionaires in Chicago and Houston. If there were 3,600 more millionaires in Chicago than Houston, find the number of millionaires in each city that year. (*Source:* The Economist.)

59. Burning Calories During strenuous exercise, an athlete can burn 9 calories per minute on a stationary bicycle and 11.5 calories per minute on a stair climber. In a 30-minute workout, an athlete burns 300 calories. How many minutes does the athlete spend on each type of exercise equipment? (*Source:* Runner's World.)

60. Distance Running An athlete runs at 9 mph and then at 12 mph, covering 10 miles in 1 hour. How long does the athlete run at each speed?

61. River Current A riverboat takes 8 hours to travel 64 miles downstream and 16 hours for the return trip. Find the speed of the current and the speed of the riverboat in still water?

62. Airplane Speed An airplane travels 3000 miles with the wind in 5 hours and takes 6 hours for the return trip into the wind. Find the speed of the wind and the speed of the airplane without any wind?

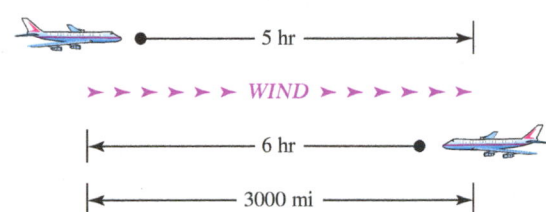

63. Investments A total of $5000 is invested at 3% and 5% annual interest. After 1 year, the total interest equals $210. How much money is invested at each interest rate?

64. Mixing Antifreeze A car radiator holds 2 gallons of fluid and initially is empty. If a mixture of water and antifreeze contains 70% antifreeze and another mixture contains 15% antifreeze, how much of each should be combined to fill the radiator with a 50% antifreeze mixture?

65. Number Problem The sum of two integers is -17, and their difference is -69. Find the two integers.

66. Supplementary Angles The measures of two supplementary angles differ by 74°. Find the two angles.

67. Picture Dimensions The figure below shows a red graph that gives possible dimensions for a rectangular picture frame with perimeter 120 inches. The blue graph shows possible dimensions for a rectangular frame whose length L is twice its width W.

Picture Dimensions

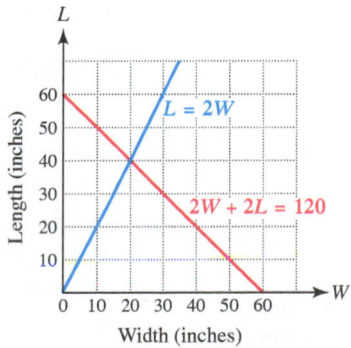

(a) Use the figure to determine the dimensions of a frame with a perimeter of 120 inches and a length that is twice the width.

(b) Solve this problem symbolically.

68. Sales of DVDs and CDs A company sells DVDs *d* and CDs *c*. The figure shows a red graph of $d + c = 2000$. The blue graph shows a revenue of $15,000 received from selling *d* DVDs at $12 each and *c* CDs at $6 each.

Digital Media

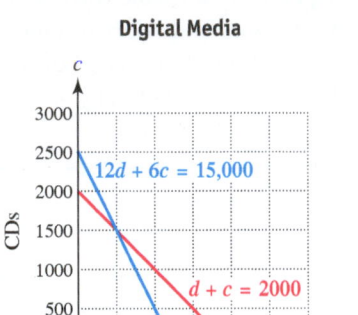

(a) If the total number of DVDs and CDs sold is 2000, determine how many of each were sold to obtain a revenue of $15,000.

(b) Solve this problem symbolically.

WRITING ABOUT MATHEMATICS

69. Suppose that a system of linear equations is solved symbolically, numerically, and graphically. How do the solutions from each method compare? Explain your answer.

70. When you are solving a system of linear equations by elimination, how can you recognize that the system has no solutions?

Group Activity

Exercises 1–4: **Facebook Apps** In early 2014, the two Facebook apps with the highest daily active users were Candy Crush Saga and Farm Heroes Saga. The total number of daily active users for the two apps was 73.8 million. The number of Candy Crush Saga users exceeded the number of Farm Heroes Saga users by 40.2 million. (*Source:* Inside Facebook.)

1. Set up a system of equations whose solution gives the number of daily active users in millions for each app. Identify what each variable represents.

2. Use substitution to solve this system. Interpret the result.

3. Use elimination to solve this system.

4. Solve this system graphically. Do all your answers agree?

4.4 Systems of Linear Inequalities

Objectives

1. Finding Solutions to a Linear Inequality in Two Variables

2. Writing and Graphing a Linear Inequality in Two Variables

3. Finding Solutions to Systems of Linear Inequalities in Two Variables

4. Solving Applications Involving Systems of Linear Inequalities

1 Finding Solutions to a Linear Inequality in Two Variables

 Connecting Concepts with Your Life Suppose that a college student works at both the library and a department store. The library pays $10 per hour, and the department store pays $8 per hour. The equation $A = 10L + 8D$ calculates the amount of money earned from working *L* hours at the library and *D* hours at the department store. If the cost of one college credit is $80, then solutions to the equation

$$10L + 8D = 80$$

are ordered pairs (L, D) that result in the student earning (exactly) enough to pay for one credit. The equation's graph is the line shown in **FIGURE 4.12(a)**. The point $(4, 5)$ lies on this line, which indicates that, if the student works 4 hours at the library and 5 hours at the department store, then the pay is $80.

NEW VOCABULARY

☐ Test point
☐ Linear inequality in two variables
☐ System of linear inequalities in two variables

STUDY TIP

If you have not been studying with other students, consider getting together with classmates so that you can work on your math homework together.

Student Work Hours

Test Points

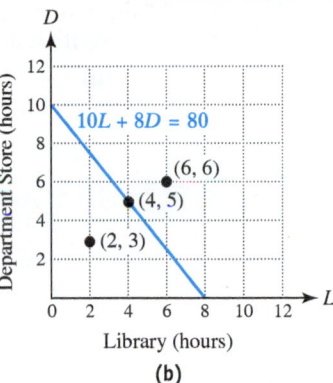

FIGURE 4.12

There are many situations in which the student can make more than $80. For example, the **test point** $(6, 6)$ lies above the line in **FIGURE 4.12(b)**, indicating that, if the student works 6 hours at both the library and the department store, then the pay is more than $80. In fact, any point *above* the line results in pay *greater than* $80. The region **above** the line is described by the inequality

$$10L + 8D > 80.$$

The test point $(6, 6)$ represents earnings of $108 and *satisfies* this inequality because

$$10(6) + 8(6) > 80$$

is a true statement. Similarly, any point *below* the line gives an ordered pair (L, D) that results in earnings *less than* $80. The point $(2, 3)$ in **FIGURE 4.12(b)** lies below the line and represents earnings of $44. That is,

$$10(2) + 8(3) < 80.$$

The region **below** the line is described by the inequality

$$10L + 8D < 80.$$

2 Writing and Graphing a Linear Inequality in Two Variables

Any linear equation in two variables can be written in standard form as

$$Ax + By = C,$$

where A, B, and C are constants. When the equals sign is replaced with $<$, $>$, $\leq$, or $\geq$, a **linear inequality in two variables** results. Examples of linear *equations* in two variables include

$$2x + 3y = 10 \quad \text{and} \quad y = \frac{1}{2}x - 5,$$

and so examples of linear *inequalities* in two variables include

$$2x + 3y < 10 \quad \text{and} \quad y \geq \frac{1}{2}x - 5.$$

A *solution* to a linear inequality in two variables is an ordered pair (x, y) that makes the inequality a true statement. The *solution set* is the set of all solutions to the inequality. The solution set to an inequality in two variables is typically a region in the *xy*-plane, which means that there are infinitely many solutions.

EXAMPLE 1   ## Writing a linear inequality

Write a linear inequality that describes each shaded region.

(a)   **(b)**

    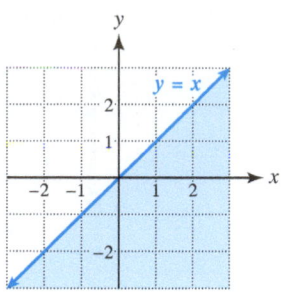

Solution
(a) The shaded region is bounded by the line $x = 2$. The *dashed* line indicates that the line is not included in the solution set. Only points with x-coordinates **greater than** 2 are shaded. Thus every point in the shaded region satisfies $x > 2$.
(b) The solution set includes all points that are **on or below** the line $y = x$. An inequality that describes this region is $y \le x$, which can also be written as $-x + y \le 0$.

Now Try Exercises 27, 29

TECHNOLOGY NOTE

Shading an Inequality
Graphing calculators can be used to shade a solution set to an inequality. The left-hand screen shows how to enter the equation from Example 1(b), and the right-hand screen shows the resulting graph.

CALCULATOR HELP

To shade an inquality, see Appendix A (pages AP-6 and AP-7).

    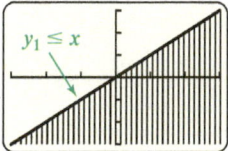

READING CHECK 1

- How do you know whether to graph a solid line or a dashed line for an inequality?

GRAPHING A LINEAR INEQUALITY IN TWO VARIABLES

STEP 1: Replace the inequality symbol with an equals sign and graph the resulting line. If the inequality is $<$ or $>$, use a dashed line, and if it is $\le$ or $\ge$, use a solid line.

STEP 2: Pick a *test point* that does *not* lie on the line. Substitute this point in the given inequality. Determine whether the resulting statement is true or false.

STEP 3: If the statement is true, shade the region containing the test point. If the statement is false, shade the region that does not contain the test point.

EXAMPLE 2   ## Graphing a linear inequality

Shade the solution set for each inequality.
(a) $y \le 1$ **(b)** $x + y < 3$ **(c)** $-x + 2y \ge 2$

Solution
(a) STEP 1: Graph the horizontal line $y = 1$, where a solid line indicates that the line is included in the solution set. See **FIGURE 4.13(a)**.

STEP 2: Select a test point either above or below the horizontal line. For example, the point $(0, 0)$ satisfies the inequality $y \le 1$, because $0 \le 1$ is a true statement.

STEP 3: Shade the region that contains $(0, 0)$, **below** the line. See **FIGURE 4.13(a)**.

(b) STEP 1: Graph the line $x + y = 3$, as shown in **FIGURE 4.13(b)**. Because the inequality is $<$ and not $\le$, the line is not included and is dashed rather than solid.

STEP 2: Select a test point in either region. For example, the point $(0, 0)$ satisfies the inequality $x + y < 3$, because $0 + 0 < 3$ is a true statement.

STEP 3: Shade the region that contains $(0, 0)$, **below** the line. See **FIGURE 4.13(b)**.

(c) STEP 1: Graph $-x + 2y = 2$ as a solid line, as shown in **FIGURE 4.13(c)**.

STEP 2: To decide whether to shade above or below the solid line, use the test point $(0, 0)$ again. (Note that other test points can be used.) The point $(0, 0)$ does not satisfy the inequality $-x + 2y \ge 2$ because $-0 + 2(0) \ge 2$ is a false statement.

STEP 3: Shade the region that does not contain $(0, 0)$, **above** the line. See **FIGURE 4.13(c)**.

READING CHECK 1

- How is a test point used when shading a solution set for an inequality?

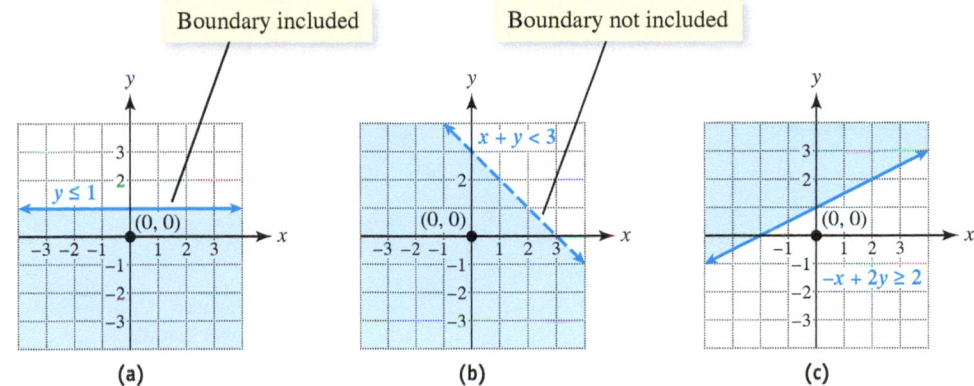

FIGURE 4.13

Now Try Exercises 35, 41, 43

NOTE: In all three parts of Example 2, the test point $(0, 0)$ was used to determine which region should be shaded. However, *any point that is not on* the solid or dashed line can be used as a test point. When the line does not pass through the origin, it is often convenient to use $(0, 0)$ as a test point because substituting 0 for both x and y in an inequality results in a very simple computation. ∎

See the Concept 1 | **GRAPHING A LINEAR INEQUALITY**

To graph $3x - 2y \le 6$:

A Solve the inequality for y.

$3x - 2y \le 6$: Given inequality

$-2y \le -3x + 6$ Subtract $3x$ from each side

$y \ge \dfrac{3}{2}x - 3$ Divide each side by -2.

B Graph the *equation* $y = \dfrac{3}{2}x - 3$.

(Use a **solid** line for $\ge$ or $\le$. Use a **dashed** line for $>$ or $<$.)

C Choose a test point that is not on the line.

D Substitute the test point $(0, 0)$ in the given inequality. Since $3(0) - 2(0) \le 6$ is true, shade the region containing the test point.

WATCH VIDEO IN MML.

3 Finding Solutions to Systems of Linear Inequalities in Two Variables

Sometimes a solution set must satisfy two inequalities in a **system of linear inequalities in two variables**. For example, the point $(2, 1)$ satisfies both of the inequalities in the system of inequalities given by

$$x > 1$$
$$y < 2.$$

FIGURE 4.14(a) shows the solution set to $x > 1$ in blue, and **FIGURE 4.14(b)** shows the solution set to $y < 2$ in red. The two regions are shaded together in **FIGURE 4.14(c)**, where the blue and red regions intersect to form a purple region. The solution set to the system of inequalities includes all points in the purple region, as shown in **FIGURE 4.14(d)**. Points in this region satisfy *both* inequalities.

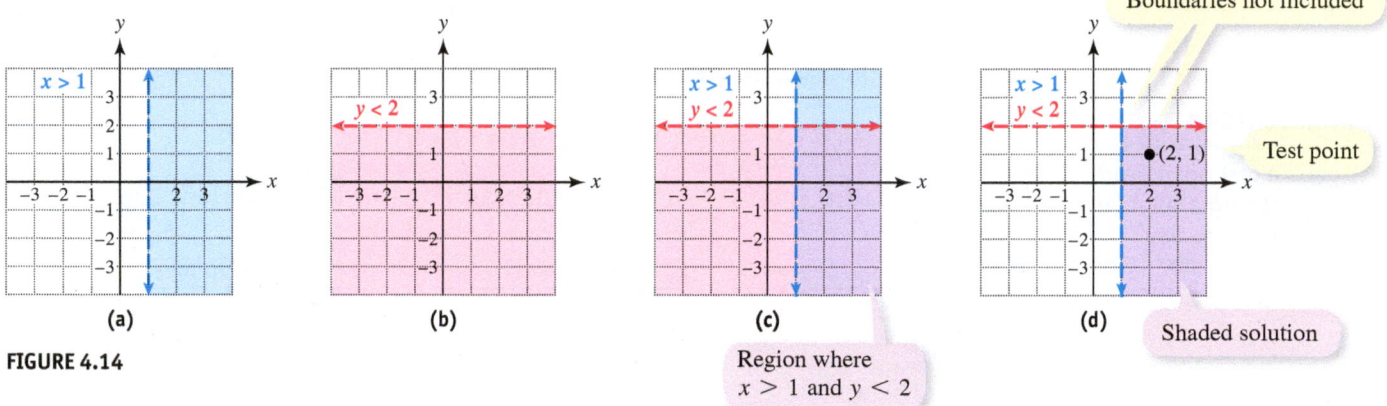

FIGURE 4.14

Region where $x > 1$ and $y < 2$

Boundaries not included

Test point

Shaded solution

EXAMPLE 3   **Graphing a system of linear inequalities**

Shade the solution set to the system of inequalities.

$$x > -1$$
$$x + y \leq 1$$

Solution

Start by graphing the solution set to each inequality. The solution set to $x > -1$ is the blue region to the **right** of the dashed vertical line $x = -1$, as shown in **FIGURE 4.15(a)**. The solution set to $x + y \leq 1$ includes the solid line $x + y = 1$ and the red region that lies **below** it, as shown in **FIGURE 4.15(b)**. For a point to satisfy the *system* of inequalities, it must satisfy *both* inequalities. Therefore the solution set is the *intersection* of the blue and red regions, shown as the purple region in **FIGURE 4.15(c)**. Note that the test point $(0, 0)$, located in the shaded region, satisfies both inequalities.

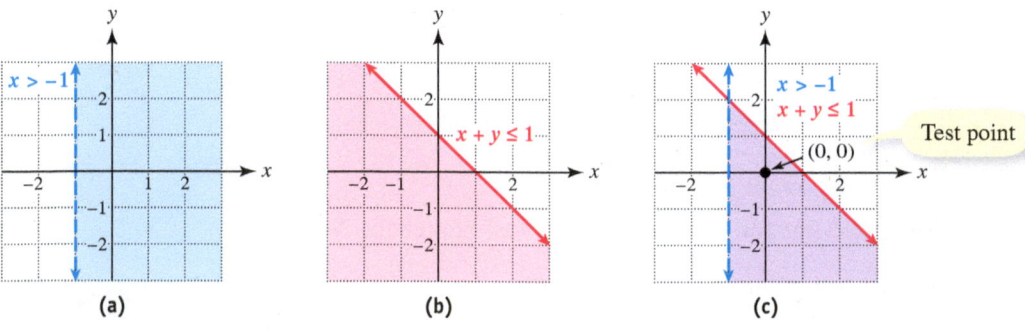

FIGURE 4.15

Test point

Now Try Exercise 67

GRAPHING A SYSTEM OF LINEAR INEQUALITIES IN TWO VARIABLES

STEP 1: For each inequality in the system, perform Step 1 from Graphing a Linear Inequality in Two Variables. See the box just before Example 2 in this section.

STEP 2: Pick a test point from one region and substitute it in the given inequalities.

STEP 3: If both of the resulting statements are true, shade the region containing the test point. If not, pick a test point from a different region and substitute it in the given inequalities. Repeat this step until the region to be shaded is found.

EXAMPLE 4 Graphing a system of linear inequalities

Shade the solution set to the system of inequalities.

$$x + 2y < -2$$
$$2x + y \geq 2$$

Solution

STEP 1: Start by graphing the dashed line $x + 2y = -2$ and the solid line $2x + y = 2$, as shown in **FIGURE 4.16(a)**. Note that these two lines divide the xy-plane into 4 regions, numbered **1, 2, 3,** and **4**.

STEP 2: If we let $(0, 0)$ be a test point, it does not satisfy either inequality. Therefore we do not shade region **2**, which contains $(0, 0)$. However, there are still 3 possible regions. If we try the test point $(4, -4)$ in region **4**, it satisfies both the given inequalities.

$$4 + 2(-4) < -2 \checkmark \quad \text{A true statement}$$
$$2(4) + (-4) \geq 2 \checkmark \quad \text{A true statement}$$

STEP 3: Shade region 4, as shown in **FIGURE 4.16(b)**. Once a test point is found that makes both inequalities true, there is no need to check test points in other regions.

CRITICAL THINKING 1

Does the solution set in **FIGURE 4.16(b)** include the point of intersection, $(2, -2)$? Explain your reasoning.

Boundaries and Regions

(a)

Shaded Solution

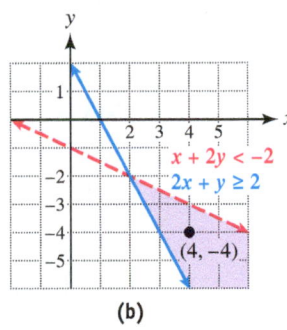

(b)

FIGURE 4.16

Now Try Exercise 63

MAKING CONNECTIONS 1

Solving for y and shading inequalities

Another way to solve the system of inequalities in Example 4 that does not involve test points is to solve each inequality for y to obtain

$$y < -\frac{1}{2}x - 1$$
$$y \geq -2x + 2.$$

The solution set is the region in **FIGURE 4.16(b)** that lies **below** the line $y = -\frac{1}{2}x - 1$ and **above and including** the line $y = -2x + 2$.

4 Solving Applications Involving Systems of Linear Inequalities

🌐 *Math in Context* The next example illustrates an application from business and manufacturing that involves systems of inequalities.

EXAMPLE 5   **Manufacturing MP3 players and digital video players**

A business manufactures MP3 players and digital video players. Because every digital video player contains an MP3 player, it must produce at least as many MP3 players as digital video players. In addition, the total number of MP3 players and digital video players produced each day cannot exceed 50 because of limited resources. Shade the region that shows the numbers of MP3 players M and digital video players V that can be produced within these restrictions. Label the horizontal axis M and the vertical axis V.

Solution
Because the company must produce *at least* as many MP3 players M as digital video players V, we have $M \geq V$, which can also be written as $V \leq M$. The total number of MP3 players and digital video players *cannot exceed* 50, so $M + V \leq 50$. To shade the solution set for

$$V \leq M$$

$$M + V \leq 50,$$

we first graph the lines $V = M$ and $M + V = 50$, as shown in **FIGURE 4.17(a)**. Because the number of MP3 players and digital video players cannot be negative, the graph includes only quadrant I. These lines divide this quadrant into four regions, and we can determine the correct region to shade by selecting one test point from each region. The region containing the test point satisfying both inequalities is the one to be shaded. For example, the test point $(20, 10)$ with $M = 20$, $V = 10$ satisfies both inequalities.

$$10 \leq 20 \ \checkmark \quad \text{A true statement; } V \leq M$$

$$20 + 10 \leq 50 \ \checkmark \quad \text{A true statement; } M + V \leq 50$$

The solution set is shaded in **FIGURE 4.17(b)**.

NOTE: An alternative solution is to write the inequalities as $V \leq M$ and $V \leq -M + 50$. Then the solution set lies *below both lines*. This region is shaded in **FIGURE 4.17(b)**. ∎

The Boundaries

(a)

The Solution Set

The numbers of MP3 players and digital video players that can be produced must fall in this region.

(b)

FIGURE 4.17

Now Try Exercise 73

🌐 *Math in ~~Context~~* Although there is no *ideal* weight for a person, government agencies and insurance companies sometimes recommend a *range* of weights for various heights. *Inequalities* are used with these recommendations.

EXAMPLE 6

Finding weight–height combinations

FIGURE 4.18 shows a shaded region containing recommended weights w for heights h. (Source: Department of Agriculture.)

FIGURE 4.18

(a) What does this graph indicate about a 68-inch person who weighs 150 pounds?

(b) The shaded region in **FIGURE 4.18** is determined by the following system of inequalities.

$$25h - 7w \leq 800$$

$$5h - w \geq 170$$

Verify that $h = 68$ and $w = 150$ satisfies the system of inequalities.

(c) What do ordered pairs (w, h) to the left of the shaded region indicate?

Solution

(a) The point $(150, 68)$ lies in the shaded region. Therefore someone who is 68 inches tall and weighs 150 pounds falls within the recommended guidelines.

(b) Both inequalities are satisfied by $h = 68$, $w = 150$.

$$25(68) - 7(150) = 650 \leq 800 ✓$$

$$5(68) - 150 = 190 \geq 170 ✓$$

(c) To the left of the shaded region are ordered pairs (w, h) that represent smaller weights and larger heights. These ordered pairs correspond to people who weigh less than recommended.

Now Try Exercise 77

4.4 Putting It All Together

CONCEPT	EXPLANATION	EXAMPLES
Linear Inequality in Two Variables	An inequality that can be written as $$Ax + By < C,$$ where $<$ can also be $\leq$, $>$, or $\geq$	$3x + y \geq 10$, $-x + 3y < 5$, $y \leq 5 - x$, and $x > 5$

continued on next page

continued from previous page

CONCEPT	EXPLANATION	EXAMPLES
Solution	A solution (x, y) makes the inequality a true statement.	The ordered pair $(0, 0)$ satisfies $$2x - y < 2,$$ so it is a solution to the inequality.
Solution Set	The set of all solutions Usually a region in the xy-plane	The solution set to $x + y > 2$ is all points above the line $x + y = 2$.
System of Linear Inequalities in Two Variables	Solutions to systems must satisfy both inequalities. The solution set usually includes infinitely many solutions.	The ordered pair $(0, 0)$ is a solution to $$x + y \leq 2$$ $$2x - y > -4,$$ because both inequalities are true when $x = 0$ and $y = 0$.

4.4 Exercises

MyMathLab®

CONCEPTS AND VOCABULARY

1. When the equals sign in $Ax + By = C$ is replaced with $<, >, \leq,$ or $\geq$, a linear _____ in two variables results.

2. A solution to a linear inequality in two variables is a(n) _____ that makes the inequality a true statement.

3. Describe the graph of the solution set to $y \leq k$ for some number k.

4. Describe the graph of the solution set to $x > k$ for some number k.

5. Describe the graph of the solution set to $y \geq x$.

6. When graphing the solution set to a linear inequality, one way to determine which region to shade is to use a(n) _____ point.

7. When graphing a linear inequality containing either $<$ or $>$, use a (dashed/solid) line.

8. When graphing a linear inequality containing either $\leq$ or $\geq$, use a (dashed/solid) line.

9. When graphing the linear inequality $Ax + By < C$, a first step is to graph the line _____.

10. A solution to a system of two inequalities must make (both inequalities/one inequality) true.

11. If two shaded regions represent the solution sets for two inequalities in a system, then the solution set for the system is where these two shaded regions _____.

12. If a test point is found that satisfies both inequalities in a system, do other test points still need to be checked?

TESTING SOLUTIONS TO ONE LINEAR INEQUALITY

Exercises 13–24: Determine whether the test point is a solution to the linear inequality.

13. $(3, 1), x > 2$

14. $(-3, 4), x \leq -3$

15. $(0, 0), y \geq 2$

16. $(0, 0), y < -3$

17. $(5, 4), y \geq x$

18. $(-1, 2), y < x$

19. $(3, 0), y < x - 1$

20. $(0, 5), y > 2x + 4$

21. $(-2, 6), x + y \leq 4$

22. $(2, -4), x - y \geq 7$

23. $(-1, -1), 2x + y \geq -1$

24. $(0, 1), -x - 5y \geq -1$

WRITING AND GRAPHING A LINEAR INEQUALITY

Exercises 25–32: Write a linear inequality that describes the shaded region.

25.

26.

27.

28.

29.

30.

31.

32.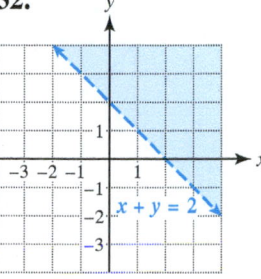

Exercises 33–44: Shade the solution set to the inequality.

33. $x \leq -1$

34. $x > 3$

35. $y < -2$

36. $y \geq 0$

37. $y > x$

38. $y \leq x$

39. $y \geq 3x$

40. $y < -2x$

41. $x + y \leq 1$

42. $x + y \geq -2$

43. $2x - y > 2$

44. $-x - y < 1$

TESTING SOLUTIONS TO SYSTEMS OF LINEAR INEQUALITIES

Exercises 45–50: Determine if the test point is a solution to the system of linear inequalities.

45. $(3, 1)$
$x - y < 3$
$x + y > 3$

46. $(0, 0)$
$x - 2y < 1$
$2x - y > -1$

47. $(-2, 3)$
$3x - 2y \geq 1$
$-x + 3y > 3$

48. $(1, 2)$
$2x - 2y < 5$
$x - y > -1$

49. $(4, -2)$
$x - 2y \geq 8$
$-2x - 5y > 0$

50. $(-1, -2)$
$x + y < 0$
$-2x - 3y \leq -1$

Exercises 51–54: The graphs of two equations are shown with four test points labeled. Use these points to decide which region should be shaded to solve the given system of inequalities.

51. $x \le 2$
    $x + y \ge 2$

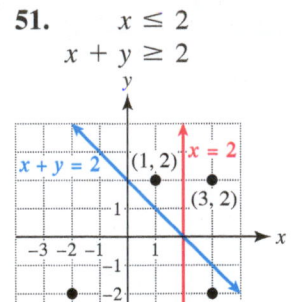

52. $y \ge 1$
    $2x - y \ge 3$

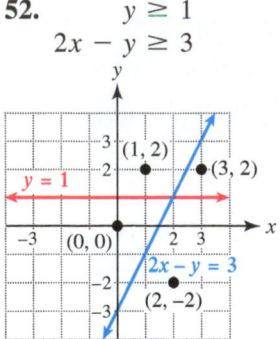

53. $x + y \le 3$
    $y \le 2x$

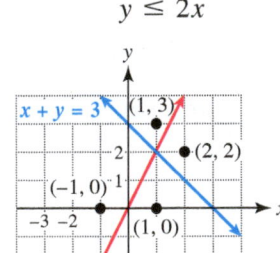

54. $y \le \ \ x$
    $y \ge -x$

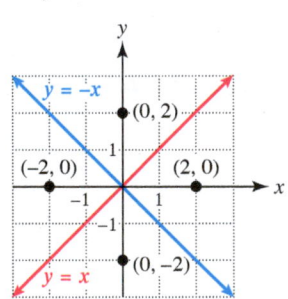

GRAPHING SOLUTIONS TO SYSTEMS OF LINEAR INEQUALITIES

Exercises 55–72: Shade the solution set to the system of inequalities.

55. $x > 2$
   $y < 3$

56. $x \le -1$
   $y \ge \ \ 3$

57. $x \le -2$
   $y < 2x$

58. $y > 2$
   $y \ge -x$

59. $y \le \ \ x$
   $y > -x$

60. $y \le \frac{1}{2}x$
   $y \ge -2x$

61. $\ \ x + y \le 3$
   $-x + y \le 1$

62. $x + y > 2$
   $x - y < 2$

63. $2x + y > -3$
   $\ \ x + y \le -1$

64. $-x + y \ge \ \ 3$
   $2x - y \ge -2$

65. $2x + y \ge -3$
   $\ \ x + y > -1$

66. $-x + y \ge \ \ 2$
   $3x - y \ge -2$

67. $\ \ \ \ \ \ y > -2$
   $x + 2y \le -4$

68. $\ \ x \ge 2$
   $3y < x - 3$

69. $\ \ x + 2y > -4$
   $2x + \ \ y \le \ \ 3$

70. $\ \ x + 3y \ge 3$
   $3x - 2y \ge 6$

71. $3x + 4y \le 12$
   $5x + 3y \ge 15$

72. $-2x + \ \ y \ge \ \ 6$
   $\ \ \ \ x - 2y \ge -4$

APPLICATIONS INVOLVING SYSTEMS OF INEQUALITIES

73. Manufacturing (Refer to Example 5.) A business manufactures at least two MP3 players for each digital video player. The total number of MP3 players and digital video players must be less than 90. Shade the region that represents the number of MP3 players *M* and digital video players *V* that can be produced within these restrictions. Put *V* on the horizontal axis.

74. Working on Two Projects An employee must spend more time on project X than on project Y. The employee can work at most 40 hours on the two projects. Shade the region in the *xy*-plane that represents the number of hours the employee can spend on each project.

75. Maximum Heart Rate When exercising, people often try to maintain heart rates that are a percentage of their maximum heart rate. Maximum heart rate *R* is given by $R = 220 - A$, where *A* is the person's age in years and *R* is the heart rate in beats per minute.
   (a) Find *R* for a person 20 years old; 70 years old.
   (b) Sketch a graph of $R \le 220 - A$. Assume that *A* is between 20 and 70 and put *A* on the horizontal axis of your graph.
   (c) Interpret your graph.

76. Target Heart Rate (Refer to the preceding exercise.) A target heart rate *T* that is half a person's maximum heart rate is given by $T = 110 - \frac{1}{2}A$, where *A* is a person's age in years.
   (a) What is *T* for a person 30 years old? 50 years old?
   (b) Sketch a graph of the system of inequalities.

$$T \ge 110 - \frac{1}{2}A$$
$$T \le 220 - A$$

   Assume that *A* is between 20 and 60.
   (c) Interpret your graph.

77. Height and Weight Use Figure 4.18 in Example 6 to determine the range of recommended weights for a person who is 74 inches tall.

78. Height and Weight Use Figure 4.18 in Example 6 to determine the range of recommended heights for a person who weighs 150 pounds.

79. Candy and Coffee Sales A small business sells freshly ground coffee for $6 per pound and candy for $4 per pound.
   (a) Let *x* be the number of pounds of coffee sold and *y* be the number of pounds of candy sold. Sketch a line that represents all possible values of *x* and *y* that result in total sales of $240. Label your graph and assume that $x \ge 0$ and $y \ge 0$. (*Hint:* Find the *x*- and *y*-intercepts.)
   (b) Shade the region in your graph that represents all sales (x, y) that result in total sales of $240 or less.

80. Candy and Coffee Sales (Continuation of Exercise 79) Suppose that the business sells at least as many pounds of candy as coffee and that total sales are $240 or less. Shade the region in your graph that represents all sales (x, y) that satisfy these two conditions.

Exercises 81–86: **Plant Growth** *The following figure illustrates the relationships among forests, grasslands, and deserts, suggested by annual Temperature T in degrees Fahrenheit and precipitation P in inches.*

Climate and Vegetation

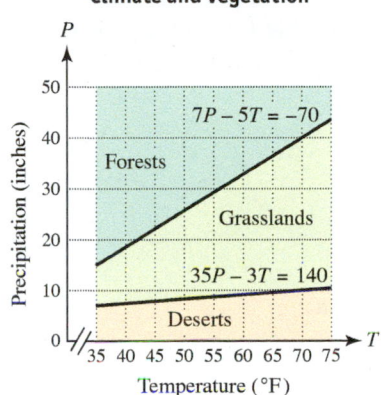

81. Use the graph to describe in terms of precipitation and temperature the regions where forests grow.

82. Use the graph to describe in terms of precipitation and temperature the regions where deserts exist.

83. The equation of the line that separates grasslands and forests is

$$7P - 5T = -70.$$

Write an inequality that describes temperatures and amounts of precipitation that correspond to forested regions. (Include the line.)

84. The equation of the line that separates grasslands and deserts is

$$35P - 3T = 140.$$

Write an inequality that describes temperatures and amounts of precipitation that correspond to desert regions. (Include the line.)

85. Using the information from Exercises 83 and 84, write a system of inequalities that describes temperatures and amounts of precipitation that correspond to grassland regions. (Include the lines.)

86. Cheyenne, Wyoming, has an average annual temperature of about 50 °F and an average annual precipitation of about 14 inches. Use the graph to predict the type of plant growth you might expect near Cheyenne. Then check to determine whether $T = 50$ and $P = 14$ satisfy the proper inequalities.

WRITING ABOUT MATHEMATICS

87. What is the solution set to the following system of inequalities? Explain your reasoning.

$$y > x$$
$$y < x - 1$$

88. Write a system of linear inequalities whose solution set is the entire xy-plane. Explain your reasoning.

SECTIONS 4.3 and 4.4 — **Checking Basic Concepts**

1. Use elimination to solve the system of equations.

$$2x + 3y = 5$$
$$x - 7y = -6$$

2. Use elimination to solve each system of equations. How many solutions are there in each case?
(a)
$$x + y = -1$$
$$x - 2y = 2$$
(b)
$$5x - 6y = 4$$
$$-5x + 6y = 1$$
(c)
$$x - 3y = 0$$
$$2x - 6y = 0$$

3. Solve the given system of equations symbolically, graphically, and numerically. How many solutions are there?

$$-2x + y = 0$$
$$y = 2x$$

4. Shade the solution set to each inequality.
(a) $y < -1$ (b) $x + y < 1$

5. Shade the solution set to the following system of inequalities.

$$x \le -1$$
$$-2x + y > -3$$

6. Large U.S. Cities The combined population of New York and Chicago was 11.1 million people in 2013. The population of New York exceeded the population of Chicago by 5.7 million people.
(a) Let x be the population of New York and y be the population of Chicago. Write a system of equations whose solution gives the population of each city in 2013.
(b) Solve the system of equations.

CHAPTER 4   Summary

System of Linear Equations

Solution	An ordered pair (x, y) that satisfies *both* equations
Solution Set	The set of all solutions
Graphical Solution	Graph each equation. A point of intersection is a solution. (Sometimes determining the exact answer when estimating from a graph may be difficult.)
Numerical Solution	Solve each equation for y and make a table for each equation. A solution occurs when two y-values are equal for a given x-value.

Example: The ordered pair $(3, 1)$ is the solution to the following system.

$$x + y = 4 \qquad 3 + 1 = 4 \text{ is a true statement.}$$
$$x - y = 2 \qquad 3 - 1 = 2 \text{ is a true statement.}$$

A Graphical Solution

The point of intersection $(3, 1)$ is the solution to the system of equations.

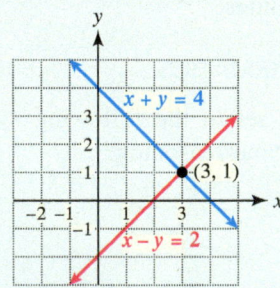

A Numerical Solution

The ordered pair $(3, 1)$ is the solution. When $x = 3$, both y-values equal 1.

x	1	2	**3**	4
$y = 4 - x$	3	2	**1**	0
$y = x - 2$	−1	0	**1**	2

Types of Systems of Linear Equations A system of linear equations can have no solutions, one solution, or infinitely many solutions.

No Solutions	**One Solution**	**Infinitely Many Solutions**
Inconsistent System	Consistent System Independent Equations	Consistent System Dependent Equations

Method of Substitution This method can be used to solve a system of equations symbolically and always gives the exact solution, provided one exists.

Example: $-2x + y = -3$

$x + y = 3$

STEP 1: Solve one of the equations for a convenient variable.

$$x + y = 3 \quad \text{becomes} \quad y = 3 - x.$$

STEP 2: Substitute this result in the other equation and then solve.

$$-2x + (3 - x) = -3 \quad \text{Substitute } (3 - x) \text{ for } y.$$
$$-3x = -6 \quad \text{Combine like terms; subtract 3.}$$
$$x = 2 \quad \text{Divide each side by } -3.$$

STEP 3: Find the value of the other variable. Because $y = 3 - x$ and $x = 2$, it follows that $y = 3 - 2 = 1$.

STEP 4: Check to determine that $(2, 1)$ is the solution.

$$-2(2) + (1) \stackrel{?}{=} -3 \; \checkmark \quad \text{A true statement}$$
$$2 + 1 \stackrel{?}{=} 3 \; \checkmark \quad \text{A true statement}$$

The solution $(2, 1)$ checks.

Recognizing Types of Systems

No solutions	The final equation is always false, such as $0 = 1$.
One solution	The final equation has one solution, such as $x = 1$.
Infinitely many solutions	The final equation is always true, such as $0 = 0$.

SECTION 4.3 ■ SOLVING SYSTEMS OF LINEAR EQUATIONS BY ELIMINATION

Method of Elimination This method can be used to solve a system of linear equations symbolically and always gives the exact solution, provided one exists.

Example:
$$\begin{aligned} x + 3y &= 1 \\ -x + y &= 3 \\ \hline 4y &= 4 \quad \text{or} \quad y = 1 \quad \text{Add and solve for } y. \end{aligned}$$

Substitute $y = 1$ in either of the given equations: $x + 3(1) = 1$ implies that $x = -2$, so $(-2, 1)$ is the solution.

NOTE: To eliminate a variable, it may be necessary to multiply one or both equations by a constant before adding. ■

Recognizing Types of Systems

No solutions	The final equation is always false, such as $0 = 1$.
One solution	The final equation has one solution, such as $x = 1$.
Infinitely many solutions	The final equation is always true, such as $0 = 0$.

SECTION 4.4 ■ SYSTEMS OF LINEAR INEQUALITIES

Graphing a Linear Inequality in Two Variables
1. Replace the inequality symbol with an equals sign and graph the resulting line. If the inequality is $<$ or $>$, use a dashed line; if it is $\leq$ or $\geq$, use a solid line.
2. Pick a *test point* that does *not* lie on the line. Substitute this point in the given inequality. Determine whether the resulting statement is true or false.
3. If the statement is true, shade the region containing the test point. If the statement is false, shade the region that does not contain the test point.

Solving a System of Linear Inequalities in Two Variables
1. For each inequality in the system, perform Step 1 from Graphing a Linear Inequality in Two Variables (above).
2. Pick a test point from one region and substitute it in the given inequalities.
3. If both of the resulting statements are true, shade the region containing the test point. If not, pick a test point from a different region and substitute it in the given inequalities. Repeat this step until the region to be shaded is found.

Example: $\quad x \le 1$

$$x + y \le 2$$

Graph the lines $x = 1$ and $x + y = 2$. Then pick a test point, such as $(0, 0)$, and substitute it in each inequality.

$$0 \le 1 \ \checkmark \quad \text{A true statement}$$

$$0 + 0 \le 2 \ \checkmark \quad \text{A true statement}$$

Because $(0, 0)$ satisfies *both* inequalities, shade the region containing $(0, 0)$. See the graph.

NOTE: When shading the solution set to a *system* of inequalities, you may need to try more than one test point. ■

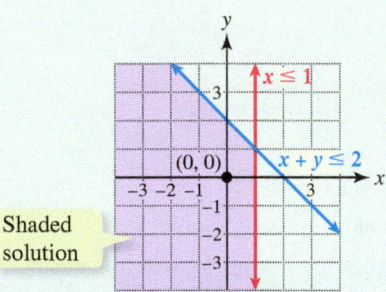

Shaded solution

NOTE: An alternative way to determine the solution set is to shade the region that is both to the **left** of the line $x = 1$ (because $x \le 1$) and **below** the line $y = -x + 2$ (because $y \le -x + 2$). ■

CHAPTER 4 Review Exercises

SECTION 4.1

Exercises 1 and 2: Determine graphically the x-value for the equation when $y = 3$.

1. $y = 2x - 3$ **2.** $y = \frac{3}{2}x$

Exercises 3–6: The graphs of two equations are shown.

  (a) *State the number of solutions to the system of equations.*

  (b) *Is the system consistent or inconsistent? If the system is consistent, state whether the equations are dependent or independent.*

3.

4.

5.

6.

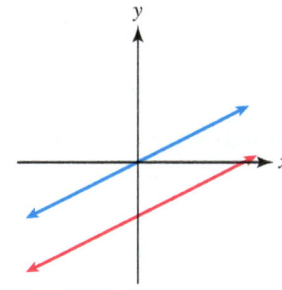

Exercises 7–10: Determine which ordered pair is a solution to the system of equations.

7. $(0, 1), (1, 2)$
$$x + 2y = 5$$
$$x - y = -1$$

8. $(5, 2), (4, 0)$
$$2x - y = 8$$
$$x + 3y = 11$$

9. $(2, 2), (4, 3)$
$$\tfrac{1}{2}x = y - 1$$
$$2x = 3y - 1$$

10. $(2, -4), (-1, 2)$
$$5x - 2y = 18$$
$$y = -2x$$

Exercises 11 and 12: The graphs for two equations are shown. Use the intersection-of-graphs method to identify the solution to both equations. Then check your result.

11.

12.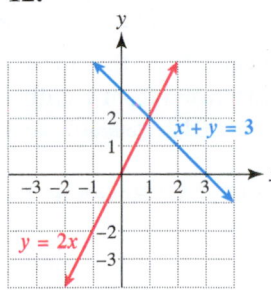

Exercises 13 and 14: A table for two equations is given. Identify the solution to both equations.

13.

x	1	2	3	4
$y = 3x$	3	6	9	12
$y = 6$	6	6	6	6

14.

x	-1	0	1	2
$y = 2x - 1$	-3	-1	1	3
$y = 2 - x$	3	2	1	0

Exercises 15–20: Solve the system of linear equations graphically.

15.
$$y = -3$$
$$x + y = 1$$

16.
$$x = 1$$
$$x - y = -1$$

17.
$$2x + y = 3$$
$$-x + y = 0$$

18.
$$y = 2x$$
$$2x + y = 4$$

19.
$$x + 2y = 3$$
$$2x + y = 3$$

20.
$$-3x - y = 7$$
$$2x + 3y = -7$$

SECTION 4.2

Exercises 21–28: Use the method of substitution to solve the system of linear equations. These systems may have no solutions, one solution, or infinitely many solutions.

21.
$$x + y = 8$$
$$y = 3x$$

22.
$$x - 2y = 22$$
$$y = -5x$$

23.
$$x + 3y = 1$$
$$-2x + 2y = 6$$

24.
$$3x - 2y = -4$$
$$2x - y = -4$$

25.
$$x + y = 2$$
$$y = -x$$

26.
$$x + y = -2$$
$$x + y = 3$$

27.
$$-x + 2y = 2$$
$$x - 2y = -2$$

28.
$$-x - y = -2$$
$$2x - y = 1$$

SECTION 4.3

Exercises 29 and 30: Use the graph to solve the system of equations. Then use the elimination method to verify your answer.

29.
$$x + y = 3$$
$$x - y = 1$$

30.
$$2x + 3y = 4$$
$$x - 2y = -5$$

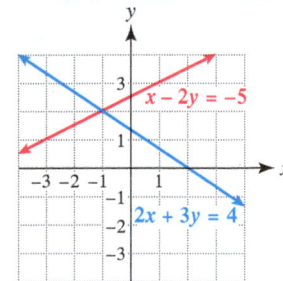

Exercises 31–38: Use the elimination method to solve the system of equations.

31.
$$x + y = 10$$
$$x - y = 12$$

32.
$$2x - y = 2$$
$$3x + y = 3$$

33.
$$-2x + 2y = -1$$
$$x - 3y = -3$$

34.
$$2x - 5y = 0$$
$$2x + 4y = 9$$

35.
$$2a + b = 3$$
$$-3a - 2b = -1$$

36.
$$a - 3b = 2$$
$$3a + b = 26$$

37.
$$5r + 3t = -1$$
$$-2r - 5t = -11$$

38.
$$5r + 2t = 5$$
$$3r - 7t = 3$$

Exercises 39 and 40: Solve the system of equations (a) symbolically, (b) graphically, and (c) numerically.

39.
$$3x + y = 6$$
$$x - y = -2$$

40.
$$2x + y = 3$$
$$-x + 2y = -4$$

Exercises 41–44: Use elimination to determine whether the system of equations has no solutions, one solution, or infinitely many solutions.

41.
$$x - y = 5$$
$$-x + y = -5$$

42.
$$3x - 3y = 0$$
$$-x + y = 0$$

43.
$$-2x + y = 3$$
$$2x - y = 3$$

44.
$$-2x + y = 2$$
$$3x - y = 3$$

SECTION 4.4

Exercises 45–48: Determine whether the test point is a solution to the linear inequality.

45. $(5, -3)$ $y \le 2$

46. $(-1, 3)$ $x > -1$

47. $(1, 2)$ $x + y < -2$

48. $(1, -4)$ $2x - 3y \ge 2$

Exercises 49 and 50: Write a linear inequality that describes the shaded region.

49.

50.

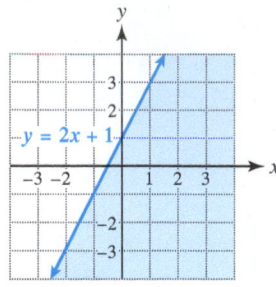

Exercises 51–56: Shade the solution set for the inequality.

51. $x \le 1$

52. $y > 2$

53. $y > 3x$

54. $x \ge 2y$

55. $y < x + 1$

56. $2x + y \ge -2$

Exercises 57 and 58: Determine whether the test point is a solution to the system of linear inequalities.

57. $(1, -2)$
$$x - 2y > 3$$
$$2x + y < 3$$

58. $(4, -3)$
$$x - y \ge 1$$
$$4x + 3y \le 4$$

Exercises 59 and 60: The graphs of two equations are shown with four test points labeled. Use these points to decide which region should be shaded to solve the system of inequalities.

59. $y \le 1$
$$2x + y \ge -1$$

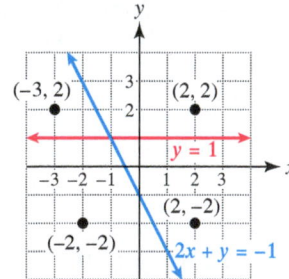

60. $y \ge x$
$$x + y \ge 2$$

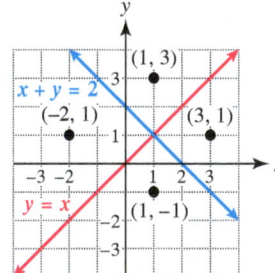

Exercises 61–66: Shade the solution set for the system of inequalities.

61. $x > -1$
$y < -2$

62. $y \le x$
$y \ge -2x$

63. $x + y \le 3$
$\quad y \ge -x$

64. $2x + y < 3$
$\qquad y > x$

65. $\frac{1}{2}x + y \ge 2$
$\quad x - 2y \le 0$

66. $2x - y < 3$
$4x + 2y > -6$

APPLICATIONS

67. Traffic Fatalities The number of traffic fatalities increased by a factor of 10.9 from 1912 to 2012. There were 30,690 more deaths in 2012 than in 1912. Find the number of traffic fatalities in each of the two years. Note that the number of vehicles on the road increased from 1 million to 253 million between 1912 and 2012. (*Source: Department of Transportation*)

68. Social Ad Revenue By 2017, U.S. total social ad revenue reached $11.8 billion. Social display ad revenue was $1.8 billion more than social native ad revenue. What was the revenue for social display ads and for social native ads?

69. Renting a Car A rental car costs $40 plus $0.20 per mile that it is driven.
   (a) Write an equation that gives the cost C of driving the car x miles.
   (b) Use the intersection-of-graphs method to determine the number of miles that the car is driven if the rental cost is $90.
   (c) Solve part (b) numerically with a table of values.

70. Supplementary Angles The smaller of two angles that sum to $180°$ is $30°$ less than the measure of the larger angle. Find each angle.

71. Triangle In a triangle, the measures of the two smaller angles are equal and their sum is $40°$ more than the larger angle.
   (a) Let x be the measure of each of the two smaller angles and y be the measure of the larger angle. Write a system of linear equations whose solution gives the measures of these angles.
   (b) Use the method of substitution to solve the system.
   (c) Use the method of elimination to solve the system.

72. Garden Dimensions A rectangular garden has 88 feet of fencing around it. The garden is 4 feet longer than it is wide. Find the dimensions of the garden.

73. Room Prices Ten rooms are rented at rates of $80 and $120 per night. The total collected for the 10 rooms is $920.
   (a) Write a system of linear equations whose solution gives the number of each type of room rented. Be sure to state what each variable represents.
   (b) Solve the system of equations.

74. Mixture Problem One type of candy sells for $2 per pound, and another type sells for $3 per pound. An order for 18 pounds of candy costs $47. How much of each type of candy was bought?

75. Burning Calories An athlete burns 9 calories per minute on a stationary bicycle and 11 calories per minute on a stair climber. In a 60-minute workout, the athlete burns 590 calories. How many minutes does the athlete spend on each type of exercise equipment? (*Source:* Runner's World.)

76. River Current A riverboat travels 140 miles downstream in 10 hours, and the return trip takes 14 hours. What is the speed of the current?

77. Garage Dimensions The blue graph shown in the figure gives possible dimensions for a rectangular garage with perimeter 80 feet. The red graph shows possible dimensions for a garage that has width W two-thirds of its length L.

Garage Dimensions

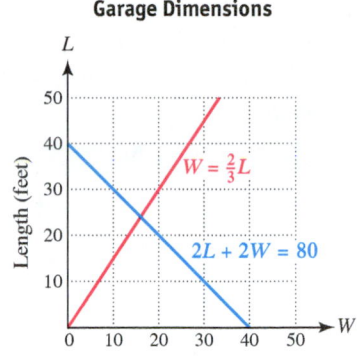

(a) Use the graph to estimate the dimensions of a garage with perimeter 80 feet and width two-thirds its length.

(b) Solve this problem symbolically.

78. Wheels and Trailers A business manufactures at least two wheels for each trailer it makes. The total number of trailers and wheels manufactured cannot exceed 30 per week. Shade the region that represents numbers of wheels W and trailers T that can be produced each week within these restrictions. Label the horizontal axis W and the vertical axis T.

CHAPTER 4   **Test**   MyMathLab® You Tube

1. Determine which ordered pair is a solution to the system of equations.

$$(3, -1), (1, 2)$$
$$3x + 2y = 7$$
$$2x - y = 0$$

2. The graphs for two equations are shown. Use the intersection-of-graphs method to identify the solution to both equations. Then check your solution.

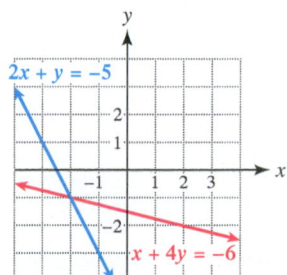

3. A table for two equations is given. Identify the solution to both equations.

x	-2	-1	0	1
$y = 2x$	-4	-2	0	2
$y = 3x + 1$	-5	-2	1	4

4. Solve the system of equations graphically.

$$x + 2y = 4$$
$$x + y = 1$$

5. Use the method of substitution to solve the system of linear equations.

$$3x + 2y = 9$$
$$y = 3x$$

6. Use the method of substitution to solve the system of linear equations. How many solutions are there? Is the system consistent or inconsistent?

(a) $x + 3y = 5$
$3x - 2y = 4$

(b) $-x + \frac{1}{2}y = 12$
$2x - y = -4$

Exercises 7 and 8: The graphs of two equations are shown.

(a) State the number of solutions to the system of equations.

(b) Is the system consistent or inconsistent? If the system is consistent, state whether the equations are dependent or independent.

7. **8.**

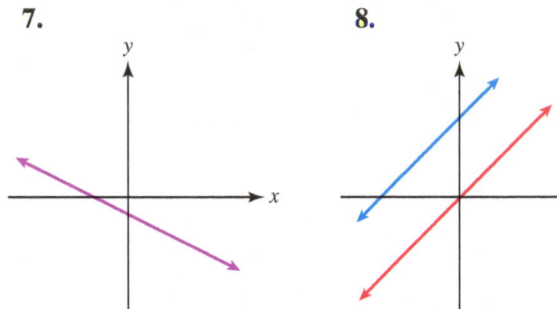

Exercises 9–12: Use the elimination method to solve the system of equations. Note that these systems may have no solutions, one solution, or infinitely many solutions.

9. $x + 2y = 5$
$3x - 2y = -17$

10. $2x - 2y = 3$
$-x + y = 5$

11. $x - 2y = 3$
$-3x + 6y = -9$

12. $4x + 3y = 5$
$3x - 2y = -9$

13. Write a linear inequality that describes the shaded region.

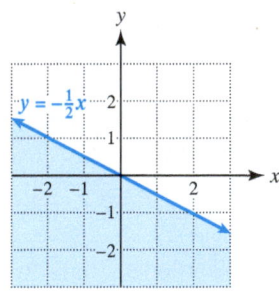

14. Determine whether the test point $(4, -3)$ is a solution to the system of linear inequalities.

$$2x + y > 3$$
$$x - y \geq 7$$

Exercises 15 and 16: Shade the solution set for the given inequality.

15. $x \leq 4$

16. $x + y > 2$

17. Determine whether region **1**, **2**, **3**, or **4** should be shaded to solve the system of inequalities.

$$x + y \leq 5$$
$$x - y \geq 1$$

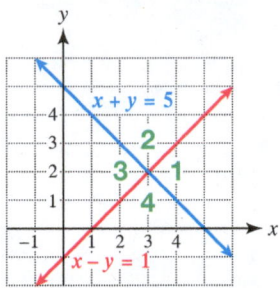

Exercises 18 and 19: Shade the solution set for the given system of inequalities.

18. $x > 2$
$y < 2x$

19. $2x + y \leq 3$
$x - y \geq 0$

20. Music Subscriptions The combined number of paid music subscriptions in 2012 and 2013 was 9.5 million. There were 2.7 million more subscriptions in 2013 than in 2012. How many paid music subscriptions were there in each year?

21. Mixture Problem A chemist has 20% and 60% solutions of acid available. How many liters of each solution should be mixed to obtain 10 liters of a 30% acid solution?

22. Jogging Speed An athlete jogs at 6 miles per hour and at 9 miles per hour for a total time of 1 hour, covering a distance of 7 miles. How long does the athlete jog at each speed?

CHAPTERS 1–4 Cumulative Review Exercises

1. Write 120 as a product of prime numbers.

2. Evaluate each expression by hand.
   (a) $2^3 \div \frac{5 + 7}{9 - 3}$
   (b) $-\frac{2}{5} \cdot (5 - 25)$

3. Classify each number as rational or irrational.
   (a) -6.9 (b) $\sqrt{14}$

4. Insert $>$ or $<$ to make each statement true.
   (a) -5 _____ $|-5|$ (b) $|7|$ _____ $|-1|$

5. Simplify each expression.
   (a) $5x^2 - x^2$ (b) $3 - 2x + 7x - 5$

Exercises 6 and 7: Solve the equation.

6. $5(2x + 1) = 7 + x$

7. $1 - (x + 1) = x - 1$

8. Determine whether the linear equation given by $2(5x + 1) = 10x - 3$ has no solutions, one solution, or infinitely many solutions.

9. Find four consecutive integers whose sum is 50.

10. Find the area of a rectangle that has a 36-inch length and a 1-foot width.

11. Solve the formula $W = 3x - 7y$ for x.

12. Solve the inequality $3 - (2x - 7) \le 8x$.

13. Use the table of xy-values to make a line graph.

x	-2	-1	0	1	2
y	-2	2	0	3	-1

14. Graph each equation.
    (a) $y = -2x + 2$ (b) $3y + 2x = 6$

15. Find the x- and y-intercepts for the graph of the equation $4x - y = 8$.

16. Sketch a line with the given slope that passes through the given point.
    (a) $m = -2$, $(1, 1)$ (b) $m = \frac{4}{3}$, $(-3, 2)$

17. Write the slope–intercept form for each line.
    (a)   (b)

     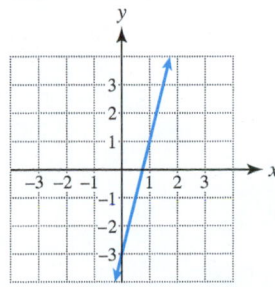

18. Find the slope–intercept form of the line passing through $(4, 2)$ and $(-4, 4)$.

19. Write the slope–intercept form of the line perpendicular to $2x - 6y = 7$, passing through $(1, 1)$.

20. Determine whether $(3, 1)$ or $(4, 4)$ is a solution to the system of equations.
$$3x - y = 8$$
$$2x + y = 12$$

Exercises 21–24: Solve the system of equations. Note that these systems may have no solutions, one solution, or infinitely many solutions.

21. $x - y = 4$ 22. $2x + 3y = 4$
       $-2x - y = 1$ $-4x - 6y = 7$

23. $3x - 4y = 8$ 24. $7x + 2y = -3$
      $-15x + 20y = -40$ $-5x - 3y = -1$

Exercises 25 and 26: Shade the solution set for the system of inequalities.

25. $y < x + 1$ 26. $3x + y \ge 6$
       $x + y \le 3$ $x - 3y \le 3$

27. **Rods to Feet** There are 16.5 feet in 1 rod. Write a formula that converts R rods to F feet.

28. **Temperature Change** Find the temperature change for a package of frozen carrots that is removed from a freezer at $-11\,°C$ and placed in water at $83\,°C$.

29. **Cost of a Digital Camera** A 7% sales tax on a digital camera amounts to $17.15. Find the cost of the digital camera.

30. **Tuition Increase** If tuition is currently $145 per credit and it is going to be increased by 9%, what will the new tuition per credit be?

31. **Gasoline Consumption** Write an equation in slope–intercept form that models the number of gallons G of gas in a truck's tank after x hours if the tank initially contains 30 gallons of gas and the truck uses 3 gallons every hour.

32. **Bank Loans** An individual has two loans totaling $2400. One loan charges 5% interest and the other charges 6% interest. If the interest for one year is $132, how much money is borrowed at each rate?

5 Polynomials and Exponents

Digital images were first sent between New York and London by cable in the early 1920s. Unfortunately, the transmission time was 3 hours and the quality was poor. Digital photography was developed further by NASA in the 1960s because ordinary pictures were subject to interference when transmitted through space. Today, digital pictures remain crystal clear even if they travel millions of miles.

Whether they are taken with a webcam, with a smartphone, or by a Mars rover, digital images consist of tiny units called pixels, which are represented by numbers. As a result, mathematics plays an important role in digital images. In this chapter, we illustrate some of the ways mathematics is used to describe digital pictures (see Example 4 and Exercise 80 in the Special Products section of this chapter). We also discuss how mathematics is used to model things such as heart rate, computer sales, motion of the planets, and interest on money. *Source:* NASA.

5.1 Rules for Exponents

Objectives

1. Reviewing Bases and Exponents
2. Computing Zero Exponents
3. Applying the Product Rule
4. Applying Power Rules

1 Reviewing Bases and Exponents

The expression 5^3 is an exponential expression with *base* **5** and *exponent* **3**. Its value is

$$5 \cdot 5 \cdot 5 = 125.$$

In general, b^n is an exponential expression with base b and exponent n. If n is a natural number, it indicates the number of times the base b is to be multiplied with itself.

$$\text{Exponent} \searrow$$
$$b^n = \underbrace{b \cdot b \cdot b \cdot \cdots \cdot b}_{n \text{ factors of } b}$$
$$\text{Base} \nearrow$$

TABLE 5.1 contains several examples of exponential expressions.

Exponential Expressions

Equal Expressions	Base	Exponent
$2 \cdot 2 \cdot 2 = 2^3$	2	3
$6 \cdot 6 \cdot 6 \cdot 6 = 6^4$	6	4
$7 = 7^1$	7	1
$0.5 \cdot 0.5 = 0.5^2$	0.5	2
$x \cdot x \cdot x = x^3$	x	3

TABLE 5.1

STUDY TIP

Exponents occur throughout mathematics. Because exponents are so important, this section is essential for your success in mathematics. It takes practice, so set aside some extra time.

EVALUATING EXPRESSIONS

When evaluating expressions, use the following order of operations. First perform all calculations within parentheses and absolute values, or above and below a fraction bar.

1. Evaluate exponents.
2. Perform negation.
3. Do multiplication and division from left to right.
4. Do addition and subtraction from left to right.

NOTE: Evaluate exponents *before* performing negation. See parts (c) and (d) in the next example. ∎

EXAMPLE 1 **Evaluating exponential expressions**

Evaluate each expression.

(a) $1 + \dfrac{2^4}{4}$ **(b)** $3\left(\dfrac{1}{3}\right)^2$ **(c)** -2^4 **(d)** $(-2)^4$

Solution

(a) Evaluate the exponent first.

$$1 + \frac{2^4}{4} = 1 + \frac{\overbrace{2 \cdot 2 \cdot 2 \cdot 2}^{\text{4 factors}}}{4} = 1 + \frac{16}{4} = 1 + 4 = 5$$

2 factors

(b) $3\left(\dfrac{1}{3}\right)^2 = 3\left(\dfrac{1}{3} \cdot \dfrac{1}{3}\right) = 3 \cdot \dfrac{1}{9} = \dfrac{3}{9} = \dfrac{1}{3}$

(c) Because exponents are evaluated before negation is performed,

Base is 2.   **4 factors**

$$-2^4 = -(2 \cdot 2 \cdot 2 \cdot 2) = -16.$$

Base is -2.   **4 factors**

(d) $(-2)^4 = (-2)(-2)(-2)(-2) = 16$

Now Try Exercises 9, 11, 13, 15

READING CHECK 1

• Simplify -4^2 and $(-4)^2$.

NOTE: Parts (c) and (d) of Example 1 appear to be very similar. However, the negation sign is inside the parentheses in part (d), which means that the base for the exponential expression is -2. In part (c), no parentheses are used, indicating that the base of the exponential expression is 2. ∎

TECHNOLOGY NOTE

Evaluating Exponents
Exponents can often be evaluated on calculators by using the ∧ key. The four expressions from Example 1 are evaluated with a calculator and the results are shown in the following two figures. When evaluating the last two expressions on your calculator, remember to use the negation key rather than the subtraction key.

CALCULATOR HELP

To evaluate exponents, see Appendix A (page AP-1).

```
1+2^4/4
              5
3(1/3)²▶Frac
            1/3
```

```
-2^4
            -16
(-2)^4
             16
```

2 Computing Zero Exponents

So far we have discussed natural number exponents. What if an exponent is 0? What does 2^0 equal? To answer these questions, consider **TABLE 5.2**, which shows values for decreasing powers of 2. Note that each time the power of 2 decreases by 1, the resulting value is divided by 2. For this pattern to continue, we need to define 2^0 to be 1 because dividing 2 by 2 results in 1.

Powers of 2

Power of 2	Value
2^3	8
2^2	4
2^1	2
2^0	?

Decrease exponent by 1 (from 2^3 to 2^2) — Divide by 2
Decrease exponent by 1 (from 2^2 to 2^1) — Divide by 2
Decrease exponent by 1 (from 2^1 to 2^0) — Divide by 2

TABLE 5.2

This discussion suggests that $2^0 = 1$, and is generalized as follows.

ZERO EXPONENT

For any nonzero real number b,

$$b^0 = 1.$$

The expression 0^0 is undefined.

EXAMPLE 2 **Evaluating zero exponents**

Evaluate each expression. Assume that all variables represent nonzero numbers.

(a) 7^0 (b) $3\left(\dfrac{4}{9}\right)^0$ (c) $\left(\dfrac{x^2y^5}{3z}\right)^0$

Solution
(a) Since 7 is a nonzero base, $7^0 = 1$.
(b) $3\left(\dfrac{4}{9}\right)^0 = 3(1) = 3$. (Note that the exponent 0 does not apply to 3.)
(c) All variables are nonzero, so the expression inside the parentheses is also nonzero.
Thus $\left(\dfrac{x^2y^5}{3z}\right)^0 = 1$.

Now Try Exercises 17, 19, 21

---

3 Applying the Product Rule

We can use a special rule to calculate products of exponential expressions *provided their bases are the same*. For example,

$$4^3 \cdot 4^2 = \underbrace{(4 \cdot 4 \cdot 4) \cdot (4 \cdot 4)}_{\text{5 factors of 4}} = 4^5.$$

The expression $4^3 \cdot 4^2$ has $3 + 2 = 5$ factors of 4, so the result is $4^{3+2} = 4^5$. To multiply exponential expressions with the *same base*, add exponents and keep the base.

READING CHECK 2

• Use the product rule to simplify the expression $3^2 \cdot 3^3$.

THE PRODUCT RULE

For any real number a and natural numbers m and n,

$$a^m \cdot a^n = a^{m+n}.$$

NOTE: The product $2^4 \cdot 3^5$ cannot be simplified by using the product rule because the exponential expressions have different bases: **2** and **3**. ∎

EXAMPLE 3 **Using the product rule**

Multiply and simplify.
(a) $2^3 \cdot 2^2$ (b) x^4x^5 (c) $2x^2 \cdot 5x^6$ (d) $x^3(2x + 3x^2)$

> Product rule: Add exponents.

Solution
(a) $2^3 \cdot 2^2 = 2^{3+2} = 2^5 = 32$ (b) $x^4x^5 = x^{4+5} = x^9$

(c) Begin by applying the commutative property of multiplication to write the product in a more convenient order.

$$2x^2 \cdot 5x^6 = 2 \cdot 5 \cdot x^2 \cdot x^6 = 10x^{2+6} = 10x^8$$

(d) To simplify this expression, first apply the distributive property.

$$x^3(2x + 3x^2) = x^3 \cdot 2x + x^3 \cdot 3x^2 = 2x^4 + 3x^5$$

Exponent is 1.

Now Try Exercises 25, 27, 31, 75

NOTE: If an exponent does not appear in an expression, it is assumed to be 1. For example, x can be written as x^1 and $(x + y)$ can be written as $(x + y)^1$. ∎

EXAMPLE 4 **Applying the product rule**

Multiply and simplify.
(a) $x \cdot x^3$ **(b)** $(a + b)(a + b)^4$

Solution
(a) Begin by writing x as x^1. Then $x^1 \cdot x^3 = x^{1+3} = x^4$.
(b) First write $(a + b)$ as $(a + b)^1$. Then

$$(a + b)^1 \cdot (a + b)^4 = (a + b)^{1+4} = (a + b)^5.$$

Now Try Exercises 23, 67

---

4 Applying Power Rules

How should $(4^3)^2$ be evaluated? To answer this question, consider how the product rule can be used in evaluating

$$3 + 3 = 3 \cdot 2$$
$$(4^3)^2 = \underbrace{4^3 \cdot 4^3}_{\text{Product rule}} = 4^{3+3} = 4^6.$$

Similarly,

$$5 + 5 + 5 = 5 \cdot 3$$
$$(a^5)^3 = \underbrace{a^5 \cdot a^5 \cdot a^5}_{\text{Product rule}} = a^{5+5+5} = a^{15}.$$

This discussion suggests that to raise a power to a power, we multiply the exponents.

READING CHECK 3

- Use the raising a power to a power rule to simplify the expression $(2^2)^3$.

RAISING A POWER TO A POWER

For any real number a and natural numbers m and n,

$$(a^m)^n = a^{mn}.$$

EXAMPLE 5 **Raising a power to a power**

Simplify each expression.
(a) $(3^2)^4$ **(b)** $(a^3)^2$

> Power rule: Multiply exponents

Solution
(a) $(3^2)^4 = 3^{2 \cdot 4} = 3^8$ **(b)** $(a^3)^2 = a^{3 \cdot 2} = a^6$

Now Try Exercises 35, 37

To decide how to simplify the expression $(2x)^3$, consider

$$(2x)^3 = \underbrace{2x \cdot 2x \cdot 2x}_{\text{3 factors}} = \underbrace{(2 \cdot 2 \cdot 2)}_{\text{3 factors}} \cdot \underbrace{(x \cdot x \cdot x)}_{\text{3 factors}} = 2^3 x^3.$$

To raise a product to a power, we raise each factor to the power.

READING CHECK 4

- Use the raising a product to a power rule to rewrite the expression $(4x)^3$.

RAISING A PRODUCT TO A POWER

For any real numbers a and b and natural number n,

$$(ab)^n = a^n b^n.$$

EXAMPLE 6 **Raising a product to a power**

Simplify each expression.
(a) $(3z)^2$ **(b)** $(-2x^2)^3$ **(c)** $4(x^2 y^3)^5$ **(d)** $(-2^2 a^5)^3$

Solution
(a) $(3z)^2 = 3^2 z^2 = 9z^2$
(b) $(-2x^2)^3 = (-2)^3 (x^2)^3 = -8x^6$
(c) $4(x^2 y^3)^5 = 4(x^2)^5 (y^3)^5 = 4x^{10} y^{15}$
(d) $(-2^2 a^5)^3 = (-4a^5)^3 = (-4)^3 (a^5)^3 = -64a^{15}$

Now Try Exercises 41, 43, 47, 49

The following equation illustrates another power rule.

$$\left(\frac{2}{3}\right)^4 = \underbrace{\frac{2}{3} \cdot \frac{2}{3} \cdot \frac{2}{3} \cdot \frac{2}{3}}_{\text{4 factors}} = \frac{2 \cdot 2 \cdot 2 \cdot 2}{3 \cdot 3 \cdot 3 \cdot 3} = \frac{2^4}{3^4}$$

To raise a quotient to a power, raise both the numerator and the denominator to the power.

READING CHECK 5

- Use the raising a quotient to a power rule to rewrite the expression $\left(\frac{x}{2}\right)^5$.

RAISING A QUOTIENT TO A POWER

For any real numbers a and b and natural number n,

$$\left(\frac{a}{b}\right)^n = \frac{a^n}{b^n}. \quad b \neq 0$$

EXAMPLE 7 **Raising a quotient to a power**

Simplify each expression.

(a) $\left(\dfrac{2}{3}\right)^3$ **(b)** $\left(\dfrac{a}{b}\right)^9$ **(c)** $\left(\dfrac{a+b}{5}\right)^2$

Solution

(a) $\left(\dfrac{2}{3}\right)^3 = \dfrac{2^3}{3^3} = \dfrac{8}{27}$ **(b)** $\left(\dfrac{a}{b}\right)^9 = \dfrac{a^9}{b^9}$

(c) Because the numerator is an expression with more than one term, we must place parentheses around it before raising it to the power 2.

$$\left(\frac{a+b}{5}\right)^2 = \frac{(a+b)^2}{5^2} = \frac{(a+b)^2}{25}$$

Now Try Exercises 51, 53, 57

MAKING CONNECTIONS 1

Raising a Sum or Difference to a Power

Although there are power rules for products and quotients, similar rules for sums and differences do not exist. In general,

$$(a+b)^n \neq a^n + b^n \qquad \text{and} \qquad (a-b)^n \neq a^n - b^n.$$

$(3+4)^2 = 7^2 = 49$, but
$3^2 + 4^2 = 9 + 16 = 25$.

$(4-1)^3 = 3^3 = 27$, but
$4^3 - 1^3 = 64 - 1 = 63$.

The five rules for exponents discussed in this section are summarized as follows.

RULES FOR EXPONENTS

The following rules hold for real numbers a and b, and natural numbers m and n.

Description	Rule	Example
Zero Exponent	$b^0 = 1$, for $b \neq 0$	$(-13)^0 = 1$
The Product Rule	$a^m \cdot a^n = a^{m+n}$	$5^4 \cdot 5^3 = 5^{4+3} = 5^7$
Power to a Power	$(a^m)^n = a^{m \cdot n}$	$(y^2)^5 = y^{2 \cdot 5} = y^{10}$
Product to a Power	$(ab)^n = a^n b^n$	$(pq)^7 = p^7 q^7$
Quotient to a Power	$\left(\dfrac{a}{b}\right)^n = \dfrac{a^n}{b^n}$, for $b \neq 0$	$\left(\dfrac{x}{y}\right)^3 = \dfrac{x^3}{y^3}$, for $y \neq 0$

Simplification of some expressions may require the application of more than one rule of exponents. This is demonstrated in the next example.

EXAMPLE 8 **Combining rules for exponents**

Simplify each expression.

(a) $(2a)^2(3a)^3$ **(b)** $\left(\dfrac{a^2 b^3}{c}\right)^4$ **(c)** $(2x^3 y)^2(-4x^2 y^3)^3$

Solution

(a) $(2a)^2(3a)^3 = 2^2a^2 \cdot 3^3a^3$ Raising a product to a power

$\quad\quad\quad\quad\quad = 4 \cdot 27 \cdot a^2 \cdot a^3$ Evaluate powers; commutative property

$\quad\quad\quad\quad\quad = 108a^5$ Product rule

(b) $\left(\dfrac{a^2b^3}{c}\right)^4 = \dfrac{(a^2)^4(b^3)^4}{c^4}$ Raising a quotient to a power; raising a product to a power

$\quad\quad\quad\quad\quad = \dfrac{a^8b^{12}}{c^4}$ Raising a power to a power

(c) $(2x^3y)^2(-4x^2y^3)^3 = 2^2(x^3)^2y^2(-4)^3(x^2)^3(y^3)^3$ Raising a product to a power

$\quad\quad\quad\quad\quad = 4x^6y^2(-64)x^6y^9$ Raising a power to a power

$\quad\quad\quad\quad\quad = 4(-64)x^6x^6y^2y^9$ Commutative property

$\quad\quad\quad\quad\quad = -256x^{12}y^{11}$ Product rule

Now Try Exercises 59, 61, 65

🌐 *Math in Context* **Financial** Exponents occur frequently in calculations involving yearly percent increases, such as the increase in property value illustrated in the next example.

EXAMPLE 9 **Calculating growth in property value**

If a parcel of property increases in value by about 11% each year for 20 years, then its value will double three times.

(a) Write an exponential expression that represents "doubling three times."

(b) If the property is initially worth $25,000, how much will it be worth after it doubles three times?

Solution

(a) Doubling three times is represented by 2^3.

(b) $2^3(25,000) = 8(25,000) = \$200,000$

Now Try Exercise 89

5.1   Putting It All Together

CONCEPT	EXPLANATION	EXAMPLES
Bases and Exponents	In the expression b^n, b is the base and n is the exponent. If n is a natural number, then $$b^n = \underbrace{b \cdot b \cdot \,\cdots\, \cdot b}_{n \text{ factors}}.$$	2^3 has base 2 and exponent 3. $9^1 = 9$, $3^2 = 3 \cdot 3 = 9$, $4^3 = 4 \cdot 4 \cdot 4 = 64$, and $-6^2 = -(6 \cdot 6) = -36$
Zero Exponents	For any nonzero number b, $b^0 = 1$.	$5^0 = 1$, $x^0 = 1$, and $(xy^3)^0 = 1$

continued on next page

continued from previous page

CONCEPT	EXPLANATION	EXAMPLES
The Product Rule	For any real number a and natural numbers m and n, $$a^m \cdot a^n = a^{m+n}.$$	$2^4 \cdot 2^3 = 2^{4+3} = 2^7,$ $x \cdot x^2 \cdot x^6 = x^{1+2+6} = x^9,$ and $(x+1) \cdot (x+1)^2 = (x+1)^3$
Raising a Power to a Power	For any real number a and natural numbers m and n, $$(a^m)^n = a^{mn}.$$	$(2^4)^2 = 2^{4 \cdot 2} = 2^8,$ $(x^2)^5 = x^{2 \cdot 5} = x^{10},$ and $(a^4)^3 = a^{4 \cdot 3} = a^{12}$
Raising a Product to a Power	For any real numbers a and b and natural number n, $$(ab)^n = a^n b^n.$$	$(3x)^3 = 3^3 x^3 = 27x^3,$ $(x^2 y)^4 = (x^2)^4 y^4 = x^8 y^4,$ and $(-xy)^6 = (-x)^6 y^6 = x^6 y^6$
Raising a Quotient to a Power	For any real numbers a and b and natural number n, $$\left(\frac{a}{b}\right)^n = \frac{a^n}{b^n}. \quad b \neq 0$$	$\left(\dfrac{x}{y}\right)^5 = \dfrac{x^5}{y^5}$ and $\left(\dfrac{a^2 b}{d^3}\right)^4 = \dfrac{(a^2)^4 b^4}{(d^3)^4} = \dfrac{a^8 b^4}{d^{12}}$

5.1 Exercises

MyMathLab®

CONCEPTS AND VOCABULARY

1. In the expression b^n, b is the _____ and n is the _____.

2. The expression $b^0 = $ _____ for any nonzero number b.

3. $a^m \cdot a^n = $ _____

4. $(a^m)^n = $ _____

5. $(ab)^n = $ _____

6. $\left(\dfrac{a}{b}\right)^n = $ _____

REVIEWING BASES AND EXPONENTS

Exercises 7–16: Evaluate the expression.

7. 8^2

8. 4^3

9. $(-2)^3$

10. $(-3)^4$

11. -2^3

12. -3^4

13. $3 + \dfrac{4^2}{2}$

14. $6 - \left(\dfrac{-4}{2}\right)^2$

15. $4\left(\dfrac{1}{2}\right)^3$

16. $16\left(\dfrac{1}{4}\right)^2$

COMPUTING ZERO EXPONENTS

Exercises 17–22: Simplify the expression. Assume that all variables represent nonzero numbers.

17. 6^0

18. $(-0.5)^0$

19. $5(-4)^0$

20. $-9(45)^0$

21. $\left(\dfrac{xy^3}{z^2}\right)^0$

22. $(x^2 y^3)^0$

APPLYING THE PRODUCT RULE

Exercises 23–34: Simplify the expression. Assume that all variables represent nonzero numbers.

23. $3 \cdot 3^2$

24. $5^3 \cdot 5^3$

25. $4^2 \cdot 4^6$

26. $10^4 \cdot 10^3$

27. $x^3 x^6$

28. $a^5 a^2$

29. $x^2 x^2 x^2$

30. $y^7 y^3 y^0$

31. $4x^2 \cdot 5x^5$

32. $-2y^6 \cdot 5y^2$

33. $3(-xy^3)(x^2 y)$

34. $(a^2 b^3)(-ab^2)$

APPLYING POWER RULES

Exercises 35–58: Simplify the expression. Assume that all variables represent nonzero numbers.

35. $(2^3)^2$

36. $(10^3)^4$

37. $(n^3)^4$

38. $(z^7)^3$

39. $x(x^3)^2$

40. $(z^3)^2(5z^5)$

41. $(-7b)^2$

42. $(-4z)^3$

43. $(ab)^3$

44. $(xy)^8$

45. $(2x^2)^0$

46. $(3a^2)^4$

47. $(-4b^2)^3$

48. $(-3r^4t^3)^2$

49. $(x^2y^3)^7$

50. $(rt^2)^5$

51. $\left(\dfrac{1}{3}\right)^3$

52. $\left(\dfrac{5}{2}\right)^2$

53. $\left(\dfrac{a}{b}\right)^5$

54. $\left(\dfrac{x}{2}\right)^4$

55. $\left(\dfrac{x-y}{3}\right)^3$

56. $\left(\dfrac{4}{x+y}\right)^2$

57. $\left(\dfrac{5}{a+b}\right)^2$

58. $\left(\dfrac{a-b}{2}\right)^3$

COMBINING RULES

Exercises 59–78: Simplify the expression. Assume that all variables represent nonzero numbers.

59. $(y^3)^2(x^4y)^3$

60. $(ab^3)^2(ab)^3$

61. $(a^2b)^2(a^2b^2)^3$

62. $(x^3y)(x^2y^4)^2$

63. $\left(\dfrac{2x}{5}\right)^3$

64. $\left(\dfrac{3y}{2}\right)^4$

65. $\left(\dfrac{3x^2}{5y^4}\right)^3$

66. $\left(\dfrac{a^2b^3}{3}\right)^5$

67. $(x+y)(x+y)^3$

68. $(a-b)^2(a-b)$

69. $(a+b)^2(a+b)^3$

70. $(x-y)^5(x-y)^4$

71. $6(x^4y^6)^0$

72. $\left(\dfrac{xy}{z^2}\right)^0$

73. $a(a^2+2b^2)$

74. $x^3(3x-5y^4)$

75. $3a^3(4a^2+2b)$

76. $2x^2(5-4y^3)$

77. $(x-y)(x^2y^3)$

78. $(r+t)(rt)$

79. Thinking Generally Students sometimes mistakenly apply the "rule" $a^m \cdot b^n \overset{?}{=} (ab)^{m+n}$. In general, this equation is *not true*. Find values for a, b, m, and n with $a \neq b$ and $m \neq n$ that will make this equation true.

80. Thinking Generally Students sometimes mistakenly apply the "rule" $(a+b)^n \overset{?}{=} a^n + b^n$. In general, this equation is *not true*. Find values for a, b, and n with $a \neq b$ that will make this equation true.

APPLICATIONS INVOLVING EXPONENTS

Exercises 81–84: Write a simplified expression for the area of each figure at the top of the next column.

81.

82.

83.

84.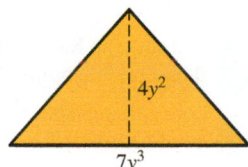

Exercises 85 and 86: Write a simplified expression for the volume of the given figure.

85.

86.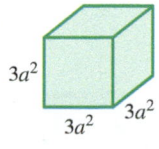

87. Compound Interest If P dollars are deposited in an account that pays 5% annual interest, then the amount of money in the account after 3 years is $P(1+0.05)^3$. Find the amount when $P = \$1000$.

88. Compound Interest If P dollars are deposited in an account that pays 9% annual interest, then the amount of money in the account after 4 years is $P(1+0.09)^4$. Find the amount when $P = \$500$.

89. Investment Growth If an investment increases in value by about 10% each year for 22 years, then its value will triple two times.
   (a) Write an exponential expression that represents "tripling two times."
   (b) If the investment has an initial value of $8000, how much will it be worth if it triples two times?

90. Stock Value If a stock decreases in value by about 23% each year for 9 years, then its value will be halved three times.
   (a) Write an exponential expression that represents "halved three times."
   (b) If the stock is initially worth $88 per share, how much will it be worth if it is halved three times?

WRITING ABOUT MATHEMATICS

91. Are the expressions $(4x)^2$ and $4x^2$ equal in value? Explain your answer.

92. Are the expressions $3^3 \cdot 2^3$ and 6^6 equal in value? Explain your answer.

5.2   Addition and Subtraction of Polynomials

1   Introducing Nonlinear Data and Polynomials

Connecting Concepts with Your Life If you have ever exercised strenuously and then taken your pulse immediately afterward, you may have discovered that your pulse slowed quickly at first and then gradually leveled off. A typical scatterplot of this phenomenon is shown in **FIGURE 5.1(a)**. These data points cannot be modeled accurately with a line because the data are nonlinear. A new expression, called a *polynomial*, is needed to model them. A graph of a polynomial that models these data is shown in **FIGURE 5.1(b)** and discussed in Exercise 75. (*Source:* V. Thomas, *Science and Sport.*)

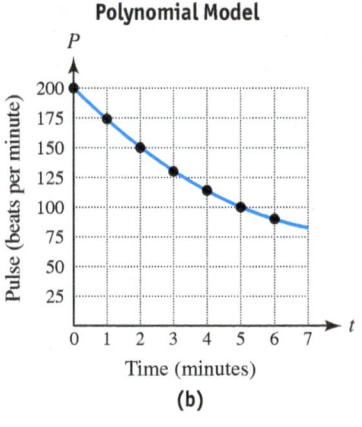

(a)   (b)

FIGURE 5.1

NEW VOCABULARY

☐ Monomial
☐ Degree of a monomial
☐ Coefficient of a monomial
☐ Polynomial
☐ Polynomial in one variable
☐ Binomial
☐ Trinomial
☐ Degree of a polynomial
☐ Like terms

READING CHECK 1

• What is the degree of the monomial $-5x^2y^4$?

2   Recognizing Monomials and Polynomials

A **monomial** is a number, a variable, or a product of numbers and variables raised to natural number powers. Examples of monomials include

$$-3, \quad xy^2, \quad 5a^2, \quad -z^3, \quad \text{and} \quad -\frac{1}{2}xy^3.$$ Monomials

A monomial may contain more than one variable, but monomials do not contain division by variables. For example, the expression $\frac{3}{z}$ is not a monomial. If an expression contains addition or subtraction signs, it is *not* a monomial.

The **degree of a monomial** is the sum of the exponents of the variables. If the monomial has only one variable, its degree is the exponent of that variable. Remember, when a variable does not have a written exponent, the exponent is implied to be 1. A nonzero number has degree 0, and the number 0 has *undefined* degree. The number in a monomial is called the **coefficient of the monomial**. **TABLE 5.3** contains the degree and coefficient of several monomials.

Properties of Monomials

Monomial	-5	$6a^3b$	$-xy$	$7y^3$
Degree	0	4	2	3
Coefficient	-5	6	-1	7

TABLE 5.3

A **polynomial** is a monomial or the sum of two or more monomials. Each monomial is called a *term* of the polynomial. Addition or subtraction signs separate terms. For example, the expression $2x^2 - 3x + 5$ is a **polynomial in one variable** with three terms. Other examples of polynomials in one variable include

$$\underset{\text{1 term}}{-2x}, \quad \underset{\text{2 terms}}{3x + 1}, \quad \underset{\text{3 terms}}{4y^2 - y + 7}, \quad \text{and} \quad \underset{\text{4 terms}}{x^5 - 3x^3 + x - 7.} \qquad \boxed{\text{Polynomials}}$$

These polynomials have 1, 2, 3, and 4 terms, respectively. A polynomial with *two terms* is called a **binomial**, and a polynomial with *three terms* is called a **trinomial**.

A polynomial can have more than one variable, as in

$$x^2y^2, \quad 2xy^2 + 5x^2y - 1, \quad \text{and} \quad a^2 + 2ab + b^2. \qquad \boxed{\text{2-variable polynomials}}$$

Note that all variables in a polynomial are raised to natural number powers. The **degree of a polynomial** is the degree of the term (or monomial) with greatest degree. See **TABLE 5.4**.

Properties of Polynomials

Polynomial	$3x^2 + 7$	$x^3 + 3x^2 + 5x + 9$	$6xy - x^2y^2$
Degree	2	3	4
Number of terms	2	4	2
Number of variables	1	1	2

TABLE 5.4

READING CHECK 2

• What is the degree of the polynomial $-2 + x - x^2 - 4x^5$?

EXAMPLE 1   **Identifying properties of polynomials**

Determine whether the expression is a polynomial. If it is, state how many terms and variables the polynomial contains and give its degree.

(a) $7x^2 - 3x + 1$ **(b)** $5x^3 - 3x^2y^3 + xy^2 - 2y^3$ **(c)** $4x^2 + \dfrac{5}{x + 1}$

Solution
(a) The expression $7x^2 - 3x + 1$ is a polynomial with three terms and one variable. The first term $7x^2$ has degree 2 because the exponent on the variable is 2. The second term $-3x$ has degree 1 because the exponent on the variable is implied to be 1. The third term 1 has degree 0 because it is a nonzero number. The term with greatest degree is $7x^2$, so the polynomial has degree **2**.
(b) The expression $5x^3 - 3x^2y^3 + xy^2 - 2y^3$ is a polynomial with four terms and two variables, x and y. The first term has degree 3 because the exponent on the variable is 3. The second term has degree 5 because the *sum* of the exponents on the variables x and y is 5. Likewise, the third term has degree 3 and the fourth term has degree 3. The term with greatest degree is $-3x^2y^3$, so the polynomial has degree $\mathbf{2} + \mathbf{3} = 5$.
(c) The expression $4x^2 + \frac{5}{x+1}$ is not a polynomial because it contains division by the polynomial $x + 1$.

Now Try Exercises 21, 23, 27

3 Adding Polynomials
Suppose that we have 2 identical rectangles with length l and width w. Then the area of one rectangle is lw and the total area is

$$lw + lw.$$

This area is equivalent to 2 times lw, which can be expressed as $2lw$. See **FIGURE 5.2**.

Adding Monomials: $lw + lw = 2lw$.

FIGURE 5.2

If two monomials contain the same variables raised to the same powers, we call them **like terms**. We can add or subtract (combine) *like* terms but cannot combine *unlike* terms. The terms lw and $2lw$ are like terms and can be combined geometrically, as shown in **FIGURE 5.3**. If we joined one of the small rectangles with area lw and a larger rectangle with area $2lw$, then the total area is $3lw$.

READING CHECK 3

Are $5xy^2$ and $5x^2y$ like terms? Are $-3xy^2$ and $9xy^2$ like terms?

Combining Like Terms: $lw + 2lw = 3lw$.

FIGURE 5.3

STUDY TIP

Do you want to know what material will be covered on your next exam? Often, the best place to look is on previously completed assignments and quizzes.

The *distributive property* justifies combining like terms.

$$1lw + 2lw = (1 + 2)lw = 3lw$$

The rectangles shown in **FIGURE 5.4** have areas of ab and xy. Together, their area is the sum, $ab + xy$. However, because these monomials are unlike terms, they cannot be combined into one term.

Unlike Terms Cannot Be Combined: $ab + xy$

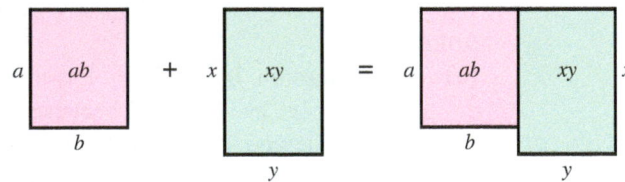

FIGURE 5.4

EXAMPLE 2   **Adding like terms**

State whether each pair of expressions contains like terms or unlike terms. If they are like terms, add them.

(a) $5x^2, -x^2$ **(b)** $7a^2b, 10ab^2$ **(c)** $4rt^2, \frac{1}{2}rt^2$

Solution
(a) The terms $5x^2$ and $-x^2$ have the same variable raised to the same power, so they are like terms. To add like terms, add their coefficients. Note that the coefficient of $-x^2$ is -1.

$$5x^2 + (-x^2) = (5 + (-1))x^2 \quad \text{Distributive property}$$
$$= 4x^2 \quad \text{Add.}$$

(b) The terms $7a^2b$ and $10ab^2$ have the same variables, but these variables are not raised to the same powers. They are unlike terms and cannot be added.

(c) The terms $4rt^2$ and $\frac{1}{2}rt^2$ have the same variables raised to the same powers, so they are like terms. We add them as follows.

$$4rt^2 + \frac{1}{2}rt^2 = \left(4 + \frac{1}{2}\right)rt^2 \qquad \text{Distributive property}$$

$$= \frac{9}{2}rt^2 \qquad\qquad \text{Add.}$$

Now Try Exercises 29, 31, 33

To add two polynomials, combine like terms, as illustrated in the next example.

EXAMPLE 3 **Adding polynomials**

Add by combining like terms.
(a) $(3x + 4) + (-4x + 2)$
(b) $(y^2 - 2y + 1) + (3y^2 + y + 11)$

Solution
(a) $(3x + 4) + (-4x + 2) = 3x + (-4x) + 4 + 2$

$$= (3 - 4)x + (4 + 2)$$

$$= -x + 6$$

(b) $(y^2 - 2y + 1) + (3y^2 + y + 11) = y^2 + 3y^2 - 2y + y + 1 + 11$

$$= (1 + 3)y^2 + (-2 + 1)y + (1 + 11)$$

$$= 4y^2 - y + 12$$

Now Try Exercises 37, 39

Recall that the commutative and associative properties of addition allow us to rearrange a sum in any order. For example, if we write each subtraction in $2x - 5 - 4x + 10$ as addition of the opposite, we have

$$2x - 5 - 4x + 10 = 2x + (-5) + (-4x) + 10,$$

and the terms can be rearranged as

$$2x + (-4x) + (-5) + 10 = 2x - 4x - 5 + 10 = -2x + 5.$$

If we pay attention to the sign in front of each term in a polynomial, the like terms can be combined without rearranging the terms, as demonstrated in the next example.

EXAMPLE 4 **Adding polynomials**

Add $(x^3 - 3x^2 + 7x - 4) + (4x^3 - 5x + 9)$ by combining like terms.

Solution
Remove parentheses and identify like terms with their signs as shown.

$$x^3 - 3x^2 + 7x - 4 + 4x^3 - 5x + 9$$

When like terms (of the same color) are added, the resulting sum is

$$5x^3 - 3x^2 + 2x + 5.$$

Now Try Exercise 41

Polynomials can also be added vertically, as demonstrated in the next example.

EXAMPLE 5 **Adding polynomials vertically**

Simplify $(3x^2 - 3x + 5) + (-x^2 + x - 6)$.

Solution
Write the polynomials in a vertical format and then add each column of like terms.

$$
\begin{array}{r}
3x^2 - 3x + 5 \\
-x^2 + x - 6 \\
\hline
2x^2 - 2x - 1
\end{array}
$$
Add like terms in each column.

Regardless of the method used, the same answer should be obtained. However, adding vertically requires that *like terms be placed in the same column.*

Now Try Exercise 47

4 Subtracting Polynomials

To subtract one integer from another, add the first integer and the *additive inverse* or *opposite* of the second integer. For example, $3 - 5$ is evaluated as follows.

$$3 - 5 = 3 + (-5) \quad \text{Add the opposite.}$$
$$= -2 \quad \text{Simplify.}$$

READING CHECK 4

• How do you subtract one polynomial from another?

Similarly, to subtract one polynomial from another, add the first polynomial and the *opposite* of the second polynomial. To find the opposite of a polynomial, simply negate each term. **TABLE 5.5** lists some polynomials and their opposites.

CRITICAL THINKING 1

What is the result when a polynomial and its opposite are added?

Opposites of Polynomials

Polynomial	Opposite
$2x - 4$	$-2x + 4$
$-x^2 - 2x + 9$	$x^2 + 2x - 9$
$6x^3 - 12$	$-6x^3 + 12$
$-3x^4 - 2x^2 - 8x + 3$	$3x^4 + 2x^2 + 8x - 3$

TABLE 5.5

EXAMPLE 6 **Subtracting polynomials**

Simplify each expression.
(a) $(3x - 4) - (5x + 1)$
(b) $(5x^2 + 2x - 3) - (6x^2 - 7x + 9)$
(c) $(6x^3 + x^2) - (-3x^3 - 9)$

Solution
(a) To subtract $(5x + 1)$ from $(3x - 4)$, we add the binomial $(3x - 4)$ and the opposite of the binomial $(5x + 1)$, or $(-5x - 1)$.

$$(3x - 4) - (5x + 1) = (3x - 4) + (-5x - 1)$$
$$= (3 - 5)x + (-4 - 1)$$
$$= -2x - 5$$

> Add the opposite.

(b) The opposite of $(6x^2 - 7x + 9)$ is $(-6x^2 + 7x - 9)$.

> Add the opposite.

$$(5x^2 + 2x - 3) - (6x^2 - 7x + 9) = (5x^2 + 2x - 3) + (-6x^2 + 7x - 9)$$
$$= (5 - 6)x^2 + (2 + 7)x + (-3 - 9)$$
$$= -x^2 + 9x - 12$$

> Add the opposite.

(c) The opposite of $(-3x^3 - 9)$ is $(3x^3 + 9)$.

$$(6x^3 + x^2) - (-3x^3 - 9) = (6x^3 + x^2) + (3x^3 + 9)$$
$$= (6 + 3)x^3 + x^2 + 9$$
$$= 9x^3 + x^2 + 9$$

Now Try Exercises 57, 59, 61

NOTE: Some students prefer to subtract one polynomial from another by noting that a subtraction sign in front of parentheses changes the signs of all of the terms within the parentheses. For example, part (a) of the previous example could be worked as follows.

$$(3x - 4) - (5x + 1) = 3x - 4 - 5x - 1$$
$$= (3 - 5)x + (-4 - 1)$$
$$= -2x - 5 \ \blacksquare$$

EXAMPLE 7   **Subtracting polynomials vertically**

Simplify $(5x^2 - 2x + 7) - (-3x^2 + 3)$.

Solution
To subtract one polynomial from another vertically, simply add the first polynomial and the opposite of the second polynomial. No x-term occurs in the second polynomial, so insert $0x$.

$$5x^2 - 2x + 7$$
$$\underline{3x^2 + 0x - 3} \quad \text{The opposite of } -3x^2 + 3 \text{ is } 3x^2 - 3 \text{ or } 3x^2 + 0x - 3.$$
$$8x^2 - 2x + 4 \quad \text{Add like terms in each column.}$$

Now Try Exercise 69

5 Evaluating Polynomial Expressions

Frequently, monomials and polynomials represent formulas that may be evaluated. We illustrate such applications in the next two examples.

EXAMPLE 8   **Writing and evaluating a monomial**

Write the monomial that represents the volume of the box having a square bottom, as shown in **FIGURE 5.5**. Find the volume of the box if $x = 3$ feet and $y = 2$ feet.

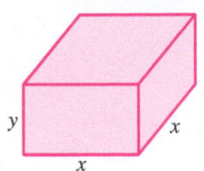

FIGURE 5.5

Solution
The volume of a box is found by multiplying the length, width, and height together. Because the length and width are both x and the height is y, the monomial xxy represents the volume

of the box. This can be written as x^2y. To calculate the volume, let $x = 3$ and $y = 2$ in the monomial x^2y.

$$x^2y = 3^2 \cdot 2 = 9 \cdot 2 = 18 \text{ cubic feet}$$

Now Try Exercise 77

🌐 *Math in Context*   *Business* Polynomials can be used to model trends that occur in business. In the next example, a polynomial is used to model computer sales.

| EXAMPLE 9 | **Modeling sales of personal computers** |

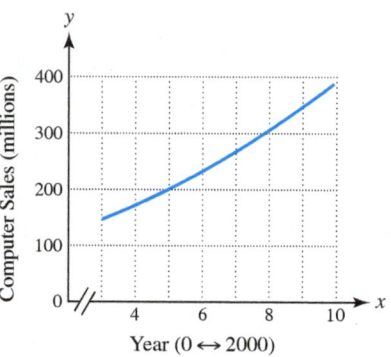

Worldwide Computer Sales

FIGURE 5.6

Worldwide sales of personal computers increased dramatically during the first 10 years of the 21st century, as illustrated in **FIGURE 5.6**. The polynomial

$$0.7868x^2 + 16.72x + 122.58$$

approximates the number of computers sold in millions, where $x = 0$ corresponds to 2000, $x = 1$ to 2001, and so on. Estimate the number of personal computers sold in 2008 by using both the graph and the polynomial. (*Source:* International Data Corporation.)

Solution
From the graph shown in **FIGURE 5.7**, it appears that personal computer sales were slightly more than 300 million, or about 310 million, in 2008.

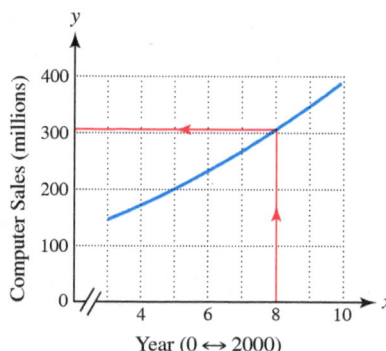

Worldwide Computer Sales

FIGURE 5.7

The year 2008 corresponds to $x = 8$ in the given polynomial, so substitute **8** for x and evaluate the resulting expression.

$$0.7868x^2 + 16.72x + 122.58 = 0.7868(8)^2 + 16.72(8) + 122.58$$
$$\approx 307 \text{ million}$$

The graph and the polynomial give similiar results.

Now Try Exercise 75

5.2   Putting It All Together

CONCEPT	EXPLANATION	EXAMPLES
Monomial	A number, variable, or product of numbers and variables raised to natural number powers The degree is the sum of the exponents. The coefficient is the number in a monomial.	$4x^2y$ Degree: 3; coefficient: 4 $-6x^2$ Degree: 2; coefficient: -6 $-a^4$ Degree: 4; coefficient: -1 x Degree: 1; coefficient: 1 -8 Degree: 0; coefficient: -8
Polynomial	A monomial or the sum of two or more monomials	$4x^2 + 8xy^2 + 3y^2$ Trinomial $-9x^4 + 100$ Binomial $-3x^2y^3$ Monomial
Like Terms	Monomials containing the same variables raised to the same powers	$10x$ and $-2x$, $4x^2$ and $3x^2$ $5ab^2$ and $-ab^2$, $5z$ and $\frac{1}{2}z$
Addition of Polynomials	To add polynomials, combine like terms.	$(x^2 + 3x + 1) + (2x^2 - 2x + 7)$ $= (1 + 2)x^2 + (3 - 2)x + (1 + 7)$ $= 3x^2 + x + 8$ $3xy + 5xy = (3 + 5)xy = 8xy$
Opposite of a Polynomial	To obtain the opposite of a polynomial, negate each term.	*Polynomial* *Opposite* $-2x^2 + x - 6$ $2x^2 - x + 6$ $a^2 - b^2$ $-a^2 + b^2$ $-3x - 18$ $3x + 18$
Subtraction of Polynomials	To subtract one polynomial from another, add the first polynomial and the opposite of the second polynomial.	$(x^2 + 3x) - (2x^2 - 5x)$ $= (x^2 + 3x) + (-2x^2 + 5x)$ $= (1 - 2)x^2 + (3 + 5)x$ $= -x^2 + 8x$
Evaluating a Polynomial	To evaluate a polynomial in x, substitute a value for x in the expression and simplify.	To evaluate the polynomial $\qquad 3x^2 - 2x + 1$ for $x = 2$, substitute 2 for x and simplify. $\qquad 3(2)^2 - 2(2) + 1 = 9$

5.2 Exercises

CONCEPTS AND VOCABULARY

1. A(n) _____ is a number, a variable, or a product of numbers and variables raised to a natural number power.

2. A(n) ____ is a monomial or a sum of monomials.

3. The ____ of a monomial is the sum of the exponents of the variables.

4. The _____ of a polynomial is the degree of the term with the greatest degree.

5. A polynomial with two terms is called a(n) ____.

6. A polynomial with three terms is called a(n) ____.

7. Two monomials with the same variables raised to the same powers are ____ terms.

8. To add two polynomials, combine ____ terms.

9. To subtract two polynomials, add the first polynomial to the ____ of the second polynomial.

10. Polynomials can be added horizontally or ____.

RECOGNIZING MONOMIALS AND POLYNOMIALS

Exercises 11–18: Identify the degree and coefficient of the monomial.

11. $3x^2$

12. y

13. $-ab$

14. $-2xy$

15. $-5rt$

16. $8x^2y^5$

17. 6

18. $-\frac{1}{2}$

Exercises 19–28: Determine whether the expression is a polynomial. If it is, state how many terms and variables the polynomial contains. Then state its degree.

19. $-x$

20. $7z$

21. $4x^2 - 5x + 9$

22. $x^3 - 9$

23. $x + \dfrac{1}{x}$

24. $\dfrac{5}{xy + 1}$

25. $3x^{-2}y^{-3}$

26. $5^2a^3b^4$

27. $-2^3a^4bc + b^2c$

28. $-7y^{-1}z^{-3}$

Exercises 29–36: State whether the given pair of expressions are like terms. If they are like terms, add them.

29. $5x, -4x$

30. $x^2, 8x^2$

31. $x^3, -6x^3$

32. $4xy, -9xy$

33. $9x, -xy$

34. $5x^2y, -3xy^2$

35. ab, ba

36. $rt^2, -2t^2r$

ADDING POLYNOMIALS

Exercises 37–46: Add the polynomials.

37. $(3x + 5) + (-4x + 4)$

38. $(-x + 5) + (2x - 5)$

39. $(3x^2 + 4x + 1) + (x^2 + 4x)$

40. $(-x^2 - x) + (2x^2 + 3x - 1)$

41. $(y^3 + 3y^2 - 5) + (3y^3 + 4y - 4)$

42. $(4z^4 + z^2 - 10) + (-z^4 + 4z - 5)$

43. $(-xy + 5) + (5xy - 4)$

44. $(2a^2 + b^2) + (3a^2 - 5b^2)$

45. $(a^3b^2 + a^2b^3) + (a^2b^3 - a^3b^2)$

46. $(a^2 + ab + b^2) + (a^2 - ab + b^2)$

Exercises 47–50: Add the polynomials vertically.

47. $\begin{aligned} 4x^2 - 2x + 1 \\ 5x^2 + 3x - 7 \end{aligned}$

48. $\begin{aligned} 8x^2 + 3x + 5 \\ -x^2 - 3x - 9 \end{aligned}$

49. $\begin{aligned} -x^2 + x \\ 2x^2 - 8x - 1 \end{aligned}$

50. $\begin{aligned} a^3 - 3a^2b + 3ab^2 - b^3 \\ a^3 + 3a^2b + 3ab^2 + b^3 \end{aligned}$

SUBTRACTING POLYNOMIALS

Exercises 51–56: Write the opposite of the polynomial.

51. $5x^2$

52. $17x + 12$

53. $3a^2 - a + 4$

54. $-b^3 + 3b$

55. $-2t^2 - 3t + 4$

56. $7t^2 + t - 10$

Exercises 57–66: Subtract the polynomials.

57. $(3x + 1) - (-x + 3)$

58. $(-2x + 5) - (x + 7)$

59. $(-x^2 + 6x) - (2x^2 + x - 2)$

60. $(2y^2 + 3y - 2) - (y^2 - y)$

61. $(z^3 - 2z^2 - z) - (4z^2 + 5z + 1)$

62. $(3z^4 - z) - (-z^4 + 4z^2 - 5)$

63. $(4xy + x^2y^2) - (xy - x^2y^2)$

64. $(a^2 + b^2) - (-a^2 + b^2)$

65. $(ab^2) - (ab^2 + a^3b)$

66. $(x^2 + 3xy + 4y^2) - (x^2 - xy + 4y^2)$

Exercises 67–70: Subtract the polynomials vertically.

67. $(x^2 + 2x - 3) - (2x^2 + 7x + 1)$

68. $(5x^2 - 9x - 1) - (x^2 - x + 3)$

69. $(3x^3 - 2x) - (5x^3 + 4x + 2)$

70. $(a^2 + 3ab + 2b^2) - (a^2 - 3ab + 2b^2)$

EVALUATING POLYNOMIAL EXPRESSIONS

Exercises 71–74: Evaluate the polynomial expression for each value of x.

71. $x^2 - x + 3$ for $x = -2, 3$

72. $2x^2 - x$ for $x = -1, 2$

73. $4x^3 + x^2 - 2x$ for $x = -3, 1$

74. $-2x^4 - x^3$ for $x = -2, 3$

APPLICATIONS INVOLVING POLYNOMIALS

75. Exercise and Heart Rate The polynomial given by $1.6t^2 - 28t + 200$ calculates the heart rate shown in **FIGURE 5.1(b)** in this section, where t represents the elapsed time in minutes since exercise stopped.
   (a) What is the heart rate when the athlete first stops exercising?
   (b) What is the heart rate after 5 minutes?
   (c) Describe what happens to the heart rate after exercise stops.

76. Music Streaming The proportion of total streaming music revenues after 2008 can be modeled by the polynomial given by $0.7x^2 - 0.4x + 4.2$, where $x = 0$ corresponds to 2008, $x = 1$ to 2009, and so on. The graph at the top of the next column illustrates this growth.

Streaming Music Revenues

   (a) Use the graph to estimate the proportion of streaming music revenues in 2012.
   (b) Use the polynomial to estimate the proportion of streaming music revenues in 2012.
   (c) Do your answers from parts (a) and (b) agree?

77. Areas of Squares Write a monomial that equals the sum of the areas of the squares. Then calculate this sum for $z = 10$ inches.

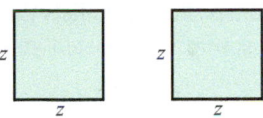

78. Areas of Rectangles Find a monomial that equals the sum of the areas of the three rectangles. Find this sum for $a = 5$ yards and $b = 3$ yards.

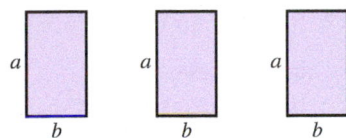

79. Area of a Figure Find a polynomial that equals the area of the figure. Calculate its area for $x = 6$ feet.

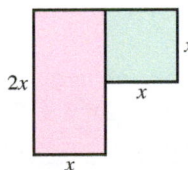

80. Area of a Rectangle Write a polynomial that gives the area of the rectangle. Calculate its area for $x = 3$ feet.

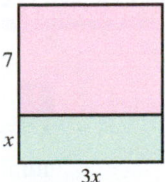

81. Areas of Circles Write a polynomial that gives the sum of the areas of two circles, one with radius x and the other with radius y. Find this sum for $x = 2$ feet and $y = 3$ feet. Leave your answer in terms of π.

82. Squares and Circles Write a polynomial that gives the sum of the areas of a square having sides of length x and a circle having diameter x. Approximate this sum to the nearest hundredth of a square foot for $x = 6$ feet.

83. World Population The table lists world population P in billions for selected years t.

t	1974	1987	1999	2012
P	4	5	6	7

Source: U.S. Census Bureau.

(a) Find the slope of each line segment connecting consecutive data points in the table. Can these data be modeled with a line? Explain.

(b) Does the polynomial $0.077t - 148$ give good estimates for the world population in year t? Explain how you decided.

84. Price of a Stamp The table lists the price P of a first-class postage stamp for selected years t.

t	1963	1975	1987	2002	2007	2011
P	5¢	13¢	25¢	37¢	41¢	44¢

Source: U.S. Postal Service.

(a) Does the polynomial $0.835t - 1635$ model the data in the table exactly?

(b) Does it give approximations that are within 1.5¢ of the actual values?

WRITING ABOUT MATHEMATICS

85. Explain what the terms monomial, binomial, trinomial, and polynomial mean. Give an example of each.

86. Explain how to determine the degree of a polynomial that has one variable. Give an example.

87. Explain how to obtain the opposite of a polynomial. Give an example.

88. Explain how to subtract two polynomials. Give an example.

SECTIONS 5.1 and 5.2

Checking Basic Concepts

1. Evaluate each expression.
(a) -5^2
(b) $3^2 - 2^3$

2. Simplify each expression.
(a) $10^3 \cdot 10^5$
(b) $(3x^2)(-4x^5)$
(c) $(a^3b)^2$
(d) $\left(\dfrac{x}{z^3}\right)^4$

3. Simplify each expression.
(a) $(4y^3)^0$
(b) $(x^3)^2(3x^4)^2$
(c) $2a^2(5a^3 - 7)$

4. State the number of terms and variables in the polynomial $5x^3y - 2x^2y + 5$. What is its degree?

5. A box has a rectangular bottom twice as long as it is wide.
(a) If the bottom has width w and the box has height h, write a monomial that gives the volume of the box.
(b) Find the volume of the box for $w = 12$ inches and $h = 10$ inches.

6. Simplify each expression.
(a) $(2a^2 + 3a - 1) + (a^2 - 3a + 7)$
(b) $(4z^3 + 5z) - (2z^3 - 2z + 8)$
(c) $(x^2 + 2xy + y^2) - (x^2 - 2xy + y^2)$

5.3 Multiplication of Polynomials

Objectives

1 Multiplying Monomials
2 Reviewing the Distributive Property
3 Multiplying Monomials and Polynomials
4 Multiplying Polynomials

1 Multiplying Monomials

A monomial is a number, a variable, or a product of numbers and variables raised to natural number powers. To multiply monomials, we often use the product rule for exponents.

READING CHECK 1

• Which rule for exponents is commonly used to multiply monomials?

EXAMPLE 1   **Multiplying monomials**

Multiply.
(a) $-5x^2 \cdot 4x^3$ **(b)** $(7xy^4)(x^3y^2)$

Solution
(a) $-5x^2 \cdot 4x^3 = (-5)(4)x^2x^3$ Commutative property

$= -20x^{2+3}$ The product rule

$= -20x^5$ Simplify.

(b) $(7xy^4)(x^3y^2) = 7xx^3y^4y^2$ Commutative property

$= 7x^{1+3}y^{4+2}$ The product rule

$= 7x^4y^6$ Simplify.

Now Try Exercises 9, 13

2 Reviewing the Distributive Properties

Distributive Property
$3(x + 2) = \mathbf{3x + 6}$

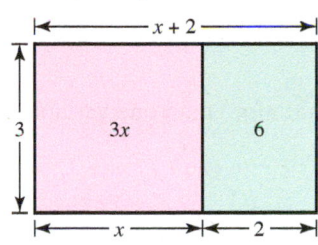

FIGURE 5.8 Area: 3x + 6

Distributive properties are used frequently for multiplying monomials and polynomials. For all real numbers a, b, and c,

$$a(b + c) = ab + ac \quad \text{and} \quad a(b - c) = ab - ac.$$

The first distributive property above can be visualized geometrically. For example,

$$3(x + 2) = \mathbf{3x + 6}$$

is illustrated in **FIGURE 5.8**. The dimensions of the large rectangle are 3 by $x + 2$, and its area is $3(x + 2)$. The areas of the two small rectangles, $3x$ and 6, equal the area of the large rectangle. Therefore $3(x + 2) = 3x + 6$.

In the next example, we use the distributive properties to multiply expressions.

EXAMPLE 2   **Using distributive properties**

Multiply.
(a) $2(3x + 4)$ **(b)** $(3x^2 + 4)5$ **(c)** $-x(3x - 6)$

Solution

(a) $2(3x + 4) = 2 \cdot 3x + 2 \cdot 4 = 6x + 8$

(b) $(3x^2 + 4)5 = 3x^2 \cdot 5 + 4 \cdot 5 = 15x^2 + 20$

(c) $-x(3x - 6) = -x \cdot 3x + x \cdot 6 = -3x^2 + 6x$

Now Try Exercises 15, 19, 21

READING CHECK 2

• What properties are commonly used to multiply a monomial and a polynomial?

3 Multiplying Monomials and Polynomials

A monomial consists of one term, whereas a polynomial consists of one or more terms separated by + or − signs. To multiply a monomial by a polynomial, we apply the distributive properties and the product rule.

EXAMPLE 3 **Multiplying monomials and polynomials**

Multiply.
(a) $9x(2x^2 - 3)$ **(b)** $(5x - 8)x^2$
(c) $-7(2x^2 - 4x + 6)$ **(d)** $4x^3(x^4 + 9x^2 - 8)$

Solution

(a)
$$9x(2x^2 - 3) = 9x \cdot 2x^2 - 9x \cdot 3 \qquad \text{Distributive property}$$
$$= 18x^3 - 27x \qquad \text{The product rule}$$

(b)
$$(5x - 8)x^2 = 5x \cdot x^2 - 8 \cdot x^2 \qquad \text{Distributive property}$$
$$= 5x^3 - 8x^2 \qquad \text{The product rule}$$

(c) $-7(2x^2 - 4x + 6) = -7 \cdot 2x^2 + 7 \cdot 4x - 7 \cdot 6 \qquad \text{Distributive property}$
$$= -14x^2 + 28x - 42 \qquad \text{Simplify.}$$

(d) $4x^3(x^4 + 9x^2 - 8) = 4x^3 \cdot x^4 + 4x^3 \cdot 9x^2 - 4x^3 \cdot 8 \qquad \text{Distributive property}$
$$= 4x^7 + 36x^5 - 32x^3 \qquad \text{The product rule}$$

Now Try Exercises 23, 25, 27, 29

We can also multiply monomials and polynomials that contain more than one variable.

EXAMPLE 4 **Multiplying monomials and polynomials**

Multiply.
(a) $2xy(7x^2y^3 - 1)$ **(b)** $-ab(a^2 - b^2)$

Solution

(a)
$$2xy(7x^2y^3 - 1) = 2xy \cdot 7x^2y^3 - 2xy \cdot 1 \qquad \text{Distributive property}$$
$$= 14xx^2yy^3 - 2xy \qquad \text{Commutative property}$$
$$= 14x^3y^4 - 2xy \qquad \text{The product rule}$$

(b)
$$-ab(a^2 - b^2) = -ab \cdot a^2 + ab \cdot b^2 \qquad \text{Distributive property}$$
$$= -aa^2b + abb^2 \qquad \text{Commutative property}$$
$$= -a^3b + ab^3 \qquad \text{The product rule}$$

Now Try Exercises 31, 35

4 Multiplying Polynomials

Monomials, binomials, and trinomials are examples of polynomials. Recall that a monomial has one term, a binomial has two terms, and a trinomial has three terms. In the next example, we multiply two binomials, using both geometric and symbolic techniques.

EXAMPLE 5   **Multiplying binomials**

Multiply $(x + 4)(x + 2)$
(a) geometrically   and   **(b)** symbolically.

Solution

(a) To multiply $(x + 4)(x + 2)$ geometrically, draw a rectangle $x + 4$ long and $x + 2$ wide, as shown in **FIGURE 5.9(a)**. The area of this rectangle equals length times width, or $(x + 4)(x + 2)$. The large rectangle can be divided into four smaller rectangles, which have areas of $x^2, 4x, 2x$, and 8, as shown in **FIGURE 5.9(b)**. Thus

$$(x + 4)(x + 2) = x^2 + 4x + 2x + 8$$
$$= x^2 + 6x + 8.$$

$(x + 2)(x + 4) = x^2 + 6x + 8$

(a) Area $= (x + 4)(x + 2)$

(b) Area $= x^2 + 4x + 2x + 8$

FIGURE 5.9

(b) To multiply $(x + 4)(x + 2)$ symbolically, apply the distributive property two times.

$$(x + 4)(x + 2) = (x + 4)(x) + (x + 4)(2)$$
$$= x \cdot x + 4 \cdot x + x \cdot 2 + 4 \cdot 2$$
$$= x^2 + 4x + 2x + 8$$
$$= x^2 + 6x + 8$$

Now Try Exercise 39

MAKING CONNECTIONS 1

Multiplying Binomials Using FOIL

The distributive properties used in part (b) of the previous example show that if we want to multiply $(x + 4)$ by $(x + 2)$, we should multiply *every* term in the binomial $x + 4$ by *every* term in the binomial $x + 2$.

$$(x + 4)(x + 2) = x^2 + 2x + 4x + 8$$
$$= x^2 + 6x + 8$$

When multiplying *binomials*, this process is sometimes called *FOIL*. This acronym may be used to remind us to multiply the first terms (F), outside terms (O), inside terms (I), and last terms (L). The *FOIL* process is a shortcut for the process used in Example 5(b).

Multiply the *First terms* to obtain x^2. $(x + 4)(x + 2)$

Multiply the *Outside terms* to obtain $2x$. $(x + 4)(x + 2)$

Multiply the *Inside terms* to obtain $4x$. $(x + 4)(x + 2)$

Multiply the *Last terms* to obtain 8. $(x + 4)(x + 2)$

NOTE: The *FOIL* acronym is useful for remembering how to multiply two *binomials*, but it is not helpful, in general, for multiplying any two *polynomials*. ∎

The following statement summarizes how to multiply two polynomials in general.

MULTIPLYING POLYNOMIALS

The product of two polynomials may be found by multiplying every term in the first polynomial by every term in the second polynomial and then combining like terms.

EXAMPLE 6 **Multiplying binomials**

Multiply. Draw arrows to show how each term is found.
(a) $(3x + 2)(x + 1)$ **(b)** $(1 - x)(1 + 2x)$ **(c)** $(4x - 3)(x^2 - 2x)$

Solution

(a) $(3x + 2)(x + 1) = 3x \cdot x + 3x \cdot 1 + 2 \cdot x + 2 \cdot 1$
$$= 3x^2 + 3x + 2x + 2$$
$$= 3x^2 + 5x + 2$$

(b) $(1 - x)(1 + 2x) = 1 \cdot 1 + 1 \cdot 2x - x \cdot 1 - x \cdot 2x$
$$= 1 + 2x - x - 2x^2$$
$$= 1 + x - 2x^2$$

(c) $(4x - 3)(x^2 - 2x) = 4x \cdot x^2 - 4x \cdot 2x - 3 \cdot x^2 + 3 \cdot 2x$
$$= 4x^3 - 8x^2 - 3x^2 + 6x$$
$$= 4x^3 - 11x^2 + 6x$$

Now Try Exercises 51, 53, 59

The *FOIL* process cannot be used for every product of polynomials. In the next example, the general process for multiplying polynomials is used.

EXAMPLE 7 **Multiplying polynomials**

Multiply.
(a) $(2x + 3)(x^2 + x - 1)$ **(b)** $(a - b)(a^2 + ab + b^2)$
(c) $(x^4 + 2x^2 - 5)(x^2 + 1)$

Solution
(a) Multiply every term in $(2x + 3)$ by every term in $(x^2 + x - 1)$.

$$(2x + 3)(x^2 + x - 1) = 2x \cdot x^2 + 2x \cdot x - 2x \cdot 1 + 3 \cdot x^2 + 3 \cdot x - 3 \cdot 1$$
$$= 2x^3 + 2x^2 - 2x + 3x^2 + 3x - 3$$
$$= 2x^3 + 5x^2 + x - 3$$

(b) $(a - b)(a^2 + ab + b^2) = a \cdot a^2 + a \cdot ab + a \cdot b^2 - b \cdot a^2 - b \cdot ab - b \cdot b^2$
$$= a^3 + a^2b + ab^2 - a^2b - ab^2 - b^3$$
$$= a^3 - b^3$$

(c) $(x^4 + 2x^2 - 5)(x^2 + 1) = x^4 \cdot x^2 + x^4 \cdot 1 + 2x^2 \cdot x^2 + 2x^2 \cdot 1 - 5 \cdot x^2 - 5 \cdot 1$
$$= x^6 + x^4 + 2x^4 + 2x^2 - 5x^2 - 5$$
$$= x^6 + 3x^4 - 3x^2 - 5$$

Now Try Exercises 63, 67, 69

STUDY TIP

Even if you know exactly how to do a math problem correctly, a simple computational error will often cause you to get an incorrect answer. Be sure to take your time on simple calculations.

Polynomials can be multiplied vertically in a manner similar to multiplication of real numbers. For example, multiplication of 123 times 12 is performed as follows.

$$
\begin{array}{r}
1\ 2\ 3 \\
\times\ 1\ 2 \\
\hline
2\ 4\ 6 \\
1\ 2\ 3 \\
\hline
1\ 4\ 7\ 6
\end{array}
$$

A similar method can be used to multiply polynomials vertically.

EXAMPLE 8   **Multiplying polynomials vertically**

Multiply $2x^2 - 4x + 1$ and $x + 3$ vertically.

Solution
Write the polynomials vertically. Then multiply every term in the first polynomial by each term in the second polynomial. Arrange the results so that *like terms are in the same column.*

$$
\begin{array}{r}
2x^2 - 4x + 1 \\
x + 3 \\
\hline
6x^2 - 12x + 3 \qquad \text{Multiply top polynomial by 3.} \\
2x^3 - 4x^2 + x \qquad \text{Multiply top polynomial by } x. \\
\hline
2x^3 + 2x^2 - 11x + 3 \qquad \text{Add each column.}
\end{array}
$$

Now Try Exercise 71

MAKING CONNECTIONS 2

Vertical and Horizontal Formats

Whether you decide to add, subtract, or multiply polynomials vertically or horizontally, remember that the *same* answer is obtained either way.

EXAMPLE 9   **Finding the volume of a box**

A box has a width 3 inches less than its height and a length 4 inches more than its height.
(a) If h represents the height of the box, write a polynomial that represents the volume of the box.
(b) Use this polynomial to calculate the volume of the box if $h = 10$ inches.

Solution
(a) If h is the height, then $h - 3$ is the width and $h + 4$ is the length, as illustrated in **FIGURE 5.10**. Its volume equals the product of these three expressions.

$$
h(h - 3)(h + 4) = (h^2 - 3h)(h + 4)
$$

$$
= h^2 \cdot h + h^2 \cdot 4 - 3h \cdot h - 3h \cdot 4
$$

$$
= h^3 + 4h^2 - 3h^2 - 12h
$$

$$
= h^3 + h^2 - 12h
$$

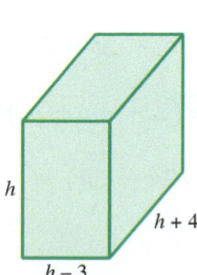

h

$h + 4$

$h - 3$

FIGURE 5.10

(b) If $h = 10$, then the volume is

$$
10^3 + 10^2 - 12(10) = 1000 + 100 - 120 = 980 \text{ cubic inches.}
$$

Now Try Exercises 79

5.3   Putting It All Together

CONCEPT	EXPLANATION	EXAMPLES
Distributive Properties	For all real numbers a, b, and c, $$a(b + c) = ab + ac \quad \text{and}$$ $$a(b - c) = ab - ac.$$	$5(x + 3) = 5x + 15$, $3(x - 6) = 3x - 18$, and $-2x(3 - 5x^3) = -6x + 10x^4$
Multiplying Polynomials	The product of two polynomials may be found by multiplying every term in the first polynomial by every term in the second polynomial and then combining like terms.	$3x(5x^2 + 2x - 7)$ $= 3x \cdot 5x^2 + 3x \cdot 2x - 3x \cdot 7$ $= 15x^3 + 6x^2 - 21x$ $(x + 2)(7x - 3)$ $= x \cdot 7x - x \cdot 3 + 2 \cdot 7x - 2 \cdot 3$ $= 7x^2 - 3x + 14x - 6$ $= 7x^2 + 11x - 6$

5.3   Exercises

MyMathLab®

CONCEPTS AND VOCABULARY

1. The equation $x^2 \cdot x^3 = x^5$ illustrates what rule of exponents?

2. The equation $3(x - 2) = 3x - 6$ illustrates what property?

3. The product of two polynomials may be found by multiplying every _____ in the first polynomial by every _____ in the second polynomial and then combining like terms.

4. Polynomials can be multiplied horizontally or _____.

MULTIPLYING MONOMIALS

Exercises 5–14: Multiply.

5. $x^2 \cdot x^5$

6. $-a \cdot a^5$

7. $-3a \cdot 4a$

8. $7x \cdot 5x$

9. $4x^3 \cdot 5x^2$

10. $6b^6 \cdot 3b^5$

11. $xy^2 \cdot 4xy$

12. $3ab \cdot ab^2$

13. $(-3xy^2)(4x^2y)$

14. $(-r^2t^2)(-r^3t)$

MULTIPLYING MONOMIALS AND POLYNOMIALS

Exercises 15–36: Multiply and simplify the expression.

15. $3(x + 4)$

16. $-7(4x - 1)$

17. $-5(9x + 1)$

18. $10(1 - 6x)$

19. $(4 - z)z$

20. $(2y - 8)2y$

21. $-y(5 + 3y)$

22. $3z(1 - 5z)$

23. $3x(5x^2 - 4)$

24. $-6x(2x^3 + 1)$

25. $(6x - 6)x^2$

26. $(1 - 2x^2)3x^2$

27. $-8(4t^2 + t + 1)$

28. $7(3t^2 - 2t - 5)$

29. $n^2(-5n^2 + n - 2)$

30. $6n^3(2 - 4n + n^2)$

31. $xy(x + y)$

32. $ab(2a - 3b)$

33. $x^2(x^2y - xy^2)$

34. $2y^2(xy - 5)$

35. $-ab(a^3 - 2b^3)$

36. $5rt(r^2 + 2rt + t^2)$

MULTIPLYING POLYNOMIALS

Exercises 37–42: (Refer to Example 5.) Multiply the given expression **(a)** *geometrically and* **(b)** *symbolically.*

37. $x(x + 3)$

38. $2x(x + 5)$

39. $(x + 2)(x + 2)$

40. $(x + 1)(x + 3)$

41. $(x + 3)(x + 6)$

42. $(x + 5)(x + 2)$

Exercises 43–70: Multiply and simplify the expression.

43. $(x + 3)(x + 5)$

44. $(x - 4)(x - 7)$

45. $(x - 8)(x - 9)$

46. $(x + 10)(x + 10)$

47. $(3z - 2)(2z - 5)$

48. $(z + 6)(2z - 1)$

49. $(8b - 1)(8b + 1)$

50. $(3t + 2)(3t - 2)$

51. $(10y + 7)(y - 1)$

52. $(y + 6)(2y + 7)$

53. $(5 - 3a)(1 - 2a)$

54. $(4 - a)(5 + 3a)$

55. $(1 - 3x)(1 + 3x)$

56. $(10 - x)(5 - 2x)$

57. $(x - 1)(x^2 + 1)$

58. $(x + 2)(x^2 - x)$

59. $(x^2 + 4)(4x - 3)$

60. $(3x^2 - 1)(3x^2 + 1)$

61. $(2n + 1)(n^2 + 3)$

62. $(2 - n^2)(1 + n^2)$

63. $(m + 1)(m^2 + 3m + 1)$

64. $(m - 2)(m^2 - m + 5)$

65. $(3x - 2)(2x^2 - x + 4)$

66. $(5x + 4)(x^2 - 3x + 2)$

67. $(x + 1)(x^2 - x + 1)$

68. $(x - 2)(x^2 + 4x + 4)$

69. $(4b^2 + 3b + 7)(b^2 + 3)$

70. $(-3a^2 - 2a + 1)(3a^2 - 3)$

Exercises 71–76: Multiply the polynomials vertically.

71. $(x + 2)(x^2 - 3x + 1)$

72. $(2y - 3)(3y^2 - 2y - 2)$

73. $(a - 2)(a^2 + 2a + 4)$

74. $(b - 3)(b^2 + 3b + 9)$

75. $(3x^2 - x + 1)(2x^2 + 1)$

76. $(2x^2 - 3x - 5)(2x^2 + 3)$

77. Thinking Generally If a polynomial with m terms and a polynomial with n terms are multiplied, how many terms are there in the product before like terms are combined?

78. Thinking Generally When a polynomial with m terms is multiplied by a second polynomial, the product contains k terms before like terms are combined. How many terms does the second polynomial contain?

APPLICATIONS INVOLVING MULTIPLYING POLYNOMIALS

79. Volume of a Box (Refer to Example 9.) A box has a width 4 inches less than its height and a length 2 inches more than its height.

   (a) If h is the height of the box, write a polynomial that represents the volume of the box.

   (b) Use this polynomial to calculate the volume for $h = 25$ inches.

80. Surface Area of a Box Use the drawing of the box to write a polynomial that represents each of the following.

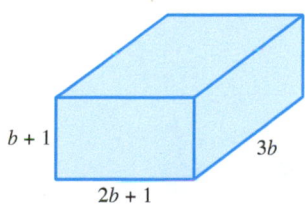

   (a) The area of its bottom

   (b) The area of its front

   (c) The area of its right side

   (d) The total area of its six sides

81. Perimeter of a Pen A rectangular pen for a pet has a perimeter of 100 feet. If one side of the pen has length x, then its area is given by $x(50 - x)$.

   (a) Multiply this expression.

   (b) Evaluate the expression obtained in part (a) for $x = 25$.

82. Rectangular Garden A rectangular garden has a perimeter of 500 feet.

   (a) If one side of the garden has length x, then write a polynomial expression that gives its area. Multiply this expression completely.

   (b) Evaluate the expression for $x = 50$ and interpret your answer.

83. Surface Area of a Cube Write a polynomial that represents the total area of the six sides of the cube having edges with length $x + 1$.

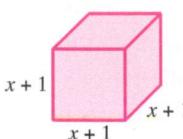

84. Surface Area of a Sphere The surface area of a sphere with radius r is $4\pi r^2$. Write a polynomial that gives the surface area of a sphere with radius $x + 2$. Leave your answer in terms of π.

85. Toy Rocket A toy rocket is shot straight up into the air. Its height h in feet above the ground after t seconds is represented by the expression $t(64 - 16t)$.
(a) Multiply this expression.
(b) Evaluate the expression obtained in part (a) and the given expression for $t = 2$.
(c) Are your answers in part (b) the same? Should they be the same?

86. Toy Rocket on the Moon (Refer to the preceding exercise.) If the same toy rocket were flown on the moon, then its height h in feet after t seconds would be $t\left(64 - \frac{5}{2}t\right)$.
(a) Multiply this expression.
(b) Evaluate the expression obtained in part (a) and the given expression for $t = 2$. Did the rocket go higher on the moon?

WRITING ABOUT MATHEMATICS

87. Explain how the acronym FOIL relates to multiplying two binomials, such as $x + 3$ and $2x + 1$.

88. Does the FOIL method work for multiplying a binomial and a trinomial? Explain.

89. Explain in words how to multiply any two polynomials. Give an example.

90. Give two properties of real numbers that are used for multiplying $3x(5x^2 - 3x + 2)$. Explain your answer.

Group Activity

Exercises 1 and 2: **Constructing a Box** *A box is constructed from a rectangular piece of metal by cutting squares from the corners and folding up the sides. The square, cutout corners are x inches by x inches.*

1. Suppose that the dimensions of the metal piece are 20 inches by 30 inches.
(a) Write a polynomial that gives the volume of the box.
(b) Find the volume of the box for $x = 4$ inches.

2. Suppose that the metal piece is square with sides of length 25 inches.
(a) Write a polynomial expression that gives the outside surface area of the box. (Assume that the box does not have a top.)
(b) Find this area for $x = 3$ inches.

5.4 Special Products

Objectives

1 Finding the Product of a Sum and Difference
2 Squaring Binomials
3 Cubing Binomials

STUDY TIP

In mathematics, there are often several correct ways to perform a particular process. If your instructor does not require you to use a specified method, choose the one that works best for you.

READING CHECK 1

• Multiply $(x - 4)(x + 4)$.

1 Finding the Product of a Sum and Difference

Products of the form $(a + b)(a - b)$ occur frequently in mathematics. Other examples include the products

$$(x + y)(x - y) \quad \text{and} \quad (2r + 3t)(2r - 3t).$$

These products can always be multiplied by using the techniques discussed in the previous section. However, there is a faster way to multiply these special products.

$$(a + b)(a - b) = a \cdot a - a \cdot b + b \cdot a - b \cdot b$$
$$= a^2 - ab + ba - b^2$$
$$= a^2 - b^2$$

In words, the product of a sum of two numbers and their difference equals the difference of their squares. We generalize this method as follows.

> **PRODUCT OF A SUM AND DIFFERENCE**
>
> For any real numbers a and b,
>
> $$(a + b)(a - b) = a^2 - b^2.$$

EXAMPLE 1 **Finding products of sums and differences**

Multiply.
(a) $(x + y)(x - y)$ **(b)** $(z - 2)(z + 2)$
(c) $(2r + 3t)(2r - 3t)$ **(d)** $(5m^2 - 4n^2)(5m^2 + 4n^2)$

Solution
(a) If we let $a = x$ and $b = y$, then we can apply the rule

$$(a + b)(a - b) = a^2 - b^2.$$

Thus

$$(x + y)(x - y) = (x)^2 - (y)^2$$
$$= x^2 - y^2.$$

(b) Because the expressions $(z + 2)(z - 2)$ and $(z - 2)(z + 2)$ are equal by the commutative property, we can apply the formula for the product of a sum and difference.

$$(z - 2)(z + 2) = (z)^2 - (2)^2$$
$$= z^2 - 4$$

(c) Let $a = 2r$ and $b = 3t$. Then the product can be evaluated as follows.

$$(2r + 3t)(2r - 3t) = (2r)^2 - (3t)^2$$
$$= 4r^2 - 9t^2$$

(d) $(5m^2 - 4n^2)(5m^2 + 4n^2) = (5m^2)^2 - (4n^2)^2 = 25m^4 - 16n^4$

Now Try Exercises 7, 9, 13, 17

The next example demonstrates how the product of a sum and difference can be used to find some products of numbers mentally.

EXAMPLE 2 **Finding a product**

Use the product of a sum and difference to find $22 \cdot 18$.

Solution
Because $22 = 20 + 2$ and $18 = 20 - 2$, rewrite and evaluate $22 \cdot 18$ as follows.

$$22 \cdot 18 = (20 + 2)(20 - 2) \quad \text{Rewrite 22 as } 20 + 2 \text{ and 18 as } 20 - 2.$$
$$= 20^2 - 2^2 \quad \text{Product of a sum and difference}$$
$$= 400 - 4 \quad \text{Evaluate exponents.}$$
$$= 396 \quad \text{Subtract.}$$

Now Try Exercise 21

Area $= (a + b)^2$

FIGURE 5.11

2 Squaring Binomials

Because each side of the square shown in **FIGURE 5.11** has length $(a + b)$, its area equals

$$(a + b)(a + b),$$

which can be written as $(a + b)^2$. We can multiply this expression as follows.

$$(a + b)^2 = (a + b)(a + b)$$
$$= a^2 + ab + ba + b^2$$
$$= a^2 + 2ab + b^2$$

This result is illustrated geometrically in **FIGURE 5.12**, where the area of the large square is $(a + b)^2$. This area can also be found by adding the areas of the four small rectangles.

$$a^2 + ab + ba + b^2 = a^2 + 2ab + b^2$$

The geometric and symbolic results are the same. Note that to obtain the middle term, $2ab$, we can multiply the two terms, a and b, in the binomial and *double* the result.

A similar product that is also the square of a binomial can be calculated as

$$(a - b)^2 = (a - b)(a - b)$$
$$= a^2 - ab - ba + b^2$$
$$= a^2 - 2ab + b^2.$$

These results are summarized as follows.

$(a + b)^2 = a^2 + 2ab + b^2$

FIGURE 5.12

READING CHECK 2

• Multiply $(a - b)^2$.

SQUARING A BINOMIAL

For any real numbers a and b,

$$(a + b)^2 = a^2 + 2ab + b^2 \quad \text{and}$$
$$(a - b)^2 = a^2 - 2ab + b^2.$$

That is, the square of a binomial equals the square of the first term, plus (or minus) twice the product of the two terms, plus the square of the last term.

NOTE: $(a + b)^2 \neq a^2 + b^2$. Don't forget the middle term when squaring a binomial. ∎

EXAMPLE 3   ### Squaring a binomial

Multiply.
(a) $(x + 3)^2$ (b) $(2x - 5)^2$ (c) $(1 - 5y)^2$ (d) $(7a^2 + 3b)^2$

Solution
(a) Let $a = x$ and $b = 3$ in the formula $(a + b)^2 = a^2 + 2ab + b^2$.

$$(x + 3)^2 = (x)^2 + 2(x)(3) + (3)^2$$
$$= x^2 + 6x + 9$$

(b) Apply the formula $(a - b)^2 = a^2 - 2ab + b^2$ with $a = 2x$ and $b = 5$.

$$(2x - 5)^2 = (2x)^2 - 2(2x)(5) + (5)^2$$
$$= 4x^2 - 20x + 25$$

(c) $(1 - 5y)^2 = (1)^2 - 2(1)(5y) + (5y)^2 = 1 - 10y + 25y^2$
(d) $(7a^2 + 3b)^2 = (7a^2)^2 + 2(7a^2)(3b) + (3b)^2 = 49a^4 + 42a^2b + 9b^2$

Now Try Exercises 27, 29, 35, 39

MAKING CONNECTIONS 1

Multiplying Binomials and Special Products

If you forget these special products, you can still multiply polynomials by using earlier techniques. For example, the binomial in Example 3(b) can be multiplied as

$$(2x - 5)^2 = (2x - 5)(2x - 5)$$
$$= 2x \cdot 2x - 2x \cdot 5 - 5 \cdot 2x + 5 \cdot 5$$
$$= 4x^2 - 10x - 10x + 25$$
$$= 4x^2 - 20x + 25.$$

Digital Photo

FIGURE 5.13

🌐 *Math in Context* Digital   NASA first developed digital photos because they were easy to transmit through space and because they provided clear images. A digital image from outer space is shown in **FIGURE 5.13**.

Today, digital cameras are readily available, and the Internet uses digital images exclusively. The next example shows how polynomials relate to digital pictures.

EXAMPLE 4   ### Calculating the size of a digital picture

A digital picture is made up of tiny square units called *pixels*. Shading individual pixels creates a picture. A simplified version of a digital picture of the letter T is shown in **FIGURE 5.14**. This picture includes an image of the letter T that measures 3 pixels by 3 pixels and a 1-pixel border.
(a) If a square digital image measures x pixels by x pixels, then a picture that includes the image and a 1-pixel border will measure $x + 2$ pixels by $x + 2$ pixels. Find a polynomial that gives the total number of pixels in the picture, including the border.
(b) Let $x = 3$ and evaluate the polynomial. Does it agree with **FIGURE 5.14**?

FIGURE 5.14

Solution
(a) The total number of pixels equals $(x + 2)$ times $(x + 2)$, or $(x + 2)^2$.

$$(x + 2)^2 = x^2 + 4x + 4$$

(b) For $x = 3$, the polynomial $x^2 + 4x + 4$ evaluates to $3^2 + 4 \cdot 3 + 4 = 25$, the total number of pixels. This result agrees with **FIGURE 5.14** on the previous page, which has a total of $5 \cdot 5 = 25$ pixels with a 3 pixel by 3 pixel image of the letter T inside.

Now Try Exercise 79

3 Cubing Binomials

Volume = $(x + 2)^3$

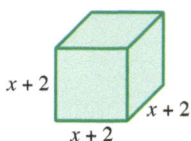

$x + 2$
$x + 2$
$x + 2$

FIGURE 5.15

To calculate the volume of the cube shown in **FIGURE 5.15**, we find the product of its length, width, and height. Because all sides have the same measure, its volume is $(x + 2)^3$. That is, the volume equals the *cube* of $x + 2$.

To multiply the expression $(x + 2)^3$, we proceed as follows.

$$(x + 2)^3 = (x + 2)(x + 2)^2 \qquad\qquad a^3 = a \cdot a^2$$

$$= (x + 2)(x^2 + 4x + 4) \qquad\qquad \text{Square the binomial.}$$

$$= x \cdot x^2 + x \cdot 4x + x \cdot 4 + 2 \cdot x^2 + 2 \cdot 4x + 2 \cdot 4 \quad \text{Multiply the polynomials.}$$

$$= x^3 + 4x^2 + 4x + 2x^2 + 8x + 8 \qquad \text{Simplify terms.}$$

$$= x^3 + 6x^2 + 12x + 8 \qquad\qquad \text{Combine like terms.}$$

EXAMPLE 5 **Cubing a binomial**

Multiply $(2z - 3)^3$.

Solution

$$(2z - 3)^3 = (2z - 3)(2z - 3)^2 \qquad\qquad a^3 = a \cdot a^2$$

$$= (2z - 3)(4z^2 - 12z + 9) \qquad\qquad \text{Square the binomial.}$$

$$= 8z^3 - 24z^2 + 18z - 12z^2 + 36z - 27 \quad \text{Multiply the polynomials.}$$

$$= 8z^3 - 36z^2 + 54z - 27 \qquad\qquad \text{Combine like terms.}$$

Now Try Exercise 47

CRITICAL THINKING 1

Suppose that a student is convinced that the expressions

$(x + y)^3$ and $x^3 + y^3$

are equal. How could you convince the student otherwise?

NOTE: In Example 5, it is not correct to find the result by simply cubing each term in the binomial. That is,

$$(2z - 3)^3 \ne (2z)^3 - (3)^3 = 8z^3 - 27. \quad \blacksquare$$

Business

🌐 *Math in ~Context* Powers of expressions often occur in business when annual interest is calculated. The next example illustrates how this is done.

EXAMPLE 6 **Calculating interest**

If a savings account pays x percent annual interest, where x is expressed as a decimal, then after 3 years a sum of money will grow by a factor of $(1 + x)^3$.

(a) Multiply this expression.

(b) Evaluate the expression for $x = 0.10$ (or 10%), and interpret the result.

Solution

(a)
$$(1 + x)^3 = (1 + x)(1 + x)^2 \qquad a^3 = a \cdot a^2$$
$$= (1 + x)(1 + 2x + x^2) \qquad \text{Square the binomial.}$$
$$= 1 + 2x + x^2 + x + 2x^2 + x^3 \qquad \text{Multiply the polynomials.}$$
$$= 1 + 3x + 3x^2 + x^3 \qquad \text{Combine like terms.}$$

(b) Let $x = 0.1$ in the expression $1 + 3x + 3x^2 + x^3$.

$$1 + 3(0.1) + 3(0.1)^2 + (0.1)^3 = 1.331$$

The sum of money will increase by a factor of 1.331. For example, $1000 deposited in this account will grow to $1331 in 3 years.

Now Try Exercise 75

5.4 Putting It All Together

CONCEPT	EXPLANATION	EXAMPLES
Product of a Sum and Difference	For any real numbers x and y, $$(x + y)(x - y) = x^2 - y^2.$$	$(x + 6)(x - 6) = x^2 - 36,$ $(2x - 3)(2x + 3) = 4x^2 - 9,$ and $(x^2 + y^2)(x^2 - y^2) = x^4 - y^4$
Squaring a Binomial	For all real numbers x and y, $$(x + y)^2 = x^2 + 2xy + y^2 \quad \text{and}$$ $$(x - y)^2 = x^2 - 2xy + y^2.$$	$(x + 4)^2 = x^2 + 8x + 16,$ $(5x - 2)^2 = 25x^2 - 20x + 4,$ and $(x^2 + y^2)^2 = x^4 + 2x^2y^2 + y^4$
Cubing a Binomial	Multiply the binomial by its square.	$(x + 3)^3$ $= (x + 3)(x + 3)^2$ $= (x + 3)(x^2 + 6x + 9)$ $= x^3 + 6x^2 + 9x + 3x^2 + 18x + 27$ $= x^3 + 9x^2 + 27x + 27$

5.4 Exercises

MyMathLab®

CONCEPTS AND VOCABULARY

1. $(a + b)(a - b) = $ _____

2. $(a + b)^2 = $ _____

3. $(a - b)^2 = $ _____

4. $(a + b)^3 = (a + b) \cdot$ _____

5. (True or False?) The two expressions $(x + y)^2$ and $x^2 + y^2$ are equal for all real numbers x and y.

6. (True or False?) The two expressions $(r - t)^2$ and $r^2 - t^2$ are equal for all real numbers r and t.

FINDING THE PRODUCT OF A SUM AND DIFFERENCE

Exercises 7–20: Multiply.

7. $(x - 3)(x + 3)$ **8.** $(x + 6)(x - 6)$

9. $(4x - 1)(4x + 1)$ **10.** $(10x + 3)(10x - 3)$

11. $(1 + 2a)(1 - 2a)$ **12.** $(4 - 9b)(4 + 9b)$

13. $(2x + 3y)(2x - 3y)$ **14.** $(5r - 6t)(5r + 6t)$

15. $(ab - 5)(ab + 5)$ **16.** $(2xy + 7)(2xy - 7)$

17. $(a^2 - b^2)(a^2 + b^2)$ **18.** $(3x^2 + y^2)(3x^2 - y^2)$

19. $(x^3 - y^3)(x^3 + y^3)$ **20.** $(2a^4 + b^4)(2a^4 - b^4)$

Exercises 21–26: (Refer to Example 2.) Use the product of a sum and a difference to evaluate the expression.

21. $101 \cdot 99$ **22.** $52 \cdot 48$

23. $23 \cdot 17$ **24.** $29 \cdot 31$

25. $90 \cdot 110$ **26.** $38 \cdot 42$

SQUARING BINOMIALS

Exercises 27–40: Multiply.

27. $(a - 2)^2$ **28.** $(x - 7)^2$

29. $(2x + 3)^2$ **30.** $(7x - 2)^2$

31. $(3b + 5)^2$ **32.** $(7t + 10)^2$

33. $\left(\frac{3}{4}a - 4\right)^2$ **34.** $\left(\frac{1}{5}a + 1\right)^2$

35. $(1 - b)^2$ **36.** $(1 - 4a)^2$

37. $(5 + y^3)^2$ **38.** $(9 - 5x^2)^2$

39. $(a^2 + b)^2$ **40.** $(x^3 - y^3)^2$

CUBING BINOMIALS

Exercises 41–48: Multiply.

41. $(a + 1)^3$ **42.** $(b + 4)^3$

43. $(x - 2)^3$ **44.** $(y - 7)^3$

45. $(2x + 1)^3$ **46.** $(4z + 3)^3$

47. $(6u - 1)^3$ **48.** $(5v + 3)^3$

MULTIPLYING POLYNOMIALS

Exercises 49–66: Multiply, using any appropriate method.

49. $4(5x + 9)$ **50.** $(2x + 1)(3x - 5)$

51. $(x - 5)(x + 7)$ **52.** $(x + 10)(x + 10)$

53. $(3x - 5)^2$ **54.** $(x - 3)(x + 9)$

55. $(5x + 3)(5x + 4)$ **56.** $-x^3(x^2 - x + 1)$

57. $(4b - 5)(4b + 5)$ **58.** $(x + 5)^3$

59. $-5x(4x^2 - 7x + 2)$ **60.** $(4x^2 - 5)(4x^2 + 5)$

61. $(4 - a)^3$ **62.** $2x(x - 3)^3$

63. $x(x + 3)^2$ **64.** $(x - 1)^2(x + 1)$

65. $(x + 2)(x - 2)(x + 1)(x - 1)$

66. $(x - y)(x + y)(x^2 + y^2)$

67. Thinking Generally Multiply $(a^n + b^n)(a^n - b^n)$.

68. Thinking Generally Multiply $(a^n + b^n)^2$.

APPLICATIONS INVOLVING SPECIAL PRODUCTS

Exercises 69–72: Do each part and verify that your answers are the same.

   (a) *Find the area of the large square by multiplying its length and width.*

   (b) *Find the sum of the areas of the smaller rectangles inside the large square.*

69.

70.

71.

72.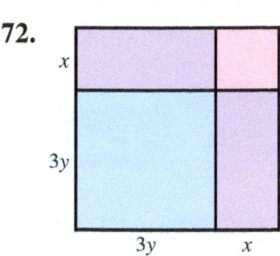

Exercises 73 and 74: Find a polynomial that represents the following.

   (a) *The outside surface area given by the six sides of the cube*

   (b) *The volume of the cube*

73.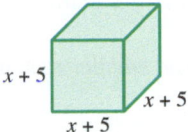

$x + 5$
$x + 5$
$x + 5$

74.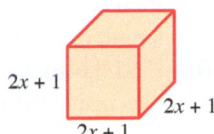

$2x + 1$
$2x + 1$
$2x + 1$

75. Compound Interest (Refer to Example 6.) If a sum of money is deposited in a savings account that is paying x percent annual interest (expressed as a decimal), then this sum of money increases by a factor of $(1 + x)^2$ after 2 years.

   (a) Multiply this expression.

   (b) Evaluate the polynomial expression found in part (a) for an annual interest rate of 10%, or $x = 0.10$, and interpret the answer.

76. Compound Interest If a sum of money is deposited in a savings account that is paying x percent annual interest, then this sum of money increases by a factor of $\left(1 + \frac{1}{100}x\right)^3$ after 3 years.

   (a) Multiply this expression.

   (b) Evaluate the polynomial expression in part (a) for an annual interest rate of 8%, or $x = 8$, and interpret the answer.

77. Probability If there is an x percent chance of rain on each of two consecutive days, then the expression $(1 - x)^2$ gives the percent chance that neither day will have rain. Assume that all percentages are expressed as decimals.

   (a) Multiply this expression.

   (b) Evaluate the polynomial expression in part (a) for a 50% chance of rain, or $x = 0.50$, and interpret the answer.

78. Probability If there is an x percent chance of rolling a 6 with one die, then the expression $(1 - x)^3$ gives the percent chance of not rolling a 6 with three dice. Assume that all percentages are expressed as decimals or fractions.

   (a) Multiply this expression.

   (b) Evaluate the polynomial expression found in part (a) for a 16.$\overline{6}$% chance of rolling a 6, or $x = \frac{1}{6}$, and interpret the answer.

79. Swimming Pool A square swimming pool has an 8-foot-wide sidewalk around it.

   (a) If the sides of the pool have length z, as shown in the accompanying figure, find a polynomial that gives the area of the sidewalk.

   (b) Evaluate the polynomial in part (a) for $z = 60$ and interpret the answer.

80. Digital Picture (Refer to Example 4.) Suppose that a digital picture, including its border, is $x + 2$ pixels by $x + 2$ pixels and that the actual image inside the border is $x - 2$ pixels by $x - 2$ pixels, as shown in the following figure.

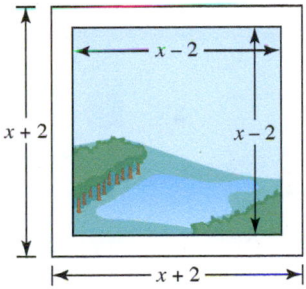

   (a) Find a polynomial that gives the number of pixels in the border.

   (b) Evaluate the polynomial in part (a) for $x = 5$.

   (c) Sketch a digital picture of the letter H with $x = 5$. Does the picture agree with the answer in part (b)?

   (d) Digital pictures typically have large values for x. If a picture has $x = 500$, find the total number of pixels in its border.

WRITING ABOUT MATHEMATICS

81. Explain why $(a + b)^2$ does not equal $a^2 + b^2$ in general for real numbers a and b.

82. Explain how to find the cube of a binomial.

Checking Basic Concepts

1. Multiply each expression.
   (a) $(-3xy^4)(5x^2y)$ (b) $-x(6 - 4x)$
   (c) $3ab(a^2 - 2ab + b^2)$

2. Multiply each expression.
   (a) $(x + 3)(4x - 3)$
   (b) $(x^2 - 1)(2x^2 + 2)$
   (c) $(x + y)(x^2 - xy + y^2)$

3. Multiply each expression.
   (a) $(5x + 2)(5x - 2)$ (b) $(x + 3)^2$
   (c) $(2 - 7x)^2$ (d) $(t + 2)^3$

4. Complete each part and verify that your answers are the same.
   (a) Find the area of the large square by squaring the length of one of its sides.
   (b) Find the sum of the areas of the smaller rectangles inside the large square.

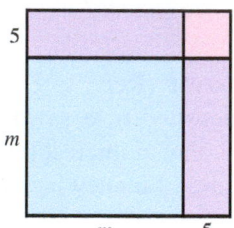

5.5 Integer Exponents and the Quotient Rule

Objectives

1. Using Negative Integers as Exponents
2. Applying the Quotient Rule
3. Applying Other Rules for Exponents
4. Reading and Writing Scientific Notation

NEW VOCABULARY

☐ Scientific notation

STUDY TIP

Mathematics often builds on concepts that have already been studied. Try to get in the regular habit of reviewing topics from earlier parts of the text.

READING CHECK 1

- Simplify 3^{-2}.

1 ▸ Using Negative Integers as Exponents

So far, we have defined exponents that are whole numbers. For example,

$$5^0 = 1 \quad \text{and} \quad 2^3 = 2 \cdot 2 \cdot 2 = 8.$$

What if an exponent is a negative integer? To answer this question, consider **TABLE 5.6**, which shows values for decreasing powers of 2. Note that each time the exponent on 2 decreases by 1, the resulting value is divided by 2.

Powers of 2

Power of 2	Value
2^1	2
2^0	1
2^{-1}	$\frac{1}{2} = \frac{1}{2^1}$
2^{-2}	$\frac{1}{4} = \frac{1}{2^2}$

Decrease exponent by 1 ⟶ (between 2^1 and 2^0)   Divide by 2
Decrease exponent by 1 ⟶ (between 2^0 and 2^{-1})   Divide by 2
Decrease exponent by 1 ⟶ (between 2^{-1} and 2^{-2})   Divide by 2

TABLE 5.6

TABLE 5.6 shows that $2^{-1} = \frac{1}{2^1}$ and $2^{-2} = \frac{1}{2^2}$. In other words, if the exponent on the base 2 is negative, then the expression is equal to the reciprocal of the corresponding expression with a positive exponent on the base 2. This discussion suggests the following definition for negative integer exponents.

NEGATIVE INTEGER EXPONENTS

Let a be a nonzero real number and n be a positive integer. Then

$$a^{-n} = \frac{1}{a^n}.$$

That is, a^{-n} is the reciprocal of a^n.

EXAMPLE 1   **Evaluating negative exponents**

Simplify each expression.
(a) 2^{-3} (b) 7^{-1} (c) x^{-2} (d) $(x + y)^{-8}$

Solution

(a) Because $a^{-n} = \frac{1}{a^n}$, $2^{-3} = \frac{1}{2^3} = \frac{1}{2 \cdot 2 \cdot 2} = \frac{1}{8}$.

(b) $7^{-1} = \frac{1}{7^1} = \frac{1}{7}$ (c) $x^{-2} = \frac{1}{x^2}$ (d) $(x + y)^{-8} = \frac{1}{(x + y)^8}$

Now Try Exercises 7, 19, 25(b)

> **TECHNOLOGY NOTE**
>
> **Negative Exponents**
> Calculators can evaluate negative exponents. The figure shows how a calculator evaluates the expressions in parts (a) and (b) of Example 1.
>
> ```
> 2^(-3)▶Frac
>              1/8
> 7^(-1)▶Frac
>              1/7
> ```
>
> **CALCULATOR HELP**
>
> To use the fraction feature (Frac), see Appendix A (pages AP-1 and AP-2).

The rules for exponents discussed in this chapter so far also apply to expressions that have negative exponents. For example, we can apply the product rule, $a^m \cdot a^n = a^{m+n}$, as follows.

$$2^{-3} \cdot 2^2 = 2^{-3+2} = 2^{-1} = \frac{1}{2}$$

We can check this result by evaluating the expression without using the product rule.

$$2^{-3} \cdot 2^2 = \frac{1}{2^3} \cdot 2^2 = \frac{1}{8} \cdot 4 = \frac{4}{8} = \frac{1}{2} \checkmark$$

EXAMPLE 2   **Using the product rule with negative exponents**

Evaluate each expression.
(a) $5^2 \cdot 5^{-4}$ (b) $3^{-2} \cdot 3^{-1}$

Solution

Product rule: Add exponents.

(a) $5^2 \cdot 5^{-4} = 5^{2+(-4)} = 5^{-2} = \frac{1}{5^2} = \frac{1}{25}$

(b) $3^{-2} \cdot 3^{-1} = 3^{-2+(-1)} = 3^{-3} = \frac{1}{3^3} = \frac{1}{27}$

Now Try Exercise 9

EXAMPLE 3 **Using the rules of exponents**

Simplify the expression. Write the answer using positive exponents.
(a) $x^2 \cdot x^{-5}$ **(b)** $(y^3)^{-4}$ **(c)** $(rt)^{-5}$ **(d)** $(ab)^{-3}(a^{-2}b)^3$

Solution
(a) Using the product rule, $a^m \cdot a^n = a^{m+n}$, gives

$$x^2 \cdot x^{-5} = x^{2+(-5)} = x^{-3} = \frac{1}{x^3}.$$

(b) Using the power rule, $(a^m)^n = a^{mn}$, gives

$$(y^3)^{-4} = y^{3(-4)} = y^{-12} = \frac{1}{y^{12}}.$$

(c) Using the power rule, $(ab)^n = a^n b^n$, gives

$$(rt)^{-5} = r^{-5}t^{-5} = \frac{1}{r^5} \cdot \frac{1}{t^5} = \frac{1}{r^5 t^5}.$$

This expression could also be simplified as follows.

$$(rt)^{-5} = \frac{1}{(rt)^5} = \frac{1}{r^5 t^5}$$

(d) $(ab)^{-3}(a^{-2}b)^3 = a^{-3}b^{-3}a^{-6}b^3$
$$= a^{-3+(-6)}b^{-3+3}$$
$$= a^{-9}b^0$$
$$= \frac{1}{a^9} \cdot 1$$
$$= \frac{1}{a^9}$$

Now Try Exercises 21, 27, 29(a)

2 Applying the Quotient Rule

Consider the division problem

$$\frac{3^4}{3^2} = \frac{3 \cdot 3 \cdot 3 \cdot 3}{3 \cdot 3} = \frac{3}{3} \cdot \frac{3}{3} \cdot 3 \cdot 3 = 1 \cdot 1 \cdot 3^2 = 3^2.$$

Because there are two more 3s in the numerator than in the denominator, the result is

$$\overset{\text{Subtract}}{3^{4-2}} = 3^2.$$

That is, to divide exponential expressions that have the *same base*, subtract the exponent of the denominator from the exponent of the numerator and keep the same base. This rule is called the *quotient rule*, which we express in symbols as follows.

READING CHECK 2

Use the quotient rule to simplify $\frac{5^2}{5^4}$.

THE QUOTIENT RULE

For any nonzero number a and integers m and n,

$$\frac{a^m}{a^n} = a^{m-n}.$$

EXAMPLE 4 **Using the quotient rule**

Simplify each expression. Write the answer using positive exponents.

(a) $\dfrac{4^3}{4^5}$ (b) $\dfrac{6a^7}{3a^4}$ (c) $\dfrac{xy^7}{x^2y^5}$

Solution

> Quotient rule: Subtract exponents.

Subtract

(a) $\dfrac{4^3}{4^5} = 4^{3-5} = 4^{-2} = \dfrac{1}{4^2} = \dfrac{1}{16}$

(b) $\dfrac{6a^7}{3a^4} = \dfrac{6}{3} \cdot \dfrac{a^7}{a^4} = 2a^{7-4} = 2a^3$

(c) $\dfrac{xy^7}{x^2y^5} = \dfrac{x^1}{x^2} \cdot \dfrac{y^7}{y^5} = x^{1-2}y^{7-5} = x^{-1}y^2 = \dfrac{y^2}{x}$

Now Try Exercises 13(b), 31(b), 33(a)

MAKING CONNECTIONS 1

The Quotient Rule and Simplifying Quotients

Some quotients can be simplified mentally. Because

$$\frac{x^5}{x^3} = \frac{x \cdot x \cdot x \cdot x \cdot x}{x \cdot x \cdot x},$$

the quotient $\frac{x^5}{x^3}$ has five factors of x in the numerator and three factors of x in the denominator. There are two more factors of x in the numerator than in the denominator, so this expression simplifies to $\frac{x^2}{1}$. or simply x^2. Similarly,

$$\frac{x^3}{x^5} = \frac{x \cdot x \cdot x}{x \cdot x \cdot x \cdot x \cdot x}$$

has two more factors of x in the denominator than in the numerator. This quotient $\frac{x^3}{x^5}$ simplifies to $\frac{1}{x^2}$.

---

3 Applying Other Rules for Exponents

Other rules can be used to simplify expressions with negative exponents.

QUOTIENTS AND NEGATIVE EXPONENTS

The following three rules hold for any nonzero real numbers a and b and positive integers m and n.

1. $\dfrac{1}{a^{-n}} = a^n$ **2.** $\dfrac{a^{-n}}{b^{-m}} = \dfrac{b^m}{a^n}$ **3.** $\left(\dfrac{a}{b}\right)^{-n} = \left(\dfrac{b}{a}\right)^n$

We demonstrate the validity of the first rule as follows.

1. $\dfrac{1}{a^{-n}} = \dfrac{1}{\dfrac{1}{a^n}} = 1 \cdot \dfrac{a^n}{1} = a^n$

We demonstrate the validity of rules 2 and 3 from the previous page as follows.

$$2.\ \frac{a^{-n}}{b^{-m}} = \frac{\dfrac{1}{a^n}}{\dfrac{1}{b^m}} = \frac{1}{a^n} \cdot \frac{b^m}{1} = \frac{b^m}{a^n}$$

$$3.\ \left(\frac{a}{b}\right)^{-n} = \frac{a^{-n}}{b^{-n}} = \frac{\dfrac{1}{a^n}}{\dfrac{1}{b^n}} = \frac{1}{a^n} \cdot \frac{b^n}{1} = \frac{b^n}{a^n} = \left(\frac{b}{a}\right)^n$$

EXAMPLE 5 **Working with quotients and negative exponents**

Simplify each expression. Write the answer using positive exponents.

(a) $\dfrac{1}{2^{-5}}$ (b) $\dfrac{3^{-3}}{4^{-2}}$ (c) $\dfrac{5x^{-4}y^2}{10x^2y^{-4}}$ (d) $\left(\dfrac{2}{z^2}\right)^{-4}$

Solution

(a) $\dfrac{1}{2^{-5}} = 2^5 = 2 \cdot 2 \cdot 2 \cdot 2 \cdot 2 = 32$ (b) $\dfrac{3^{-3}}{4^{-2}} = \dfrac{4^2}{3^3} = \dfrac{16}{27}$

(c) $\dfrac{5x^{-4}y^2}{10x^2y^{-4}} = \dfrac{y^2y^4}{2x^2x^4} = \dfrac{y^6}{2x^6}$ (d) $\left(\dfrac{2}{z^2}\right)^{-4} = \left(\dfrac{z^2}{2}\right)^4 = \dfrac{z^8}{2^4} = \dfrac{z^8}{16}$

Now Try Exercises 15(b), 17, 37, 47

The rules for natural number exponents that are summarized in the first section of this chapter also hold for integer exponents. Additional rules for integer exponents are summarized as follows.

RULES FOR INTEGER EXPONENTS

The following rules hold for nonzero real numbers a and b, and positive integers m and n.

Description	Rule	Example
Negative Exponents	$a^{-n} = \dfrac{1}{a^n}$	$9^{-2} = \dfrac{1}{9^2} = \dfrac{1}{81}$
The Quotient Rule	$\dfrac{a^m}{a^n} = a^{m-n}$	$\dfrac{2^3}{2^{-2}} = 2^{3-(-2)} = 2^5$
Negative Exponents	$\dfrac{1}{a^{-n}} = a^n$	$\dfrac{1}{7^{-5}} = 7^5$
Negative Exponents	$\dfrac{a^{-n}}{b^{-m}} = \dfrac{b^m}{a^n}$	$\dfrac{4^{-3}}{2^{-5}} = \dfrac{2^5}{4^3}$
Negative Exponents	$\left(\dfrac{a}{b}\right)^{-n} = \left(\dfrac{b}{a}\right)^n$	$\left(\dfrac{1}{5}\right)^{-2} = \left(\dfrac{5}{1}\right)^2 = 25$

READING CHECK 3

• What kinds of numbers are often expressed in scientific notation?

TECHNOLOGY NOTE

Powers of 10
Calculators make use of scientific notation, as illustrated in the accompanying figure. The letter E denotes a power of 10. That is,
$2.5\text{E}13 = 2.5 \times 10^{13}$ and $5\text{E}{-}6 = 5 \times 10^{-6}$.

```
25000000000000
              2.5E13
.000005
              5E-6
```

NOTE: The calculator has been set in *scientific mode*. ∎

4 Reading and Writing Scientific Notation

Powers of 10 are important because they are used in science to express numbers that are either very small or very large in absolute value. **TABLE 5.7** lists some powers of 10. Note that if the power of 10 decreases by 1, the result decreases by a factor of $\frac{1}{10}$, or equivalently, the decimal point is moved one place to the left. **TABLE 5.8** shows the names of some important powers of 10.

Powers of 10

Power of 10	Value
10^3	1000
10^2	100
10^1	10
10^0	1
10^{-1}	$\frac{1}{10} = 0.1$
10^{-2}	$\frac{1}{100} = 0.01$
10^{-3}	$\frac{1}{1000} = 0.001$

TABLE 5.7

Important Powers of 10

Power of 10	Name
10^{12}	Trillion
10^9	Billion
10^6	Million
10^3	Thousand
10^{-1}	Tenth
10^{-2}	Hundredth
10^{-3}	Thousandth
10^{-6}	Millionth

TABLE 5.8

Recall that numbers written in decimal notation are sometimes said to be in *standard form*. Decimal numbers that are either very large or very small in absolute value can be expressed in *scientific notation*.

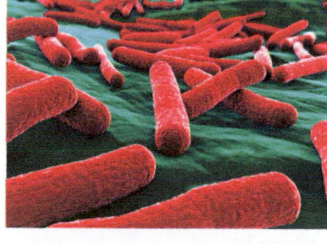

Math in Context (Science) The distance to the planet WASP-17b, discovered in 2009, is about 5,880,000,000,000,000 miles. This distance can be written in scientific notation as 5.88×10^{15} because

$$5{,}880{,}000{,}000{,}000{,}000 = 5.88 \times 10^{15}.$$

15 decimal places

The 10^{15} indicates that the decimal point in 5.88 should be moved **15** places to the **right**.

A typical virus is about 0.00000468 inch in diameter, which can be written in scientific notation as 4.68×10^{-6} because

$$0.00000468 = 4.68 \times 10^{-6}.$$

6 decimal places

The 10^{-6} indicates that the decimal point in 4.68 should be moved **6** places to the **left**. The following definition provides a more complete explanation of scientific notation.

SCIENTIFIC NOTATION

A real number a is in **scientific notation** when a is written in the form $b \times 10^n$, where $1 \le |b| < 10$ and n is an integer.

EXAMPLE 6 **Converting scientific notation to standard form**

Write each number in standard form.
(a) 5.23×10^4 **(b)** 8.1×10^{-3} **(c)** 6×10^{-2}

Solution
(a) The positive exponent 4 indicates that the decimal point in 5.23 is to be moved **4** places to the **right**.

$$5.23 \times 10^4 = 5.2\,3\,0\,0. = 52{,}300$$
$$\qquad\qquad\quad 1\,2\,3\,4$$

(b) The negative exponent -3 indicates that the decimal point in 8.1 is to be moved **3** places to the **left**.

$$8.1 \times 10^{-3} = 0.\,0\,0\,8.\,1 = 0.0081$$
$$\qquad\qquad\quad 1\,2\,3$$

(c) $6 \times 10^{-2} = 0.\,0\,6. = 0.06$
$$\qquad\qquad\; 1\,2$$

Now Try Exercises 61, 63

The following steps can be used for writing a positive number a in scientific notation.

WRITING A POSITIVE NUMBER IN SCIENTIFIC NOTATION

For a positive, rational number a expressed as a decimal, if $1 \leq a < 10$, then $a = a \times 10^0$. Otherwise, use the following process to write a in scientific notation.

1. Move the decimal point in a until it becomes a number b such that $1 \leq b < 10$.
2. Let the positive integer n be the number of places the decimal point was moved.
3. Write a in scientific notation as follows.
   - If $a \geq 10$, then $a = b \times 10^n$.
   - If $a < 1$, then $a = b \times 10^{-n}$.

NOTE: The scientific notation for a negative number a is the opposite of the scientific notation of $|a|$. For example, $450 = 4.5 \times 10^2$ and $-450 = -4.5 \times 10^2$. ■

EXAMPLE 7 **Writing a number in scientific notation**

Write each number in scientific notation.
(a) 321,000,000 (U.S. population in 2015)
(b) 0.001 (Approximate time in seconds for sound to travel one foot)

Solution
(a) Move the assumed decimal point in 321,000,000 eight places to obtain 3.21.

$$3.\,2\,1\,0\,0\,0\,0\,0\,0.$$
$$\quad 1\,2\,3\,4\,5\,6\,7\,8$$

Since $321{,}000{,}000 \geq 10$, the scientific notation is 3.21×10^8.

(b) Move the decimal point in 0.001 three places to obtain 1.

$$0.\underset{1\,2\,3}{0\,0\,1,}$$

Since $0.001 < 1$, the scientific notation is 1×10^{-3}.

Now Try Exercises 75, 79

Numbers in scientific notation can be multiplied by applying properties of real numbers and properties of exponents.

$$
\begin{aligned}
(6 \times 10^4) \cdot (3 \times 10^3) &= (6 \cdot 3) \times (10^4 \cdot 10^3) && \text{Properties of real numbers} \\
&= 18 \times 10^7 && \text{Product rule} \\
&= 1.8 \times 10^8 && \text{Scientific notation}
\end{aligned}
$$

Division can also be performed with scientific notation.

$$
\begin{aligned}
\frac{6 \times 10^4}{3 \times 10^3} &= \frac{6}{3} \times \frac{10^4}{10^3} && \text{Property of fractions} \\
&= 2 \times 10^1 && \text{Quotient rule}
\end{aligned}
$$

These results are supported in **FIGURE 5.16**, where the calculator is in scientific mode.

```
(6*10^4)(3*10^3)
               1.8E8
(6*10^4)/(3*10^3
)
               2E1
```

FIGURE 5.16

CALCULATOR HELP

To display numbers in scientific notation, see Appendix A (page AP-2).

🌐 Math in Context *(Advertising)* In the next example, we show how to use scientific notation to calculate the per person spending on advertising in the United States.

EXAMPLE 8 **Analyzing the cost of advertising**

In 2016, a total of $\$1.969 \times 10^{11}$ was spent on advertising in the United States. At that time, the population of the United States was 3.26×10^8. Determine how much was spent per person on advertising. *(Source: eMarketer.)*

CRITICAL THINKING 1

Estimate the number of seconds that you have been alive. Write your answer in scientific notation.

Solution

To determine the amount spent per person, divide $\$1.969 \times 10^{11}$ by 3.26×10^8.

$$\frac{1.969 \times 10^{11}}{3.26 \times 10^8} = \frac{1.969}{3.26} \times 10^{11-8} \approx 0.604 \times 10^3 = 604$$

In 2016, about $604 per person was spent on advertising in the United States.

Now Try Exercise 97

5.5 Putting It All Together

For the rules for integer exponents in this table, assume that a and b are nonzero real numbers and that m and n are integers.

CONCEPT	EXPLANATION	EXAMPLES
Negative Integer Exponents	$a^{-n} = \dfrac{1}{a^n}$	$2^{-4} = \dfrac{1}{2^4} = \dfrac{1}{16}$, $\quad a^{-8} = \dfrac{1}{a^8}$, $\quad$ and $(xy)^{-2} = \dfrac{1}{(xy)^2} = \dfrac{1}{x^2 y^2}$

continued on next page

continued from previous page

CONCEPT	EXPLANATION	EXAMPLES
Quotient Rule	$\dfrac{a^m}{a^n} = a^{m-n}$	$\dfrac{7^2}{7^4} = 7^{2-4} = 7^{-2} = \dfrac{1}{7^2} = \dfrac{1}{49}$ and $\dfrac{x^6}{x^3} = x^{6-3} = x^3$
Quotients and Negative Integer Exponents	1. $\dfrac{1}{a^{-n}} = a^n$ 2. $\dfrac{a^{-n}}{b^{-m}} = \dfrac{b^m}{a^n}$ 3. $\left(\dfrac{a}{b}\right)^{-n} = \left(\dfrac{b}{a}\right)^n$	1. $\dfrac{1}{5^{-2}} = 5^2 = 25$ 2. $\dfrac{x^{-4}}{y^{-2}} = \dfrac{y^2}{x^4}$ 3. $\left(\dfrac{2}{3}\right)^{-3} = \left(\dfrac{3}{2}\right)^3 = \dfrac{3^3}{2^3} = \dfrac{27}{8}$
Scientific Notation	Write a as $b \times 10^n$, where $1 \le |b| < 10$ and n is an integer.	$23{,}500 = 2.35 \times 10^4$, $0.0056 = 5.6 \times 10^{-3}$, and $1000 = 1 \times 10^3$

5.5 Exercises

MyMathLab®

CONCEPTS AND VOCABULARY

Exercises 1–5: Complete the given rule for integer exponents m and n, where a and b are nonzero real numbers.

1. $a^{-n} = $ _____

2. $\dfrac{1}{a^{-n}} = $ _____

3. $\dfrac{a^m}{a^n} = $ _____

4. $\dfrac{a^{-n}}{b^{-m}} = $ _____

5. $\left(\dfrac{a}{b}\right)^{-n} = $ _____

6. To write a positive number a in scientific notation as $b \times 10^n$, the number b must satisfy _____.

USING NEGATIVE INTEGERS AS EXPONENTS

Exercises 7–18: Simplify the expression.

7. (a) 4^{-1} (b) $\left(\dfrac{1}{3}\right)^{-2}$

8. (a) 6^{-2} (b) 2.5^{-1}

9. (a) $2^3 \cdot 2^{-2}$ (b) $10^{-1} \cdot 10^{-2}$

10. (a) $3^{-4} \cdot 3^2$ (b) $10^4 \cdot 10^{-2}$

11. (a) $3^{-2} \cdot 3^{-1} \cdot 3^{-1}$ (b) $(2^3)^{-1}$

12. (a) $2^{-3} \cdot 2^5 \cdot 2^{-4}$ (b) $(3^{-2})^{-2}$

13. (a) $(3^2 4^3)^{-1}$ (b) $\dfrac{4^5}{4^2}$

14. (a) $(2^{-2} 3^2)^{-2}$ (b) $\dfrac{5^5}{5^3}$

15. (a) $\dfrac{1^9}{1^7}$ (b) $\dfrac{1}{4^{-3}}$

16. (a) $\dfrac{-6^4}{6}$ (b) $\dfrac{1}{6^{-2}}$

17. (a) $\dfrac{5^{-2}}{5^{-4}}$ (b) $\left(\dfrac{2}{7}\right)^{-2}$

18. (a) $\dfrac{7^{-3}}{7^{-1}}$ (b) $\left(\dfrac{3}{4}\right)^{-3}$

APPLYING RULES FOR INTEGER EXPONENTS

Exercises 19–50: Simplify the expression. Write the answer using positive exponents.

19. (a) x^{-1} (b) a^{-4}

20. (a) y^{-2} (b) z^{-7}

21. (a) $x^{-2} \cdot x^{-1} \cdot x$ (b) $a^{-5} \cdot a^{-2} \cdot a^{-1}$

22. (a) $y^{-3} \cdot y^4 \cdot y^{-5}$ (b) $b^5 \cdot b^{-3} \cdot b^{-6}$

23. (a) $x^2 y^{-3} x^{-5} y^6$ (b) $(xy)^{-3}$

24. (a) $a^{-2} b^{-6} b^3 a^{-1}$ (b) $(ab)^{-1}$

25. (a) $(2t)^{-4}$ (b) $(x+1)^{-7}$

26. (a) $(8c)^{-2}$ (b) $(a+b)^{-9}$

27. (a) $(a^{-2})^{-4}$ (b) $(rt^3)^{-2}$

28. (a) $(4x^3)^{-3}$ (b) $(xy^{-3})^{-2}$

29. (a) $(ab)^2 (a^2)^{-3}$ (b) $\dfrac{x^4}{x^2}$

30. (a) $(x^3)^{-2} (xy)^4 y^{-5}$ (b) $\dfrac{y^9}{y^5}$

31. (a) $\dfrac{a^{10}}{a^{-3}}$ (b) $\dfrac{4z}{2z^4}$

32. (a) $\dfrac{b^5}{b^{-2}}$ (b) $\dfrac{12x^2}{24x^7}$

33. (a) $\dfrac{-4xy^5}{6x^3y^2}$ (b) $\dfrac{x^{-4}}{x^{-1}}$

34. (a) $\dfrac{12a^6b^2}{8ab^3}$ (b) $\dfrac{y^{-2}}{y^{-7}}$

35. (a) $\dfrac{10b^{-4}}{5b^{-5}}$ (b) $\left(\dfrac{a}{b}\right)^3$

36. (a) $\dfrac{8a^{-2}}{2a^{-3}}$ (b) $\left(\dfrac{2x}{y}\right)^5$

37. (a) $\dfrac{6x^2y^{-4}}{18x^{-5}y^4}$ (b) $\dfrac{16a^{-3}b^{-5}}{4a^{-8}b}$

38. (a) $\dfrac{m^2n^4}{3m^{-5}n^4}$ (b) $\dfrac{7x^{-3}y^{-5}}{x^{-3}y^{-2}}$

39. (a) $\dfrac{1}{y^{-5}}$ (b) $\dfrac{4}{2t^{-3}}$

40. (a) $\dfrac{1}{z^{-6}}$ (b) $\dfrac{5}{10b^{-5}}$

41. (a) $\dfrac{3a^4}{(2a^{-2})^3}$ (b) $\dfrac{(2b^5)^{-3}}{4b^{-6}}$

42. (a) $\dfrac{(2x^4)^{-2}}{5x^{-2}}$ (b) $\dfrac{2y^5}{(3y^{-4})^{-2}}$

43. (a) $\dfrac{1}{(xy)^{-2}}$ (b) $\dfrac{1}{(a^2b)^{-3}}$

44. (a) $\dfrac{1}{(ab)^{-1}}$ (b) $\dfrac{1}{(rt^4)^{-2}}$

45. (a) $\dfrac{(3m^4n)^{-2}}{(2mn^{-2})^3}$ (b) $\dfrac{(-4x^4y)^2}{(xy^{-5})^{-3}}$

46. (a) $\dfrac{(x^4y^2)^2}{(-2x^2y^{-2})^3}$ (b) $\dfrac{(m^2n^{-6})^{-2}}{(4m^2n^{-4})^{-3}}$

47. (a) $\left(\dfrac{a}{b}\right)^{-2}$ (b) $\left(\dfrac{u}{4v}\right)^{-1}$

48. (a) $\left(\dfrac{2x}{y}\right)^{-3}$ (b) $\left(\dfrac{5u}{3v}\right)^{-2}$

49. (a) $\left(\dfrac{3a^4b}{2ab^{-2}}\right)^{-2}$ (b) $\left(\dfrac{4m^4n}{5m^{-3}n^2}\right)^2$

50. (a) $\left(\dfrac{2x^4y^2}{3x^3y^{-3}}\right)^3$ (b) $\left(\dfrac{a^{-5}b^2}{2ab^{-2}}\right)^{-2}$

51. Thinking Generally For positive integers m and n, show that $\dfrac{a^n}{a^m} = \dfrac{1}{a^{m-n}}$.

52. Thinking Generally For positive integers m and n, show that $\dfrac{a^{-n}}{a^{-m}} = a^{m-n}$.

READING AND WRITING SCIENTIFIC NOTATION

Exercises 53–58: (Refer to Table 5.8.) Write the value of the power of 10 in words.

53. 10^3 **54.** 10^6

55. 10^9 **56.** 10^{-1}

57. 10^{-2} **58.** 10^{-6}

Exercises 59–70: Write the expression in standard form.

59. -2×10^{-3} **60.** -5×10^{-2}

61. 4.5×10^4 **62.** 7.1×10^6

63. 8×10^{-3} **64.** 9×10^{-1}

65. 4.56×10^{-4} **66.** 9.4×10^{-2}

67. 3.9×10^7 **68.** 5.27×10^6

69. -5×10^5 **70.** -9.5×10^3

Exercises 71–82: Write the number in scientific notation.

71. 2000   **72.** 11,000

73. −567   **74.** −9300

75. 12,000,000   **76.** 600,000

77. 0.004   **78.** 0.0008

79. 0.000895   **80.** 0.0123

81. −0.05   **82.** −0.934

Exercises 83–90: Evaluate the expression. Write the answer in standard form.

83. $(5 \times 10^3)(3 \times 10^2)$

84. $(2.1 \times 10^2)(2 \times 10^4)$

85. $(-3 \times 10^{-3})(5 \times 10^2)$

86. $(4 \times 10^2)(1 \times 10^3)(5 \times 10^{-4})$

87. $\dfrac{4 \times 10^5}{2 \times 10^2}$ **88.** $\dfrac{9 \times 10^2}{3 \times 10^6}$

89. $\dfrac{8 \times 10^{-6}}{4 \times 10^{-3}}$ **90.** $\dfrac{6.3 \times 10^2}{2 \times 10^{-3}}$

APPLICATIONS INVOLVING INTEGER EXPONENTS

91. Light-year The distance that light travels in 1 year is called a *light-year*. Light travels at 1.86×10^5 miles per second, and there are about 3.15×10^7 seconds in 1 year.
  (a) Estimate the number of miles in 1 light-year.
  (b) Except for the sun, Alpha Centauri is the nearest star, and its distance is 4.27 light-years from Earth. Estimate its distance in miles. Write your answer in scientific notation.

92. Milky Way It takes 2×10^8 years for the sun to make one orbit around the Milky Way galaxy. Write this number in standard form.

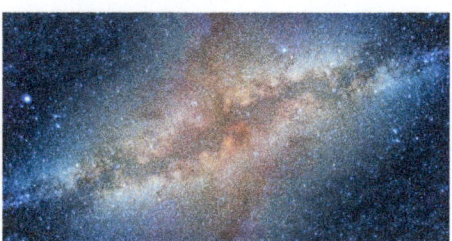

93. Speed of the Sun (Refer to the two previous exercises.) Assume that the sun's orbit in the Milky Way galaxy is circular with a diameter of 10^5 light-years. Estimate how many miles the sun travels in 1 year.

94. Distance to the Moon The moon is about 240,000 miles from Earth.
  (a) Write this number in scientific notation.
  (b) If a rocket traveled at 4×10^4 miles per hour, how long would it take for it to reach the moon?

95. Online Exploration In 1997, the creators of the Internet search engine BackRub renamed it Google. This new name is a play on the word *googol*, which is a very large number. Look up a googol and write it in scientific notation.

96. Online Exploration An astronomical unit (AU) is based on the distance from Earth to the sun. Look up the distance in kilometers from Earth to the sun.
  (a) Write an astronomical unit in standard form to the nearest million kilometers.
  (b) Convert your rounded answer from part (a) to scientific notation.

97. Gross Domestic Product The gross domestic product (GDP) is the total national output of goods and services valued at market prices *within* the United States. The GDP of the United States in 2013 was $15,684,000,000,000. (*Source:* World Bank.)
  (a) Write this number in scientific notation.
  (b) In 2013, the U.S. population was 3.17×10^8. On average, how many dollars of goods and services were produced by each individual?

98. Average Household Income Recently, there were 1.22×10^8 households in the United States and the average household income was $51,371. Calculate the total household income in the United States. (*Source:* U.S. Census Bureau.)

WRITING ABOUT MATHEMATICS

99. Explain what a negative exponent is and how it is different from a positive exponent. Give an example.

100. Explain why scientific notation is helpful for writing some numbers.

Group Activity

Water in a Lake East Battle Lake in Minnesota covers an area of about 1950 acres, or 8.5×10^7 square feet, and its average depth is about 3.2×10^1 feet.

(a) Estimate the cubic feet of water in the lake. (*Hint:* volume = area × average depth.)

(b) One cubic foot of water equals about 7.5 gallons. How many gallons of water are in this lake?

(c) The population of the United States is about 3.1×10^8, and the average American uses about 1.5×10^2 gallons of water per day. Could this lake supply the American population with water for 1 day?

5.6   Division of Polynomials

Objectives

1 Dividing by a Monomial
2 Dividing by a Polynomial

1   Dividing by a Monomial

To add two fractions with like denominators, we use the property

$$\frac{a}{d} + \frac{b}{d} = \frac{a + b}{d}.$$

For example, $\frac{1}{7} + \frac{3}{7} = \frac{1 + 3}{7} = \frac{4}{7}$.

To divide a polynomial by a monomial, we use the same property, only in reverse. That is,

$$\frac{a + b}{d} = \frac{a}{d} + \frac{b}{d}.$$

Note that each term in the numerator is divided by the monomial in the denominator. The next example shows how to divide a polynomial by a monomial.

EXAMPLE 1   **Dividing a polynomial by a monomial**

Divide.

(a) $\dfrac{a^5 + a^3}{a^2}$ **(b)** $\dfrac{5x^4 + 10x}{10x}$ **(c)** $\dfrac{3y^2 + 2y - 12}{6y}$

Solution

(a) $\dfrac{a^5 + a^3}{a^2} = \dfrac{a^5}{a^2} + \dfrac{a^3}{a^2} = a^{5-2} + a^{3-2} = a^3 + a$

(b) $\dfrac{5x^4 + 10x}{10x} = \dfrac{5x^4}{10x} + \dfrac{10x}{10x} = \dfrac{x^3}{2} + 1$

(c) $\dfrac{3y^2 + 2y - 12}{6y} = \dfrac{3y^2}{6y} + \dfrac{2y}{6y} - \dfrac{12}{6y} = \dfrac{y}{2} + \dfrac{1}{3} - \dfrac{2}{y}$

Now Try Exercises 17, 19, 21

MAKING CONNECTIONS 1

Division and Simplification

A common mistake made when dividing expressions is to "cancel" incorrectly. Note in Example 1(b) that

$$\frac{5x^4 + 10x}{10x} \neq 5x^4 + \frac{10x}{10x}.$$

> We cannot "cancel" the 10x terms.

The monomial must be divided into *every* term in the numerator.

When dividing two natural numbers, we can check our work by multiplying. For example, $\frac{10}{5} = 2$, and we can check this result by finding the product $5 \cdot 2 = 10$. Similarly, to check

$$\frac{a^5 + a^3}{a^2} = a^3 + a$$

in Example 1(a), we can multiply a^2 and $a^3 + a$.

$$a^2(a^3 + a) = a^2 \cdot a^3 + a^2 \cdot a \qquad \text{Distributive property}$$
$$= a^5 + a^3 \checkmark \qquad \text{It checks.}$$

EXAMPLE 2 ▸ **Dividing and checking**

Divide the expression $\frac{8x^3 - 4x^2 + 6x}{2x^2}$ and then check the result.

Solution
Be sure to divide $2x^2$ into *every* term in the numerator.

$$\frac{8x^3 - 4x^2 + 6x}{2x^2} = \frac{8x^3}{2x^2} - \frac{4x^2}{2x^2} + \frac{6x}{2x^2} = 4x - 2 + \frac{3}{x}$$

Check:

$$2x^2\left(4x - 2 + \frac{3}{x}\right) = 2x^2 \cdot 4x - 2x^2 \cdot 2 + 2x^2 \cdot \frac{3}{x}$$
$$= 8x^3 - 4x^2 + 6x \checkmark$$

Now Try Exercise 13

EXAMPLE 3 ▸ **Finding the length of a rectangle**

The rectangle in **FIGURE 5.17** has an area $A = x^2 + 2x$ and width x. Write an expression for its length l in terms of x.

$$x \quad \boxed{A = x^2 + 2x}$$

$$l$$

FIGURE 5.17

Solution

The area A of a rectangle equals length l times width w, or $A = lw$. Solving for l gives

$$l = \frac{A}{w}.$$

Thus to find the length of the given rectangle, divide the area by the width.

$$l = \frac{x^2 + 2x}{x} = \frac{x^2}{x} + \frac{2x}{x} = x + 2$$

The length of the rectangle is $x + 2$. The answer checks because $x(x + 2) = x^2 + 2x$.

Now Try Exercise 49

2 Dividing by a Polynomial

To understand division by a polynomial better, we first need to review some terminology related to long division of natural numbers. To compute $271 \div 4$, we complete long division as follows.

$$\begin{array}{r}
\text{Quotient} \longrightarrow 67\text{ R }3 \longleftarrow \text{Remainder}\\
\text{Divisor} \longrightarrow 4)\overline{271} \longleftarrow \text{Dividend}\\
\underline{24}\\
31\\
\underline{28}\\
3
\end{array}$$

To check this result, we find the product of the quotient and divisor and then add the remainder. Because $67 \cdot 4 + 3 = 271$, the answer checks. The quotient and remainder can also be expressed as $67\frac{3}{4}$. Division of polynomials is similar to long division of natural numbers.

EXAMPLE 4 **Dividing polynomials**

Divide $\frac{6x^2 + 13x + 3}{3x + 2}$.

Solution

Begin by dividing the first term of $3x + 2$ into the first term of $6x^2 + 13x + 3$. That is, divide $3x$ into $6x^2$ to obtain $2x$. Then find the product of $2x$ and $3x + 2$, or $6x^2 + 4x$, place it below $6x^2 + 13x$, and subtract. Bring down the 3.

$$\begin{array}{r}
2x\\
3x + 2)\overline{6x^2 + 13x + 3}\\
\underline{6x^2 + 4x}\\
9x + 3
\end{array}
\qquad
\begin{array}{l}
\frac{6x^2}{3x} = 2x\\[4pt]
2x(3x + 2) = 6x^2 + 4x\\[4pt]
\text{Subtract: } 13x - 4x = 9x. \text{ Bring down the 3.}
\end{array}$$

In the next step, divide $3x$ into the first term of $9x + 3$ to obtain 3. Then find the product of 3 and $3x + 2$, or $9x + 6$, place it below $9x + 3$, and subtract.

$$\begin{array}{r}
2x + 3\\
3x + 2)\overline{6x^2 + 13x + 3}\\
\underline{6x^2 + 4x}\\
9x + 3\\
\underline{9x + 6}\\
-3
\end{array}
\qquad
\begin{array}{l}
\frac{9x}{3x} = 3\\[20pt]
3(3x + 2) = 9x + 6\\[4pt]
\text{Subtract: } 3 - 6 = -3.
\end{array}$$

The quotient is $2x + 3$ with remainder -3. This result can also be written as

$$2x + 3 + \frac{-3}{3x + 2}, \qquad \text{Quotient} + \frac{\text{Remainder}}{\text{Divisor}}$$

in the same manner that 67 R 3 was written as $67\frac{3}{4}$.

Now Try Exercise 27

MAKING CONNECTIONS 2

Checking Polynomial Division

Check polynomial division by adding the remainder to the product of the divisor and the quotient. That is,

$$(\text{Divisor})(\text{Quotient}) + \text{Remainder} = \text{Dividend}.$$

For Example 4, the equation becomes

$$(3x + 2)(2x + 3) + (-3) = 3x \cdot 2x + 3x \cdot 3 + 2 \cdot 2x + 2 \cdot 3 - 3$$
$$= 6x^2 + 9x + 4x + 6 - 3$$
$$= 6x^2 + 13x + 3. \ \checkmark \quad \text{It checks.}$$

EXAMPLE 5 **Dividing polynomials having a missing term**

Simplify $(3x^3 + 2x - 4) \div (x - 2)$.

Solution
Because the dividend does not have an x^2-term, insert $0x^2$ as a "place holder." Then begin by dividing x into $3x^3$ to obtain $3x^2$.

$$
\begin{array}{r}
3x^2 \phantom{{}+0x^2+2x-4} \\
x - 2 \overline{) 3x^3 + 0x^2 + 2x - 4} \\
\underline{3x^3 - 6x^2} \phantom{{}+2x-4} \\
6x^2 + 2x \phantom{{}-4}
\end{array}
$$

$\dfrac{3x^3}{x} = 3x^2$

$3x^2(x - 2) = 3x^3 - 6x^2$

Subtract $0x^2 - (-6x^2) = 6x^2$. Bring down $2x$.

In the next step, divide x into $6x^2$.

$$
\begin{array}{r}
3x^2 + 6x \phantom{{}-4} \\
x - 2 \overline{) 3x^3 + 0x^2 + 2x - 4} \\
\underline{3x^3 - 6x^2} \phantom{{}+2x-4} \\
6x^2 + 2x \phantom{{}-4} \\
\underline{6x^2 - 12x} \phantom{{}} \\
14x - 4
\end{array}
$$

$\dfrac{6x^2}{x} = 6x$

$6x(x - 2) = 6x^2 - 12x$

Subtract: $2x - (-12x) = 14x$. Bring down -4.

Now divide x into $14x$.

$$
\begin{array}{r}
3x^2 + 6x + 14 \\
x - 2 \overline{) 3x^3 + 0x^2 + 2x - 4} \\
\underline{3x^3 - 6x^2} \phantom{{}+2x-4} \\
6x^2 + 2x \phantom{{}-4} \\
\underline{6x^2 - 12x} \phantom{{}} \\
14x - 4 \\
\underline{14x - 28} \\
24
\end{array}
$$

$\dfrac{14x}{x} = 14$

$14(x - 2) = 14x - 28$

Subtract: $-4 - (-28) = 24$.

The quotient is $3x^2 + 6x + 14$ with remainder 24. This result can also be written as

$$3x^2 + 6x + 14 + \frac{24}{x - 2}.$$

Now Try Exercise 37

EXAMPLE 6   **Dividing when the divisor is not linear**

Divide $x^3 - 3x^2 + 3x + 2$ by $x^2 + 1$.

Solution
Begin by writing $x^2 + 1$ as $x^2 + 0x + 1$.

$$
\require{enclose}
\begin{array}{r}
x - 3 \\
x^2 + 0x + 1 \enclose{longdiv}{x^3 - 3x^2 + 3x + 2} \\
\underline{x^3 + 0x^2 + x} \\
-3x^2 + 2x + 2 \\
\underline{-3x^2 + 0x - 3} \\
2x + 5
\end{array}
$$

The quotient is $x - 3$ with remainder $2x + 5$. This result can also be written as

$$x - 3 + \frac{2x + 5}{x^2 + 1}.$$

Now Try Exercise 41

5.6   Putting It All Together

CONCEPT	EXPLANATION	EXAMPLES
Division by a Monomial	Use the property $$\frac{a + b}{d} = \frac{a}{d} + \frac{b}{d}.$$ Be sure to divide the denominator into every term in the numerator.	$\dfrac{2x^3 + 4x}{2x^2} = \dfrac{2x^3}{2x^2} + \dfrac{4x}{2x^2} = x + \dfrac{2}{x}$ and $\dfrac{a^2 - 2a}{4a} = \dfrac{a^2}{4a} - \dfrac{2a}{4a} = \dfrac{a}{4} - \dfrac{1}{2}$
Division by a Polynomial	Is done similarly to the way long division of natural numbers is performed. If either the divisor or the dividend is missing a term, be sure to insert as a "place holder" the missing term with coefficient 0.	Divide $x^2 + 3x + 3$ by $x + 1$. $$\begin{array}{r} x + 2 \\ x + 1 \enclose{longdiv}{x^2 + 3x + 3} \\ \underline{x^2 + x} \\ 2x + 3 \\ \underline{2x + 2} \\ 1 \end{array}$$ The quotient is $x + 2$ with remainder 1, which can be expressed as $$x + 2 + \frac{1}{x + 1}.$$

continued on next page

continued from previous page

CONCEPT	EXPLANATION	EXAMPLES
Checking a Result	Dividend $=$ (Divisor)(Quotient) $+$ Remainder	When $x^2 + 3x + 3$ is divided by $x + 1$, the quotient is $x + 2$ with remainder 1. Thus $$(x + 1)(x + 2) + 1 = x^2 + 3x + 3,$$ and the answer checks.

5.6 Exercises

MyMathLab®

CONCEPTS AND VOCABULARY

1. $\dfrac{a + b}{d} = $ _____

2. $\dfrac{a + b - c}{d} = $ _____

3. When dividing a polynomial by a monomial, the monomial must be divided into every _____ of the polynomial.

4. (True or False?) The expressions $\dfrac{5x^2 + 2x}{2x}$ and $5x^2 + 1$ are equal.

5. (True or False?) The expressions $\dfrac{5x^2 + 2x}{2x}$ and $\dfrac{5x^2}{2x}$ are equal.

6. Because $\dfrac{37}{9} = 4$ with remainder 1, it follows that $37 = $ _____ $\cdot$ _____ $+$ _____.

7. Because $2x^3 - x + 5$ divided by $x + 1$ equals $2x^2 - 2x + 1$ with remainder 4, it follows that $2x^3 - x + 5 = $ _____ $\cdot$ _____ $+$ _____.

8. When dividing $2x^3 + 3x - 1$ by $x - 1$, insert ___ into the dividend as a "place holder" for the missing x^2-term.

DIVIDING BY A MONOMIAL

Exercises 9–16: Divide and check.

9. $\dfrac{6x^2}{3x}$

10. $\dfrac{-5x^2}{10x^4}$

11. $\dfrac{z^4 + z^3}{z}$

12. $\dfrac{t^3 - t}{t}$

13. $\dfrac{a^5 - 6a^3}{2a^3}$

14. $\dfrac{b^4 - 4b}{4b^2}$

15. $\dfrac{y + 6y^2}{3y^3}$

16. $\dfrac{8z^2 - z}{4z^2}$

Exercises 17–26: Divide.

17. $\dfrac{4x - 7x^4}{x^2}$

18. $\dfrac{1 + 6x^4}{3x^3}$

19. $\dfrac{6y^2 + 3y}{3y^3}$

20. $\dfrac{5z^2 - 10z^3}{5z^4}$

21. $\dfrac{9x^4 - 3x + 6}{3x}$

22. $\dfrac{y^3 - 4y + 6}{y}$

23. $\dfrac{12y^4 - 3y^2 + 6y}{3y^2}$

24. $\dfrac{2x^2 - 6x + 9}{12x}$

25. $\dfrac{15m^4 - 10m^3 + 20m^2}{5m^2}$

26. $\dfrac{n^8 - 8n^6 + 4n^4}{2n^5}$

DIVIDING BY A POLYNOMIAL

Exercises 27–34: Divide and check.

27. $\dfrac{2x^2 - 3x + 1}{x - 2}$

28. $\dfrac{4x^2 - x + 3}{x + 2}$

29. $\dfrac{x^2 + 2x + 1}{x + 1}$

30. $\dfrac{4x^2 - 4x + 1}{2x - 1}$

31. $\dfrac{x^3 - x^2 + x - 2}{x - 1}$

32. $\dfrac{2x^3 + 3x^2 + 3x - 1}{2x + 1}$

33. $\dfrac{x^3 + x^2 - 7x + 2}{x - 2}$

34. $\dfrac{x^3 + x^2 - 2x + 12}{x + 3}$

Exercises 35–46: Divide.

35. $\dfrac{4x^3 - 3x^2 + 7x + 3}{4x + 1}$

36. $\dfrac{10x^3 - x^2 - 17x - 7}{5x + 2}$

37. $\dfrac{x^3 - x + 2}{x - 2}$

38. $\dfrac{6x^3 + 8x^2 + 4}{3x + 4}$

39. $(3x^3 + 2) \div (x - 1)$

40. $(-3x^3 + 8x^2 + x) \div (3x + 4)$

41. $(x^3 + 3x^2 + 1) \div (x^2 + 1)$

42. $(x^4 - x^3 + x^2 - x + 1) \div (x^2 - 1)$

43. $\dfrac{x^3 + 1}{x^2 - x + 1}$

44. $\dfrac{4x^3 + 3x + 2}{2x^2 - x + 1}$

45. $\dfrac{x^3 + 8}{x + 2}$

46. $\dfrac{x^4 - 16}{x - 2}$

47. Thinking Generally If the quotient in a polynomial division problem is an integer, what must be true about the degrees of the dividend and divisor?

48. Thinking Generally If the quotient in a polynomial division problem is a polynomial of degree 1, what must be true about the degrees of the dividend and divisor?

APPLICATIONS INVOLVING DIVISION OF POLYNOMIALS

Exercises 49 and 50: **Area of a Rectangle** *The area of a rectangle and its width are given. Find an expression for the length l.*

49.

50.

51. **Volume of a Box** The volume V of a box is $2x^3 + 4x^2$, and the area of its bottom is $2x^2$. Find the height of the box in terms of x. Make a possible sketch of the box, and label the length of each side.

52. **Area of a Triangle** A triangle has height h and area $A = 2h^2 - 4h$. Find its base b in terms of h. Make a possible sketch of the triangle, and label the height and base. (*Hint:* $A = \frac{1}{2}bh$.)

WRITING ABOUT MATHEMATICS

53. Suppose that one polynomial is divided into another polynomial and the remainder is 0. What does the product of the divisor and quotient equal? Explain.

54. A student simplifies the expression $\dfrac{4x^3 - 1}{4x^2}$ to $x - 1$. Explain the student's error.

<table>
<tr><td>**SECTIONS 5.5 and 5.6**</td><td>## Checking Basic Concepts</td></tr>
</table>

1. Simplify each expression. Write the result with positive exponents.

  (a) 9^{-2} **(b)** $\dfrac{3x^{-3}}{6x^4}$ **(c)** $(4ab^{-4})^{-2}$

2. Simplify each expression. Write the result with positive exponents.

  (a) $\dfrac{1}{z^{-5}}$ **(b)** $\dfrac{x^{-3}}{y^{-6}}$ **(c)** $\left(\dfrac{3}{x^2}\right)^{-3}$

3. Write each number in scientific notation.
  (a) 45,000   **(b)** 0.000234   **(c)** 0.01

4. Write each expression in standard form.
  (a) 4.71×10^4 **(b)** 6×10^{-3}

5. Simplify $\dfrac{25a^4 - 15a^3}{5a^3}$.

6. Divide $3x^2 - x - 4$ by $x - 1$. State the quotient and remainder.

7. Divide $x^4 + 2x^3 - 2x^2 - 5x - 2$ by $x^2 - 3$. State the quotient and remainder.

8. **Distance to the Sun** The distance to the sun is approximately 93 million miles.
  (a) Write this distance in scientific notation.
  (b) Light travels at 1.86×10^5 miles per second. How long does it take for the sun's light to reach Earth?

CHAPTER 5 Summary

SECTION 5.1 ■ RULES FOR EXPONENTS

Bases and Exponents The expression b^n has base b and exponent n and equals the expression $\underbrace{b \cdot b \cdot b \cdot \cdots \cdot b}_{n \text{ times}}$, when n is a natural number.

Example: 2^3 has base 2 and exponent 3 and equals $2 \cdot 2 \cdot 2 = 8$.

Evaluating Expressions When evaluating expressions, use the following order of operations. First perform all calculations within parentheses and absolute values, or above and below a fraction bar.

1. Evaluate exponents.
2. Perform negation.
3. Do multiplication and division from left to right.
4. Do addition and subtraction from left to right.

Example: $-3^2 + 3 \cdot 4 = -9 + 3 \cdot 4 = -9 + 12 = 3$

Zero Exponents For any nonzero number b, $b^0 = 1$. Note that 0^0 is undefined.

Examples: $5^0 = 1$ and $\left(\dfrac{x}{y}\right)^0 = 1$, where x and y are nonzero.

Product Rule For any real number a and natural numbers m and n,

$$a^m \cdot a^n = a^{m+n}.$$

Examples: $3^4 \cdot 3^2 = 3^6$ and $x^3 x^2 x^4 = x^9$

Power Rules For any real numbers a and b and natural numbers m and n,

$$(a^m)^n = a^{mn}, \quad (ab)^n = a^n b^n, \quad \text{and} \quad \left(\frac{a}{b}\right)^n = \frac{a^n}{b^n}, \quad b \neq 0.$$

Examples: $(x^2)^3 = x^6$, $(3x)^4 = 3^4 x^4 = 81 x^4$, and $\left(\dfrac{2}{y}\right)^3 = \dfrac{2^3}{y^3} = \dfrac{8}{y^3}$

SECTION 5.2 ■ ADDITION AND SUBTRACTION OF POLYNOMIALS

Terms Related to Polynomials

Monomial	A number, variable, or product of numbers and variables raised to natural number powers
Degree of a Monomial	Sum of the exponents of the variables
Coefficient of a Monomial	The number in a monomial

Example: The monomial $-3x^2y^3$ has degree 5 and coefficient -3.

Polynomial	A monomial or the sum of two or more monomials
Term of a Polynomial	Each monomial is a term of the polynomial.
Binomial	A polynomial with two terms
Trinomial	A polynomial with three terms
Degree of a Polynomial	The degree of the term with highest degree
Opposite of a Polynomial	The opposite is found by negating each term.

Example: $2x^3 - 4x + 5$ is a trinomial with degree 3. Its opposite is $-2x^3 + 4x - 5$.

Like Terms	Two monomials with the same variables raised to the same powers

Examples: $3xy^2$ and $-xy^2$ are like terms.

$5x^3$ and $3x^3$ are like terms.

$5x^2$ and $5x$ are *unlike* terms.

Addition of Polynomials Combine like terms, using the distributive property.

Example: $(2x^2 - 4x) + (-x^2 - x) = (2 - 1)x^2 + (-4 - 1)x$
$$= x^2 - 5x$$

Subtraction of Polynomials Add the first polynomial to the opposite of the second polynomial.

Example: $(4x^4 - 5x) - (7x^4 + 6x) = (4x^4 - 5x) + (-7x^4 - 6x)$
$$= (4 - 7)x^4 + (-5 - 6)x$$
$$= -3x^4 - 11x$$

SECTION 5.3 ■ MULTIPLICATION OF POLYNOMIALS

Multiplication of Monomials Use the commutative property and the product rule.

Examples: $-2x^3 \cdot 3x^2 = -2 \cdot 3 \cdot x^3 \cdot x^2 = -6x^5$

$(2xy^2)(3x^2y^3) = 2 \cdot 3 \cdot x \cdot x^2 \cdot y^2 \cdot y^3 = 6x^3y^5$

⎵—Assumed exponent of 1

Distributive Properties

$$a(b + c) = ab + ac \quad \text{and} \quad a(b - c) = ab - ac$$

Examples: $4x(3x + 6) = 4x \cdot 3x + 4x \cdot 6 = 12x^2 + 24x$

$ab(a^2 - b^2) = ab \cdot a^2 - ab \cdot b^2 = a^3b - ab^3$

Multiplication of Monomials and Polynomials Apply the distributive properties. Be sure to multiply every term in the polynomial by the monomial.

Example: $-2x^2(4x^2 - 5x - 3) = -8x^4 + 10x^3 + 6x^2$

Multiplication of Polynomials The product of two polynomials may be found by multiplying every term in the first polynomial by every term in the second polynomial. Be sure to combine like terms.

Examples: $(x + 3)(2x - 5) = 2x^2 - 5x + 6x - 15$
$$= 2x^2 + x - 15$$

$(2x + 1)(x^2 - 5x + 2) = 2x^3 - 10x^2 + 4x + x^2 - 5x + 2$
$$= 2x^3 - 9x^2 - x + 2$$

SECTION 5.4 ■ SPECIAL PRODUCTS

Product of a Sum and Difference

$$(a + b)(a - b) = a^2 - b^2$$

Examples: $(x + 4)(x - 4) = x^2 - 16$

$(2r - 3t)(2r + 3t) = (2r)^2 - (3t)^2 = 4r^2 - 9t^2$

Squaring Binomials

$$(a + b)^2 = a^2 + 2ab + b^2 \quad \text{and} \quad (a - b)^2 = a^2 - 2ab + b^2$$

Examples: $(2x + 1)^2 = (2x)^2 + 2(2x)1 + 1^2 = 4x^2 + 4x + 1$

$(z^2 - 2)^2 = (z^2)^2 - 2z^2(2) + 2^2 = z^4 - 4z^2 + 4$

Cubing Binomials To multiply $(a + b)^3$, write it as $(a + b)(a + b)^2$.

Example: $(x + 4)^3 = (x + 4)(x + 4)^2$

$\qquad = (x + 4)(x^2 + 8x + 16)$ Square the binomial.

$\qquad = x^3 + 8x^2 + 16x + 4x^2 + 32x + 64$ Distributive property

$\qquad = x^3 + 12x^2 + 48x + 64$ Combine like terms.

SECTION 5.5 ■ INTEGER EXPONENTS AND THE QUOTIENT RULE

Negative Integers as Exponents For any nonzero real number a and positive integer n,

$$a^{-n} = \frac{1}{a^n}.$$

Examples: $5^{-2} = \dfrac{1}{5^2}$ and $x^{-4} = \dfrac{1}{x^4}$

The Quotient Rule For any nonzero real number a and integers m and n,

$$\frac{a^m}{a^n} = a^{m-n}.$$

Examples: $\dfrac{6^4}{6^2} = 6^{4-2} = 6^2 = 36$ and $\dfrac{xy^3}{x^4y^2} = x^{1-4}y^{3-2} = x^{-3}y^1 = \dfrac{y}{x^3}$

Other Rules For any nonzero real numbers a and b and positive integers m and n,

$$\frac{1}{a^{-n}} = a^n, \quad \frac{a^{-n}}{b^{-m}} = \frac{b^m}{a^n}, \quad \text{and} \quad \left(\frac{a}{b}\right)^{-n} = \left(\frac{b}{a}\right)^n.$$

Examples: $\dfrac{1}{4^{-3}} = 4^3$, $\dfrac{x^{-3}}{y^{-2}} = \dfrac{y^2}{x^3}$,   and $\left(\dfrac{4}{5}\right)^{-2} = \left(\dfrac{5}{4}\right)^2$

Scientific Notation A real number a written as $b \times 10^n$, where $1 \leq |b| < 10$ and n is an integer

Examples: $2.34 \times 10^3 = 2340$ Move the decimal point 3 places to the right.

$2.34 \times 10^{-3} = 0.00234$ Move the decimal point 3 places to the left.

SECTION 5.6 ■ DIVISION OF POLYNOMIALS

Division of a Polynomial by a Monomial Divide the monomial into *every* term of the polynomial.

Example: $\dfrac{5x^3 - 10x^2 + 15x}{5x} = \dfrac{5x^3}{5x} - \dfrac{10x^2}{5x} + \dfrac{15x}{5x} = x^2 - 2x + 3$

Division of a Polynomial by a Polynomial Division of polynomials is performed similarly to long division of natural numbers.

Example: Divide $2x^3 + 4x^2 - 3x + 1$ by $x + 1$.

$$
\begin{array}{r}
2x^2 + 2x - 5 \\
x + 1 \overline{\smash{)}2x^3 + 4x^2 - 3x + 1} \\
\underline{2x^3 + 2x^2} \\
2x^2 - 3x \\
\underline{2x^2 + 2x} \\
-5x + 1 \\
\underline{-5x - 5} \\
6
\end{array}
$$

The quotient is $2x^2 + 2x - 5$ with remainder 6, which can be written as

$$2x^2 + 2x - 5 + \frac{6}{x + 1}.$$

CHAPTER 5   Review Exercises

SECTION 5.1

Exercises 1–6: Evaluate the expression.

1. 5^3
2. -3^4
3. $4(-2)^0$
4. $3 + 3^2 - 3^0$
5. $\dfrac{-5^2}{5}$
6. $\left(\dfrac{-5}{5}\right)^2$

Exercises 7–24: Simplify the expression.

7. $6^2 \cdot 6^3$
8. $10^5 \cdot 10^7$
9. $z^4 \cdot z^5$
10. $y^2 \cdot y \cdot y^3$
11. $5x^2 \cdot 6x^7$
12. $(ab^3)(a^3b)$
13. $(2^5)^2$
14. $(m^4)^5$
15. $(ab)^3$
16. $(x^2y^3)^4$
17. $(xy)^3(x^2y^4)^2$
18. $(a^2b^9)^0$
19. $(r - t)^4(r - t)^5$
20. $(a + b)^2(a + b)^4$
21. $\left(\dfrac{3}{x - y}\right)^2$
22. $\left(\dfrac{x + y}{2}\right)^3$
23. $2x^2(3x - 5)$
24. $3x(4x + x^3)$

SECTION 5.2

Exercises 25 and 26: Identify the degree and coefficient of the monomial.

25. $6x^7$
26. $-x^2y^3$

Exercises 27–30: Determine whether the expression is a polynomial. If it is, state how many terms and variables the polynomial contains. Then state its degree.

27. $8y$
28. $8x^3 - 3x^2 + x - 5$
29. $a^2 + 2ab + b^2$
30. $\dfrac{1}{xy}$

31. Add the polynomials vertically.
$$
\begin{array}{r}
3x^2 + 4x + 8 \\
2x^2 - 5x - 5
\end{array}
$$

32. Write the opposite of $6x^2 - 3x - 7$.

Exercises 33–40: Simplify.

33. $(4x - 3) + (-x + 7)$
34. $(3x^2 - 1) - (5x^2 + 12)$
35. $(x^2 + 5x + 6) - (3x^2 - 4x + 1)$
36. $(x^2 + 3x - 5) + (2x^2 - 5x - 1)$

37. $(a^3 + 4a^2) + (a^3 - 5a^2 + 7a)$

38. $(4x^3 - 2x + 6) - (4x^3 - 6)$

39. $(xy + y^2) + (4y^2 - 4xy)$

40. $(7x^2 + 2xy + y^2) - (7x^2 - 2xy + y^2)$

SECTION 5.3

Exercises 41–54: Multiply and simplify.

41. $-x^2 \cdot x^3$

42. $-(r^2t^3)(rt)$

43. $-3(2t - 5)$

44. $2y(1 - 6y)$

45. $6x^3(3x^2 + 5x)$

46. $-x(x^2 - 2x + 9)$

47. $-ab(a^2 - 2ab + b^2)$

48. $(a - 2)(a + 5)$

49. $(8x - 3)(x + 2)$

50. $(2x - 1)(1 - x)$

51. $(y^2 + 1)(2y + 1)$

52. $(y^2 - 1)(2y^2 + 1)$

53. $(z + 1)(z^2 - z + 1)$

54. $(4z - 3)(z^2 - 3z + 1)$

Exercises 55 and 56: Multiply the expression
   (a) *geometrically and*
   (b) *symbolically.*

55. $z(z + 1)$

56. $2x(x + 2)$

SECTION 5.4

Exercises 57–72: Multiply.

57. $(z + 2)(z - 2)$

58. $(5z - 9)(5z + 9)$

59. $(1 - 3y)(1 + 3y)$

60. $(5x + 4y)(5x - 4y)$

61. $(rt + 1)(rt - 1)$

62. $(2m^2 - n^2)(2m^2 + n^2)$

63. $(x + 1)^2$

64. $(4x + 3)^2$

65. $(y - 3)^2$

66. $(2y - 5)^2$

67. $(4 + a)^2$

68. $(4 - a)^2$

69. $(x^2 + y^2)^2$

70. $(xy - 2)^2$

71. $(z + 5)^3$

72 $(2z - 1)^3$

Exercises 73 and 74: Use the product of a sum and a difference to evaluate the expression.

73. $59 \cdot 61$

74. $22 \cdot 18$

SECTION 5.5

Exercises 75–82: Simplify the expression.

75. 9^{-1}

76. 3^{-2}

77. $4^3 \cdot 4^{-2}$

78. $10^{-6} \cdot 10^3$

79. $\dfrac{1}{6^{-2}}$

80. $\dfrac{5^7}{5^9}$

81. $(3^{-1}2^2)^{-2}$

82. $(2^{-4}5^3)^0$

Exercises 83–98: Simplify the expression. Write the answer using positive exponents.

83. z^{-2}

84. y^{-4}

85. $a^{-4} \cdot a^2$

86. $x^2 \cdot x^{-5} \cdot x$

87. $(2t)^{-2}$

88. $(ab^2)^{-3}$

89. $(xy)^{-2}(x^{-2}y)^{-1}$

90. $\dfrac{x^6}{x^2}$

91. $\dfrac{4x}{2x^4}$

92. $\dfrac{20x^5y^3}{30xy^6}$

93. $\left(\dfrac{a}{b}\right)^5$

94. $\dfrac{4}{t^{-4}}$

95. $\dfrac{(3m^3n)^{-2}}{(2m^2n^{-3})^3}$

96. $\left(\dfrac{x^{-4}y^2}{3xy^{-3}}\right)^{-2}$

97. $\left(\dfrac{x}{y}\right)^{-2}$

98. $\left(\dfrac{3u}{2v}\right)^{-1}$

Exercises 99–102: Write the expression in standard form.

99. 6×10^2

100. 5.24×10^4

101. -3.7×10^{-3}

102. -6.234×10^{-2}

Exercises 103–106: Write the number in scientific notation.

103. $10,000$

104. $56,100,000$

105. 0.000054

106. 0.001

Exercises 107 and 108: Evaluate the expression. Write the result in standard form.

107. $(4 \times 10^2)(6 \times 10^4)$ **108.** $\dfrac{8 \times 10^3}{4 \times 10^4}$

SECTION 5.6

Exercises 109–116: Divide and check.

109 $\dfrac{5x^2 + 3x}{3x}$

110. $\dfrac{6b^4 - 4b^2 + 2}{2b^2}$

111. $\dfrac{3x^2 - x + 2}{x - 1}$

112. $\dfrac{9x^2 - 6x - 2}{3x + 2}$

113. $\dfrac{4x^3 - 11x^2 - 7x - 1}{4x + 1}$

114. $\dfrac{2x^3 - x^2 - 1}{2x - 1}$ **115.** $\dfrac{x^3 - x^2 - x + 1}{x^2 + 1}$

116. $\dfrac{x^4 + 3x^3 + 8x^2 + 7x + 5}{x^2 + x + 1}$

APPLICATIONS

117. Heart Rate An athlete starts running and continues for 10 seconds. The polynomial $\frac{1}{2}t^2 + 60$ calculates the heart rate of the athlete in beats per minute t seconds after beginning the run, where $t \leq 10$.
(a) What is the athlete's heart rate when the athlete first starts to run?
(b) What is the athlete's heart rate after 10 seconds?
(c) What happens to the athlete's heart rate while the athlete is running?

118. Areas of Rectangles Find a monomial equal to the sum of the areas of the rectangles. Calculate this sum for $x = 3$ feet and $y = 4$ feet.

119. Area of a Rectangle Write a polynomial that gives the area of the rectangle. Calculate its area for $z = 6$ inches.

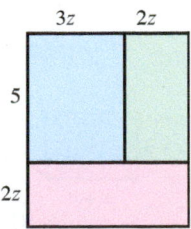

120. Area of a Square Find the area of the square whose sides have length x^2y.

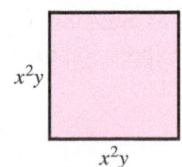

121. Media Consumption Recently, there were 240 million people in the United States over the age of 18. They spent an average of 11 hours per day with electronic media. Use scientific notation to estimate the total number of hours of media consumed by this group.

122. Volume of a Sphere The expression for the volume of a sphere with radius r is $\frac{4}{3}\pi r^3$. Find a polynomial that gives the volume of a sphere with radius $x + 2$. Leave your answer in terms of π.

123. Height Reached by a Baseball A baseball is hit straight up. Its height h in feet above the ground after t seconds is given by $t(96 - 16t)$.
(a) Multiply this expression.
(b) Evaluate both the expression in part (a) and the given expression for $t = 2$. Interpret the result.

124. Rectangular Building A rectangular building has a perimeter of 1200 feet.
(a) If one side of the building has length L, write a polynomial expression that gives its area. (Be sure to multiply your expression.)
(b) Evaluate the expression in part (a) for $L = 50$ and interpret the answer.

125. Geometry Complete each part and verify that your answers are equal.
(a) Find the area of the large square by multiplying its length and width.
(b) Find the sum of the areas of the smaller rectangles inside the large square.

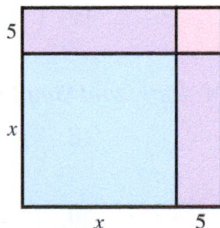

126. Digital Picture A digital picture, including its border, is $x + 4$ pixels by $x + 4$ pixels, and the actual picture inside the border is $x - 4$ pixels by $x - 4$ pixels.

(a) Find a polynomial that gives the number of pixels in the border.
(b) Let $x = 100$ and evaluate the polynomial.

1. Simplify each expression.
   (a) -5^0 (b) -9^2

2. Evaluate each expression by hand.
   (a) $-4^2 + 10$ (b) 8^{-2}
   (c) $\dfrac{1}{2^{-3}}$ (d) $-3x^0$

3. State how many terms and variables the polynomial $5x^2 - 3xy - 7y^3$ contains. Then state its degree.

4. Write the opposite of $-x^3 + 4x - 8$.

Exercises 5–8: Simplify.

5. $(-3x + 4) + (7x + 2)$

6. $(y^3 - 2y + 6) - (4y^3 + 5)$

7. $(5x^2 - x + 3) - (4x^2 - 2x + 10)$

8. $(a^3 + 5ab) + (3a^3 - 3ab)$

Exercises 9–16: Write the given expression with positive exponents.

9. $6y^4 \cdot 4y^7$ 10. $(a^2b^3)^2(ab^2)$

11. $x^7 \cdot x^{-3}$ 12. $(a^{-1}b^2)^{-3}$

13. $ab(a^2 - b^2)$ 14. $\left(\dfrac{3a^2}{2b^{-3}}\right)^{-2}$

15. $\dfrac{12xy^4}{6x^2y}$ 16. $\left(\dfrac{2}{a + b}\right)^4$

Exercises 17–22: Multiply and simplify.

17. $3x^2(4x^3 - 6x + 1)$ 18. $(z - 3)(2z + 4)$

19. $(7y^2 - 3)(7y^2 + 3)$ 20. $(3x - 2)^2$

21. $(m + 3)^3$

22. $(y + 2)(y^2 - 2y + 3)$

23. Evaluate $78 \cdot 82$ using the product of a sum and a difference.

24. Write 6.1×10^{-3} in standard form.

25. Write 5410 in scientific notation.

Exercises 26 and 27: Divide.

26. $\dfrac{9x^3 - 6x^2 + 3x}{3x^2}$ 27. $\dfrac{x^3 + x^2 - x + 1}{x + 2}$

28. **Concert Tickets** Tickets for a concert are sold for $20 each.
   (a) Write a polynomial that gives the revenue from selling t tickets.
   (b) Putting on the concert costs management $2000 to hire the band plus $2 for each ticket sold. What is the total cost of the concert if t tickets are sold?
   (c) Subtract the polynomial that you found in part (b) from the polynomial that you found in part (a). What does the resulting polynomial represent?

29. **Areas of Rectangles** Find a polynomial representing the sum of the areas of two identical rectangles that have width $2x$ and length $3x$. Calculate this sum for $x = 10$ feet.

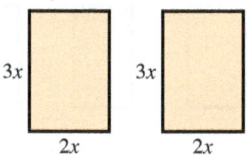

30. **Volume of a Box** Write a polynomial that represents the volume of the box. Be sure to multiply your answer completely.

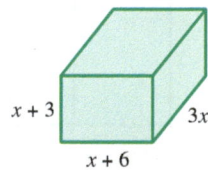

31. **Height Reached by a Golf Ball** When a golf ball is hit into the air, its height in feet above the ground after t seconds is given by $t(88 - 16t)$.
   (a) Multiply this expression.
   (b) Evaluate the expression in part (a) for $t = 3$. Interpret the result.

Cumulative Review Exercises

Exercises 1 and 2: Evaluate each expression by hand.

1. (a) $18 - 2 \cdot 5$ **(b)** $42 \div 7 + 2$

2. (a) $21 - (-8)$ **(b)** $-\frac{7}{3} \div \left(-\frac{14}{9}\right)$

Exercises 3 and 4: Solve the equation. Note that these equations may have no solutions, one solution, or infinitely many solutions.

3. (a) $(x - 3) + x = 4 + x$
   (b) $2(5x - 4) = 1 + 10x$

4. (a) $2 + 6x = 2(3x + 1)$
   (b) $11x - 9 = -31$

5. Find the average speed of a car that travels 306 miles in 4 hours 30 minutes.

6. Write each value as a fraction in lowest terms.
   (a) 42% **(b)** 0.076

7. Graph the equation $4x - 5y = 20$.

8. Sketch a line with slope $-\frac{2}{3}$ that passes through the point $(1, 1)$.

9. Write the slope–intercept form for the line shown.

10. Find the x-and y-intercepts for the graph of the equation $2y = 3x - 6$.

Exercises 11 and 12: Write the slope–intercept form of a line that satisfies the given information.

11. Parallel to $3x - 6y = 7$, passing through $(2, -3)$

12. Passing through $(-2, -5)$ and $(1, 4)$

Exercises 13–16: Solve the system of equations. Note that these systems may have no solutions, one solution, or infinitely many solutions.

13. $4x + 3y = -6$
    $8x + 6y = 12$

14. $x - 3y = 5$
    $3x + y = 5$

15. $\quad x + 4y = -8$
    $-3x - 12y = 24$

16. $x - 5y = 30$
    $2x + y = -6$

Exercises 17 and 18: Shade the solution set for the system of inequalities.

17. $x + y < 3$
    $y \geq x + 2$

18. $\quad x - 2y > 4$
    $3x + y < 6$

19. Simplify the expression.
   (a) $3x^2 \cdot 5x^3$ **(b)** $(x^3 y)^2 (x^4 y^5)$

20. Simplify.
   (a) $(5x^2 - 3x + 4) - (3x^2 - 2x + 1)$
   (b) $(7a^3 - 4a^2 - 5) + (5a^3 + 4a^2 + a)$

21. Multiply and simplify.
   (a) $(2x + 3)(x - 7)$
   (b) $(y + 3)(y^2 - 3y - 1)$
   (c) $(4x + 7)(4x - 7)$
   (d) $(5a + 3)^2$

22. Simplify the expression. Write the answer using positive exponents.

   (a) $x^{-5} \cdot x^3 \cdot x$ **(b)** $\left(\dfrac{2}{x^3}\right)^{-3}$

   (c) $\dfrac{3x^2 y^{-1}}{6x^{-2} y}$ **(d)** $(xy^{-2})^3 (x^{-2} y)^{-2}$

23. Write 24,000,000,000 in scientific notation.

24. Write 4.71×10^{-7} in standard form.

25. Divide.

   (a) $\dfrac{8x^3 - 2x}{2x}$ **(b)** $\dfrac{2x^2 + x - 14}{x + 3}$

26. Price Decrease If the price of a computer is reduced from \$1200 to \$900, find the percent change.

27. Mixing an Acid Solution How many milliliters of a 3% acid solution should be added to 400 milliliters of a 6% acid solution to dilute it to a 5% acid solution?

28. Surface Area of a Box Use the drawing of the box to write a polynomial that represents the area of each of the following.

   (a) The bottom
   (b) The front
   (c) The right side
   (d) All six sides

6

Factoring Polynomials and Solving Equations

Most cars are designed so that their exteriors are curved and smooth. This characteristic is especially true for solar cars. In fact, a side view of a solar car often resembles the cross section of an airplane wing, as illustrated in the accompanying figure. This design reduces drag from air resistance and increases fuel efficiency. As a result, solar cars are capable of traveling at speeds as high as 100 kilometers per hour.

Views of a Solar Car and a Standard Car

Mathematics plays an important role in the design of cars. A special type of third-degree polynomial is used extensively by engineers to model smooth shapes for new cars. In this chapter, we introduce many concepts necessary for understanding polynomials and polynomial equations. *Source:* R. Burden and J. Faires. *Numerical Analysis.*

6.1   Introduction to Factoring

Objectives

1   Introducing Factoring
2   Finding Common Factors
3   Finding the Greatest Common Factor
4   Factoring by Grouping

NEW VOCABULARY

☐ Factoring a polynomial
☐ Greatest common factor (GCF)

1 Introducing Factoring

When two or more numbers are multiplied, each number is called a *factor*. For example, in the equation $4 \cdot 6 = 24$, the numbers 4 and 6 are factors and we say that 24 can be *factored* into the product $4 \cdot 6$. Other ways to factor 24 are shown below.

Factors of 24

$$2 \cdot 12, \quad 3 \cdot 8, \quad 2 \cdot 2 \cdot 6, \quad \text{and} \quad 2 \cdot 2 \cdot 2 \cdot 3$$

When we factor a positive number, we reverse the multiplication process and write the number as a product of two or more smaller numbers. Similarly, we **factor a polynomial** by writing the polynomial as a product of two or more *lower-degree* polynomials. For example, possible ways to factor the polynomial $12x^5 + 6x^4$ are shown below.

Factors of $12x^5 + 6x^4$

$$2x(6x^4 + 3x^3), \quad x^2(12x^3 + 6x^2), \quad \text{and} \quad 6x^4(2x + 1)$$

The second of these factorizations shows that the **fifth**-degree polynomial $12x^5 + 6x^4$ can be written as the product of the **second**-degree polynomial x^2 and the **third**-degree polynomial $12x^3 + 6x^2$, where each factor has a lower degree than the given polynomial. Similar statements can be made about the first and third factorizations.

2 Finding Common Factors

When factoring a polynomial, we first look for factors that are common to each term. By applying the distributive property, we can often write a polynomial as a product. For example, each term in the polynomial $8x^2 + 6x$ has a factor of $2x$ because

$$8x^2 = 2x \cdot 4x \quad \text{and} \quad 6x = 2x \cdot 3.$$

Therefore by the distributive property,

$$8x^2 + 6x = 2x(4x + 3).$$

Thus the product $2x(4x + 3)$ equals $8x^2 + 6x$. We check this result by multiplying.

$$2x(4x + 3) = 2x \cdot 4x + 2x \cdot 3$$
$$= 8x^2 + 6x \ \checkmark \qquad \text{It checks.}$$

This factorization is shown visually in **FIGURE 6.1**, where possible dimensions for a rectangle with an area of $8x^2 + 6x$ are $2x$ by $4x + 3$.

READING CHECK 1

• What property allows us to write a polynomial as a product?

Geometric Illustration of Factoring:
$8x^2 + 6x = 2x(4x + 3)$

FIGURE 6.1

CALCULATOR HELP

To make a table of values, see Appendix A (pages AP-2 and AP-3).

The expressions $8x^2 + 6x$ and $2x(4x + 3)$ are equal for all values of x. **FIGURE 6.2** illustrates this fact with a partial table of values for each expression.

Numerical Illustration: $8x^2 + 6x = 2x(4x + 3)$

(a)   (b)

FIGURE 6.2

EXAMPLE 1   **Factoring an expression**

Factor the expression and sketch a rectangle that illustrates the factorization.
(a) $10x + 6$ **(b)** $6x^2 + 15x$

Solution
(a) Each term in the polynomial $10x + 6$ has a factor of 2 because

$$10x = \mathbf{2 \cdot 5x} \quad \text{and} \quad 6 = \mathbf{2 \cdot 3}. \qquad \text{Common factor is } \mathbf{2}.$$

By the distributive property, $10x + 6 = \mathbf{2(5x + 3)}$. This factorization is illustrated visually in **FIGURE 6.3(a)**.

(b) Each term in the polynomial $6x^2 + 15x$ has a factor of $3x$ because

$$6x^2 = \mathbf{3x \cdot 2x} \quad \text{and} \quad 15x = \mathbf{3x \cdot 5}. \qquad \text{Common factor is } \mathbf{3x}.$$

By the distributive property, $6x^2 + 15x = \mathbf{3x(2x + 5)}$. This factorization is illustrated visually in **FIGURE 6.3(b)**.

Geometric Illustrations of Factoring:

$10x + 6 = 2(5x + 3)$ $6x^2 + 15x = 3x(2x + 5)$

(a)   (b)

FIGURE 6.3

Now Try Exercises 9, 11

EXAMPLE 2   **Finding common factors**

Factor.
(a) $15x^2 + 10x$ **(b)** $6y^3 - 2y^2$ **(c)** $3z^3 + 9z^2 - 6z$ **(d)** $2x^2y^2 + 4xy^3$

Solution
(a) In the expression $15x^2 + 10x$, the terms $15x^2$ and $10x$ contain a common factor of $5x$.

$$15x^2 = \mathbf{5x \cdot 3x} \quad \text{and} \quad 10x = \mathbf{5x \cdot 2}$$

Therefore this polynomial can be factored as

$$15x^2 + 10x = \mathbf{5x(3x + 2)}.$$

(b) In the expression $6y^3 - 2y^2$, the terms $6y^3$ and $2y^2$ contain a common factor of $2y^2$.

$$6y^3 = \mathbf{2y^2} \cdot \mathbf{3y} \quad \text{and} \quad 2y^2 = \mathbf{2y^2} \cdot \mathbf{1}$$

Therefore this polynomial can be factored as

$$6y^3 - 2y^2 = \mathbf{2y^2}(\mathbf{3y} - \mathbf{1}).$$

(c) In $3z^3 + 9z^2 - 6z$, the terms $3z^3$, $9z^2$, and $6z$ contain a common factor of $3z$.

$$3z^3 = \mathbf{3z} \cdot \mathbf{z^2}, \quad 9z^2 = \mathbf{3z} \cdot \mathbf{3z}, \quad \text{and} \quad 6z = \mathbf{3z} \cdot \mathbf{2}$$

Therefore this polynomial can be factored as

$$3z^3 + 9z^2 - 6z = \mathbf{3z}(\mathbf{z^2} + \mathbf{3z} - \mathbf{2}).$$

(d) In $2x^2y^2 + 4xy^3$, the terms $2x^2y^2$ and $4xy^3$ contain a common factor of $2xy^2$.

$$2x^2y^2 = \mathbf{2xy^2} \cdot \mathbf{x} \quad \text{and} \quad 4xy^3 = \mathbf{2xy^2} \cdot \mathbf{2y}$$

Thus $2x^2y^2 + 4xy^3 = \mathbf{2xy^2}(\mathbf{x} + \mathbf{2y})$.

Now Try Exercises 15, 17, 19, 21

READING CHECK 2

- What is the greatest common factor of $18b^4 + 12b^3$?

STUDY TIP

In general, the ability to factor polynomials is important in solving many types of equations. Factoring out the GCF is a first step in preparing a polynomial for other factoring techniques discussed in this chapter.

3 Finding the Greatest Common Factor

In most situations, we factor out the *greatest common factor* (GCF). For example, the polynomial $12b^3 + 8b^2$ has a common factor of $2b$. We could factor this polynomial as

$$12b^3 + 8b^2 = 2b(6b^2 + 4b).$$

However, we can factor out $4b^2$ instead.

$$12b^3 + 8b^2 = 4b^2(3b + 2)$$

Because $4b^2$ is the common factor with the greatest (integer) coefficient and highest degree, we say that $4b^2$ is the **greatest common factor** of $12b^3 + 8b^2$. In Examples 1 and 2, we factored out the greatest common factor for each expression.

The following See the Concept illustrates a method for factoring the greatest common factor (GCF) from a polynomial.

See the Concept 1   **FACTORING OUT THE GCF**

To factor the polynomial $12x^4 + 9xy^3$:

Ⓐ Complete factorization

$$12x^4 = 2 \cdot 2 \cdot 3 \cdot x \cdot x \cdot x \cdot x$$

$$= \mathbf{3x} \cdot 4x^3$$

Common factors   Noncommon factors

Ⓐ Complete factorization

$$9xy^3 = 3 \cdot \mathbf{3} \cdot x \cdot y \cdot y \cdot y$$

$$= \mathbf{3x} \cdot 3y^3$$

Common factors   Noncommon factors

Ⓑ GCF times the sum of the noncommon factors

$$12x^4 + 9xy^3 = \mathbf{3x}(4x^3 + 3y^3)$$

Ⓐ Completely factor each term of the polynomial by writing its coefficient as the product of prime numbers and writing any powers of variables as repeated multiplication.

Ⓑ The given polynomial can be factored into the product of the common factors (GCF) times the sum (or difference) of the noncommon factors.

WATCH VIDEO IN MML .

| EXAMPLE 3 | **Finding the greatest common factor** |

Find the greatest common factor for each expression. Then factor the expression.
(a) $9x^2 + 6x$ **(b)** $4z^4 + 8z^2$ **(c)** $8a^2b^3 - 16a^3b^2$

Solution
(a) Because

$$9x^2 = 3 \cdot 3 \cdot x \cdot x \quad \text{and} \quad 6x = 3 \cdot 2 \cdot x,$$

the terms have common factors of 3 and x. The GCF is the product of these two factors, or $3 \cdot x = 3x$. Thus the expression $9x^2 + 6x$ can be factored as $3x(3x + 2)$.
(b) Because

$$4z^4 = 2 \cdot 2 \cdot z \cdot z \cdot z \cdot z \quad \text{and}$$

$$8z^2 = 2 \cdot 2 \cdot 2 \cdot z \cdot z,$$

the terms have common factors of 2, 2, z, and z. The GCF is the product of these four factors, or $2 \cdot 2 \cdot z \cdot z = 4z^2$. Thus $4z^4 + 8z^2$ can be factored as $4z^2(z^2 + 2)$.
(c) Because

$$8a^2b^3 = 2 \cdot 2 \cdot 2 \cdot a \cdot a \cdot b \cdot b \cdot b \quad \text{and}$$

$$16a^3b^2 = 2 \cdot 2 \cdot 2 \cdot 2 \cdot a \cdot a \cdot a \cdot b \cdot b,$$

the terms have common factors of 2, 2, 2, a, a, b, and b. The GCF is the product of these seven factors, or $2 \cdot 2 \cdot 2 \cdot a \cdot a \cdot b \cdot b = 8a^2b^2$. Here, $8a^2b^3 - 16a^3b^2$ can be factored as $8a^2b^2(b - 2a)$.

Now Try Exercises 23, 25, 37

NOTE: With practice, you may find that you can determine the GCF mentally without factoring each term, as was done in Example 3. ■

| MAKING CONNECTIONS 1 |

Checking Common Factors with Multiplication

When factoring, we can check our work by multiplying. For example, if we are uncertain whether the equation

$$6y^3 - 2y^2 = 2y^2(3y - 1)$$

is correct, we can apply the distributive property to the right side of the above equation to obtain

$$2y^2(3y - 1) = 2y^2 \cdot 3y - 2y^2 \cdot 1$$
$$= 6y^3 - 2y^2. \checkmark \qquad \text{It checks.}$$

🌐 Math in Context (Sports) In the next example, we factor an expression that occurs in an application involving the flight of a golf ball.

| EXAMPLE 4 | **Modeling the flight of a golf ball** |

If a golf ball is hit upward at 66 feet per second (45 miles per hour), then its height in feet after t seconds is approximated by $66t - 16t^2$. Factor this expression.

Solution
The GCF for $66t$ and $16t^2$ is $2t$ because

$$66t = 2t \cdot 33 \quad \text{and} \quad 16t^2 = 2t \cdot 8t.$$

Therefore this polynomial can be factored as

$$66t - 16t^2 = 2t(33 - 8t).$$

Now Try Exercise 81

4 Factoring by Grouping

Factoring by grouping is a technique that makes use of the associative and distributive properties. The next example illustrates one step in this factoring technique.

EXAMPLE 5 **Factoring out binomials**

Factor.
(a) $5x(x + 3) + 6(x + 3)$ (b) $x^2(2x - 5) - 4x(2x - 5)$

Solution
(a) Each term in the expression $5x(x + 3) + 6(x + 3)$ contains the binomial $(x + 3)$. Therefore the distributive property can be used to factor this expression.

$$5x(x + 3) + 6(x + 3) = (5x + 6)(x + 3)$$

(b) Each term in $x^2(2x - 5) - 4x(2x - 5)$ contains the binomial $(2x - 5)$. Therefore the distributive property can be used to factor this expression.

$$x^2(2x - 5) - 4x(2x - 5) = (x^2 - 4x)(2x - 5)$$
$$= x(x - 4)(2x - 5) \qquad \text{Factor out the GCF.}$$

Now Try Exercises 41, 45

Now consider the polynomial

> The middle arithmetic symbol is $+$, and it separates the polynomial into two binomials.

$$x^3 + x^2 + 2x + 2.$$

We can factor this polynomial by first grouping it into two binomials. The first binomial has a common factor of x^2, and the second binomial has a common factor of 2.

$$(x^3 + x^2) + (2x + 2) \qquad \text{Associative property}$$
$$x^2(x + 1) + 2(x + 1) \qquad \text{Factor out common factors.}$$
$$(x^2 + 2)(x + 1) \qquad \text{Factor out } (x + 1).$$

When factoring by grouping, we factor out a common factor more than once.

The first step in factoring a *four-term* polynomial by grouping requires the use of the associative property to write the polynomial as the *sum* of two binomials. However, this property must be applied carefully to avoid sign errors. The next two examples illustrate that the middle arithmetic symbol ($+$ or $-$) in a four-term polynomial determines how the associative property is applied.

READING CHECK 3

• What factoring technique is sometimes used to factor four-term polynomials?

EXAMPLE 6   **Factoring by grouping when the middle symbol is $(+)$**

Factor each polynomial.
(a) $2x^3 - 4x^2 + 3x - 6$ (b) $3x + 3y + ax + ay$

Solution
(a) Use the associative property to write the polynomial as the *sum* of two binomials.

$$2x^3 - 4x^2 + 3x - 6 = (2x^3 - 4x^2) + (3x - 6) \quad \text{Associative property}$$

The middle arithmetic symbol is +.
$$= 2x^2(x - 2) + 3(x - 2) \quad \text{Factor out common factors.}$$
$$= (2x^2 + 3)(x - 2) \quad \text{Factor out } (x - 2).$$

(b) Group the polynomial into the *sum* of two binomials.

$$3x + 3y + ax + ay = (3x + 3y) + (ax + ay) \quad \text{Associative property}$$

The middle arithmetic symbol is +.
$$= 3(x + y) + a(x + y) \quad \text{Factor out common factors.}$$
$$= (3 + a)(x + y) \quad \text{Factor out } (x + y).$$

Now Try Exercises 47, 65

EXAMPLE 7   **Factoring by grouping when the middle symbol is $(-)$**

Factor each polynomial.
(a) $3y^3 - y^2 - 9y + 3$ (b) $z^3 + 4z^2 - 5z - 20$

Solution
(a) Begin by changing the middle subtraction to addition by adding the opposite of $9y$. Then apply the associative property to write the result as the *sum* of two binomials.

$$3y^3 - y^2 - 9y + 3 = 3y^3 - y^2 + (-9y) + 3 \quad \text{Add the opposite of } 9y.$$

The middle arithmetic symbol is −.
$$= (3y^3 - y^2) + (-9y + 3) \quad \text{Associative property}$$
$$= y^2(3y - 1) - 3(3y - 1) \quad \text{Factor out common factors.}$$
$$= (y^2 - 3)(3y - 1) \quad \text{Factor out } (3y - 1).$$

Note that in the third step, -3 was factored from the second binomial.
(b) Begin by changing the middle subtraction to addition by adding the opposite of $5z$. Then apply the associative property to write the result as the *sum* of two binomials.

$$z^3 + 4z^2 - 5z - 20 = z^3 + 4z^2 + (-5z) - 20 \quad \text{Add the opposite of } 5z.$$

The middle arithmetic symbol is −.
$$= (z^3 + 4z^2) + (-5z - 20) \quad \text{Associative property}$$
$$= z^2(z + 4) - 5(z + 4) \quad \text{Factor out common factors.}$$
$$= (z^2 - 5)(z + 4) \quad \text{Factor out } (z + 4).$$

Factor out -5.

Now Try Exercises 55, 61

When factoring some polynomials, it may be necessary to factor out the greatest common factor before completing other factoring techniques such as factoring by grouping. In the next example, the GCF is factored out before grouping is applied.

EXAMPLE 8

Factoring out the GCF before grouping

Completely factor each polynomial.
(a) $6x^3 - 12x^2 - 3x + 6$
(b) $2x^5 - 8x^4 + 6x^3 - 24x^2$

Solution

(a) The GCF of $6x^3 - 12x^2 - 3x + 6$ is 3, so factor out **3** before factoring the remaining polynomial by grouping.

$$
\begin{aligned}
6x^3 - 12x^2 - 3x + 6 &= \mathbf{3}(2x^3 - 4x^2 - x + 2) && \text{Factor out the GCF.}\\
&= 3\big(2x^3 - 4x^2 + (-x) + 2\big) && \text{Add the opposite of } x.\\
&= 3\big((2x^3 - 4x^2) + (-x + 2)\big) && \text{Associative property}\\
&= 3\big(\mathbf{2x^2}(\mathbf{x - 2}) - \mathbf{1}(\mathbf{x - 2})\big) && \text{Factor out common factors.}\\
&= 3(\mathbf{2x^2 - 1})(\mathbf{x - 2}) && \text{Factor out } (x - 2).
\end{aligned}
$$

Factor out -1.

(b) The GCF of $2x^5 - 8x^4 + 6x^3 - 24x^2$ is $2x^2$, so factor out $\mathbf{2x^2}$ before factoring the remaining polynomial by grouping.

$$
\begin{aligned}
2x^5 - 8x^4 + 6x^3 - 24x^2 &= \mathbf{2x^2}(x^3 - 4x^2 + 3x - 12) && \text{Factor out the GCF.}\\
&= 2x^2\big((x^3 - 4x^2) + (3x - 12)\big) && \text{Associative property}\\
&= 2x^2\big(\mathbf{x^2}(\mathbf{x - 4}) + \mathbf{3}(\mathbf{x - 4})\big) && \text{Factor out common factors.}\\
&= 2x^2(\mathbf{x^2 + 3})(\mathbf{x - 4}) && \text{Factor out } (x - 4).
\end{aligned}
$$

Now Try Exercises 67, 71

Sometimes we need to rearrange the terms before we can apply factoring by grouping as shown in the next example.

EXAMPLE 9

Rearranging terms before factoring by grouping

Factor the polynomial $2x^3 - 15 - 3x + 10x^2$.

Solution

The terms of the first binomial, $2x^3 - 15$, have no factors in common. However, if we rearrange the terms as $2x^3 + 10x^2 - 3x - 15$, we can factor as follows.

$$
\begin{aligned}
2x^3 + 10x^2 - 3x - 15 &= 2x^3 + 10x^2 + (-3x) - 15\\
&= (2x^3 + 10x^2) + (-3x - 15)\\
&= \mathbf{2x^2}(\mathbf{x + 5}) - \mathbf{3}(\mathbf{x + 5})\\
&= (\mathbf{2x^2 - 3})(\mathbf{x + 5})
\end{aligned}
$$

Now Try Exercise 57

6.1   Putting It All Together

CONCEPT	EXPLANATION	EXAMPLES
Common Factor	Factor out a monomial common to each term in a polynomial.	$6z^2 - 6z = 6z(z - 1)$ $4y^3 - 6y^2 = 2y^2(2y - 3)$ $5x^3 - 10x^2 + 15x = 5x(x^2 - 2x + 3)$ $2a^3b^3 - 4a^2b^3 = 2a^2b^3(a - 2)$
Greatest Common Factor (GCF)	The common factor with the greatest (integer) coefficient and highest degree	The GCF of $10x^4 + 15x^2$ is $5x^2$. Common factors include 1, 5, x, $5x$, x^2, and $5x^2$. However, $5x^2$ is *the greatest* common factor.
Factoring by Grouping	Factoring by grouping is a method that can be used to factor some *four-term* polynomials into a product of two binomials. It makes use of the associative and distributive properties.	$2x^3 + 3x^2 + 2x + 3$ $\quad = (2x^3 + 3x^2) + (2x + 3)$ $\quad = x^2(2x + 3) + 1(2x + 3)$ $\quad = (x^2 + 1)(2x + 3)$ $4x^3 - 24x^2 - 3x + 18$ $\quad = 4x^3 - 24x^2 + (-3x) + 18$ $\quad = (4x^3 - 24x^2) + (-3x + 18)$ $\quad = 4x^2(x - 6) - 3(x - 6)$ $\quad = (4x^2 - 3)(x - 6)$

6.1   Exercises

MyMathLab®

CONCEPTS AND VOCABULARY

1. To _____ a polynomial, write it as a product of two or more lower degree polynomials.

2. A common factor in the expression $ab + ac$ is _____.

3. When factoring, we can check our work by _____.

4. When finding the GCF for a polynomial, it is often helpful to completely factor each term by writing its coefficient as the product of _____ numbers and writing any powers of variables as repeated _____ .

5. The _____ of a polynomial is the common factor with the greatest (integer) coefficient and highest degree.

6. Factoring by _____ can be used to factor some four-term polynomials into a product of two binomials by using the associative and distributive properties.

7. Identify four common factors of $2x^2 + 4x$.

8. Identify the greatest common factor (GCF) of the expression $2x^2 + 4x$.

FINDING COMMON FACTORS

Exercises 9–14: Factor the expression. Then make a sketch of a rectangle that illustrates this factorization.

9. $2x + 4$

10. $6 + 3x$

11. $z^2 + 4z$

12. $a^2 + 5a$

13. $3y^2 + 12y$

14. $2z^2 + 10z$

Exercises 15–22: Factor the expression.

15. $3x^2 + 9x$

16. $10y^2 + 2y$

17. $4y^3 - 2y^2$

18. $6x^4 + 9x^2$

19. $2z^3 + 8z^2 - 4z$

20. $5x^4 - 15x^3 - 10x^2$

21. $6x^2y - 3xy^2$

22. $7x^3y^3 + 14x^2y^2$

FINDING THE GREATEST COMMON FACTOR

Exercises 23–40: Identify the greatest common factor. Then factor the expression.

23. $6x - 18x^2$

24. $16x^2 - 24x^3$

25. $8y^3 - 12y^2$

26. $12y^3 - 8y^2 + 4y$

27. $6z^3 + 3z^2 + 9z$

28. $16z^3 - 24z^2 - 36z$

29. $x^4 - 5x^3 - 4x^2$

30. $2x^4 + 8x^2$

31. $5y^5 + 10y^4 - 15y^3 + 10y^2$

32. $7y^4 - 14y^3 - 21y^2 + 7y$

33. $xy + xz$

34. $ab - bc$

35. $ab^2 - a^2b$

36. $4x^2y + 6xy^2$

37. $5x^2y^4 + 10x^3y^3$

38. $3r^3t^3 - 6r^4t^2$

39. $a^2b + ab^2 + ab$

40. $6ab^2 - 9ab + 12b^2$

FACTORING BY GROUPING

Exercises 41–46: Factor.

41. $x(x + 1) - 2(x + 1)$

42. $5x(3x - 2) + 2(3x - 2)$

43. $(z + 5)z + (z + 5)4$

44. $3y^2(y - 2) + 5(y - 2)$

45. $4x^3(x - 5) - 2x(x - 5)$

46. $8x^2(x + 3) + (x + 3)$

Exercises 47–66: Factor by grouping.

47. $x^3 + 2x^2 + 3x + 6$

48. $x^3 + 6x^2 + x + 6$

49. $2y^3 + y^2 + 2y + 1$

50. $4y^3 + 10y^2 + 2y + 5$

51. $15z^3 - 5z^2 + 6z - 2$

52. $2z^3 - 6z^2 + 5z - 15$

53. $4t^3 - 20t^2 + 3t - 15$

54. $4t^3 - 12t^2 + 3t - 9$

55. $9r^3 + 6r^2 - 6r - 4$

56. $3r^3 + 12r^2 - 2r - 8$

57. $7x^3 - 6 - 2x + 21x^2$

58. $6x^3 - 5 - 10x + 3x^2$

59. $2y^3 - 7y^2 - 4y + 14$

60. $y^3 - 5y^2 - 3y + 15$

61. $z^3 - 4z^2 - 7z + 28$

62. $12z^3 - 18z^2 - 10z + 15$

63. $2x^4 - 3x^3 + 4x - 6$

64. $x^4 + x^3 + 5x + 5$

65. $ax + bx + ay + by$

66. $ax - bx + ay - by$

Exercises 67–78: Completely factor the polynomial.

67. $3x^3 + 6x^2 + 3x + 6$

68. $5x^3 - 5x^2 + 5x - 5$

69. $6y^4 - 24y^3 - 2y^2 + 8y$

70. $6x^4 - 12x^3 + 3x^2 - 6x$

71. $x^5 + 2x^4 - 3x^3 - 6x^2$

72. $y^6 + 3y^5 - 2y^4 - 6y^3$

73. $4x^5 + 2x^4 - 12x^3 - 6x^2$

74. $18y^5 + 27y^4 + 12y^3 + 18y^2$

75. $x^3y + x^2y^2 - 2x^2y - 2xy^2$

76. $6x^3y - 3x^2y^2 + 18x^2y - 9xy^2$

77. $2x^3y^3 - 2x^4y^2 + 4x^2y^3 - 4x^3y^2$

78. $4x^2y^6 - 4x^2y^5 - 8xy^7 + 8xy^6$

79. **Thinking Generally** Factor a from the polynomial expression $ax^2 + bx + c$.

80. **Thinking Generally** Factor c from the polynomial expression $ax^2 + bx + c$.

APPLICATIONS INVOLVING FACTORING

81. **Flight of a Golf Ball** The height of a golf ball in feet after t seconds is given by $80t - 16t^2$.
(a) Identify the greatest common factor.
(b) Factor this expression.

82. **Flight of a Golf Ball** Repeat the previous exercise if the height of a golf ball in feet after t seconds is given by $128t - 16t^2$.

Exercises 83–86: **Geometry** *Use the information in the figure to write a polynomial expression that represents the area of the shaded region. Then factor the expression.*

83.

84.

85.

86.

87. Volume of a Box A box is constructed by cutting out square corners of a rectangular piece of cardboard and folding up the sides. If the cutout corners have sides with length x, then the volume of the box is given by the polynomial $4x^3 - 60x^2 + 200x$.

(a) Determine the volume of the box when $x = 3$ inches.

(b) Factor out the greatest common factor for this expression.

88. Volume of a Box (Refer to the preceding exercise.) A box is constructed from a square piece of metal that is 20 inches on a side.

(a) If the square corners of length x are cut out, write a polynomial that gives the volume of the box.

(b) Find the volume when $x = 4$ inches.

(c) Factor out the greatest common factor for this polynomial expression.

WRITING ABOUT MATHEMATICS

89. Use an example to explain the difference between a common factor and the greatest common factor.

90. Use an example to explain how to factor a polynomial by grouping. What two properties of real numbers did you use?

6.2 Factoring Trinomials I ($x^2 + bx + c$)

Objectives

1 Reviewing Binomial Multiplication

2 Factoring Trinomials with Leading Coefficient 1

3 Relating Geometry and Visual Factoring

NEW VOCABULARY

☐ Leading coefficient
☐ Prime polynomial

1 Reviewing Binomial Multiplication

Recall that a *trinomial* is a polynomial that has three terms. We begin by reviewing products of binomials that result in trinomials. For example, we multiply the binomials $(x + 2)$ and $(x + 3)$ as follows.

$$(x + 2)(x + 3) = x \cdot x + x \cdot 3 + 2 \cdot x + 2 \cdot 3$$
$$= x^2 + 5x + 6$$

Note that the first term, x^2, in the trinomial results from multiplying the *first* terms of each binomial. The middle term, $5x$, results from adding the product of the *outside* terms and the product of the *inside* terms. Finally, the last term, 6, results from multiplying the *last* terms of each binomial. We discussed this method of multiplying binomials, called FOIL, in Section 3 of the previous chapter and illustrate it as follows.

$$\overset{F\quad L}{(x + 2)(x + 3)} = x^2 + 5x + 6$$

$$I \to \lfloor 2x \rfloor$$
$$O \, \underline{+3x}$$
$$\overline{5x} \checkmark \longleftarrow \text{The middle term checks.}$$

2 Factoring Trinomials with Leading Coefficient 1

Any trinomial of degree 2 in the variable x can be written in *standard form* as $ax^2 + bx + c$, where a, b, and c are constants. The constant a is called the **leading coefficient** of the trinomial. In this section, we focus on trinomials where $a = 1$ and b and c are integers.

Recall that the binomials $(x + m)$ and $(x + n)$ are multiplied as follows.

$$(x + m)(x + n) = x^2 + nx + mx + mn$$
$$= x^2 + (m + n)x + mn$$

Note that the coefficient of the x-term is the sum of m and n and that the constant (or third) term is the product of m and n. Thus to factor a trinomial in the form $x^2 + bx + c$, we start

READING CHECK 1

• How are the coefficient of the x-term b and the constant term c related to the numbers m and n?

by finding two numbers, m and n, such that when they are multiplied, $m \cdot n = c$, and when they are added, $m + n = b$. In the next example, we find integer pairs that have a specified product and sum.

EXAMPLE 1 | **Finding integer pairs with a given product and sum**

For each of the following, find an integer pair that has the given product and sum.
(a) Product: 18; Sum: 11   **(b)** Product: -20; Sum: 1

Solution

(a) For two integers to have a product of (positive) 18, the integers must have the same sign. Because the specified sum is positive and any sum of two negative integers is negative, the two integers must be positive. **TABLE 6.1** lists positive integer factor pairs for **18** along with the corresponding sum for each pair.

Factor Pairs for 18

Factors	1, 18	2, 9	3, 6
Sum	19	11	9

TABLE 6.1

From the table, we see that the integers **2** and **9** have a product of **18** and a sum of **11**.

(b) For two integers to have a negative product, they must have unlike signs. Integer factor pairs for -20 and the corresponding sum for each pair are listed in **TABLE 6.2**.

Factor Pairs for -20

Factors	1, -20	2, -10	4, -5	-1, 20	-2, 10	-4, 5
Sum	-19	-8	-1	19	8	1

TABLE 6.2

Here, we see that the integers -4 and **5** have a product of -20 and a sum of **1**.

Now Try Exercises 11, 15

STUDY TIP

Spend extra time practicing the process illustrated in Example 1. Pay special attention to factor pairs in which at least one of the integers is negative.

The ability to find integer pairs as demonstrated in Example 1 is essential for factoring trinomials of the form $x^2 + bx + c$. To illustrate the factoring process, we will find an integer pair that can be used to factor $x^2 + 6x + 8$.

Standard Form	*Example*
$x^2 + bx + c$	$x^2 + 6x + 8$
$m \cdot n = c$	$m \cdot n = 8$
$m + n = b$	$m + n = 6$

Factor Pairs for 8

Factors	1, 8	2, 4
Sum	9	6

TABLE 6.3

$m \cdot n = 8$

$m + n = 6$

To determine possible values for m and n, we list factor pairs for 8 and search for a pair whose sum is 6, as in **TABLE 6.3**.

Because $2 \cdot 4 = 8$ and $2 + 4 = 6$, we let $m = 2$ and $n = 4$. We then factor the given trinomial as

$$x^2 + 6x + 8 = (x + 2)(x + 4).$$

Note that, if you can find this factor pair mentally, making a table is not necessary. We check the result by multiplying the two binomials.

$$(x + 2)(x + 4) = x^2 + 6x + 8$$

$$\begin{array}{l} 2x \\ +4x \\ \hline 6x \checkmark \end{array} \quad \longleftarrow \text{ The middle term checks.}$$

FACTORING $x^2 + bx + c$

To factor the trinomial $x^2 + bx + c$, find two numbers m and n that satisfy

$$m \cdot n = c \quad \text{and} \quad m + n = b.$$

Then $x^2 + bx + c = (x + m)(x + n)$.

EXAMPLE 2 **Factoring a trinomial with only positive coefficients**

Factor each trinomial.
(a) $x^2 + 7x + 12$ **(b)** $x^2 + 13x + 30$ **(c)** $z^2 + 9z + 20$

Solution
(a) To factor $x^2 + 7x + 12$, we need to find a factor pair for 12 whose sum is 7. To do so, we make **TABLE 6.4**.

Factor Pairs for 12

Factors	1, 12	2, 6	3, 4
Sum	13	8	7

TABLE 6.4

The required factor pair is **3** and **4** because $3 \cdot 4 = 12$ and $3 + 4 = 7$. Therefore the given trinomial can be factored as

$$x^2 + 7x + 12 = (x + \mathbf{3})(x + \mathbf{4}).$$

(b) To factor $x^2 + 13x + 30$, we need to find a factor pair for 30 whose sum is 13. The required pair is **3** and **10**. Thus

$$x^2 + 13x + 30 = (x + \mathbf{3})(x + \mathbf{10}).$$

(c) To factor $z^2 + 9z + 20$, we need to find a factor pair for 20 whose sum is 9. The required pair is **4** and **5**. Thus

$$z^2 + 9z + 20 = (z + \mathbf{4})(z + \mathbf{5}).$$

Now Try Exercises 25, 27, 29

In the next example, the coefficients of the middle terms are negative.

EXAMPLE 3   **Factoring trinomials with a negative middle coefficient**

Factor each trinomial.
(a) $x^2 - 7x + 10$ **(b)** $x^2 - 8x + 15$ **(c)** $y^2 - 9y + 18$

Solution
(a) To factor $x^2 - 7x + 10$, we need to find a factor pair for 10 whose sum equals -7. To have a positive product *and* a negative sum, *both* numbers must be negative, as shown in **TABLE 6.5**.

Factor Pairs for 10

Factors	$-1, -10$	$-2, -5$
Sum	-11	-7

TABLE 6.5

The required pair is -2 and -5 because $-2 \cdot (-5) = 10$ and $-2 + (-5) = -7$. Therefore the given trinomial can be factored as

$$x^2 - 7x + 10 = (x - 2)(x - 5).$$

(b) To factor $x^2 - 8x + 15$, we need to find a factor pair for 15 whose sum is -8. The required pair is -3 and -5. Thus

$$x^2 - 8x + 15 = (x - 3)(x - 5).$$

(c) To factor $y^2 - 9y + 18$, we need to find a factor pair for 18 whose sum is -9. The required pair is -3 and -6. Thus

$$y^2 - 9y + 18 = (y - 3)(y - 6).$$

Now Try Exercises 33, 35, 37

In Examples 2 and 3, the coefficient of the last term was always positive. In the next example, this coefficient is negative and the coefficient of the middle term is either positive or negative.

EXAMPLE 4   **Factoring trinomials with a negative constant term**

Factor each trinomial.
(a) $x^2 - 3x - 4$ **(b)** $x^2 + 7x - 8$ **(c)** $t^2 - 2t - 24$

Solution
(a) To factor $x^2 - 3x - 4$, we need to find a factor pair for -4 whose sum is -3. To have a negative product, one factor must be positive and the other factor must be negative, as shown in **TABLE 6.6**.

Factor Pairs for -4

Factors	$-1, 4$	$1, -4$	$-2, 2$
Sum	3	-3	0

TABLE 6.6

The required pair is 1 and -4 because $1 \cdot (-4) = -4$ and $1 + (-4) = -3$. Therefore the given trinomial can be factored as

$$x^2 - 3x - 4 = (x + 1)(x - 4),$$

which can be checked by multiplying $(x + 1)(x - 4)$.

(b) To factor $x^2 + 7x - 8$, we need to find a factor pair for -8 whose sum is 7. The required pair is -1 and 8. Thus

$$x^2 + 7x - 8 = (x - 1)(x + 8).$$

(c) To factor $t^2 - 2t - 24$, we need to find a factor pair for -24 whose sum is -2. The required pair is -6 and 4. Thus

$$t^2 - 2t - 24 = (t - 6)(t + 4).$$

Now Try Exercises 43, 55, 59

READING CHECK 2

• What is a prime polynomial?

A polynomial with integer coefficients that cannot be factored by using integer coefficients is called a **prime polynomial**. The next example illustrates that some trinomials of the form $x^2 + bx + c$ cannot be factored into the product of two binomials.

EXAMPLE 5   **Discovering that a trinomial is prime**

Factor each trinomial, if possible.
(a) $x^2 + 9x + 12$ **(b)** $x^2 + 5x - 4$

Solution

(a) To factor $x^2 + 9x + 12$, we need to find a factor pair for 12 whose sum is 9. **TABLE 6.7** reveals that no such factor pair exists.

Factor Pairs for 12

Factors	1, 12	2, 6	3, 4
Sum	13	8	7

TABLE 6.7

The trinomial $x^2 + 9x + 12$ is prime.

(b) At first glance, it may appear that the required factor pair is 4 and 1 because $4 \cdot 1 = 4$ and $4 + 1 = 5$. However, it is important to pay close attention to the signs of the coefficients. To factor $x^2 + 5x - 4$, we need to find a factor pair for -4 whose sum is 5. No such factor pair exists. The trinomial $x^2 + 5x - 4$ is prime.

Now Try Exercises 23, 47

MAKING CONNECTIONS 1

The Signs in the Binomial Factors

If a trinomial of the form $x^2 + bx + c$ can be factored, the signs of the coefficients in the trinomial can be used to determine the signs in the binomial factors. If b and c represent positive numbers, this can be summarized as follows.

Form of the Trinomial	Signs in the Binomial Factors	Examples
$x^2 + bx + c$	$(+)(+)$	$x^2 + 3x + 2 = (x + 1)(x + 2)$
$x^2 - bx + c$	$(-)(-)$	$x^2 - 3x + 2 = (x - 1)(x - 2)$
$x^2 + bx - c$	$(-)(+)$	$x^2 + 4x - 5 = (x - 1)(x + 5)$
$x^2 - bx - c$	$(-)(+)$	$x^2 - 4x - 5 = (x - 5)(x + 1)$

When factoring some trinomials, it may be necessary to factor out the greatest common factor before attempting to factor the trinomial into the product of two binomials. The next example illustrates this process.

EXAMPLE 6 **Factoring out the GCF before factoring further**

Factor each trinomial completely.
(a) $7x^2 + 35x + 42$ (b) $2x^4 - 4x^3 - 6x^2$

Solution
(a) Because the GCF of $7x^2 + 35x + 42$ is 7, factor out **7** first.

$$7x^2 + 35x + 42 = \mathbf{7}(x^2 + 5x + 6)$$

Now, to factor $x^2 + 5x + 6$, we need to find a factor pair for 6 whose sum is 5. The required pair is 2 and 3. Thus

$$7x^2 + 35x + 42 = 7(x + 2)(x + 3).$$

(b) Because the GCF of $2x^4 - 4x^3 - 6x^2$ is $2x^2$, factor out $\mathbf{2x^2}$ first.

$$2x^4 - 4x^3 - 6x^2 = \mathbf{2x^2}(x^2 - 2x - 3)$$

Now, to factor $x^2 - 2x - 3$, we need to find a factor pair for -3 whose sum is -2. The required pair is -3 and 1. Thus

$$2x^4 - 4x^3 - 6x^2 = 2x^2(x - 3)(x + 1).$$

Now Try Exercises 65, 77

CRITICAL THINKING 1

A cube has a surface area of $6x^2 + 24x + 24$. What is the length of each side?

3 Relating Geometry and Visual Factoring

Trinomials of the form $x^2 + bx + c$ can be factored visually using rectangles, as illustrated in the next example and in the following Making Connections.

EXAMPLE 7 **Finding the dimensions of a rectangle**

Find one possibility for the dimensions of a rectangle that has an area of $x^2 + 6x + 5$.

Solution
The area of a rectangle equals length times width. If we can factor $x^2 + 6x + 5$, then the factors can represent its length and width. Because

$$x^2 + 6x + 5 = (x + 1)(x + 5),$$

one possibility for the rectangle's dimensions is width $x + 1$ and length $x + 5$, as illustrated in **FIGURE 6.4**.

Now Try Exercise 93

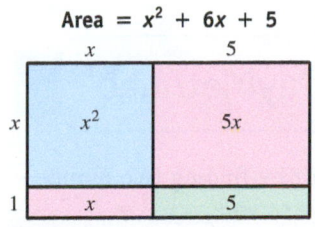

Area = $x^2 + 6x + 5$

FIGURE 6.4

MAKING CONNECTIONS 2

Factoring Trinomials Visually

A special grid can be used to factor trinomials visually. Like an xy-plane, the grid has four quadrants. But this grid is not an xy-plane because the axes are not labeled and the grid is not used to plot ordered pairs. The area of any region located in quadrants I or III represents a positive term and the area of any region in quadrants II or IV represents a negative term.

$x^2 + 10x + 24$
factors as
$(x + 4)(x + 6)$

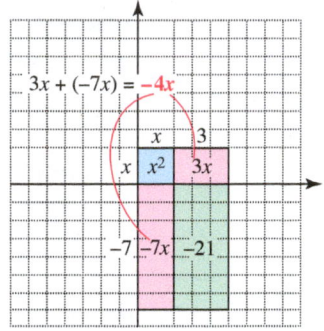

$x^2 - 4x - 21$
factors as
$(x - 7)(x + 3)$

See Exercises 99–104.

6.2   Putting It All Together

CONCEPT	EXPLANATION	EXAMPLES
Factoring Trinomials of the Form $$x^2 + bx + c$$	Find two numbers m and n that satisfy $mn = c$ and $m + n = b$. Then $$x^2 + bx + c = (x + m)(x + n).$$	$x^2 + 9x + 20 = (x + 4)(x + 5)$ because $4 \cdot 5 = 20$ and $4 + 5 = 9$. $x^2 + x - 6 = (x - 2)(x + 3)$ because $-2 \cdot 3 = -6$ and $-2 + 3 = 1$. $x^2 - 8x + 12 = (x - 6)(x - 2)$ because $-6 \cdot (-2) = 12$ and $-6 + (-2) = -8$.

6.2   Exercises

MyMathLab®

CONCEPTS AND VOCABULARY

1. In the trinomial $x^2 + bx + c$, the leading coefficient is _____.

2. Multiply $(x + m)(x + n)$. What is the coefficient of the x-term? What is the constant term?

3. To factor $x^2 + bx + c$, start by finding two numbers m and n that satisfy _____ $= c$ and _____ $= b$.

4. A trinomial with integer coefficients that cannot be factored using integer coefficients is _____.

REVIEWING BINOMIAL MULTIPLICATION

Exercises 5–10: Multiply and simplify.

5. $(x + 2)(x + 5)$

6. $(x + 1)(x + 7)$

7. $(x - 3)(x - 6)$

8. $(x - 5)(x - 2)$

9. $(x + 8)(x - 7)$

10. $(x - 4)(x + 9)$

FINDING INTEGER PAIRS

Exercises 11–18: Find the integer pair that has the given product and sum.

11. Product: 28   Sum: 11

12. Product: 35   Sum: 12

13. Product: −30   Sum: −7

14. Product: −15   Sum: −2

15. Product: −50   Sum: 5

16. Product: −100   Sum: 21

17. Product: 28   Sum: −11

18. Product: 80   Sum: −42

FACTORING TRINOMIALS

Exercises 19–64: Factor the trinomial. If the trinomial cannot be factored, write "prime."

19. $x^2 + 3x + 2$

20. $x^2 + 5x + 4$

21. $y^2 + 4y + 4$

22. $y^2 + 8y + 7$

23. $z^2 + 3z + 7$

24. $z^2 + 4z + 5$

25. $x^2 + 8x + 15$

26. $x^2 + 9x + 14$

27. $m^2 + 13m + 36$

28. $m^2 + 15m + 36$

29. $n^2 + 20n + 100$

30. $n^2 + 52n + 100$

31. $x^2 - 6x + 5$

32. $x^2 - 6x + 8$

33. $y^2 - 7y + 12$

34. $y^2 - 12y + 27$

35. $z^2 - 13z + 40$

36. $z^2 - 15z + 54$

37. $a^2 - 16a + 63$

38. $a^2 - 82a + 81$

39. $y^2 - 6y + 10$

40. $y^2 - 2y + 3$

41. $b^2 - 30b + 125$

42. $b^2 - 19b + 90$

43. $x^2 + 13x - 90$

44. $x^2 + 15x - 100$

45. $m^2 + 4m - 45$

46. $m^2 + 4m - 60$

47. $a^2 + 16a - 63$

48. $a^2 + 13a - 42$

49. $n^2 + 10n - 200$

50. $n^2 + 2n - 120$

51. $x^2 + 22x - 23$

52. $x^2 + 18x - 19$

53. $a^2 + 4a - 32$

54. $a^2 + 9a - 36$

55. $b^2 - b - 20$

56. $b^2 - b - 12$

57. $m^2 - 14m - 22$

58. $m^2 - 11m - 24$

59. $x^2 - x - 72$

60. $x^2 - 2x - 80$

61. $y^2 - 15y - 34$

62. $y^2 - 10y - 39$

63. $z^2 - 5z - 66$

64. $z^2 - 6z - 55$

Exercises 65–80: Factor the trinomial completely.

65. $5x^2 - 10x - 40$

66. $2x^2 + 8x - 10$

67. $-3m^2 - 9m + 12$

68. $-4n^2 + 20n - 24$

69. $y^3 - 7y^2 + 10y$

70. $z^3 + 9z^2 + 20z$

71. $-x^3 - 2x^2 + 15x$

72. $-y^3 + 9y^2 - 14y$

73. $3a^3 + 21a^2 + 18a$

74. $5b^3 - 5b^2 - 60b$

75. $-2x^3 + 6x^2 - 8x$

76. $-4y^3 + 20y^2 - 32y$

77. $2m^4 - 10m^3 - 28m^2$

78. $6n^4 - 18n^3 + 12n^2$

79. $-3x^4 + 3x^3 + 6x^2$

80. $-5y^4 + 25y^3 - 30y^2$

Exercises 81–84: Factor the trinomial. (Hint: Write the expression in standard form.)

81. $5 + 6x + x^2$

82. $8 + 6x + x^2$

83. $3 - 4x + x^2$

84. $10 - 7x + x^2$

Exercises 85–88: Factor the trinomial. (Hint: Write $(m - x)(n + x)$ and find m and n.)

85. $12 + 4x - x^2$

86. $28 + 3x - x^2$

87. $32 - 4x - x^2$

88. $40 - 3x - x^2$

89. Thinking Generally Factor the trinomial expression $x^2 + (k + 1)x + k$.

90. Thinking Generally Factor the trinomial expression $x^2 + (k - 2)x - 2k$.

RELATING GEOMETRY AND VISUAL FACTORING

91. A square has an area of $x^2 + 2x + 1$. Find the length of a side. Make a sketch of the square.

92. A square has an area of $x^2 + 6x + 9$. Find the length of a side. Make a sketch of the square.

93. A rectangle has an area of $x^2 + 3x + 2$. Find one possibility for its width and length. Make a sketch of the rectangle.

94. A rectangle has an area of $x^2 + 9x + 8$. Find one possibility for its width and length. Make a sketch of the rectangle.

95. A cube has a surface area of $6x^2 + 12x + 6$. Find the length of a side. (*Hint:* First factor out the GCF.)

96. A cube has a surface area of $6x^2 + 36x + 54$. Find the length of a side.

97. Write a polynomial in factored form that represents the total area of the figure.

98. Write a polynomial in factored form that represents the total area of the figure.

Exercises 99–104: Factor the trinomial visually, as is discussed in Making Connections 2 at the end of this section.

99. $x^2 + 5x + 6$ **100.** $x^2 + 9x + 20$

101. $x^2 - 11x + 30$ **102.** $x^2 - 3x - 10$

103. $x^2 + 4x - 12$ **104.** $x^2 - 8x + 16$

WRITING ABOUT MATHEMATICS

105. Explain how to determine whether a trinomial has been factored correctly. Give an example.

106. Factoring $x^2 + bx + c$ involves finding two integers m and n such that $mn = c$ and $m + n = b$. Is it better first to determine values of m and n so that the product is c or the sum is b? Explain your reasoning.

SECTIONS 6.1 and 6.2   Checking Basic Concepts

1. What is the greatest common factor for the expression $8x^3 - 12x^2 + 24x$?

2. Factor $12z^3 - 18z^2$.

3. Factor each expression completely.
   (a) $6y(y - 2) + 5(y - 2)$
   (b) $2x^3 + x^2 + 10x + 5$
   (c) $4z^3 - 12z^2 + 4z - 12$

4. Factor each trinomial completely, if possible.
   (a) $x^2 + 6x + 8$ **(b)** $x^2 - x - 42$

(c) $a^2 + 3a - 5$ **(d)** $4a^3 + 20a^2 + 24a$

5. Write a polynomial in factored form that represents the total area of the figure.

6.3   Factoring Trinomials II ($ax^2 + bx + c$)

Objectives

1   Factoring Trinomials by Grouping

2   Factoring with FOIL in Reverse

1   Factoring Trinomials by Grouping

To factor the trinomial $x^2 + 6x + 5$, we find two numbers, m and n, such that $mn = 5$ and $m + n = 6$. For this trinomial, we let $m = 1$ and $n = 5$, which gives

$$x^2 + 6x + 5 = (x + 1)(x + 5).$$

READING CHECK 1

• What are the values of m and n for $6x^2 + 11x + 3$?

To factor the trinomial $2x^2 + 7x + 3$, which has a leading coefficient of 2, we must find two numbers m and n such that $mn = 2 \cdot 3 = 6$ and $m + n = 7$. One solution is $m = 1$ and $n = 6$. Using grouping, we can now factor this trinomial by writing $7x$ as the sum $1x + 6x$.

$$
\begin{aligned}
2x^2 + 7x + 3 &= 2x^2 + \overset{\lceil 7x \rceil}{x + 6x} + 3 && \text{Write } 7x \text{ as } x + 6x. \\
&= (2x^2 + x) + (6x + 3) && \text{Associative property} \\
&= x(2x + 1) + 3(2x + 1) && \text{Factor out common factors.} \\
&= (x + 3)(2x + 1) && \text{Factor out } (2x + 1).
\end{aligned}
$$

This technique of factoring trinomials by grouping is summarized as follows.

FACTORING $ax^2 + bx + c$ BY GROUPING

To factor $ax^2 + bx + c$, perform the following steps. (Assume that a, b, and c are integers and have no common factors.)

1. Find two numbers, m and n, such that $mn = ac$ and $m + n = b$.
2. Write the trinomial as $ax^2 + mx + nx + c$.
3. Use grouping to factor this expression into two binomials.

STUDY TIP

Be sure that you understand that the values of m and n are used in different ways, depending on whether or not the leading coefficient is 1.

MAKING CONNECTIONS 1

The Values of m and n

The values of m and n found when factoring $ax^2 + bx + c$, where $a \neq 1$, are not used in the same way as the values of m and n found when factoring $x^2 + bx + c$.

Factoring $x^2 + bx + c$	*Factoring $ax^2 + bx + c$, where $a \neq 1$*
$mn = c$ and $m + n = b$	$mn = ac$ and $m + n = b$
$x^2 + bx + c = (x + m)(x + n)$	$ax^2 + bx + c = ax^2 + mx + nx + c$

Factor the resulting four-term polynomial by grouping.

EXAMPLE 1   Factoring $ax^2 + bx + c$ by grouping

Factor each trinomial.
(a) $2x^2 + 5x + 2$ (b) $3z^2 + z - 2$ (c) $10t^2 - 11t + 3$

Solution
(a) To factor $2x^2 + 5x + 2$, we need to find m and n that satisfy $mn = 2 \cdot 2 = 4$ and $m + n = 5$. **TABLE 6.8** shows that two such numbers are $m = 1$ and $n = 4$.

$$
\begin{aligned}
2x^2 + 5x + 2 &= 2x^2 + \overset{\lceil 5x \rceil}{x + 4x} + 2 && \text{Write } 5x \text{ as } x + 4x. \\
&= (2x^2 + x) + (4x + 2) && \text{Associative property} \\
&= x(2x + 1) + 2(2x + 1) && \text{Factor out common factors.} \\
&= (x + 2)(2x + 1) && \text{Factor out } (2x + 1).
\end{aligned}
$$

Factor Pairs for 4

Factors	1, 4	2, 2
Sum	5	4

TABLE 6.8

(b) To factor $3z^2 + 1z - 2$, we need to find m and n that satisfy $mn = 3 \cdot (-2) = -6$ and $m + n = 1$. Two such numbers are $m = 3$ and $n = -2$.

$$
\begin{aligned}
3z^2 + z - 2 &= 3z^2 + 3z - 2z - 2 && \text{Write } z \text{ as } 3z - 2z. \\
&= (3z^2 + 3z) + (-2z - 2) && \text{Associative property} \\
&= 3z(z + 1) - 2(z + 1) && \text{Factor out common factors.} \\
&= (3z - 2)(z + 1) && \text{Factor out } (z + 1).
\end{aligned}
$$

(c) To factor $10t^2 - 11t + 3$, we need to find m and n that satisfy $mn = 10 \cdot 3 = 30$ and $m + n = -11$. Two such numbers are $m = -5$ and $n = -6$.

$$
\begin{aligned}
10t^2 - 11t + 3 &= 10t^2 - 5t - 6t + 3 && \text{Write } -11t \text{ as } -5t - 6t. \\
&= (10t^2 - 5t) + (-6t + 3) && \text{Associative property} \\
&= 5t(2t - 1) - 3(2t - 1) && \text{Factor out common factors.} \\
&= (5t - 3)(2t - 1) && \text{Factor out } (2t - 1).
\end{aligned}
$$

Now Try Exercises 15, 25, 37

MAKING CONNECTIONS 2

Different Ways to Factor by Grouping

In Example 1(c), we could have written $-11t$ as $-6t - 5t$, rather than $-5t - 6t$. Then the factoring could have been written as

$$
\begin{aligned}
10t^2 - 11t + 3 &= 10t^2 - 6t - 5t + 3 \\
&= (10t^2 - 6t) + (-5t + 3) \\
&= 2t(5t - 3) - 1(5t - 3) \\
&= (2t - 1)(5t - 3),
\end{aligned}
$$

which gives the same result.

In the previous section, we showed that some trinomials of the form $x^2 + bx + c$ are prime and cannot be factored into the product of two binomials with integer coefficients. The next example illustrates that some trinomials of the form $ax^2 + bx + c$, with $a \neq 1$, may also be prime.

EXAMPLE 2   **Discovering that a trinomial is prime**

Factor each trinomial.
(a) $3x^2 + 5x + 4$ **(b)** $2x^2 - 8x - 3$

Solution
(a) To factor $3x^2 + 5x + 4$, we need to find integers m and n such that $mn = 3 \cdot 4 = 12$ and $m + n = 5$. Because the middle term is positive, we consider only positive factors of 12. **TABLE 6.9** reveals that no such integers exist.

Factor Pairs for 12

Factors	1, 12	2, 6	3, 4	
Sum	13	8	7	No sum equals 5.

TABLE 6.9

The trinomial $3x^2 + 5x + 4$ is prime.

(b) To factor $2x^2 - 8x - 3$, we must find integers m and n such that $mn = 2 \cdot (-3) = -6$ and $m + n = -8$. **TABLE 6.10** reveals that no such integers exist.

Factor Pairs for −6

Factors	−1, 6	−2, 3	2, −3	1, −6	
Sum	5	1	−1	−5	No sum equals −8.

TABLE 6.10

The trinomial $2x^2 - 8x - 3$ is prime.

Now Try Exercises 17, 41

Although some trinomials may look as though they can be factored using the process discussed in Example 1, it is important to remember to first factor out the greatest common factor whenever possible. In the next example, we factor out the GCF before factoring the trinomial further.

EXAMPLE 3   **Factoring out the GCF before factoring further**

Factor each trinomial completely.
(a) $15x^2 - 50x - 40$ **(b)** $4x^3 - 22x^2 + 30x$

Solution
(a) Because the GCF of $15x^2 - 50x - 40$ is 5, factor out **5** before factoring the remaining trinomial.

$$15x^2 - 50x - 40 = \mathbf{5}(3x^2 - 10x - 8)$$

To factor $3x^2 - 10x - 8$, we need numbers m and n such that $mn = 3 \cdot (-8) = -24$ and $m + n = -10$. Two such numbers are $-\mathbf{12}$ and $\mathbf{2}$.

$$
\begin{aligned}
15x^2 - 50x - 40 &= \mathbf{5}(3x^2 \mathbf{- 12x + 2x} - 8) && \text{Write } -10x \text{ as } -12x + 2x. \\
&= 5\big((3x^2 - 12x) + (2x - 8)\big) && \text{Associative property} \\
&= 5\big(3x(x - 4) + 2(x - 4)\big) && \text{Factor out common factors.} \\
&= 5(3x + 2)(x - 4) && \text{Factor out } (x - 4).
\end{aligned}
$$

(b) Because the GCF of $4x^3 - 22x^2 + 30x$ is $2x$, factor out $\mathbf{2x}$ before factoring the remaining trinomial.

$$4x^3 - 22x^2 + 30x = \mathbf{2x}(2x^2 - 11x + 15)$$

To factor the trinomial $2x^2 - 11x + 15$, we need to find numbers m and n such that $mn = 2 \cdot 15 = 30$ and $m + n = -11$. Two such numbers are $-\mathbf{6}$ and $-\mathbf{5}$.

$$
\begin{aligned}
4x^3 - 22x^2 + 30x &= \mathbf{2x}(2x^2 \mathbf{- 6x - 5x} + 15) && \text{Write } -11x \text{ as } -6x - 5x. \\
&= 2x\big((2x^2 - 6x) + (-5x + 15)\big) && \text{Associative property} \\
&= 2x\big(2x(x - 3) - 5(x - 3)\big) && \text{Factor out common factors.} \\
&= 2x(2x - 5)(x - 3) && \text{Factor out } (x - 3).
\end{aligned}
$$

Now Try Exercises 55, 59

2 Factoring with FOIL in Reverse

Rather than factoring a trinomial by grouping, we can use FOIL in reverse to determine its binomial factors. For example, the factors of $2x^2 + 5x + 2$ are two binomials:

$$2x^2 + 5x + 2 = (\underline{\quad} + \underline{\quad})(\underline{\quad} + \underline{\quad}),$$

where the expressions to be placed in the four blanks are yet to be found. By the FOIL method, we know that the product of the first terms of these two binomials is $2x^2$. Because $2x^2 = 2x \cdot x$, we can write

$$2x^2 + 5x + 2 = (\underline{2x} + \underline{\quad})(x + \underline{\quad}).$$

By FOIL, the product of the last terms in these binomials must be 2. Because $2 = 1 \cdot 2$, we could put 1 and 2 in the blanks. However, we must be sure to place them correctly so that the product of the outside terms plus the product of the inside terms is $5x$.

If we had interchanged the 1 and 2, we would have obtained an incorrect result.

MAKING CONNECTIONS 3

The Signs in the Binomial Factors

Let a, b, and c represent positive integers. If a trinomial of the form $ax^2 + bx + c$ can be factored, the signs in the binomial factors can be summarized as follows.

Form of the Trinomial	Signs in the Binomial Factors	Examples
$ax^2 + bx + c$	$(\,+\,)(\,+\,)$	$2x^2 + 5x + 3 = (2x + 3)(x + 1)$
$ax^2 - bx + c$	$(\,-\,)(\,-\,)$	$2x^2 - 5x + 3 = (2x - 3)(x - 1)$
$ax^2 + bx - c$	$(\,-\,)(\,+\,)$	$3x^2 + 5x - 2 = (3x - 1)(x + 2)$
$ax^2 - bx - c$	$(\,-\,)(\,+\,)$	$3x^2 - 2x - 5 = (3x - 5)(x + 1)$

In the next example, we factor expressions of the form $ax^2 + bx + c$, where $a \neq 1$. We use FOIL in reverse and *trial and error* to find the correct factors.

EXAMPLE 4   **Factoring the form $ax^2 + bx + c$**

Factor each trinomial.
(a) $3x^2 + 5x + 2$ **(b)** $6x^2 + 7x - 3$ **(c)** $6 + 4y^2 - 11y$

Solution

(a) To factor $3x^2 + 5x + 2$, we start by finding factors of $3x^2$, which include **$3x$** and **x**, so we write

$$3x^2 + 5x + 2 = (\mathbf{3x} + \underline{\quad})(\mathbf{x} + \underline{\quad}).$$

READING CHECK 2

- Does $6x^2 + 17x + 5$ factor as $(3x + 1)(2x + 5)$ or as $(3x + 5)(2x + 1)$?

The factors of the last term 2 are **1** and **2**. (Because the middle term is positive, we do not consider -1 and -2.) If we place values of 1 and 2 in the binomial as follows, the middle term becomes $7x$ rather than $5x$.

$$(\mathbf{3x + 1})(x + \mathbf{2}) = 3x^2 + 7x + 2$$
$$\underset{7x \; \text{✗}}{\underset{+6x}{\underset{1x}{}}} \quad \longleftarrow \text{Middle term is } not \; 5x.$$

By reversing the positions of 1 and 2, we obtain a correct factorization.

$$(\mathbf{3x + 2})(x + \mathbf{1}) = 3x^2 + 5x + 2$$
$$\underset{5x \; \text{✓}}{\underset{+3x}{\underset{2x}{}}} \quad \longleftarrow \text{Middle term checks.}$$

(b) To factor $6x^2 + 7x - 3$, we start by finding factors of $6x^2$, which include $2x$ and $3x$ or $6x$ and x. Factors of the last term, -3, include -1 and 3 or 1 and -3. The following two factorizations give incorrect results.

$$(2x + 1)(3x - 3) = 6x^2 - 3x - 3 \qquad (6x + 1)(x - 3) = 6x^2 - 17x - 3$$
$$\underset{-3x \; \text{✗}}{\underset{-6x}{\underset{3x}{}}} \longleftarrow \text{Middle term is } not \; 7x. \qquad \underset{-17x \; \text{✗}}{\underset{-18x}{\underset{1x}{}}} \longleftarrow \text{Middle term is } not \; 7x.$$

To obtain a middle term of $7x$, we use the following arrangement.

$$(3x - 1)(2x + 3) = 6x^2 + 7x - 3$$
$$\underset{7x \; \text{✓}}{\underset{+9x}{\underset{-2x}{}}} \quad \longleftarrow \text{Middle term checks.}$$

To find the correct factorization, we often need to try more than one arrangement.

(c) To factor $6 + 4y^2 - 11y$, we start by writing the trinomial in standard form as $4y^2 - 11y + 6$. Then we find possible factors of the first term, $4y^2$, which include $2y$ and $2y$ or $4y$ and y. Factors of the last term, 6, include either -1 and -6 or -2 and -3. (Because the middle term is negative, we do not use the positive factors of 1 and 6 or 2 and 3.) To obtain a middle term of $-11y$, we use the following arrangement.

$$(4y - 3)(y - 2) = 4y^2 - 11y + 6$$
$$\underset{-11y \; \text{✓}}{\underset{-8y}{\underset{-3y}{}}} \quad \longleftarrow \text{Middle term checks.}$$

Now Try Exercises 19, 33, 69

READING CHECK 3

- What is a good first step when factoring $-x^2 - 2x + 3$, which has a negative leading coefficient?

In the next example, we demonstrate how to factor trinomials that have a negative leading coefficient. This task is sometimes accomplished by first factoring out -1.

EXAMPLE 5 **Factoring trinomials that have a negative leading coefficient**

Factor each trinomial.
(a) $-6x^2 + 17x - 5$ **(b)** $1 - x - 2x^2$

Solution
(a) One way to factor $-6x^2 + 17x - 5$ is to start by factoring out -1. Then we can apply FOIL in reverse.

Opposite of $-6x^2 + 17x - 5$

$$-6x^2 + 17x - 5 = -1(6x^2 - 17x + 5) \qquad \text{Factor out } -1.$$
$$= -1(3x - 1)(2x - 5) \qquad \text{Factor the trinomial.}$$
$$= -(3x - 1)(2x - 5) \qquad \text{Rewrite.}$$

(b) To factor the trinomial $1 - x - 2x^2$, write it in standard form, factor out -1, and then apply FOIL in reverse.

$$1 - x - 2x^2 = -2x^2 - x + 1 \qquad \text{Standard form}$$
$$= -1(2x^2 + x - 1) \qquad \text{Factor out } -1.$$
$$= -(2x - 1)(x + 1) \qquad \text{Factor the trinomial.}$$

Now Try Exercises 71, 75

6.3 Putting It All Together

CONCEPT	EXPLANATION	EXAMPLES
Factoring Trinomials by Grouping	To factor $ax^2 + bx + c$, find two numbers, m and n, such that $mn = ac$ and $m + n = b$. Then write $$ax^2 + mx + nx + c.$$ Use grouping to factor this expression into two binomials.	For $3x^2 + 10x + 8$, let $m = 6$ and $n = 4$ because $mn = 24$ and $m + n = 10$. $$3x^2 + 10x + 8 = 3x^2 + 6x + 4x + 8$$ $$= (3x^2 + 6x) + (4x + 8)$$ $$= 3x(x + 2) + 4(x + 2)$$ $$= (3x + 4)(x + 2)$$
Factoring Trinomials by Using FOIL in Reverse	To factor $ax^2 + bx + c$, find factors of ax^2 and of c. Choose and arrange these factors in two binomials so that the middle term is bx.	For $3x^2 + 10x + 8$, the factors of $3x^2$ are $3x$ and x. The positive factors of 8 are either 2 and 4, or 1 and 8. A middle term of $10x$ can be obtained as follows. $$(3x + 4)(x + 2) = 3x^2 + 10x + 8$$ $4x$ $+6x$ $10x$ ✓ ← Middle term checks.
Choosing Signs When Factoring $ax^2 + bx + c$	For $ax^2 + bx + c$ with $a > 0$, **1.** If $c > 0$ and $b > 0$, use $(\_ + \_)(\_ + \_)$. **2.** If $c > 0$ and $b < 0$, use $(\_ - \_)(\_ - \_)$. **3.** If $c < 0$, use $(\_ - \_)(\_ + \_)$ or $(\_ + \_)(\_ - \_)$.	**1.** $2x^2 + 7x + 3 = (2x + 1)(x + 3)$ $(c = 3, b = 7)$ **2.** $2x^2 - 7x + 3 = (2x - 1)(x - 3)$ $(c = 3, b = -7)$ **3.** $2x^2 + 5x - 3 = (2x - 1)(x + 3)$ $(c = -3, b = 5)$ $2x^2 - 5x - 3 = (2x + 1)(x - 3)$ $(c = -3, b = -5)$

6.3 Exercises

MyMathLab®

CONCEPTS AND VOCABULARY

1. To factor the polynomial $ax^2 + bx + c$ by grouping, you first find two numbers, m and n, such that $mn =$ _____ and $m + n =$ _____.

2. To factor the polynomial $ax^2 + bx + c$ with FOIL in reverse, you first find possible factors for _____ and for _____.

Exercises 3–6: Insert the symbol + or − in each binomial factor to make the equation true.

3. $3x^2 + 5x + 2 = (3x\underline{}2)(x\underline{}1)$.

4. $3x^2 - x - 2 = (3x\underline{}2)(x\underline{}1)$.

5. $3x^2 - 5x + 2 = (3x\underline{}2)(x\underline{}1)$.

6. $3x^2 + x - 2 = (3x\underline{}2)(x\underline{}1)$.

Exercises 7–14: Fill in the blank in each binomial factor to make the equation true.

7. $4x^2 + 11x + 6 = (4x + \underline{})(\underline{} + 2)$.

8. $4x^2 - 5x - 6 = (\underline{} - 2)(\underline{} + 3)$.

9. $4x^2 + 4x - 3 = (2x - \underline{})(2x + \underline{})$.

10. $4x^2 - 8x + 3 = (2x - \underline{})(\underline{} - 3)$.

11. $6x^2 + 11x - 7 = (\underline{} + 7)(2x - \underline{})$.

12. $6x^2 - 31x + 28 = (\underline{} - 7)(\underline{} - 4)$.

13. $6x^2 - x - 15 = (3x - \underline{})(2x + \underline{})$.

14. $6x^2 - 53x + 40 = (\underline{} - 5)(x - \underline{})$.

FACTORING TRINOMIALS

Exercises 15–54: Factor the trinomial. If the trinomial cannot be factored, write "prime."

15. $2x^2 + 7x + 3$

16. $2x^2 + 3x + 1$

17. $3y^2 + 2y + 4$

18. $2y^2 + 5y + 1$

19. $3x^2 + 4x + 1$

20. $3x^2 + 10x + 3$

21. $6x^2 + 11x + 3$

22. $6x^2 + 17x + 5$

23. $5x^2 - 11x + 2$

24. $7x^2 - 8x + 1$

25. $2y^2 - 7y + 5$

26. $2y^2 - 11y + 12$

27. $3m^2 - 11m - 6$

28. $5m^2 - 7m - 2$

29. $7z^2 - 37z + 10$

30. $3z^2 - 11z + 6$

31. $3t^2 - 7t - 6$

32. $8t^2 - 6t - 9$

33. $15r^2 + r - 6$

34. $12r^2 + r - 6$

35. $24m^2 - 23m - 12$

36. $24m^2 + 29m - 4$

37. $25x^2 + 5x - 2$

38. $30x^2 + 7x - 2$

39. $6x^2 + 11x - 2$

40. $12x^2 + 28x - 5$

41. $15y^2 - 7y + 2$

42. $14y^2 - 5y + 1$

43. $21n^2 + 4n - 1$

44. $21n^2 + 10n + 1$

45. $14y^2 + 23y + 3$

46. $28y^2 + 25y + 3$

47. $28z^2 - 25z + 3$

48. $15z^2 - 19z + 6$

49. $30x^2 - 29x + 6$

50. $50x^2 - 55x + 12$

51. $20a^2 + 18a - 5$

52. $40a^2 + 21a - 2$

53. $18t^2 + 23t - 6$

54. $33t^2 + 7t - 10$

Exercises 55–64: Factor the trinomial completely.

55. $12a^2 + 12a - 9$

56. $21b^2 - 14b - 56$

57. $10z^3 + 19z^2 + 6z$

58. $12y^3 - 11y^2 + 2y$

59. $24x^3 - 30x^2 + 9x$

60. $8y^3 - 16y^2 + 6y$

61. $8x^4 - 6x^3 + 2x^2$

62. $10y^3 + 15y^2 - 5y$

63. $28x^4 + 56x^3 + 21x^2$

64. $20y^4 + 42y^3 - 20y^2$

65. **Thinking Generally** Factor the trinomial expression $3x^2 + (3k + 1)x + k$.

66. **Thinking Generally** Factor the trinomial expression $3x^2 + (3k - 2)x - 2k$.

Exercises 67–76: Factor the trinomial.

67. $2 + 15x + 7x^2$

68. $3 + 16x + 5x^2$

69. $2 - 5x + 2x^2$

70. $5 - 6x + x^2$

71. $3 - 2x - 8x^2$

72. $5 - 3x - 2x^2$

73. $-2x^2 - 7x + 15$

74. $-5x^2 - 19x + 4$

75. $-5x^2 + 14x + 3$

76. $-6x^2 + 17x + 14$

RELATING GEOMETRY AND VISUAL FACTORING

77. A rectangle has an area of $6x^2 + 7x + 2$. Find possible dimensions for this rectangle. Make a sketch of the rectangle.

78. A rectangle has an area of $2x^2 + 5x + 3$. Find possible dimensions for the rectangle. Make a sketch of the rectangle.

79. Write a polynomial in factored form that represents the total area of the figure.

80. Write a polynomial in factored form that represents the total area of the figure.

$3x^2$	$2x$
$6x$	4

WRITING ABOUT MATHEMATICS

81. Explain how the sign of the third term in the trinomial $ax^2 + bx + c$ affects how it is factored.

82. Explain the steps to be used to factor $ax^2 + bx + c$ by grouping.

6.4   Special Types of Factoring

Objectives

1 Factoring the Difference of Two Squares
2 Factoring Perfect Square Trinomials
3 Factoring the Sum and Difference of Two Cubes
4 Using Special Methods in General Factoring

NEW VOCABULARY

☐ Perfect square trinomial

STUDY TIP

The special types of factoring discussed in this section are helpful for factoring some (but not all) types of polynomials. Don't forget to review factoring techniques that were discussed earlier in this chapter.

1   Factoring the Difference of Two Squares

In Section 4 of the previous chapter, we showed that $(a - b)(a + b) = a^2 - b^2$. If we rewrite this equation as

$$a^2 - b^2 = (a - b)(a + b),$$

then we can use it to factor a difference of two squares. For example, to factor the expression $x^2 - 25$, we first write it in the form $a^2 - b^2$, where $a = x$ and $b = 5$. Doing so results in expressing $x^2 - 25$ as $x^2 - 5^2$, and the equation

$$a^2 - b^2 = (a - b)(a + b) \qquad \text{General rule}$$

becomes

$$x^2 - 5^2 = (x - 5)(x + 5). \qquad \text{Substitute: } a = x, b = 5$$

This discussion suggests the following rule for factoring a difference of two squares.

DIFFERENCE OF TWO SQUARES

For any real numbers a and b,

$$a^2 - b^2 = (a - b)(a + b).$$

In the next example, we apply this method to other expressions.

EXAMPLE 1   Factoring the difference of two squares

Factor each difference of two squares.
(a) $x^2 - 36$ **(b)** $4x^2 - 9$ **(c)** $100 - 16t^2$ **(d)** $49y^2 - 64z^2$

Solution
(a) Write $x^2 - 36$ as $x^2 - 6^2$ and then substitute x for a and 6 for b. The equation

$$a^2 - b^2 = (a - b)(a + b) \qquad \text{General rule}$$

becomes

$$x^2 - 6^2 = (x - 6)(x + 6).$$ Substitute: $a = x, b = 6$

(b) The expression $4x^2 - 9$ can be written as $(2x)^2 - 3^2$. Thus

$$4x^2 - 9 = (2x - 3)(2x + 3).$$ $a = 2x, b = 3$

(c) The expression $100 - 16t^2$ can be written as $(10)^2 - (4t)^2$. Thus

$$100 - 16t^2 = (10 - 4t)(10 + 4t).$$ $a = 10, b = 4t$

(d) The expression $49y^2 - 64z^2$ can be written as $(7y)^2 - (8z)^2$. Thus

$$49y^2 - 64z^2 = (7y - 8z)(7y + 8z).$$ $a = 7y, b = 8z$

Now Try Exercises 17, 19, 25, 29

MAKING CONNECTIONS 1

Sum of Squares versus Difference of Squares

The sum of two squares, $a^2 + b^2$, cannot be factored by using real numbers. However, the difference of two squares, $a^2 - b^2$, can be factored.

$$x^2 + 4$$ $$x^2 - 4 = (x - 2)(x + 2)$$

Sum of squares cannot be factored.   Difference of squares can be factored.

READING CHECK 1

• If possible, factor $4x^2 + 9$.

---

2   Factoring Perfect Square Trinomials

In Section 4 of the previous chapter, we also showed how to compute $(a + b)^2$ and $(a - b)^2$ as

$$(a + b)^2 = a^2 + 2ab + b^2 \quad \text{and}$$
$$(a - b)^2 = a^2 - 2ab + b^2.$$

The expressions $a^2 + 2ab + b^2$ and $a^2 - 2ab + b^2$ are called **perfect square trinomials**. If we can recognize a perfect square trinomial, we can use the following formulas to factor it.

PERFECT SQUARE TRINOMIALS

For any real numbers a and b,

$$a^2 + 2ab + b^2 = (a + b)^2 \quad \text{and}$$
$$a^2 - 2ab + b^2 = (a - b)^2.$$

READING CHECK 2

• Is either $4x^2 + 4x + 1$ or $4x^2 + 9x + 16$ a perfect square trinomial?

When factoring a trinomial as a perfect square trinomial, we must first verify that the middle term is correct. For example, to factor $x^2 + 14x + 49$, we start by rewriting the expression as $x^2 + 14x + 7^2$. In order to factor this trinomial as a perfect square trinomial, the middle term $14x$ must be equal to twice the product of x and 7.

$$x^2 + 14x + 7^2$$

$$2(x)(7)$$ ✓ The middle term checks.

When $a = x$ and $b = 7$, the equation $a^2 + 2ab + b^2 = (a + b)^2$ allows us to factor the given polynomial as $x^2 + 14x + 49 = (x + 7)^2$.

EXAMPLE 2   **Factoring perfect square trinomials**

If possible, factor each trinomial as a perfect square trinomial.
(a) $x^2 + 10x + 25$ **(b)** $4x^2 - 4x + 1$ **(c)** $9z^2 + 18z + 4$ **(d)** $x^2 - 4xy + 4y^2$

Solution
(a) Start by writing $x^2 + 10x + 25$ as $x^2 + 10x + 5^2$ and then check the middle term.

$$x^2 + 10x + 5^2$$
$$\downarrow$$
$$2(x)(5) \quad \checkmark \quad \text{The middle term checks.}$$

When $a = x$ and $b = 5$, the equation $a^2 + 2ab + b^2 = (a + b)^2$ allows us to factor the given polynomial as $x^2 + 10x + 25 = (x + 5)^2$.

(b) The polynomial $4x^2 - 4x + 1$ can be written as $(2x)^2 - 4x + 1^2$.

$$(2x)^2 - 4x + 1^2$$
$$\downarrow$$
$$2(2x)(1) \quad \checkmark \quad \text{The middle term checks.}$$

When $a = 2x$ and $b = 1$, the equation $a^2 - 2ab + b^2 = (a - b)^2$ allows us to factor the given polynomial as $4x^2 - 4x + 1 = (2x - 1)^2$.

(c) Write $9z^2 + 18z + 4$ as $(3z)^2 + 18z + 2^2$ and then check the middle term.

$$(3z)^2 + 18z + 2^2$$
$$\downarrow$$
$$2(3z)(2) \quad \text{✗} \quad \text{The middle term is } 12z \text{ and does not equal } 18z.$$

The expression $9z^2 + 18z + 4$ cannot be factored as a perfect square trinomial.

(d) The polynomial $x^2 - 4xy + 4y^2$ can be written as $x^2 - 4xy + (2y)^2$.

$$x^2 - 4xy + (2y)^2$$
$$\downarrow$$
$$2(x)(2y) \quad \checkmark \quad \text{The middle term checks.}$$

When $a = x$ and $b = 2y$, the equation $a^2 - 2ab + b^2 = (a - b)^2$ allows us to factor the given polynomial as $x^2 - 4xy + 4y^2 = (x - 2y)^2$.

Now Try Exercises 37, 41, 43, 51

MAKING CONNECTIONS 2

Special Factoring and General Techniques

If you do not recognize a polynomial as the difference of two squares or a perfect square trinomial, you can still factor the polynomial by using the methods discussed in earlier sections.

---

3 Factoring the Sum and Difference of Two Cubes

The sum or difference of two cubes may be factored—a result of the two equations

$$(a + b)(a^2 - ab + b^2) = a^3 + b^3 \quad \text{and}$$
$$(a - b)(a^2 + ab + b^2) = a^3 - b^3.$$

These equations can be verified by multiplying the left side to obtain the right side. For example, multiplying the polynomials on the left side of the first equation results in

$$(a + b)(a^2 - ab + b^2) = a \cdot a^2 - a \cdot ab + a \cdot b^2 + b \cdot a^2 - b \cdot ab + b \cdot b^2$$
$$= a^3 - a^2b + ab^2 + a^2b - ab^2 + b^3$$
$$= a^3 + b^3.$$

SUM AND DIFFERENCE OF TWO CUBES

For any real numbers a and b,

Opposite Signs

$$a^3 + b^3 = (a + b)(a^2 - ab + b^2) \quad \text{and}$$
$$a^3 - b^3 = (a - b)(a^2 + ab + b^2).$$

Opposite Signs

Any binomial whose terms can be expressed as cubes can be factored as a sum or difference of cubes. We demonstrate this method in the next example.

EXAMPLE 3 **Factoring the sum and difference of two cubes**

Factor each polynomial.
(a) $z^3 + 8$ (b) $x^3 - 27$ (c) $8x^3 - 1$

Solution
(a) To factor $z^3 + 8$, we let $a^3 = z^3$ and $b^3 = 2^3$ so that $a = z$ and $b = 2$. Then

$$a^3 + b^3 = (a + b)(a^2 - ab + b^2) \qquad \text{General rule}$$

becomes

$$z^3 + 2^3 = (z + 2)(z^2 - z \cdot 2 + 2^2) \qquad \begin{array}{l}\text{Substitute:}\\ a = z, b = 2\end{array}$$
$$= (z + 2)(z^2 - 2z + 4).$$

(b) To factor $x^3 - 27$, we let $a^3 = x^3$ and $b^3 = 3^3$ so that $a = x$ and $b = 3$. Then

$$a^3 - b^3 = (a - b)(a^2 + ab + b^2) \qquad \text{General rule}$$

becomes

$$x^3 - 3^3 = (x - 3)(x^2 + x \cdot 3 + 3^2) \qquad \begin{array}{l}\text{Substitute:}\\ a = x, b = 3\end{array}$$
$$= (x - 3)(x^2 + 3x + 9).$$

(c) To factor $8x^3 - 1$, we let $a^3 = (2x)^3$ and $b^3 = 1^3$ so that $a = 2x$ and $b = 1$. Be sure to write $2x$ in *parentheses* when substituting for a in the term a^2.

Write $2x$ in parentheses.

$$(2x)^3 - 1^3 = (2x - 1)\big((2x)^2 + 2x \cdot 1 + 1^2\big) \qquad a = 2x, b = 1$$
$$= (2x - 1)(4x^2 + 2x + 1)$$

Now Try Exercises 57, 59, 63

---

4 Using Special Methods in General Factoring

In this section, we have discussed special methods for factoring polynomials that can be identified as the difference of two squares, perfect square trinomials, the sum of two cubes, or the difference of two cubes. The next example demonstrates how to recognize and factor such polynomials.

EXAMPLE 4   **Recognizing polynomials that can be factored with special methods**

Factor each polynomial.
(a) $8x^3 + 27$ (b) $4y^2 - 20y + 25$ (c) $9z^2 - 64$

Solution
(a) Because this polynomial has only two terms, it cannot be factored as a perfect square trinomial. Since it is not a difference, it cannot be factored as the difference of two squares. We will try to factor the polynomial as the sum of two cubes. To factor $8x^3 + 27$, we note that $8x^3 = (2x)^3$ and $27 = 3^3$. Then

$$8x^3 + 27 = (2x)^3 + 3^3 \qquad \text{Sum of cubes}$$
$$= (2x + 3)\left((2x)^2 - 2x \cdot 3 + 3^2\right)$$
$$= (2x + 3)(4x^2 - 6x + 9).$$

(b) Because this polynomial has three terms, we will try to factor it as a perfect square trinomial. Begin by writing $4y^2 - 20y + 25$ as $(2y)^2 - 20y + 5^2$ and then verify that the middle term checks.

$$(2y)^2 - 20y + 5^2 \qquad \text{Perfect square trinomial}$$
$$2(2y)(5) \qquad \checkmark \quad \text{The middle term checks.}$$

When $a = 2y$ and $b = 5$, the equation $a^2 - 2ab + b^2 = (a - b)^2$ allows us to factor the given polynomial as $4y^2 - 20y + 25 = (2y - 5)^2$.

(c) Because this polynomial is a difference of two terms that appear to be square terms, we will try to factor the polynomial as the difference of two squares. The expression $9z^2 - 64$ can be written as $(3z)^2 - 8^2$. Thus

$$\text{Difference of squares} \qquad 9z^2 - 64 = (3z - 8)(3z + 8).$$

Now Try Exercises 31, 47, 67

When using the special factoring methods discussed in this section, it is important to remember to first factor out the greatest common factor whenever possible. In the next example, we factor out the GCF before factoring further.

EXAMPLE 5   **Factoring out the GCF before factoring further**

Factor each polynomial completely.
(a) $27x^3 + 72x^2 + 48x$ (b) $18a^3 - 8ab^2$

Solution
(a) Because the GCF of $27x^3 + 72x^2 + 48x$ is $3x$, factor out $3x$ before factoring the remaining trinomial.

$$27x^3 + 72x^2 + 48x = 3x(9x^2 + 24x + 16)$$

The expression $9x^2 + 24x + 16$ is a perfect square trinomial that can be written as $(3x)^2 + 24x + 4^2$.

$$(3x)^2 + 24x + 4^2 \qquad \text{Perfect square trinomial}$$
$$2(3x)(4) \qquad \checkmark \quad \text{The middle term checks.}$$

Thus $9x^2 + 24x + 16 = (3x + 4)^2$. As a result, the given polynomial can be factored as

$$27x^3 + 72x^2 + 48x = 3x(3x + 4)^2.$$

(b) Because the GCF of $18a^3 - 8ab^2$ is $2a$, factor out $2a$ before factoring the remaining polynomial.

$$18a^3 - 8ab^2 = 2a(9a^2 - 4b^2) \qquad \boxed{\text{Difference of squares}}$$

The expression $9a^2 - 4b^2$ is the difference of two squares and can be written as $(3a)^2 - (2b)^2$. Thus

$$18a^3 - 8ab^2 = 2a(3a - 2b)(3a + 2b).$$

Now Try Exercises 71, 79

6.4   Putting It All Together

FACTORING	EXPLANATION	EXAMPLES
Difference of Two Squares	$a^2 - b^2 = (a - b)(a + b)$ **NOTE:** The *sum* of two squares, $a^2 + b^2$, cannot be factored by using real numbers. ∎	$y^2 - 49 = (y - 7)(y + 7)$ $81 - z^2 = (9 - z)(9 + z)$ $4r^2 - 25t^2 = (2r - 5t)(2r + 5t)$ $16a^2 + b^2$ cannot be factored.
Perfect Square Trinomial	$a^2 + 2ab + b^2 = (a + b)^2$ $a^2 - 2ab + b^2 = (a - b)^2$ Be sure to verify that the given middle term equals $2ab$ before factoring.	$m^2 + 2m + 1 = (m + 1)^2$ $25y^2 - 30y + 9 = (5y - 3)^2$ $36r^2 + 12rt + t^2 = (6r + t)^2$ $x^2 + 5x + 4$ is *not* a perfect square trinomial because $2ab = 2 \cdot x \cdot 2 = 4x \neq 5x.$
Sum and Difference of Two Cubes	$a^3 + b^3 = (a + b)(a^2 - ab + b^2)$ $a^3 - b^3 = (a - b)(a^2 + ab + b^2)$	$y^3 + 27 = (y + 3)(y^2 - y \cdot 3 + 3^2)$ $\qquad = (y + 3)(y^2 - 3y + 9)$ $27r^3 - 64t^3$ $= (3r - 4t)\big((3r)^2 + 3r \cdot 4t + (4t)^2\big)$ $= (3r - 4t)(9r^2 + 12rt + 16t^2)$

6.4 Exercises

MyMathLab®

CONCEPTS AND VOCABULARY

1. $a^2 - b^2 =$ _____

2. If the expression $36x^2 - 49y^2$ is written in the form $a^2 - b^2$, then $a =$ _____ and $b =$ _____.

3. (True or False?) The expression $a^2 + b^2$ can be factored by using real numbers.

4. (True or False?) Using only integer coefficients, the expression $3x^2 - 5y^2$ can be factored as a difference of two squares.

5. $a^2 + 2ab + b^2 =$ _____

6. $a^2 - 2ab + b^2 =$ _____

7. $x^2 +$ _____ $+ 9$ is a perfect square trinomial.

8. $4r^2 -$ _____ $+ 25t^2$ is a perfect square trinomial.

9. $a^3 + b^3 =$ _____

10. $a^3 - b^3 =$ _____

11. If the expression $8x^3 + 27y^3$ is written in the form $a^3 + b^3$, then $a =$ ____ and $b =$ _____.

12. If the expression $x^3 - 1$ is written in the form $a^3 - b^3$, then $a =$ _____ and $b =$ _____.

13. $y^3 - 8 = (y$___$2)(y^2$___$2y + 4)$

14. $64z^3 + 27 = (4z$___$3)(16z^2$___$12z + 9)$

FACTORING THE DIFFERENCE OF TWO SQUARES

Exercises 15–32: Factor.

15. $x^2 - 1$
16. $x^2 - 16$

17. $z^2 - 100$
18. $z^2 - 81$

19. $4y^2 - 1$
20. $9y^2 - 16$

21. $36z^2 - 25$
22. $49z^2 - 64$

23. $9 - x^2$
24. $25 - x^2$

25. $1 - 9y^2$
26. $49 - 16y^2$

27. $4a^2 - 9b^2$
28. $16a^2 - b^2$

29. $36m^2 - 25n^2$
30. $49m^2 - 100n^2$

31. $81r^2 - 49t^2$
32. $625r^2 - 121t^2$

FACTORING PERFECT SQUARE TRINOMIALS

Exercises 33–54: Factor as a perfect square trinomial whenever possible.

33. $x^2 + 8x + 16$
34. $x^2 + 4x + 4$

35. $z^2 + 12z + 25$
36. $z^2 - 18z + 36$

37. $x^2 - 6x + 9$
38. $x^2 - 10x + 25$

39. $9y^2 + 6y + 1$
40. $16y^2 + 8y + 1$

41. $4z^2 - 4z + 1$
42. $25z^2 - 12z + 1$

43. $9t^2 + 16t + 4$
44. $4t^2 + 12t + 9$

45. $9x^2 + 30x + 25$
46. $25x^2 + 60x + 36$

47. $4a^2 - 36a + 81$
48. $9a^2 - 60a + 100$

49. $x^2 + 2xy + y^2$
50. $x^2 - 6xy + 9y^2$

51. $r^2 - 10rt + 25t^2$
52. $15r^2 + 10rt + t^2$

53. $4y^2 - 10yz + 9z^2$
54. $25y^2 - 20yz + 4z^2$

FACTORING THE SUM AND DIFFERENCE OF TWO CUBES

Exercises 55–68: Factor.

55. $z^3 + 1$
56. $z^3 + 8$

57. $x^3 + 64$
58. $x^3 + 125$

59. $y^3 - 8$
60. $y^3 - 27$

61. $n^3 - 1$
62. $n^3 - 64$

63. $8x^3 + 1$
64. $27x^3 - 1$

65. $m^3 - 64n^3$

66. $m^3 + 8n^3$

67. $8x^3 + 125y^3$

68. $27x^3 + 64y^3$

USING SPECIAL METHODS IN GENERAL FACTORING

Exercises 69–86: Factor the expression completely.

69. $4x^2 - 16$
70. $12x^2 - 60x + 75$

71. $2y^2 - 28y + 98$ **72.** $y^3 - 9y$

73. $5z^3 + 40$ **74.** $4z^3 + 36z^2 + 100z$

75. $x^3y - xy^3$ **76.** $8m^3 - 8$

77. $2m^3 - 10m^2 + 18m$ **78.** $2a^3b - 18ab^3$

79. $700x^4 - 63x^2y^2$ **80.** $135r^3 - 5t^3$

81. $16a^3 + 2b^3$ **82.** $192x^2y^2 - 3y^4$

83. $4b^4 + 24b^3 + 36b^2$ **84.** $2y^4 + 24y^3 + 72y^2$

85. $500r^3 - 32t^3$ **86.** $8r^3 - 64t^3$

RELATING GEOMETRY AND VISUAL FACTORING

87. A square has an area of $4x^2 + 12x + 9$. Find the length of a side. Make a sketch of the square.

88. A square has an area of $9x^2 + 30x + 25$. Find the length of a side. Make a sketch of the square.

89. Difference of Two Squares Write the difference in the areas between the large square with sides of length x and a smaller square with sides of length 5, using factored form.

 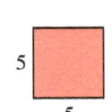

90. Difference of Two Cubes Write the difference in the volumes between a large cube with sides of length 2 and a smaller cube with sides of length x, using factored form.

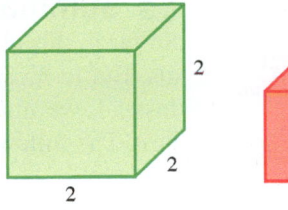

WRITING ABOUT MATHEMATICS

91. Explain how factoring $x^3 + y^3$ is different from factoring $x^3 - y^3$.

92. Using the techniques discussed in this section, can you factor the expression $4x^2 + 9y^2$ into two binomials? Explain your reasoning.

SECTIONS 6.3 and 6.4   **Checking Basic Concepts**

1. Factor each trinomial.
   (a) $2x^2 - 5x - 12$ **(b)** $6x^2 + 17x - 14$

2. Factor completely when possible.
   (a) $3y^2 + 4y - 2$ **(b)** $6y^3 - 10y^2 - 4y$

3. Write a polynomial in factored form that represents the total area of the figure.

4. Factor each polynomial.
   (a) $z^2 - 64$ **(b)** $9r^2 - 4t^2$

5. Factor each trinomial.
   (a) $x^2 + 12x + 36$ **(b)** $9a^2 - 12ab + 4b^2$

6. Factor.
   (a) $m^3 - 27$ **(b)** $125n^3 + 27$

7. Factor completely.
   (a) $16x^2 - 4$ **(b)** $3y^4 + 24y$

6.5 Summary of Factoring

Objectives

1 Summarizing General Guidelines for Factoring Polynomials

2 Factoring Polynomials

STUDY TIP

This section provides an opportunity to practice the material from several sections. In general, it is a good idea to regularly review topics that have been covered earlier.

1 Summarizing General Guidelines for Factoring Polynomials

So far in this chapter, we have discussed several useful techniques for factoring polynomials. But in most factoring problems, the specific method that should be used is not stated. Instead, we must look carefully at each factoring problem and decide which approach is best. The following guidelines can be used to factor polynomials in general.

FACTORING POLYNOMIALS

STEP 1: Factor out the greatest common factor (GCF), if possible.

STEP 2: A. If the polynomial has *four terms*, try factoring by grouping.

      B. If the polynomial is a *binomial*, try one of the following.

        1. $a^2 - b^2 = (a - b)(a + b)$ Difference of two squares

        2. $a^3 - b^3 = (a - b)(a^2 + ab + b^2)$ Difference of two cubes

        3. $a^3 + b^3 = (a + b)(a^2 - ab + b^2)$ Sum of two cubes

      C. If the polynomial is a *trinomial*, check for a perfect square.

        1. $a^2 + 2ab + b^2 = (a + b)^2$ Perfect square trinomial

        2. $a^2 - 2ab + b^2 = (a - b)^2$ Perfect square trinomial

      Otherwise, try to factor the trinomial by grouping or apply FOIL in reverse, as described in Sections 2 and 3 of this chapter.

STEP 3: Check to make sure that the polynomial is *completely* factored.

READING CHECK 1

• Why is it important to first factor out the greatest common factor?

NOTE: Always perform Step 1 first. Factoring out the greatest common factor usually makes it easier to factor the resulting polynomial. After a polynomial has been factored, remember to perform Step 3 so that you are sure the given polynomial is completely factored. ■

2 Factoring Polynomials

In the first example, we apply Step 1 to a polynomial with a common factor.

EXAMPLE 1   **Factoring out a common factor**

Factor $5x^3 - 10x^2 + 15x$.

Solution

STEP 1: The greatest common factor is $5x$.

                                         Prime trinomial

$$5x^3 - 10x^2 + 15x = 5x(x^2 - 2x + 3)$$

STEP 2C: The trinomial $x^2 - 2x + 3$ is prime and cannot be factored further.

STEP 3: The completely factored polynomial is $5x(x^2 - 2x + 3)$.

Now Try Exercise 11

When factoring polynomials completely, it is often necessary to apply more than one factoring technique. In several of the next examples, we factor polynomials that require more than one method of factoring.

EXAMPLE 2   **Factoring a difference of squares**

Factor $3x^4 - 48x^2$.

Solution

STEP 1: The greatest common factor is $3x^2$.

$$3x^4 - 48x^2 = 3x^2(x^2 - 16)$$

> Difference of squares

STEP 2B: The binomial $x^2 - 16$ can be factored as a difference of two squares.

$$3x^2(x^2 - 16) = 3x^2(x - 4)(x + 4)$$

STEP 3: The completely factored polynomial is $3x^2(x - 4)(x + 4)$.

Now Try Exercise 39

MAKING CONNECTIONS 1

Factoring Polynomials

We can often determine how a polynomial should be factored by considering the number of terms in the polynomial. This is summarized as follows.

Type of Polynomial	*Factoring Technique*
4-term Polynomial	Grouping
Trinomial	Perfect square trinomial FOIL in reverse or grouping
Binomial	Difference of two squares Sum or difference of two cubes

---

EXAMPLE 3   **Factoring a perfect square trinomial**

Factor $36y^3 - 24y^2 + 4y$.

Solution

STEP 1: The greatest common factor is $4y$.

$$36y^3 - 24y^2 + 4y = 4y(9y^2 - 6y + 1)$$

> Perfect square trinomial

STEP 2C: We can factor $9y^2 - 6y + 1$ as a perfect square trinomial.

$$4y(9y^2 - 6y + 1) = 4y(3y - 1)(3y - 1)$$

STEP 3: The completely factored polynomial is $4y(3y - 1)^2$.

Now Try Exercise 41

EXAMPLE 4   **Factoring a sum of cubes**

Factor $27z^3 + 64$.

Solution

STEP 1: There are no common factors.

STEP 2B: The binomial $27z^3 + 64$ can be written as $(3z)^3 + 4^3$ and factored as a sum of two cubes.

> Sum of cubes

$$27z^3 + 64 = (3z + 4)\big((3z)^2 - 3z \cdot 4 + 4^2\big)$$
$$= (3z + 4)(9z^2 - 12z + 16)$$

NOTE: The trinomial $9z^2 - 12z + 16$ is prime and cannot be factored further. ■

STEP 3: The completely factored polynomial is $(3z + 4)(9z^2 - 12z + 16)$.

Now Try Exercise 15

EXAMPLE 5   **Factoring a trinomial**

Factor $14x^4 + 7x^3 - 42x^2$.

Solution

STEP 1: The greatest common factor is $7x^2$.

> Factor this trinomial further.

$$14x^4 + 7x^3 - 42x^2 = 7x^2(2x^2 + x - 6)$$

STEP 2C: We can factor $2x^2 + x - 6$ using FOIL in reverse.

$$7x^2(2x^2 + x - 6) = 7x^2(2x - 3)(x + 2)$$

STEP 3: The completely factored polynomial is $7x^2(2x - 3)(x + 2)$.

Now Try Exercise 35

EXAMPLE 6   **Factoring by grouping**

Factor $15x^3 + 10x^2 - 60x - 40$.

Solution

STEP 1: The greatest common factor is 5.

> Factor by grouping.

$$15x^3 + 10x^2 - 60x - 40 = 5(3x^3 + 2x^2 - 12x - 8)$$

STEP 2A: Because the resulting polynomial has four terms, we apply grouping.

$$
\begin{aligned}
5(3x^3 + 2x^2 - 12x - 8) &= 5\big((3x^3 + 2x^2) + (-12x - 8)\big) &&\text{Associative property} \\
&= 5\big(x^2(3x + 2) - 4(3x + 2)\big) &&\text{Factor out common factors.} \\
&= 5(x^2 - 4)(3x + 2) &&\text{Factor out } (3x + 2).
\end{aligned}
$$

STEP 2B: The binomial $x^2 - 4$ can now be factored as a difference of two squares.

$$5(x^2 - 4)(3x + 2) = 5(x - 2)(x + 2)(3x + 2)$$

> Difference of squares

STEP 3: The completely factored polynomial is $5(x - 2)(x + 2)(3x + 2)$.

Now Try Exercise 33

EXAMPLE 7   **Factoring a polynomial that has two variables**

Factor $18x^3y - 8xy^3$.

Solution

STEP 1: The greatest common factor is $2xy$.

$$18x^3y - 8xy^3 = 2xy(9x^2 - 4y^2)$$

STEP 2B: The binomial $9x^2 - 4y^2$ can be written as $(3x)^2 - (2y)^2$ and can be factored as a difference of two squares.

Difference of squares $9x^2 - 4y^2 = (3x - 2y)(3x + 2y)$

STEP 3: The completely factored polynomial is $2xy(3x - 2y)(3x + 2y)$.

Now Try Exercise 55

EXAMPLE 8   **Applying several techniques**

Factor $3x^5 - 3x^3 - 24x^2 + 24$.

Solution

STEP 1: The greatest common factor is 3.

$$3x^5 - 3x^3 - 24x^2 + 24 = 3(x^5 - x^3 - 8x^2 + 8)$$

STEP 2A: The resulting four-term polynomial can be factored by grouping.

$$3(x^5 - x^3 - 8x^2 + 8) = 3\big((x^5 - x^3) + (-8x^2 + 8)\big) \quad \text{Associative property}$$
$$= 3\big(x^3(x^2 - 1) - 8(x^2 - 1)\big) \quad \text{Factor out common factors.}$$

Factor by grouping.

$$= 3(x^3 - 8)(x^2 - 1) \quad \text{Factor out } x^2 - 1.$$

STEP 2B: Both binomials in this expression can be factored further. The binomial $x^3 - 8$ can be factored as a difference of two cubes, and the binomial $x^2 - 1$ can be factored as a difference of two squares.

$$3(x^3 - 8)(x^2 - 1) = 3(x - 2)(x^2 + 2x + 4)(x - 1)(x + 1)$$

Difference of cubes   Difference of squares

NOTE: The trinomial $x^2 + 2x + 4$ is prime and cannot be factored further. ■

STEP 3: The completely factored form is $3(x - 2)(x^2 + 2x + 4)(x - 1)(x + 1)$.

Now Try Exercise 45

6.5   Putting It All Together

CONCEPT	EXPLANATION	EXAMPLES
Greatest Common Factor	Factor out the greatest common factor, or monomial, that occurs in each term.	$2x^2 - 4x + 10 = 2(x^2 - 2x + 5)$ $3x^3 + 6x = 3x(x^2 + 2)$ $7xy - x^2y = xy(7 - x)$
Factoring by Grouping	Use the associative and distributive properties to factor a polynomial with four terms.	$x^3 - 3x^2 + 2x - 6 = (x^3 - 3x^2) + (2x - 6)$ $= x^2(x - 3) + 2(x - 3)$ $= (x^2 + 2)(x - 3)$
Factoring Binomials	Use the difference of two squares, the difference of two cubes, or the sum of two cubes.	$9x^2 - 4 = (3x - 2)(3x + 2)$ $x^3 - 27 = (x - 3)(x^2 + 3x + 9)$ $x^3 + 27 = (x + 3)(x^2 - 3x + 9)$

continued on next page

continued from previous page

CONCEPT	EXPLANATION	EXAMPLES
Factoring Trinomials	Use FOIL in reverse or grouping.	$x^2 + 5x - 6 = (x + 6)(x - 1)$ Check middle term: $-x + 6x = 5x.$ ✓ $4x^2 + 4x - 3 = (4x^2 - 2x) + (6x - 3)$ $= 2x(2x - 1) + 3(2x - 1)$ $= (2x + 3)(2x - 1)$

6.5   Exercises

MyMathLab®

CONCEPTS AND VOCABULARY

1. What do the letters GCF mean?

2. A good first step for factoring polynomials is to factor out the _____ .

3. If a polynomial has four terms, what factoring method might be appropriate?

4. If a polynomial is a binomial, we can try to factor it as a difference of two _____, a difference of two _____, or a sum of two _____.

5. Can $x^2 + 1$ be factored? Explain.

6. Can $x^3 + 1$ be factored? Explain.

7. If a polynomial is a trinomial, we can try to factor it as a perfect _____ trinomial. Otherwise, try factoring it by _____ or apply _____ in reverse.

8. The last step for factoring is to be sure that the polynomial is _____ factored.

PERFORMING BASIC FACTORING

Exercises 9–24: Factor completely, if possible.

9. $4x - 2$

10. $x^2 + 3x$

11. $2y^2 - 4y + 4$

12. $5y^2 - 25y + 10$

13. $z^2 - 4$

14. $9z^2 - 25$

15. $a^3 + 8$

16. $8a^3 - 1$

17. $4b^2 - 12b + 9$

18. $b^2 + 4b + 4$

19. $m^2 + 9$

20. $4m^2 + 49$

21. $x^3 - x^2 + 5x - 5$

22. $3x^3 + 6x^2 + x + 2$

23. $y^2 - 5y + 4$

24. $y^2 - 3y - 10$

PERFORMING GENERAL FACTORING

Exercises 25–62: Factor completely.

25. $x^3 + 4x^2 - 9x - 36$

26. $6x^2 - 19x + 15$

27. $8a^3 - 64$

28. $ab^2 - 4a$

29. $12x^4 - 18x^3 + 4x^2 - 6x$

30. $3x^2y + 24xy + 48y$

31. $54t^4 + 16t$

32. $3t^3 + 18t^2 - 48t$

33. $2r^3 + 6r^2 - 2r - 6$

34. $3r^4 + 3r^3 - 24r - 24$

35. $6z^4 - 21z^3 - 45z^2$

36. $3x^4y + 24xy^4$

37. $12b^4 - 10b^3 + 2b^2$

38. $6a^4b + 4a^3b + 18a^2b + 12ab$

39. $6z - 24z^3$

40. $6y^3z - 48z^4$

41. $3x^2y - 30xy + 75y$

42. $8x^3 + y^3$

43. $27m^3 - 8n^3$

44. $45m^3 - 69m^2 + 12m$

45. $3x^5 - 12x^3 - 3x^2 + 12$

46. $8x^3 - 8$

47. $5a^2 - 27a - 18$

48. $2a^2 - 6ab + 3a - 9b$

49. $3rt^2 + 33rt + 90r$

50. $9t^2 + 24t + 16$

51. $9b^3 + 6b^2 + 12b + 8$

52. $5b^3 - 55b^2 - 60b$

53. $6n^3 + 2n^2 - 10n$

54. $7n^4 + 28n^3 - 63n^2$

55. $4x^2 - 36y^2$

56. $64x^2 - 25y^2$

57. $2a^3 - 16a^2 + 32a$

58. $24a^3 + 72a^2 + 54a$

59. $32xy^3 + 4x$

60. $24x^3 - 4x^2 - 160x$

61. $8b^4 + 24b^3 - 2b^2 - 6b$

62. $3z^3 - 6z^2 - 27z + 54$

RELATING GEOMETRY AND VISUAL FACTORING

63. Dimensions of a Square If three identical squares have a total area of $27x^2 + 18x + 3$, find the length of one side of one of the squares.

64. Dimensions of a Cube If three identical cubes have a total volume of $3x^3 + 18x^2 + 36x + 24$, find the length of one side of one of the cubes.

WRITING ABOUT MATHEMATICS

65. Explain how the number of terms in a polynomial can help determine what method should be used to factor it.

66. Describe a method for determining whether a polynomial has been factored correctly.

Group Activity

Exercises 1–6: **Difference of Two Squares** *The difference of two squares can be factored by using*

$$a^2 - b^2 = (a - b)(a + b).$$

This equation can also be used in some situations where an expression may not appear to be the difference of two squares. For example, because $(\sqrt{3})^2 = 3$, *the binomial* $x^2 - 3$ *can be written and then factored as*

$$x^2 - 3 = x^2 - (\sqrt{3})^2$$
$$= (x - \sqrt{3})(x + \sqrt{3}).$$

Use this concept to factor the following expressions as the difference of two squares.

1. $x^2 - 5$

2. $y^2 - 7$

3. $3z^2 - 25$

4. $7t^2 - 2$

5. $x - 4$ for $x \geq 0$ *(Hint:* $(\sqrt{x})^2 = x.$)

6. $x - 7$ for $x \geq 0$

6.6 Solving Equations by Factoring I (Quadratics)

Objectives

1 **Applying the Zero-Product Property**

2 **Solving Quadratic Equations**

3 **Solving Applications Involving Quadratic Equations**

NEW VOCABULARY

☐ Zero-product property
☐ Zeros (of a polynomial)
☐ Quadratic polynomial
☐ Quadratic equation
☐ Standard form

1 Applying the Zero-Product Property

To solve equations, we often use the **zero-product property**, which states that if the product of two numbers (or expressions) is 0, then at least one of the numbers (or expressions) must equal 0.

ZERO-PRODUCT PROPERTY

For all real numbers a and b, if $ab = 0$, then $a = 0$ or $b = 0$ (or both).

NOTE: The zero-product property works only for 0. If $ab = 1$, then it does *not* follow that $a = 1$ or $b = 1$. For example, $a = \frac{1}{3}$ and $b = 3$ satisfy the equation $ab = 1$. ∎

After factoring an expression, we can use the zero-product property to solve an equation. The left side of the equation $3t^2 - 9t = 0$ may be factored to obtain

$$3t(t - 3) = 0.$$

Note that the product of $3t$ and $t - 3$ is 0. By the zero-product property, either

$$3t = 0 \quad \text{or} \quad t - 3 = 0.$$

Solving each equation for t results in

> Zero-product property

$$t = 0 \quad \text{or} \quad t = 3.$$

STUDY TIP

The zero-product property is used extensively in mathematics for solving many types of equations. Be sure that you learn how to apply this important property correctly.

These values can be checked by substituting them in the given equation $3t^2 - 9t = 0$.

$$3(0)^2 - 9(0) = 0 \checkmark \quad \text{Let } t = 0. \text{ It checks.}$$
$$3(3)^2 - 9(3) = 0 \checkmark \quad \text{Let } t = 3. \text{ It checks.}$$

The t-values of 0 and 3 are called **zeros** of the polynomial $3t^2 - 9t$, because when either is substituted in this polynomial, the result is 0.

READING CHECK 1

• Which of the following are zeros of $4x^2 - 8x$: $-8, -4, -2, 0, 2, 4, 8$?

EXAMPLE 1 **Applying the zero-product property**

Solve each equation.
(a) $x(x - 1) = 0$
(b) $2z^2 = 0$
(c) $(t + 3)(t + 2) = 0$
(d) $x(x - 2)(2x + 1) = 0$

Solution
(a) By the zero-product property, $x(x - 1) = 0$ when $x = 0$ or $x - 1 = 0$. The solutions are 0 and 1.
(b) $2z^2 = 2 \cdot z \cdot z$ and $2 \neq 0$, so $2z^2 = 0$ when $z = 0$.
(c) $(t + 3)(t + 2) = 0$ implies that $t + 3 = 0$ or $t + 2 = 0$. The solutions to the given equation are -3 and -2.
(d) We can apply the zero-product property to $x(x - 2)(2x + 1) = 0$. Thus $x = 0$, or $x - 2 = 0$, or $2x + 1 = 0$. The solutions are $-\frac{1}{2}$, 0, and 2.

Now Try Exercises 13, 15, 19, 23

2 Solving Quadratic Equations

Any **quadratic polynomial** in the variable x can be written as $ax^2 + bx + c$ with $a \neq 0$. Any **quadratic equation** in the variable x can be written as $ax^2 + bx + c = 0$ with $a \neq 0$. This form of quadratic equation is called the **standard form** of a quadratic equation. **TABLE 6.11** shows examples of quadratic polynomials along with related quadratic equations that can be expressed in standard form.

Quadratic Polynomials and Equations

Quadratic Polynomial	Quadratic Equation	Standard Form
$x^2 - 2x + 1$	$x^2 - 2x + 1 = 9$	$x^2 - 2x - 8 = 0$
$3x^2 + 7x$	$3x^2 + 7x = 4$	$3x^2 + 7x - 4 = 0$
$x^2 - 9$	$x^2 - 9 = 2$	$x^2 - 11 = 0$
$6x - x^2 + 2$	$6x - x^2 + 2 = 1$	$-x^2 + 6x + 1 = 0$

TABLE 6.11

READING CHECK 2

• What does it mean for a polynomial to be written so that it contains descending powers of x?

NOTE: When a quadratic polynomial in the variable x is in standard form and we read it from left to right, the terms contain *descending powers* of x. In other words, the first term contains x^2, the second term contains x, and the third term is a constant (the exponent on x is 0). ∎

To solve a quadratic equation, we often use factoring and the zero-product property. This method is summarized by the following steps. Although it is not necessary to label each step in the solution to a quadratic equation, it is important to keep these steps in mind.

SOLVING QUADRATIC EQUATIONS

To solve a quadratic equation by factoring, follow these steps.

STEP 1: If necessary, write the equation in standard form as $ax^2 + bx + c = 0$.

STEP 2: Factor the left side of the equation using any method.

STEP 3: Apply the zero-product property.

STEP 4: Solve each of the resulting equations. Check any solutions.

EXAMPLE 2 Solving equations by factoring

Solve each quadratic equation. Check your answers.
(a) $x^2 + 2x = 0$ **(b)** $y^2 = 16$ **(c)** $z^2 - 3z + 2 = 0$ **(d)** $2x^2 = 5 - 9x$

Solution
(a) Because $x^2 + 2x = 0$ is in standard form, Step 1 is unnecessary. We begin by factoring out the GCF, x.

$$x^2 + 2x = 0 \qquad \text{Given equation}$$
$$x(x + 2) = 0 \qquad \text{Factor out } x. \text{ (Step 2)}$$
$$x = 0 \quad \text{or} \quad x + 2 = 0 \qquad \text{Zero-product property (Step 3)}$$
$$x = 0 \quad \text{or} \quad x = -2 \qquad \text{Solve for } x. \text{ (Step 4)}$$

Zero-product property

To check these values, substitute -2 and 0 for x in the given equation.

$$(-2)^2 + 2(-2) \overset{?}{=} 0 \qquad (0)^2 + 2(0) \overset{?}{=} 0 \qquad \text{Substitute } -2 \text{ and } 0.$$
$$0 = 0 \checkmark \qquad\qquad 0 = 0 \checkmark \qquad \text{Both answers check.}$$

Therefore the solutions are -2 and 0.

(b) To write $y^2 = 16$ in standard form, we begin by subtracting 16 from each side to obtain 0 on the right side.

$$y^2 = 16 \qquad \text{Given equation}$$
$$y^2 - 16 = 0 \qquad \text{Subtract 16. (Step 1)}$$
$$(y - 4)(y + 4) = 0 \qquad \text{Difference of squares (Step 2)}$$
$$y - 4 = 0 \quad \text{or} \quad y + 4 = 0 \qquad \text{Zero-product property (Step 3)}$$
$$y = 4 \quad \text{or} \quad y = -4 \qquad \text{Solve for } y. \text{ (Step 4)}$$

To check these values, substitute -4 and 4 for y in the given equation.

$$(-4)^2 \overset{?}{=} 16 \qquad (4)^2 \overset{?}{=} 16 \qquad \text{Substitute } -4 \text{ and } 4.$$
$$16 = 16 \checkmark \qquad 16 = 16 \checkmark \qquad \text{Both answers check.}$$

The solutions are -4 and 4.

(c) Step 1 is unnecessary because the equation is already in standard form. We begin by factoring the left side of the equation, $z^2 - 3z + 2$.

$$z^2 - 3z + 2 = 0 \qquad \text{Given equation}$$
$$(z - 1)(z - 2) = 0 \qquad \text{Factor. (Step 2)}$$
$$z - 1 = 0 \quad \text{or} \quad z - 2 = 0 \qquad \text{Zero-product property (Step 3)}$$
$$z = 1 \quad \text{or} \quad z = 2 \qquad \text{Solve for } z. \text{ (Step 4)}$$

To check these values, substitute 1 and 2 for z in the given equation.

$$1^2 - 3(1) + 2 \overset{?}{=} 0 \qquad 2^2 - 3(2) + 2 \overset{?}{=} 0 \qquad \text{Substitute 1 and 2.}$$
$$0 = 0 \checkmark \qquad\qquad\qquad 0 = 0 \checkmark \qquad \text{Both answers check.}$$

The solutions are 1 and 2.

(d) We write $2x^2 = 5 - 9x$ in standard form by adding -5 and $9x$ to each side.

$$2x^2 = 5 - 9x \qquad \text{Given equation}$$
$$2x^2 + 9x - 5 = 0 \qquad \text{Add } -5 \text{ and } 9x. \text{ (Step 1)}$$
$$(2x - 1)(x + 5) = 0 \qquad \text{Factor. (Step 2)}$$
$$2x - 1 = 0 \quad \text{or} \quad x + 5 = 0 \qquad \text{Zero-product property (Step 3)}$$
$$x = \frac{1}{2} \quad \text{or} \quad x = -5 \qquad \text{Solve for } x. \text{ (Step 4)}$$

To check these values, substitute -5 and $\frac{1}{2}$ for x in the given equation.

$$2(-5)^2 \overset{?}{=} 5 - 9(-5) \qquad 2\left(\frac{1}{2}\right)^2 \overset{?}{=} 5 - 9\left(\frac{1}{2}\right) \qquad \text{Substitute } -5 \text{ and } \tfrac{1}{2}.$$

$$50 = 50 \checkmark \qquad\qquad \frac{1}{2} = \frac{1}{2} \checkmark \qquad \text{Both answers check.}$$

The solutions are -5 and $\frac{1}{2}$.

Now Try Exercises 27, 35, 43, 49

EXAMPLE 3   **Solving an equation by factoring**

Solve $6x^2 - x = 12$.

Solution

We *cannot* solve the equation $6x^2 - x = 12$ by factoring out the common factor of x in $6x^2 - x$ and setting each factor equal to 12. Instead, we apply the zero-product property by first writing the given equation in the standard form as $ax^2 + bx + c = 0$.

$$6x^2 - x = 12 \qquad \text{Given equation}$$
$$6x^2 - x - 12 = 0 \qquad \text{Subtract 12. (Step 1)}$$
$$(2x - 3)(3x + 4) = 0 \qquad \text{Factor. (Step 2)}$$
$$2x - 3 = 0 \quad \text{or} \quad 3x + 4 = 0 \qquad \text{Zero-product property (Step 3)}$$
$$x = \frac{3}{2} \quad \text{or} \quad x = -\frac{4}{3} \qquad \text{Solve for } x. \text{ (Step 4)}$$

The solutions are $-\frac{4}{3}$ and $\frac{3}{2}$.

Now Try Exercise 51

READING CHECK 3

Which of the following are expressions: $x^2 - 4x + 8$, $\sqrt{x + 1} = x$, $\frac{x + 4}{5} - x^2$, $x + 5 = x^2$?

MAKING CONNECTIONS 1

Equations and Expressions

The words "equation" and "expression" occur frequently in mathematics. However, they are *not* interchangeable. We often want to *solve* equations to find the values of the variable that make the equation true. We *factor* and *simplify* expressions. An equation is a statement that two expressions are equal. For example, $2x^2 - 3 = 5x + 1$ is an equation where $2x^2 - 3$ and $5x + 1$ are each expressions.

3 ▸ Solving Applications Involving Quadratic Equations

 Math in Context *Sports* If a golf ball is hit upward at 132 feet per second, or 90 miles per hour, then its height h in feet above the ground after t seconds is given by $h = 132t - 16t^2$. The expression $132t - 16t^2$ is an example of a *quadratic polynomial*. To determine the elapsed time between when the ball is hit and when it strikes the ground (or when $h = 0$), we solve the quadratic equation

$$132t - 16t^2 = 0.$$

One method for solving this equation is by factoring.

EXAMPLE 4   **Modeling the flight of a golf ball**

If a golf ball is hit upward at 132 feet per second, or 90 miles per hour, then its height h in feet after t seconds is $h = 132t - 16t^2$. Find the time when the golf ball strikes the ground.

Solution
The golf ball strikes the ground when its height is 0.

$132t - 16t^2 = 0$	Let $h = 0$.
$4t(33 - 4t) = 0$	Factor out $4t$.
$4t = 0$ or $33 - 4t = 0$	Zero-product property
$t = 0$ or $-4t = -33$	Divide by 4; subtract 33.
$t = 0$ or $t = \dfrac{33}{4}$	Solve for t.

The ball strikes the ground after $\frac{33}{4} = 8.25$ seconds. The solution of 0 is not used in this problem because it corresponds to the time when the ball is hit from ground level.

Now Try Exercise 67 (a)

Math in Context *Highway Safety* When you try to stop a car, the faster you are driving, the further the stopping distance will be. In fact, if you drive twice as fast, the braking distance will be about four times as much, and if you drive three times faster, the braking distance will be about nine times as much.

EXAMPLE 5   **Modeling braking distance**

The braking distance D in feet required to stop a car traveling at x miles per hour on dry, level pavement can be approximated by $D = \frac{1}{11}x^2$. (*Source*: L. Haefner.)
(a) Calculate the braking distance for a car traveling 70 miles per hour.
(b) If the braking distance is 44 feet, calculate the speed of the car.
(c) If you have a calculator available, use it to solve part (b) numerically with a table of values.

Solution

(a) If $x = \mathbf{70}$, then $D = \frac{1}{11}(\mathbf{70})^2 = \frac{4900}{11} \approx 445$ feet.

(b) **Symbolic Solution** Let $D = 44$ in the given equation and solve.

$$\frac{1}{11}x^2 = 44 \qquad \text{Let } D = 44.$$

$$\frac{1}{11}x^2 - 44 = 0 \qquad \text{Subtract 44.}$$

$$x^2 - 484 = 0 \qquad \text{Multiply by 11.}$$

$$(x - 22)(x + 22) = 0 \qquad \text{Difference of two squares } (22^2 = 484)$$

$$x - 22 = 0 \quad \text{or} \quad x + 22 = 0 \qquad \text{Zero-product property}$$

$$x = 22 \quad \text{or} \quad x = -22 \qquad \text{Solve for } x.$$

The car is traveling at (approximately) 22 miles per hour. (Note that $x = -22$ has no physical meaning in this problem.)

(c) **Numerical Solution** Let $Y_1 = X^2/11$ and make a table of values. Scroll through the table, as shown in **FIGURE 6.5**, to where $x = 22$ when $y_1 = 44$. Thus the solution is 22 miles per hour.

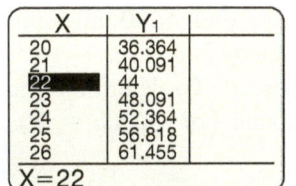

X	Y₁
20	36.364
21	40.091
22	44
23	48.091
24	52.364
25	56.818
26	61.455

X=22

FIGURE 6.5

Now Try Exercise 69

EXAMPLE 6   **Finding the dimensions of a digital photograph**

A small digital photograph is 20 pixels longer than it is wide, as illustrated in **FIGURE 6.6**. It has a total of 2400 pixels. Find the dimensions of this photograph.

Dimensions of a Photograph

x

2400 pixels

$x + 20$

FIGURE 6.6

Solution

From **FIGURE 6.6** the rectangular photograph has an area of 2400 pixels.

$$x(x + 20) = 2400 \qquad \text{Area} = \text{width} \times \text{length}$$

$$x^2 + 20x = 2400 \qquad \text{Distributive property}$$

$$x^2 + 20x - 2400 = 0 \qquad \text{Subtract 2400.}$$

$$(x - 40)(x + 60) = 0 \qquad \text{Factor.}$$

$$x - 40 = 0 \quad \text{or} \quad x + 60 = 0 \qquad \text{Zero-product property}$$

$$x = 40 \quad \text{or} \quad x = -60 \qquad \text{Solve for } x.$$

The only valid solution is 40. Thus the dimensions of the photograph are 40 pixels by $40 + 20 = 60$ pixels.

Now Try Exercise 71

CRITICAL THINKING 1

Are the two solutions to $x(2x + 1) = 1$ found by setting both factors equal to 1? Explain your reasoning. What are the solutions to this equation?

6.6 Putting It All Together

CONCEPT	EXPLANATION	EXAMPLES
Zero-Product Property	If the product of two or more expressions is 0, then at least one of the expressions must equal 0.	$ab = 0$ implies that $a = 0$ or $b = 0$. $x(x + 1) = 0$ implies that $x = 0$ or $x + 1 = 0$. $z(z - 1)(z + 2) = 0$ implies that $z = 0$ or $z - 1 = 0$ or $z + 2 = 0$.
Solving Quadratic Equations by Factoring	**1.** Write the equation as $$ax^2 + bx + c = 0.$$ **2.** Factor the left side of this equation. **3.** Apply the zero-product property. **4.** Solve each resulting equation. Check any solutions.	$2x^2 + 11x = 6$ $2x^2 + 11x - 6 = 0$ Step 1 $(2x - 1)(x + 6) = 0$ Step 2 $2x - 1 = 0$ or $x + 6 = 0$ Step 3 $x = \dfrac{1}{2}$ or $x = -6$ Step 4

6.6 Exercises

MyMathLab®

CONCEPTS AND VOCABULARY

1. If $ab = 0$, then either $a =$ _____ or $b =$ _____.

2. Can the zero-product property be used to state that if $(x - 1)(x - 2) = 3$, then either $x - 1 = 3$ or $x - 2 = 3$? Explain your answer.

3. If $2x(x + 6) = 0$, then either _____ or _____.

4. What is a good first step when you are solving the equation $4x^2 + 1 = 4x$ by factoring?

5. What is the next step when you are solving the equation $(x + 5)(x - 4) = 0$?

6. Factoring is an important method for ____ equations.

7. Because $2(4) - 8 = 0$, the value 4 is called a(n) _____ of the polynomial $2x - 8$.

8. What is the zero of the polynomial $3x - 6$?

9. Any quadratic equation in the variable x can be written in standard form as ____.

10. Standard form for $x^2 + 1 = 6x$ is ____.

11. When written in standard form and read from left to right, a quadratic polynomial in the variable x contains ____ powers of x.

12. For the constant term in a quadratic polynomial in the variable x, the exponent on x is _____.

APPLYING THE ZERO-PRODUCT PROPERTY

Exercises 13–24: Solve the equation.

13. $x^2 = 0$

14. $5m^2 = 0$

15. $2x(x + 8) = 0$

16. $x(x + 10) = 0$

17. $(y - 1)(y - 2) = 0$

18. $(y + 4)(y - 3) = 0$

19. $(6z + 5)(z - 7) = 0$

20. $(2z - 1)(4z - 3) = 0$

21. $(1 - 3n)(3 - 7n) = 0$

22. $(5 - n)(5 + n) = 0$

23. $x(x - 5)(x - 8) = 0$

24. $x(x + 1)(x - 6) = 0$

SOLVING QUADRATIC EQUATIONS

Exercises 25–60: Solve and check.

25. $x^2 - x = 0$

26. $2x^2 + 4x = 0$

27. $z^2 - 5z = 0$

28. $6z^2 - 3z = 0$

29. $10y^2 + 15y = 0$

30. $2y^2 + 3y = 0$

31. $x^2 - 1 = 0$

32. $x^2 - 9 = 0$

33. $4n^2 - 1 = 0$

34. $9n^2 - 4 = 0$

35. $z^2 + 3z + 2 = 0$

36. $z^2 - 2z - 3 = 0$

37. $x^2 - 12x + 35 = 0$

38. $x^2 - x - 20 = 0$

39. $2b^2 + 3b - 2 = 0$

40. $3b^2 + b - 2 = 0$

41. $6y^2 + 19y + 10 = 0$

42. $4y^2 - 25y - 21 = 0$

43. $x^2 = 25$

44. $x^2 = 81$

45. $t^2 = 5t$

46. $10t^2 = -5t$

47. $3m^2 = -9m$

48. $4m^2 = 9$

49. $x^2 = 5x + 6$

50. $12z^2 = 5 - 4z$

51. $2x^2 + 3x = 14$

52. $12z^2 + 11z = 15$

53. $t(t + 1) = 2$

54. $t(t - 7) = -12$

55. $x(2x + 5) = 3$

56. $x(3x + 2) = 5$

57. $12x^2 + 12x = -3$

58. $18x^2 + 2 = 12x$

59. $30y^2 + 50y + 20 = 0$

60. $30y^2 - 25y + 5 = 0$

RELATING GEOMETRY AND EQUATIONS

61. Dimensions of a Square A square has an area of 144 square feet. Find the length of a side.

62. Dimensions of a Cube A cube has a surface area of 96 square feet. Find the length of a side.

63. Radius of a Circle The numerical difference between the area and the circumference of a circle is 8π. Find the radius of the circle. (*Hint:* First factor out π in your equation.)

64. Dimensions of a Rectangle A rectangle is 5 feet longer than it is wide and has an area of 126 square feet. What are its dimensions?

Exercise 65 and 66: **The Pythagorean Theorem** *Suppose that a right triangle has legs a and b and hypotenuse c, as illustrated in the figure. Then these values satisfy* $a^2 + b^2 = c^2$.

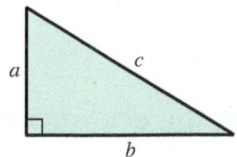

Use the Pythagorean theorem to find the value of x in the figure.

65.

66.

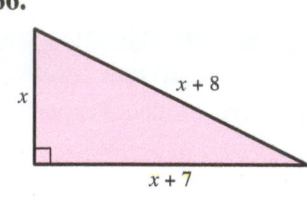

APPLICATIONS INVOLVING QUADRATIC EQUATIONS

67. Flight of a Golf Ball (Refer to Example 4.) The height h in feet of a golf ball after t seconds is given by $h = 96t - 16t^2$.
   (a) How long does it take for the golf ball to hit the ground?
   (b) Make a table of h for $t = 0, 1, 2, \ldots, 6$. After how many seconds does the golf ball reach its maximum height?

68. Flight of a Baseball The height h in feet of a baseball after t seconds is given by $h = -16t^2 + 88t + 3$. At what values of t is the height of the baseball 75 feet?

69. Braking Distance (Refer to Example 5.) The braking distance D in feet required to stop a car traveling x miles per hour on dry, level pavement can be approximated by $D = \frac{1}{11}x^2$.
   (a) Calculate the braking distance for 30 miles per hour and 60 miles per hour. How do your answers compare?
   (b) If the braking distance is 33 feet, estimate the speed of the car.
   (c) If you have a calculator, use it to solve part (b) numerically. Do your answers agree?

70. Braking Distance The braking distance D in feet required to stop a car traveling x miles per hour on wet, level pavement is approximated by $D = \frac{1}{9}x^2$.
   (a) Calculate the braking distance for 36 miles per hour and 72 miles per hour. How do your answers compare?
   (b) If the braking distance is 49 feet, estimate the speed of the car.
   (c) If you have a calculator, use it to solve part (b) numerically. Do your answers agree?

71. Digital Photographs (Refer to Example 6.) A digital photograph is 10 pixels longer than it is wide and has a total area of 2000 pixels. Find the dimensions of this rectangular photograph.

72. Dimensions of a Building The rectangular floor of a shed has a length 4 feet longer than its width, and its area is 140 square feet. Let x be the width of the floor.
  (a) Write a quadratic equation whose solution gives the width of the floor.
  (b) Solve this equation.

WRITING ABOUT MATHEMATICS

73. List four steps for solving a quadratic equation by factoring.

74. Explain why factoring is important.

SECTIONS 6.5 and 6.6   Checking Basic Concepts

1. Factor out the greatest common factor.
  (a) $9a^2 - 18a + 27$ **(b)** $7xy^2 + 28x$

2. Factor completely.
  (a) $6z^4 - 28z^3 + 16z^2$
  (b) $2r^2t^2 - 18r^2$

3. Factor completely.
  (a) $36x^3 - 48x^2 + 16x$
  (b) $24b^3 - 81$

4. Solve each quadratic equation.
  (a) $4y^2 - 6y = 0$ **(b)** $5z^2 + 2z = 3$

5. Solve $x^2 + 2x - 3 = 0$ symbolically and numerically with a table of values.

6. If a golf ball is hit upward at 60 miles per hour, then its height h in feet after t seconds is given by $h = 88t - 16t^2$. Use factoring to determine when the golf ball strikes the ground.

6.7 Solving Equations by Factoring II (Higher Degree)

Objectives

1 Recognizing Nonlinear Data and Polynomial Models
2 Factoring Higher-Degree Polynomials
3 Solving Polynomial Equations
4 Solving Applications Involving Polynomial Equations

1 Recognizing Nonlinear Data and Polynomial Models

In this section, we discuss factoring polynomials that have higher degree. Polynomials of degree 2 or higher are often used in applications. For example, the polynomial

$$0.0013x^3 - 0.085x^2 + 1.6x + 12$$

models natural gas consumption in the United States in trillions of cubic feet, where $x = 0$ corresponds to 1960, $x = 10$ corresponds to 1970, and so on until $x = 40$ corresponds to 2000, as shown in **FIGURE 6.7**. (This trend did not continue after 2000.) In Exercises 81–84 we discuss further the consumption of natural gas. (*Source:* Department of Energy.)

Natural Gas Consumption

FIGURE 6.7

2 Factoring Higher-Degree Polynomials

The first step in factoring a polynomial is to factor out the greatest common factor (GCF). Once this is accomplished, the resulting expression should be factored further by using any of the factoring methods discussed earlier in this chapter.

READING CHECK 1

• What is a good first step in factoring a polynomial?

EXAMPLE 1 **Factoring trinomials with common factors**

Factor each trinomial completely.
(a) $-4x^2 + 28x - 40$ **(b)** $10x^3 + 28x^2 - 6x$

Solution
(a) Each term in $-4x^2 + 28x - 40$ has a factor of -4, so we start by factoring out -4. (In this example, we factor out the *opposite* of the GCF to obtain a positive leading coefficient in the resulting expression.)

$$-4x^2 + 28x - 40 = -4(x^2 - 7x + 10) \quad \text{Factor out } -4.$$

$$= -4(x - 5)(x - 2) \quad \text{Factor the trinomial.}$$

(b) We start by factoring out the GCF for $10x^3 + 28x^2 - 6x$, which is $2x$.

$$10x^3 + 28x^2 - 6x = 2x(5x^2 + 14x - 3) \quad \text{Factor out } 2x.$$

$$= 2x(5x - 1)(x + 3) \quad \text{Factor the trinomial.}$$

Now Try Exercises 13, 21

EXAMPLE 2 **Factoring higher-degree polynomials**

Factor each polynomial completely.
(a) $x^4 - 16$ **(b)** $y^4 + 5y^2 + 4$
(c) $x^4 + 2x^2y^2 + y^4$ **(d)** $r^4 - t^4$

Solution
(a) We view this polynomial as the difference of two squares, where $x^4 = (x^2)^2$ and $16 = 4^2$. Then we factor twice.

$$x^4 - 16 = (x^2)^2 - 4^2 \quad\quad\quad\quad \text{Rewrite.}$$

$$= (x^2 - 4)(x^2 + 4) \quad\quad\quad \text{Difference of two squares}$$

$$= (x - 2)(x + 2)(x^2 + 4) \quad \text{Difference of two squares}$$

Note that $x^2 + 4$ does not factor.
(b) Because the trinomial $a^2 + 5a + 4$ factors as $(a + 1)(a + 4)$, we let $a = y^2$ and then factor the given trinomial.

$$y^4 + 5y^2 + 4 = (y^2)^2 + 5(y^2) + 4$$

$$= (y^2 + 1)(y^2 + 4)$$

Note that neither $y^2 + 1$ nor $y^2 + 4$ can be factored further.
(c) Because the perfect square trinomial $a^2 + 2ab + b^2$ factors as $(a + b)^2$, we let $a = x^2$ and $b = y^2$ and then factor the given trinomial.

$$x^4 + 2x^2y^2 + y^4 = (x^2)^2 + 2x^2y^2 + (y^2)^2$$

$$= (x^2 + y^2)^2$$

(d) Because the difference of squares $a^2 - b^2$ factors as $(a - b)(a + b)$, we let $a = r^2$ and $b = t^2$ and then factor the given binomial.

$$r^4 - t^4 = (r^2)^2 - (t^2)^2$$

$$= (r^2 - t^2)(r^2 + t^2)$$

$$= (r - t)(r + t)(r^2 + t^2)$$

Note that $r^2 + t^2$ cannot be factored.

Now Try Exercises 29, 37, 43, 45

3 Solving Polynomial Equations

READING CHECK 2

• When solving a polynomial equation by factoring, to which factors can we apply the zero-product property?

Many equations involving higher-degree polynomials can be solved using the four-step process discussed in the previous section. It is important to remember that the zero-product property applies to *all* factors that contain the variable. In the next example, we apply the zero-product property to three factors, resulting in three solutions.

EXAMPLE 3   **Solving higher-degree polynomial equations**

Solve each equation.
(a) $x^3 - x^2 - 6x = 0$ (b) $4x^4 + 10x^3 = 6x^2$

Solution
(a) We start by factoring out the GCF, which is x.

$$
\begin{array}{ll}
x^3 - x^2 - 6x = 0 & \text{Given equation} \\
x(x^2 - x - 6) = 0 & \text{Factor out } x. \\
x(x - 3)(x + 2) = 0 & \text{Factor the trinomial.} \\
x = 0 \quad \text{or} \quad x - 3 = 0 \quad \text{or} \quad x + 2 = 0 & \text{Zero-product property} \\
x = 0 \quad \text{or} \quad x = 3 \quad \text{or} \quad x = -2 & \text{Solve for } x.
\end{array}
$$

The solutions are $-2, 0,$ and 3.

(b) We start by subtracting $6x^2$ from each side to obtain 0 on one side of the equation.

STUDY TIP

If you are having difficulty with your studies, you may be able to find help at the student support services office on your campus.

$$
\begin{array}{ll}
4x^4 + 10x^3 = 6x^2 & \text{Given equation} \\
4x^4 + 10x^3 - 6x^2 = 0 & \text{Subtract } 6x^2. \\
2x^2(2x^2 + 5x - 3) = 0 & \text{Factor out the GCF, } 2x^2. \\
2x^2(2x - 1)(x + 3) = 0 & \text{Factor the trinomial.} \\
2x \cdot x = 0 \quad \text{or} \quad 2x - 1 = 0 \quad \text{or} \quad x + 3 = 0 & \text{Zero-product property} \\
x = 0 \quad \text{or} \quad x = \dfrac{1}{2} \quad \text{or} \quad x = -3 & \text{Solve for } x.
\end{array}
$$

The solutions are $-3, 0,$ and $\frac{1}{2}$.

Now Try Exercises 59, 65

EXAMPLE 4   **Solving a higher-degree equation**

Solve $x^5 - 81x = 0$.

Solution
We start by factoring out the common factor of x.

$$
\begin{array}{ll}
x^5 - 81x = 0 & \text{Given equation} \\
x(x^4 - 81) = 0 & \text{Factor out } x. \\
x(x^2 - 9)(x^2 + 9) = 0 & \text{Difference of two squares} \\
x(x - 3)(x + 3)(x^2 + 9) = 0 & \text{Difference of two squares} \\
x = 0 \text{ or } x - 3 = 0 \text{ or } x + 3 = 0 \text{ or } x^2 + 9 = 0 & \text{Zero-product property} \\
x = 0 \quad \text{or} \quad x = 3 \quad \text{or} \quad x = -3 & \text{Solve for } x.
\end{array}
$$

Note that $x^2 + 9 = 0$ has no real-number solutions because the square of a number plus 9 is never 0. The solutions are $-3, 0,$ and 3.

Now Try Exercise 71

4 Solving Applications Involving Polynomial Equations

CRITICAL THINKING 1

Calculate the outside surface area *A* of the box shown in **FIGURE 6.8** two different ways.

🌐 *Math in Context* ^Construction The corners of a square piece of metal are cut out to form a box, as shown in **FIGURE 6.8**. This square piece of metal has sides with length 10 inches, and the cut-out corners are squares with length *x*. The outside surface area *A* of this box, including the bottom and the sides but *not* the top, is $A = 100 - 4x^2$. (See Critical Thinking 1 in the margin.)

Construction of a Box

10 in.

$10 - 2x$

$10 - 2x$

(a)

(b)

FIGURE 6.8

EXAMPLE 5 **Finding a dimension of a box**

Find the value of *x* in **FIGURE 6.8**, if the outside surface area *A* is 84 square inches.

Solution
Let $A = 84$ in the equation $A = 100 - 4x^2$ and solve for *x*.

$$100 - 4x^2 = 84 \qquad \text{Let } A = 84.$$
$$16 - 4x^2 = 0 \qquad \text{Subtract 84.}$$
$$4(4 - x^2) = 0 \qquad \text{Factor out 4.}$$
$$4(2 - x)(2 + x) = 0 \qquad \text{Difference of two squares}$$
$$2 - x = 0 \quad \text{or} \quad 2 + x = 0 \qquad \text{Zero-product property}$$
$$x = 2 \quad \text{or} \quad x = -2 \qquad \text{Solve for } x.$$

If squares measuring 2 inches on a side are cut out of each corner, the surface area of the box is 84 square inches. (Note that $x = -2$ has no physical meaning in this problem.)

Now Try Exercise 77

6.7 Putting It All Together

CONCEPT	EXPLANATION	EXAMPLES
Common Factors	A first step when factoring polynomials is to factor out the GCF. Once this is done, factor the resulting expression, if possible.	$x^3 - 4x^2 - 5x = x(x^2 - 4x - 5)$ $= x(x - 5)(x + 1)$ $4x^4 - 16x^2 = 4x^2(x^2 - 4)$ $= 4x^2(x - 2)(x + 2)$

CONCEPT	EXPLANATION	EXAMPLES
Factoring Higher Degree Polynomials	Some higher degree polynomials can be factored by using the same methods that we use to factor quadratic polynomials.	$x^4 + 6x^2 + 5 = (x^2 + 5)(x^2 + 1)$ $4y^4 - 25 = (2y^2 - 5)(2y^2 + 5)$ $2z^4 - 32 = 2(z^4 - 16)$ $\qquad = 2(z^2 - 4)(z^2 + 4)$ $\qquad = 2(z - 2)(z + 2)(z^2 + 4)$
Solving Equations by Factoring	Use factoring and the zero-product property to solve polynomial equations.	$3x^3 - 12x^2 + 9x = 0$ $3x(x^2 - 4x + 3) = 0$ $3x(x - 3)(x - 1) = 0$ $3x = 0 \ \text{ or } \ x - 3 = 0 \ \text{ or } \ x - 1 = 0$ $x = 0 \ \text{ or } \ x = 3 \ \text{ or } \ x = 1$ The solutions are 0, 1, and 3.

6.7 Exercises

MyMathLab®

CONCEPTS AND VOCABULARY

1. When you are factoring polynomials, a good first step is to factor out the _____.

2. When you are solving an equation by factoring, the _____ property is used.

3. The zero-product property applies to all _____ containing the variable.

4. Because $a^2 - 2ab + b^2 = (a - b)^2$, it follows that $x^4 - 2x^2y^2 + y^4 =$ _____.

5. Because $x^2 + 3x + 2 = (x + 1)(x + 2)$, it follows that $z^4 + 3z^2 + 2 =$ _____.

6. If $x^4 - 1$ is factored as $(x^2 - 1)(x^2 + 1)$, is it factored *completely*? Explain.

7. If $x^3 - x^2 + x - 1$ is factored as $(x - 1)(x^2 + 1)$, is it factored *completely*? Explain.

8. When you are solving the equation $x^3 = x$, what is a good first step?

9. When you are solving $x^4 - x^2 = 0$, what is a good first step?

10. How many real-number solutions does the equation $(x - 6)(x^2 + 4) = 0$ have?

FACTORING HIGHER-DEGREE POLYNOMIALS

Exercises 11–48: Factor the polynomial completely.

11. $5x^2 - 5x - 30$

12. $3x^2 - 15x + 12$

13. $-4y^2 - 32y - 48$

14. $-7y^2 + 14y + 21$

15. $-20z^2 - 110z - 50$

16. $-12z^2 - 54z + 30$

17. $60 - 64t - 28t^2$

18. $18 - 45t - 27t^2$

19. $r^3 - r$

20. $r^3 + 2r^2 - 3r$

21. $3x^3 + 3x^2 - 18x$

22. $6x^3 - 26x^2 - 20x$

23. $72z^3 + 12z^2 - 24z$

24. $6z^3 - 4z^2 - 42z$

25. $x^4 - 4x^2$

26. $4x^4 - 36x^2$

27. $t^4 + t^3 - 2t^2$

28. $t^4 + 5t^3 - 24t^2$

29. $x^4 - 5x^2 + 6$

30. $x^4 - 3x^2 - 10$

31. $2x^4 + 7x^2 + 3$

32. $3x^4 - 8x^2 + 5$

33. $y^4 + 6y^2 + 9$

34. $y^4 - 10y^2 + 25$

35. $x^4 - 9$

36. $x^4 - 25$

37. $x^4 - 81$

38. $4x^4 - 64$

39. $z^5 + 2z^4 + z^3$

40. $6z^5 - 47z^4 + 35z^3$

41. $2x^2 + xy - y^2$

42. $2x^2 + 5xy + 2y^2$

43. $a^4 - 2a^2b^2 + b^4$

44. $a^3 + 2a^2b + ab^2$

45. $x^3 - xy^2$

46. $2x^2y - 2y^3$

47. $4x^3 + 4x^2y + xy^2$

48. $x^2y - 6xy^2 + 9y^3$

SOLVING POLYNOMIAL EQUATIONS

Exercises 49–54: Do the following.

49. (a) Factor $x^3 - 4x$.
   (b) Solve $x^3 - 4x = 0$.

50. (a) Factor $4x^3 - 16x$.
   (b) Solve $4x^3 - 16x = 0$.

51. (a) Factor $2y^3 - 6y^2 - 36y$.
   (b) Solve $2y^3 - 6y^2 - 36y = 0$.

52. (a) Factor $z^4 - 13z^2 + 36$.
   (b) Solve $z^4 - 13z^2 + 36 = 0$.

53. (a) Factor $x^3 - x^2 + 4x - 4$.
   (b) Solve $x^3 - x^2 + 4x - 4 = 0$.

54. (a) Factor $y^4 - 8y^2 + 16$.
   (b) Solve $y^4 - 8y^2 + 16 = 0$.

Exercises 55–76: Solve.

55. $3x^2 + 33x + 72 = 0$

56. $4x^2 - 16x - 20 = 0$

57. $25x^2 = 50x + 75$

58. $10x^2 = 20x + 80$

59. $y^3 - 3y^2 - 4y = 0$

60. $y^3 - 3y^2 + 2y = 0$

61. $3z^3 + 6z^2 = 72z$

62. $4z^3 = 4z^2 + 24z$

63. $x^4 - 36x^2 = 0$

64. $4x^4 = 100x^2$

65. $r^4 + 6r^3 = 7r^2$

66. $r^4 + 30r^2 = 11r^3$

67. $x^4 - 13x^2 = -36$

68. $x^4 - 17x^2 + 16 = 0$

69. $x^4 + 1 = 2x^2$

70. $x^4 - 8x^2 + 16 = 0$

71. $a^5 = 81a$

72. $b^3 = -8$

73. $x^3 - 2x^2 - x + 2 = 0$

74. $x^3 - x^2 + 4x - 4 = 0$

75. $x^3 - 5x^2 + x - 5 = 0$

76. $3x^3 + 2x^2 - 27x - 18 = 0$

APPLICATIONS

77. Dimensions of a Box (Refer to Example 5) A box is made from a rectangular piece of metal with length 20 inches and width 15 inches. The box has no top.
   (a) What are the limitations on the size of x? Explain your answer.
   (b) Write an expression that gives the outside surface area of the box. (*Hint:* Consider the size of the metal sheet and how much was cut out.)
   (c) If the outside surface area of the box is 275 square inches, find x.

78. Dimensions of a Box Refer to the previous exercise.
   (a) Find a polynomial that gives the volume of the box for a given x.
   (b) Factor your polynomial.
   (c) What are the zeros of your polynomial? What do they represent in this problem?

79. Numbers If a positive number n is increased by 3, its square equals 121. Find n.

80. Sidewalk around a Pool A 5-foot-wide side walk around a rectangle swimming pool has a total area of 900 square feet. Find the dimensions of the swimming pool if the pool is 20 feet longer than it is wide.

Exercises 81–84: **U.S. Natural Gas Consumption** *The polynomial*

$$0.0013x^3 - 0.085x^2 + 1.6x + 12$$

models natural gas consumption in trillions of cubic feet, where $x = 0$ corresponds to 1960, $x = 1$ corresponds to 1961, and so on until $x = 40$ corresponds to 2000.

81. How much natural gas was consumed in 1990?

82. In which year between 1970 and 1990 was natural gas consumption about 20.4 trillion cubic feet?

83. Explain any difficulties encountered when you try to solve the equation

$$0.0013x^3 - 0.085x^2 + 1.6x + 12 = 23.2.$$

84. How might you solve the equation in Exercise 83 without factoring? If you were to find the solution to this equation, what would it represent?

86. Suppose that a polynomial can be factored. Explain how its factors can be used to find the zeros of the polynomial. Give an example.

WRITING ABOUT MATHEMATICS

85. Compare factoring the polynomial $x^2 + 6x + 5$ with factoring the polynomial $z^4 + 6z^2 + 5$.

SECTION 6.7 — **Checking Basic Concepts**

1. Factor the trinomial completely.
(a) $3x^2 - 6x - 24$ (b) $-10y^2 + 5y + 5$

2. Factor the binomial completely.
(a) $z^4 - 25$ (b) $7t^4 - 7$

3. Factor completely.
(a) $x^4 - 8x^2 + 16$ (b) $2y^3 + 17y^2 - 30y$

4. Solve $t^4 + t^3 = 12t^2$.

5. Solve $x^3 - 3x^2 + 2x - 6 = 0$.

CHAPTER 6   Summary

SECTION 6.1 ■ INTRODUCTION TO FACTORING

Terms Related to Factoring Polynomials

Factoring	Writing a polynomial as a product, usually of lower-degree polynomials
Common Factor	An expression that is a factor of each term in a polynomial

Example: Some common factors of $4x^4 + 8x^2$ are $2x$, x^2, and $4x^2$.

Greatest Common Factor (GCF)   The common factor with the greatest (integer) coefficient and the highest degree

Example: The GCF of $4x^4 + 8x^2$ is $4x^2$.

$$4x^4 + 8x^2 = 4x^2(x^2 + 2)$$

Factoring by Grouping   Used to factor a four-term polynomial into two binomials

Example: $x^3 + 5x^2 + 3x + 15 = x^2(x + 5) + 3(x + 5)$
$$= (x^2 + 3)(x + 5)$$

SECTION 6.2 ■ FACTORING TRINOMIALS I $(x^2 + bx + c)$

Review of the FOIL Method   A method used for multiplying two binomials

First Terms	$(2x + 3)(5x + 4)$:	$2x \cdot 5x = 10x^2$
Outside Terms	$(2x + 3)(5x + 4)$:	$2x \cdot 4 = 8x$
Inside Terms	$(2x + 3)(5x + 4)$:	$3 \cdot 5x = 15x$
Last Terms	$(2x + 3)(5x + 4)$:	$3 \cdot 4 = 12$

The product is the sum of these four terms:

$$(2x + 3)(5x + 4) = 10x^2 + 8x + 15x + 12 = 10x^2 + 23x + 12.$$

Factoring Trinomials with a Leading Coefficient of 1 To factor the trinomial $x^2 + bx + c$, find two numbers, m and n, that satisfy

$$m \cdot n = c \quad \text{and} \quad m + n = b.$$

Then $x^2 + bx + c = (x + m)(x + n)$.

Example: Because $-3 \cdot 5 = -15$ and $-3 + 5 = 2$,

$$x^2 + 2x - 15 = (x - 3)(x + 5).$$

SECTION 6.3 ■ FACTORING TRINOMIALS II ($ax^2 + bx + c$)

Factoring Trinomials by Grouping To factor $ax^2 + bx + c$, perform the following steps. (Assume that a, b, and c are integers and have no factor in common.)

1. Find two numbers, m and n, such that $mn = ac$ and $m + n = b$.
2. Write the trinomial as $ax^2 + mx + nx + c$.
3. Use grouping to factor this expression into two binomials.

Example: To factor $3x^2 + 10x - 8$, find two numbers whose product is -24 and whose sum is 10. These two numbers are $m = 12$ and $n = -2$, so write $10x$ as $12x - 2x$.

$$
\begin{aligned}
3x^2 + 10x - 8 &= 3x^2 + 12x - 2x - 8 && \text{Write } 10x \text{ as } 12x - 2x. \\
&= (3x^2 + 12x) + (-2x - 8) && \text{Associative property} \\
&= 3x(x + 4) - 2(x + 4) && \text{Factor out common factors.} \\
&= (3x - 2)(x + 4) && \text{Factor out } (x + 4).
\end{aligned}
$$

Factoring with FOIL in Reverse Use trial and error and FOIL in reverse to find the factors of a trinomial.

Example: To factor $3x^2 + 10x - 8$, first find factors of $3x^2$.

$$(3x + \underline{\quad})(x + \underline{\quad})$$

Then place factors of -8 so that the resulting middle term is $10x$.

$$(3x + -2) \cdot (x + 4) = 3x^2 + 10x - 8$$

$$-2x$$
$$+12x$$
$$\overline{10x} \checkmark \longleftarrow \text{Middle term checks.}$$

SECTION 6.4 ■ SPECIAL TYPES OF FACTORING

Difference of Two Squares

$$a^2 - b^2 = (a - b)(a + b)$$

Examples: $x^2 - 16 = (x - 4)(x + 4) \qquad (a = x, b = 4)$

$\quad\quad\quad\quad\; 4r^2 - 9t^2 = (2r - 3t)(2r + 3t) \quad (a = 2r, b = 3t)$

Perfect Square Trinomials

$$a^2 + 2ab + b^2 = (a + b)^2 \quad \text{and} \quad a^2 - 2ab + b^2 = (a - b)^2$$

Examples: $4x^2 + 4x + 1 = (2x)^2 + 2(2x)1 + 1^2 = (2x + 1)^2 \quad (a = 2x, b = 1)$

$\quad\quad\quad\quad\; x^2 - 10x + 25 = x^2 - 2 \cdot x \cdot 5 + 5^2 = (x - 5)^2 \quad (a = x, b = 5)$

Sums and Differences of Two Cubes

$$a^3 + b^3 = (a + b)(a^2 - ab + b^2) \quad \text{and} \quad a^3 - b^3 = (a - b)(a^2 + ab + b^2)$$

Examples: $x^3 + 8 = (x + 2)(x^2 - 2x + 4)$ $\qquad (a = x, b = 2)$

$\qquad\qquad\quad 27x^3 - 1 = (3x - 1)(9x^2 + 3x + 1)$ $\quad (a = 3x, b = 1)$

SECTION 6.5 ■ SUMMARY OF FACTORING

Guidelines for Factoring Polynomials The following guidelines can be used to factor polynomials in general.

STEP 1: Factor out the greatest common factor, if possible.

STEP 2: A. If the polynomial has *four terms*, try factoring by grouping.

  B. If the polynomial is a *binomial*, try one of the following.

    1. $a^2 - b^2 = (a - b)(a + b)$ Difference of two squares

    2. $a^3 - b^3 = (a - b)(a^2 + ab + b^2)$ Difference of two cubes

    3. $a^3 + b^3 = (a + b)(a^2 - ab + b^2)$ Sum of two cubes

  C. If the polynomial is a *trinomial*, check for a perfect square.

    1. $a^2 + 2ab + b^2 = (a + b)^2$ Perfect square trinomial

    2. $a^2 - 2ab + b^2 = (a - b)^2$ Perfect square trinomial

  Otherwise, try to factor the trinomial by grouping or apply FOIL in reverse, as described in Sections 2 and 3 of this chapter.

STEP 3: Check to make sure that the polynomial is *completely* factored.

Examples: $12x^3 - 12x^2 + 3x = 3x(4x^2 - 4x + 1) = 3x(2x - 1)^2$ Steps 1, 2C, and 3

$\qquad\qquad\quad 9x^3 - 6x^2 + 18x - 12 = 3(3x^3 - 2x^2 + 6x - 4) = 3(x^2 + 2)(3x - 2)$ Steps 1, 2A, and 3

$\qquad\qquad\quad 16x^3 - 100x = 4x(4x^2 - 25) = 4x(2x - 5)(2x + 5)$ Steps 1, 2B, and 3

SECTION 6.6 ■ SOLVING EQUATIONS BY FACTORING I (QUADRATICS)

Zero-Product Property
For any real numbers a and b, if $ab = 0$, then $a = 0$ or $b = 0$ (or both).
The zero-product property is used to solve equations.

Examples: $xy = 0$ implies that $x = 0$ or $y = 0$.

$\qquad\qquad\quad (x + 5)(x - 3) = 0$ implies $x + 5 = 0$ or $x - 3 = 0$.

Zero of a Polynomial A number a is a zero of a polynomial if the result is 0 when a is substituted in that polynomial.

Example: The number -2 is a zero of $x^2 - 4$ because $(-2)^2 - 4 = 0$.

Solving Quadratic Equations by Factoring To solve a quadratic equation by factoring, follow these steps.

STEP 1: If necessary, write the equation in standard form as $ax^2 + bx + c = 0$.

STEP 2: Factor the left side of the equation using any method.

STEP 3: Apply the zero-product property.

STEP 4: Solve each of the resulting equations. Check any solutions.

Example:

$$x^2 + 7x = 8 \quad \text{Given equation}$$
$$x^2 + 7x - 8 = 0 \quad \text{Step 1}$$
$$(x + 8)(x - 1) = 0 \quad \text{Step 2}$$
$$x + 8 = 0 \quad \text{or} \quad x - 1 = 0 \quad \text{Step 3}$$
$$x = -8 \quad \text{or} \quad x = 1 \quad \text{Step 4}$$

SECTION 6.7 ■ SOLVING EQUATIONS BY FACTORING II (HIGHER DEGREE)

Factoring Polynomials of Higher Degree The distributive property and the techniques for factoring quadratic polynomials can also be applied to polynomials of higher degree.

Examples: $10r^3 + 15r = 5r(2r^2 + 3)$ (To check, multiply the right side.)

Because $2x^2 + x - 1 = (x + 1)(2x - 1)$,

it follows that $2z^4 + z^2 - 1 = (z^2 + 1)(2z^2 - 1)$.

Solving Equations by Factoring Use algebra to obtain 0 on one side of the equation. Factor the other side and apply the zero-product property.

Example:

$$x^3 = 4x \quad \text{Given equation}$$
$$x^3 - 4x = 0 \quad \text{Subtract } 4x.$$
$$x(x^2 - 4) = 0 \quad \text{Factor out the GCF, } x.$$
$$x(x - 2)(x + 2) = 0 \quad \text{Difference of two squares}$$
$$x = 0 \quad \text{or} \quad x - 2 = 0 \quad \text{or} \quad x + 2 = 0 \quad \text{Zero-product property}$$
$$x = 0 \quad \text{or} \quad x = 2 \quad \text{or} \quad x = -2 \quad \text{Solve for } x.$$

CHAPTER 6   Review Exercises

SECTION 6.1

Exercises 1–6: Identify the greatest common factor for the expression and then factor the expression.

1. $5x - 15$

2. $y^2 + 2y$

3. $8z^3 - 4z^2$

4. $6x^4 + 3x^3 - 12x^2$

5. $9xy + 15yz^2$

6. $a^2b^3 + a^3b^2$

Exercises 7–14: Use grouping to factor the given polynomial completely.

7. $x(x + 2) - 3(x + 2)$

8. $y^2(x - 5) + 3y(x - 5)$

9. $z^3 - 2z^2 + 5z - 10$ **10.** $t^3 + t^2 + 8t + 8$

11. $x^3 - 3x^2 + 6x - 18$ **12.** $ax + bx - ay - by$

13. $x^5 + 3x^4 - 2x^3 - 6x^2$

14. $2y^4 + 6y^3 + 2y^2 + 6y$

SECTION 6.2

Exercises 15–18: Find an integer pair that has the given product and sum.

15. Product: 20   Sum: 9

16. Product: −21   Sum: 4

17. Product: 36   Sum: −13

18. Product: −100   Sum: −21

Exercises 19–32: Factor the trinomial completely. If the trinomial cannot be factored, write "prime."

19. $x^2 - x - 12$ **20.** $x^2 + 10x + 24$

21. $x^2 + 6x - 16$ **22.** $x^2 - x - 42$

23. $x^2 + 4x - 6$ **24.** $x^2 - 5x + 8$

25. $x^2 + 2x - 3$ **26.** $x^2 + 22x + 120$

27. $2x^3 + 6x^2 - 20x$ **28.** $x^4 - 3x^3 - 28x^2$

29. $10 - 7x + x^2$

30. $24 + 2x - x^2$

31. $-2x^2 - 4x + 30$

32. $-x^3 - 9x^2 + 10x$

SECTION 6.3

Exercises 33–44: Factor the trinomial completely. If the trinomial cannot be factored, write "prime."

33. $9x^2 + 3x - 2$

34. $2x^2 + 3x - 5$

35. $3x^2 + 14x + 15$

36. $35x^2 - 2x - 1$

37. $3x^2 + 4x - 5$

38. $4x^2 - 12x - 5$

39. $24x^2 - 7x - 5$

40. $4x^2 + 33x - 27$

41. $12x^3 + 48x^2 + 21x$

42. $8x^4 + 14x^3 - 30x^2$

43. $12 - 5x - 2x^2$

44. $1 + 3x - 10x^2$

SECTION 6.4

Exercises 45–58: Factor completely.

45. $z^2 - 4$

46. $9z^2 - 64$

47. $36 - y^2$

48. $100a^2 - 81b^2$

49. $x^2 + 14x + 49$

50. $x^2 - 10x + 25$

51. $4x^2 - 12x + 9$

52. $9x^2 + 48x + 64$

53. $8t^3 - 1$

54. $27r^3 + 8t^3$

55. $2x^3 - 50x$

56. $24x^3 + 81$

57. $2x^3 + 28x^2 + 98x$

58. $2x^4 - 128x$

SECTION 6.5

Exercises 59–68: Factor completely.

59. $12x - 8$

60. $6x^3 + 9x^2$

61. $9y^2 - 6y + 6$

62. $yz^2 - 9y$

63. $x^4 + 7x^3 - 4x^2 - 28x$

64. $12x^3 + 36x^2 + 27x$

65. $3ab^3 - 24a$

66. $5x^3 + 20x$

67. $24x^3 - 6xy^2$

68. $x^3y + 27y$

SECTION 6.6

Exercises 69–80: Solve the equation.

69. $mn = 0$

70. $y^2 = 0$

71. $(4x - 3)(x + 9) = 0$

72. $(1 - 4x)(6 + 5x) = 0$

73. $z(z - 1)(z - 2) = 0$

74. $z^2 - 7z = 0$

75. $y^2 - 64 = 0$

76. $y^2 + 9y + 14 = 0$

77. $x^2 = x + 6$

78. $10x^2 + 11x = 6$

79. $t(t - 14) = 72$

80. $t(2t - 1) = 10$

SECTION 6.7

Exercises 81–90: Factor completely.

81. $5x^2 - 15x - 50$

82. $-3x^2 - 6x + 45$

83. $y^3 - 4y$

84. $3y^3 + 6y^2 - 9y$

85. $2z^4 + 14z^3 + 20z^2$

86. $8z^4 - 32z^2$

87. $x^4 - 6x^2 + 9$

88. $2x^4 - 15x^2 - 27$

89. $a^2 + 10ab + 25b^2$

90. $x^3 - xy^2$

Exercises 91–98: Solve.

91. $16x^2 - 72x - 40 = 0$

92. $2x^3 - 11x^2 + 15x = 0$

93. $t^3 = 25t$

94. $t^4 - 7t^3 + 12t^2 = 0$

95. $z^4 + 16 = 8z^2$

96. $z^4 - 256 = 0$

97. $y^3 = -64$

98. $y^3 - y^2 - y + 1 = 0$

APPLICATIONS

99. A square has area $9x^2 + 42x + 49$. Find the length of a side. Make a sketch of the square.

100. A rectangle has area $x^2 + 6x + 5$. Find possible dimensions for the rectangle. Make a sketch of the rectangle.

101. A cube has surface area $6x^2 + 12x + 6$. Find the length of a side.

102. Write a polynomial in factored form that represents the total area of the rectangle.

103. Write a polynomial in factored form that represents the total area of the rectangle.

104. Radius of a Circle The area and the circumference of a circle are numerically equal. Find the radius of the circle.

105. Dimensions of a Shed The floor of a rectangular shed is 7 feet longer than it is wide and has an area of 120 square feet. What are its dimensions?

106. Flight of a Ball A ball is hit upward. Its height h in feet after t seconds is $h = -16t^2 + 80t + 4$. At what times is the ball 100 feet in the air?

107. Stopping Distance The distance D in feet that it takes to stop a car traveling x miles per hour on wet, level pavement can be approximated by $D = \frac{1}{9}x^2 + \frac{11}{3}x$.
   (a) Estimate the distance required for the car to stop when it is traveling 45 miles per hour.
   (b) If the stopping distance is 80 feet, what is the speed of the car?
 (c) If you have a calculator, use it to solve part (b) numerically with a table of values. Do your answers agree?

108. Revenue A company makes tops for the boxes of pickup trucks. The total revenue R in dollars from selling the tops for p dollars each is given by $R = p(200 - p)$, where $p \le 200$.
   (a) Find R when $p = \$100$.
   (b) Find p when $R = \$7500$.
   (c) If you have a calculator, use it to solve part (b) numerically with a table of values. Do your answers agree?

109. Dimensions of a Box A box is made from a rectangular piece of metal with length 50 inches and width 40 inches by cutting out square corners of length x and folding up the sides.
   (a) Write an expression that gives the surface area of the inside of the box.
   (b) If the surface area of the box is 1900 square inches, find x.

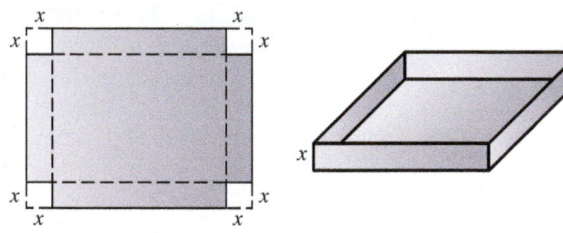

110. Dimensions of a Cube If two identical cubes have a total surface area of $12x^2 + 48x + 48$, find the length of one side of one of the cubes.

CHAPTER 6   Test   MyMathLab® YouTube™

Exercises 1 and 2: Identify the greatest common factor for the expression. Then factor the expression.

1. $4x^2y - 20xy^2 + 12xy$ **2.** $9a^3b^2 + 3a^2b^2$

Exercises 3 and 4: Factor by grouping.

3. $ay + by + az + bz$ **4.** $3x^3 + x^2 - 15x - 5$

Exercises 5–10: Factor the trinomial completely. If the trinomial cannot be factored, write "prime."

5. $y^2 + 4y - 12$ **6.** $4x^2 + 20x + 25$

7. $4z^2 - 19z + 12$ **8.** $21 - 17t + 2t^2$

9. $x^2 + 7x - 10$ **10.** $3y^2 + 4y + 2$

Exercises 11–16: Factor completely.

11. $6x^3 + 3x^2 - 3x$ **12.** $2z^4 - 12z^2 - 54$

13. $36y^3 - 100y$ **14.** $7x^4 + 56x$

15. $16a^4 + 24a^3 + 9a^2$ **16.** $2b^4 - 32$

Exercises 17–22: Solve the equation.

17. $x^2 - 16 = 0$ **18.** $y^2 = y + 20$

19. $9z^2 + 16 = 24z$ **20.** $x(x - 5) = 66$

21. $y^3 = 9y$ **22.** $x^4 - 5x^2 + 4 = 0$

23. A square has area $9x^2 + 30x + 25$. Find the length of a side in terms of x.

24. Write a polynomial in factored form that represents the total area of the rectangle.

25. Braking Distance The braking distance D in feet required for a car traveling at x miles per hour to stop on dry, level pavement can be modeled by $D = \frac{1}{11}x^2$.
  (a) Calculate the distance required for the car to stop when it is traveling 55 miles per hour.
  (b) If the braking distance is 99 feet, estimate the speed of the car.

26. Flight of a Ball A ball is hit upward. Its height h in feet after t seconds is given by $h = -16t^2 + 48t + 4$. At what times is the ball 36 feet in the air?

CHAPTERS 1–6   Cumulative Review Exercises

Exercises 1 and 2: Evaluate by hand and then simplify to lowest terms.

1. $\frac{3}{5} \cdot \frac{15}{21}$

2. $\frac{4}{5} - \frac{1}{10}$

Exercises 3 and 4: Evaluate by hand.

3. $26 - 3 \cdot 6 \div 2$

4. $-2^2 + \frac{3+2}{8+2}$

5. Complete the table. Then use the table to solve the equation $2x + 3 = 5$.

x	-2	-1	0	1	2
$2x + 3$					

6. Translate the sentence "Triple a number decreased by 5 equals the number decreased by 7" into an equation using the variable n. Then solve the equation.

7. Convert 5.7% to fraction and decimal notation.

8. Solve $P = 2W + 2L$ for W.

9. Solve $5 - 3z < -1$.

10. Make a scatterplot having the following five points: $(-2, 3)$, $(-1, 2)$, $(0, -1)$, $(1, 1)$, and $(2, 2)$.

Exercises 11 and 12: Graph the given equation. Determine any intercepts.

11. $y = 3x - 2$

12. $y = -2$

Exercises 13 and 14: Find the slope–intercept form for the line satisfying the given conditions.

13. Perpendicular to $2x - 3y = -6$, passing through the point $(1, 2)$

14. Passing through the points $(-2, 1)$ and $(1, 5)$

15. Identify the x-intercept and the y-intercept. Then write the slope–intercept form of the line.

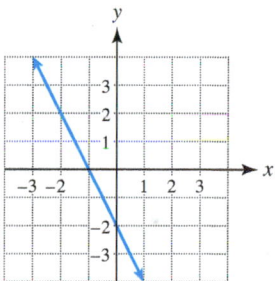

16. The graphs of two equations are shown. Use the graphs to identify the solution to the system of equations. Then check your answer.

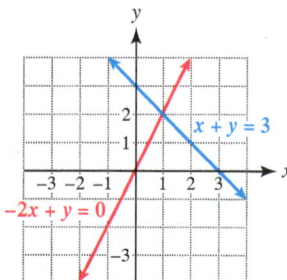

Exercises 17 and 18: Solve the system of equations.

17. $\begin{aligned} y &= -1 \\ 2x + y &= 1 \end{aligned}$

18. $\begin{aligned} 5x + y &= -5 \\ -x + 2y &= 12 \end{aligned}$

Exercises 19 and 20: Shade the solution set to the system of inequalities.

19. $\begin{aligned} x &> -1 \\ y &< x \end{aligned}$

20. $\begin{aligned} 2x - y &\leq 4 \\ x + 2y &\geq 2 \end{aligned}$

Exercises 21–24: Simplify the expression.

21. -2^4

22. $(xy)^0$

23. $(xy)^4(x^3y^{-4})^2$

24. $7x^3(-2x^2 + 3x)$

Exercises 25–28: Simplify and write the expression using positive exponents.

25. $a^{-4} \cdot a^2$

26. $(2t^3)^{-2}$

27. $(xy)^{-3}(x^{-1}y^2)^{-1}$

28. $\left(\dfrac{2x}{y^{-2}}\right)^5$

Exercises 29 and 30: Divide and check.

29. $\dfrac{6x^3 + 12x^2}{3x}$

30. $\dfrac{3x^3 - x + 1}{x^2 + 1}$

Exercises 31–36: Factor completely.

31. $x^2 + 3x - 28$

32. $6y^2 + y - 12$

33. $25x^2 - 4y^2$

34. $64x^2 - 16x + 1$

35. $27t^3 - 8$

36. $-4x^2 + 4x + 24$

Exercises 37–40: Solve the equation.

37. $y^4 = 25y^2$

38. $8z^2 + 8z - 16 = 0$

39. $4z^3 = 49z$

40. $x^4 - 18x^2 + 81 = 0$

APPLICATIONS

41. Shoveling a Driveway Two people are shoveling snow from a driveway. The first person shovels 10 square feet per minute, while the second person shovels 8 square feet per minute.
   (a) Write and simplify an expression that gives the total square feet that the two people shovel in x minutes.
   (b) How many minutes would it take for them to clear a driveway with an area of 900 square feet?

42. Running Distance A person runs the following distances over three days: $1\frac{3}{4}$ miles, $2\frac{1}{2}$ miles, and $2\frac{2}{3}$ miles. How far does the person run altogether?

43. Burning Calories An athlete can burn 12 calories per minute while cross-country skiing and 9 calories per minute while running at 5 miles per hour. If the athlete burns 615 calories in 60 minutes, how long is spent on each activity?

44. Renting a Car A rental car costs $20 plus $0.25 per mile driven.
   (a) Write an equation that gives the cost C of driving the car x miles.
   (b) Determine the number of miles that the car is driven if the rental cost is $100.

45. Triangle In an isosceles triangle, the measures of the two smaller angles are equal and their sum is 20° more than the larger angle.
   (a) Let x be the measure of one of the two smaller angles and y be the measure of the larger angle. Write a system of linear equations whose solution gives the measures of these angles.
   (b) Solve your system.

46. Areas of Rectangles Find a monomial equal to the sum of the areas of the rectangles. Calculate this sum for $x = 2$ yards and $y = 3$ yards.

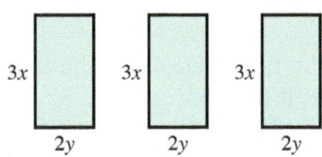

47. Area of a Square A square has area $x^2 + 12x + 36$. Find the length of a side.

48. Flight of a Golf Ball A golf ball is hit upward. Its height h in feet after t seconds is given by the formula $h = 64t - 16t^2$.
   (a) How long does it take for the ball to hit the ground?
   (b) At what times is the ball 48 feet in the air?

7

Rational Expressions

One of the most significant problems facing the U.S. transportation system is chronic highway congestion. According to a newly released "Highway Statistics," Americans drove about 4.6 trillion miles in 2014. Our ability to keep traffic moving smoothly and safely is key to keeping our economy strong, and traffic congestion costs motorists more than $121 billion annually in wasted fuel. A recent study found that, collectively, Americans spend as many as 5.5 billion hours stuck in traffic each year.

If the amount of traffic doubles on a highway, the time spent waiting in traffic may more than double. At times, only a slight increase in the traffic rate can result in a dramatic increase in the time spent waiting. (See Example 6 in the first section of this chapter.) Mathematics can describe this effect by using rational expressions. In this chapter, we introduce rational expressions and some of their applications. *Source:* Texas A&M Transportation Institute.

433

7.1   Introduction to Rational Expressions

Objectives

1 **Evaluating Rational Expressions**
   • Undefined Expressions
2 **Simplifying Rational Expressions**
3 **Solving Applications Involving Rational Expressions**
   • An Application Involving Probability

NEW VOCABULARY

☐ Rational expression
☐ Lowest terms
☐ Vertical asymptote
☐ Probability

1 Evaluating Rational Expressions

Recall that a *rational number* is any number that can be expressed as a ratio of two integers $\frac{p}{q}$, where $q \neq 0$. In this chapter, we discuss *rational expressions*, which can be written as the ratio of two polynomials. Because examples of polynomials include

$$3, \quad 2x, \quad x^2 + 4, \quad \text{and} \quad x^3 - 1, \quad \boxed{\text{Polynomials}}$$

it follows that examples of rational expressions include

$$\frac{3}{2x}, \quad \frac{2x}{x^2 + 4}, \quad \frac{x^2 + 4}{3}, \quad \text{and} \quad \frac{x^3 - 1}{x^2 + 4}. \quad \boxed{\text{Rational expressions}}$$

RATIONAL EXPRESSION

A **rational expression** can be written as $\frac{P}{Q}$, where P and Q are polynomials. A rational expression is defined whenever $Q \neq 0$.

We can evaluate polynomials for different values of a variable. For example, for $x = 2$, the polynomial $x^2 - 3x + 1$ evaluates to

$$(2)^2 - 3(2) + 1 = -1.$$

Rational expressions can be evaluated similarly.

EXAMPLE 1   Evaluating rational expressions

If possible, evaluate each expression for the given value of the variable.

(a) $\dfrac{1}{x + 1}$ $x = 2$ (b) $\dfrac{y^2}{2y - 1}$ $y = -4$

(c) $\dfrac{5z + 8}{z^2 - 2z + 1}$ $z = 1$ (d) $\dfrac{2 - x}{x - 2}$ $x = -3$

Solution

(a) If $x = 2$, then $\frac{1}{x + 1} = \frac{1}{2 + 1} = \frac{1}{3}$.

(b) If $y = -4$, then $\frac{y^2}{2y - 1} = \frac{(-4)^2}{2(-4) - 1} = -\frac{16}{9}$.

(c) If $z = 1$, then $\frac{5z + 8}{z^2 - 2z + 1} = \frac{5(1) + 8}{1^2 - 2(1) + 1}$, or $\frac{13}{0}$, which is undefined because division by 0 is not possible.

(d) If $x = -3$, then $\frac{2 - x}{x - 2} = \frac{2 - (-3)}{-3 - 2} = \frac{5}{-5} = -1$.

Now Try Exercises 7, 11, 13, 17

READING CHECK 1

• When is a rational expression undefined?

UNDEFINED EXPRESSIONS Division by 0 is undefined. As a result, rational expressions are different from polynomials because they are *undefined* whenever their *denominators are 0*. For example, the expression in Example 1(c) is undefined when $z = 1$.

EXAMPLE 2   **Determining when a rational expression is undefined**

Find all values of the variable for which each expression is undefined.

(a) $\dfrac{1}{x}$ **(b)** $\dfrac{4t}{t-3}$ **(c)** $\dfrac{1-6r}{r^2-4}$ **(d)** $\dfrac{4}{x^2+1}$

Solution
(a) A rational expression is undefined when its denominator is 0. Thus $\frac{1}{x}$ is undefined when $x = 0$.
(b) The expression $\frac{4t}{t-3}$ is undefined when its denominator, $t-3$, is 0, or when $t = 3$.
(c) The expression $\frac{1-6r}{r^2-4}$ is undefined when its denominator, r^2-4, is 0. Here,

$$r^2 - 4 = (r-2)(r+2) = 0$$

implies that the denominator is 0 when $r = -2$ or $r = 2$.
(d) In the expression $\frac{4}{x^2+1}$, the denominator $x^2 + 1$ is never 0 because any real number squared plus 1 is always greater than or equal to 1. Thus this rational expression is defined for all real numbers x.

Now Try Exercises 25, 27, 31, 33

STUDY TIP

You may want to review your notes on fractions. Many of the mathematical concepts that apply to fractions also apply to rational expressions.

2 Simplifying Rational Expressions

Earlier in this text, we used the *basic principle of fractions*,

$$\frac{a \cdot c}{b \cdot c} = \frac{a}{b}.$$

For example, this basic principle allows us to simplify the fraction $\frac{8}{12}$ as

$$\frac{8}{12} = \frac{2 \cdot 4}{3 \cdot 4} = \frac{2}{3}.$$

EXAMPLE 3   **Simplifying fractions**

Simplify each fraction by applying the basic principle of fractions.

(a) $\dfrac{5}{10}$ **(b)** $-\dfrac{36}{48}$

Solution

(a) $\dfrac{5}{10} = \dfrac{1 \cdot 5}{2 \cdot 5} = \dfrac{1}{2}$ **(b)** $-\dfrac{36}{48} = -\dfrac{3 \cdot 12}{4 \cdot 12} = -\dfrac{3}{4}$

Now Try Exercises 39, 43

We can also apply this basic principle to rational expressions. For example,

$$\frac{x(x-1)}{4(x-1)} = \frac{x}{4},$$

provided that $x \neq 1$.

NOTE: The above simplification is not valid when $x = 1$ because the expression is undefined for this x-value. When simplifying a rational expression, *we assume that values of the variable that make the rational expression undefined are excluded*, unless stated otherwise. ■

BASIC PRINCIPLE OF RATIONAL EXPRESSIONS

The following property can be used to simplify rational expressions, where P, Q, and R are polynomials.

$$\frac{P \cdot R}{Q \cdot R} = \frac{P}{Q} \qquad Q \text{ and } R \text{ are nonzero.}$$

NOTE: $\frac{P \cdot R}{Q \cdot R} = \frac{P}{Q} \cdot \frac{R}{R} = \frac{P}{Q} \cdot 1 = \frac{P}{Q}$, provided that $Q \neq 0$ and $R \neq 0$. ∎

Like fractions, rational expressions can be written in *lowest terms*. For example, the rational expression $\frac{x^2 - 1}{x^2 + 2x + 1}$ can be written in lowest terms by factoring the numerator and the denominator and then applying the basic principle of rational expressions.

READING CHECK 2

• How do you know when a rational expression is written in lowest terms?

$$\frac{x^2 - 1}{x^2 + 2x + 1} = \frac{(x - 1)(x + 1)}{(x + 1)(x + 1)} \qquad \text{Factor the numerator and the denominator.}$$

$$= \frac{x - 1}{x + 1} \qquad \text{Apply } \frac{PR}{QR} = \frac{P}{Q} \text{ with } R = x + 1.$$

Because the basic principle of rational expressions cannot be applied further to $\frac{x - 1}{x + 1}$, we say that this expression is written in **lowest terms**.

EXAMPLE 4 **Simplifying rational expressions**

Simplify each expression.

(a) $\dfrac{8y}{4y^2}$ **(b)** $\dfrac{2x + 6}{3x + 9}$ **(c)** $\dfrac{(z + 1)(z - 5)}{(z - 5)(z + 3)}$ **(d)** $\dfrac{x^2 - 9}{2x^2 + 7x + 3}$

Solution
(a) Factor out the greatest common factor, $4y$, in the numerator and the denominator.

$$\frac{8y}{4y^2} = \frac{2 \cdot 4y}{y \cdot 4y} = \frac{2}{y} \qquad \text{Apply } \frac{PR}{QR} = \frac{P}{Q} \text{ with } R = 4y.$$

(b) Start by factoring the numerator and denominator.

$$\frac{2x + 6}{3x + 9} = \frac{2(x + 3)}{3(x + 3)} = \frac{2}{3} \qquad \text{Apply } \frac{PR}{QR} = \frac{P}{Q} \text{ with } R = x + 3.$$

(c) The commutative property allows us to write $\frac{PR}{RQ}$ as $\frac{PR}{QR}$.

$$\frac{(z + 1)(z - 5)}{(z - 5)(z + 3)} = \frac{(z + 1)(z - 5)}{(z + 3)(z - 5)} = \frac{z + 1}{z + 3} \qquad \text{Apply } \frac{PR}{RQ} = \frac{P}{Q} \text{ with } R = z - 5.$$

(d) Start by factoring the numerator and the denominator.

$$\frac{x^2 - 9}{2x^2 + 7x + 3} = \frac{(x - 3)(x + 3)}{(2x + 1)(x + 3)} = \frac{x - 3}{2x + 1} \qquad \text{Apply } \frac{PR}{QR} = \frac{P}{Q} \text{ with } R = x + 3.$$

Now Try Exercises 51, 55, 61, 79

READING CHECK 3

- How do rational expressions and rational equations differ?

MAKING CONNECTIONS 1

Rational Expressions and Equations

$$\frac{x}{x + 4} \quad \text{and} \quad \frac{2}{x} \qquad\qquad \frac{x}{x + 4} = \frac{2}{x}$$

Expressions do not contain an equals sign.

An equation must contain an equals sign.

We simplify and evaluate expressions. We solve equations by finding x-values that make the equation true.

A negative sign can be placed in a fraction in a number of ways.

$$-\frac{5}{7} = \frac{-5}{7} = \frac{5}{-7}$$

Three equal fractions

This property about negative signs can also be applied to rational expressions, as demonstrated in the next example.

EXAMPLE 5 **Distributing a negative sign**

Simplify each expression.

(a) $\dfrac{-x - 6}{2x + 12}$ **(b)** $\dfrac{10 - z}{z - 10}$ **(c)** $-\dfrac{5 - x}{x - 5}$

Solution
(a) Factor -1 out of the numerator and 2 out of the denominator.

$$\frac{-x - 6}{2x + 12} = \frac{-1(x + 6)}{2(x + 6)} = -\frac{1}{2}$$

(b) Factor -1 out of the numerator.

$$\frac{10 - z}{z - 10} = \frac{-1(-10 + z)}{z - 10} = \frac{-1(z - 10)}{z - 10} = -1$$

(c) Rewrite the expression with the negative sign in the numerator and then apply the distributive property. Be sure to include parentheses around the numerator.

$$-\frac{5 - x}{x - 5} = \frac{-(5 - x)}{x - 5} = \frac{-5 + x}{x - 5} = \frac{x - 5}{x - 5} = 1$$

The same answer can be obtained by distributing the negative sign in the denominator.

$$-\frac{5 - x}{x - 5} = \frac{5 - x}{-(x - 5)} = \frac{5 - x}{-x + 5} = \frac{5 - x}{5 - x} = 1$$

Now Try Exercises 63, 67, 71

NOTE: The result for Example 5(b) becomes more obvious if we substitute a number for z. For example, if we let $z = 6$, then

$$\frac{10 - z}{z - 10} = \frac{10 - 6}{6 - 10} = \frac{4}{-4} = -1. \quad \blacksquare$$

MAKING CONNECTIONS 2

Negative Signs and Rational Expressions

In general, $(b - a)$ equals $-1(a - b)$. Thus if $a \neq b$, then $\frac{b - a}{a - b} = -1$.

3 Solving Applications Involving Rational Expressions

Highway Traffic

 Math in Context Have you ever been moving smoothly in traffic, only to come to a sudden halt? Mathematics shows that in certain conditions, if the number of cars on a road increases even slightly, then the movement of traffic can slow dramatically. To understand why this occurs, we will consider how *rational equations* can be used to model traffic flow.

EXAMPLE 6 **Modeling traffic flow**

Suppose that 10 cars per minute can pass through a construction zone. If traffic arrives *randomly* at an average rate of x cars per minute, the average time T in minutes spent waiting in line and passing through the construction zone is given by

$$T = \frac{1}{10 - x},$$

where $x < 10$. (**Source:** N. Garber and L. Hoel, *Traffic and Highway Engineering.*)

(a) Complete **TABLE 7.1** by finding T for each value of x.

Waiting in Traffic

x (cars/minute)	5	7	9	9.5	9.9	9.99
T (minutes)						

TABLE 7.1

(b) Interpret the results.

Solution

(a) When $x = 5$ cars per minute, $T = \frac{1}{10 - 5} = \frac{1}{5}$ minute. Other values are found similarly and are shown in **TABLE 7.2**.

Waiting in Traffic

x (cars/minute)	5	7	9	9.5	9.9	9.99
T (minutes)	$\frac{1}{5}$	$\frac{1}{3}$	1	2	10	100

TABLE 7.2

(b) As the average traffic rate increases from 9 cars per minute to 9.9 cars per minute, the time needed to pass through the construction zone increases from 1 minute to 10 minutes. As x nears 10 cars per minute, a small increase in x increases the waiting time dramatically.

Now Try Exercise 103

This nonlinear effect for traffic congestion in Example 6 is shown in **FIGURE 7.1**, where points from **TABLE 7.2** have been plotted and a curve passing through them has been sketched. A vertical dashed line has also been sketched at $x = 10$. This dashed line is called a **vertical asymptote** and indicates that the rational expression is undefined at this value of x. Near the left side of the vertical asymptote, the waiting time T increases dramatically for small increases in x. The graph of T does not intersect or cross this vertical asymptote.

FIGURE 7.1

TECHNOLOGY NOTE

Making Tables
TABLE 7.2 can also be created with a graphing calculator by using the Ask feature, as illustrated in the following displays.

CALCULATOR HELP

To make a table of values, see Appendix A (pages 2 and 3).

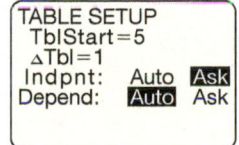 **Math in Context** When a new species of animal is introduced into an area that it did not previously inhabit, its population may grow quickly at first, and then level off over time. Rational equations can be used to model such situations, as demonstrated in the next example.

EXAMPLE 7   ### Modeling a fish population

Suppose that a small fish species is introduced into a pond that had not previously held this type of fish, and that its population P in thousands is modeled by the equation

$$P = \frac{3x + 1}{x + 4},$$

where $x \geq 0$ represents time in months.
(a) Complete **TABLE 7.3** by finding P for each value of x. Round to 3 decimal places.

Fish Population

x (months)	0	6	12	36	72
P (thousands)					

TABLE 7.3

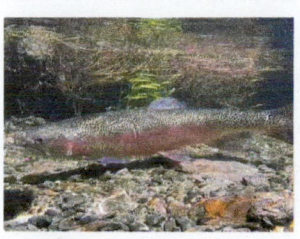

(b) How many fish were initially introduced into the pond?
(c) Interpret the results shown in your completed table.

Solution

(a) If we evaluate $P = \frac{3x + 1}{x + 4}$ when $x = 0$, the population is $P = \frac{3(0) + 1}{0 + 4} = \frac{1}{4} = 0.25$ thousand fish. The other values are found similarly and are shown in **TABLE 7.4**.

Fish Population

x (months)	0	6	12	36	72
P (thousands)	0.25	1.9	2.313	2.725	2.855

TABLE 7.4

(b) **TABLE 7.4** shows that initially (when $x = 0$) there were 0.25 thousand, or 250 fish.

(c) The fish population increased quickly at first but then leveled off. This population growth is shown graphically in **FIGURE 7.2**.

Fish Population

The horizontal dashed line $P = 3$ is called a **horizontal asymptote**. The fish population levels off at 3 thousand fish.

FIGURE 7.2

Now Try Exercise 105

AN APPLICATION INVOLVING PROBABILITY If 10 marbles, one blue and nine red, are placed in a jar, then the *probability*, or *likelihood*, of picking the blue marble at random is 1 *chance* in 10, or $\frac{1}{10}$. The probability of drawing a red marble at random is 9 chances in 10, or $\frac{9}{10}$. **Probability** is a real number from 0 to 1. A probability of 0, or 0%, indicates that an event is impossible, whereas a probability of 1, or 100%, indicates that an event is certain. Rational expressions are often used to describe probability.

EXAMPLE 8 Calculating probability

Suppose that n balls, numbered 1 to n, are placed in a container and only one ball has the winning number.

(a) What is the probability of drawing the winning ball at random?

(b) Calculate this probability for $n = 100$, 1000, and 10,000.

(c) What happens to the probability of drawing the winning ball as the number of balls increases?

Solution

(a) There is 1 chance in n of drawing the winning ball, so the probability is $\frac{1}{n}$.

(b) For $n = 100$, 1000, and 10,000, the probabilities are $\frac{1}{100}$, $\frac{1}{1000}$, and $\frac{1}{10,000}$.

(c) As the number of balls increases, the probability of picking the winning ball decreases.

Now Try Exercise 113

7.1 Putting It All Together

CONCEPT	EXPLANATION	EXAMPLES
Rational Expression	An expression of the form $\frac{P}{Q}$, where P and Q are polynomials with $Q \neq 0$	$\frac{1}{x}$, $\frac{x-3}{2x^2-1}$, $\frac{2x+9}{5x}$, and $\frac{x^2+3x-5}{1}$
Undefined Rational Expressions	A rational expression is undefined for any value of the variable that makes the *denominator* equal to 0.	$\frac{1}{x-3}$ is undefined when $x = 3$. $\frac{5y}{y^2-1}$ is undefined when $y = 1$ or when $y = -1$.
Basic Principle of Rational Expressions	Factor the numerator and the denominator completely. Then apply $$\frac{P \cdot R}{Q \cdot R} = \frac{P}{Q}.$$	$\frac{4xy^2}{6xy^3} = \frac{2(2xy^2)}{3y(2xy^2)} = \frac{2}{3y}$ $\frac{4x(x-4)}{(x+1)(x-4)} = \frac{4x}{x+1}$

7.1 Exercises

MyMathLab®

CONCEPTS AND VOCABULARY

1. A rational expression can be written as ____, where P and Q are ____ with $Q \neq 0$.

2. Is $\frac{x}{2x^2+1}$ a rational expression? Why or why not?

3. A rational expression is undefined whenever the ____ is equal to 0.

4. The rational expression $\frac{1}{x-a}$ is undefined whenever $x =$ ____.

5. The basic principle of fractions states that a fraction can be simplified by using $\frac{a \cdot c}{b \cdot c} =$ ____ .

6. The basic principle of rational expressions can be used to simplify ____ expressions.

EVALUATING RATIONAL EXPRESSIONS

Exercises 7–20: If possible, evaluate the expression for the given value of the variable.

7. $\frac{3}{x}$; $x = -7$

8. $\frac{3}{x+3}$; $x = 0$

9. $-\frac{x}{x-5}$; $x = -4$

10. $-\frac{4x}{5x+1}$; $x = 1$

11. $\frac{y+1}{y^2}$; $y = -2$

12. $\frac{3y-1}{y^2+1}$; $y = -1$

13. $\frac{7z}{z^2-4}$; $z = -2$

14. $\frac{5}{z^2-3z+2}$; $z = -1$

15. $\frac{5}{3t+6}$; $t = -2$

16. $\frac{4t}{2t+5}$; $t = -\frac{5}{2}$

17. $\frac{4-x}{x-4}$; $x = -2$

18. $\frac{x-7}{7-x}$; $x = 4$

19. $-\frac{6-x}{x-6}$; $x = 0$

20. $\frac{8-2x}{2x-8}$; $x = -5$

Exercises 21–24: Complete the table for the given expression. If a value is undefined, place a dash in the table.

21.

x	-2	-1	0	1	2
$\frac{x}{x+1}$					

22.

x	-2	-1	0	1	2
$\frac{2x}{3x-1}$					

23.

x	-2	-1	0	1	2
$\dfrac{3x}{2x^2 + 1}$					

24.

x	-2	-1	0	1	2
$\dfrac{2x - 1}{x^2 - 1}$					

DETERMINING UNDEFINED EXPRESSIONS

Exercises 25–38: Find any values of the variable that make the expression undefined.

25. $-\dfrac{8}{x}$

26. $\dfrac{7}{x + 1}$

27. $\dfrac{4}{z - 3}$

28. $\dfrac{7 - z}{z - 7}$

29. $\dfrac{4y}{5y + 4}$

30. $\dfrac{3 + y}{3y - 7}$

31. $\dfrac{5t + 2}{t^2 + 1}$

32. $\dfrac{8t}{t^2 + 25}$

33. $\dfrac{8x}{x^2 - 25}$

34. $\dfrac{x + 4}{x^2 - 36}$

35. $\dfrac{x^2 + 3x + 2}{x^2 + 5x + 6}$

36. $\dfrac{2x - 1}{x^2 - 7x + 10}$

37. $\dfrac{8z^2 + z + 1}{2z^2 - 7z + 5}$

38. $\dfrac{4n^2 + 17n - 15}{3n^2 - 8n + 4}$

SIMPLIFYING RATIONAL EXPRESSIONS

Exercises 39–46: Simplify the fraction to lowest terms.

39. $\dfrac{12}{18}$

40. $\dfrac{24}{32}$

41. $\dfrac{24}{48}$

42. $\dfrac{8}{22}$

43. $-\dfrac{6}{15}$

44. $-\dfrac{22}{33}$

45. $-\dfrac{25}{75}$

46. $-\dfrac{36}{42}$

Exercises 47–50: First simplify the fraction in part (a), then simplify the rational expression in part (b).

47. (a) $\dfrac{8}{16}$ (b) $\dfrac{x + 2}{2x + 4}$

48. (a) $\dfrac{6}{9}$ (b) $\dfrac{4x + 12}{6x + 18}$

49. (a) $\dfrac{7 - 3}{3 - 7}$ (b) $\dfrac{7 - x}{x - 7}$

50. (a) $-\dfrac{8 - 5}{5 - 8}$ (b) $-\dfrac{x - 5}{5 - x}$

Exercises 51–88: Simplify the expression.

51. $\dfrac{5x^4}{10x^6}$

52. $\dfrac{6y^2}{9y}$

53. $\dfrac{8xy^3}{6x^2y^2}$

54. $\dfrac{36x^2y^5}{6x^5y}$

55. $\dfrac{x + 4}{2x + 8}$

56. $\dfrac{5x - 10}{x - 2}$

57. $\dfrac{3z - 9}{5z - 15}$

58. $\dfrac{4z + 8}{10 + 5z}$

59. $\dfrac{(x + 1)(x - 1)}{(x + 6)(x - 1)}$

60. $\dfrac{(2x + 1)(x + 9)}{(4x - 3)(x + 9)}$

61. $\dfrac{(5y + 3)(2y - 1)}{(2y - 1)(y + 2)}$

62. $\dfrac{(4y - 1)(5y + 7)}{(5y + 7)(1 - 4y)}$

63. $\dfrac{x - 7}{7 - x}$

64. $\dfrac{5 - x}{x - 5}$

65. $\dfrac{a - b}{b - a}$

66. $\dfrac{2t - 3r}{3r - 2t}$

67. $\dfrac{-6 - x}{18 + 3x}$

68. $\dfrac{-2x - 6}{x + 3}$

69. $\dfrac{x + 1}{-2x - 2}$

70. $\dfrac{3x + 21}{-7 - x}$

71. $-\dfrac{9 - x}{x - 9}$

72. $-\dfrac{4 - 2x}{x - 2}$

73. $\dfrac{(3x + 5)(x - 1)}{(3x - 5)(1 - x)}$

74. $\dfrac{(2 - x)(x - 2)}{(x - 2)(2 - x)}$

75. $\dfrac{n^2 - n}{n^2 - 5n}$

76. $\dfrac{3n^2 - 4n}{n^2 + 4n}$

77. $\dfrac{x^2 - 3x}{6x - 18}$

78. $\dfrac{4x^2 + 16x}{5x^2 + 20x}$

79. $\dfrac{z^2 - 3z + 2}{z^2 - 4z + 3}$

80. $\dfrac{z^2 - 3z - 10}{z^2 - 2z - 8}$

81. $\dfrac{2x^2 + 7x - 4}{6x^2 + x - 2}$

82. $\dfrac{5x^2 + 3x - 2}{5x^2 + 13x - 6}$

83. $\dfrac{x - 3}{3x^2 - 11x + 6}$

84. $\dfrac{2x - 1}{4x^2 + 6x - 4}$

85. $-\dfrac{a - 9}{9 - a}$

86. $-\dfrac{-b - 6}{b + 6}$

87. $\dfrac{-2x - 1}{4x + 2}$

88. $\dfrac{4x + 3}{-8x - 6}$

Exercises 89 and 90: **Thinking Generally** *Complete the statement involving a rational expression.*

89. The expression $\frac{x-a}{a-x}$ simplifies to _____.

90. The expression $-\frac{x-a}{x+a}$ simplifies to $\frac{?}{x+a}$.

Exercises 91–98: Refer to Making Connections 1 in this section.

    (a) *Decide whether you are given an expression or an equation.*

    (b) *If you are given an expression, simplify it. If you are given an equation, solve it.*

91. $x + 1 = 7$

92. $x^2 - 4 = 0$

93. $\dfrac{x}{x(x+1)}$

94. $\dfrac{x-2}{(x-2)(x-8)}$

95. $\dfrac{x^2 - 4}{x + 2}$

96. $\dfrac{x-4}{8-2x}$

97. $\dfrac{x}{2(1+3)} = 1$

98. $\dfrac{x}{2} + 2 = \dfrac{8}{2}$

Exercises 99–102: **Graphing Rational Equations** *Complete the following.*

    (a) *Use the given equation to complete the table of values for y.*

x	-4	-3	-2	-1	0	1	2	3	4
y									

    (b) *Determine any x-value that will make the expression undefined.*

    (c) *Sketch a dashed, vertical line (asymptote) in the xy-plane at any undefined values of x.*

    (d) *Plot the points from the table.*

    (e) *Sketch a graph of the equation. Do not let your graph cross the vertical, dashed line.*

99. $y = \dfrac{1}{x-1}$

100. $y = \dfrac{1}{x+1}$

101. $y = \dfrac{x}{x+1}$

102. $y = \dfrac{x}{x-1}$

APPLICATIONS INVOLVING RATIONAL EXPRESSIONS

103. **Traffic Flow** (Refer to Example 6.) Five vehicles per minute can pass through a construction zone. If the traffic arrives randomly at an average rate of x vehicles per minute, the average time T in minutes spent waiting in line and passing through the construction zone is given by $T = \frac{1}{5-x}$ for $x < 5$. (*Source:* N. Garber.)

    (a) Evaluate T for $x = 3$ and interpret the result.

    (b) Complete the table and interpret the results.

x	2	4	4.5	4.9	4.99
T					

104. **Standing in Line** A worker at a poolside store can serve 20 customers per hour. If children arrive randomly at an average rate of x per hour, then the average number of children N waiting in line is given by $N = \frac{x^2}{400 - 20x}$ for $x < 20$. (*Source:* N. Garber.)

    (a) Complete the table.

x	5	10	18	19	20
N					

    (b) Compare the number of children waiting in line if the average rate increases from 18 to 19 children per hour.

105. **Frog Population** Suppose that a frog species is introduced into a wetland area and its population in hundreds is modeled by

$$P = \frac{7x + 3}{x + 6},$$

where $x \geq 0$ is time in months.

    (a) Complete the table by finding P for each given value of x. Round to 2 decimal places.

x (months)	0	12	36	72
P (hundreds)				

    (b) What was the initial frog population?

    (c) Interpret the results in your completed table.

106. **Insect Population** Suppose that an insect population in thousands per acre is modeled by

$$P = \frac{5x + 2}{x + 1},$$

where $x \geq 0$ is time in months.

    (a) Complete the table by finding P for each given value of x. Round to 3 decimal places.

x (months)	0	12	36	60
P (thousands)				

    (b) What was the initial insect population?

    (c) Interpret the results in your completed table.

107. **Distance and Time** A car is traveling at 60 miles per hour.

    (a) How long does it take the car to travel 360 miles?

    (b) Write a rational expression that gives the time that it takes the car to travel M miles.

108. Distance and Time A bicyclist rides uphill at 10 miles per hour for 5 miles and then rides downhill at 20 miles per hour for 5 miles. What is the bicyclist's average speed? (*Hint:* Average speed equals distance divided by time.)

Exercises 109 and 110: **Traffic Flow** *(Refer to Example 6.) The figure shows a graph of the waiting time T in minutes at a construction zone when cars are arriving randomly at an average rate of x cars per minute.*

(a) *Give the equation of the vertical asymptote.*

(b) *Explain how the graph relates to traffic flow.*

109.

110.

APPLICATIONS INVOLVING PROBABILITY

111. Probability Suppose that a coin is flipped. What is the probability that a head appears?

112. Probability A die shows the numbers 1, 2, 3, 4, 5, and 6. If each number has an equal chance of appearing on any given roll, what is the probability that a 2 or 4 appears?

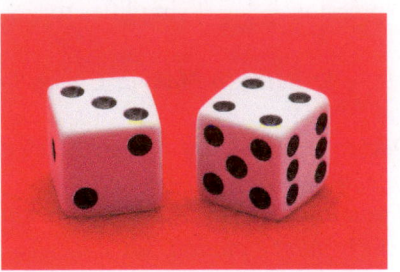

113. Probability (Refer to Example 8.) Suppose that there are n balls in a container and that three balls have a winning number. If a ball is drawn randomly, do each of the following.

(a) Write a rational expression that gives the probability of drawing a winning ball.

(b) Write a rational expression that gives the probability of not drawing a winning ball. Evaluate your expression for $n = 100$ and interpret the result.

114. Probability Suppose that there are n balls numbered 1 to n in a container and that every even-numbered ball is green while every odd-numbered ball is red. If a ball is drawn randomly, answer each of the following.

(a) If n is an odd number, are there more green balls or more red balls?

(b) If n is 39, what is the probability of drawing a green ball?

WRITING ABOUT MATHEMATICS

115. What is a rational expression? When is a rational expression undefined?

116. Does the rational expression $\frac{5x + 2}{10x + 4}$ equal $\frac{5x}{10x} + \frac{2}{4}$? Explain your answer.

Group Activity

Students per Computer In the early years of personal computers, school districts could not afford to buy a computer for every student. As the price of computers decreased, more and more school districts were able to move toward this goal. The following table lists numbers of students per computer during these early years.

Year	1983	1985	1987	1989
Students/Computer	125	50	32	22

Year	1991	1993	1995	1997
Students/Computer	18	14	10	6

Source: Quality Education Data, Inc.

(a) Make a scatterplot of the data in the table. Would a straight line model the data accurately? Explain.

(b) Discuss how well the formula

$$S = \frac{125}{1 + 0.7(x - 1983)}, \quad x \geq 1983$$

models these data, where S represents the students per computer and x represents the year.

(c) In what year does the formula suggest that there were about 17 students per computer?

7.2   Multiplication and Division of Rational Expressions

Objectives

1. Reviewing Multiplication and Division of Fractions
2. Multiplying Rational Expressions
3. Dividing Rational Expressions

1   Reviewing Multiplication and Division of Fractions

To multiply two fractions, we use the property

$$\frac{a}{b} \cdot \frac{c}{d} = \frac{ac}{bd}.$$

In the next example, we review multiplication of fractions.

EXAMPLE 1   Multiplying fractions

Multiply and simplify your answers to lowest terms.

(a) $\frac{3}{7} \cdot \frac{4}{5}$ **(b)** $2 \cdot \frac{3}{4}$ **(c)** $\frac{4}{21} \cdot \frac{7}{8}$

Solution

(a) $\frac{3}{7} \cdot \frac{4}{5} = \frac{12}{35}$ **(b)** $2 \cdot \frac{3}{4} = \frac{2}{1} \cdot \frac{3}{4} = \frac{6}{4} = \frac{3}{2}$

(c) $\frac{4}{21} \cdot \frac{7}{8} = \frac{7 \cdot 4}{21 \cdot 8} = \frac{7}{21} \cdot \frac{4}{8} = \frac{1}{3} \cdot \frac{1}{2} = \frac{1}{6}$

Now Try Exercises 5, 7, 9

To divide two fractions, we change the division problem to a multiplication problem by using the property

$$\frac{a}{b} \div \frac{c}{d} = \frac{a}{b} \cdot \frac{d}{c}.$$

Invert $\frac{c}{d}$ to $\frac{d}{c}$ and multiply.

EXAMPLE 2 **Dividing fractions**

Divide and simplify your answers to lowest terms.

(a) $\frac{1}{3} \div \frac{5}{7}$ (b) $\frac{4}{5} \div 8$ (c) $\frac{8}{9} \div \frac{10}{3}$

Solution

> Invert $\frac{5}{7}$ to $\frac{7}{5}$ and multiply.

(a) $\frac{1}{3} \div \frac{5}{7} = \frac{1}{3} \cdot \frac{7}{5} = \frac{7}{15}$

(b) $\frac{4}{5} \div 8 = \frac{4}{5} \cdot \frac{1}{8} = \frac{4}{40} = \frac{1}{10}$

(c) $\frac{8}{9} \div \frac{10}{3} = \frac{8}{9} \cdot \frac{3}{10} = \frac{24}{90} = \frac{4 \cdot 6}{15 \cdot 6} = \frac{4}{15}$

Now Try Exercises 13, 15, 17

2 Multiplying Rational Expressions

Multiplying rational expressions is similar to multiplying fractions.

PRODUCTS OF RATIONAL EXPRESSIONS

To multiply two rational expressions, multiply the numerators and multiply the denominators. That is,

$$\frac{A}{B} \cdot \frac{C}{D} = \frac{AC}{BD},$$

where B and D are nonzero.

READING CHECK 1

- How do we multiply rational expressions?

EXAMPLE 3 **Multiplying rational expressions**

Multiply and simplify to lowest terms. Leave your answers in factored form.

(a) $\frac{3}{x} \cdot \frac{2x - 5}{x - 1}$

(b) $\frac{x - 1}{4x} \cdot \frac{x + 3}{x - 1}$

(c) $\frac{x^2 - 4}{x + 3} \cdot \frac{x + 3}{x + 2}$

(d) $\frac{4}{x^2 + 3x + 2} \cdot \frac{x^2 + 2x + 1}{8}$

Solution

(a) $\frac{3}{x} \cdot \frac{2x - 5}{x - 1} = \frac{3(2x - 5)}{x(x - 1)}$ Multiply the numerators and the denominators.

(b) $\frac{x - 1}{4x} \cdot \frac{x + 3}{x - 1} = \frac{(x - 1)(x + 3)}{4x(x - 1)}$ Multiply the numerators and the denominators.

$= \frac{x + 3}{4x}$ Simplify.

(c) $\frac{x^2 - 4}{x + 3} \cdot \frac{x + 3}{x + 2} = \frac{(x - 2)(x + 2)}{x + 3} \cdot \frac{x + 3}{x + 2}$ Factor.

$= \frac{(x - 2)(x + 2)(x + 3)}{(x + 3)(x + 2)}$ Multiply the numerators and the denominators.

$= x - 2$ Simplify.

STUDY TIP

Check with your instructor about your progress in the class to be sure that your work is complete and your grades are up-to-date.

(d) $\dfrac{4}{x^2 + 3x + 2} \cdot \dfrac{x^2 + 2x + 1}{8} = \dfrac{4(x^2 + 2x + 1)}{8(x^2 + 3x + 2)}$ Multiply the numerators and the denominators.

$$= \dfrac{\mathbf{4}(x + 1)(x + 1)}{\mathbf{8}(x + 2)(x + 1)}$$ Factor.

$$= \dfrac{x + 1}{2(x + 2)}$$ Simplify.

Now Try Exercises 29, 31, 41, 45

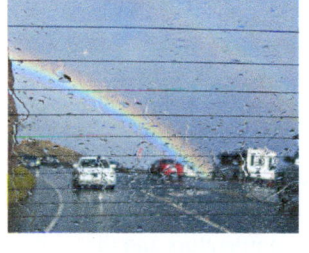

Highway Safety
🌐 *Math in Context* Stopping distance for a car can vary depending on the road conditions. If the road is slippery, more distance is needed to stop. Also, cars that are traveling downhill require additional stopping distance. Rational expressions are frequently used by highway engineers to estimate the stopping distance of a car on slippery surfaces or on hills.

EXAMPLE 4   **Estimating stopping distance**

If a car is traveling at 60 miles per hour on a slippery road, then its stopping distance D in feet can be calculated by

$$D = \dfrac{3600}{30} \cdot \dfrac{1}{x},$$

where x is the coefficient of friction between the tires and the road and $0 < x \le 1$. The more slippery the road is, the smaller the value of x. (*Source:* L. Haefner, *Introduction to Transportation Systems.*)
(a) Multiply and simplify the formula for D.
(b) Compare the stopping distance on an icy road with $x = 0.1$ to the stopping distance on dry pavement with $x = 0.4$.

Solution
(a) Because

$$\dfrac{3600}{30} \cdot \dfrac{1}{x} = \dfrac{3600}{30x} = \dfrac{120 \cdot \mathbf{30}}{x \cdot \mathbf{30}} = \dfrac{120}{x},$$

it follows that $D = \dfrac{120}{x}$.

(b) When $x = \mathbf{0.1}$, $D = \dfrac{120}{\mathbf{0.1}} = 1200$ feet, and when $x = \mathbf{0.4}$, $D = \dfrac{120}{\mathbf{0.4}} = 300$ feet. A car traveling at 60 miles per hour on an icy road requires a stopping distance that is 4 times that of a car traveling at the same speed on dry pavement.

Now Try Exercise 73

3 Dividing Rational Expressions
Dividing rational expressions is similar to dividing fractions.

READING CHECK 2

- When we divide the rational expressions $\frac{A}{B} \div \frac{C}{D}$, what do we multiply $\frac{A}{B}$ by?

QUOTIENTS OF RATIONAL EXPRESSIONS

To divide two rational expressions, multiply by the reciprocal of the divisor. That is,

$$\dfrac{A}{B} \div \dfrac{C}{D} = \dfrac{A}{B} \cdot \dfrac{D}{C},$$

where B, C, and D are nonzero.

NOTE: When we "multiply by the reciprocal of the divisor" in a division problem, we will sometimes refer to this step as "invert and multiply." ∎

EXAMPLE 5   **Dividing rational expressions**

Divide and simplify to lowest terms.

(a) $\dfrac{5}{2x} \div \dfrac{10}{x-4}$ (b) $\dfrac{x^2-9}{x^2+4} \div (x-3)$ (c) $\dfrac{x^2-x}{x^2-x-2} \div \dfrac{x}{x-2}$

Solution

(a)
$$\frac{5}{2x} \div \frac{10}{x-4} = \frac{5}{2x} \cdot \frac{x-4}{10}$$
Multiply by the reciprocal of the divisor.

Invert $\frac{10}{x-4}$ to $\frac{x-4}{10}$ and multiply.
$$= \frac{5(x-4)}{20x}$$
Multiply the numerators and the denominators.

$$= \frac{x-4}{4x}$$
Simplify. Note that $\frac{5}{20} = \frac{1}{4}$.

(b)
$$\frac{x^2-9}{x^2+4} \div (x-3) = \frac{x^2-9}{x^2+4} \cdot \frac{1}{x-3}$$
Multiply by the reciprocal of the divisor.

Invert $\frac{x-3}{1}$ to $\frac{1}{x-3}$ and multiply.
$$= \frac{x^2-9}{(x^2+4)(x-3)}$$
Multiply the numerators and the denominators.

$$= \frac{(x+3)(x-3)}{(x^2+4)(x-3)}$$
Factor the numerator.

$$= \frac{x+3}{x^2+4}$$
Simplify.

(c)
$$\frac{x^2-x}{x^2-x-2} \div \frac{x}{x-2} = \frac{x^2-x}{x^2-x-2} \cdot \frac{x-2}{x}$$
Multiply by the reciprocal of the divisor.

Invert $\frac{x}{x-2}$ to $\frac{x-2}{x}$ and multiply.
$$= \frac{(x^2-x)(x-2)}{(x^2-x-2)x}$$
Multiply the numerators and the denominators.

$$= \frac{x(x-1)(x-2)}{x(x+1)(x-2)}$$
Factor the numerator and the denominator.

$$= \frac{x-1}{x+1}$$
Simplify.

Now Try Exercises 49, 57, 65

7.2   Putting It All Together

CONCEPT	EXPLANATION	EXAMPLES
Multiplication of Rational Expressions	To multiply two rational expressions, multiply the numerators and multiply the denominators. $\dfrac{A}{B} \cdot \dfrac{C}{D} = \dfrac{AC}{BD}$ B and D are nonzero Then simplify the result to lowest terms.	$\dfrac{4x}{x-1} \cdot \dfrac{x-1}{x+1} = \dfrac{4x(x-1)}{(x-1)(x+1)}$ $= \dfrac{4x}{x+1}$

CONCEPT	EXPLANATION	EXAMPLES
Division of Rational Expressions	To divide two rational expressions, multiply the first expression by the reciprocal of the second expression. $$\frac{A}{B} \div \frac{C}{D} = \frac{A}{B} \cdot \frac{D}{C} \quad B, C, \text{ and } D \text{ are nonzero}$$ Then simplify the result to lowest terms.	$$\frac{x+1}{x-3} \div \frac{x+1}{x-5} = \frac{x+1}{x-3} \cdot \frac{x-5}{x+1}$$ $$= \frac{(x+1)(x-5)}{(x-3)(x+1)}$$ $$= \frac{x-5}{x-3}$$

7.2 Exercises

MyMathLab®

CONCEPTS AND VOCABULARY

1. To multiply rational expressions, multiply the _____ and multiply the _____ .

2. To divide two rational expressions, multiply the first expression by the _____ of the second expression.

3. $\dfrac{A}{B} \cdot \dfrac{C}{D} = $ _____

4. $\dfrac{A}{B} \div \dfrac{C}{D} = $ _____

REVIEWING MULTIPLICATION AND DIVISION OF FRACTIONS

Exercises 5–20: Multiply or divide as indicated. Simplify to lowest terms.

5. $\dfrac{1}{2} \cdot \dfrac{4}{5}$

6. $\dfrac{6}{7} \cdot \dfrac{7}{18}$

7. $\dfrac{3}{7} \cdot 4$

8. $5 \cdot \dfrac{4}{5}$

9. $\dfrac{5}{4} \cdot \dfrac{8}{15}$

10. $\dfrac{3}{10} \cdot \dfrac{5}{9}$

11. $\dfrac{1}{3} \cdot \dfrac{2}{3} \cdot \dfrac{9}{11}$

12. $\dfrac{2}{5} \cdot \dfrac{10}{11} \cdot \dfrac{1}{4}$

13. $\dfrac{2}{3} \div \dfrac{1}{6}$

14. $\dfrac{5}{7} \div \dfrac{5}{8}$

15. $\dfrac{8}{9} \div 3$

16. $\dfrac{7}{3} \div 6$

17. $8 \div \dfrac{4}{5}$

18. $7 \div \dfrac{14}{5}$

19. $\dfrac{4}{5} \div \dfrac{2}{3} \div \dfrac{1}{2}$

20. $\dfrac{1}{2} \div \dfrac{5}{4} \div \dfrac{2}{5}$

SIMPLIFYING RATIONAL EXPRESSIONS

Exercises 21–28: Simplify the expression.

21. $\dfrac{x+5}{x+5}$

22. $\dfrac{2x-3}{2x-3}$

23. $\dfrac{(z+1)(z+2)}{(z+4)(z+2)}$

24. $\dfrac{(2z-7)(z+5)}{(3z+5)(z+5)}$

25. $\dfrac{8y(y+7)}{12y(y+7)}$

26. $\dfrac{6(y+1)}{12(y+1)}$

27. $\dfrac{x(x+2)(x+3)}{x(x-2)(x+3)}$

28. $\dfrac{2(x+1)(x-1)}{4(x+1)(x-1)}$

MULTIPLYING RATIONAL EXPRESSIONS

Exercises 29–48: Multiply and simplify to lowest terms. Leave your answers in factored form.

29. $\dfrac{8}{x} \cdot \dfrac{x+1}{x}$

30. $\dfrac{7}{2x} \cdot \dfrac{x}{x-1}$

31. $\dfrac{8+x}{x} \cdot \dfrac{x-3}{x+8}$

32. $\dfrac{5x^2+x}{2x-1} \cdot \dfrac{1}{x}$

33. $\dfrac{z+3}{z+4} \cdot \dfrac{z+4}{z-7}$

34. $\dfrac{2z+1}{3z} \cdot \dfrac{3z}{z+2}$

35. $\dfrac{5x+1}{3x+2} \cdot \dfrac{3x+2}{5x+1}$

36. $\dfrac{x+1}{x+3} \cdot \dfrac{x+3}{x+1}$

37. $\dfrac{(t+1)^2}{t+2} \cdot \dfrac{(t+2)^2}{t+1}$

38. $\dfrac{(t-1)^2}{(t+5)^2} \cdot \dfrac{t+5}{t-1}$

39. $\dfrac{x^2}{x^2+4} \cdot \dfrac{x+4}{x}$

40. $\dfrac{x-1}{x^2} \cdot \dfrac{x^2}{x^2+1}$

41. $\dfrac{z^2-1}{z^2-4} \cdot \dfrac{z-2}{z+1}$

42. $\dfrac{z^2-9}{z-5} \cdot \dfrac{z-5}{z+3}$

43. $\dfrac{y^2-2y}{y^2-1} \cdot \dfrac{y+1}{y-2}$

44. $\dfrac{y^2-4y}{y+1} \cdot \dfrac{y+1}{y-4}$

45. $\dfrac{2x^2-x-3}{3x^2-8x-3} \cdot \dfrac{3x+1}{2x-3}$

46. $\dfrac{6x^2+11x-2}{3x^2+11x-4} \cdot \dfrac{3x-1}{6x-1}$

47. $\dfrac{(x-3)^3}{x^2-2x+1} \cdot \dfrac{x-1}{(x-3)^2}$

48. $\dfrac{x^2+4x+4}{x^2-2x+1} \cdot \dfrac{(x-1)^2}{(x+2)^2}$

DIVIDING RATIONAL EXPRESSIONS

Exercises 49–70: Divide and simplify to lowest terms. Leave your answers in factored form.

49. $\dfrac{2}{x} \div \dfrac{2x+3}{x}$

50. $\dfrac{6}{2x} \div \dfrac{x+2}{2x}$

51. $\dfrac{x-2}{3x} \div \dfrac{2-x}{6x}$

52. $\dfrac{x+1}{2x-1} \div \dfrac{x+1}{x}$

53. $\dfrac{z+2}{z+1} \div \dfrac{z+2}{z-1}$

54. $\dfrac{z+7}{z-4} \div \dfrac{z+7}{z-4}$

55. $\dfrac{3y+4}{2y+1} \div \dfrac{3y+4}{y+2}$

56. $\dfrac{y+5}{y-2} \div \dfrac{y}{y+3}$

57. $\dfrac{t^2-1}{t^2+1} \div \dfrac{t+1}{4}$

58. $\dfrac{4}{2t^3} \div \dfrac{8}{t^2}$

59. $\dfrac{y^2-9}{y^2-25} \div \dfrac{y+3}{y+5}$

60. $\dfrac{y+1}{y-4} \div \dfrac{y^2-1}{y^2-16}$

61. $\dfrac{2x^2-4x}{2x-1} \div \dfrac{x-2}{2x-1}$

62. $\dfrac{x-4}{x^2+x} \div \dfrac{5}{x+1}$

63. $\dfrac{2z^2-5z-3}{z^2+z-20} \div \dfrac{z-3}{z-4}$

64. $\dfrac{z^2+12z+27}{z^2-5z-14} \div \dfrac{z+3}{z+2}$

65. $\dfrac{t^2-1}{t^2+5t-6} \div (t+1)$

66. $\dfrac{t^2-2t-3}{t^2-5t-6} \div (t-3)$

67. $\dfrac{a-b}{a+b} \div \dfrac{a-b}{2a+3b}$

68. $\dfrac{x^3-y^3}{x^2-y^2} \div \dfrac{x^2+xy+y^2}{x-y}$

69. $\dfrac{x-y}{x^2+2xy+y^2} \div \dfrac{1}{(x+y)^2}$

70. $\dfrac{a^2-b^2}{4a^2-9b^2} \div \dfrac{a-b}{2a+3b}$

71. Thinking Generally Simplify $\dfrac{a-b}{b-c} \cdot \dfrac{c-b}{b-a}$.

72. Thinking Generally Simplify $\dfrac{a-b}{b-c} \div \dfrac{b-a}{a-b}$.

APPLICATIONS INVOLVING RATIONAL EXPRESSIONS

73. Stopping on Slippery Roads (Refer to Example 4.) If a car is traveling at 30 miles per hour on a slippery

road, then its stopping distance D in feet can be calculated by

$$D = \frac{900}{30} \cdot \frac{1}{x},$$

where x is the coefficient of friction between the tires and the road and $0 < x \le 1$. (*Source:* L. Haefner.)
  (a) Multiply and simplify the formula for D.
  (b) Compare the stopping distance on an icy road with $x = 0.1$ and on dry pavement with $x = 0.4$.

74. Stopping on Hills If a car is traveling at 50 miles per hour on a hill with wet pavement, then its stopping distance D is given by

$$D = \frac{2500}{30} \cdot \frac{1}{x+0.3},$$

where x equals the slope of the hill. (*Source:* L. Haefner.)
  (a) Multiply and simplify the formula for D.
  (b) Compare the stopping distance for an uphill slope of $x = 0.1$ to a downhill slope of $x = -0.1$.

75. Probability Suppose that one jar holds n balls and that a second jar holds $n+1$ balls. Each jar contains one winning ball.
  (a) The probability, or chance, of drawing the winning ball from the first jar and *not* drawing it from the second jar is

$$\frac{1}{n} \cdot \frac{n}{n+1}.$$

  Simplify this expression.
  (b) Find this probability for $n = 99$.

76. U.S. AIDS Cases The cumulative number of AIDS cases C in the United States from 1982 to 1994 can be modeled by $C = 3200x^2 + 1586$, and the cumulative number of AIDS deaths D from 1982 to 1994 can be modeled by $D = 1900x^2 + 619$. In these equations, $x = 0$ corresponds to 1982, $x = 1$ corresponds to 1983, and so on until $x = 12$ corresponds to 1994. (*Source:* U.S. Department of Health.)
  (a) Write the rational expression $\frac{D}{C}$ in terms of x.
  (b) Evaluate your expression for $x = 4, 7$, and 10. Round your answers to the nearest thousandth. Interpret the results.
  (c) Explain what the rational expression $\frac{D}{C}$ represents.

77. Google and Facebook Revenues The equation given by $G = 350(x-3)^2 + 50$ gives Google's revenue x years after its startup, and $F = 206(x-3)^2 + 150$ gives Facebook's revenue x years after its startup. Results are in millions of dollars and $x \ge 3$.
  (a) Write a formula for $R = \frac{G}{F}$.
  (b) Evaluate R when $x = 6$ and interpret your result.

78. Google and Yahoo Revenues The equation given by $G = 350(x - 3)^2 + 50$ gives Google's revenue x years after its startup, and $Y = 117(x - 3)^2 + 50$ gives Yahoo's revenue x years after its startup, Results are in millions of dollars and $x \geq 3$.
(a) Write a formula for $R = \frac{G}{Y}$.
(b) Evaluate R when $x = 5$ and interpret your result.

79. Total Production Cost The average cost per item in dollars for a company to produce x digital cameras is given by

$$C = \frac{50x + 20,000}{x}.$$

(a) If the company sells 5000 cameras, evaluate C when $x = 5000$ and interpret the result.
(b) Write an expression T that gives the *total* cost of producing x cameras.
(c) Evaluate T when $x = 5000$ and interpret the result.

80. Total Production Cost The average cost per item in dollars for a company to produce x flash drives is given by

$$C = \frac{18x + 36,000}{x}.$$

(a) The company expects to sell 9000 flash drives. Evaluate C when $x = 9000$ and interpret the result.
(b) Write an expression for T that gives the *total* cost of producing x flash drives. Simplify this expression.
(c) Evaluate T when $x = 9000$ and interpret the result.

WRITING ABOUT MATHEMATICS

81. Explain how to multiply two rational expressions.

82. Explain how to divide two rational expressions.

SECTIONS 7.1 and 7.2 **Checking Basic Concepts**

1. If possible, evaluate the expression $\frac{3}{x^2 - 1}$ for $x = -1$ and $x = 3$.

2. Simplify to lowest terms.
(a) $\frac{6x^3y^2}{15x^2y^3}$ (b) $\frac{5x - 15}{x - 3}$
(c) $\frac{x^2 - x - 6}{x^2 + x - 12}$

3. Multiply and simplify to lowest terms.
(a) $\frac{4}{3x} \cdot \frac{2x}{6}$ (b) $\frac{2x + 4}{x^2 - 1} \cdot \frac{x + 1}{x + 2}$

4. Divide and simplify to lowest terms.
(a) $\frac{7}{3z^2} \div \frac{14}{5z^3}$ (b) $\frac{x^2 + x}{x - 3} \div \frac{x}{x - 3}$

5. Waiting in Line Customers are waiting in line at a department store. They arrive randomly at an average rate of x per minute. If the clerk can wait on 2 customers per minute, then the average time in minutes spent waiting in line is given by $T = \frac{1}{2 - x}$ for $x < 2$. (*Source:* N. Garber, *Traffic and Highway Engineering.*)
(a) Complete the table.

x	0.5	1.0	1.5	1.9
T				

(b) What happens to the waiting time as x increases but remains less than 2?

7.3 Addition and Subtraction with Like Denominators

1 ▸ Reviewing Addition and Subtraction of Fractions

Previously, we demonstrated how the property

$$\frac{a}{c} + \frac{b}{c} = \frac{a+b}{c}$$

can be used to add fractions with like denominators. For example,

$$\frac{3}{7} + \frac{2}{7} = \frac{3+2}{7} = \frac{5}{7}. \qquad \text{Add the numerators.}$$

Like denominators

STUDY TIP

By this time in the semester, it is likely that you know some of your classmates. Have you started or joined a study group? Be sure not to miss the opportunity to study math with your classmates.

To subtract two fractions with like denominators, the property

$$\frac{a}{c} - \frac{b}{c} = \frac{a-b}{c}$$

is used. For example,

$$\frac{2}{5} - \frac{4}{5} = \frac{2-4}{5} = -\frac{2}{5}. \qquad \text{Subtract the numerators.}$$

Like denominators

EXAMPLE 1 Adding and subtracting fractions with like denominators

Simplify each expression to lowest terms.

(a) $\frac{3}{8} + \frac{4}{8}$ **(b)** $\frac{5}{9} + \frac{1}{9}$ **(c)** $\frac{12}{5} - \frac{7}{5}$ **(d)** $\frac{23}{20} - \frac{13}{20}$

Solution

(a) $\frac{3}{8} + \frac{4}{8} = \frac{3+4}{8} = \frac{7}{8}$

(b) $\frac{5}{9} + \frac{1}{9} = \frac{5+1}{9} = \frac{6}{9} = \frac{2}{3}$

(c) $\frac{12}{5} - \frac{7}{5} = \frac{12-7}{5} = \frac{5}{5} = 1$

(d) $\frac{23}{20} - \frac{13}{20} = \frac{23-13}{20} = \frac{10}{20} = \frac{1}{2}$

Now Try Exercises 7, 9, 11, 13

TECHNOLOGY NOTE

CALCULATOR HELP

To express a result as a fraction on a calculator, see Appendix A (pages AP-1 and AP-2).

Arithmetic of Fractions

Many calculators have the capability to perform addition and subtraction of fractions, as illustrated in the following figures. Compare these results with those from Example 1.

```
(3/8)+(4/8)▶Frac
               7/8
(5/9)+(1/9)▶Frac
               2/3
```

```
(12/5)−(7/5)▶Fra
c
               1
(23/20)−(13/20)▶
Frac
             1/2
```

2 Adding and Subtracting Rational Expressions with Like Denominators

Addition and subtraction of rational expressions with like denominators are similar to addition and subtraction of fractions. The following property can be used to add two rational expressions with like denominators.

READING CHECK 1

• How do we add rational expressions with like denominators?

SUMS OF RATIONAL EXPRESSIONS

To add two rational expressions that have like denominators, add their numerators. Keep the same denominator.

$$\frac{A}{C} + \frac{B}{C} = \frac{A + B}{C} \qquad C \text{ is nonzero.}$$

When we add rational expressions with like denominators, we add the numerators. Then we combine like terms and simplify the resulting expression by applying the basic principle of rational expressions. For example, we can add $\frac{2x}{x+1}$ and $\frac{1-x}{x+1}$ as follows.

$$\frac{2x}{x+1} + \frac{1-x}{x+1} = \frac{2x + 1 - x}{x+1} \qquad \text{Add the numerators.}$$

Like denominators

$$= \frac{2x - x + 1}{x+1} \qquad \text{Commutative property}$$

$$= \frac{x+1}{x+1} \qquad \text{Combine like terms.}$$

$$= 1 \qquad \text{Simplify.}$$

It is important to understand that the expressions $\frac{2x}{x+1} + \frac{1-x}{x+1}$ and 1 are *equivalent expressions*. That is, they are equal for *every* value of x except -1, for which the first expression is undefined.

In the next example, we add rational expressions with like denominators and simplify the result to lowest terms.

EXAMPLE 2   **Adding rational expressions with like denominators**

Add and simplify to lowest terms.

(a) $\dfrac{3}{b} + \dfrac{2}{b}$

(b) $\dfrac{z}{z+2} + \dfrac{2}{z+2}$

(c) $\dfrac{x-1}{x^2+x} + \dfrac{1}{x^2+x}$

(d) $\dfrac{t^2+t}{t-1} + \dfrac{1-3t}{t-1}$

Solution

(a) $\dfrac{3}{b} + \dfrac{2}{b} = \dfrac{3+2}{b} = \dfrac{5}{b}$ 　　　　　Add the numerators.

(b) $\dfrac{z}{z+2} + \dfrac{2}{z+2} = \dfrac{z+2}{z+2}$ 　　　　Add the numerators.

$\qquad\qquad\qquad = 1$ 　　　　　Simplify.

(c) $\dfrac{x-1}{x^2+x} + \dfrac{1}{x^2+x} = \dfrac{x-1+1}{x^2+x}$ 　　Add the numerators.

Add numerator; keep the same denominator.

$\qquad\qquad\qquad = \dfrac{x}{x(x+1)}$ 　　　Factor the denominator.

$\qquad\qquad\qquad = \dfrac{1}{x+1}$ 　　　Simplify.

(d) $\dfrac{t^2 + t}{t - 1} + \dfrac{1 - 3t}{t - 1} = \dfrac{t^2 + t + 1 - 3t}{t - 1}$ Add the numerators.

$\qquad\qquad\qquad\qquad = \dfrac{t^2 - 2t + 1}{t - 1}$ Combine like terms.

$\qquad\qquad\qquad\qquad = \dfrac{(t - 1)(t - 1)}{t - 1}$ Factor the numerator.

$\qquad\qquad\qquad\qquad = t - 1$ Simplify.

Now Try Exercises 19, 25, 33, 35

EXAMPLE 3   **Adding rational expressions with two variables**

Add and simplify to lowest terms.

(a) $\dfrac{4}{xy} + \dfrac{5}{xy}$ **(b)** $\dfrac{a}{a^2 - b^2} + \dfrac{b}{a^2 - b^2}$ **(c)** $\dfrac{1}{x - y} + \dfrac{-1}{y - x}$

Solution

(a) $\dfrac{4}{xy} + \dfrac{5}{xy} = \dfrac{4 + 5}{xy} = \dfrac{9}{xy}$ Add the numerators.

(b) $\dfrac{a}{a^2 - b^2} + \dfrac{b}{a^2 - b^2} = \dfrac{a + b}{a^2 - b^2}$ Add the numerators.

$\qquad\qquad\qquad\qquad\quad = \dfrac{a + b}{(a - b)(a + b)}$ Factor the denominator.

$\qquad\qquad\qquad\qquad\quad = \dfrac{1}{a - b}$ Simplify.

(c) First write $\dfrac{1}{x - y} + \dfrac{-1}{y - x}$ with a common denominator. Note that if we multiply the second term by 1, written in the form $\frac{-1}{-1}$, it becomes

$$\dfrac{-1}{y - x} \cdot \dfrac{-1}{-1} = \dfrac{(-1)(-1)}{(y - x)(-1)} = \dfrac{1}{-y + x} = \dfrac{1}{x - y}.$$

Thus the given sum can be simplified as follows.

$$\dfrac{1}{x - y} + \dfrac{-1}{y - x} = \dfrac{1}{x - y} + \dfrac{1}{x - y} \qquad \text{Rewrite the second term.}$$

$$\qquad\qquad\qquad\quad = \dfrac{2}{x - y} \qquad \text{Add the numerators.}$$

Now Try Exercises 51, 53, 55

Next we consider subtraction of rational expressions with like denominators.

READING CHECK 2

- How do we subtract rational expressions with like denominators?

DIFFERENCES OF RATIONAL EXPRESSIONS

To subtract two rational expressions that have like denominators, subtract their numerators. Keep the same denominator.

$$\dfrac{A}{C} - \dfrac{B}{C} = \dfrac{A - B}{C} \qquad C \text{ is nonzero.}$$

Subtraction of rational expressions with like denominators is similar to addition except that instead of adding numerators, we subtract them. For example, the expressions $\frac{3x}{x-4}$ and $\frac{2x}{x-4}$ have like denominators and can be subtracted as follows.

$$\frac{3x}{x-4} - \frac{2x}{x-4} = \frac{3x-2x}{x-4} \qquad \text{Subtract the numerators.}$$

Like denominators

$$= \frac{x}{x-4} \qquad \text{Combine like terms.}$$

In the next example, we subtract rational expressions with like denominators and simplify the result to lowest terms.

EXAMPLE 4   **Subtracting rational expressions with like denominators**

Subtract and simplify to lowest terms.

(a) $\dfrac{a+1}{a} - \dfrac{1}{a}$ **(b)** $\dfrac{2y}{3y-1} - \dfrac{3y}{3y-1}$ **(c)** $\dfrac{1+x}{2x^2+5x-3} - \dfrac{-2}{2x^2+5x-3}$

Solution

(a) $\dfrac{a+1}{a} - \dfrac{1}{a} = \dfrac{a+1-1}{a} = \dfrac{a}{a} = 1$

(b) $\dfrac{2y}{3y-1} - \dfrac{3y}{3y-1} = \dfrac{2y-3y}{3y-1} = \dfrac{-y}{3y-1}$ or $-\dfrac{y}{3y-1}$

(c) $\dfrac{1+x}{2x^2+5x-3} - \dfrac{-2}{2x^2+5x-3} = \dfrac{1+x-(-2)}{2x^2+5x-3}$

Subtract numerators; keep the same denominator.

$$= \frac{x+3}{(2x-1)(x+3)}$$

$$= \frac{1}{2x-1}$$

Now Try Exercises 21, 27, 67

If the numerator of the second fraction in a difference has more than one term, it is important to put parentheses around the second numerator.

$$\frac{x+6}{2x+1} - \frac{3-x}{2x+1} = \frac{x+6-(3-x)}{2x+1} \qquad \text{Subtract the numerators; insert parentheses.}$$

$$= \frac{x+6-3-(-x)}{2x+1} \qquad \text{Distributive property}$$

$$= \frac{x+6-3+x}{2x+1} \qquad \text{Double negative property}$$

$$= \frac{2x+3}{2x+1} \qquad \text{Combine like terms.}$$

NOTE: If parentheses were not inserted in the previous calculation, the numerator would be

$$x+6-3-x=3,$$

which would give an incorrect result. ∎

EXAMPLE 5   **Subtracting rational expressions with like denominators**

Subtract and simplify to lowest terms.

(a) $\dfrac{2x}{x+1} - \dfrac{x-1}{x+1}$ (b) $\dfrac{x+y}{3y} - \dfrac{x-y}{3y}$

Solution

(a) $\dfrac{2x}{x+1} - \dfrac{x-1}{x+1} = \dfrac{2x-(x-1)}{x+1}$ Subtract the numerators.

$= \dfrac{2x-x+1}{x+1}$ Distributive property

$= \dfrac{x+1}{x+1}$ Simplify the numerator.

$= 1$ Simplify.

(b) $\dfrac{x+y}{3y} - \dfrac{x-y}{3y} = \dfrac{x+y-(x-y)}{3y}$ Subtract the numerators.

$= \dfrac{x+y-x+y}{3y}$ Distributive property

$= \dfrac{2y}{3y}$ Simplify the numerator.

$= \dfrac{2}{3}$ Simplify.

Now Try Exercises 31, 59

Quality Control

🌐 **Math in Context** When companies manufacture a large number of items, *quality control* is important. For example, suppose that a company makes computer flash drives. Because it is not practical to check every flash drive to make sure that it works properly, inspectors often check a random sample. By using mathematics and rational expressions, this technique helps determine the likelihood that all the flash drives are good.

EXAMPLE 6   **Analyzing quality control**

A container holds a mixture of 8-GB and 16-GB computer flash drives. In this container, there is a total of n flash drives, including 2 defective 8-GB flash drives and 4 defective 16-GB flash drives. If a flash drive is picked at random by a quality control inspector, then the probability, or chance, of one of the defective flash drives being chosen is given by the expression $\dfrac{2}{n} + \dfrac{4}{n}$.
(a) Simplify this expression.   (b) Interpret the result.

Solution
(a) Because the denominators are the same, we simply add the numerators.

$$\frac{2}{n} + \frac{4}{n} = \frac{2+4}{n} = \frac{6}{n}$$ Add the numerators.

(b) There are 6 chances in n that a defective flash drive is chosen.

Now Try Exercise 73

7.3   Putting It All Together

CONCEPT	EXPLANATION	EXAMPLES
Addition of Rational Expressions	For polynomials A, B, and C, where C is nonzero, $$\frac{A}{C} + \frac{B}{C} = \frac{A + B}{C}.$$	$\dfrac{x}{x + 1} + \dfrac{1 - x}{x + 1} = \dfrac{x + 1 - x}{x + 1} = \dfrac{1}{x + 1}$ $\dfrac{2x}{x^2 - 1} + \dfrac{x}{x^2 - 1} = \dfrac{2x + x}{x^2 - 1} = \dfrac{3x}{x^2 - 1}$
Subtraction of Rational Expressions	For polynomials A, B, and C, where C is nonzero, $$\frac{A}{C} - \frac{B}{C} = \frac{A - B}{C}.$$ If B consists of more than one term, put parentheses around B and apply the distributive property.	$\dfrac{2x}{x^2 - 4} - \dfrac{x + 2}{x^2 - 4} = \dfrac{2x - (x + 2)}{x^2 - 4}$ $= \dfrac{2x - x - 2}{x^2 - 4}$ $= \dfrac{x - 2}{(x + 2)(x - 2)}$ $= \dfrac{1}{x + 2}$

7.3   Exercises

MyMathLab®

CONCEPTS AND VOCABULARY

1. When adding two rational expressions that have like denominators, add their _____ . The _____ do not change.

2. When subtracting two rational expressions that have like denominators, subtract their _____ . The _____ do not change.

3. $\dfrac{A}{C} + \dfrac{B}{C} =$ _____

4. $\dfrac{A}{C} - \dfrac{B}{C} =$ _____

ADDING AND SUBTRACTING FRACTIONS THAT HAVE LIKE DENOMINATORS

Exercises 5–18: Simplify to lowest terms.

5. $\frac{1}{2} + \frac{1}{2}$

6. $\frac{3}{7} + \frac{2}{7}$

7. $\frac{4}{5} + \frac{2}{5}$

8. $\frac{3}{11} + \frac{5}{11}$

9. $\frac{1}{6} + \frac{5}{6}$

10. $\frac{3}{10} + \frac{5}{10}$

11. $\frac{4}{7} - \frac{1}{7}$

12. $\frac{5}{13} - \frac{7}{13}$

13. $\frac{7}{8} - \frac{3}{8}$

14. $\frac{9}{16} - \frac{5}{16}$

15. $\frac{11}{12} - \frac{5}{12}$

16. $\frac{7}{24} - \frac{3}{24}$

17. $\frac{7}{15} + \frac{4}{15} - \frac{1}{15}$

18. $\frac{11}{36} - \frac{5}{36} + \frac{1}{36}$

ADDING AND SUBTRACTING RATIONAL EXPRESSIONS THAT HAVE LIKE DENOMINATORS

Exercises 19–70: Simplify to lowest terms.

19. $\dfrac{2}{x} + \dfrac{1}{x}$

20. $\dfrac{9}{x} - \dfrac{7}{x}$

21. $\dfrac{7 + 2x}{4x} - \dfrac{7}{4x}$

22. $\dfrac{x - 1}{5x} + \dfrac{2x + 1}{5x}$

23. $\dfrac{y + 3}{y - 3} + \dfrac{2y - 12}{y - 3}$

24. $\dfrac{5 - y}{y + 2} + \dfrac{3y - 1}{y + 2}$

25. $\dfrac{x}{x - 3} + \dfrac{-3}{x - 3}$

26. $\dfrac{2x}{2x + 1} + \dfrac{1}{2x + 1}$

27. $\dfrac{5z}{4z + 3} - \dfrac{z}{4z + 3}$

28. $\dfrac{z}{2z + 1} - \dfrac{1 - z}{2z + 1}$

29. $\dfrac{t + 5}{t + 6} + \dfrac{t + 7}{t + 6}$

30. $\dfrac{4t - 13}{t - 4} + \dfrac{1 - t}{t - 4}$

31. $\dfrac{5x}{2x + 3} - \dfrac{3x - 3}{2x + 3}$

32. $\dfrac{x}{5 - x} - \dfrac{2x - 5}{5 - x}$

33. $\dfrac{x - 4}{x^2 - x} + \dfrac{4}{x^2 - x}$

34. $\dfrac{2x - 2}{4x^2 - 1} + \dfrac{1}{4x^2 - 1}$

35. $\dfrac{z^2 - 1}{z - 2} + \dfrac{3 - 3z}{z - 2}$

36. $\dfrac{x^2 + 2}{x + 1} + \dfrac{3x}{x + 1}$

37. $\dfrac{x^2 + 4x - 1}{4x + 2} - \dfrac{x^2 - 4x - 5}{4x + 2}$

38. $\dfrac{2x^2 - x + 5}{x^2 - 9} - \dfrac{x^2 - x + 14}{x^2 - 9}$

39. $\dfrac{3y}{5} + \dfrac{2y - 5}{5}$

40. $\dfrac{3y - 22}{11} + \dfrac{8y}{11}$

41. $\dfrac{x + y}{4} + \dfrac{x - y}{4}$

42. $\dfrac{x + y}{4} - \dfrac{x - y}{4}$

43. $\dfrac{z^2 + 4}{z - 2} - \dfrac{4z}{z - 2}$

44. $\dfrac{z^2 + 2z}{z + 1} + \dfrac{1}{z + 1}$

45. $\dfrac{2x^2 - 5x}{2x + 1} - \dfrac{3}{2x + 1}$

46. $\dfrac{2x^2}{x + 2} + \dfrac{9x + 10}{x + 2}$

47. $\dfrac{3n}{2n^2 - n + 5} + \dfrac{4n}{2n^2 - n + 5}$

48. $\dfrac{n}{n^2 + n + 1} - \dfrac{1}{n^2 + n + 1}$

49. $\dfrac{1}{x + 3} + \dfrac{2}{x + 3} + \dfrac{3}{x + 3}$

50. $\dfrac{x}{2x - 5} - \dfrac{1}{2x - 5} + \dfrac{2x + 1}{2x - 5}$

51. $\dfrac{8}{ab} + \dfrac{1}{ab}$

52. $\dfrac{6}{xy} + \dfrac{9}{xy}$

53. $\dfrac{x}{(x + y)^2} + \dfrac{y}{(x + y)^2}$

54. $\dfrac{x - 2y}{x^2 - y^2} + \dfrac{y}{x^2 - y^2}$

55. $\dfrac{5}{x - y} + \dfrac{-5}{y - x}$

56. $\dfrac{-4}{y - x} + \dfrac{4}{x - y}$

57. $\dfrac{8}{a - b} + \dfrac{8}{b - a}$

58. $\dfrac{6}{x - y} + \dfrac{6}{y - x}$

59. $\dfrac{a + b}{4a} - \dfrac{a - b}{4a}$

60. $\dfrac{x - y}{5x} - \dfrac{x + 9y}{5x}$

61. $\dfrac{x}{x + y} + \dfrac{y}{x + y}$

62. $\dfrac{x}{x - y} - \dfrac{y}{x - y}$

63. $\dfrac{a^2}{a + b} - \dfrac{b^2}{a + b}$

64. $\dfrac{a^2}{a + b} + \dfrac{2ab + b^2}{a + b}$

65. $\dfrac{2x - 5}{2x^2 + 5x + 2} + \dfrac{6}{2x^2 + 5x + 2}$

66. $\dfrac{4x}{3x^2 + 5x - 2} + \dfrac{2 - 3x}{3x^2 + 5x - 2}$

67. $\dfrac{3x + 7}{3x^2 - 2x - 5} - \dfrac{2x + 6}{3x^2 - 2x - 5}$

68. $\dfrac{5x - 4}{2x^2 - 7x + 6} - \dfrac{3x - 1}{2x^2 - 7x + 6}$

69. $\dfrac{4x^2}{2x + 3y} - \dfrac{9y^2}{2x + 3y}$

70. $\dfrac{x^3}{x^2 + xy + y^2} - \dfrac{y^3}{x^2 + xy + y^2}$

71. **Thinking Generally** If $\dfrac{2}{3 + x} + \dfrac{3}{3 + x}$ equals $\dfrac{5}{10}$, what must be true about x?

72. **Thinking Generally** If $\dfrac{8}{6 + x} - \dfrac{4}{3 + y}$ equals 0, what must be true about x and y?

APPLICATIONS INVOLVING RATIONAL EXPRESSIONS

73. **Quality Control** (Refer to Example 6.) A container holds a total of $n + 1$ batteries. In this container, there are 6 defective AA batteries, 5 defective C batteries, and 3 defective D batteries. If a battery is chosen at random by a quality control inspector, the probability, or chance, of one of the defective batteries being chosen is

$$\dfrac{6}{n + 1} + \dfrac{5}{n + 1} + \dfrac{3}{n + 1}.$$

(a) Simplify this expression.

(b) Evaluate the simplified expression for $n = 99$ and interpret the result.

74. **Intensity of a Light Bulb** The farther a person is from a light bulb, the less intense its light. The equation $I = \dfrac{19}{4d^2}$ approximates the light intensity from a 60-watt light bulb at a distance of d meters, where I is measured in watts per square meter. (*Source:* R. Weidner.)

(a) Find I for $d = 2$ meters and interpret the result.

(b) The intensity of light from a 100-watt light bulb is about $I = \dfrac{32}{4d^2}$. Find an expression for the sum of the intensities of light from a 100-watt bulb and a 60-watt bulb.

75. **Average Cost** The average cost per item in dollars to make x video games is given by

$$M = \dfrac{5x + 1{,}000{,}000}{x}$$

and the average cost per item in dollars to package the same video game is given by

$$P = \dfrac{2x + 500{,}000}{x}.$$

Write an equation C that gives the total cost per item of making and packaging x video games.

76. Total Cost Use the result from the previous exercise to determine the total cost per item when 50,000 video games are made and packaged.

WRITING ABOUT MATHEMATICS

77. Explain how to add two rational expressions with like denominators. Give an example.

78. Explain how to subtract two rational expressions with like denominators. Give an example.

7.4 Addition and Subtraction with Unlike Denominators

Objectives

1 **Finding Least Common Multiples**
   • Using a Step Diagram to Find the LCD

2 **Reviewing Fractions with Unlike Denominators**

3 **Adding and Subtracting Rational Expressions with Unlike Denominators**

4 **Solving an Application Involving Rational Expressions**

1 Finding Least Common Multiples

Connecting Concepts with Your Life Two friends work part-time at a store. The first person works every sixth day, and the second person works every eighth day. If they both work today, how many days will pass before they work on the same day again?

We can answer this question by listing the days that each person works.

Days the First Person Works: 0, 6, 12, 18, **24**, 30, 36, 42, **48**, 54
(every sixth day)

> Both worked today.

> Both will work on day 24.

> Both will work on day 48.

Days the Second Person Works: 0, 8, 16, **24**, 32, 40, **48**, 56, 64, 72
(every eighth day)

After **24** days, the two friends work on the same day. The next time is after **48** days. The numbers 24 and 48 are *common multiples* of 6 and 8. (Find another.) However, 24 is the *least* common multiple (LCM) of 6 and 8 because it is the smallest common multiple.

Another way to find the least common multiple of 6 and 8 is to factor each number into prime numbers.

> Prime factors of 6

> Prime factors of 8

$$6 = 2 \cdot \mathbf{3} \quad \text{and} \quad 8 = \mathbf{2 \cdot 2 \cdot 2}$$

STUDY TIP

The methods discussed earlier in this text for finding the LCM of two numbers can also be applied to find the LCM of two polynomials.

To find the least common multiple, first list each factor the greatest number of times that it occurs in either factorization. Then find the product of these numbers. For this example, the factor 2 occurs three times in the factorization of 8 and only once in the factorization of 6, so list 2 three times. The factor 3 appears only once in the factorization of 6, so list it once:

$$\mathbf{2}, \mathbf{2}, \mathbf{2}, \mathbf{3}.$$

The least common multiple is their product: $\mathbf{2 \cdot 2 \cdot 2 \cdot 3} = 24$.

This same procedure can also be used to find the least common multiple of two or more polynomials.

READING CHECK 1

• What is the first step in finding the least common multiple of two polynomials?

FINDING THE LEAST COMMON MULTIPLE

The least common multiple (LCM) of two or more polynomials can be found as follows.

STEP 1: Factor each polynomial completely.

STEP 2: List each factor the greatest number of times that it occurs in any factorization.

STEP 3: Find the product of this list of factors. The result is the LCM.

EXAMPLE 1   **Finding least common multiples**

Find the least common multiple of each pair of expressions.
(a) $2x, 5x^2$ **(b)** $x^2 - x, x - 1$
(c) $x + 2, x - 3$ **(d)** $x^2 + 2x + 1, x^2 + 3x + 2$

Solution
(a) **STEP 1:** Factor $2x$ and $5x^2$ completely.

$$2x = \mathbf{2} \cdot x \quad \text{and} \quad 5x^2 = \mathbf{5} \cdot x \cdot x$$

STEP 2: In either factorization, the factor 2 occurs at most once, the factor 5 occurs at most once, and the factor x occurs at most twice. The list of factors is $\mathbf{2}, \mathbf{5}, x, x$.
STEP 3: The LCM equals the product

$$\mathbf{2} \cdot \mathbf{5} \cdot x \cdot x = 10x^2.$$

(b) **STEP 1:** Factor $x^2 - x$ and $x - 1$ completely. Note that $x - 1$ cannot be factored.

$$x^2 - x = x(x - 1) \quad \text{and} \quad x - 1 = x - 1$$

STEP 2: Both factors, x and $x - 1$, occur at most once in either factorization. The list of factors is $x, (x - 1)$.
STEP 3: The LCM is the product $x(x - 1)$, or $x^2 - x$.
(c) **STEP 1:** Neither $x + 2$ nor $x - 3$ can be factored.
STEP 2: The list of factors is $(x + 2), (x - 3)$.
STEP 3: The LCM is the product $(x + 2)(x - 3)$, or $x^2 - x - 6$.
(d) **STEP 1:** Factor $x^2 + 2x + 1$ and $x^2 + 3x + 2$ completely.

$$x^2 + 2x + 1 = (x + 1)(x + 1) \quad \text{and} \quad x^2 + 3x + 2 = (x + 1)(x + 2)$$

STEP 2: In either factorization, the factor $(x + 1)$ occurs at most twice and the factor $(x + 2)$ occurs at most once. The list is $(x + 1), (x + 1), (x + 2)$.
STEP 3: The LCM is the product $(x + 1)^2(x + 2)$.

Now Try Exercises 15, 19, 27, 29

USING A STEP DIAGRAM TO FIND THE LCM The least common multiple for two polynomials can be found using a step diagram similar to that used earlier in this text for finding the least common multiple of two numbers. The next example illustrates this process.

EXAMPLE 2   **Using a step diagram to find the LCM**

Use a step diagram to find the LCM of $2x^3 - 8x^2$ and $2x^3 - 2x^2 - 24x$.

Solution
Start by factoring $2x^3 - 8x^2$ and $2x^3 - 2x^2 - 24x$ completely.

$$2x^3 - 8x^2 = 2x^2(x - 4) \quad \text{and} \quad 2x^3 - 2x^2 - 24x = 2x(x + 3)(x - 4)$$

Write the factored form of each expression on the top step of the diagram. We find the expressions in each of the following steps of the diagram by factoring out *any* common factor from the expressions in the previous step. The process continues until no common factor can be found, as shown in **FIGURE 7.3**.

Factor Step Diagram for Finding the LCM

FIGURE 7.3

We find the LCM by multiplying the expressions along the side and at the bottom of the diagram. The LCM is $2 \cdot x \cdot (x - 4) \cdot x \cdot (x + 3) = 2x^2(x - 4)(x + 3)$.

Now Try Exercise 31

2 Reviewing Fractions with Unlike Denominators

Before we can find the sum $\frac{1}{2} + \frac{1}{3}$ by hand, we need to rewrite these fractions by using their least common denominator. The *least common denominator* (LCD) of $\frac{1}{2}$ and $\frac{1}{3}$ corresponds to the *least common multiple* of the denominators 2 and 3, which is 6. As a result, we can rewrite these fractions as

$$\frac{1}{2} \cdot \frac{3}{3} = \frac{3}{6} \quad \text{and} \quad \frac{1}{3} \cdot \frac{2}{2} = \frac{2}{6}.$$

> Multiplying a fraction by 1 in the form of $\frac{2}{2}$ or $\frac{3}{3}$ does not change its value.

Their sum is $\frac{1}{2} + \frac{1}{3} = \frac{3}{6} + \frac{2}{6} = \frac{5}{6}$.

EXAMPLE 3 **Adding and subtracting fractions with unlike denominators**

Simplify each expression.

(a) $\dfrac{3}{10} + \dfrac{4}{15}$ (b) $\dfrac{7}{8} - \dfrac{1}{6}$

Solution

(a) The LCD for $\frac{3}{10}$ and $\frac{4}{15}$ equals the LCM of 10 and 15, which is 30. We rewrite these fractions as

$$\frac{3}{10} \cdot \frac{3}{3} = \frac{9}{30} \quad \text{and} \quad \frac{4}{15} \cdot \frac{2}{2} = \frac{8}{30}.$$

Their sum is

$$\frac{3}{10} + \frac{4}{15} = \frac{9}{30} + \frac{8}{30} = \frac{17}{30}.$$

(b) The LCD for $\frac{7}{8}$ and $\frac{1}{6}$ is the LCM of 8 and 6, which is 24. We rewrite these fractions as

$$\frac{7}{8} \cdot \frac{3}{3} = \frac{21}{24} \quad \text{and} \quad \frac{1}{6} \cdot \frac{4}{4} = \frac{4}{24}.$$

Their difference is

$$\frac{7}{8} - \frac{1}{6} = \frac{21}{24} - \frac{4}{24} = \frac{17}{24}.$$

Now Try Exercises 33, 35

---

3 Adding and Subtracting Rational Expressions with Unlike Denominators

The first step in adding or subtracting rational expressions with unlike denominators is to rewrite each expression by using the least common denominator. Then the sum or difference can be found by using the techniques discussed in Section 3 of this chapter.

NOTE: The LCD of two or more rational expressions equals the LCM of their denominators. ■

To add or subtract rational expressions, we often rewrite a rational expression with a different denominator. This technique is demonstrated in the next example and is used in future examples.

EXAMPLE 4 **Rewriting rational expressions**

Rewrite each rational expression so it has the given denominator D.

(a) $\dfrac{3}{2x}$, $D = 8x^2$ (b) $\dfrac{1}{x + 1}$, $D = x^2 - 1$

Solution

(a) We need to write $\frac{3}{2x}$ so that it is equivalent to $\frac{?}{8x^2}$.

$\times\, 4x$

Since we multiply $2x$ by **$4x$** to get $8x^2$, we multiply 3 by **$4x$** also.

$$\frac{3}{2x} = \frac{?}{8x^2}$$

$\times\, 4x$

Because $8x^2 = 2x \cdot 4x$, we can multiply $\frac{3}{2x}$ by 1 in the form $\frac{4x}{4x}$ as follows.

$$\frac{3}{2x} \cdot \frac{4x}{4x} = \frac{12x}{8x^2} \qquad \text{Multiply rational expressions.}$$

(b) We must write $\frac{1}{x+1}$ so that it is equivalent to $\frac{?}{x^2-1}$.

$\times\, (x-1)$

Since we multiply $(x+1)$ by $(x-1)$ to get (x^2-1), we multiply 1 by $(x-1)$ also.

$$\frac{1}{x+1} = \frac{?}{x^2-1}$$

$\times\, (x-1)$

Because $x^2 - 1 = (x+1)(x-1)$, we can multiply $\frac{1}{x+1}$ by 1 in the form $\frac{x-1}{x-1}$ as follows.

$$\frac{1}{x+1} \cdot \frac{x-1}{x-1} = \frac{x-1}{x^2-1} \qquad \text{Multiply rational expressions.}$$

Now Try Exercises 45, 47

EXAMPLE 5   **Adding rational expressions with unlike denominators**

Find each sum and leave your answer in factored form.

(a) $\dfrac{5}{8y} + \dfrac{7}{4y^2}$ **(b)** $\dfrac{1}{x-1} + \dfrac{1}{x+1}$ **(c)** $\dfrac{x}{x^2+2x+1} + \dfrac{1}{x+1}$

Solution

(a) First find the LCM of $8y$ and $4y^2$.

$$8y = 2 \cdot 2 \cdot 2 \cdot y \quad \text{and} \quad 4y^2 = 2 \cdot 2 \cdot y \cdot y$$

Thus the LCM is $2 \cdot 2 \cdot 2 \cdot y \cdot y = 8y^2$. Now, because

$$8y^2 = 8y \cdot y \quad \text{and} \quad 8y^2 = 4y^2 \cdot 2,$$

we multiply the first expression by $\frac{y}{y}$ and the second expression by $\frac{2}{2}$.

$$\frac{5}{8y} + \frac{7}{4y^2} = \frac{5}{8y} \cdot \frac{y}{y} + \frac{7}{4y^2} \cdot \frac{2}{2} \qquad \text{Rewrite by using the LCD.}$$

The LCD is $8y^2$.

$$= \frac{5y}{8y^2} + \frac{14}{8y^2} \qquad \text{Multiply rational expressions.}$$

$$= \frac{5y+14}{8y^2} \qquad \text{Add the numerators.}$$

(b) The LCM for $x-1$ and $x+1$ is their product, $(x-1)(x+1)$.

$$\frac{1}{x-1} + \frac{1}{x+1} = \frac{1}{x-1} \cdot \frac{x+1}{x+1} + \frac{1}{x+1} \cdot \frac{x-1}{x-1} \qquad \text{Rewrite by using the LCD.}$$

The LCD is $(x-1)(x+1)$.

$$= \frac{x+1}{(x-1)(x+1)} + \frac{x-1}{(x+1)(x-1)} \qquad \text{Multiply rational expressions.}$$

$$= \frac{x + 1 + x - 1}{(x - 1)(x + 1)}$$ Add the numerators.

$$= \frac{2x}{(x - 1)(x + 1)}$$ Simplify the numerator.

(c) First find the LCM for $x^2 + 2x + 1$ and $x + 1$. Because

$$x^2 + 2x + 1 = (x + 1)(x + 1),$$

their LCM is $(x + 1)(x + 1) = (x + 1)^2$.

$$\frac{x}{x^2 + 2x + 1} + \frac{1}{x + 1} = \frac{x}{(x + 1)^2} + \frac{1}{x + 1} \cdot \frac{x + 1}{x + 1}$$ Rewrite by using the LCD.

The LCD is $(x + 1)^2$.

$$= \frac{x}{(x + 1)^2} + \frac{x + 1}{(x + 1)^2}$$ Multiply rational expressions.

$$= \frac{2x + 1}{(x + 1)^2}$$ Add the numerators.

Now Try Exercises 53, 65, 71

Subtraction of rational expressions is performed in a manner similar to addition and is illustrated in the next example.

EXAMPLE 6   **Subtracting rational expressions with unlike denominators**

Simplify each expression. Write your answer in lowest terms and leave it in factored form.

(a) $\dfrac{5}{z} - \dfrac{z}{z - 1}$ **(b)** $\dfrac{5}{x + 1} - \dfrac{1}{x^2 - 1}$

(c) $\dfrac{x}{x^2 - 2x} - \dfrac{1}{x^2 + 2x}$ **(d)** $\dfrac{3}{x - 1} - \dfrac{3}{x^2 - x} + \dfrac{5}{x}$

Solution

(a) The LCD is $z(z - 1)$.

$$\frac{5}{z} - \frac{z}{z - 1} = \frac{5}{z} \cdot \frac{z - 1}{z - 1} - \frac{z}{z - 1} \cdot \frac{z}{z}$$ Rewrite by using the LCD.

$$= \frac{5(z - 1)}{z(z - 1)} - \frac{z^2}{z(z - 1)}$$ Multiply rational expressions.

$$= \frac{5(z - 1) - z^2}{z(z - 1)}$$ Subtract the numerators.

$$= \frac{-z^2 + 5z - 5}{z(z - 1)}$$ Simplify the numerator.

(b) The LCD is $(x - 1)(x + 1)$. Note that $x^2 - 1 = (x - 1)(x + 1)$.

$$\frac{5}{x + 1} - \frac{1}{x^2 - 1} = \frac{5}{x + 1} \cdot \frac{x - 1}{x - 1} - \frac{1}{(x - 1)(x + 1)}$$ Rewrite by using the LCD.

$$= \frac{5(x - 1) - 1}{(x - 1)(x + 1)}$$ Subtract the numerators.

$$= \frac{5x - 5 - 1}{(x - 1)(x + 1)}$$ Distributive property

$$= \frac{5x - 6}{(x - 1)(x + 1)}$$ Simplify the numerator.

(c) Start by factoring each denominator in $\frac{x}{x^2 - 2x} - \frac{1}{x^2 + 2x}$. Because

$$x^2 - 2x = x(x - 2) \quad \text{and} \quad x^2 + 2x = x(x + 2),$$

the LCD is $x(x - 2)(x + 2)$.

$$\frac{x}{x^2 - 2x} - \frac{1}{x^2 + 2x} = \frac{x}{x(x - 2)} \cdot \frac{x + 2}{x + 2} - \frac{1}{x(x + 2)} \cdot \frac{x - 2}{x - 2} \qquad \text{Rewrite by using the LCD.}$$

$$= \frac{x(x + 2)}{x(x - 2)(x + 2)} - \frac{x - 2}{x(x - 2)(x + 2)} \qquad \text{Multiply rational expressions.}$$

$$= \frac{x(x + 2) - (x - 2)}{x(x - 2)(x + 2)} \qquad \boxed{\text{Insert parentheses.}} \quad \text{Subtract the numerators.}$$

$$= \frac{x^2 + 2x - x + 2}{x(x - 2)(x + 2)} \qquad \text{Distributive property}$$

$$= \frac{x^2 + x + 2}{x(x - 2)(x + 2)} \qquad \text{Simplify the numerator.}$$

(d) The given expression $\frac{3}{x - 1} - \frac{3}{x^2 - x} + \frac{5}{x}$ contains three rational expressions. Begin by finding the LCM of the three denominators: $x - 1, x^2 - x,$ and x. Because $x^2 - x = x(x - 1)$, the LCM is $x(x - 1)$.

$$\frac{3}{x - 1} - \frac{3}{x^2 - x} + \frac{5}{x} = \frac{3}{x - 1} \cdot \frac{x}{x} - \frac{3}{x(x - 1)} + \frac{5}{x} \cdot \frac{x - 1}{x - 1} \qquad \text{Rewrite by using the LCD.}$$

$$= \frac{3x}{x(x - 1)} - \frac{3}{x(x - 1)} + \frac{5(x - 1)}{x(x - 1)} \qquad \text{Multiply rational expressions.}$$

$$= \frac{3x - 3 + 5x - 5}{x(x - 1)} \qquad \text{Combine the expressions.}$$

$$= \frac{8x - 8}{x(x - 1)} \qquad \text{Simplify the numerator.}$$

$$= \frac{8(x - 1)}{x(x - 1)} \qquad \text{Factor the numerator.}$$

$$= \frac{8}{x} \qquad \text{Simplify to lowest terms.}$$

Now Try Exercises 63, 77, 79, 97

4 Solving an Application Involving Rational Expressions

Parallel Wiring

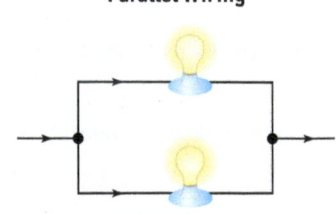

FIGURE 7.4

🌐 *Math in Context* *Electrical* Sums of rational expressions occur in applications involving electricity. Suppose that two light bulbs are wired in parallel so that electricity can flow through either light bulb, as illustrated in **FIGURE 7.4**. If the individual resistances of the light bulbs are R and S, then the total resistance of the circuit is found by adding the expression

$$\frac{1}{R} + \frac{1}{S},$$

and then taking the *reciprocal*. (*Source:* R. Weidner and R. Sells, *Elementary Classical Physics*, vol. 2.)

EXAMPLE 7   **Modeling electrical resistance**

Add $\frac{1}{R} + \frac{1}{S}$, and then find the reciprocal of the result.

Solution
The LCD for $\frac{1}{R}$ and $\frac{1}{S}$ is **RS**.

$$\frac{1}{R} + \frac{1}{S} = \frac{1}{R} \cdot \frac{S}{S} + \frac{1}{S} \cdot \frac{R}{R} \qquad \text{Rewrite by using the LCD.}$$

$$= \frac{S}{RS} + \frac{R}{RS} \qquad \text{Multiply rational expressions.}$$

$$= \frac{S + R}{RS} \qquad \text{Add the numerators.}$$

CRITICAL THINKING 1

Find the reciprocal of the sum $x + \frac{1}{x}$.

In general, the reciprocal of $\frac{a}{b}$ is $\frac{b}{a}$, so the reciprocal of $\frac{S + R}{RS}$ is

$$\frac{RS}{S + R}.$$

This final expression can be used to find the total resistance of the circuit.

Now Try Exercise 101

7.4   Putting It All Together

CONCEPT	EXPLANATION	EXAMPLES
Least Common Multiple (LCM) of Polynomials	1. Factor each polynomial completely. 2. List each factor the greatest number of times that it occurs in any factorization. 3. The LCM is the product of this list.	For $x^2 - 2x$ and $x^2 - 4x + 4$, 1. $x^2 - 2x = x(x - 2)$ $x^2 - 4x + 4 = (x - 2)(x - 2)$ 2. $x, (x - 2), (x - 2)$ 3. LCM $= x(x - 2)(x - 2)$.
Least Common Denominator (LCD)	The LCD of two or more rational expressions equals the least common multiple (LCM) of their denominators.	The LCD of $\frac{2}{x^2 - 2x}$ and $\frac{3}{x^2 - 4x + 4}$ is $x(x - 2)(x - 2)$ because the LCM of $x^2 - 2x$ and $x^2 - 4x + 4$ is $x(x - 2)(x - 2)$, as shown above.
Addition and Subtraction of Rational Expressions with Unlike Denominators	First rewrite each expression by using the LCD. Then add or subtract the expressions. Finally, write your answer in lowest terms. It is often helpful to leave the result in factored form.	The LCM of x and $x + 2$ is $x(x + 2)$. $\frac{2}{x} + \frac{5}{x + 2} = \frac{2}{x} \cdot \frac{x + 2}{x + 2} + \frac{5}{x + 2} \cdot \frac{x}{x}$ $= \frac{2x + 4}{x(x + 2)} + \frac{5x}{x(x + 2)}$ $= \frac{7x + 4}{x(x + 2)}$

7.4 Exercises

MyMathLab®

CONCEPTS AND VOCABULARY

1. Give a common multiple of 6 and 9 that is not the *least* common multiple.

2. The LCM of x and y is ___ .

3. To rewrite $\frac{3}{4}$ with denominator 12, multiply $\frac{3}{4}$ by 1 written as the fraction ___ .

4. To rewrite $\frac{4}{x-1}$ with denominator $x^2 - 1$, multiply $\frac{4}{x-1}$ by 1 written as the rational expression ___ .

FINDING LEAST COMMON MULTIPLES

Exercises 5–12: Find the least common multiple.

5. 4, 6

6. 6, 9

7. 2, 3

8. 5, 4

9. 10, 15

10. 8, 12

11. 24, 36

12. 32, 40

Exercises 13–32: Find the least common multiple. Leave your answer in factored form.

13. $4x, 6x$

14. $6x, 9x$

15. $5x, 10x^2$

16. $4x^2, 12x$

17. $x, x + 1$

18. $4x, x - 1$

19. $2x + 1, x + 3$

20. $5x + 3, x + 9$

21. $x^2 - x, x^2 + x$

22. $x^2 + 2x, x^2$

23. $(x - 8)^2, (x - 8)(x + 1)$

24. $(2x - 1)^3, (2x - 1)(x + 3)$

25. $4x^2 - 1, 2x + 1$

26. $x^2 + 4x + 3, x + 3$

27. $x^2 - 1, x + 1$

28. $x^2 - 4, x - 2$

29. $2x^2 + 7x + 6, x^2 + 5x + 6$

30. $x^2 - 3x + 2, x^2 + 2x - 3$

31. $3y^2 + 6y, 3y^3 + 3y^2 - 6y$

32. $y^2 + 3y + 2, y^4 - 4y^2$

REVIEWING FRACTIONS WITH UNLIKE DENOMINATORS

Exercises 33–40: Evaluate and simplify. Write your answer in lowest terms.

33. $\frac{4}{5} + \frac{1}{2}$

34. $\frac{3}{8} + \frac{1}{4}$

35. $\frac{5}{9} - \frac{1}{3}$

36. $\frac{7}{10} - \frac{3}{15}$

37. $\frac{4}{25} + \frac{2}{5}$

38. $\frac{7}{9} - \frac{1}{3}$

39. $\frac{1}{5} + \frac{3}{4} - \frac{1}{2}$

40. $\frac{6}{7} - \frac{8}{9} + \frac{2}{3}$

ADDING AND SUBTRACTING RATIONAL EXPRESSIONS WITH UNLIKE DENOMINATORS

Exercises 41–52: (Refer to Example 4.) Rewrite the rational expression so it has the given denominator D.

41. $\frac{1}{3}, D = 9$

42. $\frac{3}{4}, D = 24$

43. $\frac{5}{7}, D = 21$

44. $\frac{4}{5}, D = 30$

45. $\frac{1}{4x}, D = 8x^3$

46. $\frac{5}{3x}, D = 9x^2$

47. $\frac{1}{x + 2}, D = x^2 - 4$

48. $\frac{3}{x - 3}, D = x^2 - 9$

49. $\frac{1}{x + 1}, D = x^2 + x$

50. $\frac{3}{x - 3}, D = x^2 - 3x$

51. $\frac{2x}{x + 1}, D = x^2 + 2x + 1$

52. $\frac{x}{2x - 1}, D = 2x^2 + 11x - 6$

Exercises 53–100: Simplify the expression. Write your answer in lowest terms and leave it in factored form.

53. $\frac{1}{3x} + \frac{3}{4x}$

54. $\frac{4}{2x^2} + \frac{7}{3x}$

55. $\frac{5}{z^2} - \frac{7}{z^3}$

56. $\frac{8}{z} - \frac{3}{2z}$

57. $\frac{1}{x} - \frac{1}{y}$

58. $\frac{1}{xy} - \frac{4}{y}$

59. $\frac{a}{b} + \frac{b}{a}$

60. $\frac{3}{x} - \frac{4}{y}$

61. $\frac{1}{2x + 4} + \frac{3}{x + 2}$

62. $\frac{1}{5x - 10} - \frac{x}{x - 2}$

63. $\frac{2}{t - 2} - \frac{1}{t}$

64. $\frac{7}{2t} + \frac{1}{t + 5}$

65. $\frac{5}{n - 1} + \frac{n}{n + 1}$

66. $\frac{4n}{3n - 2} + \frac{n}{n + 1}$

67. $\dfrac{3}{x-3} + \dfrac{6}{3-x}$

68. $\dfrac{x}{x-8} + \dfrac{x}{8-x}$

69. $\dfrac{1}{5k-1} + \dfrac{1}{1-5k}$

70. $\dfrac{4}{4-3k} + \dfrac{3k}{3k-4}$

71. $\dfrac{2x}{(x-1)^2} + \dfrac{4}{x-1}$

72. $\dfrac{5}{(x+5)} - \dfrac{x}{(x+5)^2}$

73. $\dfrac{2y}{y(2y-1)} + \dfrac{1}{2y-1}$

74. $\dfrac{5y}{y(y+1)} - \dfrac{5}{y+1}$

75. $\dfrac{1}{x+2} - \dfrac{1}{x^2+2x}$

76. $\dfrac{1}{x-3} - \dfrac{2}{x^2-3x}$

77. $\dfrac{3}{x-2} - \dfrac{1}{x^2-4}$

78. $\dfrac{x}{9-x^2} - \dfrac{1}{3-x}$

79. $\dfrac{2}{x^2-3x} - \dfrac{1}{x^2+3x}$

80. $\dfrac{3}{x^2+4x} - \dfrac{2}{x^2-4x}$

81. $\dfrac{1}{x-2} - \dfrac{1}{x+2} + \dfrac{1}{x}$

82. $\dfrac{1}{x^2} - \dfrac{2}{x} + \dfrac{2}{x-1}$

83. $\dfrac{x}{x^2+4x+4} + \dfrac{1}{x+2}$

84. $\dfrac{1}{x^2-3x-4} - \dfrac{1}{x+1}$

85. $\dfrac{x}{(x+1)(x+2)} - \dfrac{1}{(x+2)(x+3)}$

86. $\dfrac{2x}{(x-1)(x-2)} - \dfrac{5}{x-2}$

87. $\dfrac{1}{a+b} - \dfrac{1}{a-b}$

88. $\dfrac{x}{x^2-y^2} - \dfrac{1}{x+y}$

89. $\dfrac{r}{r-t} + \dfrac{t}{t-r} - 1$

90. $\dfrac{1}{a-b} - \dfrac{a}{a^2-b^2}$

91. $\dfrac{1}{x} + \dfrac{2}{x^2-2x} + \dfrac{5}{x-2}$

92. $\dfrac{1}{b} + \dfrac{1}{b+1} + \dfrac{1}{b+2}$

93. $\dfrac{2}{x-y} + \dfrac{3}{y-x} + \dfrac{1}{x-y}$

94. $\dfrac{a}{a-b} + \dfrac{b}{b-a} + \dfrac{3}{b}$

95. $\dfrac{3}{x-3} - \dfrac{3}{x^2-3x} - \dfrac{6}{x(x-3)}$

96. $\dfrac{3}{2a-4} + \dfrac{5}{2a} - \dfrac{3}{a^2-2a}$

97. $x + \dfrac{1}{x-1} - \dfrac{1}{x+1}$

98. $\dfrac{x}{x^2-4} - \dfrac{1}{x^2+4x+4}$

99. $\dfrac{2x+1}{x-1} - \dfrac{3}{x+1} + \dfrac{x}{x-1}$

100. $\dfrac{1}{x-3} - \dfrac{2}{x+3} + \dfrac{x}{x^2-9}$

APPLICATIONS INVOLVING RATIONAL EXPRESSIONS

101. Electricity In Example 7, we demonstrated that the expressions

$$\frac{1}{R} + \frac{1}{S} \quad \text{and} \quad \frac{S+R}{RS}$$

are equivalent. Evaluate both expressions by using $R = 120$ and $S = 200$. Are your answers the same?

102. Intensity of a Light Bulb The formula $I = \dfrac{32}{4d^2}$ approximates the intensity of light from a 100-watt light bulb at a distance of d meters, where I is in watts per square meter. For light from a 40-watt bulb, the equation for intensity becomes $I = \dfrac{16}{5d^2}$. (*Source:* R. Weidner.)

(a) Find an expression for the sum of the intensities of light from the two light bulbs.

(b) Find the combined intensity of their light at $d = 5$ meters.

103. Photography A lens in a camera has a focal length, which is important for focusing the camera. If an object is at a distance D from a lens that has a focal length F, then to be in focus, the distance S between the lens and the film should satisfy the equation

$$\frac{1}{S} = \frac{1}{F} - \frac{1}{D},$$

as illustrated in the figure. Write the difference $\frac{1}{F} - \frac{1}{D}$ as one term.

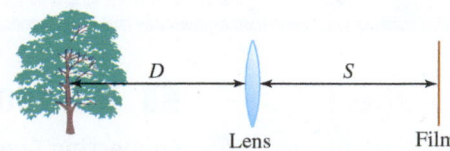

104. Geometry Find the sum of the areas of the two rectangles shown in the figure. Write your answer in factored form.

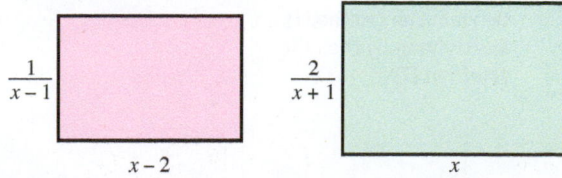

105. Planet Alignment Suppose that Saturn and Jupiter are in alignment with respect to the sun. If Jupiter orbits the sun every 12 years and Saturn orbits the sun every 30 years, how many years will it be before these two planets are back in the same location with the same alignment again?

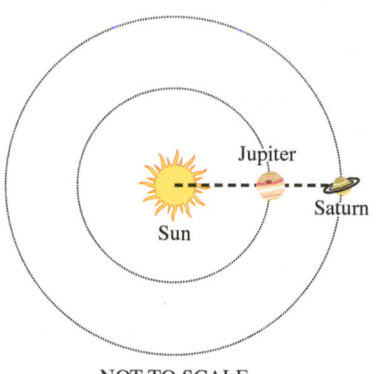

NOT TO SCALE

106. Working Shifts Two friends work part-time at a store. That first person works every sixth day and the second person works every tenth day. If they are both working today, how many days pass before they both work on the same day again?

WRITING ABOUT MATHEMATICS

107. Explain how to find the least common multiple of two polynomials.

108. Explain how to subtract two rational expressions with unlike denominators.

SECTIONS 7.3 and 7.4   Checking Basic Concepts

1. Simplify each expression.

(a) $\dfrac{x}{x+2} + \dfrac{2}{x+2}$ (b) $\dfrac{2}{3x} - \dfrac{x}{3x}$

(c) $\dfrac{z^2+z}{z+2} + \dfrac{z}{z+2}$

2. Find the least common multiple of each pair of expressions.

(a) $3x, 5x$ (b) $4x, x^2 + x$

(c) $x+1, x-1$

3. Simplify each expression.

(a) $\dfrac{1}{x+1} + \dfrac{5}{x}$

(b) $\dfrac{5}{x-3} + \dfrac{1}{3-x}$

(c) $\dfrac{-4}{4x+2} - \dfrac{x+2}{2x+1}$

4. Simplify the expression $\dfrac{a}{a-b} - \dfrac{b}{a+b}$.

7.5   Complex Fractions

Objectives

1 Introducing Complex Fractions

2 Simplifying Complex Fractions
- Simplifying the Numerator and Denominator (Method I)
- Multiplying by the LCD (Method II)

1 Introducing Complex Fractions

Connecting Concepts with Your Life If two and a half pizzas are cut so that each piece equals one-eighth of a pizza, then there are 20 pieces of pizza. The quotient two and a half divided by one-eighth can be written as

$$\dfrac{2\frac{1}{2}}{\frac{1}{8}} \quad \text{or} \quad \dfrac{2+\frac{1}{2}}{\frac{1}{8}}.$$

Complex fractions

NEW VOCABULARY

☐ Complex fraction
☐ Basic complex fraction

These expressions are examples of *complex* fractions. Typically, we want to simplify a complex fraction by rewriting it as a standard fraction in the form $\frac{a}{b}$.

A **complex fraction** is a rational expression that contains fractions in its numerator, denominator, or both. Examples of complex fractions include

$$\frac{1 + \frac{1}{2}}{1 - \frac{1}{3}}, \quad \frac{5z}{\frac{7}{z} - \frac{4}{2z}}, \quad \text{and} \quad \frac{\frac{2x}{3} + \frac{1}{x}}{x - \frac{1}{x+1}}.$$

Complex fractions

STUDY TIP

Two methods for simplifying complex fractions are discussed in this section. If your instructor does not require you to use a specific method, choose the one that works best for you.

The expression $\frac{a}{b} \div \frac{c}{d}$ can be written as a **basic complex fraction**, where both the numerator and denominator of the complex fraction are single fractions.

$$\frac{\frac{a}{b}}{\frac{c}{d}}$$

Basic complex fraction

Invert $\frac{c}{d}$ to $\frac{d}{c}$ and multiply.

Because

$$\frac{a}{b} \div \frac{c}{d} = \frac{a}{b} \cdot \frac{d}{c},$$

we can simplify basic complex fractions by multiplying the numerator and the reciprocal of the denominator. This strategy is summarized as follows.

READING CHECK 1

• Does the complex fraction $\frac{\frac{a}{b}}{\frac{c}{d}}$ equal $\frac{ad}{bc}$?

> **SIMPLIFYING BASIC COMPLEX FRACTIONS**
>
> For any real numbers a, b, c, and d,
>
> $$\frac{\frac{a}{b}}{\frac{c}{d}} = \frac{a}{b} \cdot \frac{d}{c},$$
>
> where b, c, and d are nonzero.

2   Simplifying Complex Fractions

There are two methods of simplifying a complex fraction. Method I is to simplify both the numerator and the denominator and then divide the resulting two fractions. Method II is to multiply the numerator and denominator by the least common denominator of all fractions within the complex fraction.

SIMPLIFYING THE NUMERATOR AND DENOMINATOR (METHOD I) The following steps outline Method I for simplifying complex fractions.

> **SIMPLIFYING COMPLEX FRACTIONS (METHOD I)**
>
> To simplify a complex fraction, perform the following steps.
>
> **STEP 1:** Write the numerator as a single fraction; write the denominator as a single fraction.
>
> **STEP 2:** Divide the denominator into the numerator by multiplying the numerator and the reciprocal of the denominator. (Invert and multiply.)
>
> **STEP 3:** Simplify the result to lowest terms.

READING CHECK 2

• What is the first step when using Method I to simplify complex fractions?

Sometimes a complex fraction already has both its numerator and denominator written as single fractions. As a result, we can start with Step 2. This situation is shown in Example 1.

EXAMPLE 1 **Simplifying basic complex fractions (Method I)**

Simplify each basic complex fraction.

(a) $\dfrac{\frac{3}{4}}{\frac{9}{8}}$ (b) $\dfrac{3\frac{3}{4}}{1\frac{1}{2}}$ (c) $\dfrac{\frac{x}{4}}{\frac{3}{2y}}$ (d) $\dfrac{\frac{(x-1)^2}{4}}{\frac{x-1}{8}}$

Solution

(a) We can skip Step 1. To simplify $\frac{3}{4} \div \frac{9}{8}$, multiply $\frac{3}{4}$ by the reciprocal of $\frac{9}{8}$, or $\frac{8}{9}$.

$$\dfrac{\frac{3}{4}}{\frac{9}{8}} = \frac{3}{4} \cdot \frac{8}{9} \qquad \text{Multiply by the reciprocal of } \tfrac{9}{8} \text{ (Step 2).}$$

Invert $\frac{9}{8}$ to $\frac{8}{9}$ and multiply.

$$= \frac{24}{36} \qquad \text{Multiply the fractions.}$$

$$= \frac{2}{3} \qquad \text{Simplify (Step 3).}$$

(b) Start by writing $3\frac{3}{4}$ and $1\frac{1}{2}$ as improper fractions.

$$\dfrac{3\frac{3}{4}}{1\frac{1}{2}} = \dfrac{\frac{15}{4}}{\frac{3}{2}} \qquad \text{Write as improper fractions (Step 1).}$$

Invert $\frac{3}{2}$ to $\frac{2}{3}$ and multiply.

$$= \frac{15}{4} \cdot \frac{2}{3} \qquad \text{Multiply by the reciprocal of } \tfrac{3}{2} \text{ (Step 2).}$$

$$= \frac{5}{2} \qquad \text{Multiply and simplify (Step 3).}$$

(c) We can skip Step 1. To simplify $\frac{x}{4} \div \frac{3}{2y}$, multiply $\frac{x}{4}$ by the reciprocal of $\frac{3}{2y}$, or $\frac{2y}{3}$.

$$\dfrac{\frac{x}{4}}{\frac{3}{2y}} = \frac{x}{4} \cdot \frac{2y}{3} \qquad \text{Multiply by the reciprocal of } \tfrac{3}{2y} \text{ (Step 2).}$$

Invert $\frac{3}{2y}$ to $\frac{2y}{3}$ and multiply.

$$= \frac{2xy}{12} \qquad \text{Multiply the fractions.}$$

$$= \frac{xy}{6} \qquad \text{Simplify (Step 3).}$$

(d) We can skip Step 1. Multiply $\frac{(x-1)^2}{4}$ by the reciprocal of $\frac{x-1}{8}$, or $\frac{8}{x-1}$.

$$\dfrac{\frac{(x-1)^2}{4}}{\frac{x-1}{8}} = \frac{(x-1)^2}{4} \cdot \frac{8}{x-1} \qquad \text{Multiply by the reciprocal of } \tfrac{x-1}{8} \text{ (Step 2).}$$

Invert $\frac{x-1}{8}$ to $\frac{8}{x-1}$ and multiply.

$$= \frac{8(x-1)(x-1)}{4(x-1)} \qquad \text{Multiply the fractions.}$$

$$= 2(x-1) \qquad \text{Simplify (Step 3).}$$

Now Try Exercises 13, 15, 19, 29

EXAMPLE 2 **Simplifying complex fractions (Method I)**

Simplify. Write your answer in lowest terms.

(a) $\dfrac{\dfrac{1}{a}+\dfrac{1}{b}}{\dfrac{1}{a}-\dfrac{1}{b}}$ (b) $\dfrac{x-\dfrac{4}{x}}{x+\dfrac{4}{x}}$ (c) $\dfrac{\dfrac{1}{x}+\dfrac{2}{x-1}}{\dfrac{2}{x}-\dfrac{1}{x-1}}$ (d) $\dfrac{\dfrac{1}{x}+\dfrac{1}{y}}{\dfrac{1}{x^2}-\dfrac{1}{y^2}}$

Solution
(a) In Step 1, write the numerator as one fraction by using the LCD, ab.

$$\frac{1}{a}+\frac{1}{b}=\frac{1}{a}\cdot\frac{b}{b}+\frac{1}{b}\cdot\frac{a}{a}=\frac{b}{ab}+\frac{a}{ab}=\frac{b+a}{ab}$$

Write the denominator as one fraction by using the LCD, ab.

$$\frac{1}{a}-\frac{1}{b}=\frac{1}{a}\cdot\frac{b}{b}-\frac{1}{b}\cdot\frac{a}{a}=\frac{b}{ab}-\frac{a}{ab}=\frac{b-a}{ab}$$

Finally, use these results to simplify the given complex fraction.

$$\frac{\dfrac{1}{a}+\dfrac{1}{b}}{\dfrac{1}{a}-\dfrac{1}{b}}=\frac{\dfrac{b+a}{ab}}{\dfrac{b-a}{ab}}$$ Simplify the numerator and the denominator (Step 1).

$$=\frac{b+a}{ab}\cdot\frac{ab}{b-a}$$ Multiply by the reciprocal of $\frac{b-a}{ab}$ (Step 2).

$$=\frac{ab(b+a)}{ab(b-a)}$$ Multiply.

$$=\frac{b+a}{b-a}$$ Simplify (Step 3).

(b) The LCD for both the numerator and the denominator is x.

$$\frac{x-\dfrac{4}{x}}{x+\dfrac{4}{x}}=\frac{\dfrac{x^2}{x}-\dfrac{4}{x}}{\dfrac{x^2}{x}+\dfrac{4}{x}}$$ Write by using the LCD (Step 1).

$$=\frac{\dfrac{x^2-4}{x}}{\dfrac{x^2+4}{x}}$$ Combine terms.

$$=\frac{x^2-4}{x}\cdot\frac{x}{x^2+4}$$ Multiply by the reciprocal of $\frac{x^2+4}{x}$ (Step 2).

$$=\frac{x(x^2-4)}{x(x^2+4)}$$ Multiply.

$$=\frac{x^2-4}{x^2+4}$$ Simplify (Step 3).

(c) The LCD for the numerator and the denominator is $x(x - 1)$.

$$\dfrac{\dfrac{1}{x} + \dfrac{2}{x - 1}}{\dfrac{2}{x} - \dfrac{1}{x - 1}} = \dfrac{\dfrac{x - 1}{x(x - 1)} + \dfrac{2x}{x(x - 1)}}{\dfrac{2(x - 1)}{x(x - 1)} - \dfrac{x}{x(x - 1)}} \qquad \text{Write by using the LCD (Step 1).}$$

$$= \dfrac{\dfrac{x - 1 + 2x}{x(x - 1)}}{\dfrac{2(x - 1) - x}{x(x - 1)}} \qquad \text{Combine terms.}$$

$$= \dfrac{\dfrac{3x - 1}{x(x - 1)}}{\dfrac{x - 2}{x(x - 1)}} \qquad \text{Simplify.}$$

$$= \dfrac{3x - 1}{x(x - 1)} \cdot \dfrac{x(x - 1)}{x - 2} \qquad \begin{array}{l} \text{Multiply by the reciprocal of } \frac{x - 2}{x(x - 1)} \\ \text{(Step 2).} \end{array}$$

$$= \dfrac{3x - 1}{x - 2} \qquad \text{Multiply and simplify (Step 3).}$$

(d) The LCD for the numerator is xy, and the LCD for the denominator is $x^2 y^2$.

$$\dfrac{\dfrac{1}{x} + \dfrac{1}{y}}{\dfrac{1}{x^2} - \dfrac{1}{y^2}} = \dfrac{\dfrac{y}{xy} + \dfrac{x}{xy}}{\dfrac{y^2}{x^2 y^2} - \dfrac{x^2}{x^2 y^2}} \qquad \text{Write by using the LCD (Step 1).}$$

$$= \dfrac{\dfrac{y + x}{xy}}{\dfrac{y^2 - x^2}{x^2 y^2}} \qquad \text{Combine terms.}$$

$$= \dfrac{y + x}{xy} \cdot \dfrac{x^2 y^2}{y^2 - x^2} \qquad \begin{array}{l} \text{Multiply by the reciprocal of } \frac{y^2 - x^2}{x^2 y^2} \\ \text{(Step 2).} \end{array}$$

$$= \dfrac{x^2 y^2 (y + x)}{xy(y^2 - x^2)} \qquad \text{Multiply.}$$

$$= \dfrac{x^2 y^2 (y + x)}{xy(y - x)(y + x)} \qquad \text{Factor the denominator (Step 3).}$$

$$= \dfrac{xy}{y - x} \qquad \text{Simplify.}$$

Now Try Exercises 31, 37, 41, 45

MULTIPLYING BY THE LCD (METHOD II) In Method II for simplifying a complex fraction, we multiply the numerator and denominator of the complex fraction by the least common denominator of all the fractions within the complex fraction.

The following steps outline Method II for simplifying complex fractions.

READING CHECK 3

- When using Method II, by what do we multiply the numerator and denominator of the complex fraction?

SIMPLIFYING COMPLEX FRACTIONS (METHOD II)

To simplify a complex fraction, perform the following steps.

STEP 1: Find the LCD of all fractions within the complex fraction.

STEP 2: Multiply the numerator and the denominator of the complex fraction by the LCD.

STEP 3: Simplify the result to lowest terms.

We can use this method to simplify the complex fraction in Example 2(a). Because the LCD for the numerator and the denominator is ab, we multiply the complex fraction by 1, expressed in the form $\frac{ab}{ab}$. This method is equivalent to multiplying the numerator and the denominator by ab.

See Example 2(a).

$$\frac{\dfrac{1}{a} + \dfrac{1}{b}}{\dfrac{1}{a} - \dfrac{1}{b}} = \frac{\left(\dfrac{1}{a} + \dfrac{1}{b}\right)ab}{\left(\dfrac{1}{a} - \dfrac{1}{b}\right)ab} \qquad \text{Multiply by } \frac{ab}{ab} = 1.$$

$$= \frac{\dfrac{ab}{a} + \dfrac{ab}{b}}{\dfrac{ab}{a} - \dfrac{ab}{b}} \qquad \text{Distributive property}$$

$$= \frac{b + a}{b - a} \qquad \text{Simplify the fractions.}$$

EXAMPLE 3   Simplifying complex fractions (Method II)

Simplify.

(a) $\dfrac{2z}{\dfrac{4}{z} + \dfrac{3}{z}}$ **(b)** $\dfrac{\dfrac{1}{x-3}}{\dfrac{1}{x} + \dfrac{3}{x-3}}$ **(c)** $\dfrac{\dfrac{1}{a} - \dfrac{1}{b}}{\dfrac{1}{2b^2} - \dfrac{1}{2a^2}}$

Solution

(a) The LCD for all fractions within the complex fraction is z, so multiply the numerator and denominator of the complex fraction by z (Step 1).

$$\frac{2z}{\dfrac{4}{z} + \dfrac{3}{z}} = \frac{(2z)\,z}{\left(\dfrac{4}{z} + \dfrac{3}{z}\right)z} \qquad \text{Multiply by } \frac{z}{z} = 1 \text{ (Step 2).}$$

$$= \frac{2z^2}{\dfrac{4z}{z} + \dfrac{3z}{z}} \qquad \text{Distributive property}$$

$$= \frac{2z^2}{4 + 3} \qquad \text{Simplify the fractions (Step 3).}$$

$$= \frac{2z^2}{7} \qquad \text{Add.}$$

(b) The LCD for all fractions within the complex fraction is the product $x(x-3)$ (Step 1).

$$\frac{\dfrac{1}{x-3}}{\dfrac{1}{x}+\dfrac{3}{x-3}} = \frac{\dfrac{1}{x-3}}{\left(\dfrac{1}{x}+\dfrac{3}{x-3}\right)} \cdot \frac{x(x-3)}{x(x-3)}$$

Multiply by $\frac{x(x-3)}{x(x-3)} = 1$ (Step 2).

$$= \frac{\dfrac{x(x-3)}{x-3}}{\dfrac{x(x-3)}{x}+\dfrac{3x(x-3)}{x-3}}$$

Distributive property

$$= \frac{x}{(x-3)+3x}$$

Simplify the fractions (Step 3).

$$= \frac{x}{4x-3}$$

Simplify the denominator.

(c) The LCD for the numerator *and* the denominator is $2a^2b^2$ (Step 1).

$$\frac{\dfrac{1}{a}-\dfrac{1}{b}}{\dfrac{1}{2b^2}-\dfrac{1}{2a^2}} = \frac{\dfrac{1}{a}-\dfrac{1}{b}}{\dfrac{1}{2b^2}-\dfrac{1}{2a^2}} \cdot \frac{2a^2b^2}{2a^2b^2}$$

Multiply by $\frac{2a^2b^2}{2a^2b^2}=1$ (Step 2).

$$= \frac{\dfrac{2a^2b^2}{a}-\dfrac{2a^2b^2}{b}}{\dfrac{2a^2b^2}{2b^2}-\dfrac{2a^2b^2}{2a^2}}$$

Distributive property

$$= \frac{2ab^2-2a^2b}{a^2-b^2}$$

Simplify the fractions (Step 3).

$$= \frac{2ab(b-a)}{(a-b)(a+b)}$$

Factor.

$$= -\frac{2ab}{a+b}$$

Simplify.

Now Try Exercises 33, 35, 49

CRITICAL THINKING 1

Are the expressions $\dfrac{\frac{a}{b}}{\frac{a}{b}+1}$ and $\dfrac{1}{1+1}$ equal? Explain.

Are the expressions $\dfrac{\frac{a}{b}+1}{\frac{a}{b}}$ and $1+\dfrac{b}{a}$ equal? Explain.

7.5 Putting It All Together

CONCEPT	EXPLANATION	EXAMPLES
Complex Fraction	A rational expression that contains fractions in its numerator, denominator, or both	$\dfrac{3+\dfrac{1}{x+1}}{3-\dfrac{1}{x+1}}$ and $\dfrac{\dfrac{x}{y}-\dfrac{y}{x}}{\dfrac{x}{y}+\dfrac{y}{x}}$
Simplifying Basic Complex Fractions	When b, c, and d are nonzero, $$\dfrac{\dfrac{a}{b}}{\dfrac{c}{d}} = \dfrac{a}{b}\cdot\dfrac{d}{c}.$$	$\dfrac{\dfrac{2}{x}}{\dfrac{4}{x-1}} = \dfrac{2}{x}\cdot\dfrac{x-1}{4} = \dfrac{x-1}{2x}$

CONCEPT	EXPLANATION	EXAMPLES
Method I: Simplifying the Numerator and Denominator	Combine the terms in the numerator, combine the terms in the denominator, and then multiply the numerator by the reciprocal of the denominator.	$\dfrac{\dfrac{1}{x}+\dfrac{3}{x}}{\dfrac{5}{y}-\dfrac{4}{y}}=\dfrac{\dfrac{4}{x}}{\dfrac{1}{y}}=\dfrac{4}{x}\cdot\dfrac{y}{1}=\dfrac{4y}{x}$
Method II: Multiplying the Numerator and Denominator by the LCD	Multiply the numerator *and* the denominator by the LCD of *all* fractions within the expression.	$\dfrac{\dfrac{2}{x}+\dfrac{1}{y}}{\dfrac{4}{y}-\dfrac{1}{x}}=\dfrac{\left(\dfrac{2}{x}+\dfrac{1}{y}\right)xy}{\left(\dfrac{4}{y}-\dfrac{1}{x}\right)xy}$ $=\dfrac{2y+x}{4x-y}$ Note that the LCD is xy.

7.5 Exercises

MyMathLab®

CONCEPTS AND VOCABULARY

1. $\dfrac{\frac{1}{2}}{\frac{3}{4}}=$ _____

2. $\dfrac{\frac{a}{b}}{\frac{c}{d}}=$ _____

3. A complex fraction is a rational expression that contains _____ in its numerator, denominator, or both.

4. Write the expression $\frac{x}{2} \div \frac{1}{x-1}$ as a complex fraction.

5. Write the expression $\frac{a}{b} \div \frac{c}{d}$ as a complex fraction.

6. What is the LCD for $\frac{1}{x+2}$ and $\frac{1}{x}$?

SIMPLIFYING COMPLEX FRACTIONS

Exercises 7–12: For the complex fraction, determine the LCD of all the fractions appearing within the expression.

7. $\dfrac{\dfrac{x}{5}-\dfrac{1}{6}}{\dfrac{2}{15}-3x}$

8. $\dfrac{\dfrac{1}{2}-\dfrac{1}{x}}{\dfrac{1}{2}+\dfrac{1}{x}}$

9. $\dfrac{\dfrac{2}{x+1}-x}{\dfrac{2}{x-1}+x}$

10. $\dfrac{\dfrac{1}{4x}-\dfrac{4}{x}}{\dfrac{1}{2x}+\dfrac{1}{3x}}$

11. $\dfrac{\dfrac{1}{2x-1}-\dfrac{1}{2x+1}}{\dfrac{x+1}{x}}$

12. $\dfrac{\dfrac{1}{4x^2}-\dfrac{1}{2x^3}}{\dfrac{1}{x-1}+\dfrac{1}{x-1}}$

Exercises 13–52: Simplify the complex fraction.

13. $\dfrac{\frac{2}{3}}{\frac{5}{6}}$

14. $\dfrac{\frac{8}{9}}{\frac{5}{4}}$

15. $\dfrac{2\frac{1}{2}}{1\frac{3}{4}}$

16. $\dfrac{1\frac{2}{3}}{3\frac{1}{2}}$

17. $\dfrac{1\frac{1}{2}}{2\frac{1}{3}}$

18. $\dfrac{2\frac{1}{5}}{2\frac{1}{10}}$

19. $\dfrac{\frac{r}{t}}{\frac{2r}{t}}$

20. $\dfrac{\frac{8}{p}}{\frac{4}{p}}$

21. $\dfrac{\frac{6}{x}}{\frac{2}{y}}$

22. $\dfrac{\frac{3}{14x}}{\frac{6}{7x}}$

23. $\dfrac{\frac{6}{m-2}}{\frac{2}{m-2}}$

24. $\dfrac{\frac{3}{n+1}}{\frac{6}{n+1}}$

25. $\dfrac{\frac{p+1}{p}}{\frac{p+2}{p}}$

26. $\dfrac{\frac{2p}{2p+5}}{\frac{1}{4p+10}}$

27. $\dfrac{\frac{5}{z^2-1}}{\frac{z}{z^2-1}}$

28. $\dfrac{\frac{z}{z-2}}{\frac{z}{z-2}}$

29. $\dfrac{\dfrac{y}{y^2-9}}{\dfrac{1}{y+3}}$

30. $\dfrac{\dfrac{2y}{2y-1}}{\dfrac{1}{4y^2-1}}$

31. $\dfrac{4-\dfrac{1}{x}}{4+\dfrac{1}{x}}$

32. $\dfrac{x-\dfrac{1}{x}}{x+\dfrac{1}{x}}$

33. $\dfrac{x}{\dfrac{2}{x}+\dfrac{1}{x}}$

34. $\dfrac{5x}{1+\dfrac{1}{x}}$

35. $\dfrac{\dfrac{3}{x+1}}{\dfrac{4}{x+1}-\dfrac{1}{x+1}}$

36. $\dfrac{\dfrac{5}{2x-3}-\dfrac{4}{2x-3}}{\dfrac{7}{2x-3}+\dfrac{8}{2x-3}}$

37. $\dfrac{\dfrac{1}{m^2n}+\dfrac{1}{mn^2}}{\dfrac{1}{m^2n}-\dfrac{1}{mn^2}}$

38. $\dfrac{\dfrac{1}{x^2y}-\dfrac{1}{xy^2}}{\dfrac{1}{x^2y}+\dfrac{1}{xy^2}}$

39. $\dfrac{\dfrac{1}{2x}+\dfrac{1}{y}}{\dfrac{1}{y}-\dfrac{1}{2x}}$

40. $\dfrac{\dfrac{3}{x}-\dfrac{2}{y}}{\dfrac{3}{x}+\dfrac{2}{y}}$

41. $\dfrac{\dfrac{1}{ab}+\dfrac{1}{a}}{\dfrac{1}{ab}-\dfrac{1}{b}}$

42. $\dfrac{\dfrac{1}{a}+\dfrac{2}{3b}}{\dfrac{1}{a}-\dfrac{5}{2b}}$

43. $\dfrac{\dfrac{2}{q}-\dfrac{1}{q+1}}{\dfrac{1}{q+1}}$

44. $\dfrac{\dfrac{5}{p}+\dfrac{4}{p-5}}{\dfrac{5}{p}-\dfrac{5}{p-5}}$

45. $\dfrac{\dfrac{1}{x+1}+\dfrac{1}{x+2}}{\dfrac{1}{x+1}-\dfrac{1}{x+2}}$

46. $\dfrac{\dfrac{1}{x-3}-\dfrac{1}{x+3}}{1-\dfrac{1}{x^2-9}}$

47. $\dfrac{\dfrac{1}{2x-1}-\dfrac{1}{2x+1}}{\dfrac{x+1}{x}}$

48. $\dfrac{\dfrac{1}{4x^2}-\dfrac{1}{x^3}}{\dfrac{1}{x-1}+\dfrac{1}{x-1}}$

49. $\dfrac{\dfrac{1}{ab^2}-\dfrac{1}{a^2b}}{\dfrac{1}{b}-\dfrac{1}{a}}$

50. $\dfrac{\dfrac{1}{x^2}-\dfrac{1}{y^2}}{\dfrac{1}{x}-\dfrac{1}{y}}$

51. $\dfrac{1}{a^{-1}+b^{-1}}$

52. $\dfrac{a^2-b^2}{a^{-2}-b^{-2}}$

APPLICATIONS INVOLVING COMPLEX FRACTIONS

53. Annuity If P dollars are deposited every 2 weeks in an account paying an annual interest rate r expressed as a decimal, then the amount in the account after 2 years can be approximated by

$$\left(P\left(1+\frac{r}{26}\right)^{52}-P\right)\div\frac{r}{26}.$$

Write this expression as a complex fraction.

54. Annuity (Continuation of the preceding exercise) Use a calculator to evaluate the expression when $r=0.026$ (2.6%) and $P=\$100$. Interpret the result.

55. Resistance in Electricity Light bulbs are often wired so that electricity can flow through either bulb, as illustrated in the accompanying figure.

In this way, if one bulb burns out, the other bulb still works. If two light bulbs have resistances T and S, their combined resistance R is

$$R=\dfrac{1}{\dfrac{1}{T}+\dfrac{1}{S}}.$$

Simplify this formula.

56. Resistance in Electricity (Refer to the preceding exercise.) Evaluate the formula

$$R=\dfrac{1}{\dfrac{1}{T}+\dfrac{1}{S}}$$

when $T=100$ and $S=200$.

WRITING ABOUT MATHEMATICS

57. A student simplifies a complex fraction as shown below. Explain the student's mistake and how you would simplify the complex fraction correctly.

$$\dfrac{\dfrac{1}{x}+1}{\dfrac{1}{x}}\overset{?}{=}\dfrac{\dfrac{1}{x}}{\dfrac{1}{x}}+1=2$$

58. Explain one method for simplifying a complex fraction.

7.6 Rational Equations and Formulas

Objectives

1 **Solving Rational Equations**

2 **Recognizing Rational Expressions and Equations**
  - **Expressions**
  - **Equations**

3 **Finding Graphical and Numerical Solutions**

4 **Solving an Equation for a Variable**

5 **Solving Applications Involving Rational Equations**

NEW VOCABULARY

☐ Rational equation
☐ Basic rational equation
☐ Extraneous solutions

1 Solving Rational Equations

🌐 *Math in Context* ^Traffic In the first section of this chapter, we demonstrated that if cars enter a construction zone at random at an average rate of x cars per minute and if 10 cars per minute can pass through the zone, then the average time T that a driver spends waiting in line and passing through the construction zone is

$$T = \frac{1}{10 - x},$$

where T is in minutes and $x < 10$. If the highway department wants to limit the average wait for a car to $\frac{1}{2}$ minute, then mathematics can be used to determine the corresponding value of x. (*Source*: N. Garber and L. Hoel, *Traffic and Highway Engineering*.)

If an equation contains one or more rational expressions, it is called a **rational equation**. Rational equations occur in mathematics whenever a rational expression is set equal to a constant. For example, the rational expression

$$\frac{1}{10 - x} \quad \text{Rational expression}$$

can be used to estimate the average time that drivers wait to get through a construction site. If the wait is $\frac{1}{2}$ minute, we determine x by solving the *rational equation*

$$\frac{1}{10 - x} = \frac{1}{2}. \quad \text{Rational equation}$$

To solve this equation, we multiply each side by the LCD: $2(10 - x)$.

$$\frac{1 \cdot 2(10 - x)}{10 - x} = \frac{1 \cdot 2(10 - x)}{2} \qquad \text{Multiply by the LCD.}$$

$$2 = 10 - x \qquad \text{Simplify.}$$

$$x = 8 \qquad \text{Add } x \text{; subtract 2.}$$

The average wait is $\frac{1}{2}$ minute when cars arrive randomly at an average rate of 8 cars per minute. In general, if a rational equation is in the form

$$\frac{a}{b} = \frac{c}{d}, \quad \text{Basic rational equation}$$

we can multiply each side of this equation by the common denominator bd to obtain

$$\frac{a(bd)}{b} = \frac{c(bd)}{d}, \qquad \text{Multiply each side by the common denominator } bd.$$

which simplifies to $ad = cb$. This technique can be used to solve some types of **basic rational equations**, which have a single rational expression on each side of the equals sign.

SOLVING BASIC RATIONAL EQUATIONS

The equations

$$\frac{a}{b} = \frac{c}{d} \quad \text{and} \quad ad = bc$$

are equivalent, provided that b and d are nonzero. Note that converting the first equation to the second equation is sometimes called *cross multiplying*.

READING CHECK 1

- When do we cross multiply?

EXAMPLE 1 **Solving rational equations**

Solve each equation.

(a) $\dfrac{5}{3} = \dfrac{4}{x}$ (b) $\dfrac{x+1}{5} = \dfrac{3x}{2}$ (c) $x = \dfrac{4}{3x-4}$ (d) $\dfrac{1}{x} + \dfrac{2}{x} = \dfrac{3}{7}$

Solution

(a)

$$\dfrac{5}{3} = \dfrac{4}{x} \qquad \text{Given equation}$$

$$5x = 12 \qquad \text{Cross multiply.}$$

$$x = \dfrac{12}{5} \qquad \text{Divide by 5.}$$

The solution is $\frac{12}{5}$.

(b)

$$\dfrac{x+1}{5} = \dfrac{3x}{2} \qquad \text{Given equation}$$

$$2(x+1) = 15x \qquad \text{Cross multiply.}$$

$$2x + 2 = 15x \qquad \text{Distributive property}$$

$$2 = 13x \qquad \text{Subtract } 2x.$$

$$\dfrac{2}{13} = x \qquad \text{Divide by 13.}$$

The solution is $\frac{2}{13}$.

(c)

$$\dfrac{x}{1} = \dfrac{4}{3x-4} \qquad \text{Write } x \text{ as } \tfrac{x}{1}.$$

$$x(3x-4) = 4 \cdot 1 \qquad \text{Cross multiply.}$$

$$3x^2 - 4x = 4 \qquad \text{Distributive property}$$

$$3x^2 - 4x - 4 = 0 \qquad \text{Subtract 4.}$$

$$(3x+2)(x-2) = 0 \qquad \text{Factor.}$$

$$3x + 2 = 0 \quad \text{or} \quad x - 2 = 0 \qquad \text{Zero-product property}$$

$$x = -\dfrac{2}{3} \quad \text{or} \quad x = 2 \qquad \text{Solve each equation.}$$

The solutions are $-\frac{2}{3}$ and 2.

(d)

$$\dfrac{1}{x} + \dfrac{2}{x} = \dfrac{3}{7} \qquad \text{Given equation}$$

> Add before cross multiplying; the equation must be in basic form.

$$\dfrac{3}{x} = \dfrac{3}{7} \qquad \text{Basic form}$$

$$21 = 3x \qquad \text{Cross multiply.}$$

$$7 = x \qquad \text{Divide by 3.}$$

The solution is 7.

Now Try Exercises 9, 15, 31, 41

Another technique for solving rational equations is to multiply each side by the least common denominator. Unlike cross multiplying, this technique can *always* be used.

EXAMPLE 2   **Multiplying each side by the LCD**

Solve each equation. Check your answer.

(a) $\dfrac{1}{x-1} - \dfrac{1}{x} = \dfrac{1}{9x}$

(b) $\dfrac{1}{x-1} + \dfrac{1}{x+1} = \dfrac{12}{x^2-1}$

Solution

(a) Start by multiplying each term by the LCD, $9x(x-1)$.

$$\dfrac{1}{x-1} - \dfrac{1}{x} = \dfrac{1}{9x} \qquad \text{Given equation}$$

$$\dfrac{1 \cdot 9x(x-1)}{x-1} - \dfrac{1 \cdot 9x(x-1)}{x} = \dfrac{1 \cdot 9x(x-1)}{9x} \qquad \text{Multiply each term by the LCD.}$$

$$9x - 9(x-1) = x - 1 \qquad \text{Simplify each rational expression.}$$

$$9 = x - 1 \qquad \text{Distributive property; simplify.}$$

$$10 = x \qquad \text{Add 1.}$$

Check:

$$\dfrac{1}{10-1} - \dfrac{1}{10} \overset{?}{=} \dfrac{1}{9(10)} \qquad \text{Substitute 10 for } x.$$

$$\dfrac{10}{90} - \dfrac{9}{90} \overset{?}{=} \dfrac{1}{90} \qquad \text{The LCD is 90.}$$

$$\dfrac{1}{90} = \dfrac{1}{90} \ \checkmark \qquad \text{The answer checks.}$$

(b) Start by multiplying each term by the LCD, $x^2 - 1 = (x-1)(x+1)$.

$$\dfrac{1}{x-1} + \dfrac{1}{x+1} = \dfrac{12}{x^2-1} \qquad \text{Given equation}$$

$$\dfrac{1(x-1)(x+1)}{x-1} + \dfrac{1(x-1)(x+1)}{x+1} = \dfrac{12(x-1)(x+1)}{(x-1)(x+1)} \qquad \text{Multiply each term by the LCD.}$$

$$(x+1) + (x-1) = 12 \qquad \text{Simplify.}$$

$$2x = 12 \qquad \text{Combine like terms.}$$

$$x = 6 \qquad \text{Divide by 2.}$$

Check:

$$\dfrac{1}{6-1} + \dfrac{1}{6+1} \overset{?}{=} \dfrac{12}{6^2-1} \qquad \text{Substitute 6 for } x.$$

$$\dfrac{1}{5} + \dfrac{1}{7} \overset{?}{=} \dfrac{12}{35} \qquad \text{Simplify.}$$

$$\dfrac{7}{35} + \dfrac{5}{35} \overset{?}{=} \dfrac{12}{35} \qquad \text{The LCD is 35.}$$

$$\dfrac{12}{35} = \dfrac{12}{35} \ \checkmark \qquad \text{The answer checks.}$$

Now Try Exercises 49, 55

In Example 2, we checked our answers. Although the answer checked in both cases, it may not always check. When multiplying a rational equation by the LCD, it is possible to obtain **extraneous solutions** that do not satisfy the *given* equation. This situation is demonstrated in Example 3. Before solving Example 3, we present a step-by-step strategy for solving rational equations.

STEPS FOR SOLVING A RATIONAL EQUATION

STEP 1: Find the LCD of the terms in the equation.

STEP 2: Multiply each side of the equation by the LCD.

STEP 3: Simplify each term.

STEP 4: Solve the resulting equation.

STEP 5: Check each answer in the *given* equation. Any value that makes a denominator equal 0 should be rejected because it is an extraneous solution.

READING CHECK 2

• What is the first step in solving a rational equation?

NOTE: If the rational equation is in the form $\frac{a}{b} = \frac{c}{d}$, or "fraction equals fraction," cross multiplying to obtain $ad = bc$ may be helpful. However, be sure to check your answers. ∎

EXAMPLE 3 **Solving an equation that has an extraneous solution**

If possible, solve $\dfrac{1}{x-2} + \dfrac{1}{x+2} = \dfrac{4}{x^2-4}$.

Solution
The LCD is $x^2 - 4 = (x-2)(x+2)$ (Step 1).

$$\frac{1}{x-2} + \frac{1}{x+2} = \frac{4}{x^2-4} \qquad \text{Given equation}$$

$$\frac{1(x-2)(x+2)}{x-2} + \frac{1(x-2)(x+2)}{x+2} = \frac{4(x^2-4)}{x^2-4} \qquad \text{Multiply by the LCD (Step 2).}$$

$$(x+2) + (x-2) = 4 \qquad \text{Simplify each term (Step 3).}$$

$$2x = 4 \qquad \text{Combine like terms (Step 4).}$$

$$x = 2 \qquad \text{Divide by 2.}$$

Check:

$$\frac{1}{2-2} + \frac{1}{2+2} \stackrel{?}{=} \frac{4}{2^2-4} \quad \textcolor{red}{✗} \qquad \text{Substitute 2 for } x \text{ (Step 5).}$$

$$\underbrace{\qquad\qquad}_{\text{Undefined terms}}$$

Note that terms on both sides of the equation are undefined because it is not possible to divide by 0. Therefore 2 is not a solution; rather, it is an *extraneous solution*. That is, there are no solutions to the *given* equation.

Now Try Exercise 53

---

2 Recognizing Rational Expressions and Equations

EXPRESSIONS Rational expressions and rational equations are not the same concepts.

$$\frac{3}{x} \quad \text{and} \quad \frac{x}{x^2-x} \qquad \text{Two rational expressions}$$

Neither expression contains an equals sign. We often simplify or evaluate rational expressions. By factoring the denominator and applying the basic principle of rational expressions, we can simplify the second expression.

$$\frac{x}{x^2 - x} \quad \text{is equivalent to} \quad \frac{x}{x(x - 1)} \quad \text{and}$$

$$\frac{x}{x(x - 1)} \quad \text{is equivalent to} \quad \frac{1}{x - 1} \quad (\text{if } x \neq 0).$$

Thus the expressions $\frac{x}{x^2 - x}$ and $\frac{1}{x - 1}$ are equal for every value of x (except 0 and 1). Expressions can also be evaluated. For example, if we replace x with 3, the expression $\frac{3}{x}$ equals $\frac{3}{3}$, or 1.

EQUATIONS When two expressions are set equal to each other, an equation is formed that is typically true for only a limited number of x-values. For example, $\frac{3}{x} = \frac{1}{x - 1}$ is a rational equation that is true for only one x-value.

$$\frac{3}{x} = \frac{1}{x - 1} \qquad \text{Given rational equation}$$

$$3(x - 1) = x \cdot 1 \qquad \text{Cross multiply.}$$

$$3x - 3 = x \qquad \text{Distributive property}$$

$$2x = 3 \qquad \text{Add 3; subtract } x.$$

> When solving an equation, we find values of x that make the equation true.

$$x = \frac{3}{2} \qquad \text{Divide each side by 2.}$$

The only solution is $\frac{3}{2}$. (Check this solution.) If we replace x with any value other than $\frac{3}{2}$, the equation $\frac{3}{x} = \frac{1}{x - 1}$ is a false statement.

EXAMPLE 4 **Identifying expressions and equations**

Determine whether you are given an expression or an equation. If it is an expression, *simplify* it and then *evaluate* it for $x = 5$. If it is an equation, *solve* it.

(a) $\dfrac{x - 1}{x + 1} = \dfrac{x}{x + 3}$ **(b)** $\dfrac{x^2 - 2x}{x - 1} + \dfrac{1}{x - 1}$

Solution
(a) There is an equals sign, so it is an equation that can be solved.

$$\frac{x - 1}{x + 1} = \frac{x}{x + 3} \qquad \text{Given equation}$$

$$(x - 1)(x + 3) = x(x + 1) \qquad \text{Cross multiply.}$$

$$x^2 + 2x - 3 = x^2 + x \qquad \text{Multiply.}$$

$$2x - 3 = x \qquad \text{Subtract } x^2.$$

$$x = 3 \qquad \text{Subtract } x; \text{ add 3.}$$

Check:

$$\frac{3 - 1}{3 + 1} \overset{?}{=} \frac{3}{3 + 3} \qquad \text{Replace } x \text{ with 3 in given equation.}$$

$$\frac{1}{2} = \frac{1}{2} \checkmark \qquad \text{The answer checks.}$$

The only solution to the equation is 3.

(b) There is no equals sign, so $\frac{x^2 - 2x}{x - 1} + \frac{1}{x - 1}$ is an expression that can be simplified. The common denominator is $x - 1$, so add the numerators.

$$\underbrace{\frac{x^2 - 2x}{x - 1} + \frac{1}{x - 1}}_{\text{Given expression}} = \frac{x^2 - 2x + 1}{x - 1}$$
$$= \frac{(x - 1)(x - 1)}{x - 1}$$
$$= x - 1$$

These expressions are all equivalent to the given expression when $x \neq 1$.

The given expression is equivalent to, or simplifies to, $x - 1$ when $x \neq 1$. When x equals **5**, the expression evaluates to **5** $-$ 1, or 4.

Now Try Exercises 63 and 67

MAKING CONNECTIONS 1

Expressions versus Equations

1. If a problem does not contain an equals sign, you are probably adding, subtracting, multiplying, dividing, or otherwise simplifying an expression. Your answer will be an expression, *not* a value for x.
2. If the problem has an equals sign, it is an equation to be solved. Your answer will be a value (or values) for x that makes the equation a true statement. Check all your answers.

---

3   Finding Graphical and Numerical Solutions

Like other types of equations, rational equations can also be solved graphically and numerically. Graphs of rational expressions are not lines. They are typically curves that can be graphed by plotting several points and then sketching a graph or by using a graphing calculator.

EXAMPLE 5   **Solving a rational equation graphically and numerically**

Solve $\frac{2}{x} = x + 1$ graphically and numerically.

Solution
Graphical Solution To solve $\frac{2}{x} = x + 1$, graph $y_1 = \frac{2}{x}$ and $y_2 = x + 1$. The graph of y_2 is a line with slope 1 and y-intercept $(0, 1)$. To graph y_1, make a table of values, as shown in **TABLE 7.5**. Then plot the points and connect them with a smooth curve. Note that y_1 is undefined for $x = 0$, so $x = 0$ is a vertical asymptote. Generally, it is a good idea to plot at least three points on each side of an asymptote. These points are plotted and the curves sketched in **FIGURE 7.5**. Note that the graphs of y_1 and y_2 intersect at $(-2, -1)$ and $(1, 2)$. The solutions to the equation are -2 and **1**. Check these solutions.

x	$\frac{2}{x}$
-3	$-\frac{2}{3}$
-2	-1
-1	-2
0	—
1	2
2	1
3	$\frac{2}{3}$

$x = 0$ (y-axis) is an asymptote.

TABLE 7.5

Graphical Solution

The solutions are -2 and **1**.

FIGURE 7.5

CRITICAL THINKING 1

If you solve Example 5 symbolically, what will be the result? Verify your answer.

Numerical Solution Make a table of values for $y_1 = \frac{2}{x}$ and $y_2 = x + 1$, as shown in **TABLE 7.6**. Note that $\frac{2}{x} = x + 1$ when $x = -2$ or $x = 1$.

Numerical Solution

Solutions

x	-3	-2	-1	0	1	2	3
$y_1 = \frac{2}{x}$	$-\frac{2}{3}$	-1	-2	—	2	1	$\frac{2}{3}$
$y_2 = x + 1$	-2	-1	0	1	2	3	4

TABLE 7.6

Now Try Exercise 75

A graphing calculator can be used to create **FIGURE 7.5** and **TABLE 7.6**, as illustrated in **FIGURE 7.6**.

CALCULATOR HELP

To make a graph or table, see Appendix A (pages 2, 3, and 4).

(a)

(b)

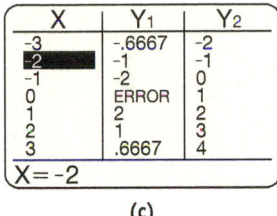
(c)

FIGURE 7.6

In the following Making Connections, we solve the rational equation $\frac{x + 3}{2x} = x$ using symbolic, numerical, and graphical methods.

MAKING CONNECTIONS 2

Symbolic, Numerical, and Graphical Solutions to a Rational Equation

STUDY TIP

It can be helpful to talk and study with other students. However, make sure that others are not "doing the work for you."

Symbolic Solution

$$\frac{x + 3}{2x} = 2$$

$$2x \cdot \frac{x + 3}{2x} = 2 \cdot 2x$$

$$x + 3 = 4x$$

$$3 = 3x$$

$$x = 1$$

The solution is **1**.

Numerical Solution

x	$(x + 3)/(2x)$
-2	-0.25
-1	-1
0	—
1	2
2	1.25

When $x = 1$, $\frac{x + 3}{2x} = 2$.

Graphical Solution

The graphs intersect at $(1, 2)$.

4 Solving an Equation for a Variable

Math in Context Equations and formulas in applications often involve both the use of rational expressions and the need to solve an equation for a variable. This idea is illustrated in the next two examples, which involve formulas from geometry, physics, and chemistry.

EXAMPLE 6 **Finding time in a distance problem**

If a person travels at a speed (rate) r for time t, then the distance d traveled is $d = rt$.
(a) How far does a person travel in 2 hours when traveling at 60 miles per hour?
(b) Solve the formula $d = rt$ for t.
(c) How long does it take a person to go 250 miles when traveling at 40 miles per hour?

Solution
(a) A person traveling at **60** miles per hour for **2** hours travels

$$d = rt = \mathbf{60} \cdot \mathbf{2} = 120 \text{ miles.}$$

(b) To solve $d = rt$ for t, divide each side by r.

$$d = r\mathbf{t} \qquad \text{Given equation}$$

$$\frac{d}{r} = \mathbf{t} \qquad \text{Divide each side by } r.$$

(c) At **40** miles per hour, a person can travel **250** miles in

$$t = \frac{d}{r} = \frac{\mathbf{250}}{\mathbf{40}} = \frac{25}{4} = 6.25 \text{ hours.}$$

Note that 6.25 hours equals 6 hours and 15 minutes.

Now Try Exercise 87

EXAMPLE 7 **Solving a formula for a variable**

Solve each equation for the specified variable.
(a) $A = \frac{bh}{2}$ for b **(b)** $P = \frac{nRT}{V}$ for V **(c)** $S = 2\pi rh + \pi r^2$ for h

Solution
(a) First, multiply each side by 2.

$$A = \frac{\mathbf{b}h}{2} \qquad \text{Given equation}$$

$$2A = \mathbf{b}h \qquad \text{Multiply each side by 2.}$$

$$\frac{2A}{h} = \mathbf{b} \qquad \text{Divide each side by } h.$$

(b) Begin by multiplying each side by V.

$$P = \frac{nRT}{\mathbf{V}} \qquad \text{Given equation}$$

$$P\mathbf{V} = nRT \qquad \text{Multiply each side by } V.$$

$$\mathbf{V} = \frac{nRT}{P} \qquad \text{Divide each side by } P.$$

(c) Begin by subtracting πr^2 from each side.

$$S = 2\pi r\mathbf{h} + \pi r^2 \qquad \text{Given equation}$$

$$S - \pi r^2 = 2\pi r\mathbf{h} \qquad \text{Subtract } \pi r^2 \text{ from each side.}$$

$$\frac{S - \pi r^2}{2\pi r} = \mathbf{h} \qquad \text{Divide each side by } 2\pi r.$$

Now Try Exercises 89, 91, 97

5 Solving Applications Involving Rational Equations

Speed and Time

🌐 *Math in ^Context* Rational equations sometimes occur in time and rate problems, as demonstrated in the next two examples.

EXAMPLE 8 | **Mowing a lawn**

Two people are mowing a large lawn. One person has a riding mower, and the other person has a push mower. The person with the riding mower can cut the lawn alone in 4 hours, and the person with the push mower can cut the lawn alone in 9 hours. How long does it take them, working together, to cut the lawn?

Solution

The first person can cut the entire lawn in 4 hours, so this person can cut $\frac{1}{4}$ of the lawn in 1 hour, $\frac{2}{4}$ of the lawn in 2 hours, and in general, $\frac{t}{4}$ of the lawn in t hours. The second person can cut the lawn in 9 hours, so (using similar reasoning) this person can cut $\frac{t}{9}$ of the lawn in t hours. Together, they can cut

$$\frac{t}{4} + \frac{t}{9}$$

of the lawn in t hours. The job is complete when the fraction of the lawn cut reaches **1**. To find out how long this task takes, solve the equation

$$\frac{t}{4} + \frac{t}{9} = 1.$$

Begin by multiplying each side by the LCD, or **36**.

$$\frac{t}{4} + \frac{t}{9} = 1 \qquad \text{Equation to be solved}$$

$$\frac{36t}{4} + \frac{36t}{9} = 36(1) \qquad \text{Multiply each term by 36.}$$

$$9t + 4t = 36 \qquad \text{Simplify.}$$

$$13t = 36 \qquad \text{Combine like terms.}$$

$$t = \frac{36}{13} \qquad \text{Divide each side by 13.}$$

Working together they can cut the lawn in $\frac{36}{13} \approx 2.8$ hours.

Now Try Exercise 103

CRITICAL THINKING 2

If one person can mow a lawn in x hours and another person can mow it in y hours, how long does it take them, working together, to mow the lawn?

EXAMPLE 9 | **Solving a distance problem**

Suppose that the winner of a 600-mile car race finishes 12 minutes ahead of the second place finisher. If the winner averages 5 miles per hour faster than the second racer, find the average speed of each racer.

Solution

Here, we apply the four-step method for solving an application problem.

STEP 1: *Identify any variables.*

x: speed of slower car in miles per hour

$x + 5$:   speed of faster car in miles per hour

STEP 2: *Write an equation.* To determine the time required for each car to finish the race, we use the equation $t = \frac{d}{r}$. Because the race is 600 miles, the time for the slower car is $\frac{600}{x}$ and the time for the faster car is $\frac{600}{x+5}$. The difference between these times is 12 minutes, or $\frac{12}{60} = \frac{1}{5}$ hour. Thus to determine x, we solve

$$\frac{600}{x} - \frac{600}{x+5} = \frac{1}{5}.$$

NOTE: Speeds are in miles per *hour*, so time must be in hours, not minutes. ∎

STEP 3: *Solve the equation.* Start by multiplying by the LCD, $5x(x+5)$.

$\dfrac{600}{x} - \dfrac{600}{x+5} = \dfrac{1}{5}$	Equation to be solved
$\dfrac{600 \cdot 5x(x+5)}{x} - \dfrac{600 \cdot 5x(x+5)}{x+5} = \dfrac{1 \cdot 5x(x+5)}{5}$	Multiply each term by the LCD.
$3000(x+5) - 3000x = x(x+5)$	Simplify.
$3000x + 15{,}000 - 3000x = x^2 + 5x$	Distributive property
$15{,}000 = x^2 + 5x$	Combine like terms.
$x^2 + 5x - 15{,}000 = 0$	Rewrite the equation.
$(x - 120)(x + 125) = 0$	Factor.
$x - 120 = 0 \quad \text{or} \quad x + 125 = 0$	Zero-product property
$x = 120 \quad \text{or} \quad x = -125$	Solve.

The slower car travels at 120 miles per hour, and the faster car travels 5 miles per hour faster, or 125 miles per hour. (The solution -125 has no meaning in this problem.)

STEP 4: *Check your answer.* The slower car travels 600 miles at 120 miles per hour, which takes $\frac{600}{120} = 5$ hours. The faster car travels 600 miles at 125 miles per hour, which requires $\frac{600}{125} = 4.8$ hours. The difference between their times is $5 - 4.8 = 0.2$ hour, or $0.2 \times 60 = 12$ minutes, so the answer checks.

Now Try Exercise 105

TECHNOLOGY NOTE

CALCULATOR HELP

To make a table, see Appendix A (pages 2 and 3).

Solving Rational Equations Numerically
Tables can be used to find solutions to rational equations. The following displays show the positive solution of 120 from Example 9. Note the use of parentheses for entering the formula for Y_1.

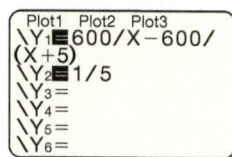

7.6   Putting It All Together

CONCEPT	EXPLANATION	EXAMPLES
Basic Rational Equations	To solve an equation of the form $$\frac{a}{b} = \frac{c}{d}, \quad b \neq 0, d \neq 0$$ cross multiply to obtain $ad = bc$.	$\frac{3}{6} = \frac{5}{2x}$ is equivalent to $6x = 30$. $\frac{x}{2} = \frac{8}{x}$ is equivalent to $x^2 = 16$. (Cross multiplication works only for equations with *one* rational expression on each side: "fraction equals fraction.")
Rational Equations	To solve a rational equation, use the following steps. **STEP 1:** Find the LCD. **STEP 2:** Multiply each side of the equation by the LCD. **STEP 3:** Simplify each term. **STEP 4:** Solve the resulting equation. **STEP 5:** Check each possible solution.	Solve $\frac{1}{x} - \frac{2}{3x} = \frac{5}{6}$. **STEP 1:** The LCD is $6x$. **STEP 2:** $\frac{1(6x)}{x} - \frac{2(6x)}{3x} = \frac{5(6x)}{6}$ **STEP 3:** $6 - 4 = 5x$ **STEP 4:** $\frac{2}{5} = x$ **STEP 5:** $\frac{1}{\frac{2}{5}} - \frac{2}{3\left(\frac{2}{5}\right)} = \frac{5}{6}$ ✓ It checks.

7.6   Exercises

MyMathLab®

CONCEPTS AND VOCABULARY

1. If an equation contains one or more rational expressions, it is called a(n) _____ equation.

2. Give an example of a rational expression and an example of a rational equation.

3. The equation $\frac{a}{b} = \frac{c}{d}$ is equivalent to _____, provided that _____ and _____ are nonzero.

4. Are the equations $\frac{2}{x-1} = 5$ and $5(x-1) = 2$ equivalent provided that $x \neq 1$?

5. One way to solve the equation $\frac{5}{3x} + \frac{3}{4x} = 1$ is to multiply each side by the LCD, which is _____.

6. To solve the equation $T = \frac{R}{SV}$ for V, multiply each side by the variable _____ and then divide each side by the variable _____.

SOLVING RATIONAL EQUATIONS

Exercises 7–62: Solve and check your answer.

7. $\frac{x}{2} = \frac{3}{4}$

8. $\frac{2x}{3} = \frac{2}{5}$

9. $\frac{3}{z} = \frac{6}{5}$

10. $\frac{2}{7} = \frac{1}{z}$

11. $\frac{3y}{4} = \frac{7y}{2}$

12. $\frac{y}{6} = \frac{5y}{3}$

13. $\frac{2}{3} = \frac{1}{2x+1}$

14. $\frac{1}{x+4} = \frac{3}{5}$

15. $\frac{5}{2x} = \frac{8}{x+2}$

16. $\frac{1}{x-1} = \frac{5}{3x}$

17. $\frac{1}{z-1} = \frac{2}{z+1}$

18. $\frac{4}{z+3} = \frac{2}{z-2}$

19. $\dfrac{3}{n+5} = \dfrac{2}{n-5}$

20. $\dfrac{4}{3n+2} = \dfrac{1}{n-1}$

21. $\dfrac{m}{m-1} = \dfrac{5}{4}$

22. $\dfrac{5m}{2m-1} = \dfrac{3}{2}$

23. $\dfrac{5x}{5-x} = \dfrac{1}{3}$

24. $\dfrac{x+2}{3x} = \dfrac{4}{3}$

25. $\dfrac{6}{5-2x} = 2$

26. $\dfrac{x+1}{x} = 6$

27. $\dfrac{2x}{2x+1} = \dfrac{-1}{2x+1}$

28. $\dfrac{x}{x-4} = \dfrac{4}{x-4}$

29. $\dfrac{1}{1-x} = \dfrac{3}{1+x}$

30. $\dfrac{2x}{1-2x} = \dfrac{1}{2}$

31. $\dfrac{1}{z+2} = -z$

32. $\dfrac{1}{z-2} = \dfrac{z}{3}$

33. $\dfrac{-1}{2x+5} = \dfrac{x}{3}$

34. $\dfrac{x}{2} = \dfrac{1}{3x+5}$

35. $\dfrac{x}{2} + \dfrac{x}{4} = 3$

36. $\dfrac{x}{4} - \dfrac{x}{3} = 1$

37. $\dfrac{3x}{4} - \dfrac{x}{2} = 1$

38. $\dfrac{2x}{3} + \dfrac{x}{3} = 6$

39. $\dfrac{4}{t+1} + \dfrac{1}{t+1} = -1$

40. $\dfrac{2}{t-5} - \dfrac{5}{t-5} = 3$

41. $\dfrac{1}{x} + \dfrac{2}{x} = \dfrac{1}{2}$

42. $\dfrac{1}{2x} - \dfrac{2}{x} = -3$

43. $\dfrac{2}{x-1} + 1 = \dfrac{4}{x^2-1}$

44. $\dfrac{1}{x} + 2 = \dfrac{1}{x^2+x}$

45. $\dfrac{1}{x+2} = \dfrac{4}{4-x^2} - 1$

46. $\dfrac{1}{x-3} + 1 = \dfrac{6}{x^2-9}$

47. $\dfrac{5}{4z} - \dfrac{2}{3z} = 1$

48. $\dfrac{3}{z+1} - \dfrac{1}{z+1} = 2$

49. $\dfrac{4}{y-1} + \dfrac{1}{y} = \dfrac{6}{5}$

50. $\dfrac{6}{y+1} + \dfrac{6}{y} = 5$

51. $\dfrac{1}{2x} - \dfrac{1}{x+3} = 0$

52. $\dfrac{2}{x} - \dfrac{6}{2x-1} = -1$

53. $\dfrac{1}{x-1} + \dfrac{1}{x+1} = \dfrac{2}{x^2-1}$

54. $\dfrac{1}{2x+1} + \dfrac{1}{2x-1} = \dfrac{2}{4x^2-1}$

55. $\dfrac{1}{x-2} + \dfrac{1}{x+2} = \dfrac{6}{x^2-4}$

56. $\dfrac{2}{x+3} - \dfrac{1}{x-3} = \dfrac{1}{x^2-9}$

57. $\dfrac{1}{p+1} + \dfrac{1}{p+2} = \dfrac{1}{p^2+3p+2}$

58. $\dfrac{1}{p-1} - \dfrac{1}{p+3} = \dfrac{1}{p^2+2p-3}$

59. $\dfrac{1}{x-2} + \dfrac{3}{2x-4} = \dfrac{6}{3x-6}$

60. $\dfrac{4}{x+1} - \dfrac{4}{2x+2} = \dfrac{1}{(x+1)^2}$

61. $\dfrac{1}{r^2-r-2} + \dfrac{2}{r^2-2r} = \dfrac{1}{r^2+r}$

62. $\dfrac{3}{r^2-1} + \dfrac{1}{r^2+r} = \dfrac{3}{r^2-r}$

RECOGNIZING RATIONAL EXPRESSIONS AND EQUATIONS

Exercises 63–70: (Refer to Example 4.) Determine if you are given an expression or an equation. If it is an expression, simplify (if possible) and evaluate it for $x = 2$. If it is an equation, solve it.

63. $\dfrac{1}{x} - \dfrac{1-x}{x}$

64. $\dfrac{1}{x} - x = 0$

65. $\dfrac{1}{2x} - \dfrac{1}{4x} = \dfrac{1}{8}$

66. $\dfrac{1}{2x} - \dfrac{1}{4x}$

67. $\dfrac{x+1}{x-1} = \dfrac{2x-3}{2x-5}$

68. $\dfrac{2x-1}{4x+1} = \dfrac{x+1}{2x-1}$

69. $\dfrac{4x+4}{x+2} + \dfrac{x^2}{x+2}$

70. $\dfrac{x^2-2}{x-2} - \dfrac{x}{x-2}$

FINDING GRAPHICAL AND NUMERICAL SOLUTIONS

Exercises 71–74: Use the graph to solve the given equation. Check your answers.

71. $\dfrac{1}{x} = 4x$

72. $\dfrac{x}{x-1} = x$

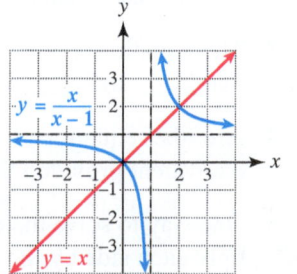

73. $\dfrac{x-1}{x} = -2x$ **74.** $\dfrac{3}{x^2-1} = 1$

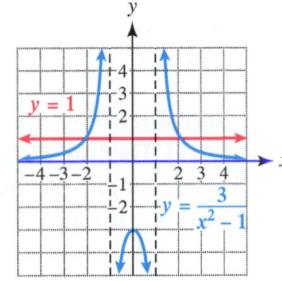

Exercises 75–82: (Refer to Example 5.) Solve the equation
  (a) *graphically and*
  (b) *numerically.*

75. $\dfrac{3}{x} = x + 2$ **76.** $-\dfrac{2}{x} = 1 - x$

77. $\dfrac{3x}{2} = \dfrac{1}{2}x - 1$ **78.** $\dfrac{x}{3} = 2 - \dfrac{2}{3}x$

79. $\dfrac{3}{x-1} = 3$ **80.** $\dfrac{2}{x+5} = 1$

81. $\dfrac{4}{x^2} = 1$ **82.** $\dfrac{-18}{x^2} = -2$

Exercises 83–86: Solve the rational equation graphically to the nearest thousandth.

83. $\dfrac{1}{\pi x - 2} + \dfrac{\sqrt{2}}{2.1x} = 1.3$ **84.** $\dfrac{6}{x^2-4} + \dfrac{\pi}{2x} = \dfrac{3}{2}$

85. $\dfrac{1}{x^3} - \dfrac{\pi}{4} = 0$ **86.** $\dfrac{1}{2\pi x^2} - \dfrac{x}{2} = 1$

SOLVING AN EQUATION FOR A VARIABLE

Exercises 87–98: (Refer to Examples 6 and 7.) Solve the equation for the specified variable.

87. $m = \dfrac{F}{a}$ for a **88.** $m = \dfrac{2K}{v^2}$ for K

89. $I = \dfrac{V}{R+r}$ for r **90.** $\dfrac{1}{T} = \dfrac{r}{R-r}$ for R

91. $h = \dfrac{2A}{b}$ for b **92.** $h = \dfrac{2A}{b_1 + b_2}$ for b_1

93. $\dfrac{3}{k} = \dfrac{z}{z+5}$ for z **94.** $\dfrac{5}{r} = \dfrac{t+r}{t}$ for t

95. $T = \dfrac{ab}{a+b}$ for b **96.** $A = \dfrac{2b}{a-b}$ for b

97. $\dfrac{3}{k} = \dfrac{1}{x} - \dfrac{2}{y}$ for x **98.** $\dfrac{1}{R} = \dfrac{1}{R_1} + \dfrac{1}{R_2}$ for R_1

APPLICATIONS INVOLVING RATIONAL EQUATIONS

99. Waiting in Traffic (Refer to the introduction to this section.) Solve the equation

$$\frac{1}{10-x} = 1$$

to determine the traffic rate x in cars per minute corresponding to an average waiting time of 1 minute.

100. Waiting in Line At a post office, customers arrive at random at an average rate of x people per minute. The clerk can wait on 4 customers per minute. The average time T in minutes spent waiting in line is given by

$$T = \frac{1}{4-x},$$

where $x < 4$.
  (a) Evaluate T for $x = 3$, 3.9, and 3.99. What happens to waiting time as the arrival rate nears 4 people per minute?
  (b) Find x when the waiting time is 5 minutes.

101. Shoveling a Sidewalk It takes a less experienced employee 4 hours to shovel the snow from a sidewalk, but a more experienced employee can shovel the same sidewalk in 3 hours. How long will it take them to clear the walk if they work together?

102. Pumping Water One pump can empty a pool in 5 days, whereas a second pump can empty the pool in 7 days. How long will it take the two pumps, working together, to empty the pool?

103. Painting a House One painter can paint a house in 8 days, yet a more experienced painter can paint the house in 4 days. How long will it take the two painters, working together, to paint the house?

104. Highway Curves To make a highway curve safe, highway engineers often bank it, as shown in the figure. If a curve is designed for a speed of 50 miles per hour and is banked with positive slope m, then a minimum radius R in feet for the curve is given by

$$R = \frac{2500}{15m + 2}.$$

(*Source:* N. Garber and L. Hoel, *Traffic and Highway Engineering.*)

(a) Find R for $m = 0.1$. Interpret the result.
(b) If $R = 500$, find m. Interpret the result.

105. Bicycle Race The winner of a 6-mile bicycle race finishes 2 minutes ahead of a teammate and travels, on average, 2 miles per hour faster than the teammate. Find the average speed of each racer.

106. Freeway Travel Two drivers travel 150 miles on a freeway and then stop at a wayside rest area. The first driver travels 5 miles per hour faster and arrives $\frac{1}{7}$ hour ahead of the second. Find the average speed of each car.

107. Braking Distance If a car is traveling *downhill* at 30 miles per hour on wet pavement, then the braking distance B in feet for this car is given by

$$B = \frac{30}{0.3 + m},$$

where $m < 0$ is the slope of the hill. (*Source*: L. Haefner, *Introduction to Transportation Systems.*)
(a) Find the braking distance for $m = -0.05$ and interpret the result.
(b) Find m if the braking distance is 150 feet.

108. Slippery Roads If a car is traveling at 30 miles per hour on a level road, then its braking distance in feet is $\frac{30}{x}$, where x is the coefficient of friction between the road and the tires. The variable x is positive and satisfies $x \le 1$. The closer the value of x is to 0, the more slippery the road is. (*Source*: L. Haefner.)
(a) Evaluate the expression for $x = 1$, 0.5, and 0.1. Interpret the results.
(b) Find x for a braking distance of 150 feet.

109. River Current A boat can travel 36 miles upstream in the same time that it can travel 54 miles downstream. If the speed of the current is 3 miles per hour, find the speed of the boat without a current.

110. Airplane Speed An airplane can travel 380 miles into the wind in the same time that it can travel 420 miles with the wind. If the wind speed is 10 miles per hour, find the speed of the airplane without any wind.

111. Airplane Speed An airplane can travel 450 miles into the wind in the same time that it can travel 750 miles with the wind. If the wind speed is 50 miles per hour, find the speed of the airplane without any wind

112. River Current A boat can travel 114 miles upstream in the same time that it can travel 186 miles downstream. If the speed of the current is 6 miles per hour, find the speed of the boat without a current.

113. Running and Jogging An athlete runs 10 miles and then jogs home. The trip home takes 1 hour longer than it took to run that distance. If the athlete runs 5 miles per hour faster than she jogs, what are her average running and jogging speeds?

114. Speed Limit A person drives 390 miles on a stretch of road. Half the distance is driven traveling 5 miles per hour below the speed limit, and half the distance is driven traveling 5 miles per hour above the speed limit. If the time spent traveling at the slower speed exceeds the time spent traveling at the faster speed by 24 minutes, find the speed limit.

115. Time Spent in Line If a parking lot attendant can wait on 5 vehicles per minute and vehicles are leaving the lot randomly at an average rate of x vehicles per minute, then the average time T in minutes spent waiting in line *and* paying the attendant is given by

$$T = \frac{1}{5 - x},$$

where $x < 5$. (*Source*: N. Garber.)
(a) Evaluate T when $x = 4$ and interpret the result.
(b) A graph of T is shown in the figure. Interpret the graph as x increases from 0 to 5. Does this result agree with your intuition?

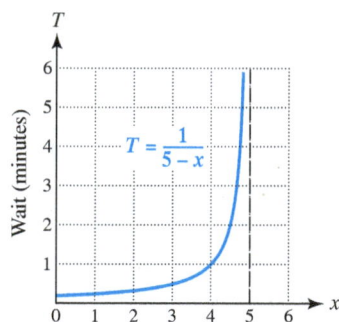

(c) Find x if the waiting time is 3 minutes.

116. People Waiting in Line At a post office, workers can wait on 50 people per hour. If people arrive randomly at an average rate of x per hour, then the average number of people N waiting in line is given by

$$N = \frac{x^2}{2500 - 50x},$$

where $x < 50$. (*Source*: N. Garber.)
(a) Evaluate N when $x = 30$ and interpret the result.

(b) A graph of N is shown in the figure. Interpret the graph as x increases from 0 to 50. Does this result agree with your intuition?

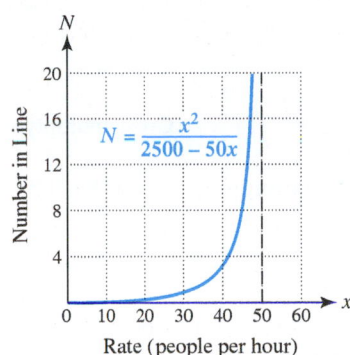

$$N = \frac{x^2}{2500 - 50x}$$

Number in Line (vertical axis)
Rate (people per hour) (horizontal axis)

(c) Find x if $N = 8$.

WRITING ABOUT MATHEMATICS

117. Do all rational equations have solutions? Explain.

118. Why is it important to check your answer when solving rational equations? Explain.

SECTIONS 7.5 and 7.6 — **Checking Basic Concepts**

1. Simplify each complex fraction.

(a) $\dfrac{\dfrac{x}{3}}{\dfrac{2x}{5}}$

(b) $\dfrac{\dfrac{2}{2x} - \dfrac{1}{3x}}{6x}$

(c) $\dfrac{\dfrac{1}{a} - \dfrac{1}{b}}{\dfrac{1}{a} + \dfrac{1}{b}}$

(d) $\dfrac{\dfrac{1}{r^2} - \dfrac{1}{t^2}}{\dfrac{2}{r} - \dfrac{2}{t}}$

2. Solve each equation. Check your answer.

(a) $\dfrac{1}{2x} = \dfrac{3}{x + 1}$

(b) $\dfrac{x}{2x + 3} = \dfrac{4}{5}$

(c) $\dfrac{1}{2x} + \dfrac{3}{2x} = 1$

(d) $\dfrac{3}{x + 1} - \dfrac{2}{x} = -2$

(e) $\dfrac{1}{x - 1} = \dfrac{2}{x^2 - 1} - \dfrac{1}{2}$

3. Solve each equation for the specified variable.

(a) $\dfrac{ax}{2} - 3y = b$ for x

(b) $\dfrac{1}{2m - 1} = \dfrac{k}{m}$ for m

4. Braking Distance If a car is traveling *uphill* at 60 miles per hour on wet pavement, then the braking distance D in feet for this car is given by

$$D = \frac{120}{0.3 + m},$$

where $m > 0$ is the slope of the hill. (*Source:* L. Haefner, *Introduction to Transportation Systems.*)

(a) Find D for $m = 0.1$ and interpret the result.

(b) Find the slope of the road if D is 200 feet. Interpret the result.

7.7 Proportions and Variation

Objectives

1 Solving Proportions • Similar Figures

2 Solving Direct Variation Problems

3 Solving Inverse Variation Problems

4 Analyzing Data

5 Solving Joint Variation Problems

1 Solving Proportions

Connecting Concepts with Your Life A **ratio** is a comparison of two quantities. For example, a math class might have 7 boys for every 8 girls. Thus the boy–girl ratio in this class is *7 to 8*, or $\frac{7}{8}$. In mathematics, ratios are typically expressed as fractions.

Ratios and proportions are sometimes used to find how much space remains for music on a compact disc (CD). A 700-megabyte CD can store about 80 minutes of music. Suppose that some music has been recorded on the CD and that 256 megabytes are still available. Using ratios and proportions, we can find how many more minutes of music could be recorded. A **proportion** is a statement that two *ratios* are equal.

NEW VOCABULARY

☐ Ratio
☐ Proportion
☐ Directly proportional (varies directly)
☐ Constant of proportionality/variation
☐ Inversely proportional (varies inversely)
☐ Varies jointly

Let x represent the number of minutes available on a CD. Then **80** minutes are to **700** megabytes as x minutes are to **256** megabytes. By setting the *ratios* $\frac{80}{700}$ and $\frac{x}{256}$ equal to each other, we obtain the *proportion*

$$\frac{\mathbf{80}}{\mathbf{700}} = \frac{x}{\mathbf{256}}.$$

$$\frac{\text{Minutes}}{\text{Megabytes}} = \frac{\text{Minutes}}{\text{Megabytes}}$$

Solving this equation for x gives

$$700x = 80(256) \qquad \text{Cross multiply.}$$

$$x = \frac{80 \cdot 256}{700} \approx 29.3 \text{ minutes.} \qquad \text{Divide by 700.}$$

About 29 minutes are available to record on the CD.

 Math in Weather Context Flooding in Fargo, North Dakota, has been an annual problem in recent years. With snow drifts reaching 10 to 15 feet high in some areas, an accurate estimate of the water content in the snow is essential for predicting spring flood levels. In the next example, we estimate the water content in snow.

EXAMPLE 1

Calculating the water content in snow

Six inches of light, fluffy snow are equivalent to about half an inch of rain in terms of water content. If 15 inches of this type of snow fall, estimate the water content.

Solution

Let x be the equivalent amount of rain. Then **6** inches of snow are to $\frac{1}{2}$ inch of rain as **15** inches of snow are to x inches of rain, which can be written as the proportion

$$\frac{\mathbf{6}}{\frac{1}{2}} = \frac{\mathbf{15}}{x}.$$

$$\frac{\text{Snow}}{\text{Rain}} = \frac{\text{Snow}}{\text{Rain}}$$

Solving this equation gives

$$6x = \frac{15}{2} \quad \text{or} \quad x = \frac{15}{12} = 1.25.$$

Thus 15 inches of light, fluffy snow are equivalent to about 1.25 inches of rain.

Now Try Exercise 65

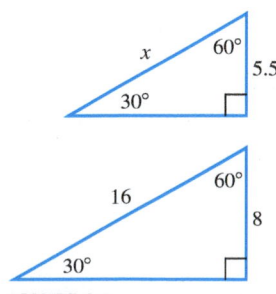

Similar Triangles

FIGURE 7.7

SIMILAR FIGURES Proportions frequently occur in geometry when we work with similar figures. Two triangles are similar if the measures of their corresponding angles are equal. Corresponding sides of similar triangles are proportional. **FIGURE 7.7** shows two right triangles that are similar because each has angles of 30°, 60°, and 90°.

We can find the length of side x by using proportions. Side x is to 16 as 5.5 is to 8, which can be written as the proportion

$$\frac{x}{16} = \frac{5.5}{8}.$$

$$\frac{\text{Hypotenuse}}{\text{Hypotenuse}} = \frac{\text{Shorter leg}}{\text{Shorter leg}}$$

Solving yields the following equation.

$$8x = 5.5(16) \qquad \text{Cross multiply.}$$

$$x = 11 \qquad \text{Divide each side by 8.}$$

NOTE: Proportions can be set up in different ways and still produce the correct result. For example, we could say that x is to 5.5 in the smaller triangle as 16 is to 8 in the larger triangle.

$$\frac{x}{5.5} = \frac{16}{8}$$

$$\frac{\text{Hypotenuse}}{\text{Shorter leg}} = \frac{\text{Hypotenuse}}{\text{Shorter leg}}$$

Solving, we obtain $8x = 5.5(16)$, or $x = 11$, which is the same answer. ∎

READING CHECK 1

• Solve the proportion $\frac{4}{x} = \frac{6}{7}$.

EXAMPLE 2   **Calculating the height of a tree**

A 6-foot-tall person casts a 4-foot-long shadow. If a nearby tree casts a 36-foot-long shadow, estimate the height of the tree. See **FIGURE 7.8**.

Forming a Proportion from Heights and Shadows

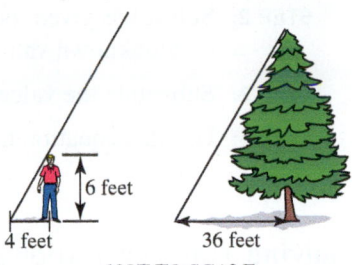

NOT TO SCALE

FIGURE 7.8

Similar Triangles

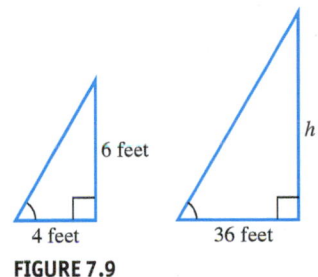

FIGURE 7.9

Solution

The triangles shown in **FIGURE 7.9** represent the situation shown in **FIGURE 7.8**. These triangles are similar because the measures of the corresponding angles are equal. Therefore their sides are proportional. Let h be the height of the tree.

$$\frac{h}{6} = \frac{36}{4} \qquad \frac{\text{Height}}{\text{Height}} = \frac{\text{Shadow length}}{\text{Shadow length}}$$

$$4h = 6(36) \qquad \text{Cross multiply.}$$

$$h = \frac{6(36)}{4} \qquad \text{Divide each side by 4.}$$

$$h = 54 \qquad \text{Simplify.}$$

The tree is 54 feet tall.

Now Try Exercise 69

2 Solving Direct Variation Problems

 Connecting Concepts with Your Life If your wage is \$12 per hour, then your total pay P is proportional to the number of hours H you work, which can be represented by the equation

$$\frac{P}{H} = \frac{12}{1}, \qquad \frac{\text{Pay}}{\text{Hours}} = \frac{\text{Pay}}{\text{Hours}}$$

or, equivalently,

$$P = 12H.$$

We say that your pay P is *directly proportional* to the number of hours H that you work. The *constant of proportionality* is 12.

DIRECT VARIATION

Let x and y denote two quantities. Then y is **directly proportional** to x, or y **varies directly** with x, if there is a nonzero number k such that

$$y = kx.$$

The number k is called the **constant of proportionality**, or the **constant of variation**.

The following 4-step process is helpful when solving variation applications. This process is used in Examples 3 and 4 and can also be used for other types of variation problems.

READING CHECK 2

- What is the first step in solving a variation problem?

SOLVING A VARIATION APPLICATION

When solving a variation problem, the following steps can be used.

STEP 1: Write the general equation for the type of variation problem that you are solving.

STEP 2: Substitute given values in this equation so the constant of variation k is the only unknown value in the equation. Solve for k.

STEP 3: Substitute the value of k in the general equation in Step 1.

STEP 4: Use this equation to find the requested quantity.

EXAMPLE 3 **Solving a direct variation problem**

Let y be directly proportional to x, or vary directly with x. Suppose $y = 7$ when $x = 5$. Find y when $x = 11$.

Solution
STEP 1: The general equation for direct variation is $y = kx$
STEP 2: Substitute **7** for y and **5** for x in $y = k\boldsymbol{x}$. Solve for k.

$$7 = k(\mathbf{5}) \qquad \text{Let } y = 7 \text{ and } x = 5.$$

$$\frac{7}{5} = k \qquad \text{Divide each side by 5.}$$

STEP 3: Replace k with $\frac{7}{5}$ in the equation $y = kx$ to obtain $y = \frac{7}{5}x$.
STEP 4: To find y when $x = 11$, let $x = \mathbf{11}$ in $y = \frac{7}{5}\boldsymbol{x}$.

$$y = \frac{7}{5}(\mathbf{11}) = \frac{77}{5} = 15.4$$

Now Try Exercise 33

EXAMPLE 4 **Solving a direct variation application**

The amount of weight that a beam of wood can support varies directly with its width. A beam that is 2.5 inches wide can support 800 pounds. How much weight can a similar beam support if its width is 3.2 inches?

Solution
STEP 1: The general equation for direct variation is $y = kx$, where y is the weight and x is the width.
STEP 2: Substitute **800** for y and **2.5** for x in $y = k\boldsymbol{x}$. Solve for k.

$$\mathbf{800} = k(\mathbf{2.5}) \qquad \text{Let } y = 800 \text{ and } x = 2.5.$$

$$\frac{\mathbf{800}}{\mathbf{2.5}} = k \qquad \text{Divide each side by 2.5.}$$

$$k = 320 \qquad \text{Simplify; rewrite the equation.}$$

STEP 3: Replace k with **320** in $y = k\boldsymbol{x}$ to obtain $y = \mathbf{320}x$.
STEP 4: To find the weight y when $x = 3.2$, substitute **3.2** for x in $y = 320\boldsymbol{x}$.

$$y = 320(\mathbf{3.2}) = 1024 \text{ pounds}$$

Now Try Exercise 67

The following See the Concept shows how the graph of a direct variation equation relates to the constant of variation k.

See the Concept 1   GRAPHS AND DIRECT VARIATION

Direct Variation: $y = kx$

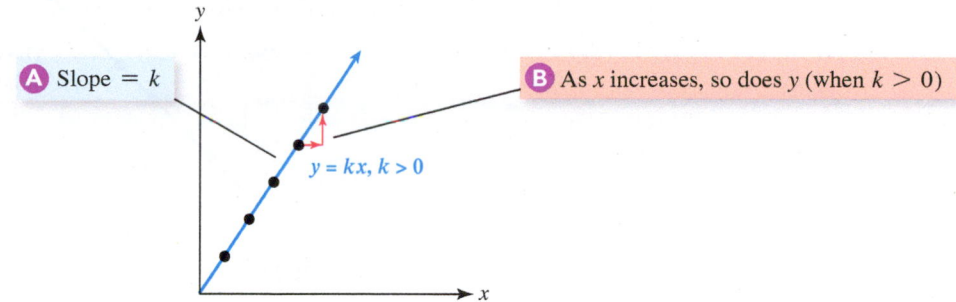

A Slope = k

B As x increases, so does y (when $k > 0$)

$y = kx, k > 0$

READING CHECK 3

- How does the constant of proportionality for data that represent direct variation compare to the slope of a line that passes through the data?

A The slope of the line equals k, which is the constant of variation. (The value of k could be negative.)

B These data lie on the line $y = kx$, so y varies directly with x.

WATCH VIDEO IN MML.

EXAMPLE 5   Modeling college tuition

TABLE 7.7 lists the tuition for taking various numbers of credits.

(a) A scatterplot of the data is shown in **FIGURE 7.10**. Could the data be modeled using a line?

Credits	Tuition
3	$189
5	$315
8	$504
11	$693
17	$1071

TABLE 7.7

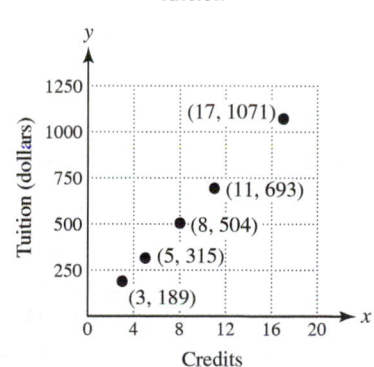

Tuition

FIGURE 7.10

(b) Explain why tuition is directly proportional to the number of credits taken.
(c) Find the constant of proportionality. Interpret your result.
(d) Predict the cost of taking 16 credits.

Solution
(a) The data are linear and suggest a line passing through the origin.
(b) Because the data can be modeled by a line passing through the origin, tuition is directly proportional to the number of credits taken. Hence doubling the credits will double the tuition and tripling the credits will triple the tuition.
(c) The slope of the line equals the constant of proportionality k. If we use the first and last data points (3, 189) and (17, 1071), the slope is

$$k = \frac{1071 - 189}{17 - 3} = 63.$$

That is, tuition is $63 per credit. If we graph the line $y = 63x$, it models the data, as shown in **FIGURE 7.11** on the next page.

The data lie on the line $y = 63x$, which indicates direct variation with $k = 63$.

FIGURE 7.11

(d) If y represents tuition and x represents the credits taken, **16** credits would cost

$$y = 63\,(\mathbf{16}) = \$1008.$$

Now Try Exercise 73

MAKING CONNECTIONS 1

Ratios and the Constant of Proportionality

The constant of proportionality in Example 5 can be found by calculating the ratios $\frac{y}{x}$.

x	3	5	8	11	17
y	189	315	504	693	1071
$\frac{y}{x}$	63	63	63	63	63

Each ratio $\frac{y}{x}$ equals 63, which indicates direct variation and that $k = 63$.

---

3 Solving Inverse Variation Problems

 Connecting Concepts with Your Life When two quantities vary inversely, an increase in one quantity results in a decrease in the second quantity. For example, at 30 miles per hour, a car travels 120 miles in 4 hours, whereas at 60 miles per hour, the car travels 120 miles in 2 hours. Doubling the speed (or rate) decreases the travel time by half. Distance equals rate times time, so $d = rt$. Thus

$$120 = rt, \quad \text{or equivalently,} \quad t = \frac{120}{r}.$$

We say that the time t to travel 120 miles is *inversely proportional* to the speed or rate r. The constant of proportionality or constant of variation is 120.

READING CHECK 4

• When two quantities, x and y, are inversely proportional with positive k, how does an increase in x affect y?

INVERSE VARIATION

Let x and y denote two quantities. Then y is **inversely proportional** to x, or y **varies inversely** with x, if there is a nonzero number k such that

$$y = \frac{k}{x}.$$

The following See the Concept shows how the graph of an inverse variation equation relates to points (x, y) on the graph and to the constant of variation k.

See the Concept 2 — GRAPHS AND INVERSE VARIATION

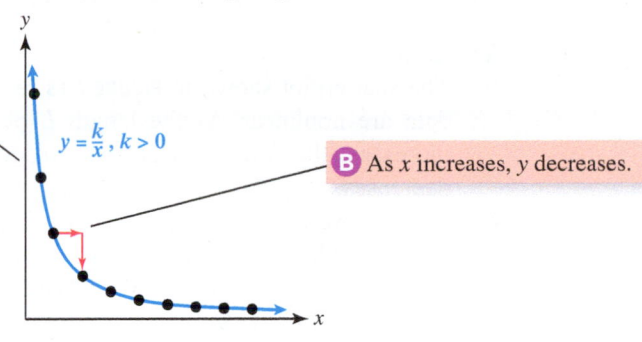

Inverse Variation: $y = \frac{k}{x}$

Ⓐ For any data point (x, y), $xy = k$

$y = \frac{k}{x}, k > 0$

Ⓑ As x increases, y decreases.

Ⓐ If (x, y) is any point on the curve, then the product xy always equals k, which is the constant of variation.

Ⓑ These data lie on the curve $y = \frac{k}{x}$, so y varies inversely with x.

WATCH VIDEO IN MML .

EXAMPLE 6 **Solving an inverse variation problem**

Let y be inversely proportional to x, or vary inversely with x. Suppose $y = 5$ when $x = 6$. Find y when $x = 21$.

Solution

STEP 1: The general equation for inverse variation is $y = \frac{k}{x}$.

STEP 2: Because $y = 5$ when $x = 6$, substitute **5** for y and **6** for x in $y = \frac{k}{x}$. Solve for k.

$$5 = \frac{k}{6} \quad \text{Let } y = 5 \text{ and } x = 6.$$

$$30 = k \quad \text{Multiply each side by 6.}$$

STEP 3: Replace k with **30** in the equation $y = \frac{k}{x}$ to obtain $y = \frac{30}{x}$.

STEP 4: To find y, let $x = \mathbf{21}$. Then $y = \frac{30}{21} = \frac{10}{7}$.

Now Try Exercise 39

 Connecting Concepts with Your Life A wrench is commonly used to loosen a nut on a bolt. See **FIGURE 7.12**. If the nut is difficult to loosen, a wrench with a longer handle is often helpful.

Loosening a Nut with a Wrench

Inverse variation: increasing the length of the wrench decreases the force required to loosen the nut.

FIGURE 7.12

EXAMPLE 7 **Illustrating inverse variation with a wrench**

TABLE 7.8 on the next page lists the force F necessary to loosen a particular nut with wrenches of different lengths L.

L (inches)	F (pounds)
6	12
8	9
12	6
18	4
24	3

TABLE 7.8

(a) Make a scatterplot of the data and discuss the graph. Are the data linear?
(b) Explain why the force F is inversely proportional to the handle length L. Find k so that $F = \frac{k}{L}$ models the data.
(c) Predict the force needed to loosen the nut with a 15-inch wrench.

Solution

(a) The scatterplot shown in **FIGURE 7.13** reveals that the data are nonlinear. As the length L of the wrench increases, the force F necessary to loosen the nut decreases.

(b) If F is inversely proportional to L, then $F = \frac{k}{L}$, or $FL = k$. That is, the product of F and L equals the constant of proportionality k. In **TABLE 7.8**, the product of F and L always equals 72 for each data point. Thus F is inversely proportional to L with constant of proportionality $k = 72$, so $F = \frac{72}{L}$.

(c) If $L = 15$, then $F = \frac{72}{15} = 4.8$. A wrench with a 15-inch handle requires a force of 4.8 pounds to loosen the nut.

Now Try Exercises 75

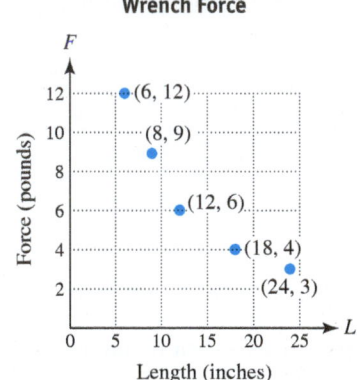

Wrench Force

FIGURE 7.13

TECHNOLOGY NOTE

CALCULATOR HELP

To make a scatterplot, see Appendix A (pages 3 and 4).

Scatterplots and Graphs

A graphing calculator can be used to create scatterplots and graphs. A scatterplot of the data in **TABLE 7.8** is shown in the first figure. In the second figure, the data and the equation $y = \frac{72}{x}$ are graphed. Note that each tick mark represents 5 units.

4 Analyzing Data

So far in this section, we have discussed *direct* and *inverse* variation. **TABLE 7.9** gives a summary of these two types of variation.

Direct and Inverse Variation

Type of Variation	Equation	Constant of Variation
y varies *directly* with x	$y = kx$	$k = \frac{y}{x}$
y varies *inversely* with x	$y = \frac{k}{x}$	$k = xy$

TABLE 7.9

READING CHECK 5

• How can the constant of variation be used to determine whether data are directly or inversely proportional?

The last column in **TABLE 7.9** shows that a set of data represents direct variation when the quotients $\frac{y}{x}$ equal a constant, and it represents inverse variation when the products xy equal a constant. In the next example, we determine if tables of data represent direct variation, inverse variation, or neither.

EXAMPLE 8 **Analyzing data**

Determine whether the data in each table represent direct variation, inverse variation, or neither.

(a)

x	4	5	10	20
y	40	32	16	8

(b)

x	2	5	9	11
y	18	45	81	99

(c)

x	2	4	6	8
y	8	20	30	56

Solution

(a) As x increases, y decreases. Because $xy = 160$ for each data point in the table, the equation $y = \frac{160}{x}$ models the data. The data represent inverse variation.

(b) Because $\frac{y}{x} = 9$ for each data point in the table, the equation $y = 9x$ models the data in the table. These data represent direct variation.

(c) Neither the products xy nor the ratios $\frac{y}{x}$ are constant for the data in the table. Therefore these data represent neither direct variation nor inverse variation.

Now Try Exercises 51(a), 53(a), 55(a)

5 Solving Joint Variation Problems

🌐 *Math in Context* Construction In many applications, a quantity depends on more than one variable. In *joint variation*, a quantity varies with the product of more than one variable. For example, the formula for the area A of a rectangle is given by

$$A = WL,$$

where W and L are the width and length, respectively. Thus the area of a rectangle varies jointly with the width and length.

Cylinder
$V = \pi r^2 h,$

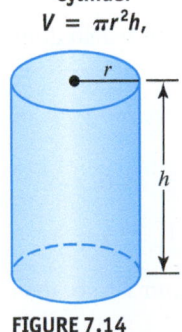

FIGURE 7.14

> **JOINT VARIATION**
>
> Let x, y, and z denote three quantities. Then z **varies jointly** with x and y if there is a nonzero number k such that
>
> $$z = kxy.$$

Sometimes joint variation can involve a power of a variable. For example, the volume V of a cylinder is given by $V = \pi r^2 h$, where r is its radius and h is its height, as illustrated in **FIGURE 7.14**. In this case, we say that the volume varies jointly with the height and the *square* of the radius. The constant of variation is $k = \pi$.

EXAMPLE 9 **Finding the strength of a rectangular beam**

The strength S of a rectangular beam varies jointly with its width w and the square of its thickness t. See **FIGURE 7.15**. If a beam 3 inches wide and 5 inches thick supports 750 pounds, how much can a similar beam 2 inches wide and 6 inches thick support?

FIGURE 7.15

Solution

STEP 1: The strength of a beam is modeled by $S = kwt^2$, where k is a constant of variation.

STEP 2: We can find k by substituting $S = 750$, $w = 3$, and $t = 5$ in the formula.

$$750 = k \cdot 3 \cdot 5^2 \quad \text{Substitute in } S = kwt^2.$$

$$k = \frac{750}{3 \cdot 5^2} \quad \text{Solve for } k; \text{ rewrite.}$$

$$= 10 \quad \text{Simplify.}$$

STEP 3: The equation $S = 10wt^2$ models the strength of this type of beam.

STEP 4: When $w = 2$ and $t = 6$, the beam can support

$$S = 10 \cdot 2 \cdot 6^2 = 720 \text{ pounds.}$$

Now Try Exercises 83

CRITICAL THINKING 1

Compare the increased strength of a beam if the width doubles and if the thickness doubles. What happens to the strength of a beam if both the width and thickness triple?

7.7 Putting It All Together

CONCEPT	EXPLANATION	EXAMPLES
Proportion	A statement that two ratios are equal	$\dfrac{8}{17} = \dfrac{49}{x}$ and $\dfrac{x}{6} = \dfrac{3}{14}$
Direct Variation	Two quantities x and y vary according to the equation $y = kx$, where k is a nonzero constant. The constant of proportionality (or variation) is k.	$y = 4x$ or $\dfrac{y}{x} = 4$ <table><tr><td>x</td><td>1</td><td>2</td><td>4</td></tr><tr><td>y</td><td>4</td><td>8</td><td>16</td></tr></table> Note that if x doubles, then y also doubles.
Inverse Variation	Two quantities x and y vary according to the equation $y = \frac{k}{x}$, where k is a nonzero constant. The constant of proportionality (or variation) is k.	$y = \dfrac{3}{x}$ or $xy = 3$ <table><tr><td>x</td><td>1</td><td>3</td><td>6</td></tr><tr><td>y</td><td>3</td><td>1</td><td>½</td></tr></table> Note that if x doubles from 3 to 6, then y decreases by half.
Joint Variation	Three quantities x, y, and z vary according to the equation $z = kxy$, where k is a nonzero constant.	The area A of a triangle varies jointly with b and h according to the equation $A = \frac{1}{2}bh$, where b is its base and h is its height. The constant of variation is $k = \frac{1}{2}$.

7.7 Exercises

MyMathLab®

CONCEPTS AND VOCABULARY

1. What is a proportion?

2. If 5 is to 6 as x is to 7, write a proportion that allows you to find x.

3. Suppose that y is directly proportional to x. If x doubles, what happens to y?

4. Suppose that y is inversely proportional to x. If x doubles, what happens to y?

5. If y varies directly with x, then $\frac{y}{x}$ equals a(n) _____.

6. If y varies inversely with x, then xy equals a(n) _____.

7. Would the food bill B generally vary directly or inversely with the number of people N being fed? Explain your reasoning.

8. Would the time T needed to paint a building vary directly or inversely with the number of painters N working on the job? Explain your reasoning.

9. If xy equals a constant for every data point (x, y) in a table, then the data represent _____ variation.

10. If $\frac{y}{x}$ equals a constant for every data point (x, y) in a table, then the data represent _____ variation.

SOLVING PROPORTIONS

Exercises 11–22: Solve the proportion.

11. $\dfrac{x}{24} = \dfrac{5}{8}$

12. $\dfrac{x}{5} = \dfrac{3}{7}$

13. $\dfrac{14}{x} = \dfrac{2}{3}$

14. $\dfrac{4}{9} = \dfrac{9}{x}$

15. $\dfrac{3}{16} = \dfrac{h}{256}$

16. $\dfrac{20}{a} = \dfrac{15}{4}$

17. $\dfrac{3}{4} = \dfrac{2x}{7}$

18. $\dfrac{7}{3z} = \dfrac{5}{4}$

19. $\dfrac{x}{6} = \dfrac{8}{3x}$

20. $\dfrac{4}{x} = \dfrac{4x}{9}$

21. $\dfrac{x}{7} = \dfrac{7}{4x}$

22. $\dfrac{2}{3x} = \dfrac{27x}{8}$

23. **Thinking Generally** Solve $\frac{a}{b} = \frac{c}{d}$ for b.

24. **Thinking Generally** Solve $\frac{a+b}{c^2} = \frac{1}{2}$ for b.

Exercises 25–32: Do the following.
(a) Write a proportion that models the situation.
(b) Solve the proportion for x.

25. 5 is to 8 as 9 is to x.

26. x is to 11 as 7 is to 4.

27. A triangle has sides of 4, 7, and 10. In a similar triangle, the shortest side is 8 and the longest side is x.

28. A rectangle has sides of 5 and 12. In a similar rectangle, the longer side is 10 and the shorter side is x.

29. If you earn \$98 in 7 hours, then you can earn x dollars in 11 hours.

30. If 14 gallons of gasoline contain 1.4 gallons of ethanol, then 22 gallons of gasoline contain x gallons of ethanol.

31. If 3 MP3 players can hold 750 songs, then 7 similar MP3 players can hold x songs.

32. If a gas pump fills a 25-gallon tank in 6 minutes, it can fill a 14-gallon tank in x minutes.

SOLVING DIRECT VARIATION PROBLEMS

Exercises 33–38: Suppose that y is directly proportional to x.
(a) Use the given information to find the constant of proportionality k.
(b) Then use $y = kx$ to find y for $x = 6$.

33. $y = 4$ when $x = 2$

34. $y = 5$ when $x = 10$

35. $y = 3$ when $x = 2$

36. $y = 11$ when $x = 55$

37. $y = -60$ when $x = 8$

38. $y = -17$ when $x = 68$

SOLVING INVERSE VARIATION PROBLEMS

Exercises 39–44: Suppose that y is inversely proportional to x.
(a) Use the given information to find the constant of proportionality k.
(b) Then use $y = \frac{k}{x}$ to find y for $x = 8$.

39. $y = 6$ when $x = 4$

40. $y = 2$ when $x = 24$

41. $y = 80$ when $x = \frac{1}{2}$ **42.** $y = \frac{1}{4}$ when $x = 32$

43. $y = 20$ when $x = 20$ **44.** $y = \frac{8}{3}$ when $x = 12$

SOLVING JOINT VARIATION PROBLEM

Exercises 45–50: Let z vary jointly with x and y.
  (a) *Find the constant of variation k.*
  (b) *Use z = kxy to find z when x = 5 and y = 7.*

45. $z = 6$ when $x = 3$ and $y = 8$

46. $z = 135$ when $x = 2.5$ and $y = 9$

47. $z = 5775$ when $x = 25$ and $y = 21$

48. $z = 1530$ when $x = 22.5$ and $y = 4$

49. $z = 25$ when $x = \frac{1}{2}$ and $y = 5$

50. $z = 12$ when $x = \frac{1}{4}$ and $y = 12$

ANALYZING DATA

Exercises 51–56: (Refer to Example 8.)
  (a) *Determine whether the data represent direct variation, inverse variation, or neither.*
  (b) *If the data represent either direct or inverse variation, find an equation that models the data.*
  (c) *Graph the equation and the data when possible.*

51.

x	2	3	4	5
y	3	4.5	6	7.5

52.

x	1	5	9	15
y	6	30	54	90

53.

x	3	6	9	12
y	12	6	4	3

54.

x	2	6	10	14
y	105	35	21	15

55.

x	4	6	12	20
y	10	20	30	40

56.

x	10	20	30	40
y	12	6	5	4

Exercises 57–62: Use the graph to determine whether the data represent direct variation, inverse variation, or neither. Find the constant of variation whenever possible.

57.

58.

59.

60.

61.

62.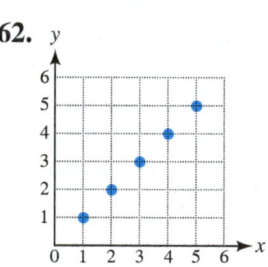

APPLICATIONS INVOLVING PROPORTIONS AND VARIATION

63. Recording Music A 750-megabyte CD can record 85 minutes of music. How many minutes can be recorded on 420 megabytes?

64. Making Fudge If $2\frac{2}{3}$ cups of sugar can make 14 pieces of fudge, how much sugar is needed to make 49 pieces of fudge?

65. Water Content in Snow (Refer to Example 1.) Eight inches of heavy, wet snow are equivalent to an inch of rain. Estimate the water content in 13 inches of heavy, wet snow.

66. Wages If a person working for an hourly wage earns \$143 in 13 hours, how much will that person earn in 15 hours?

67. Strength of a Beam (Refer to Example 4.) The strength of a metal beam varies directly with its width. A beam that is 6.2 inches wide can support 2800 pounds. How much weight can a similar beam support if it is 4.7 inches wide?

68. Strength of a Beam The strength of a wood beam varies inversely with its length. A beam that is 33 feet long can support 1200 pounds. How much weight can a similar beam support if it is 23 feet long?

69. Height of a Tree (Refer to Example 2.) A 5-foot-tall person casts an 8-foot-long shadow, and a nearby tree casts a 30-foot-long shadow. See the figure below. Estimate the height of the tree.

70. Making Coffee If 6 tablespoons of coffee grounds make 10 cups of coffee, how many tablespoons of coffee grounds are needed to make 35 cups of coffee?

71. Rolling Resistance of Cars If you were to try to push a car, you would experience *rolling resistance*. This resistance equals the force necessary to keep the car moving slowly in neutral gear. The following table shows the rolling resistance R for passenger cars of different gross weights W. (*Source:* N. Garber and L. Hoel, *Traffic and Highway Engineering.*)

W (pounds)	2000	2500	3000	3500
R (pounds)	24	30	36	42

(a) Do the data represent direct or inverse variation? Explain.
(b) Find an equation that models the data. Graph the equation with the data.
(c) Estimate the rolling resistance of a 3200-pound car.

72. Transportation Costs The use of a particular toll bridge varies inversely according to the toll. If the toll is $0.50, then 8000 vehicles use the bridge. Estimate the number of users if the toll is $0.80. (*Source:* N. Garber.)

73. Flow of Water The gallons of water G flowing in 1 minute through a hose with a cross-sectional area A are shown in the table.

A (square inch)	0.2	0.3	0.4	0.5
G (gallons)	5.4	8.1	10.8	13.5

(a) Do the data represent direct or inverse variation? Explain.
(b) Find an equation that models the data. Graph the equation with the data.
(c) Interpret the constant of variation k.

74. Hooke's Law The table shows the distance D that a spring stretches when a weight W is hung on it.

W (pounds)	2	6	9	15
D (inches)	1.6	4.8	7.2	12

(a) Do the data represent direct or inverse variation? Explain.
(b) Find an equation that models the data.
(c) How far will the spring stretch if an 11-pound weight is hung on it?

75. Tightening Lug Nuts (Refer to Example 7.) When a tire is mounted on a car, the lug nuts should not be over-tightened. The table below shows the maximum force used with wrenches of different lengths.

L (inches)	8	10	16
F (pounds)	150	120	75

Source: Tires plus.

(a) Model the data, using the equation $F = \frac{k}{L}$.
(b) How much force should be used with a wrench 15 inches long?

76. Ozone and UV Radiation Ozone in the atmosphere filters out approximately 90% of the harmful ultraviolet (UV) rays from the sun. Depletion of the ozone layer increases the amount of UV radiation reaching Earth's surface. An increase in UV radiation is associated with skin cancer. The following graph shows the percentage increase y in UV radiation for a decrease in the ozone layer of x percent. (*Source:* R. Turner, D. Pearce, and I. Bateman, *Environmental Economics.*)

continued on next page

(a) Does this graph represent direct or inverse variation?

(b) Find an equation for the line in the graph.

(c) Estimate the percentage increase in UV radiation if the ozone layer decreases by 5%.

77. **Air Temperature and Altitude** In the first 6 miles of Earth's atmosphere, air cools as the altitude increases. The following graph shows the temperature change y in degrees Fahrenheit at an altitude of x miles. (*Source:* A. Miller and R. Anthes, *Meteorology.*)

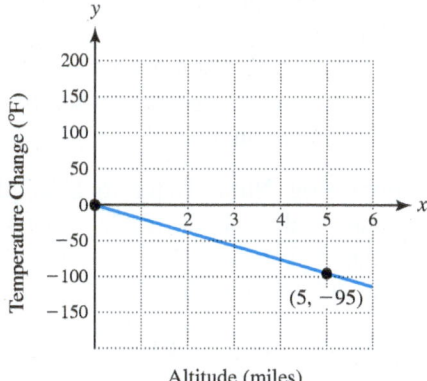

Altitude (miles)

(a) Does this graph represent direct variation or inverse variation?

(b) Find an equation that models the data in the graph.

(c) Is the constant of proportionality k positive or negative? Interpret k.

(d) Find the change in air temperature 2.5 miles high.

78. **Cost of Tuition** (Refer to Example 5.) The cost of tuition is directly proportional to the number of credits taken. If 6 credits cost $483, find the cost of 13 credits. What does the constant of proportionality represent?

79. **Electrical Resistance** The electrical resistance of a wire is directly proportional to its length. If a 30-foot-long wire has a resistance of 3 ohms, find the resistance of an 18-foot-long wire.

80. **Resistance and Current** The current that flows through an electrical circuit is inversely proportional to the resistance. When the resistance R is 180 ohms, the current I is 0.6 amp. Find the current when the resistance is 54 ohms.

81. **Joint Variation** The variable z varies jointly with the second power of x and the third power of y. Write a formula for z if $z = 31.9$ when $x = 2$ and $y = 2.5$.

82. **Wind Power** The electrical power generated by a windmill varies jointly with the square of the diameter of the area swept out by the blades and the cube of the wind velocity. If a windmill with a 10-foot diameter and a 16-mile-per-hour wind generates 15,392 watts, how much power would be generated if the blades swept out an area 12 feet in diameter and the wind speed was 15 miles per hour?

83. **Strength of a Beam** (Refer to Example 9.) If a wood beam 5 inches wide and 3 inches thick supports 300 pounds, how much can a similar beam 5 inches wide and 2 inches thick support?

84. **Carpeting** The cost of carpet for a rectangular room varies jointly with its width and length. If a room 11 feet wide and 14 feet long costs $539 to carpet, find the cost to carpet a room 17 feet by 19 feet. Interpret the constant of variation k.

85. **Weight on the Moon** The weight of a person on the moon is directly proportional to the weight of the person on Earth. If a 175-pound person weighs 28 pounds on the moon, how much will a 220-pound person weigh on the moon?

86. **Weight Near Earth** The weight W of a person near Earth is inversely proportional to the square of the person's distance d from the *center* of Earth. If a person weighs 200 pounds when $d = 4000$ miles, how much does the same person weigh when $d = 7000$ miles? (*Note:* The radius of Earth is about 4000 miles.)

87. **Ohm's Law** The voltage V in an electrical circuit varies jointly with the amperage I and resistance R. If $V = 220$ when $I = 10$ and $R = 22$, find V when $I = 15$ and $R = 50$.

88. Revenue The revenue R from selling x items at price p varies jointly with x and p. If $R = \$24{,}000$ when $x = 3000$ and $p = \$8$, find the number of items x sold when $R = \$30{,}000$ and $p = \$6$.

89. Whales and Diving At a depth of 1800 feet, a sperm whale experiences water pressure equal to 800 pounds per square inch. Find the pressure at 2700 feet. (*Hint:* Water pressure is proportional to water depth.)

90. Eagle Population Twenty-two bald eagles are tagged and released into the wilderness. Later, an observed sample of 56 bald eagles contains 7 eagles that are tagged. Estimate the bald eagle population in this wilderness area. (*Hint:* The number of tagged birds in the sample is proportional to the number of tagged birds in the population)

WRITING ABOUT MATHEMATICS

91. Explain what it means for a quantity y to be directly proportional to a quantity x.

92. Explain what it means for a quantity y to be inversely proportional to a quantity x.

SECTION 7.7   Checking Basic Concepts

1. Solve each proportion.

(a) $\dfrac{x}{9} = \dfrac{2}{5}$ (b) $\dfrac{4}{3} = \dfrac{5}{b}$

2. Write a proportion that models each situation. Then solve it.

(a) 4 is to 6 as 8 is to x.

(b) If 2 compact discs can record 148 minutes of music, then 5 compact discs can record x minutes of music.

3. Suppose that y is inversely proportional to x. If $y = 4$ when $x = 15$, find the constant of proportionality k. Find y when $x = 10$.

4. Decide whether the data in the table represent direct or inverse variation. Explain your reasoning. Find the constant of variation.

(a)

x	2	4	6	8
y	3	6	9	12

(b)

x	2	4	6	8
y	12	6	4	3

5. Wages If a person working for an hourly wage earns \$272 in 17 hours, how much will that person earn in 10 hours?

CHAPTER 7   Summary

Rational Expression

A rational expression can be written in the form $\frac{P}{Q}$, where P and Q are polynomials, and is defined whenever $Q \neq 0$.

Example: $\frac{x^2}{x-5}$ is a rational expression that is defined for all real numbers except $x = 5$.

Undefined Rational Expression A rational expression is undefined for any value of the variable that makes the denominator equal to 0.

Examples: $\frac{1}{x-5}$ is undefined when $x = 5$.

$\frac{3y}{y^2 - 4}$ is undefined when $y = -2$ or when $y = 2$.

Simplifying a Rational Expression To simplify a rational expression, factor the numerator and the denominator. Then apply the basic principle of rational expressions,

$$\frac{PR}{QR} = \frac{P}{Q}. \quad Q \text{ and } R \text{ nonzero}$$

Examples: $\dfrac{x^2 - 4}{x^2 - 3x + 2} = \dfrac{(x+2)(x-2)}{(x-1)(x-2)} = \dfrac{x+2}{x-1}, \quad \dfrac{x-3}{3-x} = -\dfrac{x-3}{x-3} = -1$

Multiplying Rational Expressions

To multiply two rational expressions, multiply the numerators and multiply the denominators.

$$\frac{A}{B} \cdot \frac{C}{D} = \frac{AC}{BD} \quad B \text{ and } D \text{ nonzero}$$

Example: $\dfrac{3}{x-1} \cdot \dfrac{4}{x+1} = \dfrac{12}{(x-1)(x+1)}$

Dividing Rational Expressions To divide two rational expressions, multiply by the reciprocal of the divisor.

$$\frac{A}{B} \div \frac{C}{D} = \frac{A}{B} \cdot \frac{D}{C} = \frac{AD}{BC} \quad B, C, \text{ and } D \text{ nonzero}$$

Example: $\dfrac{x+2}{x^2 + 3x} \div \dfrac{x+2}{x} = \dfrac{x+2}{x(x+3)} \cdot \dfrac{x}{x+2} = \dfrac{x(x+2)}{x(x+3)(x+2)} = \dfrac{1}{x+3}$

Addition of Rational Expressions That Have Like Denominators To add two rational expressions that have like denominators, add their numerators. Keep the same denominator.

$$\frac{A}{C} + \frac{B}{C} = \frac{A+B}{C} \quad C \text{ nonzero}$$

Example: $\dfrac{5x}{x+4} + \dfrac{1}{x+4} = \dfrac{5x+1}{x+4}$

Subtraction of Rational Expressions That Have Like Denominators To subtract two rational expressions that have like denominators, subtract their numerators. Keep the same denominator.

$$\frac{A}{C} - \frac{B}{C} = \frac{A - B}{C} \qquad C \text{ nonzero}$$

Example: $\dfrac{6}{2x + 1} - \dfrac{2}{2x + 1} = \dfrac{4}{2x + 1}$

SECTION 7.4 ■ ADDITION AND SUBTRACTION WITH UNLIKE DENOMINATORS

Finding Least Common Multiples The least common multiple (LCM) of two or more polynomials can be found as follows.

STEP 1: Factor each polynomial completely.

STEP 2: List each factor the greatest number of times that it occurs in any factorization.

STEP 3: Find the product of this list of factors. It is the LCM.

Example: $4x^2(x + 1) = 2 \cdot 2 \cdot x \cdot x \cdot (x + 1)$
$2x(x^2 - 1) = 2 \cdot x \cdot (x + 1) \cdot (x - 1)$

Listing each factor the greatest number of times and multiplying give the LCM.

$$2 \cdot 2 \cdot x \cdot x \cdot (x + 1)(x - 1) = 4x^2(x^2 - 1)$$

Finding the Least Common Denominator The least common denominator (LCD) is the least common multiple (LCM) of the denominators.

Example: From the preceding example, the LCD for $\dfrac{1}{4x^2(x + 1)}$ and $\dfrac{1}{2x(x^2 - 1)}$ is $4x^2(x^2 - 1)$.

Addition and Subtraction of Rational Expressions That Have Unlike Denominators First write each rational expression by using the LCD. Then add or subtract the resulting rational expressions. Finally, write your answer in lowest terms.

Example: $\dfrac{1}{x - 1} - \dfrac{1}{x} = \dfrac{x}{x(x - 1)} - \dfrac{x - 1}{x(x - 1)} = \dfrac{x - (x - 1)}{x(x - 1)} = \dfrac{1}{x(x - 1)}$

Note that the LCD is $x(x - 1)$.

SECTION 7.5 ■ COMPLEX FRACTIONS

Complex Fractions A complex fraction is a rational expression that contains fractions in its numerator, denominator, or both. The following equation can be used to simplify basic complex fractions.

$$\frac{\dfrac{a}{b}}{\dfrac{c}{d}} = \frac{a}{b} \cdot \frac{d}{c} \qquad b, c, \text{ and } d \text{ nonzero}$$

Example: $\dfrac{\dfrac{x}{3}}{\dfrac{x}{x - 1}} = \dfrac{x}{3} \cdot \dfrac{x - 1}{x} = \dfrac{x(x - 1)}{3x} = \dfrac{x - 1}{3}$

Simplifying Complex Fractions

Method I Combine terms in the numerator, combine terms in the denominator, and then multiply the numerator by the reciprocal of the denominator.

Method II Multiply the numerator and denominator by the LCD of all fractions within the expression and simplify the resulting expression.

Example: *Method I*
$$\frac{\dfrac{1}{a} - \dfrac{1}{b}}{\dfrac{1}{a} + \dfrac{1}{b}} = \frac{\dfrac{b-a}{ab}}{\dfrac{b+a}{ab}} = \frac{b-a}{ab} \cdot \frac{ab}{b+a} = \frac{b-a}{b+a}$$

Method II The least common denominator is ab.
$$\frac{\dfrac{1}{a} - \dfrac{1}{b}}{\dfrac{1}{a} + \dfrac{1}{b}} = \frac{\left(\dfrac{1}{a} - \dfrac{1}{b}\right)ab}{\left(\dfrac{1}{a} + \dfrac{1}{b}\right)ab} = \frac{\dfrac{ab}{a} - \dfrac{ab}{b}}{\dfrac{ab}{a} + \dfrac{ab}{b}} = \frac{b-a}{b+a}$$

SECTION 7.6 ■ RATIONAL EQUATIONS AND FORMULAS

Solving Rational Equations One way to solve the equation $\frac{a}{b} = \frac{c}{d}$ is to cross multiply to obtain $ad = bc$. (Check each answer.) A general way to solve rational equations is to use the following steps.

STEP 1: Find the LCD of the terms in the equation.

STEP 2: Multiply each side of the equation by the LCD.

STEP 3: Simplify each term.

STEP 4: Solve the resulting equation.

STEP 5: Check each answer in the *given* equation. Reject any value that makes a denominator equal 0.

Examples: $\frac{1}{2x} = \frac{2}{x+3}$ implies that $x + 3 = 4x$, or $x = 1$. This answer checks.

To solve $\frac{5}{x} - \frac{1}{3x} = \frac{7}{3}$, multiply each term by the LCD, $3x$:
$$\frac{5(3x)}{x} - \frac{1 \cdot (3x)}{3x} = \frac{7(3x)}{3},$$

which simplifies to $15 - 1 = 7x$, or $x = 2$. This answer checks.

Solving for a Variable Many formulas contain more than one variable. To solve for a particular variable, use the rules of algebra to isolate the variable.

Example: To solve $S = \frac{2\pi}{r}$ for r, multiply each side by r to obtain $Sr = 2\pi$ and then divide each side by S to obtain $r = \frac{2\pi}{S}$.

SECTION 7.7 ■ PROPORTIONS AND VARIATION

Proportions A proportion is a statement that two ratios are equal.

Example: $\frac{5}{x} = \frac{4}{7}$

Similar Triangles Two triangles are similar if the measures of their corresponding angles are equal. Corresponding sides of similar triangles are proportional.

Example: The following triangles are similar.

$$\frac{3}{6} = \frac{4}{x} \qquad \frac{\text{Shorter leg}}{\text{Shorter leg}} = \frac{\text{Longer leg}}{\text{Longer leg}}$$

By solving this proportion, we see that $x = 8$.

Direct Variation A quantity y is *directly proportional* to a quantity x, or y *varies directly* with x, if there is a nonzero constant k such that $y = kx$. The number k is called the *constant of proportionality* or the *constant of variation*.

Example: If y varies directly with x, then the ratios $\frac{y}{x}$ always equal k. The following data satisfy $\frac{y}{x} = 4$, so the constant of variation is 4. Thus $y = 4x$.

x	1	2	3	4
y	4	8	12	16

Inverse Variation A quantity y is *inversely proportional* to a quantity x, or y *varies inversely* with x, if there is a nonzero constant k such that $y = \frac{k}{x}$.

Example: If y varies inversely with x, then the products xy always equal k. The following data satisfy $xy = 12$, so the constant of variation is 12. Thus $y = \frac{12}{x}$.

x	1	2	4	6
y	12	6	3	2

Joint Variation The quantity z *varies jointly* with x and y if $z = kxy$, $k \neq 0$.

Example: The area A of a rectangle varies jointly with the width W and length L because $A = LW$. Note that $k = 1$ in this example.

CHAPTER 7	**Review Exercises**

SECTION 7.1

Exercises 1–4: If possible, evaluate the expression for the given value of x.

1. $\dfrac{3}{x - 3}$ $x = -2$

2. $\dfrac{4x}{5 - x^2}$ $x = 3$

3. $\dfrac{-x}{7 - x}$ $x = 7$

4. $\dfrac{4x}{x^2 - 3x + 2}$ $x = 2$

5. Complete the table for the rational expression. If a value is undefined, place a dash in the table.

x	-2	-1	0	1	2
$\dfrac{3x}{x - 1}$					

6. Find the x-values that make $\dfrac{8}{x^2 - 4}$ undefined.

Exercises 7–10: Simplify to lowest terms.

7. $\dfrac{25x^3y^4}{15x^5y}$

8. $\dfrac{x^2 - 36}{x + 6}$

9. $\dfrac{2x^2 + 5x - 3}{2x^2 + x - 1}$

10. $\dfrac{3x^2 + 10x - 8}{3x^2 + x - 2}$

Exercises 11 and 12: Do the following.
   (a) *Decide whether you are given an expression or an equation.*
   (b) *If you are given an expression, simplify it. If you are given an equation, solve it.*

11. $\dfrac{x + 1}{(x - 3)(x + 1)}$

12. $\dfrac{x}{3(4 - 1)} = 2$

SECTION 7.2

Exercises 13–16: Multiply and write in lowest terms.

13. $\dfrac{x - 3}{x + 1} \cdot \dfrac{2x + 2}{x - 3}$

14. $\dfrac{2x + 5}{(x + 5)(x - 1)} \cdot \dfrac{x - 1}{2x + 5}$

15. $\dfrac{z + 3}{z - 4} \cdot \dfrac{z - 4}{(z + 3)^2}$

16. $\dfrac{x^2}{x^2 - 4} \cdot \dfrac{x + 2}{x}$

Exercises 17–20: Divide and write in lowest terms.

17. $\dfrac{x + 1}{2x} \div \dfrac{3x + 3}{5x}$

18. $\dfrac{4}{x^3} \div \dfrac{x + 1}{2x^2}$

19. $\dfrac{x - 5}{x + 2} \div \dfrac{2x - 10}{x + 2}$

20. $\dfrac{x^2 - 6x + 5}{x^2 - 25} \div \dfrac{x - 1}{x + 5}$

SECTION 7.3

Exercises 21–28: Add or subtract and write in lowest terms.

21. $\dfrac{2}{x + 10} + \dfrac{8}{x + 10}$

22. $\dfrac{9}{x - 1} - \dfrac{8}{x - 1}$

23. $\dfrac{x + 2y}{2x} + \dfrac{x - 2y}{2x}$

24. $\dfrac{x}{x - 3} - \dfrac{3}{x - 3}$

25. $\dfrac{x}{x^2 - 1} - \dfrac{1}{x^2 - 1}$

26. $\dfrac{2x}{x^2 - 25} + \dfrac{10}{x^2 - 25}$

27. $\dfrac{3}{xy} - \dfrac{1}{xy}$

28. $\dfrac{x + y}{2y} + \dfrac{x - y}{2y}$

SECTION 7.4

Exercises 29–32: Find the least common multiple for the expressions. Leave your answer in factored form.

29. $3x,\ 5x$

30. $5x^2,\ 10x$

31. $x,\ x - 5$

32. $10x^2,\ x^2 - x$

Exercises 33–36: Rewrite the rational expression by using the given denominator D.

33. $\dfrac{3}{8}, D = 24$

34. $\dfrac{4}{3x}, D = 12x$

35. $\dfrac{3x}{x - 2}, D = x^2 - 4$

36. $\dfrac{2x}{2x - 3}, D = 2x^2 + x - 6$

Exercises 37–46: Simplify the expression. Write your answer in lowest terms.

37. $\dfrac{5}{8} + \dfrac{1}{6}$

38. $\dfrac{3}{4x} + \dfrac{1}{x}$

39. $\dfrac{5}{9x} - \dfrac{2}{3x}$

40. $\dfrac{7}{x - 1} - \dfrac{3}{x}$

41. $\dfrac{1}{x + 1} + \dfrac{1}{x - 1}$

42. $\dfrac{4}{3x^2} - \dfrac{3}{2x}$

43. $\dfrac{1 + x}{3x} - \dfrac{3}{2x}$

44. $\dfrac{x}{x^2 - 1} - \dfrac{1}{x - 1}$

45. $\dfrac{2}{x - y} - \dfrac{3}{x + y}$

46. $\dfrac{x}{y - x} + \dfrac{y}{x - y}$

SECTION 7.5

Exercises 47–54: Simplify the complex fraction.

47. $\dfrac{\frac{3}{4}}{\frac{7}{11}}$

48. $\dfrac{\frac{x}{5}}{\frac{2x}{7}}$

49. $\dfrac{\frac{m}{n}}{\frac{2m}{n^2}}$

50. $\dfrac{\frac{3}{p - 1}}{\frac{1}{p + 1}}$

51. $\dfrac{\frac{1}{2x} - \frac{1}{3x}}{\frac{2}{3x} - \frac{1}{6x}}$

52. $\dfrac{\frac{2}{xy} - \frac{1}{y}}{\frac{2}{xy} + \frac{1}{y}}$

53. $\dfrac{\dfrac{1}{x} - \dfrac{1}{x+1}}{\dfrac{x}{x+1}}$

54. $\dfrac{\dfrac{2}{x-1} - \dfrac{1}{x+1}}{\dfrac{1}{x^2-1}}$

SECTION 7.6

Exercises 55–58: Solve and check your answer.

55. $\dfrac{x}{5} = \dfrac{4}{7}$

56. $\dfrac{4}{x} = \dfrac{3}{2}$

57. $\dfrac{3}{z+1} = \dfrac{1}{2z}$

58. $\dfrac{x+2}{x} = \dfrac{3}{5}$

Exercises 59–70: If possible, solve. Check your answer.

59. $\dfrac{1}{5x} + \dfrac{3}{5x} = \dfrac{1}{5}$

60. $\dfrac{1}{x-1} + \dfrac{2x}{x-1} = 1$

61. $\dfrac{1}{x} + \dfrac{2}{3x} = \dfrac{1}{3}$

62. $\dfrac{1}{x+3} + \dfrac{2x}{x+3} = \dfrac{3}{2}$

63. $\dfrac{5}{x} - \dfrac{3}{x+1} = \dfrac{1}{2}$

64. $\dfrac{1}{x-1} - \dfrac{1}{x+1} = \dfrac{1}{4}$

65. $\dfrac{4}{p} - \dfrac{5}{p+2} = 0$

66. $\dfrac{1}{x-3} - \dfrac{1}{x+3} = \dfrac{1}{x^2-9}$

67. $\dfrac{1}{x+1} = \dfrac{-x}{x+1}$

68. $\dfrac{2}{x} = \dfrac{2}{x^2+x} - 4$

69. $\dfrac{2}{x^2-2x} + \dfrac{1}{x^2-4} = \dfrac{1}{x^2+2x}$

70. $\dfrac{3}{x^2-3x} - \dfrac{1}{x^2-9} = \dfrac{1}{x^2+3x}$

Exercises 71 and 72: Do the following.
(a) Decide whether you are given an expression or an equation.
(b) If you are given an expression, simplify it. If you are given an equation, solve it.

71. $\dfrac{4}{x} - x = 0$

72. $\dfrac{x^2}{x-3} - \dfrac{9}{x-3}$

Exercises 73 and 74: Solve for the specified variable.

73. $\dfrac{1}{a} + \dfrac{2}{b} = \dfrac{3}{c}$ for b

74. $y = \dfrac{x}{x-1}$ for x

SECTION 7.7

Exercises 75 and 76: Solve the proportion.

75. $\dfrac{x}{6} = \dfrac{1}{3}$

76. $\dfrac{5}{x} = \dfrac{7}{3}$

*Exercises 77 and 78: **Proportions** Do the following.*
(a) Write a proportion that models the situation.
(b) Solve the proportion for x.

77. A rectangle has sides of 6 and 13. In a similar rectangle, the longer side is 20 and the shorter side is x.

78. If you earn \$341 in 11 hours, then you can earn x dollars in 8 hours.

*Exercises 79 and 80: **Direct Variation** Suppose that y is directly proportional to x.*
(a) Use the given information to find the constant of proportionality k.
(b) Then use $y = kx$ to find y for $x = 5$.

79. $y = 8$ when $x = 2$

80. $y = 21$ when $x = 7$

*Exercises 81 and 82: **Inverse Variation** Suppose that y is inversely proportional to x.*
(a) Use the given information to find the constant of proportionality k.
(b) Then use $y = \frac{k}{x}$ to find y for $x = 5$.

81. $y = 2.5$ when $x = 4$

82. $y = 7$ when $x = 3$

83. Joint Variation Suppose that z varies jointly with x and y. If $z = 483$ when $x = 23$ and $y = 7$, find the constant of variation k.

84. Joint Variation Suppose that z varies jointly with x and the square of y. If $z = 891$ when $x = 22$ and $y = 3$, find z when $x = 10$ and $y = 4$.

Exercises 85 and 86: Do the following.
(a) Determine whether the data represent direct or inverse variation.
(b) Find an equation that models the data.
(c) Graph the data and your equation.

85.

x	2	3	4	5
y	30	20	15	12

86.

x	2	4	6	8
y	6	12	18	24

Exercises 87 and 88: Use the graph to determine whether the data represent direct or inverse variation. Find the constant of variation.

87.

88.

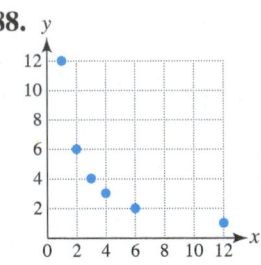

APPLICATIONS

89. Modeling Traffic Flow Fifteen vehicles per minute can pass through an intersection. If vehicles arrive randomly at an average rate of x per minute, the average waiting time T in minutes is given by $T = \frac{1}{15 - x}$ for $x < 15$. (*Source: N. Garber, Traffic and Highway Engineering.*)
   (a) Find T when $x = 10$ and interpret the result.
   (b) Complete the table.

x	5	10	13	14	14.9
T					

   (c) What happens to the waiting time as the traffic rate x approaches 15 vehicles per minute?

90. Distance and Time A car traveled at 50 miles per hour for 150 miles and then traveled at 75 miles per hour for 150 miles. What was the car's average speed?

91. Emptying a Swimming Pool A large pump can empty a swimming pool in 100 hours, whereas a small pump can empty the pool in 160 hours. How long will it take to empty the pool if both pumps are used?

92. Running Two athletes run 10 miles. One of the athletes runs 2 miles per hour faster and finishes 10 minutes ahead of the other athlete. Find the average speed of each athlete.

93. River Current A boat can travel 16 miles upstream in the same time that it can travel 48 miles downstream. If the speed of the current is 4 miles per hour, find the boat's speed.

94. Height of a Tree A 5-foot-tall person has a 6-foot-long shadow, and a nearby tree has a 40-foot-long shadow. Estimate the tree's height.

95. Transportation Costs Use of a toll road varies inversely with the toll. If the toll is $0.25, then 400 vehicles use the road. Estimate the number of users for a toll of $0.50.

96. Cost of Carpet The cost of carpet is directly proportional to the amount of carpet purchased. If 17 square yards cost $612, find the cost of 13 square yards.

97. Tightening a Bolt The torque exerted on a nut by a wrench is inversely proportional to the length of the wrench's handle. Suppose that a 12-inch wrench can be used to tighten a nut by using 30 pounds of force. How much force is necessary to tighten the same nut by using a 10-inch wrench?

98. Water Content in Snow Twenty inches of extremely dry snow are equivalent to an inch of rain. Estimate the water content in 32 inches of this type of snow.

99. Polar Plunge When a person swims in extremely cold water, the water removes body heat 25 times faster than air at the same temperature. To be safe, a person has between 30 and 90 seconds to get out of the water. How long could a person safely remain in air at the same temperature?

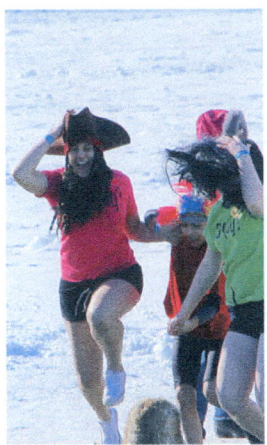

100. Strength of a Beam The strength of a wood beam varies inversely with its length. A beam that is 18 feet long can support 900 pounds. How much weight can a similar beam support if its length is 21 feet?

101. Wind Power The electric power generated by a windmill varies jointly with the square of the diameter of the area swept out by the blades and the cube of the wind velocity. If a windmill with 6-foot-diameter blades and a 20-mile-per-hour wind generates 10,823 watts, how much power would be generated if the blades were 10 feet in diameter and the wind speed were 12 miles per hour?

102. Strength of a Beam The strength of a beam varies jointly with its width w and the square of its thickness t. If a beam 8 inches wide and 5 inches thick supports 650 pounds, how much can a similar beam 6 inches wide and 6 inches thick support?

CHAPTER 7   Test   MyMathLab® You Tube

1. Evaluate the expression $\frac{3x}{2x - 1}$ for $x = 3$.

2. Find any x-value that makes $\frac{x - 1}{x + 2}$ undefined.

Exercises 3 and 4: Simplify the expression.

3. $\dfrac{x^2 - 25}{x - 5}$

4. $\dfrac{3x^2 - 15x}{3x}$

Exercises 5–10: Simplify the expression. Write your answer in lowest terms.

5. $\dfrac{x - 2}{x + 4} \cdot \dfrac{3x + 12}{x - 2}$

6. $\dfrac{z + 1}{z + 3} \cdot \dfrac{2z + 6}{z + 1}$

7. $\dfrac{x + 1}{5x} \div \dfrac{2x + 2}{x - 1}$

8. $\dfrac{2}{x^2} \div \dfrac{x + 3}{3x}$

9. $\dfrac{x}{x + 4} + \dfrac{3x + 1}{x + 4}$

10. $\dfrac{4t + 1}{2t - 3} - \dfrac{3t - 6}{2t - 3}$

11. Find the least common multiple for
$$6x^2 \quad \text{and} \quad 3x^2 - 3x.$$

12. Rewrite the rational expression $\frac{4}{7x}$ by using the denominator $7x^2 - 7x$.

Exercises 13 and 14: Simplify the expression. Write your answer in lowest terms.

13. $\dfrac{1}{y^2 + y} - \dfrac{y - 1}{y^2 - y}$

14. $\dfrac{1}{xy} + \dfrac{x}{y} - \dfrac{1}{y^2}$

Exercises 15 and 16: Simplify the complex fraction.

15. $\dfrac{\dfrac{a}{3b}}{\dfrac{5a}{b^2}}$

16. $\dfrac{1 + \dfrac{1}{p - 1}}{1 - \dfrac{1}{p - 1}}$

Exercises 17–24: Solve the equation and check your answer.

17. $\dfrac{2}{7} = \dfrac{5}{x}$

18. $\dfrac{x + 3}{2x} = 1$

19. $\dfrac{1}{2x} + \dfrac{2}{5x} = \dfrac{9}{10}$

20. $\dfrac{1}{x - 1} + \dfrac{2}{x + 2} = \dfrac{3}{2}$

21. $\dfrac{1}{x^2 - 1} - \dfrac{4}{x + 1} = \dfrac{3}{x - 1}$

22. $\dfrac{1}{x^2 - 4x} + \dfrac{2}{x^2 - 16} = \dfrac{2}{x^2 + 4x}$

23. $\dfrac{x}{2x - 1} = \dfrac{1 - x}{2x - 1}$

24. $\dfrac{x}{x - 5} + \dfrac{x}{x + 5} = \dfrac{10x}{x^2 - 25}$

Exercises 25 and 26: Solve the equation for the specified variable.

25. $y = \dfrac{2}{3x - 5}$ for x

26. $\dfrac{a + b}{ab} = 1$ for b

27. Suppose that y is directly proportional to x.
   (a) If $y = 14$ when $x = 4$, find k so that $y = kx$.
   (b) Use $y = kx$ to find y for $x = 6$.

28. Use the table to determine whether y varies directly or inversely with x. Find the constant of variation.

x	2	4	8	16
y	16	8	4	2

29. **Emptying a Swimming Pool** It takes a large pump 40 hours to empty a swimming pool, whereas a small pump can empty the pool in 60 hours. How long will it take to empty the pool if both pumps are used?

30. **Height of a Building** A 5-foot-tall post has a 4-foot-long shadow, and a nearby building has a 54-foot-long shadow. Estimate the height of the building.

31. **Standing in Line** A department store clerk can wait on 30 customers per hour. If people arrive randomly at an average rate of x per hour, then the average number of customers N waiting in line is given by

$$N = \frac{x^2}{900 - 30x},$$

for $x < 30$. Evaluate the expression for $x = 24$ and interpret your result.

CHAPTERS 1–7   Cumulative Review Exercises

1. Evaluate $\pi r^2 h$ when $r = 2$ and $h = 6$.

2. Translate the phrase "two less than twice a number" into an algebraic expression using the variable x.

Exercises 3 and 4: Evaluate and simplify to lowest terms.

3. $\frac{1}{2} \div \frac{5}{4}$

4. $\frac{5}{8} + \frac{1}{8}$

Exercises 5 and 6: Simplify the expression.

5. $-2 + 7x + 4 - 5x$

6. $-4(4 - y) + (5 - 3y)$

7. Solve $-2x + 11 = 13$ and check the solution.

8. Solve $-3x + 1 \geq x$.

Exercises 9 and 10: Graph the equation. Determine any intercepts.

9. $2x - 3y = 6$

10. $x = 1$

11. Sketch a line with slope $m = 3$ passing through $(-2, -1)$. Write its slope–intercept form.

12. The table lists points located on a line. Write the slope–intercept form of the line.

x	-2	-1	0	1
y	-5	-3	-1	1

Exercises 13 and 14: Find the slope–intercept form for the line satisfying the given conditions.

13. Parallel to $y = -\frac{2}{3}x + 1$, passing through $(2, -1)$

14. Passing through $(-1, 2)$ and $(2, 4)$

15. Determine which ordered pair is a solution to the system of equations: $(2, -6)$ or $(1, -2)$.

$$4x + y = 2$$
$$x - 4y = 9$$

16. Solve the system of equations.

$$-3r - t = 2$$
$$2r + t = -4$$

Exercises 17 and 18: Determine if the system of linear equations has no solutions, one solution, or infinitely many solutions.

17. $2x - y = 5$
    $-2x + y = -5$

18. $4x - 6y = 12$
    $-6x + 9y = 18$

Exercises 19–22: Simplify the expression.

19. $3z^2 \cdot 5z^6$

20. $(ab)^3$

21. $(2y - 3)(5y + 2)$

22. $(x^2 - y^2)^2$

Exercises 23 and 24: Simplify and write the expression using positive exponents.

23. $(3x^2)^{-3}$

24. $\dfrac{4x^2}{2x^4}$

25. Write 0.00123 in scientific notation.

26. Divide and check: $\dfrac{2x^2 - x + 3}{x - 1}$.

Exercises 27–30: Factor completely.

27. $6 + 13x - 5x^2$

28. $9z^2 - 4$

29. $t^2 + 16t + 64$

30. $x^3 - 16x$

Exercises 31 and 32: Solve the equation.

31. $y^2 + 5y - 14 = 0$

32. $x^3 = 4x$

Exercises 33 and 34: Simplify.

33. $\dfrac{3x}{4y} \cdot \dfrac{y}{9x^2}$

34. $\dfrac{x}{x^2 - 4} \div \dfrac{2x}{x - 2}$

35. Simplify $\dfrac{1 + \frac{2}{x}}{1 - \frac{2}{x}}$.

36. Solve $z = 3x - 2y$ for x.

Exercises 37 and 38: Solve.

37. $\dfrac{4}{3x} - \dfrac{3}{4x} = 1$

38. $\dfrac{1}{x - 1} + \dfrac{2}{x + 2} = \dfrac{3}{2}$

39. Suppose that y is directly proportional to x and that $y = 7$ when $x = 14$. Find y when $x = 11$.

40. Determine whether the data in the table represent direct or inverse variation. Find an equation that models the data.

x	1	2	4	10
y	20	10	5	2

APPLICATIONS

41. Shoveling a Driveway Two people are shoveling snow from a driveway. The first person shovels 12 square feet per minute, while the second person shovels 9 square feet per minute.
  (a) Write a simplified expression that gives the total square feet that the two people shovel together in x minutes.
  (b) How long would it take them to clear a driveway with 1890 square feet?

42. Burning Calories An athlete can burn 10 calories per minute while running and 4 calories per minute while walking. If the athlete burns 450 calories in 60 minutes, how long is spent on each activity?

43. Triangle In an isosceles triangle, the measures of the two smaller angles are equal and their sum is 32° more than the largest angle.
  (a) Let x be the measure of one of the two smaller angles and y be the measure of the largest angle. Write a system of linear equations whose solution gives the measures of these angles.
  (b) Solve your system.

44. Strength of a Beam The strength of a wood beam varies inversely with its length. A beam that is 10 feet long can support 1100 pounds. How much weight can a similar beam support if it is 22 feet long?

8 Introduction to Functions

The trend today is to use more green building material. As a result, demand for green building material is expected to increase dramatically, particularly bamboo. Bamboo doesn't need pesticides or much water, it pulls large amounts of carbon dioxide out of the air, and it can be harvested every 5–10 years. One essential reason why bamboo is valuable is because it grows at an astonishing rate, as much as 2 inches per hour.

In Examples 6 and 7, in the second section of this chapter, the mathematical concept of a *function* is used to describe these facts about bamboo and green building material. However, functions can also be used to describe other things like human behavior and social networks, which are discussed in Example 10, in the fourth section of this chapter.

8.1 Functions and Their Representations

NEW VOCABULARY

☐ Function
☐ Function notation
☐ Input/output
☐ Name of the function
☐ Dependent variable
☐ Independent variable
☐ Verbal representation
☐ Numerical representation
☐ Symbolic representation
☐ Graphical representation
☐ Diagrams/diagrammatic representation
☐ Relation
☐ Domain/range
☐ Nonlinear functions
☐ Vertical line test

1 Introducing Functions

Connecting Concepts with Your Life Functions are used to calculate many important quantities. For example, suppose that a person works for $7 per hour. Then we could use a function *named f* to calculate the amount of money the person earns after working x hours simply by multiplying the *input x* by 7. The result y is called the *output*. This concept is shown visually in the following diagram.

Uses input x to
compute a value of y

For each valid input x, a function computes *exactly one* output y. This can be represented by the ordered pair (x, y). If the input is 5 hours, f outputs $7 \cdot 5 = \$35$; if the input is 8 hours, f outputs $7 \cdot 8 = \$56$. These results can be represented by the ordered pairs $(5, 35)$ and $(8, 56)$. Sometimes an input may not be valid. For example, if $x = -3$, there is no reasonable output because a person cannot work -3 hours. The set of all valid inputs is called the **domain** of the function and the set of corresponding outputs is called the **range** of the function.

We say that *y is a function of x* because the output y is determined by and *depends* on the input x. As a result, y is called the *dependent variable* and x is the *independent variable*. To emphasize that y is a function of x, we use the notation $y = f(x)$. The symbol $f(x)$ does not represent multiplication of a variable f and a variable x. The notation $y = f(x)$ is called *function notation*, is read "y equals f of x," and means that function f with input x produces output y. For example, if $x = 3$ hours, $y = f(3) = 7 \cdot 3 = \$21$.

The following See the Concept summarizes these basic concepts about a function.

See the Concept 1   **FUNCTION NOTATION**

"y equals f of x"

Ⓐ Name
$y = f(x)$
Ⓒ Output   Ⓑ Input

Ⓐ Common function names are f, g, and h.

Ⓑ The set of input values is the *domain*. The variable x is the *independent variable*.

Ⓒ The set of output values is the *range*. The variable y is the *dependent variable*.

WATCH VIDEO IN MML.

NOTE: Functions can be given *meaningful* names and variables. For example, function f could have been defined by $P(h) = 7h$, where function P calculates the pay after working h hours for $7 per hour. ∎

Connecting Concepts with Your Life Functions can be used to compute a variety of quantities. For example, suppose that a boy has a sister who is exactly 5 years older than he is. If the age of the boy is x, then a function g can calculate the age of his sister by adding 5 to x. Thus $g(4) = 4 + 5 = 9$, $g(10) = 10 + 5 = 15$, and in general $g(x) = x + 5 = y$. That is, function g adds 5 to input x to obtain the output $y = g(x)$.

Functions can be represented by an input–output machine, as illustrated in **FIGURE 8.1**. This machine represents function g and receives input $x = 4$, adds 5 to this value, and then outputs $g(4) = 4 + 5 = 9$.

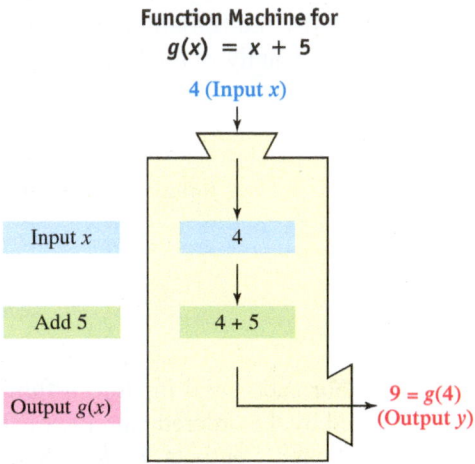

Function Machine for
$g(x) = x + 5$

FIGURE 8.1

2 Recognizing Representations of Functions

 Connecting Concepts with Your Life A function f forms a relation between inputs x and outputs y that can be represented verbally, numerically, symbolically, and graphically. Functions can also be represented with diagrams. We begin by considering a function f that converts yards to feet.

VERBAL REPRESENTATION (WORDS) To convert x yards to y feet, we multiply x by 3. Therefore, if function f computes the number of feet in x yards, a **verbal representation** of f is "Multiply the input x in yards by 3 to obtain the output y in feet."

NUMERICAL REPRESENTATION (TABLE OF VALUES) A function f that converts yards to feet is shown in **TABLE 8.1**, where $y = f(x)$.

A *table of values* is called a **numerical representation** of a function. Many times it is impossible to list all valid inputs x in a table. On the one hand, if a table does not contain every x-input, it is a *partial* numerical representation. On the other hand, a *complete* numerical representation includes *all* valid inputs. **TABLE 8.1** is a partial numerical representation of f because many valid inputs, such as $x = 10$ or $x = 5.3$, are not shown in it. Note that for each valid input x, there is exactly one output y. *For a function, inputs are not listed more than once in a table.*

SYMBOLIC REPRESENTATION (FORMULA) A *formula* can be used to provide a **symbolic representation** of a function. The computation performed by f to convert x yards to y feet is expressed by $y = 3x$. A formula for f is $f(x) = 3x$, where $y = f(x)$. We say that function f is *defined by* or *given by* $f(x) = 3x$. Thus $f(2) = 3 \cdot 2 = 6$.

GRAPHICAL REPRESENTATION (GRAPH) A **graphical representation**, or **graph**, visually associates an x-input with a y-output. The ordered pairs

$$(1, 3), (2, 6), (3, 9), (4, 12), (5, 15), (6, 18), \text{ and } (7, 21)$$

Yards to Feet

x (yards)	y (feet)
1	3
2	6
3	9
4	12
5	15
6	18
7	21

TABLE 8.1

from **TABLE 8.1** are plotted in **FIGURE 8.2(a)**. This scatterplot suggests a line for the graph of f. For each real number x, there is exactly one real number y determined by $y = 3x$. If we restrict inputs to $x \geq 0$ and plot all ordered pairs $(x, 3x)$, then a line with no breaks will appear, as shown in **FIGURE 8.2(b)**.

STUDY TIP

Be sure that you understand what verbal, numerical, graphical, and symbolic representations are.

Yards to Feet

FIGURE 8.2

Because $f(1) = 3$, it follows that the point $(1, 3)$ lies on the graph of f, as shown in **FIGURE 8.3**. Graphs can sometimes be used to define a function f. For example, because the point $(1, 3)$ lies on the graph of f in **FIGURE 8.3**, we can conclude that $f(1) = 3$. That is, each point on the graph of f defines an input–output pair for f. Because there are infinitely many points on the graph of f, there are also infinitely many input-output pairs for function f.

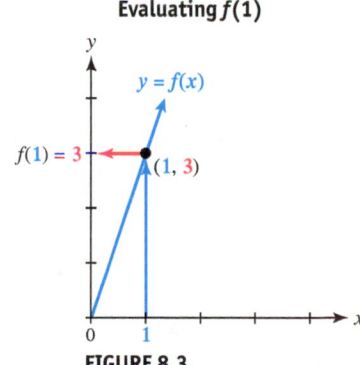

FIGURE 8.3

MAKING CONNECTIONS 1

Functions, Points, and Graphs

If $f(a) = b$, then the point (a, b) lies on the graph of f. Conversely, if the point (a, b) lies on the graph of f, then $f(a) = b$. See **FIGURE 8.4(a)**. Thus each point on the graph of f can be written in the form $(a, f(a))$. See **FIGURE 8.4(b)**.

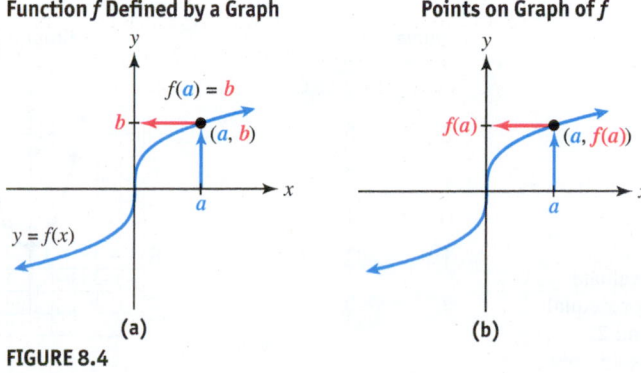

FIGURE 8.4

In the next example, we find a formula and then sketch a graph of a function.

EXAMPLE 1   **Finding representations of a function**

Let function f square the input x and then subtract 1 to obtain the output y.
(a) Write a formula, or symbolic representation, for f.
(b) Make a table of values, or numerical representation, for f. Use $x = -2, -1, 0, 1, 2$.
(c) Sketch a graph, or graphical representation, of f.

Solution
(a) **Symbolic Representation** If we square x and then subtract 1, we obtain $x^2 - 1$. Thus a formula for f is $f(x) = x^2 - 1$.
(b) **Numerical Representation** Make a table of values for $f(x)$, as shown in **TABLE 8.2**. For example,

$$f(-2) = (-2)^2 - 1 = 4 - 1 = \textbf{3}.$$

(c) **Graphical Representation** To obtain a graph of $f(x) = x^2 - 1$, plot the points from **TABLE 8.2** and then connect them with a smooth curve, as shown in **FIGURE 8.5**. Note that we need to plot enough points so that we can determine the overall shape of the graph.

$f(x) = x^2 - 1$

x	$f(x)$
-2	**3**
-1	0
0	-1
1	0
2	3

TABLE 8.2

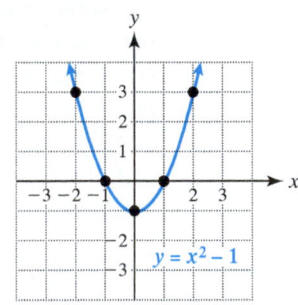

FIGURE 8.5

Now Try Exercise 63

READING CHECK 1

• Give a verbal and a symbolic representation of a function that calculates the number of days in a given number of weeks. Choose meaningful variables.

The following See the Concept illustrates four basic representations of a function.

See the Concept 2   **FOUR REPRESENTATIONS OF A FUNCTION**

Symbolic	Numerical		Graphical	Verbal

Symbolic
$f(x) = x^2 + 1$

Numerical

x	y
-2	5
-1	2
0	1
1	2
2	5

To get the table, evaluate $f(x) = x^2 + 1$ for x equal to $-2, -1, 0, 1,$ and 2.

Graphical

Plot the points from the table, and sketch a smooth curve.

Verbal
Function f squares the input x and then adds 1 to produce the output y.

WATCH VIDEO IN MML.

Yards to Feet

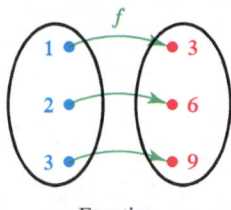

Function

FIGURE 8.6

DIAGRAMMATIC REPRESENTATION (DIAGRAM) Functions may be represented by **diagrams**. **FIGURE 8.6** is a diagram of a function where an arrow is used to identify the output y associated with input x. For example, an arrow is drawn from input **2** to output **6**, which is written in function notation as $f(2) = 6$. That is, **2** yards are equivalent to **6** feet.

   FIGURE 8.7(a) shows a (different) function f even though $f(1) = 4$ and $f(2) = 4$. Although two inputs for f have the same output, each valid input has exactly one output. In contrast, **FIGURE 8.7(b)** is *not* a function because input 2 results in two different outputs, 5 and 6.

Function

Not a Function

(a)   (b)

FIGURE 8.7

EVALUATING REPRESENTATIONS OF FUNCTIONS In the next three examples, we evaluate several representations of functions.

EXAMPLE 2   **Evaluating symbolic representations (formulas)**

Evaluate each function f at the given value of x.
(a) $f(x) = 3x - 7$ $\quad\quad x = -2$

(b) $f(x) = \dfrac{x}{x + 2}$ $\quad\quad x = 0.5$

(c) $f(x) = \sqrt{x - 1}$ $\quad\quad x = 10$

Solution
(a) $f(-2) = 3(-2) - 7 = -6 - 7 = -13$

(b) $f(0.5) = \dfrac{0.5}{0.5 + 2} = \dfrac{0.5}{2.5} = 0.2$

(c) $f(10) = \sqrt{10 - 1} = \sqrt{9} = 3$

Now Try Exercises 21, 23, 31

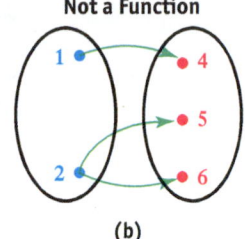 *Math in Context* In the next example, we calculate sales tax by evaluating different representations of a function.

EXAMPLE 3   **Calculating sales tax**

Let a function f compute a sales tax of 7% on a purchase of x dollars. Use the given representation to evaluate $f(2)$.
(a) **Verbal Representation** Multiply a purchase of x dollars by 0.07 to obtain a sales tax of y dollars.
(b) **Numerical Representation (partial)** Shown in **TABLE 8.3** on the next page
(c) **Symbolic Representation** $f(x) = 0.07x$
(d) **Graphical Representation** Shown in **FIGURE 8.8** on the next page
(e) **Diagrammatic Representation** Shown in **FIGURE 8.9** on the next page

Sales Tax of 7%

x	$f(x)$
$1.00	$0.07
$2.00	$0.14
$3.00	$0.21
$4.00	$0.28

TABLE 8.3

Sales Tax of 7%

(Graph: Sales Tax (dollars) vs Purchase (dollars))

FIGURE 8.8

Sales Tax of 7%

(Diagram: mapping f from {1,2,3,4} to {0.07, 0.14, 0.21, 0.28})

FIGURE 8.9

Solution

(a) Verbal To evaluate $f(2)$, multiply the input 2 by 0.07 to obtain 0.14. The sales tax on a $2.00 purchase is $0.14.

(b) Numerical From **TABLE 8.3**, $f(2) = \$0.14$.

(c) Symbolic Because $f(x) = 0.07x$, $f(2) = 0.07(2) = 0.14$, or $0.14.

(d) Graphical To evaluate $f(2)$ with a graph, first find **2** on the x-axis in **FIGURE 8.10**. Then move vertically upward until you reach the graph of f. The point on the graph may be estimated as $(2, 0.14)$, meaning that $f(2) = 0.14$. Note that it may not be possible to find the exact answer from a graph. For example, one might estimate $f(2)$ to be 0.13 or 0.15 instead of 0.14.

Evaluating a Function

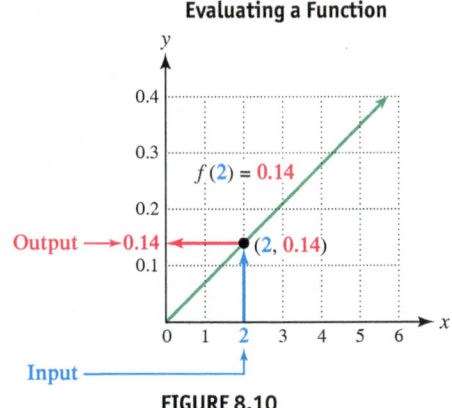

FIGURE 8.10

(e) Diagrammatic In **FIGURE 8.9**, follow the arrow from **2** to **0.14**. Thus $f(2) = 0.14$.

Now Try Exercises 25, 33, 53, 59, 61

The following See the Concept summarizes how to evaluate a function graphically.

See the Concept 3 **EVALUATING A FUNCTION GRAPHICALLY**

If $f(x) = x^2 - 2x$, evaluate $f(-1)$ graphically.

B Move vertically to the graph of f.

C Move horizontally to the y-axis.

A Find -1 on the x-axis.

A To evaluate $f(-1)$ graphically, begin by finding $x = -1$ on the x-axis.

B Next move vertically until the graph of f is reached.

C Finally, move horizontally to the y-axis. The y-value corresponding to an x-value of -1 is **3**. Thus $f(-1) = 3$.

WATCH VIDEO IN MML .

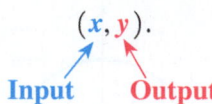 *Math in* ^Medical *Context* There are many examples of functions. To give more meaning to a function, sometimes we change both its name and its input variable. For instance, if we know the radius r of a circle, we can calculate its circumference by using $C(r) = 2\pi r$. The next example illustrates how functions are used in physical therapy.

EXAMPLE 4   ### Computing crutch length

People who sustain leg injuries often require crutches. A proper crutch length can be estimated without using trial and error. The function L, given by $L(t) = 0.72t + 2$, outputs an appropriate crutch length L in inches for a person t inches tall. (*Source: Journal of the American Physical Therapy Association.*)

(a) Find $L(60)$ and interpret the result.

(b) If one person is 70 inches tall and another person is 71 inches tall, what should be the difference in their crutch lengths?

Solution

(a) $L(60) = 0.72(60) + 2 = $ **45.2**. Thus a person 60 inches tall needs crutches that are about 45.2 inches long.

(b) From the formula $L(t) = 0.72t + 2$, we can see that each 1-inch increase in t results in a 0.72-inch increase in $L(t)$. For example,

$$L(71) - L(70) = 53.12 - 52.4 = 0.72.$$

Now Try Exercise 75

---

3 ▸ Defining a Function

RELATIONS Sometimes there is a relationship between two quantities or variables, such as the cost y of parking x hours. We can represent this relationship by using a set of ordered pairs. For example, it might cost $2 to park 1 hour, $4 to park 2 hours, and $8 to park from 3 to 5 hours. Thus, the ordered pairs $(1, 2)$, $(2, 4)$, $(3, 8)$, $(4, 8)$, and $(5, 8)$ are elements or members of this relation.

A **relation** is a set of ordered pairs. The sets C and S are examples of relations.

$$C = \{(1, 2), (2, 4), (3, 8), (4, 8), (5, 8)\} \quad \text{and} \quad S = \{(1, 1), (3, 4), (3, 5)\}$$

The *domain* of a relation is the set of x-values and the *range* of a relation is the set of y-values. For example, the domain of S is $\{1, 3\}$ and the range is $\{1, 4, 5\}$.

FUNCTIONS A function is a fundamental concept in mathematics. Its definition should allow for all representations of a function. *A function receives an input x and produces exactly one output y*, which can be expressed as an ordered pair:

$$(x, y).$$

↑ ↑
Input   **Output**

A *relation* is a set of ordered pairs, and a function is a special type of relation.

FUNCTION

A **function** f is a set of ordered pairs (x, y), where each x-value corresponds to exactly one y-value.

The *domain* of f is the set of all x-values, and the *range* of f is the set of all y-values. For example, a function f that converts 1, 2, 3, and 4 yards to feet could be expressed as

$$f = \{(1, 3), (2, 6), (3, 9), (4, 12)\}.$$

The domain of f is $D = \{1, 2, 3, 4\}$, and the range of f is $R = \{3, 6, 9, 12\}$.

MAKING CONNECTIONS 2

Relations and Functions

A relation can be thought of as a set of input–output pairs. A function is a special type of relation whereby each input results in *exactly one output*.

🌐 *Math in Context* ^Education   In the next example, we see how education can improve a person's chances for earning a higher income.

EXAMPLE 5   **Computing average income**

A function f computes the median individual income in dollars in relation to educational attainment. This function is defined by $f(N) = 25{,}376$, $f(H) = 34{,}736$, $f(B) = 57{,}252$, and $f(M) = 68{,}952$, where N denotes no diploma, H a high school diploma, B a bachelor's degree, and M a master's degree. (*Source:* Bureau of Labor.)
(a) Write f as a set of ordered pairs.
(b) Give the domain and range of f.
(c) Discuss the relationship between education and income.

Solution
(a) $f = \{(N, 25376), (H, 34736), (B, 57252), (M, 68952)\}$
(b) The domain of function f is given by $D = \{N, H, B, M\}$, and the range of function f is given by $R = \{25376, 34736, 57252, 68952\}$.
(c) Education pays—the greater the educational attainment, the greater annual earnings are.

Now Try Exercise 101

DOMAIN AND RANGE   We can determine the domain and range of a function from either a graph or a formula, as shown in the next two examples.

EXAMPLE 6   **Finding the domain and range graphically**

Use the graphs of f shown in **FIGURES 8.11** and **8.12** to find each function's domain and range.

(a)

FIGURE 8.11

(b)

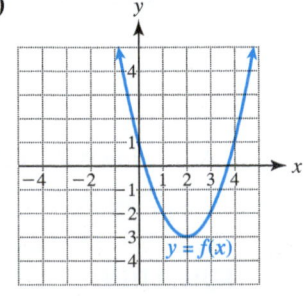

FIGURE 8.12

READING CHECK 2

• Use the graph in **FIGURE 8.12** to evaluate $f(3)$.

Solution

(a) The domain is the set of all x-values that correspond to points on the graph of f. **FIGURE 8.13** shows that the domain D includes all x-values satisfying $-3 \leq x \leq 3$. (Recall that the symbol $\leq$ is read "*less than or equal to*.") Because the graph is a semi-circle with no breaks, the domain includes all real numbers between and including -3 and 3. The range R is the set of y-values that correspond to points on the graph of f. Thus R includes all y-values satisfying $0 \leq y \leq 3$.

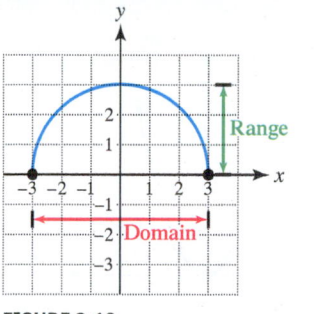

FIGURE 8.13   **FIGURE 8.14**

(b) The arrows on the ends of the graph in **FIGURE 8.12** indicate that the graph extends indefinitely left and right, as well as upward. Thus D includes **all real numbers**. See **FIGURE 8.14**. The smallest y-value on the graph is $y = -3$, which occurs when $x = 2$. Thus the range R is $y \geq -3$. (Recall that the symbol $\geq$ is read "*greater than or equal to*.")

Now Try Exercises 77, 81

CRITICAL THINKING 1

Suppose that a car travels at 50 miles per hour to a city that is 250 miles away. Sketch a graph of a function f that gives the distance y traveled after x hours. Identify the domain and range of f.

The domain of a function is the set of all valid inputs. To determine the domain of a function from a formula, we must find x-values for which the formula is defined. To do this, we must determine if we can substitute any real number in the formula for $f(x)$. If we can, then the domain of f is *all real numbers*. However, there are situations in which we must limit the domain of f. For example, the domain must often be limited when there is either division or a square root in the formula for f. When division occurs, we must be careful to avoid values of the variable that result in division by 0, which is undefined. When a square root occurs, we must be careful to avoid values of the variable that result in the square root of a negative number, which is not a real number. This concept is demonstrated in the next example.

EXAMPLE 7   **Finding the domain of a function**

Use $f(x)$ to find the domain of f.

(a) $f(x) = 5x$ (b) $f(x) = \dfrac{1}{x-2}$ (c) $f(x) = \sqrt{x}$

Solution

(a) Because we can always multiply a real number x by 5, $f(x) = 5x$ is defined for all real numbers. Thus the domain of f includes all real numbers.

(b) Because we cannot divide by 0, input $x = 2$ is not valid for $f(x) = \frac{1}{x-2}$. The expression for $f(x)$ is defined for all other values of x. Thus the domain of f includes all real numbers except 2, or $x \neq 2$.

(c) Because square roots of negative numbers are not real numbers, inputs for $f(x) = \sqrt{x}$ cannot be negative. Thus the domain of f includes all nonnegative numbers, or $x \geq 0$.

Now Try Exercises 87, 91, 95

Symbolic, numerical, and graphical representations of three common functions are shown in **FIGURE 8.15**. Note that their graphs are not lines. For this reason, they are called **nonlinear functions**. Use the graphs to find the domain and range of each function.

Absolute Value: $f(x) = |x|$

x	-2	-1	0	1	2
$|x|$	2	1	0	1	2

Square: $f(x) = x^2$

x	-2	-1	0	1	2
x^2	4	1	0	1	4

Square Root: $f(x) = \sqrt{x}$

x	0	1	4	9
$\sqrt{x}$	0	1	2	3

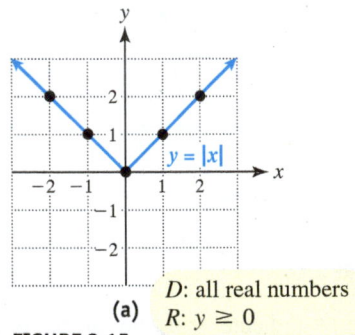

(a) D: all real numbers
     R: $y \geq 0$

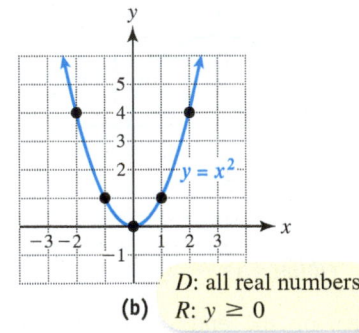

(b) D: all real numbers
     R: $y \geq 0$

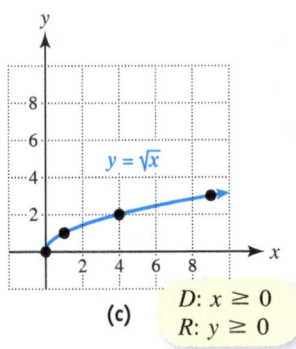

(c) D: $x \geq 0$
     R: $y \geq 0$

FIGURE 8.15

4 Identifying a Function

ORDERED PAIRS AND TABLES AS FUNCTIONS Recall that for a function each valid input x produces exactly one output y. In the next three examples, we demonstrate techniques for identifying a function.

EXAMPLE 8 **Determining whether a set of ordered pairs is a function**

The set S of ordered pairs (x, y) represents the number of mergers and acquisitions y in a recent year for selected technology companies x.

$$S = \{(\text{IBM}, 12), (\text{HP}, 7), (\text{Oracle}, 5), (\text{Apple}, 5), (\text{Microsoft}, 0)\}$$

Determine if S is a function. (*Source:* cbinsights.)

Solution
The input x is the name of the technology company, and the output y is the number of mergers and acquisitions associated with that company. The set S *is* a function because each company x is associated with exactly one number y. Note that even though there were 5 mergers and acquisitions corresponding to both Oracle and Apple, S is nonetheless a function.

Now Try Exercise 131

EXAMPLE 9 **Determining whether a table of values represents a function**

x	y
1	-4
2	8
3	2
1	5
4	-6

TABLE 8.4

Determine whether the relation in **TABLE 8.4** represents a function.

Solution
The table does not represent a function because input $x = 1$ produces two outputs: -4 and 5. That is, the following two ordered pairs both belong to this relation.

Same input x

$(1, -4) \qquad (1, 5) \qquad \leftarrow$ **Not a function**

Different outputs y

Now Try Exercise 133

VERTICAL LINE TEST To determine whether a graph represents a function, we must be convinced that it is impossible for an input x to have two or more outputs y. If two distinct points have the *same* x-coordinate on a graph, then the graph cannot represent a function. For example, the ordered pairs $(-1, 1)$ and $(-1, -1)$ could not lie on the graph of a function because input -1 results in *two* outputs: 1 and -1. When the points $(-1, 1)$ and $(-1, -1)$ are plotted, they lie on the same vertical line, as shown in **FIGURE 8.16(a)**. A graph passing through these points intersects the vertical line twice, as illustrated in **FIGURE 8.16(b)**.

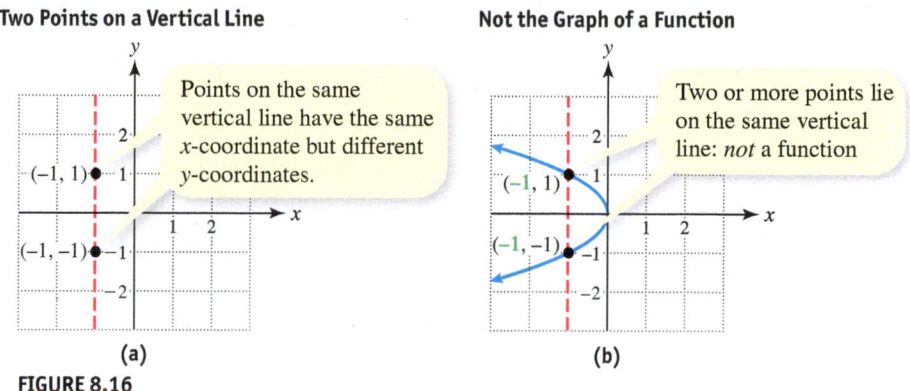

Two Points on a Vertical Line

Points on the same vertical line have the same x-coordinate but different y-coordinates.

(a)

Not the Graph of a Function

Two or more points lie on the same vertical line: *not* a function

(b)

FIGURE 8.16

To determine whether a graph represents a function, visualize vertical lines moving across the xy-plane. If each vertical line intersects the graph *at most once*, then it is a graph of a function. This test is called the **vertical line test**. Note that the graph in **FIGURE 8.16(b)** fails the vertical line test and therefore does not represent a function.

READING CHECK 3

• What is the vertical line test used for?

VERTICAL LINE TEST

If every vertical line intersects a graph at no more than one point, then the graph represents a function.

EXAMPLE 10

Determining whether a graph represents a function

Determine whether the graphs shown in **FIGURE 8.17** represent functions.

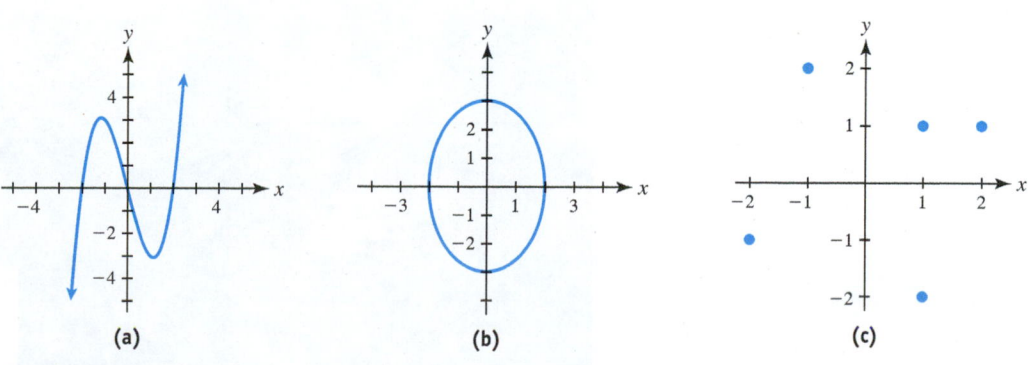

(a)   **(b)**   **(c)**

FIGURE 8.17

Solution

(a) Visualize vertical lines moving across the xy-plane from left to right. Any (red) vertical line will intersect the graph at most once, as depicted in **FIGURE 8.18(a)** on the next page. Therefore the graph *does* represent a function.

FIGURE 8.18

(b) Visualize vertical lines moving across the xy-plane from left to right. The graph *does not* represent a function because there exist (red) vertical lines that can intersect the graph twice. One such line is shown in **FIGURE 8.18(b)**.

(c) Visualize vertical lines moving across the xy-plane from left to right. The graph is a scatterplot and *does not* represent a function because there exists one (red) vertical line that intersects two points: $(1, 1)$ and $(1, -2)$. These points have the same x-coordinate but different y-coordinates. See **FIGURE 8.18(c)**.

Now Try Exercises 119, 121, 127

5 Using Graphing Calculators

Graphing calculators provide several features beyond those found on scientific calculators. For example, graphing calculators have additional keys that can be used to create tables, scatterplots, and graphs.

 Connecting Concepts with Your Life The **viewing rectangle**, or **window**, on a graphing calculator is similar to the viewfinder in a camera. A camera cannot take a picture of an entire scene. The camera must be centered on some object and can photograph only a portion of the available scenery. A camera can capture different views of the same scene by zooming in and out, as can graphing calculators. The xy-plane is infinite, but the calculator screen can show only a finite, rectangular region of the xy-plane. The viewing rectangle must be specified by setting minimum and maximum values for both the x- and y-axes before a graph can be drawn.

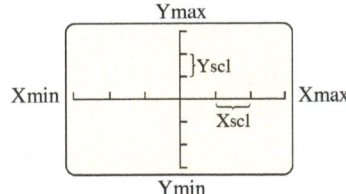

FIGURE 8.19

We use the following terminology regarding the size of a viewing rectangle. **Xmin** is the minimum x-value along the x-axis, and **Xmax** is the maximum x-value. Similarly, **Ymin** is the minimum y-value along the y-axis, and **Ymax** is the maximum y-value. Most graphs show an x-scale and a y-scale with tick marks on the respective axes. Sometimes the distance between consecutive tick marks is 1 unit, but at other times it might be 5 or 10 units. The distance represented by consecutive tick marks on the x-axis is called **Xscl**, and the distance represented by consecutive tick marks on the y-axis is called **Yscl** (see **FIGURE 8.19**).

If we want to represent this information about the viewing rectangle, we can write [Xmin, Xmax, Xscl] by [Ymin, Ymax, Yscl]. For example, $[-10, 10, 1]$ by $[-10, 10, 1]$ indicates that Xmin $= -10$, Xmax $= 10$, Xscl $= 1$, Ymin $= -10$, Ymax $= 10$, and Yscl $= 1$. This setting is referred to as the **standard viewing rectangle**. The window in **FIGURE 8.19** is $[-3, 3, 1]$ by $[-3, 3, 1]$.

EXAMPLE 11   ### Setting the viewing rectangle

Show the viewing rectangle $[-2, 3, 0.5]$ by $[-100, 200, 50]$ on your calculator.

Solution
The window setting and viewing rectangle are displayed in **FIGURE 8.20**. Note that in **FIGURE 8.20(b)**, there are 6 tick marks on the positive *x*-axis because its length is 3 units and the distance between consecutive tick marks is 0.5 unit.

CALCULATOR HELP

To set a viewing rectangle, see Appendix A (page AP-3).

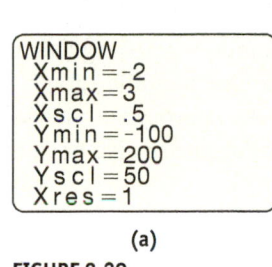
(a)

$[-2, 3, 0.5]$ by $[-100, 200, 50]$

(b)

FIGURE 8.20

Now Try Exercise 141

SCATTERPLOTS Many graphing calculators have the capability to create scatterplots and line graphs. The next example illustrates how to make a scatterplot with a graphing calculator.

EXAMPLE 12   ### Making a scatterplot with a graphing calculator

Plot the points $(-2, -2)$, $(-1, 3)$, $(1, 2)$, and $(2, -3)$ in $[-4, 4, 1]$ by $[-4, 4, 1]$.

Solution
We enter the points $(-2, -2)$, $(-1, 3)$, $(1, 2)$, and $(2, -3)$ as shown in **FIGURE 8.21(a)** using the STAT EDIT feature. The variable L1 represents the list of *x*-values, and the variable L2 represents the list of *y*-values. In **FIGURE 8.21(b)**, we set the graphing calculator to make a scatterplot with the STATPLOT feature, and in **FIGURE 8.21(c)**, the points have been plotted. If you have a different model of calculator, you may need to consult your owner's manual.

CALCULATOR HELP

To make a scatterplot, see Appendix A (pages AP-3 and AP-4).

(a)

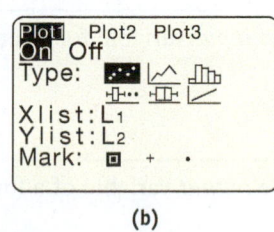
(b)

$[-4, 4, 1]$ by $[-4, 4, 1]$

(c)

FIGURE 8.21

Now Try Exercise 145

GRAPHS AND TABLES We can use graphing calculators to create graphs and tables, usually more efficiently and reliably than with pencil-and-paper techniques. However, a graphing calculator uses the same techniques that we might use to sketch a graph.

For example, one way to sketch a graph of $y = 2x - 1$ is first to make a table of values, as shown in **TABLE 8.5**. We can plot these points in the xy-plane, as shown in **FIGURE 8.22**. Next we might connect the points, as shown in **FIGURE 8.23**.

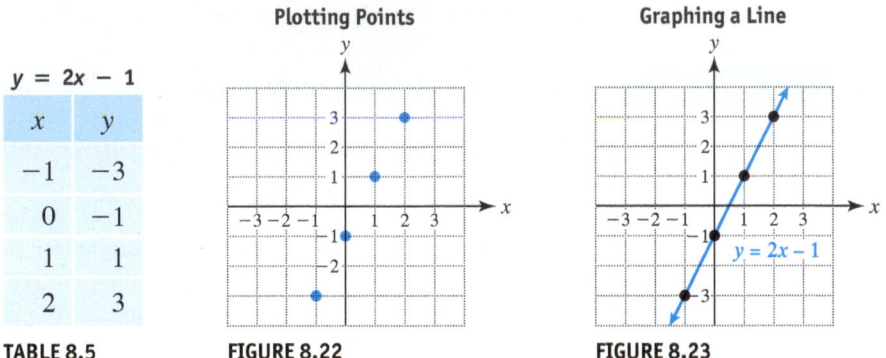

Plotting Points   **Graphing a Line**

$y = 2x - 1$

x	y
-1	-3
0	-1
1	1
2	3

TABLE 8.5   **FIGURE 8.22**   **FIGURE 8.23**

In a similar manner, a graphing calculator plots numerous points and connects them to make a graph. To create a similar graph with a graphing calculator, we enter the formula $Y_1 = 2X - 1$, set an appropriate viewing rectangle, and graph as shown in **FIGURES 8.24** and **8.25**. A table of values can also be generated, as illustrated in **FIGURE 8.26**.

CALCULATOR HELP

To make a graph, see Appendix A (page AP-3). To make a table, see Appendix A (pages AP-2 and AP-3).

$[-10, 10, 1]$ by $[-10, 10, 1]$

FIGURE 8.24   **FIGURE 8.25**   **FIGURE 8.26**

8.1   Putting It All Together

CONCEPT	EXPLANATION	EXAMPLES
Function	A set of ordered pairs (x, y), where each x-value corresponds to exactly one y-value	$f = \{(1, 3), (2, 3), (3, 1)\}$ $f(x) = 2x$ A graph of $y = x + 2$ A table of values for $y = 4x$
Independent Variable	The *input* variable for a function	*Function* *Independent Variable* $\quad f(x) = 2x$ $\qquad\qquad\qquad x$ $\quad A(r) = \pi r^2$ $\qquad\qquad\quad r$ $\quad V(s) = s^3$ $\qquad\qquad\qquad s$
Dependent Variable	The *output* variable of a function There is exactly one output for each valid input.	*Function* *Dependent Variable* $\quad y = f(x)$ $\qquad\qquad\qquad\quad y$ $\quad T = F(r)$ $\qquad\qquad\qquad\quad T$ $\quad V = g(r)$ $\qquad\qquad\qquad\quad V$
Domain and Range of a Function	The domain D is the set of all valid inputs. The range R is the set of all outputs.	For $S = \{(-1, 0), (3, 4), (5, 0)\}$, $D = \{-1, 3, 5\}$ and $R = \{0, 4\}$. For $f(x) = \frac{1}{x}$, the domain includes all real numbers except 0, or $x \neq 0$.

CONCEPT	EXPLANATION	EXAMPLES
Vertical Line Test	If every vertical line intersects a graph at no more than one point, the graph represents a function.	This graph does *not* pass this test and thus does not represent a function. Two points lie on the same vertical line: *not* a function

A function can be represented verbally, symbolically, numerically, and graphically.

REPRESENTATION	EXPLANATION	COMMENTS
Verbal	Precise word description of what is computed	May be oral or written Must be stated *precisely*
Symbolic	Mathematical formula	Efficient and concise way of representing a function (e.g., $f(x) = 2x - 3$)
Numerical	List of specific inputs and their outputs	May be in the form of a table or an explicit set of ordered pairs
Graphical, diagrammatic	Shows inputs and outputs visually	No words, formulas, or tables Many types of graphs and diagrams are possible.

8.1 Exercises

MyMathLab®

CONCEPTS AND VOCABULARY

1. The notation $y = f(x)$ is called _____ notation.

2. The notation $y = f(x)$ is read _____.

3. The notation $f(x) = x^2 + 1$ is a(n) _____ representation of a function.

4. A table of values is a(n) _____ representation of a function.

5. The set of valid inputs for a function is the _____.

6. The set of outputs for a function is the _____.

7. A function computes _____ output for each valid input.

8. (True or False?) The vertical line test is used to identify graphs of relations.

9. (True or False?) Four ways to represent functions are verbal, numerical, symbolic, and graphical.

10. If $f(3) = 4$, the point _____ is on the graph of f. If $(3, 6)$ is on the graph of f, then $f(\_\_\_\_) = \_\_\_\_$.

11. **Thinking Generally** If $f(a) = b$, the point _____ is on the graph of f.

12. **Thinking Generally** If (c, d) is on the graph of g, then $g(c) = \_\_\_\_$.

13. **Thinking Generally** If a is in the domain of f, then $f(a)$ represents how many outputs?

14. **Thinking Generally** If $f(x) = x$ for every x in the domain of f, then the domain and range of f are _____.

RECOGNIZING FUNCTIONS

Exercises 15–20: Determine whether the phrase describes a function.

15. Calculating the square of a number

16. Calculating the low temperature for a day

17. Listing the students who passed a given math exam

18. Listing the children of parent x

19. Finding sales tax on a purchase

20. Naming the people in your class

REPRESENTING AND EVALUATING FUNCTIONS

Exercises 21–32: Evaluate $f(x)$ at the given values of x.

21. $f(x) = 4x - 2$ $x = -1, 0$

22. $f(x) = 5 - 3x$ $x = -4, 2$

23. $f(x) = \sqrt{x}$ $x = 0, \frac{9}{4}$

24. $f(x) = \sqrt[3]{x}$ $x = -1, 27$

25. $f(x) = x^2$ $x = -5, \frac{3}{2}$

26. $f(x) = x^3$ $x = -2, 0.1$

27. $f(x) = 3$ $x = -8, \frac{7}{3}$

28. $f(x) = 100$ $x = -\pi, \frac{1}{3}$

29. $f(x) = 5 - x^3$ $x = -2, 3$

30. $f(x) = x^2 + 5$ $x = -\frac{1}{2}, 6$

31. $f(x) = \dfrac{2}{x + 1}$ $x = -5, 4$

32. $f(x) = \dfrac{x}{x - 4}$ $x = -3, 1$

Exercises 33–38: Do the following.

   (a) *Write a formula for the function described.*
   (b) *Evaluate the function for input* 10 *and interpret the result.*

33. Function I computes the number of inches in x yards.

34. Function A computes the area of a circle with radius r.

35. Function M computes the number of miles in x feet.

36. Function C computes the circumference of a circle with radius r.

37. Function A computes the square feet in x acres. (*Hint:* There are 43,560 square feet in one acre.)

38. Function K computes the number of kilograms in x pounds. (*Hint:* There are about 2.2 pounds in one kilogram.)

Exercises 39–42: Write each function f as a set of ordered pairs. Give the domain and range of f.

39. $f(1) = 3, f(2) = -4, f(3) = 0$

40. $f(-1) = 4, f(0) = 6, f(1) = 4$

41. $f(a) = b, f(c) = d, f(e) = a, f(d) = b$

42. $f(a) = 7, f(b) = 7, f(c) = 7, f(d) = 7$

Exercises 43–52: Sketch a graph of f.

43. $f(x) = -x + 3$ 44. $f(x) = -2x + 1$

45. $f(x) = 2x$ 46. $f(x) = \frac{1}{2}x - 2$

47. $f(x) = 4 - x$ 48. $f(x) = 6 - 3x$

49. $f(x) = x^2$ 50. $f(x) = \sqrt{x}$

51. $f(x) = \sqrt{x + 1}$ 52. $f(x) = \frac{1}{2}x^2 - 1$

Exercises 53–58: Use the graph of f to evaluate the given expressions.

53. $f(0)$ and $f(2)$ 54. $f(-2)$ and $f(2)$

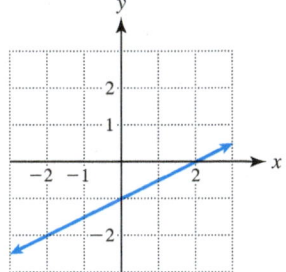

55. $f(-2)$ and $f(1)$ 56. $f(-1)$ and $f(0)$

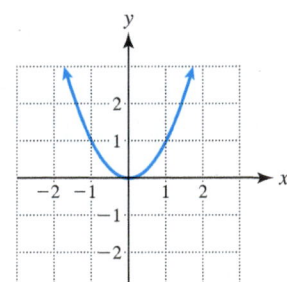

57. $f(1)$ and $f(2)$ 58. $f(-1)$ and $f(4)$

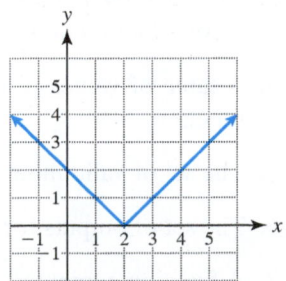

Exercises 59 and 60: Use the table to evaluate the given expressions.

59. $f(0)$ and $f(2)$

x	0	1	2	3	4
$f(x)$	5.5	4.3	3.7	2.5	1.9

60. $f(-10)$ and $f(5)$

x	-10	-5	0	5	10
$f(x)$	23	96	-45	-33	23

Exercises 61 and 62: Use the diagram to evaluate $f(1990)$. Interpret your answer.

61. Fuel Efficiency The function f computes average fuel efficiency of new U.S. passenger cars in miles per gallon during year x. (*Source: Department of Transportation.*)

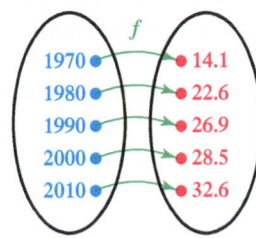

62. Cost of Tuition The function f computes average cost of tuition at public colleges and universities during academic year x. (*Source: The College Board.*)

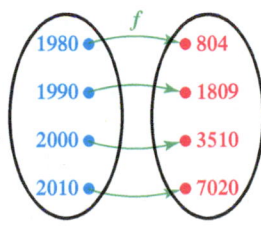

Exercises 63–66: Express the verbal representation for the function f numerically, symbolically, and graphically. Let $x = -3, -2, -1, \ldots, 3$ for the numerical representation (table), and let $-3 \le x \le 3$ for the graph.

63. Add 5 to the input x to obtain the output y.

64. Square the input x to obtain the output y.

65. Multiply the input x by 5 and then subtract 2 to obtain the output y.

66. Divide the input x by 2 and then add 3 to obtain the output y.

Exercises 67–72: Give a verbal representation for $f(x)$.

67. $f(x) = x - \frac{1}{2}$ **68.** $f(x) = \frac{3}{4}x$

69. $f(x) = \frac{x}{3}$ **70.** $f(x) = x^2 + 1$

71. $f(x) = \sqrt{x - 1}$ **72.** $f(x) = 1 - 3x$

73. Cost of Driving In 2015, the average cost of driving a new car in the United States was about 60 cents per mile. Symbolically, graphically, and numerically represent a function f that computes the cost in dollars of driving x miles. For the numerical representation (table), let $x = 10, 20, 30, \ldots, 70$. (*Source: AAA.*)

74. Federal Income Taxes In 2015, the lowest U.S. income tax rate was 10 percent. Symbolically, graphically, and numerically represent a function f that computes the tax on a taxable income of x dollars. For the numerical representation (table) let $x = 1000, 2000, 3000, \ldots, 7000$. For the graphical representation let $0 \le x \le 10{,}000$. (*Source: Internal Revenue Service.*)

75. Google Web Searches The number of Google searches S in billions during year x can be approximated by $S(x) = 120x - 240{,}280$ from 2009 to 2015. Evaluate $S(2014)$ and interpret the result. (*Source: DMR.*)

76. Price of Android Smartphones The average price P in dollars for an Android smartphone during year x is projected to be $P(x) = -9.75x + 19{,}890.5$, where $2014 \le x \le 2019$. (*Source: Computerworld.*)
   (a) Evaluate $P(2018)$ and interpret your answer.
   (b) What does function P indicate about how the price of Android smartphones is changing?

IDENTIFYING DOMAINS AND RANGES

Exercises 77–84: Use the graph of f to identify its domain and range.

77.

78.

79.

80.

81.

82.

83. **84.**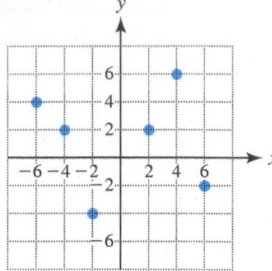

Exercises 85 and 86: Use the diagram to find the domain and range of f.

85. **86.**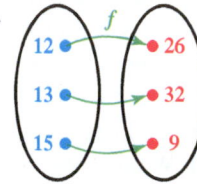

Exercises 87–100: Find the domain.

87. $f(x) = 10x$

88. $f(x) = 5 - x$

89. $f(x) = x^2 - 3$

90. $f(x) = \frac{1}{2}x^2$

91. $f(x) = \frac{3}{x - 5}$

92. $f(x) = \frac{x}{x + 1}$

93. $f(x) = \frac{2x}{x^2 + 1}$

94. $f(x) = \frac{6}{1 - x}$

95. $f(x) = \sqrt{x - 1}$

96. $f(x) = |x|$

97. $f(x) = |x - 5|$

98. $f(x) = \sqrt{2 - x}$

99. $f(x) = \frac{1}{x}$

100. $f(x) = 1 - 3x^2$

101. Humpback Whales The number of humpback whales W sighted in Maui's annual whale census for year x is given by $W(2011) = 1607$, $W(2012) = 1054$, $W(2013) = 1126$, $W(2014) = 1331$, and finally $W(2015) = 1488$. (*Source: Pacific Whale Foundation.*)
(a) Evaluate $W(2015)$ and interpret the result.
(b) Identify the domain and range of *W*.
(c) Describe the pattern in the data.

102. Global Digital Revenues Revenue R in billions of dollars from global sales of digital music during year x is given by $R(2010) = 4.6$, $R(2011) = 5.1$, $R(2012) = 5.6$, $R(2013) = 6.4$, and $R(2014) = 6.9$. (*Source: IFPI.*)
(a) Evaluate $R(2014)$ and interpret the result.
(b) Identify the domain and range of function R.
(c) Describe the pattern in the data.

103. Cost of Tuition Suppose that a student can take from 1 to 20 credits at a college and that each credit costs \$200. If function C calculates the cost of taking x credits, determine the domain and range of C.

104. Falling Ball Suppose that a ball is dropped from a window that is 64 feet above the ground and that the ball strikes the ground after 2 seconds. If function H calculates the height of the ball after t seconds, determine a domain and range for H, while the ball is falling.

IDENTIFYING RELATIONS AND FUNCTIONS

Exercises 105–112: Identify whether set S is a relation, function, or both. If it is neither, state so. Give its domain and range, if possible.

105. $S = \{(2, 3), (3, 5), (2, -4)\}$

106. $S = \{(-1, 4), (-1, 6), (-1, 7)\}$

107. $S = \{(4, 5), (-7, 5), (3, 5)\}$

108. $S = \{(-5, 0), (0, 0), (5, 0)\}$

109. $S = \{1, 2, 3, 4\}$

110. $S = \{a, b, c, d, e\}$

111. $S = \{(-5, -5), (-2, -2), (0, 0), (3, 3)\}$

112. $S = \{(-1, 3), (1, 2), (1, 7), (-1, -2)\}$

IDENTIFYING A FUNCTION

Exercises 113–116: Determine whether the diagram could represent a function.

113. **114.**

115. **116.**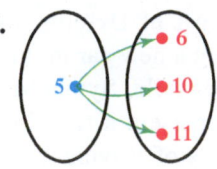

117. Average Precipitation The table lists the monthly average precipitation P in Las Vegas, Nevada, where $x = 1$ corresponds to January and $x = 12$ corresponds to December.

x (month)	1	2	3	4	5	6
P (inches)	0.5	0.4	0.4	0.2	0.2	0.1

x (month)	7	8	9	10	11	12
P (inches)	0.4	0.5	0.3	0.2	0.4	0.3

Source: J. Williams.

(a) Determine the value of P during May.
(b) Is P a function of x? Explain.
(c) If $P = 0.4$, find x.

118. Wind Speeds The table lists the monthly average wind speed W in Louisville, Kentucky, where $x = 1$ corresponds to January and $x = 12$ corresponds to December.

x (month)	1	2	3	4	5	6
W (mph)	10.4	12.7	10.4	10.4	8.1	8.1

x (month)	7	8	9	10	11	12
W (mph)	6.9	6.9	6.9	8.1	9.2	9.2

Source: J. Williams.

(a) Determine the month with the highest average wind speed.
(b) Is W a function of x? Explain.
(c) If $W = 6.9$, find x.

Exercises 119–130: Determine whether the graph represents a function. If it does, identify the domain and range.

119.

120.

121.

122.

123.

124.

125.

126.

127.   **128.**

129.   **130.**

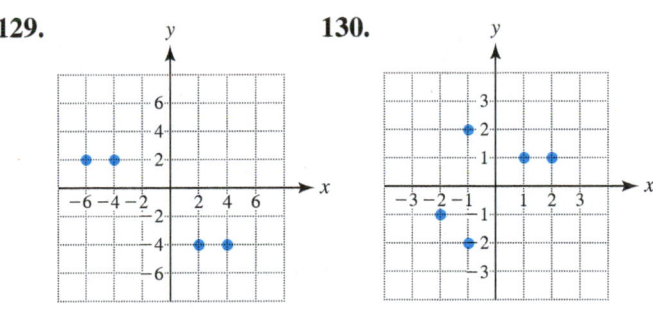

Exercises 131–134: Determine whether S is a function.

131. $S = \{(1, 2), (4, 5), (7, 8), (5, 4), (2, 2)\}$

132. $S = \{(4, 7), (-2, 1), (3, 8), (4, 9)\}$

133. S is given by the table.

x	5	10	5
y	2	1	0

134. S is given by the table.

x	-3	-2	-1
y	10	10	10

INTERPRETING GRAPHS

Exercises 135 and 136: The graph represents the distance that a person is from home while walking on a straight path. The x-axis represents time and the y-axis represents distance. Interpret the graph.

135.   **136.**

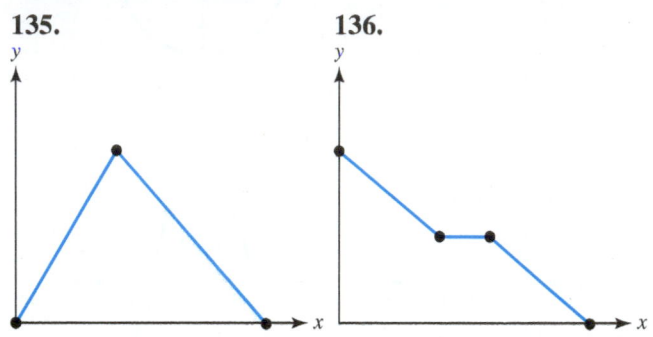

137. Texting The average 18- to 24-year-old person texts about 2000 messages per month. Sketch a graph that shows the total number of text messages sent over a period of 4 months. Assume that the same number of texts is sent each day. (*Source:* The Nielsen Company.)

138. Computer Viruses In 2000, there were about 50 thousand computer viruses. In 2015, there were about 27 million computer viruses. Sketch a graph of this increase from 2000 to 2015. Answers may vary. (*Source:* Symantec.)

USING GRAPHING CALCULATORS

 Exercises 139–144: Show the given viewing rectangle on your graphing calculator. Predict the number of tick marks on the positive x-axis and the positive y-axis.

139. Standard viewing rectangle

140. $[-12, 12, 2]$ by $[-8, 8, 2]$

141. $[0, 100, 10]$ by $[-50, 50, 10]$

142. $[-30, 30, 5]$ by $[-20, 20, 5]$

143. $[1980, 1995, 1]$ by $[12000, 16000, 1000]$

144. $[1900, 1990, 10]$ by $[1700, 2800, 100]$

 Exercises 145–150: Use your calculator to make a scatterplot of the relation after determining an appropriate viewing rectangle.

145. $\{(4, 3), (-2, 1), (-3, -3), (5, -2)\}$

146. $\{(5, 5), (2, 0), (-2, 7), (2, -8), (-1, -5)\}$

147. $\{(20, 40), (-25, -15), (-20, 25), (15, -25)\}$

148. $\{(-13, 12), (3, 10), (-15, -4), (12, -9)\}$

149. $\{(100, -100), (50, 200), (-150, -140), (-30, 80)\}$

150. $\{(-125, 75), (45, 65), (-53, -67), (150, -80)\}$

Exercises 151–154: Make a table and graph of $y = f(x)$. Let $x = -3, -2, -1, \ldots, 3$ for your table and use the standard window for your graph.

151. $f(x) = \sqrt{x + 3}$ **152.** $f(x) = x^3 - \frac{1}{2}x^2$

153. $f(x) = \dfrac{5 - x}{5 + x}$ **154.** $f(x) = |2 - x| + \sqrt[3]{x}$

WRITING ABOUT MATHEMATICS

155. Give an example of a function. Identify the domain and range of your function.

156. Explain in your own words what a function is. How is a function different from other relations?

157. Explain how to evaluate a function by using a graph. Give an example.

158. Give one difficulty that may occur when you use a table of values to evaluate a function.

Group Activity

Digital Music Album Downloads The following table lists the number of digital music album downloads in the United States for selected years.

Year	Downloads (millions)
2004	5
2007	45
2009	75
2012	115

(*Source:* Statista 2015.)

(a) Make a scatterplot of the data. Let x represent the number of years after 2004. Discuss any trend in numbers of album downloads.

(b) Estimate the slope of a line that could be used to model the data.

(c) Find an equation of a line $y = mx + b$ that models the data.

(d) Interpret the slope as a rate of change.

(e) Use your results to estimate the number of album downloads in 2014. Compare your answer with the true value of 118 million. Comment on the prediction made by your equation.

8.2 Linear Functions

STUDY TIP

Be sure you understand how functions can be represented.

1 Recognizing Representations of Linear Functions

Connecting Concepts with Your Life Suppose that the air conditioner is turned on when the temperature inside a house is 80 °F. The resulting temperatures are listed in **TABLE 8.6** for various elapsed times. Note that for each 1-hour increase in elapsed time, the temperature decreases by 2 °F.

House Temperature

1-hour increase

Elapsed Time (hours)	0	1	2	3	4	5
Temperature (°F)	80	78	76	74	72	70

TABLE 8.6

2 °F decrease

We want to determine a function f that models, or calculates, the temperature inside the house after x hours. To do this, we will find numerical, graphical, verbal, and symbolic representations of f.

NUMERICAL REPRESENTATION (TABLE OF VALUES) We can think of **TABLE 8.6** as a numerical representation (table of values) for the function f. A similar numerical representation that uses x and $f(x)$ is shown in **TABLE 8.7**.

Numerical Representation of $f(x)$

x	0	1	2	3	4	5
$f(x)$	80	78	76	74	72	70

TABLE 8.7

GRAPHICAL REPRESENTATION (GRAPH) To graph $y = f(x)$, we begin by plotting the points in **TABLE 8.7**, as shown in **FIGURE 8.27**. This scatterplot suggests that a line models these data, as shown in **FIGURE 8.28**. We call f a *linear function* because its graph is a *line*.

FIGURE 8.27 A Scatterplot

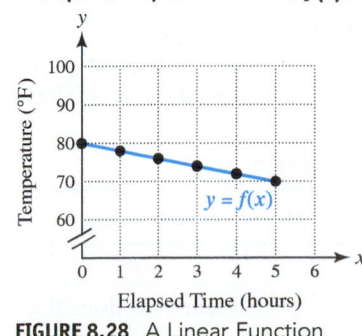

FIGURE 8.28 A Linear Function

FIGURE 8.29

Another graph of $y = f(x)$ with a different y-scale is shown in **FIGURE 8.29**. Because the y-values always decrease by the same amount for each 1-hour increase on the x-axis, we say that function f has a *constant rate of change*. In this example, the constant rate of change is -2 °F per hour.

VERBAL REPRESENTATION (WORDS) Over a 5-hour period, the air conditioner lowers the initial temperature of 80 °F by 2 °F for each elapsed hour x. Thus a description of how to calculate the temperature is written as follows.

Verbal representation of $f(x)$

"Multiply x by -2 °F and then add 80 °F."

SYMBOLIC REPRESENTATION (FORMULA) Our verbal representation of $f(x)$ makes it straightforward for us to write a formula.

Symbolic representation of $f(x)$

Elapsed Hours

$$f(x) = -2x + 80$$

Rate of Change   Initial Temperature

For example,

$$f(2.5) = -2(2.5) + 80 = 75$$

means that the temperature is **75** °F after the air conditioner has run for **2.5** hours. In this instance, it might be appropriate to *limit the domain* of f to x-values between 0 and 5, inclusive.

LINEAR FUNCTION

A function f defined by $f(x) = mx + b$, where m and b are constants, is a **linear function**.

For $f(x) = -2x + 80$, we have $m = -2$ and $b = 80$. The constant m represents the rate at which the air conditioner cools the building, and the constant b represents the initial temperature.

NOTE: The value of m represents the slope of the graph of $f(x) = mx + b$, and $(0, b)$ is the y-intercept. ∎

 Connecting Concepts with Your Life In general, a linear function defined by $f(x) = mx + b$ changes by m units for each unit increase in x. This **rate of change** is an increase if $m > 0$ and a decrease if $m < 0$. For example, if new carpet costs $20 per square yard, then the linear function defined by $C(x) = 20x$ gives the cost of buying x square yards of carpet. The value of $m = 20$ gives the cost (rate of change) for each additional square yard of carpet. For function C, the value of b is 0 because it costs $0 to buy 0 square yards of carpet.

READING CHECK 1

• Explain what a linear function is and what its graph looks like.

NOTE: If f is a linear function, then $f(0) = m(0) + b = b$. Thus b can be found by evaluating $f(x)$ at $x = 0$. ∎

IDENTIFYING LINEAR FUNCTIONS We can easily identify the graph of a linear function because it is always a line. The next two examples illustrate how to identify a linear function from its formula and from a table of values.

EXAMPLE 1 **Identifying linear functions from formulas**

Determine whether f is a linear function. If f is a linear function, find values for m and b so that $f(x) = mx + b$.
(a) $f(x) = 4 - 3x$ **(b)** $f(x) = 8$ **(c)** $f(x) = 2x^2 + 8$

Solution
(a) Let $m = -3$ and $b = 4$. Then $f(x) = -3x + 4$, and f is a linear function.
(b) Let $m = 0$ and $b = 8$. Then $f(x) = 0x + 8$, and f is a linear function.
(c) Function f is not linear because its formula contains x^2. The formula for a linear function cannot contain a variable with an exponent other than 1.

Now Try Exercises 11, 13, 15

EXAMPLE 2 **Identifying linear functions from tables of values**

Use each table of values to determine whether $f(x)$ could represent a linear function. If f could be linear, write a formula for f in the form $f(x) = mx + b$.

(a)

x	0	1	2	3
$f(x)$	10	15	20	25

(b)

x	-2	0	2	4
$f(x)$	4	2	0	-2

(c)

x	0	1	2	3
$f(x)$	1	2	4	7

(d)

x	-2	0	3	5
$f(x)$	7	7	7	7

Solution
(a) For each unit increase in x, $f(x)$ increases by 5 units, so $f(x)$ could be linear with $m = 5$. Because $f(0) = 10$, $b = 10$. Thus $f(x) = 5x + 10$.
(b) For each 2-unit increase in x, $f(x)$ decreases by 2 units. Equivalently, each unit increase in x results in a 1-unit decrease in $f(x)$, so $f(x)$ could be linear with $m = -1$. Because $f(0) = 2$, $b = 2$. Thus $f(x) = -x + 2$.
(c) Each unit increase in x does not result in a constant change in $f(x)$. Thus $f(x)$ does *not* represent a linear function.
(d) For any change in x, $f(x)$ does *not* change, so $f(x)$ could be linear with $m = 0$. Because $f(0) = 7$, let $b = 7$. Thus $f(x) = 0x + 7$, or $f(x) = 7$. (When $m = 0$, we say that f is a *constant function*. See Example 8.)

Now Try Exercises 23, 25, 27, 31

2 Recognizing Graphs of Linear Functions

The graph of a linear function is a line. To graph a linear function f, we can start by making a table of values and then plotting three or more points. We can then sketch the graph of f by drawing a line through these points, as demonstrated in the next example.

EXAMPLE 3 **Graphing a linear function by hand**

Sketch a graph of $f(x) = x - 1$. Use the graph to evaluate $f(-2)$.

Solution
Begin by making a table of values containing at least three points. Pick convenient values of x, such as $x = -1, 0, 1$.

$$f(-1) = -1 - 1 = -2$$
$$f(0) = 0 - 1 = -1$$
$$f(1) = 1 - 1 = 0$$

Display the results, as shown in **TABLE 8.8** on the next page.

Plot the points $(-1, -2)$, $(0, -1)$, and $(1, 0)$. Sketch a line through these points to obtain the graph of f. A graph of a line results when *infinitely* many points are plotted, as shown in **FIGURE 8.30**.

To evaluate $f(-2)$, first find $x = -2$ on the x-axis. See **FIGURE 8.31**. Then move downward to the graph of f. By moving across to the y-axis, we see that the corresponding y-value is -3. Thus $f(-2) = -3$.

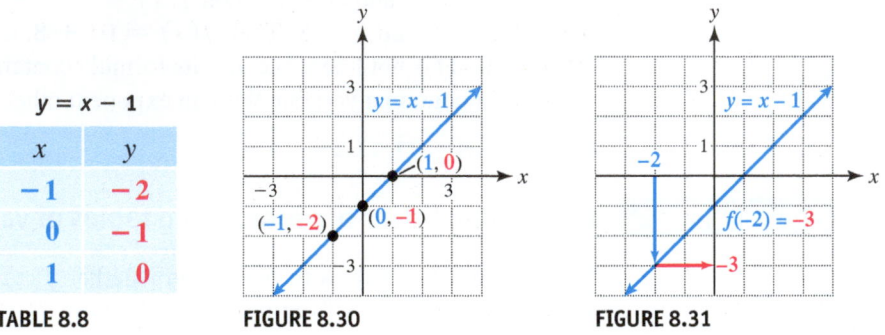

$y = x - 1$

x	y
-1	-2
0	-1
1	0

TABLE 8.8   FIGURE 8.30   FIGURE 8.31

Now Try Exercises 39, 57

In the next example, a graphing calculator is used to create a graph and table.

EXAMPLE 4   **Using a graphing calculator**

Give numerical and graphical representations of $f(x) = \frac{1}{2}x - 2$.

Solution

Numerical Representation To make a numerical representation, construct the table for $Y_1 = 0.5X - 2$, starting at $x = -3$ and incrementing by 1, as shown in **FIGURE 8.32(a)**. (Other tables are possible.)

Graphical Representation Graph Y_1 in the standard viewing rectangle, as shown in **FIGURE 8.32(b)**. (Other viewing rectangles may be used.)

CALCULATOR HELP

To make a table, see Appendix A (pages AP-2 and AP-3). To make a graph, see Appendix A (page AP-3).

Table of values

$[-10, 10, 1]$ by $[-10, 10, 1]$

$y = \frac{1}{2}x - 2$ Graph

(a)   (b)

FIGURE 8.32

Now Try Exercise 75

CRITICAL THINKING 1

Two points determine a line. Why is it a good idea to plot at least three points when graphing a linear function by hand?

EXAMPLE 5 **Representing a linear function**

A linear function is given by $f(x) = -3x + 2$.
(a) Give a verbal representation (description) of f.
(b) Make a numerical representation (table) of f by letting $x = -1, 0, 1$.
(c) Plot the points listed in the table from part (b). Then sketch a graph of $y = f(x)$.

Solution
(a) Verbal Representation Multiply the input x by -3 and add 2 to obtain the output.
(b) Numerical Representation Evaluate the formula $f(x) = -3x + 2$ at $x = -1, 0, 1$, which results in **TABLE 8.9**. Note that $f(-1) = 5$, $f(0) = 2$, and $f(1) = -1$.
(c) Graphical Representation To make a graph of f by hand without a graphing calculator, plot the points $(-1, 5)$, $(0, 2)$, and $(1, -1)$ from **TABLE 8.9**. Then draw a line passing through these points, as shown in **FIGURE 8.33**.

$y = -3x + 2$

x	$f(x)$
-1	5
0	2
1	-1

TABLE 8.9

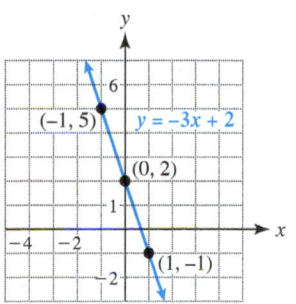

FIGURE 8.33

Now Try Exercise 71

NOTE: To graph $y = -3x + 2$ in Example 5, we could also graph a line with slope -3 and y-intercept $(0, 2)$. ■

MAKING CONNECTIONS 1

Mathematics in Newspapers

Think of the mathematics that you see in newspapers or in online publications. Often, percentages are described *verbally*, numbers are displayed in *tables*, and data are shown in *graphs*. Seldom are *formulas* given, which is an important reason to study verbal, numerical, and graphical representations.

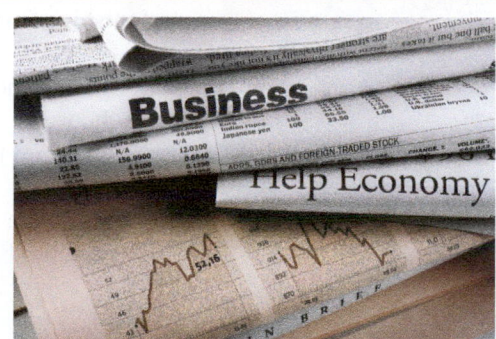

---

3 **Modeling Data with Linear Functions**

🌐 *Math in Context* (Wireless) A distinguishing feature of a linear function is that when the input x increases by 1 unit, the output $f(x) = mx + b$ always changes by an amount equal to m. For example, the percentage of wireless-only households during year x from 2005 to 2014 can be modeled by the linear function

$$f(x) = 4x - 8013,$$

where x is the year. The value of $m = 4$ indicates that the percentage of wireless-only households has increased, on average, by 4% per year. (*Source*: National Center for Health Statistics.)

The following are other examples of quantities that are modeled by linear functions. Try to determine the value of the constant m.

- The wages earned by an individual working x hours at $8 per hour
- The distance traveled by a jet airliner in x hours if the speed of the jet is 500 miles per hour
- The cost of tuition and fees when registering for x credits if each credit costs $200 and the fees are fixed at $300

When we are modeling data with a linear function defined by $f(x) = mx + b$, the following concepts are helpful to determine m and b.

MODELING DATA WITH A LINEAR FUNCTION

The formula $f(x) = mx + b$ may be interpreted as follows.

$$f(x) \quad = \quad mx \quad + \quad b$$

$$(\text{New amount}) = (\text{Change}) + (\text{Fixed amount})$$

When x represents time, *change* equals (rate of change) $\times$ (time).

$$f(x) \quad = \quad m \quad \times \quad x \quad + \quad b$$

$$(\text{Future amount}) = (\text{Rate of change}) \times (\text{Time}) + (\text{Initial amount})$$

🌐 *Math in Context* ^Biology These concepts are applied in the next example where a linear function is used to model the height of a fast-growing bamboo plant.

EXAMPLE 6   **Modeling growth of bamboo**

Bamboo is gaining popularity as a *green* building material because of its fast-growing, regenerative characteristics. Under ideal conditions, some species of bamboo grow at an astonishing 2 inches per hour. Suppose a bamboo plant is initially 6 inches tall. (*Source:* Cali Bamboo.)

(a) Find a function H that models the plant's height in inches under ideal conditions after t hours.

(b) Find $H(3)$ and interpret the result.

Solution

(a) The initial height is **6** inches and the rate of change is **2** inches per hour.

$$H(t) \quad = \quad 2 \quad \times \quad t \quad + \quad 6,$$

$$(\text{Future height}) = (\text{Rate of change}) \times (\text{Time}) + (\text{Initial height})$$

or $H(t) = 2t + 6$.

(b) $H(3) = 2(3) + 6 = 12$. After **3** hours, the bamboo plant is **12** inches tall.

Now Try Exercise 117

🌐 *Math in Context* ^Green People are becoming more energy conscious, so there is an increase in the demand for green building material. In the next example, this increase in demand is modeled with a linear function.

| EXAMPLE 7 | Modeling demand for green building material |

TABLE 8.10 lists actual and projected sales of green building material in billions of dollars.

Green Building Material Sales ($ billions)

Year	2015	2016	2017	2018
Sales	72	79	86	93

TABLE 8.10 *Source: Freedonia Group, Green Building Material.*

(a) Make a scatterplot of the data and sketch the graph of a function f that models these data. Let x represent years after 2015. That is, let $x = 0$ correspond to 2015, $x = 1$ to 2016, and so on.

(b) What were sales in 2015? What was the annual increase in sales each year?

(c) Find a formula for $f(x)$.

(d) Use your formula to estimate sales in 2019.

READING CHECK 2

• How can you determine whether data in a table can be modeled by a linear function?

Solution

(a) In **FIGURE 8.34** the scatterplot suggests that a linear function models the data. A line has been sketched with the data.

Green Building Material Sales

FIGURE 8.34 A Linear Model

(b) From **TABLE 8.10**, sales for green building material totaled $72 billion in 2015, with sales increasing at a *constant rate* of $7 billion per year.

(c) From part (b) initial sales $(x = 0)$ were $72 billion, and sales increased by $7 billion per year. Thus

$$f(x) \quad = \quad 7 \quad \times \quad x \quad + \quad 72$$

$$(\text{Future sales}) = (\text{Rate of change in sales}) \times (\text{Time}) + (\text{Initial sales})$$

or $f(x) = 7x + 72$.

(d) Because $x = 4$ corresponds to 2019, evaluate $f(4)$.

$$f(4) = 7(4) + 72 = 100$$

This model projects sales of green building material to be $100 billion in 2019.

Now Try Exercise 119

🌐 *Math in Context* Driving In the next example, we consider a simple function that models the speed of a car.

| EXAMPLE 8 | **Modeling with a constant function** |

A car travels on a freeway with its speed recorded at regular intervals, as listed in **TABLE 8.11**.

Speed of a Car					
Elapsed Time (hours)	0	1	2	3	4
Speed (miles per hour)	70	70	70	70	70

TABLE 8.11

(a) Discuss the speed of the car during this time interval.
(b) Find a formula for a function *f* that models these data.
(c) Sketch a graph of *f* together with the data.

Solution
(a) The speed of the car appears to be constant at 70 miles per hour.
(b) Because the speed is constant, the rate of change is 0. Thus

$$f(x) \quad = \quad 0x \quad + \quad 70$$

(Future speed) = (Change in speed) + (Initial speed)

and $f(x) = 70$. We call *f* a *constant function.*
(c) Because $y = f(x)$, graph $y = 70$ with the data points

$$(0, 70), (1, 70), (2, 70), (3, 70), \text{ and } (4, 70)$$

to obtain **FIGURE 8.35**.

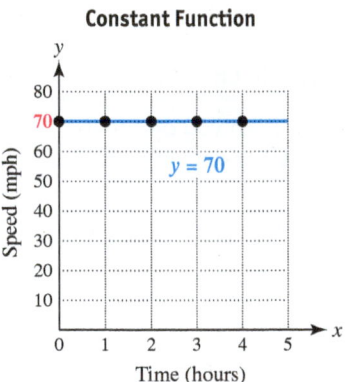

FIGURE 8.35 Speed of a Car

Now Try Exercise 113

The function defined by $f(x) = 70$ is an example of a *constant function.* A **constant function** *is a linear function with $m = 0$* and can be written as $f(x) = b$. Regardless of the input, a constant function always outputs the same value, *b*. Its graph is a horizontal line. In general, its domain is all real numbers and its range is $R = \{b\}$.

CRITICAL THINKING 2

Find a formula for a function *D* that calculates the *distance* traveled by the car in Example 8 after *x* hours. What is the rate of change for $D(x)$?

 Math in ∧Context (Real-World) The following are three applications of constant functions.

- A thermostat calculates a constant function regardless of the weather outside by maintaining a set temperature.
- A cruise control in a car calculates a constant function by maintaining a fixed speed, regardless of the type of road or terrain.
- A constant function calculates the 1250-foot height of the Empire State Building.

4 Applying the Midpoint Formula

 Connecting Concepts with Your Life A common way to make estimations is to average data items. For example, in 2000 the average cost of tuition and fees at public two-year colleges was about $1700, and in 2014 it was about $3300. (*Source:* The College Board.) To estimate tuition and fees in 2007, we could average the 2000 and 2014 amounts.

$$\frac{1700 + 3300}{2} = \$2500 \quad \text{Finding the average}$$

This technique predicts that the average cost of tuition and fees was $2500 in 2007 and is referred to as finding the *midpoint.*

FIGURE 8.36

MIDPOINT FORMULA ON THE REAL NUMBER LINE The **midpoint** of a line segment is the unique point on the line segment that is an equal distance from the endpoints. For example, in **FIGURE 8.36**, the midpoint M of -3 and 5 on the real number line is 1.

We can calculate the value of M as follows.

$$M = \frac{x_1 + x_2}{2} = \frac{-3 + 5}{2} = \frac{2}{2} = 1$$

Average the x-values to find the midpoint.

MIDPOINT FORMULA IN THE xy-PLANE The midpoint of a line segment in the xy-plane can be found in a similar way. **FIGURE 8.37(a)** shows the midpoint on the line segment connecting the points (x_1, y_1) and (x_2, y_2). The x-coordinate of M is equal to the average of x_1 and x_2, and the y-coordinate of M is equal to the average of y_1 and y_2. For example, the line segment shown in **FIGURE 8.37(b)** has endpoints $(-2, 1)$ and $(4, -3)$. The coordinates of the midpoint are

$$M = \left(\frac{-2 + 4}{2}, \frac{1 + (-3)}{2} \right) = (1, -1).$$

(a)

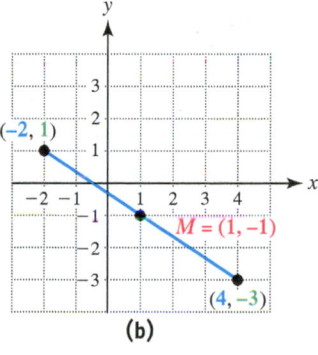

(b)

FIGURE 8.37

This discussion is summarized as follows.

MIDPOINT FORMULA IN THE xy-PLANE

The midpoint of the line segment with endpoints (x_1, y_1) and (x_2, y_2) in the xy-plane is

$$\left(\frac{x_1 + x_2}{2}, \frac{y_1 + y_2}{2} \right).$$

EXAMPLE 9   **Finding the midpoint**

Find the midpoint of the line segment connecting the points $(-3, -2)$ and $(4, 1)$.

Solution

In the midpoint formula, let $(-3, -2)$ be (x_1, y_1) and $(4, 1)$ be (x_2, y_2).

$$M = \left(\frac{x_1 + x_2}{2}, \frac{y_1 + y_2}{2} \right) \qquad \text{Midpoint formula}$$

$$= \left(\frac{-3 + 4}{2}, \frac{-2 + 1}{2} \right) \qquad \text{Substitute.}$$

$$= \left(\frac{1}{2}, -\frac{1}{2} \right) \qquad \text{Simplify.}$$

The midpoint of the line segment is $\left(\frac{1}{2}, -\frac{1}{2} \right)$.

Now Try Exercise 93

🌐 *Math in Context* **Social Science** In the next example, we use the midpoint formula to estimate the divorce rate in the United States in 2007.

EXAMPLE 10 **Estimating the U.S. divorce rate**

The divorce rate per 1000 people in 2000 was 4.2, and in 2014 it was 3.6. (*Source:* Statistical Abstract of the United States.)
(a) Use the midpoint formula to estimate the divorce rate in 2007.
(b) Could the midpoint formula be used to estimate the divorce rate in 2003? Explain.

Solution
(a) In the midpoint formula, let $(2000, 4.2)$ be (x_1, y_1) and let $(2014, 3.6)$ be (x_2, y_2).

$$M = \left(\frac{x_1 + x_2}{2}, \frac{y_1 + y_2}{2}\right) \qquad \text{Midpoint formula}$$

$$= \left(\frac{2000 + 2014}{2}, \frac{4.2 + 3.6}{2}\right) \qquad \text{Substitute.}$$

$$= (2007, 3.9) \qquad \text{Simplify.}$$

The midpoint formula estimates that the divorce rate was **3.9** per 1000 people in **2007**.
(b) No, the midpoint formula can only be used to estimate data that are exactly halfway between two given data points. Because the year 2003 is not exactly halfway between 2000 and 2014, the midpoint formula cannot be used.

Now Try Exercise 107

NOTE: An estimate obtained from the midpoint formula is equal to an estimate obtained from a linear function whose graph passes through the endpoints of the line segment. This fact is illustrated in the next example. ∎

EXAMPLE 11 **Relating midpoints to linear functions**

The graph of the linear function f shown in **FIGURE 8.38** is a straight line passing through the points $(-1, 3)$ and $(2, -3)$.
(a) Find a formula for $f(x)$.
(b) Evaluate $f\left(\frac{1}{2}\right)$. Does your answer agree with the graph?
(c) Find the midpoint M of the line segment connecting the points $(-1, 3)$ and $(2, -3)$. Comment on your result.

Solution
(a) The graph of f is a line that passes through $(-1, 3)$ and $(2, -3)$. The slope m of the line is

$$m = \frac{-3 - 3}{2 - (-1)} = -\frac{6}{3} = -2,$$

and from the graph, the y-intercept is $(0, 1)$. Thus $f(x) = -2x + 1$.
(b) $f\left(\frac{1}{2}\right) = -2\left(\frac{1}{2}\right) + 1 = 0$. Yes, they agree because the point $\left(\frac{1}{2}, 0\right)$ lies on the graph of $y = f(x)$ in **FIGURE 8.38**.

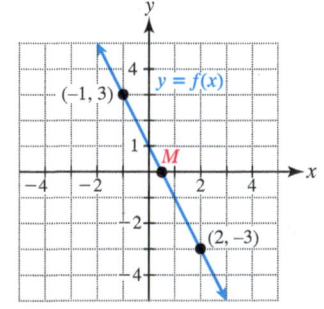

FIGURE 8.38

(c) The midpoint of the line segment connecting $(-1, 3)$ and $(2, -3)$ is

$$M = \left(\frac{-1 + 2}{2}, \frac{3 + (-3)}{2}\right) = \left(\frac{1}{2}, 0\right).$$

Finding the midpoint $M = \left(\frac{1}{2}, 0\right)$ of the line segment with endpoints $(-1, 3)$ and $(2, -3)$ is equivalent to evaluating the linear function f, whose graph passes through $(-1, 3)$ and $(2, -3)$, at $x = \frac{1}{2}$.

Now Try Exercise 103

8.2 Putting It All Together

CONCEPT	EXPLANATION	EXAMPLES
Linear Function	Can be represented by $f(x) = mx + b$. Its graph is a line with slope m and y-intercept $(0, b)$.	$f(x) = 2x - 6$, $m = 2$ and $b = -6$ $f(x) = 10$, $m = 0$ and $b = 10$
Constant Function	Can be represented by $f(x) = b$. Its graph is a horizontal line.	$f(x) = -7$, $b = -7$ $f(x) = 22$, $b = 22$
Rate of Change for a Linear Function	The output of a linear function changes by a constant amount for each unit increase in the input.	$f(x) = -3x + 8$ decreases 3 units for each unit increase in x. $f(x) = 5$ neither increases nor decreases. The rate of change is 0.
Midpoint Formula	The midpoint of the line segment connecting (x_1, y_1) and (x_2, y_2) is $$\left(\frac{x_1 + x_2}{2}, \frac{y_1 + y_2}{2}\right).$$	The midpoint of the line segment connecting $(-2, 3)$ and $(4, 5)$ is $$\left(\frac{-2 + 4}{2}, \frac{3 + 5}{2}\right) = (1, 4).$$

REPRESENTATION	COMMENTS	EXAMPLE
Symbolic	Mathematical formula in the form $f(x) = mx + b$	$f(x) = 2x + 1$, where $m = 2$ and $b = 1$
Verbal	Multiply the input x by m and add b.	Multiply the input x by 2 and then add 1 to obtain the output.
Numerical (table of values)	For each unit increase in x in the table, the output of $f(x) = mx + b$ changes by an amount equal to m.	1-unit increase | x	0	1	2
-----	---	---	---
$f(x)$	1	3	5	 2-unit increase
Graphical	The graph of a linear function is a line. Plot at least 3 points and then sketch the line. If $f(x) = mx + b$, then the graph of f has slope m and y-intercept $(0, b)$.	$y = 2x + 1$

8.2 Exercises

CONCEPTS AND VOCABULARY

1. The formula for a linear function is $f(x) =$ ___.

2. The formula for a constant function is $f(x) =$ ___.

3. The graph of a linear function is a(n) ___.

4. The graph of a constant function is a(n) ___ line.

5. If $f(x) = 7x + 5$, each time x increases by 1 unit, $f(x)$ increases by ___ units.

6. If $f(x) = 5$, each time x increases by 1 unit, $f(x)$ increases by ___ units.

7. (True or False?) Every constant function is a linear function.

8. (True or False?) Every linear function is a constant function.

9. If $C(x) = 2x$ calculates the cost in dollars of buying x square feet of carpet, what does 2 represent in the formula? Interpret the fact that the point $(10, 20)$ lies on the graph of C.

10. If $G(x) = 100 - 4x$ calculates the number of gallons of water in a tank after x minutes, what does -4 represent in the formula? Interpret the fact that the point $(5, 80)$ lies on the graph of G.

IDENTIFYING LINEAR FUNCTIONS

Exercises 11–18: Determine whether f is a linear function. If f is linear, give values for m and b so that f may be expressed as $f(x) = mx + b$.

11. $f(x) = \dfrac{1}{2}x - 6$

12. $f(x) = x$

13. $f(x) = \dfrac{5}{2} - x^2$

14. $f(x) = \sqrt{x} + 3$

15. $f(x) = -9$

16. $f(x) = 1.5 - 7.3x$

17. $f(x) = -9x$

18. $f(x) = \dfrac{1}{x}$

Exercises 19–22: Determine whether the graph represents a linear function.

19.

20.

21.

22.

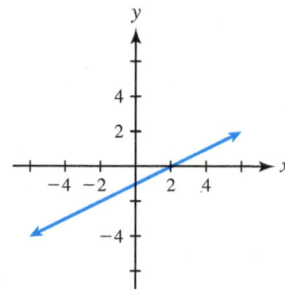

Exercises 23–32: (Refer to Example 2.) Use the table to determine whether $f(x)$ could represent a linear function. If it could, write $f(x)$ in the form $f(x) = mx + b$.

23.

x	0	1	2	3
$f(x)$	−6	−3	0	3

24.

x	0	2	4	6
$f(x)$	−2	2	6	10

25.

x	−2	0	2	4
$f(x)$	6	3	0	−3

26.

x	0	3	6	9
$f(x)$	8	4	2	1

27.

x	−2	−1	0	1
$f(x)$	−5	0	20	40

28.

x	−2	−1	0	1
$f(x)$	6	3	0	−3

29.

x	0	2	3	4
$f(x)$	0	4	6	8

30.

x	1	2	3	4
$f(x)$	0	1	3	7

31.

x	-1	0	1	2
$f(x)$	-4	-4	-4	-4

32.

x	2	5	6	8
$f(x)$	5	5	5	5

EVALUATING LINEAR FUNCTIONS

Exercises 33–38: Evaluate $f(x)$ at the given values of x.

33. $f(x) = 4x$ $\qquad$ $x = -4, 5$

34. $f(x) = -2x + 1$ $\qquad$ $x = -2, 3$

35. $f(x) = 5 - x$ $\qquad$ $x = -\frac{2}{3}, 3$

36. $f(x) = \frac{1}{2}x - \frac{1}{4}$ $\qquad$ $x = 0, \frac{1}{2}$

37. $f(x) = -22$ $\qquad$ $x = -\frac{3}{4}, 13$

38. $f(x) = 9x - 7$ $\qquad$ $x = -1.2, 2.8$

Exercises 39–44: Use the graph of f to evaluate the given expressions.

39. $f(-1)$ and $f(0)$ $\qquad$ **40.** $f(-2)$ and $f(2)$

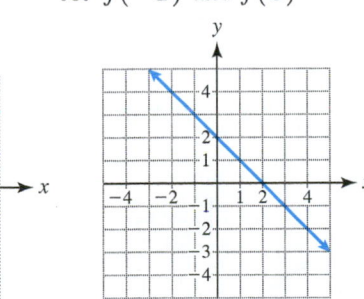

41. $f(-2)$ and $f(4)$ $\qquad$ **42.** $f(0)$ and $f(3)$

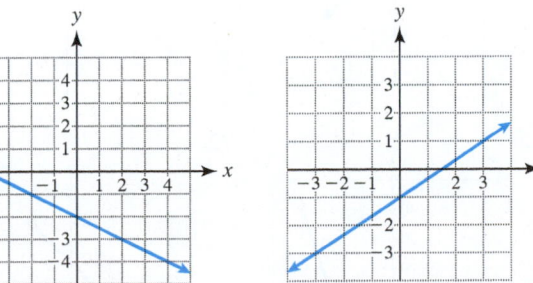

43. $f(-3)$ and $f(1)$ $\qquad$ **44.** $f(1.5)$ and $f(0.5\pi)$

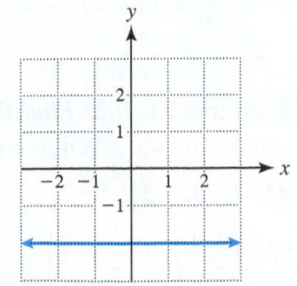

Exercises 45–48: Use the verbal description to write a formula for $f(x)$. Then evaluate $f(3)$.

45. Multiply the input by 6.

46. Multiply the input by -3 and add 7.

47. Divide the input by 6 and subtract $\frac{1}{2}$.

48. Output 8.7 for every input.

REPRESENTING LINEAR FUNCTIONS

Exercises 49–52: Match $f(x)$ with its graph (a–d).

49. $f(x) = 3x$ $\qquad$ **50.** $f(x) = -2x$

51. $f(x) = x - 2$ $\qquad$ **52.** $f(x) = 2x + 1$

a. $\qquad$ **b.**

c. $\qquad$ **d.**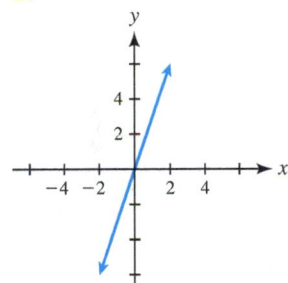

Exercises 53–62: Sketch a graph of $y = f(x)$.

53. $f(x) = 2$ $\qquad$ **54.** $f(x) = -1$

55. $f(x) = -\frac{1}{2}x$ $\qquad$ **56.** $f(x) = 2x$

57. $f(x) = x + 1$ $\qquad$ **58.** $f(x) = x - 2$

59. $f(x) = 3x - 3$ $\qquad$ **60.** $f(x) = -2x + 1$

61. $f(x) = 3 - x$ $\qquad$ **62.** $f(x) = \frac{1}{4}x + 2$

Exercises 63–68: Write a symbolic representation (formula) for a linear function f that calculates the following.

63. The number of pounds in x ounces

64. The number of dimes in x dollars

65. The distance traveled by a car moving at 65 miles per hour for t hours

66. The international phone bill *in dollars* for calling t minutes at 10 cents per minute and a fixed fee of $4.95

67. The total number of hours in a day during day x

68. The total cost of downhill skiing x times with a $500 season pass

69. Thinking Generally For each 1-unit increase in x with $y = ax + b$ and $a > 0$, y increases by ____ units.

70. Thinking Generally For each 1-unit decrease in x with $y = cx + d$ and $c < 0$, y increases by ____ units.

 Exercises 71–74: Do the following.

    (a) *Give a verbal representation of f.*
    (b) *Make a numerical representation (table) of f for $x = -2, 0, 2$.*
    (c) *Plot the points listed in the table from part (b), then sketch a graph of f.*

71. $f(x) = -2x + 1$ **72.** $f(x) = 1 - x$

73. $f(x) = \frac{1}{2}x - 1$ **74.** $f(x) = \frac{3}{4}x$

 Exercises 75–78: Do the following.

    (a) *Make a numerical representation (table) of f for $x = -3, -2, -1, \ldots, 3$.*
    (b) *Graph f in the window $[-6, 6, 1]$ by $[-4, 4, 1]$.*

75. $f(x) = \frac{1}{3}x + \sqrt{2}$ **76.** $f(x) = -\frac{2}{3}x - \sqrt{3}$

77. $f(x) = \dfrac{x + 2}{5}$ **78.** $f(x) = \dfrac{2 - 3x}{7}$

MODELING WITH LINEAR FUNCTIONS

Exercises 79–82: Match the situation with the graph (a–d) that models it best, where x-values represent time from 2000 to 2010.

79. The cost of college tuition

80. The cost of 1 gigabyte of computer memory

81. The distance between Chicago and Denver

82. The total distance traveled by a satellite that is orbiting Earth if the satellite was launched in 2000

a.

b.

c.

d.

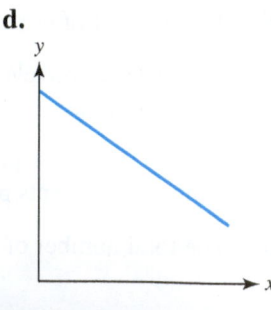

83. Online Exploration Look up the fuel efficiency E in miles per gallon for one of your favorite cars. (Answers will vary.)

    (a) Find a function G that calculates the number of gallons required to travel x miles.
    (b) If the cost of gasoline is $3 per gallon, find function C that calculates the cost of fuel to travel x miles.

84. Online Exploration Suppose that you would like to drive to Miami for spring break (if it is possible) in the car that you chose in Exercise 83. Calculate the gallons of gasoline needed for the trip.

APPLYING THE MIDPOINT FORMULA

Exercises 85–92: Find the midpoint of the line segment shown.

85.

86.

87.

88.

89.

90.

91.

92.

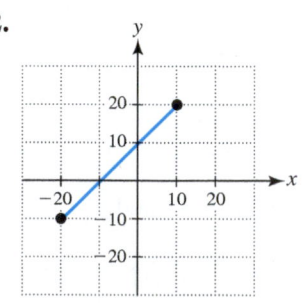

Exercises 93–102: Find the midpoint of the line segment connecting the given points.

93. $(-9, -3), (-7, 1)$ **94.** $(7, -2), (-5, 8)$

95. $\left(\frac{1}{2}, \frac{1}{3}\right), \left(-\frac{5}{2}, -\frac{2}{3}\right)$ **96.** $\left(-\frac{3}{5}, -\frac{1}{4}\right), \left(\frac{1}{10}, \frac{1}{2}\right)$

97. $(-0.3, 0.1)$, $(0.7, 0.4)$

98. $(0.8, -0.4)$, $(0.9, -0.1)$

99. $(2000, 5)$, $(2010, 13)$ **100.** $(2005, 9)$, $(2011, 3)$

101. Thinking Generally $(a, -b)$, $(3a, 5b)$

102. Thinking Generally $(-a, b)$, $(a, -b)$

Exercises 103–106: (Refer to Example 11.) The graph of a linear function f passes through the two given points.

  (a) *Find a formula for $f(x)$. Determine $f(2)$.*

  (b) *Determine $f(2)$ by finding the midpoint of the line segment connecting the given points.*

  (c) *Compare your answers for parts (a) and (b).*

103. $(0, 5)$, $(4, -3)$

104. $(0, 2)$, $(4, 10)$

105. $(-3, -1)$, $(7, 3)$

106. $(-1, 3)$, $(5, -5)$

Exercises 107–112: Use the midpoint formula to make the requested estimation.

107. U.S. Life Expectancy The life expectancy of a female born in 1990 was 78.8 years, and the life expectancy of a female born in 2010 rose to 80.8 years. Estimate the life expectancy of a female born in 2000. (*Source:* Centers for Disease Control and Prevention.)

108. U.S. Life Expectancy The life expectancy of a male born in 1990 was 71.8 years, and the life expectancy of a male born in 2010 rose to 75.6 years. Estimate the life expectancy of a male born in 2000. (*Source:* Centers for Disease Control and Prevention.)

109. U.S. Population The population of the United States in 1970 was 205 million, and in 2010 it was 308 million. Estimate the population in 1990. (*Source:* U.S. Census Bureau.)

110. Distance Traveled A car is moving at a constant speed on an interstate highway. After 1 hour, the car passes the 103-mile marker, and after 5 hours, the car passes the 391-mile marker. What mile marker does the car pass after 3 hours?

111. U.S. Median Income In 2009, the median family income was \$49,800, and in 2013, it was \$51,900. Estimate the median family income in 2011.

112. Estimating Fish Populations In 2008, there were approximately 3200 large-mouth bass in a lake. This number increased to 3800 in 2012. Estimate the number of large-mouth bass in the lake in 2010.

APPLICATIONS INVOLVING LINEAR FUNCTIONS

113. Thermostat Let $y = f(x)$ describe the temperature y of a room that is kept at $70\,°F$ for x hours.

  (a) Represent f symbolically and graphically over a 24-hour period for $0 \le x \le 24$.

  (b) Construct a table of f for $x = 0, 4, 8, 12, \ldots, 24$.

  (c) What type of function is f?

114. Cruise Control Let $y = f(x)$ describe the speed y of an automobile after x minutes if the cruise control is set at 60 miles per hour.

  (a) Represent f symbolically and graphically over a 15-minute period for $0 \le x \le 15$.

  (b) Construct a table of f for $x = 0, 1, 2, \ldots, 6$.

  (c) What type of function is f?

115. Distance A car is initially 50 miles south of the Minnesota—Iowa border, traveling south on Interstate 35. Distances D between the car and the border are recorded in the table for various elapsed times t. Find a linear function D that models these data.

t (hours)	0	2	3	5
D (miles)	50	170	230	350

116. Estimating the Weight of a Bass Sometimes the weight of a fish can be estimated by measuring its length. The table lists typical weights of bass that have various lengths.

Length (inches)	12	14	16	18	20	22
Weight (pounds)	1.0	1.7	2.5	3.6	5.0	6.6

Source: Minnesota Department of Natural Resources.

  (a) Let x be the length and y be the weight. Make a line graph of the data.

  (b) Could the data be modeled accurately with a linear function? Explain your answer.

117. Texting In 2014, the average American between the ages of 18 and 24 sent approximately 130 texts per day, whereas the average adult over age 65 sent approximately 5 texts per day.

  (a) Find a formula for a function K that calculates the number of texts sent in x days by the average person between the ages of 18 and 24.

  (b) Find a formula for a function A that calculates the number of texts sent in x days by the average person over age 65.

  (c) Evaluate $K(365)$ and $A(365)$. Interpret your results.

118. Rain Forests Rain forests are forests that grow in regions receiving more than 70 inches of rain per year. The world is losing about 49 million acres of rain forest each year. (*Source:* New York Times Almanac.)

  (a) Find a linear function f that calculates the change in the acres of rain forest in millions in x years.

  (b) Evaluate $f(7)$ and interpret the result.

119. Historical Car Sales The table shows the number of U.S. Toyota vehicles sold in millions for past years.

Year	2000	2001	2002	2003	2004
Vehicles	1.6	1.7	1.8	1.9	2.0

Source: Autodata.

(a) What were the sales in 2000?

(b) What was the annual increase in sales?

(c) Find a linear function f that models these data. Let $x = 0$ correspond to 2000, $x = 1$ to 2001, and so on.

(d) Use f to estimate sales in 2006.

120. Tuition and Fees Suppose tuition costs $300 per credit and that student fees are fixed at $100.

(a) Find a formula for a linear function T that models the cost of tuition and fees for x credits.

(b) Evaluate $T(16)$ and interpret the result.

121. Skype Users The number of active Skype users S in millions x years after 2010 can be modeled by the formula $S(x) = 11x + 145$.

(a) How many active users were there in 2014?

(b) What does the number 145 indicate in the formula?

(c) What does the number 11 indicate in the formula?

122. Temperature and Volume If a sample of a gas such as helium is heated, it will expand. The formula $V(T) = 0.147T + 40$ calculates the volume V in cubic inches of a sample of gas at temperature T in degrees Celsius.

(a) Evaluate $V(0)$ and interpret the result.

(b) If the temperature increases by $10\,°C$, by how much does the volume increase?

(c) What is the volume of the gas when the temperature is $100\,°C$?

123. Temperature and Volume (Refer to the preceding exercise.) A sample of gas at $0\,°C$ has a volume V of 137 cubic centimeters, which increases in volume by 0.5 cubic centimeter for every $1\,°C$ increase in temperature T.

(a) Write a formula $V(T) = aT + b$ that gives the volume of the gas at temperature T.

(b) Find the volume of the gas when $T = 50\,°C$.

124. Cost To make a music video, it costs $750 to rent a studio plus $5 for each copy produced.

(a) Write a formula $C(x) = ax + b$ that calculates the cost of producing x videos.

(b) Find the cost of producing 2500 videos.

125. Weight Lifting Lifting weights can increase a person's muscle mass. Each additional pound of muscle burns an extra 40 calories per day. Write a linear function that models the number of calories burned each day by x pounds of muscle. By burning an extra 3500 calories, a person can lose 1 pound of fat. How many pounds of muscle are needed to burn 1 pound of fat in 30 days? (*Source:* Runner's World.)

126. Wireless-Only Households The percentage P of wireless-only households x years after 2005 can be modeled by the formula $P(x) = 4x + 7$, where $0 \le x \le 7$.

(a) Evaluate $P(0)$ and $P(3)$. Interpret your results.

(b) Explain the meaning of 4 and 7 in the formula.

127. Smartphone Users The table lists the percentage P of people with cell phones who have smartphones.

Year	2015	2016	2017
Percentage	43%	46%	49%

(a) What was this percentage in 2015?

(b) By how much did this percentage change each year?

(c) Write a function P that models these data. Let x be years after 2015.

(d) Estimate this percentage in 2019.

128. Past Walmart Sales The table shows Walmart's share as a percentage of overall U.S. retail sales for past years. (This percentage excludes restaurants and motor vehicles.)

Year	1998	1999	2000	2001	2002
Share(%)	6	6.5	7	7.5	8

Source: Commerce Department, Walmart.

(a) What was Walmart's share in 1998?

(b) By how much (percent) did Walmart's share increase each year?

(c) Find a linear function f that models these data. Let $x = 0$ correspond to 1998.

(d) Use f to estimate Walmart's share in 2014. Compare your answer to the actual value of 11.3%.

WRITING ABOUT MATHEMATICS

129. Explain how you can determine whether a function is linear by using its

(a) symbolic representation,

(b) graphical representation, and

(c) numerical representation.

130. Describe one way to determine whether data can be modeled by a linear function.

Checking Basic Concepts

1. Find a formula and sketch a graph for a function f that squares the input x and then subtracts 1.

2. Use the graph of f to do the following.
   (a) Find the domain and range of f.
   (b) Evaluate $f(0)$ and $f(2)$.
   (c) Is f a linear function? Explain.

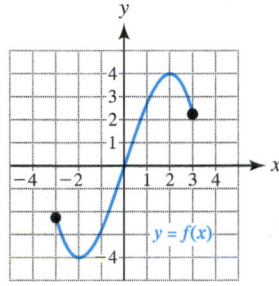

3. Determine whether f is a linear function.
   (a) $f(x) = 4x - 2$
   (b) $f(x) = 2\sqrt{x} - 5$
   (c) $f(x) = -7$
   (d) $f(x) = 9 - 2x + 5x$

4. Graph $f(x) = 4 - 3x$. Evaluate $f(-2)$.

5. Find a formula for a linear function that models the data.

x	0	1	2	3	4
$f(x)$	-1	$-\frac{1}{2}$	0	$\frac{1}{2}$	1

6. The median age in the United States from 1970 to 2015 can be approximated by

$$f(x) = 0.2x + 27.7,$$

where $x = 0$ corresponds to 1970, $x = 1$ to 1971, and so on.
   (a) Evaluate $f(45)$ and interpret the result.
   (b) Interpret the numbers 0.2 and 27.7.

7. Find the midpoint of the line segment connecting the points $(-3, 4)$ and $(5, -6)$.

8.3 Compound Inequalities and Piecewise-Defined Functions

Objectives

1. Introducing Compound Inequalities
2. Finding Symbolic Solutions and Graphing Them on a Number Line
3. Solving Inequalities and Writing Solutions in Interval Notation
4. Finding Numerical and Graphical Solutions
5. Introducing Piecewise-Defined Functions

1 Introducing Compound Inequalities

A person weighing 143 pounds and needing to purchase a life vest for white-water rafting is not likely to find one designed exactly for this weight. Life vests are manufactured to support a range of body weights. A vest approved for weights between 100 and 160 pounds might be appropriate for this person. In other words, if a person's weight is w, this life vest is safe if $w \geq 100$ *and* $w \leq 160$. This example illustrates the concept of a *compound inequality*.

A **compound inequality** consists of two inequalities joined by the words *and* or *or*. The following are two examples of compound inequalities.

$2x \geq -3$ **and** $2x < 5$ $x + 2 \geq 3$ **or** $x - 1 < -5$

First compound inequality   Second compound inequality

If a compound inequality contains the word *and*, a solution must satisfy *both* inequalities. For example, 1 is a solution to the first compound inequality because

Both parts are true so this compound inequality containing "and" is also true.

$2(1) \geq -3$ and $2(1) < 5$ First compound inequality
    True   True   with $x = 1$

are *both* true statements.

If a compound inequality contains the word *or*, a solution must satisfy *at least one* of the two inequalities. Thus 5 is a solution to the second compound inequality, because the first statement is true.

> One part is true so this compound inequality containing "or" is also true.

$$5 + 2 \geq 3 \quad \text{or} \quad 5 - 1 < -5$$
$$\text{True} \qquad\qquad \text{False}$$

Second compound inequality with $x = 5$

Note that 5 does not need to satisfy both statements for this compound inequality to be true.

| EXAMPLE 1 | **Determining solutions to compound inequalities** |

NEW VOCABULARY

- ☐ Compound inequality
- ☐ Intersection
- ☐ Three-part inequality
- ☐ Union
- ☐ Interval notation
- ☐ Infinity
- ☐ Negative infinity
- ☐ Piecewise-defined function
- ☐ Piecewise-defined linear function

Determine whether the given x-values are solutions to the compound inequalities.

(a) $x + 1 < 9$ and $2x - 1 > 8$ $\qquad x = 5, -5$
(b) $5 - 2x \leq -4$ or $5 - 2x \geq 4$ $\qquad x = 2, -3$

Solution
(a) Substitute $x = 5$ in the compound inequality $x + 1 < 9$ and $2x - 1 > 8$.

$$5 + 1 < 9 \quad \text{and} \quad 2(5) - 1 > 8$$
$$\text{True} \qquad\qquad \text{True}$$

> Check to see if 5 satisfies *both* inequalities.

Both inequalities are true, so 5 is a solution. Now substitute $x = -5$.

$$-5 + 1 < 9 \quad \text{and} \quad 2(-5) - 1 > 8$$
$$\text{True} \qquad\qquad \text{False}$$

> Check to see if -5 satisfies *both* inequalities.

To be a solution, both inequalities must be true, so -5 is *not* a solution.
(b) Substitute $x = 2$ into the compound inequality $5 - 2x \leq -4$ or $5 - 2x \geq 4$.

$$5 - 2(2) \leq -4 \quad \text{or} \quad 5 - 2(2) \geq 4$$
$$\text{False} \qquad\qquad \text{False}$$

> Check to see if 2 satisfies *at least one* inequality.

Neither inequality is true, so 2 is *not* a solution. Now substitute $x = -3$.

$$5 - 2(-3) \leq -4 \quad \text{or} \quad 5 - 2(-3) \geq 4$$
$$\text{False} \qquad\qquad \text{True}$$

> Check to see if -3 satisfies *at least one* inequality.

At least one of the two inequalities is true, so -3 is a solution.

Now Try Exercises 7, 9

2 ▸ Finding Symbolic Solutions and Graphing Them on a Number Line

We can use a number line to graph solutions to compound inequalities, such as

$$x \leq 6 \quad \text{and} \quad x > -4.$$

The solution set for $x \leq 6$ is shaded to the left of 6, with a bracket placed at $x = 6$, as shown in **FIGURE 8.39** on the next page. The solution set for $x > -4$ can be shown by shading a different number line to the right of -4 and placing a left parenthesis at -4. Because the inequalities are connected by *and*, the solution set consists of all numbers that are shaded on *both* number lines. The final number line represents the intersection of the two solution sets. That is, the solution set includes real numbers where the graphs "overlap." For any two sets A and B, the **intersection** of A and B, denoted $A \cap B$, is defined by

$$A \cap B = \{x \mid x \text{ is an element of } A \text{ and an element of } B\}.$$

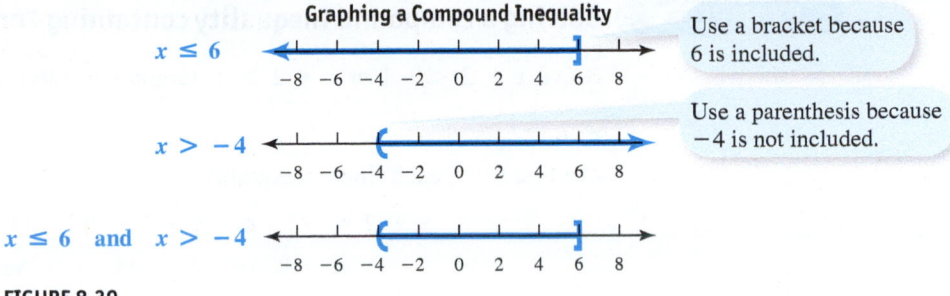

Graphing a Compound Inequality

$x \leq 6$ — Use a bracket because 6 is included.

$x > -4$ — Use a parenthesis because -4 is not included.

$x \leq 6$ and $x > -4$

FIGURE 8.39

NOTE: A bracket, either [or], is used when an inequality contains $\leq$ or $\geq$. A parenthesis, either (or), is used when an inequality contains $<$ or $>$. This notation makes clear whether or not an endpoint is included in the inequality. ■

EXAMPLE 2   **Solving a compound inequality containing "and"**

Solve $2x + 4 > 8$ and $5 - x < 9$. Graph the solution set.

Solution
First solve each linear inequality separately.

$$2x + 4 > 8 \quad \text{and} \quad 5 - x < 9$$

Subtract 4 and divide by 2. $2x > 4 \quad \text{and} \quad -x < 4$ Subtract 5 and multiply by -1.

$$x > 2 \quad \text{and} \quad x > -4$$

Graph $x > 2$ and $x > -4$ on two different number lines. On a third number line, shade solutions that appear on both of the first two number lines. As shown in **FIGURE 8.40**, the solution set written in set-builder notation is $\{x \mid x > 2\}$.

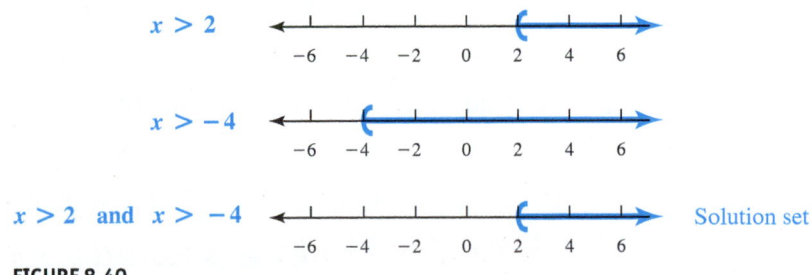

$x > 2$

$x > -4$

$x > 2$ and $x > -4$ Solution set

FIGURE 8.40

Now Try Exercise 43

We can also solve compound inequalities containing the word *or*. To write the solution to such an inequality, we sometimes use *union* notation. For any two sets A and B, the **union** of A and B, denoted $A \cup B$, is defined by

$$A \cup B = \{x \mid x \text{ is an element of } A \text{ or an element of } B\}.$$

If the solution to an inequality is $\{x \mid x < 1\}$ or $\{x \mid x \geq 3\}$, then it can also be written as

$$\{x \mid x < 1\} \cup \{x \mid x \geq 3\}.$$

That is, we can replace the word *or* with the $\cup$ symbol.

EXAMPLE 3 **Solving a compound inequality containing "or"**

Solve $x + 2 < -1$ or $x + 2 > 1$. Graph the solution set.

Solution
We first solve each linear inequality.

$$x + 2 < -1 \quad \text{or} \quad x + 2 > 1 \qquad \text{Given compound inequality}$$

$$x < -3 \quad \text{or} \quad x > -1 \qquad \text{Subtract 2.}$$

We can graph the simplified inequalities on different number lines, as shown in **FIGURE 8.41**. A solution must satisfy at least one of the two inequalities. Thus the solution set for the compound inequality results from taking the *union* of the first two number lines. We can write the solution, using set-builder notation, as $\{x \mid x < -3\} \cup \{x \mid x > -1\}$ or as $\{x \mid x < -3 \text{ or } x > -1\}$.

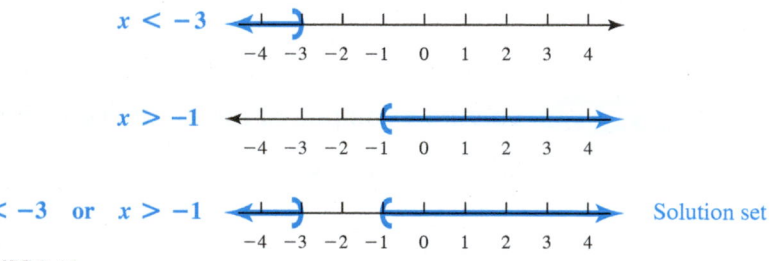

FIGURE 8.41

Now Try Exercise 47

3 Solving Inequalities and Writing Solutions in Interval Notation

READING CHECK 1

• Write the compound inequality $x < 2$ and $x \ge -1$ as a three-part inequality.

Sometimes a compound inequality containing the word *and* can be combined into a three-part inequality. For example, rather than writing

$$x > 5 \quad \text{and} \quad x \le 10,$$

we could write the **three-part inequality**

$$5 < x \le 10.$$

This three-part inequality is represented by the number line shown in **FIGURE 8.42**.

FIGURE 8.42

A convenient notation for number line graphs is called **interval notation**. Instead of drawing the entire number line as in **FIGURE 8.42**, the solution set can be expressed as $(5, 10]$ in interval notation. The solution set does not include 5, but does include 10. To denote this, a parenthesis is used next to 5 and a bracket is used next to 10, as is done on the number line. A solution set that includes all real numbers satisfying $-2 < x < 3$ can be expressed as $(-2, 3)$. Parentheses indicate that the endpoints are *not* included. The interval $0 \le x \le 4$ is represented by $[0, 4]$.

MAKING CONNECTIONS 1

Points and Intervals

The expression $(1, 2)$ may represent a point in the *xy*-plane or the interval $1 < x < 2$. To alleviate confusion, phrases such as "the point $(1, 2)$" or "the interval $(1, 2)$" are used.

TABLE 8.12 provides examples of interval notation. The symbol ∞ refers to **infinity**, and it does not represent a real number. The interval $(5, \infty)$ represents $x > 5$, which has no maximum x-value, so ∞ is used for the right endpoint. The symbol $-\infty$ may be used similarly and denotes **negative infinity**. Real numbers are denoted $(-\infty, \infty)$.

Interval Notation

Inequality	Interval Notation	Number Line Graph
$-1 < x < 3$	$(-1, 3)$	
$-3 < x \le 2$	$(-3, 2]$	
$-2 \le x \le 2$	$[-2, 2]$	
$x < -1$ or $x > 2$	$(-\infty, -1) \cup (2, \infty)$ ($\cup$ is the union symbol.)	
$x > -1$	$(-1, \infty)$	
$x \le 2$	$(-\infty, 2]$	

TABLE 8.12

EXAMPLE 4   **Writing inequalities in interval notation**

Write each expression in interval notation.
(a) $-2 \le x < 5$ (b) $x \ge 3$ (c) $x < -5$ or $x \ge 2$
(d) $\{x \mid x > 0 \text{ and } x \le 3\}$ (e) $\{x \mid x \le 1 \text{ or } x \ge 3\}$

Solution
(a) $[-2, 5)$ (b) $[3, \infty)$ (c) $(-\infty, -5) \cup [2, \infty)$
(d) $(0, 3]$ (e) $(-\infty, 1] \cup [3, \infty)$

Now Try Exercises 13, 17, 23, 27, 37

EXAMPLE 5   **Solving three-part inequalities**

Solve each inequality and graph each solution set. Write the solution set in interval notation.
(a) $4 < t + 2 \le 8$ (b) $-3 \le 3z \le 6$ (c) $-\dfrac{5}{2} < \dfrac{1 - m}{2} < 4$

Solution
(a) To solve a three-part inequality, isolate the variable by applying properties of inequalities to each part of the inequality.

$$4 < t + 2 \le 8 \qquad \text{Given three-part inequality}$$
$$4 - \mathbf{2} < t + 2 - \mathbf{2} \le 8 - \mathbf{2} \qquad \text{Subtract 2 from each part.}$$
$$2 < t \le 6 \qquad \text{Simplify each part.}$$

The solution set is $(2, 6]$. See **FIGURE 8.43**.

$2 < t \le 6$ Solution set

FIGURE 8.43

CRITICAL THINKING

Graph the following
inequalities and discuss
your results.
1. $x < 2$ and $x > 5$
2. $x > 2$ or $x < 5$

(b) To simplify, divide each part by 3.

$$-3 \le 3z \le 6 \qquad \text{Given three-part inequality}$$

$$\frac{-3}{3} \le \frac{3z}{3} \le \frac{6}{3} \qquad \text{Divide each part by 3.}$$

$$-1 \le z \le 2 \qquad \text{Simplify each part.}$$

The solution set is $[-1, 2]$. See **FIGURE 8.44**.

$-1 \le z \le 2$ Solution set

FIGURE 8.44

(c) Multiply each part by 2 to clear (eliminate) fractions.

$$-\frac{5}{2} < \frac{1 - m}{2} < 4 \qquad \text{Given three-part inequality}$$

$$2 \cdot \left(-\frac{5}{2}\right) < 2 \cdot \left(\frac{1 - m}{2}\right) < 2 \cdot 4 \qquad \text{Multiply each part by 2.}$$

$$-5 < 1 - m < 8 \qquad \text{Simplify each part.}$$

$$-5 - 1 < 1 - m - 1 < 8 - 1 \qquad \text{Subtract 1 from each part.}$$

$$-6 < -m < 7 \qquad \text{Simplify each part.}$$

$$-1 \cdot (-6) > -1 \cdot (-m) > -1 \cdot 7 \qquad \text{Multiply each part by } -1;$$
$$\text{\textit{reverse} inequality symbols.}$$

$$6 > m > -7 \qquad \text{Simplify each part.}$$

$$-7 < m < 6 \qquad \text{Rewrite inequality.}$$

The solution set is $(-7, 6)$. See **FIGURE 8.45**.

$-7 < m < 6$ Solution set

FIGURE 8.45

STUDY TIP

When simplifying a three-part inequality, be sure to perform the same step on each of the three parts.

NOTE: Either $6 > m > -7$ or $-7 < m < 6$ is a correct way to write a three-part inequality. However, we usually write the smaller number on the left side and the larger number on the right side. ∎

Now Try Exercises 59, 63, 79

🌐 *Math in Context* ^Meteorology Three-part inequalities occur frequently in applications. In the next example, we find altitudes at which the air temperature is within a certain range.

EXAMPLE 6   **Solving a three-part inequality**

If the ground-level temperature is 80 °F, the air temperature x miles above Earth's surface is cooler and can be modeled by $T(x) = 80 - 19x$. Find the altitudes at which the air temperature ranges from 42 °F down to 23 °F. (*Source:* A. Miller and R. Anthes, *Meteorology.*)

Solution

We write and solve the three-part inequality $23 \leq T(x) \leq 42$.

$$23 \leq 80 - 19x \leq 42 \qquad \text{Substitute for } T(x).$$

$$-57 \leq -19x \leq -38 \qquad \text{Subtract 80 from each part.}$$

$$\frac{-57}{-19} \geq x \geq \frac{-38}{-19} \qquad \text{Divide by } -19; \textit{ reverse} \text{ inequality symbols.}$$

$$3 \geq x \geq 2 \qquad \text{Simplify.}$$

$$2 \leq x \leq 3 \qquad \text{Rewrite inequality.}$$

The air temperature ranges from 42 °F to 23 °F for altitudes between 2 and 3 miles. In interval notation, the solution set can be expressed as $[2, 3]$.

Now Try Exercise 123

MAKING CONNECTIONS 2

Writing Three-Part Inequalities

The inequality $-2 < x < 1$ means that $x > -2$ *and* $x < 1$. A three-part inequality should *not* be used when *or* connects a compound inequality. Writing $x < -2$ or $x > 1$ as $1 < x < -2$ is **incorrect** because it states that x must be both greater than 1 *and* less than -2. It is impossible for any value of x to satisfy this statement.

CRITICAL THINKING 1

Carbon dioxide is emitted when human beings breathe. In one study of college students, the amount of carbon dioxide exhaled in grams per hour was measured during both lectures and exams. The average amount exhaled during lectures L satisfied $25.33 \leq L \leq 28.17$, whereas the average amount exhaled during exams E satisfied $36.58 \leq E \leq 40.92$. What do these results indicate? Explain. (*Source:* T. Wang, *ASHRAE Trans.*)

EXAMPLE 7   ### Solving an inequality

Solve $2x + 1 \leq -1$ or $2x + 1 \geq 3$. Write the solution set in interval notation.

Solution

First solve each inequality.

$$2x + 1 \leq -1 \quad \text{or} \quad 2x + 1 \geq 3 \qquad \text{Given compound inequality}$$

$$2x \leq -2 \quad \text{or} \quad 2x \geq 2 \qquad \text{Subtract 1.}$$

$$x \leq -1 \quad \text{or} \quad x \geq 1 \qquad \text{Divide by 2.}$$

The solution set may be written in interval notation as $(-\infty, -1] \cup [1, \infty)$.

Now Try Exercise 55

---

4 ▸ Finding Numerical and Graphical Solutions

Compound inequalities can also be solved graphically and numerically, as illustrated in the next example.

EXAMPLE 8 **Estimating numbers of Internet users**

The number of U.S. Internet users in millions during year x can be modeled by the formula $f(x) = 11.6(x - 2000) + 124$. Estimate the years when the number of users was between 240 and 275 million. (*Source:* The Nielsen Company.)

Solution

Numerical Solution Let $y_1 = 11.6(x - 2000) + 124$. Make a table of values, as shown in **FIGURE 8.46(a)**. In 2010, the number of Internet users was 240 million, and in 2013, this number was about 275 million. Thus from 2010 to about 2013, the number of Internet users was between 240 million and 275 million.

Graphical Solution The graph of $y_1 = 11.6(x - 2000) + 124$ is shown between the graphs of $y_2 = 240$ and $y_3 = 275$ in **FIGURES 8.46(b)** and **8.46(c)** from 2010 to about 2013, or when $2010 \le x \le 2013$.

CALCULATOR HELP

To find a point of intersection, see Appendix A (page AP-6).

(a)

$[2006, 2015, 1]$ by $[150, 350, 50]$
(b)

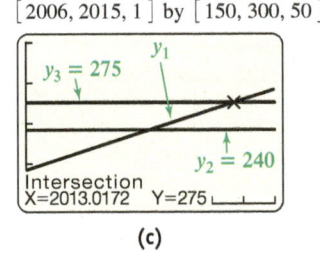
$[2006, 2015, 1]$ by $[150, 300, 50]$
(c)

FIGURE 8.46

Now Try Exercise 117(a) and (b)

5 Introducing Piecewise-Defined Functions

When a function f models data, there may not be one formula for $f(x)$ that works. In this case, the function is sometimes defined on pieces of its domain and is therefore called a **piecewise-defined function**. If each piece is linear, the function is a **piecewise-defined linear function**.

An example of a piecewise-defined function is the *Fujita scale*, which classifies tornadoes by intensity. If a tornado has wind speeds between 40 and 72 miles per hour, it is an F1 tornado. Tornadoes with wind speeds *greater than* 72 miles per hour but not more than 112 miles per hour are F2 tornadoes. The Fujita scale is represented by the following function F, where the input x represents the maximum wind speed of a tornado and the output is the F-scale number from 1 to 5.

Fujita Scale (F1 → F5)

F is defined in five pieces over intervals of its domain.

$$F(x) = \begin{cases} 1 & \text{if } 40 \le x \le 72 \\ 2 & \text{if } 72 < x \le 112 \\ 3 & \text{if } 112 < x \le 157 \\ 4 & \text{if } 157 < x \le 206 \\ 5 & \text{if } 206 < x \le 260 \end{cases}$$

$F(180) = 4$

For example, if the maximum wind speed is 180 miles per hour, then $F(180) = 4$ because **180** is between **157** and **206**; that is, $157 < 180 \le 206$. Thus a tornado with a maximum wind speed of 180 miles per hour is an F4 tornado.

A graph of $y = f(x)$ is shown in **FIGURE 8.47**. It is composed of horizontal line segments. Because each piece is constant, F is sometimes called a **piecewise-constant function** or **step function**. A solid dot occurs at the point $(72, 1)$ and an open circle occurs at the point $(72, 2)$, because technically a tornado with 72-mile-per-hour winds is an F1 tornado, not an F2 tornado.

Tornado Intensities

FIGURE 8.47

EXAMPLE 9 **Evaluating and graphing a piecewise-defined function**

Use $f(x)$ to complete the following.

$$f(x) = \begin{cases} x - 1 & \text{if } -4 \le x < 2 \\ -2x & \text{if } \ \ 2 \le x \le 4 \end{cases}$$

(a) What is the domain of f?
(b) Evaluate $f(-3), f(2), f(4)$, and $f(5)$.
(c) Sketch a graph of f.
(d) What is the range of f?

Solution

(a) Function f is defined for x-values satisfying either $-4 \le x < 2$ or $2 \le x \le 4$. Thus the domain of f is $-4 \le x \le 4$.

(b) For x-values satisfying $-4 \le x < 2$, $f(x) = x - 1$ and so $f(-3) = -4$. Similarly, if $2 \le x \le 4$, then $f(x) = -2x$. Thus $f(2) = -4$ and $f(4) = -8$. The expression $f(5)$ is undefined because 5 is not in the domain of f.

(c) Because each piece of $f(x)$ is linear, the graph of $y = f(x)$ consists of two line segments. Therefore we can find the endpoints of each line segment and then sketch the two line segments together to form the graph of f.

A **Graph the First Piece** Evaluate $y_1 = x - 1$ at $x = -4$ and $x = 2$. Place a dot at $(-4, -5)$ and an open circle at $(2, 1)$, because 2 is not included in $-4 \le x < 2$. Sketch a line segment between these points. See **FIGURE 8.48**.

B **Graph the Second Piece** Evaluate $y_2 = -2x$ at $x = 2$ and $x = 4$. Place dots at $(2, -4)$ and $(4, -8)$. Sketch a line segment between these points. See **FIGURE 8.49**.

Graph the First Piece

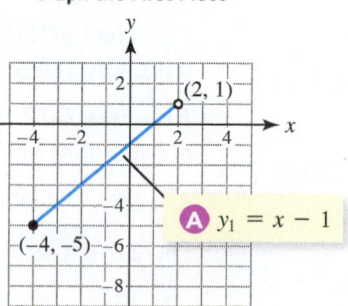

FIGURE 8.48

Graph the Second Piece

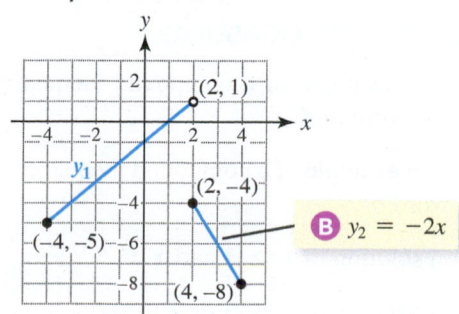

FIGURE 8.49

(d) The y-values in **FIGURE 8.49** vary from $y = -8$ up to but *not* including $y = 1$. The range is $-8 \le x < 1$.

Now Try Exercise 111

8.3   Putting It All Together

CONCEPT	EXPLANATION	EXAMPLES
Compound Inequality	Two inequalities joined by *and* or *or*	$2x \geq 10$ and $x + 2 < 16$; $x < -1$ or $x > 2$
Three-Part Inequality	Can be used to write some types of compound inequalities involving *and*	$x > -2$ and $x \leq 3$ is equivalent to $-2 < x \leq 3$.
Interval Notation	Notation used to write sets of real numbers rather than using number lines or inequalities	$-2 \leq z \leq 4$ is equivalent to $[-2, 4]$. $x < 4$ is equivalent to $(-\infty, 4)$. $x \leq -2$ or $x > 0$ is equivalent to $(-\infty, -2] \cup (0, \infty)$.
Piecewise-Defined Function	A function that is defined by using different formulas on different intervals of its domain	$f(x) = \begin{cases} x^2 & \text{if } x \leq 2 \\ 2x & \text{if } x > 2 \end{cases}$

TYPE OF INEQUALITY	METHOD USED TO SOLVE THE INEQUALITY
Solving a Compound Inequality Containing *and*	**STEP 1:** First solve each inequality individually. **STEP 2:** The solution set includes values that satisfy *both* inequalities from Step 1.
Solving a Compound Inequality Containing *or*	**STEP 1:** First solve each inequality individually. **STEP 2:** The solution set includes values that satisfy *at least one* of the inequalities from Step 1.
Solving a Three-Part Inequality	Work on all three parts at the same time. Be sure to perform the same step on each part. Continue until the variable is isolated in the middle part.

8.3   Exercises

MyMathLab®

CONCEPTS AND VOCABULARY

1. Give an example of a compound inequality containing the word *and*.

2. Give an example of a compound inequality containing the word *or*.

3. Is 1 a solution to $x > 3$ and $x \leq 5$?

4. Is 1 a solution to $x < 3$ or $x \geq 5$?

5. Is the compound inequality $x \geq -5$ and $x \leq 5$ equivalent to $-5 \leq x \leq 5$?

6. Name three ways to solve a compound inequality.

TESTING SOLUTIONS TO COMPOUND INEQUALITIES

Exercises 7–12: Determine whether the given values of x are solutions to the compound inequality.

7. $x - 1 < 5$ and $2x > 3$ $x = 2, x = 6$

8. $2x + 1 \geq 4$ and $1 - x \leq 3$ $x = -2, x = 3$

9. $3x < -5$ or $2x \geq 3$ $x = 0, x = 3$

10. $x + 1 \leq -4$ or $x + 1 \geq 4$ $x = -5, x = 2$

11. $2 - x > -5$ and $2 - x \leq 4$ $x = -3, x = 0$

12. $x + 5 \geq 6$ or $3x \leq 3$ $x = -1, x = 1$

WRITING INTERVAL NOTATION

Exercises 13–38: Write the inequality in interval notation.

13. $2 \le x \le 10$

14. $-1 < x < 5$

15. $5 < x \le 8$

16. $-\frac{1}{2} \le x \le \frac{5}{6}$

17. $x < 4$

18. $x \le -3$

19. $x > -2$

20. $x \ge 6$

21. $x \ge -2$ and $x < 5$

22. $x \le 6$ and $x \ge 2$

23. $x \le 8$ and $x > -8$

24. $x \ge -4$ and $x < 3$

25. $x \ge 6$ or $x > 3$

26. $x \le -4$ or $x < -3$

27. $x \le -2$ or $x \ge 4$

28. $x \le -1$ or $x > 6$

29. $x < 1$ or $x \ge 5$

30. $x < -3$ or $x > 3$

31.

32.

33.

34.

35. $\{x \mid x < 4\}$

36. $\{x \mid -1 \le x < 4\}$

37. $\{x \mid x < 1 \text{ or } x > 2\}$

38. $\{x \mid -\infty < x < \infty\}$

FINDING SYMBOLIC SOLUTIONS

Exercises 39–48: Solve the compound inequality. Graph the solution set on a number line.

39. $x \le 3$ and $x \ge -1$

40. $x \ge 5$ and $x > 6$

41. $2x < 5$ and $2x > -4$

42. $2x + 1 < 3$ and $x - 1 \ge -5$

43. $x + 2 > 5$ and $3 - x < 10$

44. $x + 2 > 5$ or $3 - x < 10$

45. $x \le -1$ or $x \ge 2$

46. $2x \le -6$ or $x \ge 6$

47. $5 - x > 1$ or $x + 3 \ge -1$

48. $1 - 2x > 3$ or $2x - 4 \ge 4$

Exercises 49–58: Solve the compound inequality. Write your answer in interval notation.

49. $x - 3 \le 4$ and $x + 5 \ge -1$

50. $2z \ge -10$ and $z < 8$

51. $3t - 1 > -1$ and $2t - \frac{1}{2} > 6$

52. $2(x + 1) < 8$ and $-2(x - 4) > -2$

53. $x - 4 \ge -3$ or $x - 4 \le 3$

54. $1 - 3n \ge 6$ or $1 - 3n \le -4$

55. $-x < 1$ or $5x + 1 < -10$

56. $7x - 6 > 0$ or $-\frac{1}{2}x \le 6$

57. $1 - 7x < -48$ and $3x + 1 \le -9$

58. $3x - 4 \le 8$ or $4x - 1 \le 13$

Exercises 59–80: Solve the three-part inequality. Write your answer in interval notation.

59. $-2 \le t + 4 < 5$

60. $5 < t - 7 < 10$

61. $-\frac{5}{8} \le y - \frac{3}{8} < 1$

62. $-\frac{1}{2} < y - \frac{3}{2} < \frac{1}{2}$

63. $-27 \le 3x \le 9$

64. $-4 < 2y < 22$

65. $\frac{1}{2} < -2y \le 8$

66. $-16 \le -4x \le 8$

67. $-4 < 5z + 1 \le 6$

68. $-3 \le 3z + 6 < 9$

69. $3 \le 4 - n \le 6$

70. $-1 < 3 - n \le 1$

71. $-1 < 2z - 1 < 3$

72. $2 \le 4z + 5 \le 6$

73. $-2 \le 5 - \frac{1}{3}m < 2$

74. $-\frac{3}{2} < 4 - 2m < \frac{7}{2}$

75. $100 \le 10(5x - 2) \le 200$

76. $-15 < 5(x - 1990) < 30$

77. $-3 < \dfrac{3z + 1}{4} < 1$

78. $-3 < \dfrac{z - 1}{2} < 5$

79. $-\dfrac{5}{2} \le \dfrac{2 - m}{4} \le \dfrac{1}{2}$

80. $\dfrac{4}{5} \le \dfrac{4 - 2m}{10} \le 2$

FINDING NUMERICAL AND GRAPHICAL SOLUTIONS

Exercises 81–84: Use the table to solve the three-part inequality. Write your answer in interval notation.

81. $-3 \le 3x \le 6$

82. $-5 \le 2x - 1 \le 1$

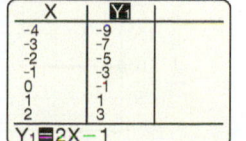

83. $-1 < 1 - x < 2$ **84.** $-2 \leq -2x < 4$

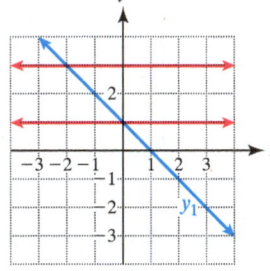

Exercises 85–88: Use the graph to solve the compound inequality. Write your answer in interval notation.

85. $-2 \leq y_1 \leq 2$ **86.** $1 \leq y_1 < 3$

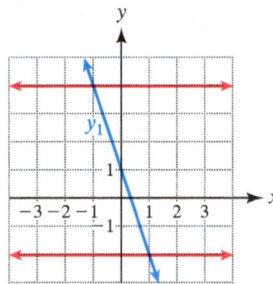

87. $y_1 < -2$ or $y_1 > 2$ **88.** $y_1 \leq -2$ or $y_1 \geq 4$

89. Distance The function f, shown in the figure, gives the distance y in miles between a car and Omaha, Nebraska, after x hours, where $0 \leq x \leq 6$.
  (a) Is the car moving toward or away from Omaha? Explain.
  (b) Determine the times when the car is 100 miles or 200 miles from Omaha.
  (c) When is the car from 100 to 200 miles from Omaha?
  (d) When is the car's distance from Omaha greater than or equal to 200 miles?

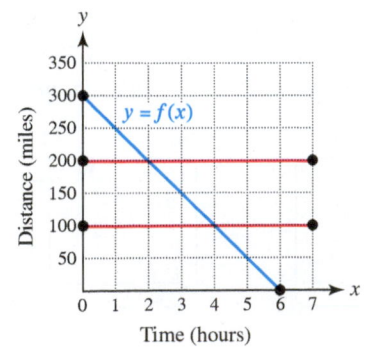

90. Distance The function g, shown in the figure, gives the distance y in miles between a train and Seattle after x hours, where $0 \leq x \leq 5$.

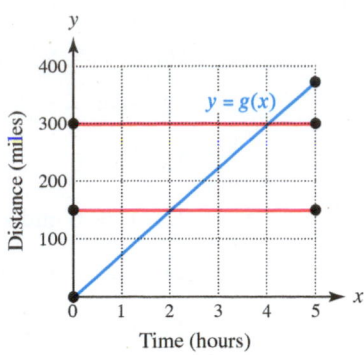

  (a) Is the train moving toward or away from Seattle? Explain.
  (b) Determine the times when the train is 150 miles or 300 miles from Seattle.
  (c) When is the train from 150 to 300 miles from Seattle?
  (d) When is the train's distance from Seattle less than or equal to 150 miles?

91. Use the figure to solve each equation or inequality. Let the domains of y_1, y_2, and y_3 be $0 \leq x \leq 8$.
  (a) $y_1 = y_2$ **(b)** $y_2 = y_3$
  (c) $y_1 \leq y_2 \leq y_3$ **(d)** $y_2 < y_1$

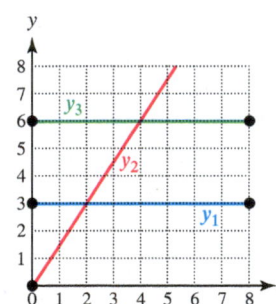

92. Use the figure to solve each equation or inequality. Let the domains of y_1, y_2, and y_3 be $0 \leq x \leq 5$.
  (a) $y_1 = y_2$ **(b)** $y_2 = y_3$
  (c) $y_1 \leq y_2 \leq y_3$ **(d)** $y_2 < y_3$

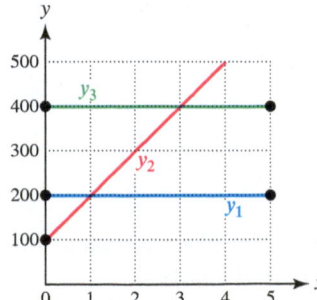

Exercises 93–98: Solve numerically or graphically. Write your answer in interval notation.

93. $-2 \le 2x - 4 \le 4$ **94.** $-1 \le 1 - x \le 3$

95. $x + 1 < -1$ or $x + 1 > 1$

96. $2x - 1 < -3$ or $2x - 1 > 5$

97. $95 \le 25(x - 2000) + 45 \le 295$

98. $42 \le -13(x - 2005) + 120 \le 94$

USING MORE THAN ONE METHOD

Exercises 99–104: Solve symbolically, graphically, and numerically. Write the solution set in interval notation.

99. $4 \le 5x - 1 \le 14$ **100.** $-4 < 2x < 4$

101. $4 - x \ge 1$ or $4 - x < 3$

102. $x + 3 \ge -2$ or $x + 3 \le 1$

103. $2x + 1 < 3$ or $2x + 1 \ge 7$

104. $3 - x \le 4$ or $3 - x > 8$

105. Thinking Generally Solve $c < x + b \le d$ for x.

106. Thinking Generally Solve $c \le ax + b \le d$ for x, if $a < 0$.

WORKING WITH PIECEWISE-DEFINED FUNCTIONS

Exercises 107–110: Graph each piecewise-defined function.

107. $f(x) = \begin{cases} x - 1 & \text{if } x \le 3 \\ 2 & \text{if } x > 3 \end{cases}$

108. $f(x) = \begin{cases} 6 - x & \text{if } x \le 3 \\ 3x - 6 & \text{if } x > 3 \end{cases}$

109. $f(x) = \begin{cases} 4 - x & \text{if } x < 2 \\ 1 + 2x & \text{if } x \ge 2 \end{cases}$

110. $f(x) = \begin{cases} 2x + 1 & \text{if } x \ge 0 \\ x & \text{if } x < 0 \end{cases}$

Exercises 111–116: Complete the following for $f(x)$.

(a) *Determine the domain of f.*
(b) *Evaluate $f(-2)$, $f(0)$, and $f(3)$.*
(c) *Graph f.*
(d) *What is the range of f?*

111. $f(x) = \begin{cases} 2 & \text{if } -5 \le x \le -1 \\ x + 3 & \text{if } -1 < x \le 5 \end{cases}$

112. $f(x) = \begin{cases} 2x + 1 & \text{if } -3 \le x < 0 \\ x - 1 & \text{if } 0 \le x \le 3 \end{cases}$

113. $f(x) = \begin{cases} 3x & \text{if } -1 \le x < 1 \\ x + 1 & \text{if } 1 \le x \le 2 \end{cases}$

114. $f(x) = \begin{cases} -2 & \text{if } -6 \le x < -2 \\ 0 & \text{if } -2 \le x < 0 \\ 3x & \text{if } 0 \le x \le 4 \end{cases}$

115. $f(x) = \begin{cases} x & \text{if } -3 \le x \le -1 \\ 1 & \text{if } -1 < x < 1 \\ 2 - x & \text{if } 1 \le x \le 3 \end{cases}$

116. $f(x) = \begin{cases} 3 & \text{if } -4 \le x \le -1 \\ x - 2 & \text{if } -1 < x \le 2 \\ 0.5x & \text{if } 2 < x \le 4 \end{cases}$

APPLICATIONS INVOLVING COMPOUND INEQUALITIES

117. Worldwide Gambling Losses Global betting losses in billions of dollars can be estimated and projected to the year 2018 by $L(x) = 20x + 250$, where x represents years after 2003. Use each of the following methods to estimate when losses ranged from $450 billion to $530 billion. (*Source:* The Economist.)
   (a) Numerical
   (b) Graphical
   (c) Symbolic

118. College Tuition From 1980 to 2000, college tuition and fees at private colleges could be modeled by the linear function $f(x) = 575(x - 1980) + 3600$. (Since 2000, the increase in tuition and fees has been more than linear growth.) Use each method to estimate when the average tuition and fees ranged from $8200 to $10,500. (*Source:* The College Board.)
   (a) Numerical
   (b) Graphical
   (c) Symbolic

119. Altitude and Dew Point If the dew point D on the ground is $60\,°F$, then the dew point x miles high is given by $D(x) = 60 - 5.8x$. Find the altitudes where the dew point ranges from $57.1\,°F$ to $51.3\,°F$. (*Source:* A. Miller.)

120. Cigarette Consumption Worldwide cigarette consumption in trillions from 1950 to 2015 can be modeled by $C(x) = 0.09x - 173.8$, where x is the year. Estimate the years when cigarette consumption was between 5.3 and 6.2 trillion. (*Source:* Department of Agriculture.)

121. Geometry For what values of x is the perimeter of the rectangle from 40 to 60 feet?

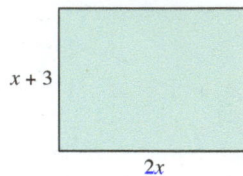

122. Geometry A rectangle is three times as long as it is wide. If the perimeter ranges from 100 to 160 inches, what values for the width are possible?

123. Altitude and Temperature If the air temperature at ground level is 70 °F, the air temperature x miles high is given by $T(x) = 70 - 19x$. Determine the altitudes at which the air temperature is from 41.5 °F to 22.5 °F. (*Source:* A. Miller and R. Anthes, Meteorology.)

124. Distance A car's distance in miles from a rest stop after x hours is given by $f(x) = 70x + 50$.
(a) Make a table for $f(x)$ for $x = 4, 5, 6, \ldots, 10$ and then use the table to solve the inequality $470 \le f(x) \le 680$. Explain your result.
(b) Solve the inequality in part (a) symbolically.

125. Medicare Costs In 2000, Medicare cost taxpayers $250 billion, and in 2010, it cost $500 billion. (*Source:* Department of Health and Human Services.)
(a) Find a linear function M that models these data x years after 2000.
(b) Estimate when Medicare costs might be from $600 billion to $700 billion.

126. Temperature Conversion Water freezes at 32 °F, or 0 °C, and boils at 212 °F, or 100 °C.
(a) Find a linear function $C(F)$ that converts Fahrenheit temperature to Celsius temperature.
(b) The greatest temperature ranges on Earth are recorded in Siberia, where temperature has varied from about -70 °C to 35 °C. Find this temperature range in Fahrenheit.

127. Speed Limits The graph of $y = f(x)$ in the next column gives the speed limit y along a rural highway x miles from its starting point.
(a) What are the maximum and minimum speed limits along this stretch of highway?

(b) Estimate the miles of highway with a speed limit of 55 miles per hour.
(c) Evaluate $f(4), f(12)$, and $f(18)$.

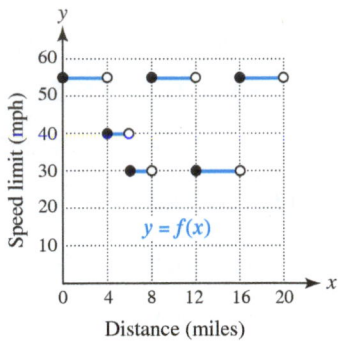

128. First-Class Mail Recently, the retail flat rate in dollars for first-class mail weighing up to 5 ounces could be computed by the piecewise-constant function P, where x is the number of ounces.

$$P(x) = \begin{cases} 0.90 & \text{if } 0 < x \le 1 \\ 1.10 & \text{if } 1 < x \le 2 \\ 1.30 & \text{if } 2 < x \le 3 \\ 1.50 & \text{if } 3 < x \le 4 \\ 1.70 & \text{if } 4 < x \le 5 \end{cases}$$

(a) Evaluate $P(1.5)$ and $P(3)$. Interpret your results.
(b) Sketch a graph of P. What is the domain of P?

WRITING ABOUT MATHEMATICS

129. Suppose that the solution set for a compound inequality can be written as $x < -3$ or $x > 2$. A student writes it as $2 < x < -3$. Is the student's three-part inequality correct? Explain your answer.

130. How can you determine whether an x-value is a solution to a compound inequality containing the word *and*? Give an example. Repeat the question for a compound inequality containing the word *or*.

8.4   Other Functions and Their Properties

Objectives

1   Introducing Nonlinear Functions
2   Expressing Domain and Range in Interval Notation
3   Recognizing Absolute Value Functions and Their Properties
4   Recognizing Polynomial Functions and Their Properties
5   Recognizing Rational Functions and Their Properties
6   Performing Operations on Functions

NEW VOCABULARY

☐ Absolute value function
☐ Polynomial function of one variable
☐ Linear function
☐ Quadratic function
☐ Cubic function
☐ Rational function

1   Introducing Nonlinear Functions

Many quantities in applications cannot be modeled with linear functions and equations. If data points do not lie on a line, we say that the data are *nonlinear*. For example, a scatterplot of the *cumulative* number of AIDS deaths from 1981 through 2015 is nonlinear, as shown in **FIGURE 8.50**. To model such data, we often use *nonlinear functions*, whose graphs are *not* lines. Because scatterplots of nonlinear data can have a variety of shapes, mathematicians use many different types of nonlinear functions, such as polynomial functions, which we discuss in this section. See Exercise 133. (*Source:* U.S. Department of Health and Human Services.)

U.S. AIDS Deaths

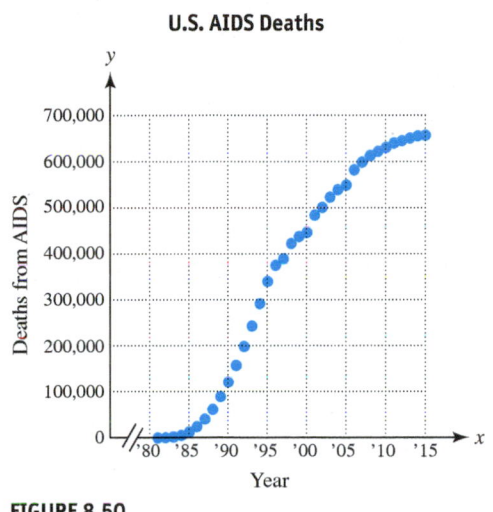

FIGURE 8.50

2   Expressing Domain and Range in Interval Notation

The set of all valid inputs for a function is called the *domain*, and the set of all outputs from a function is called the *range*. For example, all real numbers are valid inputs for $f(x) = x^2$. Rather than writing "the set of all real numbers" for the domain of f, we can use *interval notation* to express the domain as $(-\infty, \infty)$. The symbol ∞ represents infinity and is not a real number. Because $x^2 \geq 0$ for every real number x, the output from $f(x) = x^2$ is never negative. Therefore the range of f is $[0, \infty)$, which denotes all nonnegative real numbers, or $x \geq 0$. Note that 0 is in the range of f because $f(0) = 0$, and a bracket "$[$" is used to indicate that 0 is included in the range of f.

EXAMPLE 1   **Writing domains in interval notation**

Write the domain for each function in interval notation.
(a) $f(x) = 4x$ **(b)** $g(t) = \sqrt{t - 1}$ **(c)** $h(v) = \dfrac{1}{v + 3}$

Solution
(a) The expression $4x$ is defined for all real numbers x. Thus the domain of f is $(-\infty, \infty)$.
(b) The square root $\sqrt{t - 1}$ is defined only when $t - 1$ is *not* negative. Thus the domain of g includes all real numbers satisfying $t - 1 \geq 0$ or $t \geq 1$. In interval notation this inequality is written as $[1, \infty)$.
(c) The expression $\frac{1}{v + 3}$ is defined except when $v + 3 = 0$ or $v = -3$. Thus the domain of h includes all real numbers except -3 and can be written as $(-\infty, -3) \cup (-3, \infty)$. Parentheses are used because -3 is not included in the domain of h.

Now Try Exercises 13, 21, 25

In the next example, we determine the domain and range of a function from its graph. Note that dots placed at each end of a graph indicate that the endpoints are included.

EXAMPLE 2 **Writing the domain and range in interval notation**

Use the graph of f in **FIGURE 8.51** to write its domain and range in interval notation.

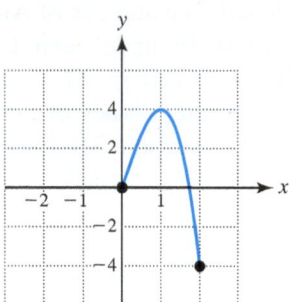

FIGURE 8.51

Solution
Because dots are placed at $(0, 0)$ and $(2, -4)$, the endpoints are included in the graph of f. Thus the graph in **FIGURE 8.51** includes x-values from $x = 0$ to $x = 2$. In interval notation, the domain of f is $[0, 2]$. The range of f includes y-values from -4 to 4 and can be expressed in interval notation as $[-4, 4]$.

Now Try Exercise 39

3 Recognizing Absolute Value Functions and Their Properties

In an earlier chapter, we discussed the absolute value of a number. We can define a function called the **absolute value function** as $f(x) = |x|$. We evaluate f as follows.

$$f(11) = |11| = 11, \quad f(-4) = |-4| = 4, \quad \text{and} \quad f(-\pi) = |-\pi| = \pi$$

To graph $y = |x|$, we begin by making a table of values, as shown in **TABLE 8.13**. Next we plot these points and sketch the graph, as shown in **FIGURE 8.52**. Note that the graph is V-shaped and never lies below the x-axis because the absolute value of a number cannot be negative.

$y = |x|$

x	$|x|$
-2	2
-1	1
0	0
1	1
2	2

TABLE 8.13

FIGURE 8.52

Because the input for $f(x) = |x|$ is any real number, the domain of f is all real numbers, or $(-\infty, \infty)$. The graph of the absolute value function shows that the output y (range) is any real number greater than or equal to 0. That is, the range is $[0, \infty)$.

4 Recognizing Polynomial Functions and Their Properties

In a previous chapter, we introduced polynomials and defined their degrees.

Polynomials of One Variable

$$1 - 5x, \quad 3t^2 - 5t + 1, \quad \text{and} \quad z^3 + 5$$

(The exponents on variables in polynomials must be nonnegative integers.) Recall that the *degree* of a polynomial of one variable equals the largest exponent on the variable. Thus the degree of $1 - 5x$ is 1, the degree of $3t^2 - 5t + 1$ is 2, and the degree of $z^3 + 5$ is 3.

Polynomial Functions of One Variable

$$f(x) = 1 - 5x, \quad g(t) = 3t^2 - 5t + 1, \quad \text{and} \quad h(z) = z^3 + 5$$

The equations shown above define three **polynomial functions of one variable**. Function f is a **linear function** because it has degree 1, function g is a **quadratic function** because it has degree 2, and function h is a **cubic function** because it has degree 3.

NOTE: The domain of *every* polynomial function is *all* real numbers, $(-\infty, \infty)$. ∎

EXAMPLE 3   **Identifying polynomial functions**

Determine whether $f(x)$ represents a polynomial function. If possible, identify the type of polynomial function and its degree.
(a) $f(x) = 5x^3 - x + 10$
(b) $f(x) = x^{-2.5} + 1$
(c) $f(x) = 1 - 2x$
(d) $f(x) = \dfrac{3}{x - 1}$

Solution
(a) The expression $5x^3 - x + 10$ is a cubic polynomial, so $f(x)$ represents a cubic polynomial function. It has degree 3.
(b) $f(x) = x^{-2.5} + 1$ does not represent a polynomial function because the variables in a polynomial must have *nonnegative integer* exponents.
(c) $f(x) = 1 - 2x$ represents a polynomial function that is linear. It has degree 1.
(d) $f(x) = \frac{3}{x-1}$ does not represent a polynomial function because $\frac{3}{x-1}$ is not a polynomial.

Now Try Exercises 43, 45, 47, 51

Frequently, polynomials represent functions or formulas that can be evaluated. This situation is illustrated in the next two examples.

EXAMPLE 4   **Evaluating a polynomial function graphically and symbolically**

A graph of $f(x) = 4x - x^3$ is shown in **FIGURE 8.53**, where $y = f(x)$. Evaluate $f(-1)$ graphically and check your result symbolically.

Solution
Graphical Evaluation To evaluate $f(-1)$ graphically, find -1 on the x-axis and move down until the graph of f is reached. Then move horizontally to the y-axis, as shown in **FIGURE 8.54** on the next page. Thus when $x = -1$, $y = -3$ and $f(-1) = -3$.

Symbolic Evaluation When $x = -1$, evaluation of $f(x) = 4x - x^3$ is performed as follows.

$$f(-1) = 4(-1) - (-1)^3 = -4 - (-1) = -3$$

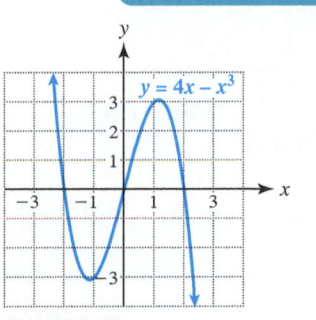

FIGURE 8.53

Evaluating $f(-1) = -3$

Find -1 on the x-axis. Move downward to the graph of f and then across to the y-axis.

FIGURE 8.54

Now Try Exercise 73

EXAMPLE 5 **Evaluating a polynomial function symbolically**

Evaluate $f(x)$ at the given value of x.
(a) $f(x) = -3x^4 - 2$, $x = 2$ **(b)** $f(x) = -2x^3 - 4x^2 + 5$, $x = -3$

Solution
(a) Be sure to evaluate exponents before multiplying.

$$f(2) = -3(2)^4 - 2 = -3 \cdot 16 - 2 = -50$$

(b) $f(-3) = -2(-3)^3 - 4(-3)^2 + 5 = -2(-27) - 4(9) + 5 = 23$

Now Try Exercises 61, 63

 Math in Context A well-conditioned athlete's heart rate can reach 200 beats per minute during strenuous physical activity. Upon quitting, a typical heart rate decreases rapidly at first and then more gradually after a few minutes, as illustrated in the next example.

EXAMPLE 6 **Modeling heart rate of an athlete**

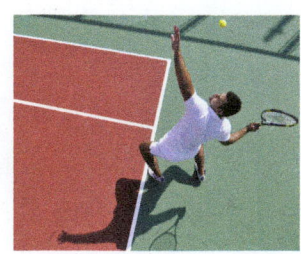

Let $P(t) = 1.875t^2 - 30t + 200$ model an athlete's heart rate (or pulse P) in beats per minute (bpm) t minutes after strenuous exercise has stopped, where $0 \le t \le 8$. (*Source:* V. Thomas, *Science and Sport.*)
(a) What is the initial heart rate when the athlete stops exercising?
(b) What is the heart rate after 8 minutes?
(c) A graph of P is shown in **FIGURE 8.55**. Interpret this graph.

Athlete's Heart Rate

$P(t) = 1.875t^2 - 30t + 200$

Time (minutes)

FIGURE 8.55

Solution
(a) To find the initial heart rate, evaluate $P(t)$ at $t = 0$, or

$$P(0) = 1.875(0)^2 - 30(0) + 200 = 200.$$

When the athlete stops exercising, the heart rate is **200** beats per minute. (This result agrees with the graph.)

(b) $P(8) = 1.875(8)^2 - 30(8) + 200 = 80$ beats per minute.

(c) The heart rate does not drop at a constant rate; rather, it drops rapidly at first and then gradually begins to level off.

Now Try Exercise 127

5 Recognizing Rational Functions and Their Properties

A rational expression is formed when a polynomial is divided by a polynomial.

Rational Expressions

$$\frac{2x-1}{x}, \quad \frac{5}{x^2-1}, \quad \text{and} \quad \frac{2x-5}{x^2-9}$$ Polynomial divided by a polynomial

Rational expressions can be used to define *rational functions*.

RATIONAL FUNCTIONS

Let $p(x)$ and $q(x)$ be polynomials. Then a **rational function** is given by

$$f(x) = \frac{p(x)}{q(x)}.$$

The domain of f includes all x-values such that $q(x) \neq 0$.

Using Rational Expressions to Define Rational Functions

$$f(x) = \frac{2x-1}{x}, \quad g(x) = \frac{5}{x^2+1}, \quad \text{and} \quad h(x) = \frac{2x-5}{x^2-9}$$

The domain of f includes all real numbers except 0, the domain of g includes all real numbers because $x^2 + 1 \neq 0$ for any x-value, and the domain of h includes all real numbers except ± 3.

Formulas for linear and polynomial functions are *defined* for all x-values. However, formulas for a rational function are *undefined* for x-values that make the denominator equal to 0. (Division by 0 is undefined.) For example, if $f(x) = \frac{1}{x-2}$, then $f(2) = \frac{1}{2-2} = \frac{1}{0}$ is undefined because the denominator equals 0. A graph of $y = f(x)$ is shown in **FIGURE 8.56**. The graph does not cross the dashed vertical line $x = 2$, because $f(x)$ is *undefined* at $x = 2$. The red vertical dashed line $x = 2$ is called a *vertical asymptote*, and is used as an aid for sketching a graph of f. It is not actually part of the graph of the function.

READING CHECK 1

- How can you determine the domain of a rational function?

A Rational Function

FIGURE 8.56

EXAMPLE 7 **Identifying the domains of rational functions**

Write the domain of each function in interval notation.

(a) $f(x) = \dfrac{1}{x + 2}$ **(b)** $g(x) = \dfrac{2x}{x^2 - 3x + 2}$ **(c)** $h(t) = \dfrac{4}{t^3 - t}$

Solution
(a) The domain of f includes all x-values except when the denominator equals 0.

$$x + 2 = 0 \qquad \text{Set the denominator equal to 0.}$$
$$x = -2 \qquad \text{Subtract 2.}$$

Thus $f(-2)$ is undefined and -2 must be excluded from the domain of f. In interval notation, the domain of f is $(-\infty, -2) \cup (-2, \infty)$.

(b) The domain of g includes all real numbers except when $x^2 - 3x + 2 = 0$.

$$x^2 - 3x + 2 = 0 \qquad \text{Set the denominator equal to 0.}$$
$$(x - 1)(x - 2) = 0 \qquad \text{Factor.}$$
$$x = 1 \quad \text{or} \quad x = 2 \qquad \text{Zero-product property}$$

Because $g(1)$ and $g(2)$ are both undefined, 1 and 2 must be excluded from the domain of g. In interval notation, the domain of g is $(-\infty, 1) \cup (1, 2) \cup (2, \infty)$.

(c) The domain of h includes all real numbers except when $t^3 - t = 0$.

$$t^3 - t = 0 \qquad \text{Set the denominator equal to 0.}$$
$$t(t^2 - 1) = 0 \qquad \text{Factor out } t.$$
$$t(t - 1)(t + 1) = 0 \qquad \text{Difference of squares}$$
$$t = 0 \quad \text{or} \quad t = 1 \quad \text{or} \quad t = -1 \qquad \text{Zero-product property}$$

In interval notation, the domain of h is $(-\infty, -1) \cup (-1, 0) \cup (0, 1) \cup (1, \infty)$.

Now Try Exercises 27, 33, 35

To graph a rational function by hand, we usually start by making a table of values, as demonstrated in the next example. Because the graphs of rational functions are typically nonlinear, it is a good idea to plot at least 3 points on each side of an x-value where the formula is undefined—that is, where the denominator equals 0.

EXAMPLE 8 **Graphing a rational function**

Graph $f(x) = \dfrac{1}{x}$. State the domain of f.

Solution
Make a table of values for $f(x) = \frac{1}{x}$, as shown in **TABLE 8.14**. Notice that $x = 0$ is not in the domain of f, and a dash can be used to denote this undefined value. The domain of f is all real numbers such that $x \neq 0$. Start by picking three x-values on each side of 0.

x	-2	-1	$-\frac{1}{2}$	0	$\frac{1}{2}$	1	2
$\frac{1}{x}$	$-\frac{1}{2}$	-1	-2	—	2	1	$\frac{1}{2}$

A table of values to help graph $y = \frac{1}{x}$

TABLE 8.14

⬆ ———— **Undefined**

Plot the points shown in **TABLE 8.14** and then connect the points with a smooth curve, as shown in **FIGURE 8.57**. Because $f(0)$ is undefined, the graph of $f(x) = \frac{1}{x}$ does not cross the line $x = 0$ (the y-axis). The line $x = 0$ is a vertical asymptote.

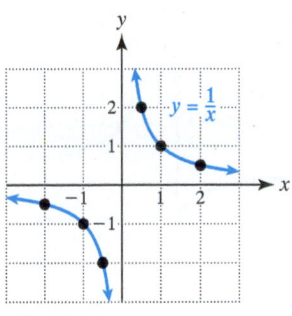

FIGURE 8.57

Now Try Exercise 89

In the next example, we evaluate a rational function in three ways.

EXAMPLE 9 **Evaluating a rational function**

Use **TABLE 8.15**, the formula for $f(x)$, and **FIGURE 8.58** to evaluate $f(-1), f(1),$ and $f(2)$.

(a)

x	$f(x)$
-3	$\frac{3}{2}$
-2	$\frac{4}{3}$
-1	1
0	0
1	—
2	4
3	3

TABLE 8.15

(b) $f(x) = \dfrac{2x}{x-1}$

(c)

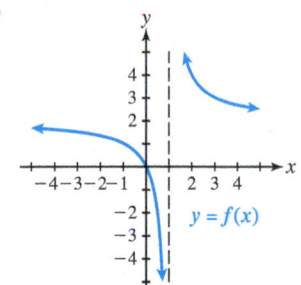

FIGURE 8.58

Solution

(a) **Numerical Evaluation** **TABLE 8.15** shows that

$$f(-1) = 1, \quad f(1) \text{ is undefined}, \quad \text{and} \quad f(2) = 4.$$

(b) **Symbolic Evaluation** Let $f(x) = \dfrac{2x}{x-1}$.

$$f(-1) = \frac{2(-1)}{-1-1} = 1$$

$$f(1) = \frac{2(1)}{1-1} = \frac{2}{0}, \text{ which is undefined. Input 1 is } not \text{ in the domain of } f.$$

$$f(2) = \frac{2(2)}{2-1} = 4$$

(c) **Graphical Evaluation** To evaluate $f(-1)$ graphically, find $x = -1$ on the x-axis and move upward to the graph of f. The y-value is **1** at the point of intersection, so

$f(-1) = 1$, as shown in **FIGURE 8.59(a)**. In **FIGURE 8.59(b)**, the red, dashed vertical line $x = 1$ is a vertical asymptote. Because the graph of f does not intersect this line, $f(1)$ is undefined. **FIGURE 8.59(c)** reveals that $f(2) = 4$.

Evaluating a Rational Function Graphically

FIGURE 8.59

Now Try Exercises 67, 77, 79

The following See the Concept explains more about vertical asymptotes and graphs of rational functions.

See the Concept 1   RATIONAL FUNCTIONS AND VERTICAL ASYMPTOTES

Ⓐ A vertical asymptote often occurs at x-values in the graph of a rational function $y = f(x)$ for which the denominator equals 0 and f is undefined.

Ⓑ The graph of a rational function never crosses a vertical asymptote, so a vertical asymptote is not part of the graph of f. However, it can be used as a guide when graphing a rational function.

Ⓒ On either side of a vertical asymptote, the y-values on the graph of a rational function typically become very large (approach ∞) or very small (approach $-\infty$).

WATCH VIDEO IN MML.

🔬 **Connecting Concepts with Your Life** You may have noticed that a relatively small percentage of people do the vast majority of postings on social networks, such as Facebook and Twitter. This phenomenon is called *participation inequality*. That is, a vast majority of the population falls under the category of "lurkers," who are on the network but are not posting material. This characteristic of a social network can be modeled approximately by a rational function, as illustrated in the next example. (*Source:* Wu, Michael, *The Economics of 90–9–1.*)

EXAMPLE 10

Modeling social network participation

The rational function given by

$$f(x) = \frac{100}{101 - x}, \quad 5 \le x \le 100,$$

models participation inequality in a social network. In this formula, $f(x)$ outputs the percentage of the postings done by the least active (bottom) x percent of the population.
(a) Evaluate $f(95)$. Interpret your answer.
(b) A graph of $y = f(x)$ is shown in **FIGURE 8.60**. Interpret the graph.

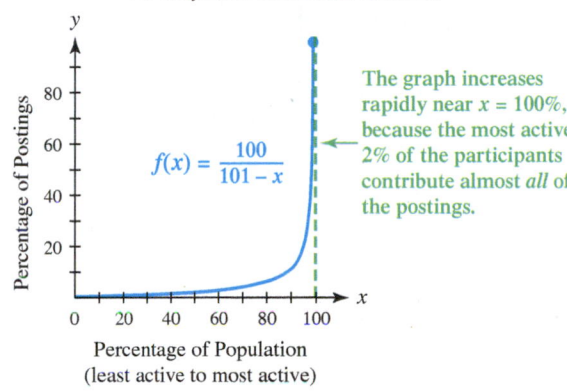

Participation in a Social Network

The graph increases rapidly near $x = 100\%$, because the most active 2% of the participants contribute almost *all* of the postings.

$f(x) = \dfrac{100}{101 - x}$

Percentage of Population (least active to most active)

FIGURE 8.60 Participation Inequality

(c) Solve the rational equation $\frac{100}{101 - x} = 9$. Interpret your answer.

Solution

(a) $f(\mathbf{95}) = \dfrac{100}{101 - \mathbf{95}} = \dfrac{100}{6} \approx 16.7\%$ Let $x = 95$.

This means that the least active 95% of the population contributes only 16.7% of the postings, so the most active 5% of the population is responsible for the remaining 83.3% of the postings.

(b) The graph shows participation inequality visually. The graph remains at a relatively low percentage until $x = 90\%$. This means that the bottom 90% of the population does very few postings. For $x \ge 90\%$, the graph rises rapidly because the top 10% contributes a vast majority of the postings.

(c) To solve this equation, we begin by multiplying each side by $(\mathbf{101 - x})$.

$$\frac{100}{101 - x} = 9 \qquad \text{Given equation}$$

$$(\mathbf{101 - x}) \cdot \frac{100}{\mathbf{101 - x}} = 9(\mathbf{101 - x}) \qquad \text{Multiply by } (101 - x).$$

$$100 = 9(101 - x) \qquad \text{Simplify left side.}$$

$$100 = 909 - 9x \qquad \text{Distributive property}$$

$$\mathbf{9x} + 100 - \mathbf{100} = 909 - \mathbf{100} + \mathbf{9x} - 9x \qquad \text{Add } 9x. \text{ Subtract } 100.$$

$$9x = 809 \qquad \text{Simplify.}$$

$$x = \frac{809}{9} \approx 90\% \qquad \text{Simplify.}$$

This result indicates that the least active 90% of the population contributes only 9% of the postings.

Now Try Exercise 129

CRITICAL THINKING 1

Suppose that a social network had *participation equality*, in which every member contributed an equal number of postings. Sketch a graph similar to **FIGURE 8.60** that describes this social network.

TECHNOLOGY NOTE

Asymptotes, Dot Mode, and Decimal Windows

When a rational function is graphed on a graphing calculator in connected mode, pseudo-asymptotes often occur because the calculator is simply connecting dots to draw a graph. The accompanying figures show the graph of $y = \frac{2}{x - 2}$ in connected mode, in dot mode, and with a *decimal*, or *friendly*, window. In dot mode, pixels in the calculator screen are not connected. With dot mode (and sometimes with a decimal window) pseudo-asymptotes do not appear. To learn more about these features, consult your owner's manual.

CALCULATOR HELP

To set a calculator in dot mode or to set a decimal window, see Appendix A (pages AP-10 and AP-11).

$[-6, 6, 1]$ by $[-4, 4, 1]$

Connected Mode

$[-6, 6, 1]$ by $[-4, 4, 1]$

Dot Mode

$[-4.7, 4.7, 1]$ by $[-3.1, 3.1, 1]$

Decimal Window

6 Performing Operations on Functions

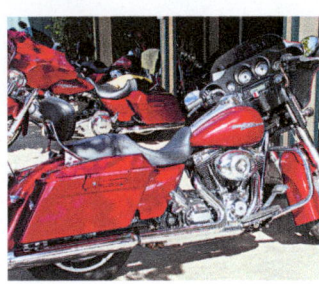

🌐 *Math in Context* **Business** A business incurs a *cost* to make its product and then it receives *revenue* from selling this product. For example, suppose a small business reconditions motorcycles. The graphs of its cost and of its revenue for reconditioning and selling x motorcycles are shown in **FIGURE 8.61**.

FIGURE 8.61

In general, *profit equals revenue minus cost*. In **FIGURE 8.61**, **profit** is shown visually as the length of the vertical green arrow between the graphs of **revenue** and **cost**. For any x-value, the distance by which revenue is above cost is called the profit for reconditioning and selling x motorcycles. *Maximum profit* for the company occurs at the x-value where the *length of the vertical green arrow is greatest.*

If we let $C(x), R(x)$, and $P(x)$ be functions that calculate the cost, revenue, and profit, respectively, for reconditioning and selling x motorcycles, then

$$P(x) = R(x) - C(x).$$

Profit *equals* **Revenue** *minus* **Cost**.

This example helps explain why we subtract functions in the real world. See Exercise 137. Functions can be added, multiplied, and divided in a similar manner.

Given two functions f and g, we define the sum $f + g$, difference $f - g$, product fg, and quotient $\frac{f}{g}$, as follows.

OPERATIONS ON FUNCTIONS

If $f(x)$ and $g(x)$ are both defined, then the sum, difference, product, and quotient of two functions f and g are defined by

$$(f + g)(x) = f(x) + g(x) \qquad \text{Sum}$$
$$(f - g)(x) = f(x) - g(x) \qquad \text{Difference}$$
$$(fg)(x) = f(x) \cdot g(x) \qquad \text{Product}$$
$$\left(\frac{f}{g}\right)(x) = \frac{f(x)}{g(x)}, \text{where } g(x) \neq 0. \quad \text{Quotient}$$

EXAMPLE 11   **Performing arithmetic on functions**

Use $f(x) = x^2$ and $g(x) = 2x - 4$ to evaluate each of the following.

(a) $(f + g)(3)$ (b) $(fg)(-1)$ (c) $\left(\frac{f}{g}\right)(0)$ (d) $(f/g)(2)$

Solution

(a) $(f + g)(3) = f(3) + g(3) = 3^2 + (2 \cdot 3 - 4) = 9 + 2 = 11$

(b) $(fg)(-1) = f(-1) \cdot g(-1) = (-1)^2 \cdot (2 \cdot (-1) - 4) = 1 \cdot (-6) = -6$

(c) $\left(\frac{f}{g}\right)(0) = \frac{f(0)}{g(0)} = \frac{0^2}{2 \cdot 0 - 4} = \frac{0}{-4} = 0$

(d) Note that $(f/g)(2)$ is equivalent to $\left(\frac{f}{g}\right)(2)$.

$$(f/g)(2) = \frac{f(2)}{g(2)} = \frac{2^2}{2 \cdot 2 - 4} = \frac{4}{0},$$

which is not possible because division by 0 is undefined. Thus $(f/g)(2)$ is *undefined*.

Now Try Exercise 101

In the next example, we find the sum, difference, product, and quotient of two functions for a general x.

EXAMPLE 12   **Performing arithmetic on functions**

Use $f(x) = 4x - 5$ and $g(x) = 3x + 1$ to evaluate each of the following.

(a) $(f + g)(x)$ (b) $(f - g)(x)$ (c) $(fg)(x)$ (d) $\left(\frac{f}{g}\right)(x)$

Solution

(a) $(f + g)(x) = f(x) + g(x) = (4x - 5) + (3x + 1) = 7x - 4$

(b) $(f - g)(x) = f(x) - g(x) = (4x - 5) - (3x + 1) = x - 6$

(c) $(fg)(x) = f(x) \cdot g(x) = (4x - 5)(3x + 1) = 12x^2 - 11x - 5$

(d) $\left(\frac{f}{g}\right)(x) = \frac{f(x)}{g(x)} = \frac{4x - 5}{3x + 1}$

Now Try Exercise 105

8.4 Putting It All Together

CONCEPT	COMMENTS	EXAMPLES
Writing Domain and Range in Interval Notation	Interval notation can be used to specify the domain and range of a function.	If $f(x) = x^2 + 1$, the domain of f is $(-\infty, \infty)$, and the range of f is $[1, \infty)$.
Absolute Value Function	Defined by $$f(x) = |x|$$ and has a V-shaped graph	$f(-5) = |-5| = 5$ $f(0) = |0| = 0$ $f(4) = |4| = 4$
Polynomial Function of One Variable	Can be defined by a polynomial; its degree equals the largest exponent of the variable.	Because $x^3 - 4x^2 + 6$ is a polynomial with degree 3, $$f(x) = x^3 - 4x^2 + 6$$ defines a polynomial function of degree 3 and is called a cubic function.
Rational Function	A rational function can be written as $$f(x) = \frac{p(x)}{q(x)},$$ where $p(x)$ and $q(x)$ are polynomials. Note that $q(x) \neq 0$.	Because $2x - 3$ and $x + 1$ are polynomials, $$f(x) = \frac{2x - 3}{x + 1}$$ defines a rational function. Because $f(x)$ is undefined at $x = -1$, the domain of f is $(-\infty, -1) \cup (-1, \infty)$
Operations on Functions	$(f + g)(x) = f(x) + g(x)$ Sum $(f - g)(x) = f(x) - g(x)$ Difference $(fg)(x) = f(x)g(x)$ Product $\left(\dfrac{f}{g}\right)(x) = \dfrac{f(x)}{g(x)}, g(x) \neq 0$ Quotient	Let $f(x) = x^2$ and $g(x) = 1 - x^2$. $(f + g)(x) = f(x) + g(x)$ $\qquad = x^2 + (1 - x^2)$ $\qquad = 1$ $(f - g)(x) = f(x) - g(x)$ $\qquad = x^2 - (1 - x^2)$ $\qquad = 2x^2 - 1$ $(fg)(x) = f(x)g(x)$ $\qquad = x^2(1 - x^2)$ $\qquad = x^2 - x^4$ $\left(\dfrac{f}{g}\right)(x) = \dfrac{f(x)}{g(x)}$ $\qquad = \dfrac{x^2}{1 - x^2}, x \neq -1, x \neq 1$

8.4 Exercises

MyMathLab®

CONCEPTS AND VOCABULARY

1. The set of all valid inputs for a function is called its _____.

2. The set of all outputs for a function is called its _____.

3. The set of all real numbers can be written in interval notation as _____.

4. If the domain of a function includes all real numbers except 5, then its domain can be written in interval notation as _____.

5. The graph of the _____ function is V-shaped.

6. The degree of a polynomial of one variable equals the largest _____ of the variable.

7. A quadratic function has degree _____.

8. If a function is linear, then its degree is _____.

9. If $f(x) = \frac{x}{2x + 1}$, then f is a(n) _____ function.

10. If $f(x) = \frac{x}{2x + 1}$, then the domain of f includes all real numbers except _____.

11. Which of the following expressions (a.–d.) is not a rational function?

    a. $f(x) = \frac{1}{x}$ **b.** $f(x) = x^2 + 1$

    c. $f(x) = \sqrt{x}$ **d.** $f(x) = \frac{2x^2}{x - 1}$

12. Which (a.–d.) is the domain of $f(x) = \frac{2x}{2x - 1}$?

    a. $\{x \mid x \neq \frac{1}{2}\}$ **b.** $\{x \mid x \neq 1\}$

    c. $\{x \mid x \neq 0\}$ **d.** $\{x \mid x = 1\}$

FINDING DOMAIN AND RANGE OF FUNCTIONS

Exercises 13–24: Write the domain and the range of the function in interval notation. (Hint: You may want to consider the graph of the function.)

13. $f(x) = -2x$ 14. $f(x) = -\frac{1}{4}x + 1$

15. $g(t) = \frac{2}{3}t - 3$ 16. $g(t) = 9t$

17. $h(z) = z^2 + 2$ 18. $h(z) = z^2 - 1$

19. $f(z) = -z^2$ 20. $f(z) = -\frac{1}{4}z^2$

21. $g(x) = \sqrt{x + 1}$ 22. $g(x) = \sqrt{x - 2}$

23. $h(x) = |x - 1|$ 24. $h(x) = |2x|$

Exercises 25–36: Write the domain of the rational function in interval notation.

25. $f(x) = \frac{1}{x - 1}$ 26. $f(x) = \frac{6}{x}$

27. $f(x) = \frac{x}{6 - 3x}$ 28. $f(x) = \frac{3x}{2x - 4}$

29. $g(t) = \frac{2}{t^2 - 4}$ 30. $g(t) = \frac{5}{1 - t^2}$

31. $g(t) = \frac{5t}{t^2 - 2t}$ 32. $g(t) = \frac{-t}{2t^2 - 3t}$

33. $h(z) = \frac{2 - z}{z^3 - 1}$ 34. $h(z) = \frac{z + 1}{z^3 - z^2}$

35. $f(x) = \frac{4}{x^2 - 2x - 3}$

36. $f(x) = \frac{1}{x^2 + 4x - 5}$

Exercises 37–42: A graph of a function is shown. Write the domain and range of the function in interval notation.

37.

38.

39.

40.

41.

42.

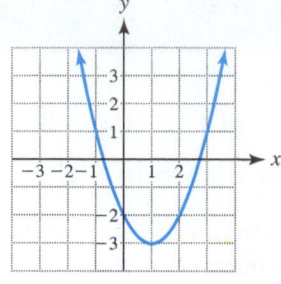

68. $g(t) = -\dfrac{2 - t}{4t}$ $t = 4, t = -1$

69. $g(t) = \dfrac{t^2}{t^2 - t}$ $t = -5, t = 1$

70. $g(t) = \dfrac{t - 3}{t^2 - 3t + 2}$ $t = -2, t = 1$

RECOGNIZING POLYNOMIAL FUNCTIONS

Exercises 43–54: Determine whether $f(x)$ represents a polynomial function. If possible, identify the degree and type of polynomial function.

43. $f(x) = 5x - 11$ **44.** $f(x) = 9 - x$

45. $f(x) = x^3$ **46.** $f(x) = x^2 + 3$

47. $f(x) = \dfrac{6}{x + 5}$ **48.** $f(x) = |x|$

49. $f(x) = 1 + 2x - x^2$ **50.** $f(x) = \frac{1}{4}x^3 - x$

51. $f(x) = 5x^{-2}$ **52.** $f(x) = x^2 + x^{-1}$

53. $f(x) = x^4 + 2x^2$ **54.** $f(x) = x^5 - 3x^3$

EVALUATING NONLINEAR FUNCTIONS

Exercises 55–70: If possible, evaluate $g(t)$ for the given values of t.

55. $g(t) = |4t|$ $t = 3, t = 0$

56. $g(t) = |t + 12|$ $t = 18, t = -15$

57. $g(t) = |t - 2|$ $t = 1, t = -\frac{3}{4}$

58. $g(t) = |2t + 1|$ $t = 2, t = -\frac{1}{2}$

59. $g(t) = t^2 - t - 6$ $t = 3, t = -3$

60. $g(t) = 3t^2 - 2t$ $t = -2, t = 4$

61. $g(t) = -2t^3 + t$ $t = 2, t = -2$

62. $g(t) = \frac{1}{3}t^3$ $t = 1, t = -3$

63. $g(t) = t^2 - 2t - 6$ $t = 0, t = -3$

64. $g(t) = 2t^3 - t^2 + 4$ $t = 2, t = -1$

65. $g(t) = \dfrac{1}{t}$ $t = 11, t = -7$

66. $g(t) = \dfrac{2}{3 - t}$ $t = 10, t = 3$

67. $g(t) = -\dfrac{t}{t + 1}$ $t = 5, t = -1$

Exercises 71–78: If possible, use the graph to evaluate each expression. Then use the formula for $f(x)$ to check your results.

71. $f(0)$ and $f(1)$ **72.** $f(-1)$ and $f(2)$

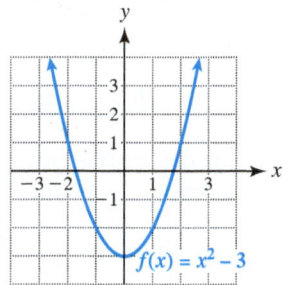

73. $f(-1)$ and $f(2)$ **74.** $f(0)$ and $f(-2)$

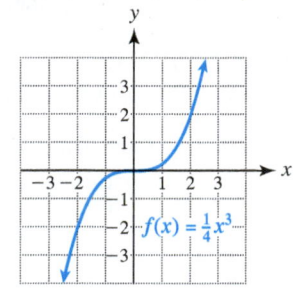

75. $f(-2)$ and $f(2)$ **76.** $f(-1)$ and $f(0)$

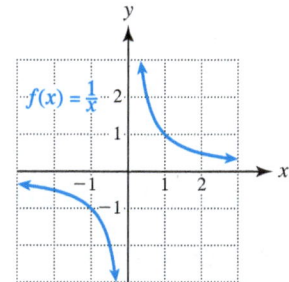

77. $f(-3)$ and $f(-1)$ **78.** $f(0)$ and $f(1)$

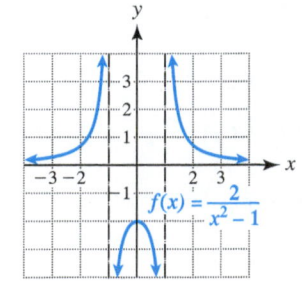

Exercises 79 and 80: Complete the table. Then evaluate $f(1)$.

79.

x	-2	-1	0	1	2
$f(x) = \frac{1}{x-1}$					

80.

x	-2	-1	0	1	2
$f(x) = \frac{2x}{x+2}$					

GRAPHING NONLINEAR FUNCTIONS

Exercises 81–100: Graph $y = f(x)$.

81. $f(x) = |2x|$ **82.** $f(x) = \left|\frac{1}{2}x\right|$

83. $f(x) = |x + 2|$ **84.** $f(x) = |x - 2|$

85. $f(x) = 1 - 2x^2$ **86.** $f(x) = \frac{1}{2}x^2 + 1$

87. $f(x) = \frac{1}{2}x^2$ **88.** $f(x) = x^2 - 2$

89. $f(x) = \dfrac{1}{x-1}$ **90.** $f(x) = \dfrac{1}{x+1}$

91. $f(x) = \dfrac{1}{2x}$ **92.** $f(x) = \dfrac{2}{x}$

93. $f(x) = \dfrac{1}{x+2}$ **94.** $f(x) = \dfrac{1}{x-2}$

95. $f(x) = \dfrac{4}{x^2+1}$ **96.** $f(x) = \dfrac{6}{x^2+2}$

97. $f(x) = \dfrac{3}{2x-3}$ **98.** $f(x) = \dfrac{1}{3x+2}$

99. $f(x) = \dfrac{1}{x^2-1}$ **100.** $f(x) = \dfrac{4}{4-x^2}$

PERFORMING OPERATIONS ON FUNCTIONS

Exercises 101–104: Use $f(x)$ *and* $g(x)$ *to evaluate each of the following.*

  (a) $(f + g)(3)$ **(b)** $(f - g)(-2)$
  (c) $(fg)(5)$ **(d)** $(f/g)(0)$

101. $f(x) = 5x, g(x) = x + 1$

102. $f(x) = x^2 + 2, g(x) = -2x$

103. $f(x) = 2x - 1, g(x) = 4x^2$

104. $f(x) = x^2 - 1, g(x) = x + 2$

Exercises 105–108: Use $f(x)$ *and* $g(x)$ *to find each of the following.*

  (a) $(f + g)(x)$ **(b)** $(f - g)(x)$
  (c) $(fg)(x)$ **(d)** $(f/g)(x)$

105. $f(x) = x + 1, g(x) = x + 2$

106. $f(x) = -3x, g(x) = x - 1$

107. $f(x) = 1 - x, g(x) = x^2$

108. $f(x) = x^2 + 4, g(x) = 6x$

109. **Thinking Generally** If $f(x) = x^2 - 2x$, then it follows that $f(a) = $ _____.

110. **Thinking Generally** If $f(x) = 2x - 1$, then it follows that $f(a + 2) = $ _____.

WORKING WITH PIECEWISE-DEFINED FUNCTIONS

Exercises 111–114: Match each piecewise-defined function with its graph (a–d).

111. $f(x) = \begin{cases} x^2 - 4 & \text{if } x \geq 0 \\ -x + 5 & \text{if } x < 0 \end{cases}$

112. $g(x) = \begin{cases} |x - 2| & \text{if } x \geq -1 \\ -x^2 & \text{if } x < -1 \end{cases}$

113. $h(x) = \begin{cases} 4 & \text{if } x \geq 0 \\ -4 & \text{if } x < 0 \end{cases}$

114. $k(x) = \begin{cases} \sqrt{x} & \text{if } x \geq 0 \\ -x^2 & \text{if } x < 0 \end{cases}$

a.

b.

c.

d.

Exercises 115–120: Evaluate $f(x)$ at the given values of x.

115. $x = -2$ and 1

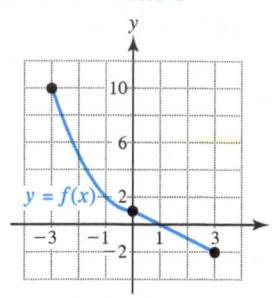

116. $x = -1, 0,$ and 3

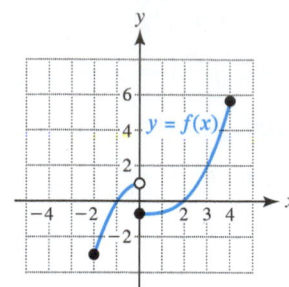

117. $x = -1, 1,$ and 2

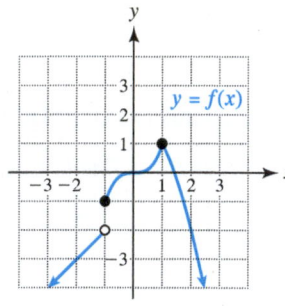

118. $x = -2, 0,$ and 2

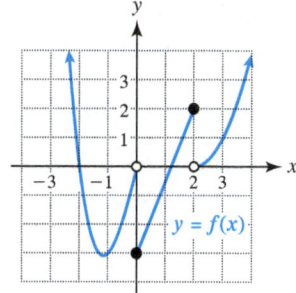

119. $x = -3, 1,$ and 4

$$f(x) = \begin{cases} x^3 - 4x^2 & \text{if } x \le -3 \\ 3x^2 & \text{if } -3 < x < 4 \\ x^3 - 54 & \text{if } x \ge 4 \end{cases}$$

120. $x = -4, 0,$ and 4

$$f(x) = \begin{cases} -4x & \text{if } x \le -4 \\ x^3 + 2 & \text{if } -4 < x \le 2 \\ 4 - x^2 & \text{if } x > 2 \end{cases}$$

Exercises 121–122: Graph the piecewise-defined function and state its domain in interval notation.

121. $f(x) = \begin{cases} |x| & \text{if } -3 \le x \le 0 \\ x^2 & \text{if } 0 < x \le 3 \end{cases}$

122. $f(x) = \begin{cases} x^2 & \text{if } -2 \le x < 0 \\ x + 1 & \text{if } 0 \le x \le 2 \end{cases}$

APPLICATIONS INVOLVING NONLINEAR FUNCTIONS

Exercises 123–126: **Graphical Interpretation** *Match the physical situation with the graph of the rational function (a.–d.) in the next column that models it best.*

123. A population of fish that increases and then levels off

124. An insect population that dies out

125. The length of a ticket line as the rate at which people arrive in line increases

126. The wind speed during a day that is initially calm, becomes windy, and then is calm again

a.

b.

c.

d.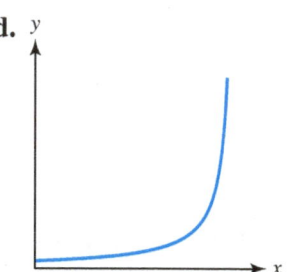

127. Heart Rate of an Athlete The following table lists the heart rate of an athlete running a 100-meter race. The race lasts 10 seconds.

Time (minutes)	0	2	4	6	8	10
Heart Rate (bpm)	90	100	113	127	143	160

(a) Does $P(t) = 0.2t^2 + 5t + 90$ model the data in the table exactly? Explain.

(b) Does P provide a reasonable model for the athlete's heart rate?

(c) Does $P(12)$ have significance in this situation? What should be the domain of P?

128. Heart Rate of an Athlete The following table lists an athlete's heart rate after the athlete finishes exercising strenuously.

Time (minutes)	0	2	4	6
Heart Rate (bpm)	180	137	107	90

(a) Does $P(t) = \frac{5}{3}t^2 - 25t + 180$ model the data in the table exactly? Explain.

(b) Does P provide a reasonable model for the athlete's heart rate?

(c) Does $P(12)$ have significance in this situation? What should be the domain of P?

129. Time Spent in Line If a parking lot attendant can wait on 5 vehicles per minute and vehicles are leaving the lot randomly at an average rate of x vehicles per minute, then the average time T in minutes spent waiting in line *and* paying the attendant is given by

$$T(x) = \frac{1}{5 - x},$$

where $x < 5$. (*Source:* N. Garber.)

(a) Evaluate $T(4)$ and interpret the result.
(b) A graph of T is shown in the figure. Interpret the graph as x increases from 0 to 5. Does this result agree with your intuition?

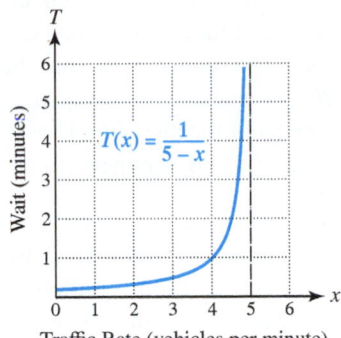

Traffic Rate (vehicles per minute)

(c) Find x if the waiting time is 3 minutes.

130. **People Waiting in Line** At a post office, workers can wait on 50 people per hour. If people arrive randomly at an average rate of x per hour, then the average number of people N waiting in line is given by

$$N(x) = \frac{x^2}{2500 - 50x},$$

where $x < 50$. (*Source:* N. Garber.)
(a) Evaluate $N(30)$ and interpret the result.
(b) A graph of N is shown in the figure. Interpret the graph as x increases from 0 to 50. Does this result agree with your intuition?

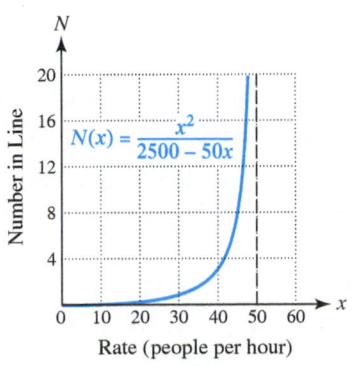

Rate (people per hour)

(c) Find x if $N = 8$.

131. **Uphill Highway Grade** The *grade* x of a hill is a measure of its steepness and corresponds to the slope of the road. For example, if a road rises 10 feet for every 100 feet of horizontal distance, it has an uphill grade of $x = \frac{10}{100}$, or 10%, as illustrated in the figure.

The braking distance for a car traveling 30 miles per hour on a wet, *uphill* grade x is given by

$$D(x) = \frac{900}{10.5 + 30x}.$$

(*Source:* N. Garber.)
(a) Evaluate $D(0.05)$ and interpret the result.
(b) If the braking distance for this car is 60 feet, find the uphill grade x.

132. **Downhill Highway Grade** (See Exercise 131.) The braking distance for a car traveling 30 miles per hour on a wet, *downhill* grade x is given by

$$S(x) = \frac{900}{10.5 - 30x}.$$

(a) Evaluate $S(0.05)$ and interpret the result.
(b) Make a table for $D(x)$ from Exercise 131 and $S(x)$, starting at $x = 0$ and incrementing by 0.05.
(c) How do the braking distances for uphill and downhill grades compare? Does this result agree with your driving experience?

133. **U.S. AIDS Deaths Historically** The following scatterplot shows the cumulative number of reported AIDS deaths. The data may be modeled x years after 1980 by $f(x) = 2.4x^2 - 14x + 23$, where the output is in thousands of deaths. (Note that after 1994, the number of deaths started to level off because of new drug therapy.)

Year

(a) Use $f(x)$ to estimate the cumulative total of AIDS deaths in 1990. Compare it with the actual value of 121.6 thousand.
(b) In 1997, the cumulative number of AIDS deaths was 390 thousand. What estimate does $f(x)$ give? Discuss your result.

134. **A PC for All?** Worldwide sales of computers have climbed as prices have continued to drop. The function $f(x) = 0.29x^2 + 8x + 19$ models the number of personal computers sold in millions during year x, where $x = 0$ corresponds to 1990, $x = 1$ corresponds to 1991, and so on until $x = 25$ corresponds to 2015. Estimate the number of personal computers

sold in 2010, using both the graph and the polynomial. (*Source:* eTForcasts.)

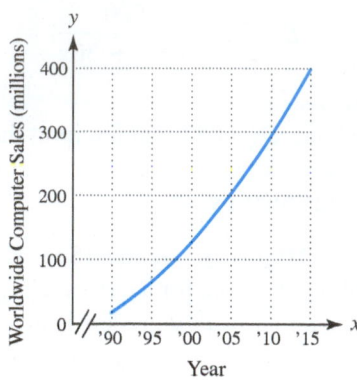

Exercises 135 and 136: **Remembering What You Learn**
After a test, students often forget what they learned. The rational function

$$R(x) = \frac{100}{1.2x + 1}, 0 \le x \le 5,$$

gives an estimate of the percentage of the material a student remembers x days after a test.

135. Evaluate $R(1)$ and $R(3)$. Interpret your results.

136. If a student takes notes in class, these percentages increase by 30% for $1 \le x \le 5$. Write another function $N(x)$ that models this result. Evaluate $N(3)$.

137. Profit A company makes and sells notebook computers. The company's cost function in thousands of dollars is $C(x) = 0.3x + 100$, and the revenue function in thousands of dollars is $R(x) = 0.75x$, where x is the number of notebook computers.
(a) Evaluate and interpret $C(100)$.
(b) Interpret the y-intercepts on the graphs of C and R.
(c) Give the profit function $P(x)$.
(d) How many computers need to be sold to make a profit?

138. Profit A company makes and sells sailboats. The company's cost function in thousands of dollars is $C(x) = 2x + 20$, and the revenue function in thousands of dollars is $R(x) = 4x$, where x is the number of sailboats.
(a) Evaluate and interpret $C(5)$.
(b) Interpret the y-intercepts on the graphs of C and R.
(c) Give the profit function $P(x)$.
(d) How many sailboats need to be sold to break even?

WRITING ABOUT MATHEMATICS

139. Name two functions. Give their formulas, sketch their graphs, and state their domains and ranges.

140. Explain the difference between the domain and the range of a function.

**SECTIONS
8.3 and 8.4**   Checking Basic Concepts

1. (a) Is 3 a solution to the compound inequality $x + 2 < 4$ or $2x - 1 \ge 3$?
(b) Is 3 a solution to the compound inequality $x + 2 < 4$ and $2x - 1 \ge 3$?

2. Solve the following compound inequalities. Write your answers in interval notation.
(a) $-5 \le 2x + 1 \le 3$
(b) $1 - x \le -2$ or $1 - x \ge 2$
(c) $-2 < \dfrac{4 - 3x}{2} \le 6$

3. Write the domain of each function in interval notation.
(a) $f(x) = x^2$
(b) $g(t) = \dfrac{1}{t - 1}$
(c) $h(z) = \sqrt{z}$

4. Use the graph of f to do the following.
(a) Write the domain and range of f in interval notation.
(b) Evaluate $f(0)$ and $f(-2)$.

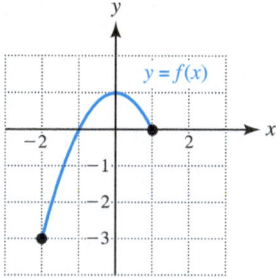

5. Graph $f(x) = |x - 3|$.

8.5 Absolute Value Equations and Inequalities

NEW VOCABULARY

☐ Absolute value equation
☐ Absolute value inequality

1 Solving Absolute Value Equations

SOLVING $|x| = k$ An equation that contains an absolute value expression is called an **absolute value equation**.

Absolute Value Equations

$$|x| = 2, \quad |2x - 1| = 5, \quad |5 - 3x| - 3 = 1$$

Consider the absolute value equation $|x| = 2$. This equation has *two* solutions, -2 and 2, because $|-2| = 2$ and $|2| = 2$. We can also demonstrate this result with a table of values or a graph. In **TABLE 8.16**, $|x| = 2$ when $x = -2$ or $x = 2$. In **FIGURE 8.62**, the graph of $y_1 = |x|$ intersects the graph of $y_2 = 2$ at the points $(-2, 2)$ and $(2, 2)$ The x-values of these points of intersection correspond to the solutions -2 and 2.

$y = |x|$

x	$|x|$
-2	2
-1	1
0	0
1	1
2	2

Solutions

TABLE 8.16

Solving $|x| = 2$ Graphically

FIGURE 8.62

We generalize this discussion in the following manner.

SOLVING $|x| = k$

1. If $k > 0$, then $|x| = k$ is equivalent to $x = k$ or $x = -k$.
2. If $k = 0$, then $|x| = k$ is equivalent to $x = 0$.
3. If $k < 0$, then $|x| = k$ has no solutions.

EXAMPLE 1 Solving absolute value equations

Solve each equation.
(a) $|x| = 20$　　(b) $|x| = -5$　　(c) $|x| = 0$

Solution
(a) The solutions are -20 and 20.
(b) There are no solutions because $|x|$ is never negative.
(c) The only solution is 0.

Now Try Exercises 21, 23, 24

EXAMPLE 2   Graphing an absolute value equation

Graph the equation $y = |3x + 3|$.

Solution
Begin by making a table of values for $y = |3x + 3|$, as shown in **TABLE 8.17**. Plot these points in the xy-plane, and sketch the resulting V-shaped graph, as shown in **FIGURE 8.63**.

$y = |3x + 3|$

x	$|3x + 3|$
-3	6
-2	3
-1	0
0	3
1	6

TABLE 8.17

Graph of Absolute Value Equation

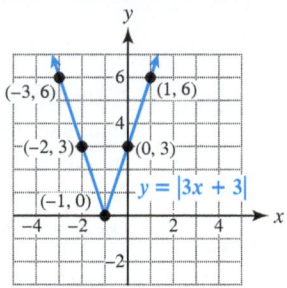

FIGURE 8.63

Now Try Exercise 15

MAKING CONNECTIONS 1

Graphs and the Absolute Value

The graph of $y = |ax + b|$ is a reflection of the line $y = ax + b$ across the x-axis whenever $ax + b < 0$, which is shown in red. Otherwise, when $ax + b \geq 0$, their graphs are identical, which is shown in blue. Note that the graph of $y = |ax + b|$ is *always* V-shaped and *never* goes below the x-axis, whereas the graph of $y = ax + b$ is a line with slope $m = a$.

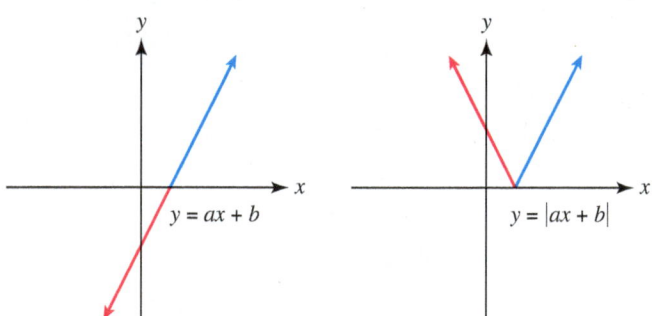

SOLVING $|ax + b| = k$ The next example illustrates how to solve an absolute value equation symbolically, numerically, and graphically.

EXAMPLE 3   Solving an absolute value equation

Solve $|2x - 5| = 3$
(a) symbolically,   **(b)** numerically, and   **(c)** graphically.

Solution
(a) Symbolic Solution If $|2x - 5| = 3$, then either $2x - 5 = 3$ or $2x - 5 = -3$. Solve each equation separately.

$$2x - 5 = 3 \quad \text{or} \quad 2x - 5 = -3 \quad \text{Equations to be solved}$$
$$2x = 8 \quad \text{or} \quad 2x = 2 \quad \text{Add 5.}$$
$$x = 4 \quad \text{or} \quad x = 1 \quad \text{Divide by 2.}$$

The solutions are **1** and **4**.

(b) Numerical Solution A table of values can be used to solve the equation $|2x - 5| = 3$. **TABLE 8.18**, shows that $|2x - 5| = 3$ when $x = 1$ or $x = 4$.

$$|2x - 5| = 3$$

x	0	1	2	3	4	5	6
$|2x - 5|$	5	3	1	1	3	5	7

TABLE 8.18   Solutions are **1** and **4**.

(c) Graphical Solution The equation $|2x - 5| = 3$ can be solved by graphing the equations $y_1 = |2x - 5|$ and $y_2 = 3$. To graph y_1, first plot some of the points from **TABLE 8.18**. Its graph is V-shaped, as shown in **FIGURE 8.64**. Note that the x-coordinate of the "point" or vertex of the V is found by solving the equation $2x - 5 = 0$ to obtain $\frac{5}{2}$. The graph of y_1 intersects the graph of y_2 at the points $(1, 3)$ and $(4, 3)$, giving the solutions **1** and **4**. The graphical solutions agree with the numerical and symbolic solutions.

Graphical Solution

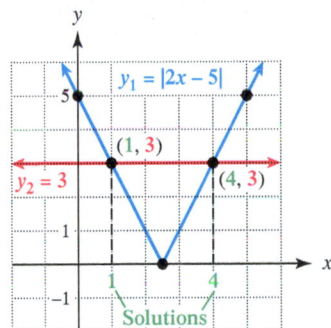

FIGURE 8.64

Now Try Exercises 29, 89 (a), 91

This discussion leads to the following result.

ABSOLUTE VALUE EQUATIONS

If $k > 0$, then

$$|ax + b| = k$$

is equivalent to

$$ax + b = k \quad \text{or} \quad ax + b = -k.$$

EXAMPLE 4 **Solving absolute value equations**

Solve.
(a) $|5 - x| - 2 = 8$ **(b)** $\left|\frac{1}{2}(x - 6)\right| = \frac{3}{4}$

Solution
(a) Start by adding 2 to each side to obtain $|5 - x| = 10$. This new equation is satisfied by the solution from either of the following equations.

$$5 - x = 10 \quad \text{or} \quad 5 - x = -10 \quad \text{Equations to be solved}$$
$$-x = 5 \quad \text{or} \quad -x = -15 \quad \text{Subtract 5.}$$
$$x = -5 \quad \text{or} \quad x = 15 \quad \text{Multiply by } -1.$$

The solutions are -5 and 15.

(b) The solutions to $\left|\frac{1}{2}(x - 6)\right| = \frac{3}{4}$ are found by solving the following equations.

$$\frac{1}{2}(x - 6) = \frac{3}{4} \quad \text{or} \quad \frac{1}{2}(x - 6) = -\frac{3}{4} \qquad \text{Equations to be solved}$$

$$4 \cdot \frac{1}{2}(x - 6) = 4 \cdot \frac{3}{4} \quad \text{or} \quad 4 \cdot \frac{1}{2}(x - 6) = 4\left(-\frac{3}{4}\right) \qquad \text{Multiply by 4 to clear fractions.}$$

$$2(x - 6) = 3 \quad \text{or} \quad 2(x - 6) = -3 \qquad \text{Simplify.}$$

$$2x - 12 = 3 \quad \text{or} \quad 2x - 12 = -3 \qquad \text{Distributive property}$$

$$2x = 15 \quad \text{or} \quad 2x = 9 \qquad \text{Add 12.}$$

$$x = \frac{15}{2} \quad \text{or} \quad x = \frac{9}{2} \qquad \text{Divide by 2.}$$

The solutions are $\frac{9}{2}$ and $\frac{15}{2}$.

Now Try Exercises 31, 35

EXAMPLE 5   **Solving absolute value equations that have no solutions or one solution**

Solve.
(a) $|2x - 1| = -2$ **(b)** $|4 - 2x| = 0$

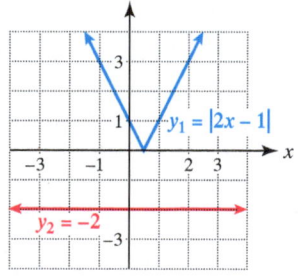

FIGURE 8.65

Solution
(a) Because an absolute value is never negative, there are no solutions. **FIGURE 8.65** shows that the graph of $y_1 = |2x - 1|$ never intersects the graph of $y_2 = -2$.
(b) If $|y| = 0$, then $y = 0$. Thus $|4 - 2x| = 0$ when $4 - 2x = 0$ or when $x = 2$. The only solution is 2.

Now Try Exercises 33, 37

SOLVING $|ax + b| = |cx + d|$ Sometimes an equation can have an absolute value on each side. An example would be $|2x| = |x - 3|$. In this situation, either $2x = x - 3$ (the two expressions are equal) or $2x = -(x - 3)$ (the two expressions are opposites).
These concepts are summarized as follows.

SOLVING $|ax + b| = |cx + d|$

Let a, b, c, and d be constants. Then $|ax + b| = |cx + d|$ is equivalent to
$$ax + b = cx + d \quad \text{or} \quad ax + b = -(cx + d).$$

EXAMPLE 6   **Solving an absolute value equation**

Solve $|2x| = |x - 3|$.

Solution
Solve the following equations.

$$2x = x - 3 \quad \text{or} \quad 2x = -(x - 3)$$
$$x = -3 \quad \text{or} \quad 2x = -x + 3$$
$$3x = 3$$
$$x = 1$$

The solutions are -3 and 1.

Now Try Exercise 41

FIGURE 8.66

Example 6 is solved graphically in **FIGURE 8.66**. The graphs of $y_1 = |2x|$ and $y_2 = |x - 3|$ are V-shaped and intersect at $(-3, 6)$ and $(1, 2)$. The solutions are -3 and 1.

2 Solving Absolute Value Inequalities

FINDING NUMERICAL AND GRAPHICAL SOLUTIONS We can solve absolute value inequalities numerically and graphically. For example, to solve $|x| < 3$ graphically, let $y_1 = |x|$ and $y_2 = 3$ (see **FIGURE 8.67**). Their graphs intersect at $(-3, 3)$ and $(3, 3)$. The V-shaped graph of y_1 is *below* the graph of y_2 for x-values *between*, but not including, $x = -3$ and $x = 3$. The solution set for $|x| < 3$ is $(-3, 3)$ in interval notation and is shaded on the x-axis.

$$|x| < 3$$

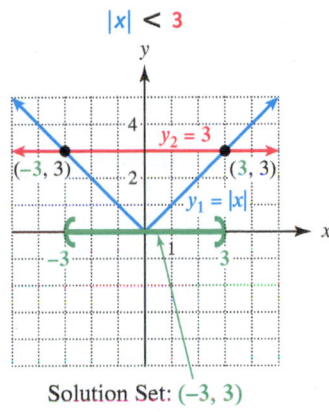

Solution Set: $(-3, 3)$
FIGURE 8.67

In the next example, we solve absolute value inequalities both numerically and graphically.

EXAMPLE 7 **Solving inequalities numerically and graphically**

Solve each inequality numerically and graphically.
(a) $|2x - 1| < 3$ **(b)** $|2x - 1| > 3$

Solution
(a) Numerical Solution **TABLE 8.19** shows a table of values for $y_1 = |2x - 1|$. Note that when $x = -1$ and $x = 2$, the absolute value $|2x - 1| = 3$. From this table, we see that $|2x - 1| < 3$ when $-1 < x < 2$. The solution set is $(-1, 2)$.

$$y = |2x - 1|$$

x	$\lvert 2x - 1 \rvert$	
-3	7	$\left.\right\}\leftarrow \lvert 2x - 1 \rvert > 3$
-2	5	
-1	3	$\leftarrow \lvert 2x - 1 \rvert = 3$
0	1	$\left.\right\}\leftarrow \lvert 2x - 1 \rvert < 3$
1	1	
2	3	$\leftarrow \lvert 2x - 1 \rvert = 3$
3	5	$\left.\right\}\leftarrow \lvert 2x - 1 \rvert > 3$
4	7	

TABLE 8.19

Graphical Solution In **FIGURE 8.68**, the V-shaped graph of $y_1 = |2x - 1|$ lies *below* the horizontal line $y_2 = 3$ **between $x = -1$ and $x = 2$**, or when $-1 < x < 2$. Thus the solution set to $|2x - 1| < 3$ is $(-1, 2)$ in interval notation, and the graphical and numerical solutions agree.

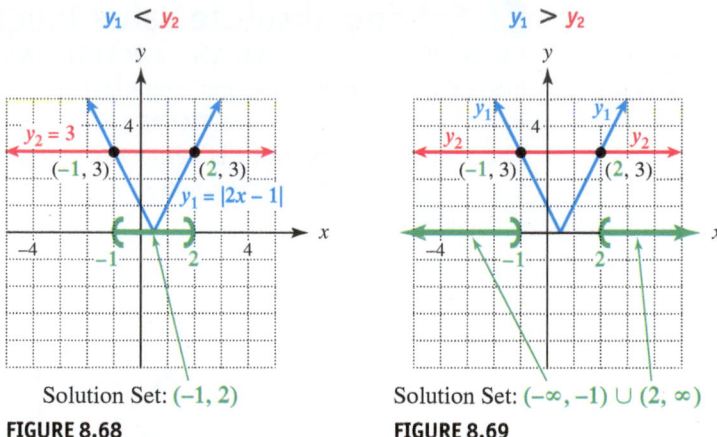

Solution Set: $(-1, 2)$

FIGURE 8.68

Solution Set: $(-\infty, -1) \cup (2, \infty)$

FIGURE 8.69

(b) Numerical Solution From **TABLE 8.19** on the previous page, we see that $|2x - 1| > 3$ when $x < -1$ or when $x > 2$. Thus the solution set is $(-\infty, -1) \cup (2, \infty)$.

Graphical Solution In **FIGURE 8.69**, the V-shaped graph of $y_1 = |2x - 1|$ lies *above* the horizontal line $y_2 = 3$ to the **left of $x = -1$** or to the **right of $x = 2$**. Thus the solution set to $|2x - 1| > 3$ is $(-\infty, -1) \cup (2, \infty)$ in interval notation, and the graphical and numerical solutions agree.

Now Try Exercises 107 (b), 107 (c)

SYMBOLIC SOLUTIONS BASED ON VISUALIZATION (METHOD I) By understanding graphical solutions to absolute value inequalities, we can use visualization to solve absolute value inequalities symbolically. If we find the solutions to $|ax + b| = k$, then we can write the solutions to $|ax + b| < k$ and $|ax + b| > k$ directly without further work. This technique is summarized as follows.

ABSOLUTE VALUE INEQUALITIES

Let the solutions to $|ax + b| = k$ be c and d, where $c < d$ and $k > 0$.

1. $|ax + b| < k$ is equivalent to $c < x < d$.
2. $|ax + b| > k$ is equivalent to $x < c$ or $x > d$.

Similar statements can be made for inequalities involving $\leq$ or $\geq$.

EXAMPLE 8 **Solve absolute value equations and inequalities**

Solve each absolute value equation or inequality.
(a) $|2 - 3x| = 4$ **(b)** $|2 - 3x| < 4$ **(c)** $|2 - 3x| > 4$

Solution
(a) The given equation is equivalent to the following equations.

$$2 - 3x = 4 \quad \text{or} \quad 2 - 3x = -4 \quad \text{Equations to be solved}$$

$$-3x = 2 \quad \text{or} \quad -3x = -6 \quad \text{Subtract 2.}$$

$$x = -\frac{2}{3} \quad \text{or} \quad x = 2 \quad \text{Divide by } -3.$$

The solutions are $-\frac{2}{3}$ and **2**.

STUDY TIP

Be sure you understand
how to write the solution
to Example 8(c).

(b) Solutions to $|2 - 3x| < 4$ include x-values **between**, but not including, $-\frac{2}{3}$ and **2**. Thus the solution set is $\left\{x \mid -\frac{2}{3} < x < 2\right\}$, or in interval notation, $\left(-\frac{2}{3}, 2\right)$.

(c) Solutions to $|2 - 3x| > 4$ include x-values to the **left** of $x = -\frac{2}{3}$ **or** to the **right** of $x = 2$. Thus the solution set is $\left\{x \mid x < -\frac{2}{3} \text{ or } x > 2\right\}$, or in interval notation, $\left(-\infty, -\frac{2}{3}\right) \cup (2, \infty)$.

Now Try Exercise 55

FIGURE 8.70(a) can be used to visualize the solution to $|2 - 3x| = 4$ in Example 8(a). Similarly, **FIGURES 8.70(b)** and **8.70(c)** can be used to visualize the solutions to $|2 - 3x| < 4$ and $|2 - 3x| > 4$ in parts (b) and (c) of Example 8.

Visualizing Solutions to Equations and Inequalities

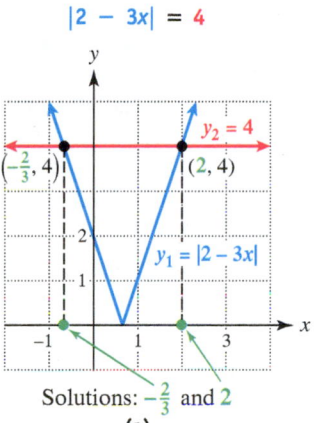
$|2 - 3x| = 4$
Solutions: $-\frac{2}{3}$ and 2
(a)

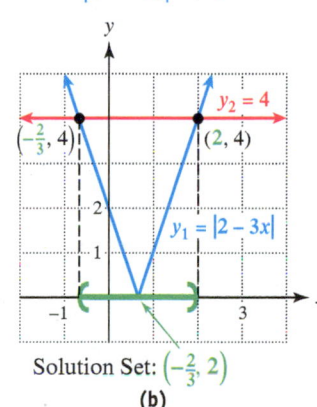
$|2 - 3x| < 4$
Solution Set: $\left(-\frac{2}{3}, 2\right)$
(b)

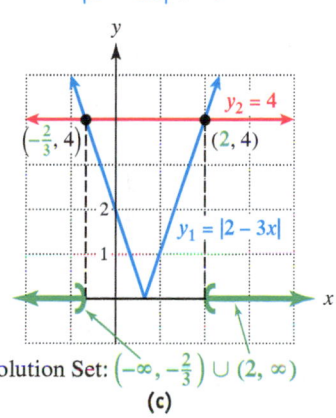
$|2 - 3x| > 4$
Solution Set: $\left(-\infty, -\frac{2}{3}\right) \cup (2, \infty)$
(c)

FIGURE 8.70

EXAMPLE 9   Solving an absolute value inequality

Solve $\left|\frac{2x - 5}{3}\right| > 3$. Write the solution set in interval notation.

Solution
Start by solving $\left|\frac{2x - 5}{3}\right| = 3$ as follows.

$$\frac{2x - 5}{3} = 3 \quad \text{or} \quad \frac{2x - 5}{3} = -3 \qquad \text{Equations to be solved}$$

$$2x - 5 = 9 \quad \text{or} \quad 2x - 5 = -9 \qquad \text{Multiply by 3.}$$

$$2x = 14 \quad \text{or} \qquad 2x = -4 \qquad \text{Add 5.}$$

$$x = 7 \quad \text{or} \qquad x = -2 \qquad \text{Divide by 2.}$$

Because the inequality symbol is $>$, the solution set is given by $x < -2$ or $x > 7$, or in interval notation, $(-\infty, -2) \cup (7, \infty)$.

Now Try Exercise 83

FINDING SYMBOLIC SOLUTIONS (METHOD II) The absolute value of a number gives its distance from the origin on a number line. For example $|3| = 3$ and $|-3| = 3$ because both 3 and -3 are located 3 units from the origin, as shown in **FIGURE 8.71**.

Absolute Values of -3 and 3

FIGURE 8.71

Now suppose we want to find the solution set to $|x| < 3$. An x-value that satisfies this inequality has a distance that is less than 3 from the origin. For example, both $x = 1$ and $x = -2$ are solutions to $|x| < 3$ because $|1| < 3$ and $|-2| < 3$. In general, solutions to $|x| < 3$ satisfy $-3 < x < 3$, as shown in **FIGURE 8.72**.

Next suppose we want to find the solution set to $|x| > 3$. An x-value that satisfies this inequality has a distance that is greater than 3 from the origin. For example, both $x = 4$ and $x = -4$ are solutions to $|x| > 3$ because $|4| > 3$ and $|-4| > 3$. In general, solutions to $|x| > 3$ satisfy $x < -3$ or $x > 3$, as shown in **FIGURE 8.73**.

Distance from Origin is Less Than 3: $|x| < 3$
$-3 < x < 3$

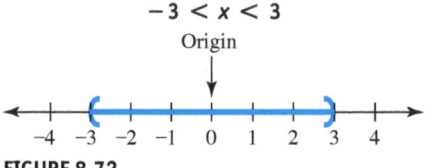

FIGURE 8.72

Distance from Origin is Greater Than 3: $|x| > 3$
$x < -3$ or $x > 3$

FIGURE 8.73

From the preceding discussion, it follows that if $|ax + b| < 3$, then $-3 < ax + b < 3$. Similarly, if $|ax + b| > 3$, then either $ax + b < -3$ or $ax + b > 3$. This method for solving absolute value inequalities symbolically is summarized as follows.

CRITICAL THINKING 1

How many solutions are there to $|2x + 1| < 4$? How many solutions are there to $|2x + 1| < -4$?

ABSOLUTE VALUE INEQUALITIES

Let k be a positive number.

1. $|ax + b| < k$ is equivalent to $-k < ax + b < k$.
2. $|ax + b| > k$ is equivalent to $ax + b < -k$ or $ax + b > k$.

Similar statements can be made for inequalities involving $\leq$ or $\geq$.

The next two examples, use this symbolic technique to solve absolute value inequalities.

EXAMPLE 10   **Solving absolute value inequalities symbolically**

Solve each absolute value inequality. Write your answer in interval notation.
(a) $|4 + x| < 5$ **(b)** $|4 + x| > 5$

Solution
(a) The solution set to $|4 + x| < 5$ is found by solving the following three-part inequality.

$$-5 < 4 + x < 5 \qquad \text{Three-part inequality}$$
$$-9 < x < 1 \qquad \text{Subtract 4 from each part.}$$

Thus the solution set in interval notation is $(-9, 1)$.

(b) The solution set to $|4 + x| > 5$ is found by solving the following compound inequality.

$$4 + x < -5 \quad \text{or} \quad 4 + x > 5 \qquad \text{Compound inequality}$$
$$x < -9 \quad \text{or} \qquad x > 1 \qquad \text{Subtract 4 from each inequality.}$$

Thus the solution set in interval notation is $(-\infty, -9) \cup (1, \infty)$.

Now Try Exercises 53 (b), 53 (c)

EXAMPLE 11   **Solving absolute value inequalities symbolically**

Solve each absolute value inequality. Write your answer in interval notation.
(a) $|4 - 5x| \leq 3$ (b) $|-4x - 6| > 2$

Solution
(a) $|4 - 5x| \leq 3$ is equivalent to the following three-part inequality.

$$-3 \leq 4 - 5x \leq 3 \qquad \text{Equivalent inequality}$$
$$-7 \leq -5x \leq -1 \qquad \text{Subtract 4 from each part.}$$
$$\frac{7}{5} \geq x \geq \frac{1}{5} \qquad \text{Divide each part by } -5;$$
$$\text{reverse the inequality.}$$

In interval notation, the solution is $\left[\frac{1}{5}, \frac{7}{5}\right]$.

(b) $|-4x - 6| > 2$ is equivalent to the following compound inequality.

$$-4x - 6 < -2 \quad \text{or} \quad -4x - 6 > 2 \qquad \text{Equivalent compound inequality}$$
$$-4x < 4 \quad \text{or} \quad -4x > 8 \qquad \text{Add 6 to each side.}$$
$$x > -1 \quad \text{or} \quad x < -2 \qquad \text{Divide each by } -4;$$
$$\text{reverse the inequality.}$$

In interval notation, the solution set is $(-\infty, -2) \cup (-1, \infty)$.

Now Try Exercises 65, 67

 Math in Context *Manufacturing* Absolute value inequalities often occur in real life when error in manufacturing occurs. The next example illustrates the fact that a circular cover cannot be made with an *exact* diameter. There is always some error.

EXAMPLE 12   **Analyzing error**

An engineer is designing a circular cover for a container. The diameter d of the cover is to be 4.25 inches and must be accurate to within 0.01 inch. Write an absolute value inequality that gives acceptable values for d.

Solution
The diameter d must satisfy $4.24 \leq d \leq 4.26$. Subtracting 4.25 from each part gives

$$-0.01 \leq d - 4.25 \leq 0.01,$$

which is equivalent to $|d - 4.25| \leq 0.01$. The "distance" or difference between 4.25 and the diameter is less than or equal to 0.01.

Now Try Exercise 127

CALCULATOR HELP

To graph an absolute value, see Appendix A (page AP-8).

Connecting Concepts with Your Life Monthly average temperatures can vary greatly from one month to another, whereas yearly average temperatures remain fairly constant from one year to the next. In Boston, Massachusetts, the yearly average temperature is 50 °F, but monthly average temperatures can vary from 28 °F to 72 °F. Because **50 °F** − 28 °F = **22 °F** and 72 °F − **50 °F** = **22 °F**, the monthly average temperatures are always within 22 °F of the yearly average temperature. If T represents a monthly average temperature, we can model this situation by using the *absolute value inequality*

$$|T - \mathbf{50}| \leq \mathbf{22}.$$

The absolute value is necessary because a monthly average temperature T can be either greater than or less than 50 °F by as much as 22 °F.

> **EXAMPLE 13**   **Modeling temperature in Boston**
>
> The inequality $|T - 50| \le 22$ models the range for the monthly average temperatures T in Boston.
> **(a)** Solve this inequality and interpret the result.
> **(b)** Give graphical support for part (a).

$[0, 100, 10]$ by $[0, 70, 10]$

(a)

$[0, 100, 10]$ by $[0, 70, 10]$

(b)

FIGURE 8.74

Solution
(a) Symbolic Solution To solve $|T - 50| \le 22$, we can solve the three-part inequality $-22 \le T - 50 \le 22$.

$$-22 \le T - 50 \le 22 \quad \text{Inequality to be solved}$$
$$28 \le T \le 72 \quad \text{Add 50 to each part.}$$

Thus the solution set to $|T - 50| \le 22$ is $\{T \,|\, 28 \le T \le 72\}$. Monthly average temperatures in Boston vary from 28 °F to 72 °F.
(b) Graphical Solution The graphs of $y_1 = |x - 50|$ and $y_2 = 22$ intersect at $(28, 22)$ and $(72, 22)$, as shown in **FIGURES 8.74(a)** and **(b)**. The V-shaped graph of y_1 intersects the horizontal graph of y_2, or is below it, when $28 \le x \le 72$. Thus the solution set is $\{T \,|\, 28 \le T \le 72\}$. This result agrees with the symbolic result.

Now Try Exercise 119

Sometimes the solution set to an absolute value inequality can be either empty or the set of all real numbers. These two situations are illustrated in the next example.

> **EXAMPLE 14**   **Solving absolute value inequalities**
>
> Solve if possible.
> **(a)** $|2x - 5| > -1$ **(b)** $|5x - 1| + 3 \le 2$

Solution
(a) Because the absolute value of an expression cannot be negative, $|2x - 5|$ is greater than -1 for every x-value. The solution set is all real numbers, or $(-\infty, \infty)$.
(b) Subtracting 3 from each side results in $|5x - 1| \le -1$. Because the absolute value is always greater than or equal to 0, no x-values satisfy this inequality. There are no solutions.

Now Try Exercises 79, 81

8.5   Putting It All Together

PROBLEM	SYMBOLIC SOLUTION	GRAPHICAL SOLUTION
$\lvert ax + b \rvert = k, k > 0$	Solve the equations $$ax + b = k$$ and $$ax + b = -k.$$	Graph $y_1 = \lvert ax + b \rvert$ and $y_2 = k$. Find the x-values of the two points of intersection.

PROBLEM	SYMBOLIC SOLUTION	GRAPHICAL SOLUTION
$\lvert ax + b\rvert < k,\ k > 0$	**Method I:** If the solutions to $$\lvert ax + b\rvert = k$$ are c and d, $c < d$, then the solutions to $$\lvert ax + b\rvert < k$$ satisfy $$c < x < d.$$ **Method II:** The solutions to $\lvert ax + b\rvert < k$ can be found by solving $$-k < ax + b < k.$$	Graph $y_1 = \lvert ax + b\rvert$ and $y_2 = k$. Find the x-values of the two points of intersection. The solutions are between these x-values on the number line, where the graph of y_1 lies below the graph of y_2.
$\lvert ax + b\rvert > k,\ k > 0$	**Method I:** If the solutions to $$\lvert ax + b\rvert = k$$ are c and d, $c < d$, then the solutions to $$\lvert ax + b\rvert > k$$ satisfy $$x < c \quad \text{or} \quad x > d.$$ **Method II:** The solutions to $\lvert ax + b\rvert > k$ can be found by solving $$ax + b < -k \text{ or } ax + b > k.$$	Graph $y_1 = \lvert ax + b\rvert$ and $y_2 = k$. Find the x-values of the two points of intersection. The solutions are "outside" (left or right of) these x-values on the number line, where the graph of y_1 is above the graph of y_2.

8.5 Exercises

MyMathLab®

CONCEPTS AND VOCABULARY

1. Give an example of an absolute value equation.

2. Give an example of an absolute value inequality.

3. Is -3 a solution to $\lvert x\rvert = 3$?

4. Is -4 a solution to $\lvert x\rvert > 3$?

5. Is the solution set to $\lvert x\rvert = 5$ equal to $\{-5, 5\}$?

6. Is $\lvert x\rvert < 3$ equivalent to $x < -3$ or $x > 3$? Explain.

7. How many times does the graph of $y = \lvert 2x - 1\rvert$ intersect the graph of $y = 5$?

8. How many times does the graph of $y = \lvert 2x - 1\rvert$ intersect the graph of $y = -5$?

Exercises 9–14: Determine whether the given values of x are solutions to the absolute value equation or inequality.

9. $\lvert 2x - 5\rvert = 1$ $x = -3, x = 3$

10. $\lvert 5 - 6x\rvert = 1$ $x = 1, x = 0$

11. $\lvert 7 - 4x\rvert \le 5$ $x = -2, x = 2$

12. $\lvert 2 + x\rvert < 2$ $x = -4, x = -1$

13. $\lvert 7x + 4\rvert > -1$ $x = -\frac{4}{7}, x = 2$

14. $\lvert 12x + 3\rvert \ge 3$ $x = -\frac{1}{4}, x = 2$

GRAPHING ABSOLUTE VALUE EQUATIONS

Exercises 15–20: Graph the absolute value equation.

15. $y = \lvert 2x + 2\rvert$

16. $y = \lvert x + 1\rvert$

17. $y = \lvert -x - 3\rvert$

18. $y = \lvert 3x - 3\rvert$

19. $y = \lvert 3 - x\rvert$

20. $y = \lvert 4 - 2x\rvert$

FINDING SYMBOLIC SOLUTIONS

Exercises 21–46: Solve the absolute value equation.

21. $\lvert x\rvert = 7$

22. $\lvert x\rvert = 4$

23. $\lvert x\rvert = -6$

24. $\lvert x\rvert = 0$

25. $|4x| = 9$

26. $|-3x| = 7$

27. $|-2x| - 6 = 2$

28. $|5x| + 1 = 5$

29. $|2x + 1| = 11$

30. $|1 - 3x| = 4$

31. $|-2x + 3| + 3 = 4$

32. $|6x + 2| - 2 = 6$

33. $|3 - 4x| = 0$

34. $|5x - 3| = -1$

35. $\left|\frac{1}{2}x - 1\right| = 5$

36. $\left|6 - \frac{3}{4}x\right| = 3$

37. $|2x - 6| = -7$

38. $\left|1 - \frac{2}{3}x\right| = 0$

39. $\left|\frac{2}{3}z - 1\right| - 3 = 8$

40. $|1 - 2z| + 5 = 10$

41. $|z - 1| = |2z|$

42. $|2z + 3| = |2 - z|$

43. $|3t + 1| = |2t - 4|$

44. $\left|\frac{1}{2}t - 1\right| = \left|3 - \frac{3}{2}t\right|$

45. $\left|\frac{1}{4}x\right| = \left|3 + \frac{1}{4}x\right|$

46. $|2x - 1| = |2x + 2|$

Exercises 47–56: Solve each equation or inequality. Use interval notation for inequalities.

47. $|x| = 3$

48. $|x| = 5$

49. $|x| < 3$

50. $|x| < 5$

51. $|x| > 3$

52. $|x| > 5$

53. (a) $|2x| = 8$ (b) $|2x| < 8$
(c) $|2x| > 8$

54. (a) $|3x - 9| = 6$
(b) $|3x - 9| \le 6$
(c) $|3x - 9| \ge 6$

55. (a) $|5 - 4x| = 3$
(b) $|5 - 4x| \le 3$
(c) $|5 - 4x| \ge 3$

56. (a) $\left|\frac{x - 5}{2}\right| = 2$ (b) $\left|\frac{x - 5}{2}\right| < 2$
(c) $\left|\frac{x - 5}{2}\right| > 2$

Exercises 57–88: Solve the absolute value inequality. Write your answer in interval notation.

57. $|x| \le 3$

58. $|x| < 2$

59. $|k| > 4$

60. $|k| \ge 5$

61. $|t| \le -3$

62. $|t| < -1$

63. $|z| > 0$

64. $|2z| \ge 0$

65. $|2x| > 7$

66. $|-12x| < 30$

67. $|-4x + 4| < 16$

68. $|-5x - 8| > 2$

69. $2|x + 5| \ge 8$

70. $-3|x - 1| \ge -9$

71. $|8 - 6x| - 1 \le 2$

72. $4 - \left|\frac{2x}{3}\right| < -7$

73. $5 + \left|\frac{2 - x}{3}\right| \le 9$

74. $\left|\frac{x + 3}{5}\right| \le 12$

75. $|2x - 1| \le -3$

76. $|x + 6| \ge -5$

77. $|x + 1| - 1 > -3$

78. $-2|1 - 7x| \ge 2$

79. $|2z - 4| + 2 \le 1$

80. $|4 - z| \le 0$

81. $|3z - 1| > -3$

82. $|2z| \ge -2$

83. $\left|\frac{2 - t}{3}\right| \ge 5$

84. $\left|\frac{2t + 3}{5}\right| \ge 7$

85. $|t - 1| \le 0.1$

86. $|t - 2| \le 0.01$

87. $|b - 10| > 0.5$

88. $|b - 25| \ge 1$

FINDING NUMERICAL AND GRAPHICAL SOLUTIONS

Exercises 89 and 90: Use the table of $y = |ax + b|$ to solve each equation or inequality. Write your answers in interval notation for parts (b) and (c).

89. (a) $y = 2$ (b) $y < 2$ (c) $y > 2$

x	-2	-1	0	1	2	3	4
y	3	2	1	0	1	2	3

90. (a) $y = 6$ (b) $y \le 6$ (c) $y \ge 6$

x	-12	-6	0	6	12	18	24
y	9	6	3	0	3	6	9

Exercises 91 and 92: Use the graph to solve the equation.

91. $|x - 2| = 2$

92. $|2x + 1| = 3$

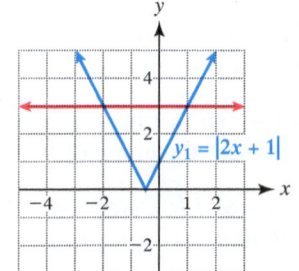

Exercises 93 and 94: Use the graph of y_1 to solve each equation or inequality. Write your answers in interval notation for parts (b) and (c).

93. (a) $|2x + 1| = 1$

   (b) $|2x + 1| \leq 1$

   (c) $|2x + 1| \geq 1$

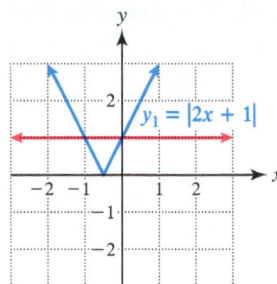

94. (a) $|x - 1| = 3$ **(b)** $|x - 1| < 3$

   (c) $|x - 1| > 3$

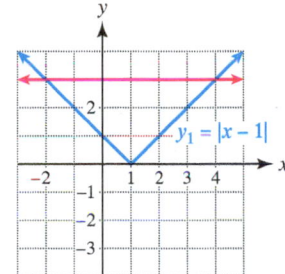

Exercises 95–104: Solve the inequality graphically. Write your answer in interval notation.

95. $|x| \geq 1$ **96.** $|x| < 2$

97. $|x - 1| \leq 3$ **98.** $|x + 5| \geq 2$

99. $|4 - 2x| > 2$ **100.** $|1.5x - 3| \geq 6$

101. $|10 - 3x| < 4$ **102.** $|7 - 4x| \leq 2.5$

103. $|8.1 - x| > -2$ **104.** $\left|\dfrac{5x - 9}{2}\right| \leq -1$

USING MORE THAN ONE METHOD

Exercises 105–108: Solve the absolute value inequality

   (a) *symbolically,*

   (b) *graphically, and*

   (c) *numerically.*

Write your answer in set-builder notation.

105. $|3x| \leq 9$ **106.** $|5 - x| \geq 3$

107. $|2x - 5| > 1$ **108.** $|-8 - 4x| < 6$

WRITING ABSOLUTE VALUE INEQUALITIES

Exercises 109–116: Write each compound inequality as an absolute value inequality. Do not simplify Exercises 113–116.

109. $-4 \leq x \leq 4$ **110.** $-0.1 < y < 0.1$

111. $y < -2$ or $y > 2$ **112.** $-0.1 \leq x \leq 0.1$

113. $-0.3 \leq 2x + 1 \leq 0.3$

114. $4x < -5$ or $4x > 5$

115. $\pi x \leq -7$ or $\pi x \geq 7$

116. $-0.9 \leq x - \sqrt{2} \leq 0.9$

117. Thinking Generally If $a \neq 0$ and $k > 0$, then the graph of $y = |ax + b|$ intersects the graph of $y = k$ at _____ points.

118. Thinking Generally If a and k are positive, then the solution set to $|ax + b| < k$ is _____.

APPLICATIONS INVOLVING ABSOLUTE VALUES

Exercises 119–122: **Average Temperatures** *(Refer to Example 13.) The given inequality models the range for the monthly average temperatures T in degrees Fahrenheit at the location specified.*

   (a) *Solve the inequality.*

   (b) *Give a possible interpretation of the inequality.*

119. $|T - 43| \leq 24$, Marquette, Michigan

120. $|T - 62| \leq 19$, Memphis, Tennessee

121. $|T - 10| \leq 36$, Chesterfield, Canada

122. $|T - 61.5| \leq 12.5$, Buenos Aires, Argentina

123. Highest Elevations The table lists the highest elevation on each continent.

Continent	Elevation (feet)
Asia	29,028
S. America	22,834
N. America	20,320
Africa	19,340
Europe	18,510
Antarctica	16,066
Australia	7,310

Source: National Geographic.

   (a) Calculate the average A of these elevations.

   (b) Which continents have their highest elevations within 1000 feet of A?

continued on next page

continued from previous page

(c) Which continents have their highest elevations within 5000 feet of A?

(d) Let E be an elevation. Write an absolute value inequality that says E is within 5000 feet of A.

124. **Distance** Suppose that two cars, both traveling at a constant speed of 60 miles per hour, approach each other on a straight highway.

(a) If the two cars are initially 4 miles apart, sketch a graph of the distance between the two cars after x minutes, where $0 \le x \le 4$. (*Hint:* 60 miles per hour = 1 mile per minute.)

(b) Write an absolute value equation whose solution gives the times when the cars are 2 miles apart.

(c) Solve your equation from part (b).

125. **Error in Measurements** (Refer to Example 12.) The maximum error in the diameter of a container is restricted to 0.002 inch, so an acceptable diameter d must satisfy the absolute value inequality

$$|d - 2.5| \le 0.002.$$

Solve this inequality for d and interpret the result.

126. **Error in Measurements** Suppose that a person can operate a stopwatch accurately to within 0.02 second. If Byron Dyce's time in the 800-meter race is recorded as 105.30 seconds, write an absolute value inequality that gives the possible values for the actual time t.

127. **Error in Measurements** A circular lid is being designed for a container. The diameter d of the lid is to be 3.8 inches and must be accurate to within 0.03 inch. Write an absolute value inequality that gives acceptable values for d.

128. **Manufacturing a Tire** An engineer is designing a tire for a truck. The diameter d of the tire is to be 36 inches and the *circumference* must be accurate to within 0.1 inch. Write an absolute value inequality that gives acceptable values for d.

129. **Relative Error** If a quantity is measured to be x and the true value is t, then the relative error in the measurement is $\left|\frac{x - t}{t}\right|$. If the true measurement is $t = 20$ and you want the relative error to be less than 0.05 (5%), what values for x are possible?

130. **Relative Error** (Refer to the preceding exercise.) The volume V of a box is 50 cubic inches. How accurately must you measure the volume of the box for the relative error to be less than 3%?

WRITING ABOUT MATHEMATICS

131. If $a \ne 0$, how many solutions are there to the equation $|ax + b| = k$ when
(a) $k > 0$,   (b) $k = 0$, and   (c) $k < 0$?
Explain each answer.

132. Suppose that you know two solutions to the equation $|ax + b| = k$. How can you use these solutions to solve the inequalities $|ax + b| < k$ and $|ax + b| > k$? Give an example.

SECTION 8.5

Checking Basic Concepts

Write answers in interval notation whenever possible.

1. Solve $\left|\frac{3}{4}x - 1\right| - 3 = 5$.

2. Solve the absolute value equation and inequalities.

(a) $|3x - 6| = 8$

(b) $|3x - 6| < 8$

(c) $|3x - 6| > 8$

3. Solve the inequality $|-2(3 - x)| < 6$, and then solve $|-2(3 - x)| \ge 6$.

4. Use the graph to solve the equation and inequalities.

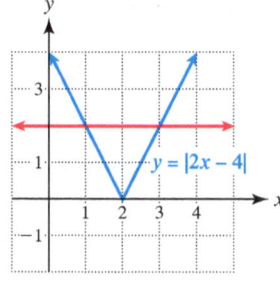

$y = |2x - 4|$

(a) $|2x - 4| = 2$ (b) $|2x - 4| \le 2$

(c) $|2x - 4| \ge 2$

CHAPTER 8   Summary

Function A function is a set of ordered pairs (x, y), where each x-value corresponds to exactly one y-value. A function takes a valid input x and computes exactly one output y, forming the ordered pair (x, y).

Domain and Range of a Function The domain D is the set of all valid inputs, or x-values, and the range R is the set of all outputs, or y-values.

Examples: $f = \{(1, 2), (2, 3), (3, 3)\}$ has $D = \{1, 2, 3\}$ and $R = \{2, 3\}$.

$f(x) = x^2$ has domain all real numbers and range $y \geq 0$. (See the graph below.)

Function Notation $y = f(x)$ and is read "y equals f of x."

Example: $f(x) = \frac{2x}{x-1}$ implies that $f(3) = \frac{2 \cdot 3}{3-1} = \frac{6}{2} = 3$. Thus the point $(3, 3)$ is on the graph of f.

Function Representations A function can be represented symbolically, numerically, graphically, or verbally.

Symbolic Representation (Formula) $f(x) = x^2$

Numerical Representation (Table)   **Graphical Representation (Graph)**

x	y
-2	4
-1	1
0	0
1	1
2	4

Verbal Representation (Words) f computes the square of the input x.

Vertical Line Test If every vertical line intersects a graph at most once, then the graph represents a function.

Linear Function A linear function can be represented by $f(x) = mx + b$. Its graph is a (straight) line, where m is the slope and $(0, b)$ is the y-intercept. For each unit increase in x, $f(x)$ changes by an amount equal to m.

Example: $f(x) = 2x - 1$ represents a linear function with $m = 2$ and $b = -1$.

Numerical Representation   **Graphical Representation**

x	$f(x)$
-1	-3
0	-1
1	1
2	3

1 unit → (between -1 and 0); 2 units ←
1 unit → (between 0 and 1); 2 units ←
1 unit → (between 1 and 2); 2 units ←

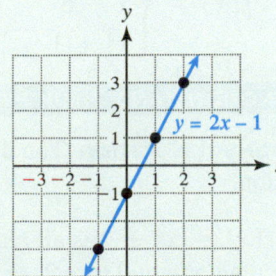

Each 1-unit increase in x results in a 2-unit increase in $f(x)$; thus $m = 2$.

NOTE: A numerical representation is a table of values of $f(x)$. ■

Modeling Data with Linear Functions When data have a constant rate of change, they can be modeled by $f(x) = mx + b$. The constant m represents the *rate of change*, and the constant b represents the *initial amount* or the value when $x = 0$. That is,

$$f(x) = (\text{Rate of change})x + (\text{Initial amount}).$$

Example: In the following table, the y-values decrease by 3 units for each 1-unit increase in x. When $x = 0$, $y = 4$. Thus the data are modeled by $f(x) = -3x + 4$.

x	-2	-1	0	1	2
y	10	7	4	1	-2

Midpoint Formula The midpoint of the line segment connecting (x_1, y_1) and (x_2, y_2) is

$$\left(\frac{x_1 + x_2}{2}, \frac{y_1 + y_2}{2} \right).$$

Example: The midpoint of the line segment connecting $(-5, 8)$ and $(9, 4)$ is

$$\left(\frac{-5 + 9}{2}, \frac{8 + 4}{2} \right) = (2, 6).$$

SECTION 8.3 ■ COMPOUND INEQUALITIES

Compound Inequality Two inequalities connected by *and* or *or*.

Examples: For $x + 1 < 3$ *or* $x + 1 > 6$, a solution satisfies *at least* one of the inequalities.

For $2x + 1 < 3$ *and* $1 - x > 6$, a solution satisfies *both* inequalities.

Three-Part Inequality A compound inequality in the form $x > a$ *and* $x < b$ can be written as the three-part inequality $a < x < b$.

Example: $1 \leq x < 7$ means $x \geq 1$ *and* $x < 7$.

Interval Notation Can be used to identify intervals on the real number line

Examples: $-2 < x \leq 3$ is equivalent to $(-2, 3]$.

$x < 5$ is equivalent to $(-\infty, 5)$.

All real numbers are denoted $(-\infty, \infty)$.

Piecewise-Defined Function A function that is defined using different formulas over different intervals of its domain

Example: $f(0) = 0 + 6 = 6$, because when $x = 0, f(x) = x + 6$.

$$f(x) = \begin{cases} x^2 + 2x + 6 & \text{if } -5 \leq x < 0 \\ x + 6 & \text{if } 0 \leq x < 2 \\ x^3 + 1 & \text{if } 2 \leq x \leq 5 \end{cases}$$

SECTION 8.4 ■ OTHER FUNCTIONS AND THEIR PROPERTIES

Domain and Range in Interval Notation The domain and range of a function can often be expressed in interval notation.

Example: The domain of $f(x) = x^2 - 2$ is all real numbers, or $(-\infty, \infty)$, and its range is real numbers greater than or equal to -2, or $[-2, \infty)$.

Absolute Value Function The domain of $f(x) = |x|$ is $(-\infty, \infty)$, and the range is $[0, \infty)$.

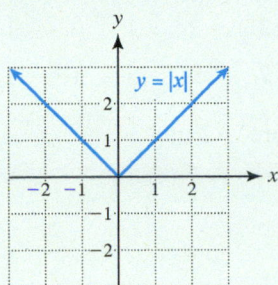

Polynomial Functions The degree of a polynomial function (of one variable) equals the largest exponent of a variable. The graphs of polynomial functions with degree greater than 1 are not lines. The domain of a polynomial function is $(-\infty, \infty)$.

Examples: $f(x) = 4x - 1$ defines a linear function with degree 1.

$g(x) = 4x^2 + x - 4$ defines a quadratic function with degree 2.

$h(x) = x^3 + 0.7x - 1$ defines a cubic function with degree 3.

Rational Functions If $f(x) = \frac{p(x)}{q(x)}$, where $p(x)$ and $q(x)$ are polynomials, f is a rational function. The domain of a rational function includes all real numbers, except x-values for which $q(x) = 0$.

Examples: $f(x) = \frac{1}{x}$ has domain $(-\infty, 0) \cup (0, \infty)$, or $x \neq 0$.

$g(x) = \frac{x}{x^2 - 9}$ has domain $(-\infty, -3) \cup (-3, 3) \cup (3, \infty)$, or $x \neq -3, x \neq 3$.

Operations on Functions If $f(x)$ and $g(x)$ are both defined, then the sum, difference, product, and quotient of two functions f and g are defined by

$$(f + g)(x) = f(x) + g(x) \qquad \text{Sum}$$

$$(f - g)(x) = f(x) - g(x) \qquad \text{Difference}$$

$$(fg)(x) = f(x) \cdot g(x) \qquad \text{Product}$$

$$\left(\frac{f}{g}\right)(x) = \frac{f(x)}{g(x)}, \text{ where } g(x) \neq 0. \quad \text{Quotient}$$

Examples: Let $f(x) = x^2 - 1$ and $g(x) = x^2 + 1$.

$$(f + g)(x) = f(x) + g(x) = (x^2 - 1) + (x^2 + 1) = 2x^2$$

$$(f - g)(x) = f(x) - g(x) = (x^2 - 1) - (x^2 + 1) = -2$$

$$(fg)(x) = f(x) \cdot g(x) = (x^2 - 1)(x^2 + 1) = x^4 - 1$$

$$\left(\frac{f}{g}\right)(x) = \frac{f(x)}{g(x)} = \frac{x^2 - 1}{x^2 + 1}$$

SECTION 8.5 ■ ABSOLUTE VALUE EQUATIONS AND INEQUALITIES

Absolute Value Equations The graph of $y = |ax + b|, a \neq 0$, is V-shaped and intersects the horizontal line $y = k$ twice if $k > 0$. In this case, there are two solutions to the equation $|ax + b| = k$ determined by $ax + b = k$ or $ax + b = -k$.

Example: The equation $|2x - 1| = 5$ has two solutions.

Symbolic Solution

$$2x - 1 = 5 \quad \text{or} \quad 2x - 1 = -5$$

$$2x = 6 \quad \text{or} \qquad 2x = -4 \quad \text{Add 1.}$$

$$x = 3 \quad \text{or} \qquad x = -2 \quad \text{Divide by 2.}$$

Graphical Solution

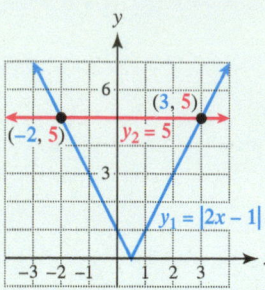

The solutions are -2 and 3.

Numerical Solution

x	-3	-2	-1	0	1	2	3
$\lvert 2x - 1\rvert$	7	5	3	1	1	3	5

The solutions are -2 and 3.

Absolute Value Inequalities

Method I If the solutions to $\lvert ax + b\rvert = k$ are c and d with $c < d$, then the solution set for $\lvert ax + b\rvert < k$ is $\{x \mid c < x < d\}$, and the solution set for $\lvert ax + b\rvert > k$ is $\{x \mid x < c \text{ or } x > d\}$.

Examples: The solutions to the equation $\lvert 2x - 1\rvert = 5$ are -2 and 3, so the solution set for $\lvert 2x - 1\rvert < 5$ is $\{x \mid -2 < x < 3\}$, and the solution set for $\lvert 2x - 1\rvert > 5$ is $\{x \mid x < -2 \text{ or } x > 3\}$.

Method II The solution set to $\lvert ax + b\rvert < k$ can be found by solving the three-part inequality

$$-k < ax + b < k.$$

The solution set to $\lvert ax + b\rvert > k$ can be found by solving the compound inequality

$$ax + b < -k \text{ or } ax + b > k.$$

Examples: $\lvert 3 - x\rvert < 5$ is equivalent to $-5 < 3 - x < 5$ and $\lvert 3 - x\rvert > 5$ is equivalent to $3 - x < -5$ or $3 - x > 5$.

CHAPTER 8   Review Exercises

SECTION 8.1

Exercises 1–4: Evaluate $f(x)$ for the given values of x.

1. $f(x) = 3x - 1$ $x = -2, \frac{1}{3}$

2. $f(x) = 5 - 3x^2$ $x = -3, 1$

3. $f(x) = \sqrt{x} - 2$ $x = 0, 9$

4. $f(x) = 5$ $x = -5, \frac{7}{5}$

Exercises 5 and 6: Do the following.
  (a) Write a symbolic representation (formula) for the function described.
  (b) Evaluate the function for input 5 and interpret the result.

5. Function P computes the number of pints in q quarts.

6. Function f computes 3 less than 4 times a number x.

7. If $f(3) = -2$, then the point _____ lies on the graph of f.

8. If the point $(4, -6)$ lies on the graph of function f, then $f(\_\_\_\_) = \_\_\_\_$.

Exercises 9–12: Sketch a graph of f.

9. $f(x) = -2x$ **10.** $f(x) = \frac{1}{2}x - \frac{3}{2}$

11. $f(x) = x^2 - 1$ **12.** $f(x) = \sqrt{x + 1}$

Exercises 13 and 14: Use the graph of f to evaluate the given expressions.

13. $f(0)$ and $f(-3)$ **14.** $f(-2)$ and $f(1)$

    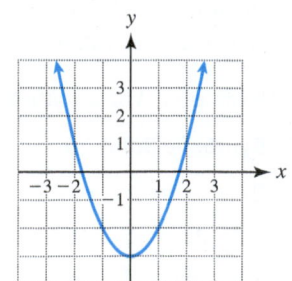

15. Use the table to evaluate $f(-1)$ and $f(3)$.

x	-1	1	3	5
$f(x)$	7	3	-1	-5

16. A function f is represented verbally by "Multiply the input x by 3 and then subtract 2." Give numerical, symbolic, and graphical representations for f. Let $x = -3, -2, -1, \ldots, 3$ in the table of values, and let $-3 \le x \le 3$ for the graph.

Exercises 17 and 18: Use the graph of f to estimate its domain and range.

17.

18.

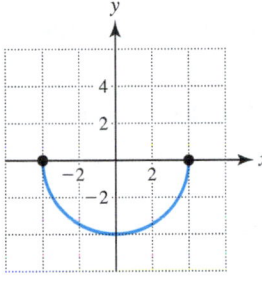

Exercises 19 and 20: Does the graph represent a function?

19.

20.

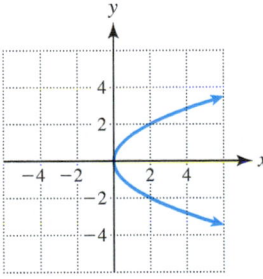

Exercises 21 and 22: Find the domain and range of S. Then state whether S defines a function.

21. $S = \{(-3, 4), (-1, 4), (2, 3), (4, -1)\}$

22. $S = \{(-1, 5), (0, 3), (1, -2), (-1, 2), (2, 4)\}$

Exercises 23–30: Find the domain.

23. $f(x) = -3x + 7$ **24.** $f(x) = \sqrt{x}$

25. $f(x) = \dfrac{3}{x}$ **26.** $f(x) = x^2 + 2$

27. $f(x) = \sqrt{5 - x}$ **28.** $f(x) = \dfrac{x}{x + 2}$

29. $f(x) = |2x + 1|$ **30.** $f(x) = x^3$

SECTION 8.2

Exercises 31 and 32: Does the graph represent a linear function?

31.

32.

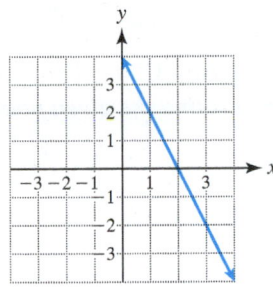

Exercises 33–36: Determine whether f is a linear function. If f is linear, give values for m and b so that f may be expressed as $f(x) = mx + b$.

33. $f(x) = -4x + 5$ **34.** $f(x) = 7 - x$

35. $f(x) = \sqrt{x}$ **36.** $f(x) = 6$

Exercises 37 and 38: Use the table to determine whether $f(x)$ could represent a linear function. If it could, write the formula for f in the form $f(x) = mx + b$.

37.

x	0	2	4	6
$f(x)$	-3	0	3	6

38.

x	-1	0	1	2
$f(x)$	-5	0	10	15

39. Evaluate $f(x) = \frac{1}{2}x + 3$ at $x = -4$.

40. Use the graph to evaluate $f(-2)$ and $f(1)$.

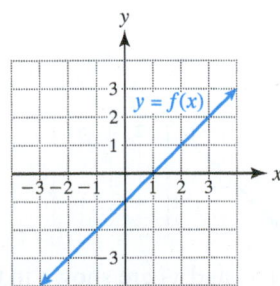

Exercises 41–44: Sketch a graph of $y = f(x)$.

41. $f(x) = x + 1$ **42.** $f(x) = 1 - 2x$

43. $f(x) = -\frac{1}{3}x$ **44.** $f(x) = -1$

45. Write a symbolic representation (formula) for a linear function H that calculates the number of hours in x days. Evaluate $H(2)$ and interpret the result.

46. Let $f(x) = \sqrt{x + 2} - x^2$.

  (a) Make a numerical representation (table) for the function f with $x = 1, 2, 3, \ldots, 7$.

  (b) Graph f in the standard window. What is the domain of f?

Exercises 47–48: Find the midpoint of the line segment connecting the given points.

47. $(-5, 3), (6, -9)$ **48.** $\left(\frac{2}{3}, -\frac{3}{4}\right), \left(\frac{1}{6}, \frac{3}{2}\right)$

SECTION 8.3

Use interval notation whenever possible for the remaining exercises.

Exercises 49–52: Solve the compound inequality. Graph the solution set on a number line.

49. $x + 1 \leq 3$ and $x + 1 \geq -1$

50. $2x + 7 < 5$ and $-2x \geq 6$

51. $5x - 1 \leq 3$ or $1 - x < -1$

52. $3x + 1 > -1$ or $3x + 1 < 10$

53. Use the table to solve $-2 \leq 2x + 2 \leq 4$.

x	−3	−2	−1	0	1	2	3
$2x + 2$	−4	−2	0	2	4	6	8

54. Use the following figure to solve each equation or inequality.

  (a) $y_1 = y_2$ (b) $y_2 = y_3$
  (c) $y_1 \leq y_2 \leq y_3$ (d) $y_2 < y_3$

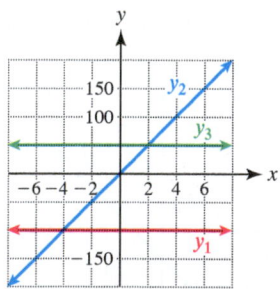

55. The graphs of y_1 and y_2 are shown in the figure. Solve each equation or inequality.

  (a) $y_1 = y_2$ (b) $y_1 < y_2$
  (c) $y_1 > y_2$

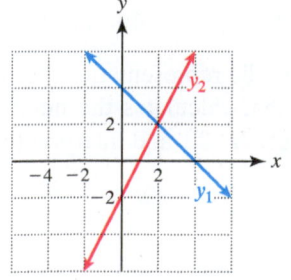

56. The graphs of three linear functions f, g, and h are shown in the following figure. Solve each equation or inequality.

  (a) $f(x) = g(x)$ (b) $g(x) = h(x)$
  (c) $f(x) < g(x) < h(x)$

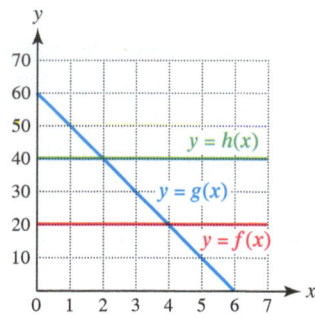

Exercises 57–62: Write the given inequality in interval notation.

57. $-3 \leq x \leq \frac{2}{3}$ **58.** $-6 < x \leq 45$

59. $x < \frac{7}{2}$ **60.** $x \geq 1.8$

61. $x > -3$ and $x < 4$ **62.** $x < 4$ or $x > 10$

Exercises 63–66: Solve the three-part inequality. Write the solution set in interval notation.

63. $-4 < x + 1 < 6$ **64.** $20 \leq 2x + 4 \leq 60$

65. $-3 < 4 - \frac{1}{3}x < 7$

66. $30 \leq \dfrac{2x - 6}{5} - 4 < 50$

Exercises 67 and 68: Let f be defined as follows.

$$f(x) = \begin{cases} 3x - 1 & \text{if } -5 \leq x < 1 \\ 4 & \text{if } 1 \leq x \leq 3 \\ 6 - x & \text{if } 3 < x \leq 5 \end{cases}$$

Use $f(x)$ to evaluate each expression.

67. $f(1)$ and $f(4)$ **68.** $f(-3)$ and $f(6)$

SECTION 8.4

Exercises 69 and 70: Write the domain and the range of the function.

69. $f(t) = \frac{1}{2}t^2$ **70.** $f(x) = |x + 2|$

71. Write the domain of $f(x) = \frac{x + 1}{2x - 8}$.

72. Write the domain and range of the function shown in the graph.

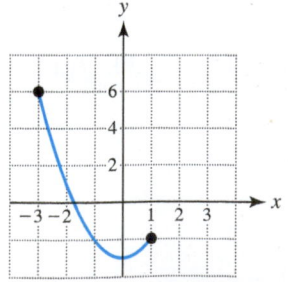

Exercises 73–76: Determine whether $f(x)$ represents a polynomial function. If possible, identify the degree and type of polynomial function.

73. $f(x) = 1 + 2x - 3x^2$

74. $f(x) = 5 + 7x$

75. $f(x) = x^3 + 2x$ **76.** $f(x) = |2x - 1|$

Exercises 77 and 78: If possible, evaluate $g(t)$ for the given values of t.

77. $g(t) = |1 - 4t|$ $t = 3, t = -\frac{1}{4}$

78. $g(t) = \dfrac{4}{4 - t^2}$ $t = 3, t = -2$

Exercises 79–82: Graph $y = f(x)$.

79. $f(x) = |x + 3|$ **80.** $f(x) = x^2 + 1$

81. $f(x) = \dfrac{1}{x}$ **82.** $f(x) = -3x$

83. Use $f(x) = 2x^2 - 3x$ and $g(x) = 2x - 3$ to find each of the following.
(a) $(f + g)(3)$ (b) $(fg)(3)$

84. Use $f(x) = x^2 - 1$ and $g(x) = x - 1$ to find each of the following.
(a) $(f - g)(x)$ (b) $(f/g)(x)$

SECTION 8.5

Exercises 85–88: Determine whether the given values of x are solutions to the absolute value equation or inequality.

85. $|12x - 24| = 24$ $x = -3; x = 2$

86. $|5 - 3x| > 3$ $x = \frac{4}{3}; x = 0$

87. $|3x - 6| \le 6$ $x = -3; x = 4$

88. $|2 + 3x| + 4 < 11$ $x = -3; x = \frac{2}{3}$

89. Use the table to solve each equation or inequality.
(a) $y_1 = 2$ (b) $y_1 < 2$
(c) $y_1 > 2$

90. Use the graph of $y = |2x + 2|$ at the top of the next column to solve each equation or inequality.
(a) $|2x + 2| = 4$ (b) $|2x + 2| \le 4$
(c) $|2x + 2| \ge 4$

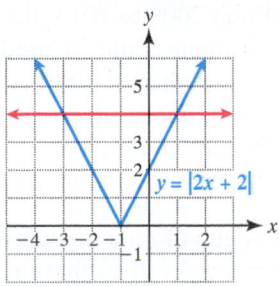

Exercises 91–96: Solve the absolute value equation.

91. $|x| = 22$ **92.** $|2x - 9| = 7$

93. $|4 - \frac{1}{2}x| = 17$ **94.** $\frac{1}{3}|3x - 1| + 1 = 9$

95. $|2x - 5| = |5 - 3x|$

96. $|-3 + 3x| = |-2x + 6|$

Exercises 97 and 98: Solve each absolute value equation or inequality.

97. (a) $|x + 1| = 7$ (b) $|x + 1| \le 7$
(c) $|x + 1| \ge 7$

98. (a) $|1 - 2x| = 6$ (b) $|1 - 2x| \le 6$
(c) $|1 - 2x| \ge 6$

Exercises 99–106: Solve the absolute value inequality.

99. $|x| > 3$ **100.** $|-5x| < 20$

101. $|4x - 2| \le 14$ **102.** $|1 - \frac{4}{5}x| \ge 3$

103. $|t - 4.5| \le 0.1$ **104.** $-2|13t - 5| \ge -4$

105. $|5 - 4x| > -5$ **106.** $|2t - 3| \le 0$

Exercises 107 and 108: Solve the inequality graphically.

107. $|2x| \ge 3$ **108.** $|\frac{1}{2}x - 1| \le 2$

Exercises 109 and 110: Write each compound inequality as an absolute value inequality.

109. $-0.05 \le x \le 0.05$

110. $5x - 1 < -4$ or $5x - 1 > 4$

APPLICATIONS

111. Age at First Marriage The median age at the first marriage for men from 1890 to 1960 can be modeled by $f(x) = -0.0492x + 119.1$, where x is the year. (*Source: National Center of Health Statistics.*)
(a) Find the median age in 1910.
(b) Graph f in $[1885, 1965, 10]$ by $[22, 26, 1]$. What happened to the median age?
(c) What is the slope of the graph of f? Interpret the slope as a rate of change.

112. Marriages From 2009 to 2014, the number of U.S. marriages in millions could be modeled by the formula $f(x) = 2.1$, where x is the year.
(a) Estimate the number of marriages in 2012.
(b) What information does f give about the number of marriages from 2009 to 2014?

113. Fat Grams A cup of milk contains 8 grams of fat.
(a) Give a formula for $f(x)$ that calculates the number of fat grams in x cups of milk.
(b) What is the slope of the graph of f?
(c) Interpret the slope as a rate of change.

114. Birth Rate The U.S. birth rate per 1000 people for selected years is shown in the table.

Year	1950	1970	1990	2010
Birth Rate	24.1	18.4	16.7	13.5

Source: U.S. Census Bureau.

(a) Make a scatterplot of the data.
(b) Model the data with $f(x) = mx + b$, where x is the year. Answers may vary.
(c) Use f to estimate the birth rate in 2000.

115. Unhealthy Air Quality The Environmental Protection Agency (EPA) monitors air quality in U.S. cities. The function f gives the annual number of days with unhealthy air quality in Los Angeles, California, for selected years.

x	2006	2008	2010	2012	2014
$f(x)$	148	145	128	122	120

Source: Environmental Protection Agency.

(a) Find $f(2010)$ and interpret your result.
(b) Identify the domain and range of f.
(c) Discuss the trend of air pollution in Los Angeles.

116. Temperature Scales The table shown below gives equivalent temperatures in degrees Celsius and degrees Fahrenheit.

°C	−40	0	15	35	100
°F	−40	32	59	95	212

(a) Plot the data. Let the x-axis correspond to the Celsius temperature and the y-axis correspond to the Fahrenheit temperature. What type of relation exists between the data?

(b) Find $f(x) = mx + b$ so that f receives the Celsius temperature x as input and outputs the corresponding Fahrenheit temperature. Interpret the slope of the graph of f.
(c) If the temperature is 20 °C, what is the equivalent temperature in degrees Fahrenheit?

117. Distance between Bicyclists The following graph shows the distance between two bicyclists traveling toward each other along a straight road after x hours.
(a) After how long did the bicycle riders meet?
(b) When were they 20 miles apart?
(c) Find the times when they were less than 20 miles apart.
(d) Estimate the sum of the speeds of the bicyclists.

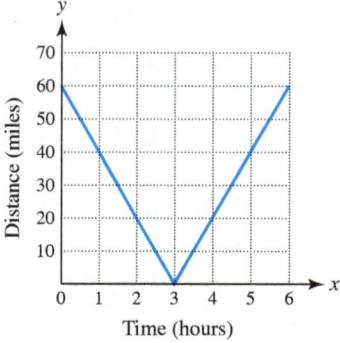

118. Violent Crimes in the U.S. The number of violent crimes reported has dropped from 1.3 million in 2009 to 1.1 million in 2013.
(a) Find a linear function $f(x) = mx + b$ that models the data x years after 2009.
(b) Use $f(x)$ to estimate the number of violent crimes in 2012.

119. Average Precipitation The average rainfall in Houston, Texas, is 3.9 inches per month. Each month's average A is within 1.7 inches of 3.9 inches.
(Source: J. Williams, The Weather Almanac)
(a) Write an absolute value inequality that models this situation.
(b) Solve the inequality.

120. Relative Error If a quantity is measured to be T and the actual value is A, then the relative error in this measurement is $\left|\frac{T-A}{A}\right|$. If $A = 35$ and the relative error is to be less than 0.08 (8%), what values for T are possible?

1. Evaluate $f(4)$ if $f(x) = 3x^2 - \sqrt{x}$. Give a point on the graph of f.

2. Write a symbolic representation (formula) for a function C that calculates the cost of buying x pounds of candy at \$4 per pound. Evaluate $C(5)$ and interpret your result.

3. Sketch a graph of f.
   (a) $f(x) = -2x + 1$ (b) $f(x) = x^2 + 1$
   (c) $f(x) = \sqrt{x + 3}$ (d) $f(x) = |x + 1|$

4. Use the graph of f to evaluate $f(-3)$ and $f(0)$. Determine the domain and range of f.

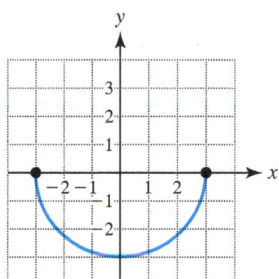

5. A function f is represented verbally by "Square the input x and then subtract 5." Give symbolic, numerical, and graphical representations of f. Let $x = -3, -2, -1, \ldots, 3$ in the numerical representation (table) and let $-3 \le x \le 3$ for the graph.

6. Determine whether the graph represents a function. Explain your reasoning.

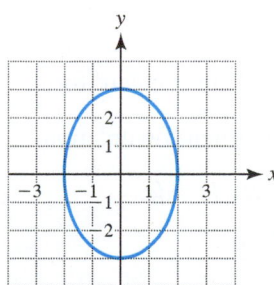

7. Find the domain of function f.
   (a) $f = \{(-2, 3), (-1, 5), (0, 3), (5, 7)\}$
   (b) $f(x) = \frac{3}{4}x - 5$
   (c) $f(x) = \sqrt{x + 4}$
   (d) $f(x) = 2x^2 - 1$
   (e) $f(x) = \frac{3x}{5 - x}$

8. Determine if $f(x) = 6 - 8x$ is a linear function. If it is, write it in the form $f(x) = mx + b$.

9. Graph the solution set to $2x + 6 < 2$ and $-3x \ge 3$ on a number line.

10. Use the given table to solve the compound inequality $-3x < -3$ or $-3x > 6$. Use interval notation.

x	-3	-2	-1	0	1	2	3
$-3x$	9	6	3	0	-3	-6	-9

11. Use the following figure to solve each equation or inequality. Write your answers for parts (c) and (d) in interval notation.
   (a) $y_1 = y_2$ (b) $y_2 = y_3$
   (c) $y_1 \le y_2 \le y_3$ (d) $y_2 < y_3$

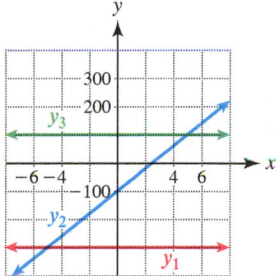

12. Solve $-2 < 2 + \frac{1}{2}x < 2$ and write the solution set in interval notation.

13. Solve the equation $|2 - \frac{1}{3}x| = 6$.

14. Solve each inequality. Write your answer in interval notation.
   (a) $|x| \le 5$ (b) $|x| > 0$

15. Determine whether $f(x) = 1 - 2x + x^3$ represents a polynomial function. If possible, identify the degree and type of polynomial function.

16. Evaluate $h(t) = -\frac{4t}{5 - t}$ at $t = -2$. Write the domain of h in interval notation.

17. Let $f(x) = x^2 + 1$ and $g(x) = 2x$. Find each of the following.
   (a) $(f - g)(-2)$ (b) $(fg)(x)$

18. **Drinking Fluids and Exercise** To determine the number of ounces of fluid that a person should drink in a day, divide his or her weight in pounds by 2 and then add 0.4 ounce for every minute of exercise.
   (a) Write a function that gives the fluid requirements for a person weighing 150 pounds and exercising x minutes a day.
   (b) If a 150-pound runner needs 89 ounces of fluid each day, determine the runner's daily minutes of exercise.

19. Heart Rate of an Athlete The table below lists the heart rate or pulse of an athlete running a 400-meter race. The race lasts 50 seconds.

Time (seconds)	0	20	30	50
Heart Rate (bpm)	100	134	150	180

(a) Does $P(t) = -\frac{1}{300}t^2 + \frac{53}{30}t + 100$ model the data in the table exactly? Explain.

(b) Does $P(60)$ have significance in this situation? What should be the domain of P?

20. Time Spent in Line Suppose that parking lot attendants can wait on 25 vehicles per minute and vehicles are arriving randomly at an average rate of x vehicles per minute. Then the average time T in minutes spent waiting in line *and* paying the attendant is given by

$$T(x) = \frac{1}{25 - x},$$

where $x < 25$. (*Source:* N. Garber.)

(a) Graph T in $[0, 25, 5]$ by $[0, 2, 0.5]$. Identify any vertical asymptotes.

(b) If the wait is 1 minute, how many vehicles are arriving on average?

CHAPTERS 1–8 Cumulative Review Exercises

1. Write 120 as a product of prime numbers.

2. Translate the sentence "Triple a number decreased by 4 equals the number" into an equation using the variable n. Solve the equation.

Exercises 3 and 4: Simplify completely.

3. $\frac{1}{4} \div \frac{3}{4} - \frac{1}{2}$

4. $12 - 3^2 \div 3 \cdot 2$

5. Solve $-3(3 - x) - 6 = 2x$.

6. Convert 0.075 to a percentage.

7. Solve $A = \frac{1}{3}(2a - b)$ for a.

8. Solve $5 - 3t < 1 - t$. Write the solution set in set-builder notation.

Exercises 9 and 10: Graph the equation.

9. $3x - 2y = -6$

10. $y = |2x - 4|$

11. Write the slope–intercept form of a line that passes through the points $(-4, 4)$ and $(2, 1)$.

12. Solve the system of equations. Write your answer as an ordered pair.

$$2x + 3y = 5$$
$$3x - 2y = 1$$

Exercises 13 and 14: Shade the solution set for the given inequality.

13. $x \geq -1$

14. $x - y \geq 2$

15. Shade the solution set for the system of inequalities.

$$x - y \geq 1$$
$$2x + y \leq 0$$

16. Simplify $(3x^2 + 2x - 4) - (4x^2 + 5)$.

Exercises 17–20: Simplify and use positive exponents.

17. $\dfrac{x^{-4}}{x^{-3}}$

18. $(3b^{-3})(2b^4)$

19. $3(2t)^3$

20. $\left(\dfrac{2x^3}{x^2y^{-1}}\right)^{-2}$

Exercises 21–24: Multiply the expression.

21. $2x^2(x^3 - 4x^2 - 5)$

22. $(5x + 1)(2x - 7)$

23. $(y - 3)(y + 3)$

24. $(x - 4y)^2$

25. Write 2.5×10^4 in standard form.

26. Write 0.028 in scientific notation.

Exercises 27 and 28: Divide.

27. $\dfrac{6x^3 - 4x^2 + 8x}{2x}$

28. $(3x^3 + 2x^2 + 1) \div (x - 1)$

Exercises 29–34: Factor completely.

29. $10x^2y^3 - 15x^3y^2$

30. $x^3 + 3x^2 - x - 3$

31. $2z^2 + z - 3$

32. $16x^2 - 25$

33. $a^3 - 8$

34. $z^4 + 7z^2 + 6$

Exercises 35–38: Solve the equation.

35. $x(x + 5) = 0$ **36.** $4x^2 = 0$

37. $2x^2 + 5x = 3$ **38.** $x^3 = x$

39. Simplify $\dfrac{x^2 + 4x + 4}{x + 2}$ to lowest terms.

40. If possible, evaluate $\dfrac{x^2 + 1}{x - 1}$ for $x = -2$ and $x = 1$.

Exercises 41 and 42: Simplify to lowest terms.

41. $\dfrac{x^2 - 3x + 2}{x + 2} \div \dfrac{x - 1}{2x + 4}$

42. $\dfrac{1}{x + 3} + \dfrac{2}{x + 1}$

43. Solve $\dfrac{x + 2}{x - 1} = \dfrac{2}{3}$.

44. Suppose that y is directly proportional to x and that $y = 5$ when $x = 10$. Find y when $x = 20$.

45. Evaluate $f(x) = x^2 - 4x$ for $x = -2$.

46. Graph $f(x) = x^2 - 2$ and identify the domain and range of f. Write your answer in interval notation.

47. Solve $x - 2 < 3$ or $x - 2 > 6$. Write your answer in interval notation.

48. Solve $|2x - 4| = 6$.

49. Solve $|2x - 4| \le 6$. Write your answer in interval notation.

50. Solve $|x - 4| > 2$. Write your answer in interval notation.

APPLICATIONS

51. Modeling Motion The table lists the distance d in miles traveled by a car for various elapsed times t in hours. Find an equation that models these data.

t (hours)	2	3	4	6
d (miles)	144	216	288	432

52. Rainfall At noon, 2 inches of rain had fallen. For the next 6 hours, rain fell at $\frac{1}{4}$ inch per hour.
  (a) Find an equation in the form $y = mx + b$ that calculates the number of inches of rain that fell x hours past noon.
  (b) What is the slope of the graph of f?
  (c) Interpret the slope as a rate of change.
  (d) How much rain had fallen by 4 P.M.?

53. Interest A total of \$5000 is deposited in two accounts paying 5% and 7% annual interest. If total interest received at the end of the year is \$308, determine how much is invested at each interest rate.

54. Height of a Building An 8-foot-tall stop sign casts a 5-foot-long shadow, while a nearby building casts a 65-foot-long shadow. Find the height of the building.

55. Shoveling the Driveway Two people are shoveling snow from a driveway. The first person shovels 9 square feet per minute, while the second person shovels 11 square feet per minute.
  (a) Write a simplified expression that gives the total number of square feet the two people can shovel in x minutes.
  (b) How many minutes would it take for them to clear a driveway with 1000 square feet?

56. Cost of a Television A 5% sales tax on a television amounts to \$82.50. Find the cost of the television.

9 Systems of Linear Equations

In 1940, a physicist named John Atanasoff at Iowa State University needed to solve 29 equations with 29 variables simultaneously. This task was too difficult to do by hand, so he and a graduate student invented the first fully electronic digital computer. Thus the desire to solve a mathematical problem led to one of the most important inventions of the twentieth century. Today, people can solve thousands of equations with thousands of variables. Solutions to such equations have resulted in better airplanes, cars, electronic devices, weather forecasts, and medical equipment.

Equations are even used in biology. The following table contains the weight W, neck size N, and chest size C for three black bears. Suppose that park rangers find a bear with a neck size of 22 inches and a chest size of 38 inches. Can they use the data in the table to estimate the bear's weight? Using systems of linear equations, they *can* estimate the bear's weight. See Example 8 and Exercise 61 in the second section of this chapter.

Sources: A. Tucker, Fundamentals of Computing; M. Triola, Elementary Statistics; Minitab, Inc.

Black Bear Measurements

W (pounds)	N (inches)	C (inches)
80	16	26
344	28	45
416	31	54
?	22	38

9.1 Systems of Linear Equations in Three Variables

Objectives

1. Writing Systems of Linear Equations and Recognizing Solutions
   • Checking for Solutions
   • Setting up a System
2. Solving Linear Systems with Substitution and Elimination
3. Modeling with Linear Systems
4. Recognizing Linear Systems That Have No Solutions
5. Recognizing Linear Systems That Have Infinitely Many Solutions

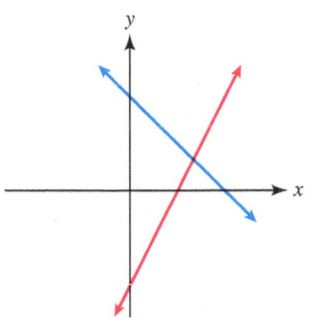

FIGURE 9.1

NEW VOCABULARY

☐ Linear system in three variables
☐ Ordered triple

1 Writing Systems of Linear Equations and Recognizing Solutions

When we solve a linear system in two variables, we can express a solution as an ordered pair (x, y). A linear equation in two variables can be represented graphically by a line. A system of two linear equations with a unique solution can be represented graphically by two lines intersecting at a point, as shown in **FIGURE 9.1**.

When solving **linear systems in three variables,** we often use the variables x, y, and z. A solution is expressed as an **ordered triple** (x, y, z), rather than an ordered pair (x, y). For example, if the ordered triple $(1, 2, 3)$ is a solution, $x = 1$, $y = 2$, and $z = 3$ satisfy each equation. A linear equation in three variables can be represented by a flat plane in space. If the solution is unique, we can represent a linear system of three equations in three variables graphically by three planes intersecting at a single point. (See **B** below.)

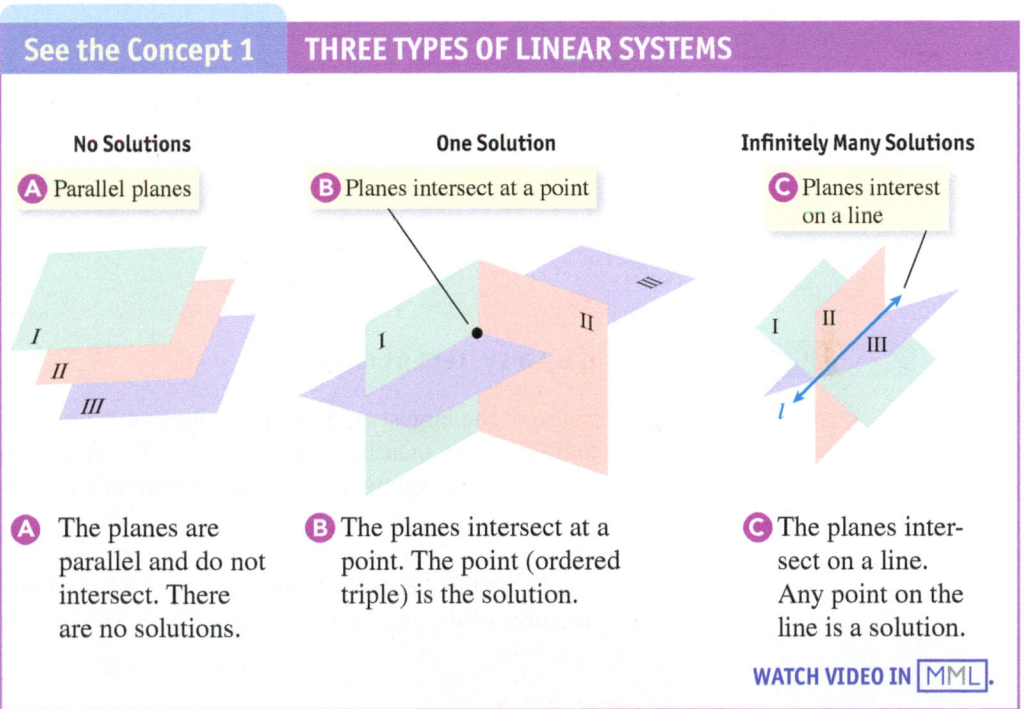

See the Concept 1 · THREE TYPES OF LINEAR SYSTEMS

No Solutions

A Parallel planes

A The planes are parallel and do not intersect. There are no solutions.

One Solution

B Planes intersect at a point

B The planes intersect at a point. The point (ordered triple) is the solution.

Infinitely Many Solutions

C Planes interest on a line

C The planes intersect on a line. Any point on the line is a solution.

WATCH VIDEO IN MML.

NOTE: The three planes in **B** above all intersect at right angles. In general, three planes can intersect at a point even if they are not at right angles to each other. ■

READING CHECK 1

• What is the possible number of solutions to a system of linear equations in three variables?

CHECKING FOR SOLUTIONS The next example shows how to check whether an ordered triple is a solution to a system of linear equations in three variables.

EXAMPLE 1 **Checking for solutions to a system of three equations**

Determine whether $(4, 2, -1)$ or $(-1, 0, 3)$ is a solution to the system.

$$2x - 3y + z = 1$$
$$x - 2y + 2z = 5$$
$$2y + z = 3$$

Solution
To check $(4, 2, -1)$, substitute $x = 4$, $y = 2$, and $z = -1$ in each equation.

$$2(4) - 3(2) + (-1) \stackrel{?}{=} 1 \checkmark \quad \text{True}$$
$$4 - 2(2) + 2(-1) \stackrel{?}{=} 5 \ \text{✗} \quad \text{False}$$
$$2(2) + (-1) \stackrel{?}{=} 3 \checkmark \quad \text{True}$$

> A solution must satisfy *all* equations.

The ordered triple $(4, 2, -1)$ *does not satisfy all three equations, so it is not a solution.* Next substitute $x = -1$, $y = 0$, and $z = 3$ in each equation.

$$2(-1) - 3(0) + 3 \stackrel{?}{=} 1 \checkmark \quad \text{True}$$
$$-1 - 2(0) + 2(3) \stackrel{?}{=} 5 \checkmark \quad \text{True}$$
$$2(0) + 3 \stackrel{?}{=} 3 \checkmark \quad \text{True}$$

The ordered triple $(-1, 0, 3)$ *satisfies all three equations, so it is a solution.*

Now Try Exercise 9

SETTING UP A SYSTEM In the next example, we set up a system of three equations in three variables that involves the angles of a triangle. You are asked to solve this system in Exercise 51.

EXAMPLE 2 **Setting up a system of equations**

The measure of the largest angle in a triangle is 40° greater than the sum of the two smaller angles and 90° more than the smallest angle. Set up a system of three linear equations in three variables whose solution gives the measure of each angle.

Solution
Let x, y, and z be the measures of the three angles from largest to smallest. Because the sum of the measures of the angles in a triangle equals 180°, we have

$$x + y + z = 180. \quad (1)$$

The measure of the largest angle x is 40° greater than the sum of the measures of the two smaller angles $y + z$, so

$$x - (y + z) = 40 \quad \text{or} \quad x - y - z = 40. \quad (2)$$

The measure of the largest angle x is 90° more than the measure of the smallest angle z, so

$$x - z = 90. \quad (3)$$

Thus the required system of equations can be written as follows.

$$x + y + z = 180 \quad (1)$$
$$x - y - z = 40 \quad (2)$$
$$x - z = 90 \quad (3)$$

Now Try Exercise 49(a)

Biology

🌐 *Math in Context* In the next example, we show how a linear system involving three equations and three variables can be used to model a real-world situation. We solve this system of equations in Example 7.

EXAMPLE 3   ### Modeling real data with a linear system

The Bureau of Land Management studies antelope populations in Wyoming. It monitors the number of adult antelope, the number of fawns each spring, and the severity of the winter. The first two columns of **TABLE 9.1** contain counts of fawns and adults for three representative winters. The third column shows the severity of each winter. The severity of the winter is measured from 1 to 5, with 1 being mild and 5 being severe.

Antelope Numbers and Winter Severity

Fawns (F)	Adults (A)	Winter (W)
405	870	3
414	848	2
272	684	5
?	750	4

TABLE 9.1

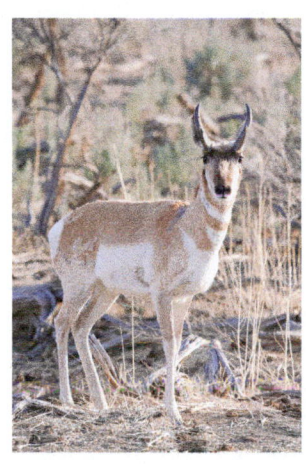

We want to use the data in the first three rows of the table to estimate the number of fawns F in the fourth row when the number of adults is 750 and the severity of the winter is 4. To do so, we use the formula

$$F = a + bA + cW,$$

where a, b, and c are constants. Write a system of linear equations whose solution gives appropriate values for a, b, and c.

Solution
From the first row in the table, when $F = 405$, $A = 870$, and $W = 3$, the formula

$$F = a + bA + cW$$

becomes

$$405 = a + b(870) + c(3). \quad (1)$$

Similarly, $F = 414$, $A = 848$, and $W = 2$ gives

$$414 = a + b(848) + c(2), \quad (2)$$

and $F = 272$, $A = 684$, and $W = 5$ yields

$$272 = a + b(684) + c(5). \quad (3)$$

To find values for a, b, and c, we can solve the following system of linear equations.

System of equations →
$$\begin{cases} 405 = a + 870b + 3c & (1) \\ 414 = a + 848b + 2c & (2) \\ 272 = a + 684b + 5c & (3) \end{cases}$$

We can also write these equations as a linear system in the following form.

$$a + 870b + 3c = 405 \quad (1)$$
$$a + 848b + 2c = 414 \quad (2)$$
$$a + 684b + 5c = 272 \quad (3)$$

Finding values for a, b, and c will allow us to use the formula $F = a + bA + cW$ to predict the number of fawns F when the number of adults A is 750 and the severity of the winter W is 4 (see Example 7).

Now Try Exercise 53(a)

NOTE: Linear systems of two equations can have no solutions, one solution, or infinitely many solutions. The same is true for larger linear systems. In the following subsection, we focus on linear systems that have one solution. ∎

2 Solving Linear Systems with Substitution and Elimination

When solving systems of linear equations with more than two variables, we usually use both substitution and elimination. However, in the next example, we use only substitution to solve a particular type of linear system in three variables.

> **EXAMPLE 4** **Using substitution to solve a linear system of equations**

Solve the following system.

$$2x - y + z = 7$$
$$3y - z = 1$$
$$z = 2$$

Solution

From the third equation, we know z is 2. Let $z = 2$ in the second equation and determine y.

	$3y - z = 1$	Second equation
Let $z = 2$ and then find y.	$3y - 2 = 1$	Substitute $z = 2$.
	$3y = 3$	Add 2 to each side.
	$y = 1$	Divide each side by 3.

Knowing that $y = 1$ and $z = 2$ allows us to find x by using the first equation.

	$2x - y + z = 7$	First equation
Let $y = 1$ and $z = 2$ and then find x.	$2x - 1 + 2 = 7$	Let $y = 1$ and $z = 2$.
	$2x = 6$	Simplify and subtract 1.
	$x = 3$	Divide each side by 2.

Thus $x = 3$, $y = 1$, and $z = 2$, and the solution is $(3, 1, 2)$.

Now Try Exercise 13

In the next example, we use elimination and substitution in a four-step method to solve a system of linear equations. This method is summarized by the following.

STUDY TIP

Be sure to follow this four-step method. It will help you get the correct answer *consistently.*

SOLVING A LINEAR SYSTEM IN THREE VARIABLES

STEP 1: Eliminate one variable, such as x, from two of the equations.

STEP 2: Use the two resulting equations in two variables to eliminate one of the variables, such as y. Solve for the remaining variable, z.

STEP 3: Substitute z in one of the two equations from Step 2. Solve for the unknown variable y.

STEP 4: Substitute values for y and z in one of the given equations and find x. The solution is (x, y, z).

EXAMPLE 5 **Solving a linear system in three variables**

Solve the following system.

$$\begin{aligned} x - y + 2z &= 6 \\ 2x + y - 2z &= -3 \\ -x - 2y + 3z &= 7 \end{aligned}$$

Solution

STEP 1: We begin by eliminating the variable x from the second and third equations. To eliminate x from the second equation, we multiply the first equation by -2 and then add it to the second equation. To eliminate x from the third equation, we add the first and third equations.

$-2x + 2y - 4z = -12$	First equation times -2		$x - y + 2z = 6$	First equation
$\underline{2x + y - 2z = -3}$	Second equation		$\underline{-x - 2y + 3z = 7}$	Third equation
$3y - 6z = -15$	Add.		$-3y + 5z = 13$	Add.

STEP 2: Take the two resulting equations from Step 1 and eliminate either variable. Here, we add the two equations to eliminate the variable y.

$$\begin{aligned} 3y - 6z &= -15 \\ \underline{-3y + 5z} &= \underline{13} \end{aligned}$$

Add to eliminate y.

$$\begin{aligned} -z &= -2 \quad \text{Add the equations.} \\ z &= 2 \quad \text{Multiply by } -1. \end{aligned}$$

STEP 3: Now we can use substitution to find the values of x and y. We let $z = 2$ in either equation used in Step 2 to find y.

$$\begin{aligned} 3y - 6z &= -15 \quad \text{Equation from Step 2} \\ 3y - 6(2) &= -15 \quad \text{Substitute } z = 2. \\ 3y - 12 &= -15 \quad \text{Multiply.} \\ 3y &= -3 \quad \text{Add 12.} \\ y &= -1 \quad \text{Divide by 3.} \end{aligned}$$

Substitute 2 for z.

STEP 4: Finally, we substitute $y = -1$ and $z = 2$ in one of the given equations to find x.

$$\begin{aligned} x - y + 2z &= 6 \quad \text{First given equation} \\ x - (-1) + 2(2) &= 6 \quad \text{Let } y = -1 \text{ and } z = 2. \\ x + 1 + 4 &= 6 \quad \text{Simplify.} \\ x &= 1 \quad \text{Subtract 5.} \end{aligned}$$

Substitute −1 for y and 2 for z.

The solution is $(1, -1, 2)$. Check this solution.

Now Try Exercise 25

🌐 *Math in Context* In the next example, we solve a system of linear equations to determine the number of tickets sold at a play.

EXAMPLE 6 **Finding the number of tickets sold**

One thousand tickets were sold for a play, which generated $3800 in revenue. The prices of the tickets were $3 for children, $4 for students, and $5 for adults. There were 100 fewer student tickets sold than adult tickets. Find the number of each type of ticket sold.

Solution

Let x be the number of tickets sold to children, y be the number of tickets sold to students, and z be the number of tickets sold to adults. The total number of tickets sold was 1000, so

$$x + y + z = 1000. \quad (1)$$

Each child's ticket cost \$3, so the revenue generated from selling x tickets was $3x$. Similarly, the revenue generated from student tickets was $4y$, and the revenue from adult tickets was $5z$. Total ticket revenue was \$3800, so

$$3x + 4y + 5z = 3800. \quad (2)$$

The equation $z - y = 100$, or

$$y - z = -100, \quad (3)$$

must also be satisfied, because 100 fewer tickets were sold to students than to adults.

To find the number of each type of ticket sold, we need to solve the following system of linear equations.

$$\left.\begin{array}{rcr} x + y + z = & 1000 & (1) \\ 3x + 4y + 5z = & 3800 & (2) \\ y - z = & -100 & (3) \end{array}\right\} \quad \text{System to be solved}$$

STEP 1: We begin by eliminating the variable x from the second equation. To do so, we multiply the first equation by -3 and add the second equation.

Add equations to eliminate x.

$$\begin{array}{ll} -3x - 3y - 3z = -3000 & \text{Equation (1) times } -3 \\ \underline{3x + 4y + 5z = 3800} & \text{Equation (2)} \\ y + 2z = 800 & \text{Add Equations to get equation (4).} \end{array}$$

STEP 2: We then multiply the resulting equation from Step 1 by -1 and add the third given equation to eliminate y.

Add equations to eliminate y.

$$\begin{array}{ll} -y - 2z = -800 & \text{Equation (4) times } -1 \\ \underline{y - z = -100} & \text{Equation (3)} \\ -3z = -900 & \text{Add Equations.} \\ z = 300 & \text{Divide by } -3 \text{ to find the value of } z. \end{array}$$

STEP 3: To find y, we can substitute $z = 300$ in one of the equations from Step 2.

$$\begin{array}{ll} y - z = -100 & \text{Equation (3)} \\ y - 300 = -100 & \text{From Step 2, let } z = 300. \\ y = 200 & \text{Add 300 to find the value of } y. \end{array}$$

STEP 4: Finally, we substitute $y = 200$ and $z = 300$ in the first given equation.

$$\begin{array}{ll} x + y + z = 1000 & \text{Equation (1)} \\ x + 200 + 300 = 1000 & \text{From Steps 2 and 3, let } y = 200 \text{ and } z = 300. \\ x = 500 & \text{Subtract 500 to find the value of } x. \end{array}$$

Thus **500** tickets were sold to children, **200** to students, and **300** to adults.

Now Try Exercise 47

3 Modeling with Linear Systems

In the next example, we solve the system of equations that we discussed in Example 3.

EXAMPLE 7　**Predicting fawns in the spring**

Solve the following linear system for a, b, and c.

$$a + 870b + 3c = 405$$
$$a + 848b + 2c = 414$$
$$a + 684b + 5c = 272$$

Then use $F = a + bA + cW$ to predict the number of fawns when there are 750 adults and the severity of the winter is 4.

Solution

STEP 1: We begin by eliminating the variable a from the second and third equations. To do so, we add the second and third equations to the first equation times -1.

$-a - 870b - 3c = -405$	$-a - 870b - 3c = -405$	First times -1
$a + 848b + 2c = \quad 414$	$a + 684b + 5c = \quad 272$	Second/third equation
$-22b - \quad c = \quad 9$	$-186b + 2c = -133$	Add.

STEP 2: We use the two resulting equations from Step 1 to eliminate c. To do, so we multiply the equation $-22b - c = 9$ by 2 and add it to the other equation.

> Add equations to eliminate c.

$$-44b - 2c = \quad 18 \quad (-22b - c = 9) \text{ times } 2$$
$$-186b + 2c = -133$$
$$-230b \qquad = -115 \quad \text{Add the equations.}$$
$$b \qquad = \quad 0.5 \quad \text{Divide by } -230.$$

STEP 3: To find c, we substitute $b = 0.5$ in either equation used in Step 2.

> Let $b = 0.5$ and find c.

$$-44b - 2c = \quad 18 \quad \text{First equation in Step 2}$$
$$-44(0.5) - 2c = \quad 18 \quad \text{Let } b = 0.5.$$
$$-22 - 2c = \quad 18 \quad \text{Multiply.}$$
$$-2c = \quad 40 \quad \text{Add 22.}$$
$$c = -20 \quad \text{Divide by } -2.$$

STEP 4: Finally, we substitute $b = 0.5$ and $c = -20$ in one of the given equations to find a.

> Let $b = 0.5$ and $c = -20$ and find a.

$$a + 870b + 3c = 405 \quad \text{First given equation}$$
$$a + 870(0.5) + 3(-20) = 405 \quad \text{Let } b = 0.5 \text{ and } c = -20.$$
$$a + 435 - 60 = 405 \quad \text{Multiply.}$$
$$a = \quad 30 \quad \text{Solve for } a.$$

The solution is $a = 30$, $b = 0.5$, and $c = -20$. Thus we may write

$$F = a + bA + cW$$
$$= 30 + 0.5A - 20W. \quad \text{Modeling equation}$$

If there are 750 adults and the winter has a severity of 4, this model predicts

$$F = 30 + 0.5(750) - 20(4)$$
$$= 325 \text{ fawns.}$$

CRITICAL THINKING 1

Give reasons why the coefficient for A is positive and the coefficient for W is negative in the formula

$$F = 30 + 0.5A - 20W.$$

Now Try Exercise 53

4 ▸ Recognizing Linear Systems That Have No Solutions

It is possible for a system of three linear equations in three variables to be inconsistent and have no solutions. If we apply substitution and elimination to this type of system, we arrive at a contradiction. This case is demonstrated in the next example.

EXAMPLE 8 **Recognizing an inconsistent system**

Solve the system, if possible.

$$\begin{aligned} x + y + z &= 4 \\ -x + y + z &= 2 \\ y + z &= 1 \end{aligned}$$

Solution

STEP 1: If we add the first two equations, we can eliminate x. The variable x is already eliminated from the third equation.

Add equations to eliminate x.

$$\begin{array}{rl} x + y + z = 4 & \text{First equation} \\ -x + y + z = 2 & \text{Second equation} \\ \hline 2y + 2z = 6 & \text{Add.} \end{array}$$

STEP 2: If we multiply the third *given* equation by -2 and add it to the resulting equation in Step 1, we arrive at a contradiction.

The equation $0 = 4$ is never true.

$$\begin{array}{rl} -2y - 2z = -2 & \text{Third equation times } -2 \\ 2y + 2z = 6 & \text{Equation from Step 1} \\ \hline 0 = 4 \; \text{✗} & \text{Add. (Contradiction)} \end{array}$$

Because $0 = 4$ is a contradiction, there are no solutions to the given system of equations.

Now Try Exercise 31

5 ▸ Recognizing Linear Systems That Have Infinitely Many Solutions

It is possible for a system of linear equations in three variables to have infinitely many solutions. If we apply substitution and elimination to this type of system, we arrive at an identity. This case is demonstrated in the next example.

EXAMPLE 9 **Solving a system with infinitely many solutions**

Solve the system.

$$\begin{aligned} x + y + z &= 2 \\ x - y + z &= 4 \\ 3x - y + 3z &= 10 \end{aligned}$$

Solution

STEP 1: To eliminate y from the second equation, add the first equation to the second. To eliminate y from the third equation, add the first equation to the third equation.

$$\begin{array}{rl} x + y + z = 2 & \text{First equation} \\ x - y + z = 4 & \text{Second equation} \\ \hline 2x + 2z = 6 & \text{Add.} \end{array} \qquad \begin{array}{rl} x + y + z = 2 & \text{First equation} \\ 3x - y + 3z = 10 & \text{Third equation} \\ \hline 4x + 4z = 12 & \text{Add.} \end{array}$$

STEP 2: If we multiply the first resulting equation in Step 1 by -2 and add it to the second resulting equation in Step 1, we arrive at an identity.

$$\begin{array}{rl} -4x - 4z = -12 & (2x + 2z = 6) \text{ times } -2 \\ \underline{4x + 4z = 12} & \text{Second equation from Step 1} \\ 0 = 0 & \text{Add. (Identity)} \end{array}$$

The equation $0 = 0$ is always true.

The variable x can be written in terms of z by solving $2x + 2z = 6$ for x.

$$\begin{array}{rl} 2x + 2z = 6 & \text{Equation from Step 1} \\ 2x = 6 - 2z & \text{Subtract } 2z. \\ x = 3 - z & \text{Divide by 2.} \end{array}$$

STEP 3: To find y in terms of z, substitute $3 - z$ for x in the first *given* equation.

$$\begin{array}{rl} x + y + z = 2 & \text{First equation} \\ (3 - z) + y + z = 2 & \text{Let } x = 3 - z. \\ y = -1 & \text{Solve for } y. \end{array}$$

All solutions have the form $(3 - z, -1, z)$, where z can be any real number. For example, if $z = 1$, then $(2, -1, 1)$ is one of infinitely many solutions to the system of equations. Note that in this particular system, y must always equal -1.

Now Try Exercise 33

9.1 Putting It All Together

In this section, we discussed how to solve a system of three linear equations in three variables. Systems of linear equations can have no solutions, one solution, or infinitely many solutions.

CONCEPT	EXPLANATION/EXAMPLE
System of Linear Equations in Three Variables	The following is a system of three linear equations in three variables. Each equation is linear. $$\begin{array}{rcrcrcr} x & - & 2y & + & z & = & 0 \\ -x & + & y & + & z & = & 4 \\ & & -y & + & 4z & = & 10 \end{array}$$
Solution to a Linear System in Three Variables	The solution to a linear system in three variables is an ordered triple, expressed as (x, y, z). The solution to the preceding system is $(1, 2, 3)$ because substituting $x = 1$, $y = 2$, and $z = 3$ in each equation results in a true statement. We can check solutions this way. $$\begin{array}{rl} (1) - 2(2) + (3) = 0 \checkmark & \text{True} \\ -(1) + (2) + (3) = 4 \checkmark & \text{True} \\ -(2) + 4(3) = 10 \checkmark & \text{True} \end{array}$$

continued on next page

continued from previous page

CONCEPT	EXPLANATION
Solving a Linear System with Substitution and Elimination	**STEP 1:** Eliminate one variable, such as x, from two of the equations.
	STEP 2: Use the two resulting equations in two variables to eliminate one of the variables, such as y. Solve for the remaining variable z.
	STEP 3: Substitute z in one of the two equations from Step 2. Solve for the unknown variable y.
	STEP 4: Substitute values for y and z in one of the given equations and find x. The solution is (x, y, z).

9.1 Exercises

MyMathLab®

CONCEPTS AND VOCABULARY

1. Can a system of three linear equations in three variables have exactly two solutions? Explain.

2. Give an example of a system of three linear equations in three variables.

3. Does the ordered triple $(1, 2, 3)$ satisfy the equation $x + y + z = 6$?

4. Does $(3, 4)$ represent a solution to the equation $x + y + z = 7$? Explain.

5. To solve uniquely for two variables, how many equations do you usually need?

6. To solve uniquely for three variables, how many equations do you usually need?

7. If a contradiction occurs while solving a system of linear equations, then there is/are _____ solution(s).

8. If an identity occurs while solving a linear system, how many solutions are there?

RECOGNIZING SOLUTIONS TO LINEAR SYSTEMS

Exercises 9–12: Determine which ordered triple is a solution to the linear system.

9. $(1, 2, 3)$, $(0, 2, 4)$
$$x + y + z = 6$$
$$x - y - z = -4$$
$$-x - y + z = 0$$

10. $(-1, 0, 2)$, $(0, 4, 4)$
$$2x + y - 3z = -8$$
$$x - 3y + 2z = -4$$
$$3x - 2y + z = -4$$

11. $(1, 0, 3)$, $(-1, 1, 2)$
$$3x - 2y + z = -3$$
$$-x + 3y - 2z = 0$$
$$x + 4y + 2z = 7$$

12. $\left(\frac{1}{2}, \frac{3}{2}, -\frac{1}{2}\right)$, $(-1, 0, -2)$
$$x + 3y - 4z = 7$$
$$-x + 5y + 3z = \frac{11}{2}$$
$$3x - 2y - 7z = 2$$

SOLVING LINEAR SYSTEMS WITH SUBSTITUTION

Exercises 13–18: (Refer to Example 4.) Use substitution to solve the system of linear equations. Check your solution.

13. $$x + y - z = 1$$
$$2y + z = -1$$
$$z = 1$$

14. $$2x + y - 3z = 1$$
$$y + 4z = 0$$
$$z = -1$$

15. $$-x - 3y + z = -2$$
$$2y + 3z = 3$$
$$z = 2$$

16. $$3x + 2y - 3z = -4$$
$$-y + 2z = 4$$
$$z = 0$$

17. $$a - b + 2c = 3$$
$$-3b + c = 4$$
$$c = -2$$

18. $$5a + 2b - 3c = 10$$
$$5b - 2c = -4$$
$$c = 3$$

SOLVING LINEAR SYSTEMS

Exercises 19–44: Solve the system, if possible.

19. $$x + y - z = 11$$
$$-x + 2y + 3z = -1$$
$$2z = 4$$

20. $$x + 2y - 3z = -7$$
$$-2x + y + z = -1$$
$$3z = 9$$

21. $$x + y - z = -2$$
$$-x + z = 1$$
$$y + 2z = 3$$

22. $$x + y - 3z = 11$$
$$-2x + y + 2z = 1$$
$$-3y + 3z = -21$$

23. $$x + y - 2z = -7$$
$$y + z = -1$$
$$-y + 3z = 9$$

24. $$2x + 3y + z = 5$$
$$y + 2z = 4$$
$$-2y + z = 2$$

25. $$x + 2y + 2z = 1$$
$$x + y + z = 0$$
$$-x - 2y + 3z = -11$$

26. $\begin{aligned} x + y - z &= 0 \\ x - 3y + z &= -2 \\ x - y + 3z &= 8 \end{aligned}$

27. $\begin{aligned} x + y + z &= 5 \\ y + z &= 6 \\ x + z &= 3 \end{aligned}$

28. $\begin{aligned} x + y + z &= 0 \\ x - y - z &= 6 \\ -x + y - z &= 4 \end{aligned}$

29. $\begin{aligned} x + 2y + 3z &= 24 \\ -x + y + 2z &= 1 \\ x + y - 2z &= 9 \end{aligned}$

30. $\begin{aligned} 5x - 15y + z &= 22 \\ -10x + 12y - 2z &= -8 \\ 4x - 2y - 3z &= 9 \end{aligned}$

31. $\begin{aligned} x + y + z &= 2 \\ x - y + z &= 1 \\ x + z &= 3 \end{aligned}$

32. $\begin{aligned} 4x - y + 3z &= 3 \\ 2x + y + z &= 2 \\ x - y + z &= 1 \end{aligned}$

33. $\begin{aligned} x + y + z &= 6 \\ x - y + z &= 2 \\ -x + 5y - z &= 6 \end{aligned}$

34. $\begin{aligned} x - y + z &= 3 \\ 2x - y + z &= 2 \\ -x - y + z &= 5 \end{aligned}$

35. $\begin{aligned} 2x + y + z &= 3 \\ 2x - y - z &= 9 \\ x + y - z &= 0 \end{aligned}$

36. $\begin{aligned} x + 3y + z &= -8 \\ x - 2y &= 11 \\ 2y - z &= -16 \end{aligned}$

37. $\begin{aligned} 2x + 6y - 2z &= 47 \\ 2x + y + 3z &= -28 \\ -x + y + z &= -\tfrac{7}{2} \end{aligned}$

38. $\begin{aligned} x + y + 2z &= 23 \\ 3x - y + 3z &= 8 \\ 2x + 2y + z &= 13 \end{aligned}$

39. $\begin{aligned} x + 3y - 4z &= \tfrac{13}{2} \\ -2x + 3y - z &= \tfrac{1}{2} \\ 3x + z &= 4 \end{aligned}$

40. $\begin{aligned} x - 2y + z &= \tfrac{9}{2} \\ 4x - y + 3z &= 9 \\ x + 2y &= -\tfrac{3}{2} \end{aligned}$

41. $\begin{aligned} 2x - 2y + z &= 4 \\ x - y + z &= 1 \\ x - y + z &= 3 \end{aligned}$

42. $\begin{aligned} x - 2y + 3z &= 1 \\ x + 3y + z &= 2 \\ 3x + 4y + 5z &= 7 \end{aligned}$

43. $\begin{aligned} x + y + z &= 5 \\ x - y + z &= 3 \\ 2x + y + 2z &= 9 \end{aligned}$

44. $\begin{aligned} x + y - 2z &= 0 \\ x + 2y - 4z &= -1 \\ y - 2z &= -1 \end{aligned}$

Exercises 45 and 46: **Thinking Generally** *Solve the system of linear equations, if $a \neq 0$.*

45. $\begin{aligned} x + y + z &= a \\ x + y + z &= 2a \\ -x + y + z &= 0 \end{aligned}$

46. $\begin{aligned} x + y + z &= a \\ -x - y - z &= -a \\ y - z &= 0 \end{aligned}$

APPLICATIONS OF LINEAR SYSTEMS

47. Finding Costs The accompanying table shows the costs of purchasing different combinations of hamburgers, fries, and soft drinks.

Hamburgers	Fries	Soft Drinks	Total Cost
1	2	4	$10
1	4	6	$15
0	3	2	$6

(a) Let x be the cost of a hamburger, y the cost of fries, and z the cost of a soft drink. Write a system of three linear equations that represents the data in the table.

(b) Solve the system and interpret your answer.

48. Cost of CDs The accompanying table shows the total cost of purchasing combinations of differently priced CDs. The types of CDs are labeled A, B, and C.

A	B	C	Total Cost
1	1	1	$37
3	2	1	$69
1	1	4	$82

(a) Let x be the cost of a CD of type A, y the cost of a CD of type B, and z the cost of a CD of type C. Write a system of three linear equations that represents the data in the table.

(b) Solve the system and interpret your answer.

49. Geometry The largest angle in a triangle is 55° more than the smallest angle. The sum of the measures of the two smaller angles is 10° more than the measure of the largest angle.

(a) Let x, y, and z be the measures of the three angles from largest to smallest. Write a system of three linear equations whose solution gives the measure of each angle.

(b) Solve the system.

(c) Check your solution.

50. Geometry The perimeter of a triangle is 90 inches. The longest side is 20 inches longer than the shortest side and 10 inches longer than the remaining side.

(a) Let x, y, and z be the lengths of the three sides from largest to smallest. Write a system of three linear equations whose solution gives the lengths of each side.

(b) Solve the system.

(c) Check your solution.

51. Geometry (Refer to Example 2.) Solve the system to find the measure of each angle.

$$\begin{aligned} x + y + z &= 180 \\ x - y - z &= 40 \\ x - z &= 90 \end{aligned}$$

52. Loan Mixture A student takes out a total of $5000 in three loans: one subsidized, one unsubsidized, and one from the parents of the student. The subsidized loan is $200 more than the combined total of the unsubsidized and parent loans. The unsubsidized loan is twice the amount of the parent loan. Find the amount of each loan.

53. Predicting Fawns (Refer to Examples 3 and 7.) The accompanying table shows counts for fawns and adult deer and the severity of the winter. These data may be modeled by the equation $F = a + bA + cW$.

Fawns (F)	Adults (A)	Winter (W)
525	600	4
365	400	2
805	900	5
?	500	3

(a) Use the first three rows to write a system of three linear equations in three variables whose solution gives values for a, b, and c.
(b) Solve the system.
(c) Predict the number of fawns when there are 500 adults and the winter has severity 3.

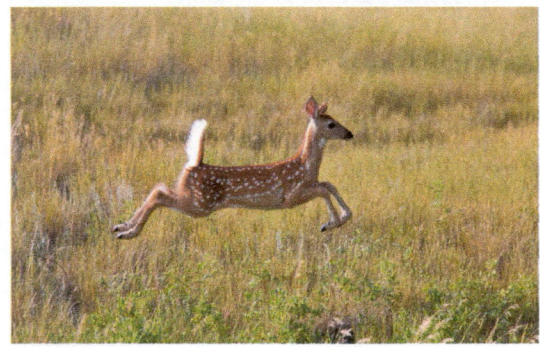

54. Business Production A business has three machines that manufacture containers. Together, they make 100 containers per day, whereas the two fastest machines together can make 80 containers per day. The fastest machine makes 34 more containers per day than the slowest machine.
(a) Let x, y, and z be the number of containers that the machines make from fastest to slowest. Write a system of three equations whose solution gives the number of containers that each machine can make.
(b) Solve the system.

55. Mixture Problem One type of lawn fertilizer consists of a mixture of nitrogen, N, phosphorus, P, and potassium, K. An 80-pound sample contains 8 more pounds of nitrogen and phosphorus than potassium. There is 9 times as much potassium as phosphorus.
(a) Write a system of three equations whose solution gives the amount of nitrogen, phosphorus, and potassium in this sample.
(b) Solve the system.

56. Predicting Home Prices Selling prices of homes can depend on several factors such as size and age. The accompanying table shows the selling price for three homes. In this table, price P is given in thousands of dollars, age A in years, and home size S in thousands of square feet. These data may be modeled by the equation $P = a + bA + cS$.

Price (P)	Age (A)	Size (S)
190	20	2
320	5	3
50	40	1
?	10	2.5

(a) Write a system of linear equations whose solution gives a, b, and c.
(b) Solve the system.
(c) Predict the price of a home that is 10 years old and has 2500 square feet.

57. Investment Mixture A sum of $30,000 was invested in three mutual funds. In one year, the first fund grew by 4%, the second by 5%, and the third by 7.5%. Total earnings were $1775. The amount invested in the third fund was $2000 less than the combined amount invested in the other two funds. Use a linear system of equations to determine the amount invested in each fund.

58. Football Tickets A total of 2500 tickets to a football game were sold. Prices were $2 for children, $3 for students, and $5 for adults. Twice as many tickets were sold to students as to children, and ticket revenues were $7250. Use a system of linear equations to determine how many of each type of ticket were sold.

WRITING ABOUT MATHEMATICS

59. Earlier in this text, we solved problems with two variables; to obtain a unique solution, we needed two linear equations. In this section, we solved problems with three variables; to obtain a unique solution, we needed three equations. Try to generalize these results. In the design of aircraft, problems commonly involve 100,000 variables. How many equations are required to solve such problems? Can such problems be solved by hand? If not, how are such problems solved? Explain your answers.

60. In Exercise 56, the price of a home was estimated by its age and size. What other factors might affect the price of a home? Explain how these factors might affect the number of variables and equations in the linear system.

9.2 Matrix Solutions of Linear Systems

Objectives

1 Representing Systems of Linear Equations with Matrices

2 Relating Matrices and Social Networks

3 Applying Gauss–Jordan Elimination

4 Using Technology to Solve Systems of Linear Equations

NEW VOCABULARY

☐ Matrix
☐ Element
☐ Dimension of a matrix
☐ Square matrix
☐ Augmented matrix
☐ Main diagonal
☐ Reduced row–echelon form
☐ Gauss–Jordan elimination
☐ Matrix row transformations

1 Representing Systems of Linear Equations with Matrices

Arrays of numbers are used frequently in many different real-world situations. Spreadsheets often make use of arrays. A **matrix** is a rectangular array of numbers. Each number in a matrix is called an **element**. The following are examples of *matrices* (plural of matrix), with their dimensions written below them.

Matrices and Their Dimensions

$$\begin{bmatrix} 2 & 0 \\ 3 & 1 \end{bmatrix} \quad \begin{bmatrix} -1.2 & 5 & 0 \\ 1 & 0 & 1 \\ 4 & -5 & 7 \end{bmatrix} \quad \begin{bmatrix} 3 & -6 & 0 & \sqrt{3} \\ 1 & 4 & 0 & 9 \\ -3 & 1 & 1 & 18 \\ -10 & -4 & 5 & -1 \end{bmatrix} \quad \begin{bmatrix} 4 & 2 \\ 0 & 1 \\ 1 & 0 \end{bmatrix} \quad \begin{bmatrix} 1 & 5 & -1 \\ 3 & 4 & 2 \end{bmatrix}$$

2×2 3×3 4×4 3×2 2×3

Rows × Columns

Connecting Concepts with Your Life The dimension of a matrix is stated much like the dimensions of a rectangular room. We might say that a room is *m* feet long and *n* feet wide. Similarly, the **dimension of a matrix** is $m \times n$ (*m* by *n*) if it has *m* rows and *n* columns. For example, the last matrix in the preceding group has a dimension of 2×3 because it has 2 rows and 3 columns. If the number of rows equals the number of columns, the matrix is a **square matrix**. The first three matrices in that group are square matrices.

Matrices can be used to represent a system of linear equations.

See the Concept 1   **REPRESENTING A LINEAR SYSTEM WITH A MATRIX**

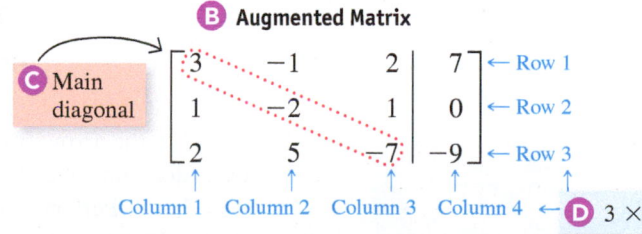

A System of Three Equations

$$\begin{aligned} 3x - y + 2z &= 7 \\ x - 2y + z &= 0 \\ 2x + 5y - 7z &= -9 \end{aligned}$$

C Main diagonal

B Augmented Matrix

$$\begin{bmatrix} 3 & -1 & 2 & 7 \\ 1 & -2 & 1 & 0 \\ 2 & 5 & -7 & -9 \end{bmatrix}$$ ← Row 1 / ← Row 2 / ← Row 3

Column 1 Column 2 Column 3 Column 4   **D** 3×4

A Write the system of equations as shown, with all variables to the left of the equals sign, constants to the right of the equals sign, and the matching variables aligned vertically in columns.

B The coefficients and constants are represented by the **augmented matrix**. The vertical line in the matrix corresponds to the position of the equals sign in each equation.

C The elements of the **main diagonal** of the augmented matrix are circled.

D The matrix has 3 rows and 4 columns, so its dimension is 3×4.

WATCH VIDEO IN MML.

EXAMPLE 1 **Representing a linear system**

Represent each linear system with an augmented matrix. State the dimension of the matrix.

(a) $x - 2y = 9$
$6x + 7y = 16$

(b) $x - 3y + 7z = 4$
$2x + 5y - z = 15$
$2x + y = 8$

Solution

(a) This system can be represented by the following 2×3 augmented matrix.

2 rows
3 columns
$$\left[\begin{array}{cc|c} 1 & -2 & 9 \\ 6 & 7 & 16 \end{array}\right]$$

(b) This system can be represented by the following 3×4 augmented matrix.

3 rows
4 columns
$$\left[\begin{array}{ccc|c} 1 & -3 & 7 & 4 \\ 2 & 5 & -1 & 15 \\ 2 & 1 & 0 & 8 \end{array}\right]$$

Now Try Exercises 11, 13

2 ▶ Relating Matrices and Social Networks

Social Networking
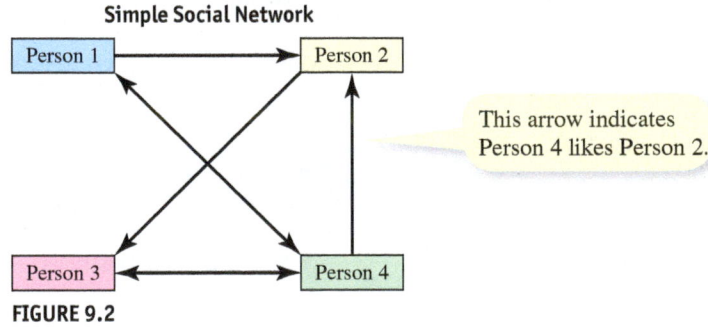 Math in Context   Today, there are several different types of online social networks, such as Facebook and Twitter. Mathematics is essential to the success of these social networks. Consider **FIGURE 9.2**, which represents a simple social network of four people.

STUDY TIP

Matrices are very important because they have a wide variety of applications in the real world.

Simple Social Network

| Person 1 | → | Person 2 |

This arrow indicates Person 4 likes Person 2.

| Person 3 | ↔ | Person 4 |

FIGURE 9.2

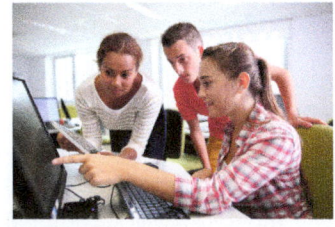

In **FIGURE 9.2**, the arrow from Person 1 to Person 2 indicates that Person 1 likes Person 2, but Person 2 does not like Person 1 because there is no arrow from Person 2 to Person 1. On the other hand, Person 3 and Person 4 like each other because there is a double arrow between them. In the next example, we can see how a matrix can be used to represent this simple social network.

EXAMPLE 2 **Representing social networks**

Use a matrix to model the social network shown in **FIGURE 9.2**.

Solution

To model a social network with four people, we use a square 4×4 matrix, where rows 1, 2, 3, and 4 represents the "likes" of each person, respectively. In this matrix, a 1 represents a connection (or "like") and a 0 represents no connection. Because Person 1 likes Person 2, we put a **1** in row 1 column 2. Because Person 1 also likes Person 4, we put a **1** in row 1 column 4. Continuing in the same manner, we arrive at the following matrix.

A Social Network

$$\begin{bmatrix} 0 & 1 & 0 & 1 \\ 0 & 0 & 1 & 0 \\ 0 & 0 & 0 & 1 \\ 1 & 1 & 1 & 0 \end{bmatrix}$$

← Person 1's "likes"
← Person 2's "likes"
← Person 3's "likes"
← Person 4's "likes"

Person 4 "likes"
Person 1

Now Try Exercise 21

3 ▸ Applying Gauss–Jordan Elimination

A convenient matrix form for representing a system of linear equations is **reduced row–echelon form**. The following augmented matrices are examples of reduced row–echelon form. Note that there are 1s on the main diagonal with 0s above and below the 1s.

Reduced Row–Echelon Form

$$\begin{bmatrix} 1 & 0 & \big| & 3 \\ 0 & 1 & \big| & -2 \end{bmatrix} \quad \begin{bmatrix} 1 & 0 & 0 & \big| & 3 \\ 0 & 1 & 0 & \big| & 1 \\ 0 & 0 & 1 & \big| & -1 \end{bmatrix} \quad \begin{bmatrix} 1 & 0 & 0 & \big| & 8 \\ 0 & 1 & 0 & \big| & 2 \\ 0 & 0 & 1 & \big| & 3 \end{bmatrix}$$

If an augmented matrix representing a linear system is in reduced row–echelon form, we can usually determine the solution easily.

EXAMPLE 3 **Determining a solution from reduced row–echelon form**

Each matrix represents a system of linear equations. Find the solution.

(a) $\begin{bmatrix} 1 & 0 & 0 & \big| & 2 \\ 0 & 1 & 0 & \big| & -3 \\ 0 & 0 & 1 & \big| & 5 \end{bmatrix}$ (b) $\begin{bmatrix} 1 & 0 & \big| & 10 \\ 0 & 1 & \big| & -4 \end{bmatrix}$

Solution
(a) The top row **1 0 0 2** represents $1x + 0y + 0z = 2$ or $x = 2$. The second and third rows tell us that $y = -3$ and $z = 5$. The solution is $(2, -3, 5)$.
(b) The system involves two equations in two variables. The solution is $(10, -4)$.

Now Try Exercise 19

In the previous section, we solved systems of three linear equations in three variables by using elimination and substitution. In real life, systems of equations often contain thousands of variables. To solve a large system of equations, we need an efficient method. Long before the invention of the computer, Carl Friedrich Gauss (1777–1855) developed a method called *Gaussian elimination* to solve systems of linear equations. Even though it was developed more than 150 years ago, it is still used today in modern computers and calculators.

We will use a numerical method called **Gauss–Jordan elimination** to solve a linear system. It makes use of the following **matrix row transformations**.

MATRIX ROW TRANSFORMATIONS

READING CHECK 1

• Why do we use matrix row transformations?

For any augmented matrix representing a system of linear equations, the following row transformations result in an equivalent system of linear equations.

1. Any two rows may be interchanged.
2. The elements of any row may be multiplied by a nonzero constant.
3. Any row may be changed by adding to (or subtracting from) its elements a nonzero multiple of the corresponding elements of another row.

Gauss–Jordan elimination can be used to transform an augmented matrix into reduced row–echelon form. Its objective is to use the matrix row transformations to obtain a matrix that has the following reduced row–echelon form, where (a, b) represents the solution.

$$\begin{bmatrix} 1 & 0 & | & a \\ 0 & 1 & | & b \end{bmatrix}$$

This method is illustrated in the next example.

EXAMPLE 4 **Transforming a matrix into reduced row–echelon form**

Use Gauss–Jordan elimination to transform the augmented matrix of the linear system into reduced row–echelon form. Find the solution.

$$x + y = 5$$
$$-x + y = 1$$

Solution
Both the linear system and the augmented matrix are shown.

Linear System

$$x + y = 5$$
$$-x + y = 1$$

Augmented Matrix

$$\begin{bmatrix} 1 & 1 & | & 5 \\ -1 & 1 & | & 1 \end{bmatrix}$$

First, we want to obtain a 0 in the second row, where the -1 is highlighted. To do so, we add row 1 to row 2 and place the result in row 2. This step is denoted $R_2 + R_1$ and eliminates the x-variable from the second equation.

$$x + \ y = 5$$
$$2y = 6 \qquad R_2 + R_1 \rightarrow \begin{bmatrix} 1 & 1 & | & 5 \\ 0 & 2 & | & 6 \end{bmatrix}$$

$$\begin{array}{r} 1\ 1\ 5 \\ -1\ 1\ 1 \\ \hline 0\ 2\ 6 \ \text{Add.} \end{array}$$

To obtain a 1 where the 2 in the second row is located, we multiply the second row by $\frac{1}{2}$, denoted $\frac{1}{2}R_2$.

$$x + y = 5$$
$$y = 3 \qquad \frac{1}{2}R_2 \rightarrow \begin{bmatrix} 1 & 1 & | & 5 \\ 0 & 1 & | & 3 \end{bmatrix}$$

Half of
0 2 6 is
0 1 3.

Next, we need to obtain a 0 where the 1 is highlighted. We do so by subtracting row 2 from row 1 and placing the result in row 1, denoted $R_1 - R_2$.

$$x = 2 \qquad R_1 - R_2 \rightarrow \begin{bmatrix} 1 & 0 & | & 2 \\ 0 & 1 & | & 3 \end{bmatrix}$$
$$y = 3$$

$$\begin{array}{r} 1\ 1\ 5 \\ 0\ 1\ 3 \\ \hline 1\ 0\ 2 \ \text{Subtract.} \end{array}$$

This matrix is in reduced row–echelon form. The solution is $(2, 3)$.

Now Try Exercise 25

NOTE: When using Gauss–Jordan elimination to solve a system, it is not necessary to write the system alongside the matrix solution. ■

In the next example, we use Gauss–Jordan elimination to solve a system with three linear equations and three variables. To do so, we transform the matrix into the following reduced row–echelon form, where (a, b, c) represents the solution.

$$\begin{bmatrix} 1 & 0 & 0 & | & a \\ 0 & 1 & 0 & | & b \\ 0 & 0 & 1 & | & c \end{bmatrix}$$

EXAMPLE 5   **Transforming a matrix into reduced row–echelon form**

Use Gauss–Jordan elimination to transform the augmented matrix of the linear system into reduced row–echelon form. Find the solution.

$$x + y + 2z = 1$$
$$-x + \quad z = -2$$
$$2x + y + 5z = -1$$

Solution

The linear system and the augmented matrix are both shown.

Linear System	Augmented Matrix

$$x + y + 2z = 1$$
$$-x + \quad z = -2$$
$$2x + y + 5z = -1$$

$$\begin{bmatrix} 1 & 1 & 2 & | & 1 \\ -1 & 0 & 1 & | & -2 \\ 2 & 1 & 5 & | & -1 \end{bmatrix}$$

First, we want to put 0s in the second and third rows, where the -1 and 2 are highlighted. To obtain a 0 in the first position of the second row, we add row 1 to row 2 and place the result in row 2, denoted $R_2 + R_1$. To obtain a 0 in the first position of the third row, we subtract 2 times row 1 from row 3 and place the result in row 3, denoted $R_3 - 2R_1$. Row 1 does not change. These steps eliminate the x-variable from the second and third equations.

$$x + y + 2z = 1$$
$$y + 3z = -1 \qquad R_2 + R_1 \rightarrow$$
$$-y + z = -3 \qquad R_3 - 2R_1 \rightarrow$$

$$\begin{bmatrix} 1 & 1 & 2 & | & 1 \\ 0 & 1 & 3 & | & -1 \\ 0 & -1 & 1 & | & -3 \end{bmatrix}$$

To eliminate the y-variable in row 1, we subtract row 2 from row 1. To eliminate the y-variable from row 3, we add row 2 to row 3.

$$x \quad - z = 2 \qquad R_1 - R_2 \rightarrow$$
$$y + 3z = -1$$
$$4z = -4 \qquad R_3 + R_2 \rightarrow$$

$$\begin{bmatrix} 1 & 0 & -1 & | & 2 \\ 0 & 1 & 3 & | & -1 \\ 0 & 0 & 4 & | & -4 \end{bmatrix}$$

To obtain a 1 in row 3, where the highlighted 4 is located, we multiply row 3 by $\frac{1}{4}$.

$$x \quad - z = 2$$
$$y + 3z = -1$$
$$z = -1 \qquad \tfrac{1}{4}R_3 \rightarrow$$

$$\begin{bmatrix} 1 & 0 & -1 & | & 2 \\ 0 & 1 & 3 & | & -1 \\ 0 & 0 & 1 & | & -1 \end{bmatrix}$$

For the matrix to be in reduced row–echelon form, we need 0s in the highlighted locations. We first add row 3 to row 1 and then subtract 3 times row 3 from row 2.

$$x = 1 \qquad R_1 + R_3 \rightarrow$$
$$y = 2 \qquad R_2 - 3R_3 \rightarrow$$
$$z = -1$$

$$\begin{bmatrix} 1 & 0 & 0 & | & 1 \\ 0 & 1 & 0 & | & 2 \\ 0 & 0 & 1 & | & -1 \end{bmatrix}$$

This matrix is now in reduced row–echelon form. The solution is $(1, 2, -1)$.

Now Try Exercise 35

CRITICAL THINKING 1

An *inconsistent* system of linear equations has no solutions, and a system of *dependent* linear equations has infinitely many solutions. Suppose that an augmented matrix row reduces to either of the following matrices. Explain what each matrix indicates about the given system of linear equations.

$$\begin{bmatrix} 1 & 0 & 0 & | & 2 \\ 0 & 1 & 0 & | & 3 \\ 0 & 0 & 0 & | & 1 \end{bmatrix} \qquad \begin{bmatrix} 1 & 0 & 1 & | & 2 \\ 0 & 1 & 2 & | & 3 \\ 0 & 0 & 0 & | & 0 \end{bmatrix}$$

In the next example, we find the amounts invested in three mutual funds.

EXAMPLE 6 **Determining investment amounts in mutual funds**

A total of \$8000 was invested in three funds that grew at a rate of 5%, 10%, and 20% over 1 year. After 1 year, the combined value of the three funds had grown by \$1200. Five times as much money was invested at 20% as at 10%. Find the amount invested in each fund.

Solution

Let x be the amount invested at 5%, y be the amount invested at 10%, and z be the amount invested at 20%. The total amount invested was \$8000, so

$$x + y + z = 8000.$$

The growth in the first mutual fund, paying 5% of x, is given by $0.05x$. Similarly, the growths in the other mutual funds are given by $0.10y$ and $0.20z$. As the total growth was \$1200, we can write

$$0.05x + 0.10y + 0.20z = 1200.$$

Multiplying each side of this equation by 20 to eliminate decimals results in

$$x + 2y + 4z = 24,000.$$

Five times as much was invested at 20% as at 10%, so $z = 5y$, or $5y - z = 0$.

These three equations can be written as a system of linear equations and as an augmented matrix.

Linear System	Augmented Matrix

$$
\begin{array}{rrrr}
x + & y + & z = & 8{,}000 \\
x + & 2y + & 4z = & 24{,}000 \\
& 5y - & z = & 0
\end{array}
\qquad
\left[\begin{array}{rrr|r}
1 & 1 & 1 & 8{,}000 \\
1 & 2 & 4 & 24{,}000 \\
0 & 5 & -1 & 0
\end{array}\right]
$$

A 0 can be obtained in the highlighted position by subtracting row 1 from row 2.

$$
\begin{array}{rrrr}
x + & y + & z = & 8{,}000 \\
& y + & 3z = & 16{,}000 \\
& 5y - & z = & 0
\end{array}
\qquad
R_2 - R_1 \rightarrow
\left[\begin{array}{rrr|r}
1 & 1 & 1 & 8{,}000 \\
0 & 1 & 3 & 16{,}000 \\
0 & 5 & -1 & 0
\end{array}\right]
$$

Zeros can be obtained in the highlighted positions by subtracting row 2 from row 1 and by subtracting 5 times row 2 from row 3.

$$
\begin{array}{rrrr}
x & - & 2z = & -8{,}000 \\
& y + & 3z = & 16{,}000 \\
& & -16z = & -80{,}000
\end{array}
\quad
\begin{array}{l}
R_1 - R_2 \rightarrow \\
\\
R_3 - 5R_2 \rightarrow
\end{array}
\left[\begin{array}{rrr|r}
1 & 0 & -2 & -8{,}000 \\
0 & 1 & 3 & 16{,}000 \\
0 & 0 & -16 & -80{,}000
\end{array}\right]
$$

To obtain a 1 in the highlighted position, multiply row 3 by $-\frac{1}{16}$.

$$
\begin{array}{rrrr}
x & - & 2z = & -8{,}000 \\
& y + & 3z = & 16{,}000 \\
& & z = & 5{,}000
\end{array}
\qquad
-\frac{1}{16}R_3 \rightarrow
\left[\begin{array}{rrr|r}
1 & 0 & -2 & -8{,}000 \\
0 & 1 & 3 & 16{,}000 \\
0 & 0 & 1 & 5{,}000
\end{array}\right]
$$

To obtain a 0 in each of the highlighted positions, add twice row 3 to row 1 and subtract three times row 3 from row 2.

$$
\begin{array}{rr}
x = & 2{,}000 \\
y = & 1{,}000 \\
z = & 5{,}000
\end{array}
\quad
\begin{array}{l}
R_1 + 2R_3 \rightarrow \\
R_2 - 3R_3 \rightarrow \\
\\
\end{array}
\left[\begin{array}{rrr|r}
1 & 0 & 0 & 2{,}000 \\
0 & 1 & 0 & 1{,}000 \\
0 & 0 & 1 & 5{,}000
\end{array}\right]
$$

Thus \$2000 was invested at 5%, \$1000 at 10%, and \$5000 at 20%.

Now Try Exercise 67

4 Using Technology to Solve Systems of Linear Equations

Examples 5 and 6 involve a lot of arithmetic. Trying to solve a large system of equations by hand is an enormous—if not impossible—task. In the real world, people use technology to solve large systems. In the next example, we solve the linear systems from Examples 4 and 5 with a graphing calculator.

EXAMPLE 7 **Using technology**

Use a graphing calculator to solve the following systems of equations.

(a) $x + y = 5$ (b) $x + y + 2z = 1$
$\;\, -x + y = 1$ $\;\, -x + + z = -2$
$\;\; 2x + y + 5z = -1$

Solution

(a) Enter the 2×3 augmented matrix from Example 4 in a graphing calculator, as shown in **FIGURE 9.3(a)**. Then transform the matrix into reduced row–echelon form (rref), as shown in **FIGURE 9.3(b)**. The solution is $(2, 3)$.

(a)　　　　　　　　(b)

FIGURE 9.3

(b) Enter the 3×4 augmented matrix from Example 5, as shown in **FIGURE 9.4(a)**. (The fourth column of A can be seen by scrolling right.) Then transform the matrix into reduced row–echelon form (rref), as shown in **FIGURE 9.4(b)**. The solution is $(1, 2, -1)$.

CALCULATOR HELP

To enter a matrix and put it in reduced row–echelon form, see Appendix A (pages AP-8 and AP-9).

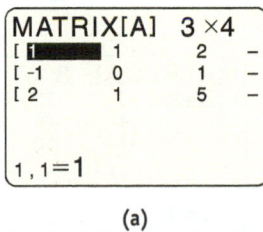

(a)　　　　　　　　(b)

FIGURE 9.4

Now Try Exercises 43, 45

Math in Context (Biology) Suppose that the size of a bear's head and its overall length are known. Can its weight be estimated from these variables? In the next example, we show that a system of linear equations can be used to make this estimate.

EXAMPLE 8 **Modeling the weight of male bears**

The data shown in **TABLE 9.2** on the next page give the weight W, head length H, and overall length L of three bears. These data can be modeled with the equation $W = a + bH + cL$, where a, b, and c are constants that we need to determine. (*Sources:* M. Triola, *Elementary Statistics;* Minitab, Inc.)

Bear Measurements

W (pounds)	H (inches)	L (inches)
362	16	72
300	14	68
147	11	52
?	13	65

TABLE 9.2

(a) Set up a system of equations whose solution gives values for constants a, b, and c.

(b) Solve the system.

(c) Predict the weight of a bear with $H = 13$ inches and $L = 65$ inches.

Solution

(a) Substitute each row of **TABLE 9.2** in the equation $W = a + bH + cL$.

$$362 = a + b(16) + c(72)$$
$$300 = a + b(14) + c(68)$$
$$147 = a + b(11) + c(52)$$

Rewrite this system as

$$a + 16b + 72c = 362$$
$$a + 14b + 68c = 300$$
$$a + 11b + 52c = 147$$

and represent it as the following augmented matrix.

$$A = \begin{bmatrix} 1 & 16 & 72 & | & 362 \\ 1 & 14 & 68 & | & 300 \\ 1 & 11 & 52 & | & 147 \end{bmatrix}$$

(b) Enter A and put it in reduced row–echelon form, as shown in **FIGURES 9.5(a)** and **(b)**, respectively. The solution is $a = -374$, $b = 19$, and $c = 6$.

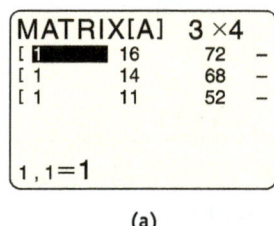

(a)

FIGURE 9.5

(c) For $W = a + bH + cL$, use

$$W = -374 + 19H + 6L$$

to predict the weight of a bear with head length $H = 13$ and overall length $L = 65$.

$$W = -374 + 19(13) + 6(65) = 263 \text{ pounds}$$

Now Try Exercise 63

9.2   Putting It All Together

A matrix is a rectangular array of numbers. An augmented matrix may be used to represent any system of linear equations. One common method for solving a system of linear equations is Gauss–Jordan elimination. Matrix row operations may be used to transform an augmented matrix to reduced row–echelon form.

CONCEPT	EXPLANATION/EXAMPLE
Augmented Matrix	A linear system can be represented by an augmented matrix. The following matrix has dimension 3×4. **Linear System** $x + 2y - z = 6$ $-2x + y - z = 7$ $2x + 3z = -11$ **Augmented Matrix** $\begin{bmatrix} 1 & 2 & -1 & 6 \\ -2 & 1 & -1 & 7 \\ 2 & 0 & 3 & -11 \end{bmatrix}$
Reduced Row–Echelon Form	The following augmented matrix is in reduced row–echelon form, which results from transforming the preceding system to reduced row–echelon form. There are 1s along the main diagonal and 0s elsewhere in the first three columns. The solution to the linear system is $(-1, 2, -3)$. $\begin{bmatrix} 1 & 0 & 0 & -1 \\ 0 & 1 & 0 & 2 \\ 0 & 0 & 1 & -3 \end{bmatrix}$

9.2   Exercises

MyMathLab®

CONCEPTS AND VOCABULARY

1. What is a matrix?

2. Give an example of a matrix and state its dimension.

3. Give an example of an augmented matrix and state its dimension.

4. If an augmented matrix is used to solve a system of three linear equations in three variables, what will be its dimension?

5. Give an example of a matrix that is in reduced row–echelon form.

6. Identify the elements on the main diagonal in the augmented matrix.

$$\begin{bmatrix} 4 & -6 & -1 & 3 \\ 6 & 2 & -2 & 9 \\ 7 & 5 & -3 & 1 \end{bmatrix}$$

REPRESENTING SYSTEMS WITH MATRICES

Exercises 7–10: State the dimension of the matrix.

7. $\begin{bmatrix} 3 & -3 & 7 \\ 2 & 6 & -2 \\ 4 & 2 & 5 \end{bmatrix}$
8. $\begin{bmatrix} -2 & 3 & 0 \\ 1 & -8 & 4 \end{bmatrix}$

9. $\begin{bmatrix} 1 & 7 \\ 0 & 2 \\ 2 & -5 \end{bmatrix}$
10. $\begin{bmatrix} 4 & 2 & -3 & -1 \\ 4 & -3 & 2 & -7 \\ 14 & 6 & 4 & 0 \end{bmatrix}$

Exercises 11–14: Represent the linear system as an augmented matrix.

11. $x - 3y = 1$
$-x + 3y = -1$

12. $4x + 2y = -5$
$5x + 8y = 2$

13. $2x - y + 2z = -4$
$x - 2y = 2$
$-x + y - 2z = -6$

14. $3x - 2y + z = 5$
$-x + 2z = -4$
$x - 2y + z = -1$

Exercises 15–20: Write the system of linear equations that the augmented matrix represents. Use the variables x, y, and z.

15. $\begin{bmatrix} 1 & 2 & | & -6 \\ 5 & -1 & | & 4 \end{bmatrix}$ **16.** $\begin{bmatrix} 1 & -5 & | & 7 \\ 0 & -3 & | & 6 \end{bmatrix}$

17. $\begin{bmatrix} 1 & -1 & 2 & | & 6 \\ 2 & 1 & -2 & | & 1 \\ -1 & 2 & -1 & | & 3 \end{bmatrix}$ **18.** $\begin{bmatrix} 3 & -1 & 2 & | & -1 \\ 2 & -2 & 2 & | & 4 \\ 1 & 7 & -2 & | & 2 \end{bmatrix}$

19. $\begin{bmatrix} 1 & 0 & 0 & | & 4 \\ 0 & 1 & 0 & | & -2 \\ 0 & 0 & 1 & | & 7 \end{bmatrix}$ **20.** $\begin{bmatrix} 1 & 0 & 0 & | & 6 \\ 0 & 1 & 0 & | & -2 \\ 0 & 0 & 1 & | & 4 \end{bmatrix}$

RELATING MATRICES AND SOCIAL NETWORKS

Exercises 21 and 24: (Refer to Example 2.) Use a 4 × 4 matrix to represent the social network.

21.

22.

23.

24.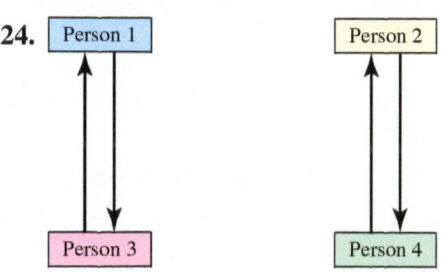

APPLYING GAUSS–JORDAN ELIMINATION

Exercises 25–42: Use Gauss–Jordan elimination to find the solution. Write the solution as an ordered pair or ordered triple and check the solution.

25. $\begin{aligned} x + y &= 4 \\ x + 3y &= 10 \end{aligned}$ **26.** $\begin{aligned} x - 3y &= -7 \\ 2x + y &= 0 \end{aligned}$

27. $\begin{aligned} 2x + 3y &= 3 \\ -2x + 2y &= 7 \end{aligned}$ **28.** $\begin{aligned} x + 3y &= -14 \\ 2x + 5y &= -24 \end{aligned}$

29. $\begin{aligned} x - y &= 5 \\ x + 3y &= -1 \end{aligned}$ **30.** $\begin{aligned} x + 4y &= 1 \\ 3x - 2y &= 10 \end{aligned}$

31. $\begin{aligned} 4x - 8y &= -10 \\ x + y &= 2 \end{aligned}$ **32.** $\begin{aligned} x - 7y &= -16 \\ 4x + 10y &= 50 \end{aligned}$

33. $\begin{aligned} x + y + z &= 6 \\ 2y - z &= 1 \\ y + z &= 5 \end{aligned}$ **34.** $\begin{aligned} x + y + z &= 3 \\ x + y - z &= 2 \\ y + z &= 2 \end{aligned}$

35. $\begin{aligned} x + 2y + 3z &= 6 \\ -x + 3y + 4z &= 0 \\ x + y - 2z &= -6 \end{aligned}$

36. $\begin{aligned} 2x - 4y + 2z &= 10 \\ -x + 3y - 4z &= -19 \\ 2x - y - 6z &= -28 \end{aligned}$

37. $\begin{aligned} x + y + z &= 0 \\ 2x + y + 2z &= -1 \\ x + y &= 0 \end{aligned}$ **38.** $\begin{aligned} x + y - 2z &= 5 \\ x + 2y - 2z &= 4 \\ -x - y + z &= -4 \end{aligned}$

39. $\begin{aligned} x + y + z &= 3 \\ -x - z &= -2 \\ x + y + 2z &= 4 \end{aligned}$ **40.** $\begin{aligned} x + 2y - z &= 3 \\ -x - y + z &= 0 \\ x + 2y &= 5 \end{aligned}$

41. $\begin{aligned} x + 2y + z &= 3 \\ 2x + y - z &= -6 \\ -x - y + 2z &= 5 \end{aligned}$ **42.** $\begin{aligned} x + y + z &= -3 \\ x - y - z &= -1 \\ -2x + y + 4z &= 4 \end{aligned}$

USING TECHNOLOGY TO SOLVE SYSTEMS

Exercises 43–52: **Technology** *Use a graphing calculator to solve the system of linear equations.*

43. $\begin{aligned} x + 4y &= 13 \\ 5x - 3y &= -50 \end{aligned}$ **44.** $\begin{aligned} 9x - 11y &= 7 \\ 5x + 6y &= 16 \end{aligned}$

45. $\begin{aligned} 2x - y + 3z &= 9 \\ -4x + 5y + 2z &= 12 \\ 2x + 7z &= 23 \end{aligned}$

46. $\begin{aligned} 3x - 2y + 4z &= 29 \\ 2x + 3y - 7z &= -14 \\ 5x - y + 11z &= 59 \end{aligned}$

47.
$$\begin{aligned} 6x + 2y + z &= 4 \\ -2x + 4y + z &= -3 \\ 2x - 8y &= -2 \end{aligned}$$

48.
$$\begin{aligned} -x - 9y + 2z &= -28.5 \\ 2x - y + 4z &= -17 \\ x - y + 8z &= -9 \end{aligned}$$

49.
$$\begin{aligned} 4x + 3y + 12z &= -9.25 \\ 15y + 8z &= -4.75 + x \\ 7z &= -5.5 - 6y \end{aligned}$$

50.
$$\begin{aligned} 5x + 4y &= 13.3 + z \\ 7y + 9z &= 16.9 - x \\ x - 3y + 4z &= -4.1 \end{aligned}$$

51.
$$\begin{aligned} 1.2x - 0.9y + 2.7z &= 5.37 \\ 3.1x - 5.1y + 7.2z &= 14.81 \\ 1.8y + 6.38 &= 3.6z - 0.2x \end{aligned}$$

52.
$$\begin{aligned} 11x + 13y - 17z &= 380 \\ 5x - 14y - 19z &= 24 \\ -21y + 46z &= -676 + 7x \end{aligned}$$

Exercises 53–58: (Refer to Critical Thinking 1 in this section.) Row-reduce the matrix associated with the given system to determine whether the system of linear equations is inconsistent or has dependent equations.

53.
$$\begin{aligned} x + 2y &= 4 \\ -2x - 4y &= -8 \end{aligned}$$
54.
$$\begin{aligned} x - 5y &= 4 \\ -2x + 10y &= 8 \end{aligned}$$

55.
$$\begin{aligned} x + y + z &= 3 \\ x + y - z &= 1 \\ x + y &= 3 \end{aligned}$$
56.
$$\begin{aligned} x + y + z &= 5 \\ x - y - z &= 8 \\ 2x + 2y + 2z &= 6 \end{aligned}$$

57.
$$\begin{aligned} x + 2y + 3z &= 14 \\ 2x - 3y - 2z &= -10 \\ 3x - y + z &= 4 \end{aligned}$$
58.
$$\begin{aligned} x + 2y + 3z &= 6 \\ -x + 3y + 4z &= 6 \\ 5y + 7z &= 12 \end{aligned}$$

Exercises 59 and 60: **Thinking Generally** *The matrix represents a linear system of equations. Solve the system, if a and b are nonzero constants.*

59. $\begin{bmatrix} a & 0 & 0 & | & 1 \\ 0 & b & 0 & | & 1 \\ 0 & 0 & ab & | & 2 \end{bmatrix}$
60. $\begin{bmatrix} a & 0 & 0 & | & 1 \\ 0 & b & 0 & | & 2 \\ 0 & 0 & 0 & | & a \end{bmatrix}$

APPLICATIONS INVOLVING MATRICES

61. Weight of a Bear Use the results of Example 8 to estimate the weight of a bear with a head length of 12 inches and an overall length of 60 inches.

62. Garbage and Household Size A larger household produces more garbage, on average, than a smaller household. If we know the amount of metal M and plastic P waste produced each week, we can estimate the household size H from $H = a + bM + cP$. The table contains representative data for three households. (*Source:* M. Triola, *Elementary Statistics.*)

H (people)	M (pounds)	P (pounds)
3	2.00	1.40
2	1.50	0.65
6	4.00	3.40

(a) Set up a system of equations whose solution gives values for the constants a, b, and c.

(b) Solve this system.

(c) Predict the size of a household that produces 3 pounds of metal waste and 2 pounds of plastic waste each week.

63. Weight of a Bear (Refer to Example 8.) Head length and overall length are not the only variables that can be used to estimate the weight of a bear. The data in the accompanying table list the weight W, neck size N, and chest size C of three bears. These data can be modeled by $W = a + bN + cC$. (*Sources:* M. Triola, *Elementary Statistics*; Minitab, Inc.)

W (pounds)	N (inches)	C (inches)
80	16	26
344	28	45
416	31	54

(a) Set up a system of equations whose solution gives values for the constants a, b, and c.

(b) Solve this system. Round each value to the nearest tenth.

(c) Predict the weight of a bear with neck size $N = 22$ inches and chest size $C = 38$ inches.

64. Old Faithful Geyser In Yellowstone National Park, Old Faithful Geyser has been a favorite attraction for decades. Although this geyser erupts about every 80 minutes, this time interval varies, as do the duration and height of the eruptions. The accompanying table shows the height H, duration D, and time interval T for three representative eruptions. (*Source:* National Park Service.)

H (feet)	D (seconds)	T (minutes)
160	276	94
125	203	84
140	245	79

continued on next page

continued from previous page

(a) Assume that these data can be modeled by $H = a + bD + cT$. Set up a system of equations whose solution gives values for the constants a, b, and c.

(b) Solve this system. Round each value to the nearest thousandth.

(c) Use this equation to estimate H when $D = 220$ and $T = 81$.

65. Jogging Speeds A runner in preparation for a marathon jogs at 5, 6, and 8 miles per hour. The runner travels a total distance of 12.5 miles in 2 hours and jogs the same length of time at 5 miles per hour and at 8 miles per hour. How long does the runner jog at each speed?

66. Mixture Problem Three types of candy that cost $2, $3, and $4 per pound are to be mixed to produce a 5-pound bag of candy that costs $14.50. If there are to be equal amounts of the $3-per-pound candy and the $4-per-pound candy, how much of each type of candy should be included in the mixture?

67. Interest and Investments (Refer to Example 6.) A total of $3000 is invested at 3%, 4%, and 6% annual interest. The interest earned after 1 year equals $145. The amount invested at 6% is triple the amount invested at 3%. Find the amount invested at each rate.

68. Geometry The measure of the largest angle in a triangle is twice the measure of the smallest angle. The remaining angle is 10° less than the largest angle. Find the measure of each angle.

69. Online Exploration Go to your Facebook profile and select three or four of your friends plus yourself. Answers will vary.

(a) Draw a social network that shows the friendships between each pair of people. Use a double arrow to represent this. (See **FIGURE 9.2.**)

(b) Use a matrix to represent your social network.

70. Online Exploration Go online and look up a diagram of the molecular structure of water (H_2O).

(a) Draw double arrows wherever the hydrogen and oxygen are connected.

(b) Use a matrix to represent the structure of water. Let the first row represent oxygen.

SOLVING AN EQUATION IN FOUR VARIABLES

71. Weight of a Bear Earlier in this section, we estimated the weight of a bear by using three variables. We may be able to make more accurate estimates by using four variables. The accompanying table shows the weight W, neck size N, overall length L, and chest size C for four bears. (*Sources:* M. Triola, *Elementary Statistics*, Minitab, Inc.)

W (pounds)	N (inches)	L (inches)	C (inches)
125	19	57.5	32
316	26	65	42
436	30	72	48
514	30.5	75	54
?	24	63	39

(a) We can model the data in the table with the equation $W = a + bN + cL + dC$, where a, b, c, and d are constants. To do so, represent a system of linear equations by a 4×5 augmented matrix whose solution gives values for a, b, c, and d.

(b) Solve the system with a graphing calculator. Round each value to the nearest thousandth.

72. Refer to the previous exercise. Predict the weight of a bear with $N = 24$, $L = 63$, and $C = 39$. Interpret the result.

WRITING ABOUT MATHEMATICS

73. Explain what the dimension of a matrix means. What is the difference between a matrix that has dimension 3×4 and one that has dimension 4×3?

74. Discuss the advantages of using technology to transform an augmented matrix to reduced row–echelon form. Are there any disadvantages? Explain.

Checking Basic Concepts

1. Determine which ordered triple is the solution to the system of equations: $(5, -4, 0)$ or $(1, 3, -1)$.

$$x - y + 7z = -9$$
$$2x - 2y + 5z = -9$$
$$-x + 3y - 2z = 10$$

2. Solve the system of equations by using elimination and substitution.

$$x - y + z = 2$$
$$2x - 3y + z = -1$$
$$-x + y + z = 4$$

3. Use an augmented matrix to represent the system of equations. Solve the system by using
(a) Gauss–Jordan elimination and
(b) technology.

$$x + 2y + z = 1$$
$$x + y + z = -1$$
$$y + z = 1$$

4. A total of $1500 is invested at 1%, 2%, and 4% annual interest. The interest after 1 year equals $46. The amount invested at 2% is double the amount invested at 1%. Find the amount invested at each rate.

Group Activity

Adjacency Matrix *A matrix A can be used to represent a map showing distances between cities. Let a_{ij} denote the number in row i and column j of a matrix A. Now consider the following map illustrating freeway distances in miles between four cities. Each city has been assigned a number. For example, there is a direct route from Denver, Colorado (city 1), to Colorado Springs, Colorado (city 2), of approximately 60 miles. Therefore $a_{12} =$ **60** in the accompanying matrix A. (Note that a_{12} is the number in row 1 and column 2.) The distance from Colorado Springs to Denver is also 60 miles, so $a_{21} =$ **60**. As there is no direct freeway connection between Las Vegas, Nevada (city 4), and Colorado Springs (city 2), we let $a_{24} = a_{42} =$ *. The matrix A is called an **adjacency matrix**.* (Source: S. Baase, *Computer Algorithms: Introduction to Design and Analysis.*)

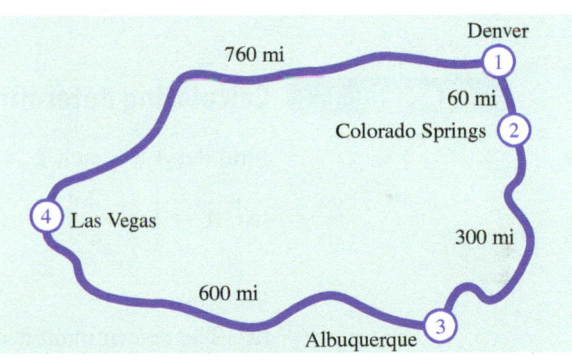

$$A = \begin{bmatrix} 0 & 60 & * & 760 \\ 60 & 0 & 300 & * \\ * & 300 & 0 & 600 \\ 760 & * & 600 & 0 \end{bmatrix}$$

1. Explain how to use A to find the freeway distance from Denver to Las Vegas.

2. Explain how to use A to find the shortest freeway distance from Denver to Albuquerque.

3. If a map shows 20 cities, what would be the dimension of the adjacency matrix? How many elements would there be in this matrix?

4. Why are there only zeros on the main diagonal of A?

5. What does $a_{14} + a_{41}$ equal?

6. What does $a_{11} + a_{44}$ equal?

9.3   Determinants

NEW VOCABULARY

☐ Determinants
☐ Expansion by minors
☐ Minors
☐ Cramer's rule

1 Calculating Determinants

The concept of determinants originated with the Japanese mathematician Seki Kowa (1642–1708), who used them to solve systems of linear equations. Later, Gottfried Leibniz (1646–1716) formally described determinants and also used them to solve systems of linear equations. (*Source: Historical Topics for the Mathematical Classroom*, NCTM.)

We begin by defining a determinant of a 2×2 matrix.

DETERMINANT OF A 2 × 2 MATRIX

The **determinant** of

$$A = \begin{bmatrix} a & b \\ c & d \end{bmatrix}$$

is a *real number* defined by

$$\det A = ad - cb.$$

EXAMPLE 1   Calculating determinants

Find $\det A$ for each 2×2 matrix.

(a) $A = \begin{bmatrix} 1 & 2 \\ 3 & 4 \end{bmatrix}$ **(b)** $A = \begin{bmatrix} -1 & -3 \\ 2 & -8 \end{bmatrix}$

Solution

(a) The determinant is calculated as follows.

$$\det A = \det \begin{bmatrix} 1 & 2 \\ 3 & 4 \end{bmatrix} = (1)(4) - (3)(2) = -2$$

(b) Similarly,

$$\det A = \det \begin{bmatrix} -1 & -3 \\ 2 & -8 \end{bmatrix} = (-1)(-8) - (2)(-3) = 14.$$

Now Try Exercises 5, 7

READING CHECK 1

• How do you calculate the determinant of a 2×2 matrix?

We can use determinants of 2×2 matrices to find determinants of 3×3 matrices. This method is called **expansion of a determinant by minors**.

DETERMINANT OF A 3 × 3 MATRIX

$$\det A = \det \begin{bmatrix} a_1 & b_1 & c_1 \\ a_2 & b_2 & c_2 \\ a_3 & b_3 & c_3 \end{bmatrix}$$

$$= a_1 \cdot \det \begin{bmatrix} b_2 & c_2 \\ b_3 & c_3 \end{bmatrix} - a_2 \cdot \det \begin{bmatrix} b_1 & c_1 \\ b_3 & c_3 \end{bmatrix} + a_3 \cdot \det \begin{bmatrix} b_1 & c_1 \\ b_2 & c_2 \end{bmatrix}$$

The 2×2 matrices in this equation are called **minors**.

EXAMPLE 2   **Calculating 3 × 3 determinants**

Evaluate det A.

(a) $A = \begin{bmatrix} 2 & 1 & -1 \\ -1 & 3 & 2 \\ 4 & -3 & -5 \end{bmatrix}$ (b) $A = \begin{bmatrix} 5 & -2 & 4 \\ 0 & 2 & 1 \\ -1 & 4 & -4 \end{bmatrix}$

Solution

(a) We evaluate the determinant as follows.

$$\det \begin{bmatrix} \mathbf{2} & 1 & -1 \\ \mathbf{-1} & 3 & 2 \\ \mathbf{4} & -3 & -5 \end{bmatrix} = \mathbf{2} \cdot \det \begin{bmatrix} 3 & 2 \\ -3 & -5 \end{bmatrix} - (\mathbf{-1}) \cdot \det \begin{bmatrix} 1 & -1 \\ -3 & -5 \end{bmatrix}$$

$$+ \, \mathbf{4} \cdot \det \begin{bmatrix} 1 & -1 \\ 3 & 2 \end{bmatrix}$$

$$= 2(-9) + 1(-8) + 4(5)$$

$$= -6$$

(b) We evaluate the determinant as follows.

$$\det \begin{bmatrix} \mathbf{5} & -2 & 4 \\ \mathbf{0} & 2 & 1 \\ \mathbf{-1} & 4 & -4 \end{bmatrix} = \mathbf{5} \cdot \det \begin{bmatrix} 2 & 1 \\ 4 & -4 \end{bmatrix} - (\mathbf{0}) \cdot \det \begin{bmatrix} -2 & 4 \\ 4 & -4 \end{bmatrix}$$

$$+ \, (\mathbf{-1}) \cdot \det \begin{bmatrix} -2 & 4 \\ 2 & 1 \end{bmatrix}$$

$$= 5(-12) - 0(-8) + (-1)(-10)$$

$$= -50$$

Now Try Exercises 13, 15

Many graphing calculators can evaluate the determinant of a matrix, as illustrated in the next example, where we evaluate the determinants from Example 2.

EXAMPLE 3   **Using technology to find determinants**

Find each determinant of A, using a graphing calculator.

(a) $A = \begin{bmatrix} 2 & 1 & -1 \\ -1 & 3 & 2 \\ 4 & -3 & -5 \end{bmatrix}$ (b) $A = \begin{bmatrix} 5 & -2 & 4 \\ 0 & 2 & 1 \\ -1 & 4 & -4 \end{bmatrix}$

Solution

(a) Begin by entering the matrix and then evaluate the determinant, as shown in **FIGURE 9.6**. The result is det $A = -6$, which agrees with our earlier calculation.

CALCULATOR HELP

To find a determinant, see Appendix A (pages AP-9 and AP-10).

```
MATRIX[A] 3×3
[ 2     1      -1   ]
[ -1    3      2    ]
[ 4     -3     -5   ]
```
(a)

```
[A]
     [[2   1   -1]
      [-1  3   2 ]
      [4   -3  -5]]
det([A])
                -6
```
(b)

FIGURE 9.6

(b) The determinant of A evaluates to -50. See **FIGURE 9.7**.

(a)	(b)

FIGURE 9.7

Now Try Exercise 21

2 Finding the Area of a Region

 Math in Context (Surveying) A determinant may be used in surveying to find the area of a triangular region. For example, if a triangle has vertices (a_1, a_2), (b_1, b_2), and (c_1, c_2), its area equals the absolute value of D, where

$$D = \frac{1}{2}\det\begin{bmatrix} a_1 & b_1 & c_1 \\ a_2 & b_2 & c_2 \\ 1 & 1 & 1 \end{bmatrix}.$$

If the vertices are entered in the columns of D counterclockwise as they appear in the xy-plane, D will be positive. (*Source:* W. Taylor, *The Geometry of Computer Graphics.*)

EXAMPLE 4   **Computing the area of a triangular parcel of land**

A triangular parcel of land is shown in **FIGURE 9.8**. If all units are miles, find the area of the parcel of land by using a determinant.

Solution

The vertices of the triangular parcel of land are $(2, 2)$, $(5, 4)$, and $(3, 8)$. The area of the triangle is

$$D = \frac{1}{2}\det\begin{bmatrix} 2 & 5 & 3 \\ 2 & 4 & 8 \\ 1 & 1 & 1 \end{bmatrix} = \frac{1}{2} \cdot 16 = 8.$$

The area of the triangle is 8 square miles.

Now Try Exercise 27

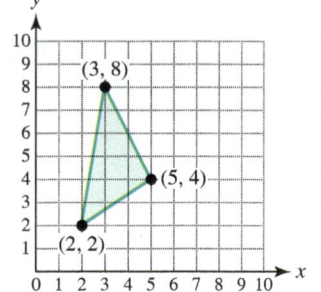

FIGURE 9.8

3 Applying Cramer's Rule

The use of determinants was expanded by Gabriel Cramer (1704–1752). His work, published in 1750, provided a method called **Cramer's rule** for solving systems of linear equations.

CRITICAL THINKING 1

Suppose that you are given three distinct points in the xy-plane and $D = 0$. What must be true about the three points?

CRAMER'S RULE FOR LINEAR SYSTEMS IN TWO VARIABLES

The solution to the system of linear equations

$$a_1 x + b_1 y = c_1$$
$$a_2 x + b_2 y = c_2$$

is given by $x = \frac{E}{D}$ and $y = \frac{F}{D}$, where

$$E = \det\begin{bmatrix} c_1 & b_1 \\ c_2 & b_2 \end{bmatrix}, \quad F = \det\begin{bmatrix} a_1 & c_1 \\ a_2 & c_2 \end{bmatrix}, \quad \text{and} \quad D = \det\begin{bmatrix} a_1 & b_1 \\ a_2 & b_2 \end{bmatrix} \neq 0.$$

NOTE: If $D = 0$, the system has either no solutions or infinitely many solutions. ■

EXAMPLE 5 **Using Cramer's rule**

Use Cramer's rule to solve the following linear systems.

(a) $3x - 4y = 18$ (b) $-4x + 9y = -24$
$\ 7x + 5y = -1$ $\ 6x + 17y = -25$

Solution

(a) $E = \det\begin{bmatrix} c_1 & b_1 \\ c_2 & b_2 \end{bmatrix} = \det\begin{bmatrix} 18 & -4 \\ -1 & 5 \end{bmatrix} = (18)(5) - (-1)(-4) = \mathbf{86}$

$\ F = \det\begin{bmatrix} a_1 & c_1 \\ a_2 & c_2 \end{bmatrix} = \det\begin{bmatrix} 3 & 18 \\ 7 & -1 \end{bmatrix} = (3)(-1) - (7)(18) = \mathbf{-129}$

$\ D = \det\begin{bmatrix} a_1 & b_1 \\ a_2 & b_2 \end{bmatrix} = \det\begin{bmatrix} 3 & -4 \\ 7 & 5 \end{bmatrix} = (3)(5) - (7)(-4) = \mathbf{43}$

Because $x = \dfrac{E}{D} = \dfrac{86}{43} = 2$ and $y = \dfrac{F}{D} = \dfrac{-129}{43} = -3$, the solution is $(2, -3)$.

(b) $E = \det\begin{bmatrix} c_1 & b_1 \\ c_2 & b_2 \end{bmatrix} = \det\begin{bmatrix} -24 & 9 \\ -25 & 17 \end{bmatrix} = (-24)(17) - (-25)(9) = \mathbf{-183}$

$\ F = \det\begin{bmatrix} a_1 & c_1 \\ a_2 & c_2 \end{bmatrix} = \det\begin{bmatrix} -4 & -24 \\ 6 & -25 \end{bmatrix} = (-4)(-25) - (6)(-24) = \mathbf{244}$

$\ D = \det\begin{bmatrix} a_1 & b_1 \\ a_2 & b_2 \end{bmatrix} = \det\begin{bmatrix} -4 & 9 \\ 6 & 17 \end{bmatrix} = (-4)(17) - (6)(9) = \mathbf{-122}$

Because $x = \dfrac{E}{D} = \dfrac{-183}{-122} = 1.5$ and $y = \dfrac{F}{D} = \dfrac{244}{-122} = -2$, the solution is $(1.5, -2)$.

Now Try Exercises 33, 35

Cramer's rule can be applied to systems that have any number of linear equations. Cramer's rule for three linear equations is discussed in Exercises 39–44.

9.3 Putting It All Together

CONCEPT	EXPLANATION
Determinant of a 2×2 Matrix	The determinant of a 2×2 matrix A is given by $$\det A = \det\begin{bmatrix} a & b \\ c & d \end{bmatrix} = ad - cb.$$
Determinant of a 3×3 Matrix	The determinant of a 3×3 matrix A is given by $$\det A = \det\begin{bmatrix} a_1 & b_1 & c_1 \\ a_2 & b_2 & c_2 \\ a_3 & b_3 & c_3 \end{bmatrix}$$ $$= a_1 \cdot \det\begin{bmatrix} b_2 & c_2 \\ b_3 & c_3 \end{bmatrix} - a_2 \cdot \det\begin{bmatrix} b_1 & c_1 \\ b_3 & c_3 \end{bmatrix}$$ $$+ a_3 \cdot \det\begin{bmatrix} b_1 & c_1 \\ b_2 & c_2 \end{bmatrix}.$$

continued on next page

continued from previous page

CONCEPT	EXPLANATION
Area of a Triangle	If a triangle has vertices (a_1, a_2), (b_1, b_2), and (c_1, c_2), its area equals the absolute value of D, where $$D = \frac{1}{2}\det\begin{bmatrix} a_1 & b_1 & c_1 \\ a_2 & b_2 & c_2 \\ 1 & 1 & 1 \end{bmatrix}.$$
Cramer's Rule for Linear Systems in Two Variables	The solution to the linear system $$a_1 x + b_1 y = c_1$$ $$a_2 x + b_2 y = c_2$$ is given by $x = \frac{E}{D}$ and $y = \frac{F}{D}$ where $$E = \det\begin{bmatrix} c_1 & b_1 \\ c_2 & b_2 \end{bmatrix}, \quad F = \det\begin{bmatrix} a_1 & c_1 \\ a_2 & c_2 \end{bmatrix}, \quad \text{and}$$ $$D = \det\begin{bmatrix} a_1 & b_1 \\ a_2 & b_2 \end{bmatrix} \neq 0.$$ **NOTE:** If $D = 0$, then the system has either no solutions or infinitely many solutions. ∎

9.3 Exercises

MyMathLab®

CONCEPTS AND VOCABULARY

1. We can find the determinant of a(n) _____ matrix.

2. If we find a determinant, the answer is a(n) _____.

3. Cramer's rule can be used to solve a(n) _____.

4. If the first column of a matrix is all 0s, then its determinant equals _____.

CALCULATING DETERMINANTS

Exercises 5–20: Evaluate det A by hand where A is the given matrix.

5. $\begin{bmatrix} 1 & -2 \\ 3 & -8 \end{bmatrix}$

6. $\begin{bmatrix} 5 & -1 \\ 3 & 7 \end{bmatrix}$

7. $\begin{bmatrix} -3 & 7 \\ 8 & -1 \end{bmatrix}$

8. $\begin{bmatrix} 0 & -7 \\ -3 & 1 \end{bmatrix}$

9. $\begin{bmatrix} 23 & 4 \\ 6 & -13 \end{bmatrix}$

10. $\begin{bmatrix} 44 & -51 \\ -9 & 32 \end{bmatrix}$

11. $\begin{bmatrix} 1 & -1 & 2 \\ 0 & 1 & -3 \\ 0 & -4 & 7 \end{bmatrix}$

12. $\begin{bmatrix} 2 & -1 & -5 \\ -1 & 4 & -2 \\ 0 & 1 & 4 \end{bmatrix}$

13. $\begin{bmatrix} 2 & -1 & 0 \\ 1 & -2 & 6 \\ 0 & 1 & 8 \end{bmatrix}$

14. $\begin{bmatrix} 0 & 1 & -4 \\ 3 & -6 & 10 \\ 4 & -2 & 7 \end{bmatrix}$

15. $\begin{bmatrix} -1 & 3 & 5 \\ 3 & -3 & 5 \\ 2 & -3 & 7 \end{bmatrix}$

16. $\begin{bmatrix} 6 & -1 & 9 \\ 7 & 0 & -3 \\ 2 & 5 & -1 \end{bmatrix}$

17. $\begin{bmatrix} 5 & 0 & 0 \\ 0 & -2 & 0 \\ 0 & 0 & 5 \end{bmatrix}$

18. $\begin{bmatrix} 1 & 2 & 3 \\ 2 & 4 & 6 \\ 3 & 6 & 9 \end{bmatrix}$

19. $\begin{bmatrix} 0 & 2 & -3 \\ 0 & 3 & -9 \\ 0 & 5 & 9 \end{bmatrix}$

20. $\begin{bmatrix} 3 & -1 & 2 \\ 0 & 5 & 7 \\ 0 & 0 & -1 \end{bmatrix}$

Exercises 21–24: Use technology to calculate det A, where A is the given matrix.

21. $\begin{bmatrix} 2 & -5 & 13 \\ 10 & 15 & -10 \\ 17 & -19 & 22 \end{bmatrix}$

22. $\begin{bmatrix} 1.6 & 3.1 & 5.7 \\ 2.1 & 6.7 & 8.1 \\ -0.4 & -0.8 & -3.1 \end{bmatrix}$

23. $\begin{bmatrix} 17 & 0 & 4 \\ -9 & 14 & 1.5 \\ 13 & 67 & -11 \end{bmatrix}$

24. $\begin{bmatrix} 121 & 45 & -56 \\ -45 & 87 & 32 \\ -14 & -34 & 67 \end{bmatrix}$

Exercises 25 and 26: **Thinking Generally** *Find det A, if a, b, and c are nonzero constants.*

25. $A = \begin{bmatrix} a & 0 & 0 \\ 0 & b & 0 \\ 0 & 0 & c \end{bmatrix}$

26. $A = \begin{bmatrix} 0 & 0 & 0 \\ a & b & 0 \\ 0 & 0 & c \end{bmatrix}$

FINDING THE AREA OF A REGION

Exercises 27–32: Find the area of the figure by using a determinant. Assume that units are feet.

27.

28.

29.

30.

31.

32.

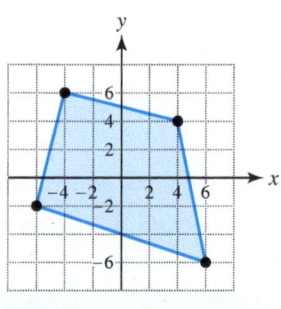

APPLYING CRAMER'S RULE IN TWO VARIABLES

Exercises 33–38: Solve the system of equations by using Cramer's rule.

33. $5x + 3y = 4$
$6x - 4y = 20$

34. $-5x + 4y = -5$
$4x + 4y = -32$

35. $7x - 5y = -3$
$-4x + 6y = -8$

36. $-4x - 9y = -17$
$8x + 4y = 9$

37. $8x = 3y - 61$
$-x = 4y - 23$

38. $15y = -188 - 22x$
$23y = -173 - 16x$

APPLYING CRAMER'S RULE IN THREE VARIABLES

Exercises 39–44: Cramer's rule can be applied to systems of three equations in three variables. For the system of equations

$$a_1x + b_1y + c_1z = d_1$$
$$a_2x + b_2y + c_2z = d_2$$
$$a_3x + b_3y + c_3z = d_3,$$

the solution can be written as follows.

$$D = \det\begin{bmatrix} a_1 & b_1 & c_1 \\ a_2 & b_2 & c_2 \\ a_3 & b_3 & c_3 \end{bmatrix}, \qquad E = \det\begin{bmatrix} d_1 & b_1 & c_1 \\ d_2 & b_2 & c_2 \\ d_3 & b_3 & c_3 \end{bmatrix}$$

$$F = \det\begin{bmatrix} a_1 & d_1 & c_1 \\ a_2 & d_2 & c_2 \\ a_3 & d_3 & c_3 \end{bmatrix}, \qquad G = \det\begin{bmatrix} a_1 & b_1 & d_1 \\ a_2 & b_2 & d_2 \\ a_3 & b_3 & d_3 \end{bmatrix}$$

If $D \neq 0$, a unique solution exists and is given by

$$x = \frac{E}{D}, \; y = \frac{F}{D}, \; z = \frac{G}{D}.$$

Use Cramer's rule to solve the system of equations.

39. $x + y + z = 6$
$2x + y + 2z = 9$
$y + 3z = 9$

40. $y + z = 1$
$2x - y - z = -1$
$x + y - z = 3$

41. $x + z = 2$
$x + y = 0$
$y + 2z = 1$

42. $x + y + 2z = 1$
$-x - 2y - 3z = -2$
$y - 3z = 5$

43. $x + 2z = 7$
$-x + y + z = 5$
$2x - y + 2z = 6$

44. $x + 2y + 3z = -1$
$2x - 3y - z = 12$
$x + 4y - 2z = -12$

WRITING ABOUT MATHEMATICS

45. Suppose that one row of a 3×3 matrix A is all 0s. What is the value of det A? Give an example and explain your answer.

46. Explain how to evaluate a determinant of a 2×2 matrix. Can you find the determinant of a 2×3 matrix? Explain your answer.

Checking Basic Concepts

1. Evaluate det A.

   (a) $A = \begin{bmatrix} -3 & 4 \\ -2 & 3 \end{bmatrix}$

   (b) $A = \begin{bmatrix} 1 & -2 & 3 \\ 5 & 1 & 1 \\ 0 & 2 & -1 \end{bmatrix}$

2. Use Cramer's rule to solve the system of linear equations.

   $$2x - y = -14$$
   $$3x - 4y = -36$$

3. Find the area of a triangle that has the vertices $(-1, 2)$, $(5, 6)$, and $(2, -3)$.

CHAPTER 9 Summary

SECTION 9.1 ■ SYSTEMS OF LINEAR EQUATIONS IN THREE VARIABLES

Solution to a System of Linear Equations in Three Variables An ordered triple (x, y, z) that satisfies all three equations

Example:
$$x - y + 2z = 3$$
$$2x - y + z = 5$$
$$x + y + z = 6$$

The solution is $(3, 2, 1)$ because these values for (x, y, z) satisfy all three equations.

$$3 - 2 + 2(1) = 3 \checkmark \quad \text{True}$$
$$2(3) - 2 + 1 = 5 \checkmark \quad \text{True}$$
$$3 + 2 + 1 = 6 \checkmark \quad \text{True}$$

Elimination and Substitution Systems of linear equations in three variables can be solved by elimination and substitution, using the following steps.

STEP 1: Eliminate one variable, such as x, from two of the given equations.

STEP 2: Use the two resulting equations in two variables to eliminate one of the variables, such as y. Solve for the remaining variable z.

STEP 3: Substitute z in one of the two equations from Step 2. Solve for the unknown variable y.

STEP 4: Substitute values for y and z in one of the given equations. Then find x. The solution is the ordered triple (x, y, z).

SECTION 9.2 ■ MATRIX SOLUTIONS OF LINEAR SYSTEMS

Matrix A rectangular array of numbers is a matrix. If a matrix has m rows and n columns, it has dimension $m \times n$.

Example: Matrix $A = \begin{bmatrix} 3 & -1 & 7 \\ 0 & 6 & -2 \end{bmatrix}$ has dimension 2×3.

Augmented Matrix Any linear system can be represented with an augmented matrix.

Linear System
$$4x - 3y = 5$$
$$x + 2y = 4$$

Augmented Matrix
$$\left[\begin{array}{cc|c} 4 & -3 & 5 \\ 1 & 2 & 4 \end{array}\right]$$

Gauss–Jordan Elimination A numerical method that uses matrix row transformations to transform a matrix into reduced row–echelon form

Example: The matrix $\begin{bmatrix} 4 & -3 & | & 5 \\ 1 & 2 & | & 4 \end{bmatrix}$ reduces to $\begin{bmatrix} 1 & 0 & | & 2 \\ 0 & 1 & | & 1 \end{bmatrix}$.

The solution to the system is $(2, 1)$.

SECTION 9.3 ■ DETERMINANTS

Determinant for a 2 × 2 Matrix A determinant is a *real number*. The determinant of a 2 × 2 matrix is

$$\det A = \det \begin{bmatrix} a & b \\ c & d \end{bmatrix} = ad - cb.$$

Example: $\det \begin{bmatrix} 2 & 3 \\ 4 & 5 \end{bmatrix} = (2)(5) - (4)(3) = -2$

Determinant for a 3 × 3 Matrix

$$\det A = \det \begin{bmatrix} a_1 & b_1 & c_1 \\ a_2 & b_2 & c_2 \\ a_3 & b_3 & c_3 \end{bmatrix}$$

$$= a_1 \cdot \det \begin{bmatrix} b_2 & c_2 \\ b_3 & c_3 \end{bmatrix} - a_2 \cdot \det \begin{bmatrix} b_1 & c_1 \\ b_3 & c_3 \end{bmatrix} + a_3 \cdot \det \begin{bmatrix} b_1 & c_1 \\ b_2 & c_2 \end{bmatrix}$$

Example: $\det \begin{bmatrix} 2 & 3 & 2 \\ 3 & 7 & -3 \\ 0 & 0 & -1 \end{bmatrix} = 2 \det \begin{bmatrix} 7 & -3 \\ 0 & -1 \end{bmatrix} - 3 \det \begin{bmatrix} 3 & 2 \\ 0 & -1 \end{bmatrix} + 0 \det \begin{bmatrix} 3 & 2 \\ 7 & -3 \end{bmatrix}$

$$= 2(-7) - 3(-3) + 0(-23) = -5$$

Cramer's rule uses determinants to solve linear systems of equations. Determinants can also be used to find areas of triangles. See Putting It All Together for this section.

CHAPTER 9 **Review Exercises**

SECTION 9.1

1. Is $(3, -4, 5)$ a solution for $x + y + z = 4$?

2. Decide whether either ordered triple is a solution: $(1, -1, 2)$ or $(1, 0, 5)$.
$$\begin{aligned} 2x - 3y + z &= 7 \\ -x - y + 3z &= 6 \\ 3x - 2y + z &= 7 \end{aligned}$$

Exercises 3–8: Use elimination and substitution to solve the system of linear equations, if possible.

3. $\begin{aligned} x - y - 2z &= -11 \\ -x + 2y + 3z &= 16 \\ 3z &= 6 \end{aligned}$

4. $\begin{aligned} x + y &= 4 \\ -2x + y + 3z &= -2 \\ x - 2y + 5z &= -26 \end{aligned}$

5. $\begin{aligned} 2x - y &= -5 \\ x + 2y + z &= 7 \\ -2x + y + z &= 7 \end{aligned}$

6. $\begin{aligned} 2x + 3y + z &= 6 \\ -x + 2y + 2z &= 3 \\ x + y + 2z &= 4 \end{aligned}$

7. $\begin{aligned} x - y + 3z &= 2 \\ 2x + y + 4z &= 3 \\ x + 2y + z &= 5 \end{aligned}$

8. $\begin{aligned} x - y + 3z &= 3 \\ x + y - z &= 1 \\ x + z &= 2 \end{aligned}$

SECTION 9.2

Exercises 9–12: Write the system of linear equations as an augmented matrix. Then use Gauss–Jordan elimination to solve the system, writing the solution as an ordered triple. Check your solution.

9.
$$\begin{aligned} x + y + z &= -6 \\ x + 2y + z &= -8 \\ y + z &= -5 \end{aligned}$$

10.
$$\begin{aligned} x + y + z &= -3 \\ -x + y &= 5 \\ y + z &= -1 \end{aligned}$$

11.
$$\begin{aligned} x + 2y - z &= 1 \\ -x + y - 2z &= 5 \\ 2y + z &= 10 \end{aligned}$$

12.
$$\begin{aligned} 2x + 2y - 2z &= -14 \\ -2x - 3y + 2z &= 12 \\ x + y - 4z &= -22 \end{aligned}$$

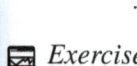 *Exercises 13 and 14: Use a graphing calculator to solve the system of linear equations.*

13.
$$\begin{aligned} 3x - 2y + 6z &= -17 \\ -2x - y + 5z &= 20 \\ 4y + 7z &= 30 \end{aligned}$$

14.
$$\begin{aligned} 19x - 13y - 7z &= 7.4 \\ 22x + 33y - 8z &= 110.5 \\ 10x - 56y + 9z &= 23.7 \end{aligned}$$

SECTION 9.3

Exercises 15–18: Evaluate det A, if A is the given matrix.

15. $\begin{bmatrix} 6 & -5 \\ -4 & 2 \end{bmatrix}$

16. $\begin{bmatrix} 0 & -6 \\ 5 & 9 \end{bmatrix}$

17. $\begin{bmatrix} 3 & -5 & -3 \\ 1 & 4 & 7 \\ 0 & -3 & 1 \end{bmatrix}$

18. $\begin{bmatrix} -2 & -1 & -7 \\ 2 & 1 & -3 \\ 3 & -5 & 8 \end{bmatrix}$

 Exercises 19 and 20: Use technology to calculate det A, where A is the given matrix.

19. $\begin{bmatrix} 22 & -45 & 3 \\ 15 & -12 & -93 \\ 5 & 81 & -21 \end{bmatrix}$

20. $\begin{bmatrix} 0.5 & -7.3 & 9.6 \\ 0.1 & 3.1 & 9.2 \\ -0.5 & -1.9 & 5.4 \end{bmatrix}$

Exercises 21–22: Use a determinant to find the area of the triangle. Assume that the units are feet.

21.

22.

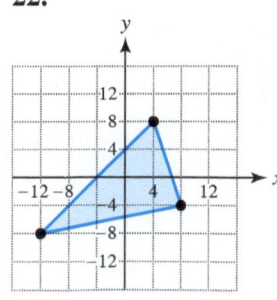

Exercises 23 and 24: Use Cramer's rule to solve the system.

23.
$$\begin{aligned} 7x + 6y &= 8 \\ 5x - 8y &= 18 \end{aligned}$$

24.
$$\begin{aligned} -2x + 5y &= 25 \\ 3x + 4y &= -3 \end{aligned}$$

APPLICATIONS

25. Pedestrian Fatalities Forty-seven percent of pedestrian fatalities occur on Friday and Saturday nights. The combined total of pedestrian fatalities in 1994 and 2004 was 10,130. There were 848 more fatalities in 1994 than in 2004. Find the number of pedestrian fatalities during each year. (*Source:* National Highway Traffic Safety Administration.)

26. Tickets Tickets for a football game cost $8 and $12. If 480 tickets were sold for total receipts of $4620, how many of each type of ticket were sold?

27. Determining Costs The accompanying table shows the costs for purchasing different combinations of malts, cones, and ice cream bars.

Malts	Cones	Bars	Total Cost
1	3	5	$14
1	2	4	$11
0	1	3	$5

(a) Let *m* be the cost of a malt, *c* the cost of a cone, and *b* the cost of an ice cream bar. Write a system of equations that represents the data in the table.

(b) Solve this system.

28. Geometry The largest angle in a triangle is 20° more than the sum of the two smaller angles. The measure of the largest angle is 85° more than the smallest angle. Find the measure of each angle in the triangle.

29. Mixture Problem Three types of candy that cost $1.50, $2.00, and $2.50 per pound are to be mixed to produce 12 pounds of candy worth $26.00. If there are to be 2 pounds more of the $2.50 candy than the $2.00 candy, how much of each type of candy should be used in the mixture?

30. Estimating the Chest Size of a Bear The accompanying table shows the chest size C, weight W, and overall length L of three bears. These data can be modeled with the formula $C = a + bW + cL$. (*Sources:* M. Triola, Elementary Statistics; Minitab, Inc.)

C (inches)	W (pounds)	L (inches)
40	202	63
50	365	70
55	446	77
?	300	68

(a) Set up a system of linear equations whose solution gives values for the constants a, b, and c.

 (b) Solve this system. Round each value to the nearest thousandth.

(c) Predict the chest size of a bear weighing 300 pounds and having a length of 68 inches.

CHAPTER 9   Test   MyMathLab®, You Tube

1. Can a system of three equations and three variables have exactly three solutions? Explain.

2. Determine which ordered triple is a solution to the linear system.

$$(-4, 3, 3), \ (1, -2, -2)$$

$$
\begin{aligned}
x - 3y + 4z &= -1 \\
2x + y - 3z &= 6 \\
x - y + z &= 1
\end{aligned}
$$

Exercises 3–6: Use substitution and elimination to solve the system, if possible.

3.
$$
\begin{aligned}
x + 3y &= 2 \\
-2x + y + z &= 5 \\
y + z &= -3
\end{aligned}
$$

4.
$$
\begin{aligned}
x + y - z &= 1 \\
2x - 3y + z &= 0 \\
x - 4y + 2z &= 2
\end{aligned}
$$

5.
$$
\begin{aligned}
x - y + 2z &= 5 \\
-x + y - 3z &= -8 \\
x - 2y + 2z &= 3
\end{aligned}
$$

6.
$$
\begin{aligned}
x - y + z &= 1 \\
x + y - z &= 1 \\
3x + y - z &= 3
\end{aligned}
$$

7. Consider the system of linear equations.

$$
\begin{aligned}
2x - 4y &= -10 \\
-3x - 2y &= 7
\end{aligned}
$$

(a) Write the system of linear equations as an augmented matrix.
(b) Use Gauss–Jordan elimination to solve the system, writing the solution as an ordered pair.

8. Consider the system of linear equations.

$$
\begin{aligned}
x + y + z &= 2 \\
x - y - z &= 3 \\
2x + 2y + z &= 6
\end{aligned}
$$

(a) Write the system as an augmented matrix.

(b) Use Gauss–Jordan elimination to solve the system, writing the solution as an ordered triple.

9. Evaluate det A if $A = \begin{bmatrix} -1 & 2 \\ -5 & 4 \end{bmatrix}$.

10. Evaluate det A if $A = \begin{bmatrix} 3 & 2 & -1 \\ 6 & 2 & -6 \\ 0 & 8 & -3 \end{bmatrix}$.

11. Solve the system of equations using Cramer's rule.

$$
\begin{aligned}
5x - 3y &= 7 \\
-4x + 2y &= 11
\end{aligned}
$$

12. Jogging Speeds A runner in preparation for a race jogs at 6, 7, and 9 miles per hour and travels a total of 7.1 miles in 1 hour. The runner jogs 12 minutes longer at 6 miles per hour than at 9 miles per hour. How many minutes did the runner jog at each speed?

13. Geometry The largest angle in a triangle is 50° more than the smallest angle. The sum of the measures of the smaller two angles is 10° more than the largest angle. Find the measure of each angle.

14. Weight of a Bear The data in the accompanying table list the weight W, neck size N, and chest size C of three bears. These data can be modeled by $W = a + bN + cC$.

W (pounds)	N (inches)	C (inches)
168	20	25
270	24	40
405	30	50

(a) Set up a system of equations whose solution gives values for the constants a, b, and c.
(b) Solve this system.
(c) Predict the weight W of a bear with neck size $N = 26$ and chest size $C = 44$.

CHAPTERS 1–9 Cumulative Review Exercises

1. Write 360 as a product of prime numbers.

2. Translate the sentence "Double a number increased by 7 equals the number decreased by 2" into an equation by using the variable n. Then solve the equation.

Exercises 3 and 4: Simplify to lowest terms.

3. $\frac{2}{3} + \frac{4}{7} \cdot \frac{21}{28}$ **4.** $\frac{3}{5} \div \frac{6}{5} - \frac{2}{3}$

Exercises 5 and 6: Evaluate by hand.

5. $30 - 4 \div 2 \cdot 6$ **6.** $\dfrac{3^2 - 2^3}{20 - 5 \cdot 2}$

7. Solve $2(x + 1) - 6x = x - 4$.

8. Solve $4 - 3x = -2$ graphically. Check your answer.

9. Convert 124% to fraction and decimal notation.

10. If $A = 30$ square miles and $h = 10$ miles, use the formula $A = \frac{1}{2}bh$ to find b.

11. Solve $6t - 1 < 3 - t$.

12. Make a scatterplot with the points $(-1, 2)$, $(1, -2)$, $(0, 3)$, $(-2, 0)$, and $(2, 3)$.

Exercises 13 and 14: Graph the equation and determine any intercepts.

13. $-3x + 4y = 12$

14. $x = -2$

15. Use the graph to identify the x-intercept and the y-intercept. Then write the slope–intercept form of the line.

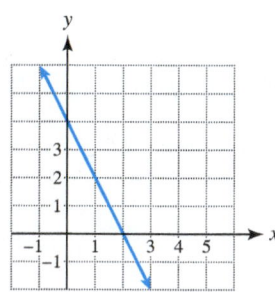

16. Write the slope–intercept form of a line that passes through the point $(-1, 3)$ with slope $m = -2$.

Exercises 17 and 18: Find the slope–intercept form of the line satisfying the given conditions.

17. Passing through the points $(-3, 5)$ and $(2, 8)$

18. Perpendicular to $x + 2y = 5$, passing through the point $(-1, 1)$

19. An insect population P is initially 4000 and increases by 500 insects per day. Write an equation in slope–intercept form that calculates P after x days.

20. Solve the system of equations. Write your answer as an ordered pair.

$$-2a + b = -5$$
$$4a - 3b = 0$$

Exercises 21 and 22: Shade the solution set.

21. $y \geq -1$ **22.** $x - y < 3$
   $2x + y \geq 6$

Exercises 23–28: Simplify and use positive exponents, when appropriate, to write the expression.

23. $5 - 3^4$ **24.** $(8t^{-3})(3t^2)(t^5)$

25. $\dfrac{2^{-4}}{4^{-2}}$ **26.** $(2t^3)^{-2}$

27. $(4a^2b^3)^2(2ab)^{-3}$ **28.** $\left(\dfrac{2a^{-1}}{ab^{-2}}\right)^{-3}$

Exercises 29–32: Multiply the expression.

29. $2a^2(a^2 - 2a + 3)$

30. $(5x + 1)(x - 7)$ **31.** $(2x + 3y)^2$

32. $(a + b)(a - b)$

Exercises 33 and 34: Divide.

33. $\dfrac{4x^3 - 8x^2 + 6x}{2x}$

34. $(x^4 - 9x^3 + 23x^2 - 17x + 11) \div (x - 5)$

Exercises 35–42: Factor completely.

35. $10ab^2 - 25a^3b^5$ **36.** $y^3 - 3y^2 + 2y - 6$

37. $6z^2 + 7z - 3$ **38.** $4z^2 - 9$

39. $4y^2 - 20y + 25$ **40.** $a^3 - 27$

41. $4z^4 - 17z^2 + 15$ **42.** $2a^3b + a^2b^2 - ab^3$

Exercises 43 and 44: Solve the equation.

43. $(x - 1)(x + 2) = 0$ **44.** $6y^2 - 7y = 3$

Exercises 45 and 46: Simplify to lowest terms.

45. $\dfrac{x^2 - 16}{x + 4}$

46. $\dfrac{2x^2 - 11x - 6}{6x^2 - 5x - 4}$

Exercises 47 and 48: Simplify and write in lowest terms.

47. $\dfrac{x^2 - 3x + 2}{x + 7} \div \dfrac{x - 2}{2x + 14}$

48. $\dfrac{x}{2x + 3} + \dfrac{x + 3}{2x + 3}$

Exercises 49 and 50: Solve the equation.

49. $\dfrac{x + 2}{5} = \dfrac{x}{4}$

50. $\dfrac{1}{3x} + \dfrac{5}{2x} = 2$

51. Suppose that y is inversely proportional to x and that $y = 25$ when $x = 4$. Find y for $x = 10$.

52. Evaluate $f(x) = 1 - 4x$ for $x = -3$.

53. Graph $f(x) = x^2 - 2$ and identify the domain and range of f. Use interval notation.

54. Use the graph to evaluate $f(0)$ and $f(-2)$.

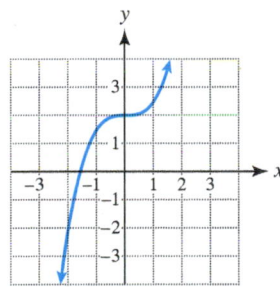

Exercises 55 and 56: Solve the inequality symbolically and write the solution set in interval notation.

55. $\dfrac{4x - 9}{6} > \dfrac{1}{2}$

56. $\dfrac{2}{3}z - 2 \le \dfrac{1}{4}z - (2z + 2)$

Exercises 57 and 58: Solve the compound inequality and graph the solution set on a number line.

57. $x + 2 > 1$ and $2x - 1 \le 9$

58. $4x + 7 < 1$ or $3x + 2 \ge 11$

Exercises 59 and 60: Solve the three-part inequality and write the solution set in interval notation.

59. $-7 \le 2x - 3 \le 5$

60. $-8 \le -\frac{1}{2}x - 3 \le 5$

61. Use the graph to solve the equation and inequalities.
   (a) $y_1 = 2$ **(b)** $y_1 \le 2$
   (c) $y_1 \ge 2$

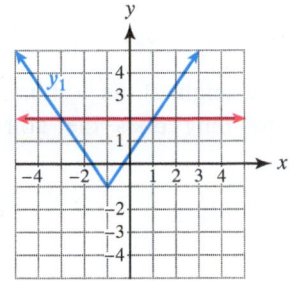

62. Solve the absolute value equation $\left|\frac{2}{3}x - 4\right| = 8$.

Exercises 63 and 64: Solve the absolute value inequality. Write the solution set in interval notation.

63. $|3x + 5| > 13$

64. $-3|2t - 11| \ge -9$

65. Use elimination and substitution to solve the system of linear equations.

$$\begin{aligned} 2x + 3y - z &= 3 \\ 3x - y + 4z &= 10 \\ 2x + y - 2z &= -1 \end{aligned}$$

66. Write the system of linear equations as an augmented matrix. Then use Gauss–Jordan elimination to solve the system. Write the solution as an ordered triple.

$$\begin{aligned} x + y - z &= 4 \\ -x - y - z &= 0 \\ x - 2y + z &= -9 \end{aligned}$$

APPLICATIONS

67. **Graphical Model** An athlete rides a bicycle on a straight road away from home at 20 miles per hour for 1.5 hours. The athlete then turns around and rides home at 15 miles per hour. Sketch a graph that depicts the athlete's distance d from home after t hours.

68. **Ticket Prices** The price of admission to a baseball game is \$15 for children and \$25 for adults. If a group of 8 people pays \$170 to enter the game, find the number of children and the number of adults in the group.

69. **Working Together** One person can shovel the snow from a sidewalk in 2 hours and another person can shovel the same sidewalk in 1.5 hours. How long will it take for them to shovel the snow from the sidewalk if they work together?

70. Flight of a Golf Ball If a golf ball is hit upward with a velocity of 88 feet per second (60 miles per hour), then its height h in feet after t seconds can be approximated by

$$h(t) = 88t - 16t^2.$$

(a) What is the height of the golf ball after 2 seconds?
(b) After how long does the golf ball strike the ground?

71. Height of a Building A 7-foot-tall stop sign casts a 4-foot-long shadow, while a nearby building casts a 35-foot-long shadow. Find the height of the building.

72. Determining Costs The accompanying table shows the costs for purchasing different combinations of burgers, fries, and malts.

Burgers	Fries	Malts	Total Cost
4	3	4	$23
1	2	1	$7
3	1	2	$13

(a) Let b be the cost of a burger, f be the cost of an order of fries, and m be the cost of a malt. Write a system of three linear equations that represents the data in the table.
(b) Solve this system and find the price of each item.

10 Radical Expressions and Functions

Wind power is rapidly becoming a larger portion of America's total energy production. Recently, the United States had a wind capacity that was able to produce 61,000 megawatts of energy, which was enough to power about 15.5 million homes. At that time, there were over 500 new wind turbines installed and their placement was critical in maximizing their efficiency. For example, suppose location A has an average wind velocity that is twice the velocity of location B. Does this mean that a wind turbine at location A will be able to produce twice the wind power as a wind turbine at location B? To better understand the mathematics behind wind energy, we need to learn more about radical expressions and functions. (See Section 6, Example 11, of this chapter.) *Source:* American Wind Energy Association.

10.1 Radical Expressions and Functions

Objectives

1. **Introducing Radical Expressions**
   - **Square Roots**
   - **Square Roots of Negative Numbers**
   - **Cube Roots**
   - *n*th Roots
   - **Absolute Value**

2. **Recognizing Square Root Functions and Their Properties**

3. **Recognizing Cube Root Functions and Their Properties**

1 Introducing Radical Expressions

SQUARE ROOTS Recall the definition of the square root of a number a.

> **SQUARE ROOT**
>
> The number b is a *square root* of a if $b^2 = a$.

For example, since

$$6^2 = 36 \quad \text{and} \quad (-6)^2 = 36,$$

we know that both 6 and -6 are square roots of 36. From this discussion, we can see that every positive number has two square roots: one positive and one negative.

EXAMPLE 1 **Finding square roots**

Find the square roots of 100.

Solution

The square roots of 100 are 10 and -10, because $10^2 = 100$ and $(-10)^2 = 100$.

Now Try Exercise 1

NEW VOCABULARY

☐ Radical sign
☐ Radicand
☐ Radical expression
☐ *n*th root
☐ Index
☐ Odd root
☐ Even root
☐ Principal *n*th root
☐ Square root function
☐ Cube root function

Recall that the *positive square root* of a positive number a is called the *principal square root* and is denoted $\sqrt{a}$. The *negative square root* is denoted $-\sqrt{a}$. To identify both square roots, we write $\pm\sqrt{a}$. The symbol $\pm$ is read "plus or minus." The symbol $\sqrt{}$ is called the **radical sign**. The expression under the radical sign is called the **radicand**, and an expression containing a radical sign is called a **radical expression**.

Radical Expression

Radical sign $\quad \sqrt{6} \quad$ Radicand

EXAMPLE 2 **Finding principal square roots**

Evaluate each square root.

(a) $\sqrt{25}$ **(b)** $\sqrt{0.49}$ **(c)** $\sqrt{\frac{4}{9}}$ **(d)** $\sqrt{c^2}, c > 0$

Solution

(a) Because $5 \cdot 5 = 25$, the principal, or *positive*, square root of 25 is $\sqrt{25} = 5$.

(b) Because $(0.7)(0.7) = 0.49$, the principal square root of 0.49 is $\sqrt{0.49} = 0.7$.

(c) Because $\frac{2}{3} \cdot \frac{2}{3} = \frac{4}{9}$, the principal square root of $\frac{4}{9}$ is $\sqrt{\frac{4}{9}} = \frac{2}{3}$.

(d) The principal square root of c^2 is $\sqrt{c^2} = c$, as c is positive.

Now Try Exercises 15, 17, 19, 21

The square roots of many real numbers, such as $\sqrt{17}$, $\sqrt{1.2}$, and $\sqrt{\frac{5}{7}}$, cannot be conveniently evaluated (or approximated) by hand. In these cases, we sometimes use a calculator to give a decimal *approximation*, as demonstrated in the next example.

EXAMPLE 3   **Approximating a square root**

Approximate $\sqrt{17}$ to the nearest thousandth.

Solution
FIGURE 10.1 shows that $\sqrt{17} \approx 4.123$, rounded to the nearest thousandth. This result means that

$$4.123 \times 4.123 \approx 17.$$

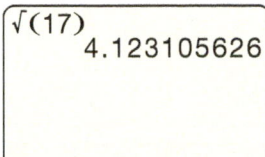

√(17)
                4.123105626

FIGURE 10.1

Now Try Exercise 39

READING CHECK 1

Are square roots of negative numbers real numbers?

SQUARE ROOTS OF NEGATIVE NUMBERS The square root of a *negative* number is *not* a real number. For example, $\sqrt{-4} \neq 2$ because $2 \cdot 2 \neq -4$ and $\sqrt{-4} \neq -2$ because $(-2)(-2) \neq -4$. (Later in this chapter, we will use the complex numbers to identify square roots of negative numbers.)

 Connecting Concepts with Your Life Have you ever noticed that if you climb up a hill or a tower, you can see farther to the horizon? This phenomenon can be described by a formula containing a square root, as demonstrated in the next example.

EXAMPLE 4   **Seeing the horizon**

A formula for calculating the distance d in miles that one can see to the horizon on a clear day is approximated by $d = 1.22\sqrt{x}$, where x is the elevation, in feet, of a person.
(a) Approximate how far a 6-foot-tall person can see to the horizon.
(b) Approximate how far a person can see from an 8000-foot mountain.
(c) If a person's elevation doubles, can this person see twice as far to the horizon?

Solution
(a) Let $x = 6$ in the formula $d = 1.22\sqrt{x}$.

$$d = 1.22\sqrt{6} \approx 3, \text{ or about 3 miles.}$$

(b) If $x = 8000$, then $d = 1.22\sqrt{8000} \approx 109$, or about 109 miles.
(c) No, this person can see $\sqrt{2} \approx 1.4$ times as far, which is less than twice as far.

Now Try Exercise 107

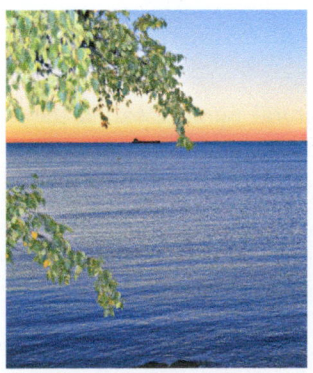

CUBE ROOTS The cube root of a number a is denoted $\sqrt[3]{a}$.

CALCULATOR HELP

To calculate a cube root, see Appendix A (page AP-1).

CUBE ROOT

The number b is a *cube root* of a if $b^3 = a$.

Although the square root of a negative number is *not* a real number, the cube root of a negative number is a negative real number. *Every real number has one real cube root.*

EXAMPLE 5 **Finding cube roots**

Evaluate the cube root. Approximate your answer to the nearest hundredth when appropriate.

(a) $\sqrt[3]{8}$ **(b)** $\sqrt[3]{-27}$ **(c)** $\sqrt[3]{\frac{1}{64}}$ **(d)** $\sqrt[3]{d^6}$ **(e)** $\sqrt[3]{16}$

Solution

(a) $\sqrt[3]{8} = 2$ because $2^3 = 2 \cdot 2 \cdot 2 = 8$.

(b) $\sqrt[3]{-27} = -3$ because $(-3)^3 = (-3)(-3)(-3) = -27$.

(c) $\sqrt[3]{\frac{1}{64}} = \frac{1}{4}$ because $\left(\frac{1}{4}\right)^3 = \frac{1}{4} \cdot \frac{1}{4} \cdot \frac{1}{4} = \frac{1}{64}$.

(d) $\sqrt[3]{d^6} = d^2$ because $(d^2)^3 = d^2 \cdot d^2 \cdot d^2 = d^{2+2+2} = d^6$.

(e) $\sqrt[3]{16}$ is not an integer. **FIGURE 10.2** shows that $\sqrt[3]{16} \approx 2.52$.

Now Try Exercises 23, 25, 27, 31, 41

```
³√(16)
        2.5198421
```

FIGURE 10.2

NOTE: $\sqrt[3]{-b} = -\sqrt[3]{b}$ for any real number b. That is, the cube root of a negative number is the negative of the cube root of the number. For example, $\sqrt[3]{-8} = -\sqrt[3]{8} = -2$. ∎

TABLE 10.1 illustrates how to evaluate both square roots and cube roots. If the radical expression is undefined, a dash is used.

Evaluating Radical Expressions

Expression	$\sqrt{64}$	$-\sqrt{64}$	$\sqrt{-64}$	$\sqrt[3]{64}$	$-\sqrt[3]{64}$	$\sqrt[3]{-64}$
Evaluated	8	−8	—	4	−4	−4

TABLE 10.1   Undefined

***n*th ROOTS** We can generalize square roots and cube roots to include *n*th roots of a number a. The number b is an ***n*th root** of a if $b^n = a$, where n is a positive integer. For example, $2^5 = 32$, so the **5**th root of **32** is **2** and can be written as $\sqrt[5]{32} = 2$.

THE NOTATION $\sqrt[n]{a}$

The equation $\sqrt[n]{a} = b$ means that $b^n = a$, where n is a natural number called the **index**. If n is odd, we are finding an **odd root** and if n is even, we are finding an **even root**.

1. If $a > 0$, then $\sqrt[n]{a}$ is a positive number. $\sqrt[4]{16} = 2$ and is positive.

2. If $a < 0$ and

   (a) n is odd, then $\sqrt[n]{a}$ is a negative number. $\sqrt[3]{-8} = -2$ and is negative.

   (b) n is even, then $\sqrt[n]{a}$ is *not* a real number. $\sqrt[4]{-8}$ is not a real number.

If $a > 0$ and n is even, then a has two real *n*th roots: one positive and one negative. The positive root is denoted $\sqrt[n]{a}$ and called the **principal *n*th root** of a. For example, $(-3)^4 = 81$ *and* $3^4 = 81$, but $\sqrt[4]{81} = 3$ in the same way *principal square roots* are calculated.

EXAMPLE 6   **Finding *n*th roots**

Find each root, if possible.
(a) $\sqrt[4]{16}$ (b) $\sqrt[5]{-32}$ (c) $\sqrt[4]{-81}$ (d) $-\sqrt[4]{81}$

Solution
(a) $\sqrt[4]{16} = 2$ because $2^4 = 2 \cdot 2 \cdot 2 \cdot 2 = 16$.
(b) $\sqrt[5]{-32} = -2$ because $(-2)^5 = (-2)(-2)(-2)(-2)(-2) = -32$.
(c) An *even* root of a *negative* number is *not* a real number.
(d) $-\sqrt[4]{81} = -3$ because $\sqrt[4]{81} = 3$.

Now Try Exercises 33, 35, 37

ABSOLUTE VALUE Consider the calculations

$$\sqrt{3^2} = \sqrt{9} = 3, \quad \sqrt{(-4)^2} = \sqrt{16} = 4, \quad \text{and} \quad \sqrt{(-6)^2} = \sqrt{36} = 6.$$

In general, the expression $\sqrt{x^2}$ equals $|x|$. See **FIGURE 10.3**.

CRITICAL THINKING 1

Evaluate $\sqrt[6]{(-2)^6}$ and $\sqrt[3]{(-2)^3}$.

Identical Graphs

(a)

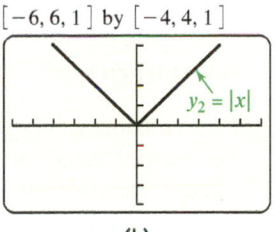
(b)

The graphs of
$y_1 = \sqrt{x^2}$
and
$y_2 = |x|$
are identical.

FIGURE 10.3

THE EXPRESSION $\sqrt{x^2}$

For every real number x, $\sqrt{x^2} = |x|$.

EXAMPLE 7   **Simplifying expressions**

Write each expression in terms of an absolute value.
(a) $\sqrt{(-3)^2}$ (b) $\sqrt{(x + 1)^2}$ (c) $\sqrt{z^2 - 4z + 4}$

Solution
(a) $\sqrt{x^2} = |x|$, so $\sqrt{(-3)^2} = |-3|$
(b) $\sqrt{(x + 1)^2} = |x + 1|$
(c) $\sqrt{z^2 - 4z + 4} = \sqrt{(z - 2)^2} = |z - 2|$

Now Try Exercises 45, 49, 51

2 Recognizing Square Root Functions and Their Properties

The important properties of the **square root function** are shown in the following See the Concept.

See the Concept 1 — THE SQUARE ROOT FUNCTION

Formula	Table of Values	Graph

$$f(x) = \sqrt{x}, \; x \geq 0$$

x	$\sqrt{x}$
0	0
1	1
4	2

A Domain: $x \geq 0$

$f(x) = \sqrt{x}$

B No graph left of origin

A The domain of the square root function is $[0, \infty)$.

B $f(x) = \sqrt{x}$ is defined for only nonnegative real numbers, so its graph does not appear to the left of the origin.

WATCH VIDEO IN MML.

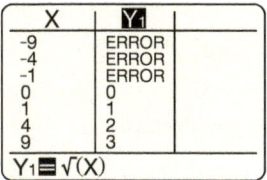

X	Y₁
-9	ERROR
-4	ERROR
-1	ERROR
0	0
1	1
4	2
9	3

Y₁ = √(X)

TECHNOLOGY NOTE

Square Roots of Negative Numbers

If a table of values for $y_1 = \sqrt{x}$ includes both negative and positive values for x, then many calculators give error messages when x is negative, as shown in the figure in the margin.

EXAMPLE 8 — Evaluating functions involving square roots

If possible, evaluate $f(1)$ and $f(-2)$ for each $f(x)$.

(a) $f(x) = \sqrt{2x - 1}$ **(b)** $f(x) = \sqrt{4 - x^2}$

Solution

(a) $f(1) = \sqrt{2(1) - 1} = \sqrt{1} = 1$

$f(-2) = \sqrt{2(-2) - 1} = \sqrt{-5}$, which does not equal a real number.

(b) $f(1) = \sqrt{4 - (1)^2} = \sqrt{3}$

$f(-2) = \sqrt{4 - (-2)^2} = \sqrt{0} = 0$

Now Try Exercises 61, 63

🌐 *Math in Context*   Sports   A good punter can kick a football so that the ball has a long *hang time*. Hang time is the length of time that the ball is in the air, and a long hang time gives the kicking team time to run down the field and stop the punt return. By using a function involving a square root, we can estimate hang time.

EXAMPLE 9 — Calculating hang time

If a football is kicked x feet high, then the time T in seconds that the ball is in the air is given by the function

$$T(x) = \frac{1}{2}\sqrt{x}.$$

(a) Find the hang time if the ball is kicked 50 feet into the air.

(b) Does the hang time double if the ball is kicked 100 feet in the air?

CRITICAL THINKING 2

How high would a football have to be kicked to have twice the hang time of a football kicked 50 feet into the air?

Solution

(a) The hang time is $T(50) = \frac{1}{2}\sqrt{50} \approx 3.5$ seconds.

(b) The hang time is $T(100) = \frac{1}{2}\sqrt{100} = 5$ seconds. The time does *not* double when the height doubles.

Now Try Exercise 105

EXAMPLE 10   **Finding the domain of a square root function**

Let $f(x) = \sqrt{x-1}$.

(a) Find the domain of f. Write your answer in interval notation.

(b) Graph $y = f(x)$ and compare it to the graph of $y = \sqrt{x}$.

Solution

(a) For $f(x)$ to be defined, $x - 1$ cannot be negative. Thus valid inputs for x must satisfy

$$x - 1 \geq 0 \quad \text{or} \quad x \geq 1.$$

To solve, add 1 to each side.

The domain is $[1, \infty)$.

(b) **TABLE 10.2** lists points that lie on the graph of $y = \sqrt{x-1}$. Note in **FIGURE 10.4** that the graph appears only when $x \geq 1$. Compared to the graph of $y = \sqrt{x}$, the graph of $y = \sqrt{x-1}$ is shifted one unit to the right.

$y = \sqrt{x-1}$

x	$\sqrt{x-1}$
1	0
2	1
5	2

TABLE 10.2

Shift the graph of $y = \sqrt{x}$ one unit to the right.

FIGURE 10.4

Now Try Exercises 75, 89

EXAMPLE 11   **Finding the domains of square root functions**

Find the domain of each function. Write your answer in interval notation.

(a) $f(x) = \sqrt{4 - 2x}$ (b) $g(x) = \sqrt{x^2 + 1}$

Solution

(a) To determine when $f(x) = \sqrt{4 - 2x}$ is defined, we must solve $4 - 2x \geq 0$.

The radicand must be nonnegative for f to be defined.

$4 - 2x \geq 0$	Inequality to be solved
$4 \geq 2x$	Add $2x$ to each side.
$2 \geq x$	Divide each side by 2.

The domain is $(-\infty, 2]$.

(b) Regardless of the value of x in $g(x) = \sqrt{x^2 + 1}$, the expression $x^2 + 1$ is always positive because $x^2 \geq 0$. Thus $g(x)$ is defined for all real numbers, and its domain is $(-\infty, \infty)$.

Now Try Exercises 83, 85

MAKING CONNECTIONS 1

Domains of Functions and Their Graphs

The graphs of f and g from Example 11 are shown below and can be used to find their domains.

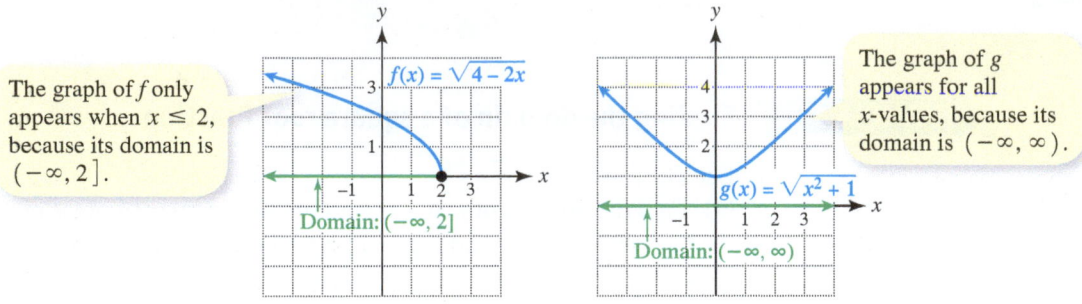

The graph of f only appears when $x \leq 2$, because its domain is $(-\infty, 2]$.

Domain: $(-\infty, 2]$

The graph of g appears for all x-values, because its domain is $(-\infty, \infty)$.

Domain: $(-\infty, \infty)$

🌐 *Math in Context* Business ^ Suppose you start a summer painting business, and initially you are the only employee. Because business is good, you hire another employee and productivity goes up, more than enough to pay for the new employee. As time goes on, you hire more employees. Eventually, productivity starts to level off because there are too many employees to keep busy. This situation is common in business and can be modeled by using a function involving a square root. In the next example, we analyze the benefit of adding employees to a small business.

EXAMPLE 12

Increasing productivity with more employees

The function $R(x) = 108\sqrt{x}$ gives the total revenue per year in thousands of dollars generated by a small business that has x employees. A graph of $y = R(x)$ is shown in **FIGURE 10.5**.

Revenue as a Function of Employees

$R(x) = 108\sqrt{x}$

Revenue ($ thousands)

Number of Employees

FIGURE 10.5

(a) Approximate values for $R(4)$, $R(8)$, and $R(12)$ by using both the formula and the graph.

(b) Evaluate $R(12) - R(8)$ and $R(8) - R(4)$ using the formula. Interpret your answer.

Solution

(a) $R(4) = 108\sqrt{4} = \$216$ thousand, $R(8) = 108\sqrt{8} \approx \$305$ thousand, and finally $R(12) = 108\sqrt{12} \approx \$374$ thousand.

We can graphically evaluate $R(4) \approx \$215$ thousand, $R(8) \approx \$305$ thousand, and $R(12) \approx \$375$ thousand, as shown in **FIGURE 10.6**.

Revenue as a Function of Employees

FIGURE 10.6

(b) $R(8) - R(4) \approx 305 - 216 = \$89$ thousand
$R(12) - R(8) \approx 374 - 305 = \$69$ thousand

There is more revenue gained when increasing from 4 to 8 employees than there is when increasing from 8 to 12 employees. Because the graph of $R(x) = 108\sqrt{x}$ starts to level off, there is a limited benefit to adding employees.

Now Try Exercise 109

3 Recognizing Cube Root Functions and Their Properties

The important properties of the cube root function are shown in the following See the Concept. Note that, unlike the square root function, the cube root function is defined for negative numbers.

See the Concept 2	THE CUBE ROOT FUNCTION

Formula	Table of Values	Graph
$f(x) = \sqrt[3]{x}$		

x	$\sqrt[3]{x}$
-8	-2
-1	-1
0	0
1	1
8	2

A Domain: $-\infty < x < \infty$

$f(x) = \sqrt[3]{x}$

B Graph is both left and right of the origin

A The domain of the cube root function is $(-\infty, \infty)$.

B $f(x) = \sqrt[3]{x}$ is defined for all real numbers, so its graph appears both to the left and to the right of the origin.

WATCH VIDEO IN MML.

EXAMPLE 13   **Evaluating functions involving cube roots**

Evaluate $f(1)$ and $f(-3)$ for each $f(x)$.

(a) $f(x) = \sqrt[3]{x^2 - 1}$ **(b)** $f(x) = \sqrt[3]{2 - x^2}$

Solution

(a) $f(1) = \sqrt[3]{1^2 - 1} = \sqrt[3]{0} = 0$; $f(-3) = \sqrt[3]{(-3)^2 - 1} = \sqrt[3]{8} = 2$

(b) $f(1) = \sqrt[3]{2 - 1^2} = \sqrt[3]{1} = 1$; $f(-3) = \sqrt[3]{2 - (-3)^2} = \sqrt[3]{-7}$ or $-\sqrt[3]{7}$

Now Try Exercises 65, 69

10.1 Putting It All Together

CONCEPT	EXPLANATION	EXAMPLES
nth Root of a Real Number	An nth root of a real number a is b if $b^n = a$, and the (principal) nth root is denoted $\sqrt[n]{a}$. If $a < 0$ and n is even, $\sqrt[n]{a}$ is not a real number.	The square roots of 25 are 5 and -5. The principal square root is $\sqrt{25} = 5$. $\sqrt[3]{-125} = -5$ because $(-5)^3 = -125$. $\sqrt[4]{-9}$ is not a real number.
Square Root and Cube Root Functions	$$f(x) = \sqrt{x} \quad \text{and} \quad g(x) = \sqrt[3]{x}$$ The cube root function g is defined for all inputs, whereas the square root function f is defined only for nonnegative inputs. **Square Root Function** **Cube Root Function**	$f(64) = \sqrt{64} = 8$ $f(-64) = \sqrt{-64}$ is *not* a real number. $g(64) = \sqrt[3]{64} = 4$ $g(-64) = \sqrt[3]{-64} = -4$

10.1 Exercises

MyMathLab®

CONCEPTS AND VOCABULARY

1. What are the square roots of 9?

2. What is the principal square root of 9?

3. What is the cube root of 8?

4. Does every real number have a cube root?

5. If $b^n = a$ and $b > 0$, then $\sqrt[n]{a} =$ _____.

6. What is $\sqrt{x^2}$ equal to?

7. Evaluate $\sqrt{-25}$, if possible.

8. Evaluate $\sqrt[3]{-27}$, if possible.

9. Which of the following (a–d) equals -4?
   a. $\sqrt{16}$ b. $\sqrt{-16}$
   c. $-\sqrt[3]{16}$ d. $\sqrt[3]{-64}$

10. Which of the following (a–d) equals $|2x + 1|$?
    a. $\sqrt{(2x + 1)^2}$ b. $\sqrt[3]{(2x + 1)^3}$
    c. $\left(\sqrt{2x + 1}\right)^2$ d. $\sqrt{(2x)^2 + 1}$

11. Sketch a graph of the square root function.

12. Sketch a graph of the cube root function.

13. What is the domain of the square root function?

14. What is the domain of the cube root function?

EVALUATING RADICAL EXPRESSIONS

Exercises 15–38: Evaluate the expression by hand, if possible. Variables represent any real number.

15. $\sqrt{9}$ 16. $\sqrt{121}$

17. $\sqrt{0.36}$ 18. $\sqrt{0.64}$

19. $\sqrt{\frac{16}{25}}$ 20. $\sqrt{\frac{9}{49}}$

21. $\sqrt{x^2}, x > 0$ 22. $\sqrt{(x - 1)^2}, x > 1$

23. $\sqrt[3]{27}$ 24. $\sqrt[3]{64}$

25. $\sqrt[3]{-64}$ 26. $-\sqrt[3]{-1}$

27. $\sqrt[3]{\frac{8}{27}}$ 28. $\sqrt[3]{-\frac{1}{125}}$

29. $-\sqrt[3]{x^9}$ 30. $\sqrt[3]{(x + 1)^6}$

31. $\sqrt[3]{27x^3}$ 32. $\sqrt[3]{(2x)^6}$

33. $\sqrt[4]{81}$ 34. $\sqrt[5]{-1}$

35. $\sqrt[5]{-243}$ 36. $\sqrt[4]{625}$

37. $\sqrt[4]{-16}$ 38. $\sqrt[6]{-64}$

Exercises 39–44: Approximate to the nearest hundredth.

39. $\sqrt{11}$ 40. $-\sqrt{5}$

41. $\sqrt[3]{5}$ 42. $\sqrt[3]{-13}$

43. $\sqrt[5]{-7}$ 44. $\sqrt[4]{6}$

CONNECTING ABSOLUTE VALUE AND RADICALS

Exercises 45–58: Simplify the expression. Assume that all variables are real numbers.

45. $\sqrt{(-4)^2}$ 46. $\sqrt{9^2}$

47. $\sqrt{y^2}$ 48. $\sqrt{z^4}$

49. $\sqrt{(x - 5)^2}$ 50. $\sqrt{(2x - 1)^2}$

51. $\sqrt{x^2 - 2x + 1}$ 52. $\sqrt{4x^2 + 4x + 1}$

53. $\sqrt[4]{y^4}$ 54. $\sqrt[4]{z^4}$

55. $\sqrt[4]{x^{12}}$ 56. $\sqrt[6]{x^6}$

57. $\sqrt[5]{x^5}$ 58. $\sqrt[5]{32(x + 4)^5}$

EVALUATING SQUARE AND CUBE ROOT FUNCTIONS

Exercises 59–74: If possible, evaluate the function at the given value(s) of the variable.

59. $f(x) = \sqrt{x - 1}$ $x = 10, 0$

60. $f(x) = \sqrt{4 - 3x}$ $x = -4, 1$

61. $f(x) = \sqrt{3 - 3x}$ $x = -1, 5$

62. $f(x) = \sqrt{x - 5}$ $x = -1, 5$

63. $f(x) = \sqrt{x^2 - x}$ $x = -4, 3$

64. $f(x) = \sqrt{2x^2 - 3}$ $x = -1, 2$

65. $f(x) = \sqrt[3]{x^2 - 8}$ $x = -3, 4$

66. $f(x) = \sqrt[3]{2x^2}$ $x = -2, 2$

67. $f(x) = \sqrt[3]{5x - 2}$ $x = -5, 2$

68. $f(x) = \sqrt[3]{x - 9}$ $x = 1, 10$

69. $f(x) = \sqrt[3]{3 - x^2}$ $x = -2, 3$

70. $f(x) = \sqrt[3]{-1 - x^2}$ $x = 0, 3$

71. $T(h) = \frac{1}{2}\sqrt{h}$ $h = 64$

72. $L(k) = 2\sqrt{k + 2}$ $k = 23$

73. $f(x) = \sqrt{x + 5} + \sqrt{x}$ $x = 4$

74. $f(x) = \dfrac{\sqrt{x - 5} - \sqrt{x}}{2}$ $x = 9$

FINDING THE DOMAIN OF ROOT FUNCTIONS

Exercises 75–88: Find the domain of f. Write your answer in interval notation.

75. $f(x) = \sqrt{x + 2}$ 76. $f(x) = \sqrt{x - 1}$

77. $f(x) = \sqrt{x - 2}$ 78. $f(x) = \sqrt{x + 1}$

79. $f(x) = \sqrt{2x - 4}$ 80. $f(x) = \sqrt{4x + 2}$

81. $f(x) = \sqrt{1 - x}$ 82. $f(x) = \sqrt{6 - 3x}$

83. $f(x) = \sqrt{8 - 5x}$ 84. $f(x) = \sqrt{3 - 2x}$

85. $f(x) = \sqrt{3x^2 + 4}$ 86. $f(x) = \sqrt{1 + 2x^2}$

87. $f(x) = \dfrac{1}{\sqrt{2x + 1}}$ 88. $f(x) = \dfrac{1}{\sqrt{x - 1}}$

APPLYING TRANSLATIONS OF GRAPHS

Exercises 89–94: Graph the function and give its domain in interval notation. Compare the graph of $y = f(x)$ to either $y = \sqrt{x}$ or $y = \sqrt[3]{x}$.

89. $f(x) = \sqrt{x+2}$ **90.** $f(x) = \sqrt{x-1}$

91. $f(x) = \sqrt{x} + 2$ **92.** $f(x) = \sqrt[3]{x} + 2$

93. $f(x) = \sqrt[3]{x+2}$ **94.** $f(x) = \sqrt[3]{x} - 1$

REPRESENTING ROOT FUNCTIONS

Exercises 95–102: Give symbolic, numerical, and graphical representations for the function f.

95. Function f takes the square root of x and then adds 1 to the result.

96. Function f takes the square root of x and then subtracts 2 from the result.

97. Function f takes the square root of the quantity three times x.

98. Function f takes the square root of the quantity x plus 1.

99. Function f takes the cube root of x and then multiplies the result by 2.

100. Function f takes the cube root of the quantity 4 times x.

101. Function f takes the cube root of the quantity x minus 1.

102. Function f takes the cube root of x and then adds 1.

FINDING AREA OF A TRIANGLE

Exercises 103 and 104: **Heron's Formula** *Suppose the lengths of the sides of a triangle are a, b, and c as illustrated in the figure.*

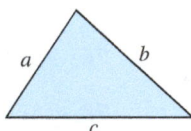

If the **semiperimeter** *(half of the perimeter) of the triangle is $s = \frac{1}{2}(a + b + c)$, then the area of the triangle is*

$$A = \sqrt{s(s-a)(s-b)(s-c)}.$$

Find the area A of the triangle with the given sides.

103. $a = 3, b = 4, c = 5$

104. $a = 5, b = 9, c = 10$

APPLICATIONS INVOLVING RADICAL EXPRESSIONS

105. **Jumping** (Refer to Example 9.) If a person jumps 4 feet off the ground, estimate how long the person is in the air.

106. **Hang Time** (Refer to Example 9.) Find the hang time for a golf ball hit 80 feet into the air.

107. **Distance to the Horizon** (Refer to Example 4.) Use the formula $d = 1.22\sqrt{x}$ to estimate how many miles a person can see from a jet airliner at 10,000 feet.

108. **Distance to the Horizon** (Refer to Example 4.) Use the formula $d = 1.22\sqrt{x}$ to estimate how many miles a 5-foot-tall person can see standing on the deck of a ship that is 50 feet above the ocean.

109. **Increasing Productivity** (Refer to Example 12.) Use $R(x) = 108\sqrt{x}$ to evaluate $R(16) - R(15)$. If the salary for the sixteenth employee is $25,000, is it a good decision to hire the sixteenth employee?

110. **Increasing Productivity** (Refer to Example 12.) Use $R(x) = 108\sqrt{x}$ to evaluate $R(2) - R(1)$. If the salary for the second employee is $25,000, is it a good decision to hire the second employee?

111. **Productivity of Workers** If workers are given more equipment, or *physical capital*, they are often more productive. For example, a carpenter definitely needs a hammer to be productive, but probably does not need 20 hammers. There is a leveling off in a worker's productivity as more is spent on equipment. The function $P(x) = 400\sqrt{x} + 8000$ approximates the worth of the goods produced by a typical U.S. worker in dollars when x dollars are spent on equipment per worker.
   (a) Evaluate $P(25{,}000)$ and interpret the result.
   (b) Sketch a graph of $y = P(x)$.
   (c) Use the graph from (b) and the formula to evaluate $P(50{,}000) - P(25{,}000)$ and also to evaluate $P(75{,}000) - P(50{,}000)$. Interpret the result.

112. **Design of Open Channels** To protect cities from flooding during heavy rains, open channels are sometimes constructed to handle runoff. The rate R at which water flows through the channel is modeled by $R = k\sqrt{m}$, where m is the slope of the channel and k is a constant determined by the shape of the channel. (*Source:* N. Garber and L. Hoel, *Traffic and Highway Design.*)
   (a) Suppose that a channel has a slope of $m = 0.01$ (or 1%) and a runoff rate of $R = 340$ cubic feet per second (cfs). Find k.
   (b) If the slope of the channel increases to $m = 0.04$ (or 4%), what happens to R? Be specific.

113. **Genetically Engineered Soybeans** The following graph shows the percentage y of planted acres of genetically engineered soybeans from 2000 to 2012, where x is the number of years after 2000. This growth is modeled by $y = 13\sqrt{x} + 50$.

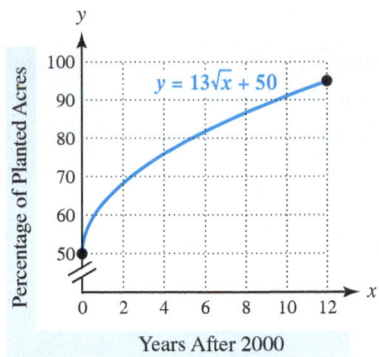

Years After 2000

(a) Find y when $x = 0, 6$, and 12.

(b) Interpret your result when $x = 12$.

(c) What happens if you use this equation to estimate this percentage in 1990?

114. **Women in Pharmacy** From 1975 to 2005, the percentage y of pharmacy degrees granted to women x years after 1975 could be modeled by $y = 7.5\sqrt{x} + 30$.

(a) Find y when $x = 0$ and 30.

(b) Interpret each answer.

(c) What happens if you use this equation to estimate this percentage in 1970?

WRITING ABOUT MATHEMATICS

115. Try to calculate $\sqrt{-7}$, $\sqrt[4]{-56}$, and $\sqrt[6]{-10}$ with a calculator. Describe what happens when you evaluate an even root of a negative number. Does the same difficulty occur when you evaluate an odd root of a negative number? Try to evaluate $\sqrt[3]{-7}$, $\sqrt[5]{-56}$, and $\sqrt[7]{-10}$. Explain.

116. Explain the difference between a root and a positive integer power of a number. Give examples.

10.2 Rational Exponents

Objectives

1 **Using Rational Numbers as Exponents**
- The Expression $a^{1/n}$
- The Expression $a^{m/n}$
- The Expression $a^{-m/n}$

2 **Applying Properties of Rational Exponents**

1 Using Rational Numbers as Exponents

THE EXPRESSION $a^{1/n}$ When m and n are integers, the product rule states that $a^m a^n = a^{m+n}$. This rule can be extended to include exponents that are fractions. For example,

$$4^{1/2} \cdot 4^{1/2} = 4^{1/2 + 1/2} = 4^1 = 4.$$

That is, if we multiply the expression $4^{1/2}$ by itself, the result is 4. Because we also know that $\sqrt{4} \cdot \sqrt{4} = 4$, this discussion suggests that $4^{1/2} = \sqrt{4}$ and leads to the following definition.

THE EXPRESSION $a^{1/n}$

If n is an integer greater than 1 and a is a real number, then

$$a^{1/n} = \sqrt[n]{a}. \qquad 7^{1/5} = \sqrt[5]{7}$$

NOTE: If $a < 0$ and n is an even positive integer, then $a^{1/n}$ is *not* a real number. ∎

In the next two examples, we show how to interpret rational exponents.

| EXAMPLE 1 | **Interpreting rational exponents** |

Write each expression in radical notation. Then evaluate the expression and round to the nearest hundredth when appropriate.

(a) $36^{1/2}$ (b) $23^{1/5}$ (c) $x^{1/3}$ (d) $(5x)^{1/2}$

Solution

(a) The exponent $\frac{1}{2}$ indicates a square root. Thus $36^{1/2} = \sqrt{36}$, which evaluates to 6.

(b) The exponent $\frac{1}{5}$ indicates a fifth root. Thus $23^{1/5} = \sqrt[5]{23}$, which is not an integer. **FIGURE 10.7** shows this expression approximated in both exponential and radical notation. In either case, $23^{1/5} \approx 1.87$.

(c) The exponent $\frac{1}{3}$ indicates a cube root, so $x^{1/3} = \sqrt[3]{x}$.

(d) The exponent $\frac{1}{2}$ indicates a square root, so $(5x)^{1/2} = \sqrt{5x}$.

Now Try Exercises 17, 45, 59, 63

```
23^(1/5)
          1.872171231
5×√(23)
          1.872171231
```

FIGURE 10.7

CALCULATOR HELP

To calculate other roots, see Appendix A (pages AP-7 and AP-8).

THE EXPRESSION $a^{m/n}$ Suppose that we want to define the expression $8^{2/3}$. On the one hand, using properties of exponents, we have

$$8^{1/3} \cdot 8^{1/3} = 8^{1/3 + 1/3} = 8^{2/3}.$$

On the other hand, we have

$$8^{1/3} \cdot 8^{1/3} = \sqrt[3]{8} \cdot \sqrt[3]{8} = 2 \cdot 2 = 4.$$

Thus $8^{2/3} = 4$, and that value is obtained whether we interpret $8^{2/3}$ as either

$$8^{2/3} = (8^{1/3})^2 = (\sqrt[3]{8})^2 = 2^2 = 4$$

or

$$8^{2/3} = (8^2)^{1/3} = \sqrt[3]{8^2} = \sqrt[3]{64} = 4.$$

This result suggests the following definition.

STUDY TIP

Be sure that you understand how rational exponents relate to radical notation.

THE EXPRESSION $a^{m/n}$

If m and n are positive integers with $\frac{m}{n}$ in lowest terms, then

$$a^{m/n} = \sqrt[n]{a^m} = (\sqrt[n]{a})^m. \qquad 2^{3/4} = \sqrt[4]{2^3} = (\sqrt[4]{2})^3$$

NOTE: If $a < 0$ and n is an even integer, then $a^{m/n}$ is *not* a real number. ∎

The exponent $\frac{m}{n}$ indicates that we either take the nth root and then calculate the mth power of the result or calculate the mth power and then take the nth root. For example, $4^{3/2}$ means that we can either take the **square root** of 4 and then **cube** the result or we can **cube** 4 and then take the **square root** of the result. In either case, the result is the same: $4^{3/2} = 8$. This concept is illustrated in the next example.

| EXAMPLE 2 | **Interpreting rational exponents** |

Write each expression in radical notation. Evaluate the expression by hand when possible.

(a) $(-27)^{2/3}$ (b) $12^{3/5}$

Solution

(a) The exponent $\frac{2}{3}$ indicates either that we take the **cube root** of -27 and then **square** it or that we **square** -27 and then take the **cube root**. Thus

$$(-27)^{2/3} = (\sqrt[3]{-27})^2 = (-3)^2 = 9$$

or

$$(-27)^{2/3} = \sqrt[3]{(-27)^2} = \sqrt[3]{729} = 9.$$

Same result

(b) The exponent $\frac{3}{5}$ indicates either that we take the **fifth root** of 12 and then **cube** it or that we **cube** 12 and then take the **fifth root**. Thus

$$12^{3/5} = (\sqrt[5]{12})^3 \quad \text{or} \quad 12^{3/5} = \sqrt[5]{12^3}.$$

This result cannot be evaluated by hand.

Now Try Exercises 47, 51

TECHNOLOGY NOTE

Rational Exponents
When evaluating expressions with rational (fractional) exponents, be sure to put parentheses around the fraction. For example, most calculators will evaluate 8^(2/3) and 8^2/3 differently. The accompanying figure shows that evaluating $8^{2/3}$ correctly as 8^(2/3) results in 4, but evaluating $8^{2/3}$ incorrectly as 8^2/3 results in $\frac{8^2}{3} = 21.\overline{3}$.

Correct →
Incorrect →

```
8^(2/3)
                4
8^2/3
        21.33333333
```

🌐 *Math in Context* Online Video In today's society, people are accustomed to having a lot of choices when it comes to online videos. In fact, if a typical online video has 1,000,000 viewers, then about 200,000 of them, or 20%, have abandoned the video in 10 seconds or less. This average viewer abandonment rate can be modeled by a new function A, given by

$$A(x) = 7.3x^{7/16},$$

where A computes the percentage of viewers who abandon an online video after x seconds. (*Source: Business insider.*)

EXAMPLE 3 **Calculating abandonment rates for online videos**

The function $A(x) = 7.3x^{7/16}$ computes the percentage of viewers who abandon an online video after x seconds.
(a) Find $A(10)$ and $A(90)$. Interpret your answers.
(b) Use radical notation to write $A(x)$.

Solution
(a) To approximate these expressions, we will use a calculator, as shown in **FIGURE 10.8**.

$$A(\mathbf{10}) = 7.3(\mathbf{10})^{7/16} \approx \mathbf{20\%}$$
$$A(\mathbf{90}) = 7.3(\mathbf{90})^{7/16} \approx \mathbf{52\%}$$

After **10** seconds, about **20%** of the viewers have abandoned the online video, and after **90** seconds, over half, or **52%**, of the viewers have abandoned the online video.

(b) The exponent $\frac{7}{16}$ means we raise x to the **seventh** power and then take the **sixteenth** root. That is, $A(x) = 7.3\sqrt[16]{x^7}$.

```
7.3*10^(7/16)
          19.99046333
7.3*90^(7/16)
          52.27619364
```

FIGURE 10.8

Now Try Exercise 97

THE EXPRESSION $a^{-m/n}$ From properties of exponents, we know that $a^{-n} = \frac{1}{a^n}$, where n is a positive integer. We now define this property for negative rational exponents.

READING CHECK 1

• Write the expression $8^{-2/3}$ in radical notation and then evaluate it.

> ### THE EXPRESSION $a^{-m/n}$
>
> If m and n are positive integers with $\frac{m}{n}$ in lowest terms, then
>
> $$a^{-m/n} = \frac{1}{a^{m/n}}, \qquad a \neq 0. \qquad 2^{-3/4} = \frac{1}{2^{3/4}}$$

Examples of changing rational exponents to radical expressions are shown in **TABLE 10.3**.

Converting Rational Exponents to Radical Notation

Expression	$a^{3/4}$	$5^{1/7}$	$x^{-5/3}$
Radical Form	$\sqrt[4]{a^3}$	$\sqrt[7]{5}$	$\dfrac{1}{\sqrt[3]{x^5}}$

TABLE 10.3

EXAMPLE 4 **Interpreting negative rational exponents**

Write each expression in radical notation and then evaluate.
(a) $64^{-1/3}$ **(b)** $81^{-3/4}$

Solution

(a) $64^{-1/3} = \dfrac{1}{64^{1/3}} = \dfrac{1}{\sqrt[3]{64}} = \dfrac{1}{4}.$

(b) $81^{-3/4} = \dfrac{1}{81^{3/4}} = \dfrac{1}{(\sqrt[4]{81})^3} = \dfrac{1}{3^3} = \dfrac{1}{27}.$

Now Try Exercises 53, 55

EXAMPLE 5 **Converting to rational exponents**

Use rational exponents to write each radical expression.
(a) $\sqrt[5]{x^4}$ **(b)** $\dfrac{1}{\sqrt{b^5}}$ **(c)** $\sqrt[3]{(x-1)^2}$ **(d)** $\sqrt[3]{a^2 + b^2}$

Solution
(a) $\sqrt[5]{x^4} = x^{4/5}$

(b) $\dfrac{1}{\sqrt{b^5}} = b^{-5/2}$

(c) $\sqrt[3]{(x-1)^2} = (x-1)^{2/3}$

(d) $\sqrt[3]{a^2 + b^2} = (a^2 + b^2)^{1/3}$ Note that $\sqrt[3]{a^2 + b^2} \neq a^{2/3} + b^{2/3}$.

Now Try Exercises 35, 37, 41, 43

🌐 *Math in Context* Biology In the next example, we use a formula from biology that involves a rational exponent.

EXAMPLE 6   **Analyzing stepping frequency**

When smaller (four-legged) animals walk, they tend to take faster, shorter steps, whereas larger animals tend to take slower, longer steps. If an animal is h feet high at the shoulder, then the number N of steps per second that the animal takes *while walking* can be estimated by $N(h) = 1.6h^{-1/2}$. (*Source:* C. Pennycuick, *Newton Rules Biology.*)

(a) Use radical notation to write $N(h)$.

(b) Estimate the stepping frequency for an adult African elephant that is 10 feet high at the shoulders and a baby elephant that is only 33 inches high.

Solution

(a) We can rewrite the formula as $N(h) = \dfrac{1.6}{\sqrt{h}}$.

(b) We can evaluate N as follows. Note that 33 inches is $\dfrac{33}{12} = 2.75$ feet.

$$N(10) = \frac{1.6}{\sqrt{10}} \approx 0.51 \quad \text{and} \quad N(2.75) = \frac{1.6}{\sqrt{2.75}} \approx 0.96$$

While walking, the adult elephant takes about $\frac{1}{2}$ step every second, or 1 step every 2 seconds, and the baby elephant takes about 1 step every second.

Now Try Exercise 99

2 Applying Properties of Rational Exponents

Any rational number can be written as a ratio of two integers. That is, if p is a rational number, then $p = \frac{m}{n}$, where m and n are integers. Properties for integer exponents also apply to rational exponents, with one exception. If n is even in the expression $a^{m/n}$ and $\frac{m}{n}$ is written in lowest terms, then a must be nonnegative for the result to be a real number.

For example, the expression $(-8)^{3/2}$ is not a real number because -8 is negative and 2 is even. That is $(-8)^{3/2} = (\sqrt{-8})^3$ is not a real number because the square root of a negative number is not a real number.

PROPERTIES OF EXPONENTS

Let p and q be rational numbers written in lowest terms. For all real numbers a and b for which the expressions are real numbers, the following properties hold.

1. $a^p \cdot a^q = a^{p+q}$ Product rule for exponents $3^{1/7} \cdot 3^{2/7} = 3^{3/7}$

2. $a^{-p} = \dfrac{1}{a^p},\ \dfrac{1}{a^{-p}} = a^p$ Negative exponents $7^{-1/4} = \dfrac{1}{7^{1/4}},\ \dfrac{1}{7^{-1/4}} = 7^{1/4}$

3. $\left(\dfrac{a}{b}\right)^{-p} = \left(\dfrac{b}{a}\right)^p$ Negative exponents for quotients $\left(\dfrac{3}{7}\right)^{-1/3} = \left(\dfrac{7}{3}\right)^{1/3}$

4. $\dfrac{a^p}{a^q} = a^{p-q}$ Quotient rule for exponents $\dfrac{5^{4/7}}{5^{1/7}} = 5^{3/7}$

5. $(a^p)^q = a^{pq}$ Power rule for exponents $(2^{1/3})^{1/2} = 2^{1/6}$

6. $(ab)^p = a^p b^p$ Power rule for products $(5x)^{2/3} = 5^{2/3}x^{2/3}$

7. $\left(\dfrac{a}{b}\right)^p = \dfrac{a^p}{b^p}$ Power rule for quotients $\left(\dfrac{5}{3}\right)^{3/4} = \dfrac{5^{3/4}}{3^{3/4}}$

In the next two examples, we apply these properties.

EXAMPLE 7 **Applying properties of exponents**

Write each expression using rational exponents and simplify. Write the answer with a positive exponent. Assume that all variables are positive numbers.

(a) $\sqrt{x} \cdot \sqrt[3]{x}$ (b) $\sqrt[3]{27x^2}$ (c) $\dfrac{\sqrt[4]{16x}}{\sqrt[3]{x}}$ (d) $\left(\dfrac{x^2}{81}\right)^{-1/2}$

Solution

(a) $\sqrt{x} \cdot \sqrt[3]{x} = x^{1/2} \cdot x^{1/3}$ Use rational exponents.

$= x^{1/2 + 1/3}$ Product rule for exponents

$= x^{5/6}$ Simplify: $\frac{1}{2} + \frac{1}{3} = \frac{3}{6} + \frac{2}{6} = \frac{5}{6}$.

(b) $\sqrt[3]{27x^2} = (27x^2)^{1/3}$ Use rational exponents.

$= 27^{1/3}(x^2)^{1/3}$ Power rule for products

$= 3x^{2/3}$ Simplify; power rule for exponents.

(c) $\dfrac{\sqrt[4]{16x}}{\sqrt[3]{x}} = \dfrac{(16x)^{1/4}}{x^{1/3}}$ Use rational exponents.

$= \dfrac{16^{1/4}x^{1/4}}{x^{1/3}}$ Power rule for products

$= 16^{1/4}x^{1/4 - 1/3}$ Quotient rule for exponents

$= 2x^{-1/12}$ Simplify: $\frac{1}{4} - \frac{1}{3} = \frac{3}{12} - \frac{4}{12} = -\frac{1}{12}$.

$= \dfrac{2}{x^{1/12}}$ Negative exponents

(d) $\left(\dfrac{x^2}{81}\right)^{-1/2} = \left(\dfrac{81}{x^2}\right)^{1/2}$ Negative exponents for quotients

$= \dfrac{(81)^{1/2}}{(x^2)^{1/2}}$ Power rule for quotients

$= \dfrac{9}{x}$ Simplify; power rule for exponents.

Now Try Exercises 73, 79, 87, 93

EXAMPLE 8 **Applying properties of exponents**

Write each expression with positive rational exponents and simplify, if possible.

(a) $\sqrt[3]{\sqrt{x+1}}$ (b) $\sqrt[5]{c^{15}}$ (c) $\dfrac{y^{-1/2}}{x^{-1/3}}$ (d) $\sqrt{x}(\sqrt{x} - 1)$

Solution

(a) $\sqrt[3]{\sqrt{x+1}} = \left((x+1)^{1/2}\right)^{1/3} = (x+1)^{1/6}$

(b) $\sqrt[5]{c^{15}} = c^{15/5} = c^3$

(c) $\dfrac{y^{-1/2}}{x^{-1/3}} = \dfrac{x^{1/3}}{y^{1/2}}$

(d) $\sqrt{x}(\sqrt{x} - 1) = x^{1/2}(x^{1/2} - 1) = x^{1/2}x^{1/2} - x^{1/2} = x - x^{1/2}$

Now Try Exercises 81, 85, 91

10.2 Putting It All Together

CONCEPT	EXPLANATION	EXAMPLES
Rational Exponents	If m and n are positive integers with $\frac{m}{n}$ in lowest terms, $$a^{m/n} = \sqrt[n]{a^m} = \left(\sqrt[n]{a}\right)^m.$$ If $a < 0$ and n is even, $a^{m/n}$ is not a real number.	$8^{4/3} = \left(\sqrt[3]{8}\right)^4 = 2^4 = 16$ $(-27)^{3/4} = \left(\sqrt[4]{-27}\right)^3$ is *not* a real number.
Properties of Exponents	Let p and q be rational numbers. 1. $a^p \cdot a^q = a^{p+q}$ 2. $a^{-p} = \frac{1}{a^p}, \frac{1}{a^{-p}} = a^p$ 3. $\left(\frac{a}{b}\right)^{-p} = \left(\frac{b}{a}\right)^p$ 4. $\frac{a^p}{a^q} = a^{p-q}$ 5. $(a^p)^q = a^{pq}$ 6. $(ab)^p = a^p b^p$ 7. $\left(\frac{a}{b}\right)^p = \frac{a^p}{b^p}$	1. $2^{1/3} \cdot 2^{2/3} = 2^{1/3 + 2/3} = 2^1 = 2$ 2. $2^{-1/2} = \frac{1}{2^{1/2}}, \frac{1}{3^{-1/4}} = 3^{1/4}$ 3. $\left(\frac{3}{4}\right)^{-4/5} = \left(\frac{4}{3}\right)^{4/5}$ 4. $\frac{7^{2/3}}{7^{1/3}} = 7^{2/3-1/3} = 7^{1/3}$ 5. $\left(8^{2/3}\right)^{1/2} = 8^{(2/3)\cdot(1/2)} = 8^{1/3} = 2$ 6. $(2x)^{1/3} = 2^{1/3} x^{1/3}$ 7. $\left(\frac{x}{y}\right)^{1/6} = \frac{x^{1/6}}{y^{1/6}}$

10.2 Exercises

MyMathLab®

CONCEPTS AND VOCABULARY

1. Simplify $4^{1/2}$.

2. Simplify $8^{1/3}$.

3. Simplify $4^{-1/2}$.

4. Simplify $8^{-1/3}$.

5. Use a rational exponent to write $\sqrt{x}$.

6. Use a rational exponent to write $\sqrt[3]{a^4}$.

7. Write $a^{1/n}$ in radical notation.

8. Write $a^{m/n}$ in radical notation.

Exercises 9–16: Match the given expression to the expression (a.–h.) that it equals.

9. $\sqrt{x^3}$

10. $\sqrt[3]{8^2}$

11. $25^{-1/2}$

12. $27^{-2/3}$

13. $x^{1/5}$

14. $(x^{1/3})^2$

15. $\sqrt{x} \cdot \sqrt[3]{x}$

a. $\sqrt[6]{x^5}$

c. $x^{3/2}$

e. $\sqrt[3]{x^2}$

g. $\frac{1}{5}$

16. $\frac{x^{1/2}}{x^{1/3}}$

b. $\sqrt[6]{x}$

d. $\sqrt[5]{x}$

f. 4

h. $\frac{1}{9}$

EVALUATING RATIONAL EXPONENTS

Exercises 17–24: Approximate to the nearest hundredth.

17. $16^{1/5}$

18. $7^{1/4}$

19. $5^{1/3}$

20. $11^{1/2}$

21. $9^{3/5}$

22. $13^{5/4}$

23. $4^{-3/7}$

24. $2^{-3/4}$

CONVERTING BETWEEN RATIONAL EXPONENTS AND RADICAL NOTATION

Exercises 25–34: Use radical notation to write each expression.

25. $7^{1/2}$

26. $10^{1/2}$

27. $a^{1/3}$

28. $y^{1/4}$

29. $x^{5/6}$

30. $x^{3/2}$

31. $(x + y)^{1/2}$

32. $(x + y)^{2/3}$

33. $b^{-2/3}$

34. $b^{-3/4}$

Exercise 35–44: Use rational exponents to write each expression.

35. $\sqrt{t}$

36. $\sqrt[3]{t}$

37. $\sqrt[3]{(x + 1)}$

38. $\sqrt{y - 3}$

39. $\dfrac{1}{\sqrt{x + 1}}$

40. $\dfrac{1}{\sqrt[3]{x - 3}}$

41. $\sqrt{a^2 - b^2}$

42. $\sqrt[3]{a^3 + b^3}$

43. $\dfrac{1}{\sqrt[3]{x^7}}$

44. $\dfrac{1}{\sqrt[3]{y^4}}$

Exercises 45–64: Write each expression in radical notation. Evaluate the expression by hand when possible.

45. $9^{1/2}$

46. $100^{1/2}$

47. $8^{1/3}$

48. $(-8)^{1/3}$

49. $\left(\frac{4}{9}\right)^{1/2}$

50. $\left(\frac{64}{27}\right)^{1/3}$

51. $(-8)^{2/3}$

52. $(16)^{3/2}$

53. $\left(\frac{1}{8}\right)^{-1/3}$

54. $\left(\frac{1}{81}\right)^{-1/4}$

55. $16^{-3/4}$

56. $(32)^{-2/5}$

57. $(4^{1/2})^{-3}$

58. $(8^{1/3})^{-2}$

59. $z^{1/4}$

60. $b^{1/3}$

61. $y^{-2/5}$

62. $z^{-3/4}$

63. $(3x)^{1/3}$

64. $(5x)^{1/2}$

USING PROPERTIES OF RATIONAL EXPONENTS

Exercises 65–94: Use positive rational exponents to simplify the expression. Assume that all variables are positive.

65. $\sqrt{x} \cdot \sqrt{x}$

66. $\sqrt[3]{x} \cdot \sqrt[3]{x}$

67. $\sqrt[3]{8x^2}$

68. $\sqrt[3]{27z}$

69. $\dfrac{\sqrt{49x}}{\sqrt[3]{x^2}}$

70. $\dfrac{\sqrt[3]{8x}}{\sqrt[4]{x^3}}$

71. $(x^2)^{3/2}$

72. $(y^4)^{1/2}$

73. $\sqrt[3]{x^3 y^6}$

74. $\sqrt{16x^4}$

75. $\sqrt{y^3} \cdot \sqrt[3]{y^2}$

76. $\left(\dfrac{x^6}{81}\right)^{1/4}$

77. $\left(\dfrac{x^6}{27}\right)^{2/3}$

78. $\left(\dfrac{1}{x^8}\right)^{-1/4}$

79. $\left(\dfrac{x^2}{y^6}\right)^{-1/2}$

80. $\dfrac{\sqrt{x}}{\sqrt[3]{27x^6}}$

81. $\sqrt{\sqrt{y}}$

82. $\sqrt{\sqrt[3]{(3x)^2}}$

83. $(a^{-1/2})^{4/3}$

84. $(x^{-3/2})^{2/3}$

85. $\dfrac{(k^{1/2})^{-3}}{(k^2)^{1/4}}$

86. $\dfrac{(b^{3/4})^4}{(b^{4/5})^{-5}}$

87. $\sqrt{b} \cdot \sqrt[4]{b}$

88. $\sqrt[3]{t} \cdot \sqrt[5]{t}$

89. $p^{1/2}(p^{3/2} + p^{1/2})$

90. $d^{3/4}(d^{1/4} - d^{-1/4})$

91. $\sqrt[3]{x}(\sqrt{x} - \sqrt[3]{x^2})$

92. $\frac{1}{2}\sqrt{x}(\sqrt{x} + \sqrt[4]{x^2})$

93. $\dfrac{\sqrt[3]{27x}}{\sqrt{x}}$

94. $\dfrac{\sqrt[5]{32x^2}}{\sqrt[3]{8x}}$

Exercises 95 and 96: **Thinking Generally** *Simplify the expression. Assume that a and b are positive integers.*

95. $(b^{1/a} b^{2/a})^a$

96. $b^{(b-1)/b} \cdot \sqrt[b]{b}$

APPLICATIONS INVOLVING RATIONAL EXPONENTS

97. **Online Videos** (Refer to Example 3.) The function given by $A(x) = 7.3x^{7/16}$ computes the percentage of viewers who abandon an online video after x seconds for $0 \le x \le 120$.

   (a) Make a table of values for $x = 0, 20, 40, 60,$ and 80. Round values to the nearest percent.

   (b) Interpret the table in terms of how people watch online videos.

98. **Online Videos** (Refer to Example 3.) Suppose that the function given by $A(x) = 7.3x^{4/3}$ computes the percentage of viewers who abandon an online video after x seconds. Evaluate $A(10)$. Is this function realistic?

99. **Animal Stepping Frequency** (Refer to Example 6.) Use the formula $N(h) = 1.6h^{-1/2}$ to estimate the stepping frequency of a dog that is 2.5 feet high at the shoulders. (*Source:* C. Pennycuick.)

100. Animal Pulse Rate According to one model, an animal's heart rate varies according to its weight. The formula $N(w) = 885w^{-1/2}$ gives an estimate for the average number N of beats per minute for an animal that weighs w pounds. Use the formula to estimate the heart rate for a horse that weighs 800 pounds. (*Source:* C. Pennycuick.)

101. Bird Wings Heavier birds tend to have larger wings than lighter birds do. For some birds, the relationship between the surface area A of the bird's wings in square inches and its weight W in pounds can be modeled by $A = 100\sqrt[3]{W^2}$. (*Source:* C. Pennycuick, *Newton Rules Biology.*)
  (a) Find the area of the wings when the weight is 8 pounds.
  (b) Write this formula with rational exponents.

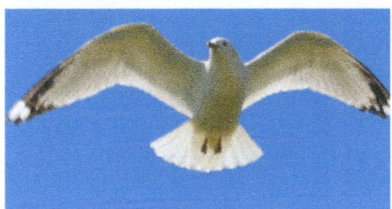

102. Planet Orbits and Distance Johannes Kepler (1571–1630) discovered a relationship between a planet's distance D from the sun and the time T it takes to orbit the sun. This formula is $T = \sqrt{D^3}$, where T is in Earth years and $D = 1$ corresponds to the distance between Earth and the sun, or 93,000,000 miles.
  (a) The planet Neptune is 30 times farther from the sun than Earth $(D = 30)$. Estimate the number of years required for Neptune to orbit the sun.
  (b) Write this formula with rational exponents.

103. Baby's Head Size If a female baby's head circumference is 35.2 centimeters at birth, then the function given by $H(x) = 35.2x^{3/40}$ gives the infant's head circumference at x months for $1 \leq x \leq 36$.
  (a) Evaluate $H(12)$ and $H(24)$.
  (b) Does an infant's head circumference increase the most in the first year or the second year?

104. Musical Tones One octave on a piano contains 12 keys (including both the black and white keys). The frequency of each successive key increases by a factor of $2^{1/12}$. For example, middle C is two keys below the first D above it. Therefore the frequency of this D is

$$2^{1/12} \cdot 2^{1/12} = 2^{1/6} \approx 1.12$$

times the frequency of middle C.
  (a) If two tones are one octave apart, how do their frequencies compare?
  (b) The A tone below middle C on a piano has a frequency of 220 cycles per second. Middle C is 3 keys above this A note. Estimate the frequency of middle C.

105. Online Exploration Go online and look up the shoulder height of three animals of different sizes. Use the formula in Exercise 99 to calculate their stepping frequencies while walking.

106. Online Exploration Go online and look up the weight of three animals of different sizes. Use the formula in Exercise 100 to calculate their heart rates.

WRITING ABOUT MATHEMATICS

107. Explain the meaning of the rational exponent in $x^{3/5}$.

108. Explain the meaning of each expression: $\sqrt[3]{x^5}$ and $(\sqrt[3]{x})^5$. Are these two expressions equal?

SECTIONS 10.1 and 10.2 **Checking Basic Concepts**

1. Find the following.
  (a) The square roots of 49
  (b) The principal square root of 49

2. Evaluate.
  (a) $\sqrt[3]{-8}$ **(b)** $-\sqrt[4]{81}$

3. Write the expression in radical notation.
  (a) $x^{3/2}$ **(b)** $x^{2/3}$
  (c) $x^{-2/5}$

4. Simplify $\sqrt{(x-1)^2}$ for any real number x.

5. Evaluate each function at the given value of x.
  (a) $f(x) = \sqrt{x}$, $x = 9$
  (b) $g(x) = \sqrt[3]{x}$, $x = 125$
  (c) $h(x) = x^{7/12}$, $x = \frac{12}{7}$

10.3 Simplifying Radical Expressions

Objectives

1 Applying the Product Rule for Radical Expressions
   - Multiplying Radical Expressions
   - Simplifying Radicals
2 Applying the Quotient Rule for Radical Expressions

NEW VOCABULARY

☐ Perfect nth power
☐ Perfect square
☐ Perfect cube

1 Applying the Product Rule for Radical Expressions

MULTIPLYING RADICAL EXPRESSIONS Consider the following examples of multiplying radical expressions. The equations

$$\sqrt{4} \cdot \sqrt{25} = 2 \cdot 5 = 10 \quad \text{and} \quad \sqrt{4 \cdot 25} = \sqrt{100} = 10$$

imply that

$$\sqrt{4} \cdot \sqrt{25} = \sqrt{4 \cdot 25} \quad \text{(see FIGURE 10.9(a)).}$$

Similarly, the equations

$$\sqrt[3]{8} \cdot \sqrt[3]{27} = 2 \cdot 3 = 6 \quad \text{and} \quad \sqrt[3]{8 \cdot 27} = \sqrt[3]{216} = 6$$

imply that

$$\sqrt[3]{8} \cdot \sqrt[3]{27} = \sqrt[3]{8 \cdot 27} \quad \text{(see FIGURE 10.9(b)).}$$

Illustrating $\sqrt[n]{a} \cdot \sqrt[n]{b} = \sqrt[n]{a \cdot b}$

$\sqrt{(4)} * \sqrt{(25)}$	
	10
$\sqrt{(4*25)}$	
	10

$\sqrt[3]{(8)} * \sqrt[3]{(27)}$	
	6
$\sqrt[3]{(8*27)}$	
	6

(a)   (b)

FIGURE 10.9

These examples suggest that *the product of the roots is equal to the root of the product*.

PRODUCT RULE FOR RADICAL EXPRESSIONS

Let a and b be real numbers, where $\sqrt[n]{a}$ and $\sqrt[n]{b}$ are both defined. Then

$$\sqrt[n]{a} \cdot \sqrt[n]{b} = \sqrt[n]{a \cdot b}. \qquad \sqrt[4]{2} \cdot \sqrt[4]{8} = \sqrt[4]{2 \cdot 8}$$

READING CHECK 1

- Can the product rule be used to simplify a product of radicals with different indexes?

NOTE: The product rule works only when the radicals have the *same index*. For example, the product $\sqrt{2} \cdot \sqrt[3]{4}$ cannot be simplified because the indexes are 2 and 3. (However, by using rational exponents, we can simplify this product. See Example 6(b).) ∎

We apply the product rule in the next two examples.

EXAMPLE 1 **Multiplying radical expressions**

Multiply each radical expression.

(a) $\sqrt{5} \cdot \sqrt{20}$ (b) $\sqrt[3]{-3} \cdot \sqrt[3]{9}$ (c) $\sqrt[4]{\frac{1}{3}} \cdot \sqrt[4]{\frac{1}{9}} \cdot \sqrt[4]{\frac{1}{3}}$

Solution

(a) $\sqrt{5} \cdot \sqrt{20} = \sqrt{5 \cdot 20} = \sqrt{100} = 10$

(b) $\sqrt[3]{-3} \cdot \sqrt[3]{9} = \sqrt[3]{-3 \cdot 9} = \sqrt[3]{-27} = -3$

(c) The product rule can also be applied to three or more factors. Thus

$$\sqrt[4]{\frac{1}{3}} \cdot \sqrt[4]{\frac{1}{9}} \cdot \sqrt[4]{\frac{1}{3}} = \sqrt[4]{\frac{1}{3} \cdot \frac{1}{9} \cdot \frac{1}{3}} = \sqrt[4]{\frac{1}{81}} = \frac{1}{3},$$

because $\frac{1}{3} \cdot \frac{1}{3} \cdot \frac{1}{3} \cdot \frac{1}{3} = \frac{1}{81}$.

Now Try Exercises 13, 15, 21

EXAMPLE 2 **Multiplying radical expressions containing variables**

Multiply each radical expression. Assume that all variables are positive.

(a) $\sqrt{x} \cdot \sqrt{x^3}$ **(b)** $\sqrt[3]{2a} \cdot \sqrt[3]{5a}$ **(c)** $\sqrt{11} \cdot \sqrt{xy}$ **(d)** $\sqrt[5]{\frac{2x}{y}} \cdot \sqrt[5]{\frac{16y}{x}}$

Solution

(a) $\sqrt{x} \cdot \sqrt{x^3} = \sqrt{x \cdot x^3} = \sqrt{x^4} = x^2$

(b) $\sqrt[3]{2a} \cdot \sqrt[3]{5a} = \sqrt[3]{2a \cdot 5a} = \sqrt[3]{10a^2}$

(c) $\sqrt{11} \cdot \sqrt{xy} = \sqrt{11xy}$

(d) $\sqrt[5]{\frac{2x}{y}} \cdot \sqrt[5]{\frac{16y}{x}} = \sqrt[5]{\frac{2x}{y} \cdot \frac{16y}{x}}$ Product rule

$$= \sqrt[5]{\frac{32xy}{xy}}$$ Multiply fractions.

$$= \sqrt[5]{32}$$ Simplify.

$$= 2$$ $2^5 = 32$

Now Try Exercises 23, 51, 57, 61

SIMPLIFYING RADICALS An integer a is a **perfect nth power** if there exists an integer b such that $b^n = a$. Thus 36 is a **perfect square** because $6^2 = 36$. Similarly, 8 is a **perfect cube** because $2^3 = 8$, and 81 is a *perfect fourth power* because $3^4 = 81$. **TABLE 10.4** lists examples of perfect squares and perfect cubes.

First Six Perfect Squares and Cubes

Perfect Squares	1	4	9	16	25	36
Perfect Cubes	1	8	27	64	125	216

TABLE 10.4

To simplify a square root, we sometimes need to recognize the *largest* perfect square factor of the radicand. For example, 50 has several factors.

1, 2, 5, 10, **25**, 50   Factors of 50

From **TABLE 10.4**, the perfect square factors of 50 are **1** and **25**. The *largest* perfect square factor of 50 is **25**.

The product rule for radicals can be used to simplify radical expressions. For example, because the largest perfect square factor of 50 is 25, the expression $\sqrt{50}$ can be simplified as

$$\sqrt{50} = \sqrt{25 \cdot 2} = \sqrt{25} \cdot \sqrt{2} = 5\sqrt{2}.$$

This procedure is generalized as follows.

READING CHECK 2

• Simplify the expression $\sqrt{40}$.

SIMPLIFYING RADICALS (nth ROOTS)

STEP 1: Determine the largest perfect nth power factor of the radicand.

STEP 2: Use the product rule to factor out and simplify this perfect nth power.

EXAMPLE 3 **Simplifying radical expressions**

Simplify each expression.

(a) $\sqrt{300}$ (b) $\sqrt[3]{16}$ (c) $\sqrt{54}$ (d) $\sqrt[4]{512}$

Solution

(a) First note that $300 = 100 \cdot 3$ and that 100 is the largest perfect square factor of 300.

$$\sqrt{300} = \sqrt{100 \cdot 3} = \sqrt{100} \cdot \sqrt{3} = 10\sqrt{3}$$

(b) The largest perfect cube factor of 16 is 8. (See **TABLE 10.4** on the previous page.)

$$\sqrt[3]{16} = \sqrt[3]{8} \cdot \sqrt[3]{2} = 2\sqrt[3]{2}$$

(c) $\sqrt{54} = \sqrt{9} \cdot \sqrt{6} = 3\sqrt{6}$

(d) $\sqrt[4]{512} = \sqrt[4]{256} \cdot \sqrt[4]{2} = 4\sqrt[4]{2}$ because $4^4 = 256$.

Now Try Exercises 75, 77, 79, 81

 Math in Context *Driving* When a car stops suddenly, it often leaves skid marks. If the car is involved in an accident, authorities often measure the length of the skid marks M and use this information to determine the minimum speed S of the car. One formula that is sometimes used when the road surface is both wet and level is given by $S = \sqrt{9M}$. In the next example, we simplify this radical expression and estimate the speed of a car based on the length of its skid marks.

EXAMPLE 4 **Calculating minimum speed**

A car is in an accident and leaves skid marks on wet, level pavement that are 121 feet long. The formula $S = \sqrt{9M}$ calculates the minimum speed S in miles per hour for a car that leaves skid marks M feet long.

(a) Simplify this formula.   (b) Determine the minimum speed of the car.

Solution

(a) Because 9 is a perfect square, we can factor it out.

$$\sqrt{9M} = \sqrt{9} \cdot \sqrt{M} = 3\sqrt{M}$$

Thus $S = 3\sqrt{M}$.

(b) $S = 3\sqrt{121} = 3 \cdot 11 = 33$ miles per hour.

Now Try Exercise 111

READING CHECK 3

• Simplify $\sqrt[3]{-54}$.

NOTE: To simplify a cube root of a negative number, we factor out the negative of the largest perfect cube factor. For example, $-16 = -8 \cdot 2$, so $\sqrt[3]{-16} = \sqrt[3]{-8} \cdot \sqrt[3]{2} = -2\sqrt[3]{2}$. This procedure can be used with any odd root of a negative number. (See Example 5(c).) ■

EXAMPLE 5 **Simplifying radical expressions**

Simplify each expression. Assume that all variables are positive.

(a) $\sqrt{25x^4}$ (b) $\sqrt{32n^3}$ (c) $\sqrt[3]{-16x^3y^5}$ (d) $\sqrt[3]{2a} \cdot \sqrt[3]{4a^2b}$

Solution

(a) $\sqrt{25x^4} = 5x^2$ Perfect square: $(5x^2)^2 = 25x^4$

(b) $\sqrt{32n^3} = \sqrt{(16n^2)2n}$ $16n^2$ is the largest perfect square factor.

           $= \sqrt{16n^2} \cdot \sqrt{2n}$ Product rule

           $= 4n\sqrt{2n}$ $(4n)^2 = 16n^2$

(c)
$$\sqrt[3]{-16x^3y^5} = \sqrt[3]{(-8x^3y^3)2y^2} \qquad 8x^3y^3 \text{ is the largest perfect cube factor.}$$
$$= \sqrt[3]{-8x^3y^3} \cdot \sqrt[3]{2y^2} \qquad \text{Product rule}$$
$$= -2xy\sqrt[3]{2y^2} \qquad (-2xy)^3 = -8x^3y^3$$

(d)
$$\sqrt[3]{2a} \cdot \sqrt[3]{4a^2b} = \sqrt[3]{(2a)(4a^2b)} \qquad \text{Product rule}$$
$$= \sqrt[3]{(8a^3)b} \qquad 8a^3 \text{ is the largest perfect cube factor.}$$
$$= \sqrt[3]{8a^3} \cdot \sqrt[3]{b} \qquad \text{Product rule}$$
$$= 2a\sqrt[3]{b} \qquad (2a)^3 = 8a^3$$

Now Try Exercises 45, 85, 89, 91

The product rule for radical expressions cannot be used if the radicals do not have the same indexes. In this case, we use rational exponents, as illustrated in the next example.

EXAMPLE 6 **Multiplying radicals with different indexes**

Simplify each expression. Write your answer in radical notation.

(a) $\sqrt{5} \cdot \sqrt[4]{5}$ **(b)** $\sqrt{2} \cdot \sqrt[3]{4}$ **(c)** $\sqrt[3]{x} \cdot \sqrt[4]{x}$

Solution

(a) Because $\sqrt{5} = 5^{1/2}$ and $\sqrt[4]{5} = 5^{1/4}$,
$$\sqrt{5} \cdot \sqrt[4]{5} = 5^{1/2} \cdot 5^{1/4} = 5^{1/2+1/4} = 5^{3/4}.$$
In radical notation, $5^{3/4} = \sqrt[4]{5^3} = \sqrt[4]{125}$.

(b) Because $\sqrt[3]{4} = \sqrt[3]{2^2} = 2^{2/3}$,
$$\sqrt{2} \cdot \sqrt[3]{4} = 2^{1/2} \cdot 2^{2/3} = 2^{1/2+2/3} = 2^{7/6}.$$
In radical notation, $2^{7/6} = \sqrt[6]{2^7} = \sqrt[6]{2^6 \cdot 2^1} = \sqrt[6]{2^6} \cdot \sqrt[6]{2} = 2\sqrt[6]{2}$.

(c) $\sqrt[3]{x} \cdot \sqrt[4]{x} = x^{1/3} \cdot x^{1/4} = x^{7/12} = \sqrt[12]{x^7}$

Now Try Exercises 101, 103, 107

2 ▶ Applying the Quotient Rule for Radical Expressions

Consider the following examples of dividing radical expressions. The equations
$$\sqrt{\frac{4}{9}} = \sqrt{\frac{2}{3} \cdot \frac{2}{3}} = \frac{2}{3} \quad \text{and} \quad \frac{\sqrt{4}}{\sqrt{9}} = \frac{2}{3}$$
imply that
$$\sqrt{\frac{4}{9}} = \frac{\sqrt{4}}{\sqrt{9}} \qquad \text{(see FIGURE 10.10)}.$$

These examples suggest that *the root of a quotient is equal to the quotient of the roots.*

FIGURE 10.10

```
√(4/9)▶Frac
              2/3
√(4)/√(9)▶Frac
              2/3
```

CALCULATOR HELP

To use the Frac feature, see Appendix A (pages AP-1 and AP-2).

QUOTIENT RULE FOR RADICAL EXPRESSIONS

Let a and b be real numbers, where $\sqrt[n]{a}$ and $\sqrt[n]{b}$ are both defined and $b \neq 0$. Then
$$\sqrt[n]{\frac{a}{b}} = \frac{\sqrt[n]{a}}{\sqrt[n]{b}}. \qquad \sqrt[3]{\frac{23}{5}} = \frac{\sqrt[3]{23}}{\sqrt[3]{5}}$$

EXAMPLE 7 **Simplifying quotients**

Simplify each radical expression. Assume that all variables are positive.

(a) $\sqrt[3]{\dfrac{5}{8}}$ (b) $\sqrt[4]{\dfrac{x}{16}}$ (c) $\sqrt{\dfrac{16}{y^2}}$

Solution

(a) $\sqrt[3]{\dfrac{5}{8}} = \dfrac{\sqrt[3]{5}}{\sqrt[3]{8}} = \dfrac{\sqrt[3]{5}}{2}$ (b) $\sqrt[4]{\dfrac{x}{16}} = \dfrac{\sqrt[4]{x}}{\sqrt[4]{16}} = \dfrac{\sqrt[4]{x}}{2}$

(c) $\sqrt{\dfrac{16}{y^2}} = \dfrac{\sqrt{16}}{\sqrt{y^2}} = \dfrac{4}{y}$ because $y > 0$.

Now Try Exercises 25, 27, 29

EXAMPLE 8 **Simplifying quotients**

Simplify each radical expression. Assume that all variables are positive.

(a) $\dfrac{\sqrt{40}}{\sqrt{10}}$ (b) $\dfrac{\sqrt[3]{2}}{\sqrt[3]{16}}$ (c) $\dfrac{\sqrt{x^2y}}{\sqrt{y}}$

Solution

(a) $\dfrac{\sqrt{40}}{\sqrt{10}} = \sqrt{\dfrac{40}{10}} = \sqrt{4} = 2$

(b) $\dfrac{\sqrt[3]{2}}{\sqrt[3]{16}} = \sqrt[3]{\dfrac{2}{16}} = \sqrt[3]{\dfrac{1}{8}} = \dfrac{1}{2}$ because $\dfrac{1}{2} \cdot \dfrac{1}{2} \cdot \dfrac{1}{2} = \dfrac{1}{8}$.

(c) $\dfrac{\sqrt{x^2y}}{\sqrt{y}} = \sqrt{\dfrac{x^2y}{y}} = \sqrt{x^2} = x$ because $x > 0$.

Now Try Exercises 33, 39, 41

MAKING CONNECTIONS 1

Rules for Radical Expressions and Rational Exponents

The rules for radical expressions are a result of the properties of rational exponents.

$$\sqrt[n]{a \cdot b} = \sqrt[n]{a} \cdot \sqrt[n]{b} \quad \text{is equivalent to} \quad (a \cdot b)^{1/n} = a^{1/n} \cdot b^{1/n}.$$

$$\sqrt[n]{\dfrac{a}{b}} = \dfrac{\sqrt[n]{a}}{\sqrt[n]{b}} \qquad \text{is equivalent to} \quad \left(\dfrac{a}{b}\right)^{1/n} = \dfrac{a^{1/n}}{b^{1/n}}.$$

READING CHECK 4

• Use rational exponents to show that $\sqrt[3]{x} \cdot \sqrt[3]{x} \cdot \sqrt[3]{x} = x$.

EXAMPLE 9 **Simplifying radical expressions**

Simplify each radical expression. Assume that all variables are positive.

(a) $\sqrt[4]{\dfrac{16x^3}{y^4}}$ (b) $\sqrt{\dfrac{5a^2}{8}} \cdot \sqrt{\dfrac{5a^3}{2}}$

Solution

(a) To simplify this expression, we first use the quotient rule for radical expressions and then apply the product rule for radical expressions.

$$\sqrt[4]{\frac{16x^3}{y^4}} = \frac{\sqrt[4]{16x^3}}{\sqrt[4]{y^4}} \qquad \text{Quotient rule}$$

$$= \frac{\sqrt[4]{16}\,\sqrt[4]{x^3}}{\sqrt[4]{y^4}} \qquad \text{Product rule}$$

$$= \frac{2\sqrt[4]{x^3}}{y} \qquad \text{Evaluate 4th roots.}$$

(b) To simplify this expression, we use both the product and quotient rules.

$$\sqrt{\frac{5a^2}{8}} \cdot \sqrt{\frac{5a^3}{2}} = \sqrt{\frac{5a^2 \cdot 5a^3}{8 \cdot 2}} \qquad \text{Product rule}$$

$$= \sqrt{\frac{25a^5}{16}} \qquad \text{Multiply.}$$

$$= \frac{\sqrt{25a^5}}{\sqrt{16}} \qquad \text{Quotient rule}$$

$$= \frac{\sqrt{25a^4 \cdot a}}{\sqrt{16}} \qquad \text{Factor out largest perfect square.}$$

$$= \frac{\sqrt{25a^4} \cdot \sqrt{a}}{\sqrt{16}} \qquad \text{Product rule}$$

$$= \frac{5a^2\sqrt{a}}{4} \qquad (5a^2)^2 = 25a^4$$

Now Try Exercises 95, 97

EXAMPLE 10 **Simplifying products and quotients of roots**

Simplify each expression. Assume all radicands are positive.

(a) $\sqrt{x-3} \cdot \sqrt{x+3}$ **(b)** $\dfrac{\sqrt[3]{x^2+3x+2}}{\sqrt[3]{x+1}}$

Solution

(a) Start by applying the product rule for radical expressions.

$$\sqrt{x-3} \cdot \sqrt{x+3} = \sqrt{(x-3)(x+3)} \qquad \text{Product rule}$$
$$= \sqrt{x^2-9} \qquad \text{Multiply binomials.}$$

NOTE: The expression $\sqrt{x^2-9}$ does *not* simplify further. It is important to realize that
$$\sqrt{x^2-9} \neq \sqrt{x^2} - \sqrt{9} = x - 3.$$
For example, $\sqrt{5^2-3^2} = \sqrt{16} = 4$, but $\sqrt{5^2} - \sqrt{3^2} = 5 - 3 = 2.$ ■

(b) Start by applying the quotient rule for radical expressions.

$$\frac{\sqrt[3]{x^2+3x+2}}{\sqrt[3]{x+1}} = \sqrt[3]{\frac{x^2+3x+2}{x+1}} \qquad \text{Quotient rule}$$

$$= \sqrt[3]{\frac{(x+1)(x+2)}{x+1}} \qquad \text{Factor trinomial.}$$

$$= \sqrt[3]{x+2} \qquad \text{Simplify quotient.}$$

Now Try Exercises 63, 67

10.3   Putting It All Together

CONCEPT	EXPLANATION	EXAMPLES
Product Rule for Radical Expressions	Let a and b be real numbers, where $\sqrt[n]{a}$ and $\sqrt[n]{b}$ are both defined. Then $$\sqrt[n]{a} \cdot \sqrt[n]{b} = \sqrt[n]{a \cdot b}.$$	$$\sqrt{2} \cdot \sqrt{32} = \sqrt{64} = 8$$ $$\sqrt{500} = \sqrt{100} \cdot \sqrt{5} = 10\sqrt{5}$$
Simplifying Radicals (nth roots)	**STEP 1:** Find the largest perfect nth power factor of the radicand. **STEP 2:** Factor out and simplify this perfect nth power.	$$\sqrt{12} = \sqrt{4 \cdot 3} = \sqrt{4} \cdot \sqrt{3} = 2\sqrt{3}$$ $$\sqrt[3]{81} = \sqrt[3]{27 \cdot 3} = \sqrt[3]{27} \cdot \sqrt[3]{3} = 3\sqrt[3]{3}$$ $$\sqrt[4]{x^5} = \sqrt[4]{x^4 \cdot x} = \sqrt[4]{x^4} \cdot \sqrt[4]{x} = x\sqrt[4]{x},$$ provided $x \geq 0$.
Quotient Rule for Radical Expressions	Let a and b be real numbers, where $\sqrt[n]{a}$ and $\sqrt[n]{b}$ are both defined and $b \neq 0$. Then $$\sqrt[n]{\dfrac{a}{b}} = \dfrac{\sqrt[n]{a}}{\sqrt[n]{b}}.$$	$$\frac{\sqrt{60}}{\sqrt{15}} = \sqrt{\frac{60}{15}} = \sqrt{4} = 2$$ $$\sqrt[3]{\frac{x^2}{-27}} = \frac{\sqrt[3]{x^2}}{\sqrt[3]{-27}} = \frac{\sqrt[3]{x^2}}{-3} = -\frac{\sqrt[3]{x^2}}{3}$$

10.3   Exercises

MyMathLab®

CONCEPTS AND VOCABULARY

1. Does $\sqrt{2} \cdot \sqrt{3}$ equal $\sqrt{6}$?

2. Does $\sqrt{5} \cdot \sqrt[3]{5}$ equal 5?

3. $\sqrt[3]{a} \cdot \sqrt[3]{b} = $ _____   4. $\dfrac{\sqrt{a}}{\sqrt{b}} = \sqrt{?}$

5. $\dfrac{\sqrt[n]{a}}{\sqrt[n]{b}} = \sqrt[n]{?}$ 6. $\dfrac{\sqrt{3}}{\sqrt{27}} = $ _____

7. Does $\sqrt{50} = \sqrt{25} + \sqrt{25}$?

8. Does $\sqrt{50} = 5\sqrt{2}$?

9. Is $\sqrt[3]{3}$ equal to 1? Explain.

10. Is 64 a perfect cube? Explain.

MULTIPLYING AND DIVIDING RADICAL EXPRESSIONS

Exercises 11–62: Simplify the expression. Assume that all variables are positive.

11. $\sqrt{3} \cdot \sqrt{3}$ 12. $\sqrt{2} \cdot \sqrt{18}$

13. $\sqrt{2} \cdot \sqrt{50}$ 14. $\sqrt[3]{-2} \cdot \sqrt[3]{-4}$

15. $\sqrt[3]{4} \cdot \sqrt[3]{16}$ 16. $\sqrt[3]{x} \cdot \sqrt[3]{x^2}$

17. $\sqrt{\dfrac{9}{25}}$ 18. $\sqrt[3]{\dfrac{x}{8}}$

19. $\sqrt{\dfrac{1}{2}} \cdot \sqrt{\dfrac{1}{8}}$ 20. $\sqrt{\dfrac{5}{3}} \cdot \sqrt{\dfrac{1}{3}}$

21. $\sqrt[3]{\dfrac{2}{3}} \cdot \sqrt[3]{\dfrac{4}{3}} \cdot \sqrt[3]{\dfrac{1}{3}}$ 22. $\sqrt[4]{\dfrac{8}{3}} \cdot \sqrt[4]{\dfrac{4}{9}} \cdot \sqrt[4]{\dfrac{8}{3}}$

23. $\sqrt{x^3} \cdot \sqrt{x^3}$ 24. $\sqrt{z} \cdot \sqrt{z^7}$

25. $\sqrt[3]{\dfrac{7}{27}}$ 26. $\sqrt[3]{\dfrac{9}{64}}$

27. $\sqrt[4]{\dfrac{x}{81}}$ 28. $\sqrt[4]{\dfrac{16x}{81}}$

29. $\sqrt{\dfrac{x^2}{81}}$ 30. $\sqrt{\dfrac{9}{z^2}}$

31. $\sqrt{\dfrac{x}{2}} \cdot \sqrt{\dfrac{x}{8}}$ 32. $\sqrt{\dfrac{4}{y}} \cdot \sqrt{\dfrac{y}{5}}$

33. $\dfrac{\sqrt{45}}{\sqrt{5}}$ 34. $\dfrac{\sqrt{7}}{\sqrt{28}}$

35. $\sqrt[3]{-4} \cdot \sqrt[3]{-16}$ 36. $\sqrt[3]{9} \cdot \sqrt[3]{3}$

37. $\sqrt[4]{9} \cdot \sqrt[4]{9}$ 38. $\sqrt[5]{16} \cdot \sqrt[5]{-2}$

39. $\dfrac{\sqrt[5]{64}}{\sqrt[5]{-2}}$ 40. $\dfrac{\sqrt[4]{324}}{\sqrt[4]{4}}$

41. $\dfrac{\sqrt{4xy^2}}{\sqrt{x}}$ 42. $\dfrac{\sqrt{a^2b}}{\sqrt{b}}$

43. $\dfrac{\sqrt[3]{54}}{\sqrt[3]{2}}$ 44. $\dfrac{\sqrt[3]{x^3y^7}}{\sqrt[3]{y^4}}$

45. $\sqrt{4x^4}$ **46.** $\sqrt[3]{-8y^3}$

47. $\sqrt[3]{-5a^6}$ **48.** $\sqrt{9x^2y}$

49. $\sqrt[4]{16x^4y}$ **50.** $\sqrt[3]{8xy^3}$

51. $\sqrt{6x^5} \cdot \sqrt{6x}$ **52.** $\sqrt{3x} \cdot \sqrt{12x}$

53. $\sqrt{16x^4y^6}$ **54.** $\sqrt[3]{8x^6y^3z^9}$

55. $\sqrt[4]{\frac{3}{4}} \cdot \sqrt[4]{\frac{27}{4}}$ **56.** $\sqrt[5]{-\frac{4}{9}} \cdot \sqrt[5]{-\frac{8}{27}}$

57. $\sqrt[3]{12} \cdot \sqrt[3]{ab}$ **58.** $\sqrt{5x} \cdot \sqrt{5z}$

59. $\sqrt[4]{25z} \cdot \sqrt[4]{25z}$ **60.** $\sqrt[5]{3z^2} \cdot \sqrt[5]{7z}$

61. $\sqrt[5]{\frac{7a}{b^2}} \cdot \sqrt[5]{\frac{b^2}{7a^6}}$ **62.** $\sqrt[3]{\frac{8m}{n}} \cdot \sqrt[3]{\frac{n^4}{m^2}}$

Exercises 63–68: Use properties of polynomials to simplify the expression. Assume all radicands are positive.

63. $\sqrt{x+4} \cdot \sqrt{x-4}$

64. $\sqrt[3]{x-1} \cdot \sqrt[3]{x^2+x+1}$

65. $\sqrt[3]{a+1} \cdot \sqrt[3]{a^2-a+1}$

66. $\sqrt{b-1} \cdot \sqrt{b+1}$

67. $\dfrac{\sqrt{x^2+2x+1}}{\sqrt{x+1}}$

68. $\dfrac{\sqrt{x^2-4x+4}}{\sqrt{x-2}}$

SIMPLIFYING RADICAL EXPRESSIONS

Exercises 69–74: Complete the equation.

69. $\sqrt{500} = \underline{\quad} \sqrt{5}$ **70.** $\sqrt{28} = \underline{\quad} \sqrt{7}$

71. $\sqrt{8} = \underline{\quad} \sqrt{2}$ **72.** $\sqrt{99} = \underline{\quad} \sqrt{11}$

73. $\sqrt{45} = \underline{\quad} \sqrt{5}$ **74.** $\sqrt{243} = \underline{\quad} \sqrt{3}$

Exercises 75–98: Simplify the radical expression by factoring out the largest perfect nth power. Assume that all variables are positive.

75. $\sqrt{200}$ **76.** $\sqrt{72}$

77. $\sqrt[3]{81}$ **78.** $\sqrt[3]{256}$

79. $\sqrt[4]{64}$ **80.** $\sqrt[5]{27 \cdot 81}$

81. $\sqrt[5]{-64}$ **82.** $\sqrt[3]{-81}$

83. $\sqrt{b^5}$ **84.** $\sqrt{t^3}$

85. $\sqrt{8n^3}$ **86.** $\sqrt{32a^2}$

87. $\sqrt{12a^2b^5}$ **88.** $\sqrt{20a^3b^2}$

89. $\sqrt[3]{-125x^4y^5}$ **90.** $\sqrt[3]{-81a^5b^2}$

91. $\sqrt[3]{5t} \cdot \sqrt[3]{125t}$ **92.** $\sqrt[4]{4bc^3} \cdot \sqrt[4]{64ab^3c^2}$

93. $\sqrt[4]{\frac{9t^5}{r^8}} \cdot \sqrt[4]{\frac{9r}{5t}}$ **94.** $\sqrt[5]{\frac{4t^6}{r}} \cdot \sqrt[5]{\frac{8t}{r^6}}$

95. $\sqrt[3]{\frac{27x^2}{y^3}}$ **96.** $\sqrt[4]{\frac{32x^8}{z^4}}$

97. $\sqrt{\frac{7a^2}{27}} \cdot \sqrt{\frac{7a}{3}}$ **98.** $\sqrt{\frac{8a}{125}} \cdot \sqrt{\frac{2a}{5}}$

Exercises 99 and 100: Thinking Generally Let m and n be positive integers and assume each expression exists.

99. Simplify $\left(\sqrt[mn]{a^m b^m} \right)^n$ so that it equals ab.

100. Simplify $\left(\sqrt[mn]{a^n b^m} \right) \cdot \left(\sqrt[mn]{a^m b^n} \right)$ so that it equals
$\sqrt[m]{ab} \cdot \sqrt[n]{ab}$.

Exercises 101–110: Simplify the expression. Let all variables be positive and write your answer in radical notation.

101. $\sqrt{3} \cdot \sqrt[3]{3}$ **102.** $\sqrt{5} \cdot \sqrt[3]{5}$

103. $\sqrt[4]{8} \cdot \sqrt[3]{4}$ **104.** $\sqrt[5]{16} \cdot \sqrt{2}$

105. $\sqrt[4]{27} \cdot \sqrt[3]{9} \cdot \sqrt{3}$ **106.** $\sqrt[5]{16} \cdot \sqrt[3]{16}$

107. $\sqrt[4]{x^3} \cdot \sqrt[3]{x}$ **108.** $\sqrt[4]{x^3} \cdot \sqrt{x}$

109. $\sqrt[4]{rt} \cdot \sqrt[3]{r^2t}$ **110.** $\sqrt[3]{a^3b^2} \cdot \sqrt{a^2b}$

APPLICATIONS INVOLVING RADICAL EXPRESSIONS

111. Minimum Speed (Refer to Example 4.) If the pavement is dry cement, then $S = \sqrt{25M}$ is an estimate for a car's speed S in miles per hour when it leaves skid marks M feet long.
(a) Simplify this formula.
(b) Determine the minimum speed of the car if the skid marks are 100 feet.

112. Minimum Speed (Refer to Example 4.) If the road surface is gravel, then $S = \sqrt{15M}$ is an estimate for a car's speed S in miles per hour when it leaves skid marks M feet long in the gravel. Determine the minimum speed of the car if the skid marks are 30 feet.

113. Baseball Ticket Prices The formula $y = \sqrt{36x}$ gives the percent price markup of a Major League Baseball ticket x days prior to game day. (*Source: StubHub.com*)
(a) Find y when $x = 4$ and interpret your result.
(b) Simplify the equation $y = \sqrt{36x}$.
(c) Find y when $x = 4$ using your equation from part (b). Do you get the same answer as in part (a)?

114. Earnings for Women The equation $I = 710\sqrt{25x}$ gives an estimate for the median annual earnings in thousands of dollars for *all* women x years after 1975. This equation has been adjusted to 2009 dollars. (*Source:* Current Populations Survey.)

(a) Find I when $x = 25$ and interpret your result.

(b) Simplify the equation $I = 710\sqrt{25x}$

(c) Find I when $x = 25$ using your equation from part (b). Do you get the same answer as in part (a)?

115. Hang Time of a Football A good punter can kick a football so that it stays in the air for a relatively long time. This time is called *hang time*. If a football is kicked h feet high, then its hang time T in seconds is

$$T = \sqrt{\frac{h}{4}}.$$

(a) Calculate the exact hang time for a football kicked 81 feet into the air.

(b) Use the quotient rule to simplify the formula.

116. Orbits and Distance Johannes Kepler (1571–1630) discovered a relationship between a planet's distance

D from the sun and the time T it takes the planet to orbit the sun. This formula is

$$T = \sqrt{D} \cdot \sqrt{D} \cdot \sqrt{D},$$

where T is in Earth years and $D = 1$ corresponds to the average distance between Earth and the sun, or 93,000,000 miles.

(a) If $D = 5.2$ for Jupiter, estimate the number of years required for Jupiter to orbit the sun

(b) Use the product rule to simplify the formula.

WRITING ABOUT MATHEMATICS

117. Explain what it means for a positive integer to be a *perfect square* or a *perfect cube*. List the positive integers that are perfect squares and less than 101. List the positive integers that are perfect cubes and less than 220.

118. Explain how the product and quotient rules for radical expressions are the result of properties of rational exponents.

Group Activity

Designing a Paper Cup A paper drinking cup is being designed in the shape shown in the accompanying figure. The amount of paper needed to manufacture the cup is determined by the surface area S of the cup, which is given by

$$S = \pi r\sqrt{r^2 + h^2},$$

where r is the radius and h is the height.

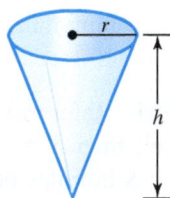

(a) Approximate S to the nearest hundredth when $r = 1.5$ inches and $h = 4$ inches.

(b) Could the formula for S be simplified as follows?

$$\pi r\sqrt{r^2 + h^2} \stackrel{?}{=} \pi r(\sqrt{r^2} + \sqrt{h^2})$$
$$\stackrel{?}{=} \pi r(r + h)$$

Try evaluating this second formula with $r = 1.5$ inches and $h = 4$ inches. Is your answer the same as in part (a)? Explain.

(c) Discuss why evaluating real-world formulas correctly is important.

(d) In general, does $\sqrt{a + b}$ equal $\sqrt{a} + \sqrt{b}$? Justify your answer by completing the following table. Approximate answers to the nearest hundredth when appropriate.

a	b	$\sqrt{a + b}$	$\sqrt{a} + \sqrt{b}$
0	4		
4	0		
5	4		
9	7		
4	16		
25	100		

10.4 Operations on Radical Expressions

Objectives

1 Adding and Subtracting Radical Expressions
  • Adding Like Radicals
  • Finding Like Radicals and Adding
  • Subtracting Like Radicals
2 Multiplying Radical Expressions
3 Rationalizing the Denominator

1 Adding and Subtracting Radical Expressions

ADDING LIKE RADICALS We can add $2x^2$ and $5x^2$ to obtain $7x^2$ because they are *like* terms. That is,

$$2x^2 + 5x^2 = (2 + 5)\,x^2 = 7x^2.$$

Like terms

We can also add and subtract *like radicals*. **Like radicals** *have the same index and the same radicand.* For example, we can add $3\sqrt{2}$ and $5\sqrt{2}$ because they are like radicals.

$$3\sqrt{2} + 5\sqrt{2} = (3 + 5)\sqrt{2} = 8\sqrt{2}$$

Like radicals

EXAMPLE 1   **Adding like radicals**

If possible, add the expressions and simplify.
(a) $10\sqrt{11} + 4\sqrt{11}$ **(b)** $5\sqrt[3]{6} + \sqrt[3]{6}$
(c) $4 + 5\sqrt{3}$ **(d)** $\sqrt{7} + \sqrt{11}$

Solution
(a) These terms are like radicals because they have the same index, 2, and the same radicand, 11.

$$10\sqrt{11} + 4\sqrt{11} = (10 + 4)\sqrt{11} = 14\sqrt{11}$$

READING CHECK 1

• What are like radicals?

(b) These terms are like radicals because they have the same index, 3, and the same radicand, 6. Note that the coefficient on the second term is understood to be 1.

$$5\sqrt[3]{6} + 1\sqrt[3]{6} = (5 + 1)\sqrt[3]{6} = 6\sqrt[3]{6}$$

(c) The expression $4 + 5\sqrt{3}$ can be written as $4\sqrt{1} + 5\sqrt{3}$. These terms cannot be added because they are not like radicals.

NOTE: $4 + 5\sqrt{3} \neq 9\sqrt{3}$ because multiplication is performed before addition. ∎

(d) The expression $\sqrt{7} + \sqrt{11}$ contains unlike radicals that *cannot* be simplified.

Now Try Exercises 19, 21, 23, 25

NEW VOCABULARY

☐ Like radicals
☐ Rationalizing the denominator
☐ Conjugate

FINDING LIKE RADICALS AND ADDING Sometimes two radical expressions that are not alike can be added by changing them to like radicals. For example, $\sqrt{20}$ and $\sqrt{5}$ are unlike radicals. However,

because $\sqrt{20} = \sqrt{4 \cdot 5} = \sqrt{4} \cdot \sqrt{5} = 2\sqrt{5},$

we have $\sqrt{20} + \sqrt{5} = 2\sqrt{5} + 1\sqrt{5} = 3\sqrt{5}.$

We cannot combine $x + x^2$ because they are unlike terms. Similarly, we cannot combine $\sqrt{2} + \sqrt{5}$ because they are unlike radicals. When combining radicals, the first step is to see if we can write pairs of terms as like radicals, as demonstrated in the next example.

EXAMPLE 2   **Finding like radicals**

Write each pair of terms as like radicals, if possible.
(a) $\sqrt{45}, \sqrt{20}$ **(b)** $\sqrt{27}, \sqrt{5}$ **(c)** $5\sqrt[3]{16}, 4\sqrt[3]{54}$

Solution

(a) The expressions $\sqrt{45}$ and $\sqrt{20}$ are unlike radicals. However, they can be changed to like radicals, as follows.

> Write as like radicals.

$$\sqrt{45} = \sqrt{9 \cdot 5} = \sqrt{9} \cdot \sqrt{5} = 3\sqrt{5} \quad \text{and}$$
$$\sqrt{20} = \sqrt{4 \cdot 5} = \sqrt{4} \cdot \sqrt{5} = 2\sqrt{5}$$

The expressions $3\sqrt{5}$ and $2\sqrt{5}$ are like radicals.

(b) Because $\sqrt{27} = \sqrt{9 \cdot 3} = \sqrt{9} \cdot \sqrt{3} = 3\sqrt{3}$, the given expressions $\sqrt{27}$ and $\sqrt{5}$ are unlike radicals and cannot be written as like radicals.

(c) $5\sqrt[3]{16} = 5\sqrt[3]{8 \cdot 2} = 5\sqrt[3]{8} \cdot \sqrt[3]{2} = 5 \cdot 2 \cdot \sqrt[3]{2} = 10\sqrt[3]{2}$ and

$$4\sqrt[3]{54} = 4\sqrt[3]{27 \cdot 2} = 4\sqrt[3]{27} \cdot \sqrt[3]{2} = 4 \cdot 3 \cdot \sqrt[3]{2} = 12\sqrt[3]{2}$$

The expressions $10\sqrt[3]{2}$ and $12\sqrt[3]{2}$ are like radicals.

Now Try Exercises 9, 11, 13

We use these techniques to add radical expressions in the next two examples.

EXAMPLE 3 **Adding radical expressions**

Add the expressions and simplify.

(a) $\sqrt{12} + 7\sqrt{3}$ (b) $\sqrt[3]{16} + \sqrt[3]{2}$ (c) $3\sqrt{2} + \sqrt{8} + \sqrt{18}$

Solution

(a)
$$\sqrt{12} + 7\sqrt{3} = \sqrt{4 \cdot 3} + 7\sqrt{3}$$
$$= \sqrt{4} \cdot \sqrt{3} + 7\sqrt{3}$$
$$= 2\sqrt{3} + 7\sqrt{3} \quad \text{Like radicals}$$
$$= 9\sqrt{3}$$

(b)
$$\sqrt[3]{16} + \sqrt[3]{2} = \sqrt[3]{8 \cdot 2} + \sqrt[3]{2}$$
$$= \sqrt[3]{8} \cdot \sqrt[3]{2} + \sqrt[3]{2}$$
$$= 2\sqrt[3]{2} + 1\sqrt[3]{2} \quad \text{Like radicals}$$
$$= 3\sqrt[3]{2}$$

(c)
$$3\sqrt{2} + \sqrt{8} + \sqrt{18} = 3\sqrt{2} + \sqrt{4 \cdot 2} + \sqrt{9 \cdot 2}$$
$$= 3\sqrt{2} + \sqrt{4} \cdot \sqrt{2} + \sqrt{9} \cdot \sqrt{2}$$
$$= 3\sqrt{2} + 2\sqrt{2} + 3\sqrt{2} \quad \text{Like radicals}$$
$$= 8\sqrt{2}$$

Now Try Exercises 29, 31, 43

READING CHECK 2

Which pairs of terms can be written as like radicals.

- $\sqrt{45}, \sqrt{80}$
- $\sqrt{5}, \sqrt{20}$
- $\sqrt{18}, \sqrt{6}$
- $\sqrt{2}, \sqrt{32}$

EXAMPLE 4 **Adding radical expressions**

Add the expressions and simplify. Assume that all variables are positive.

(a) $\sqrt[4]{32} + 3\sqrt[4]{2}$ (b) $-2\sqrt{4x} + \sqrt{x}$ (c) $3\sqrt{3k} + 5\sqrt{12k} + 9\sqrt{48k}$

Solution

(a) Because $\sqrt[4]{32} = \sqrt[4]{16 \cdot 2} = \sqrt[4]{16} \cdot \sqrt[4]{2} = 2\sqrt[4]{2}$, we can add and simplify as follows.

$$\sqrt[4]{32} + 3\sqrt[4]{2} = 2\sqrt[4]{2} + 3\sqrt[4]{2} = 5\sqrt[4]{2}$$

(b) Note that $\sqrt{4x} = \sqrt{4} \cdot \sqrt{x} = 2\sqrt{x}$.

$$-2\sqrt{4x} + \sqrt{x} = -2(2\sqrt{x}) + \sqrt{x} = -4\sqrt{x} + 1\sqrt{x} = -3\sqrt{x}$$

(c) Note that $\sqrt{12k} = \sqrt{4} \cdot \sqrt{3k} = 2\sqrt{3k}$ and that $\sqrt{48k} = \sqrt{16} \cdot \sqrt{3k} = 4\sqrt{3k}$.

$$3\sqrt{3k} + 5\sqrt{12k} + 9\sqrt{48k} = 3\sqrt{3k} + 5(2\sqrt{3k}) + 9(4\sqrt{3k})$$
$$= (3 + 10 + 36)\sqrt{3k}$$
$$= 49\sqrt{3k}$$

Now Try Exercises 45, 47, 49

SUBTRACTING LIKE RADICALS Subtraction of radical expressions is similar to addition, as illustrated in the next four examples.

EXAMPLE 5   **Subtracting like radicals**

Simplify the expressions.

(a) $5\sqrt{7} - 3\sqrt{7}$ (b) $8\sqrt[3]{5} - 3\sqrt[3]{5} + \sqrt[3]{11}$ (c) $5\sqrt{z} + \sqrt[3]{z} - 2\sqrt{z}$

Solution

(a) $5\sqrt{7} - 3\sqrt{7} = (5 - 3)\sqrt{7} = 2\sqrt{7}$

(b) $8\sqrt[3]{5} - 3\sqrt[3]{5} + \sqrt[3]{11} = (8 - 3)\sqrt[3]{5} + \sqrt[3]{11} = 5\sqrt[3]{5} + \sqrt[3]{11}$

(c) $5\sqrt{z} + \sqrt[3]{z} - 2\sqrt{z} = 5\sqrt{z} - 2\sqrt{z} + \sqrt[3]{z}$ Commutative property

$\qquad\qquad\qquad\qquad = (5 - 2)\sqrt{z} + \sqrt[3]{z}$ Distributive property

$\qquad\qquad\qquad\qquad = 3\sqrt{z} + \sqrt[3]{z}$ Subtract.

NOTE: We cannot combine $3\sqrt{z} + \sqrt[3]{z}$ because their indexes are different. That is, one term is a square root and the other is a cube root. ∎

Now Try Exercises 33, 35, 39

EXAMPLE 6   **Subtracting radical expressions**

Subtract and simplify. Assume that all variables are positive.

(a) $3\sqrt[3]{xy^2} - 2\sqrt[3]{xy^2}$ (b) $\sqrt{16x^3} - \sqrt{x^3}$ (c) $\sqrt[3]{\dfrac{5x}{27}} - \dfrac{\sqrt[3]{5x}}{6}$

Solution

(a) $3\sqrt[3]{xy^2} - 2\sqrt[3]{xy^2} = (3 - 2)\sqrt[3]{xy^2} = \sqrt[3]{xy^2}$

(b) $\sqrt{16x^3} - \sqrt{x^3} = \sqrt{16x^2} \cdot \sqrt{x} - \sqrt{x^2} \cdot \sqrt{x}$ Factor out perfect squares.

$\qquad\qquad\quad = 4x\sqrt{x} - x\sqrt{x}$ Simplify.

$\qquad\qquad\quad = (4x - x)\sqrt{x}$ Distributive property

$\qquad\qquad\quad = 3x\sqrt{x}$ Subtract.

(c) $\sqrt[3]{\dfrac{5x}{27}} - \dfrac{\sqrt[3]{5x}}{6} = \dfrac{\sqrt[3]{5x}}{\sqrt[3]{27}} - \dfrac{\sqrt[3]{5x}}{6}$ Quotient rule for radical expressions

$\qquad\qquad\qquad = \dfrac{\sqrt[3]{5x}}{3} - \dfrac{\sqrt[3]{5x}}{6}$ Evaluate $\sqrt[3]{27} = 3$.

$\qquad\qquad\qquad = \dfrac{2\sqrt[3]{5x}}{6} - \dfrac{\sqrt[3]{5x}}{6}$ Find a common denominator.

$\qquad\qquad\qquad = \dfrac{2\sqrt[3]{5x} - \sqrt[3]{5x}}{6}$ Subtract numerators.

$\qquad\qquad\qquad = \dfrac{\sqrt[3]{5x}}{6}$ Simplify.

Now Try Exercises 55, 63, 65

Math in ~~Context~~ *Business* Our productivity as a nation has increased significantly during the past 70 years because of new technology. For example, if a typical business bought $10,000 in equipment per worker in 2005, it would have generated about $48,000 per worker in revenue. This same investment by a business in 1935 would have generated only $23,000 in revenue (adjusted to 2005 dollars). This situation can be described by the following two functions, which approximate the increase in revenue resulting from investing x dollars in equipment (per worker).

$$N(x) = 400\sqrt{x} + 8000 \quad \text{and} \quad O(x) = 195\sqrt{x} + 3500$$

2005 technology (new)   **1935 technology (old)**

What does the difference of these two functions represent? In the next example, we learn how to simplify and interpret this difference.

EXAMPLE 7   **Increasing productivity with new technology**

For the functions given by

$$N(x) = 400\sqrt{x} + 8000 \quad \text{and} \quad O(x) = 195\sqrt{x} + 3500,$$

2005 technology (new)   **1935 technology (old)**

complete each of the following.
(a) Find their difference $D(x) = N(x) - O(x)$. Simplify your answer.
(b) Evaluate $D(40,000)$ and interpret the result.

Solution
(a) $D(x) = N(x) - O(x)$

$$\begin{aligned}
&= (400\sqrt{x} + 8000) - (195\sqrt{x} + 3500) && \text{Substitute.} \\
&= 400\sqrt{x} - 195\sqrt{x} + 8000 - 3500 && \text{Rewrite terms.} \\
&= (400 - 195)\sqrt{x} + 4500 && \text{Distributive property} \\
&= 205\sqrt{x} + 4500 && \text{Simplify.}
\end{aligned}$$

The difference is given by $D(x) = 205\sqrt{x} + 4500$.
(b) $D(40,000) = 205\sqrt{40,000} + 4500 = \$45,500$. An investment of $40,000 per worker resulted in $45,500 *more* revenue per worker in 2005 than in 1935 (adjusted to 2005 dollars).

Now Try Exercises 81, 135

EXAMPLE 8   **Subtracting radical expressions**

Subtract and simplify. Assume that all variables are positive.
(a) $\dfrac{5\sqrt{2}}{3} - \dfrac{2\sqrt{2}}{4}$ **(b)** $\sqrt[4]{81a^5b^6} - \sqrt[4]{16ab^2}$ **(c)** $3\sqrt[3]{\dfrac{n^5}{27}} - 2\sqrt[3]{n^2}$

Solution

(a)
$$\begin{aligned}
\frac{5\sqrt{2}}{3} - \frac{2\sqrt{2}}{4} &= \frac{5\sqrt{2}}{3} \cdot \frac{4}{4} - \frac{2\sqrt{2}}{4} \cdot \frac{3}{3} && \text{LCD is 12.} \\
&= \frac{20\sqrt{2}}{12} - \frac{6\sqrt{2}}{12} && \text{Multiply fractions.} \\
&= \frac{14\sqrt{2}}{12} && \text{Subtract numerators.} \\
&= \frac{7\sqrt{2}}{6} && \text{Simplify.}
\end{aligned}$$

(b) $\sqrt[4]{81a^5b^6} - \sqrt[4]{16ab^2} = \sqrt[4]{81a^4b^4} \cdot \sqrt[4]{ab^2} - \sqrt[4]{16} \cdot \sqrt[4]{ab^2}$ Factor out perfect powers.

$= 3ab\sqrt[4]{ab^2} - 2\sqrt[4]{ab^2}$ Simplify.

$= (3ab - 2)\sqrt[4]{ab^2}$ Distributive property

(c) $3\sqrt[3]{\dfrac{n^5}{27}} - 2\sqrt[3]{n^2} = 3\sqrt[3]{\dfrac{n^3}{27}} \cdot \sqrt[3]{n^2} - 2\sqrt[3]{n^2}$ Factor out perfect cube.

$= \dfrac{3\sqrt[3]{n^3}}{\sqrt[3]{27}} \cdot \sqrt[3]{n^2} - 2\sqrt[3]{n^2}$ Quotient rule

$= n\sqrt[3]{n^2} - 2\sqrt[3]{n^2}$ Simplify: $\sqrt[3]{n^3} = n$ and $\sqrt[3]{27} = 3$.

$= (n - 2)\sqrt[3]{n^2}$ Distributive property

Now Try Exercises 69, 77, 79

Radicals often occur in geometry. In the next example, we find the perimeter of a triangle by adding radical expressions.

FIGURE 10.11

EXAMPLE 9   **Finding the perimeter of a triangle**

Find the *exact* perimeter of the right triangle shown in **FIGURE 10.11**. Then approximate your answer to the nearest hundredth of a foot.

Solution

The sum of the lengths of the sides of the triangle is

$$\sqrt{18} + \sqrt{32} + \sqrt{50} = 3\sqrt{2} + 4\sqrt{2} + 5\sqrt{2} = 12\sqrt{2}.$$

The perimeter is $12\sqrt{2} \approx 16.97$ feet.

Now Try Exercise 125

2 Multiplying Radical Expressions

Some types of radical expressions can be multiplied like binomials. For example, because

$$(x + 1)(x + 2) = x^2 + 3x + 2,$$

we have

$$(\sqrt{x} + 1)(\sqrt{x} + 2) = (\sqrt{x})^2 + 2\sqrt{x} + 1\sqrt{x} + 2 = x + 3\sqrt{x} + 2,$$

provided that $x \geq 0$. The next example demonstrates this technique.

EXAMPLE 10   **Multiplying radical expressions**

Multiply and simplify.
(a) $(\sqrt{b} - 4)(\sqrt{b} + 5)$ **(b)** $(4 + \sqrt{3})(4 - \sqrt{3})$

Solution

(a) This expression can be multiplied and then simplified.

$$(\sqrt{b} - 4)(\sqrt{b} + 5) = \sqrt{b} \cdot \sqrt{b} + 5\sqrt{b} - 4\sqrt{b} - 4 \cdot 5$$

> Multiply in the same way as we multiply $(b - 4)(b + 5)$.

$$= b + \sqrt{b} - 20$$

Compare this product with $(b - 4)(b + 5) = b^2 + b - 20$.

(b) $(4 + \sqrt{3})(4 - \sqrt{3}) = (4)^2 - (\sqrt{3})^2$

> Multiply as a sum and difference:
> $(a + b)(a - b) = a^2 - b^2$.

$= 16 - 3$

$= 13$

Now Try Exercises 85, 87

NOTE: Example 10(b) illustrates a special case for multiplying radicals. In general,

$$(\sqrt{a} + \sqrt{b})(\sqrt{a} - \sqrt{b}) = (\sqrt{a})^2 - (\sqrt{b})^2 = a - b,$$

provided a and b are nonnegative. ∎

3 Rationalizing the Denominator

In mathematics, it is common to write expressions without radicals in the denominator. Quotients containing radical expressions in the numerator or denominator can appear to be different but actually be equal. For example, $\frac{1}{\sqrt{3}}$ and $\frac{\sqrt{3}}{3}$ represent the same real number even though they look as though they are unequal. To show that they are equal, we multiply the first quotient by **1** in the form $\frac{\sqrt{3}}{\sqrt{3}}$.

$$\frac{1}{\sqrt{3}} \cdot \frac{\sqrt{3}}{\sqrt{3}} = \frac{1 \cdot \sqrt{3}}{\sqrt{3} \cdot \sqrt{3}} = \frac{\sqrt{3}}{3}$$

READING CHECK 3

Rationalize the denominator of $\frac{1}{\sqrt{5}}$.

If the denominator of a quotient contains only one term with one square root, then we can rationalize the denominator by multiplying the numerator and denominator by this square root. For example, the denominator of $\frac{1}{\sqrt{3}}$ contains one term, which is $\sqrt{3}$. Therefore, we multiplied $\frac{1}{\sqrt{3}}$ by $\frac{\sqrt{3}}{\sqrt{3}}$ to rationalize the denominator.

NOTE: $\sqrt{b} \cdot \sqrt{b} = \sqrt{b^2} = b$ for any *positive* number b. ∎

One way to *standardize* radical expressions is to remove any radical expressions from the denominator. This process is called **rationalizing the denominator**. Exercise 139 suggests one reason why people rationalized denominators before calculators were invented. The next example demonstrates how to rationalize the denominator of several quotients.

EXAMPLE 11   **Rationalizing the denominator**

Rationalize each denominator. Assume that all variables are positive.

(a) $\frac{1}{\sqrt{2}}$ **(b)** $\frac{3}{5\sqrt{3}}$ **(c)** $\sqrt{\frac{x}{24}}$ **(d)** $\frac{xy}{\sqrt{y^3}}$

Solution

(a) We start by multiplying this expression by 1 in the form $\frac{\sqrt{2}}{\sqrt{2}}$.

$$\frac{1}{\sqrt{2}} \cdot \frac{\sqrt{2}}{\sqrt{2}} = \frac{\sqrt{2}}{\sqrt{4}} = \frac{\sqrt{2}}{2}$$

Note that the expression $\frac{\sqrt{2}}{2}$ does not have a radical in the denominator.

(b) We multiply $\frac{3}{5\sqrt{3}}$ by **1** in the form $\frac{\sqrt{3}}{\sqrt{3}}$.

$$\frac{3}{5\sqrt{3}} \cdot \frac{\sqrt{3}}{\sqrt{3}} = \frac{3\sqrt{3}}{5\sqrt{9}} = \frac{3\sqrt{3}}{5 \cdot 3} = \frac{\sqrt{3}}{5}$$

(c) Because $\sqrt{24} = \sqrt{4} \cdot \sqrt{6} = 2\sqrt{6}$, we start by simplifying $\sqrt{\dfrac{x}{24}}$.

$$\sqrt{\dfrac{x}{24}} = \dfrac{\sqrt{x}}{\sqrt{24}} = \dfrac{\sqrt{x}}{2\sqrt{6}}$$

To rationalize the denominator, we multiply this expression by **1** in the form $\dfrac{\sqrt{6}}{\sqrt{6}}$.

$$\dfrac{\sqrt{x}}{2\sqrt{6}} = \dfrac{\sqrt{x}}{2\sqrt{6}} \cdot \dfrac{\sqrt{6}}{\sqrt{6}} = \dfrac{\sqrt{6x}}{12}$$

(d) Because $\sqrt{y^3} = \sqrt{y^2} \cdot \sqrt{y} = y\sqrt{y}$, we start by simplifying $\dfrac{xy}{\sqrt{y^3}}$.

$$\dfrac{xy}{\sqrt{y^3}} = \dfrac{xy}{y\sqrt{y}} = \dfrac{x}{\sqrt{y}}$$

To rationalize the denominator, we multiply by **1** in the form $\dfrac{\sqrt{y}}{\sqrt{y}}$.

$$\dfrac{x}{\sqrt{y}} \cdot \dfrac{\sqrt{y}}{\sqrt{y}} = \dfrac{x\sqrt{y}}{y}$$

Now Try Exercises 97, 101, 103, 105

FIGURE 10.12

EXAMPLE 12

Rationalizing a denominator in geometry

A square with a diagonal of length x has sides of length $\dfrac{x}{\sqrt{2}}$. Find the perimeter and rationalize the denominator. See **FIGURE 10.12**.

Solution
The perimeter is 4 times the length of one side.

$$4 \cdot \dfrac{x}{\sqrt{2}} = \dfrac{4x}{\sqrt{2}}$$

To rationalize the denominator, we multiply the ratio by **1** in the form $\dfrac{\sqrt{2}}{\sqrt{2}}$.

$$\dfrac{4x}{\sqrt{2}} \cdot \dfrac{\sqrt{2}}{\sqrt{2}} = \dfrac{4x\sqrt{2}}{2} = 2x\sqrt{2}$$

The perimeter is $2x\sqrt{2}$.

Now Try Exercise 127

When the denominator is either a sum or difference containing a square root, we rationalize the denominator by multiplying the numerator and denominator by the *conjugate* of the denominator. In this case, the **conjugate** of the denominator is found by changing a $+$ sign to a $-$ sign or vice versa. **TABLE 10.5** lists examples of conjugates.

Expressions and Their Conjugates

Expression	$1 + \sqrt{2}$	$\sqrt{3} - 2$	$\sqrt{x} + 7$	$\sqrt{a} - \sqrt{b}$
Conjugate	$1 - \sqrt{2}$	$\sqrt{3} + 2$	$\sqrt{x} - 7$	$\sqrt{a} + \sqrt{b}$

TABLE 10.5

In the next two examples, we use this method to rationalize a denominator.

EXAMPLE 13 **Using a conjugate to rationalize the denominator**

Rationalize the denominator of $\dfrac{1}{1 + \sqrt{2}}$.

Solution

From **TABLE 10.5**, on the previous page, the conjugate of $1 + \sqrt{2}$ is $\mathbf{1 - \sqrt{2}}$.

$$\frac{1}{1 + \sqrt{2}} = \frac{1}{1 + \sqrt{2}} \cdot \frac{(\mathbf{1 - \sqrt{2}})}{(\mathbf{1 - \sqrt{2}})} \qquad \text{Multiply numerator and denominator by the conjugate.}$$

$$= \frac{1 - \sqrt{2}}{(1)^2 - (\sqrt{2})^2} \qquad (a + b)(a - b) = a^2 - b^2$$

$$= \frac{1 - \sqrt{2}}{1 - 2} \qquad \text{Simplify.}$$

$$= \frac{1 - \sqrt{2}}{-1} \qquad \text{Subtract.}$$

$$= \frac{1}{-1} - \frac{\sqrt{2}}{-1} \qquad \frac{a - b}{c} = \frac{a}{c} - \frac{b}{c}$$

$$= -1 + \sqrt{2} \qquad \text{Simplify.}$$

Now Try Exercise 107

READING CHECK 4

- Rationalize the denominator of $\dfrac{1}{1 - \sqrt{3}}$.

EXAMPLE 14 **Rationalizing the denominator**

Rationalize the denominator.

(a) $\dfrac{3 + \sqrt{5}}{2 - \sqrt{5}}$ **(b)** $\dfrac{\sqrt{x}}{\sqrt{x} - 2}$

Solution

(a) The conjugate of the denominator is $\mathbf{2 + \sqrt{5}}$.

$$\frac{3 + \sqrt{5}}{2 - \sqrt{5}} = \frac{(3 + \sqrt{5})}{(2 - \sqrt{5})} \cdot \frac{(\mathbf{2 + \sqrt{5}})}{(\mathbf{2 + \sqrt{5}})} \qquad \text{Multiply by 1.}$$

$$= \frac{6 + 3\sqrt{5} + 2\sqrt{5} + (\sqrt{5})^2}{(2)^2 - (\sqrt{5})^2} \qquad \text{Multiply.}$$

$$= \frac{11 + 5\sqrt{5}}{4 - 5} \qquad \text{Combine terms.}$$

$$= -11 - 5\sqrt{5} \qquad \text{Simplify.}$$

(b) The conjugate of the denominator is $\mathbf{\sqrt{x} + 2}$.

$$\frac{\sqrt{x}}{\sqrt{x} - 2} = \frac{\sqrt{x}}{(\sqrt{x} - 2)} \cdot \frac{(\mathbf{\sqrt{x} + 2})}{(\mathbf{\sqrt{x} + 2})} \qquad \text{Multiply by 1.}$$

$$= \frac{x + 2\sqrt{x}}{x - 4} \qquad \text{Multiply.}$$

Now Try Exercises 111, 115

EXAMPLE 15 **Rationalizing a denominator that has a cube root**

Rationalize the denominator of $\dfrac{5}{\sqrt[3]{x}}$.

Solution

The expression $\dfrac{5}{\sqrt[3]{x}}$ is equal to $\dfrac{5}{x^{1/3}}$. To rationalize the denominator, $x^{1/3}$, we can multiply it by $x^{2/3}$ because $x^{1/3} \cdot x^{2/3} = x^{1/3+2/3} = x^1$.

$$\dfrac{5}{x^{1/3}} = \dfrac{5}{x^{1/3}} \cdot \dfrac{x^{2/3}}{x^{2/3}} \qquad \text{Multiply by 1.}$$

$$= \dfrac{5x^{2/3}}{x^{1/3+2/3}} \qquad \text{Product rule}$$

$$= \dfrac{5\sqrt[3]{x^2}}{x} \qquad \text{Add; write in radical notation.}$$

Now Try Exercise 121

10.4   Putting It All Together

CONCEPT	EXPLANATION	EXAMPLES
Like Radicals	Like radicals have the same index and the same radicand.	$7\sqrt{5}$ and $3\sqrt{5}$ are like radicals. $5\sqrt[3]{ab}$ and $\sqrt[3]{ab}$ are like radicals. $\sqrt[3]{5}$ and $\sqrt[3]{4}$ are unlike radicals. $\sqrt[3]{7}$ and $\sqrt{7}$ are unlike radicals.
Adding and Subtracting Radical Expressions	Combine like radicals when adding or subtracting. We cannot combine unlike radicals such as $\sqrt{2}$ and $\sqrt{5}$. But sometimes we can rewrite radicals and then combine them.	$6\sqrt{13} + \sqrt{13} = (6+1)\sqrt{13} = 7\sqrt{13}$ $$\begin{aligned}\sqrt{40} - \sqrt{10} &= \sqrt{4} \cdot \sqrt{10} - \sqrt{10}\\ &= 2\sqrt{10} - \sqrt{10}\\ &= \sqrt{10}\end{aligned}$$
Multiplying Radical Expressions	Radical expressions can sometimes be multiplied like binomials.	$(\sqrt{a} - 5)(\sqrt{a} + 5) = a - 25$ and $(\sqrt{x} - 3)(\sqrt{x} + 1) = x - 2\sqrt{x} - 3$
Conjugate	The conjugate is found by changing a $+$ sign to a $-$ sign or vice versa.	*Expression* *Conjugate* $\sqrt{x} + 7$ $\sqrt{x} - 7$ $\sqrt{a} - 2\sqrt{b}$ $\sqrt{a} + 2\sqrt{b}$
Rationalizing a Denominator that has One Term	Write the quotient without a radical expression in the denominator.	To rationalize $\dfrac{5}{\sqrt{7}}$, multiply the expression by 1 in the form $\dfrac{\sqrt{7}}{\sqrt{7}}$. $$\dfrac{5}{\sqrt{7}} \cdot \dfrac{\sqrt{7}}{\sqrt{7}} = \dfrac{5\sqrt{7}}{\sqrt{49}} = \dfrac{5\sqrt{7}}{7}$$
Rationalizing a Denominator that has Two Terms	Multiply the numerator and denominator by the conjugate of the denominator.	$$\dfrac{1}{2 - \sqrt{3}} = \dfrac{1}{2 - \sqrt{3}} \cdot \dfrac{(2 + \sqrt{3})}{(2 + \sqrt{3})}$$ $$= \dfrac{2 + \sqrt{3}}{4 - 3} = 2 + \sqrt{3}$$

10.4   Exercises

CONCEPTS AND VOCABULARY

1. $\sqrt{a} + \sqrt{a} =$ _____

2. $\sqrt[3]{b} + \sqrt[3]{b} + \sqrt[3]{b} =$ _____

3. You cannot simplify $\sqrt[3]{4} + \sqrt[3]{7}$ because they are not _____ radicals.

4. Can you simplify $4\sqrt{15} - 3\sqrt{15}$? Explain.

5. Does $6 + 3\sqrt{5}$ equal $9\sqrt{5}$?

6. To rationalize the denominator of $\frac{2}{\sqrt{7}}$, multiply this expression by _____.

7. What is the conjugate of $\sqrt{t} - 5$?

8. To rationalize the denominator of $\frac{1}{5 - \sqrt{2}}$, multiply this expression by _____.

WRITING LIKE RADICALS

Exercises 9–18: (Refer to Example 2.) Write the terms as like radicals, if possible. Assume that all variables are positive.

9. $\sqrt{12}, \sqrt{24}$

10. $\sqrt{18}, \sqrt{27}$

11. $\sqrt{7}, \sqrt{28}, \sqrt{63}$

12. $\sqrt{200}, \sqrt{300}, \sqrt{500}$

13. $\sqrt[3]{16}, \sqrt[3]{-54}$

14. $\sqrt[3]{80}, \sqrt[3]{10}$

15. $\sqrt{x^2 y}, \sqrt{4y^2}$

16. $\sqrt{x^5 y^3}, \sqrt{9xy}$

17. $\sqrt[3]{8xy}, \sqrt[3]{x^4 y^4}$

18. $\sqrt[3]{64x^4}, \sqrt[3]{-8x}$

ADDING AND SUBSTRACTING RADICAL EXPRESSIONS

Exercises 19–80: If possible, simplify the expression. Assume that all variables are positive.

19. $2\sqrt{3} + 7\sqrt{3}$

20. $8\sqrt{7} + 2\sqrt{7}$

21. $4\sqrt[3]{5} + 2\sqrt[3]{5}$

22. $\sqrt[3]{13} + 3\sqrt[3]{13}$

23. $7 + 4\sqrt{7}$

24. $8 - 4\sqrt{3}$

25. $2\sqrt{3} + 3\sqrt{2}$

26. $\sqrt{6} + \sqrt{17}$

27. $\sqrt{3} + \sqrt[3]{3}$

28. $\sqrt{6} + \sqrt[4]{6}$

29. $\sqrt[3]{16} + 3\sqrt[3]{2}$

30. $\sqrt[3]{24} + \sqrt[3]{81}$

31. $\sqrt{2} + \sqrt{18} + \sqrt{32}$

32. $2\sqrt{3} + \sqrt{12} + \sqrt{27}$

33. $11\sqrt{11} - 5\sqrt{11}$

34. $9\sqrt{5} + \sqrt{2} - \sqrt{5}$

35. $\sqrt{x} + \sqrt{x} - \sqrt{y}$

36. $\sqrt{xy^2} - \sqrt{x}$

37. $\sqrt[3]{z} + \sqrt[3]{z}$

38. $\sqrt[3]{y} - \sqrt[3]{y}$

39. $2\sqrt[3]{6} - 7\sqrt[3]{6}$

40. $18\sqrt[3]{3} + 3\sqrt[3]{3}$

41. $\sqrt[3]{y^6} - \sqrt[3]{y^3}$

42. $2\sqrt{20} + 7\sqrt{5} + 3\sqrt{2}$

43. $3\sqrt{28} + 3\sqrt{7}$

44. $9\sqrt{18} - 2\sqrt{8}$

45. $\sqrt[4]{48} + 4\sqrt[4]{3}$

46. $\sqrt[4]{32} + \sqrt[4]{16}$

47. $\sqrt{9x} + \sqrt{16x}$

48. $-3\sqrt{x} + 5\sqrt{x}$

49. $3\sqrt{2k} + \sqrt{8k} + \sqrt{18k}$

50. $3\sqrt{k} + 2\sqrt{4k} + \sqrt{9k}$

51. $\sqrt{44} - 4\sqrt{11}$

52. $\sqrt[4]{5} + 2\sqrt[4]{5}$

53. $2\sqrt[3]{16} + \sqrt[3]{2} - \sqrt{2}$

54. $5\sqrt[3]{x} - 3\sqrt[3]{x}$

55. $\sqrt[3]{xy} - 2\sqrt[3]{xy}$

56. $3\sqrt{x^3} - \sqrt{x}$

57. $\sqrt{4x + 8} + \sqrt{x + 2}$

58. $\sqrt{2a + 1} + \sqrt{8a + 4}$

59. $\sqrt{9x + 18} - \sqrt{4x + 8}$

60. $\sqrt{25x - 25} - \sqrt{4x - 4}$

61. $\sqrt{x^3 + x^2} - \sqrt{x + 1}, x \geq 0$

62. $\sqrt{x^3 - x^2} - \sqrt{4x - 4}, x \geq 1$

63. $\sqrt{25x^3} - \sqrt{x^3}$

64. $\sqrt{36x^5} - \sqrt{25x^5}$

65. $\sqrt[3]{\frac{7x}{8}} - \frac{\sqrt[3]{7x}}{3}$

66. $\sqrt[3]{\frac{8x^2}{27}} - \sqrt[3]{\frac{x^2}{8}}$

67. $\frac{4\sqrt{3}}{3} + \frac{\sqrt{3}}{6}$

68. $\frac{8\sqrt{5}}{7} + \frac{4\sqrt{5}}{2}$

69. $\frac{15\sqrt{8}}{4} - \frac{2\sqrt{2}}{5}$

70. $\frac{23\sqrt{11}}{2} - \frac{\sqrt{44}}{8}$

71. $2\sqrt[4]{64} - \sqrt[4]{324} + \sqrt[4]{4}$

72. $2\sqrt[3]{16} - 5\sqrt[3]{54} + 10\sqrt[3]{2}$

73. $5\sqrt[4]{x^5} - \sqrt[4]{x}$

74. $20\sqrt[3]{b^4} - 4\sqrt[3]{b}$

75. $\sqrt{64x^3} - \sqrt{x} + 3\sqrt{x}$

76. $2\sqrt{3z} + 3\sqrt{12z} + 3\sqrt{48z}$

77. $\sqrt[4]{81a^5 b^5} - \sqrt[4]{ab}$

78. $\sqrt[4]{xy^5} - \sqrt[4]{x^5 y}$

79. $5\sqrt[3]{\frac{n^4}{125}} - 2\sqrt[3]{n}$

80. $\sqrt[3]{\frac{8x}{27}} - \frac{2\sqrt[3]{x}}{3}$

PERFORMING OPERATIONS ON RADICAL FUNCTIONS

Exercises 81–84: Find $(f + g)(x)$ *and* $(f - g)(x)$.

81. $f(x) = 5\sqrt{x} - 2$, $g(x) = -2\sqrt{x} + 3$

82. $f(x) = 3\sqrt{4x - 4}$, $g(x) = 5\sqrt{x - 1}$

83. $f(x) = \sqrt[3]{8x} + 1$, $g(x) = 2\sqrt[3]{x} - 1$

84. $f(x) = \sqrt[3]{27x}$, $g(x) = \sqrt[3]{64x}$

MULTIPLYING BINOMIALS CONTAINING RADICALS

Exercises 85–96: Multiply and simplify.

85. $(\sqrt{x} - 3)(\sqrt{x} + 2)$

86. $(2\sqrt{x} + 1)(\sqrt{x} + 4)$

87. $(3 + \sqrt{7})(3 - \sqrt{7})$

88. $(5 - \sqrt{5})(5 + \sqrt{5})$

89. $(11 - \sqrt{2})(11 + \sqrt{2})$

90. $(6 + \sqrt{3})(6 - \sqrt{3})$

91. $(\sqrt{x} + 8)(\sqrt{x} - 8)$

92. $(\sqrt{ab} - 3)(\sqrt{ab} + 3)$

93. $(\sqrt{ab} - \sqrt{c})(\sqrt{ab} + \sqrt{c})$

94. $(\sqrt{2x} + \sqrt{3y})(\sqrt{2x} - \sqrt{3y})$

95. $(\sqrt{x} - 7)(\sqrt{x} + 8)$

96. $(\sqrt{ab} - 1)(\sqrt{ab} - 2)$

RATIONALIZING THE DENOMINATOR

Exercises 97–124: Rationalize the denominator.

97. $\frac{1}{\sqrt{7}}$

98. $\frac{1}{\sqrt{23}}$

99. $\frac{4}{\sqrt{3}}$

100. $\frac{8}{\sqrt{2}}$

101. $\frac{5}{3\sqrt{5}}$

102. $\frac{6}{11\sqrt{3}}$

103. $\sqrt{\frac{b}{12}}$

104. $\sqrt{\frac{5b}{72}}$

105. $\frac{rt}{2\sqrt{r^3}}$

106. $\frac{m^2n}{2\sqrt{m^5}}$

107. $\frac{1}{3 - \sqrt{2}}$

108. $\frac{1}{\sqrt{3} - 2}$

109. $\frac{\sqrt{2}}{\sqrt{5} + 2}$

110. $\frac{\sqrt{3}}{\sqrt{3} + 2}$

111. $\frac{\sqrt{7} - 2}{\sqrt{7} + 2}$

112. $\frac{\sqrt{3} - 1}{\sqrt{3} + 1}$

113. $\frac{1}{\sqrt{7} - \sqrt{6}}$

114. $\frac{1}{\sqrt{8} - \sqrt{7}}$

115. $\frac{\sqrt{z}}{\sqrt{z} - 3}$

116. $\frac{2\sqrt{z}}{2 - \sqrt{z}}$

117. $\frac{\sqrt{a} + \sqrt{b}}{\sqrt{a} - \sqrt{b}}$

118. $\frac{\sqrt{x} - 2\sqrt{y}}{\sqrt{x} + 2\sqrt{y}}$

119. $\frac{1}{\sqrt{x + 1} - \sqrt{x}}$

120. $\frac{1}{\sqrt{a + 1} + \sqrt{a}}$

121. $\frac{3}{\sqrt[3]{x}}$

122. $\frac{6}{5\sqrt[3]{x}}$

123. $\frac{1}{\sqrt[3]{x^2}}$

124. $\frac{2}{\sqrt[3]{(x - 2)^2}}$

RELATING GEOMETRY AND RADICALS

125. Perimeter (Refer to Example 9.) Find the exact perimeter of the right triangle. Then approximate your answer.

126. Perimeter Find the exact perimeter of the rectangle. Then approximate your answer.

127. Geometry (Refer to Example 12.) A square has a diagonal with length $\sqrt{3}$. Find the perimeter and rationalize the denominator.

128. Geometry A rectangle has sides of length $\sqrt{8}$ and $\sqrt{32}$. Find the perimeter of the rectangle.

129. Geometry (Refer to Example 12.) A square has a diagonal that is 60 feet long. Find the exact perimeter of the rectangle and simplify your answer.

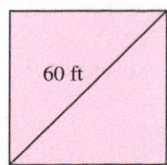

130. Geometry A square has a diagonal that is 10 feet long. Find the exact perimeter of the rectangle and simplify your answer.

131. Geometry A square has an area of x square feet. Find the length of the diagonal of the square.

132. Geometry A square has an area of 16 square feet. Find the length of the diagonal of the square.

APPLICATIONS INVOLVING RADICAL EXPRESSIONS

Exercises 133 and 134. **Flood Channels** *The rate R at which water flows through a channel is given by the formula $R = k\sqrt{m}$, where m is the slope of the channel and k is a constant determined by the shape of the channel. Suppose that two open flood channels have flow rates R_1 and R_2 given by*

$$R_1 = 1500\sqrt{m_1} \quad \text{and} \quad R_2 = 2000\sqrt{m_2},$$

where R_1 and R_2 are in cubic feet per second and m_1 and m_2 are the slopes of each channel, respectively.

133. Find $R_1 + R_2$ if $m_1 = 0.09$ and $m_2 = 0.04$.

134. Find $R_1 + R_2$ if both channels have slope m.

Exercises 135–138: **Drug Law Violations** *The number of adults in thousands incarcerated in state or federal prisons for drug law violations x years after 1984 can be estimated by the formula $P = 70\sqrt{x} + 45$. Similarly, the formula $J = 20\sqrt{x} + 25$ gives the number of adults in thousands incarcerated in jails for drug law violations x years after 1984.*

135. Evaluate P and J for $x = 16$. Interpret each.

136. Add your results from Exercise 135. Interpret the result.

137. Simplify $T = (70\sqrt{x} + 45) + (20\sqrt{x} + 25)$.

138. What does T represent?

WRITING ABOUT MATHEMATICS

139. Suppose that a student knows that $\sqrt{3} \approx 1.73205$ and does not have a calculator. Which of the following expressions would be easier to evaluate by hand? Why?

$$\frac{1}{\sqrt{3}} \quad \text{or} \quad \frac{\sqrt{3}}{3}$$

140. A student simplifies an expression *incorrectly:*

$$\sqrt{8} + \sqrt[3]{16} \overset{?}{=} \sqrt{4 \cdot 2} + \sqrt[3]{8 \cdot 2}$$
$$\overset{?}{=} \sqrt{4} \cdot \sqrt{2} + \sqrt[3]{8} \cdot \sqrt[3]{2}$$
$$\overset{?}{=} 2\sqrt{2} + 2\sqrt[3]{2}$$
$$\overset{?}{=} 4\sqrt{4}$$
$$\overset{?}{=} 8.$$

Explain any errors that the student made. What would you do differently?

Checking Basic Concepts

1. Simplify each expression. Assume that all variables are positive.
   (a) $(64^{-3/2})^{1/3}$ **(b)** $\sqrt{5} \cdot \sqrt{20}$
   (c) $\sqrt[3]{-8x^4y}$ **(d)** $\sqrt{\dfrac{4b}{5}} \cdot \sqrt{\dfrac{4b^3}{5}}$

2. Simplify $\sqrt[3]{7} \cdot \sqrt{7}$

3. Simplify each expression.
   (a) $\sqrt{3} \cdot \sqrt{12}$ **(b)** $\dfrac{\sqrt[3]{81}}{\sqrt[3]{3}}$
   (c) $\sqrt{36x^6}, x > 0$

4. Simplify each expression.
   (a) $5\sqrt{6} + 2\sqrt{6} + \sqrt{7}$
   (b) $8\sqrt[3]{x} - 3\sqrt[3]{x}$
   (c) $\sqrt{9x} - \sqrt{4x}$

5. Simplify each expression.
   (a) $\sqrt[3]{xy^4} - \sqrt[3]{x^4y}$
   (b) $(4 - \sqrt{2})(4 + \sqrt{2})$

6. Rationalize the denominator of $\dfrac{6}{2\sqrt{6}}$.

7. Rationalize the denominator of $\dfrac{2}{\sqrt{5} - 1}$.

10.5 More Radical Functions

Objectives

1 Recognizing Root Functions and Their Properties

2 Recognizing Power Functions and Their Properties

3 Modeling with Power Functions

1 Recognizing Root Functions and Their Properties

Earlier in this chapter, we discussed the square root and cube root functions. There are also functions for higher roots. The following See the Concept shows graphs of four common root functions.

See the Concept 1   THE ROOT FUNCTIONS

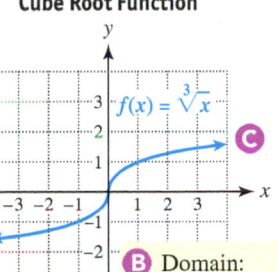

Square Root Function

$f(x) = \sqrt{x}$

Ⓒ

Ⓐ Domain: $x \geq 0$

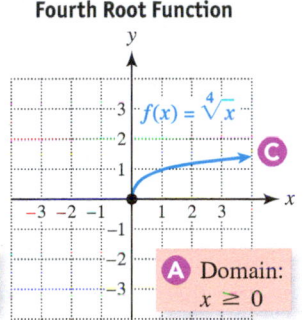

Cube Root Function

$f(x) = \sqrt[3]{x}$

Ⓒ

Ⓑ Domain: $-\infty < x < \infty$

Fourth Root Function

$f(x) = \sqrt[4]{x}$

Ⓒ

Ⓐ Domain: $x \geq 0$

Fifth Root Function

$f(x) = \sqrt[5]{x}$

Ⓒ

Ⓑ Domain: $-\infty < x < \infty$

Ⓐ The domain for an even root function is $[0, \infty)$.

Ⓑ The domain for an odd root function is $(-\infty, \infty)$.

Ⓒ The higher the root, the more slowly the graph increases for $x \geq 1$.

WATCH VIDEO IN MML.

NEW VOCABULARY

☐ Root function
☐ Power function

EXAMPLE 1   Evaluating root functions

If possible, evaluate each root function f at the given x-values.
(a) $f(x) = \sqrt{x + 1}, \quad x = -5, x = 3$
(b) $f(x) = \sqrt[3]{1 - x}, \quad x = -7, x = 28$
(c) $f(x) = \sqrt[4]{x - 7}, \quad x = 5, x = 88$
(d) $f(x) = \sqrt[5]{x}, \qquad x = -1, x = 32$

Solution
(a) $f(-5) = \sqrt{-5 + 1} = \sqrt{-4}$, which is *not* a real number.

$f(3) = \sqrt{3 + 1} = \sqrt{4} = 2$ because $2^2 = 4$.

(b) $f(-7) = \sqrt[3]{1 - (-7)} = \sqrt[3]{8} = 2$ because $2^3 = 8$.

$f(28) = \sqrt[3]{1 - 28} = \sqrt[3]{-27} = -3$ because $(-3)^3 = -27$.

(c) For $f(x) = \sqrt[4]{x - 7}$, $f(5) = \sqrt[4]{5 - 7} = \sqrt[4]{-2}$, which is not a real number.

$f(88) = \sqrt[4]{88 - 7} = \sqrt[4]{81} = 3$ because $3^4 = 81$.

(d) For $f(x) = \sqrt[5]{x}$, $f(-1) = \sqrt[5]{-1} = -1$ because $(-1)^5 = -1$.

$f(32) = \sqrt[5]{32} = 2$ because $2^5 = 32$.

Now Try Exercises 9, 11, 13, 15

2 Recognizing Power Functions and Their Properties

Power functions are a generalization of root functions.

STUDY TIP

Power functions are important in many areas of study. Be sure to become familiar with them.

Power Functions

$$f(x) = x^{1/2}, g(x) = x^{2/3}, \quad \text{and} \quad h(x) = x^{-3/5}.$$

The exponent for a power function is frequently a rational number p, written in lowest terms. That is, we usually write $f(x) = x^{1/2}$, rather than $f(x) = x^{3/6}$. The power function $f(x) = x^{m/n}$ can also be written as $f(x) = \sqrt[n]{x^m}$. A power function f is defined for all real numbers when n is odd and defined for only nonnegative numbers when n is even.

POWER FUNCTION

If a function f can be represented by

$$f(x) = x^p,$$

where p is a rational number, then f is a **power function**. If $p = \frac{1}{n}$, where $n \geq 2$ is an integer, then f is also a **root function**, which is given by

$$f(x) = \sqrt[n]{x}.$$

EXAMPLE 2 **Evaluating power functions**

If possible, evaluate $f(x)$ at the given value of x.
(a) $f(x) = x^{0.75}$ at $x = 16$ (b) $f(x) = x^{1/4}$ at $x = -81$

Solution
(a) $0.75 = \frac{3}{4}$, so $f(x) = x^{3/4}$. Thus $f(16) = 16^{3/4} = (16^{1/4})^3 = 2^3 = 8$.

NOTE: $16^{1/4} = \sqrt[4]{16} = 2$. ∎

(b) $f(-81) = (-81)^{1/4} = \sqrt[4]{-81}$, which is not a real number. There is no real number a such that $a^4 = -81$ because a^4 is never negative.

Now Try Exercises 23, 27

🌐 *Math in Context* ^Biology The surface area of the skin covering the human body is influenced by both the height and weight of a person. A taller person tends to have a larger surface area, as does a heavier person. In the next example, we use a power function from biology to model this situation. Also, see the Group Activity at the end of this section.

EXAMPLE 3 **Modeling surface area of the human body**

The surface area of a person who is 66 inches tall and weighs w pounds can be estimated by $S(w) = 327w^{0.425}$, where S is in square inches. (*Source:* H. Lancaster, *Quantitative Methods in Biological and Medical Sciences.*)

(a) Find S if this person weighs 130 pounds.

(b) If the person gains 20 pounds, by how much does the person's surface area increase?

(c) A graph of $y = S(w)$ is shown in **FIGURE 10.13**. Suppose that a person's weight is more than 50 pounds and it doubles. Use the graph to determine if the person's surface area also doubles.

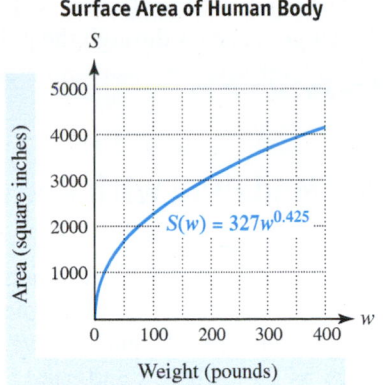

Surface Area of Human Body

$S(w) = 327w^{0.425}$

FIGURE 10.13

Solution

(a) $S(130) = 327(130)^{0.425} \approx 2588$ square inches

(b) Because $S(150) = 327(150)^{0.425} \approx 2750$ square inches, the surface area of the person increases by about $2750 - 2588 = 162$ square inches.

(c) For $w \geq 50$, the graph increases more slowly. This means that the person's surface area (S-values) will not double even if his or her weight doubles (w-values). For example, $S(100) \approx 2300$ square inches and $S(200) \approx 3100$ square inches. The surface area did not double.

Now Try Exercise 57

In the next example, we investigate the graph of $y = x^p$ for different values of p.

EXAMPLE 4   ### Graphing power functions

The graphs of three power functions,

$$f(x) = x^{1/3}, \ g(x) = x^{0.75}, \quad \text{and} \quad h(x) = x^{1.4},$$

are shown in **FIGURE 10.14**, where the two decimal exponents have been written as fractions. Discuss how the value of p affects the graph of $y = x^p$ when $x > 1$ and when $0 < x < 1$.

Power Functions

(a)

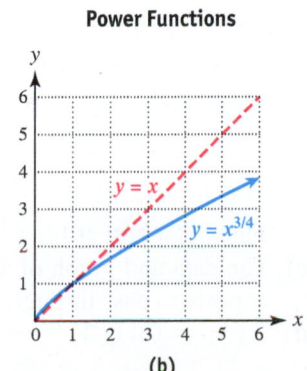

(b)

(c)

FIGURE 10.14

Solution

The graphs of these power functions have the following characteristics.

1. If $x > 1$, then $x^{1/3} < x^{3/4} < x^{7/5}$. Larger exponents result in graphs that increase (rise) *faster* for $x > 1$.
2. If $0 < x < 1$, then $x^{7/5} < x^{3/4} < x^{1/3}$. Smaller exponents result in graphs that have larger y-values for $0 < x < 1$.
3. All graphs pass through the points $(0, 0)$ and $(1, 1)$.

Now Try Exercise 35

3 Modeling with Power Functions

Biology

Math in Context Allometry is the study of the relative sizes of different characteristics of an organism. For example, the weight of a bird is related to the surface area of its wings; heavier birds tend to have larger wings. Allometric relations are often modeled with $f(x) = kx^p$, where k and p are constants. (*Source:* C. Pennycuick, *Newton Rules Biology.*)

EXAMPLE 5

Modeling surface area of wings

Weights w of various birds and the corresponding surface area of their wings A are shown in **TABLE 10.6**.

Weight and Wing Size

w (kilograms)	0.5	2.0	3.5	5.0
A (square meters)	0.069	0.175	0.254	0.325

TABLE 10.6

CALCULATOR HELP

To make a scatterplot, see Appendix A (pages AP-3 and AP-4).

(a) Make a scatterplot of the data. Discuss any trends in the data.
(b) Biologists modeled the data with $A(w) = kw^{2/3}$, where k is a constant. Find k.
(c) Graph A and the data in the same viewing rectangle.
(d) Estimate the area of the wings of a 3-kilogram bird.

$[0, 6, 1]$ by $[0, 0.4, 0.1]$

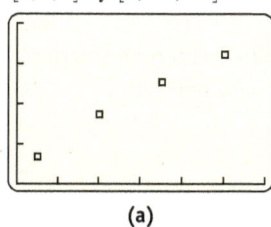

(a)

$[0, 6, 1]$ by $[0, 0.4, 0.1]$

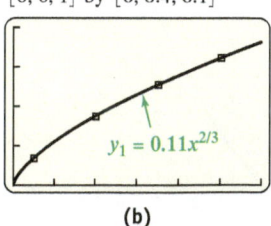

$y_1 = 0.11x^{2/3}$

(b)

FIGURE 10.15

Solution

(a) A scatterplot of the data is shown in **FIGURE 10.15(a)**. As the weight of a bird increases so does the surface area of its wings.
(b) To determine k, substitute one of the data points into $A(w)$. We use $(2.0, 0.175)$.

$$A(w) = kw^{2/3} \qquad \text{Given formula}$$

$$0.175 = k(2)^{2/3} \qquad \text{Let } w = 2 \text{ and } A(w) = 0.175 \text{ (any data point could be used).}$$

$$k = \frac{0.175}{2^{2/3}} \qquad \text{Solve for } k; \text{ rewrite equation.}$$

$$k \approx 0.11 \qquad \text{Approximate } k.$$

Thus $A(w) = 0.11w^{2/3}$.

(c) The data and graph of $y_1 = 0.11x^{2/3}$ are shown in **FIGURE 10.15(b)**. Note that the graph appears to pass through each data point.
(d) $A(3) = 0.11(3)^{2/3} \approx 0.23$ square meter

Now Try Exercise 65

10.5 Putting It All Together

FUNCTION	EXPLANATION	EXAMPLES
Root	Odd root functions are defined for all real numbers, and even root functions are defined for nonnegative real numbers.	$f(x) = \sqrt[3]{x}, \quad g(x) = \sqrt[4]{x}, \quad \text{and} \quad h(x) = \sqrt[5]{x}$
Power	$f(x) = x^p$, where p is a rational number, is a power function. If $p = \frac{1}{n}$, where $n \geq 2$ is an integer, f is also a root function, given by $f(x) = \sqrt[n]{x}$.	$f(x) = x^{5/3}$ Power function $g(x) = x^{1/4} = \sqrt[4]{x}$ Root function

10.5 Exercises

MyMathLab®

CONCEPTS AND VOCABULARY

1. Sketch a graph of $y = x^{1/2}$.

2. Sketch a graph of $y = x^{1/3}$.

3. What is the domain of $f(x) = x^{1/2}$?

4. What is the domain of $f(x) = x^{1/3}$?

5. Give a symbolic representation for a power function.

6. Give a symbolic representation for a root function.

7. What is the domain of $f(x) = \sqrt[4]{x}$?

8. What is the domain of $f(x) = \sqrt[5]{x}$?

EVALUATING ROOT FUNCTIONS

Exercises 9–16: If possible, evaluate each root function f at the given x-values. When the result is not an integer, approximate it to the nearest hundredth.

9. $f(x) = \sqrt{x^2 - 1}, \quad x = -2, x = 0$

10. $f(x) = \sqrt{2x - 5}, \quad x = 0, x = 3$

11. $f(x) = \sqrt[4]{1 - x}, \quad x = -5, x = 2$

12. $f(x) = \sqrt[4]{x + 1}, \quad x = -15, x = 15$

13. $f(x) = \sqrt[5]{4 - 3x}, \quad x = -3, x = 1$

14. $f(x) = \sqrt[5]{1 - x^2}, \quad x = 3, x = 1$

15. $f(x) = \sqrt[3]{1 - x}, \quad x = -5, x = 2$

16. $f(x) = \sqrt[3]{(x + 1)^2}, \quad x = -2, x = 7$

WORKING WITH POWER FUNCTIONS

Exercises 17–22: Use radical notation to write f(x).

17. $f(x) = x^{1/2}$ 18. $f(x) = x^{1/3}$

19. $f(x) = x^{2/3}$ 20. $f(x) = x^{3/4}$

21. $f(x) = x^{-1/5}$ 22. $f(x) = x^{-2/5}$

Exercises 23–30: If possible, evaluate f(x) at the given values of x. When appropriate, approximate the answer to the nearest hundredth.

23. $f(x) = x^{5/2} \quad\quad x = 4, x = 5$

24. $f(x) = x^{-3/4} \quad\quad x = 1, x = 3$

25. $f(x) = x^{-7/5} \quad\quad x = -32, x = 10$

26. $f(x) = x^{4/3} \quad\quad x = -8, x = 27$

27. $f(x) = x^{1/4} \quad\quad x = 256, x = -10$

28. $f(x) = x^{3/4} \quad\quad x = 16, x = -1$

29. $f(x) = x^{2/5} \quad\quad x = 32, x = -32$

30. $f(x) = x^{5/6} \quad\quad x = -5, x = 64$

Exercises 31–34: Graph y = f(x). Write the domain of f in interval notation.

31. $f(x) = x^{1/4}$ 32. $f(x) = x^{1/5}$

33. $f(x) = x^{2/3}$ 34. $f(x) = (x + 1)^{1/4}$

Exercises 35–38: Graph f and g in the window $[0, 6, 1]$ by $[0, 6, 1]$. Which function is greater when $x > 1$?

35. $f(x) = x^{1/5}, g(x) = x^{1/3}$

36. $f(x) = x^{4/5}, g(x) = x^{5/4}$

37. $f(x) = x^{1.2}, g(x) = x^{0.45}$

38. $f(x) = x^{-1.4}, g(x) = x^{1.4}$

Exercises 39 and 40: **Thinking Generally** *Let $0 < q < p$, where p and q are rational numbers.*

39. For $x > 1$, is $x^p > x^q$ or $x^p < x^q$?

40. For $0 < x < 1$, is $x^p > x^q$ or $x^p < x^q$?

PERFORMING OPERATIONS ON ROOT FUNCTIONS

Exercises 41 and 42: For the given $f(x)$ and $g(x)$, evaluate each expression.

    (a) $(f + g)(2)$ **(b)** $(f - g)(x)$
    (c) $(fg)(x)$ **(d)** $(f/g)(x)$

41. $f(x) = \sqrt{8x}, g(x) = \sqrt{2x}$

42. $f(x) = \sqrt{9x + 18}, g(x) = \sqrt{4x + 8}$

RECOGNIZING GRAPHS OF POWER FUNCTIONS

Exercises 43–46: **Graphical Interpretation** *Match the situation with the graph of the power function (a–d) that models it best.*

43. Amount of water in a barrel that is initially full and has a hole near the bottom

44. The average weight of a type of bird as its wing span increases (*Hint:* The exponent is greater than 1.)

45. Money made after x hours by a person who works for a fixed hourly wage

46. Rapid growth of an insect population that eventually slows down

a. y

b. y

c. y

d. y
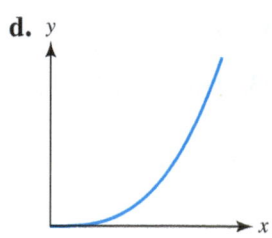

Exercises 47–50: **Graphical Recognition** *Match the equation with its graph (a–d). Do not use a calculator.*

47. $y = x^{7/4}$ **48.** $y = x^{4/3}$

49. $y = x^{1/3}$ **50.** $y = x^{1/2}$

a.

b.

c.

d.
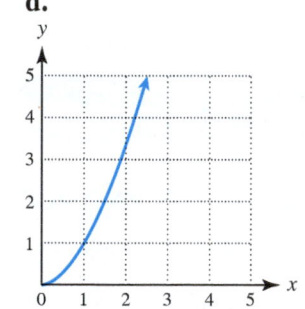

USING ZEROS TO WRITE A FORMULA FOR A FUNCTION

Exercises 51–56: A quadratic function f has the given zeros and leading coefficient 1.

    (a) *Write the formula for $f(x)$ in factored form.*
    (b) *Write the expanded form for $f(x)$ by multiplying the factor form.*

(Hint: If -5 and 5 are the two zeros of $f(x)$, then $f(x)$ can be written in factored form using the formula given by $f(x) = (x + 5)(x - 5)$.)

51. $-2, 2$

52. $-3, 3$

53. $-\sqrt{7}, \sqrt{7}$

54. $-\sqrt{3}, \sqrt{3}$

55. $-\sqrt{6}, \sqrt{6}$

56. $-\sqrt{11}, \sqrt{11}$

APPLICATIONS INVOLVING ROOT AND POWER FUNCTIONS

Exercises 57 and 58: **Surface Area** *(Refer to Example 3.) The surface area of the skin of a person who is 70 inches tall and weighs w pounds can be estimated by $S(w) = 342w^{0.425}$, where S is in square inches. Evaluate each of the following.*

57. $S(150)$ **58.** $S(200)$

Exercises 59–62: **Learning Curves** *The graph shows the percentage P of times that individuals completed a simple task correctly after practicing it x times.*

$P(x) = 0.156x^{7/20}$

Number of Trials

59. If a group of people does 10 practice trials each, what percentage of the time is the task done correctly?

60. How many trials need to be performed to reach 50% accuracy?

61. If a person doubles the number of practice trials, does the accuracy always double? Explain.

62. What happens to a person's accuracy after a large number of trials?

63. **Aging More Slowly** In his theory of relativity, Albert Einstein showed that if a person travels at nearly the speed of light, then time slows down significantly. Suppose that there are twins; one remains on Earth and the other leaves in a very fast spaceship that has velocity v. If the twin on Earth ages T_0 years, then according to Einstein, the twin in the spaceship ages T years, where

$$T(v) = T_0\sqrt{1 - (v/c)^2}.$$

In this formula, c represents the speed of light, which is 186,000 miles per second.
(a) Evaluate T when $v = 0.8c$ (eight-tenths the speed of light) and $T_0 = 10$ years. (*Hint:* Simplify $\frac{v}{c}$ without using 186,000 miles per second.)
(b) Interpret your result.

64. **Increasing Your Weight** (Refer to Exercise 63.) Albert Einstein also showed that the weight (mass) of an object increases when traveling near the speed of light. If a person's weight on Earth is W_0, then the same person's weight W in a spaceship traveling at velocity v is

$$W(v) = \frac{W_0}{\sqrt{1 - (v/c)^2}}.$$

(a) Evaluate function W when $v = 0.6c$ (six-tenths the speed of light) and $W_0 = 220$ pounds (100 kilograms).
(b) Interpret your result.

65. **Modeling Wing Size** (Refer to Example 5.) The surface area of wings for a species of bird with weight w is shown in the table.

w (kilograms)	2	3	4
A (square meters)	0.254	0.333	0.403

(a) Make a scatterplot of the data.
(b) The data can be modeled with $A(w) = kw^{2/3}$. Find k.
(c) Graph A and the data in the same viewing rectangle. Does the graph of A pass through the data points?
(d) Estimate the area of the wings of a 2.5-kilogram bird.

66. **Pulse Rate in Animals** The following table lists typical pulse rates R in beats per minute (bpm) for animals with various weights W in pounds. (*Source:* C. Pennycuick.)

W (pounds)	20	150	500	1500
R (beats per minute)	198	72	40	23

(a) Describe what happens to the pulse rate as the weight of the animal increases.
(b) Plot the data in $[0, 1600, 400]$ by $[0, 220, 20]$.
(c) If $R = kW^{-1/2}$, find k.
(d) Find R when $W = 700$, and interpret the result.

67. **Modeling Wing Span** (Refer to Example 5.) Biologists have found that the weight W of a bird and the length L of its wing span are related by $L = kW^{1/3}$, where k is a constant. The following table lists L and W for one species of bird. (*Source:* C. Pennycuick.)

L

W (kilograms)	0.1	0.4	0.8	1.1
L (meters)	0.422	0.670	0.844	0.938

(a) Use the data to approximate the value of k.
(b) Graph L and the data in the same viewing rectangle. What happens to L as W increases?
(c) Find the wing span of a 0.7-kilogram bird.
(d) Find L when $W = 0.65$, and interpret the result.

68. Orbits and Distances The time T in years for a planet to orbit the sun is given by $T = D^{3/2}$, where D is the planet's distance from the sun and $D = 1$ corresponds to Earth's distance from the sun. If a planet is twice the distance from the sun as Earth, does the time to orbit the sun also double?

WRITING ABOUT MATHEMATICS

69. Explain why a root function is an example of a power function.

70. Discuss the shape of the graph of $y = x^p$ as p increases. Assume that p is a positive rational number and that x is a positive real number.

Group Activity

Area of Skin The surface area of the skin covering the human body is a function of more than one variable. Both height and weight influence the surface area of a person's body. Hence a taller person tends to have a larger surface area, as does a heavier person. A formula to determine the area of a person's skin in square inches is $S = 15.7w^{0.425}h^{0.725}$, where w is weight in pounds and h is height in inches. (**Source:** H. Lancaster, *Quantitative Methods in Biological and Medical Sciences.*)

(a) Use S to estimate the area of skin covering a person who is 65 inches tall and weighs 154 pounds.

(b) If a person's weight doubles, what happens to the area of the person's skin? Explain.

(c) If a person's height doubles, what happens to the area of the person's skin? Explain.

10.6   Equations Involving Radical Expressions

Objectives

1 **Solving Radical Equations**
 • **Power Rule**
 • **Extraneous Solutions**
 • **Squaring Twice**
 • **Equations Involving Cube Roots**

2 **Deriving and Applying the Distance Formula**
 • **Pythagorean Theorem**
 • **Distance Between Two Points**

3 **Solving Equations of the Form $x^n = k$**

NEW VOCABULARY

☐ Extraneous solutions
☐ Pythagorean theorem
☐ Distance

1 Solving Radical Equations

POWER RULE Many times, equations contain either radical expressions or rational exponents. Examples include

$$\sqrt{x} = 6, \quad 5x^{1/2} = 1, \quad \text{and} \quad \sqrt[3]{x-1} = 3. \qquad \boxed{\text{Radical equations}}$$

One strategy for solving an equation containing a square root is to isolate the square root (if necessary) and then square each side of the equation. This technique is an example of the *power rule for solving equations*.

POWER RULE FOR SOLVING EQUATIONS

If each side of an equation is raised to the same positive integer power, then any solutions to the given equation are among the solutions to the new equation. That is, the solutions to the equation $a = b$ are among the solutions to $a^n = b^n$.

NOTE: After applying the power rule, the new equation can have solutions that do not satisfy the given equation, which are sometimes called **extraneous solutions**. We must *always* check our answers in the *given* equation after applying the power rule. ■

READING CHECK 1

• When you apply the power rule to solve an equation, what must you do with your solutions?

We illustrate the power rule in the next example.

EXAMPLE 1

Solving a radical equation symbolically

Solve $\sqrt{2x - 1} = 3$. Check your solution.

Solution

Begin by squaring each side of the equation. That is, apply the power rule with $n = 2$.

$$
\begin{aligned}
\sqrt{2x - 1} &= 3 &&\text{Given equation} \\
(\sqrt{2x - 1})^2 &= 3^2 &&\text{Square each side.} \\
2x - 1 &= 9 &&\text{Simplify.} \\
2x &= 10 &&\text{Add 1.} \\
x &= 5 &&\text{Divide by 2.}
\end{aligned}
$$

To check this answer, we substitute $x = 5$ in the given equation.

$$
\begin{aligned}
\sqrt{2(5) - 1} &\overset{?}{=} 3 \\
3 &= 3 \quad \checkmark \quad \text{It checks.}
\end{aligned}
$$

Now Try Exercise 21

NOTE: To simplify $(\sqrt{2x - 1})^2$ in Example 1, we used the fact that

$$(\sqrt{a})^2 = \sqrt{a} \cdot \sqrt{a} = a. \quad \blacksquare$$

The following steps can be used to solve a radical equation.

SOLVING A RADICAL EQUATION

STEP 1: Isolate a radical term on one side of the equation.

STEP 2: Apply the power rule by raising each side of the equation to the power equal to the index of the isolated radical term.

STEP 3: Solve the equation. If it still contains a radical, repeat Steps 1 and 2.

STEP 4: Check your answers by substituting each result in the *given* equation.

MAKING CONNECTIONS 1

Expressions and Equations

An equation is a statement that two expressions are equal. For example,

$$\sqrt{x + 1} \quad \text{and} \quad \sqrt{5 - x}$$

Expressions do not have an equals sign.

are two different expressions, and

$$\sqrt{x + 1} = \sqrt{5 - x}$$

Equations contain an equals sign.

is an equation. We often *solve equations*, whereas we *simplify* and *evaluate* expressions.

In the next example, we apply these steps to solve a radical equation.

EXAMPLE 2

Isolating the radical term before applying the power rule

Solve $\sqrt{4 - x} + 5 = 8$.

Solution

STEP 1: To isolate the radical term, we subtract 5 from each side of the equation.

$$\sqrt{4 - x} + 5 = 8 \qquad \text{Given equation}$$
$$\sqrt{4 - x} = 3 \qquad \text{Subtract 5.}$$

STEP 2: The isolated term involves a square root, so we must **square** each side.

$$(\sqrt{4 - x})^2 = 3^2 \qquad \text{Square each side.}$$

STEP 3: Next we solve the resulting equation. (It is not necessary to repeat Steps 1 and 2 because the resulting equation does not contain any radical expressions.)

$$4 - x = 9 \qquad \text{Simplify.}$$
$$-x = 5 \qquad \text{Subtract 4.}$$
$$x = -5 \qquad \text{Multiply by } -1.$$

STEP 4: To check this answer, we substitute $x = -5$ in the given equation.

$$\sqrt{4 - (-5)} + 5 \stackrel{?}{=} 8$$
$$\sqrt{9} + 5 \stackrel{?}{=} 8$$
$$8 = 8 \checkmark \qquad \text{It checks.}$$

Now Try Exercise 23

EXTRANEOUS SOLUTIONS The next example shows that we must check our answers to identify extraneous solutions when squaring each side of an equation.

EXAMPLE 3   **Solving a radical equation**

Solve $\sqrt{3x + 3} = 2x - 1$ symbolically. Check your results and then solve the equation graphically.

Solution

Symbolic Solution Begin by squaring each side of the equation.

$$\sqrt{3x + 3} = 2x - 1 \qquad \text{Given equation}$$
$$(\sqrt{3x + 3})^2 = (2x - 1)^2 \qquad \text{Square each side.}$$
$$3x + 3 = 4x^2 - 4x + 1 \qquad \text{Multiply.}$$
$$0 = 4x^2 - 7x - 2 \qquad \text{Subtract } 3x + 3.$$
$$0 = (4x + 1)(x - 2) \qquad \text{Factor.}$$
$$x = -\frac{1}{4} \quad \text{or} \quad x = 2 \qquad \text{Solve for } x.$$

To check these values, start by substituting $x = -\frac{1}{4}$ in the given equation.

$$\sqrt{3\left(-\frac{1}{4}\right) + 3} \stackrel{?}{=} 2\left(-\frac{1}{4}\right) - 1$$
$$\sqrt{2.25} \stackrel{?}{=} -1.5$$
$$1.5 \neq -1.5 \; \textbf{X} \qquad \text{It does not check.}$$

Thus $-\frac{1}{4}$ is an *extraneous solution*. Next substitute $x = 2$ in the given equation.

$$\sqrt{3(2) + 3} \stackrel{?}{=} 2(2) - 1$$
$$\sqrt{9} \stackrel{?}{=} 3$$
$$3 = 3 \checkmark \qquad \text{It checks.}$$

The only solution is 2. Next we find a graphical solution.

$[-5, 5, 1]$ by $[-5, 5, 1]$

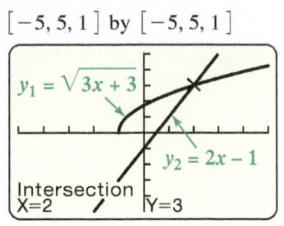

FIGURE 10.16

CALCULATOR HELP

To find a point of intersection, see Appendix A (page AP-6).

Graphical Solution The solution **2** to the equation $\sqrt{3x + 3} = 2x - 1$ is supported graphically in **FIGURE 10.16**, where the graphs of $y_1 = \sqrt{3x + 3}$ and $y_2 = 2x - 1$ intersect at the point $(2, 3)$. *Note that the graphical solution does not give an extraneous solution.*

Now Try Exercises 27, 79(a), (b)

Example 3 demonstrates that *checking a solution is essential when you are squaring each side of an equation.* Squaring may introduce extraneous solutions, which are solutions to the resulting equation but are not solutions to the given equation.

CRITICAL THINKING 1

Will a numerical solution give extraneous solutions?

SQUARING TWICE When an equation contains two or more terms with square roots, it may be necessary to square each side of the equation more than once. In these situations, isolate one of the square roots and then square each side of the equation. If a radical term remains after simplifying, repeat these steps. We apply this technique in the next example.

EXAMPLE 4   **Solving a radical equation by squaring twice**

Solve $\sqrt{2x} - 1 = \sqrt{x + 1}$.

Solution
Begin by squaring each side of the equation.

$$\sqrt{2x} - 1 = \sqrt{x + 1} \qquad \text{Given equation (Step 1)}$$
$$(\sqrt{2x} - 1)^2 = (\sqrt{x + 1})^2 \qquad \text{Square each side. (Step 2)}$$
$$(\sqrt{2x})^2 - 2(\sqrt{2x})(1) + 1^2 = x + 1 \qquad (a - b)^2 = a^2 - 2ab + b^2$$
$$2x - 2\sqrt{2x} + 1 = x + 1 \qquad \text{Simplify.}$$
$$2x - 2\sqrt{2x} = x \qquad \text{Subtract 1.}$$
$$x = 2\sqrt{2x} \qquad \text{Subtract } x \text{ and add } 2\sqrt{2x}. \text{ (Step 1)}$$
$$x^2 = 4(2x) \qquad \text{Square each side again. (Step 2)}$$
$$x^2 - 8x = 0 \qquad \text{Subtract } 8x.$$
$$x(x - 8) = 0 \qquad \text{Factor.}$$
$$x = 0 \quad \text{or} \quad x = 8 \qquad \text{Solve. (Step 3)}$$

To check these answers, substitute $x = 0$ and $x = 8$ in the given equation (Step 4).

$$\sqrt{2(0)} - 1 \overset{?}{=} \sqrt{0 + 1} \qquad\qquad \sqrt{2(8)} - 1 \overset{?}{=} \sqrt{8 + 1}$$
$$-1 \neq 1 \; \text{✗} \quad \text{It does not check.} \qquad\qquad 3 = 3 \; \text{✓} \quad \text{It checks.}$$

Because 0 is an extraneous solution, the only solution is 8.

Now Try Exercise 43

EQUATIONS INVOLVING CUBE ROOTS In the next example, we apply the power rule to an equation that contains a cube root.

EXAMPLE 5 **Solving an equation containing a cube root**

Solve $\sqrt[3]{4x - 7} = 4$.

Solution

STEP 1: The cube root term is already isolated, so we proceed to Step 2.

STEP 2: Because the index is 3, we **cube** each side of the equation.

$$\sqrt[3]{4x - 7} = 4 \qquad \text{Given equation}$$
$$(\sqrt[3]{4x - 7})^3 = (4)^3 \qquad \text{Cube each side.}$$

STEP 3: We solve the resulting equation.

$$4x - 7 = 64 \qquad \text{Simplify.}$$
$$4x = 71 \qquad \text{Add 7 to each side.}$$
$$x = \frac{71}{4} \qquad \text{Divide each side by 4.}$$

STEP 4: To check this answer, we substitute $x = \frac{71}{4}$ in the given equation.

$$\sqrt[3]{4\left(\frac{71}{4}\right) - 7} \stackrel{?}{=} 4$$
$$\sqrt[3]{64} \stackrel{?}{=} 4$$
$$4 = 4 \checkmark \qquad \text{It checks.}$$

Now Try Exercise 31

🌐 *Math in Context* Biology Larger birds tend to have larger wings. This relationship can sometimes be modeled by $A(W) = 100\sqrt[3]{W^2}$, where W is the weight in pounds of the bird and A is the area of the wings in square inches. In the next example, we use the formula to determine the weight of a bird that has wings with a surface area of 600 square inches. (*Source:* C. Pennycuick, *Newton Rules Biology.*)

EXAMPLE 6 **Finding the weight of a bird**

Solve the equation $600 = 100\sqrt[3]{W^2}$ to determine the weight in pounds of a bird that has wings with an area of 600 square inches.

Solution

Begin by dividing each side of the equation $600 = 100\sqrt[3]{W^2}$ by 100 to isolate the radical term on the right side of the equation.

$$\frac{600}{100} = \sqrt[3]{W^2} \qquad \text{Divide each side by 100.}$$
$$(6)^3 = (\sqrt[3]{W^2})^3 \qquad \text{Cube each side.}$$
$$216 = W^2 \qquad \text{Simplify.}$$
$$\sqrt{216} = W \qquad \text{Take principal square root, } W > 0.$$
$$W \approx 14.7 \qquad \text{Approximate.}$$

The weight of the bird is approximately 14.7 pounds.

Now Try Exercise 117

TECHNOLOGY NOTE

Graphing Radical Expressions

The equation in Example 6 can be solved graphically. Sometimes it is more convenient to use rational exponents than radical notation. Thus $y = 100\sqrt[3]{W^2}$ can be entered as $y_1 = 100x^{2/3}$. The accompanying figure shows y_1 intersecting the line $y_2 = 600$ near the point $(14.7, 600)$, which supports our symbolic result.

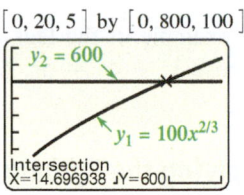

EXAMPLE 7 **Solving an equation with rational exponents graphically**

Solve $x^{2/3} = 3 - x^2$ graphically.

Solution

Graph $y_1 = x^{2/3}$ and $y_2 = 3 - x^2$. The graphs intersect near $(-1.34, 1.21)$ and $(1.34, 1.21)$, as shown in **FIGURE 10.17**. Thus the solutions are given by $x \approx \pm 1.34$.

(a)

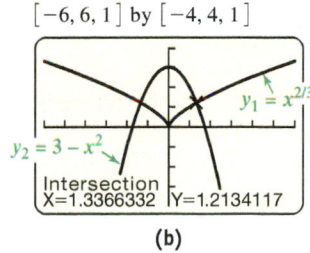
(b)

FIGURE 10.17

Now Try Exercise 73

The following See the Concept illustrates how the equation $\sqrt{x + 2} = x$ can be solved symbolically, numerically, and graphically.

See the Concept 1 **WAYS TO SOLVE AN EQUATION**

Solve the equation $\sqrt{x + 2} = x$ symbolically, numerically, and graphically.

Symbolic Solution

$$\sqrt{x + 2} = x$$
$$x + 2 = x^2$$
$$x^2 - x - 2 = 0$$
$$(x - 2)(x + 1) = 0$$
$$x = 2 \quad \text{or} \quad x = -1$$

$Check$: $\sqrt{2 + 2} = 2$ ✓

$\sqrt{-1 + 2} \neq -1$ ✗

The only solution is **2**.

Numerical Solution

X	Y₁	Y₂
-2	0	-2
-1	1	-1
0	1.4142	0
1	1.7321	1
2	2	2
3	2.2361	3
4	2.4495	4

X=2

$y_1 = y_2$ when $x = 2$, so **2** is a solution. Note that -1 is *not* a solution.

Graphical Solution

$[-4.7, 4.7, 1]$ by $[-3.1, 3.1, 1]$

The graphs intersect at $(2, 2)$. Note that there is no point of intersection when $x = -1$.

Note: The solution is **2** for each method. ∎

WATCH VIDEO IN MML.

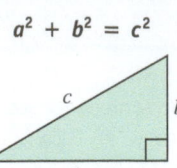

$$a^2 + b^2 = c^2$$

FIGURE 10.18

2 Deriving and Applying the Distance Formula

PYTHAGOREAN THEOREM One of the most famous theorems in mathematics is the **Pythagorean theorem**. It states that if a right triangle has legs a and b with hypotenuse c (see **FIGURE 10.18**), then

$$a^2 + b^2 = c^2. \qquad \text{Pythagorean theorem}$$

For example, if the legs of a right triangle are $a = 3$ and $b = 4$, the hypotenuse is $c = 5$ because $3^2 + 4^2 = 5^2$. Also, if the sides of a triangle satisfy $a^2 + b^2 = c^2$, it is a right triangle.

EXAMPLE 8   **Applying the Pythagorean theorem**

An HDTV has a width of 60 inches and a height of 32 inches. Find the diagonal of the television. Why is it called a 68-inch television?

Solution

In **FIGURE 10.19**, let $a = \mathbf{60}$ and $b = \mathbf{32}$. Then the diagonal of the television corresponds to the hypotenuse of a right triangle with legs of 60 inches and 32 inches.

$$
\begin{aligned}
c^2 &= a^2 + b^2 & &\text{Pythagorean theorem} \\
c &= \sqrt{a^2 + b^2} & &\text{Take the principal square root, } c > 0. \\
c &= \sqrt{\mathbf{60}^2 + \mathbf{32}^2} & &\text{Substitute } a = 60 \text{ and } b = 32. \\
c &= 68 & &\text{Simplify.}
\end{aligned}
$$

A 68-inch television has a diagonal of 68 inches.

Now Try Exercise 125

32 in.

60 in.

FIGURE 10.19

DISTANCE BETWEEN TWO POINTS The Pythagorean theorem can be used to determine the distance between two points, as illustrated in the following See the Concept.

See the Concept 2   DISTANCE BETWEEN TWO POINTS

A The length of the vertical leg is $y_2 - y_1$.

B The length of the horizontal leg is $x_2 - x_1$.

C By the Pythagorean theorem:
$$d^2 = (x_2 - x_1)^2 + (y_2 - y_1)^2$$

Thus, $d = \sqrt{(x_2 - x_1)^2 + (y_2 - y_1)^2}$.

Distance is nonnegative.   **WATCH VIDEO IN** $\boxed{\text{MML}}$.

DISTANCE FORMULA

The **distance** d between the points (x_1, y_1) and (x_2, y_2) in the xy-plane is

$$d = \sqrt{(x_2 - x_1)^2 + (y_2 - y_1)^2}.$$

EXAMPLE 9 **Finding distance between points**

Find the distance between the points $(-2, 3)$ and $(1, -4)$.

Solution
Start by letting $(x_1, y_1) = (-2, 3)$ and $(x_2, y_2) = (1, -4)$. Then substitute these values into the distance formula.

$$
\begin{aligned}
d &= \sqrt{(x_2 - x_1)^2 + (y_2 - y_1)^2} && \text{Distance formula} \\
&= \sqrt{(1 - (-2))^2 + (-4 - 3)^2} && \text{Substitute.} \\
&= \sqrt{9 + 49} && \text{Simplify.} \\
&= \sqrt{58} && \text{Add.} \\
&\approx 7.62 && \text{Approximate.}
\end{aligned}
$$

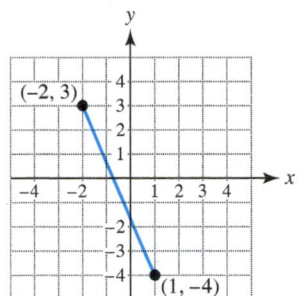

FIGURE 10.20

The distance between the points, as shown in **FIGURE 10.20**, is exactly $\sqrt{58}$ units, or about 7.62 units. Note that we would obtain the same result if we let $(x_1, y_1) = (1, -4)$ and $(x_2, y_2) = (-2, 3)$.

Now Try Exercise 109

NOTE: In Example 9, $\sqrt{9 + 49} \neq \sqrt{9} + \sqrt{49} = 3 + 7 = 10$. In general, for any a and b, $\sqrt{a^2 + b^2} \neq a + b$. ∎

3 ▸ Solving Equations of the Form $x^n = k$

The equation $x^n = k$, where n is a *positive integer*, can be solved by taking the nth root of each side of the equation. The following technique allows us to find all *real* solutions to this equation.

SOLVING THE EQUATION $x^n = k$

Take the nth root of each side of $x^n = k$ to obtain $\sqrt[n]{x^n} = \sqrt[n]{k}$.

1. If n is odd, then $\sqrt[n]{x^n} = x$ and the equation becomes $x = \sqrt[n]{k}$.
2. If n is *even* and $k > 0$, then $\sqrt[n]{x^n} = |x|$ and the equation becomes $|x| = \sqrt[n]{k}$. (If $k < 0$, there are no real solutions.)

To understand this technique better, consider the following examples. First let $x^3 = 8$ so that n is odd. Taking the cube root of each side gives

$$\sqrt[3]{x^3} = \sqrt[3]{8}, \text{ which is equivalent to } x = \sqrt[3]{8} \text{ or } x = 2.$$

Next let $x^2 = 4$ so that n is even. Taking the square root of each side gives

$$\sqrt{x^2} = \sqrt{4}, \text{ which is equivalent to } |x| = \sqrt{4} \text{ or } |x| = 2.$$

The solutions to $|x| = 2$ are -2 or 2, which can be written as ± 2.

EXAMPLE 10 **Solving equations of the form $x^n = k$**

Solve each equation.
(a) $x^3 = -64$ (b) $x^2 = 12$ (c) $2(x - 1)^4 = 32$

Solution
(a) Taking the cube root of each side of $x^3 = -64$ gives

$$\sqrt[3]{x^3} = \sqrt[3]{-64} \text{ or } x = -4.$$

(b) Taking the square root of each side of $x^2 = 12$ gives

$$\sqrt{x^2} = \sqrt{12} \quad \text{or} \quad |x| = \sqrt{12}.$$

The equation $|x| = \sqrt{12}$ is equivalent to $x = \pm\sqrt{12}$.

(c) First divide each side of $2(x-1)^4 = 32$ by 2 to isolate the power of $x - 1$.

$(x-1)^4 = 16$	Divide each side by 2.
$\sqrt[4]{(x-1)^4} = \sqrt[4]{16}$	Take the 4th root of each side.
$\|x - 1\| = 2$	Simplify: $\sqrt[4]{y^4} = \|y\|$.
$x - 1 = -2 \quad \text{or} \quad x - 1 = 2$	Solve the absolute value equation.
$x = -1 \quad \text{or} \quad x = 3$	Add 1 to each side.

Now Try Exercises 47, 55, 67

$\blacktriangleright$ *Math in Context* ~Wind Energy~ In some parts of the United States, wind power is used to generate electricity. Suppose that the diameter of the circular path created by the blades of a wind-powered generator is 8 feet. Then the wattage W generated by a wind velocity of v miles per hour is modeled by

$$W(v) = 2.4v^3.$$

If the wind blows at 10 miles per hour, the generator can produce about

$$W(\mathbf{10}) = 2.4 \cdot \mathbf{10}^3 = 2400 \text{ watts.}$$

(*Source: Conquering the Sciences*, Sharp Electronics.)

EXAMPLE 11   **Modeling a wind generator**

The formula $W(v) = 2.4v^3$ is used to calculate the watts generated when there is a wind velocity of v miles per hour.
(a) Find a function f that calculates the wind velocity when W watts are being produced.
(b) If the wattage doubles, has the wind velocity also doubled? Explain.

Solution
(a) Given W we need a formula to find v, so solve $W = 2.4v^3$ for v.

$W = 2.4v^3$	Given formula
$\dfrac{W}{2.4} = v^3$	Divide by 2.4.
$\sqrt[3]{\dfrac{W}{2.4}} = \sqrt[3]{v^3}$	Take the cube root of each side.
$v = \sqrt[3]{\dfrac{W}{2.4}}$	Simplify and rewrite equation.

Thus $f(W) = \sqrt[3]{\dfrac{W}{2.4}}$.

(b) Suppose that the power generated is 1000 watts. Then the wind speed is

$$f(1000) = \sqrt[3]{\frac{1000}{2.4}} \approx 7.5 \text{ miles per hour.}$$

If the power doubles to 2000 watts, then the wind speed is

$$f(2000) = \sqrt[3]{\frac{2000}{2.4}} \approx 9.4 \text{ miles per hour.}$$

Thus for the wattage to double, the wind speed does not need to double.

Now Try Exercise 133

10.6   Putting It All Together

CONCEPT	EXPLANATION	EXAMPLES
Power Rule for Solving Equations	If each side of an equation is raised to the same positive integer power, any solutions to the given equation are among the solutions to the new equation.	$\sqrt{2x} = x$ $2x = x^2$ Square each side. $x^2 - 2x = 0$ Rewrite equation. $x = 0$ or $x = 2$ Factor and solve. *Be sure to check any solutions.*
Pythagorean Theorem	If c is the hypotenuse of a right triangle and a and b are its legs, then $a^2 + b^2 = c^2$. 	If the sides of the right triangle are $a = 5$, $b = 12$, and $c = 13$, then they satisfy $a^2 + b^2 = c^2$ or $5^2 + 12^2 = 13^2$.
Distance Formula	The distance d between the points (x_1, y_1) and (x_2, y_2) is $$d = \sqrt{(x_2 - x_1)^2 + (y_2 - y_1)^2}.$$	The distance between the points $(2, 3)$ and $(-3, 4)$ is $$d = \sqrt{(-3 - 2)^2 + (4 - 3)^2}$$ $$= \sqrt{(-5)^2 + (1)^2} = \sqrt{26}.$$
Solving the Equation $x^n = k$, Where n is a Positive Integer	Take the nth root of each side to obtain $\sqrt[n]{x^n} = \sqrt[n]{k}$. Then 1. $x = \sqrt[n]{k}$, if n is odd. 2. $x = \pm\sqrt[n]{k}$, if n is even and $k \geq 0$.	1. n odd: If $x^5 = 32$, then $x = \sqrt[5]{32} = 2$. 2. n even: If $x^4 = 81$, then $x = \pm\sqrt[4]{81} = \pm 3$.

10.6   Exercises

MyMathLab®

CONCEPTS AND VOCABULARY

1. What is a good first step for solving $\sqrt{4x - 1} = 5$?

2. What is a good first step for solving $\sqrt[3]{x + 1} = 6$?

3. Can an equation involving rational exponents have more than one solution?

4. When you square each side of an equation to solve for an unknown, what must you do with any answers?

5. What is the Pythagorean theorem used for?

6. If the legs of a right triangle are 3 and 4, what is the length of the hypotenuse?

7. What formula can you use to find the distance d between two points?

8. Write the equation $\sqrt{x} + \sqrt[4]{x^3} = 2$ with rational exponents.

SIMPLIFYING RADICALS

Exercises 9–16: Simplify. Assume radicands of square roots are positive.

9. $\sqrt{2} \cdot \sqrt{2}$

10. $\sqrt{5} \cdot \sqrt{5}$

11. $\sqrt{x} \cdot \sqrt{x}$

12. $\sqrt{2x} \cdot \sqrt{2x}$

13. $(\sqrt{2x + 1})^2$

14. $(\sqrt{7x})^2$

15. $(\sqrt[3]{5x^2})^3$

16. $(\sqrt[3]{2x - 5})^3$

FINDING SYMBOLIC SOLUTIONS

Exercises 17–46: Solve the equation symbolically. Check your results.

17. $\sqrt{x} = 8$

18. $\sqrt{3z} = 6$

19. $\sqrt[4]{x} = 3$

20. $\sqrt[3]{x - 4} = 2$

21. $\sqrt{2t + 4} = 4$

22. $\sqrt{y + 4} = 3$

23. $\sqrt{x + 1} - 3 = 4$ **24.** $\sqrt{2x + 5} + 2 = 5$

25. $2\sqrt{x - 2} + 1 = 5$ **26.** $-\sqrt{x + 7} - 1 = -7$

27. $\sqrt{x + 6} = x$ **28.** $\sqrt{z} + 6 = z$

29. $\sqrt[3]{x} = 3$ **30.** $\sqrt[3]{x + 10} = 4$

31. $\sqrt[3]{2z - 4} = -2$ **32.** $\sqrt[3]{z - 1} = -3$

33. $\sqrt[4]{t + 1} = 2$ **34.** $\sqrt[4]{5t} = 5$

35. $\sqrt{5z - 1} = \sqrt{z + 1}$ **36.** $y = \sqrt{y + 1} + 1$

37. $\sqrt{1 - x} = 1 - x$ **38.** $\sqrt[3]{4x} = x$

39. $\sqrt{b^2 - 4} = b - 2$ **40.** $\sqrt{b^2 - 2b + 1} = b$

41. $\sqrt{1 - 2x} = x + 7$ **42.** $\sqrt{4 - y} = y - 2$

43. $\sqrt{x} = \sqrt{x - 5} + 1$ **44.** $\sqrt{x - 1} = \sqrt{x + 4} - 1$

45. $\sqrt{2t - 2} + \sqrt{t} = 7$ **46.** $\sqrt{x + 1} - \sqrt{x - 6} = 1$

SOLVING THE EQUATION $x^n = k$

Exercises 47–68: (Refer to Example 10.) Solve.

47. $x^2 = 49$ **48.** $x^2 = 9$

49. $2z^2 = 200$ **50.** $3z^2 = 48$

51. $(t + 1)^2 = 16$ **52.** $(t - 5)^2 = 81$

53. $(3x - 6)^2 = 25$ **54.** $(4 - 2x)^2 = 100$

55. $b^3 = 64$ **56.** $a^3 = 1000$

57. $2t^3 = -128$ **58.** $3t^3 = -81$

59. $(x + 1)^3 = 8$ **60.** $(4 - x)^3 = -1$

61. $(2 - 5z)^3 = -125$ **62.** $(2x + 4)^3 = 125$

63. $x^4 = 16$ **64.** $x^4 = 7$

65. $x^5 = 12$ **66.** $x^5 = -32$

67. $2(x + 2)^4 = 162$ **68.** $\frac{1}{2}(x - 1)^5 = 16$

FINDING GRAPHICAL SOLUTIONS

Exercises 69–78: Solve graphically. Approximate solutions to the nearest hundredth when appropriate.

69. $\sqrt[3]{x + 5} = 2$ **70.** $\sqrt[3]{x} + \sqrt{x} = 3.43$

71. $\sqrt{2x - 3} = \sqrt{x} - \frac{1}{2}$ **72.** $x^{4/3} - 1 = 2$

73. $x^{5/3} = 2 - 3x^2$ **74.** $x^{3/2} = \sqrt{x + 2} - 2$

75. $z^{1/3} - 1 = 2 - z$ **76.** $z^{3/2} - 2z^{1/2} - 1 = 0$

77. $\sqrt{x + 1} - \sqrt{x - 1} = 4$

78. $\sqrt{y + 2} + \sqrt{3y + 2} = 2$

USING MORE THAN ONE METHOD TO SOLVE EQUATIONS

Exercises 79–82: Solve the equation
  (a) *symbolically,*   (b) *graphically, and*
  (c) *numerically.*

79. $2\sqrt{x} = 8$ **80.** $\sqrt[3]{5 - x} = 2$

81. $\sqrt{6z - 2} = 8$ **82.** $\sqrt{y + 4} = \dfrac{y}{3}$

SOLVING AN EQUATION FOR A VARIABLE

Exercises 83–86: Solve for the indicated variable.

83. $T = 2\pi\sqrt{\dfrac{L}{32}}$ for L

84. $Z = \sqrt{L^2 + R^2}$ for R

85. $r = \sqrt{\dfrac{A}{\pi}}$ for A

86. $F = \dfrac{1}{2\pi\sqrt{LC}}$ for C

APPLYING THE PYTHAGOREAN THEOREM

Exercises 87–94: If the sides of a triangle are a, b, and c and they satisfy $a^2 + b^2 = c^2$, the triangle is a right triangle. Determine whether the triangle with the given sides is a right triangle.

87. $a = 6$ $b = 8$ $c = 10$

88. $a = 5$ $b = 12$ $c = 13$

89. $a = \sqrt{5}$ $b = \sqrt{9}$ $c = \sqrt{14}$

90. $a = 4$ $b = 5$ $c = 7$

91. $a = 7$ $b = 24$ $c = 25$

92. $a = 1$ $b = \sqrt{3}$ $c = 2$

93. $a = 8$ $b = 8$ $c = 16$

94. $a = 11$ $b = 60$ $c = 61$

Exercises 95–98: Find the length of the missing side in the right triangle.

95.

96.

97.

98.
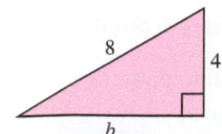

Exercises 99–104: A right triangle has legs a and b with hypotenuse c. Find the length of the missing side.

99. $a = 3, b = 4$

100. $a = 4, b = 7$

101. $a = \sqrt{3}, c = 8$

102. $a = \sqrt{6}, c = \sqrt{10}$

103. $b = 48, c = 50$

104. $b = 10, c = 26$

APPLYING THE DISTANCE FORMULA

Exercises 105–108: Find the length of the line segment.

105.

106.

107.

108.

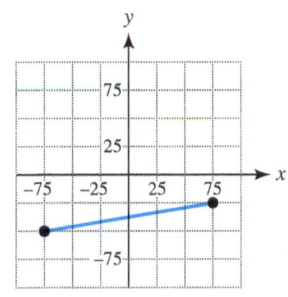

Exercises 109–112: Find the distance between the points.

109. $(-1, 2), (4, 10)$

110. $(5, -40), (-6, 20)$

111. $(0, -3), (4, 0)$

112. $(3, 9), (-4, 2)$

Exercises 113–116: Find x if the distance between the points is d. Assume that $x \geq 0$.

113. $(x, 3), (0, 6)$ $d = 5$

114. $(x, -1), (6, 11)$ $d = 13$

115. $(x, -5), (62, 6)$ $d = 61$

116. $(x, 3), (12, -4)$ $d = 25$

APPLICATIONS INVOLVING RADICAL EQUATIONS

Exercises 117 and 118: **Weight of a Bird** *(Refer to Example 6.) Estimate the weight W of a bird that has wings with area A. Let $A = 100\sqrt[3]{W^2}$.*

117. $A = 400$ square inches

118. $A = 1000$ square inches

Exercises 119 and 120: **Abandonment Rate** *In a fast-paced society, people do not spend time watching videos that they find boring. For example, in Example 3 of the second section of this chapter, we learned how the function*

$$A(x) = 7.3\sqrt[16]{x^7}$$

computes the percentage of viewers who abandon an online video after x seconds. Use $A(x) = 7.3\sqrt[16]{x^7}$ to find the average number of seconds x that it takes for A percent of the viewers to abandon an online movie.

119. $A = 50$

120. $A = 20$

Exercises 121–124: **Distance to the Horizon** *Because of Earth's curvature, a person can see a limited distance to the horizon. The higher the location of the person, the farther that person can see. The distance D in miles to the horizon can be estimated by $D(h) = 1.22\sqrt{h}$, where h is the height of the person above the ground in feet.*

121. Find D for a 6-foot-tall person standing on level ground.

122. Find D for a person on top of Mount Everest with a height of 29,028 feet.

123. What height allows a person to see 20 miles?

124. How high does a plane need to fly for the pilot to be able to see 100 miles?

125. **Diagonal of a Television** (Refer to Example 8.) A rectangular television screen is 11.4 inches by 15.2 inches. Find the diagonal of the television screen.

126. **Dimensions of a Television** The height of a television with a 13-inch diagonal is $\frac{3}{4}$ of its width. Find the width and height of the television set.

127. DVD and Picture Dimensions If the picture shown on a television set is h units high and w units wide, the *aspect ratio* of the picture is $\frac{w}{h}$ (see the accompanying figure). Digital video discs (DVDs) support the newer aspect ratio of $\frac{16}{9}$ rather than the older ratio of $\frac{4}{3}$. If the width of a picture with an aspect ratio of $\frac{16}{9}$ is 29 inches, approximate the height and diagonal of the rectangular picture. (*Source:* J. Taylor, *DVD Demystified.*)

128. Flood Control The spillway capacity of a dam is important in flood control. Spillway capacity Q in cubic feet of water per second flowing over the spillway depends on the width W and the depth D of the spillway, as illustrated in the accompanying figure. If W and D are measured in feet, capacity can be modeled by $Q = 3.32WD^{3/2}$. (*Source:* D. Callas, Project Director, *Snapshots of Applications in Mathematics.*)

(a) Find the capacity of a spillway with $W = 20$ feet and $D = 5$ feet.

(b) A spillway with a width of 30 feet is to have a capacity of $Q = 2690$ cubic feet per second. Estimate to the nearest foot the appropriate depth of the spillway.

129. Sky Diving When sky divers initially fall from an airplane, their velocity v in miles per hour after free falling d feet can be approximated by $v = \frac{60}{11}\sqrt{d}$. (Because of air resistance, they will eventually reach a terminal velocity.) How far do sky divers need to fall to attain the following velocities? (These values for d represent minimum distances.)

(a) 60 miles per hour

(b) 100 miles per hour

130. Guy Wire A guy wire attached to the top of a 30-foot-long pole is anchored 10 feet from the base of the pole, as illustrated in the figure at the top of the next column. Find the length of the guy wire to the nearest tenth of a foot.

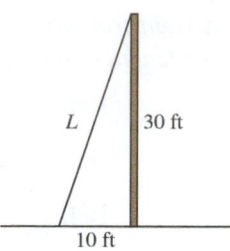

131. Skid Marks Vehicles involved in accidents often leave skid marks, which can be used to determine how fast a vehicle was traveling. To determine this speed, officials often use a test vehicle to compare skid marks on the same section of road. Suppose that a vehicle in a crash left skid marks D feet long and that a test vehicle traveling at v miles per hour leaves skid marks d feet long. Then the speed V of the vehicle involved in the crash is given by

$$V = v\sqrt{\frac{D}{d}}.$$

(*Source:* N. Garber and L. Hoel, *Traffic and Highway Engineering.*)

(a) Find V if $v = 30$ mph, $D = 285$ feet, and $d = 178$ feet. Interpret your result.

(b) A test vehicle traveling at 45 mph leaves skid marks 255 feet long. How long would the skid marks be for a vehicle traveling 60 miles per hour?

132. Highway Curves If a circular highway curve without any banking has a radius of R feet, the speed limit L in miles per hour for the curve is $L = 1.5\sqrt{R}$. (*Source:* N. Garber.)

(a) Find the speed limit for a curve that has a radius of 400 feet.

(b) If the radius of a curve doubles, what happens to the speed limit?

(c) A curve with a 40-mile-per-hour speed limit is being designed. What should be its radius?

133. Wind Power (Refer to Example 11.) If a wind-powered generator has blades that create a circular path with a diameter of 10 feet, then the wattage W generated by a wind velocity of v miles per hour is modeled by $W(v) = 3.8v^3$.

(a) If the wind velocity doubles, what happens to the wattage generated?

(b) Solve $W = 3.8v^3$ for v.

(c) If the wind generator is producing 30,400 watts, find the wind speed.

134. Height and Weight Suppose that the weight of a person is directly proportional to the cube of the person's height. If one person weighs twice as much as a (similarly proportioned) second person, by what factor is the heavier person's height greater than the shorter person's height?

135. 45°–45° Right Triangle Suppose that the legs of a right triangle with angles of 45° and 45° both have length a, as depicted in the accompanying figure. Find the length of the hypotenuse.

136. 30°–60° Right Triangle In a right triangle with angles of 30° and 60°, the shortest side is half the length of the hypotenuse (see the accompanying figure). If the hypotenuse has length c, find the length of the other two sides, a and b, in terms of c.

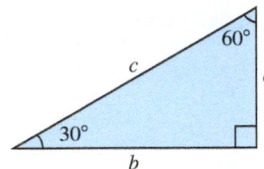

137. Minimizing Cost A natural gas line running along a river is to be connected from point A to a cabin on the other bank located at point D, as illustrated in the figure. The width of the river is 500 feet, and the distance from point A to point C is 1000 feet. The cost of running the pipe along the shoreline is $30 per foot, and the cost of running it underwater is $50 per foot. The cost of connecting the gas line from A to D is to be minimized.

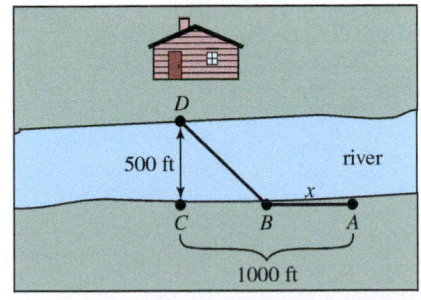

(a) Write an expression that gives the cost of running the line from A to B if the distance between these points is x feet.

(b) Find the distance from B to D in terms of x.

(c) Write an expression that gives the cost of running the line from B to D.

(d) Use your answer from parts (a) and (c) to write an expression that gives the cost of running the line from A to B to D.

(e) Graph your expression from part (d) in the window $[0, 1000, 100]$ by $[40000, 60000, 5000]$ to determine the value of x that minimizes the cost of the line going from A to D. What is the minimum cost?

138. Minimizing Cost Repeat Exercise 137 if the width of the river is 600 feet and the distance from point A to point C is 1500 feet. For the window in part (e), use $[0, 1500, 500]$ by $[50000, 90000, 10000]$.

WRITING ABOUT MATHEMATICS

139. A student solves an equation *incorrectly* as follows.

$$\sqrt{3 - x} = \sqrt{x} - 1$$
$$(\sqrt{3 - x})^2 \overset{?}{=} (\sqrt{x})^2 - (1)^2$$
$$3 - x \overset{?}{=} x - 1$$
$$-2x \overset{?}{=} -4$$
$$x \overset{?}{=} 2$$

(a) How could you convince the student that the answer is wrong?

(b) Discuss where any errors were made.

140. When each side of an equation is squared, you must check your results. Explain why.

SECTIONS 10.5 and 10.6   Checking Basic Concepts

1. Sketch a graph of each function and then evaluate $f(-1)$, if possible.
(a) $f(x) = \sqrt{x}$
(b) $f(x) = \sqrt[3]{x}$
(c) $f(x) = \sqrt{x^2}$

2. Evaluate $f(x) = 0.2x^{2/3}$ when $x = 64$.

3. Find the domain of $f(x) = \sqrt{x - 4}$. Write your answer in interval notation.

4. Solve each equation. Check your answers.
(a) $\sqrt{2x - 4} = 2$
(b) $\sqrt[3]{x - 1} = 3$
(c) $\sqrt{3x} = 1 + \sqrt{x + 1}$

5. Find the distance between $(-3, 5)$ and $(2, -7)$.

6. A 16-inch diagonal television set has a rectangular picture with a width of 12.8 inches. Find the height of the picture.

7. Solve $(x + 1)^4 = 16$.

10.7 Complex Numbers

1 Introducing the Imaginary Unit and Standard Form

A graph of $y = x^2 + 1$ is shown in **FIGURE 10.21**. There are no x-intercepts, so the quadratic equation $x^2 + 1 = 0$ has no real number solutions.

If we try to solve $x^2 + 1 = 0$ by subtracting 1 from each side, the result is $x^2 = -1$. Because $x^2 \geq 0$ for any real number x, there are no real solutions. However, mathematicians have invented (or discovered) solutions.

$$x^2 = -1$$
$$x = \pm\sqrt{-1} \quad \text{Solve for } x.$$

We now define a number called the **imaginary unit**, denoted i.

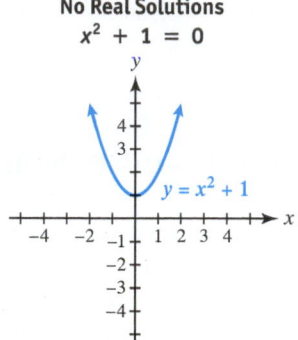

No Real Solutions
$x^2 + 1 = 0$

FIGURE 10.21

PROPERTIES OF THE IMAGINARY UNIT i

$$i = \sqrt{-1} \quad \text{and} \quad i^2 = -1$$

By creating the number i, the solutions to the equation $x^2 + 1 = 0$ are i and $-i$. Using the real numbers and the imaginary unit i, we can define a new set of numbers called the *complex numbers*. A **complex number** can be written in **standard form**, as $a + bi$, where a and b are real numbers. The **real part** is a and the **imaginary part** is b. Every real number a is also a complex number because it can be written $a + 0i$. A complex number $a + bi$ with $b \neq 0$ is a **nonreal complex number**, or **imaginary number**. A complex number $a + bi$ with $a = 0$ and $b \neq 0$ is sometimes called a **pure imaginary number**. Examples of pure imaginary numbers include $4i$ and $-2i$. **TABLE 10.7** lists several complex numbers with their real and imaginary parts.

Real and Imaginary Parts of Complex Numbers

Complex Number: $a + bi$	$-3 + 2i$	5	$-3i$	$-1 + 7i$	$-5 - 2i$	$4 + 6i$
Real Part: a	-3	5	0	-1	-5	4
Imaginary Part: b	2	0	-3	7	-2	6

TABLE 10.7

The following See the Concept illustrates how different sets of numbers are related.

See the Concept 1   SETS OF NUMBERS

The complex numbers include all the real numbers.

Complex Numbers
$a + bi$, a and b real

Real Numbers
$a + bi$, $b = 0$

Nonreal Complex Numbers
$a + bi$, $b \neq 0$

Rational Numbers
$-3, \frac{2}{3}, 0,$ and $-\frac{1}{2}$

Irrational Numbers
$\sqrt{3}, \pi,$ and $\sqrt[3]{-11}$

Pure Imaginary Numbers
$a + bi$, $a = 0, b \neq 0$

WATCH VIDEO IN MML.

Using the imaginary unit i, we may write the square root of a negative number as a complex number. For example, $\sqrt{-2} = i\sqrt{2}$, and $\sqrt{-4} = i\sqrt{4} = 2i$. This method is summarized as follows.

CALCULATOR HELP

To set your calculator in $a + bi$ mode, see Appendix A (page AP-7).

THE EXPRESSION $\sqrt{-a}$

If $a > 0$, then $\sqrt{-a} = i\sqrt{a}$.

NOTE: Although it is standard for a complex number to be expressed as $a + bi$, we often write $\sqrt{2}i$ as $i\sqrt{2}$ so that it is clear i is not under the square root. Similarly, $\frac{1}{2}\sqrt{2}i$ is sometimes written as $\frac{1}{2}i\sqrt{2}$ or $\frac{i\sqrt{2}}{2}$. ∎

EXAMPLE 1 Writing the square root of a negative number

Write each square root using the imaginary unit i.
(a) $\sqrt{-25}$ (b) $\sqrt{-7}$ (c) $\sqrt{-20}$

Solution
(a) $\sqrt{-25} = i\sqrt{25} = 5i$
(b) $\sqrt{-7} = i\sqrt{7}$
(c) $\sqrt{-20} = i\sqrt{20} = i\sqrt{4}\sqrt{5} = 2i\sqrt{5}$

Now Try Exercises 13, 15, 19

---

2 Adding, Subtracting, and Multiplying Complex Numbers

Addition can be defined for complex numbers in a manner similar to how we add binomials. For example,

$$(-3 + 2x) + (2 - x) = (-3 + 2) + (2x - x) = -1 + x.$$

ADDITION AND SUBTRACTION To add the complex numbers $(-3 + 2i)$ and $(2 - i)$, add the real parts and then add the imaginary parts.

$$(-3 + 2i) + (2 - i) = (-3 + 2) + (2i - i)$$
$$= (-3 + 2) + (2 - 1)i$$
$$= -1 + i$$

This same process works for subtraction.

$$(6 - 3i) - (2 + 5i) = (6 - 2) + (-3i - 5i)$$
$$= (6 - 2) + (-3 - 5)i$$
$$= 4 - 8i$$

This method is summarized as follows.

SUM OR DIFFERENCE OF COMPLEX NUMBERS

READING CHECK 1

Add $(1 - 2i) + (3 + i)$.

Let $a + bi$ and $c + di$ be two complex numbers. Then

$$(a + bi) + (c + di) = (a + c) + (b + d)i \quad \text{Sum}$$

and

$$(a + bi) - (c + di) = (a - c) + (b - d)i. \quad \text{Difference}$$

EXAMPLE 2　**Adding and subtracting complex numbers**

Write each sum or difference in standard form.
(a) $(-7 + 2i) + (3 - 4i)$　　(b) $3i - (5 - i)$

Solution
(a) $(-7 + 2i) + (3 - 4i) = (-7 + 3) + (2 - 4)i = -4 - 2i$
(b) $3i - (5 - i) = 3i - 5 + i = -5 + (3 + 1)i = -5 + 4i$

Now Try Exercises 23, 29

CALCULATOR HELP

To access the imaginary unit i, see Appendix A (page AP-7).

MULTIPLICATION We multiply two complex numbers in the same way that we multiply binomials and then we apply the property $i^2 = -1$. For example,

$$(2 - 3x)(1 + 4x) = 2 + 8x - 3x - 12x^2 = 2 + 5x - 12x^2.$$

In the next example, we find the product of $2 - 3i$ and $1 + 4i$ in a similar manner.

EXAMPLE 3　**Multiplying complex numbers**

Write each product in standard form.
(a) $(2 - 3i)(1 + 4i)$　　(b) $(5 - 2i)(5 + 2i)$

Solution
(a) Multiply the complex numbers like binomials.

Complex Multiplication

```
(2-3i)(1+4i)
            14+5i
(5-2i)(5+2i)
               29
```

FIGURE 10.22

$$(2 - 3i)(1 + 4i) = (2)(1) + (2)(4i) - (3i)(1) - (3i)(4i)$$
$$= 2 + 8i - 3i - 12i^2$$
$$= 2 + 5i - 12(-1) \quad i^2 = -1$$
$$= 14 + 5i$$

(b)
$$(5 - 2i)(5 + 2i) = (5)(5) + (5)(2i) - (2i)(5) - (2i)(2i)$$
$$= 25 + 10i - 10i - 4i^2$$
$$= 25 - 4(-1) \quad i^2 = -1$$
$$= 29$$

TECHNOLOGY NOTE

Complex Numbers
Many calculators can perform arithmetic with complex numbers. The figure shows a calculator display for the results in Example 2.

```
(-7+2i)+(3-4i)
             -4-2i
3i-(5-i)
             -5+4i
```

These results are supported in **FIGURE 10.22**.

Now Try Exercises 31, 35

3　Calculating Powers of i

An interesting pattern appears when powers of i are calculated.

$$i^1 = i$$
$$i^2 = -1$$
$$i^3 = i^2 \cdot i = -1 \cdot i = -i$$
$$i^4 = i^2 \cdot i^2 = (-1)(-1) = 1$$
$$i^5 = i^4 \cdot i = (1)i = i$$
$$i^6 = i^4 \cdot i^2 = (1)(-1) = -1$$
$$i^7 = i^4 \cdot i^3 = (1)(-i) = -i$$
$$i^8 = i^4 \cdot i^4 = (1)(1) = 1$$

Note the pattern:
$i = i^5, i^2 = i^6,$
and so on.

The powers of i cycle with the pattern i, -1, $-i$, and 1. These examples suggest the following method for calculating powers of i.

POWERS OF i

The value of i^n can be found by dividing n (a positive integer) by 4. If the remainder is r, then

$$i^n = i^r.$$

Note that $i^0 = 1$, $i^1 = i$, $i^2 = -1$, and $i^3 = -i$.

EXAMPLE 4 Calculating powers of i

Evaluate each expression.
(a) i^9 (b) i^{19} (c) i^{40}

Solution
(a) When 9 is divided by 4, the result is 2 with remainder **1**. Thus $i^9 = i^1 = i$.
(b) When 19 is divided by 4, the result is 4 with remainder **3**. Thus $i^{19} = i^3 = -i$.
(c) When 40 is divided by 4, the result is 10 with remainder **0**. Thus $i^{40} = i^0 = 1$.

Now Try Exercises 49, 51, 55

4 Finding Complex Conjugates

The **complex conjugate** of $a + bi$ is $a - bi$. To find the conjugate, we change the sign of the imaginary part b. **TABLE 10.8** contains some complex numbers and their conjugates.

Complex Conjugates

Number	$2 + 5i$	$6 - 3i$	$-2 + 7i$	$-1 - i$	5	$-5i$
Conjugate	$2 - 5i$	$6 + 3i$	$-2 - 7i$	$-1 + i$	5	$5i$

TABLE 10.8

EXAMPLE 5 Finding complex conjugates

Find the complex conjugate of each number.
(a) $5 + 3i$ (b) $-2 - i$ (c) $-4i$ (d) -6

Solution
(a) The conjugate is found by changing the imaginary part of $5 + 3i$ to its opposite. The conjugate is $5 - 3i$.
(b) The conjugate is $-2 + i$.
(c) $-4i$ can be written as $0 - 4i$. The conjugate is $0 + 4i$, or $4i$.
(d) -6 can be written as $-6 + 0i$. The conjugate is $-6 - 0i$, or -6. The conjugate of a real number is simply the same real number.

Now Try Exercises 57, 59, 61, 63

NOTE: The product of a complex number and its conjugate is a *real* number. That is,

$$(a + bi)(a - bi) = a^2 + b^2.$$

For example, $(3 + 4i)(3 - 4i) = 3^2 + 4^2 = 25$, which is a real number. ∎

5 Dividing Complex Numbers

The property of complex conjugates in the NOTE above is used to divide two complex numbers. To convert the quotient $\frac{2 + 3i}{3 - i}$ into standard form $a + bi$, we multiply the numerator and the denominator by the complex conjugate of the *denominator*, which is $3 + i$. The next example illustrates this method.

EXAMPLE 6 Dividing complex numbers

Write each quotient in standard form.

(a) $\dfrac{2 + 3i}{3 - i}$ (b) $\dfrac{4}{2i}$

Solution

(a) Multiply the numerator and denominator by $3 + i$.

$$\frac{2 + 3i}{3 - i} = \frac{(2 + 3i)(3 + i)}{(3 - i)(3 + i)}$$

The conjugate of $3 - i$ is $3 + i$.   Multiply by 1.

$$= \frac{2(3) + (2)(i) + (3i)(3) + (3i)(i)}{(3)(3) + (3)(i) - (i)(3) - (i)(i)}$$ Multiply.

$$= \frac{6 + 2i + 9i + 3i^2}{9 + 3i - 3i - i^2}$$ Simplify.

$$= \frac{6 + 11i + 3(-1)}{9 - (-1)}$$ $i^2 = -1$

$$= \frac{3 + 11i}{10}$$ Simplify.

$$= \frac{3}{10} + \frac{11}{10}i$$ $\frac{a + bi}{c} = \frac{a}{c} + \frac{b}{c}i$

(b) Multiply the numerator and denominator by $-2i$.

$$\frac{4}{2i} = \frac{(4)(-2i)}{(2i)(-2i)}$$

The conjugate of $2i$ is $-2i$.   Multiply by 1.

$$= \frac{-8i}{-4i^2}$$ Simplify.

$$= \frac{-8i}{-4(-1)}$$ $i^2 = -1$

$$= \frac{-8i}{4}$$ Simplify.

$$= -2i$$ Divide.

Complex Division

```
(2+3i)/(3-i)►Fra
c
        3/10+11/10i
4/(2i)
              -2i
```

FIGURE 10.23

These results are supported in **FIGURE 10.23**.

Now Try Exercises 71, 73

10.7 Putting It All Together

CONCEPT	EXPLANATION	EXAMPLES
Complex Numbers	A complex number can be expressed as $a + bi$, where a and b are real numbers. The imaginary unit i satisfies $i = \sqrt{-1}$ and $i^2 = -1$. As a result, we can write $\sqrt{-a} = i\sqrt{a}$ if $a > 0$.	The real part of $5 - 3i$ is 5 and the imaginary part is -3. $$\sqrt{-13} = i\sqrt{13} \quad \text{and} \quad \sqrt{-9} = 3i$$
Addition, Subtraction, and Multiplication	To add (subtract) complex numbers, add (subtract) the real parts and then add (subtract) the imaginary parts.	$(3 + 6i) + (-1 + 2i)$ Sum $\quad = (3 + (-1)) + (6 + 2)i$ $\quad = 2 + 8i$ $(2 - 5i) - (1 + 4i)$ Difference $\quad = (2 - 1) + (-5 - 4)i$ $\quad = 1 - 9i$
	Multiply complex numbers in a manner similar to the way *FOIL* is used to multiply binomials. Then apply the property $i^2 = -1$.	$(-1 + 2i)(3 + i)$ Product $= (-1)(3) + (-1)(i) + (2i)(3) + (2i)(i)$ $= -3 - i + 6i + 2i^2$ $= -3 + 5i + 2(-1)$ $= -5 + 5i$
Complex Conjugates	The conjugate of $a + bi$ is $a - bi$.	The conjugate of $3 - 5i$ is $3 + 5i$. The conjugate of $2i$ is $-2i$.
Division	To simplify a quotient, multiply the numerator and denominator by the complex conjugate of the *denominator*. Then simplify the expression and write it in standard form as $a + bi$.	$\dfrac{10}{1 + 2i} = \dfrac{10(1 - 2i)}{(1 + 2i)(1 - 2i)}$ Quotient $\quad = \dfrac{10 - 20i}{5}$ $\quad = 2 - 4i$

10.7 Exercises

MyMathLab®

CONCEPTS AND VOCABULARY

1. Give an example of a complex number that is not a real number.

2. Can you give an example of a real number that is not a complex number? Explain.

3. $\sqrt{-1} =$ _____

4. $i^2 =$ _____

5. $\sqrt{-a} =$ _____, if $a > 0$.

6. The complex conjugate of $10 + 7i$ is _____.

7. The standard form for a complex number is _____.

8. Write $\frac{2 + 4i}{2}$ in standard form.

9. The real part of $4 - 5i$ is _____.

10. The imaginary part of $4 - 5i$ is _____.

11. The imaginary part of -7 is _____.

12. The number $7i$ is a(n) _____ imaginary number.

SIMPLIFYING SQUARE ROOTS OF NEGATIVE NUMBERS

Exercises 13–22: Use i to write the expression.

13. $\sqrt{-5}$ 14. $\sqrt{-21}$ 15. $\sqrt{-100}$

16. $\sqrt{-49}$ 17. $\sqrt{-144}$ 18. $\sqrt{-64}$

19. $\sqrt{-12}$

20. $\sqrt{-8}$

21. $\sqrt{-18}$

22. $\sqrt{-48}$

ADDING, SUBTRACTING, AND MULTIPLYING COMPLEX NUMBERS

Exercises 23–46: Write the expression in standard form.

23. $(5 + 3i) + (-2 - 3i)$

24. $(1 - i) + (5 - 7i)$

25. $2i + (-8 + 5i)$ **26.** $-3i + 5i$

27. $(2 - 7i) - (1 + 2i)$ **28.** $(1 + 8i) - (3 + 9i)$

29. $5i - (10 - 2i)$ **30.** $-3(4 - 3i)$

31. $(3 + 2i)(-1 + 5i)$ **32.** $(1 + i) - (1 - i)$

33. $4(5 - 3i)$ **34.** $(1 + 2i)(-6 - i)$

35. $(5 + 4i)(5 - 4i)$ **36.** $(3 + 5i)(3 - 5i)$

37. $(-4i)(5i)$ **38.** $(-6i)(-4i)$

39. $3i + (2 - 3i) - (1 - 5i)$

40. $4 - (5 - 7i) + (3 + 7i)$

41. $(2 + i)^2$ **42.** $(-1 + 2i)^2$

43. $2i(-3 + i)$ **44.** $5i(1 - 9i)$

45. $i(1 + i)^2$ **46.** $2i(1 - i)^2$

Exercises 47 and 48: **Thinking Generally** *Write in standard form.*

47. $(a + 3bi)(a - 3bi)$ **48.** $(a + bi) - (a - bi)$

SIMPLIFYING POWERS OF *i*

Exercises 49–56: (Refer to Example 4.) Simplify.

49. i^{11} **50.** i^{50} **51.** i^{21}

52. i^{103} **53.** i^{58} **54.** i^{61}

55. i^{64} **56.** i^{28}

FINDING COMPLEX CONJUGATES

Exercises 57–64: Write the complex conjugate.

57. $3 + 4i$ **58.** $1 - 4i$

59. $-6i$ **60.** -10

61. $5 - 4i$ **62.** $7 + 2i$

63. -1 **64.** $19i$

DIVIDING COMPLEX NUMBERS

Exercises 65–78: Write the expression in standard form.

65. $\dfrac{2}{1 + i}$ **66.** $\dfrac{-6}{2 - i}$

67. $\dfrac{3i}{5 - 2i}$ **68.** $\dfrac{-8}{2i}$

69. $\dfrac{8 + 9i}{5 + 2i}$ **70.** $\dfrac{3 - 2i}{1 + 4i}$

71. $\dfrac{5 + 7i}{1 - i}$ **72.** $\dfrac{-7 + 4i}{3 - 2i}$

73. $\dfrac{2 - i}{i}$ **74.** $\dfrac{3 + 2i}{-i}$

75. $\dfrac{1}{i} + \dfrac{1}{2i}$ **76.** $\dfrac{3}{4i} + \dfrac{2}{i}$

77. $\dfrac{1}{-1 + i} - \dfrac{2}{i}$ **78.** $-\dfrac{3}{2i} - \dfrac{2}{1 + i}$

USING ZEROS TO WRITE A FORMULA FOR A FUNCTION

Exercises 79–84: A quadratic function f has the given complex zeros with leading coefficient 1.

   (a) *Write the formula for $f(x)$ in factored form.*
   (b) *Write the expanded form for $f(x)$ by multiplying the factor form.*

(Hint: If $-5i$ and $5i$ are the two zeros of $f(x)$, then $f(x)$ can be written in factored form using the formula given by $f(x) = (x + 5i)(x - 5i)$.)

79. $-i, i$ **80.** $-3i, 3i$

81. $-7i, 7i$ **82.** $-9i, 9i$

83. $-i\sqrt{6}, i\sqrt{6}$ **84.** $-i\sqrt{11}, i\sqrt{11}$

APPLICATIONS INVOLVING COMPLEX NUMBERS

Exercises 85 and 86: **Corrosion in Airplanes** *Corrosion in the metal surface of an airplane can be difficult to detect visually. One test used to locate it involves passing an alternating current through a small area on the plane's surface. If the current varies from one region to another, it may indicate that corrosion is occurring. The impedance Z (or opposition to the flow of electricity) of the metal is related to the voltage V and current I by the equation $Z = \frac{V}{I}$, where Z, V, and I are complex numbers. Calculate Z* *for the given values of V and I.* (**Source:** Society for Industrial and Applied Mathematics.)

85. $V = 40 + 70i, I = 2 + 3i$

86. $V = 10 + 20i, I = 3 + 7i$

87. Online Exploration We graph real numbers on a number line. Can we graph complex numbers on a number line? Go online and find out how to graph complex numbers.

88. Online Exploration Go online and find out how to calculate the absolute value of a complex number. If a complex number is graphed, what does the absolute value tell us?

WRITING ABOUT MATHEMATICS

89. A student multiplies $(2 + 3i)(4 - 5i)$ *incorrectly* to obtain $8 - 15i$. What is the student's mistake?

90. A student divides the ratio $\frac{6 - 10i}{3 + 2i}$ *incorrectly* to obtain $2 - 5i$. What is the student's mistake?

SECTION 10.7 Checking Basic Concepts

1. Use i to write each expression.
(a) $\sqrt{-64}$
(b) $\sqrt{-17}$

2. Simplify each expression.
(a) $(2 - 3i) + (1 - i)$
(b) $4i - (2 + i)$
(c) $(3 - 2i)(1 + i)$
(d) $\dfrac{3}{2 - 2i}$

CHAPTER 10 Summary

SECTION 10.1 ■ RADICAL EXPRESSIONS AND FUNCTIONS

Radicals and Radical Notation

Square Root b is a square root of a if $b^2 = a$.

Principal Square Root $\sqrt{a} = b$ if $b^2 = a$ and $b \geq 0$.

Examples: $\sqrt{16} = 4, -\sqrt{9} = -3,$
and $\pm\sqrt{36} = \pm 6$

Cube Root b is a cube root of a if $b^3 = a$.

Examples: $\sqrt[3]{27} = 3, \sqrt[3]{-8} = -2$

nth Root b is an nth root of a if $b^n = a$.

Example: $\sqrt[4]{16} = 2$ because $2^4 = 16$.

NOTE: An *even* root of a *negative* number is not a real number. Also, $\sqrt[n]{a}$ denotes the *principal* nth root. ■

Absolute Value   The expressions $|x|$ and $\sqrt{x^2}$ are equivalent.

Example: $\sqrt{(x + y)^2} = |x + y|$

The Square Root Function The square root function is denoted $f(x) = \sqrt{x}$. Its domain is $\{x | x \geq 0\}$ and its graph is shown in the figure.

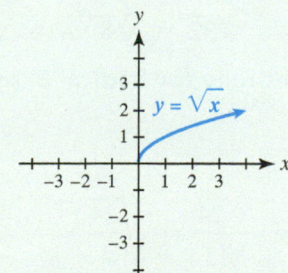

The Cube Root Function The cube root function is denoted $f(x) = \sqrt[3]{x}$. Its domain is all real numbers and its graph is shown in the figure.

$y = \sqrt[3]{x}$

SECTION 10.2 ■ RATIONAL EXPONENTS

The Expression $a^{1/n}$

$a^{1/n} = \sqrt[n]{a}$ if n is an integer greater than 1.

Examples: $5^{1/2} = \sqrt{5}$ and $64^{1/3} = \sqrt[3]{64} = 4$

The Expression $a^{m/n}$

$a^{m/n} = \sqrt[n]{a^m}$ or $a^{m/n} = (\sqrt[n]{a})^m$

Examples: $8^{2/3} = \sqrt[3]{8^2} = \sqrt[3]{64} = 4$ and
$8^{2/3} = (\sqrt[3]{8})^2 = (2)^2 = 4$

Properties of Exponents

Product Rule	$a^p a^q = a^{p+q}$	$3^{1/7} \cdot 3^{2/7} = 3^{3/7}$
Negative Exponents	$a^{-p} = \dfrac{1}{a^p}, \dfrac{1}{a^{-p}} = a^p$	$7^{-1/4} = \dfrac{1}{7^{1/4}}, \dfrac{1}{7^{-1/4}} = 7^{1/4}$
Negative Exponents for Quotients	$\left(\dfrac{a}{b}\right)^{-p} = \left(\dfrac{b}{a}\right)^{p}$	$\left(\dfrac{3}{7}\right)^{-1/3} = \left(\dfrac{7}{3}\right)^{1/3}$
Quotient Rule for Exponents	$\dfrac{a^p}{a^q} = a^{p-q}$	$\dfrac{5^{4/7}}{5^{1/7}} = 5^{3/7}$
Power Rule for Exponents	$(a^p)^q = a^{pq}$	$(2^{1/3})^{1/2} = 2^{1/6}$
Power Rule for Products	$(ab)^p = a^p b^p$	$(5x)^{2/3} = 5^{2/3} x^{2/3}$
Power Rule for Quotients	$\left(\dfrac{a}{b}\right)^{p} = \dfrac{a^p}{b^p}$	$\left(\dfrac{5}{3}\right)^{3/4} = \dfrac{5^{3/4}}{3^{3/4}}$

SECTION 10.3 ■ SIMPLIFYING RADICAL EXPRESSIONS

Product Rule for Radical Expressions Provided each expression is defined,

$$\sqrt[n]{a} \cdot \sqrt[n]{b} = \sqrt[n]{a \cdot b}.$$

Example: $\sqrt[3]{3} \cdot \sqrt[3]{9} = \sqrt[3]{27} = 3$

Perfect nth Power An integer a is a perfect nth power if $b^n = a$ for some integer b.

Examples: 25 is a perfect square, 8 is a perfect cube, and 16 is a perfect fourth power.

Simplifying Radicals (nth roots)

STEP 1: Determine the largest perfect nth power factor of the radicand.

STEP 2: Use the product rule to factor out and simplify this perfect nth power.

Examples: $\sqrt{32} = \sqrt{16 \cdot 2} = \sqrt{16} \cdot \sqrt{2} = 4\sqrt{2}$
$\sqrt[3]{32} = \sqrt[3]{8 \cdot 4} = \sqrt[3]{8} \cdot \sqrt[3]{4} = 2\sqrt[3]{4}$

Quotient Rule for Radical Expressions Provided each expression is defined,

$$\sqrt[n]{\dfrac{a}{b}} = \dfrac{\sqrt[n]{a}}{\sqrt[n]{b}}.$$

Example: $\dfrac{\sqrt[3]{24}}{\sqrt[3]{3}} = \sqrt[3]{\dfrac{24}{3}} = \sqrt[3]{8} = 2$

Addition and Subtraction Combine like radicals.

Examples: $2\sqrt[3]{4} + 3\sqrt[3]{4} = 5\sqrt[3]{4}$ and $\sqrt{5} - 2\sqrt{5} = -\sqrt{5}$

Multiplication Sometimes radical expressions can be multiplied like binomials.

Examples: $(4 - \sqrt{2})(2 + \sqrt{2}) = 8 + 4\sqrt{2} - 2\sqrt{2} - 2 = 6 + 2\sqrt{2}$

$(5 - \sqrt{3})(5 + \sqrt{3}) = (5)^2 - (\sqrt{3})^2 = 25 - 3 = 22$ because

$(a - b)(a + b) = a^2 - b^2.$

Rationalizing the Denominator One technique is to multiply the numerator and denominator by the conjugate of the denominator if the denominator is a binomial containing one or more square roots.

Examples: $\dfrac{1}{4 + \sqrt{2}} = \dfrac{1}{(4 + \sqrt{2})} \cdot \dfrac{(4 - \sqrt{2})}{(4 - \sqrt{2})} = \dfrac{4 - \sqrt{2}}{(4)^2 - (\sqrt{2})^2} = \dfrac{4 - \sqrt{2}}{14}$

$\dfrac{4}{\sqrt{7}} = \dfrac{4}{\sqrt{7}} \cdot \dfrac{\sqrt{7}}{\sqrt{7}} = \dfrac{4\sqrt{7}}{7}$

Root Functions $f(x) = x^{1/n} = \sqrt[n]{x}$

Odd root functions are defined for all real numbers, and even root functions are defined for nonnegative real numbers.

Examples: $f(x) = x^{1/3} = \sqrt[3]{x},$ $g(x) = x^{1/4} = \sqrt[4]{x},$ and $h(x) = x^{1/5} = \sqrt[5]{x}$

Power Functions If a function can be defined by $f(x) = x^p$, where p is a rational number, then it is a power function.

Examples: $f(x) = x^{4/5}$ and $g(x) = x^{2.3}$

Power Rule for Solving Radical Equations The solutions to $a = b$ are among the solutions to $a^n = b^n$, where n is a positive integer.

Example: The solutions to the equation $\sqrt{3x + 3} = 2x - 1$ are among the solutions to the equation $3x + 3 = (2x - 1)^2.$

Solving Radical Equations

STEP 1: Isolate a radical term on one side of the equation.

STEP 2: Apply the power rule by raising each side of the equation to the power equal to the index of the isolated radical term.

STEP 3: Solve the equation. If it still contains a radical, repeat Steps 1 and 2.

STEP 4: Check your answers by substituting each result in the *given* equation.

Example: To isolate the radical in $\sqrt{x + 1} + 4 = 6$, subtract 4 from each side to obtain $\sqrt{x + 1} = 2$. Next square each side, which gives $x + 1 = 4$ or $x = 3$. Checking verifies that 3 is a solution.

Pythagorean Theorem If a right triangle has legs a and b with hypotenuse c, then

$$a^2 + b^2 = c^2.$$

Example: If a right triangle has legs 8 and 15, then the hypotenuse equals

$$c = \sqrt{8^2 + 15^2} = \sqrt{289} = 17.$$

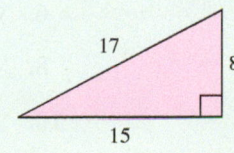

The Distance Formula The distance d between (x_1, y_1) and (x_2, y_2) is

$$d = \sqrt{(x_2 - x_1)^2 + (y_2 - y_1)^2}.$$

continued on next page

Example: The distance between $(-1, 3)$ and $(4, 5)$ is

$$d = \sqrt{(4 - (-1))^2 + (5 - 3)^2} = \sqrt{25 + 4} = \sqrt{29}.$$

Solving the Equation $x^n = k$ Let n be a positive integer.

Take the nth root of each side of $x^n = k$ to obtain $\sqrt[n]{x^n} = \sqrt[n]{k}$.

1. If n is *odd*, then $\sqrt[n]{x^n} = x$ and the equation becomes $x = \sqrt[n]{k}$.

2. If n is *even* and $k \geq 0$, then $\sqrt[n]{x^n} = |x|$ and the equation becomes $|x| = \sqrt[n]{k}$.

Examples: $x^3 = -27$ implies that $x = \sqrt[3]{-27} = -3$.

$x^4 = 81$ implies that $|x| = \sqrt[4]{81} = 3$ or $x = \pm 3$.

SECTION 10.7 ■ COMPLEX NUMBERS

Complex Numbers

Imaginary Unit	$i = \sqrt{-1}$ and $i^2 = -1$
Standard Form	$a + bi$, where a and b are real numbers
	Examples: $4 + 3i$, $5 - 6i$, 8, and $-2i$
Real Part	The real part of $a + bi$ is a.
	Example: The real part of $3 - 2i$ is 3.
Imaginary Part	The imaginary part of $a + bi$ is b.
	Example: The imaginary part of $2 - i$ is -1.
Arithmetic Operations	Arithmetic operations are similar to arithmetic operations on binomials.

Examples:

$(2 + 2i) + (3 - i) = 5 + i$ Sum

$(1 - i) - (1 - 2i) = i$ Difference

$(1 - i)(1 + i) = 1^2 - i^2 = 1 - (-1) = 2$ Product

The complex conjugate of $1 - i$ is $1 + i$, so multiply by $\frac{1+i}{1+i}$.

$\dfrac{2}{1 - i} = \dfrac{2}{1 - i} \cdot \dfrac{1 + i}{1 + i} = \dfrac{2 + 2i}{2} = 1 + i$ Quotient

Powers of i   The value of i^n equals i^r, where r is the remainder when n is divided by 4. Note that $i^0 = 1$, $i^1 = i$, $i^2 = -1$, and $i^3 = -i$.

Example: $i^{21} = i^1 = i$ because when 21 is divided by 4, the remainder is 1.

CHAPTER 10   Review Exercises

SECTION 10.1

Exercises 1–12: Simplify the expression.

1. $\sqrt{4}$
2. $\sqrt{36}$
3. $\sqrt{9x^2}$
4. $\sqrt{(x - 1)^2}$
5. $\sqrt[3]{-64}$
6. $\sqrt[3]{-125}$
7. $\sqrt[3]{x^6}$
8. $\sqrt[3]{27x^3}$
9. $\sqrt[4]{16}$
10. $\sqrt[5]{-1}$
11. $\sqrt[4]{x^8}$
12. $\sqrt[5]{(x + 1)^5}$

SECTION 10.2

Exercises 13–16: Write the expression in radical notation.

13. $14^{1/2}$
14. $(-5)^{1/3}$
15. $\left(\dfrac{x}{y}\right)^{3/2}$
16. $(xy)^{-2/3}$

Exercises 17–20: Evaluate the expression.

17. $(-27)^{2/3}$
18. $16^{1/4}$
19. $16^{3/2}$
20. $81^{3/4}$

Exercises 21–24: Simplify the expression. Assume that all variables are positive.

21. $(z^3)^{2/3}$

22. $(x^2y^4)^{1/2}$

23. $\left(\dfrac{x^2}{y^6}\right)^{3/2}$

24. $\left(\dfrac{x^3}{y^6}\right)^{-1/3}$

SECTION 10.3

Exercises 25–40: Simplify the expression. Assume that all variables are positive.

25. $\sqrt{2} \cdot \sqrt{32}$

26. $\sqrt[3]{-4} \cdot \sqrt[3]{2}$

27. $\sqrt[3]{x^4} \cdot \sqrt[3]{x^2}$

28. $\dfrac{\sqrt{80}}{\sqrt{20}}$

29. $\sqrt[3]{-\dfrac{x}{8}}$

30. $\sqrt{\dfrac{1}{3}} \cdot \sqrt{\dfrac{1}{3}}$

31. $\sqrt{48}$

32. $\sqrt{54}$

33. $\sqrt[3]{\dfrac{3}{x}} \cdot \sqrt[3]{\dfrac{9}{x^2}}$

34. $\sqrt{32a^3b^2}$

35. $\sqrt{3xy} \cdot \sqrt{27xy}$

36. $\sqrt[3]{-25z^2} \cdot \sqrt[3]{-5z^2}$

37. $\sqrt{x^2 + 2x + 1}$

38. $\sqrt[4]{\dfrac{2a^2}{b}} \cdot \sqrt[4]{\dfrac{8a^3}{b^3}}$

39. $2\sqrt{x} \cdot \sqrt[3]{x}$

40. $\sqrt[3]{rt} \cdot \sqrt[4]{r^2t^4}$

SECTION 10.4

Exercises 41–50: Simplify the expression. Assume that all variables are positive.

41. $3\sqrt{3} + \sqrt{3}$

42. $\sqrt[3]{x} + 2\sqrt[3]{x}$

43. $3\sqrt[3]{5} - 6\sqrt[3]{5}$

44. $\sqrt[4]{y} - 2\sqrt[4]{y}$

45. $2\sqrt{12} + 7\sqrt{3}$

46. $3\sqrt{18} - 2\sqrt{2}$

47. $7\sqrt[3]{16} - \sqrt[3]{2}$

48. $\sqrt{4x + 4} + \sqrt{x + 1}$

49. $\sqrt{4x^3} - \sqrt{x}$

50. $\sqrt[3]{ab^4} + 2\sqrt[3]{a^4b}$

Exercises 51–56: Multiply and simplify.

51. $(1 + \sqrt{2})(3 + \sqrt{2})$

52. $(7 - \sqrt{5})(1 + \sqrt{3})$

53. $(3 + \sqrt{6})(3 - \sqrt{6})$

54. $(10 - \sqrt{5})(10 + \sqrt{5})$

55. $(\sqrt{a} + \sqrt{2b})(\sqrt{a} - \sqrt{2b})$

56. $(\sqrt{xy} - 1)(\sqrt{xy} + 2)$

Exercises 57–62: Rationalize the denominator.

57. $\dfrac{4}{\sqrt{5}}$

58. $\dfrac{r}{2\sqrt{t}}$

59. $\dfrac{1}{\sqrt{2} + 3}$

60. $\dfrac{2}{5 - \sqrt{7}}$

61. $\dfrac{1}{\sqrt{8} - \sqrt{7}}$

62. $\dfrac{\sqrt{a} - \sqrt{b}}{\sqrt{a} + \sqrt{b}}$

SECTION 10.5

Exercises 63 and 64: Graph the equation.

63. $y = \sqrt[4]{x}$

64. $y = \sqrt[3]{x}$

Exercises 65 and 66: Write $f(x)$ in radical notation and evaluate $f(4)$.

65. $f(x) = x^{1/2}$

66. $f(x) = x^{2/7}$

Exercises 67 and 68: Graph the equation. Compare the graph to either $y = \sqrt{x}$ or $y = \sqrt[3]{x}$.

67. $y = \sqrt{x} - 2$

68. $y = \sqrt[3]{x - 1}$

Exercises 69–72: Find the domain of f. Write your answer in interval notation.

69. $f(x) = \sqrt{x - 1}$

70. $f(x) = \sqrt{6 - 2x}$

71. $f(x) = \sqrt{x^2 + 1}$

72. $f(x) = \dfrac{1}{\sqrt{x + 2}}$

SECTION 10.6

Exercises 73–78: Solve. Check your answer.

73. $\sqrt{x + 2} = x$

74. $\sqrt{2x - 1} = \sqrt{x + 3}$

75. $\sqrt[3]{x - 1} = 2$

76. $\sqrt[3]{3x} = 3$

77. $\sqrt{2x} = x - 4$

78. $\sqrt{x + 1} = \sqrt{x + 2}$

Exercises 79 and 80: Solve graphically. Approximate solutions to the nearest hundredth when appropriate.

79. $\sqrt[3]{2x - 1} = 2$

80. $x^{2/3} = 3 - x$

Exercises 81 and 82: A right triangle has legs a and b with hypotenuse c. Find the length of the missing side.

81. $a = 4, b = 7$

82. $a = 5, c = 8$

Exercises 83 and 84: Find the exact distance between the given points.

83. $(-2, 3), (2, -2)$

84. $(2, -3), (-4, 1)$

Exercises 85–94: Solve.

85. $x^2 = 121$

86. $2z^2 = 32$

87. $(x - 1)^2 = 16$

88. $x^3 = 64$

89. $(x - 1)^3 = 8$

90. $(2x - 1)^3 = 27$

91. $x^4 = 256$

92. $x^5 = -1$

93. $(x - 3)^5 = -32$

94. $3(x + 1)^4 = 3$

SECTION 10.7

Exercises 95–100: Write the given complex expression in standard form.

95. $(1 - 2i) + (-3 + 2i)$

96. $(1 + 3i) - (3 - i)$

97. $(1 - i)(2 + 3i)$

98. $\dfrac{3 + i}{1 - i}$

99. $\dfrac{i(4 + i)}{2 - 3i}$

100. $(1 - i)^2(1 + i)$

APPLICATIONS

101. Hang Time A football is punted and has a hang time T of 4.6 seconds. Use the formula $T(h) = \frac{1}{2}\sqrt{h}$ to estimate to the nearest foot the height h to which it was kicked.

102. Baseball Diamond The four bases of a baseball diamond form a square that is 90 feet on a side. Find the distance from home plate to second base.

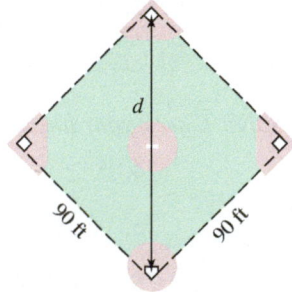

103. Falling Time The time T in seconds for an object to fall from a height of h feet is given by $T = \frac{1}{4}\sqrt{h}$. If a person steps off a 10-foot-high board into a swimming pool, how long is the person in the air?

104. Geometry A cube has sides of length $\sqrt{5}$.
(a) Find the area of one side of the cube.
(b) Find the volume of the cube.

(c) Find the length of the diagonal of one of the sides.
(d) Find the distance from A to B in the figure.

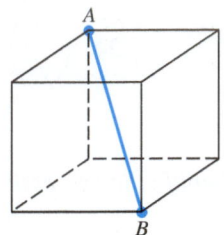

105. Population Growth In 1790, the population of the United States was 4 million, and by 2000, it had grown to 281 million. The *average* annual percentage growth in the population r (expressed as a decimal) can be determined by the polynomial equation $281 = 4(1 + r)^{210}$. Solve this equation for r and interpret the result.

106. Highway Curves If a circular highway curve is banked with a slope of $m = \frac{1}{10}$ (see the accompanying figure) and has a radius of R feet, then the speed limit L in miles per hour for the curve is given by

$$L = \sqrt{3.75R}.$$

(*Source:* N. Garber and L. Hoel, *Traffic and Highway Engineering.*)
(a) Find the speed limit if R is 500 feet.
(b) With no banking, the speed limit is given by $L = 1.5\sqrt{R}$. Find the speed limit for a curve with no banking and a radius of 500 feet. How does banking affect the speed limit? Does this result agree with your intuition?

107. Geometry Find the length of a side of a square if the square has an area of 7 square feet.

CHAPTER 10 Test MyMathLab®, YouTube

Exercises 1–6: Simplify the expression.

1. $\sqrt[3]{-27}$

2. $\sqrt{(z + 1)^2}$

3. $\sqrt{25x^4}$

4. $\sqrt[3]{8z^6}$

5. $\sqrt[4]{16x^4y^5}$; $x > 0, y > 0$

6. $(\sqrt{3} - \sqrt{2})(\sqrt{3} + \sqrt{2})$

Exercises 7 and 8: Write the given expression using radical notation.

7. $7^{2/5}$

8. $\left(\dfrac{x}{y}\right)^{-2/3}$

Exercises 9 and 10: Evaluate the expression by hand.

9. $(-8)^{4/3}$

10. $36^{-3/2}$

Exercises 11 and 12: Use rational exponents to write the expression.

11. $\sqrt[3]{x^4}$

12. $\sqrt{x} \cdot \sqrt[5]{x}$

13. Find the domain of $f(x) = \sqrt{4-x}$. Write your answer in interval notation.

14. Sketch a graph of $y = \sqrt{x+3}$.

Exercises 15–22: Simplify the expression. Assume that all variables are positive.

15. $(2z^{1/2})^3$

16. $\left(\dfrac{y^2}{z^3}\right)^{-1/3}$

17. $\sqrt{3} \cdot \sqrt{27}$

18. $\dfrac{\sqrt{y^3}}{\sqrt{4y}}$

19. $7\sqrt{7} - 3\sqrt{7} + \sqrt{5}$

20. $7\sqrt[3]{x} - \sqrt[3]{x}$

21. $4\sqrt{18} + \sqrt{8}$

22. $\dfrac{\sqrt[3]{32}}{\sqrt[3]{4}}$

23. Solve each equation.
(a) $\sqrt{x-2} = 5$
(b) $\sqrt[3]{x+1} = 2$
(c) $(x-1)^3 = 8$
(d) $\sqrt{2x+2} = x - 11$

24. Rationalize the denominator.
(a) $\dfrac{2}{3\sqrt{7}}$
(b) $\dfrac{1}{1+\sqrt{5}}$

 25. Solve $\sqrt{3x} - x + 1 = \sqrt[3]{x-1}$ graphically. Round solutions to the nearest hundredth.

26. One leg of a right triangle has length 7 and the hypotenuse has length 13. Find the length of the third side.

27. Find the distance between $(-3, 5)$ and $(-1, 7)$.

Exercises 28–31: Write the given complex expression in standard form.

28. $(-5 + i) + (7 - 20i)$

29. $(3i) - (6 - 5i)$

30. $\left(\dfrac{1}{2} - i\right)\left(\dfrac{1}{2} + i\right)$

31. $\dfrac{2i}{5 + 2i}$

32. Distance to the Horizon A formula for calculating the distance d in miles that one can see to the horizon on a clear day is approximated by the formula $d = 1.22\sqrt{x}$, where x is the elevation, in feet, of a person.
(a) If a person climbs a cliff and stops at an elevation of 200 feet, how far can the person see to the horizon?
(b) How high should the person climb to see 25 miles to the horizon?

33. Wing Span of a Bird The wing span L of a bird with weight W can sometimes be modeled by $L = 27.4W^{1/3}$, where L is in inches and W is in pounds. Use this formula to estimate the weight of a bird that has a wing span of 30 inches. (*Source:* C. Pennycuick, *Newton Rules Biology.*)

CHAPTERS 1–10   Cumulative Review Exercises

1. Evaluate $S = 4\pi r^2$ when $r = 3$.

2. Identify the domain and range of the relation given by $S = \{(-1, 2), (0, 4), (1, 2)\}$.

3. Simplify each expression. Write the result using positive exponents.
(a) $\left(\dfrac{ab^2}{b^{-1}}\right)^{-3}$
(b) $\dfrac{(x^2y)^3}{x^2(y^2)^{-3}}$
(c) $(rt)^2(r^2t)^3$

4. Write 0.00043 in scientific notation.

5. Find $f(3)$ if $f(x) = \dfrac{x}{x-2}$. What is the domain?

6. Find the domain of $f(x) = \sqrt[3]{2x}$.

Exercises 7 and 8: Graph f by hand.

7. $f(x) = 1 - 2x$

8. $f(x) = \sqrt{x+1}$

9. Sketch a graph of $f(x) = x^2 - 4$.
(a) Identify the domain and range.
(b) Evaluate $f(-2)$.
(c) Identify the x-intercepts.
(d) Solve the equation $f(x) = 0$.

10. Find the slope–intercept form of the line that passes through $(-1, 2)$ and is perpendicular to $y = -2x$.

11. Let f be a linear function. Find a formula for $f(x)$.

x	-2	-1	1	2
$f(x)$	7	4	-2	-5

12. Sketch a graph of a line passing through $(-1, 1)$ with slope $-\dfrac{1}{2}$.

13. Solve $5x - (3 - x) = \dfrac{1}{2}x$.

14. Solve $2x - 5 \le 4 - x$.

15. Solve $|x - 2| \le 3$.

16. Solve $-1 \le 1 - 2x \le 6$.

17. Solve each system symbolically, if possible.

(a) $2x - y = 4$
$x + y = 8$

(b) $3x - 4y = 2$
$x - \frac{4}{3}y = 1$

18. Shade the solution set in the xy-plane.

$$x + 2y \le 2$$
$$-x + 3y \ge 3$$

19. Solve the system.

$$x + 2y - z = 6$$
$$x - 3y + z = -2$$
$$x + y + z = 6$$

20. Calculate $\det \begin{bmatrix} 4 & 2 & -1 \\ 2 & 1 & 0 \\ 0 & -2 & 1 \end{bmatrix}$.

Exercises 21–24: Multiply the expression.

21. $4x(4 - x^3)$

22. $(x - 4)(x + 4)$

23. $(5x + 3)(x - 2)$

24. $(4x + 9)^2$

Exercises 25–30: Factor the expression.

25. $9x^2 - 16$

26. $x^2 - 4x + 4$

27. $15x^3 - 9x^2$

28. $12x^2 - 5x - 3$

29. $r^3 - 1$

30. $x^3 - 3x^2 + 5x - 15$

Exercises 31 and 32: Solve each equation.

31. $x^2 - 3x + 2 = 0$

32. $x^3 = 4x$

Exercises 33 and 34: Simplify the expression.

33. $\dfrac{x^2 + 3x + 2}{x - 3} \div \dfrac{x + 1}{2x - 6}$

34. $\dfrac{2}{x - 1} + \dfrac{5}{x}$

Exercises 35–42: Simplify the given expression. Assume all variables are positive.

35. $\sqrt{36x^2}$

36. $\sqrt[3]{64}$

37. $16^{-3/2}$

38. $\sqrt[4]{625}$

39. $\sqrt{2x} \cdot \sqrt{8x}$

40. $\sqrt{x} \cdot \sqrt[4]{x}$

41. $\dfrac{\sqrt[3]{16x^4}}{\sqrt[3]{2x}}$

42. $4\sqrt{12x} - 2\sqrt{3x}$

43. Multiply $(2x + \sqrt{3})(x - \sqrt{3})$.

44. Find the domain of $f(x) = \sqrt{1 - x}$.

45. Graph $f(x) = 3\sqrt[3]{x}$.

46. Find the distance between $(-2, 3)$ and $(1, 2)$.

47. Write $(1 - i)(2 + 3i)$ in standard form.

48. The lengths of the legs of a right triangle are 5 and 12. What is the length of the hypotenuse?

Exercises 49–52: Solve symbolically.

49. $2\sqrt{x + 3} = x$

50. $\sqrt[3]{x - 1} = 3$

51. $\sqrt{x} + 4 = 2\sqrt{x + 5}$

52. $\frac{1}{3}x^4 = 27$

Exercises 53 and 54: Solve graphically and approximate your answer to the nearest hundredth.

53. $\sqrt[3]{x^2 - 2} + x = \sqrt{x}$

54. $x - \sqrt[3]{x} = \sqrt{x + 2}$

APPLICATIONS

55. Calculating Water Flow The gallons G of water in a tank after t minutes are given by $G(t) = 300 - 15t$. Interpret the y-intercept and the slope of the graph of G as a rate of change.

56. Distance A person jogs away from home at 8 miles per hour for 1 hour, rests in a park for 2 hours, and then walks back home at 4 miles per hour. Sketch a graph that models the distance y that the person is from home after x hours.

57. Investment A total of $2000 is deposited in two accounts paying 5% and 4% annual interest. If the total interest after 1 year is $93, how much is invested at each rate?

58. Geometry A rectangular box with a square base has a volume of 256 cubic inches. If its height is 4 inches less than the length of an edge of the base, find the dimensions of the box.

59. Height of a Building A 5-foot 3-inch person casts a shadow that is 7.5 feet while a nearby building casts a 32-foot shadow. Find the height of the building.

60. Angles in a Triangle The measure of the largest angle in a triangle is 20° less than the sum of the measures of the two smaller angles. The sum of the measures of the two larger angles is 90° greater than the measure of the smaller angle. Find the measure of each angle.

11 Quadratic Functions and Equations

What size television should you buy? The farther you sit from your television, the larger it should be. For example, if you sit only 6 feet from the screen, then a 32-inch television would be adequate; if you sit 10 feet from the screen, then a 50-inch television is more appropriate.

We can use a *quadratic function* to calculate the size S of the television screen needed for a person who sits x feet from the screen. You might want to use S to determine the size of screen that is recommended for you. (See Exercises 99 and 100 in the first section of this chapter.) Quadratic functions occur frequently in applications involving economics, road construction, falling objects, geometry, and modeling real-world data. (*Source: Money*, January 2007, p. 107; hdguru.com.)

11.1 Quadratic Functions and Their Graphs

NEW VOCABULARY

☐ Quadratic function
☐ Vertex
☐ Axis of symmetry

1 Working with Graphs of Quadratic Functions

PARABOLAS Earlier in this text, we discussed how a quadratic function could be represented by a polynomial of degree 2. We now give an alternative definition of a quadratic function.

QUADRATIC FUNCTION

A **quadratic function** can be written in the form

$$f(x) = ax^2 + bx + c,$$

where a, b, and c are constants with $a \neq 0$.

NOTE: The domain of a quadratic function is all real numbers. ∎

The graph of *any* quadratic function is a *parabola*. Recall that a parabola is a ∪-shaped graph that opens either upward or downward. The following See the Concept outlines the basic characteristics of parabolas.

READING CHECK 1

• How can you identify the vertex and the axis of symmetry on the graph of a parabola?

See the Concept 1   PARABOLAS

A The graph of the simple quadratic function $f(x) = x^2$ is an important parabola.

B The **vertex** is the *lowest* point on a parabola that opens *upward*.

C The **vertex** is the *highest* point on a parabola that opens *downward*.

D The **axis of symmetry** is a vertical line that passes through the vertex, but it is not part of the parabola. The left and right sides of the parabola would match if we folded the *xy*-plane along the axis of symmetry.

WATCH VIDEO IN MML.

EXAMPLE 1   **Identifying the vertex and the axis of symmetry**

Use the graph of the quadratic function to identify the vertex, the axis of symmetry, and whether the parabola opens upward or downward.

(a)

FIGURE 11.1

(b)

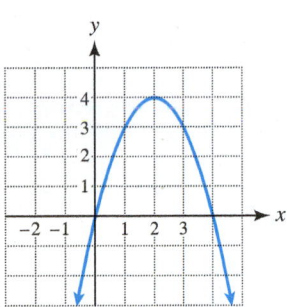

FIGURE 11.2

Solution

(a) The vertex is the lowest point on the graph shown in **FIGURE 11.1**, and its coordinates are $(1, -3)$. The axis of symmetry is the vertical line passing through the vertex, so its equation is $x = 1$. The parabola opens upward.

(b) The vertex is the highest point on the graph shown in **FIGURE 11.2**, and its coordinates are $(2, 4)$. The axis of symmetry is the vertical line passing through the vertex, so its equation is $x = 2$. The parabola opens downward.

Now Try Exercises 25, 27

THE VERTEX FORMULA   In order to graph a parabola by hand, it is helpful to know the location of the vertex. The following formula can be used to find the coordinates of the vertex for *any* parabola. This formula can be derived by *completing the square*, a technique discussed in the next section. However, we can also derive the vertex formula by using the fact that a parabola is symmetric, as follows.

The equation for a parabola is $y = ax^2 + bx + c$ and its y-intercept is $(0, c)$. See **FIGURE 11.3**. Because the parabola is symmetric about the axis of symmetry, there must be a second point where the y-value is c.

Finding the x-Value of the Vertex

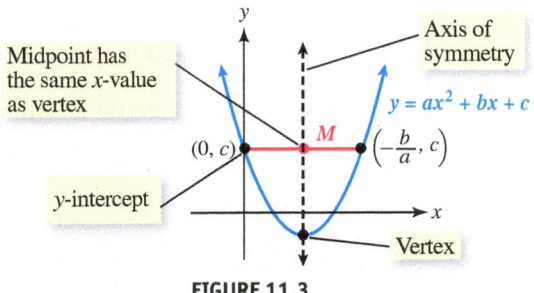

FIGURE 11.3

We can find the two x-values where $y = c$ by solving the following equation.

$$ax^2 + bx + c = c \qquad \text{Find } x\text{-values where } y = c.$$

$$ax^2 + bx = 0 \qquad \text{Subtract } c \text{ from each side.}$$

$$x(ax + b) = 0 \qquad \text{Factor out } x.$$

$$x = 0 \quad \text{or} \quad ax + b = 0 \qquad \text{Zero-product property}$$

$$x = 0 \quad \text{or} \qquad x = -\frac{b}{a} \qquad \text{Solve for } x.$$

The two points $(0, c)$ and $\left(-\frac{b}{a}, c\right)$ lie on the parabola and the x-coordinate of the vertex is the same as the x-coordinate of the midpoint M of the line segment connecting these two points.

$$x = \frac{0 + \left(-\frac{b}{a}\right)}{2} = -\frac{b}{2a}$$

> The x-value of the midpoint is the average of the x-values of the endpoints.

READING CHECK 2

- How do you use the vertex formula to find the coordinates of the vertex?

VERTEX FORMULA

The x-coordinate of the vertex of the graph of $y = ax^2 + bx + c$, $a \neq 0$, is given by

$$x = -\frac{b}{2a}.$$

To find the y-coordinate of the vertex, substitute this x-value in $y = ax^2 + bx + c$.

The following See the Concept shows how the vertex formula can be used to find the vertex (h, k) of a parabola and its axis of symmetry.

See the Concept 2   FINDING THE VERTEX AND AXIS OF SYMMETRY

$$f(x) = ax^2 + bx + c$$

 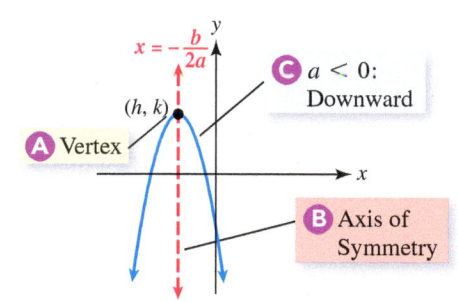

A The x-coordinate of the vertex (h, k) is $h = -\frac{b}{2a}$.

The y-coordinate of the vertex (h, k) is $k = f\left(-\frac{b}{2a}\right)$.

B The equation of the axis of symmetry is $x = -\frac{b}{2a}$.

C The parabola opens upward if $a > 0$, and downward if $a < 0$.

WATCH VIDEO IN MML.

EXAMPLE 2   Finding the vertex of a parabola

Find the vertex for the graph of $f(x) = 2x^2 - 4x + 1$. Support your answer graphically.

Solution

For $f(x) = 2x^2 - 4x + 1$, $a = 2$ and $b = -4$. The x-coordinate of the vertex is

$$x = -\frac{b}{2a} = -\frac{(-4)}{2(2)} = 1.$$

> x-coordinate of vertex

$[-4.7, 4.7, 1]$ by $[-3.1, 3.1, 1]$

FIGURE 11.4

To find the y-coordinate of the vertex, substitute $x = 1$ in the given formula.

$$f(1) = 2(1)^2 - 4(1) + 1 = -1 \qquad \text{y-coordinate of vertex}$$

Thus the vertex is located at $(1, -1)$, which is supported by **FIGURE 11.4**.

Now Try Exercise 29

GRAPHING QUADRATIC FUNCTIONS BY HAND One way to graph a quadratic function without a graphing calculator is to first apply the vertex formula. After the vertex is located, a table of values can be made to locate points on either side of the vertex. This technique is used in the next example to graph three quadratic functions.

EXAMPLE 3 **Graphing quadratic functions**

Identify the vertex and axis of symmetry on the graph of $y = f(x)$. Graph $y = f(x)$.

(a) $f(x) = x^2 - 1$ **(b)** $f(x) = -(x + 1)^2$ **(c)** $f(x) = x^2 + 4x + 3$

Solution
(a) Begin by applying the vertex formula with $a = 1$ and $b = 0$ to locate the vertex.

$$x = -\frac{b}{2a} = -\frac{0}{2(1)} = 0 \qquad \text{x-coordinate of vertex}$$

The y-coordinate of the vertex is found by evaluating $f(0)$.

$$y = f(0) = 0^2 - 1 = -1 \qquad \text{y-coordinate of vertex}$$

Thus the coordinates of the vertex are $(0, -1)$. **TABLE 11.1** is made by finding points on either side of the vertex. Plotting these points and connecting a smooth $\cup$-shaped graph results in **FIGURE 11.5**. The axis of symmetry is the vertical line passing through the vertex, so its equation is $x = 0$.

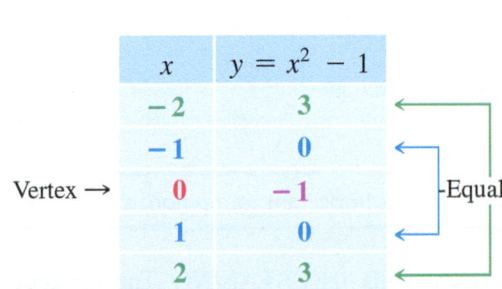

x	$y = x^2 - 1$
-2	3
-1	0
0	-1
1	0
2	3

Vertex → (at row $x = 0$)

-Equal

TABLE 11.1

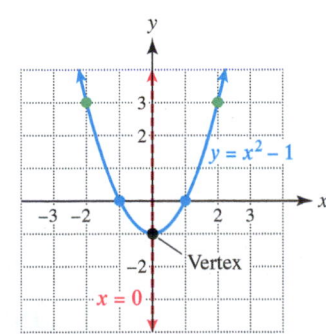

FIGURE 11.5

(b) Before we can apply the vertex formula, we need to determine values for a and b by multiplying the expression for $f(x) = -(x + 1)^2$.

$$-(x + 1)^2 = -(x^2 + 2x + 1) \qquad \text{Square the binomial.}$$
$$= -x^2 - 2x - 1 \qquad \text{Distribute the negative sign.}$$

Substitute $a = -1$ and $b = -2$ into the vertex formula.

$$x = -\frac{b}{2a} = -\frac{(-2)}{2(-1)} = -1 \qquad \text{x-coordinate of vertex}$$

The y-coordinate of the vertex is found by evaluating $f(-1)$.

$$y = f(-1) = -(-1 + 1)^2 = 0 \qquad \text{y-coordinate of vertex}$$

Thus the coordinates of the vertex are $(-1, 0)$. **TABLE 11.2** is made by finding points on either side of the vertex. Plotting these points and connecting a smooth $\cap$-shaped graph result in **FIGURE 11.6**. The axis of symmetry is the vertical line passing through the vertex, so its equation is $x = -1$.

x	$y = -(x + 1)^2$
-3	-4
-2	-1
-1	0
0	-1
1	-4

Vertex → -1

Equal

TABLE 11.2

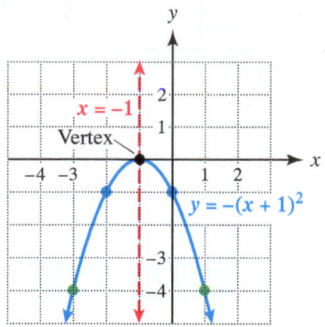

FIGURE 11.6

(c) The graph of $f(x) = x^2 + 4x + 3$ can be found in a manner similar to that used for the graphs in (a) and (b). See **TABLE 11.3** and **FIGURE 11.7**. With $a = 1$ and $b = 4$, the vertex formula can be used to show that the vertex is located at $(-2, -1)$. The equation of the axis of symmetry is $x = -2$.

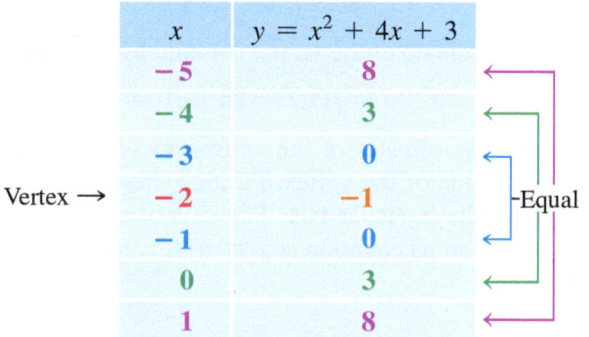

x	$y = x^2 + 4x + 3$
-5	8
-4	3
-3	0
-2	-1
-1	0
0	3
1	8

Vertex → -2

Equal

TABLE 11.3

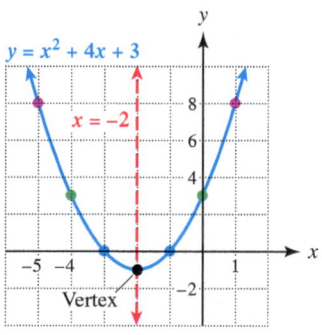

FIGURE 11.7

Now Try Exercises 41, 47, 49

STUDY TIP

The vertex formula is important to memorize because it is often used when graphing parabolas.

INCREASING AND DECREASING The concept of increasing and decreasing is frequently used when modeling data with functions. Suppose that the graph of the equation $y = x^2$ shown in **FIGURE 11.8** represents a valley. If we walk from *left to right* the valley "goes down" and then "goes up." Mathematically, we say that the graph of $y = x^2$ is *decreasing* when $x < 0$ and *increasing* when $x > 0$. In **FIGURE 11.6**, the graph of $y = -(x + 1)^2$ increases when $x < -1$ and decreases when $x > -1$.

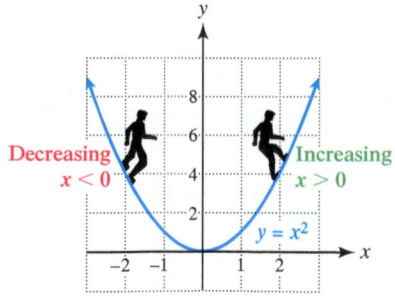

FIGURE 11.8

NOTE: When determining where a graph is increasing and where it is decreasing, we must "walk" along the graph *from left to right*. (We read English from left to right, which might help you remember.) ∎

2 Solving Min–Max Applications

🌐 Math in Context *Pizza Eating* Have you ever noticed that the first piece of pizza tastes really great and gives a person a lot of satisfaction? The second piece may be almost as good. After a few more pieces, there is often a point where the satisfaction starts to level off, and it can even go down as a person starts to eat too much. Eventually, if a person overeats, he or she might regret eating any pizza at all.

In **FIGURE 11.9**, a parabola models the total satisfaction received from eating x slices of pizza. Notice that the satisfaction level increases and then decreases. Although maximum satisfaction would vary with the individual, it always occurs at the vertex. (This is a seemingly simple model, but it is nonetheless used frequently in economics to describe consumer satisfaction from buying x identical items.)

Satisfaction from Eating Pizza

The parabola opens downward and has a maximum y-value of 100 at the vertex.

Eating too much pizza decreases satisfaction.

FIGURE 11.9

EXAMPLE 4

Finding maximum satisfaction from pizza

The graph of $S(x) = -6.25x^2 + 50x$ is shown in **FIGURE 11.9**. Use the vertex formula to determine the number of slices of pizza that results in maximum satisfaction. Does this maximum agree with what is shown in the graph of $y = S(x)$?

Solution

The parabola in **FIGURE 11.9** opens downward, so the greatest y-value occurs at the vertex. To find the x-coordinate of the vertex, we can apply the vertex formula. Substitute $a = -6.25$ and $b = 50$ into the vertex formula.

$$x = -\frac{b}{2a} = -\frac{50}{2(-6.25)} = 4 \quad \textcolor{blue}{x\text{-coordinate of vertex}}$$

Maximum satisfaction on a scale of 1 to 100 occurs when **4** slices of pizza are eaten, which agrees with **FIGURE 11.9**. (This number obviously varies from individual to individual.)

Now Try Exercise 83

READING CHECK 3

• How do you determine if a parabola has a maximum y-value or a minimum y-value?

FINDING MIN–MAX When a quadratic function f is used to model real data, the y-coordinate of the vertex represents either a maximum value of $f(x)$, when the parabola opens downward, or a minimum value of $f(x)$, when the parabola opens upward. **FIGURE 11.10** on the next page illustrates maximum and minimum y-values on a parabola.

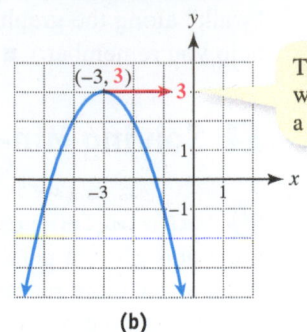

Minimum: 1

This parabola opens upward with vertex $(2, 1)$, so it has a *minimum* y-value of 1.

Maximum: 3

This parabola opens downward with vertex $(-3, 3)$, so it has a *maximum* y-value of 3.

(a)　　　　　(b)

FIGURE 11.10

EXAMPLE 5 **Finding a minimum y-value**

Find the minimum y-value on the graph of $f(x) = x^2 - 4x + 3$.

Solution

To locate the minimum y-value on a parabola, first apply the vertex formula with $a = 1$ and $b = -4$ to find the x-coordinate of the vertex.

$$x = -\frac{b}{2a} = -\frac{(-4)}{2(1)} = 2$$

x-coordinate of vertex

The minimum y-value on the graph is found by evaluating $f(2)$.

$$y = f(2) = 2^2 - 4(2) + 3 = -1$$

y-coordinate of vertex

Thus the minimum y-value on the graph of $y = f(x)$ is -1. This result is supported by **FIGURE 11.11**.

Now Try Exercise 69

Locating a Minimum y-Value

FIGURE 11.11

 Math in Context In the next example, we demonstrate finding a maximum height reached by a baseball.

EXAMPLE 6 **Finding maximum height**

A baseball is hit into the air, and its height h in feet after t seconds can be calculated by $h(t) = -16t^2 + 96t + 3$.
(a) What is the height of the baseball when it is hit?
(b) Determine the maximum height of the baseball.

Solution
(a) The baseball is hit when $t = 0$, so $h(0) = -16(0)^2 + 96(0) + 3 = 3$ feet.
(b) The graph of h opens downward because $a = -16 < 0$. Thus the maximum height of the baseball occurs at the vertex. To find the vertex, we apply the vertex formula with $a = -16$ and $b = 96$ because $h(t) = -16t^2 + 96t + 3$.

$$t = -\frac{b}{2a} = -\frac{96}{2(-16)} = 3 \text{ seconds}$$

The maximum height of the baseball occurs at $t = 3$ seconds and is

$$h(3) = -16(3)^2 + 96(3) + 3 = 147 \text{ feet}.$$

Now Try Exercise 89

🌐 *Math in Context* ^(Business) Suppose that a hotel is considering giving a group discount on room rates. The regular price is $80, but for each room rented, the price decreases by $2. On the one hand, if the hotel rents one room, it makes only $78. On the other hand, if the hotel rents 40 rooms, the rooms are all free and the hotel makes nothing. Is there a number of rooms between 1 and 40 that should be rented to maximize the revenue from the group?

In **FIGURE 11.12**, the hotel's revenue is graphed. From the graph, it is apparent that "peak" revenue occurs at the vertex of the parabola.

Hotel Revenue

FIGURE 11.12

EXAMPLE 7 **Maximizing revenue**

A hotel gives the following group discount on room rates. The regular price for a room is $80, but for each room rented the price decreases by $2. A graph of the revenue received from renting x rooms is shown in **FIGURE 11.12**.

(a) Interpret the graph.

(b) What is the maximum revenue? How many rooms should be rented to receive the maximum revenue?

(c) Write a formula for $f(x)$ whose graph is shown in **FIGURE 11.12**.

(d) Use $f(x)$ to determine symbolically the maximum revenue and the number of rooms that should be rented.

TECHNOLOGY NOTE

Locating a Vertex
Graphing calculators can locate a vertex with the MAXIMUM or MINIMUM utility. The maximum in Example 7 is found in the figure.

$[0, 50, 10]$ by $[0, 1000, 100]$

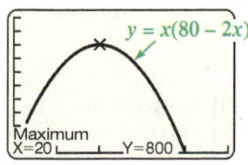

CALCULATOR HELP

To find a minimum or maximum, see Appendix A (page AP-11).

Solution

(a) The revenue increases at first, reaches a maximum (which corresponds to the vertex), and then decreases.

(b) In **FIGURE 11.12** the vertex is $(\textbf{20}, \textbf{800})$. Thus the maximum revenue of **$800** occurs when **20** rooms are rented.

(c) If x rooms are rented, the price for each room is $80 - 2x$. The revenue equals the number of rooms rented times the price of each room. Thus $f(x) = x(80 - 2x)$.

(d) First, multiply $x(80 - 2x)$ to obtain $80x - 2x^2$ and then let $f(x) = -2x^2 + 80x$. The x-coordinate of the vertex is

$$x = -\frac{b}{2a} = -\frac{80}{2(-2)} = \textbf{20}.$$

The y-coordinate is $f(\textbf{20}) = -2(\textbf{20})^2 + 80(\textbf{20}) = \textbf{800}$. These calculations verify our results in part (b).

Now Try Exercise 93

11.1 Putting It All Together

CONCEPT	EXPLANATION	EXAMPLES
Quadratic Function	Can be written as $f(x) = ax^2 + bx + c, a \neq 0$	$f(x) = x^2 + x - 2$ and $g(x) = -2x^2 + 4$ $(b = 0)$
Vertex of a Parabola	The x-coordinate of the vertex for the function $f(x) = ax^2 + bx + c$ with $a \neq 0$ is given by $$x = -\frac{b}{2a}.$$ The y-coordinate of the vertex is found by substituting this x-value in the equation. Hence the vertex is $\left(-\frac{b}{2a}, f\left(-\frac{b}{2a}\right)\right)$.	If $f(x) = -2x^2 + 8x - 7$, then $$x = -\frac{8}{2(-2)} = 2$$ and $$f(2) = -2(2)^2 + 8(2) - 7 = 1.$$ The vertex is $(2, 1)$. The graph of f opens downward because $a < 0$.
Graph of a Quadratic Function	Its graph is a parabola that opens upward ($\cup$-shaped) if $a > 0$ and downward ($\cap$-shaped) if $a < 0$. The vertex can be used to determine the maximum or minimum y-value.	The graph of $f(x) = -x^2 - 2x$ opens downward because $a = -1 < 0$. The vertex is $(-1, 1)$, so the maximum y-value is 1.

Quadratic functions can have verbal, symbolic, numerical, and graphical representations. For example, the quadratic function represented verbally as *f squares the input x and then subtracts* 1 can also be represented as follows.

Symbolic Representation

$$f(x) = x^2 - 1$$

Numerical Representation

Graphical Representation

11.1 Exercises

MyMathLab®

CONCEPTS AND VOCABULARY

1. The graph of a quadratic function is called a(n) _____.

2. If a parabola opens upward, what is the lowest point on the parabola called?

3. If a parabola is symmetric with respect to the y-axis, the y-axis is called the _____.

4. The vertex on the graph of $y = x^2$ is _____.

5. Sketch a parabola that opens downward with a vertex of $(1, 2)$.

6. If $y = ax^2 + bx + c$, the x-coordinate of the vertex is given by $x =$ _____.

7. Any quadratic function can be written in the form $f(x) =$ _____.

8. If a parabola opens downward, the point with the largest y-value is called the _____.

9. (True or False?) The axis of symmetry for the graph of $y = ax^2 + bx + c$ is given by $x = -\frac{b}{2a}$.

10. (True or False?) If the vertex of a parabola is located at (a, b), then the axis of symmetry is given by $x = b$.

11. (True or False?) If a parabola opens downward and its vertex is (a, b), then the minimum y-value on the parabola is b.

12. (True or False?) The graph of $y = -ax^2$ with $a > 0$ opens downward.

EVALUATING QUADRATIC FUNCTIONS

Exercises 13–20: Evaluate the given quadratic function as indicated.

13. Find $f(4)$ if $f(x) = x^2 + 1$.

14. Find $f(3)$ if $f(x) = x^2 - 11$.

15. Find $f(2)$ if $f(x) = x^2 - 6x$.

16. Find $f(-1)$ if $f(x) = x^2 + 5x$.

17. Find $f(-3)$ if $f(x) = 2x^2 + 4x + 1$.

18. Find $f(5)$ if $f(x) = -2x^2 + 7x - 3$.

19. Find $f(8)$ if $f(x) = -3x^2 + 16x + 72$.

20. Find $f(-6)$ if $f(x) = 4x^2 + 11x - 88$.

Exercises 21–24: Use the given graph of f to evaluate the expressions.

21. $f(-2)$ and $f(0)$

22. $f(-2)$ and $f(2)$

23. $f(-3)$ and $f(1)$

24. $f(-1)$ and $f(2)$

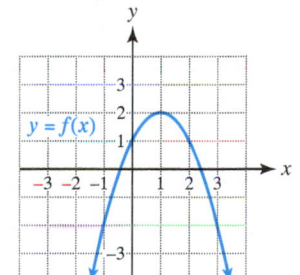

IDENIFYING CHARACTERISTICS OF PARABOLAS

Exercises 25–28: Identify the vertex, axis of symmetry, and whether the parabola opens upward or downward.

25.

26.

27.

28.

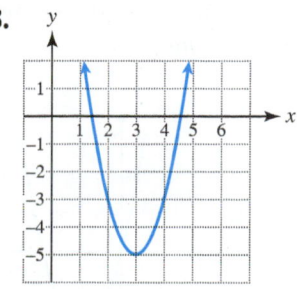

USING THE VERTEX FORMULA

Exercises 29–38: Find the vertex of the parabola.

29. $f(x) = x^2 - 4x - 2$

30. $f(x) = 2x^2 + 6x - 3$

31. $f(x) = -\frac{1}{3}x^2 - 2x + 1$

32. $f(x) = 5 - 4x + x^2$

33. $f(x) = 3 - 2x^2$

34. $f(x) = \frac{1}{4}x^2 - 3x - 2$

35. $f(x) = -0.3x^2 + 0.6x + 1.1$

36. $f(x) = 25 - 10x + 20x^2$

37. $f(x) = 6x - x^2$ **38.** $f(x) = x - \frac{1}{2}x^2$

GRAPHING QUADRATIC FUNCTIONS

Exercises 39–58: Do the following for the given $f(x)$.

   (**a**) *Identify the vertex and axis of symmetry on the graph of $y = f(x)$.*
   (**b**) *Graph $y = f(x)$.*
   (**c**) *Evaluate $f(-2)$ and $f(3)$.*

39. $f(x) = \frac{1}{2}x^2$ **40.** $f(x) = -3x^2$

41. $f(x) = x^2 - 2$ **42.** $f(x) = x^2 - 1$

43. $f(x) = -3x^2 + 1$ **44.** $f(x) = \frac{1}{2}x^2 + 2$

45. $f(x) = (x - 1)^2$ **46.** $f(x) = (x + 2)^2$

47. $f(x) = -(x + 2)^2$ **48.** $f(x) = -(x - 1)^2$

49. $f(x) = x^2 + x - 2$ **50.** $f(x) = x^2 - 2x + 2$

51. $f(x) = 2x^2 - 3$ **52.** $f(x) = 1 - 2x^2$

53. $f(x) = 2x - x^2$ **54.** $f(x) = x^2 + 2x - 8$

55. $f(x) = -2x^2 + 4x - 1$

56. $f(x) = -\frac{1}{2}x^2 + 2x - 3$

57. $f(x) = \frac{1}{4}x^2 - x + 5$ **58.** $f(x) = 3 - 6x - 4x^2$

IDENTIFYING INCREASING AND DECREASING INTERVALS

Exercises 59–62: Use the graph of $y = f(x)$ to determine the intervals where f is increasing and where f is decreasing.

59.

60.

61.

62.

Exercises 63–68: Determine the intervals where $y = f(x)$ is increasing and where it is decreasing.

63. $y = 3x^2 - 4x + 1$ **64.** $y = 2x^2 - 5x - 2$

65. $y = -x^2 - 3x$ **66.** $y = -2x^2 + 3x - 4$

67. $y = 5 - x - 4x^2$ **68.** $y = 3 + x + x^2$

FINDING MAXIMUMS AND MINIMUMS

Exercises 69–74: Find the minimum y-value on the graph of $y = f(x)$.

69. $f(x) = x^2 + 2x - 1$ **70.** $f(x) = x^2 + 6x + 2$

71. $f(x) = x^2 - 5x$ **72.** $f(x) = x^2 - 3x$

73. $f(x) = 2x^2 + 2x - 3$ **74.** $f(x) = 3x^2 - 3x + 7$

Exercises 75–80: Find the maximum y-value on the graph of $y = f(x)$.

75. $f(x) = -x^2 + 2x + 5$

76. $f(x) = -x^2 + 4x - 3$

77. $f(x) = 4x - x^2$

78. $f(x) = 6x - x^2$

79. $f(x) = -2x^2 + x - 5$

80. $f(x) = -5x^2 + 15x - 2$

81. Numbers Find two positive numbers whose sum is 20 and whose product is maximum.

82. Thinking Generally Find two positive numbers whose sum is k and whose product is maximum.

APPLICATIONS INVOLVING QUADRATIC FUNCTIONS

Exercises 83 and 84: **Eating Pizza** *(Refer to Example 4.) Let $S(x)$ denote the satisfaction, on a scale from 0 to 100, from eating x slices of pizza. Find the number of slices that gives the maximum satisfaction.*

83. $S(x) = -\frac{100}{9}x^2 + \frac{200}{3}x$

84. $S(x) = -16x^2 + 80x$

Exercises 85–88: **Quadratic Models** *Match the physical situation with the graph (a–d) that models it best.*

85. The height y of a stone thrown from ground level after x seconds

86. The number of people attending a popular movie x weeks after its opening

87. The temperature after x hours in a house when the furnace quits and then a repair person fixes it

88. U.S. population from 1800 to the present

a.

b.

c.

d.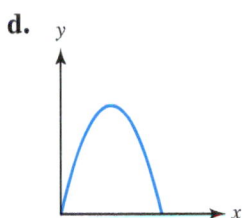

89. Height Reached by a Baseball A baseball is hit into the air, and its height h in feet after t seconds is given by $h(t) = -16t^2 + 64t + 2$.
 (a) What is the height of the baseball when it is hit?
 (b) After how many seconds does the baseball reach its maximum height?
 (c) Determine the maximum height of the baseball.

90. Height Reached by a Golf Ball A golf ball is hit into the air, and its height h in feet after t seconds is given by $h(t) = -16t^2 + 128t$.
 (a) What is the height of the golf ball when it is hit?
 (b) After how many seconds does the golf ball reach its maximum height?
 (c) Determine the maximum height of the golf ball.

91. Height Reached by a Baseball Suppose that a baseball is thrown upward with an initial velocity of 66 feet per second (45 miles per hour) and it is released 6 feet above the ground. Its height h after t seconds is given by

$$h(t) = -16t^2 + 66t + 6.$$

After how many seconds does the baseball reach a maximum height? Estimate this height.

92. Throwing a Baseball on the Moon (Refer to Exercise 91.) If the same baseball were thrown the same

way on the moon, its height h above the moon's surface after t seconds would be

$$h(t) = -2.55t^2 + 66t + 6.$$

Does the baseball go higher on the moon or on Earth? What is the difference in these two heights?

93. Concert Tickets (Refer to Example 7.) An agency is promoting concert tickets by offering a group-discount rate. The regular price is $100 and for each ticket bought the price decreases by $1. (One ticket costs $99, two tickets cost $98 *each*, and so on.)
 (a) A graph of the revenue received from selling x tickets is shown in the figure. Interpret the graph.

Tickets Sold

 (b) What is the maximum revenue? How many tickets should be sold to maximize revenue?
 (c) Write a formula for $y = f(x)$ whose graph is shown in the figure.
 (d) Use $f(x)$ to determine symbolically the maximum revenue and the number of tickets that should be sold to maximize revenue.

94. Maximizing Revenue The regular price for a round-trip ticket to Las Vegas, Nevada, charged by an airline charter company is $300. For a group rate, the company will reduce the price of each ticket by $1.50 for every passenger on the flight.
 (a) Write a formula $f(x)$ that gives the revenue from selling x tickets.
 (b) Determine how many tickets should be sold to maximize the revenue. What is the maximum revenue?

95. Monthly Facebook Visitors The number of *unique* monthly Facebook visitors in millions x years after 2010 can be modeled by

$$V(x) = -0.5x^2 + 10x + 138.$$

 (a) Evaluate $V(1)$, $V(2)$, $V(3)$, and $V(4)$. Interpret your answer.
 (b) Explain why numbers of Facebook visitors can not be modeled by a linear function.

96. Cell Phone Complexity People often enjoy having certain features on their cell phones, such as e-mail and the ability to surf the Web. However, as phones become more and more complicated to operate, the benefits from the additional complexity start to decrease. The following parabola models this general situation. Interpret why a parabola is appropriate to model this consumer experience.

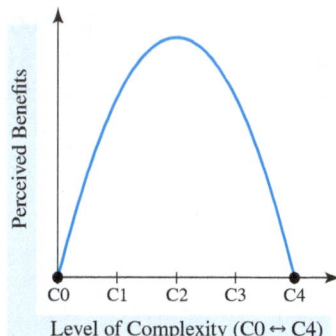

Level of Complexity (C0 ↔ C4)

97. Maximizing Area A farmer is fencing a rectangular area for cattle and uses a straight portion of a river as one side of the rectangle, as illustrated in the figure. Note that there is no fence along the river. If the farmer has 1200 feet of fence, find the dimensions for the rectangular area that give a maximum area for the cattle.

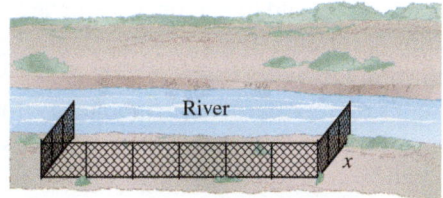

River

x

98. Maximizing Area A rectangular pen being constructed for a pet requires 60 feet of fence.
(a) Write a formula $f(x)$ that can be used to find the total area of the pen if one side of the pen has length x.
(b) Find the dimensions of the pen that give the largest area. What is the largest area?

Exercises 99 and 100: **Large-Screen Televisions** *(Refer to the introduction to this chapter.) Use the formula*

$$S(x) = -0.227x^2 + 8.155x - 8.8,$$

where $6 \leq x \leq 16$, to estimate the recommended screen size in inches when viewers sit x feet from the screen.

99. $x = 8$ feet   **100.** $x = 12$ feet

101. Carbon Emissions Past and future carbon emissions in billions of metric tons during year x can be modeled by

$$C(x) = \frac{1}{300}x^2 - \frac{199}{15}x + \frac{39,619}{3},$$

where $1990 \leq x \leq 2020$. (*Source:* U.S. Department of Energy.)
(a) Evaluate $C(1990)$. Interpret your answer.
(b) Find the expected *increase* in carbon emissions from 1990 to 2020.

102. Seedling Growth In a study of the effect of temperature on the growth of melon seedlings, the seedlings were grown at different temperatures, and their heights were measured after a fixed period of time. The findings of this study can be modeled by

$$f(x) = -0.095x^2 + 5.4x - 52.2,$$

where x is the temperature in degrees Celsius and the output $f(x)$ gives the resulting average height in centimeters. (*Source:* R. Pearl, "The growth of *Cucumis melo* seedlings at different temperatures.")
(a) Graph f in $[20, 40, 5]$ by $[0, 30, 5]$.
(b) Estimate graphically the temperature that resulted in the greatest height for the melon seedlings.
(c) Solve part (b) symbolically.

WRITING ABOUT MATHEMATICS

103. If $f(x) = ax^2 + bx + c$, explain how to find the vertex on the graph of $y = f(x)$.

104. Suppose that a quantity Q is modeled by the formula $Q(x) = ax^2 + bx + c$ with $a < 0$. Explain how to find the x-value that maximizes $Q(x)$. How do you find the maximum value of $Q(x)$?

11.2 Transformations and Translations of Parabolas

Objectives

1. Graphing Basic Transformations of $y = ax^2$
   - The Graph of $y = ax^2, a > 0$
   - The Graph of $y = ax^2, a < 0$
   - Reflections of $y = ax^2$

2. Performing Vertical and Horizontal Translations

3. Applying Vertex Form
   - Graphing with Vertex Form
   - Completing the Square to Find the Vertex

4. Modeling with Quadratic Functions

NEW VOCABULARY

☐ Reflection
☐ Translations
☐ Vertex form
☐ Completing the square

1 Graphing Basic Transformations of $y = ax^2$

THE GRAPH OF $y = ax^2, a > 0$ The graphs of $y_1 = \frac{1}{2}x^2$, $y_2 = x^2$, and $y_3 = 2x^2$ are shown in **FIGURE 11.13** and open upward. Note that $a = \frac{1}{2}$, $a = 1$, and $a = 2$, respectively, and that as the value of a increases, the resulting parabola becomes narrower. When $a > 0$, the graph of $y = ax^2$ never lies *below* the x-axis.

Open Upward: $y = ax^2, a > 0$

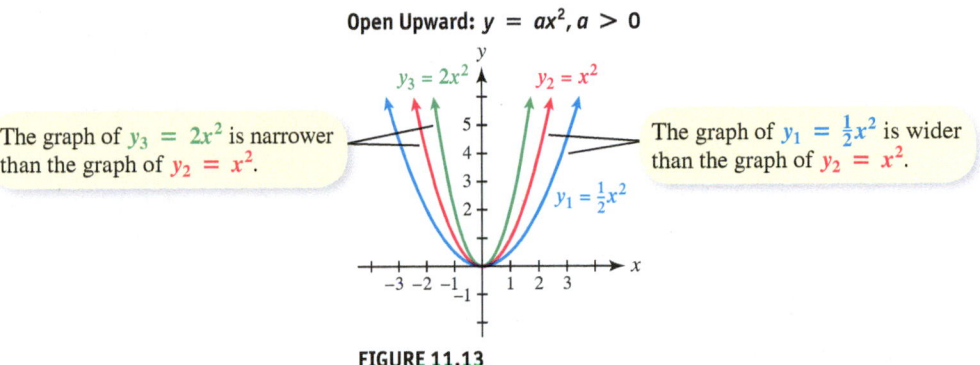

The graph of $y_3 = 2x^2$ is narrower than the graph of $y_2 = x^2$.

The graph of $y_1 = \frac{1}{2}x^2$ is wider than the graph of $y_2 = x^2$.

FIGURE 11.13

THE GRAPH OF $y = ax^2, a < 0$ The graphs of $y_4 = -\frac{1}{2}x^2$, $y_5 = -x^2$, and $y_6 = -2x^2$ are shown in **FIGURE 11.14** and open downward. When $a < 0$, the graph of $y = ax^2$ never lies *above* the x-axis.

Open Downward: $y = ax^2, a < 0$

The graph of $y_6 = -2x^2$ is narrower than the graph of $y_5 = -x^2$.

The graph of $y_4 = -\frac{1}{2}x^2$ is wider than the graph of $y_5 = -x^2$.

FIGURE 11.14

> **THE GRAPH OF** $y = ax^2$
>
> The graph of $y = ax^2$ is a parabola with the following characteristics.
>
> 1. The vertex is $(0, 0)$, and the axis of symmetry is given by $x = 0$.
> 2. It opens upward if $a > 0$ and opens downward if $a < 0$.
> 3. It is wider than the graph of $y = x^2$, if $0 < |a| < 1$. It is narrower than the graph of $y = x^2$, if $|a| > 1$.

Reflection Across the x-Axis

$y = ax^2$
$(a > 0)$

$y = -ax^2$
$(a > 0)$

FIGURE 11.15

REFLECTIONS OF $y = ax^2$ The graph of $y_1 = \frac{1}{2}x^2$ in **FIGURE 11.13** can be *transformed* into the graph of $y_4 = -\frac{1}{2}x^2$ in **FIGURE 11.14** by *reflecting* it across the x-axis. In general, the graph of $y = -ax^2$ is a **reflection** of the graph of $y = ax^2$ across the x-axis. That is, if we folded the xy-plane along the x-axis, the two graphs would match. See **FIGURE 11.15**.

EXAMPLE 1   Graphing $y = ax^2$

Compare the graph of $g(x) = -3x^2$ to the graph of $f(x) = x^2$. Then graph both functions in the same xy-plane.

Solution

Both graphs are parabolas. However, the graph of $y = g(x)$ opens downward and is narrower than the graph of $y = f(x)$. These graphs are shown in **FIGURE 11.16**.

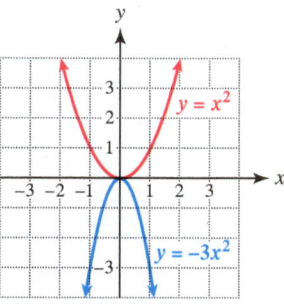

FIGURE 11.16

Now Try Exercise 11

2 Performing Vertical and Horizontal Translations

The graph of $y = x^2$ is a parabola opening upward with vertex $(0, 0)$. Suppose that we graph $y_1 = x^2$, $y_2 = x^2 + 1$, and $y_3 = x^2 - 2$ in the same xy-plane, as calculated in **TABLE 11.4** and shown in **FIGURE 11.17**. All three graphs have the same shape. However, compared to the graph of $y_1 = x^2$, the graph of $y_2 = x^2 \mathbin{\color{red}+} \mathbf{1}$ is shifted ***upward*** **1** unit and the graph of $y_3 = x^2 \mathbin{\color{green}-} \mathbf{2}$ is shifted *downward* **2** units. Such *shifts* are called **translations** because they do not change the shape of a graph—only its position.

Vertical Shifts of $y = x^2$

x	$y_2 = x^2 + 1$	$y_1 = x^2$	$y_3 = x^2 - 2$
-2	5	4	2
-1	2	1	-1
0	1	0	-2
1	2	1	-1
2	5	4	2
↑	↑		↑
The x-values do NOT change.	Add 1 to find the y_2-values.	Subtract 2 to find the y_3-values.	

TABLE 11.4

Shifted up 1 unit

Shifted down 2 units

FIGURE 11.17

Next, suppose that we graph $y_1 = x^2$ and $y_2 = (x - 1)^2$ in the same xy-plane. Compare **TABLE 11.5** and **TABLE 11.6**. Note that the y-values are equal when the x-value for y_2 is 1 unit *larger* than the x-value for y_1. For example, $y_1 = 4$ when $x = -2$ and $y_2 = 4$ when $x = -1$. Thus the graph of $y_2 = (x - 1)^2$ has the same shape as the graph of $y_1 = x^2$ except that it is translated *horizontally to the right* 1 unit, as illustrated in **FIGURE 11.18**.

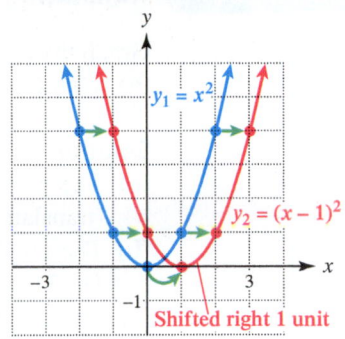

Horizontal Shift of $y = x^2$

FIGURE 11.18

x	$y_1 = x^2$
-2	4
-1	1
0	0
1	1
2	4

TABLE 11.5

x	$y_2 = (x-1)^2$
-1	4
0	1
1	0
2	1
3	4

TABLE 11.6

Add 1 to find the x-values.   The y-values do NOT change.

The graphs $y_1 = x^2$ and $y_2 = (x + 2)^2$ are shown in **FIGURE 11.19**. Note that **TABLE 11.7** and **TABLE 11.8** show their y-values to be equal when the x-value for y_2 is 2 units *smaller* than the x-value for y_1. As a result, the graph of $y_2 = (x + 2)^2$ has the same shape as the graph of $y_1 = x^2$ except that it is translated *horizontally to the left* 2 units.

Horizontal Shift of $y = x^2$

FIGURE 11.19

x	$y_1 = x^2$
-2	4
-1	1
0	0
1	1
2	4

TABLE 11.7

x	$y_2 = (x+2)^2$
-4	4
-3	1
-2	0
-1	1
0	4

TABLE 11.8

Subtract 2 to find the x-values.   The y-values do NOT change.

VERTICAL AND HORIZONTAL TRANSLATIONS OF PARABOLAS

Let h and k be positive numbers.

To graph	**shift the graph of $y = x^2$ by k units**
$y = x^2 + k$	upward.
$y = x^2 - k$	downward.

To graph	**shift the graph of $y = x^2$ by h units**
$y = (x - h)^2$	right.
$y = (x + h)^2$	left.

EXAMPLE 2   **Translating the graph** $y = x^2$

Graph the equation and identify the vertex. Compare this graph to the graph of $y = x^2$.

(a) $y = x^2 + 2$ **(b)** $y = (x + 3)^2$ **(c)** $y = (x - 2)^2 - 3$

Solution

(a) The graph of $y = x^2 + 2$ is similar to the graph of $y = x^2$ except that it is shifted (translated) *upward* 2 units, as shown in **FIGURE 11.20(a)**. The vertex is $(0, 2)$.

(b) The graph of $y = (x + 3)^2$ is similar to the graph of $y = x^2$ except that it is shifted *left* 3 units, as shown in **FIGURE 11.20(b)**. The vertex is $(-3, 0)$.

(c) The graph of $y = (x - 2)^2 - 3$ is similar to the graph of $y = x^2$ except that it is shifted *right* 2 units *and downward* 3 units. See **FIGURE 11.20(c)**. The vertex is $(2, -3)$.

Shifted Up 2 Units

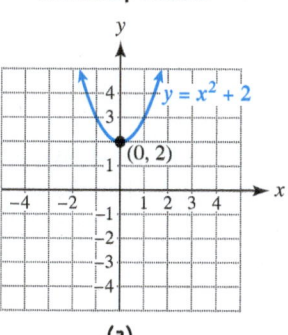

(a)

Shifted Left 3 Units

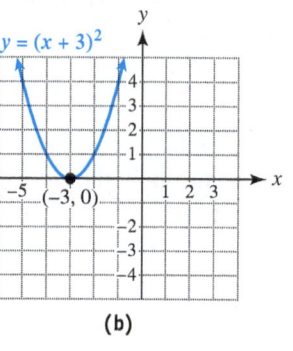

(b)

Shifted Right 2 Units and Down 3 Units

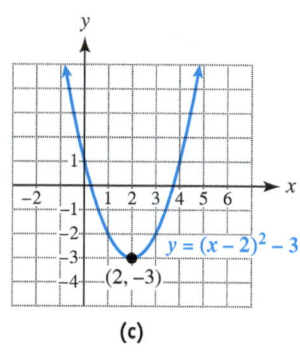

(c)

FIGURE 11.20

Now Try Exercises 23, 27, 35

3   Applying Vertex Form

GRAPHING WITH VERTEX FORM The graphs of $y = 2x^2$ and $y = 2x^2 - 12x + 20$ have *exactly* the same shape, as illustrated in the following See the Concept. However, the vertex for $y = 2x^2$ is $(0, 0)$, whereas the vertex for $y = 2x^2 - 12x + 20$ is $(3, 2)$.

See the Concept 1   **FINDING VERTEX FORM**

Vertex $(0, 0)$

Vertex $(3, 2)$

Shifting $y = 2x^2$

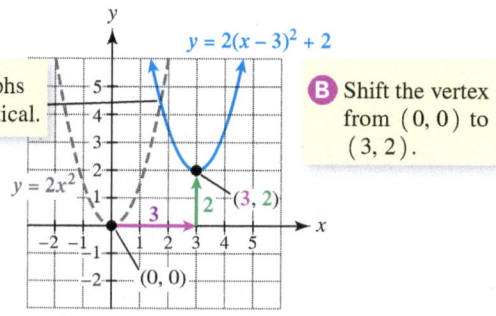

A The graphs have identical shapes.

C The graphs are identical.

B Shift the vertex from $(0, 0)$ to $(3, 2)$.

A The graphs of $y = 2x^2$ and $y = 2x^2 - 12x + 20$ have identical shapes because $a = 2$ in both equations.

B Shifting the graph of $y = 2x^2$ right 3 units and upward 2 units:

Vertex form ⟶ $y = 2(x - 3)^2 + 2.$ 2 units upward

3 units right

C The graphs of $y = 2x^2 - 12x + 20$ and $y = 2(x - 3)^2 + 2$ are *identical* because they have the same shape and the same vertex $(3, 2)$.

WATCH VIDEO IN MML.

CRITICAL THINKING 1

Expand $y = 2(x - 3)^2 + 2$ and combine like terms. What does it equal?

In general, the equation for any parabola can be written as either $y = ax^2 + bx + c$ or $y = a(x - h)^2 + k$, where the vertex is (h, k). The second form is sometimes called *vertex form.*

READING CHECK 1

• If the equation of a parabola is $y = -3(x + 4)^2 - 5$, identify the vertex.

VERTEX FORM

The **vertex form** of the equation of a parabola with vertex (h, k) is

$$y = a(x - h)^2 + k,$$

where $a \neq 0$ is a constant. If $a > 0$, the parabola opens upward; if $a < 0$, the parabola opens downward.

NOTE: The vertex form of the equation of a parabola is sometimes called **standard form for a parabola with a vertical axis.** ■

In the next three examples, we demonstrate graphing parabolas in vertex form, finding their equations, and writing vertex forms of equations.

EXAMPLE 3   Graphing parabolas in vertex form

Compare the graph of $y = f(x)$ to the graph of $y = x^2$. Then sketch a graph of $y = f(x)$ and $y = x^2$ in the same xy-plane.
(a) $f(x) = \frac{1}{2}(x - 5)^2 + 2$ (b) $f(x) = -3(x + 5)^2 - 3$

Solution
(a) Because $a = \frac{1}{2}$, the graph of $y = f(x)$ *opens upward* and is *wider* than the graph of $y = x^2$. **FIGURE 11.21(a)** describes all transformations associated with this graph.
(b) Because $a = -3$, the graph of $y = f(x)$ *opens downward* and is *narrower* than the graph of $y = x^2$. **FIGURE 11.21(b)** describes all transformations associated with this graph.

Shift the graph of $y_1 = x^2$ right 5 units and up 2 units.

Shift the graph of $y_1 = x^2$ left 5 units, reflect it across the x-axis, and shift it down 3 units.

Graphing Vertex Form

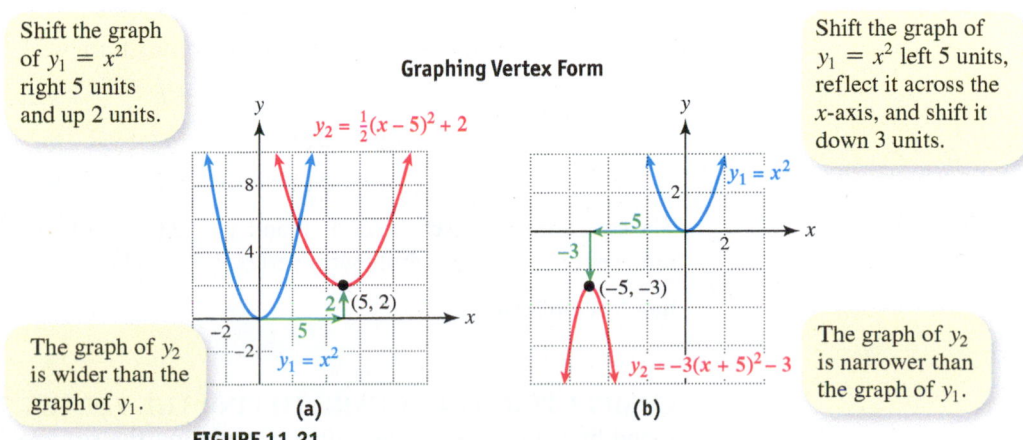

The graph of y_2 is wider than the graph of y_1.

The graph of y_2 is narrower than the graph of y_1.

(a)   (b)

FIGURE 11.21

Now Try Exercises 43, 45

EXAMPLE 4   **Finding equations of parabolas**

Write the vertex form of a parabola with $a = 2$ and vertex $(-2, 1)$. Then express this equation in the form $y = ax^2 + bx + c$.

Solution
The vertex form of a parabola is given by $y = a(x - h)^2 + k$, where the vertex is (h, k). For $a = 2$, $h = -2$, and $k = 1$, the equation becomes

$$y = 2(x - (-2))^2 + 1 \quad \text{or} \quad y = 2(x + 2)^2 + 1.$$

To write $y = 2(x + 2)^2 + 1$ in the form $y = ax^2 + bx + c$, do the following.

$$y = 2(x^2 + 4x + 4) + 1 \quad \text{Multiply } (x + 2)^2 \text{ to get } x^2 + 4x + 4.$$
$$= 2x^2 + 8x + 8 + 1 \quad \text{Distributive property}$$
$$= 2x^2 + 8x + 9 \quad \text{Add.}$$

The equivalent equation is $y = 2x^2 + 8x + 9$.

Now Try Exercise 49

In the next example, the vertex formula is used to write the equation of a parabola in vertex form.

EXAMPLE 5   **Using the vertex formula to write vertex form**

Find the vertex on the graph of $y = 3x^2 + 6x + 1$. Write this equation in vertex form.

Solution
We can use the vertex formula, $x = -\frac{b}{2a}$, to find the x-coordinate of the vertex with $a = 3$ and $b = 6$.

$$x = -\frac{b}{2a} = -\frac{6}{2(3)} = -1 \qquad x\text{-coordinate of vertex}$$

To find the y-coordinate, let $x = -1$ in $y = 3x^2 + 6x + 1$.

$$y = 3(-1)^2 + 6(-1) + 1 = -2 \qquad y\text{-coordinate of vertex}$$

The vertex is $(-1, -2)$. We now find the vertex form with $a = 3$, $h = -1$, and $k = -2$.

$$y = a(x - h)^2 + k \qquad \text{Vertex form}$$
$$y = 3(x - (-1))^2 - 2 \qquad \text{Substitute.}$$
$$y = 3(x + 1)^2 - 2 \qquad \text{Simplify.}$$

The equations $y = 3x^2 + 6x + 1$ and $y = 3(x + 1)^2 - 2$ are equivalent, and their graphs represent the same parabola with vertex $(-1, -2)$.

Now Try Exercise 61

COMPLETING THE SQUARE TO FIND THE VERTEX The vertex of a parabola can be found by using a technique called **completing the square**. To complete the square for the expression $x^2 + 4x$, consider the following See the Concept.

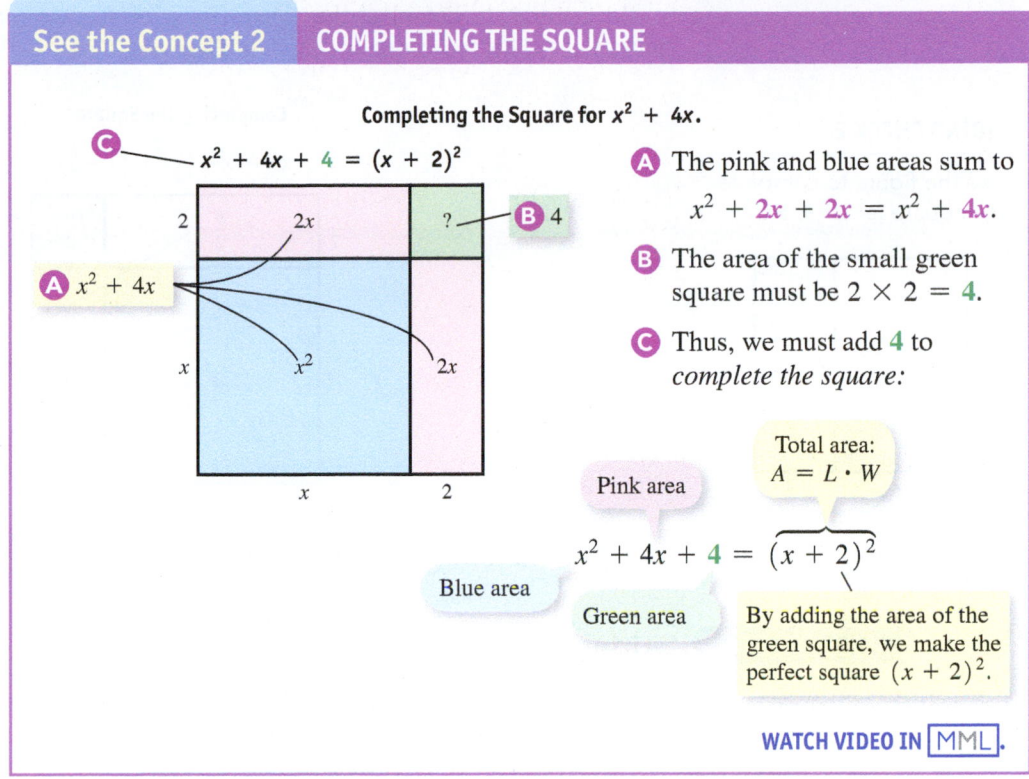

See the Concept 2   **COMPLETING THE SQUARE**

Completing the Square for $x^2 + 4x$.

C → $x^2 + 4x + 4 = (x + 2)^2$

A $x^2 + 4x$

B 4

A The pink and blue areas sum to
$$x^2 + 2x + 2x = x^2 + 4x.$$

B The area of the small green square must be $2 \times 2 = 4$.

C Thus, we must add 4 to *complete the square:*

Total area:
$A = L \cdot W$

Pink area

Blue area

Green area

$$x^2 + 4x + 4 = \overbrace{(x + 2)^2}$$

By adding the area of the green square, we make the perfect square $(x + 2)^2$.

WATCH VIDEO IN MML.

In the next example, we use the above result to complete the square and find the vertex on the graph of $y = x^2 + 4x$.

EXAMPLE 6   **Writing vertex form by completing the square**

Write the equation $y = x^2 + 4x$ in vertex form by completing the square. Identify the vertex on the graph of the equation.

Solution
As discussed in See the Concept 2, we must add 4 to $x^2 + 4x$ to complete the square. Because we are given an equation, we will add 4 *and* subtract 4 from the right side of the equation in order to keep the equation "balanced."

$$y = x^2 + 4x. \qquad \text{Given equation}$$
$$y = x^2 + 4x + 4 - 4 \qquad \text{Add and subtract 4 on the right.}$$
$$y = (x^2 + 4x + 4) - 4 \qquad \text{Associative property}$$
$$y = (x + 2)^2 - 4 \qquad \text{Perfect square trinomial}$$

The vertex form is $y = (x + 2)^2 - 4$, and the vertex is $(-2, -4)$.

Now Try Exercise 69

NOTE: Adding 4 and subtracting 4 from the right side of the equation in Example 6 is equivalent to adding 0 to the right side, which does not change the equation. ∎

In general, to complete the square for $x^2 + bx$, we must add $\left(\frac{b}{2}\right)^2$, as illustrated in **FIGURE 11.22**. This technique is shown in Example 7.

READING CHECK 2

- Use the figure to *complete the square* for $x^2 + 8x$.

Completing the Square

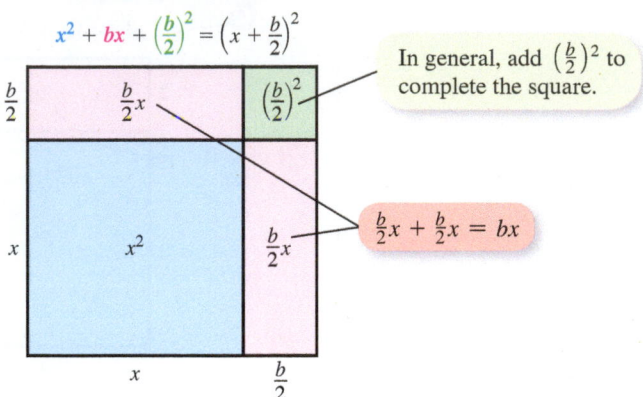

In general, add $\left(\frac{b}{2}\right)^2$ to complete the square.

$\frac{b}{2}x + \frac{b}{2}x = bx$

FIGURE 11.22

EXAMPLE 7 **Writing vertex form**

Write each equation in vertex form. Identify the vertex.
(a) $y = x^2 - 6x - 1$ **(b)** $y = x^2 + 3x + 4$ **(c)** $y = 2x^2 + 4x - 1$

Solution
(a) Because $\left(\frac{b}{2}\right)^2 = \left(\frac{-6}{2}\right)^2 = 9$, add *and* subtract 9 on the right side.

$$y = x^2 - 6x - 1 \qquad \text{Given equation}$$
$$= (x^2 - 6x + 9) - 9 - 1 \qquad \text{Add and subtract 9.}$$
$$= (x - 3)^2 - 10 \qquad \text{Perfect square trinomial}$$

The vertex is $(3, -10)$.

(b) Because $\left(\frac{b}{2}\right)^2 = \left(\frac{3}{2}\right)^2 = \frac{9}{4}$, add *and* subtract $\frac{9}{4}$ on the right side.

$$y = x^2 + 3x + 4 \qquad \text{Given equation}$$
$$= \left(x^2 + 3x + \frac{9}{4}\right) - \frac{9}{4} + 4 \qquad \text{Add and subtract } \tfrac{9}{4}.$$
$$= \left(x + \frac{3}{2}\right)^2 + \frac{7}{4} \qquad \text{Perfect square trinomial}$$

The vertex is $\left(-\frac{3}{2}, \frac{7}{4}\right)$.

(c) This equation is slightly different because the leading coefficient is 2 rather than 1. Start by factoring 2 from the first two terms on the right side.

$$y = 2x^2 + 4x - 1 \qquad \text{Given equation}$$

Now we have the form $x^2 + bx$ inside the parentheses, so we can complete the square.

$$= 2(x^2 + 2x) - 1 \qquad \text{Factor out 2.}$$
$$= 2(x^2 + 2x + 1 - 1) - 1 \qquad \left(\frac{b}{2}\right)^2 = \left(\frac{2}{2}\right)^2 = 1$$
$$= 2(x^2 + 2x + 1) - 2 - 1 \qquad \text{Distributive property: } 2 \cdot (-1)$$
$$= 2(x + 1)^2 - 3 \qquad \text{Perfect square trinomial}$$

The vertex is $(-1, -3)$.

Now Try Exercises 73, 77, 81

4 Modeling with Quadratic Functions

 Math in Context *Airline Safety* A taxiway used by an airplane to exit a runway often contains curves. A curve that is too sharp for the speed of the plane is a safety hazard. The scatterplot shown in **FIGURE 11.23** gives an appropriate radius R of a curve designed for an airplane taxiing at x miles per hour. The data are nonlinear because they do not lie on a line. In this subsection we explain how a quadratic function may be used to model such data. *(Source: FAA.)*

Safe Runway Curve Speed

FIGURE 11.23

The data shown in **FIGURE 11.23** are listed in **TABLE 11.9**.

Safe Runway Curve Speed

x (mph)	10	20	30	40	50	60
R (ft)	50	200	450	800	1250	1800

Source: Federal Aviation Administration.
TABLE 11.9

A second scatterplot of the data is shown in **FIGURE 11.24**. The data may be modeled by $R(x) = ax^2$ for some value a. To illustrate this relation, graph R for different values of a. In **FIGURES 11.25–11.27**, R has been graphed for $a = 2$, -1, and $\frac{1}{2}$, respectively. When $a > 0$, the parabola opens upward, and when $a < 0$, the parabola opens downward. Larger values of $|a|$ make a parabola narrower, whereas smaller values of $|a|$ make the parabola wider. Through trial and error, $a = \frac{1}{2}$ gives a good fit to the data, so $R(x) = \frac{1}{2}x^2$ models the data.

Finding a so $R(x) = ax^2$ Models the Data

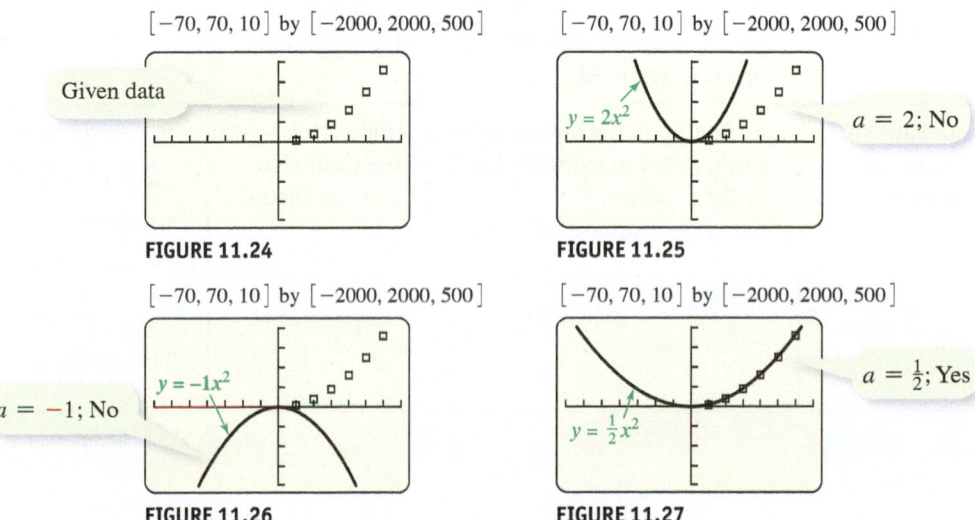

CALCULATOR HELP

To make a scatterplot, see Appendix A (pages AP-3 and AP-4).

$[-70, 70, 10]$ by $[-2000, 2000, 500]$

Given data

FIGURE 11.24

$[-70, 70, 10]$ by $[-2000, 2000, 500]$

$y = 2x^2$ ⟶ $a = 2$; No

FIGURE 11.25

$[-70, 70, 10]$ by $[-2000, 2000, 500]$

$y = -1x^2$

$a = -1$; No

FIGURE 11.26

$[-70, 70, 10]$ by $[-2000, 2000, 500]$

$a = \frac{1}{2}$; Yes

$y = \frac{1}{2}x^2$

FIGURE 11.27

This value of a can also be found *symbolically*, as demonstrated in the next example.

| EXAMPLE 8 | **Modeling safe runway curve speed** |

Find a value for the constant a so that $R(x) = ax^2$ models the data in **TABLE 11.9** on the previous page. Check your result by making a table of values for $R(x)$.

Solution
From **TABLE 11.9**, when $x = 10$ miles per hour, the curve radius is **50** feet. Therefore

$$R(10) = 50 \quad \text{or} \quad a(10)^2 = 50.$$

Solving for a gives

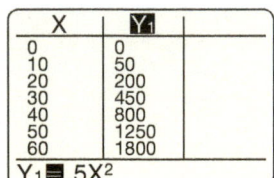

$$a = \frac{50}{10^2} = \frac{1}{2}.$$

To be sure that $R(x) = \frac{1}{2}x^2$ is correct, make a table, as shown in **FIGURE 11.28**. Its values agree with those in **TABLE 11.9**.

FIGURE 11.28

Now Try Exercise 95

11.2   Putting It All Together

CONCEPT	EXPLANATION	EXAMPLES
Translations of Parabolas	Compared to the graph $y = x^2$, the graph of $y = x^2 + k$ is shifted vertically k units and the graph of $y = (x - h)^2$ is shifted horizontally h units.	Compared to the graph of $y = x^2$, the graph of $y = x^2 - 4$ is shifted *downward* 4 units. Compared to the graph of $y = x^2$, the graph of $y = (x - 4)^2$ is shifted *right* 4 units and the graph of $y = (x + 4)^2$ is shifted *left* 4 units.
Vertex Form	The vertex form of the equation of a parabola with vertex (h, k) is $$y = a(x - h)^2 + k,$$ where $a \neq 0$ is a constant. If $a > 0$, the parabola opens upward; if $a < 0$, the parabola opens downward.	The graph of $y = 3(x + 2)^2 - 7$ has a vertex of $(-2, -7)$ and opens upward because $3 > 0$.
Completing the Square Method	To complete the square to obtain the vertex form, add *and* subtract $\left(\frac{b}{2}\right)^2$ on the right side of the equation $y = x^2 + bx + c$. Then factor the perfect square trinomial.	If $y = x^2 + 10x - 3$, then add *and* subtract $\left(\frac{b}{2}\right)^2 = \left(\frac{10}{2}\right)^2 = 25$ on the right side of this equation. $$y = (x^2 + 10x + 25) - 25 - 3$$ $$= (x + 5)^2 - 28$$ The vertex is $(-5, -28)$.

11.2 Exercises

MyMathLab®

CONCEPTS AND VOCABULARY

1. Compared to the graph of $y = x^2$, the graph of $y =$ _____ is shifted upward 2 units.

2. Compared to the graph of $y = x^2$, the graph of $y =$ _____ is shifted to the right 2 units.

3. The vertex of $y = (x - 1)^2 + 2$ is _____

4. The vertex of $y = (x + 1)^2 - 2$ is _____.

5. A quadratic function f may be written either in the form _____ or _____.

6. The vertex form of a parabola is given by _____ and its vertex is _____.

7. The graph of the equation $y = -x^2$ is a parabola that opens _____.

8. The x-coordinate of the vertex of $y = ax^2 + bx + c$ is $x =$ _____.

9. Compared to the graph of $y = x^2$, the graph of $y = x^2 + k$ with $k > 0$ is shifted k units _____.
   a. upward   b. downward
   c. left   d. right

10. Compared to the graph of $y = x^2$, the graph of $y = (x - k)^2$ with $k > 0$ is shifted k units _____.
    a. upward   b. downward
    c. left   d. right

APPLYING TRANSFORMATIONS OF GRAPHS

Exercises 11–18: Graph $f(x)$. Compare the graph of $y = f(x)$ to the graph of $y = x^2$.

11. $f(x) = -x^2$ 12. $f(x) = -2x^2$

13. $f(x) = 2x^2$ 14. $f(x) = 3x^2$

15. $f(x) = \frac{1}{4}x^2$ 16. $f(x) = \frac{1}{2}x^2$

17. $f(x) = -\frac{1}{2}x^2$ 18. $f(x) = -\frac{3}{2}x^2$

RELATING TABLES AND TRANSLATIONS

Exercises 19–22: Complete the table for each translation of $y = x^2$. State what the translation does.

19.

x	-2	-1	0	1	2
$y = x^2$					
$y = x^2 - 3$					

20.

x	-2	-1	0	1	2
$y = x^2$					
$y = x^2 + 5$					

21.

x					
$y = x^2$	4	1	0	1	4

x					
$y = (x - 3)^2$	4	1	0	1	4

22.

x					
$y = x^2$	16	4	0	4	16

x					
$y = (x + 4)^2$	16	4	0	4	16

GRAPHING PARABOLAS

Exercises 23–42: Do the following.

(a) Sketch a graph of the function.
(b) Identify the vertex.
(c) Compare the graph of $y = f(x)$ to the graph of $y = x^2$. (State any transformations used.)

23. $f(x) = x^2 - 4$ 24. $f(x) = x^2 - 1$

25. $f(x) = 2x^2 + 1$ 26. $f(x) = \frac{1}{2}x^2 + 1$

27. $f(x) = (x - 3)^2$ 28. $f(x) = (x + 1)^2$

29. $f(x) = -x^2$ 30. $f(x) = -(x + 2)^2$

31. $f(x) = 2 - x^2$ 32. $f(x) = (x - 1)^2$

33. $f(x) = (x + 2)^2$ 34. $f(x) = (x - 2)^2 - 3$

35. $f(x) = (x + 1)^2 - 2$

36. $f(x) = (x - 3)^2 + 1$

37. $f(x) = (x - 1)^2 + 2$

38. $f(x) = \frac{1}{2}(x + 3)^2 - 3$

39. $f(x) = 2(x - 5)^2 - 4$

40. $f(x) = -3(x + 4)^2 + 5$

41. $f(x) = -\frac{1}{2}(x + 3)^2 + 1$

42. $f(x) = 2(x - 5)^2 + 10$

Exercises 43–46: Compare the graph of $y = f(x)$ to the graph of $y = x^2$. Then sketch a graph of $y = f(x)$ and $y = x^2$ in the same xy-plane.

43. $f(x) = \frac{1}{2}(x - 1)^2 - 2$

44. $f(x) = 2(x + 2)^2 - 1$

45. $f(x) = -2(x + 1)^2 + 3$

46. $f(x) = -\frac{1}{2}(x - 2)^2 + 2$

Exercises 47 and 48: Graph the equation in a window that shows the vertex and all intercepts.

47. $f(x) = -0.4x^2 + 6x - 10$

48. $f(x) = 3x^2 - 40x + 50$

WRITING VERTEX FORM

Exercises 49–52: (Refer to Example 4.) Write the vertex form of a parabola that satisfies the conditions given. Then write the equation in the form $y = ax^2 + bx + c$.

49. Vertex $(3, 4)$ and $a = 3$

50. Vertex $(-1, 3)$ and $a = -5$

51. Vertex $(5, -2)$ and $a = -\frac{1}{2}$

52. Vertex $(-2, -6)$ and $a = \frac{3}{4}$

Exercises 53–56: Write the vertex form of a parabola that satisfies the conditions given. Assume that $a = \pm 1$.

53. Opens upward, vertex $(1, 2)$

54. Opens downward, vertex $(-1, -2)$

55. Opens downward, vertex $(0, -3)$

56. Opens upward, vertex $(5, -4)$

Exercises 57–60: Write the vertex form of the parabola shown in the graph. Assume that $a = \pm 1$.

57.

58.

59.

60.

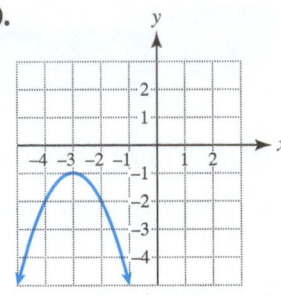

Exercises 61–66: (Refer to Example 5.) Do the following.
(a) Find the vertex on the graph of the equation.
(b) Write the equation in vertex form.

61. $y = 4x^2 - 8x + 5$

62. $y = 3x^2 - 12x + 15$

63. $y = -x^2 - 2x - 3$

64. $y = -x^2 - 4x - 5$

65. $y = -2x^2 - 4x + 1$

66. $y = 2x^2 - 16x + 27$

COMPLETING THE SQUARE

Exercises 67 and 68: Use the given figure to determine what number should be added to the expression to complete the square.

67. $x^2 + 2x$

68. $x^2 + 6x$

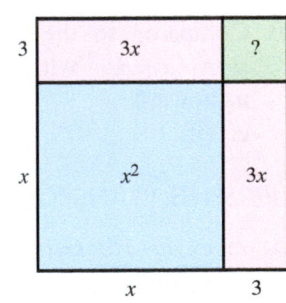

Exercises 69–86: (Refer to Example 7.) Write the equation in vertex form. Identify the vertex.

69. $y = x^2 + 2x$ **70.** $y = x^2 + 6x$

71. $y = x^2 - 4x$ **72.** $y = x^2 - 8x$

73. $y = x^2 + 2x - 3$ **74.** $y = x^2 + 4x + 1$

75. $y = x^2 - 4x + 5$ **76.** $y = x^2 - 8x + 10$

77. $y = x^2 + 3x - 2$ **78.** $y = x^2 + 5x - 4$

79. $y = x^2 - 7x + 1$ **80.** $y = x^2 - 3x + 5$

81. $y = 3x^2 + 6x - 1$ **82.** $y = 2x^2 + 4x - 9$

83. $y = 2x^2 - 3x$ **84.** $y = 3x^2 - 7x$

85. $y = -2x^2 - 8x + 5$ **86.** $y = -3x^2 + 6x + 1$

MODELING DATA

Exercises 87–90: Find a value for the constant a so that $f(x) = ax^2$ models the data. If you are uncertain about your value for a, check it by making a table of values.

87.

x	1	2	3
y	2	8	18

88.

x	−2	0	2
y	6	0	6

89.

x	2	4	6	8
y	1.2	4.8	10.8	19.2

90.

x	5	10	15	20
y	17.5	70	157.5	280

Exercises 91–94: **Modeling Quadratic Data** *Find a quadratic function expressed in vertex form that models the data in the given table. (Hint: Let the first data point in the table be the vertex.)*

91.

x	1	2	3	4
y	−3	−1	5	15

92.

x	−2	−1	0	1	2
y	5	2	−7	−22	−43

93.

x	1980	1990	2000	2010
y	6	56	206	456

94.

x	1990	1995	2000	2005
y	10	60	210	460

95. Braking Distance The table lists approximate braking distances D in feet for cars traveling at x miles per hour on dry, level pavement.

x	12	24	36	48
D	12	48	108	192

(a) Make a scatterplot of the data.
(b) Find a function given by $D(x) = ax^2$ that models these data.

96. Health-Care Costs The table lists approximate *annual* percent increases in the cost of health insurance premiums between 1992 and 2000.

Year	1992	1994	1996	1998	2000
Increase	11%	4%	1%	4%	11%

Source: Kaiser Family Foundation.

(a) Describe what happened to health-care costs from 1992 to 2000.
(b) Can a linear function model these data? Explain.

(c) What type of function might model these data? Explain.
(d) What might be a good choice for the vertex? Explain.
(e) Find a quadratic function C expressed in vertex form that models the data. (Answers may vary.)
(f) Graph C and the data.

97. Sub-Saharan Africa The table lists actual and projected real gross domestic product (GDP) per capita for selected years in sub-Saharan Africa.

Year	1980	1995	2010	2025
Real GDP	$1000	$600	$1000	$2200

Source: IMF WEO, Standard Chartered Research.

(a) Describe what happens to the real GDP.
(b) Can a linear function model these data? Explain.
(c) What type of function might model these data? Explain.
(d) What might be a good choice for the vertex? Explain.
(e) Find a quadratic function C expressed in vertex form that models the data. (Answers may vary.)
(f) Graph C and the data.

98. Tax Theory and Modeling If a government wants to generate enough revenue from income taxes, then it is important to assess the correct tax rates. On the one hand, if tax rates are too low, then enough revenue may not be generated. On the other hand, if taxes are too high, people may work less and not earn as much money. Again, enough revenue may not be generated. The following parabolic graph illustrates this phenomenon.

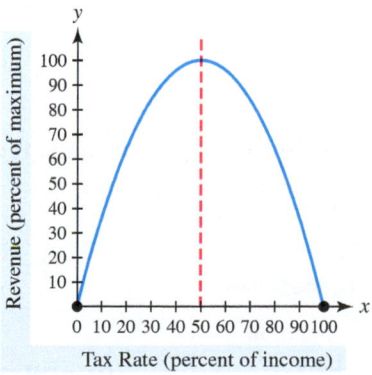

(a) The points $(0, 0)$ and $(100, 0)$ are on this graph. Explain why.
(b) Explain why a linear model is not suitable.
(c) According to *this* graph, what income tax rate maximizes revenue?

WRITING ABOUT MATHEMATICS

WRITING ABOUT MATHEMATICS

99. Explain how to find the vertex of $y = x^2 + bx + c$ by completing the square.

100. If $f(x) = a(x - h)^2 + k$, explain how the values of a, h, and k affect the graph of $y = f(x)$.

101. If $k > 0$, how does the graph of $f(x) = x^2 + k$ compare to the graph of $f(x) = x^2$?

102. A quadratic function can be written in the form $f(x) = ax^2 + bx + c$ or it can be written in the form $f(x) = a(x - h)^2 + k$. Which form is best for identifying the vertex on the graph of $y = f(x)$. Explain.

SECTIONS 11.1 and 11.2 | **Checking Basic Concepts**

1. Graph each quadratic function. Identify the vertex and axis of symmetry.
 (a) $f(x) = x^2 - 2$
 (b) $f(x) = x^2 - 2x - 2$

2. Compare the graph of $y_1 = 2x^2$ to the graph of $y_2 = -\frac{1}{2}x^2$.

3. Find the maximum y-value on the graph of the equation $y = -3x^2 + 12x - 5$.

4. Sketch a graph of $y = f(x)$. Compare this graph to the graph of $y = x^2$.
 (a) $f(x) = (x - 1)^2 + 2$
 (b) $f(x) = -(x + 3)^2$

5. Write the vertex form for each equation.
 (a) $y = x^2 + 14x - 7$
 (b) $y = 4x^2 + 8x - 2$

11.3 Quadratic Equations

Objectives

NEW VOCABULARY

☐ Quadratic equation
☐ Square root property

1 Learning the Basics of Quadratic Equations

RECOGNIZING QUADRATIC EQUATIONS Any quadratic function f can be represented by $f(x) = ax^2 + bx + c$ with $a \neq 0$.

Quadratic Functions

$$f(x) = 2x^2 - 1, \quad g(x) = -\frac{1}{3}x^2 + 2x, \quad h(x) = x^2 + 2x - 1$$

Quadratic functions can be used to write quadratic equations.

Quadratic Equations

$$2x^2 - 1 = 0, \quad -\frac{1}{3}x^2 + 2x = 0, \quad x^2 + 2x - 1 = 0$$

QUADRATIC EQUATION

A **quadratic equation** is an equation that can be written as

$$ax^2 + bx + c = 0,$$

where a, b, and c are constants with $a \neq 0$.

READING CHECK 1

• How can you identify a quadratic equation?

The following See the Concepts explains that a quadratic equation can have zero, one, or two real solutions.

See the Concept 1　　**SOLUTIONS TO QUADRATIC EQUATIONS**

No *x*-intercepts	One *x*-intercept	Two *x*-intercepts
A $f(x) = ax^2 + bx + c, \quad a < 0$	**A** $f(x) = ax^2 + bx + c, \quad a > 0$	**A** $f(x) = ax^2 + bx + c, \quad a > 0$

| **B** $ax^2 + bx + c = 0$ has no solutions. | **B** $ax^2 + bx + c = 0$ has one solution. | **B** $ax^2 + bx + c = 0$ has two solutions. |

A The graph of a quadratic function $f(x) = ax^2 + bx + c$ is either ∪-shaped or ∩-shaped and can intersect the *x*-axis zero, one, or two times.

B The *x*-coordinate of an *x*-intercept corresponds to a real solution to the quadratic equation $ax^2 + bx + c = 0$. Thus a quadratic equation can have zero, one, or two solutions.

WATCH VIDEO IN MML .

NOTE: Some quadratic equations may not have 0 on the right side of the equation, such as $4x^2 = 1$. To solve this quadratic equation graphically, we will sometimes rewrite it as $4x^2 - 1 = 0$ and graph $y = 4x^2 - 1$. Then solutions are the *x*-coordinates of the *x*-intercepts, because $y = 0$ on the *x*-axis. ∎

SOLVING QUADRATIC EQUATIONS (THREE METHODS) We have already solved quadratic equations by factoring, graphing, and constructing tables. In the next example, we apply these three techniques to quadratic equations that have no real solutions, one real solution, and two real solutions.

EXAMPLE 1　　**Solving quadratic equations**

Solve each quadratic equation. Support your results numerically and graphically.
(a) $2x^2 + 1 = 0$　　(No real solutions)
(b) $x^2 + 4 = 4x$　　(One real solution)
(c) $x^2 - 6x + 8 = 0$　　(Two real solutions)

Solution
(a) Symbolic Solution

$$2x^2 + 1 = 0 \qquad \text{Given equation}$$
$$2x^2 = -1 \qquad \text{Subtract 1 from each side.}$$
$$x^2 = -\frac{1}{2} \qquad \text{Divide each side by 2.}$$

This equation has no real-number solutions because $x^2 \geq 0$ for all real numbers *x*.

Numerical and Graphical Solution On the next page, the points in **TABLE 11.10** for $y = 2x^2 + 1$ are plotted in **FIGURE 11.29** and connected with a parabolic graph. The graph of $y = 2x^2 + 1$ has no *x*-intercepts, indicating that there are no real solutions.

No Solutions

$y = 2x^2 + 1$

x	y
-2	9
-1	3
0	1
1	3
2	9

For all x, $y \neq 0$, so there are no solutions.

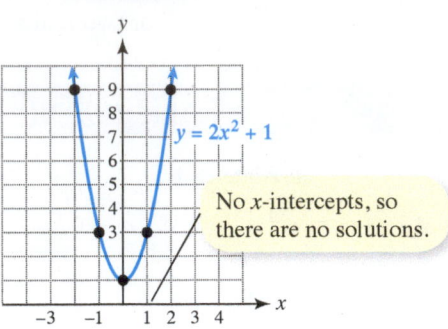

$y = 2x^2 + 1$

No x-intercepts, so there are no solutions.

TABLE 11.10   **FIGURE 11.29**

(b) Symbolic Solution Next, we solve $x^2 + 4 = 4x$.

$$x^2 + 4 = 4x \qquad \text{Given equation}$$
$$x^2 - 4x + 4 = 0 \qquad \text{Subtract } 4x \text{ from each side.}$$
$$(x - 2)(x - 2) = 0 \qquad \text{Factor.}$$
$$x - 2 = 0 \quad \text{or} \quad x - 2 = 0 \qquad \text{Zero-product property}$$
$$x = 2 \qquad \text{There is one solution.}$$

Numerical and Graphical Solution Because the given quadratic equation is equivalent to $x^2 - 4x + 4 = 0$, we let $y = x^2 - 4x + 4$. The points in **TABLE 11.11** are plotted in **FIGURE 11.30** and connected with a parabolic graph. The graph of $y = x^2 - 4x + 4$ has one x-intercept, $(2, 0)$. Note that in **TABLE 11.11**, $y = 0$ when $x = 2$, indicating that the equation has one solution.

One Solution

$y = x^2 - 4x + 4$

x	y
0	4
1	1
2	**0**
3	1
4	4

One solution: **2**

$y = x^2 - 4x + 4$

One x-intercept: $(2, 0)$ so there is one solution: **2**.

TABLE 11.11   **FIGURE 11.30**

(c) Symbolic Solution Next, we solve $x^2 - 6x + 8 = 0$.

$$x^2 - 6x + 8 = 0 \qquad \text{Given equation}$$
$$(x - 2)(x - 4) = 0 \qquad \text{Factor.}$$
$$x - 2 = 0 \quad \text{or} \quad x - 4 = 0 \qquad \text{Zero-product property}$$
$$x = 2 \quad \text{or} \quad x = 4 \qquad \text{There are two solutions.}$$

Numerical and Graphical Solution The points in **TABLE 11.12** for $y = x^2 - 6x + 8$ are plotted in **FIGURE 11.31** and connected with a parabolic graph. The graph of $y = x^2 - 6x + 8$ has two x-intercepts, $(2, 0)$ and $(4, 0)$, indicating two solutions. Note that in **TABLE 11.12** $y = 0$ when $x = 2$ or $x = 4$.

$$y = x^2 - 6x + 8$$

x	y
0	8
1	3
2	0
3	−1
4	0
5	3
6	8

Two solutions: **2** and **4**

Two Solutions

$y = x^2 - 6x + 8$

Two x-intercepts: $(\mathbf{2}, \mathbf{0})$ and $(\mathbf{4}, \mathbf{0})$, so there are two solutions: **2** and **4**.

TABLE 11.12   **FIGURE 11.31**

Now Try Exercises 29, 37, 39

Connecting Concepts with Your Life When a person throws a ball into the air, it follows a parabolic path that opens downward. See **FIGURE 11.32**.

Heights Reached by a Ball

50 feet

40 feet

24 feet

Parabolic path

FIGURE 11.32

If the ball reaches a maximum height of **40** feet, then there would be two different times when the ball is **24** feet above the ground because the ball goes up and then down. However, there would be only one time when the ball is **40** feet above the ground, and never a time when the ball is **50** feet above the ground. If the height of the ball h after t seconds is determined by $h(t) = -16t^2 + 48t + 4$, it follows that

$$-16t^2 + 48t + 4 = 24$$

has two solutions because the ball is 24 feet above the ground at two different times,

$$-16t^2 + 48t + 4 = 40$$

has one solution because the ball is 40 feet above the ground at only one point, and

$$-16t^2 + 48t + 4 = 50$$

has no solutions because the ball is never 50 feet above the ground. A quadratic equation can have zero, one, or two solutions in real life. See Exercises 127–130.

READING CHECK 2

• How many solutions can a quadratic equation have?

2 Applying the Square Root Property

The **square root property** is used to solve quadratic equations that have no x-terms. The following is an example of the square root property.

$$x^2 = 25 \text{ is equivalent to } x = \pm 5.$$

The equation $x = \pm 5$ (read "x equals plus or minus 5") indicates that either $x = 5$ or $x = -5$. Each value is a solution because $(5)^2 = 25$ and $(-5)^2 = 25$.

We can derive this result in general for $k \geq 0$.

$x^2 = k$	Given quadratic equation
$\sqrt{x^2} = \sqrt{k}$	Take the square root of each side.
$|x| = \sqrt{k}$	$\sqrt{x^2} = |x|$ for all x.
$x = \pm\sqrt{k}$	$|x| = b$ implies $x = \pm b, b \geq 0$.

This result is summarized by the *square root property*.

SQUARE ROOT PROPERTY

Let k be a nonnegative number. Then the solutions to the equation

$$x^2 = k$$

are given by $x = \pm\sqrt{k}$. If $k < 0$, then this equation has no real solutions.

EXAMPLE 2   **Using the square root property**

Solve each equation.
(a) $x^2 = 7$ (b) $16x^2 - 9 = 0$ (c) $(x - 4)^2 = 25$

Solution
(a) $x^2 = 7$ is equivalent to $x = \pm\sqrt{7}$ by the square root property. The solutions are $\sqrt{7}$ and $-\sqrt{7}$.

(b)

$16x^2 - 9 = 0$	Given equation
$16x^2 = 9$	Add 9 to each side.
$x^2 = \dfrac{9}{16}$	Divide each side by 16.
$x = \pm\sqrt{\dfrac{9}{16}}$	Square root property
$x = \pm\dfrac{3}{4}$	Simplify.

The solutions are $\frac{3}{4}$ and $-\frac{3}{4}$.

(c)

$(x - 4)^2 = 25$	Given equation
$(x - 4) = \pm\sqrt{25}$	Square root property
$x - 4 = \pm 5$	Simplify.
$x = 4 \pm 5$	Add 4 to each side.
$x = 9 \quad \text{or} \quad x = -1$	Evaluate $4 + 5$ and $4 - 5$.

The solutions are 9 and -1.

Now Try Exercises 51, 53, 57

READING CHECK 3

• Would it be more appropriate to use the square root property to solve $x^2 - x + 3 = 0$ or $x^2 - 9 = 0$? Use the square root property to solve the equation that you chose.

Falling Object

🌐 *Math in Context* If an object is dropped from a height of h feet, its distance d above the ground after t seconds is given by

$$d(t) = h - 16t^2.$$

This formula can be used to estimate the time it takes for a falling object to hit the ground.

EXAMPLE 3   **Modeling a falling object**

A toy falls from a window that is 30 feet above the ground. How long does it take for the toy to hit the ground?

Solution
The height h of the window above the ground is 30 feet, so let $d(t) = 30 - 16t^2$. The toy strikes the ground when the distance d above the ground equals 0.

$$30 - 16t^2 = 0 \qquad \text{Equation to solve for } t$$
$$-16t^2 = -30 \qquad \text{Subtract 30 from each side.}$$
$$t^2 = \frac{30}{16} \qquad \text{Divide each side by } -16.$$
$$t = \pm\sqrt{\frac{30}{16}} \qquad \text{Square root property}$$
$$t = \pm\frac{\sqrt{30}}{4} \qquad \text{Simplify.}$$

Time cannot be negative in this problem, so the appropriate solution is $t = \frac{\sqrt{30}}{4} \approx 1.4$. The toy hits the ground after about 1.4 seconds.

Now Try Exercise 125

3 ▸ Completing the Square

The *method of completing the square* can be used to solve quadratic equations. Because

$$x^2 + bx + \left(\frac{b}{2}\right)^2 = \left(x + \frac{b}{2}\right)^2,$$

we can solve a quadratic equation in the form $x^2 + bx = d$, where b and d are constants, by adding $\left(\frac{b}{2}\right)^2$ to each side and then factoring the resulting perfect square trinomial. For example, in the equation $x^2 + 6x = 7$, we have $b = 6$, so we add $\left(\frac{6}{2}\right)^2 = 9$ to each side.

$$x^2 + 6x = 7 \qquad \text{Given equation}$$
$$x^2 + 6x + 9 = 7 + 9 \qquad \text{Add 9 to each side.}$$
$$(x + 3)^2 = 16 \qquad \text{Perfect square trinomial}$$
$$x + 3 = \pm 4 \qquad \text{Square root property}$$
$$x = -3 \pm 4 \qquad \text{Add } -3 \text{ to each side.}$$
$$x = 1 \quad \text{or} \quad x = -7 \qquad \text{Simplify } -3 + 4 \text{ and } -3 - 4.$$

Now we can factor the perfect square trinomial.

The solutions are 1 and −7. Note that after completing the square, the left side of the equation is a perfect square trinomial. We show how to create one in the next example.

EXAMPLE 4   **Creating a perfect square trinomial**

Find the term that should be added to $x^2 - 10x$ to form a perfect square trinomial.

Solution
The coefficient of the x-term is -10, so we let $b = -10$. To complete the square, we divide b by 2 and then square the result.

$$\left(\frac{b}{2}\right)^2 = \left(\frac{-10}{2}\right)^2 = 25$$

If we add 25 to $x^2 - 10x$, a perfect square trinomial is formed.

$$x^2 - 10x + 25 = (x - 5)^2$$

Now Try Exercise 77

READING CHECK 4

- Use the figure to complete the square for $x^2 + 6x$.

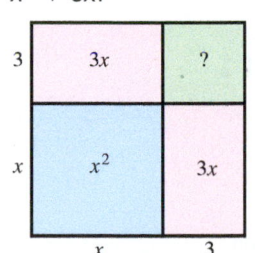

3	3x	?
x	x²	3x
	x	3

The following steps can be used to solve any quadratic equation by completing the square.

SOLVING A QUADRATIC EQUATION BY COMPLETING THE SQUARE

STEP 1: If necessary, use algebra to write the given equation in the form $x^2 + bx = d$, where the coefficient on x^2 is 1 and the constant term is on the right side of the equation.

STEP 2: Add $\left(\frac{b}{2}\right)^2$ to each side of the equation from Step 1.

STEP 3: Factor the resulting perfect square trinomial on the left side of the equation.

STEP 4: Apply the square root property to solve for x.

EXAMPLE 5   Completing the square when the leading coefficient is 1

Solve the equation $x^2 - 4x + 2 = 0$.

Solution
Start by writing the equation in the form $x^2 + bx = d$.

$$x^2 - 4x + 2 = 0 \qquad \text{Given equation}$$
$$x^2 - 4x = -2 \qquad \text{Step 1: subtract 2.}$$
$$x^2 - 4x + 4 = -2 + 4 \qquad \text{Step 2: add } \left(\tfrac{b}{2}\right)^2 = \left(\tfrac{-4}{2}\right)^2 = 4.$$
$$(x - 2)^2 = 2 \qquad \text{Step 3: perfect square trinomial; add.}$$
$$x - 2 = \pm\sqrt{2} \qquad \text{Step 4: square root property}$$
$$x = 2 \pm \sqrt{2} \qquad \text{Add 2.}$$

The solutions are $2 + \sqrt{2} \approx 3.41$ and $2 - \sqrt{2} \approx 0.59$.

Now Try Exercise 83

EXAMPLE 6   Completing the square when the leading coefficient is not 1

Solve the equation $2x^2 + 7x - 5 = 0$.

Solution

$$2x^2 + 7x - 5 = 0 \qquad \text{Given equation}$$
$$2x^2 + 7x = 5 \qquad \text{Step 1: add 5 to each side.}$$

> Be sure to first write the given equation as $x^2 + bx = d$.

$$x^2 + \frac{7}{2}x = \frac{5}{2} \qquad \text{Divide each side by 2.}$$
$$x^2 + \frac{7}{2}x + \frac{49}{16} = \frac{5}{2} + \frac{49}{16} \qquad \text{Step 2: add } \left(\tfrac{b}{2}\right)^2 = \left(\tfrac{7}{4}\right)^2 = \tfrac{49}{16}.$$
$$\left(x + \frac{7}{4}\right)^2 = \frac{89}{16} \qquad \text{Step 3: perfect square trinomial; add.}$$
$$x + \frac{7}{4} = \pm\frac{\sqrt{89}}{4} \qquad \text{Step 4: square root property}$$
$$x = -\frac{7}{4} \pm \frac{\sqrt{89}}{4} \qquad \text{Add } -\tfrac{7}{4}.$$
$$x = \frac{-7 \pm \sqrt{89}}{4} \qquad \text{Combine fractions.}$$

The solutions are $\dfrac{-7 + \sqrt{89}}{4} \approx 0.61$ and $\dfrac{-7 - \sqrt{89}}{4} \approx -4.1$.

Now Try Exercise 91

CRITICAL THINKING 1

What happens if you try to solve

$$2x^2 - 13 = 1$$

by completing the square? What method could you use to solve this problem?

---

4 Solving an Equation for a Variable

FIGURE 11.33

We often need to solve an equation or formula for a variable. For example, the formula $V = \frac{1}{3}\pi r^2 h$ calculates the volume of the cone shown in **FIGURE 11.33**. Let's say that we know the volume V is 120 cubic inches and the height h is 15 inches. We can then find the radius of the cone by solving the equation for r.

$$V = \frac{1}{3}\pi r^2 h \qquad \text{Solve the equation for } r.$$

$$3V = \pi r^2 h \qquad \text{Multiply by 3.}$$

$$\frac{3V}{\pi} = r^2 h \qquad \text{Divide by } \pi.$$

$$\frac{3V}{\pi h} = r^2 \qquad \text{Divide by } h.$$

$$r = \pm\sqrt{\frac{3V}{\pi h}} \qquad \text{Square root property; rewrite.}$$

Because $r \geq 0$, we use the positive or *principal square root*. Thus for $V = \mathbf{120}$ cubic inches and $h = \mathbf{15}$ inches,

$$r = \sqrt{\frac{3(\mathbf{120})}{\pi(\mathbf{15})}} = \sqrt{\frac{24}{\pi}} \approx 2.8 \text{ inches.}$$

EXAMPLE 7   Solving equations for variables

Solve each equation for the specified variable.
(a) $s = -\frac{1}{2}gt^2 + h$ for t **(b)** $d^2 = x^2 + y^2$ for y

Solution
(a) Begin by subtracting h from each side of the equation.

$$s = -\frac{1}{2}gt^2 + h \qquad \text{Solve the equation for } t.$$

$$s - h = -\frac{1}{2}gt^2 \qquad \text{Subtract } h.$$

$$-2(s - h) = gt^2 \qquad \text{Multiply by } -2.$$

$$\frac{2h - 2s}{g} = t^2 \qquad \text{Divide by } g; \text{ simplify.}$$

$$t = \pm\sqrt{\frac{2h - 2s}{g}} \qquad \text{Square root property; rewrite.}$$

(b) Begin by subtracting x^2 from each side of the equation.

$$d^2 = x^2 + y^2 \qquad \text{Solve the equation for } y.$$
$$d^2 - x^2 = y^2 \qquad \text{Subtract } x^2.$$
$$y = \pm\sqrt{d^2 - x^2} \qquad \text{Square root property; rewrite.}$$

Now Try Exercises 115, 117

5 Solving Applications of Quadratic Equations

🌐 *Math in Context* *Airline Safety* In the second section of this chapter, we modeled curves on airport taxiways by using $R(x) = \frac{1}{2}x^2$. In this formula, x represents the airplane's speed in miles per hour, and R represents the radius of the curve in feet. This formula may be used to determine the speed limit for a curve with a radius of **650** feet by solving the following *quadratic equation*. (*Source*: FAA.)

$$\frac{1}{2}x^2 = 650$$

In the next example, we solve this quadratic equation.

EXAMPLE 8   **Finding a safe speed limit**

Solve $\frac{1}{2}x^2 = 650$ and interpret any solutions.

Solution
Use the square root property to solve this problem.

$$\frac{1}{2}x^2 = 650 \qquad \text{Given equation}$$

$$x^2 = 1300 \qquad \text{Multiply by 2.}$$
$$x = \pm\sqrt{1300} \qquad \text{Square root property}$$

The solutions are $\sqrt{1300} \approx 36$ and $-\sqrt{1300} \approx -36$. The solution of $x \approx 36$ indicates that a safe speed limit for a curve with a radius of 650 feet is about 36 miles per hour. (The negative solution has no physical meaning in this problem.)

Now Try Exercise 123

🌐 *Math in Context* *Social Network* For many years, MySpace was the largest social network in the United States. However, during 2010 the number of unique monthly visitors fell dramatically from 70 million in January 2010 to 45 million in January 2011. **TABLE 11.13** lists the number of unique visitors V to MySpace in millions x months after January 2010.

Unique MySpace Visitors (millions)

x (months after January 2010)	0	6	12
V (unique monthly visitors)	70	63	45

Source: comScore.

TABLE 11.13

Decreased 7 million   Decreased 18 million

The number of unique visitors to MySpace decreased faster in the second half of 2010 than it did in the first half, so a linear function will *not* model these data. Instead, these data can be modeled by the *quadratic function*

$$V(x) = -\frac{25}{144}x^2 + 70.$$

EXAMPLE 9   Modeling the decline in MySpace visitors

Use the equation $V(x) = -\frac{25}{144}x^2 + 70$ given in the above discussion to determine the month when MySpace had 56 million unique visitors.

Solution
To determine when there were **56** million visitors, we must solve

$$-\frac{25}{144}x^2 + 70 = \mathbf{56}.$$

There is no x-term in the equation, so we use the square root property.

$$-\frac{25}{144}x^2 + 70 = 56. \qquad \text{Equation to be solved}$$

$$-\frac{25}{144}x^2 = -14 \qquad \text{Subtract 70 from each side.}$$

$$x^2 = \frac{2016}{25} \qquad \text{Multiply each side by } -\frac{144}{25}.$$

$$x = \pm\frac{\sqrt{2016}}{5} \qquad \text{Square root property}$$

Because x must be positive, $x = \frac{\sqrt{2016}}{5} \approx 9$. Thus 9 months after January, or in October, MySpace had 56 million unique visitors.

Now Try Exercise 127

11.3   Putting It All Together

CONCEPT	EXPLANATION	EXAMPLES
Square Root Property	If $k \geq 0$, the solutions to the equation $x^2 = k$ are $\pm\sqrt{k}$.	$x^2 = 100$ is equivalent to $x = \pm 10$ and $x^2 = 13$ is equivalent to $x = \pm\sqrt{13}$. $x^2 = -2$ has no real solutions.
Method of Completing the Square	To solve an equation in the form $x^2 + bx = d$, add $\left(\frac{b}{2}\right)^2$ to each side of the equation. Factor the resulting perfect square trinomial and solve for x by applying the square root property.	To solve $x^2 + 8x = 3$, add $\left(\frac{8}{2}\right)^2 = \mathbf{16}$ to each side. $\quad x^2 + 8x + \mathbf{16} = 3 + \mathbf{16} \qquad$ Add 16. $\quad (x + 4)^2 = 19 \qquad$ Perfect square trinomial $\quad x + 4 = \pm\sqrt{19} \qquad$ Square root property $\quad x = -4 \pm\sqrt{19} \qquad$ Add -4.

11.3   Exercises

CONCEPTS AND VOCABULARY

1. Give an example of a quadratic equation. How many real solutions can a quadratic equation have?

2. Is a quadratic equation a linear equation or a nonlinear equation?

3. Name three symbolic methods that can be used to solve a quadratic equation.

4. Sketch a graph of a quadratic function that has two x-intercepts and opens downward.

5. Sketch a graph of a quadratic function that has no x-intercepts and opens upward.

6. If the graph of $y = ax^2 + bx + c$ intersects the x-axis twice, how many solutions does the equation $ax^2 + bx + c = 0$ have? Explain.

7. Solve $x^2 = 64$. What property did you use?

8. To solve $x^2 + bx = 6$ by completing the square, what value should be added to each side of the equation?

RECOGNIZING QUADRATIC EQUATIONS

Exercises 9–16: Is the given equation quadratic?

9. $x^2 - 3x + 1 = 0$ 10. $2x^2 - 3 = 0$

11. $3x + 1 = 0$ 12. $x^3 - 3x^2 + x = 0$

13. $-3x^2 + x = 16$ 14. $x^2 - 1 = 4x$

15. $x^2 = \sqrt{x} + 1$ 16. $\dfrac{1}{x - 1} = 5$

SOLVING QUADRATIC EQUATIONS

Exercises 17–20: Approximate to the nearest hundredth.

17. (a) $1 \pm \sqrt{7}$ (b) $-2 \pm \sqrt{11}$

18. (a) $\pm\dfrac{\sqrt{3}}{2}$ (b) $\pm\dfrac{2\sqrt{5}}{7}$

19. (a) $\dfrac{3 \pm \sqrt{13}}{5}$ (b) $\dfrac{-5 \pm \sqrt{6}}{9}$

20. (a) $\dfrac{2}{5} \pm \dfrac{\sqrt{5}}{5}$ (b) $-\dfrac{3}{7} \pm \dfrac{\sqrt{3}}{7}$

Exercises 21–24: A graph of $y = ax^2 + bx + c$ is given. Use this graph to solve $ax^2 + bx + c = 0$, if possible.

21.

22.

23.

24.
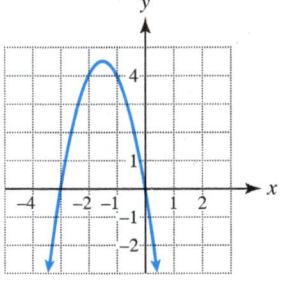

Exercises 25–28: A table of $y = ax^2 + bx + c$ is given. Use this table to solve $ax^2 + bx + c = 0$.

25.

X	Y1
-3	6
-2	0
-1	-4
0	-6
1	-6
2	-4
3	0

Y1 ■X^2−X−6

26.

X	Y1
-6	0
-4	-16
-2	-24
0	-24
2	-16
4	0
6	24

Y1 ■X^2+2X−24

27.

X	Y1
-2	9
-1.5	4
-1	1
-.5	0
0	1
.5	4
1	9

Y1 ■4X^2+4X+1

28.

X	Y1
-4	-2.5
-3	0
-2	1.5
-1	2
0	1.5
1	0
2	-2.5

Y1 ■-X^2/2−X+3/2

Exercises 29–40: Use each method to solve the equation.

(a) Symbolic   (b) Graphical   (c) Numerical

29. $x^2 - 4x - 5 = 0$ 30. $x^2 - x - 6 = 0$

31. $x^2 + 2x = 3$ 32. $x^2 + 4x = 5$

33. $x^2 = 9$ 34. $x^2 = 4$

35. $4x^2 - 4x = 3$ 36. $2x^2 + x = 1$

37. $x^2 + 2x = -1$ 38. $4x^2 - 4x + 1 = 0$

39. $x^2 + 2 = 0$ 40. $-4x^2 - 1 = 2$

Exercises 41–50: Solve by factoring.

41. $x^2 + 2x - 35 = 0$ **42.** $2x^2 - 7x + 3 = 0$

43. $6x^2 - x - 1 = 0$ **44.** $x^2 + 4x + 6 = -3x$

45. $4x^2 + 13x + 9 = x$ **46.** $9x^2 + 4 = 12x$

47. $25x^2 - 350 = 125x$ **48.** $20x^2 + 150 = 130x$

49. $2(5x^2 + 9) = 27x$ **50.** $15(3x^2 + x) = 10$

USING THE SQUARE ROOT PROPERTY

Exercises 51–62: Use the square root property to solve.

51. $x^2 = 144$ **52.** $4x^2 - 5 = 0$

53. $5x^2 - 64 = 0$ **54.** $3x^2 = 7$

55. $(x + 1)^2 = 25$ **56.** $(x + 4)^2 = 9$

57. $(x - 1)^2 = 64$ **58.** $(x - 3)^2 = 0$

59. $(2x - 1)^2 = 5$ **60.** $(5x + 3)^2 = 7$

61. $10(x - 5)^2 = 50$ **62.** $7(3x + 1)^2 = 14$

Exercises 63–72: **Complex Solutions** *Although the equation $x^2 = -1$ has no real solutions, it does have complex solutions. In the previous chapter, we defined the imaginary unit as $i = \sqrt{-1}$, or $i^2 = -1$. Thus the square root property can be used to solve $x^2 = -1$, to obtain $x = \pm\sqrt{-1}$, or $x = \pm i$. Use the square root property to find the complex solutions to each equation.*

63. $x^2 = -4$ **64.** $x^2 = -9$

65. $x^2 = -3$ **66.** $x^2 = -7$

67. $x^2 = -20$ **68.** $x^2 = -8$

69. $x^2 + 16 = 0$ **70.** $x^2 + 25 = 0$

71. $x^2 + 18 = 0$ **72.** $x^2 + 12 = 0$

COMPLETING THE SQUARE

Exercises 73–76: To solve by completing the square, what value should you add to each side of the equation?

73. $x^2 + 4x = -3$ **74.** $x^2 - 6x = 4$

75. $x^2 - 5x = 4$ **76.** $x^2 + 3x = 1$

Exercises 77–80: (Refer to Example 4.) Find the term that should be added to the expression to form a perfect square trinomial. Write the resulting perfect square trinomial in factored form.

77. $x^2 - 8x$ **78.** $x^2 - 5x$

79. $x^2 + 9x$ **80.** $x^2 + x$

Exercises 81–96: Solve by completing the square.

81. $x^2 - 2x = 24$ **82.** $x^2 - 2x + \frac{1}{2} = 0$

83. $x^2 + 6x - 2 = 0$ **84.** $x^2 - 16x = 5$

85. $x^2 - 3x = 5$ **86.** $x^2 + 5x = 2$

87. $x^2 - 5x + 1 = 0$ **88.** $x^2 - 9x + 7 = 0$

89. $x^2 - 4 = 2x$ **90.** $x^2 + 1 = 7x$

91. $2x^2 - 3x = 4$ **92.** $3x^2 + 6x - 5 = 0$

93. $4x^2 - 8x - 7 = 0$ **94.** $25x^2 - 20x - 1 = 0$

95. $36x^2 + 18x + 1 = 0$ **96.** $12x^2 + 8x - 2 = 0$

Exercises 97–106: Solve by any method.

97. $3x^2 + 12x = 36$ **98.** $6x^2 + 9x = 27$

99. $x^2 + 4x = -2$ **100.** $x^2 + 6x + 3 = 0$

101. $3x^2 - 4 = 2$ **102.** $-2x^2 + 3 = 1$

103. $-6x^2 + 70 = 16x$

104. $-15x^2 + 25x + 10 = 0$

105. $-3x(x - 8) = 6$ **106.** $-2x(4 - x) = 8$

Exercises 107 and 108: Thinking Generally *Solve for x. Assume a and c are positive.*

107. $ax^2 - c = 0$ **108.** $ax^2 + bx = 0$

SOLVING EQUATIONS BY MORE THAN ONE METHOD

Exercises 109–114: Solve the quadratic equation

   (a) *symbolically,*
   (b) *graphically, and*
   (c) *numerically.*

109. $x^2 - 3x - 18 = 0$ **110.** $\frac{1}{2}x^2 + 2x - 6 = 0$

111. $x^2 - 8x + 15 = 0$ **112.** $2x^2 + 3 = 7x$

113. $4(x^2 + 35) = 48x$ **114.** $4x(2 - x) = -5$

SOLVING AN EQUATION FOR A VARIABLE

Exercises 115–122: Solve for the specified variable.

115. $x = y^2 - 1$ for y

116. $x = 9y^2$ for y

117. $K = \frac{1}{2}mv^2$ for v

118. $c^2 = a^2 + b^2$ for b

119. $E = \dfrac{k}{r^2}$ for r **120.** $W = I^2R$ for I

121. $LC = \dfrac{1}{(2\pi f)^2}$ for f **122.** $F = \dfrac{KmM}{r^2}$ for r

APPLICATIONS INVOLVING QUADRATIC EQUATIONS

123. Safe Curve Speed (Refer to Example 8.) Find a safe speed limit x for an airport taxiway curve with the given radius R by using $R = \frac{1}{2}x^2$.
(a) $R = 450$ feet   (b) $R = 800$ feet

124. Braking Distance The braking distance y in feet that it takes for a car to stop on wet, level pavement can be estimated by $y = \frac{1}{9}x^2$, where x is the speed of the car in miles per hour. Find the speed associated with each braking distance. (*Source:* L. Haefner, *Introduction to Transportation Systems.*)
(a) 25 feet   (b) 361 feet   (c) 784 feet

125. Falling Object (Refer to Example 3.) How long does it take for a toy to hit the ground if it is dropped out of a window 60 feet above the ground? Does it take twice as long as it takes to fall from a window 30 feet above the ground?

126. Falling Object If a metal ball is thrown *downward* with an initial velocity of 22 feet per second (15 mph) from a 100-foot water tower, its height h in feet above the ground after t seconds is modeled by

$$h(t) = -16t^2 - 22t + 100.$$

(a) Determine symbolically when the height of the ball is 62 feet.
(b) Support your result in part (a) either graphically or numerically.
(c) If the ball is thrown *upward* at 22 feet per second, then its height is given by

$$h(t) = -16t^2 + 22t + 100.$$

Determine when the height of the ball is 80 feet.

Exercises 127–130: **Height of a Ball** *(Refer to the Connecting Concepts with Your Life discussion in this section.) Suppose that a person throws a ball into the air and the ball's height in feet after t seconds is given by $h(t) = -16t^2 + 48t + 4$. If possible, find the time(s) when the ball reaches the given height.*

127. 24 feet   **128.** 40 feet

129. 50 feet   **130.** 4 feet

Exercises 131 and 132: **Television Size** *(Refer to the introduction of this chapter.) The size S of the television screen recommended for a person who sits x feet from the screen $(6 \le x \le 16)$ is given by*

$$S(x) = -0.227x^2 + 8.155x - 8.8.$$

If a person buys a television set with a size S screen, how far from the screen should the person sit?

131. $S = 42$ inches   **132.** $S = 50$ inches

133. Distance Two athletes start jogging at the same time from the same location. One jogs north at 6 miles per hour while the second jogs east at 8 miles per hour. After how long are the two athletes 20 miles apart?

134. Geometry A triangle has an area of 35 square inches, and its base is 3 inches more than its height. Find the base and height of the triangle.

135. Construction A rectangular plot of land has an area of 520 square feet and is 6 feet longer than it is wide.
(a) Write a quadratic equation in the form $ax^2 + bx + c = 0$, whose solution gives the width of the rectangular plot of land.
(b) Solve the equation. What is the width?

136. Modeling Motion The height y in feet of a tennis ball after x seconds is shown in the graph. Estimate when the ball is 25 feet above the ground.

Time (seconds)

137. Seedling Growth The heights of melon seedlings grown at different temperatures are shown in the graph at the top of the next column. At what temperatures are the heights of the seedlings about 22 centimeters?

Temperature (°C)

138. U.S. Population The three tables show the population of the United States in millions from 1800 through 2010 for selected years.

Year	1800	1820	1840	1860
Population	5	10	17	31

Year	1880	1900	1920	1940
Population	50	76	106	132

Year	1960	1980	2000	2010
Population	179	226	269	308

Source: U.S. Census Bureau.

(a) Without plotting the data, how do you know that the data are nonlinear?
(b) These data are modeled (approximately) by

$$f(x) = 0.0066(x - 1800)^2 + 5.$$

Find the vertex of the graph of f and interpret it.
(c) Estimate when the U.S. population reached 85 million.

139. Facebook Visitors The number of unique monthly U.S. visitors V in millions to Facebook x years after January 2011 can be modeled by

$$V(x) = -\frac{1}{2}(x - 9)^2 + 188,$$

where $0 \le x \le 5$. Determine the year when this number of visitors was 170 million.

140. Federal Debt The federal debt D in trillions of dollars held by foreign and international investors x years after 1970 can be modeled by

$$D(x) = \frac{1}{320}x^2 - \frac{3}{80}x,$$

where $15 \le x \le 50$.
(a) Evaluate $D(40)$ and interpret the result.
(b) Determine the year when the federal debt held by foreign and international investors first reached $500 billion ($0.5 trillion).

WRITING ABOUT MATHEMATICS

141. Suppose that you are asked to solve

$$ax^2 + bx + c = 0.$$

Explain how the graph of $y = ax^2 + bx + c$ can be used to find any real solutions to the equation.

142. Explain why a quadratic equation could not have more than two solutions. (*Hint:* Consider the graph of $y = ax^2 + bx + c$.)

Group Activity

Video Games *Exercises 1 and 2: In older video games with two-dimensional graphics, the background is often translated to give the illusion that a character in the game is moving. The simple scene on the left shows a mountain and an airplane. To make it appear that the airplane is flying, the mountain can be translated to the left, as shown in the figure on the right.* (*Source:* C. Pokorny and C. Gerald, *Computer Graphics.*)

1. Suppose that the mountain in the figure on the left is modeled by $f(x) = -0.4x^2 + 4$ and that the airplane is located at the point $(1, 5)$.
(a) Graph f in $[-4, 4, 1]$ by $[0, 6, 1]$, where the units are kilometers. Plot the point $(1, 5)$ to show the location of the airplane.
(b) Assume that the airplane is moving horizontally to the right at 0.2 kilometer per second. To give a video game player the illusion that the airplane is moving, graph the image of the mountain and the position of the airplane after 10 seconds.

2. Discuss how you could create the illusion of the airplane moving to the left and gaining altitude as it passes over the mountain. Try to perform a translation of this type. Explain your reasoning.

11.4   The Quadratic Formula

NEW VOCABULARY

☐ Quadratic formula
☐ Discriminant

1   Solving Quadratic Equations

DERIVATION OF THE QUADRATIC FORMULA   Thus far, we have solved quadratic equations by factoring, the square root property, and completing the square. In this subsection, we derive a formula that can be used to solve *any* quadratic equation. To do this, we solve the general quadratic equation $ax^2 + bx + c = 0$ for x by completing the square. The resulting formula is called the **quadratic formula**. We assume that $a > 0$ and derive this formula as follows.

$ax^2 + bx + c = 0$	Quadratic equation
$ax^2 + bx = -c$	Subtract c.
$x^2 + \dfrac{b}{a}x = -\dfrac{c}{a}$	Divide by a.
$x^2 + \dfrac{b}{a}x + \dfrac{b^2}{4a^2} = -\dfrac{c}{a} + \dfrac{b^2}{4a^2}$	Add $\left(\dfrac{b/a}{2}\right)^2 = \dfrac{b^2}{4a^2}$.
$\left(x + \dfrac{b}{2a}\right)^2 = -\dfrac{c}{a} + \dfrac{b^2}{4a^2}$	Perfect square trinomial
$\left(x + \dfrac{b}{2a}\right)^2 = -\dfrac{c \cdot 4a}{a \cdot 4a} + \dfrac{b^2}{4a^2}$	Multiply $-\dfrac{c}{a}$ by $\dfrac{4a}{4a}$.
$\left(x + \dfrac{b}{2a}\right)^2 = -\dfrac{4ac}{4a^2} + \dfrac{b^2}{4a^2}$	Simplify.
$\left(x + \dfrac{b}{2a}\right)^2 = \dfrac{-4ac + b^2}{4a^2}$	Add fractions.
$\left(x + \dfrac{b}{2a}\right)^2 = \dfrac{b^2 - 4ac}{4a^2}$	Rewrite.
$x + \dfrac{b}{2a} = \pm\sqrt{\dfrac{b^2 - 4ac}{4a^2}}$	Square root property
$x = -\dfrac{b}{2a} \pm \sqrt{\dfrac{b^2 - 4ac}{4a^2}}$	Add $-\dfrac{b}{2a}$.
$x = -\dfrac{b}{2a} \pm \dfrac{\sqrt{b^2 - 4ac}}{2a}$	Property of square roots
$x = \dfrac{-b \pm \sqrt{b^2 - 4ac}}{2a}$	Combine fractions.

STUDY TIP

Memorize the quadratic formula.

QUADRATIC FORMULA

The solutions to $ax^2 + bx + c = 0$ with $a \neq 0$ are given by

$$x = \dfrac{-b \pm \sqrt{b^2 - 4ac}}{2a}.$$

SOLVING EQUATIONS WITH THE QUADRATIC FORMULA   The quadratic formula can be used to solve *any* quadratic equation. It always "works."

EXAMPLE 1   **Solving a quadratic equation that has two solutions**

Solve the equation $2x^2 - 3x - 1 = 0$. Support your results graphically.

Solution

Symbolic Solution Note that $a = 2$, $b = -3$, and $c = -1$ in $2x^2 - 3x - 1 = 0$.

$$x = \frac{-b \pm \sqrt{b^2 - 4ac}}{2a} \qquad \text{Quadratic formula}$$

$$x = \frac{-(-3) \pm \sqrt{(-3)^2 - 4(2)(-1)}}{2(2)} \qquad \text{Substitute for } a, b, \text{ and } c.$$

$$x = \frac{3 \pm \sqrt{17}}{4} \qquad \text{Simplify.}$$

The solutions are $\frac{3 + \sqrt{17}}{4} \approx 1.78$ and $\frac{3 - \sqrt{17}}{4} \approx -0.28$.

Graphical Solution The graph of $y = 2x^2 - 3x - 1$ is shown in **FIGURE 11.34**. Note that the two x-intercepts correspond to the two solutions for $2x^2 - 3x - 1 = 0$. Estimating from this graph, we see that the solutions are approximately -0.25 and 1.75, which supports our symbolic solution. (You could also use a graphing calculator to find the x-intercepts.)

Two Solutions

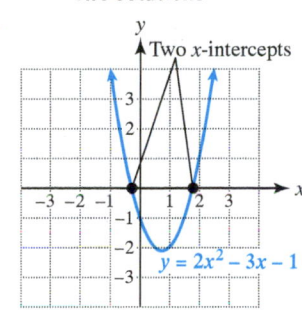

FIGURE 11.34

Now Try Exercise 9

CRITICAL THINKING 1

Use the equation and results from Example 1 to evaluate each expression mentally.

$$2\left(\frac{3 + \sqrt{17}}{4}\right)^2 - 3\left(\frac{3 + \sqrt{17}}{4}\right) - 1 \quad \text{and} \quad 2\left(\frac{3 - \sqrt{17}}{4}\right)^2 - 3\left(\frac{3 - \sqrt{17}}{4}\right) - 1 \qquad \text{Both equal 0.}$$

EXAMPLE 2   **Solving a quadratic equation that has one solution**

Solve the equation $25x^2 + 20x + 4 = 0$. Support your result graphically.

Solution

Symbolic Solution Note that $a = 25$, $b = 20$, and $c = 4$ in $25x^2 + 20x + 4 = 0$.

$$x = \frac{-b \pm \sqrt{b^2 - 4ac}}{2a} \qquad \text{Quadratic formula}$$

$$= \frac{-20 \pm \sqrt{20^2 - 4(25)(4)}}{2(25)} \qquad \text{Substitute for } a, b, \text{ and } c.$$

$$= \frac{-20 \pm \sqrt{0}}{50} \qquad \text{Simplify.}$$

$$= \frac{-20}{50} = -0.4 \qquad \sqrt{0} = 0$$

There is one solution, -0.4.

One Solution

FIGURE 11.35

Graphical Solution The graph of $y = 25x^2 + 20x + 4$ is shown in **FIGURE 11.35**. Note that the one x-intercept, $(-0.4, 0)$, corresponds to the solution to $25x^2 + 20x + 4 = 0$.

Now Try Exercise 11

EXAMPLE 3 **Recognizing a quadratic equation that has no real solutions**

Solve the equation $5x^2 - x + 3 = 0$. Support your result graphically.

Solution

Symbolic Solution Note that $a = \mathbf{5}$, $b = \mathbf{-1}$, and $c = \mathbf{3}$ in $\mathbf{5}x^2 - \mathbf{1}x + \mathbf{3} = 0$.

$$x = \frac{-b \pm \sqrt{b^2 - 4ac}}{2a} \qquad \text{Quadratic formula}$$

$$= \frac{-(\mathbf{-1}) \pm \sqrt{(\mathbf{-1})^2 - 4(\mathbf{5})(\mathbf{3})}}{2(\mathbf{5})} \qquad \text{Substitute for } a, b, \text{ and } c.$$

$$= \frac{1 \pm \sqrt{-59}}{10} \qquad \text{Simplify.}$$

No Real Solutions

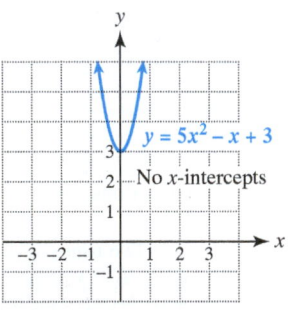

FIGURE 11.36

There are *no real solutions* to this equation because $\sqrt{-59}$ *is not a real number*. (Later in this section, we discuss how to find complex solutions to quadratic equations like this one.)

Graphical Solution The graph of $y = 5x^2 - x + 3$ is shown in **FIGURE 11.36**. There are no *x*-intercepts, indicating that the equation $5x^2 - x + 3 = 0$ has no real solutions.

Now Try Exercise 13

The following Making Connections shows three ways to solve a quadratic equation.

MAKING CONNECTIONS 1

Three Ways to Solve $x^2 - x - 2 = 0$

Symbolic Solution

Solve $x^2 - x - 2 = 0$.

$$x = \frac{-(-1) \pm \sqrt{(-1)^2 - 4(1)(-2)}}{2(1)}$$

$$x = \frac{1 \pm 3}{2}$$

$$x = 2, -1$$

Solutions are $-\mathbf{1}$ and $\mathbf{2}$.

Numerical Solution

$x^2 - x - 2 = 0$ when

$x = -\mathbf{1}$ or $\mathbf{2}$.

Graphical Solution

x-intercepts: $(-1, 0), (2, 0)$

The *x*-intercepts are $(-\mathbf{1}, 0)$ and $(\mathbf{2}, 0)$.

The solutions are $-\mathbf{1}$ and $\mathbf{2}$.

Each method gives the same solutions.

 Connecting Concepts with Your Life To model the stopping distance of a car, we need to compute two quantities. The first quantity is the *reaction distance*, which is the distance a car travels from the time a driver first recognizes a hazard until the brakes are applied. The second quantity is *braking distance*, which is the distance a car travels after a driver applies the brakes. *Stopping distance* equals the sum of the reaction distance and the braking distance. If a car is traveling x miles per hour, we can estimate the reaction distance in feet as $\frac{11}{3}x$ and the braking distance in feet as $\frac{1}{9}x^2$. See **FIGURE 11.37** at the top of the next page.

(*Source:* L. Haefoer *Introduction to Transportation Systems.*)

Stopping Distance

FIGURE 11.37

To estimate the total stopping distance d in feet, add the two expressions to obtain

$$d(x) = \frac{1}{9}x^2 + \frac{11}{3}x.$$

If a car's headlights don't illuminate the road beyond **500** feet, a safe nighttime speed limit x for the car can be determined by solving the quadratic equation

$$\frac{1}{9}x^2 \quad + \quad \frac{11}{3}x \quad = \quad 500.$$

Quadratic equation

Braking Distance + Reaction Distance = Stopping Distance

EXAMPLE 4   **Modeling stopping distance**

If a car's headlights do not illuminate the road beyond 500 feet, estimate a safe nighttime speed limit x for the car by solving $\frac{1}{9}x^2 + \frac{11}{3}x = 500$.

Solution
Begin by subtracting 500 from each side of the given equation.

$$\frac{1}{9}x^2 + \frac{11}{3}x - 500 = 0 \quad \text{Subtract 500.}$$

To eliminate fractions, multiply each side by the LCD, which is 9. (This step is not necessary, but it makes the problem easier to work.)

$$x^2 + 33x - 4500 = 0 \quad \text{Multiply by 9.}$$

Now let $a = 1$, $b = 33$, and $c = -4500$ in the quadratic formula.

$$x = \frac{-b \pm \sqrt{b^2 - 4ac}}{2a} \quad \text{Quadratic formula}$$

$$= \frac{-33 \pm \sqrt{33^2 - 4(1)(-4500)}}{2(1)} \quad \text{Substitute for } a, b, \text{ and } c.$$

$$= \frac{-33 \pm \sqrt{19{,}089}}{2} \quad \text{Simplify.}$$

The solutions are

$$\frac{-33 + \sqrt{19{,}089}}{2} \approx 52.6 \quad \text{and} \quad \frac{-33 - \sqrt{19{,}089}}{2} \approx -85.6.$$

The negative solution has no physical meaning because negative speeds are not possible. The other solution is 52.6, so an appropriate speed limit might be 50 miles per hour.

Now Try Exercise 115

STUDY TIP

The quadratic formula can be used to solve *any* quadratic equation.

2 Finding and Applying the Discriminant

The expression $b^2 - 4ac$ in the quadratic formula is called the **discriminant**. It provides information about the number of solutions to a quadratic equation.

THE DISCRIMINANT AND QUADRATIC EQUATIONS

To determine the number of solutions to the quadratic equation $ax^2 + bx + c = 0$, evaluate the discriminant $b^2 - 4ac$.

1. If $b^2 - 4ac > 0$, there are two real solutions.
2. If $b^2 - 4ac = 0$, there is one real solution.
3. If $b^2 - 4ac < 0$, there are no real solutions; there are two complex solutions.

READING CHECK 1

- How many real solutions does $ax^2 + bx + c = 0$ have, if $b^2 - 4ac = 0$?

The following See the Concept shows how the discriminant gives information about the graph of $y = ax^2 + bx + c$ and the quadratic equation $ax^2 + bx + c = 0$.

See the Concept 1   GRAPHS, EQUATIONS, AND THE DISCRIMINANT

$b^2 - 4ac < 0$: No x-intercepts

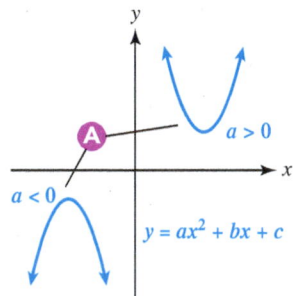

$b^2 - 4ac = 0$: One x-intercept

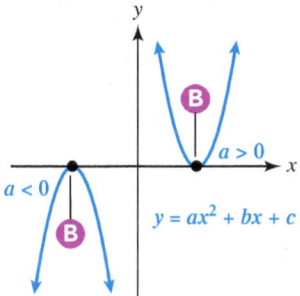

$b^2 - 4ac > 0$: Two x-intercepts

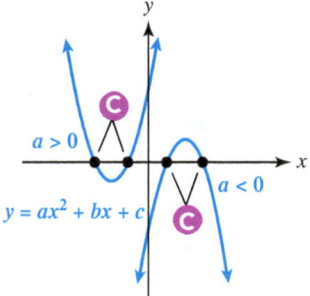

A When $b^2 - 4ac < 0$, the graph of $y = ax^2 + bx + c$ has no x-intercepts and the quadratic equation $ax^2 + bx + c = 0$ has no real solutions.

B When $b^2 - 4ac = 0$, the graph of $y = ax^2 + bx + c$ has one x-intercept and the quadratic equation $ax^2 + bx + c = 0$ has one real solution.

C When $b^2 - 4ac > 0$, the graph of $y = ax^2 + bx + c$ has two x-intercepts and the quadratic equation $ax^2 + bx + c = 0$ has two real solutions.

WATCH VIDEO IN MML.

EXAMPLE 5   Analyzing graphs of quadratic functions

A graph of $f(x) = ax^2 + bx + c$ is shown in **FIGURE 11.38**.
(a) State whether $a > 0$ or $a < 0$.
(b) Solve the equation $ax^2 + bx + c = 0$.
(c) Determine whether the discriminant is positive, negative, or zero.

FIGURE 11.38

Solution
(a) The parabola opens downward, so $a < 0$.
(b) The graph of $f(x) = ax^2 + bx + c$ intersects the x-axis at $(-3, 0)$ and $(2, 0)$. So $f(-3) = 0$ and $f(2) = 0$. The solutions to $ax^2 + bx + c = 0$ are -3 and 2.
(c) There are two solutions, so the discriminant is positive.

Now Try Exercise 43

EXAMPLE 6 **Using the discriminant**

Use the discriminant to determine the number of solutions to $4x^2 + 25 = 20x$. Then solve the equation, using the quadratic formula.

Solution
Write the equation as $4x^2 - 20x + 25 = 0$ so that $a = 4$, $b = -20$, and $c = 25$. The discriminant evaluates to

$$b^2 - 4ac = (-20)^2 - 4(4)(25) = 0.$$

Thus there is **one real solution**.

$$x = \frac{-b \pm \sqrt{b^2 - 4ac}}{2a} \qquad \text{Quadratic formula}$$

$$= \frac{-(-20) \pm \sqrt{0}}{2(4)} \qquad \text{Substitute.}$$

$$= \frac{20}{8} = 2.5 \qquad \text{Simplify.}$$

The only solution is 2.5.

Now Try Exercise 49(a), (b)

3 Using Intercepts to Graph Quadratic Functions

Instead of using vertex form to graph a quadratic function, we can first plot key points on the parabola to determine its position in the xy-plane. Key points include any intercepts and the vertex. The discriminant can be used to determine the number of x-intercepts. This method is demonstrated in the next example.

EXAMPLE 7 **Analyzing the graph of $y = ax^2 + bx + c$**

Let $f(x) = -\frac{1}{2}x^2 + x + \frac{3}{2}$.
(a) Does the graph of $y = f(x)$ open upward or downward? Is this graph wider or narrower than the graph of $y = x^2$?
(b) Find the axis of symmetry and the vertex.
(c) Find the discriminant and determine the number of x-intercepts.
(d) Find the y-intercept and any x-intercepts.
(e) Sketch a graph of $y = f(x)$.

Solution
(a) If $f(x) = -\frac{1}{2}x^2 + x + \frac{3}{2}$, then $a = -\frac{1}{2}$, $b = 1$, and $c = \frac{3}{2}$. Because $a = -\frac{1}{2} < 0$, the parabola opens downward. Also, because $0 < |a| < 1$, the graph is wider than the graph of $y = x^2$.

(b) The axis of symmetry is $x = -\frac{b}{2a} = -\frac{1}{2\left(-\frac{1}{2}\right)} = 1$, or $x = 1$. Because

$$f(1) = -\frac{1}{2}(1)^2 + (1) + \frac{3}{2} = -\frac{1}{2} + 1 + \frac{3}{2} = 2,$$

the vertex is $(1, 2)$.

(c) To find the x-intercepts, we must find the solutions to

$$-\tfrac{1}{2}x^2 + x + \tfrac{3}{2} = 0.$$

Because $a = -\tfrac{1}{2}$, $b = 1$, and $c = \tfrac{3}{2}$, the discriminant equals

$$b^2 - 4ac = (1)^2 - 4\left(-\tfrac{1}{2}\right)\left(\tfrac{3}{2}\right) = 4 > 0.$$

Thus there are two x-intercepts to plot.

(d) To find the y-intercept, evaluate $f(0)$ because $x = 0$ at the y-intercept.

$$f(0) = -\tfrac{1}{2}(0)^2 + (0) + \tfrac{3}{2} = \tfrac{3}{2}$$

The y-intercept is $\left(0, \tfrac{3}{2}\right)$. Factoring, completing the square, or the quadratic formula can be used to find the x-intercepts. We use factoring and let $y = 0$ to solve for x.

$$-\tfrac{1}{2}x^2 + x + \tfrac{3}{2} = 0 \qquad \text{Equation to be solved}$$
$$x^2 - 2x - 3 = 0 \qquad \text{Multiply by } -2; \text{ clear fractions.}$$
$$(x + 1)(x - 3) = 0 \qquad \text{Factor.}$$
$$x + 1 = 0 \quad \text{or} \quad x - 3 = 0 \qquad \text{Zero-product property}$$
$$x = -1 \quad \text{or} \quad x = 3 \qquad \text{Solve.}$$

The x-intercepts are $(-1, 0)$ and $(3, 0)$

(e) Start by plotting the vertex and intercepts, as shown in **FIGURE 11.39**. Then sketch a smooth, $\cap$-shaped graph that connects these points.

FIGURE 11.39

Now Try Exercise 57

4 Solving Quadratic Equations That Have Complex Solutions

A quadratic equation written in the form $ax^2 + bx + c = 0$ has no real solutions if the discriminant, $b^2 - 4ac$, is negative. For example, the quadratic equation given by $x^2 + 4 = 0$ has $a = 1$, $b = 0$, and $c = 4$. Its discriminant is

$$b^2 - 4ac = 0^2 - 4(1)(4) = -16 < 0,$$

so this equation has no real solutions. However, if we use complex numbers, we can solve this equation as follows.

$$x^2 + 4 = 0 \qquad \text{Given equation}$$
$$x^2 = -4 \qquad \text{Subtract 4.}$$
$$x = \pm\sqrt{-4} \qquad \text{Square root property}$$
$$x = \sqrt{-4} \quad \text{or} \quad x = -\sqrt{-4} \qquad \text{Meaning of } \pm$$
$$x = 2i \quad \text{or} \quad x = -2i \qquad \text{The expression } \sqrt{-a}$$

No x-intercepts

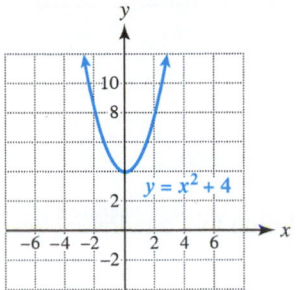

FIGURE 11.40

The solutions are $\pm 2i$. We check each solution to $x^2 + 4 = 0$ as follows.

$$(2i)^2 + 4 = (2)^2 i^2 + 4 = 4(-1) + 4 = 0 \checkmark \quad \text{It checks.}$$
$$(-2i)^2 + 4 = (-2)^2 i^2 + 4 = 4(-1) + 4 = 0 \checkmark \quad \text{It checks.}$$

The fact that the equation $x^2 + 4 = 0$ has only imaginary solutions is apparent from the graph of $y = x^2 + 4$, shown in **FIGURE 11.40**. This parabola does not intersect the x-axis, so the equation $x^2 + 4 = 0$ has no real solutions.

These results can be generalized as follows.

THE EQUATION $x^2 + k = 0$

If $k > 0$, the solutions to $x^2 + k = 0$ are given by $x = \pm i\sqrt{k}$.

NOTE: This result is a form of the *square root property* that includes complex solutions. ∎

EXAMPLE 8 **Solving a quadratic equation that has complex solutions**

Solve $x^2 + 5 = 0$.

Solution
The solutions are $\pm i\sqrt{5}$. That is, $x = i\sqrt{5}$ or $x = -i\sqrt{5}$.

Now Try Exercise 63

When $b \neq 0$, the preceding method cannot be used. Consider the quadratic equation $2x^2 + x + 3 = 0$, which has $a = 2$, $b = 1$, and $c = 3$. Its discriminant is negative.

$$b^2 - 4ac = 1^2 - 4(2)(3) = -23 < 0$$

This equation has two complex solutions, as demonstrated in the next example.

EXAMPLE 9 **Solving a quadratic equation that has complex solutions**

Solve $2x^2 + x + 3 = 0$. Write your answer in standard form: $a + bi$.

Solution
Let $a = 2$, $b = 1$, and $c = 3$.

CALCULATOR HELP

To set your calculator in $a + bi$ mode or to access the imaginary unit i, see Appendix A (page AP-7).

$$x = \frac{-b \pm \sqrt{b^2 - 4ac}}{2a} \qquad \text{Quadratic formula}$$

$$= \frac{-1 \pm \sqrt{1^2 - 4(2)(3)}}{2(2)} \qquad \text{Substitute for } a, b, \text{ and } c.$$

$$= \frac{-1 \pm \sqrt{-23}}{4} \qquad \text{Simplify.}$$

$$= \frac{-1 \pm i\sqrt{23}}{4} \qquad \sqrt{-23} = i\sqrt{23}$$

$$= -\frac{1}{4} \pm i\frac{\sqrt{23}}{4} \qquad \text{Property of fractions}$$

The solutions are $-\frac{1}{4} + i\frac{\sqrt{23}}{4}$ and $-\frac{1}{4} - i\frac{\sqrt{23}}{4}$.

Now Try Exercise 79

CRITICAL THINKING 2

Use the results of Example 9 to evaluate each expression mentally.

$$2\left(-\tfrac{1}{4} + i\tfrac{\sqrt{23}}{4}\right)^2 + \left(-\tfrac{1}{4} + i\tfrac{\sqrt{23}}{4}\right) + 3 \quad \text{and} \quad 2\left(-\tfrac{1}{4} - i\tfrac{\sqrt{23}}{4}\right)^2 + \left(-\tfrac{1}{4} - i\tfrac{\sqrt{23}}{4}\right) + 3$$

Both expressions equal 0.

Sometimes we can use properties of radicals to simplify a solution to a quadratic equation, as demonstrated in the next example.

EXAMPLE 10 **Solving a quadratic equation that has complex solutions**

Solve $\tfrac{3}{4}x^2 + 1 = x$. Write your answer in standard form: $a + bi$.

Solution
Begin by subtracting x from each side of the equation and then multiply by 4 to clear fractions. The resulting equation is $3x^2 - 4x + 4 = 0$. Substitute $a = 3$, $b = -4$, and $c = 4$ in the quadratic formula.

$$x = \frac{-b \pm \sqrt{b^2 - 4ac}}{2a} \qquad \text{Quadratic formula}$$

$$= \frac{-(-4) \pm \sqrt{(-4)^2 - 4(3)(4)}}{2(3)} \qquad \text{Substitute.}$$

$$= \frac{4 \pm \sqrt{-32}}{6} \qquad \text{Simplify.}$$

$$= \frac{4 \pm 4i\sqrt{2}}{6} \qquad \sqrt{-32} = i\sqrt{32} = i\sqrt{16}\sqrt{2} = 4i\sqrt{2}$$

$$= \frac{2}{3} \pm \frac{2}{3}i\sqrt{2} \qquad \text{Property of fractions; simplify.}$$

Now Try Exercise 85

In the next example, we use completing the square to obtain complex solutions.

EXAMPLE 11 **Completing the square to find complex solutions**

Solve $x(x + 2) = -2$ by completing the square.

Solution
After applying the distributive property, we obtain the equation $x^2 + 2x = -2$. Because $b = 2$, add $\left(\tfrac{b}{2}\right)^2 = \left(\tfrac{2}{2}\right)^2 = 1$ to each side of the equation.

$$x^2 + 2x = -2 \qquad \text{Equation to be solved}$$

$$x^2 + 2x + 1 = -2 + 1 \qquad \text{Add 1 to each side.}$$

$$(x + 1)^2 = -1 \qquad \text{Perfect square trinomial; add.}$$

$$x + 1 = \pm\sqrt{-1} \qquad \text{Square root property}$$

$$x + 1 = \pm i \qquad \sqrt{-1} = i, \text{ the imaginary unit}$$

$$x = -1 \pm i \qquad \text{Add } -1 \text{ to each side.}$$

The solutions are $-1 + i$ and $-1 - i$.

Now Try Exercise 97

11.4 Putting It All Together

CONCEPT	EXPLANATION	EXAMPLES
Quadratic Formula	The quadratic formula can be used to solve *any* quadratic equation written as $ax^2 + bx + c = 0$. The solutions are given by $$x = \frac{-b \pm \sqrt{b^2 - 4ac}}{2a}.$$	For the equation $$2x^2 - 3x + 1 = 0$$ with $a = 2$, $b = -3$, and $c = 1$, the solutions are $$\frac{-(-3) \pm \sqrt{(-3)^2 - 4(2)(1)}}{2(2)} = \frac{3 \pm \sqrt{1}}{4} = 1, \frac{1}{2}.$$
The Discriminant	The expression $b^2 - 4ac$ is called the discriminant. 1. $b^2 - 4ac > 0$ indicates two real solutions. 2. $b^2 - 4ac = 0$ indicates one real solution. 3. $b^2 - 4ac < 0$ indicates no real solutions; rather, there are two complex solutions.	For the equation $$x^2 + 4x - 1 = 0$$ with $a = 1$, $b = 4$, and $c = -1$, the discriminant is $$b^2 - 4ac = 4^2 - 4(1)(-1) = 20 > 0,$$ indicating two real solutions.
Quadratic Formula and Complex Solutions	If the discriminant is negative $(b^2 - 4ac < 0)$, the two solutions are complex numbers that are not real numbers. If $k > 0$, the solutions to $x^2 + k = 0$ are given by $x = \pm i\sqrt{k}$.	Solve $2x^2 - x + 3 = 0$. $$x = \frac{-(-1) \pm \sqrt{(-1)^2 - 4(2)(3)}}{2(2)}$$ $$= \frac{1 \pm \sqrt{-23}}{4} = \frac{1}{4} \pm i\frac{\sqrt{23}}{4}$$ $x^2 + 7 = 0$ is equivalent to $x = \pm i\sqrt{7}$.

11.4 Exercises

MyMathLab®

CONCEPTS AND VOCABULARY

1. What is the quadratic formula used for?

2. What basic algebraic technique is used to derive the quadratic formula?

3. Write the discriminant.

4. If the discriminant evaluates to 0, what does that indicate about the quadratic equation?

5. Name four symbolic techniques for solving a quadratic equation.

6. Does every quadratic equation have at least one real solution? Explain.

7. Solve $x^2 - k = 0$, if $k > 0$.

8. Solve $x^2 + k = 0$, if $k > 0$.

USING THE QUADRATIC FORMULA

Exercises 9–14: Use the quadratic formula to solve the equation. Support your result graphically. If there are no real solutions, say so.

9. $2x^2 + 11x - 6 = 0$ 10. $x^2 + 2x - 24 = 0$

11. $-x^2 + 2x - 1 = 0$ 12. $3x^2 - x + 1 = 0$

13. $-2x^2 + x - 1 = 0$ 14. $-x^2 + 4x - 4 = 0$

Exercises 15–32: Solve by using the quadratic formula. If there are no real solutions, say so.

15. $x^2 - 6x - 16 = 0$ 16. $2x^2 - 9x + 7 = 0$

17. $4x^2 - x - 1 = 0$ 18. $-x^2 + 2x + 1 = 0$

19. $-3x^2 + 2x - 1 = 0$ 20. $x^2 + x + 3 = 0$

21. $36x^2 - 36x + 9 = 0$ **22.** $4x^2 - 5.6x + 1.96 = 0$

23. $2x(x - 3) = 2$ **24.** $x(x + 1) + x = 5$

25. $(x - 1)(x + 1) + 2 = 4x$

26. $\frac{1}{2}(x - 6) = x^2 + 1$ **27.** $\frac{1}{2}x(x + 1) = 2x^2 - \frac{3}{2}$

28. $\frac{1}{2}x^2 - \frac{1}{4}x + \frac{1}{2} = x$ **29.** $2x(x - 1) = 7$

30. $3x(x - 4) = 4$ **31.** $-3x^2 + 10x - 5 = 0$

32. $-2x^2 + 4x - 1 = 0$

Exercises 33–42: Use the quadratic formula to find any x-intercepts on the graph of the function.

33. $f(x) = x^2 - 2x - 1$

34. $f(x) = x^2 + 3x + 1$

35. $f(x) = -2x^2 - x + 3$

36. $f(x) = -3x^2 - x + 4$

37. $f(x) = x^2 + x + 5$

38. $f(x) = 3x^2 - 2x + 5$

39. $f(x) = x^2 + 9$ **40.** $f(x) = x^2 + 11$

41. $f(x) = 3x^2 + 4x - 2$

42. $f(x) = 4x^2 - 2x - 3$

IDENTIFYING THE DISCRIMINANT

Exercises 43–48: A graph of $y = ax^2 + bx + c$ is shown.

   (a) *State whether $a > 0$ or $a < 0$.*
   (b) *Solve $ax^2 + bx + c = 0$, if possible.*
   (c) *Determine whether the discriminant is positive, negative, or zero.*

43.

44.

45.

46.

47.

48.
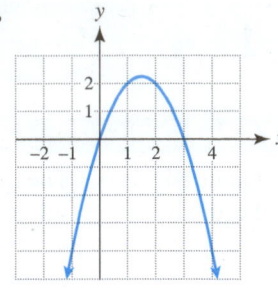

Exercises 49–56: Do the following for the given equation.

   (a) *Evaluate the discriminant.*
   (b) *How many real solutions are there?*
   (c) *Support your answer for part (b) graphically.*

49. $3x^2 + x - 2 = 0$ **50.** $5x^2 - 13x + 6 = 0$

51. $x^2 - 4x + 4 = 0$ **52.** $\frac{1}{4}x^2 + 4 = 2x$

53. $\frac{1}{2}x^2 + \frac{3}{2}x + 2 = 0$ **54.** $x - 3 = 2x^2$

55. $x(x + 3) = 3$

56. $(4x - 1)(x - 3) = -25$

GRAPHING QUADRATIC FUNCTIONS

Exercises 57–62: (Refer to Example 7.) Use the given $f(x)$ to complete the following.

   (a) *Does the graph of $y = f(x)$ open upward or downward? Is this graph wider, narrower, or the same as the graph of $y = x^2$?*
   (b) *Find the axis of symmetry and the vertex.*
   (c) *Find the discriminant and determine the number of x-intercepts.*
   (d) *Find the y-intercept and any x-intercepts.*
   (e) *Sketch a graph of $y = f(x)$.*

57. $f(x) = \frac{1}{2}x^2 + x - \frac{3}{2}$ **58.** $f(x) = -x^2 + 4x + 5$

59. $f(x) = 2x - x^2$ **60.** $f(x) = x - 2x^2$

61. $f(x) = 2x^2 + 2x - 4$ **62.** $f(x) = \frac{1}{2}x^2 - \frac{1}{2}x - 1$

FINDING COMPLEX SOLUTIONS

Exercises 63–94: Solve the equation. Write complex solutions in standard form.

63. $x^2 + 9 = 0$ **64.** $x^2 + 16 = 0$

65. $x^2 + 80 = 0$ **66.** $x^2 + 20 = 0$

67. $x^2 + \frac{1}{4} = 0$ **68.** $x^2 + \frac{9}{4} = 0$

69. $16x^2 + 9 = 0$ **70.** $25x^2 + 36 = 0$

71. $x^2 = -6$ **72.** $x^2 = -75$

73. $x^2 - 3 = 0$

74. $x^2 - 8 = 0$

75. $x^2 + 2 = 0$

76. $x^2 + 4 = 0$

77. $x^2 - x + 2 = 0$

78. $x^2 + 2x + 3 = 0$

79. $2x^2 + 3x + 4 = 0$

80. $3x^2 - x = 1$

81. $x^2 + 1 = 4x$

82. $3x^2 + 2 = x$

83. $x(x + 1) = -2$

84. $x(x - 4) = -8$

85. $5x^2 + 2x + 4 = 0$

86. $7x^2 - 2x + 4 = 0$

87. $\frac{1}{2}x^2 + \frac{3}{4}x = -1$

88. $-\frac{1}{3}x^2 + x = 2$

89. $x(x + 2) = x - 4$

90. $x - 5 = 2x(2x + 1)$

91. $x(2x - 1) = 1 + x$

92. $2x = x(3 - 4x)$

93. $x^2 = x(1 - x) - 2$

94. $2x^2 = 2x(5 - x) - 8$

COMPLETING THE SQUARE

Exercises 95–100: Solve by completing the square.

95. $x^2 + 2x + 4 = 0$

96. $x^2 - 2x + 2 = 0$

97. $x(x + 4) = -5$

98. $x(8 - x) = 25$

99. $2x^2 - 4x + 6 = 0$

100. $2x^2 + 2x + 1 = 0$

USING THE METHOD OF YOUR CHOICE

Exercises 101–114: Find exact solutions to the quadratic equation, using a method of your choice. Explain why you chose the method you did. Answers may vary.

101. $x^2 - 3x + 2 = 0$

102. $x^2 + 2x + 1 = 0$

103. $0.5x^2 - 1.75x = 1$

104. $\frac{3}{5}x^2 + \frac{9}{10}x = \frac{3}{5}$

105. $x^2 - 5x + 2 = 0$

106. $2x^2 - x - 4 = 0$

107. $2x^2 + x = -8$

108. $4x^2 = 2x - 3$

109. $4x^2 - 1 = 0$

110. $3x^2 = 9$

111. $3x^2 + 6 = 0$

112. $4x^2 + 7 = 0$

113. $9x^2 + 1 = 6x$

114. $10x^2 + 15x = 25$

APPLICATIONS INVOLVING QUADRATIC EQUATIONS

Exercises 115–118: **Modeling Stopping Distance** *(Refer to Example 4.) Use* $d = \frac{1}{9}x^2 + \frac{11}{3}x$ *to find a safe speed x for the following stopping distances d.*

115. 42 feet

116. 152 feet

117. 390 feet

118. 726 feet

119. **Groupon's Growth** In its early years, Groupon experienced a dramatic increase in value from October 2010 to March 2011. The function

$$G(x) = 0.4x^2 + 1.8x + 6$$

approximates the company's value in billions of dollars x months after October 2010.
 (a) Evaluate $G(0)$. Interpret the result.
 (b) Determine when Groupon's value was about $15 billion.

120. **Foursquare Users** Foursquare provides a service that allows your friends to know your whereabouts by "checking in." From March 2010 to March 2011, it experienced amazing growth. The function

$$F(x) = \frac{1}{18}x^2 - \frac{1}{12}x + \frac{1}{2}$$

approximates Foursquare users in millions x months after March 2010.
 (a) Evaluate $F(3)$. Interpret the result.
 (b) Determine symbolically when Foursquare had 4.25 million users.

121. **U.S. AIDS Deaths** In the early years of the HIV/ AIDS epidemic, the cumulative numbers in thousands of AIDS deaths from 1984 through 1994 may be modeled by

$$f(x) = 2.39x^2 + 5.04x + 5.1,$$

where $x = 0$ corresponds to 1984, $x = 1$ corresponds to 1985, and so on until $x = 10$ corresponds to 1994. See the accompanying graph. Use the formula for $f(x)$ to estimate the year when the total number of AIDS deaths reached 200 thousand. Compare your result with that shown in the graph.

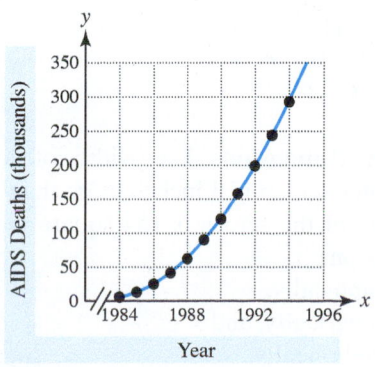

122. **Canoeing** A camper paddles a canoe 2 miles downstream in a river that has a 2-mile-per-hour current. To return to camp, the canoeist travels upstream on a different branch of the river. It is 4 miles long and has a 1-mile-per-hour current. The total trip (both ways) takes 3 hours. Find the average speed of the canoe in still water. (*Hint:* Time equals distance divided by rate.)

123. Historical Music Sales From 1989 to 2009, the global music sales S of both compact discs and tapes in billions of dollars can be modeled by $S(x) = -0.095x^2 + 1.85x + 6$, where x is years after 1989. Estimate symbolically when sales were $12 billion. (*Source: RIAA, Bain Analysis.*)

124. Monthly Facebook Visitors The number of *unique* monthly U.S. Facebook visitors in millions x years after 2011 can be modeled by

$$V(x) = -0.5x^2 + 9x + 147,$$

where $0 \le x \le 5$. Assuming current trends continue, estimate symbolically when this number was 175 million.

125. Airplane Speed A pilot flies 500 miles against a 20-mile-per-hour wind. On the next day, the pilot flies back home with a 10-mile-per-hour tail wind. The total trip (both ways) takes 4 hours. Find the speed of the airplane without a wind.

126. Distance Two cars leave an intersection, one traveling south and one traveling east, as shown in the figure. After 1 hour, the two cars are 50 miles apart and the car traveling east has traveled 10 miles farther than the car traveling south. How far did each car travel?

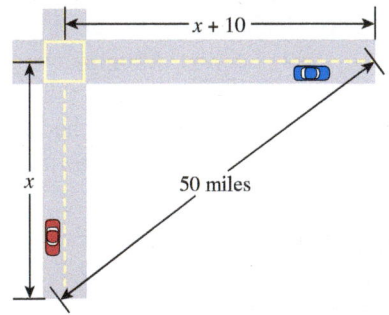

127. Screen Dimensions The width of a rectangular computer screen is 3 inches more than its height. If the area of the screen is 154 square inches, find its dimensions

 (a) graphically,
(b) numerically, and
(c) symbolically.

128. Sidewalk Dimension A rectangular flower garden in a park is 30 feet wide and 40 feet long. A sidewalk around the perimeter of the garden is being planned, as shown in the figure. The gardener has enough money to pour 624 square feet of cement sidewalk. Find the width of the sidewalk.

129. Modeling Water Flow When water runs out of a hole in a cylindrical container, the height of the water in the container can often be modeled by a quadratic function. The data in the table show the height y in centimeters of water at 30-second intervals in a metal container that has a small hole in it.

Time	0	30	60	90
Height	16	11.9	8.4	5.3

Time	120	150	180
Height	3.1	1.4	0.5

These data are modeled by

$$f(x) = 0.0004x^2 - 0.15x + 16.$$

(a) Explain why a linear function would not be appropriate for modeling these data.
(b) Use the table to estimate the time at which the height was 7 centimeters.
(c) Use the quadratic formula to solve part (b).

130. Hospitals The general trend in the number of hospitals in the United States from 1945 through 2000 is shown in the graph and can be modeled by

$$f(x) = -1.38x^2 + 84x + 5865,$$

where $x = 5$ corresponds to 1945, $x = 10$ corresponds to 1950, and so on until $x = 60$ represents 2000.

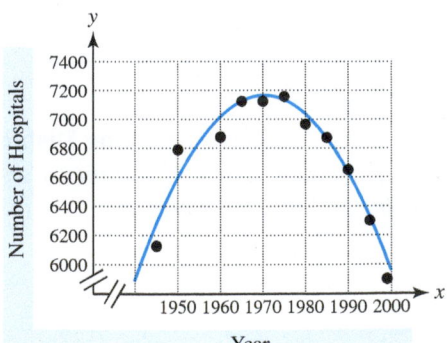

(a) Describe any trends in the numbers of hospitals from 1945 to 2000.

(b) What information does the vertex give?

(c) Use the formula for $f(x)$ to estimate the number of hospitals in 1970. Compare your result with that shown in the graph.

(d) Use the formula for $f(x)$ to estimate the year (or years) when there were 6300 hospitals. Compare your result with that shown in the graph.

WRITING ABOUT MATHEMATICS

131. Explain how the discriminant $b^2 - 4ac$ can be used to determine the number of solutions to a quadratic equation.

132. Let $f(x) = ax^2 + bx + c$ be a quadratic function. If you know the value of $b^2 - 4ac$, what information does this give you about the graph of $y = f(x)$? Explain your answer.

SECTIONS 11.3 and 11.4 | **Checking Basic Concepts**

1. Solve the quadratic equation $2x^2 - 7x + 3 = 0$ symbolically and graphically.

2. Use the square root property to solve $x^2 = 5$.

3. Complete the square to solve $x^2 - 4x + 1 = 0$.

4. Solve the equation $x^2 + y^2 = 1$ for y.

5. Use the quadratic formula to solve each equation.
(a) $2x^2 = 3x + 1$
(b) $9x^2 - 24x + 16 = 0$

6. Calculate the discriminant for each equation and give the number of *real* solutions.
(a) $x^2 - 5x + 5 = 0$
(b) $2x^2 - 5x + 4 = 0$
(c) $49x^2 - 56x + 16 = 0$

7. Solve each equation.
(a) $x^2 + 5 = 0$
(b) $x^2 + x + 3 = 0$

11.5 Quadratic Inequalities

Objectives

1 Recognizing Quadratic Inequalities

2 Finding Graphical and Numerical Solutions

3 Finding Symbolic Solutions

NEW VOCABULARY

☐ Quadratic inequality
☐ Test value

1 Recognizing Quadratic Inequalities

If the equals sign in a quadratic equation is replaced with $>$, $\geq$, $<$, or $\leq$, a **quadratic inequality** results. Examples of quadratic inequalities include

$$x^2 + 4x - 3 < 0, \quad 5x^2 \geq 5, \quad \text{and} \quad 1 - z \leq z^2.$$

Any quadratic equation can be written as

$$ax^2 + bx + c = 0, \quad a \neq 0, \qquad \text{Quadratic equation}$$

so any quadratic inequality can be written as

$$ax^2 + bx + c > 0, \quad a \neq 0, \qquad \text{Quadratic inequality}$$

where $>$ may be replaced with $\geq$, $<$, or $\leq$.

| EXAMPLE 1 | **Identifying a quadratic inequality** |

Determine whether the inequality is quadratic.

(a) $5x + x^2 - x^3 \leq 0$ **(b)** $4 + 5x^2 > 4x^2 + x$

Solution

(a) The inequality $5x + x^2 - x^3 \leq 0$ is not quadratic because it has an x^3-term.

(b) Write the inequality as follows.

$$
\begin{array}{ll}
4 + 5x^2 > 4x^2 + x & \text{Given inequality} \\
4 + 5x^2 - 4x^2 - x > 0 & \text{Subtract } 4x^2 \text{ and } x. \\
4 + x^2 - x > 0 & \text{Combine like terms.} \\
x^2 - x + 4 > 0 & \text{Rewrite.}
\end{array}
$$

Because the inequality can be written in the form $ax^2 + bx + c > 0$ with $a = 1$, $b = -1$, and $c = 4$, it is a quadratic inequality.

Now Try Exercises 7, 11

READING CHECK 1

Which of the following are quadratic inequalities?

$$2x - 5 < 2$$
$$x^2 - 1 > 0$$
$$2x^2 + 4x + 5 \leq 0$$
$$-4x^2 - 5 \geq -7x$$
$$x < 1$$

2 Finding Graphical and Numerical Solutions

Equality often is the boundary between *greater than* and *less than*, so a first step in solving an inequality is to determine the x-values where equality occurs. We begin by using this concept with graphical techniques.

The graph of $y = x^2 - x - 2$ has x-intercepts $(-1, 0)$ and $(2, 0)$. See **FIGURE 11.41**. The solutions to $x^2 - x - 2 = 0$ are given by $x = -1$ or $x = 2$. Between the x-intercepts the graph dips **below** the x-axis and the y-values are **negative**. Thus solutions to $x^2 - x - 2 < 0$ satisfy $-1 < x < 2$. To support this result, we select a **test value**. For example, 0 lies between -1 and 2. If we substitute $x = 0$ in $x^2 - x - 2 < 0$, it results in a true statement.

$$0^2 - 0 - 2 < 0 \checkmark \quad \text{True}$$

Visualizing Quadratic Inequalities

FIGURE 11.41

When $x < -1$ or $x > 2$, the graph lies **above** the x-axis and the y-values are **positive**. Thus the solutions to $x^2 - x - 2 > 0$ satisfy $x < -1$ or $x > 2$. For example, 3 is greater than 2 and -3 is less than -1. Therefore both 3 and -3 are solutions. We can verify this result by substituting 3 and -3 as test values in $x^2 - x - 2 > 0$.

$$3^2 - 3 - 2 > 0 \checkmark \quad \text{True}$$
$$(-3)^2 - (-3) - 2 > 0 \checkmark \quad \text{True}$$

In the next three examples, we use these concepts to solve quadratic inequalities.

| EXAMPLE 2 | **Solving a quadratic inequality** |

Make a table of values for $y = x^2 - 3x - 4$ and then sketch the graph. Use the table and graph to solve $x^2 - 3x - 4 \leq 0$. Write your answer in interval notation.

Solution

The points calculated for **TABLE 11.14** are plotted in **FIGURE 11.42** and connected with a smooth U-shaped graph.

Numerical Solution TABLE 11.14 shows that $x^2 - 3x - 4$ equals 0 when $x = -1$ or $x = 4$. Between these values, $x^2 - 3x - 4$ is negative, so the solution set to $x^2 - 3x - 4 \leq 0$ is given by $-1 \leq x \leq 4$ or, in interval notation, $[-1, 4]$.

Graphical Solution In FIGURE 11.42, the graph of $y = x^2 - 3x - 4$ shows that the x-intercepts are $(-1, 0)$ and $(4, 0)$. Between these values, the graph dips **below** the x-axis. Thus the solution set is $[-1, 4]$.

STUDY TIP

Learn how to use a parabola to help solve a quadratic inequality.

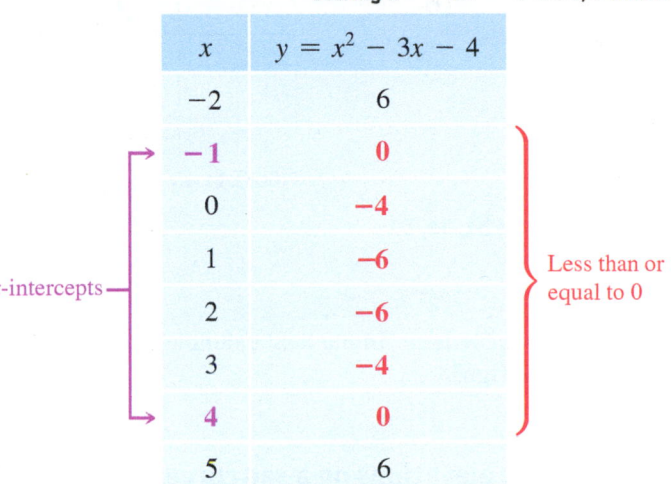

Solving $x^2 - 3x - 4 \leq 0$; x-Intercepts: $(-1, 0)$, $(4, 0)$

x	$y = x^2 - 3x - 4$
-2	6
-1	0
0	-4
1	-6
2	-6
3	-4
4	0
5	6

TABLE 11.14

FIGURE 11.42

Now Try Exercises 19, 27

EXAMPLE 3 Solving a quadratic inequality

Solve $x^2 > 1$. Write your answer in interval notation.

Solution

First, we rewrite $x^2 > 1$ as $x^2 - 1 > 0$. The graph of $y = x^2 - 1$ is shown in FIGURE 11.43 with x-intercepts $(-1, 0)$ and $(1, 0)$. The graph lies **above** the x-axis and is shaded green to the left of $x = -1$ or to the right of $x = 1$. Thus the solution set is given by $x < -1$ or $x > 1$, which can be written in interval notation as $(-\infty, -1) \cup (1, \infty)$.

Now Try Exercise 49

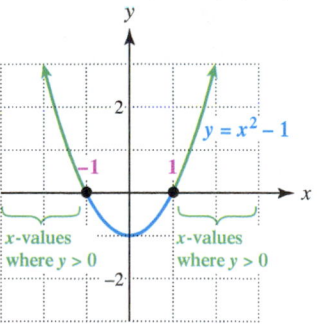

Solving $x^2 - 1 > 0$
x-Intercepts: $(-1, 0)$, $(1, 0)$

FIGURE 11.43

EXAMPLE 4 Solving some special cases

Solve each of the inequalities graphically.
(a) $x^2 + 1 > 0$ **(b)** $x^2 + 1 < 0$ **(c)** $(x - 1)^2 \leq 0$

Solution
(a) Because the graph of $y = x^2 + 1$, shown in FIGURE 11.44, is always above the x-axis, $x^2 + 1$ is always greater than 0. The solution set includes all real numbers, or $(-\infty, \infty)$.
(b) Because the graph of $y = x^2 + 1$, shown in FIGURE 11.44, never goes below the x-axis, $x^2 + 1$ is never less than 0. Thus there are no real solutions.

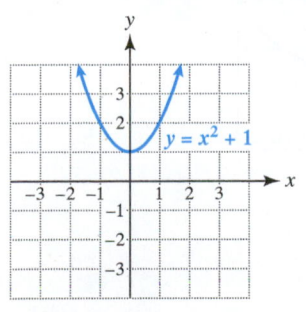

FIGURE 11.44

(c) Because the graph of $y = (x - 1)^2$, shown in **FIGURE 11.45**, never goes below the x-axis, $(x - 1)^2$ is never less than 0. When $x = 1$, $y = 0$, so **1** is the only solution to the inequality $(x - 1)^2 \le 0$.

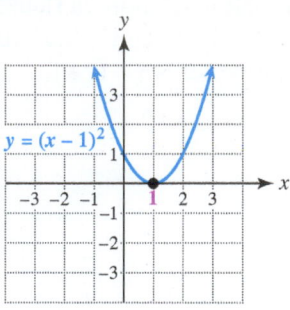

FIGURE 11.45

Now Try Exercises 51, 55, 57

 Math in Context *Highway Design* In the next example, we show how quadratic inequalities are used in highway design.

EXAMPLE 5   Determining elevations on a sag curve

Parabolas are frequently used in highway design to model hills and sags (valleys) along a proposed route. Suppose that the elevation E in feet of a sag, or *sag curve*, is given by

$$E(x) = 0.00004x^2 - 0.4x + 2000,$$

where x is the horizontal distance in feet along the sag curve and $0 \le x \le 10{,}000$. See **FIGURE 11.46**. Estimate graphically the x-values where the elevation is 1500 feet or less. (*Source:* F. Mannering and W. Kilareski, *Principles of Highway Engineering and Traffic Analysis.*)

FIGURE 11.46

Solution

Graphical Solution We must solve the quadratic inequality

$$0.00004x^2 - 0.4x + 2000 \le 1500.$$

Let $y_1 = 0.00004x^2 - 0.4x + 2000$ be the sag curve and $y_2 = 1500$ be a line with an elevation of 1500 feet. Their graphs intersect at $x \approx 1464$ and $x \approx 8536$, as shown in **FIGURE 11.47**. The elevation of the road is less than 1500 feet between these x-values. Therefore the elevation is 1500 feet or less when $1464 \le x \le 8536$ (approximately).

Determining Where the Elevation Is 1500 Feet

CALCULATOR HELP

To find a point of intersection, see Appendix A (page AP-6).

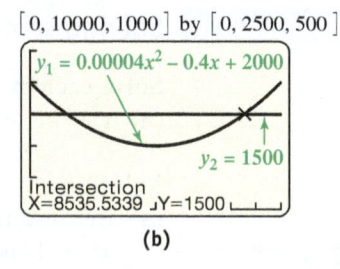

(a)   (b)

FIGURE 11.47

Now Try Exercise 65

The following chart summarizes visual solutions for several possibilities when solving quadratic equations and inequalities.

Visualizing Solutions to Quadratic Equations and Inequalities

Two *x*-Intercepts

Case 1: Upward Opening

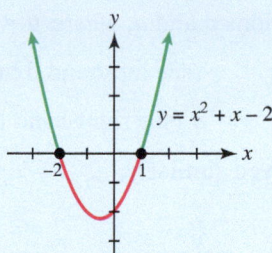

Equation/Inequality	Solutions
$x^2 + x - 2 = 0$	$x = -2, 1$
$x^2 + x - 2 < 0$	$-2 < x < 1$
$x^2 + x - 2 > 0$	$x < -2$ or $x > 1$

Case 2: Downward Opening

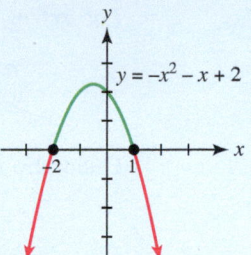

Equation/Inequality	Solutions
$-x^2 - x + 2 = 0$	$x = -2, 1$
$-x^2 - x + 2 < 0$	$x < -2$ or $x > 1$
$-x^2 - x + 2 > 0$	$-2 < x < 1$

One *x*-Intercept

Case 1: Upward Opening

Equation/Inequality	Solutions
$(x - 1)^2 = 0$	$x = 1$
$(x - 1)^2 < 0$	no solutions
$(x - 1)^2 > 0$	$x < 1$ or $x > 1$

Case 2: Downward Opening

Equation/Inequality	Solutions
$-(x - 1)^2 = 0$	$x = 1$
$-(x - 1)^2 < 0$	$x < 1$ or $x > 1$
$-(x - 1)^2 > 0$	no solutions

No *x*-Intercepts

Case 1: Upward Opening

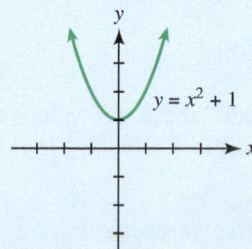

Equation/Inequality	Solutions
$x^2 + 1 = 0$	no solutions
$x^2 + 1 < 0$	no solutions
$x^2 + 1 > 0$	all real numbers

Case 2: Downward Opening

Equation/Inequality	Solutions
$-x^2 - 1 = 0$	no solutions
$-x^2 - 1 < 0$	all real numbers
$-x^2 - 1 > 0$	no solutions

3   Finding Symbolic Solutions

To solve a quadratic inequality symbolically, we first solve the corresponding equality. We can then write the solution to the inequality, using the following method.

SOLUTIONS TO QUADRATIC INEQUALITIES

Let $ax^2 + bx + c = 0$, $a > 0$, have two real solutions p and q, where $p < q$.

$ax^2 + bx + c < 0$ is equivalent to $p < x < q$ (see left-hand figure).

$ax^2 + bx + c > 0$ is equivalent to $x < p$ or $x > q$ (see right-hand figure).

Quadratic inequalities involving $\leq$ or $\geq$ can be solved similarly.

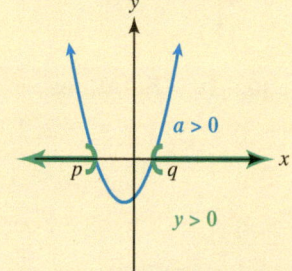

Solutions lie between p and q.   Solutions lie "outside" p and q.

One way to handle the situation where $a < 0$ is to multiply each side of the inequality by -1, in which case we must be sure to *reverse* the inequality symbol. For example, the inequality $-2x^2 + 8 \leq 0$ has $a = -2$, which is negative. If we multiply each side of this inequality by -1, we obtain $2x^2 - 8 \geq 0$ and now $a = 2$, which is positive.

EXAMPLE 6   **Solving quadratic inequalities**

Solve each inequality symbolically. Write your answer in interval notation.
(a) $6x^2 - 7x - 5 \geq 0$ **(b)** $x(3 - x) > -18$

Solution
(a) Begin by solving $6x^2 - 7x - 5 = 0$.

$6x^2 - 7x - 5 = 0$	Quadratic equation
$(2x + 1)(3x - 5) = 0$	Factor.
$2x + 1 = 0$ or $3x - 5 = 0$	Zero-product property
$x = -\dfrac{1}{2}$ or $x = \dfrac{5}{3}$	Solve.

Therefore solutions to $6x^2 - 7x - 5 \geq 0$ lie "**outside**" these two values and satisfy $x \leq -\frac{1}{2}$ or $x \geq \frac{5}{3}$. In interval notation, the solution set is $\left(-\infty, -\frac{1}{2}\right] \cup \left[\frac{5}{3}, \infty\right)$.

(b) First, rewrite the inequality as follows.

$x(3 - x) > -18$	Given inequality
$3x - x^2 > -18$	Distributive property
$3x - x^2 + 18 > 0$	Add 18.
$-x^2 + 3x + 18 > 0$	Rewrite.
$x^2 - 3x - 18 < 0$	Multiply by -1 because $a < 0$; reverse the inequality symbol.

CRITICAL THINKING 1

The graph of $y = -x^2 + x + 12$ is a parabola with x-intercepts $(-3, 0)$, and $(4, 0)$. Solve $-x^2 + x + 12 < 0$.

Next, solve $x^2 - 3x - 18 = 0$.

$$(x + 3)(x - 6) = 0 \quad \text{Factor.}$$

$$x = -3 \quad \text{or} \quad x = 6 \quad \text{Solve.}$$

Solutions to the inequality $x^2 - 3x - 18 < 0$ lie **between** these two values and satisfy $-3 < x < 6$. In interval notation, the solution set is $(-3, 6)$.

Now Try Exercise 31, 37

EXAMPLE 7 **Finding the dimensions of a building**

A rectangular building needs to be 7 feet longer than it is wide, as illustrated in **FIGURE 11.48**. The area of the building must be at least 450 square feet. What widths x are possible for this building? Support your results with a table of values.

FIGURE 11.48

Solution

Symbolic Solution If x is the width of the building, $x + 7$ is the length of the building and its area is $x(x + 7)$. The area must be at least 450 square feet, so the inequality $x(x + 7) \geq 450$ must be satisfied.

First solve the following quadratic equation.

$$
\begin{array}{ll}
x(x + 7) = 450 & \text{Quadratic equation} \\
x^2 + 7x = 450 & \text{Distributive property} \\
x^2 + 7x - 450 = 0 & \text{Subtract 450.} \\
x = \dfrac{-7 \pm \sqrt{7^2 - 4(1)(-450)}}{2(1)} & \begin{array}{l}\text{Quadratic formula: } a = 1, \\ b = 7, \text{ and } c = -450\end{array} \\
= \dfrac{-7 \pm \sqrt{1849}}{2} & \text{Simplify.} \\
= \dfrac{-7 \pm 43}{2} & \sqrt{1849} = 43 \\
= 18, -25 & \text{Evaluate.}
\end{array}
$$

Thus the solutions to $x(x + 7) \geq 450$ are $x \leq -25$ or $x \geq 18$. The width is positive, so the building width must be 18 feet or more.

Numerical Solution A table of values is shown in **FIGURE 11.49**, where $y_1 = x(x + 7)$ equals 450 when $x = 18$. For $x \geq 18$, the area is *at least* 450 square feet.

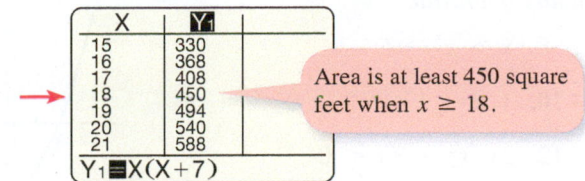

FIGURE 11.49

Now Try Exercise 63

11.5 Putting It All Together

CONCEPT	EXPLANATION
Solving a Quadratic Inequality Symbolically	Let $ax^2 + bx + c = 0, a > 0$, have two real solutions p and q, where $p < q$. $ax^2 + bx + c < 0$ is equivalent to $p < x < q$. $ax^2 + bx + c > 0$ is equivalent to $x < p$ or $x > q$. *Examples:* The solutions to $x^2 - 3x + 2 = 0$ are 1 and 2. The solutions to $x^2 - 3x + 2 < 0$ satisfy $1 < x < 2$. The solutions to $x^2 - 3x + 2 > 0$ satisfy $x < 1$ or $x > 2$.
Solving a Quadratic Inequality Graphically	Given $ax^2 + bx + c < 0$ with $a > 0$, graph $y = ax^2 + bx + c$ and locate any x-intercepts. If there are two x-intercepts, then solutions correspond to x-values between the x-intercepts. Solutions to $ax^2 + bx + c > 0$ correspond to x-values "outside" (left or right of) the x-intercepts.
Solving a Quadratic Inequality Numerically	If a quadratic inequality is expressed as $ax^2 + bx + c < 0$ with $a > 0$, then we start by solving $y = ax^2 + bx + c = 0$ with a table. If there are two solutions, then the solutions to the given inequality lie between these values. Solutions to $ax^2 + bx + c > 0$ lie before or after these values in the table.

11.5 Exercises

MyMathLab®

CONCEPTS AND VOCABULARY

1. How is a quadratic inequality different from a quadratic equation?

2. Do quadratic inequalities typically have two solutions? Explain.

3. Is 3 a solution to $x^2 < 7$?

4. Is 5 a solution to $x^2 \geq 25$?

5. The solutions to $x^2 - 2x - 8 = 0$ are -2 and 4. What are the solutions to $x^2 - 2x - 8 < 0$? Write your answer as an inequality.

6. The solutions to $x^2 + 2x - 3 = 0$ are -3 and 1. What are the solutions to $x^2 + 2x - 3 > 0$? Write your answer as an inequality.

RECOGNIZING QUADRATIC INEQUALITIES AND SOLUTIONS

Exercises 7–12: Is the inequality quadratic?

7. $x^2 + 4x + 5 < 0$
8. $x > x^3 - 5$
9. $x^2 > 19$
10. $x(x - 1) - 2 \geq 0$
11. $4x > 1 - x$
12. $2x(x^2 + 3) < 0$

Exercises 13–18: Is the given value of x a solution?

13. $2x^2 + x - 1 > 0$ $x = 3$

14. $x^2 - 3x + 2 \leq 0$ $x = 2$

15. $x^2 + 2 \leq 0$ $x = 0$

16. $2x(x - 3) \geq 0$ $x = 1$

17. $x^2 - 3x \leq 1$ $x = -3$

18. $4x^2 - 5x + 1 > 30$ $x = -2$

FINDING GRAPHICAL SOLUTIONS

Exercises 19–24: The graph of $y = ax^2 + bx + c$ is given. Solve each equation or inequality.

(a) $ax^2 + bx + c = 0$
(b) $ax^2 + bx + c < 0$
(c) $ax^2 + bx + c > 0$

19.

20.

21.   **22.**

23.   **24.**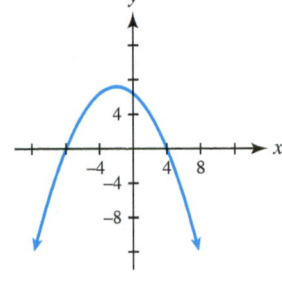

Exercises 25 and 26: Solve the inequality graphically to the nearest thousandth. Use interval notation.

25. $\pi x^2 - \sqrt{3}x \leq \frac{3}{11}$ **26.** $\sqrt{5}x^2 - \pi^2 x \geq 10.3$

FINDING NUMERICAL SOLUTIONS

Exercises 27–30: A table for $y = ax^2 + bx + c$ is given. Solve each equation or inequality.

(a) $ax^2 + bx + c = 0$
(b) $ax^2 + bx + c < 0$
(c) $ax^2 + bx + c > 0$

27. $y = x^2 - 4$

x	−3	−2	−1	0	1	2	3
y	5	0	−3	−4	−3	0	5

28. $y = x^2 - x - 2$

x	−3	−2	−1	0	1	2	3
y	10	4	0	−2	−2	0	4

29. $y = x^2 + 4x$

x	−5	−4	−3	−2	−1	0	1
y	5	0	−3	−4	−3	0	5

30. $y = -2x^2 - 2x + 1.5$

x	−2	−1.5	−1	−0.5	0	0.5	1
y	−2.5	0	1.5	2	1.5	0	−2.5

FINDING SYMBOLIC SOLUTIONS

Exercises 31–40: Solve the quadratic inequality symbolically. Write your answer in interval notation.

31. $x^2 + 10x + 21 \leq 0$ **32.** $x^2 - 7x - 18 < 0$

33. $3x^2 - 9x + 6 > 0$ **34.** $7x^2 + 34x - 5 \geq 0$

35. $x^2 < 10$ **36.** $x^2 \geq 64$

37. $x(6 - x) < 0$ **38.** $1 - x^2 \leq 0$

39. $x(4 - x) \leq 2$ **40.** $2x(1 - x) \geq 2$

USING THE METHOD OF YOUR CHOICE

Exercises 41–44: Solve the equation in part (a). Use the results to solve the inequalities in parts (b) and (c).

41. (a) $x^2 - 4 = 0$
    (b) $x^2 - 4 < 0$
    (c) $x^2 - 4 > 0$

42. (a) $x^2 - 5 = 0$
    (b) $x^2 - 5 \leq 0$
    (c) $x^2 - 5 \geq 0$

43. (a) $x^2 + x - 1 = 0$
    (b) $x^2 + x - 1 < 0$
    (c) $x^2 + x - 1 > 0$

44. (a) $x^2 + 4x - 5 = 0$
    (b) $x^2 + 4x - 5 \leq 0$
    (c) $x^2 + 4x - 5 \geq 0$

Exercises 45–62: Solve the inequality by any method. Write your answer in interval notation when appropriate.

45. $x^2 + 4x + 3 < 0$ **46.** $x^2 + x - 2 \leq 0$

47. $2x^2 - x - 15 \geq 0$ **48.** $3x^2 - 3x - 6 > 0$

49. $2x^2 \leq 8$ **50.** $x^2 < 9$

51. $x^2 > -5$ **52.** $-x^2 \geq 1$

53. $-x^2 + 3x > 0$ **54.** $-8x^2 - 2x + 1 \leq 0$

55. $x^2 + 2 \leq 0$ **56.** $x^2 + 3 \geq -5$

57. $(x - 2)^2 \leq 0$ **58.** $(x + 2)^2 \leq 0$

59. $(x + 1)^2 > 0$ **60.** $(x - 3)^2 > 0$

61. $x(1 - x) \geq -2$ **62.** $x(x - 2) < 3$

APPLICATIONS INVOLVING QUADRATIC EQUATIONS

63. Dimensions of a Pen A rectangular pen for a pet is 5 feet longer than it is wide. Give possible values for the width w of the pen if its area must be between 176 and 500 square feet, inclusively.

64. Dimensions of a Cylinder The volume of a cylindrical container is given by $V = \pi r^2 h$, where r is its radius and h is its height. See the accompanying figure. If $h = 6$ inches and the volume of the container must be 50 cubic inches or more, estimate to the nearest tenth of an inch possible values for r.

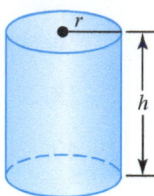

65. Highway Design (Refer to Example 5.) The elevation E of a sag curve, in feet, is given by

$$E(x) = 0.0000375x^2 - 0.175x + 1000,$$

where $0 \le x \le 4000$.
(a) Estimate graphically the x-values for which the elevation is 850 feet or less. (*Hint:* Use the window $[0, 4000, 1000]$ by $[500, 1200, 100]$.)
(b) For which x-values is the elevation 850 feet or more?

66. Early Cellular Phone Use The number of cellular subscribers in the United States in thousands from 1985 to 1991 can be modeled by

$$f(x) = 163x^2 - 146x + 205,$$

where x is the year and $x = 0$ corresponds to 1985, $x = 1$ to 1986, and so on. (*Source:* M. Paetsch, *Mobile Communication in the U.S. and Europe.*)
(a) Write a quadratic inequality whose solution set represents the years when there were 2 million subscribers or more.
(b) Solve this inequality.

67. Heart Disease Death Rates From 1960 to 2010, age-adjusted heart disease rates decreased dramatically. The number of deaths per 100,000 people can be modeled by

$$f(x) = -0.05107x^2 + 194.74x - 184{,}949,$$

where x is the year, as illustrated in the accompanying figure. (*Source:* Department of Health and Human Services.)

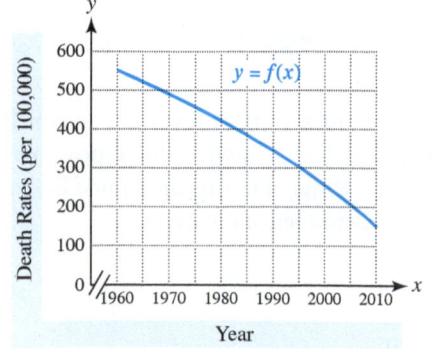

(a) Evaluate $f(1985)$ using both the formula and the graph. How do your results compare?
(b) Use the graph to estimate the years when this death rate was 500 or less.
(c) Solve part (b) by using the quadratic formula.

68. Accidental Deaths From 1910 to 2010 the number of accidental deaths per 100,000 people generally decreased and can be modeled by

$$f(x) = -0.001918x^2 + 6.93x - 6156,$$

where x is the year, as shown in the accompanying figure. Note that after 2000, these rates increased and are not modeled by $f(x)$. (*Source:* Department of Health and Human Services.)

(a) Evaluate $f(1955)$, using both the formula and the graph. How do your results compare?
(b) Use the graph to estimate when this death rate was 60 or more.
(c) Solve part (b) by using the quadratic formula.

WRITING ABOUT MATHEMATICS

69. Consider the inequality $x^2 < 0$. Discuss the solutions to this inequality and explain your reasoning.

70. Explain how the graph of $y = ax^2 + bx + c$ can be used to solve the inequality

$$ax^2 + bx + c > 0$$

when $a < 0$. Assume that the x-intercepts of the graph are $(p, 0)$ and $(q, 0)$ with $p < q$.

11.6 Equations in Quadratic Form

1 Solving Higher Degree Polynomial Equations

Sometimes a fourth-degree polynomial can be factored like a quadratic trinomial, provided it does not have an x-term or an x^3-term. Let's consider the equation $x^4 - 5x^2 + 4 = 0$.

$$x^4 - 5x^2 + 4 = 0 \qquad \text{Given equation}$$
$$(x^2)^2 - 5(x^2) + 4 = 0 \qquad \text{Properties of exponents}$$

We use the substitution $u = x^2$.

$$u^2 - 5u + 4 = 0 \qquad \text{Let } u = x^2.$$
$$(u - 4)(u - 1) = 0 \qquad \text{Factor.}$$
$$u - 4 = 0 \quad \text{or} \quad u - 1 = 0 \qquad \text{Zero-product property}$$
$$u = 4 \quad \text{or} \quad u = 1 \qquad \text{Solve each equation.}$$

Because the given equation $x^4 - 5x^2 + 4 = 0$ uses the variable x, we must give the solutions in terms of x. We substitute x^2 for u in $u = 4$ and $u = 1$ and then solve to obtain the following four solutions.

$$u = 4 \qquad \text{or} \qquad u = 1 \qquad \text{Solutions in terms of } u$$
$$x^2 = 4 \qquad \text{or} \qquad x^2 = 1 \qquad \text{Substitute } x^2 \text{ for } u.$$
$$x = \pm 2 \qquad \text{or} \qquad x = \pm 1 \qquad \text{Square root property}$$

The solutions are -2, -1, 1, and 2.

EXAMPLE 1 **Solving a sixth-degree equation by substitution**

Solve $2x^6 + x^3 = 1$.

Solution
Start by subtracting 1 from each side.

$$2x^6 + x^3 - 1 = 0 \qquad \text{Subtract 1.}$$
$$2(x^3)^2 + (x^3) - 1 = 0 \qquad \text{Properties of exponents}$$
$$2u^2 + u - 1 = 0 \qquad \text{Let } u = x^3.$$
$$(2u - 1)(u + 1) = 0 \qquad \text{Factor.}$$
$$2u - 1 = 0 \quad \text{or} \quad u + 1 = 0 \qquad \text{Zero-product property}$$
$$u = \frac{1}{2} \quad \text{or} \quad u = -1 \qquad \text{Solve.}$$

Now substitute x^3 for u, and solve for x to obtain the following two solutions.

$$x^3 = \frac{1}{2} \qquad \text{or} \quad x^3 = -1 \qquad \text{Substitute } x^3 \text{ for } u.$$
$$x = \sqrt[3]{\frac{1}{2}} \quad \text{or} \quad x = -1 \qquad \text{Take cube root of each side.}$$

Now Try Exercise 3

2 Solving Equations That Have Rational Exponents

Equations that have negative exponents are sometimes reducible to quadratic form. Consider the following example, in which two methods are presented.

EXAMPLE 2   **Solving an equation that has negative exponents**

Solve $-6m^{-2} + 13m^{-1} + 5 = 0$.

Solution
Method I Use the substitution $u = m^{-1} = \frac{1}{m}$ and $u^2 = m^{-2} = \frac{1}{m^2}$.

$$-6\boldsymbol{m^{-2}} + 13\boldsymbol{m^{-1}} + 5 = 0 \qquad \text{Given equation}$$
$$-6\boldsymbol{u}^2 + 13\boldsymbol{u} + 5 = 0 \qquad \text{Let } u = m^{-1} \text{ and } u^2 = m^{-2}.$$
$$6u^2 - 13u - 5 = 0 \qquad \text{Multiply by } -1.$$
$$(2u - 5)(3u + 1) = 0 \qquad \text{Factor.}$$
$$2u - 5 = 0 \quad \text{or} \quad 3u + 1 = 0 \qquad \text{Zero-product property}$$
$$u = \frac{5}{2} \quad \text{or} \quad u = -\frac{1}{3} \qquad \text{Solve for } u.$$

Because $u = \frac{1}{m}$, $m = \frac{1}{u}$. Thus the solutions are given by $m = \frac{2}{5}$ or $m = -3$.

Method II Another way to solve this equation is to multiply each side by the LCD, m^2.

$$-6m^{-2} + 13m^{-1} + 5 = 0 \qquad \text{Given equation}$$
$$m^2(-6m^{-2} + 13m^{-1} + 5) = m^2 \cdot 0 \quad \text{Multiply by } m^2.$$
$$-6m^2m^{-2} + 13m^2m^{-1} + 5m^2 = 0 \qquad \text{Distributive property}$$
$$-6 + 13m + 5m^2 = 0 \qquad \text{Add exponents.}$$
$$5m^2 + 13m - 6 = 0 \qquad \text{Rewrite the equation.}$$
$$(5m - 2)(m + 3) = 0 \qquad \text{Factor.}$$
$$5m - 2 = 0 \quad \text{or} \quad m + 3 = 0 \qquad \text{Zero-product property}$$
$$m = \frac{2}{5} \quad \text{or} \quad m = -3 \qquad \text{Solve.}$$

Now Try Exercise 5

In the next example, we solve an equation that has fractional exponents.

EXAMPLE 3   **Solving an equation that has fractional exponents**

Solve $x^{2/3} - 2x^{1/3} - 8 = 0$.

Solution
Use the substitution $\boldsymbol{u = x^{1/3}}$.

$$x^{2/3} - 2x^{1/3} - 8 = 0 \qquad \text{Given equation}$$
$$(\boldsymbol{x^{1/3}})^2 - 2(\boldsymbol{x^{1/3}}) - 8 = 0 \qquad \text{Properties of exponents}$$
$$\boldsymbol{u}^2 - 2\boldsymbol{u} - 8 = 0 \qquad \text{Let } u = x^{1/3}.$$
$$(u - 4)(u + 2) = 0 \qquad \text{Factor.}$$
$$u - 4 = 0 \quad \text{or} \quad u + 2 = 0 \qquad \text{Zero-product property}$$
$$u = 4 \quad \text{or} \quad u = -2 \qquad \text{Solve.}$$

Because $u = x^{1/3}$, $u^3 = (x^{1/3})^3 = x$. Thus $x = 4^3 = 64$ or $x = (\boldsymbol{-2})^3 = -8$. The solutions are -8 and 64.

Now Try Exercise 13

3 Solving Equations That Have Complex Solutions

Sometimes an equation that is reducible to quadratic form also has complex solutions. This situation is discussed in the next two examples.

EXAMPLE 4 Solving a fourth-degree equation

Find all complex solutions to $x^4 - 1 = 0$.

Solution

$$x^4 - 1 = 0 \qquad \text{Given equation}$$
$$(x^2)^2 - 1 = 0 \qquad \text{Properties of exponents}$$
$$u^2 - 1 = 0 \qquad \text{Let } u = x^2.$$
$$(u - 1)(u + 1) = 0 \qquad \text{Factor difference of squares.}$$
$$u - 1 = 0 \quad \text{or} \quad u + 1 = 0 \qquad \text{Zero-product property}$$
$$u = 1 \quad \text{or} \quad u = -1 \qquad \text{Solve for } u.$$

Now substitute x^2 for u, and solve for x.

$$u = 1 \qquad \text{or} \qquad u = -1 \qquad \text{Solutions in terms of } u$$
$$x^2 = 1 \qquad \text{or} \qquad x^2 = -1 \qquad \text{Let } x^2 = u.$$
$$x = \pm 1 \quad \text{or} \quad x = \pm i \qquad \text{Square root property}$$

There are four complex solutions: $-1, 1, -i,$ and i.

Now Try Exercise 25

EXAMPLE 5 Solving a rational equation

Find all complex solutions to $\dfrac{1}{x} + \dfrac{1}{x^2} = -1$.

Solution

This equation is a rational equation. However, if we multiply through by the LCD, x^2, we clear fractions and obtain a quadratic equation with complex solutions.

$$\frac{1}{x} + \frac{1}{x^2} = -1 \qquad \text{Given equation}$$

$$\frac{x^2}{x} + \frac{x^2}{x^2} = -1x^2 \qquad \text{Multiply each term by } x^2.$$

$$x + 1 = -x^2 \qquad \text{Simplify.}$$

$$x^2 + x + 1 = 0 \qquad \text{Add } x^2.$$

$$x = \frac{-1 \pm \sqrt{1^2 - 4(1)(1)}}{2(1)} \qquad \text{Quadratic formula}$$

$$x = \frac{-1 \pm i\sqrt{3}}{2} \qquad \sqrt{-3} = i\sqrt{3}$$

$$x = -\frac{1}{2} \pm \frac{i\sqrt{3}}{2} \qquad \frac{a \pm b}{c} = \frac{a}{c} \pm \frac{b}{c}$$

Now Try Exercise 31

11.6   Putting It All Together

CONCEPT	EXPLANATION	EXAMPLES
Higher Degree Polynomial Equations	Let $u = x^n$ for some integer n.	To solve $x^4 - 3x^2 - 4 = 0$, let $u = x^2$. This equation becomes $$u^2 - 3u - 4 = 0.$$
Equations That Have Rational Exponents	Pick a substitution that reduces the equation to quadratic form.	To solve $n^{-2} + 6n^{-1} + 9 = 0$, let $u = n^{-1}$. This equation becomes $$u^2 + 6u + 9 = 0.$$ To solve $6x^{2/5} - 5x^{1/5} - 4 = 0$, let $u = x^{1/5}$. This equation becomes $$6u^2 - 5u - 4 = 0.$$
Equations That Have Complex Solutions	Both polynomial and rational equations can have complex solutions. Use the fact that if $a > 0$, then $\sqrt{-a} = i\sqrt{a}$.	$$1 + \frac{1}{x^2} = 0$$ $$x^2 \cdot \left(1 + \frac{1}{x^2}\right) = 0 \cdot x^2$$ $$x^2 + 1 = 0$$ $$x^2 = -1$$ $$x = \pm i$$

11.6   Exercises

MyMathLab®

SOLVING EQUATIONS THAT ARE REDUCIBLE TO QUADRATIC FORM

Exercises 1–6: Use the given substitution to solve the equation.

1. $x^4 - 7x^2 + 6 = 0$ $\qquad u = x^2$

2. $2k^4 - 7k^2 + 6 = 0$ $\qquad u = k^2$

3. $3z^6 + z^3 - 10 = 0$ $\qquad u = z^3$

4. $2x^6 + 17x^3 + 8 = 0$ $\qquad u = x^3$

5. $4n^{-2} + 17n^{-1} + 15 = 0$ $\qquad u = n^{-1}$

6. $m^{-2} + 24 = 10m^{-1}$ $\qquad u = m^{-1}$

Exercises 7–24: Solve. Find all real solutions.

7. $x^4 = 8x^2 + 9$
8. $3x^4 = 10x^2 + 8$

9. $3x^6 - 5x^3 - 2 = 0$
10. $6x^6 + 11x^3 + 4 = 0$

11. $2z^{-2} + 11z^{-1} = 40$
12. $z^{-2} - 10z^{-1} + 25 = 0$

13. $x^{2/3} - 2x^{1/3} + 1 = 0$
14. $3x^{2/3} + 18x^{1/3} = 48$

15. $x^{2/5} - 33x^{1/5} = -32$
16. $x^{2/5} - 80x^{1/5} = 81$

17. $x - 13\sqrt{x} + 36 = 0$
18. $x - 17\sqrt{x} + 16 = 0$

19. $z^{1/2} - 2z^{1/4} + 1 = 0$
20. $z^{1/2} - 4z^{1/4} + 4 = 0$

21. $(x + 1)^2 - 5(x + 1) - 14 = 0$

22. $2(x - 5)^2 + 5(x - 5) + 3 = 0$

23. $(x^2 - 1)^2 - 4 = 0$

24. $(x^2 - 9)^2 - 8(x^2 - 9) + 16 = 0$

SOLVING EQUATIONS THAT HAVE COMPLEX SOLUTIONS

Exercises 25–34: Find all complex solutions.

25. $x^4 - 16 = 0$
26. $\frac{1}{3}x^4 - 27 = 0$

27. $x^3 + x = 0$
28. $4x^3 + x = 0$

29. $x^4 - 2 = x^2$
30. $x^4 - 3 = 2x^2$

31. $\dfrac{1}{x} + \dfrac{1}{x^2} = -\dfrac{1}{2}$ **32.** $\dfrac{2}{x-1} - \dfrac{1}{x} = -1$

33. $\dfrac{2}{x-2} - \dfrac{1}{x} = -\dfrac{1}{2}$ **34.** $\dfrac{1}{x} - \dfrac{1}{x^2} = \dfrac{1}{2}$

WRITING ABOUT MATHEMATICS

35. Explain how to solve $ax^4 - bx^2 + c = 0$. Assume that the left side of the equation factors.

36. Explain what it means for an equation to be reducible to quadratic form.

SECTIONS 11.5 and 11.6   **Checking Basic Concepts**

1. Solve the inequality $x^2 - x - 6 > 0$. Write your answer in interval notation.

2. Solve the inequality $3x^2 + 5x + 2 \le 0$. Write your answer in interval notation.

3. Solve $x^6 + 6x^3 - 16 = 0$.

4. Solve $x^{2/3} - 7x^{1/3} - 8 = 0$.

5. Find all complex solutions to $x^4 + 2x^2 + 1 = 0$.

CHAPTER 11   Summary

SECTION 11.1 ■ QUADRATIC FUNCTIONS AND THEIR GRAPHS

Quadratic Function Any quadratic function f can be written as

$$f(x) = ax^2 + bx + c \quad (a \ne 0).$$

Graph of a Quadratic Function Its graph is a parabola that opens upward if $a > 0$ and opens downward if $a < 0$. The y-intercept is $(0, c)$.

  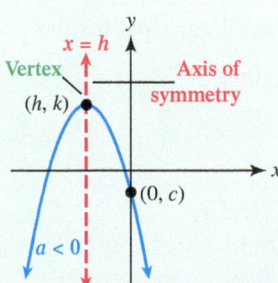

Axis of Symmetry The parabola is symmetric with respect to this vertical line. The axis of symmetry passes through the vertex. Its equation is $x = -\dfrac{b}{2a}$.

Vertex Formula The x-coordinate of the vertex is $-\dfrac{b}{2a}$.

Example: Let $y = x^2 - 4x + 1$ with $a = 1$ and $b = -4$.

$$x = -\dfrac{-4}{2(1)} = 2 \quad \text{and} \quad y = 2^2 - 4(2) + 1 = -3. \text{ The vertex is } (2, -3).$$

Min–Max Applications The graph of a quadratic function $f(x) = ax^2 + bx + c$ is a parabola that has vertex (h, k), where $h = -\dfrac{b}{2a}$ and $k = f(-\dfrac{b}{2a})$. If $a > 0$, the parabola opens upward and the minimum y-value is k. If $a < 0$, the parabola opens downward and the maximum y-value is k. These concepts can be used in certain applications to find maximum and minimum values.

Graph of $y = ax^2$ The graph of $y = ax^2$ is a parabola that has vertex $(0, 0)$. It opens upward if $a > 0$ and opens downward if $a < 0$. It is wider than the graph of $y = x^2$ if $0 < |a| < 1$, and narrower than the graph of $y = x^2$ if $|a| > 1$.

Vertical and Horizontal Translations Let h and k be positive numbers.

To graph	**shift the graph of $y = x^2$ by k units**
$y = x^2 + k$	upward.
$y = x^2 - k$	downward.

To graph	**shift the graph of $y = x^2$ by h units**
$y = (x - h)^2$	right.
$y = (x + h)^2$	left.

Example: Compared to $y = x^2$, the graph of $y = (x - 1)^2 + 2$ is translated right 1 unit and upward 2 units.

Vertex Form Any quadratic function can be expressed as $f(x) = a(x - h)^2 + k$. In this form, the point (h, k) is the vertex. A quadratic function can be put in this form by completing the square or by applying the vertex formula.

Example:
$$
\begin{aligned}
y &= x^2 + 10x - 4 && \text{Given equation}\\
&= (x^2 + 10x + 25) - 25 - 4 && \left(\tfrac{b}{2}\right)^2 = \left(\tfrac{10}{2}\right)^2 = 25;\ \text{complete the square.}\\
&= (x + 5)^2 - 29 && \text{Perfect square trinomial; subtract.}
\end{aligned}
$$
The vertex is $(-5, -29)$.

SECTION 11.3 ■ QUADRATIC EQUATIONS

Quadratic Equations Any quadratic equation can be written as $ax^2 + bx + c = 0$ and can have no real solutions, one real solution, or two real solutions. These solutions correspond to the x-coordinates of the x-intercepts for the graph of $y = ax^2 + bx + c$.

Example:
$$
\begin{aligned}
x^2 + x - 2 &= 0\\
(x + 2)(x - 1) &= 0\\
x = -2 \quad &\text{or} \quad x = 1
\end{aligned}
$$

The solutions are -2 and 1.
See the graph to the right.

x-intercepts:
$(-2, 0), (1, 0)$

$y = x^2 + x - 2$

Square Root Property Let k be a nonnegative number. Then the solutions to the equation $x^2 = k$ are given by $x = \pm\sqrt{k}$. If $k < 0$, then this equation has no real solutions.

Example: $x^2 = 5$ implies that $x = \pm\sqrt{5}$.

Completing the Square Write the equation in the form $x^2 + bx = d$. Complete the square by adding $\left(\tfrac{b}{2}\right)^2$ to each side of the equation.

Example:
$$
\begin{aligned}
x^2 - 8x &= 3 &&\\
x^2 - 8x + 16 &= 3 + 16 && \text{Add } \left(\tfrac{-8}{2}\right)^2 = 16 \text{ to each side.}\\
(x - 4)^2 &= 19 && \text{Perfect square trinomial}\\
x - 4 &= \pm\sqrt{19} && \text{Square root property}\\
x &= 4 \pm\sqrt{19} && \text{Add 4 to each side.}
\end{aligned}
$$

SECTION **11.4** ■ THE QUADRATIC FORMULA

The Quadratic Formula The solutions to $ax^2 + bx + c = 0$ $(a \neq 0)$ are given by

$$x = \frac{-b \pm \sqrt{b^2 - 4ac}}{2a}.$$

Example: Solve $2x^2 + 3x - 1 = 0$ by letting $a = 2$, $b = 3$, and $c = -1$.

$$x = \frac{-3 \pm \sqrt{3^2 - 4(2)(-1)}}{2(2)} = \frac{-3 \pm \sqrt{17}}{4} \approx 0.28, -1.78$$

The Discriminant The expression $b^2 - 4ac$ is called the discriminant. If $b^2 - 4ac > 0$, there are two real solutions; if $b^2 - 4ac = 0$, there is one real solution; and if $b^2 - 4ac < 0$, there are no real solutions—rather there are two complex solutions.

Example: For $2x^2 + 3x - 1 = 0$, the discriminant is

$$b^2 - 4ac = 3^2 - 4(2)(-1) = 17 > 0.$$

There are two real solutions to this quadratic equation, as shown in the previous example.

Quadratic Equations That Have Complex Solutions A quadratic equation sometimes has no real solutions.

Example: $x^2 + 4 = 0$

$$x^2 = -4 \qquad \text{Subtract 4 from each side.}$$

$$x = \pm 2i \qquad \text{Square root property; two complex solutions}$$

SECTION **11.5** ■ QUADRATIC INEQUALITIES

Quadratic Inequalities When the equals sign in a quadratic equation is replaced with $<, >, \leq,$ or $\geq$, a quadratic inequality results. For example,

$$3x^2 - x + 1 = 0$$

is a quadratic equation and

$$3x^2 - x + 1 > 0$$

is a quadratic inequality. Like quadratic equations, quadratic inequalities can be solved symbolically, graphically, and numerically. An important first step in solving a quadratic inequality is to solve the corresponding quadratic equation.

Examples: The solutions to $x^2 - 5x - 6 = 0$ are -1 and 6.

The solutions to $x^2 - 5x - 6 < 0$ satisfy $-1 < x < 6$.

The solutions to $x^2 - 5x - 6 > 0$ satisfy $x < -1$ or $x > 6$.

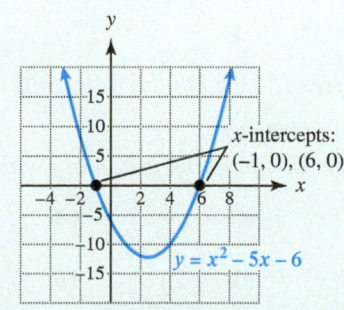

x-intercepts: $(-1, 0), (6, 0)$

$y = x^2 - 5x - 6$

SECTION **11.6** ■ EQUATIONS IN QUADRATIC FORM

Equations Reducible to Quadratic Form An equation that is not quadratic, but can be put into quadratic form by using a substitution, is reducible to quadratic form.

Example: To solve $x^{2/3} - 2x^{1/3} - 15 = 0$, let $u = x^{1/3}$. This equation becomes

$$u^2 - 2u - 15 = 0.$$

Factoring results in $(u + 3)(u - 5) = 0$, so $u = -3$ or $u = 5$.
Because $u = x^{1/3}$, $x = u^3$ and $x = (-3)^3 = -27$ or $x = (5)^3 = 125$.

CHAPTER 11   Review Exercises

SECTION 11.1

Exercises 1 and 2: Identify the vertex, axis of symmetry, and whether the parabola opens upward or downward.

1.

2.

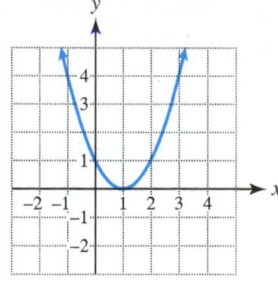

Exercises 3–6: Do the following.
  (a) *Graph f.*
  (b) *Identify the vertex and axis of symmetry.*
  (c) *Evaluate* $f(x)$ *at the given value of x.*

3. $f(x) = x^2 - 2$, $x = -1$

4. $f(x) = -x^2 + 4x - 3$, $x = 3$

5. $f(x) = -\frac{1}{2}x^2 + x + \frac{3}{2}$, $x = -2$

6. $f(x) = 2x^2 + 8x + 5$, $x = -3$

7. Find the minimum y-value located on the graph of $y = 2x^2 - 6x + 1$.

8. Find the maximum y-value located on the graph of $y = -3x^2 + 2x - 5$.

Exercises 9–12: Find the vertex of the parabola.

9. $f(x) = x^2 - 4x - 2$ **10.** $f(x) = 5 - x^2$

11. $f(x) = -\frac{1}{4}x^2 + x + 1$ **12.** $f(x) = 2 + 2x + x^2$

SECTION 11.2

Exercises 13–18: Do the following.

  (a) *Graph f.*
  (b) *Compare the graph of f with the graph of* $y = x^2$.

13. $f(x) = x^2 + 2$ **14.** $f(x) = 3x^2$

15. $f(x) = (x - 2)^2$ **16.** $f(x) = (x + 1)^2 - 3$

17. $f(x) = \frac{1}{2}(x + 1)^2 + 2$

18. $f(x) = 2(x - 1)^2 - 3$

19. Write the vertex form of a parabola with $a = -4$ and vertex $(2, -5)$.

20. Write the vertex form of a parabola that opens downward with vertex $(-4, 6)$. Assume that $a = \pm 1$.

Exercises 21–24: Write the equation in vertex form. Identify the vertex.

21. $y = x^2 + 4x - 7$ **22.** $y = x^2 - 7x + 1$

23. $y = 2x^2 - 3x - 8$ **24.** $y = 3x^2 + 6x - 2$

Exercises 25 and 26: Find a value for the constant a so that $f(x) = ax^2 - 1$ *models the data.*

25.

x	1	2	3
$f(x)$	2	11	26

26.

x	-1	0	1
$f(x)$	$-\frac{3}{4}$	-1	$-\frac{3}{4}$

Exercises 27 and 28: Write $f(x)$ *in the form given by* $f(x) = ax^2 + bx + c$. *Identify the y-intercept on the graph of* $y = f(x)$.

27. $f(x) = -5(x - 3)^2 + 4$

28. $f(x) = 3(x + 2)^2 - 4$

SECTION 11.3

Exercises 29–32: Use the graph of $y = ax^2 + bx + c$ *to solve* $ax^2 + bx + c = 0$.

29.

30.

31.

32.

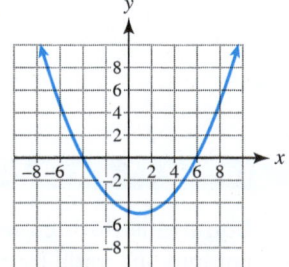

Exercises 33 and 34: A table of $y = ax^2 + bx + c$ is given. Solve $ax^2 + bx + c = 0$.

33.

34.

 Exercises 35–38: Solve the quadratic equation

   (a) *graphically and*
   (b) *numerically.*

35. $x^2 - 5x - 50 = 0$ **36.** $\frac{1}{2}x^2 + x - \frac{3}{2} = 0$

37. $\frac{1}{4}x^2 + \frac{1}{2}x = 2$ **38.** $\frac{1}{2}x + \frac{3}{4} = \frac{1}{4}x^2$

Exercises 39–42: Solve by factoring.

39. $x^2 + x - 20 = 0$ **40.** $x^2 + 11x + 24 = 0$

41. $15x^2 - 4x - 4 = 0$ **42.** $7x^2 - 25x + 12 = 0$

Exercises 43–46: Use the square root property to solve.

43. $x^2 = 100$ **44.** $3x^2 = \frac{1}{3}$

45. $4x^2 - 6 = 0$ **46.** $5x^2 = x^2 - 4$

Exercises 47–50: Solve by completing the square.

47. $x^2 + 6x = -2$ **48.** $x^2 - 4x = 6$

49. $x^2 - 2x - 5 = 0$ **50.** $2x^2 + 6x - 1 = 0$

Exercises 51 and 52: Solve for the specified variable.

51. $F = \dfrac{k}{(R + r)^2}$ for R **52.** $2x^2 + 3y^2 = 12$ for y

SECTION 11.4

Exercises 53–58: Use the quadratic formula to solve.

53. $x^2 - 9x + 18 = 0$ **54.** $x^2 - 24x + 143 = 0$

55. $6x^2 + x = 1$ **56.** $5x^2 + 1 = 5x$

57. $x(x - 8) = 5$ **58.** $2x(2 - x) = 3 - 2x$

Exercises 59–64: Solve by any method.

59. $x^2 - 4 = 0$ **60.** $4x^2 - 1 = 0$

61. $2x^2 + 15 = 11x$ **62.** $2x^2 + 15 = 13x$

63. $x(5 - x) = 2x + 1$ **64.** $-2x(x - 1) = x - \frac{1}{2}$

Exercises 65–68: Graphs of $y = ax^2 + bx + c$ are shown at the top of the next column.

   (a) *State whether $a > 0$ or $a < 0$.*
   (b) *Solve $ax^2 + bx + c = 0$.*
   (c) *Determine whether the discriminant is positive, negative, or zero.*

65.   **66.**

67.   **68.**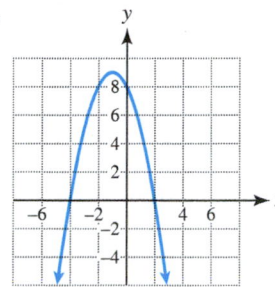

Exercises 69–72: Do the following for the given equation.

   (a) *Evaluate the discriminant.*
   (b) *How many real solutions are there?*
   (c) *Support your answer for part (b) graphically.*

69. $2x^2 - 3x + 1 = 0$ **70.** $7x^2 + 2x - 5 = 0$

71. $3x^2 + x + 2 = 0$

72. $4.41x^2 - 12.6x + 9 = 0$

Exercises 73–76: Solve. Write any complex solutions in standard form.

73. $x^2 + x + 5 = 0$ **74.** $2x^2 + 8 = 0$

75. $2x^2 = x - 1$ **76.** $7x^2 = 2x - 5$

SECTION 11.5

Exercises 77 and 78: The graph of $y = ax^2 + bx + c$ is shown. Solve each equation or inequality.

   (a) $ax^2 + bx + c = 0$
   (b) $ax^2 + bx + c < 0$
   (c) $ax^2 + bx + c > 0$

77.   **78.**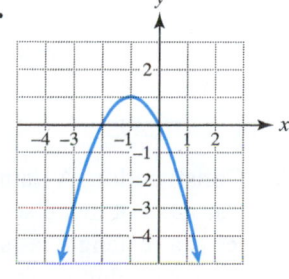

Exercises 79 and 80: A table of $y = ax^2 + bx + c$ is shown. Solve each equation or inequality.

(a) $ax^2 + bx + c = 0$
(b) $ax^2 + bx + c < 0$
(c) $ax^2 + bx + c > 0$

79. $y = x^2 - 16$

x	−6	−4	−2	0	2	4	6
y	20	0	−12	−16	−12	0	20

80. $y = x^2 + x - 2$

x	−3	−2	−1	0	1	2	3
y	4	0	−2	−2	0	4	10

Exercises 81 and 82: Solve the equation in part (a). Use the results to solve the inequalities in parts (b) and (c).

81. (a) $x^2 - 2x - 3 = 0$
(b) $x^2 - 2x - 3 < 0$
(c) $x^2 - 2x - 3 > 0$

82. (a) $2x^2 - 7x - 15 = 0$
(b) $2x^2 - 7x - 15 \le 0$
(c) $2x^2 - 7x - 15 \ge 0$

Exercises 83–88: Solve the quadratic inequality. Write your answer in interval notation.

83. $x^2 + 4x + 3 \le 0$ **84.** $5x^2 - 16x + 3 < 0$

85. $6x^2 - 13x + 2 > 0$ **86.** $x^2 \ge 5$

87. $(x - 1)^2 \ge 0$ **88.** $x^2 + 3 < 2$

SECTION 11.6

Exercises 89–92: Solve the equation.

89. $x^4 - 14x^2 + 45 = 0$

90. $2z^{-2} + z^{-1} - 28 = 0$

91. $x^{2/3} - 9x^{1/3} + 8 = 0$

92. $(x - 1)^2 + 2(x - 1) + 1 = 0$

Exercises 93 and 94: Find all complex solutions.

93. $4x^4 + 4x^2 + 1 = 0$ **94.** $\dfrac{1}{x - 2} - \dfrac{3}{x} = -1$

APPLICATIONS

95. Construction A rain gutter is being made from a flat sheet of metal so that the cross section of the gutter is a rectangle, as shown in the figure at the top of the next column. The width of the metal sheet is 12 inches.

(a) Write a formula $f(x)$ that gives the area of the cross section.
(b) To hold the greatest amount of rainwater, the cross section should have maximum area. Find the dimensions that result in this maximum.

96. Height of a Stone Suppose that a stone is thrown upward with an initial velocity of 44 feet per second (30 miles per hour) and is released 4 feet above the ground. Its height h in feet after t seconds is given by

$$h(t) = -16t^2 + 44t + 4.$$

(a) When does the stone reach a height of 32 feet?
(b) After how many seconds does the stone reach maximum height? Estimate this height.

97. Maximizing Revenue Suppose that hotel rooms cost $90 per night. However, for a group rate the management is considering reducing the cost of a room by $3 for every room rented.
(a) Write a formula $f(x)$ that gives the revenue for x rooms at the group rate.
(b) Graph f in $[0, 30, 5]$ by $[0, 800, 100]$.
(c) How many rooms should be rented to receive revenue of $600?
(d) How many rooms should be rented to maximize revenue?

98. Numbers The product of two numbers is 143. One number is 2 more than the other.
(a) Write a quadratic equation whose solution gives the smaller number x.
(b) Solve the equation.

99. Braking Distance On dry pavement, a safe braking distance d in feet for a car traveling x miles per hour is $d = \frac{x^2}{12}$. For each distance d, find x. (*Source*: F. Mannering, *Principles of Highway Engineering and Traffic Control.*)
(a) $d = 144$ feet
(b) $d = 300$ feet

100. U.S. Energy Consumption From 1950 to 1970, per capita consumption of energy in millions of Btu can be modeled by $f(x) = \frac{1}{4}(x - 1950)^2 + 220$, where x is the year. (*Source*: Department of Energy.)
(a) Find and interpret the vertex.

(b) Graph f in $[\,1950, 1970, 5\,]$ by $[\,200, 350, 25\,]$. What happened to energy consumption during this time period?

(c) Use f to predict the consumption in 2010. Actual consumption was 321 million Btu. Did f provide a good model for 2010? Explain.

101. Screens A square computer screen has an area of 123 square inches. Approximate its dimensions to the nearest tenth of an inch.

102. Flying a Kite A kite is being flown, as illustrated in the accompanying figure. If 130 feet of string have been let out, find the value of x.

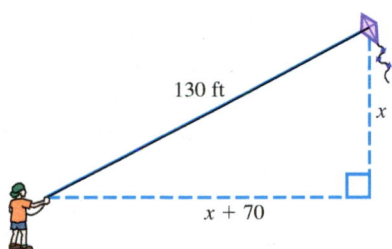

130 ft

x

$x + 70$

103. Area A uniform strip of grass is to be planted around a rectangular swimming pool, as illustrated in the accompanying figure. The swimming pool is 30 feet wide and 50 feet long. If there is only enough grass seed to cover 250 square feet, estimate the width x that the strip of grass should be.

104. Dimensions of a Cone The volume V of a cone is given by $V = \frac{1}{3}\pi r^2 h$, where r is its base radius and h is its height. See the accompanying figure. If $h = 20$ inches and the volume of the cone must be between 750 and 1700 cubic inches, inclusively, estimate, to the nearest tenth of an inch, possible values for r.

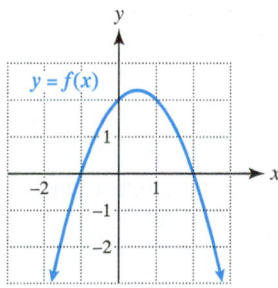

CHAPTER 11   Test   MyMathLab®, YouTube

1. Find the vertex and axis of symmetry for the graph of $f(x) = -\frac{1}{2}x^2 + x + 1$. Evaluate $f(-2)$.

2. Find the minimum y-value located on the graph of $y = x^2 + 3x - 5$.

3. Find the exact value for the constant a so that $f(x) = ax^2 + 2$ models the data in the table.

x	-2	0	2	4
$f(x)$	0	2	0	-6

4. Compare the graph of $y = f(x)$ to the graph of $y = x^2$. Then graph $y = f(x)$.
(a) $f(x) = (x - 1)^2$ **(b)** $f(x) = x^2 - 2$
(c) $f(x) = \frac{1}{2}(x - 3)^2 + 2$

5. Write $y = x^2 - 6x + 2$ in vertex form. Identify the vertex and axis of symmetry.

6. Use the graph of $f(x) = ax^2 + bx + c$ to solve $ax^2 + bx + c = 0$. Then evaluate $f(1)$.

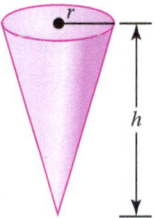

$y = f(x)$

Exercises 7 and 8: Solve the quadratic equation.

7. $3x^2 + 11x - 4 = 0$ **8.** $2x^2 = 2 - 6x^2$

9. Solve $x^2 - 8x = 1$ by completing the square.

10. Solve $x(-2x + 3) = -1$ by using the quadratic formula.

11. Solve $9x^2 - 16 = 0$.

12. Solve $F = \dfrac{Gm^2}{r^2}$ for m.

13. A graph of $y = ax^2 + bx + c$ is shown.
   (a) State whether $a > 0$ or $a < 0$.
   (b) Solve $ax^2 + bx + c = 0$.
   (c) Determine whether the discriminant is positive, negative, or zero.

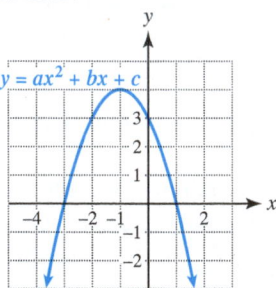

14. Complete the following for $-3x^2 + 4x - 5 = 0$.
   (a) Evaluate the discriminant.
   (b) How many real solutions are there?
   (c) Support your answer for part (b) graphically.

Exercises 15 and 16: The graph of $y = ax^2 + bx + c$ is shown. Solve each equation or inequality.
   (a) $ax^2 + bx + c = 0$
   (b) $ax^2 + bx + c < 0$
   (c) $ax^2 + bx + c > 0$

15.

16.

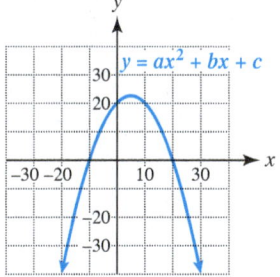

17. Solve the quadratic equation in part (a). Use the result to solve the inequalities in parts (b) and (c) and write your answer in interval notation.
   (a) $8x^2 - 2x - 3 = 0$
   (b) $8x^2 - 2x - 3 \le 0$
   (c) $8x^2 - 2x - 3 \ge 0$

18. Solve $x^2 + 2x \le 0$.

19. Solve $x^6 - 3x^3 + 2 = 0$. Find all real solutions.

20. Solve $2x^2 + 4x + 3 = 0$. Write all complex solutions in standard form.

21. Solve $\sqrt{2} - \pi x^2 = 2.12x - 0.5\pi$ graphically. Round your answers to the nearest hundredth.

22. Braking Distance On wet pavement, a safe braking distance d in feet for a car traveling x miles per hour is $d = \frac{x^2}{9}$. What speed corresponds to a braking distance of 250 feet? (*Source:* F. Mannering, *Principles of Highway Engineering and Traffic Control.*)

23. Construction A fence is being constructed along a 20-foot building, as shown in the accompanying figure. No fencing is used along the building.
   (a) If 200 feet of fence are available, find a formula $f(x)$ using only the variable x that gives the area enclosed.
   (b) What value of x gives the greatest area?

24. Height of a Stone Suppose that a stone is thrown upward with an initial velocity of 88 feet per second (60 miles per hour) and is released 8 feet above the ground. Its height h in feet after t seconds is given by
$$h(t) = -16t^2 + 88t + 8.$$
   (a) Graph h in $[0, 6, 1]$ by $[0, 150, 50]$.
   (b) When does the stone strike the ground?
   (c) After how many seconds does the stone reach maximum height? Estimate this height.

CHAPTERS 1–11   Cumulative Review Exercises

1. Evaluate $F = \dfrac{5}{z^2 + 1}$ when $z = -2$.

2. Classify each number as one or more of the following: natural number, whole number, integer, rational number, or irrational number:
$$0.\overline{4}, \ \sqrt{7}, \ 0, \ -5, \ \sqrt[3]{8}, \ -\tfrac{4}{3}.$$

3. Simplify each expression. Write the result using positive exponents.
   (a) $\left(\dfrac{x^2 y^6}{x^{-3}}\right)^2$
   (b) $\dfrac{(xy^{-3})^2}{x(y^{-2})^{-1}}$
   (c) $(a^2 b)^2 (ab^3)^{-4}$

4. Write 9,290,000 in scientific notation.

5. Find $f(-2)$, if $f(x) = \sqrt{2 - x}$. Write the domain of f?

6. If $f(2) = 5$, then what point lies on the graph of f?

Exercises 7 and 8: Graph f by hand.

7. $f(x) = x^2 + 2x$

8. $f(x) = |2x - 4|$

9. Find the slope–intercept form of the line that passes through $(4, -1)$ and is parallel to the line passing through $(0, 1)$ and $(-2, 4)$.

10. Find the equation of a vertical line that passes through the point $(-3, 4)$.

11. Solve $2x - 3(x + 2) = 6$.

Exercises 12–14: Solve the inequality. Write your answer in interval notation.

12. $7 - x > 3x$

13. $|3x - 2| \le 1$

14. $-4 \le 1 - x < 2$

15. Solve the system.
$$-x - 4y = -3$$
$$5x + y = -4$$

16. Shade the solution set in the xy-plane.
$$3x + y \le 3$$
$$x - 3y \le 3$$

17. Solve the system.
$$x + y - z = 3$$
$$x - y + z = 1$$
$$2x - y - z = 1$$

Exercises 18–20: Multiply the expression.

18. $(3x - 2)(2x + 7)$

19. $3xy(x^2 + y^2)$

20. $(\sqrt{x} + 3)(\sqrt{x} - 3)$

Exercises 21 and 22: Factor the expression.

21. $x^3 - x^2 - 2x$ **22.** $4x^2 - 25$

Exercises 23 and 24: Solve each equation.

23. $x^2 - 3 = 0$ **24.** $x^2 + 1 = 2x$

Exercises 25 and 26: Simplify the expression.

25. $\dfrac{(x + 3)^2}{x + 2} \cdot \dfrac{x + 2}{2x + 6}$ **26.** $\dfrac{1}{x + 2} - \dfrac{1}{x}$

Exercises 27–30: Simplify the expression. Assume all variables are positive.

27. $\sqrt{16x^6}$ **28.** $16^{-3/2}$

29. $\dfrac{\sqrt[3]{81x}}{\sqrt[3]{3x}}$ **30.** $\sqrt{8x} + \sqrt{2x}$

31. Graph $f(x) = \sqrt{4x}$.

32. Find the distance between $(-1, 2)$ and $(4, 3)$.

33. Write $\frac{3 - i}{2 + i}$ in standard form.

34. Solve $3\sqrt{x + 1} = 2x$.

35. Solve the equation $2x = \sqrt{2.1 - x} + \sqrt[3]{0.1x}$ to the nearest hundredth.

36. Sketch a graph of $f(x) = x^2 - 2x + 3$.
   (a) Find the vertex.
   (b) Evaluate $f(-1)$.
   (c) What is the axis of symmetry?
   (d) Where is f increasing?

37. Write $f(x) = 2x^2 - 4x - 1$ in vertex form.

38. Compare the graph of $f(x) = 4(x + 1)^2 - 2$ to the graph of $y = x^2$.

39. Solve $x^2 + 6x = 2$ by completing the square.

40. Solve $2x^2 - 3x = 1$ by using the quadratic formula.

41. Solve $x(4 - x) = 3$.

42. The graph of $y = ax^2 + bx + c$ is shown. Solve each equation or inequality.
   (a) $ax^2 + bx + c = 0$
   (b) $ax^2 + bx + c \le 0$

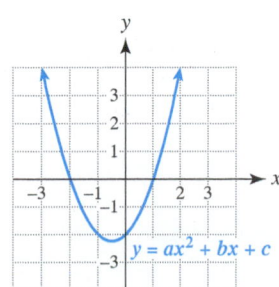

43. Solve $x^2 - 3x + 2 > 0$.

44. Solve $x^4 - 256 = 0$. Find all complex solutions.

Exercises 45–52: **Thinking Generally** *Match the graph (a–h) with its equation. Assume that a, b, and c are positive constants.*

45. $y = ax - b$

46. $y = b$

47. $y = -ax^2 + c$

48. $y = \dfrac{a}{x}$

49. $y = ax^3$

50. $y = |ax + b|$

51. $y = a\sqrt{x}$

52. $y = a\sqrt[3]{x}$

a.

b.

c.

d.

e.

f.

g.

h.

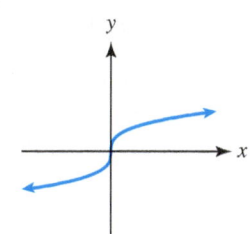

APPLICATIONS

53. Calculating Water Flow Water is being pumped out of a tank. The number of gallons G of water in the tank after t minutes is shown in the figure, where $y = G(t)$.

(a) Evaluate $G(0)$. Interpret your answer.

(b) What is the t-intercept? Interpret your answer.

(c) What is the slope of the graph of G? Interpret your answer.

(d) Find a formula for $G(t)$.

54. Investment Suppose $4000 is deposited in three accounts paying 4%, 5%, and 6% annual interest. The amount invested at 6% is $1000 more than the amount invested at 5%. The interest after 1 year is $216. How much is invested at each rate?

55. Maximizing Area There are 490 feet of fence available to surround the perimeter of a rectangular garden. On one side, there is a 10-foot gate that requires no fencing. What dimensions for the garden give the largest area?

56. Height of a Tree A 6-foot-tall person casts a shadow that is 10 feet long while a nearby tree casts a 55-foot shadow. Find the height of the tree.

12

Exponential and Logarithmic Functions

When a link is posted on a social network, the majority of hits on the link happen within the first few hours. For example, within the first 3 hours of posting a link on Facebook, the link will receive half of its hits. This pattern can continue over time so the number of hits on the link is decreasing by a factor of $\frac{1}{2}$ every 3 hours. This is sometimes referred to as the half-life of a link. The half-life is even shorter for Twitter links, 2.8 hours, but is much longer, 400 hours, for StumbleUpon. In this chapter, we use exponential functions to model these engagements with links posted on social networks. See Exercises 95–96 in Section 2 of this chapter.

12.1   Composite and Inverse Functions

NEW VOCABULARY

☐ Composite function
☐ Composition
☐ One-to-one correspondence
☐ One-to-one
☐ Horizontal line test
☐ Inverse function

1   Introducing Composition of Functions

Connecting Concepts with Your Life   Many tasks in life are performed in *sequence*, such as putting on your socks and then your shoes. These types of situations also occur in mathematics. For example, suppose that we want to calculate the number of ounces in 3 tons. Because there are 2000 pounds in 1 ton, we might first multiply **3** by 2000 to obtain **6000** pounds. There are 16 ounces in a pound, so we could multiply **6000** by 16 to obtain **96,000** ounces. This particular calculation involves a *sequence* of calculations that can be represented by the diagram shown in **FIGURE 12.1**.

Converting Tons to Ounces

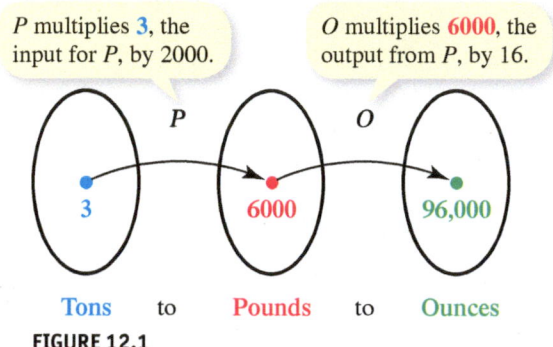

P multiplies **3**, the input for P, by 2000.

O multiplies **6000**, the output from P, by 16.

FIGURE 12.1

SYMBOLIC EVALUATION   The results shown in **FIGURE 12.1** can be calculated by using function notation. Suppose that we let $P(x) = 2000x$ convert x tons to P pounds and also let $O(x) = 16x$ convert x pounds to O ounces. We can calculate the number of ounces in 3 tons by performing the *composition of O and P*. This method can be expressed symbolically as follows.

Composition: $(O \circ P)(3)$ $= O\big(P(3)\big)$

Input 3 $= O(2000 \cdot 3)$ Let $x = 3$ in $P(x) = 2000x$.

$\downarrow$ $\downarrow$

$P(x)$ $P(3) = 6000.$

$\downarrow$

Output $P(3)$ $= O(6000)$ Simplify.
becomes input for $O(x)$. $= 16(6000)$ Let $x = 6000$ in $O(x) = 16x$.

$\downarrow$ $\downarrow$

$(O \circ P)(x) = O(P(x))$ P first multiplies 3 by 2000 and then O multiplies 6000 by 16 to get 96,000. This is written as $(O \circ P)(3)$.

$\downarrow$ $\downarrow$

Output $O(P(3))$ $= 96{,}000.$ Multiply.

COMPOSITION OF FUNCTIONS

If f and g are functions, then the **composite function** $g \circ f$, or **composition** of g and f, is defined by

$$(g \circ f)(x) = g\big(f(x)\big).$$

NOTE: We read $g\big(f(x)\big)$ as "g of f of x." ∎

READING CHECK 1

- Calculate $(f \circ g)(2)$ and $(g \circ f)(2)$ if $f(x) = x^2$ and $g(x) = 2x - 1$.

MAKING CONNECTIONS 1

$g \circ f$ vs. $f \circ g$

The compositions $g \circ f$ and $f \circ g$ represent evaluating functions f and g in different orders. When evaluating $g \circ f$, function f is evaluated first followed by function g, whereas for $f \circ g$, the function g is evaluated first followed by function f. Note that, in general

$$(g \circ f)(x) \neq (f \circ g)(x).$$

That is, *the order in which functions are applied makes a difference*, in the same way that putting on your socks and then your shoes is quite different from putting on your shoes and then your socks. See Example 2(a) and (b).

EXAMPLE 1 **Evaluating composite functions symbolically**

Evaluate $(g \circ f)(2)$ and then find a formula for $(g \circ f)(x)$.

(a) $f(x) = x^3, g(x) = 3x - 2$ (b) $f(x) = 5x, g(x) = x^2 - 3x + 1$

(c) $f(x) = \sqrt{2x}, g(x) = \dfrac{1}{x - 1}$

Solution

(a) $\begin{aligned}(g \circ f)(2) &= g\big(f(2)\big) && \text{Composition of functions} \\ &= g(8) && f(2) = 2^3 = 8 \\ &= 22 && g(8) = 3(8) - 2 = 22\end{aligned}$

Note that the output from f, which is $f(2)$, becomes the input for g.

$\begin{aligned}(g \circ f)(x) &= g\big(f(x)\big) && \text{Composition of functions} \\ &= g(x^3) && f(x) = x^3 \\ &= 3x^3 - 2 && \text{Replace } x \text{ with } x^3 \text{ in } g(x) = 3x - 2.\end{aligned}$

(b) $\begin{aligned}(g \circ f)(2) &= g\big(f(2)\big) && \text{Composition of functions} \\ &= g(10) && f(2) = 5(2) = 10 \\ &= 71 && g(10) = 10^2 - 3(10) + 1 = 71\end{aligned}$

$\begin{aligned}(g \circ f)(x) &= g\big(f(x)\big) && \text{Composition of functions} \\ &= g(5x) && f(x) = 5x \\ &= (5x)^2 - 3(5x) + 1 && \text{Replace } x \text{ with } 5x \text{ in } g(x) = x^2 - 3x + 1. \\ &= 25x^2 - 15x + 1 && \text{Simplify.}\end{aligned}$

(c) $\begin{aligned}(g \circ f)(2) &= g\big(f(2)\big) && \text{Composition of functions} \\ &= g(2) && f(2) = \sqrt{2(2)} = 2 \\ &= 1 && g(2) = \dfrac{1}{2 - 1} = 1\end{aligned}$

$\begin{aligned}(g \circ f)(x) &= g\big(f(x)\big) && \text{Composition of functions} \\ &= g(\sqrt{2x}) && f(x) = \sqrt{2x} \\ &= \dfrac{1}{\sqrt{2x} - 1} && \text{Replace } x \text{ with } \sqrt{2x} \text{ in } g(x) = \dfrac{1}{x - 1}.\end{aligned}$

Now Try Exercises 13, 15, 19

NUMERICAL AND GRAPHICAL EVALUATION Composite functions can also be evaluated numerically and graphically, as demonstrated in the next two examples.

EXAMPLE 2   **Evaluating composite functions with tables**

Use **TABLES 12.1** and **12.2** to evaluate each expression.

(a) $(f \circ g)(2)$ **(b)** $(g \circ f)(2)$ **(c)** $(f \circ f)(0)$

x	0	1	2	3
$f(x)$	3	2	0	1

TABLE 12.1

x	0	1	2	3
$g(x)$	1	3	2	0

TABLE 12.2

Solution

(a) $(f \circ g)(2) = f\big(g(2)\big)$ Composition of functions

$\qquad\qquad = f(2)$ $g(2) = 2$

$\qquad\qquad = 0$ $f(2) = 0$

(b) $(g \circ f)(2) = g\big(f(2)\big)$ Composition of functions

$\qquad\qquad = g(0)$ $f(2) = 0$

$\qquad\qquad = 1$ $g(0) = 1$

NOTE: From parts (a) and (b), we see that $(f \circ g)(2) \neq (g \circ f)(2)$. ∎

(c) $(f \circ f)(0) = f\big(f(0)\big)$ Composition of functions

$\qquad\qquad = f(3)$ $f(0) = 3$

$\qquad\qquad = 1$ $f(3) = 1$

Now Try Exercises 23, 25

EXAMPLE 3   **Evaluating composite functions graphically**

Use **FIGURE 12.2** in the margin to evaluate $(g \circ f)(2)$.

Solution

Because $(g \circ f)(2) = g\big(f(2)\big)$, start by using **FIGURE 12.2** to evaluate $f(2)$. **FIGURE 12.3(a)** shows that $f(2) = 4$, which becomes the input for g. **FIGURE 12.3(b)** reveals that $g(4) = 2$.

$$(g \circ f)(2) = g\big(f(2)\big) = g(4) = 2$$

FIGURE 12.2

Evaluate: $f(2) = 4$

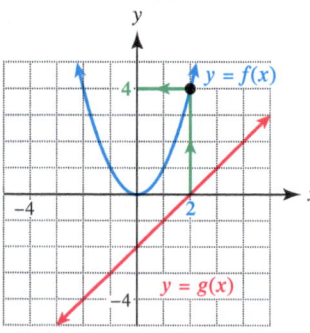

(a)

Evaluate: $g(4) = 2$

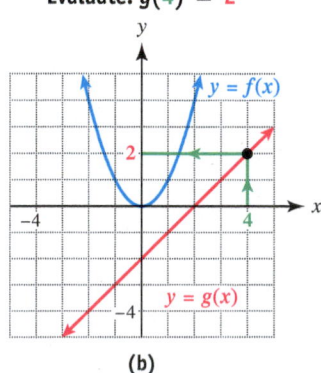

(b)

FIGURE 12.3

Now Try Exercise 29

---

2 ▸ Recognizing One-to-One Functions

🌐 *Math in Context* (Real-World) In a typical city, there is *usually* a **one-to-one correspondence** between homes and addresses. That is to say that each home has one address, and each address corresponds to one home. We can apply this concept to functions and ask the question, do *different inputs* always result in *different outputs* for every function? The answer is no. Use $f(x) = x^2 + 1$ as an example.

Different Inputs

$$f(-2) = 5 \text{ and } f(2) = 5.$$

Same Output

However, for $g(x) = 2x$, *different inputs* always result in *different outputs*. For function g, there is a one-to-one correspondence between inputs and outputs. Thus we say that g is a *one-to-one function*, whereas f is not.

ONE-TO-ONE FUNCTION

A function f is **one-to-one** if, for any c and d in the domain of f,

$$c \neq d \quad \text{implies that} \quad f(c) \neq f(d).$$

That is, different inputs always result in different outputs.

HORIZONTAL LINE TEST The following See the Concept shows how to use the horizontal line test to determine if a function is one-to-one.

See the Concept 1 — HORIZONTAL LINE TEST

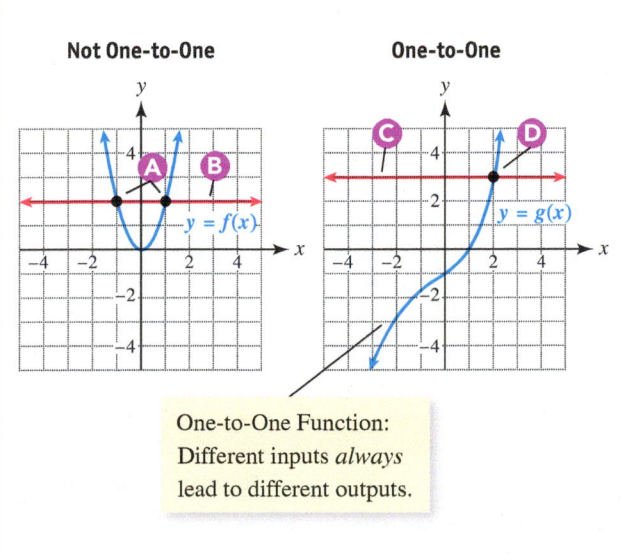

Not One-to-One

One-to-One

One-to-One Function:
Different inputs *always*
lead to different outputs.

Ⓐ The points $(-1, 2)$ and $(1, 2)$ both lie on the graph of f and so $f(-1) = f(1) = 2$. Since different inputs do *not always* lead to different outputs, the function is not one-to-one.

Ⓑ Two points with different x-values and the same y-value determine a horizontal line as shown. Thus if any horizontal line intersects the graph of f at more than one point, f is *NOT* one-to-one.

Ⓒ To determine if a function is one-to-one by its graph, apply the **horizontal line test** and visualize a horizontal line moving up and down over the graph. For a function to be one-to-one, every possible horizontal line can intersect the graph *at most once*.

Ⓓ Because every horizontal line intersects the graph at most once, g is a one-to-one function.

WATCH VIDEO IN MML.

READING CHECK 2

• How can you determine if a function is one-to-one by looking at its graph?

HORIZONTAL LINE TEST

If every horizontal line intersects the graph of a function f at most once, then f is a one-to-one function.

EXAMPLE 4   ## Using the horizontal line test

Determine whether each graph in **FIGURE 12.4** represents a one-to-one function.

(a)

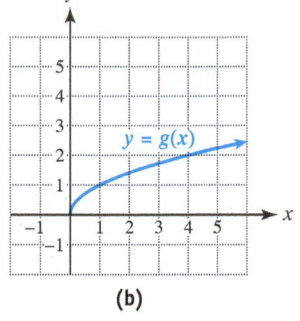
(b)

FIGURE 12.4

Solution

FIGURE 12.5(a) shows one of many horizontal lines that intersect the graph of $y = f(x)$ twice. Therefore, function f is *not* one-to-one.

Two Points of Intersection:
f Is Not One-to-One

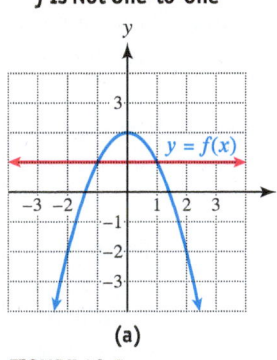
(a)

At Most One Point of
Intersection: g Is One-to-One

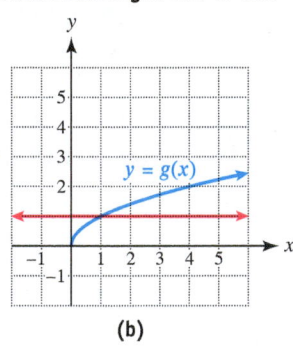
(b)

FIGURE 12.5

FIGURE 12.5(b) suggests that every horizontal line will intersect the graph of $y = g(x)$ *at most* once. Therefore, function g is one-to-one.

Now Try Exercises 37, 39

MAKING CONNECTIONS 2

STUDY TIP

Be sure that you understand when to use the vertical line test and when to use the horizontal line test.

Vertical and Horizontal Line Tests

The **vertical line test** is used to identify **functions**, whereas the **horizontal line test** is used to identify **one-to-one** functions. For example, consider the graph of $f(x) = x^2$. A vertical line never intersects the graph more than once, so f is a function. A horizontal line can intersect the graph twice, so f is *not* a one-to-one function.

Function; Not One-to-One

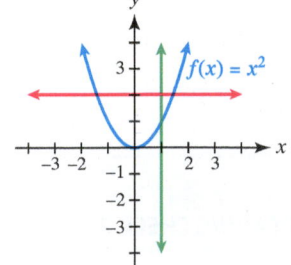

INCREASING, DECREASING, AND ONE-TO-ONE FUNCTIONS When a continuous function f is always increasing on its domain, every horizontal line will intersect the graph of f at most once. By the horizontal line test, f is a one-to-one function. For example, function f in Example 4(b) is always increasing on its domain and so it is one-to-one. Similarly, when a continuous function g is always decreasing on its domain, g is a one-to-one function.

3 Finding Inverse Functions

INVERSE OPERATIONS There are many examples of inverse operations in everyday life.

 Connecting Concepts with Your Life Suppose that you walk into a classroom, turn on the lights, and sit down at your desk. How could you undo or reverse these actions? You would stand up from the desk, turn off the lights, and walk out of the classroom. Note that you must not only perform the "inverse" of each action, but you also must do them in the *reverse order*. In mathematics, we undo an arithmetic operation by performing its inverse operation. For example, the inverse operation of addition is subtraction, and the inverse operation of multiplication is division. Adding **5** to x and subtracting **5** from x are inverse operations.

$$x + 5 - 5 = x$$

> Adding 5 and then subtracting 5 gives x back.

Similarly, multiplying x by **5** and dividing x by **5** are inverse operations.

$$\frac{5x}{5} = x$$

> Multiplying by 5 and then dividing by 5 gives x back.

EXAMPLE 5 **Finding inverse operations**

Give the inverse operations for each statement. Then write a function f for the given statement and a function g for its inverse operations.
(a) Divide x by 3. **(b)** Cube x and then add 1 to the result.

Solution
(a) The inverse of dividing x by 3 is to *multiply* x by 3. Thus

$$f(x) = \frac{x}{3} \quad \text{and} \quad g(x) = 3x.$$

(b) The inverse of cubing a number is taking a cube root, and the inverse of adding 1 is subtracting 1. The inverse operations of "cubing a number and then adding 1" are "subtracting 1 and then taking a cube root." For example, 2 cubed plus 1 is $2^3 + 1 = 9$. For the inverse operations, we *first* subtract 1 from 9 and then take the cube root to obtain 2. That is, $\sqrt[3]{9 - 1} = 2$. When there is more than one operation, we must perform the inverse operations in *reverse order*. Thus

$$f(x) = x^3 + 1 \quad \text{and} \quad g(x) = \sqrt[3]{x - 1}.$$

Now Try Exercises 43, 49

Inverse Functions f and g

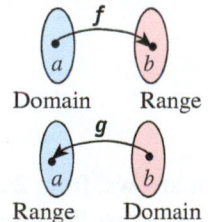

The domain of f is the range of g, and the domain of g is the range of f.

Functions f and g in each part of Example 5 are examples of *inverse functions*. Note that in part (a), if $f(x) = \frac{x}{3}$ and $g(x) = 3x$, then

$$f(\mathbf{15}) = \mathbf{5} \quad \text{and} \quad g(\mathbf{5}) = \mathbf{15}.$$

> If f outputs b with input a, g must output a with input b.

See the diagram to the left.

In general, if f and g are inverse functions, $f(\mathbf{a}) = \mathbf{b}$ implies $g(\mathbf{b}) = \mathbf{a}$. Thus

$$(g \circ f)(\mathbf{a}) = g(f(\mathbf{a})) = g(\mathbf{b}) = \mathbf{a}$$

for any $\mathbf{a}$ in the domain of f, whenever g and f are inverse functions. The composition of a function with its inverse leaves the input unchanged.

READING CHECK 3

- What happens to the input of the composition of a function with its inverse?

INVERSE FUNCTIONS

Let f be a one-to-one function. Then f^{-1} is the **inverse function** of f if

$$(f^{-1} \circ f)(x) = f^{-1}(f(x)) = x, \quad \text{for every } x \text{ in the domain of } f, \quad \text{and}$$
$$(f \circ f^{-1})(x) = f(f^{-1}(x)) = x, \quad \text{for every } x \text{ in the domain of } f^{-1}.$$

NOTE: In the expression $f^{-1}(x)$, the -1 is *not* an exponent. That is, $f^{-1}(x) \neq \frac{1}{f(x)}$. Rather, if $f(x) = \frac{x}{3}$, then $f^{-1}(x) = 3x$, and if $f(x) = x^3 + 1$, then $f^{-1}(x) = \sqrt[3]{x - 1}$. ∎

EXAMPLE 6 **Verifying inverses**

Verify that $f^{-1}(x) = 3x$ if $f(x) = \frac{x}{3}$.

Solution
We must show that $(f^{-1} \circ f)(x) = x$ and that $(f \circ f^{-1})(x) = x$.

$$
\begin{aligned}
(f^{-1} \circ f)(x) &= f^{-1}(f(x)) && \text{Composition of functions} \\
&= f^{-1}\left(\frac{x}{3}\right) && f(x) = \frac{x}{3} \\
&= 3\left(\frac{x}{3}\right) && f^{-1}(x) = 3x \\
&= x \checkmark && \text{Simplify; it checks.} \\
(f \circ f^{-1})(x) &= f(f^{-1}(x)) && \text{Composition of functions} \\
&= f(3x) && f^{-1}(x) = 3x \\
&= \frac{3x}{3} && f(x) = \frac{x}{3} \\
&= x \checkmark && \text{Simplify; it checks.}
\end{aligned}
$$

Now Try Exercise 51

The definition of inverse functions states that *f must be a one-to-one function.* To understand why, consider **FIGURE 12.6(a)**, where a one-to-one function f is represented by a diagram. To find f^{-1}, the arrows are simply reversed. For example, $f(1) = 3$ implies that $f^{-1}(3) = 1$, so the arrow from **1** to **3** for f must be redrawn in **FIGURE 12.6(b)**, from **3** to **1** for f^{-1}.

One-to-One, So Inverse Exists   **Inverse Is Also One-to-One**

No two arrows go to the same output.   Reverse each arrow to find f^{-1}.

(a) $f(1) = 3$ (b) $f^{-1}(3) = 1$

FIGURE 12.6

In **FIGURE 12.7(a)**, function g is *not* one-to-one because two different arrows, from **2** and **3**, both go to **1**. In **FIGURE 12.7(b)**, the arrows are reversed in an attempt to draw the inverse *function.* However, **two** arrows now go from input **1** to outputs **2** and **3**. This means that *one input* has *two outputs*, which is not possible if g^{-1} is a function.

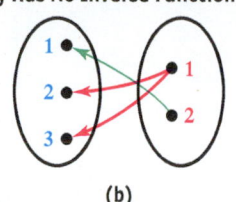

g Is Not One-to-One

Two inputs give the same output: not one-to-one.

g Has No Inverse Function

One input gives two outputs: not a function.

(a)   (b)

FIGURE 12.7

FINDING AN INVERSE The following steps can be used to find the inverse of a function symbolically. Note that for a function f to have an inverse function, f must be one-to-one.

FINDING AN INVERSE FUNCTION

To find f^{-1} for a one-to-one function f, perform the following steps.

STEP 1: Let $y = f(x)$.

STEP 2: Interchange x and y.

STEP 3: Solve the formula for y. The resulting formula is $y = f^{-1}(x)$.

EXAMPLE 7 **Finding an inverse function**

Find the inverse of each one-to-one function.
(a) $f(x) = 3x - 7$
(b) $g(x) = (x + 2)^3$

Solution
(a) STEP 1: Let $y = 3x - 7$.

STEP 2: Interchange x and y to get $x = 3y - 7$.

STEP 3: To solve for y, start by adding 7 to each side.

$$x + 7 = 3y \qquad \text{Add 7 to each side.}$$

$$\frac{x + 7}{3} = y \qquad \text{Divide each side by 3.}$$

Thus $f^{-1}(x) = \frac{x + 7}{3}$ or $f^{-1}(x) = \frac{1}{3}x + \frac{7}{3}$.

(b) STEP 1: Let $y = (x + 2)^3$.

STEP 2: Interchange x and y to get $x = (y + 2)^3$.

STEP 3: To solve for y, start by taking the cube root of each side.

$$\sqrt[3]{x} = y + 2 \qquad \text{Take the cube root of each side.}$$

$$\sqrt[3]{x} - 2 = y \qquad \text{Subtract 2 from each side.}$$

Thus $g^{-1}(x) = \sqrt[3]{x} - 2$.

Now Try Exercises 63, 71

4 Using Tables and Graphs to Express Inverse Functions

TABLES OF INVERSE FUNCTIONS Inverse functions can be represented with tables. **TABLE 12.3** on the next page shows a table of values for a function f.

A Function f

x	1	2	3	4	5
$f(x)$	3	6	9	12	15

TABLE 12.3

Because $f(1) = 3$, $f^{-1}(3) = 1$. Similarly, $f(2) = 6$ implies that $f^{-1}(6) = 2$ and so on. **TABLE 12.4** lists values for $f^{-1}(x)$.

The Inverse Function f^{-1}

x	3	6	9	12	15
$f^{-1}(x)$	1	2	3	4	5

Interchange values

TABLE 12.4

Note that the domain of f is $\{1, 2, 3, 4, 5\}$ and that the range of f is $\{3, 6, 9, 12, 15\}$, whereas the domain of f^{-1} is $\{3, 6, 9, 12, 15\}$ and the range of f^{-1} is $\{1, 2, 3, 4, 5\}$. *The domain of f is the range of f^{-1}, and the range of f is the domain of f^{-1}.* This statement is true in general for a function and its inverse.

GRAPHS OF INVERSE FUNCTIONS If $f(a) = b$, then the point (a, b) lies on the graph of f. This statement also means that $f^{-1}(b) = a$ and that the point (b, a) lies on the graph of f^{-1}. These points are shown in **FIGURE 12.8(a)** with a solid green line segment connecting them. The line $y = x$ is perpendicular to this line segment and divides it into two equal parts. As a result, the graph of f^{-1} can be sketched from the graph of f by reflecting the graph of f across the line $y = x$. See **FIGURE 12.8(b)**.

Reflections Across the Line $y = x$

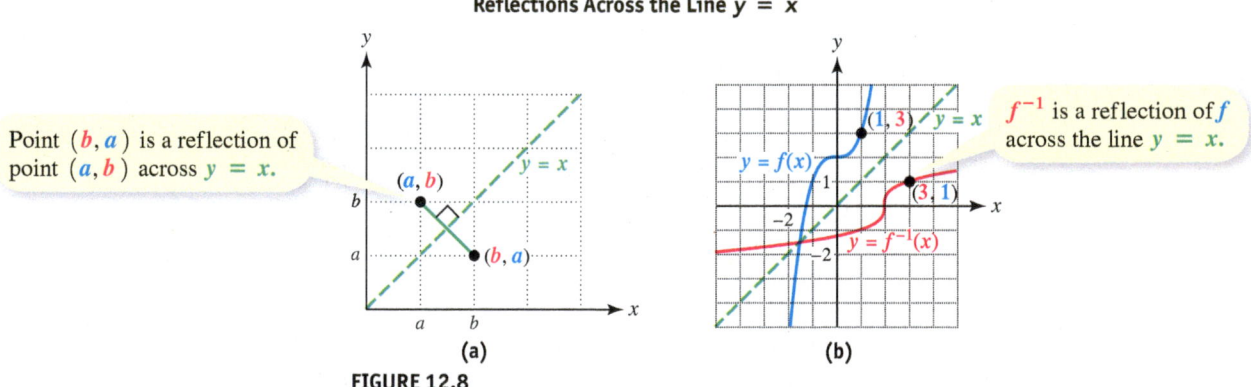

Point (b, a) is a reflection of point (a, b) across $y = x$.

f^{-1} is a reflection of f across the line $y = x$.

(a)

(b)

FIGURE 12.8

The relationship between the graph of a function and the graph of its inverse function is summarized as follows.

GRAPHS OF FUNCTIONS AND THEIR INVERSES

The graph of f^{-1} is a reflection of the graph of f across the line $y = x$.

EXAMPLE 8 Graphing an inverse function

The graph of $y = f(x)$ is shown in **FIGURE 12.9**.
Sketch a graph of $y = f^{-1}(x)$.

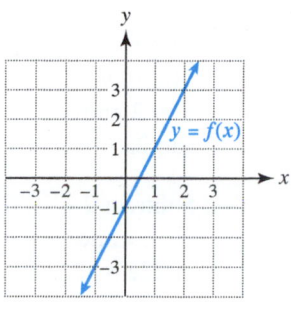

FIGURE 12.9

CRITICAL THINKING 2

The graph of a linear function f passes through the
points $(1, 2)$ and $(2, 1)$.
What two points does the
graph of f^{-1} pass through?
Find $f(x)$ and $f^{-1}(x)$.

Solution
The graph of $y = f^{-1}(x)$ is the reflection of the
graph of $y = f(x)$ across the line $y = x$ and is shown
in **FIGURE 12.10**. Note that the reflection of a line is
another line.

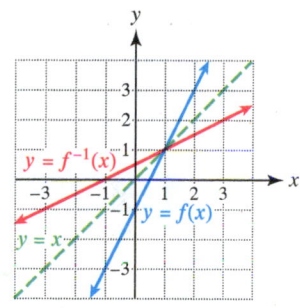

FIGURE 12.10

Now Try Exercise 83

EXAMPLE 9 Graphing an inverse function

The line graph shown in **FIGURE 12.11** represents a function f.
(a) Is f a one-to-one function?
(b) Sketch a graph of $y = f^{-1}(x)$.

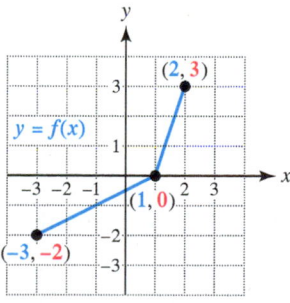

FIGURE 12.11

Solution
(a) Every horizontal line intersects the graph of f at most once. By the horizontal line test,
the graph represents a one-to-one function.
(b) The points $(-3, -2)$, $(1, 0)$, and $(2, 3)$ lie on the graph of f. It follows that the
points $(-2, -3)$, $(0, 1)$, and $(3, 2)$ lie on the graph of f^{-1}. Plot these three points
and then connect them with line segments, as shown in **FIGURE 12.12(a)**. Note that the
graph of $y = f^{-1}(x)$ is a reflection of the graph of $y = f(x)$ across the line $y = x$, as
shown in **FIGURE 12.12(b)**.

READING CHECK 4

• If we are given the
graph of $y = f(x)$,
where f is one-to-one,
how can we sketch the
graph of $y = f^{-1}(x)$?

(a)

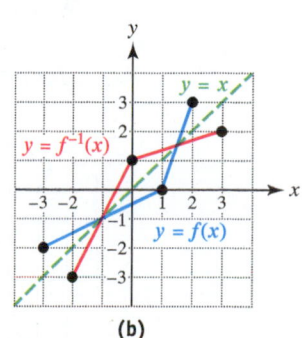

(b)

FIGURE 12.12

Now Try Exercise 81

See the Concept 2   REPRESENTING INVERSE FUNCTIONS

Verbal	Symbolic	Numerical	Graphical

f: Multiply x by 2 and add 2.

f^{-1}: Subtract 2 from x and divide by 2.

$$f(x) = 2x + 2$$
$$y = 2x + 2$$
$$x = 2y + 2$$
$$x - 2 = 2y$$
$$\frac{x - 2}{2} = y$$
$$f^{-1}(x) = \frac{x - 2}{2}$$

x	$f(x)$
-2	-2
-1	0
0	2
1	4

x	$f^{-1}(x)$
-2	-2
0	-1
2	0
4	1

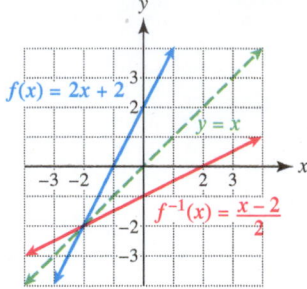

Use inverse operations in reverse order.	Interchange x and y. Solve for y.	Interchange the x- and y-values.	Reflect the graph of f across the line $y = x$.

WATCH VIDEO IN MML.

EXAMPLE 10   **Evaluating f and f^{-1} graphically**

Use the graph of f in **FIGURE 12.13** to evaluate each expression.

(a) $f(2)$ **(b)** $f^{-1}(-3)$ **(c)** $f^{-1}(2)$

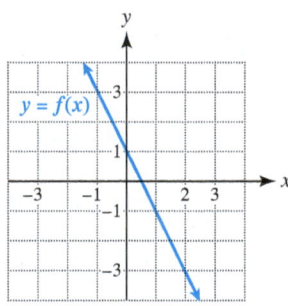

FIGURE 12.13

Solution

(a) To evaluate $f(2)$, find **2** on the x-axis, move downward to the graph of f, and then move left to the y-axis to obtain $f(2) = -3$, as shown in **FIGURE 12.14(a)**.

(b) Start by finding -3 on the y-axis, move right to the graph of f, and then move upward to the x-axis to obtain $f^{-1}(-3) = 2$, as shown in **FIGURE 12.14(b)**. Notice that $f(2) = -3$ from part (a) and $f^{-1}(-3) = 2$ here.

(c) Find **2** on the y-axis, move left to the graph of f, and then move downward to the x-axis. We can see from **FIGURE 12.14(c)** that $f^{-1}(2) = -\frac{1}{2}$.

$f(2) = -3$ $f^{-1}(-3) = 2$ $f^{-1}(2) = -\frac{1}{2}$

(a)

(b)

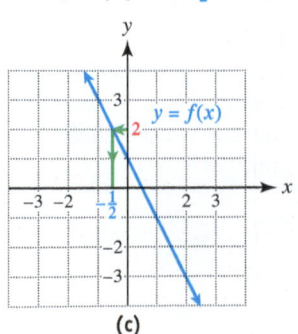

(c)

FIGURE 12.14

Now Try Exercise 87

12.1 Putting It All Together

CONCEPT	EXPLANATION	EXAMPLES
Composite Functions	The composition of g and f is $$(g \circ f)(x) = g(f(x)),$$ and represents a *new* function whose name is $g \circ f$.	If $f(x) = 1 - 4x$ and $g(x) = x^3$, then $$\begin{aligned}(g \circ f)(x) &= g(f(x)) \\ &= g(1 - 4x) \\ &= (1 - 4x)^3.\end{aligned}$$
One-to-One Functions	Function f is one-to-one if different inputs always give different outputs.	$f(x) = x^2$ is not one-to-one because $f(-4) = f(4) = 16$, whereas $g(x) = x + 1$ is one-to-one because every input has a unique output.
Horizontal Line Test	This test is used to determine whether a function is one-to-one from its graph: If every horizontal line intersects the graph of a function f at most once, then f is a one-to-one function.	$f(x) = x^2$ is not one-to-one. A horizontal line can intersect its graph more than once.
Inverse Functions	f^{-1} will undo the operations performed by f. That is, $$(f^{-1} \circ f)(x) = x \quad \text{and}$$ $$(f \circ f^{-1})(x) = x.$$	If $f(x) = x^3$, then $f^{-1}(x) = \sqrt[3]{x}$ because cubing a number x and then taking its cube root result in the number x.

12.1 Exercises

MyMathLab®

CONCEPTS AND VOCABULARY

1. $(g \circ f)(7) =$ _____

2. $(f \circ g)(x) =$ _____

3. Does $(f \circ g)(x)$ always equal $(g \circ f)(x)$?

4. If a function f is one-to-one, then different _____ always result in different _____.

5. If $f(3) = 5$ and $f(7) = 5$, could f be one-to-one?

6. If every horizontal line intersects the graph of f at most once, then f is _____.

7. The inverse operation of subtracting 10 is _____.

8. $(f^{-1} \circ f)(7) =$ _____.

9. If $f(6) = 8$, then f^{-1} (_____) $=$ _____.

10. If $f^{-1}(y) = x$, then f (_____) $=$ _____.

11. For f to have an inverse function, f must be _____.

12. The graph of f^{-1} is a(n) _____ of the graph of f across the line _____.

EVALUATING COMPOSITE FUNCTIONS SYMBOLICALLY

Exercises 13–22: For the given $f(x)$ and $g(x)$, find each of the following.

(a) $(g \circ f)(-2)$ (b) $(f \circ g)(4)$

(c) $(g \circ f)(x)$ (d) $(f \circ g)(x)$

13. $f(x) = x^2$ $g(x) = x + 3$

14. $f(x) = 4x^2$ $g(x) = 5x$

15. $f(x) = 2x$ $g(x) = x^3 - 1$

16. $f(x) = 3x + 1$ $g(x) = x^2 + 4x$

17. $f(x) = \frac{1}{2}x$ $g(x) = |x - 2|$

18. $f(x) = 6x$ $g(x) = \dfrac{2}{x - 5}$

19. $f(x) = \dfrac{1}{x}$ $g(x) = 3 - 5x$

20. $f(x) = \sqrt{x + 3}$ $g(x) = x^3 - 3$

21. $f(x) = 2x$ $g(x) = 4x^2 - 2x + 5$

22. $f(x) = 9x - \dfrac{1}{3x}$ $g(x) = \dfrac{x}{3}$

EVALUATING COMPOSITE FUNCTIONS NUMERICALLY

Exercises 23–28: For one-to-one functions f and g, evaluate each expression numerically.

x	-2	-1	0	1	2
$f(x)$	2	1	0	-1	-2

x	-2	-1	0	1	2
$g(x)$	0	1	-1	2	-2

23. (a) $(f \circ g)(0)$ **(b)** $(g \circ f)(-1)$

24. (a) $(f \circ g)(1)$ **(b)** $(g \circ f)(-2)$

25. (a) $(f \circ f)(-1)$ **(b)** $(g \circ g)(0)$

26. (a) $(g \circ g)(2)$ **(b)** $(f \circ f)(1)$

27. (a) $(f^{-1} \circ g)(-2)$ **(b)** $(g^{-1} \circ f)(2)$

28. (a) $(f \circ g^{-1})(1)$ **(b)** $(g \circ f^{-1})(-2)$

EVALUATING COMPOSITE FUNCTIONS GRAPHICALLY

Exercises 29 and 30: Evaluate each expression graphically.

29. (a) $(f \circ g)(0)$

   (b) $(g \circ f)(1)$

   (c) $(f \circ f)(-1)$

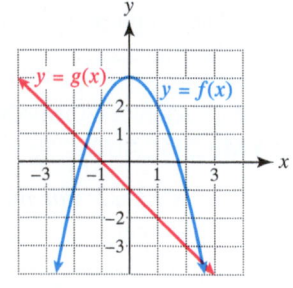

30. (a) $(f \circ g)(1)$

   (b) $(g \circ f)(-2)$

   (c) $(g \circ g)(-2)$

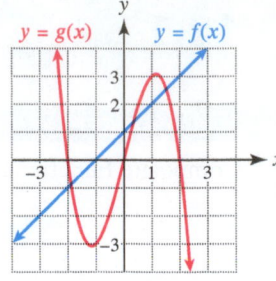

IDENTIFYING ONE-TO-ONE FUNCTIONS

Exercises 31–36: Show that f is not a one-to-one function by finding two inputs that result in the same output. Answers may vary.

31. $f(x) = 5x^2$ **32.** $f(x) = 4 - x^2$

33. $f(x) = x^4 + 100$ **34.** $f(x) = \dfrac{x^2}{x^2 + 1}$

35. $f(x) = x^4 - 3x^2$ **36.** $f(x) = \sqrt{x^2 - 1}$

APPLYING THE HORIZONTAL LINE TEST

Exercises 37–42: Use the horizontal line test to determine whether the graph represents a one-to-one function.

37.

38.

39.

40.

41.

42.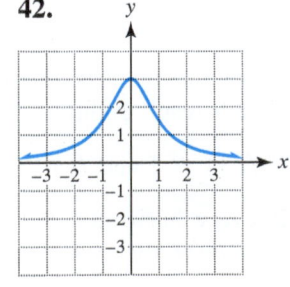

FINDING INVERSE FUNCTIONS

Exercises 43–50: (Refer to Example 5.) Give the inverse operation for the statement. Then write a function f for the given statement and a function g for its inverse.

43. Multiply x by 7.

44. Subtract 10 from x.

45. Add 5 to x and then divide the result by 2.

46. Multiply x by 6 and then add 8 to the result.

47. Multiply x by $\frac{1}{2}$ and then subtract 3 from the result.

48. Divide x by 10 and then add 20 to the result.

49. Cube the sum of x and 5.

50. Take the cube root of x and then subtract 2.

Exercises 51–58: (Refer to Example 6.) Verify that $f(x)$ and $f^{-1}(x)$ are indeed inverse functions.

51. $f(x) = 4x$ $\qquad$ $f^{-1}(x) = \dfrac{x}{4}$

52. $f(x) = \dfrac{2x}{3}$ $\qquad$ $f^{-1}(x) = \dfrac{3x}{2}$

53. $f(x) = 3x + 5$ $\qquad$ $f^{-1}(x) = \dfrac{x-5}{3}$

54. $f(x) = x + 7$ $\qquad$ $f^{-1}(x) = x - 7$

55. $f(x) = x^3$ $\qquad$ $f^{-1}(x) = \sqrt[3]{x}$

56. $f(x) = \sqrt[3]{x-4}$ $\qquad$ $f^{-1}(x) = x^3 + 4$

57. $f(x) = \dfrac{1}{x}$ $\qquad$ $f^{-1}(x) = \dfrac{1}{x}$

58. $f(x) = \dfrac{x+7}{7}$ $\qquad$ $f^{-1}(x) = 7x - 7$

Exercises 59–74: (Refer to Example 7.) Find $f^{-1}(x)$.

59. $f(x) = 12x$ $\qquad$ **60.** $f(x) = \frac{3}{4}x$

61. $f(x) = x + 8$ $\qquad$ **62.** $f(x) = x - 3$

63. $f(x) = 5x - 2$ $\qquad$ **64.** $f(x) = 3x + 4$

65. $f(x) = -\frac{1}{2}x + 1$ $\qquad$ **66.** $f(x) = \frac{3}{4}x - \frac{1}{4}$

67. $f(x) = 8 - x$ $\qquad$ **68.** $f(x) = 5 - x$

69. $f(x) = \dfrac{x+1}{2}$ $\qquad$ **70.** $f(x) = \dfrac{3-x}{5}$

71. $f(x) = \sqrt[3]{2x}$ $\qquad$ **72.** $f(x) = \sqrt[3]{x+4}$

73. $f(x) = x^3 - 8$ $\qquad$ **74.** $f(x) = (x-5)^3$

RELATING TABLES AND GRAPHS TO INVERSE FUNCTIONS

Exercises 75–78: Use the table to make a table of values for $f^{-1}(x)$. State the domain and range for f and for f^{-1}.

75.

x	0	1	2	3	4
$f(x)$	0	5	10	15	20

76.

x	−4	−2	0	2	4
$f(x)$	1	2	3	4	5

77.

x	−5	0	5	10	15
$f(x)$	4	2	0	−2	−4

78.

x	0	2	4	6	8
$f(x)$	8	6	4	2	0

Exercises 79–82: (Refer to Example 9.) Use the graph of $y = f(x)$ to sketch a graph of $y = f^{-1}(x)$.

79.

80.

81.

82.
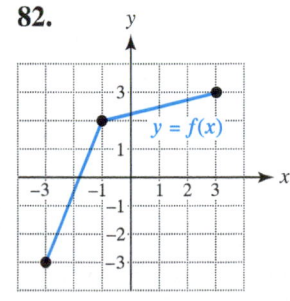

Exercises 83–86: (Refer to Example 8.) Use the graph of $y = f(x)$ to sketch a graph of $y = f^{-1}(x)$. Include the graph of f and the line $y = x$ in your graph.

83.

84.

85.

86.

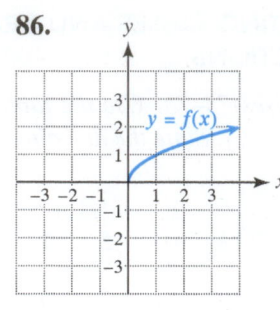

Exercises 87–90: Use the graph of $f(x)$ to evaluate the expression.

87. (a) $f^{-1}(-8)$

 (b) $f^{-1}(8)$

 (c) $f^{-1}(0)$

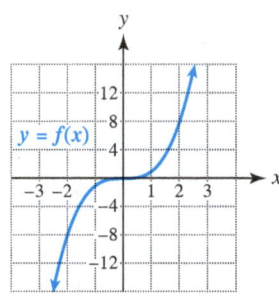

88. (a) $f^{-1}(1)$

 (b) $f^{-1}\left(\frac{1}{2}\right)$

 (c) $f^{-1}(2)$

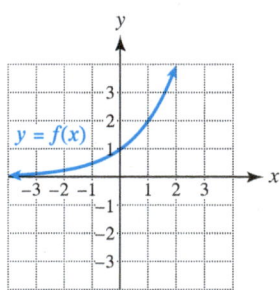

89. (a) $f^{-1}(7)$

 (b) $f^{-1}(2)$

 (c) $f^{-1}(5)$

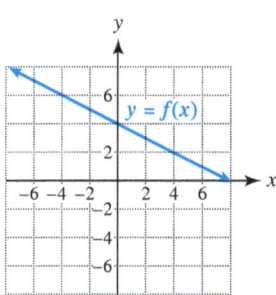

90. (a) $f^{-1}(-5)$

 (b) $f^{-1}(3)$

 (c) $f^{-1}(5)$

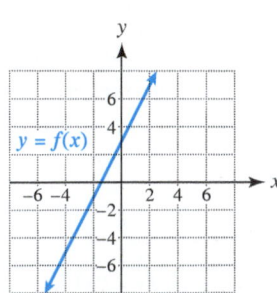

PERFORMING OPERATIONS ON FUNCTIONS

Exercises 91–94: Find each of the following for the given $f(x)$ and $g(x)$.

   (a) $(fg)(2)$ (b) $(f-g)(x)$ (c) $(f \circ g)(x)$

91. $f(x) = x^2 - 2, g(x) = x^2 + 2$

92. $f(x) = 2x^2, g(x) = 2x - 1$

93. $f(x) = \dfrac{1}{x}, g(x) = \dfrac{2}{x}$

94. $f(x) = x^3, g(x) = \sqrt[3]{x}$

APPLICATIONS INVOLVING COMPOSITION AND INVERSE FUNCTIONS

95. Circular Wave A stone is dropped in a lake, creating a circular wave. The radius r of the wave in feet after t seconds is $r(t) = 2t$.
 (a) The wave's circumference C is $C(r) = 2\pi r$. Evaluate $(C \circ r)(5)$ and interpret your result.
 (b) Find $(C \circ r)(t)$.

96. Volume of a Balloon The volume V of a spherical balloon with radius r is given by $V(r) = \frac{4}{3}\pi r^3$. Suppose that the balloon is being inflated so that the radius in inches after t seconds is $r(t) = \sqrt[3]{t}$.
 (a) Evaluate $(V \circ r)(3)$ and interpret your result.
 (b) Find $(V \circ r)(t)$.

97. College Degree The table lists the percentage P of people 25 or older who have completed 4 or more years of college during year x.

x	1960	1980	2000	2010
$P(x)$	8	16	27	29

Source: U.S Census Bureau.
 (a) Evaluate $P(1980)$ and interpret the results.
 (b) Make a table for $P^{-1}(x)$.
 (c) Evaluate $P^{-1}(16)$.

98. Skin Cancer and Ozone Ozone in the stratosphere filters out most of the harmful ultraviolet (UV) rays from the sun. However, depletion of the ozone layer affects this protection. The formula $U(x) = 1.5x$ calculates the percent increase in UV radiation for an x percent decrease in the thickness of the ozone layer. The formula $C(x) = 3.5x$ calculates the percent increase in skin cancer cases when the UV radiation increases by x percent. (*Source:* R. Turner, D. Pierce, and I. Bateman, *Environmental Economics*).

(a) Evaluate $U(2)$ and $C(3)$ and interpret each result.

(b) Find $(C \circ U)(2)$ and interpret the result.

(c) Find $(C \circ U)(x)$. What does it calculate?

99. **Temperature and Mosquitoes** Temperature can affect the number of mosquitoes observed on a summer night. Graphs of two functions, T and M, are shown. Function T calculates the temperature on a summer evening h hours past midnight, and M calculates the number of mosquitoes observed per 100 square feet when the outside temperature is T.

(a) Find $T(1)$ and $M(75)$.

(b) Evaluate $(M \circ T)(1)$ and interpret your result.

(c) What does $(M \circ T)(h)$ calculate?

(d) Find equations for the lines in each graph.

(e) Use your answers from part (d) to write a formula for $(M \circ T)(h)$.

100. **High School Grades** The table lists the percentage P of college freshmen with a high school grade average of A or A– during year x.

x	1970	1980	1990	2000	2010
$P(x)$	20	26	29	43	46

Source: Department of Education.

(a) Evaluate $P(1970)$ and interpret the results.

(b) Make a table for $P^{-1}(x)$.

(c) Evaluate $P^{-1}(43)$.

101. **Temperature** The function given by $f(x) = \frac{9}{5}x + 32$ converts x degrees Celsius to an equivalent temperature in degrees Fahrenheit.

(a) Is f a one-to-one function? Why or why not?

(b) Find $f^{-1}(x)$ and interpret what it calculates.

102. **Feet and Yards** The function $f(x) = 3x$ converts x yards to feet.

(a) Is f a one-to-one function? Why or why not?

(b) Find $f^{-1}(x)$ and interpret what it calculates.

103. **Quarts and Gallons** Write a function f that converts x gallons to quarts. Then find $f^{-1}(x)$ and interpret what it computes.

104. **Wind Speed** The table lists monthly average wind speeds at Hilo, Hawaii, in miles per hour from July through December, where x is the month.

x	July	Aug	Sept	Oct	Nov	Dec
$f(x)$	7	7	7	7	7	7

(a) Is function f one-to-one? Explain.

(b) Does f^{-1} exist?

(c) What happens if you try to make a table for f^{-1}?

(d) Could f be one-to-one if it were computed at a different location? What would have to be true about the monthly average wind speeds?

105. **Online Exploration** Go online and find how many cups there are in one gallon and how many teaspoons there are in one cup.

(a) Write a function C that converts x gallons to cups.

(b) Write a function T that converts x cups to teaspoons.

(c) Write the composite function $(T \circ C)(x)$, where x represents gallons.

(d) Evaluate $(T \circ C)(3)$ and interpret the result.

106. **Online Exploration** Go online and find the number of yuan (Chinese currency) in 1 dollar and the number of rupees (Indian currency) in 1 yuan. Answers may vary over time.

(a) Write a function Y that converts x dollars to yuan.

(b) Write a function R that converts x yuan to rupees.

(c) Write the composite function $(R \circ Y)(x)$, where x represents dollars.

(d) Evaluate $(R \circ Y)(100)$ and interpret the result.

WRITING ABOUT MATHEMATICS

107. Explain the difference between the expressions $(g \circ f)(2)$ and $(f \circ g)(2)$. Are they always equal? If f and g are inverse functions, evaluate $(g \circ f)(2)$ and $(f \circ g)(2)$.

108. Explain what it means for a function to be one-to-one.

12.2 Exponential Functions

NEW VOCABULARY

☐ Exponential function with base a and coefficient C
☐ Growth factor
☐ Decay factor
☐ Exponential growth
☐ Exponential decay
☐ Percent change
☐ Compound interest
☐ Base e
☐ Natural exponential function
☐ Continuous growth

1 Introducing Exponential Functions

 Math in Context (Biology) Suppose that an insect population (per acre) doubles each week. **TABLE 12.5** shows the size of the population after x weeks. Note that, as the population of insects becomes larger, the *increase* in population each week becomes greater. The population is increasing by 100%, or doubling, each week. When a quantity increases by a constant percentage (or constant factor) at regular intervals, its growth is *exponential*.

Insect Population

Week	0	1	2	3	4	5
Population	100	200	400	800	1600	3200

TABLE 12.5

$\times 2$ $\times 2$ $\times 2$ $\times 2$ $\times 2$

Doubles each week

We can model the data in **TABLE 12.5** by using the exponential function

$$f(x) = 100(2)^x.$$

For example,

$$f(0) = 100(2)^0 = 100 \cdot 1 = 100,$$
$$f(1) = 100(2)^1 = 100 \cdot 2 = 200,$$
$$f(2) = 100(2)^2 = 100 \cdot 4 = 400,$$

and so on. Note that the exponential function f has a *variable as an exponent*.

EXPONENTIAL FUNCTION

A function represented by

$$f(x) = Ca^x, \quad a > 0 \quad \text{and} \quad a \neq 1,$$

is an **exponential function with base a and coefficient C**. (Unless stated otherwise, we assume that $C > 0$.)

In the formula $f(x) = Ca^x$, a is called the **growth factor** when $a > 1$ and the **decay factor** when $0 < a < 1$. For an exponential function, each time x increases by 1 unit, $f(x)$ increases by a factor of a when $a > 1$ and decreases by a factor of a when $0 < a < 1$. Moreover, because

$$f(0) = Ca^0 = C \cdot 1 = C,$$

the value of C equals the value of $f(x)$ when $x = 0$. That is, $(0, C)$ is the y-intercept. If x represents time, C represents the initial value of f when time equals 0. **FIGURE 12.15** illustrates **exponential growth** and **exponential decay** for $x > 0$.

Growth ($a > 1$) for Positive x

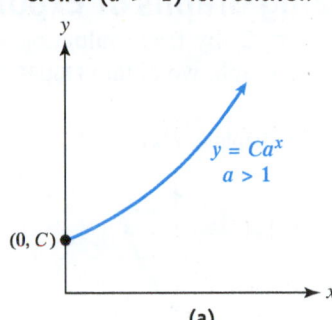

Decay ($0 < a < 1$) for Positive x

FIGURE 12.15

The set of valid inputs (domain) for an exponential function includes all real numbers. The set of corresponding outputs (range) includes all positive real numbers.

When evaluating an exponential function, we evaluate exponents *before* we multiply. For example, if $f(x) = 4(3)^x$, then

$$f(2) = 4(3)^2 = 4(9) = 36.$$

Evaluate exponents first.

The next example illustrates how exponential functions are evaluated.

EXAMPLE 1 **Evaluating exponential functions**

Evaluate $f(x)$ for the given value of x.
(a) $f(x) = 10(3)^x$ $x = 2$
(b) $f(x) = 5\left(\frac{1}{2}\right)^x$ $x = 3$
(c) $f(x) = \frac{1}{3}(2)^x$ $x = -1$

Solution
(a) $f(2) = 10(3)^2 = 10 \cdot 9 = 90$
(b) $f(3) = 5\left(\frac{1}{2}\right)^3 = 5 \cdot \frac{1}{8} = \frac{5}{8}$
(c) $f(-1) = \frac{1}{3}(2)^{-1} = \frac{1}{3} \cdot \frac{1}{2} = \frac{1}{6}$

NOTE: $f(x)$ is also defined for all *negative* inputs. ∎

Now Try Exercises 11, 13, 15

MAKING CONNECTIONS 1

The Expressions a^{-x} and $\left(\frac{1}{a}\right)^x$

Using properties of exponents, we can write 2^{-x} as

$$2^{-x} = \frac{1}{2^x} = \left(\frac{1}{2}\right)^x.$$

In general, the expressions a^{-x} and $\left(\frac{1}{a}\right)^x$ are equal for all positive values of a.

2 Recognizing Graphs of Exponential Functions

We can graph $f(x) = 2^x$ by first evaluating some points, as in **TABLE 12.6**. If we plot these points and sketch the graph, we obtain **FIGURE 12.16**.

$f(x) = 2^x$

x	2^x
-2	$\frac{1}{4}$
-1	$\frac{1}{2}$
0	1
1	2
2	4

TABLE 12.6

Graphing $f(x) = 2^x$

FIGURE 12.16

- This graph is always above the x-axis.
- The graph passes through the point $(0, 1)$.
- Negative x-values give y-values between 0 and 1.
- Positive x-values give y-values greater than 1.

See the Concept 1   GRAPHING $y = a^x$

$a > 1$

Values of a greater than 1

$y = 2.5^x$
$y = 1.7^x$
$y = 1.3^x$

$0 < a < 1$

Values of a between 0 and 1

$y = 0.5^x$
$y = 0.15^x$
$y = 0.7^x$

A As x increases, the y-values increase.

B Larger values of a $(a > 1)$ result in y-values that increases more rapidly.

C The graphs pass through $(0, 1)$ because $a^0 = 1$.

D The graphs are always above the x axis.

A As x increases, the y-values decrease.

B Smaller values of a $(0 < a < 1)$ result in y-values that decrease more rapidly.

C The graphs pass through $(0, 1)$ because $a^0 = 1$.

D The graphs are always above the x axis.

WATCH VIDEO IN MML.

EXAMPLE 2   **Transforming exponential graphs**

Explain how to obtain the graph of $g(x) = 2^{x+1}$ from the graph of $f(x) = 2^x$. Then graph both f and g in the same xy-plane.

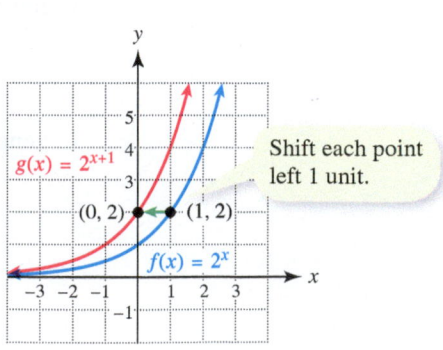

Shift each point left 1 unit.

$g(x) = 2^{x+1}$

$(0, 2)$ $(1, 2)$

$f(x) = 2^x$

Solution
If we replace x with $x + 1$ in the formula $f(x) = 2^x$, we obtain $f(x + 1) = 2^{x+1}$, which is $g(x)$. Thus if we shift the graph of $y = 2^x$ left 1 unit, we obtain the graph of g. See **FIGURE 12.17**.

FIGURE 12.17

Now Try Exercise 45

EXPONENTIAL AND LINEAR GROWTH In the next example, we show the dramatic difference between the outputs of linear and exponential functions.

EXAMPLE 3 Comparing exponential and linear functions

Compare $f(x) = 3^x$ and $g(x) = 3x$ graphically and numerically for $x \geq 0$.

Solution
Graphical Comparison The graphs of $y_1 = 3^x$ and $y_2 = 3x$ are shown in **FIGURE 12.18**. For large values of x, the graph of the exponential function y_1 increases much faster than the graph of the linear function y_2.

Numerical Comparison The tables of $y_1 = 3^x$ and $y_2 = 3x$ are shown in **FIGURE 12.19**. For large values of x, the values for y_1 increase much faster than the values for y_2.

Comparing Exponential and Linear Growth

$[0, 5, 1]$ by $[0, 120, 20]$

$y_1 = 3^x$

$y_2 = 3x$

FIGURE 12.18

X	Y₁	Y₂
0	1	0
1	3	3
2	9	6
3	27	9
4	81	12
5	243	15
6	729	18

X=0

FIGURE 12.19

Exponential growth — Linear growth

For each unit increase, linear functions increase by a fixed amount, and exponential functions increase by a constant *factor*.

Now Try Exercise 91

NOTE: The results of Example 3 are true in general: for large enough inputs, exponential functions with $a > 1$ grow far faster than any linear function. ■

MAKING CONNECTIONS 2

Exponential and Polynomial Functions

The function $f(x) = 2^x$ is an exponential function. The base **2** is a constant and the exponent x is a variable, so $f(3) = 2^3 = 8$.

The function $g(x) = x^2$ is a quadratic (polynomial) function. The base x is a variable and the exponent **2** is a constant, so $g(3) = 3^2 = 9$.

The table clearly shows that the exponential function grows much faster than the quadratic function for large values of x.

	x	0	2	4	6	8	10	12	
Exponential expression ⟶	2^x	1	4	16	64	256	1024	4096	⟵ Exponential growth
Polynomial expression ⟶	x^2	0	4	16	36	64	100	144	⟵ Quadratic growth

In the next example, we determine whether a function is linear or exponential.

EXAMPLE 4 **Finding linear and exponential functions**

For each table, determine whether f is a linear function or an exponential function. Find a formula for f.

(a)

x	0	1	2	3	4
$f(x)$	16	8	4	2	1

(b)

x	0	1	2	3	4
$f(x)$	5	7	9	11	13

(c)

x	0	1	2	3	4
$f(x)$	1	3	9	27	81

(d)

x	0	1	2	3	4
$f(x)$	2	−1	−4	−7	−10

Solution

(a) Each time x increases by 1 unit, $f(x)$ decreases by a factor of $\frac{1}{2}$. Therefore f is an exponential function with a *decay factor* of $\frac{1}{2}$. Because $f(0) = 16$, $C = \mathbf{16}$, so $f(x) = \mathbf{16}\left(\frac{1}{2}\right)^x$. This formula can also be written as $f(x) = 16(2)^{-x}$.

(b) Each time x increases by 1 unit, $f(x)$ increases by $\mathbf{2}$ units. Therefore f is a linear function, and the slope of its graph is $\mathbf{2}$. The y-intercept is $(0, \mathbf{5})$, so $f(x) = \mathbf{2}x + \mathbf{5}$.

(c) Each time x increases by 1 unit, $f(x)$ increases by a factor of $\mathbf{3}$. Therefore f is an exponential function with a *growth factor* of $\mathbf{3}$. Because $f(0) = 1$, it follows that $C = \mathbf{1}$ and $f(x) = \mathbf{1}(\mathbf{3})^x$, or $f(x) = \mathbf{3}^x$.

(d) Each time x increases by 1 unit, $f(x)$ **decreases by 3** units. Therefore f is a linear function, and the slope of its graph equals $-\mathbf{3}$. The y-intercept is $(0, \mathbf{2})$ so $f(x) = -\mathbf{3}x + \mathbf{2}$.

Now Try Exercises 47, 49, 51

MAKING CONNECTIONS 3

Linear and Exponential Functions

Linear Function

y increases by 2 for each unit increase in x. The slope is 2.

Exponential Function

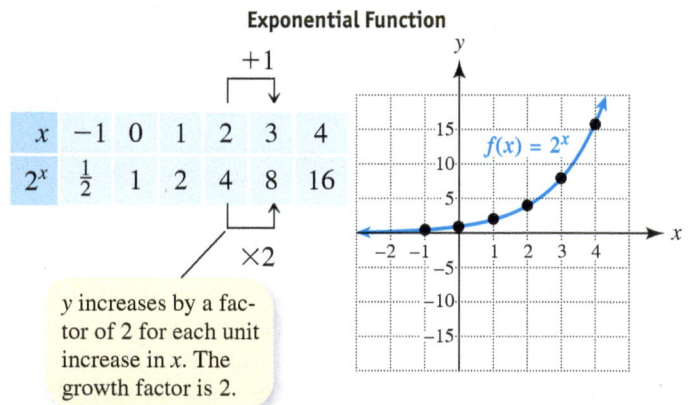

y increases by a factor of 2 for each unit increase in x. The growth factor is 2.

For a *linear function,* given by $f(x) = ax + b$, each time x increases by 1 unit, y increases (or decreases) by a units, where a equals the slope of the graph of f.

For an *exponential function,* given by $f(x) = Ca^x$, each time x increases by 1 unit, y increases by a factor of a when $a > 1$ and decreases by a factor of a when $0 < a < 1$. The constant a equals either the growth factor or the decay factor.

---

3 **Relating Percent Change and Exponential Functions**

PERCENT CHANGE When an amount A changes to a new amount B, then the **percent change** is calculated by

$$\frac{B - A}{A} \times 100. \quad \boxed{\text{Percent change formula}}$$

We multiply the ratio by 100 to change decimal form to percent form.

EXAMPLE 5 **Finding percent change**

Complete the following.
(a) Find the percent change if an account balance increases from $500 to $1000.
(b) Find the percent change if an account balance decreases from $1000 to $500.

Solution
(a) Let $A = 500$ and $B = 1000$.

$$\frac{1000 - 500}{500} \times 100 = \frac{500}{500} \times 100 \qquad \frac{B - A}{A} \times 100$$

$$= 1 \times 100 \qquad \text{Simplify.}$$

$$= 100\% \qquad \text{Multiply.}$$

The percent change (increase) is 100%; that is, the account balance doubles.

(b) Let $A = 1000$ and $B = 500$.

$$\frac{500 - 1000}{1000} \times 100 = -\frac{500}{1000} \times 100 \qquad \frac{B - A}{A} \times 100$$

$$= -\frac{1}{2} \times 100 \qquad \text{Simplify.}$$

$$= -50\% \qquad \text{Multiply.}$$

The percent change (decrease) is −50%.

Now Try Exercise 53

NOTE: In Example 5, the account did *not* increase by 100% and then decrease by 100% to return the account to its initial value of $500. In part (b), the initial amount is $A = \$1000$ and needs to decrease only by a factor of $\frac{1}{2}$, or 50%, to decrease to $500. ∎

PERCENT CHANGE AND GROWTH FACTOR Suppose a savings account A increases from $200 to $600. The percent change, $R\%$, equals

$$\frac{600 - 200}{200} \times 100 = 2 \times 100 = 200\%.$$

Note that the account balance *tripled* from $200 to 600, but the percent change is 200%, *not* 300%. Also, the *increase* in the account balance A is $400 because $600 − \$200 = \$400$. This $400 increase equals 200% of $200. If we let r represent the percent change as a decimal, then

$$\underbrace{200\% \text{ of } \$200}_{R\% \text{ of } A} = \underbrace{2.00 \times \$200}_{rA} = \underbrace{\$400.}_{\text{Increase in } A} \qquad R\% = 200\%; r = 2.00$$

In general, if the *percent increase* in an account balance A is given by r in decimal form, then the *amount of increase* in A is equal to rA and the new balance after the increase is $A + rA$. For example, if $200 increases by 200%, then $r = 2.00$, $A = \$200$, and

$$rA = 2.00(200) = \$400, \qquad \text{Increase in account balance}$$

so the balance increases by $400 as noted above. The new balance for the account is

$$A + rA = \$200 + \$400 = \$600.$$

Initial amount + Increase = Final amount

If we factor out A in the expression $A + rA$, we get

$$A + rA = A(1 + r). \quad \text{Factor out } A.$$

Initial amount $+$ Increase in A $=$ Initial amount $\times$ Growth factor

Thus if an account increases from \$200 to \$600, the increase is \$400 and the percent increase is 200%. In addition, the account balance increased by a *growth factor* equal to

$$a = 1 + r = 1 + 2.00 = 3, \quad \text{Growth factor is } a = 3.$$

which means that the account balance tripled.

EXAMPLE 6   **Analyzing the decrease in an account balance**

An account that contains \$2000 decreases in value by 20%.
(a) Find the decrease in value of the account.
(b) Find the final value of the account.
(c) By what factor a did the account decrease?

Solution
(a) Let $A = 2000$ and $r = -0.20$ (-20% in decimal form). The decrease is

$$rA = -0.20(2000) = -400. \quad \text{Decrease in } A \text{ is } rA.$$

The account decreased in value by \$400.

(b) The final value of the account is

$$A + rA = 2000 + (-400) = \$1600.$$

Initial amount $+$ Decrease $=$ New amount

(c) The account decreased by a factor of

$$a = 1 + r = 1 + (-0.20) = 0.80. \quad \text{Decay factor is } a = 0.80.$$

The account value decreased to 80% of its original value because 80% of \$2000 is \$1600.

Now Try Exercise 57

> **STUDY TIP**
>
> Be sure that you understand how a percent change relates to the growth factor of an exponential function.

NOTE: In general, a positive amount A cannot *decrease* by more than 100% because a 100% decrease would reduce A to 0. However, percent *increases* can be more than 100%. ∎

PERCENT CHANGE AND EXPONENTIAL FUNCTIONS An exponential function results when an *initial value C* is multiplied by a *growth* (or *decay*) *factor a* for each unit increase in x. For example, if the population P of a city is 100,000 people, and the city is growing at 5% per year, then $C = 100,000$ and the growth factor is

$$a = 1 + r = 1 + 0.05 = 1.05.$$

Thus the exponential function

$$P(x) = 100,000(1.05)^x \quad P(x) = Ca^x \text{ with } C = 100,000 \\ \text{and } a = 1.05.$$

models the city's population after x years. For example,

$$P(6) = 100,000(1.05)^6 \approx 134,000$$

indicates that after 6 years the city's population has grown to about 134,000 people.

PERCENT CHANGE AND EXPONENTIAL FUNCTIONS

Suppose that an amount A increases or decreases by $R\%$ (or by r expressed in decimal form) for each x-unit increase in time.

1. If $r > 0$, the *growth factor* is $a = 1 + r$ and $a > 1$.
2. If $-1 < r < 0$, the *decay factor* is $a = 1 + r$ and $0 < a < 1$.
3. If the initial amount is C, the amount A after x-units of time is given by

$$A(x) = Ca^x \quad \text{or equivalently,} \quad A(x) = C(1 + r)^x.$$

READING CHECK 1

- If your income increases by 200%, what is the growth factor?

EXAMPLE 7 **Analyzing growth of bacteria**

Initially, a laboratory culture contains 50,000 bacteria per milliliter and it is increasing in numbers by 20% per hour.
(a) Write the formula for an exponential function B that gives the number of bacteria per milliliter after x hours.
(b) Evaluate $B(3)$ and interpret your result.

Solution
(a) The initial value is $C = 50,000$ and the hourly percent increase is 20%, or $r = 0.20$ in decimal form. Thus the growth factor is

$$a = 1 + r = 1 + 0.20 = 1.20. \qquad \text{Growth factor} = 1.20$$
$$\text{with } r = 0.20.$$

Because the exponential function can be written as $B(x) = Ca^x$, it follows that

$$B(x) = 50,000(1.20)^x.$$

(b) $B(3) = 50,000(1.20)^3 = 50,000(1.728) = 86,400$. Thus after 3 hours, the culture contains 86,400 bacteria per milliliter.

Now Try Exercise 87

4 Computing Compound Interest

🌐 *Math in Context* ^Savings If a total $100 is deposited in a savings account paying 10% annual interest, the interest earned after 1 year equals $100 \times 0.10 = $10. The total amount of money in the account after 1 year is $100(1 + 0.10) = $110. Each year, the money in the account increases by a growth factor of $a = 1 + r = 1 + 0.10 = 1.10$, so after x years, there will be $100(1.10)^x$ dollars in the account. Thus **compound interest** is an example of exponential growth.

COMPOUND INTEREST

If a total of P dollars is deposited in an account and if interest is paid at the end of each year with an annual rate of interest r, expressed in decimal form, then after t years, the account will contain A dollars, where

$$A = P(1 + r)^t.$$

The growth factor is $(1 + r)$.

NOTE: The compound interest formula takes the form of an exponential function with

$$a = 1 + r. \;\blacksquare$$

EXAMPLE 8   **Calculating compound interest**

A 20-year-old worker deposits $2000 in a retirement account that pays 6% annual interest at the end of each year. How much money will be in the account when the worker is 65 years old? What is the growth factor?

Solution

Here, $P = 2000$, $r = 0.06$, and $t = 45$. The amount in the account after 45 years is

$$A = 2000 (1 + 0.06)^{45} \approx \$27{,}529.22, \quad A = P(1 + r)^t$$

which is supported by **FIGURE 12.20**. Each year, the amount of money in the account is multiplied by a factor of $(1 + 0.06)$, so the growth factor is 1.06.

Now Try Exercise 67

```
2000(1+.06)^45
       27529.22165
```

FIGURE 12.20

INTEREST PAID MORE THAN ONCE A YEAR Many times, interest is paid more than once a year. For example, suppose an account gives 8% annual interest that is paid every 3 months, or *quarterly*. It follows that $\frac{1}{4}$ of the annual 8% interest, or 2%, is paid every 3 months. The growth factor is $a = 1 + 0.02 = 1.02$ for each 3-month period. If the initial balance is $1000, then after **5** years, the **quarterly** 2% interest has been paid $4 \cdot 5 = 20$ times, and the account contains

$$\$1000 (1.02)^{20} \approx \$1485.95.$$

In general, if a total of P dollars is deposited in an account paying an *annual* interest rate r (in decimal form) and this interest rate is compounded or paid n times per year, then after t years, the account contains A dollars, given by the following formula.

Compound Interest

Amount after t years   Annual interest (decimal)

Number of years

$$A = P\left(1 + \frac{r}{n}\right)^{nt}$$

Initial deposit   Number of times interest is paid per year

EXAMPLE 9   **Calculating compound interest**

Initially, a total of $1500 is deposited in an account paying 6% annual interest, compounded monthly. What is the account balance after 5 years?

Solution

Let $P = 1500$, $r = 0.06$, $n = 12$, and $t = 5$. The balance after 5 years is

$$A = P\left(1 + \frac{r}{n}\right)^{nt} \qquad \text{Interest formula}$$

$$= 1500\left(1 + \frac{0.06}{12}\right)^{(12 \cdot 5)} \qquad \text{Substitute.}$$

$$= 1500 (1.005)^{60} \qquad \text{Evaluate.}$$

$$\approx \$2023.28. \qquad \text{Approximate.}$$

Now Try Exercise 75

NOTE: In Example 9, if the annual 6% interest had been paid only once a year, the growth factor would be $a = 1.06$, but interest would have been paid only 5 times. The balance would be $1500(1.06)^5 \approx \$2007.34$, rather than the $2023.28 that resulted from monthly compounding. ∎

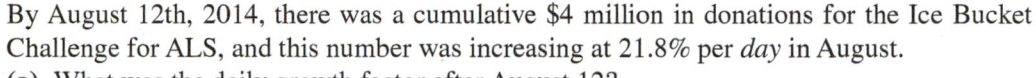

🌐 *Math in Context* The Ice Bucket Challenge for ALS became a fundraising phenomenon in August 2014. The next example shows how an exponential function can model donations.

5 Modeling with Exponential Functions

EXAMPLE 10 **Modeling donations for the Ice Bucket Challenge for ALS**

By August 12th, 2014, there was a cumulative $4 million in donations for the Ice Bucket Challenge for ALS, and this number was increasing at 21.8% per *day* in August.
(a) What was the daily growth factor after August 12?
(b) Write an exponential function $D(x) = Ca^x$ that models the cumulative donations in millions of dollars x days after August 12.
(c) Evaluate $D(16)$ and interpret the result.

Solution
(a) Because $r = 0.218$, the daily growth factor was $a = 1 + 0.218 = 1.218$.
(b) Let $C = 4$ and $a = 1.218$. Then $D(x) = 4(1.218)^x$.
(c) $D(16) = 4(1.218)^{16} \approx 93.8$. In the 16 days following August 12, ALS donations grew from $4 million to approximately $93.8 million.

Now Try Exercise 93

🌐 *Math in Context* Traffic flow at intersections can be modeled by exponential functions whenever traffic patterns occur randomly. In the next example, we model traffic at an intersection by using an exponential function.

EXAMPLE 11 **Modeling traffic flow**

On average, a particular intersection has 360 vehicles arriving randomly each hour. Traffic engineers use $f(x) = (0.905)^x$ to estimate the likelihood, or probability, that *no* vehicle will enter the intersection within an interval of x seconds. (*Source:* F. Mannering and W. Kilareski, *Principles of Highway Engineering and Traffic Analysis.*)

(a) Compute $f(5)$ and interpret the results.
(b) A graph of $y = f(x)$ is shown in **FIGURE 12.21** Discuss this graph.
(c) Is this function an example of exponential growth or decay?

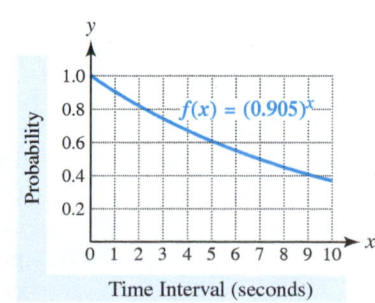

FIGURE 12.21

Solution
(a) The result $f(5) = (0.905)^5 \approx 0.61$ indicates that there is a 61% chance no vehicle will enter the intersection during any particular 5-second interval.
(b) The graph decreases, which means that as the interval of time increases, there is less chance (likelihood) that a car will *not* enter the intersection.
(c) Because the graph is decreasing and $a = 0.905 < 1$, this function is an example of exponential decay.

Now Try Exercise 101

6 Recognizing the Natural Exponential Function

A special type of exponential function that occurs often in applications is called the *natural exponential function*, expressed as $f(x) = e^x$. The **base** *e* is a special number in mathematics similar to π. The number π is approximately 3.14, whereas the number *e* is approximately 2.72. The number *e* is named for the great Swiss mathematician Leonhard Euler (1707–1783). Most calculators have a special key that can be used to compute the natural exponential function.

NATURAL EXPONENTIAL FUNCTION

The function represented by

$$f(x) = e^x$$

is the **natural exponential function**, where $e \approx 2.71828183$.

EXAMPLE 12   Sketching a graph of $y = e^x$

Sketch a graph of $y = 2^x$ and $y = 3^x$ in the same *xy*-plane. Then use these two graphs to sketch a graph of $y = e^x$. How do these graphs compare?

Solution

The graphs of $y = 2^x$, $y = 3^x$, and $y = e^x$ are shown in **FIGURE 12.22**.

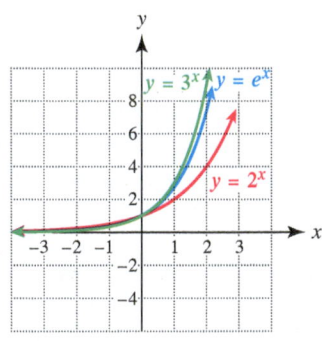

FIGURE 12.22

Because $e \approx 2.718$, the graph of $y = e^x$ lies between the graphs of $y = 2^x$ and $y = 3^x$. It is closer to the graph of $y = 3^x$.

Now Try Exercise 83

🌐 *Math in ₐContext* (*Population*) The natural exponential function is frequently used to model **continuous growth** and **continuous decay**. For example, the fact that births and deaths occur throughout the year, not just at one time during the year, must be recognized when population growth or decay is being modeled. If a population *P* is increasing or decreasing continuously at *r* percent per year, expressed as a decimal, we can model this population after *x* years by

$$P = Ce^{rx}, \quad \text{Continuous growth, } r > 0$$

$$\text{Continuous decay, } r < 0$$

where *C* is the initial population. To evaluate natural exponential functions, we use a calculator, as demonstrated in the next example.

EXAMPLE 13   **Modeling population**

In 2013, the population of Texas was 26.5 million people and was growing at a continuous rate of 1.8% per year. This population in millions x years after 2013 can be modeled by

$$f(x) = 26.5e^{0.018x}.$$

Estimate the population in 2017.

Solution

Because 2017 is 4 years after 2013, we evaluate $f(4)$ to obtain

$$f(4) = 26.5e^{0.018(4)} \approx 28.5,$$

which is supported by **FIGURE 12.23**. This model estimates the population of Texas to be about 28.5 million in 2017.

CALCULATOR HELP

To evaluate the natural exponential function, see Appendix A (page AP-8).

```
26.5e^.018(4)
           28.47836662
```

FIGURE 12.23

Now Try Exercise 103

12.2   Putting It All Together

CONCEPT	EXPLANATION	EXAMPLES
Exponential Function	The variable is an exponent. $$f(x) = Ca^x$$ Initial value when $x = 0$ Growth factor (base) Growth: $a > 1$ Decay: $0 < a < 1$	$f(x) = 3(2)^x$ models exponential **growth**, and $g(x) = 2\left(\frac{1}{3}\right)^x$ models exponential **decay**.
Graphs of Exponential Functions	• Is above the x-axis for all inputs • Increases for $a > 1$ • Decreases for $0 < a < 1$ • The y-intercept is $(0, C)$.	**Exponential Functions: $f(x) = Ca^x$** Growth → $a > 1$ Decay → $0 < a < 1$ y-intercept is $(0, C)$.
Percent Change	An amount changes from A to B: $$\text{Percent change} = \frac{B - A}{A} \times 100.$$	If \$200 increases to \$300, then $$\frac{300 - 200}{200} \times 100 = 50\%$$ is the percent change.

continued on next page

continued from previous page

CONCEPT	EXPLANATION	EXAMPLES
Constant Percent Change and Exponential Functions	Constant percent change (decimal) Time $$f(x) = C(1 + r)^x$$ Initial amount Growth factor	If 3000 bacteria increase in number by 12% daily for x days, then $$f(x) = 3000(1.12)^x$$ models these numbers after x days.
Compound Interest	Compounded annually: $$A = P(1 + r)^t$$ Compounded n times per year: $$A = P\left(1 + \frac{r}{n}\right)^{nt}$$ P: Initial deposit A: Final amount r: Annual interest rate (decimal) n: Interest paid n times per year t: Number of years	Depositing \$1000 at 3% annual interest for 4 years gives $$A = \$1000(1.03)^4$$ $$\approx 1125.51.$$ Depositing \$1000 at 3% annual interest compounded monthly for 4 years gives $$A = \$1000\left(1 + \frac{0.03}{12}\right)^{(12 \cdot 4)}$$ $$\approx \$1127.33.$$
Natural Exponential Function	$$f(x) = e^x$$ Growth factor, or base, is $e \approx 2.72$.	Using a calculator, $$f(2) = e^2 \approx 7.39.$$

12.2 Exercises

MyMathLab®

CONCEPTS AND VOCABULARY

1. Give a general formula for an exponential function f.

2. Sketch a graph of an exponential function that illustrates exponential decay.

3. Give the domain and range of an exponential function.

4. Evaluate the expressions 2^x and x^2 for $x = 5$.

5. Approximate e to the nearest thousandth.

6. Evaluate e^2 and π^2 using your calculator.

7. If a quantity y grows exponentially, then for each unit increase in x, y increases by a constant _____.

8. If $f(x) = 1.5^x$, what is the growth factor?

9. If a quantity increases from A to B, then the percent change equals _____.

10. If a quantity increases by 35% each year, then the growth factor a is _____.

EVALUATING EXPONENTIAL FUNCTIONS

Exercises 11–22: Evaluate the exponential function for the given values of x by hand when possible. Approximate answers to the nearest hundredth when appropriate.

11. $f(x) = 3^x$ $x = -2, x = 2$

12. $f(x) = 5^x$ $x = -1, x = 3$

13. $f(x) = 5(2^x)$ $x = 0, x = 5$

14. $f(x) = 3(7^x)$ $x = -2, x = 0$

15. $f(x) = \left(\frac{1}{2}\right)^x$ $x = -2, x = 3$

16. $f(x) = \left(\frac{1}{4}\right)^x$ $x = 0, x = 2$

17. $f(x) = 5(3)^{-x}$ $x = -1, x = 2$

18. $f(x) = 4\left(\frac{3}{7}\right)^x$ $x = 1, x = 4$

19. $f(x) = 1.8^x$ $x = -3, x = 1.5$

20. $f(x) = 0.91^x$ $x = 5.1, x = 10$

21. $f(x) = 3(0.6)^x$ $x = -1, x = 2$

22. $f(x) = 5(4.5)^{-x}$ $x = -2.1, x = 1.9$

Exercises 23 and 24: **Thinking Generally** *For the given exponential function, evaluate $f(0)$ and $f(-1)$.*

23. $f(x) = a^x$

24. $f(x) = (1 + r)^{2x}$

RECOGNIZING GRAPHS OF EXPONENTIAL FUNCTIONS

Exercises 25–28: Match the formula with its graph (a–d). Do not use a calculator.

25. $f(x) = 1.5^x$ **26.** $f(x) = \frac{1}{4}(2^x)$

27. $f(x) = 4\left(\frac{1}{2}\right)^x$ **28.** $f(x) = \left(\frac{1}{3}\right)^x$

a.

b.

c.

d.
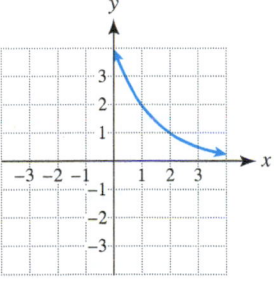

Exercises 29–32: Use the graph of $y = Ca^x$ to determine the constants C and a.

29.

30.

31.

32.
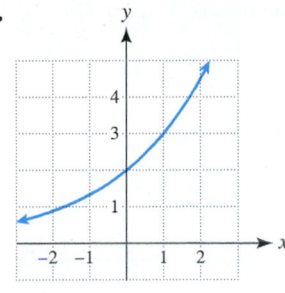

Exercises 33–44: Graph $y = f(x)$. State whether the graph depicts exponential growth or exponential decay.

33. $f(x) = 2^x$ **34.** $f(x) = 3^x$

35. $f(x) = \left(\frac{1}{4}\right)^x$ **36.** $f(x) = \left(\frac{1}{2}\right)^x$

37. $f(x) = 2^{-x}$ **38.** $f(x) = 3^{-x}$

39. $f(x) = 3^x - 1$ **40.** $f(x) = 2^x + 1$

41. $f(x) = 2^{x-1}$ **42.** $f(x) = 2^{x+1}$

43. $f(x) = 4\left(\frac{1}{3}\right)^x$ **44.** $f(x) = 3\left(\frac{1}{2}\right)^x$

Exercises 45–46: Compare the graph of each equation to the graph of $y = f(x)$.

45. $f(x) = 2^x$
     (a) $y = 2^{x-1}$
     (b) $y = 2^x - 1$
     (c) $y = 2^{-x}$
     (d) $y = -2^x$

46. $f(x) = e^x$
     (a) $y = -e^x$
     (b) $y = e^x + 3$
     (c) $y = e^{-x}$
     (d) $y = e^{x+2}$

COMPARING EXPONENTIAL AND LINEAR GROWTH

Exercises 47–52: (Refer to Example 4.) A table for a function f is given.

     (a) *Determine whether function f represents exponential growth, exponential decay, or linear growth.*
     (b) *Find a formula for f.*

47.

x	0	1	2	3	4
$f(x)$	64	16	4	1	$\frac{1}{4}$

48.

x	0	1	2	3	4
$f(x)$	$\frac{1}{2}$	1	2	4	8

49.

x	0	1	2	3	4
$f(x)$	8	11	14	17	20

50.

x	-2	-1	0	1	2
$f(x)$	4	2	1	$\frac{1}{2}$	$\frac{1}{4}$

51.

x	-2	-1	0	1	2
$f(x)$	2.56	3.2	4	5	6.25

52.

x	-2	-1	0	1	2
$f(x)$	-6	-2	2	6	10

RELATING PERCENT CHANGE AND EXPONENTIAL FUNCTIONS

Exercises 53–56: For the given amounts A and B, find each of the following.

(a) *The percent change if A changes to B*
(b) *The percent change if B changes to A*

53. $A = \$200, B = \$400$

54. $A = \$1.50, B = \$1.00$

55. $A = 150, B = 30$

56. $A = 80, B = 200$

Exercises 57–62: An account contains A dollars and increases or decreases by R%. For each A and R, answer the following.

(a) *Find the increase or decrease in the value of the account.*
(b) *Find the final value of the account.*
(c) *By what factor did the account value increase or decrease?*

57. $A = \$1000, R = 120\%$

58. $A = \$500, R = 230\%$

59. $A = \$650, R = 20\%$

60. $A = \$70, R = 35\%$

61. $A = \$800, R = -10\%$

62. $A = \$950, R = -60\%$

Exercises 63–66: For the given f(x), state the initial value C, the growth or decay factor a, and percent change R for each unit increase in x.

63. $f(x) = 9(1.07)^x$

64. $f(x) = 3(1.351)^x$

65. $f(x) = 1.5(0.45)^x$

66. $f(x) = 0.9^x$

COMPUTING COMPOUND INTEREST

Exercises 67–72: (Refer to Example 8.) The amount P is deposited in an account paying R percent annual interest. Approximate the amount in the account after x years.

67. $P = \$1500 \quad R = 9\% \quad x = 10$ years

68. $P = \$1500 \quad R = 15\% \quad x = 10$ years

69. $P = \$200 \quad R = 20\% \quad x = 50$ years

70. $P = \$5000 \quad R = 8.4\% \quad x = 7$ years

71. $P = \$560 \quad R = 1.4\% \quad x = 25$ years

72. $P = \$750 \quad R = 10\% \quad x = 13$ years

73. Thinking Generally A total of \$1000 is deposited in an account paying 8% annual interest for 10 years. If \$2000 had been deposited instead of \$1000, would there be twice the money in the account after 10 years? Explain.

74. Thinking Generally A total of \$500 is deposited in an account paying 5% annual interest for 10 years. If the interest rate had been 10% instead of 5%, would the total interest earned after 10 years be twice as much? Explain.

Exercises 75–78: (Refer to Example 9.) The amount P is deposited in an account giving R% annual interest compounded n times a year. Find the amount A in the account after t years.

75. $P = \$700, R = 4\%, n = 4, t = 3$

76. $P = \$550, R = 3\%, n = 2, t = 5$

77. $P = \$1200, R = 2.5\%, n = 12, t = 7$

78. $P = \$1500, R = 6.5\%, n = 365, t = 20$

EVALUATING AND GRAPHING THE NATURAL EXPONENTIAL FUNCTION

Exercises 79–82: Evaluate f(x) for the given value of x. Approximate answers to the nearest hundredth.

79. $f(x) = e^x \qquad x = 1.2$

80. $f(x) = 2e^x \qquad x = 2$

81. $f(x) = 1 - e^x \qquad x = -2$

82. $f(x) = 4e^{-x} \qquad x = 1.5$

Exercises 83–86: Graph $f(x)$ in $[-4, 4, 1]$ by $[0, 8, 1]$. State whether the graph illustrates exponential growth or exponential decay.

83. $f(x) = e^{0.5x}$ **84.** $f(x) = e^x + 1$

85. $f(x) = 1.5e^{-0.32x}$ **86.** $f(x) = 2e^{-x} + 1$

APPLICATIONS OF EXPONENTIAL FUNCTIONS

Exercises 87–90: **Exponential Models** *Write the formula for an exponential function, $f(x) = Ca^x$, that models the situation. Evaluate $f(4)$.*

87. A sample of 5000 bacteria decreases in number by 25% per week.

88. A sample of 700 insects increases in number by 200% per day.

89. A sample of 50 birds increases in number by 10% per month.

90. A sample of 137 fish decreases in number by 2% per week.

91. **Salary Growth** Suppose your salary is $50,000 per year. Would you rather have a 20% raise each year or a $20 raise each year?

92. **Salary Growth** Suppose your salary is $35,000 per year. Would you rather have a 10% raise each year or a $4000 raise each year? Assume that you will keep this job for 10 years.

93. **Blood Alcohol** Suppose that a person's peak blood alcohol level is 0.07 (grams per 100 mL) and that this level decreases by 40% each hour. (*Source:* National Institutes of Health.)
 (a) What is the hourly decay factor?
 (b) Write an exponential function $B(x) = Ca^x$ that models the blood alcohol level after x hours.
 (c) Evaluate $B(2)$ and interpret the result.

94. **Apple's Revenue** From 2001 to 2015, Apple's first quarter revenue grew at an annual rate of 33.7%, from $1 billion to about $58 billion. (*Source:* Apple Corp.)
 (a) What is the annual growth factor?
 (b) Write an exponential function $R(x) = Ca^x$ that models the revenue in billions of dollars x years after 2001.
 (c) Evaluate $R(14)$ and interpret the result.

95. **Facebook Link Half-Life** (Refer to the chapter opener.) The half-life for a link on Facebook is 3 hours. Write an exponential function F that gives the percentage of hits remaining on a typical Facebook link after t hours. Estimate this percentage after 4 hours.

96. **StumbleUpon Link Half-Life** (Refer to the chapter opener.) The half-life for a link on StumbleUpon is 400 hours. Write an exponential function S that gives the percentage of engagements remaining on a typical StumbleUpon link after t hours. Estimate this percentage after 250 hours.

97. **Tweets per Month** From July 2010 to July 2015, the number of tweets T per month increased dramatically and could be modeled in billions by

$$T(x) = 0.1(2.724)^x,$$

where x represents years after July 2010. (*Source:* Twitter.)
 (a) What is the yearly growth factor?
 (b) Interpret the 0.1 in the formula.
 (c) Evaluate $T(5)$ and interpret the result.

98. **Dating Artifacts** Radioactive carbon-14 is found in all living things and is used to date objects containing organic material. Suppose that an object initially contains C grams of carbon-14. After x years, it will contain A grams, where

$$A = C(0.99988)^x.$$

 (a) Let $C = 10$ and graph A over a 20,000-year period. Is this function an example of exponential growth or decay?
 (b) If $C = 10$, how many grams are left after 5700 years? What fraction of the carbon-14 is left?

99. **E. coli Bacteria** A strain of bacteria that inhabits the intestines of animals is named *Escherichia coli* (*E. coli*). These bacteria are capable of rapid growth and can be dangerous to humans—particularly children. The table shows the results of one study of the growth of *E. coli* bacteria, where concentrations are listed in *thousands* of bacteria per milliliter.

t (minutes)	0	50	100
Concentration	500	1000	2000

t (minutes)	150	200	250
Concentration	4000	8000	16,000

Source: G.S. Stent, *Molecular Biology of Bacterial Viruses.*

 (a) Find C and a so that $f(t) = Ca^{t/50}$ models these data.
 (b) Use $f(t)$ to estimate the concentration of bacteria after 170 minutes.
 (c) Discuss the growth of this strain of bacteria over a 250-minute time period.

100. Swimming Pool Maintenance Chlorine is frequently used to disinfect swimming pools. The chlorine concentration should remain between 1.5 and 2.5 parts per million (ppm). After a warm, sunny day, only 80% of the chlorine may remain in the water, with the other 20% dissipating into the air or combining with other chemicals in the water. (*Source: D. Thomas, Swimming Pool Operator's Handbook.*)

(a) Let $f(x) = 3(0.8)^x$ model the concentration of chlorine in parts per million after x days. What is the initial concentration of chlorine in the pool?

(b) If no more chlorine is added, estimate when the chlorine level first drops below 1.6 parts per million.

101. Modeling Traffic Flow (Refer to Example 11.) Construct a table of $f(x) = (0.905)^x$, starting at $x = 0$ and incrementing by 10, until $x = 50$.

(a) Evaluate $f(0)$ and interpret the result.

(b) For a time interval of what length is there only a 5% chance that no cars will enter the intersection?

102. Pros and Putts The percentage of putts P from 3 feet to 25 feet made by professional golfers can be modeled by the exponential function

$$P(x) = 99(0.872)^{x-3},$$

where x is the length of the putt.

(a) Find the percentage of putts made from 3 feet by professionals.

(b) Evaluate $P(8)$ and interpret the results.

(c) What is the decay factor? Interpret it.

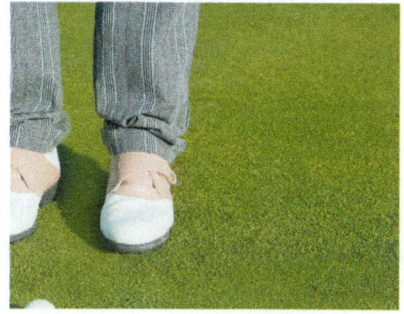

Exercises 103–106: **Population Growth** *(Refer to Example 13.) The population P in 2010 for a state is given along with R%, its annual percentage rate of continuous growth.*

(a) *Write the formula* $f(x) = Pe^{rx}$, *where r is in decimal notation, that models the population in millions x years after 2010.*

(b) *Estimate the population in 2020.*

103. Nevada: $P = 2.7$ million, $R = 1.4\%$

104. North Carolina: $P = 9.4$ million, $R = 1.36\%$

105. California: $P = 38$ million, $R = 1.02\%$

106. Arizona: $P = 6.6$ million, $R = 1.44\%$

107. Online Exploration In many states, landlords are mandated to return a tenant's security deposit plus interest. Go online and look up this interest rate for your state. Answers may vary.

(a) If you are a tenant, calculate the interest you should receive after 1 year. (If you are not, assume the security deposit is $1200.)

(b) Write a function I that calculates the interest that you should receive after x years. (Does your landlord have to pay compound interest?)

108. Online Exploration Go online and determine how long it takes the fastest growing bacteria to double in number. Answers may vary.

(a) Suppose that you start with a sample of 4 million such bacteria. Write an exponential function that gives the number N of bacteria in millions after x minutes.

(b) How many bacteria are there after 2 hours?

WRITING ABOUT MATHEMATICS

109. A student evaluates $f(x) = 4(2)^x$ at $x = 3$ and obtains 512. Did the student evaluate the function correctly? What was the student's error?

110. For a set of data, how can you distinguish between linear growth and exponential growth? Give an example of each type of data.

Checking Basic Concepts

1. If $f(x) = 2x^2 + 5x - 1$ and $g(x) = x + 1$, find each expression.
   (a) $(g \circ f)(1)$ **(b)** $(f \circ g)(x)$

2. Sketch a graph of $f(x) = x^2 - 1$.
   (a) Is f a one-to-one function? Explain.
   (b) Does f have an inverse function?

3. If $f(x) = 4x - 3$, find $f^{-1}(x)$.

4. Evaluate $f(-2)$ if $f(x) = 3(2)^x$.

5. Sketch a graph of $f(x) = \left(\frac{1}{3}\right)^x$.

6. Use the graph of $y = Ca^x$ to determine the constants C and a

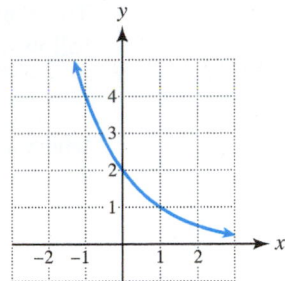

12.3   Logarithmic Functions

Objectives

1 Introducing the Common Logarithmic Function
• The Graph of the Common Logarithm Function

2 Finding the Inverse of the Common Logarithmic Function

3 Working with Logarithms that Have Other Bases
• Base 2
• Base e (The Natural Logarithm)
• Base a
• Inverses of Base a Logarithmic Functions

NEW VOCABULARY

☐ Common logarithm of a positive number x
☐ Common logarithmic function
☐ Natural logarithm
☐ Logarithm with base a of a positive number x
☐ Logarithmic function with base a

1   Introducing the Common Logarithmic Function

🌐 Math in Context (Measurement)   In applications, measurements can vary greatly in size. **TABLE 12.7** lists some examples of objects, with the approximate distances in meters across each.

Sizes of Objects

Object	Distance (meters)
Atom	10^{-9}
Protozoan	10^{-4}
Small Asteroid	10^2
Earth	10^7
Universe	10^{26}

Source: C. Ronan, *The Natural History of the Universe.*
TABLE 12.7

Each distance is listed in the form 10^k for some k. The value of k distinguishes one measurement from another. The *common logarithmic function* or *base-10 logarithmic function*, denoted log or log_{10}, outputs k if the input x can be written as 10^k for some real number k. For example, $\log 10^{-9} = -9$, $\log 10^2 = 2$, and $\log 10^{1.43} = 1.43$. For any real number k, $\log 10^k = k$. Some values for $\log x$ are given in **TABLE 12.8**.

Common Logarithms of Powers of 10

> Write x as a power of 10.

x	10^{-4}	10^{-3}	10^{-2}	10^{-1}	10^0	10^1	10^2	10^3	10^4
$\log x$	-4	-3	-2	-1	0	1	2	3	4

TABLE 12.8

> A common logarithm is the *exponent* on base 10.

We use this information to define the common logarithm.

COMMON LOGARITHM

The **common logarithm of a positive number** x, denoted log x, is calculated as follows. If x is written as $x = 10^k$, then

$$\log x = k,$$

where k is a real number. That is, $\log 10^k = k$.
The function given by

$$f(x) = \log x$$

is called the **common logarithmic function**.

NOTE: Previously, we have always used one letter, such as f or g, to name a function. The common logarithm is the *first* function for which we use *three* letters, log, to name it. Thus $f(x)$, $g(x)$, and log (x) all represent functions. Generally, log (x) is written without parentheses as log x. ∎

The equation $\log x = k$ is called logarithmic form and the equation $x = 10^k$ is called exponential form. These forms are *equivalent*. That is, if one equation is true, the other equation must also be true.

Equivalent Logarithmic and Exponential Forms

Logarithmic Form	Exponential Form
$\log 100 = 2$	$10^2 = 100$
$\log 1000 = 3$	$10^3 = 1000$
$\log \frac{1}{10} = -1$	$10^{-1} = \frac{1}{10}$
$\log x = k$	$10^k = x$

General forms

A *logarithm is an exponent*. For example, to evaluate the logarithm log 100, ask the question, "10 to what power equals 100?" The necessary *exponent* equals log 100. That is, $\log 100 = 2$ because $10^2 = 100$.

EXAMPLE 1 **Using exponential and logarithmic forms**

Change the given equation into the equivalent logarithmic or exponential form.
(a) $\log \frac{1}{100} = -2$ **(b)** $\log 100{,}000 = 5$ **(c)** $10^4 = 10{,}000$ **(d)** $10^{-3} = \frac{1}{1000}$

Solution

(a) We place the exponent -2 on 10 to get $\frac{1}{100}$, so $10^{-2} = \frac{1}{100}$
(b) We place the exponent 5 on 10 to get 100,000, so $10^5 = 100{,}000$
(c) The exponent on 10 is 4, so $\log 10{,}000 = 4$
(d) The exponent on 10 is -3, so $\log \frac{1}{1000} = -3$

Now Try Exercise 13

STEPS USED TO CALCULATE THE COMMON LOGARITHM

STEP 1: Is x positive? If not, then log x is *undefined*.

STEP 2: Write x as 10^k for some real number k. If this is not possible, then use a calculator to approximate log x.

STEP 3: If $x = 10^k$, then log $x = k$. That is, log $10^k = k$.

EXAMPLE 2 **Calculating log x**

Evaluate each expression, if possible.
(a) log (-4) **(b)** log 1000

Solution

(a) **STEP 1:** Because -4 is negative, log -4 is undefined.

 NOTE: Because 10^k is always positive, there is no k such that $10^k = -4$. The common logarithm of *any* negative number is undefined. ∎

(b) **STEP 1:** Since 1000 is positive, log 1000 is defined.
 STEP 2: Because $1000 = 10^3$, we can write log 1000 as log 10^3.
 STEP 3: Because log $10^k = k$, it follows that

$$\log 1000 = \log 10^3 = \mathbf{3}.$$

Now Try Exercises 29, 33

EXAMPLE 3 **Evaluating common logarithms**

Evaluate each common logarithm.
(a) log 100   **(b)** log $\frac{1}{10}$ **(c)** log $\sqrt{1000}$ **(d)** log 45

Solution
(a) $100 = 10^2$, so log $100 = \log 10^2 = \mathbf{2}$
(b) $\frac{1}{10} = 10^{-1}$, so log $\frac{1}{10} = \log 10^{-1} = \mathbf{-1}$
(c) log $\sqrt{1000} = \log (1000)^{1/2} = \log (10^3)^{1/2} = \log 10^{3/2} = \frac{3}{2}$
(d) It is not obvious how to write 45 as a power of 10. However, we can use a calculator to determine that log $45 \approx 1.6532$. Thus $10^{1.6532} \approx 45$.

FIGURE 12.24 supports these answers.

CALCULATOR HELP

To evaluate the common logarithmic function, see Appendix A (page AP-8).

READING CHECK 1

• What does log x equal?

```
log(100)
               2
log(1/10)
              -1
```
(a)

```
log(√(1000))
             1.5
log(45)
      1.653212514
10^1.6532
      44.99870339
```
(b)

FIGURE 12.24

Now Try Exercises 19, 25, 31, 45

THE GRAPH OF THE COMMON LOGARITHM FUNCTION The following See the Concept describes the graph of $y = \log x$.

See the Concept 1   THE GRAPH OF $y = \log x$

$y = \log x$

x	y
10^{-1}	-1
10^{0}	0
$10^{0.5}$	0.5
10^{1}	1

Four points of the graph of $y = \log x$

Range: $(-\infty, \infty)$

$y = \log x$

Domain: $(0, \infty)$

D Vertical Asymptote

A The graph passes through $(1, 0)$, so $\log(1) = 0$.

B The graph increases slowly for large values of x. For example, $\log(100) = 2$ and $\log(1000) = 3$.

C When $0 < x < 1$, $\log x$ outputs negative values.

D As x approaches 0, $\log x$ approaches $-\infty$.

WATCH VIDEO IN MML.

EXAMPLE 4   Graphing logarithmic functions

Graph each function f and compare its graph to $y = \log x$.
(a) $f(x) = \log(x - 2)$ **(b)** $f(x) = \log(x) + 1$

Solution

(a) Recall that the graph of $y = (x - 2)^2$ is similar to the graph of $y = x^2$, except that it is translated 2 units to the *right*. It follows that the graph of $y = \log(x - 2)$ is similar to the graph of $y = \log x$, except that it is also translated 2 units to the right. This translation is shown in **FIGURE 12.25**. The graph of $y = \log x$ passes through $(1, 0)$, so the graph of $y = \log(x - 2)$ passes through a point that is 2 units to the right, or $(3, 0)$. The vertical asymptote is also translated 2 units to the right to the line $x = 2$.

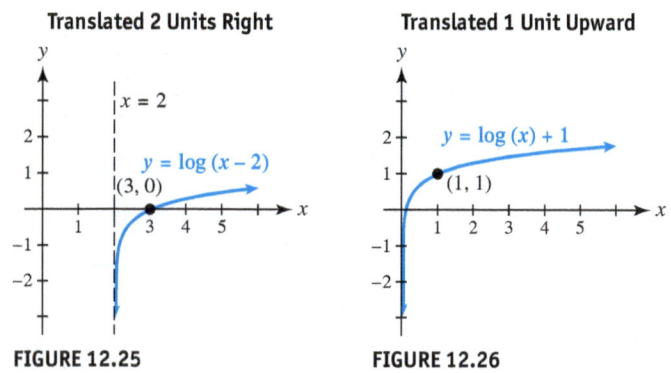

Translated 2 Units Right

$x = 2$

$y = \log(x - 2)$

$(3, 0)$

FIGURE 12.25

Translated 1 Unit Upward

$y = \log(x) + 1$

$(1, 1)$

FIGURE 12.26

NOTE: The graph of $y = \log x$ is undefined when $x < 0$ and the graph of $\log(x - 2)$ in **FIGURE 12.25** is undefined when $x < 2$. ∎

(b) The graph of $y = \log(x) + 1$ is similar to the graph of $y = \log x$, except that it is translated 1 unit *upward*. This graph is shown in **FIGURE 12.26**. Note that the graph of $y = \log(x) + 1$ passes through the point $(1, 1)$.

Now Try Exercises 49, 51

MAKING CONNECTIONS 1

The Common Logarithmic Function and the Square Root Function

Some similarities of $f(x) = \sqrt{x}$ and $f(x) = \log x$ are outlined in the table below.

Similarities	$\sqrt{x}$	$\log x$
Some calculations are easy.	$\sqrt{100} = 10$	$\log 1000 = 3$
Some calculations require a calculator.	$\sqrt{2} \approx 1.41$	$\log 45 \approx 1.65$
The implied notation is similar.	$\sqrt{x} = \sqrt[2]{x}$	$\log x = \log_{10} x$
The input x can not be negative.	$\sqrt{-3}$	$\log(-3)$

Undefined expressions

2 Finding the Inverse of the Common Logarithmic Function

The graph of $y = \log x$ is a one-to-one function because it passes the horizontal line test. Thus the common logarithmic function has an inverse function. To determine this inverse function for $\log x$, consider **TABLES 12.9** and **12.10**.

$y = 10^x$

x	-2	-1	0	1	2
10^x	10^{-2}	10^{-1}	10^0	10^1	10^2

(1)

TABLE 12.9

Inverse functions

$y = \log x$

x	10^{-2}	10^{-1}	10^0	10^1	10^2
$\log x$	-2	-1	0	1	2

(2)

TABLE 12.10

If we start with the input **2** for 10^x, we compute 10^2, as shown by (1) in **TABLE 12.9**. If we use 10^2 as the input for $\log x$, we compute **2**, as shown by (2) in **TABLE 12.10**. That is,

$$\log(10^2) = 2. \quad \text{Calculate } 10^x \text{ and then } \log x.$$

Next, suppose we find the logarithm first and then compute a power of 10. If we start with the input 10^2, we compute **2**, as shown by (2) in **TABLE 12.10**. If we use **2** as the input for 10^x, we compute 10^2, as shown by (1) in **TABLE 12.9**. That is,

$$10^{\log(10^2)} = 10^2. \quad \text{Calculate } \log x \text{ and then } 10^x.$$

In general, if $10^a = b$, then $\log b = a$, and if $\log b = a$, then $10^a = b$. Thus the *inverse function* of $f(x) = \log x$ is $f^{-1}(x) = 10^x$. Note that composition of these two functions satisfies the definition of an inverse function.

$$(f \circ f^{-1})(x) = f(f^{-1}(x)) \quad \text{and} \quad (f^{-1} \circ f)(x) = f^{-1}(f(x))$$
$$= f(10^x) \qquad\qquad\qquad = f^{-1}(\log x)$$
$$= \log 10^x \qquad\qquad\qquad = 10^{\log x}$$
$$= x \qquad\qquad\qquad\qquad = x$$

The graphs of $y = \log x$ and $y = 10^x$ are shown in **FIGURE 12.27**. Note that the graph of $y = 10^x$ is a reflection of the graph of $y = \log x$ across the dashed line $y = x$.

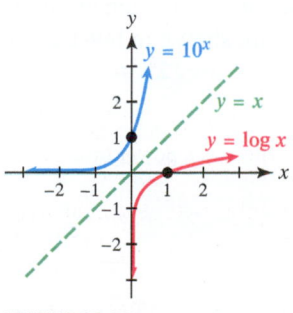

FIGURE 12.27

INVERSE PROPERTIES OF THE COMMON LOGARITHM

The following properties hold for common logarithms.

$$\log 10^x = x, \qquad \text{for any real number } x$$
$$10^{\log x} = x, \qquad \text{for any positive real number } x$$

EXAMPLE 5   Applying inverse properties

Use inverse properties to simplify each expression.

(a) $\log 10^\pi$ **(b)** $\log 10^{x^2+1}$ **(c)** $10^{\log 7}$ **(d)** $10^{\log 3x}, x > 0$

Solution
(a) Because $\log 10^x = x$ for any real number x, $\log 10^\pi = \boldsymbol{\pi}$.
(b) $\log 10^{x^2+1} = \boldsymbol{x^2 + 1}$
(c) Because $10^{\log x} = x$ for any positive real number x, $10^{\log 7} = \boldsymbol{7}$.
(d) $10^{\log 3x} = \boldsymbol{3x}$, provided x is a *positive* number.

Now Try Exercises 27, 37, 39, 43

🌐 *Math in Context* [Audiology]   Logarithms are used to model quantities that vary greatly in intensity. For example, the human ear is extremely sensitive and able to detect intensities on the eardrum ranging from 10^{-16} watts per square centimeter (w/cm^2) to 10^{-4} w/cm^2, which is usually painful. The next example illustrates modeling sound with logarithms.

EXAMPLE 6   Modeling sound levels

Sound levels in decibels (dB) can be computed by $f(x) = 160 + 10 \log x$, where x is the intensity of the sound in watts per square centimeter. Ordinary conversation has an intensity of 10^{-10} w/cm^2. What decibel level is this? (*Source:* R. Weidner and R. Sells, *Elementary Classical Physics*, Vol. 2.)

Solution

CRITICAL THINKING 1

If the sound level increases by 10 dB, by what factor does the intensity x increase?

To find the decibel level for ordinary conversation, evaluate $f(10^{-10})$.

$$\begin{aligned}
f(10^{-10}) &= 160 + 10\,\boldsymbol{\log\,(10^{-10})} && \text{Substitute } x = 10^{-10}.\\
&= 160 + 10(\boldsymbol{-10}) && \text{Evaluate } \log\,(10^{-10}).\\
&= 60 && \text{Simplify.}
\end{aligned}$$

Ordinary conversation corresponds to 60 dB.

Now Try Exercise 111

3 Working with Logarithms that Have Other Bases

BASE 2 Common logarithms are base-10 logarithms, but we can define logarithms using other bases. For example, base-2 logarithms are frequently used in computer science. Some values for the base-2 logarithmic function, denoted $f(x) = \log_2 x$, are shown in **TABLE 12.11**. If x can be expressed as $x = 2^k$ for some real number k, then $\log_2 x = \log_2 \boldsymbol{2^k} = \boldsymbol{k}$.

READING CHECK 2

• What does $\log_2 x$ mean?

Write x as a power of 2.

Base-2 Logarithms of Powers of 2

x	2^{-3}	2^{-2}	2^{-1}	2^0	2^1	2^2	2^3
$\log_2 x$	-3	-2	-1	0	1	2	3

TABLE 12.11

A base-2 logarithm is the *exponent* on base 2.

NOTE: A base-2 logarithm is an *exponent* on base 2. ∎

EXAMPLE 7   **Evaluating base-2 logarithms**

Simplify each logarithm.
(a) $\log_2 8$ (b) $\log_2 \frac{1}{4}$

Solution
(a) The logarithmic expression $\log_2 8$ represents the exponent on base 2 that gives 8. Because $8 = 2^3$, $\log_2 8 = \log_2 2^3 = 3$.
(b) Because $\frac{1}{4} = \frac{1}{2^2} = 2^{-2}$, $\log_2 \frac{1}{4} = \log_2 2^{-2} = -2$.

Now Try Exercises 87, 89

BASE e (THE NATURAL LOGARITHM) Some values of base-e logarithms are shown in TABLE 12.12. A base-e logarithm is referred to as a **natural logarithm** and is denoted either $\log_e x$ or $\ln x$. Natural logarithms are used in many areas including mathematics, science, economics, electronics, and communications.

CALCULATOR HELP

To evaluate the natural logarithmic function, see Appendix A (page AP-8).

Natural Logarithms of Powers of e

x	e^{-3}	e^{-2}	e^{-1}	e^0	e^1	e^2	e^3
$\ln x$	-3	-2	-1	0	1	2	3

TABLE 12.12

NOTE: A natural logarithm is an *exponent* on base e. ∎

To evaluate natural logarithms, we usually use a calculator.

EXAMPLE 8   **Evaluating natural logarithms**

Approximate to the nearest hundredth.
(a) $\ln 10$ (b) $\ln \frac{1}{2}$

```
ln(10)
        2.302585093
ln(1/2)
        -.6931471806
```

FIGURE 12.28

Solution
(a) **FIGURE 12.28** shows that $\ln 10 \approx 2.30$.
(b) **FIGURE 12.28** shows that $\ln \frac{1}{2} \approx -0.69$.

Now Try Exercises 65, 67

BASE a We now define base-a logarithms.

BASE-a LOGARITHMS

The **logarithm with base a of a positive number x**, denoted $\log_a x$, is calculated as follows. If x is written as $x = a^k$, then

$$\log_a x = k,$$

where $a > 0$, $a \neq 1$, and k is a real number. That is, $\log_a a^k = k$.
The function given by

$$f(x) = \log_a x$$

is called the **logarithmic function with base a**.

CRITICAL THINKING 2

Evaluate $\log_2 1$.

Remember that *a logarithm is an exponent.* The expression $\log_a x$ equals the exponent k such that $a^k = x$. The graph of $y = \log_a x$ with $a > 1$ is shown in **FIGURE 12.29**. Note that the graph passes through the point $(1, 0)$. Thus $\log_a 1 = 0$. Furthermore, the y-axis is a vertical asymptote and the expression $\log_a x$ is defined only when x is positive.

NOTE: The natural logarithm, $\ln x$, is a base-a logarithm with $a = e$. That is, $\ln x = \log_e x$. ∎

$y = \log_a x, a > 1$

$\log_a 1 = 0$ for all $a > 1$

Vertical asymptote $x = 0$

$(1, 0)$

FIGURE 12.29

EXAMPLE 9 **Evaluating base-a logarithms**

Simplify each logarithm.

(a) $\log_5 25$ **(b)** $\log_4 \frac{1}{64}$ **(c)** $\log_7 1$ **(d)** $\log_3 9^{-1}$

Solution

(a) $25 = 5^2$, so $\log_5 25 = \log_5 5^2 = 2$.

(b) $\frac{1}{64} = \frac{1}{4^3} = 4^{-3}$, so $\log_4 \frac{1}{64} = \log_4 4^{-3} = -3$.

(c) $1 = 7^0$, so $\log_7 1 = \log_7 7^0 = 0$. (The logarithm of 1 is 0, regardless of the base.)

(d) $9^{-1} = (3^2)^{-1} = 3^{-2}$, so $\log_3 9^{-1} = \log_3 3^{-2} = -2$.

Now Try Exercises 59, 81, 91, 93

INVERSES OF BASE a LOGARITHMIC FUNCTIONS The graph of $y = \log_a x$ in **FIGURE 12.29** passes the horizontal line test, so it is a one-to-one function and it has an inverse. If $f(x) = \log_a x$, then $f^{-1}(x) = a^x$. This is a generalization of the fact that $f(x) = \log x$ and $f^{-1}(x) = 10^x$ represent inverse functions. These and other properties are illustrated for $f(x) = 2^x$ and $g(x) = e^x$ in the following See the Concept.

See the Concept 2 **INVERSES OF EXPONENTIAL FUNCTIONS**

Base 2 / Base e

Ⓐ The graph of the inverse is a reflection across the line $y = x$.

Ⓑ Rapid exponential growth

Ⓒ Slow logarithmic growth

Ⓓ $\log_2 x$ and $\ln x$ are negative (below the x-axis) for x between 0 and 1.

WATCH VIDEO IN MML.

These inverse properties are summarized as follows.

INVERSE PROPERTIES OF BASE-a LOGARITHMS

The following properties hold for logarithms with base a.

$$\log_a a^x = x, \qquad \text{for any real number } x$$

$$a^{\log_a x} = x, \qquad \text{for any positive real number } x$$

NOTE: The inverse function of $f(x) = \ln x$ is $f^{-1}(x) = e^x$. The inverse function of $f(x) = \log x$ is $f^{-1}(x) = 10^x$. ∎

EXAMPLE 10 **Applying inverse properties**

Simplify each expression.
(a) $\ln e^{0.5x}$ **(b)** $e^{\ln 4}$ **(c)** $2^{\log_2 7x}$ **(d)** $10^{\log (9x-3)}$

Solution
(a) $\ln e^{0.5x} = 0.5x$ because $\ln e^k = k$ for all k.
(b) $e^{\ln 4} = 4$ because $e^{\ln k} = k$ for all positive k.
(c) $2^{\log_2 7x} = 7x$ for $x > 0$ because $a^{\log_a k} = k$ for all positive k.
(d) $10^{\log (9x-3)} = 9x - 3$ for $x > \frac{1}{3}$ because $10^{\log k} = k$ for all positive k.

Now Try Exercises 41, 61, 63, 95

🌐 *Airline Safety* **Math in Context** Logarithms occur in many applications, including calculating runway length for airplanes.

EXAMPLE 11 **Calculating runway length**

There is a mathematical relationship between an airplane's weight x and the runway length required at takeoff. For certain types of airplanes, the minimum runway length L in thousands of feet may be modeled by $L(x) = 1.3 \ln x$, where x is in thousands of pounds. (*Source:* L. Haefner, *Introduction to Transportation Systems.*)
(a) Estimate the runway length needed for an airplane that weighs 10,000 pounds.
(b) Does a 20,000-pound airplane need twice the runway length that a 10,000-pound airplane needs? Explain.

Solution
(a) Because $L(x) = 1.3 \ln x$, it follows that $L(10) = 1.3 \ln 10 \approx 3$. An airplane that weighs 10,000 pounds requires a runway (at least) 3000 feet long.
(b) Because $L(20) = 1.3 \ln 20 \approx 3.9$, a 20,000-pound airplane does not need twice the runway length needed by a 10,000-pound airplane. Rather, the heavier airplane needs roughly 3900 feet of runway, or only an extra 900 feet.

Now Try Exercise 113

12.3   Putting It All Together

Common logarithms are base-10 logarithms. If a positive number x is written as $x = 10^k$, then $\log x = k$. The value of $\log x$ represents the exponent on the base 10 that gives x.

CONCEPT	EXPLANATION	EXAMPLES
Base-a Logarithms	*Definition:* $\log_a x = k$ means $x = a^k$, where $a > 0$ and $a \neq 1$. *Domain:* all *positive* real numbers *Range:* all real numbers *Graph:* $a > 1$ (shown to the right for $\log x$ and $\ln x$); passes through $(1, 0)$; vertical asymptote: y-axis *Common Logarithm:* base-10 logarithm and denoted $\log x$ *Natural Logarithm:* base-e logarithm, where $e \approx 2.718$, and denoted $\ln x$	$\log 1000 = \log 10^3 = 3,$ $\log_2 16 = \log_2 2^4 = 4,$ and $\log_3 \dfrac{1}{81} = \log_3 3^{-4} = -4$
Inverse Properties	The following properties hold for base-a logarithms. $\log_a a^x = x,$ for any real number x $a^{\log_a x} = x,$ for any positive number x	$\log 10^{7.48} = 7.48$ $2^{\log_2 63} = 63$ $10^{\log 23} = 23$ $\ln e^4 = 4$

12.3   Exercises   MyMathLab®

CONCEPTS AND VOCABULARY

1. What is the base of the common logarithm?

2. What is the base of the natural logarithm?

3. What are the domain and range of $\log x$?

4. What are the domain and range of 10^x?

5. $\log 10^k =$ _____

6. $\ln e^k =$ _____

7. If $\log x = k$, then $10^k =$ _____.

8. If $x > 0$, then $10^{\log x} =$ _____.

9. What does k equal if $10^k = 5$?

10. What does k equal if $e^k = 5$?

11. $\log 1 =$ _____

12. $\log (-1)$ is _____.

CHANGING EXPONENTIAL AND LOGARITHMIC FORMS

Exercises 13–16: Change the given equation into the equivalent exponential or logarithmic form.

13. (a) $\log 100 = 2$
    (b) $10^{-1} = \frac{1}{10}$

14. (a) $10^5 = 100{,}000$
    (b) $\log \frac{1}{1000} = -3$

15. (a) $10^0 = 1$
    (b) $\log \frac{1}{100{,}000} = -5$

16. (a) $\log 1000 = 3$
    (b) $10^1 = 10$

EVALUATING COMMON LOGARITHMS

Exercises 17 and 18: Complete the table.

17.

x	10^{-5}	10^0	$10^{0.5}$	$10^{2.2}$
$\log x$				

18.

x	10^{-6}	10^{-1}	$10^{\frac{5}{7}}$	10^{π}
$\log x$				

Exercises 19–44: Simplify the expression, if possible.

19. $\log 10^5$ **20.** $\log 10$

21. $\log 10^{-4}$ **22.** $\log 10^{-1}$

23. $\log 1$ **24.** $\log \sqrt[3]{100}$

25. $\log \frac{1}{100}$ **26.** $\log \frac{1}{10}$

27. $\log 10^{4.7}$ **28.** $\log 10^{2x+4}$

29. $\log 10{,}000$ **30.** $\log 100{,}000$

31. $\log \sqrt{10}$ **32.** $\log 1{,}000{,}000$

33. $\log (-23)$ **34.** $\log (-8)$

35. $\log 0.001$ **36.** $\log 0.0001$

37. $10^{\log 2}$ **38.** $10^{\log 7.5}$

39. $10^{\log x^2}$ **40.** $10^{\log |x|}$

41. $10^{\log 5}$ **42.** $\log 10^{\frac{3}{4}}$

43. $\log 10^{2x-7}$ **44.** $\log 10^{8-4x}$

Exercises 45–48: Evaluate the common logarithm, using a calculator. Round values to the nearest thousandth.

45. $\log 25$ **46.** $\log 0.501$

47. $\log 1.45$ **48.** $\log \frac{1}{35}$

GRAPHING COMMON LOGARITHMS

Exercises 49–56: Graph $y = f(x)$. Compare the graph to the graph of $y = \log x$.

49. $f(x) = \log (x) - 1$ **50.** $f(x) = \log (x) + 2$

51. $f(x) = \log (x + 1)$ **52.** $f(x) = \log (x + 2)$

53. $f(x) = \log (x - 1)$ **54.** $f(x) = \log (x - 3)$

55. $f(x) = 2 \log x$ **56.** $f(x) = -\log x$

EVALUATING NATURAL LOGARITHMS

Exercises 57 and 58: Complete the table.

57.

x	$e^{0.5}$	e^{-12}	$e^{\frac{1}{2}}$	$e^{-2.4}$
$\ln x$				

58.

x	e^{-6}	e^{-1}	$e^{\frac{5}{7}}$	e^{π}
$\ln x$				

Exercises 59–64: Simplify the expression.

59. $\ln 1$ **60.** $\ln e^2$

61. $\ln e^{-5x}$ **62.** $e^{\ln 2x}$

63. $e^{\ln x^2}$ **64.** $\ln e^{7x}$

Exercises 65–68: Evaluate the natural logarithm, using a calculator. Round values to the nearest thousandth.

65. $\ln 7$ **66.** $\ln 126$

67. $\ln \frac{4}{7}$ **68.** $\ln 0.67$

GRAPHING NATURAL LOGARITHMS

Exercises 69–72: Graph f in $[-4, 4, 1]$ by $[-4, 4, 1]$. Compare this graph to the graph of $y = \ln x$. Identify the domain of f.

69. $f(x) = \ln |x|$ **70.** $f(x) = \ln (x) - 2$

71. $f(x) = \ln (x + 2)$ **72.** $f(x) = 2 \ln x$

EVALUATING BASE-*a* LOGARITHMS

Exercises 73 and 74: Complete the table.

73.

x	$5^{\frac{2}{3}}$	5^{-1}	$5^{5.5}$	$5^{-0.9}$
$\log_5 x$				

74.

x	2^{-5}	2^0	$2^{0.5}$	$2^{2.2}$
$\log_2 x$				

Exercises 75–100: Simplify the expression, if possible.

75. $\log_5 5^{6x}$ **76.** $\log_3 3^3$

77. $\log_2 \sqrt{\frac{1}{8}}$ **78.** $\log_2 2^6$

79. $\log_2 2^8$ **80.** $\log_2 2^{-5}$

81. $\log_2 \sqrt{8}$ **82.** $\log_2 \sqrt{32}$

83. $\log_2 \sqrt[3]{\frac{1}{4}}$ **84.** $\log_2 \frac{1}{64}$

85. $\log_2 (-8)$ **86.** $\log_2 (-7)$

87. $\log_2 4$ **88.** $\log_2 \frac{1}{32}$

89. $\log_2 \frac{1}{16}$ **90.** $\log_3 27$

91. $\log_3 \frac{1}{9}$ **92.** $\log_4 16$

93. $\log_5 5^{-2}$ **94.** $\log_8 64$

95. $5^{\log_5 17}$ **96.** $9^{\log_9 73}$

97. $4^{\log_4 (2x)^2}$ **98.** $b^{\log_b (x-1)}$

99. $5^{\log_5 0.6z}$ **100.** $7^{\log_7 (x-9)}$

Exercises 101 and 102: Complete the table.

101.

x	$\frac{1}{4}$	$\frac{1}{2}$	1	$\sqrt{2}$	64
$\log_2 x$					

102.

x	$\frac{1}{7}$	1	$\sqrt{7}$	7	49
$\log_7 x$					

GRAPHING BASE-*a* LOGARITHMS

Exercises 103–106: Graph $y = f(x)$.

103. $f(x) = \log_2 x$ **104.** $f(x) = \log_2 (x - 2)$

105. $f(x) = 2 + \log_2 x$ **106.** $f(x) = \log_2 (x + 2)$

Exercises 107–110: Without using a calculator match $f(x)$ with its graph (a–d).

107. $f(x) = \log x$ **108.** $f(x) = \log_3 x$

109. $f(x) = \log_3 (x) + 2$ **110.** $f(x) = \log (x + 1)$

a.

b.

c.

d.

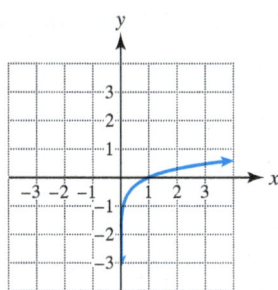

APPLICATIONS OF LOGARITHMS

111. Modeling Sound (Refer to Example 6.) At professional football games in domed stadiums, the decibel level may reach 120. The eardrum usually experiences pain when the intensity of the sound reaches 10^{-4} watts per square centimeter. How many decibels does this quantity represent? Is the noise at a football game likely to hurt some people's eardrums?

112. Hurricanes The barometric air pressure P in inches of mercury at a distance of d miles from the eye of a severe hurricane can sometimes be modeled by the formula $P(d) = 0.48 \ln (d + 1) + 27$. Average air pressure is about 30 inches of mercury. (*Source:* A. Miller and R. Anthes, *Meteorology.*)

(a) Evaluate $P(0)$ and $P(50)$. Interpret the results.

(b) A graph of $y = P(d)$ is shown in the figure. Describe how the air pressure changes as the distance from the eye of the hurricane increases.

Distance from the Eye (miles)

(c) Is the eye of the hurricane a low-pressure area or a high-pressure area?

113. Runway Length (Refer to Example 11.)

(a) A graph of $L(x) = 1.3 \ln x$ is shown in the figure, where $y = L(x)$. As the weight of the plane increases, what can be said about the length of the runway required?

Weight (pounds × 1000)

(b) Evaluate $L(50)$ and interpret the result.

114. Growth in Salary Suppose that a person's salary is initially \$40,000 and could be determined by either $f(x)$ or $g(x)$, where x represents the number of years of experience.

$$f(x) = 40{,}000(1.1)^x$$
$$g(x) = 40{,}000 \log (10 + x)$$

Would most people prefer that their salaries increase exponentially or logarithmically? Explain.

115. Magnitude of a Star The first stellar brightness scale was developed 2000 years ago by two Greek astronomers, Hipparchus and Ptolemy. The brightest star in the sky was given a magnitude of 1, and the faintest star was given a magnitude of 6. In 1856, this scale was described mathematically by the formula

$$M = 6 - 2.5 \log \frac{I}{I_0},$$

where M is the magnitude of a star with an intensity of I and I_0 is the intensity of the faintest star seen in the sky. (*Source:* M. Zeilik, *Introductory Astronomy and Astrophysics.*)

(a) Find M if $I = 10$ and $I_0 = 1$.

(b) What is the magnitude of a star that is 100 times more intense than the faintest star?

(c) If the intensity of a star increases by a factor of 10, what happens to its magnitude?

116. Population of Urban Regions While less industrialized urban regions of the world are experiencing exponential population growth, industrialized urban regions are experiencing logarithmic population growth. Population in less industrialized urban regions can be modeled by

$$f(x) = 0.338(1.035)^x,$$

whereas the population in industrialized urban regions can be modeled by

$$g(x) = 0.36 + 0.15 \ln(x + 1).$$

In these formulas, the output is in billions of people and x is in years, where $x = 0$ corresponds to 1950, $x = 10$ corresponds to 1960, and so on until $x = 80$ corresponds to 2030. (*Source:* D. Meadows, *Beyond The Limits.*)

(a) Evaluate $f(50)$ and $g(50)$. Interpret the results.

(b) Graph f and g in $[0, 80, 10]$ by $[0, 5, 1]$. Compare the two graphs.

(c) If x increases from 20 to 40, by what factor does $f(x)$ increase? By what factor does $g(x)$ increase?

117. Earthquakes The Richter scale is used to determine the intensity of earthquakes, which corresponds to the amount of energy released. If an earthquake has an intensity of x, its *magnitude*, as computed by the Richter scale, is given by $R(x) = \log \frac{x}{I_0}$, where I_0 is the intensity of a small, measurable earthquake.

(a) On July 26, 1963, an earthquake in Yugoslavia had a magnitude of 6.0 on the Richter scale, and on August 19, 1977, an earthquake in Indonesia measured 8.0. Find the intensity x for each of these earthquakes if $I_0 = 1$.

(b) How many times more intense was the Indonesian earthquake than the Yugoslavian earthquake?

118. Path Loss for Cellular Phones For cellular phones to work throughout a country, large numbers of cellular towers are necessary. How well the signal is propagated throughout a region depends on the location of these towers. One quick way to estimate the strength of a signal at x kilometers is to use the formula

$$D(x) = -121 - 36 \log x.$$

This formula computes the decrease in the signal, using decibels, so it is always negative. For example, $D(1) = -121$ means that at a distance of 1 kilometer, the signal has decreased in strength by 121 decibels. (*Source:* C. Smith, *Practical Cellular & PCS Design.*)

(a) Evaluate $D(3)$ and interpret the result.

(b) Graph D in $[1, 10, 1]$ by $[-160, -120, 10]$.

(c) What happens to the signal as x increases?

WRITING ABOUT MATHEMATICS

119. Explain what $\log_a x$ means and give an example.

120. How would you explain to a student that $\log_a 1 \neq 1$?

Group Activity

 Greenhouse Gases Carbon dioxide (CO_2) is a greenhouse gas in the atmosphere that may raise average temperatures on Earth. The burning of fossil fuels could be responsible for the increased levels of carbon dioxide. If current trends continue, future concentrations of atmospheric carbon dioxide in parts per million (ppm) could reach the levels shown in the accompanying table. The CO_2 concentration in the year 2000 was greater than it had been at any time in the previous 160,000 years.

Year	2000	2050	2100	2150	2200
CO_2 (ppm)	364	467	600	769	987

Source: R. Turner, *Environmental Economics.*

(a) Let x be in years, where $x = 0$ corresponds to 2000, $x = 1$ to 2001, and so on. Find values for C and a so that $f(x) = Ca^x$ models the data.

(b) Graph f and the data in the same viewing rectangle.

(c) Use $f(x)$ to estimate graphically the year when the carbon dioxide concentration will be double the preindustrial level of 280 ppm.

12.4 Properties of Logarithms

Objectives

1 **Introducing Basic Properties of Logarithms**
   - **The Product Rule**
   - **The Quotient Rule**
   - **The Power Rule**
   - **Applying More Than One Property**

2 **Using the Change of Base Formula**

NEW VOCABULARY

☐ Change of base formula

---

1 Introducing Basic Properties of Logarithms

In this subsection, we discuss three important properties of logarithms. The first property is the product rule for logarithms.

THE PRODUCT RULE The product rule for logarithms can be applied when computing a logarithm of a product.

> **PRODUCT RULE FOR LOGARITHMS**
>
> For positive numbers m, n, and $a \neq 1$,
> $$\log_a mn = \log_a m + \log_a n.$$

This property is illustrated in **FIGURE 12.30**, which shows that
$$\log 10 = \log (2 \cdot 5) = \log 2 + \log 5.$$

This product rule may be verified by using properties of exponents and the fact that $\log_a a^k = k$ for any real number k. Here, we verify the product property for logarithms. The two other properties presented later can be verified in a similar manner.

If m and n are positive numbers, we can write $m = a^c$ and $n = a^d$ for some real numbers c and d.

$$\log_a mn = \log_a (a^c a^d) = \log_a (a^{c+d}) = c + d \quad \text{and}$$
$$\log_a m + \log_a n = \log_a a^c + \log_a a^d = c + d$$

Thus $\log_a mn = \log_a m + \log_a n$.

```
log(10)
                    1
log(2)+log(5)
                    1
```

FIGURE 12.30

EXAMPLE 1   **Expanding logarithms as sums**

Write each expression as a sum of logarithms. Assume that x is positive.
(a) $\log 21$ (b) $\ln 5x$ (c) $\log x^3$

Solution
(a) $\log 21 = \log(3 \cdot 7) = \log 3 + \log 7$ (b) $\ln 5x = \ln(5 \cdot x) = \ln 5 + \ln x$
(c) $\log x^3 = \log(x \cdot x \cdot x) = \log x + \log x + \log x$

Now Try Exercises 13, 15, 17

EXAMPLE 2   **Combining logarithms**

Write each expression as one logarithm. Assume that x and y are positive.
(a) $\log 5 + \log 6$ (b) $\ln x + \ln xy$ (c) $\log 2x + \log 5x$

Solution
(a) $\log 5 + \log 6 = \log(5 \cdot 6) = \log 30$ (b) $\ln x + \ln xy = \ln(x \cdot xy) = \ln x^2 y$
(c) $\log 2x + \log 5x = \log(2x \cdot 5x) = \log 10x^2$

Now Try Exercises 25, 27, 29

THE QUOTIENT RULE The second property is the quotient rule for logarithms.

> **QUOTIENT RULE FOR LOGARITHMS**
>
> For positive numbers m, n, and $a \neq 1$,
>
> $$\log_a \frac{m}{n} = \log_a m - \log_a n.$$

```
log(10)
              1
log(20)-log(2)
              1
```

FIGURE 12.31

This property is illustrated in **FIGURE 12.31**, which shows that

$$\log 10 = \log \frac{20}{2} = \log 20 - \log 2.$$

EXAMPLE 3   **Expanding logarithms as differences**

Write each expression as a difference of two logarithms. Assume that variables are positive.

(a) $\log \frac{3}{2}$ (b) $\ln \frac{3x}{y}$ (c) $\log_5 \frac{x}{z^4}$

Solution

(a) $\log \frac{3}{2} = \log 3 - \log 2$ (b) $\ln \frac{3x}{y} = \ln 3x - \ln y$

(c) $\log_5 \frac{x}{z^4} = \log_5 x - \log_5 z^4$

Now Try Exercises 19, 21, 23

NOTE: $\log_a(m + n) \neq \log_a m + \log_a n;$ $\log_a(m - n) \neq \log_a m - \log_a n;$
$\log_a(mn) \neq \log_a m \cdot \log_a n;$ $\log_a\left(\dfrac{m}{n}\right) \neq \dfrac{\log_a m}{\log_a n}$ ∎

EXAMPLE 4 **Combining logarithms**

Write each expression as one term. Assume that x is positive.
(a) $\log 50 - \log 25$ **(b)** $\ln x^3 - \ln x$ **(c)** $\log 15x - \log 5x$

Solution

(a) $\log 50 - \log 25 = \log \dfrac{50}{25} = \log 2$ **(b)** $\ln x^3 - \ln x = \ln \dfrac{x^3}{x} = \ln x^2$

(c) $\log 15x - \log 5x = \log \dfrac{15x}{5x} = \log 3$

Now Try Exercises 33, 35, 37

READING CHECK 1

• Is $\log (4+5)$ the same as $\log 4 + \log 5$? Why or why not?

THE POWER RULE The third property is the power rule for logarithms. To illustrate this rule, we use the following example.

$$\log x^3 = \log (x \cdot x \cdot x) = \log x + \log x + \log x = 3 \log x$$

Thus $\log x^3 = 3 \log x$. This example is generalized in the following rule.

POWER RULE FOR LOGARITHMS

For positive numbers m and $a \neq 1$ and any real number r,

$$\log_a (m^r) = r \log_a m.$$

```
log(10^2)
                 2
2log(10)
                 2
```

FIGURE 12.32

We use a calculator and the equation

$$\log 10^2 = 2 \log 10$$

to illustrate this property. **FIGURE 12.32** shows the result.

EXAMPLE 5 **Applying the power rule**

Rewrite each expression, using the power rule.
(a) $\log 5^6$ **(b)** $\ln 0.55^{x-1}$ **(c)** $\log_5 8^{kx}$

Solution

(a) $\log 5^6 = 6 \log 5$ **(b)** $\ln 0.55^{x-1} = (x - 1) \ln 0.55$
(c) $\log_5 8^{kx} = kx \log_5 8$

Now Try Exercises 39, 41, 47

See the Concept 1   **EXPANDING VERSUS COMBINING**

EXPANDING Ⓐ

$$\log 10a = \log 10 + \log a$$

$$\log \frac{x}{5} = \log x - \log 5$$

$$\log x^3 = 3 \log x$$

Ⓑ **COMBINING**

Ⓐ Transforming the left side of the equation into the right side is called **EXPANDING**.

Ⓑ Transforming the right side of the equation into the left side is called **COMBINING**.

WATCH VIDEO IN MML.

APPLYING MORE THAN ONE PROPERTY Sometimes we use more than one property to simplify an expression. We assume that all variables are positive in the next two examples.

EXAMPLE 6   **Combining logarithms**

Write each expression as the logarithm of a single expression.
(a) $3 \log x + \log x^2$ **(b)** $2 \ln x - \ln \sqrt{x}$

Solution

(a) $3 \log x + \log x^2 = \log x^3 + \log x^2$ Power rule

$\qquad\qquad\qquad\quad = \log (x^3 \cdot x^2)$ Product rule

$\qquad\qquad\qquad\quad = \log x^5$ Properties of exponents

(b) $2 \ln x - \ln \sqrt{x} = 2 \ln x - \ln x^{1/2}$ $\sqrt{x} = x^{1/2}$

$\qquad\qquad\qquad\quad = \ln x^2 - \ln x^{1/2}$ Power rule

$\qquad\qquad\qquad\quad = \ln \dfrac{x^2}{x^{1/2}}$ Quotient rule

$\qquad\qquad\qquad\quad = \ln x^{3/2}$ Properties of exponents

Now Try Exercises 59, 67

EXAMPLE 7   **Expanding logarithms**

Write each expression in terms of logarithms of x, y, and z.
(a) $\log \dfrac{x^2 y^3}{\sqrt{z}}$ **(b)** $\ln \sqrt[3]{\dfrac{xy}{z}}$

Solution

(a) $\log \dfrac{x^2 y^3}{\sqrt{z}} = \log x^2 y^3 - \log \sqrt{z}$ Quotient rule

$\qquad\qquad\quad = \log x^2 + \log y^3 - \log z^{1/2}$ Product rule; $\sqrt{z} = z^{1/2}$

$\qquad\qquad\quad = 2 \log x + 3 \log y - \tfrac{1}{2} \log z$ Power rule

(b) $\ln \sqrt[3]{\dfrac{xy}{z}} = \ln \left(\dfrac{xy}{z} \right)^{1/3}$ $\sqrt[3]{m} = m^{1/3}$

$\qquad\qquad = \tfrac{1}{3} \ln \dfrac{xy}{z}$ Power rule

$\qquad\qquad = \tfrac{1}{3} (\ln xy - \ln z)$ Quotient rule

$\qquad\qquad = \tfrac{1}{3} (\ln x + \ln y - \ln z)$ Product rule

$\qquad\qquad = \tfrac{1}{3} \ln x + \tfrac{1}{3} \ln y - \tfrac{1}{3} \ln z$ Distributive property

Now Try Exercises 73, 75

EXAMPLE 8   **Applying properties of logarithms**

Using only properties of logarithms and the approximations $\ln 2 \approx 0.7$, $\ln 3 \approx 1.1$, and $\ln 5 \approx 1.6$, find an approximation for each expression.

(a) $\ln 8$ **(b)** $\ln 15$ **(c)** $\ln \dfrac{10}{3}$

Solution

Using $\ln 2 \approx 0.7$, $\ln 3 \approx 1.1$, and $\ln 5 \approx 1.6$, we approximate as follows.

(a) $\ln 8 = \ln 2^3 = 3 \ln 2 \approx 3(0.7) = 2.1$

(b) $\ln 15 = \ln(3 \cdot 5) = \ln 3 + \ln 5 \approx 1.1 + 1.6 = 2.7$

(c) $\ln \dfrac{10}{3} = \ln\left(\dfrac{2 \cdot 5}{3}\right) = \ln 2 + \ln 5 - \ln 3 \approx 0.7 + 1.6 - 1.1 = 1.2$

Now Try Exercises 49, 51, 55

READING CHECK 2

• How would you evaluate a logarithm with a base other than 10 or *e*?

2 ▸ Using the Change of Base Formula

Most calculators only have keys to evaluate common and natural logarithms. Occasionally, it is necessary to evaluate a logarithmic function with a base other than 10 or *e*. In these situations, we use the following **change of base formula**, which we illustrate in the next example.

CHANGE OF BASE FORMULA

Let x and $a \neq 1$ be positive real numbers. Then

$$\log_a x = \frac{\log x}{\log a} \quad \text{or} \quad \log_a x = \frac{\ln x}{\ln a}.$$

EXAMPLE 9   **Applying the change of base formula**

Approximate $\log_2 14$ to the nearest thousandth.

Solution

Using the change of base formula,

$$\log_2 14 = \frac{\log 14}{\log 2} \approx 3.807 \quad \text{or} \quad \log_2 14 = \frac{\ln 14}{\ln 2} \approx 3.807.$$

FIGURE 12.33 supports these results.

Now Try Exercise 89

```
log(14)/log(2)
        3.807354922
ln(14)/ln(2)
        3.807354922
```

FIGURE 12.33

12.4   Putting It All Together

CONCEPT	EXPLANATION	EXAMPLES
Properties of Logarithms **1.** Product Rule **2.** Quotient Rule **3.** Power Rule	The following properties hold for positive numbers m, n, and $a \neq 1$ and for any real number r. **1.** $\log_a mn = \log_a m + \log_a n$ **2.** $\log_a \frac{m}{n} = \log_a m - \log_a n$ **3.** $\log_a(m^r) = r \log_a m$	**1.** $\log(10 \cdot 2) = \log 10 + \log 2$ **2.** $\log \frac{45}{6} = \log 45 - \log 6$ **3.** $\ln x^6 = 6 \ln x$
Change of Base Formula	Let x and $a \neq 1$ be positive numbers. Then $$\log_a x = \frac{\log x}{\log a} \quad \text{and} \quad \log_a x = \frac{\ln x}{\ln a}.$$	The expression $\log_3 6$ is equivalent to either $$\frac{\log 6}{\log 3} \quad \text{or} \quad \frac{\ln 6}{\ln 3}.$$

12.4 Exercises

MyMathLab®

NOTE: Assume that all variables are positive and that all expressions are defined in this exercise set. ∎

CONCEPTS AND VOCABULARY

1. $\log 12 = \log 3 + \log (\underline{\hspace{1cm}})$

2. $\ln 5 = \ln 20 - \ln (\underline{\hspace{1cm}})$

3. $\log 8 = (\underline{\hspace{1cm}}) \log 2$

4. $\log mn = \underline{\hspace{1cm}}$

5. $\log \dfrac{m}{n} = \underline{\hspace{1cm}}$

6. $\log (m^r) = \underline{\hspace{1cm}}$

7. Does $\log x + \log y$ equal $\log (x + y)$?

8. Does $\log x - \log y$ equal $\log \left(\frac{x}{y}\right)$?

9. Does $\log (xy)$ equal $(\log x)(\log y)$?

10. Does $\log \left(\frac{x}{y}\right)$ equal $\dfrac{\log x}{\log y}$?

11. Give the change of base formula.

12. $\log_a 1 = \underline{\hspace{1cm}}$ and $\log_a a = \underline{\hspace{1cm}}$.

APPLYING PROPERTIES OF LOGARITHMS

Exercises 13–18: Write the expression as a sum of two or more logarithms.

13. $\ln (15)$

14. $\log (77)$

15. $\log xy$

16. $\ln 10z$

17. $\log y^2$

18. $\log x^2 y$

Exercises 19–24: Write the expression as a difference of two logarithms.

19. $\log \frac{7}{3}$

20. $\ln \frac{11}{13}$

21. $\ln \dfrac{x}{y}$

22. $\log \dfrac{2x}{z}$

23. $\log_2 \dfrac{45}{x}$

24. $\log_7 \dfrac{5x}{4z}$

Exercises 25–32: Write the expression as one logarithm.

25. $\log 45 + \log 5$

26. $\log 30 - \log 10$

27. $\ln x + \ln y$

28. $\ln m + \ln n - \ln n$

29. $\ln 7x^2 + \ln 2x$

30. $\ln x + \ln y - \ln z$

31. $\ln x + \ln y^2 - \ln y$

32. $\ln \sqrt{z} - \ln z^3 + \ln y^3$

Exercises 33–38: Write the expression as one term. Evaluate by hand if possible.

33. $\log 20 - \log 4$

34. $\log 900 - \log 9$

35. $\ln x^4 - \ln x^2$

36. $\ln 9x^2 - \ln 3x$

37. $\log 300x - \log 3x$

38. $\log 18x^2 - \log 2x^2$

Exercises 39–48: Rewrite using the power rule.

39. $\log 3^6$

40. $\log x^7$

41. $\ln 2^x$

42. $\ln (0.77)^{x+1}$

43. $\log_2 5^{1/4}$

44. $\log_3 \sqrt{x}$

45. $\log_4 \sqrt[3]{z}$

46. $\log_7 3^{\pi}$

47. $\log x^{y-1}$

48. $\ln a^{2b}$

Exercises 49–58: (Refer to Example 8.) Using only properties of logarithms and the approximations $\log 2 \approx 0.3$, $\log 5 \approx 0.7$, and $\log 13 \approx 1.1$, find an approximation for the expression.

49. $\log 16$

50. $\log 125$

51. $\log 65$

52. $\log 26$

53. $\log 130$

54. $\log 100$

55. $\log \frac{5}{2}$

56. $\log \frac{26}{5}$

57. $\log \frac{1}{13}$

58. $\log \frac{1}{65}$

COMBINING LOGARITHMS

Exercises 59–70: Use properties of logarithms to write the expression as the logarithm of a single expression.

59. $4 \log z - \log z^3$

60. $2 \log_5 y + \log_5 x$

61. $\log x + 2 \log x + 2 \log y$

62. $\log x^2 + 3 \log z - 5 \log y$

63. $\log x - 2 \log \sqrt{x}$

64. $\ln y^2 - 6 \ln \sqrt[3]{y}$

65. $\ln 2^{x+1} - \ln 2$

66. $\ln 8^{1/2} + \ln 2^{1/2}$

67. $\ln \sqrt[3]{x} + \ln \sqrt{x}$

68. $2 \log_3 \sqrt{x} - 3 \log_3 x$

69. $2 \log_a (x + 1) - \log_a (x^2 - 1)$

70. $\log_b (x^2 - 9) - \log_b (x - 3)$

EXPANDING LOGARITHMS

Exercises 71–82: Use properties of logarithms to write the expression in terms of logarithms of x, y, and z.

71. $\log xy^2$

72. $\log \dfrac{x^2}{y^3}$

73. $\ln \dfrac{x^4 y}{z}$

74. $\ln \dfrac{\sqrt{x}}{y}$

75. $\log \dfrac{\sqrt[3]{z}}{\sqrt{y}}$

76. $\log \sqrt{\dfrac{x}{y}}$

77. $\log (x^4 y^3)$

78. $\log (x^2 y^4 z^3)$

79. $\ln \dfrac{1}{y} - \ln \dfrac{1}{x}$

80. $\ln \dfrac{1}{xy}$

81. $\log_4 \sqrt{\dfrac{x^3 y}{z^2}}$

82. $\log_3 \left(\dfrac{x^2 \sqrt{z}}{y^3} \right)$

Exercises 83–86: Graph f and g in the window $[-6, 6, 1]$ by $[-4, 4, 1]$. If the two graphs appear to be identical, prove that they are, using properties of logarithms.

83. $f(x) = \log x^3, g(x) = 3 \log x$

84. $f(x) = \ln x + \ln 3, g(x) = \ln 3x$

85. $f(x) = \ln (x + 5), g(x) = \ln x + \ln 5$

86. $f(x) = \log (x - 2), g(x) = \log x - \log 2$

USING THE CHANGE OF BASE FORMULA

Exercises 87–92: Use the change of base formula to approximate each expression to the nearest hundredth.

87. $\log_3 5$

88. $\log_5 12$

89. $\log_2 25$

90. $\log_7 8$

91. $\log_9 102$

92. $\log_6 293$

APPLICATIONS OF LOGARITHMS

93. Modeling Sound The formula $f(x) = 10 \log (10^{16} x)$ can be used to calculate the decibel level of a sound with an intensity x. Use properties of logarithms to simplify this formula to $f(x) = 160 + 10 \log x$.

94. Cellular Phone Technology A formula used to calculate the strength of a signal for a cellular phone is

$$L = 110.7 - 19.1 \log h + 55 \log d,$$

where h is the height of the cellular phone tower and d is the distance the phone is from the tower. Use properties of logarithms to write an expression for L that contains only one logarithm. (*Source:* C. Smith, *Practical Cellular & PCS Design.*)

WRITING ABOUT MATHEMATICS

95. State the three basic properties of logarithms and give an example of each.

96. A student insists that $\log (x - y)$ is equal to the expression $\log x - \log y$. How could you convince the student otherwise?

<table>
<tr><td>SECTIONS
12.3 and 12.4</td><td>Checking Basic Concepts</td></tr>
</table>

1. Simplify each expression by hand.
  (a) $\log 10^4$ **(b)** $\ln e^x$
  (c) $\log_2 \frac{1}{8}$ **(d)** $\log_5 \sqrt{5}$

2. Sketch a graph of $f(x) = \log x$.
  (a) What are the domain and range of f?
  (b) Evaluate $f(1)$.
  (c) Can the common logarithm of a positive number be negative? Explain.
  (d) Can the common logarithm of a negative number be positive? Explain.

3. Write the expression in terms of logarithms of x, y, and z. Assume that variables are positive.
  (a) $\log xy$ **(b)** $\ln \dfrac{x}{yz}$
  (c) $\ln x^2$ **(d)** $\log \dfrac{x^2 y^3}{\sqrt{z}}$

4. Write as the logarithm of a single expression.
  (a) $\log x + \log y$
  (b) $\ln 2x - 3 \ln y$
  (c) $2 \log_2 x + 3 \log_2 y - \log_2 z$

12.5 Exponential and Logarithmic Equations

1 Recognizing Exponential Equations and Models

Recall that to solve the equation $10 + x = 100$, **we subtract 10 from each side** because addition and subtraction are inverse operations.

$$10 + x - 10 = 100 - 10 \quad \text{Subtract 10 from each side.}$$
$$x = 90$$

Similarly, to solve the equation $10x = 100$, **we divide each side by 10** because multiplication and division are inverse operations.

$$\frac{10x}{10} = \frac{100}{10} \quad \text{Divide each side by 10.}$$
$$x = 10$$

Now suppose that we want to solve the *exponential equation*

> The exponent is a variable.

$$10^x = 100. \quad \text{Exponential equation}$$

What is new about this type of equation is that the variable x is an *exponent*. The inverse operation of 10^x is $\log x$. Rather than subtracting 10 from each side or dividing each side by 10, **we take the base-10 logarithm of each side**. Doing so results in

$$\mathbf{log}\ 10^x = \mathbf{log}\ 100. \quad \text{Take base-10 logarithm of each side.}$$

Because $\log 10^x = x$ for all real numbers x, the equation becomes

$$x = \log 100 \quad \text{or, equivalently,} \quad x = 2.$$

CHANGING EXPONENTIAL FORM TO LOGARITHMIC FORM In order to solve exponential equations, we take a logarithm of each side to change the exponential form of the equation into the logarithmic form. In general, we will show this step, but you may be able to move directly from the exponential form to the logarithmic form. The table shown below gives several examples of exponential equations along with their equivalent logarithmic forms.

Equivalent Forms

Exponential Form	Logarithmic Form	
$10^x = 100$	$x = \log 100$	Take base-10 logarithm of each side.
$2^{3x} = 8$	$3x = \log_2 8$	Take base-2 logarithm of each side.
$e^x = 9$	$x = \ln 9$	Take the natural logarithm of each side.
$8^{4x} = 20$	$4x = \log_8 20$	Take base-8 logarithm of each side.
$a^{2x} = b$	$2x = \log_a b$	Take base-a logarithm of each side.

SOLVING EXPONENTIAL EQUATIONS In the next example, we solve exponential equations by using inverse operations to change them to logarithmic form.

EXAMPLE 1 Solving exponential equations

Solve and approximate to the nearest hundredth.
(a) $10^x = 150$ (b) $e^x = 40$ (c) $2^x = 50$ (d) $0.9^x = 0.5$

Solution

(a)

$$10^x = 150 \qquad \text{Given equation}$$

Change to logarithmic form.

$$\log 10^x = \log 150 \qquad \text{Take the common logarithm of each side.}$$

$$\longrightarrow x = \log 150 \approx 2.18 \qquad \text{Inverse property: } \log 10^k = k \text{ for all } k$$

(b) The inverse operation of e^x is $\ln x$, so we take the natural logarithm of each side.

$$e^x = 40 \qquad \text{Given equation}$$

Change to logarithmic form.

$$\ln e^x = \ln 40 \qquad \text{Take the natural logarithm of each side.}$$

$$\longrightarrow x = \ln 40 \approx 3.69 \qquad \text{Inverse property: } \ln e^k = k \text{ for all } k$$

(c) The inverse operation of 2^x is $\log_2 x$. Calculators do not usually have a base-2 logarithm key, so we take the common logarithm of each side and then apply the power rule.

$$2^x = 50 \qquad \text{Given equation}$$

$$\log 2^x = \log 50 \qquad \text{Take the common logarithm of each side.}$$

$$x \log 2 = \log 50 \qquad \text{Power rule: } \log m^r = r \log m$$

$$x = \frac{\log 50}{\log 2} \approx 5.64 \qquad \text{Divide by } \log 2 \text{ and approximate.}$$

(d) This time, we begin by taking the natural logarithm of each side.

$$0.9^x = 0.5 \qquad \text{Given equation}$$

$$\ln 0.9^x = \ln 0.5 \qquad \text{Take the natural logarithm of each side.}$$

$$x \ln 0.9 = \ln 0.5 \qquad \text{Power rule: } \ln m^r = r \ln m$$

$$x = \frac{\ln 0.5}{\ln 0.9} \approx 6.58 \qquad \text{Divide by } \ln 0.9 \text{ and approximate.}$$

Now Try Exercises 21, 23, 33, 35

MAKING CONNECTIONS 1

Logarithms of Quotients and Quotients of Logarithms

The solution in Example 1(c) is $\frac{\log 50}{\log 2}$. Note that

$$\frac{\log 50}{\log 2} \neq \log 50 - \log 2.$$

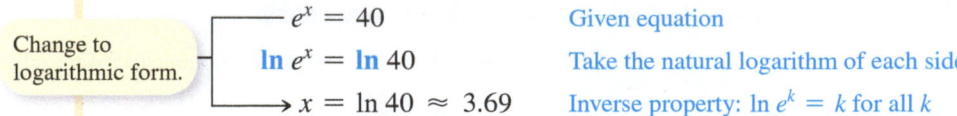

```
log(50)/log(2)
           5.64385619
log(50)-log(2)
           1.397940009
log(25)
           1.397940009
```

Rather, $\log 50 - \log 2 = \log \frac{50}{2} = \log 25$ by the quotient rule for logarithms, as shown in the figure.

EXAMPLE 2 **Solving exponential equations**

Solve each equation and approximate to the nearest hundredth.

(a) $2e^x - 1 = 5$ **(b)** $3^{x-5} = 15$ **(c)** $e^{2x} = e^{x+5}$ **(d)** $3^{2x} = 2^{x+3}$

Solution

(a) Begin by solving for e^x.

$$2e^x - 1 = 5 \qquad \text{Given equation}$$

$$2e^x = 6 \qquad \text{Add 1 to each side.}$$

$$e^x = 3 \qquad \text{Divide each side by 2.}$$

Change to logarithmic form.

$$\ln e^x = \ln 3 \qquad \text{Take the natural logarithm.}$$

$$\longrightarrow x = \ln 3 \approx 1.10 \qquad \text{Inverse property: } \ln e^k = k$$

(b) Start by taking the common logarithm of each side. (We could also take the natural logarithm of each side.)

$$3^{x-5} = 15 \qquad \text{Given equation}$$

$$\log 3^{x-5} = \log 15 \qquad \text{Take the common logarithm of each side.}$$

$$(x-5)\log 3 = \log 15 \qquad \text{Power rule for logarithms}$$

$$x - 5 = \frac{\log 15}{\log 3} \qquad \text{Divide by log 3.}$$

$$x = \frac{\log 15}{\log 3} + 5 \approx 7.46 \qquad \text{Add 5 to each side and approximate.}$$

(c) For $e^{2x} = e^{x+5}$, the bases are equal, so the exponents must also be equal. To verify this assertion, take the natural logarithm of each side.

$$e^{2x} = e^{x+5} \qquad \text{Given equation}$$

$$\ln e^{2x} = \ln e^{x+5} \qquad \text{Take the natural logarithm.}$$

$$2x = x + 5 \qquad \text{Inverse property: } \ln e^k = k$$

$$x = 5 \qquad \text{Subtract } x.$$

(d) For $3^{2x} = 2^{x+3}$, the bases are not equal. However, we can still solve the equation by taking the common logarithm of each side. A logarithm of any base could be used.

$$3^{2x} = 2^{x+3} \qquad \text{Given equation}$$

$$\log 3^{2x} = \log 2^{x+3} \qquad \text{Take the common logarithm.}$$

$$2x \log 3 = (x+3)\log 2 \qquad \text{Power rule for logarithms}$$

$$2x \log 3 = x \log 2 + 3 \log 2 \qquad \text{Distributive property}$$

$$2x \log 3 - x \log 2 = 3 \log 2 \qquad \text{Subtract } x \log 2.$$

$$x(2 \log 3 - \log 2) = 3 \log 2 \qquad \text{Factor out } x.$$

$$x = \frac{3 \log 2}{2 \log 3 - \log 2} \qquad \text{Divide by } 2 \log 3 - \log 2.$$

$$x \approx 1.38 \qquad \text{Approximate.}$$

Now Try Exercises 37, 39, 43, 51

EXAMPLE 3   **Solving an exponential equation**

Graphs for $f(x) = 0.2e^x$ and $g(x) = 4$ are shown in **FIGURE 12.34**.
(a) Use the graphs to estimate the solution to the equation $f(x) = g(x)$.
(b) Check your estimate by solving the equation symbolically.

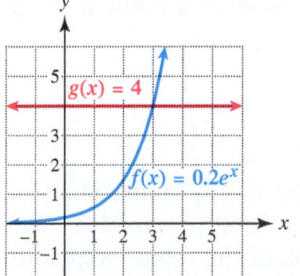

FIGURE 12.34

Solution
(a) The graphs intersect near the point $(3, 4)$. Therefore the solution is given by $x \approx 3$.
(b) We must solve the equation $0.2e^x = 4$.

$$0.2e^x = 4 \qquad \text{Given equation}$$

$$e^x = 20 \qquad \text{Divide each side by 0.2.}$$

$$\ln e^x = \ln 20 \qquad \text{Take the natural logarithm of each side.}$$

$$x = \ln 20 \qquad \text{Inverse property: } \ln e^k = k$$

$$x \approx 2.996 \qquad \text{Approximate.}$$

NOTE: The graphical estimate did not give the *exact* solution of ln 20. ∎

Now Try Exercise 53

2 Solving Applications of Exponential Equations

Math in Context   *Savings*   As discussed in Section 2 of this chapter, if a total of $1000 is deposited in a savings account paying 10% annual interest at the end of each year, the amount A in the account after x years is given by

$$A(x) = 1000(1.1)^x.$$

After 10 years, there will be

$$A(10) = 1000(1.1)^{10} \approx \$2593.74$$

in the account. To calculate how long it will take for $4000 to accrue in the account, we need to solve the exponential equation

$$1000(1.1)^x = 4000.$$

We do so in the next example.

EXAMPLE 4   **Solving an exponential equation**

Solve $1000(1.1)^x = 4000$ symbolically. Give graphical support for your answer.

Solution
Symbolic Solution Begin by dividing each side of the equation by 1000.

$1000(1.1)^x = 4000$	Given equation
$1.1^x = 4$	Divide by 1000.
$\log 1.1^x = \log 4$	Take the common logarithm of each side.
$x \log 1.1 = \log 4$	Power rule for logarithms
$x = \dfrac{\log 4}{\log 1.1} \approx 14.5$	Divide by log 1.1 and approximate.

Interest is paid at the end of the year, so it will take 15 years for $1000 earning 10% annual interest to grow to (at least) $4000.

Intersection
X=14.545082 Y=4000

FIGURE 12.35

Graphical Solution Graphical support is shown in **FIGURE 12.35**, where the graphs of $y_1 = 1000(1.1)^x$ and $y_2 = 4000$ intersect when $x \approx 14.5$.

Now Try Exercise 105

Math in Context   *Biology*   In the next example, we model the life span of a robin with an exponential function.

EXAMPLE 5   **Modeling the life span of a robin**

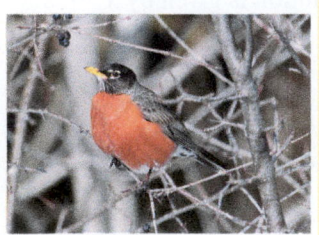

The life spans of 129 robins were monitored over a 4-year period in one study. The formula $f(x) = 10^{-0.42x}$ can be used to calculate the percentage of robins remaining after x years. For example, $f(1) \approx 0.38$ means that after 1 year, 38% of the robins were still alive. (*Source:* D. Lack, *The Life Span of a Robin.*)
(a) Evaluate $f(2)$ and interpret the result.
(b) Determine when 5% of the robins remained.

Solution
(a) $f(2) = 10^{-0.42(2)} \approx 0.145$. After **2** years, about **14.5%** of the robins were still alive.
(b) Use 5% = 0.05 and solve the following equation.

$$10^{-0.42x} = 0.05 \qquad \text{Equation to solve}$$

$$\log 10^{-0.42x} = \log 0.05 \qquad \text{Take the common logarithm of each side.}$$

$$-0.42x = \log 0.05 \qquad \text{Inverse property: } \log 10^k = k$$

$$x = \frac{\log 0.05}{-0.42} \approx 3.1 \qquad \text{Divide by } -0.42.$$

After about 3 years, only 5% of the robins were still alive.

Now Try Exercise 107

3 Recognizing Logarithmic Equations and Models

To solve an exponential equation, we use logarithms. To solve a logarithmic equation, we *exponentiate* each side of the equation. To do so, we use the fact that if $x = y$, then $a^x = a^y$ for any positive base a. For example, to solve

$$\log x = 3, \qquad \text{Logarithmic equation}$$

we exponentiate each side of the equation, using base 10.

$$10^{\log x} = 10^3 \qquad \text{If } a = b, \text{ then } 10^a = 10^b.$$

Because $10^{\log x} = x$ for all positive x, it follows that

$$x = 10^3. \qquad \text{Inverse property}$$

To solve logarithmic equations, we frequently use the inverse property

$$a^{\log_a x} = x.$$

> Bases must be equal.

Examples of this inverse property include

$$e^{\ln 2k} = 2k, \qquad 2^{\log_2 x} = x, \qquad \text{and} \qquad 10^{\log (x+5)} = x + 5.$$

> Both are base-e.   Both are base-2.   Both are base-10.

CHANGING LOGARITHMIC FORM TO EXPONENTIAL FORM In order to solve logarithmic equations, we exponentiate each side of the equation to change the logarithmic form of the equation into the exponential form. In general, we will show this step, but you may find that you can move directly from the logarithmic form to the exponential form. The table shown below gives several examples of logarithmic equations along with their equivalent exponential forms.

Equivalent Forms

Logarithmic Form	Exponential Form	
$\log x = 3$	$x = 10^3$	Exponentiate each side: base 10.
$\log_2 3x = 5$	$3x = 2^5$	Exponentiate each side: base 2.
$\ln 8x = 6$	$8x = e^6$	Exponentiate each side: base e.
$\log_9 (4 + x) = 5$	$4 + x = 9^5$	Exponentiate each side: base 9.
$\log_a x = b$	$x = a^b$	Exponentiate each side: base a.

SOLVING LOGARITHMIC EQUATIONS In the next example, we solve logarithmic equations by using inverse operations to change them to exponential form.

EXAMPLE 6 **Solving logarithmic equations**

Solve and approximate solutions to the nearest hundredth when appropriate.
(a) $2 \log x = 4$ **(b)** $\ln 3x = 5.5$ **(c)** $\log_2 (x + 4) = 7$

Solution

(a)

$2 \log x = 4$ Given equation

$\log x = 2$ Divide each side by 2.

Change to exponential form

$10^{\log x} = 10^2$ Exponentiate each side, using base 10.

$x = 100$ Inverse property: $10^{\log k} = k$

(b)

$\ln 3x = 5.5$ Given equation

Change to exponential form

$e^{\ln 3x} = e^{5.5}$ Exponentiate each side, using base e.

$3x = e^{5.5}$ Inverse property: $e^{\ln k} = k$

$x = \dfrac{e^{5.5}}{3} \approx 81.56$ Divide each side by 3 and approximate.

(c)

$\log_2 (x + 4) = 7$ Given equation

Change to exponential form

$2^{\log_2 (x+4)} = 2^7$ Exponentiate each side, using base 2.

$x + 4 = 2^7$ Inverse property: $2^{\log_2 k} = k$

$x = 2^7 - 4$ Subtract 4 from each side.

$x = 124$ Simplify.

Now Try Exercises 69, 71, 77

Because the domain of any logarithmic function includes only positive numbers, it is important to check answers, as emphasized in the next example.

EXAMPLE 7 **Solving a logarithmic equation**

Solve $\log (x + 2) + \log (x - 2) = \log 5$. Check any answers.

Solution

$\log (x + 2) + \log (x - 2) = \log 5$ Given equation

$\log ((x + 2)(x - 2)) = \log 5$ Product rule

$\log (x^2 - 4) = \log 5$ Multiply.

$10^{\log (x^2 - 4)} = 10^{\log 5}$ Exponentiate using base 10.

$x^2 - 4 = 5$ Inverse properties

$x^2 = 9$ Add 4.

$x = \pm 3$ Square root property

Check each answer.

$\log (3 + 2) + \log (3 - 2) \stackrel{?}{=} \log 5$ $\log (-3 + 2) + \log (-3 - 2) \stackrel{?}{=} \log 5$

$\log 5 + \log 1 \stackrel{?}{=} \log 5$ $\log (-1) + \log (-5) \neq \log 5$ ✗

$\log 5 + 0 \stackrel{?}{=} \log 5$ Undefined

$\log 5 = \log 5$ ✓

Although 3 is a solution, -3 is not, because both $\log (-1)$ and $\log (-5)$ are undefined expressions. *Be sure to check your answers.*

Now Try Exercise 85

4 Solving Applications of Logarithmic Equations

EXAMPLE 8   **Modeling bird populations**

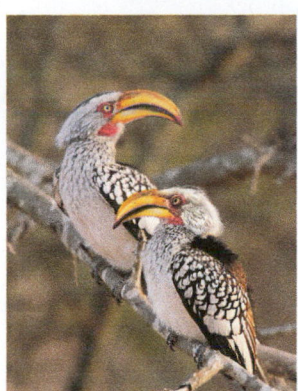

Near New Guinea, there is a relationship between the number of different species of birds and the size of an island. Larger islands tend to have a greater variety of birds. **TABLE 12.13** lists the number of species of birds y found on islands with an area of x square kilometers.

Numbers of Bird Species on Islands

$x \ (\text{km}^2)$	0.1	1	10	100	1000
$y \ (\text{species})$	10	15	20	25	30

Source: B. Freedman, Environmental Ecology.

TABLE 12.13

(a) Find values for the constants a and b so that $y = a + b \log x$ models the data.
(b) Predict the number of bird species on an island of 4000 square kilometers.

Solution
(a) Substitute $x = 1$ and $y = 15$ in the equation to find a.

$$15 = a + b \log 1 \qquad y = a + b \log x$$
$$15 = a + b \cdot 0 \qquad \log 1 = 0$$
$$15 = a \qquad \text{Simplify.}$$

Thus $y = 15 + b \log x$. To find b, substitute $x = 10$ and $y = 20$.

$$20 = 15 + b \log 10 \qquad y = 15 + b \log x$$
$$20 = 15 + b \cdot 1 \qquad \log 10 = 1$$
$$5 = b \qquad \text{Simplify.}$$

The data in **TABLE 12.13** are modeled by $y = 15 + 5 \log x$. This result is supported by **FIGURE 12.36**.

X	Y₁
.1	10
1	15
10	20
100	25
1000	30

Y₁■15+5log(X)

FIGURE 12.36

(b) To predict the number of species on an island of 4000 square kilometers, let $x = 4000$ and find y.

$$y = 15 + 5 \log 4000 \approx 33$$

The model estimates about 33 different species of birds on this island.

Now Try Exercise 113

EXAMPLE 9   **Modeling runway length**

For some types of airplanes with weight x, the minimum runway length L required at take-off is modeled by

$$L(x) = 3 \log x.$$

In this equation, L is measured in thousands of feet and x is measured in thousands of pounds. Estimate the weight of the heaviest airplane that can take off from a runway 5100 feet long. (*Source: L. Haefner, Introduction to Transportation Systems.*)

CRITICAL THINKING 1

Previously, we used $L(x) = 1.3 \ln x$ to model runway length. Are $L(x) = 1.3 \ln x$ and $L(x) = 3 \log x$ equivalent formulas? Explain.

Solution

Runway length is measured in thousands of feet, so we must solve the equation $L(x) = 5.1$.

$$3 \log x = 5.1 \qquad L(x) = 5.1$$
$$\log x = 1.7 \qquad \text{Divide each side by 3.}$$
$$10^{\log x} = 10^{1.7} \qquad \text{Exponentiate each side, using base 10.}$$
$$x = 10^{1.7} \qquad \text{Inverse property: } 10^{\log k} = k$$
$$x \approx 50.1 \qquad \text{Approximate.}$$

The largest airplane that can take off from this runway weighs about 50,000 pounds.

Now Try Exercise 109

12.5   Putting It All Together

TYPE OF EQUATION	PROCEDURE	EXAMPLE	
Exponential	Begin by solving for the exponential expression a^x. Then take a logarithm of each side.	$4e^x + 1 = 9$	Given equation
		$e^x = 2$	Solve for e^x.
		$\ln e^x = \ln 2$	Take the natural logarithm.
		$x = \ln 2$	Inverse property: $\ln e^k = k$
Logarithmic	Begin by solving for the logarithm in the equation. Then exponentiate each side of the equation, using the same base as the logarithm.	$\frac{1}{3} \log 2x = 1$	Given equation
		$\log 2x = 3$	Multiply by 3.
		$10^{\log 2x} = 10^3$	Exponentiate using base 10.
		$2x = 1000$	Inverse property: $10^{\log k} = k$
		$x = 500$	Divide by 2.

12.5   Exercises

MyMathLab®

CONCEPTS AND VOCABULARY

1. To solve $x - 5 = 50$, what should be done?

2. To solve $5x = 50$, what should be done?

3. To solve $10^x = 50$, what should be done?

4. To solve $\log x = 5$, what should be done?

5. $\log 10^x = $ _____

6. $10^{\log x} = $ _____

7. $\ln e^{2x} = $ _____

8. $e^{\ln (x+7)} = $ _____

9. Does $\frac{\log 5}{\log 4}$ equal $\log \frac{5}{4}$? Explain.

10. Does $\frac{\log 5}{\log 4}$ equal $\log 5 - \log 4$? Explain.

11. How many solutions are there to $\log x = k$, where k is any real number?

12. How many solutions are there to $10^x = k$, where k is a positive number?

WRITING EQUIVALENT EXPONENTIAL AND LOGARITHMIC FORMS

Exercises 13–16: Change each equation to its equivalent logarithmic form.

13. (a) $6^{3x} = 8$ (b) $e^{7x} = 10$

14. (a) $10^x = 50$ (b) $2^{2x} = 5$

15. (a) $5^{8x} = 12$ (b) $e^x = 3$

16. (a) $a^{3x} = b$ (b) $d^x = c$

Exercises 17–20: Change each equation to its equivalent exponential form.

17. (a) $\log_2 x = 8$ (b) $\ln 2x = 3$

18. (a) $\log(3 + x) = 5$ (b) $\log_9 x = 4$

19. (a) $\log_4 6x = 6$ (b) $\ln(x + 1) = 10$

20. (a) $\log_a(x + 2) = b$ (b) $\log_c 3x = d$

SOLVING EXPONENTIAL EQUATIONS

Exercises 21–52: Solve the given exponential equation. Approximate answers to the nearest hundredth when appropriate.

21. $10^x = 1000$ **22.** $10^x = 0.01$

23. $2^x = 64$ **24.** $3^x = 27$

25. $2^{x-3} = 8$ **26.** $3^{2x} = 81$

27. $4^x + 3 = 259$ **28.** $3(5^{2x}) = 300$

29. $10^{0.4x} = 124$ **30.** $0.75^x = 0.25$

31. $e^{-x} = 1$ **32.** $0.5^{-5x} = 5$

33. $e^x = 25$ **34.** $e^x = 0.4$

35. $0.4^x = 2$ **36.** $0.7^x = 0.3$

37. $e^x - 1 = 6$ **38.** $2e^{4x} = 15$

39. $2(10)^{x+2} = 35$ **40.** $10^{3x} + 10 = 1500$

41. $3.1^{2x} - 4 = 16$ **42.** $5.4^{x-1} = 85$

43. $e^{3x} = e^{2x-1}$ **44.** $e^{x^2} = e^{3x-2}$

45. $5^{4x} = 5^{x^2-5}$ **46.** $2^{4x} = 2^{x+3}$

47. $e^{2x} \cdot e^x = 10$ **48.** $10^{x-2} \cdot 10^x = 1000$

49. $e^x = 2^{x+2}$ **50.** $2^{2x} = 3^{x-1}$

51. $4^{0.5x} = 5^{x+2}$ **52.** $3^{2x} = 7^{x+1}$

Exercises 53–56: (Refer to Example 3.) The symbolic and graphical representations of f and g are given.

    (a) Use the graph to solve $f(x) = g(x)$.
    (b) Solve $f(x) = g(x)$ symbolically.

53. $f(x) = 0.2(10^x)$,
    $g(x) = 2$

54. $f(x) = e^x$,
    $g(x) = 7.4$

55. $f(x) = 2^{-x}$,
    $g(x) = 4$

56. $f(x) = 0.1(3^x)$,
    $g(x) = 0.9$

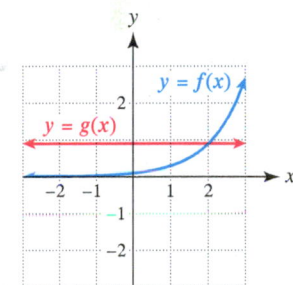

Exercises 57–64: Solve the equation symbolically. Give graphical or numerical support. Approximate answers to the nearest hundredth when appropriate.

57. $10^x = 0.1$ **58.** $2(10^x) = 2000$

59. $4e^x + 5 = 9$ **60.** $e^x + 6 = 36$

61. $4^x = 1024$ **62.** $3^x = 729$

63. $(0.55)^x + 0.55 = 2$ **64.** $5(0.9)^x = 3$

Exercises 65–68: The given equation cannot be solved symbolically. Find any solutions graphically to the nearest hundredth.

65. $e^x - x = 2$ **66.** $x \log x = 1$

67. $\ln x = e^{-x}$

68. $10^x - 2 = \log(x + 2)$

SOLVING LOGARITHMIC EQUATIONS

Exercises 69–90: Solve the given equation. Approximate answers to the nearest hundredth when appropriate.

69. $\log x = 2$ **70.** $\log x = 0.01$

71. $\ln x = 5$ **72.** $2 \ln x = 4$

73. $\log 2x = 7$ **74.** $6 \ln 4x = 12$

75. $\log_2 x = 4$

76. $\log_2 x = 32$

77. $\log_2 5x = 2.3$

78. $2 \log_3 4x = 10$

79. $2 \log x + 5 = 7.8$

80. $\ln (x - 1) = 3.3$

81. $5 \ln (2x + 1) = 55$

82. $5 - \log (x + 3) = 2.6$

83. $\log x^2 = \log x$

84. $\ln x^2 = \ln (3x - 2)$

85. $\ln x + \ln (x + 1) = \ln 30$

86. $\log (x - 1) + \log (2x + 1) = \log 14$

87. $\log_3 3x - \log_3 (x + 2) = \log_3 2$

88. $\log_4 (x^2 - 1) - \log_4 (x - 1) = \log_4 6$

89. $\log_2 (x - 1) + \log_2 (x + 1) = 3$

90. $\log_4 (x^2 + 2x + 1) - \log_4 (x + 1) = 2$

Exercises 91–96: Solve the equation symbolically. Give graphical or numerical support. Approximate answers to the nearest hundredth.

91. $\log x = 1.6$

92. $\ln x = 2$

93. $\ln (x + 1) = 1$

94. $2 \log (2x + 3) = 8$

95. $17 - 6 \log_3 x = 5$

96. $4 \log_2 x + 7 = 12$

Exercises 97–100: Two functions, f and g, are given.

   (a) *Use the graph to solve* $f(x) = g(x)$.

   (b) *Solve* $f(x) = g(x)$ *symbolically.*

97. $f(x) = \ln x$,
    $g(x) = 0.7$

98. $f(x) = \log_2 x$,
    $g(x) = 1.6$

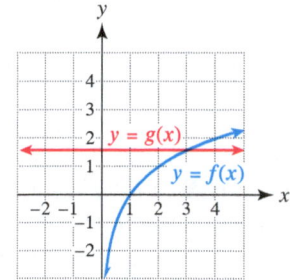

99. $f(x) = 5 \log 2x$,
    $g(x) = 3$

100. $f(x) = 2 \ln (x) - 3$,
    $g(x) = 0.9$

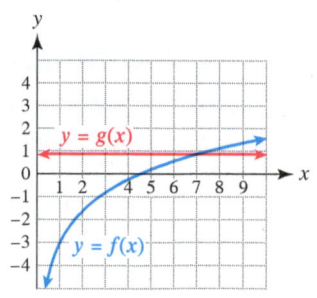

APPLICATIONS OF EXPONENTIAL AND LOGARITHMIC EQUATIONS

101. Smartwatch Market The market for smartwatches is increasing dramatically. The annual shipment forecasts in millions can be modeled by $W(x) = 8(1.47)^x$, where x represents years after 2013. (*Source:* BI Intelligence.)

   (a) Evaluate $f(5)$ and interpret the result.

   (b) Determine when the annual shipments of smartwatches will reach 120 million.

102. Risk of Down Syndrome The likelihood of a live birth having Down syndrome varies with the age of the mother. The function

$$D(x) = 0.0000816(10^{0.102x})$$

gives the percentage of live births that have Down syndrome when the mother is x years old.

   (a) Evaluate $D(35)$ and interpret the result.

   (b) Determine the mother's age when this percentage reaches 4%.

103. Bacteria Growth A sample of 5 million bacteria increases by 15% each hour. To the nearest hour, how long does it take for the sample to reach 20 million?

104. Bacteria Growth A sample of 3 million bacteria increases by 125% each day. To the nearest day, how long does it take for the bacteria sample to reach 173 million?

105. Growth of a Mutual Fund (Refer to Example 4.) An investor deposits \$2000 in a mutual fund that returns 15% at the end of 1 year. Determine the length of time required for the investment to triple its value if the annual rate of return remains the same.

106. Savings Account (Refer to Example 4.) If a savings account pays 6% annual interest at the end of each year, how many years will it take for the account to double in value?

107. Liver Transplants In the United States, the gap between available organs for liver transplants and people who need them has widened. The number of individuals in thousands waiting for liver transplants can be modeled by

$$f(x) = 120(1.025)^{x-2015},$$

where x is the year. (*Source:* United Network for Organ Sharing.)

   (a) Evaluate $f(2017)$ and interpret the result.

   (b) Determine when the number of individuals waiting for liver transplants will be 139 thousand.

108. Life Span of a Robin (Refer to Example 5.) Determine when 50% of the robins in the study were still alive.

109. Runway Length (Refer to Example 9.) Determine the weight of the heaviest airplane that can take off from a runway having a length of $\frac{3}{4}$ mile. (*Hint:* 1 mile = 5280 feet.)

110. Runway Length (Refer to Example 9.)
(a) Suppose that an airplane is 10 times heavier than another airplane. How much longer should the runway be for the heavier airplane than for the lighter airplane? (*Hint:* Let the heavier airplane have weight $10x$.)
(b) If the runway length is increased by 3000 feet, by what factor can the weight of an airplane that uses the runway be increased?

111. The Decline of Bluefin Tuna At the end of the 20th century, the number of western Atlantic bluefin tuna declined dramatically. Their numbers in thousands between 1974 and 1991 can be modeled by $f(x) = 230(10)^{-0.055x}$, where x is the number of years after 1974. See the accompanying graph.
(*Source:* B. Freedman.)

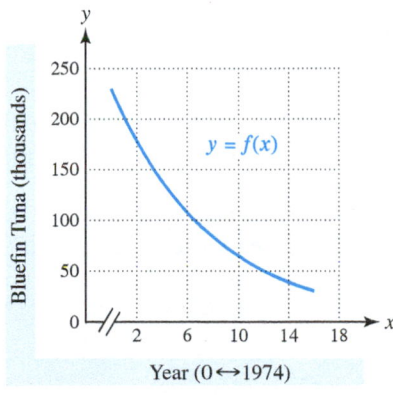

Year (0 ↔ 1974)

(a) Evaluate $f(1)$ and interpret the result.
(b) Use the graph to estimate the year in which blue-fin tuna numbered 115 thousand.
(c) Solve part (b) symbolically.

112. Insect Populations (Refer to Example 8.) The table lists numbers of species of insects y found on islands having areas of x square miles.

x (square miles)	1	2	4
y (species)	1000	1500	2000

x (square miles)	8	16	32
y (species)	2500	3000	3500

(a) Find values for the constants a and b so that $y = a + b \log_2 x$ models the data.

(b) Construct a table for y and verify that your equation models the data.
(c) Estimate the number of species of insects on an island that has an area of 12 square miles.

Exercises 113 and 114: (Refer to Example 8.) Find values for a and b so that $y = a + b \log x$ models the data.

113.

x	0.1	1	10	100
y	22	25	28	31

114.

x	0.01	1	100	1000
y	−10	−2	6	10

115. Calories Consumed and Land Ownership In developing countries, there is a relationship between the amount of land a person owns and the average number of calories that person consumes daily. This relationship is modeled by $f(x) = 645 \log(x + 1) + 1925$, where x is the amount of land owned in acres and $0 \le x \le 4$.
(*Source:* D. Grigg, *The World Food Problem.*)
(a) Estimate graphically the number of acres owned by a person consuming 2200 calories per day.
(b) Solve part (a) symbolically.

116. Population of Industrialized Urban Regions The number of people living in industrialized urban regions throughout the world has not grown exponentially. Instead, it has grown logarithmically and is modeled by

$$f(x) = 0.36 + 0.15 \ln(x - 1949).$$

In this formula, the output is billions of people and the input x is the year, where $1950 \le x \le 2030$.
(*Source:* D. Meadows, *Beyond The Limits.*)
(a) Determine graphically when this population may reach 1 billion.
(b) Solve part (a) symbolically.

117. Fitbit Sales The table lists the number of Fitbits sold in millions in year x. (*Source:* Fitbit)

x	2011	2012	2013	2014
y	0.3	1.3	4.5	10.9

(a) Are the data linear or nonlinear? Explain.
(b) The equation $f(x) = 0.345(3.33)^{x-2011}$ may be used to model the data. The growth factor is 3.33. What does this growth factor indicate about Fitbit sales during this period?
(c) If this trend continued, estimate the year when Fitbit sales was 140 million.

118. Greenhouse Gases If current trends continue, concentrations of atmospheric carbon dioxide (CO_2) in parts per million (ppm) are expected to increase. This increase in concentration of CO_2 has been accelerated by burning fossil fuels and deforestation. The exponential equation $y = 364(1.005)^x$ may be used to model CO_2 in parts per million x years after 2000. Estimate the year when the CO_2 concentration could be double the preindustrial level of 280 parts per million. (*Source:* R. Turner, *Environmental Economics.*)

119. Modeling Sound The formula

$$f(x) = 160 + 10 \log x$$

is used to calculate the decibel level of a sound with intensity x measured in watts per square centimeter. The noise level at a basketball game can reach 100 decibels. Find the intensity x of this sound.

120. Loudness of a Sound (Refer to Exercise 119.)
   (a) Show that if the intensity of a sound increases by a factor of 10 from x to $10x$, the decibel level increases by 10 decibels. *Hint:* Show that

$$160 + 10 \log 10x = 170 + 10 \log x.$$

   (b) Find the increase in decibels if the intensity x increases by a factor of 1000.
   (c) Find the increase in the intensity x if the decibel level increases by 20.

121. Hurricanes The barometric air pressure in inches of mercury at a distance of x miles from the eye of a severe hurricane is given by $f(x) = 0.48 \ln(x + 1) + 27$. How far from the eye is the pressure 28 inches of mercury? (*Source:* A. Miller and R. Anthes, *Meteorology.*)

122. Earthquakes The Richter scale is used to determine the intensity of earthquakes, which corresponds to the amount of energy released. If an earthquake has an intensity of x, its magnitude, as computed by the Richter scale, is given by $R(x) = \log \frac{x}{I_0}$, where I_0 is the intensity of a small, measurable earthquake.
   (a) If x is 1000 times greater than I_0, how large is this increase on the Richter scale?
   (b) If the Richter scale increases from 5 to 8, by what factor does the intensity x increase?

Exercises 123 and 124: **Investment Account** *If a total of x dollars is deposited every 2 weeks (26 times per year) in an account paying an annual interest rate r, expressed in decimal form, the amount A in the account after n years can be approximated by the formula*

$$A = x \left[\frac{(1 + r/26)^{26n} - 1}{(r/26)} \right].$$

123. If a total of $100 is deposited every 2 weeks in an account paying 9% interest, approximate the amount in the account after 10 years.

124. Suppose that your retirement account pays 12% annual interest. Determine how much you should deposit in this account every 2 weeks, in order to have one million dollars at age 65.

WRITING ABOUT MATHEMATICS

125. Explain the basic steps for solving the equation $a(10^x) - b = c$. Then write the solution.

126. Explain the basic steps for solving the equation $a \log 3x = b$. Then write the solution.

SECTION 12.5 **Checking Basic Concepts**

1. Solve the equation. Approximate answers to the nearest hundredth when appropriate.
   (a) $2(10^x) = 40$ **(b)** $2^{3x} + 3 = 150$
   (c) $\ln x = 4.1$ **(d)** $4 \log 2x = 12$

2. Solve $\log(x + 4) + \log(x - 4) = \log 48$. Check the answers.

3. If a total of $500 is deposited in a savings account that pays 3% annual interest at the end of each year, the amount of money A in the account after x years is given by $A = 500(1.03)^x$. Estimate the number of years required for this amount to reach $900.

CHAPTER 12   Summary

SECTION 12.1 ■ COMPOSITE AND INVERSE FUNCTIONS

Composition of Functions If f and g are functions, then the composite function $g \circ f$, or composition of g and f, is defined by $(g \circ f)(x) = g(f(x))$.

Example: If $f(x) = x - 5$ and $g(x) = 2x^2 + 4x - 6$, then $(g \circ f)(x)$ is
$$g(f(x)) = g(x - 5)$$
$$= 2(x - 5)^2 + 4(x - 5) - 6.$$

One-to-One Function A function f is one-to-one if, for any c and d in the domain of f,
$$c \neq d \quad \text{implies that} \quad f(c) \neq f(d).$$

That is, different inputs always result in different outputs.

Example: $f(x) = x^2 + 4$ is *not* one-to-one because $f(-3) = f(3) = 13$.   Different inputs; same output

Horizontal Line Test If every horizontal line intersects the graph of a function f at most once, then f is a one-to-one function.

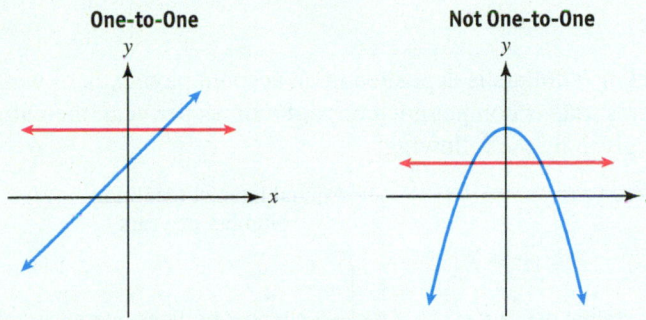

Inverse Functions If f is one-to-one, then f has an inverse function, denoted f^{-1}, that satisfies $(f^{-1} \circ f)(x) = x$ and $(f \circ f^{-1})(x) = x$.

Example: $f(x) = 7x$ and $f^{-1}(x) = \frac{x}{7}$ are inverse functions.

SECTION 12.2 ■ EXPONENTIAL FUNCTIONS

Exponential Function An exponential function is defined by $f(x) = Ca^x$, where $a > 0$, $C > 0$, and $a \neq 1$. Its domain (set of valid inputs) is all real numbers and its range (outputs) is all positive real numbers. The base is a.

Example: $f(x) = e^x$ is the natural exponential function and $e \approx 2.71828$.

Exponential Growth and Decay When $a > 1$, the graph of $f(x) = Ca^x$ models exponential growth, and when $0 < a < 1$, it models exponential decay. The base a represents either the growth factor or the decay factor. The constant C corresponds to the y-intercept, $(0, C)$.

Example: $f(x) = 1.5(2)^x$ is an exponential function with $a = 2$ and $C = 1.5$. It models exponential growth because $a > 1$. The growth factor is 2 because for each unit increase in x, the output from $f(x)$ increases by a *factor* of 2.

Percent Change If an amount A changes to a new amount B, then the percent change is given by

$$\frac{B - A}{A} \times 100.$$

Example: If $500 increases to $700, the percent change is

$$\frac{700 - 500}{500} \times 100 = 40\%.$$

Percent Change and Growth Factor If an initial amount C experiences a constant percent change r (in decimal form) for each x-unit increase in time, then the new amount A is

$$A = C(1 + r)^x,$$

where the growth factor is $a = 1 + r$.

Example: If $800 increases by 4% each year, then after 5 years, the amount is

$$A = 800(1 + 0.04)^5 \approx \$973.32.$$

The growth factor is $a = 1.04$.

Compound Interest If a total of P dollars is deposited in an account paying an *annual* interest rate r (in decimal form) and this interest rate is compounded or paid n times per year, then after t years, the account contains A dollars given by the following.

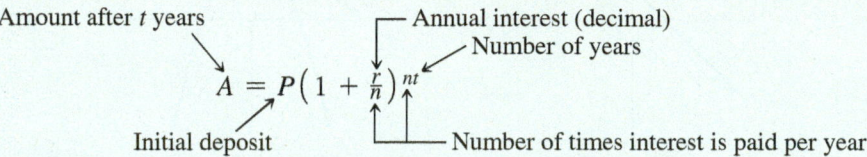

Example: If a total of $1200 is deposited at 6% annual interest, compounded quarterly, then after 8 years, the account contains

$$A = 1200\left(1 + \frac{0.06}{4}\right)^{(4 \cdot 8)} \approx \$1932.39.$$

SECTION 12.3 ■ LOGARITHMIC FUNCTIONS

Base-a Logarithms The logarithm with base a of a positive number x is denoted $\log_a x$. If $\log_a x = b$, then $x = a^b$. That is, $\log_a x$ represents the exponent on base a that results in x.

Examples: $\log_2 16 = 4$ because $16 = 2^4$ and $\log 100 = 2$ because $100 = 10^2$.

Base-a Logarithm

Domain and Range of Logarithmic Functions The domain (set of valid inputs) of a logarithmic function is the set of all positive real numbers and the range (outputs) is the set of real numbers.

Graph of a Logarithmic Function The graph of a logarithmic function passes through $(1, 0)$, as illustrated in the graph. As x becomes large, $\log_a x$ with $a > 1$ grows very slowly.

Exponential and Logarithmic Form The equation $y = a^x$ is equivalent to $\log_a y = x$.

SECTION **12.4** ■ PROPERTIES OF LOGARITHMS

Basic Properties Logarithms have several important properties. For positive numbers m, n, and $a \neq 1$ and any real number r,

1. $\log_a mn = \log_a m + \log_a n$.

2. $\log_a \frac{m}{n} = \log_a m - \log_a n$.

3. $\log_a (m^r) = r \log_a m$.

Examples: 1. $\log 5 + \log 20 = \log (5 \cdot 20) = \log 100 = 2$

2. $\log 100 - \log 5 = \log \frac{100}{5} = \log 20 \approx 1.301$

3. $\ln 2^6 = 6 \ln 2 \approx 4.159$

NOTE: $\log_a 1 = 0$ for any valid base a. Thus $\log 1 = 0$ and $\ln 1 = 0$. ■

Inverse Properties The following inverse properties are important for solving exponential and logarithmic equations.

1. $\log_a a^x = x$, for any real number x

2. $a^{\log_a x} = x$, for any positive number x

Examples: 1. $\log_2 2^\pi = \pi$ **2.** $10^{\log 2.5} = 2.5$

3. $e^{\ln 5} = 5$ **4.** $\log 10^7 = 7$

NOTE: If $f(x) = \log x$, then $f^{-1}(x) = 10^x$, and if $g(x) = \ln x$, then $g^{-1}(x) = e^x$. ■

SECTION **12.5** ■ EXPONENTIAL AND LOGARITHMIC EQUATIONS

Solving Equations When solving an exponential equation, we usually take a logarithm of each side. When solving a logarithmic equation, we usually exponentiate each side.

Examples:

1.
$2(5)^x = 22$	Exponential equation
$5^x = 11$	Divide by 2.
$\log 5^x = \log 11$	Take the common logarithm.
$x \log 5 = \log 11$	Power rule
$x = \dfrac{\log 11}{\log 5}$	Divide by $\log 5$.

2.
$\log 2x = 2$	Logarithmic equation
$10^{\log 2x} = 10^2$	Exponentiate each side.
$2x = 100$	Inverse properties
$x = 50$	Divide by 2.

CHAPTER 12   Review Exercises

SECTION 12.1

Exercises 1 and 2: Find the following.

   (a) $(g \circ f)(-2)$ *(b)* $(f \circ g)(x)$

1. $f(x) = 2x^2 - 4x$, $g(x) = 5x + 1$

2. $f(x) = \sqrt[3]{x - 6}$, $g(x) = 4x^3$

3. Use the two tables to evaluate each expression.
   (a) $(f \circ g)(2)$ **(b)** $(g \circ f)(1)$

x	0	1	2	3
$f(x)$	3	2	1	0

x	0	1	2	3
$g(x)$	1	2	3	0

4. Use the graph to evaluate each expression.
   (a) $(f \circ g)(-1)$
   (b) $(g \circ f)(2)$
   (c) $(f \circ f)(1)$

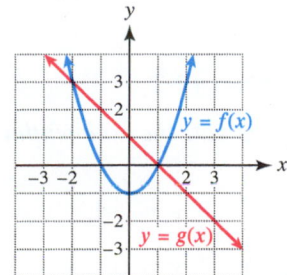

Exercises 5 and 6: Show that f is not one-to-one by finding two inputs that result in the same output. Answers may vary.

5. $f(x) = \dfrac{4}{1 + x^2}$ **6.** $f(x) = x^2 - 2x + 1$

Exercises 7 and 8: Use the horizontal line test to determine whether the graph represents a one-to-one function.

7.

8.

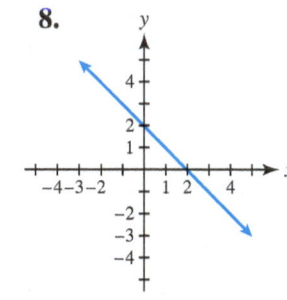

Exercises 9 and 10: Verify that $f(x)$ and $f^{-1}(x)$ are indeed inverse functions.

9. $f(x) = 2x - 9$ $f^{-1}(x) = \dfrac{x + 9}{2}$

10. $f(x) = x^3 + 1$ $f^{-1}(x) = \sqrt[3]{x - 1}$

Exercises 11–14: Find $f^{-1}(x)$.

11. $f(x) = 5x$ **12.** $f(x) = x - 11$

13. $f(x) = 2x + 7$ **14.** $f(x) = \dfrac{4}{x}$

15. Use the table to make a table of values for $f^{-1}(x)$. What are the domain and range for f^{-1}?

x	0	1	2	3
$f(x)$	10	8	7	3

16. Use the graph of $y = f(x)$ to sketch a graph of $y = f^{-1}(x)$. Include the graph of f and the line $y = x$.

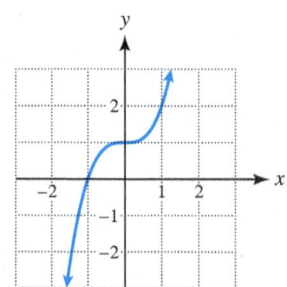

SECTIONS 12.2 AND 12.3

Exercises 17–20: Evaluate the exponential function for the given values of x.

17. $f(x) = 6^x$ $x = -1, \quad x = 2$

18. $f(x) = 5(2^{-x})$ $x = 0, \quad x = 3$

19. $f(x) = \left(\dfrac{1}{3}\right)^x$ $x = -1, \quad x = 4$

20. $f(x) = 3\left(\dfrac{1}{6}\right)^x$ $x = 0, \quad x = 1$

Exercises 21–24: Graph f. State whether the graph illustrates exponential growth, exponential decay, or logarithmic growth.

21. $f(x) = 2^x$ **22.** $f(x) = \left(\dfrac{1}{2}\right)^x$

23. $f(x) = \ln(x + 1)$ **24.** $f(x) = 3^{-x}$

Exercises 25 and 26: A table for a function f is given.

    (a) Determine whether f represents linear or exponential growth.

    (b) Find a formula for f.

25.

x	0	1	2	3	4
$f(x)$	5	10	20	40	80

26.

x	0	1	2	3	4
$f(x)$	5	10	15	20	25

27. Use the graph of $y = Ca^x$ to find C and a.

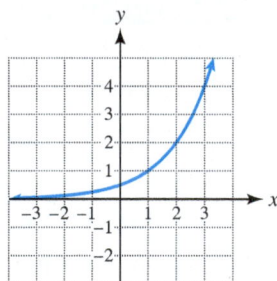

28. Use the graph of $y = k \log_2 x$ to find k.

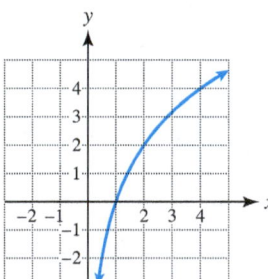

29. Find the percent change if \$150 decreases to \$120.

30. Find the growth factor if \$1500 increases by 7% each year.

Exercises 31 and 32: For the given amounts A and B, find each of the following. Round values to the nearest hundredth of a percent when appropriate.

    (a) The percent change if A changes to B

    (b) The percent change if B changes to A

31. $A = \$600, B = \$1200$

32. $A = \$2.20, B = \$1.00$

Exercises 33 and 34: An account contains A dollars and increases or decreases by R%. For each A and R, answer the following.

    (a) Find the increase or decrease in value of the account.

    (b) Find the final value of the account.

    (c) By what factor did the account value increase or decrease?

33. $A = \$500, R = 210\%$

34. $A = \$700, R = -25\%$

Exercises 35 and 36: **Exponential Models** *Write the formula for an exponential function, $f(x) = Ca^x$, that models the situation. Evaluate $f(2)$.*

35. A city's population of 20,000 decreases in number by 5% per year.

36. A sample of 1500 insects increases in number by 300% per day.

Exercises 37 and 38: If a total of P dollars is deposited in an account that pays R percent annual interest at the end of each year, approximate the amount in the account after x years.

37. $P = \$1200,\quad R = 10\%,\quad x = 9$ years

38. $P = \$900,\quad R = 18\%,\quad x = 40$ years

Exercises 39–42: Evaluate $f(x)$ for the given value of x. Approximate answers to the nearest hundredth.

39. $f(x) = 2e^x - 1$ $x = 5.3$

40. $f(x) = 0.85^x$ $x = 2.1$

41. $f(x) = 2\log x$ $x = 55$

42. $f(x) = \ln(2x + 3)$ $x = 23$

Exercises 43–46: Evaluate the logarithm by hand.

43. $\log 0.001$ **44.** $\log \sqrt{10{,}000}$

45. $\ln e^{-4}$ **46.** $\log_4 16$

Exercises 47–50: Approximate to the nearest thousandth.

47. $\log 65$ **48.** $\ln 0.85$

49. $\ln 120$ **50.** $\log \frac{2}{5}$

Exercises 51–54: Simplify, using inverse properties.

51. $10^{\log 7}$ **52.** $\log_2 2^{5/9}$

53. $\ln e^{6-x}$ **54.** $e^{2\ln x}$

SECTION 12.4

Exercises 55–60: Write the expression by using sums and differences of logarithms of x, y, and z.

55. $\ln xy$ **56.** $\log \dfrac{x}{y}$

57. $\ln(x^2 y^3)$ **58.** $\log \dfrac{\sqrt{x}}{z^3}$

59. $\log_2 \dfrac{x^2 y}{z}$ **60.** $\log_3 \sqrt[3]{\dfrac{x}{y}}$

Exercises 61–64: Write as the logarithm of one expression.

61. $\log 45 + \log 5 - \log 3$

62. $\log_4 2x + \log_4 5x$

63. $2\ln x - 3\ln y$

64. $\log x^4 - \log x^3 + \log y$

Exercises 65–68: Rewrite, using the power rule.

65. $\log 6^3$ **66.** $\ln x^2$

67. $\log_2 5^{2x}$ **68.** $\log_4 0.6^{x+1}$

SECTION 12.5

Exercises 69–78: Solve the given equation. Approximate answers to the nearest hundredth when appropriate.

69. $10^x = 100$ **70.** $2^{2x} = 256$

71. $3e^x + 1 = 28$ **72.** $0.85^x = 0.2$

73. $5\ln x = 4$ **74.** $\ln 2x = 5$

75. $2\log x = 80$ **76.** $3\log x - 5 = 1$

77. $2^{x+4} = 3^x$

78. $\ln(2x + 1) + \ln(x - 5) = \ln 13$

Exercises 79 and 80: Do the following.

    (a) Solve $f(x) = g(x)$ graphically.

    (b) Solve $f(x) = g(x)$ symbolically.

79. $f(x) = \frac{1}{2}(2^x)$,   **80.** $f(x) = \log_2 2x$,
     $g(x) = 4$ $g(x) = 3$

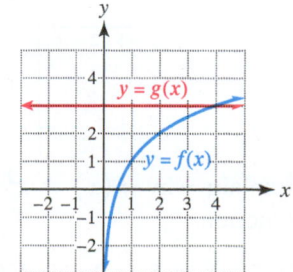

APPLICATIONS

81. Surface Area of a Balloon The surface area S of a spherical balloon with radius r is given by the formula $S(r) = 4\pi r^2$. Suppose that the balloon is being inflated so that its radius in inches after t seconds is $r(t) = \sqrt{2t}$.
(a) Evaluate $(S \circ r)(8)$ and interpret your result.
(b) Find $(S \circ r)(t)$.

82. Sales Tax Suppose that $f(x) = 0.08x$ calculates the sales tax in dollars on an item that costs x dollars.
(a) Is f a one-to-one function? Why?
(b) Find a formula for f^{-1} and interpret what it calculates.

83. Growth of a Mutual Fund An investor deposits $1500 in a mutual fund that returns 11% annually. Determine the time required for the investment to double in value.

84. Modeling Data Find values for the constants a and b so that $y = a + b \log x$ models the data in the table.

x	0.1	1	10	100	1000
y	50	100	150	200	250

85. Modeling Data Find values for the constants C and a so that $y = Ca^x$ models the data in the table.

x	0	1	2	3	4
y	3	6	12	24	48

86. Earthquakes The Richter scale, used to determine the magnitude of earthquakes, is based on the formula $R(x) = \log \frac{x}{I_0}$, where x is the measured intensity. Let $I_0 = 1$. Find the intensity x for an earthquake with $R = 7$.

87. Modeling Population In 2000, the population of Nevada was 2 million and growing continuously at an annual rate of 5.1%. The population of Nevada in millions x years after 2000 can be modeled by

$$f(x) = 2e^{0.051x}.$$

(a) Graph f in the window $[0, 10, 2]$ by $[0, 4, 1]$. Does this function represent exponential growth or decay?
(b) Estimate the population of Nevada in 2010.
(c) Estimate the year when the population of Nevada was 3 million.

88. Modeling Bacteria A colony of bacteria can be modeled by $N(t) = 1000e^{0.0014t}$, where N is measured in bacteria per milliliter and t is in minutes.
(a) Evaluate $N(0)$ and interpret the result.
(b) Estimate how long it takes for N to double.

89. Modeling Wind Speed Wind speeds vary at different heights above the ground. For a particular day, $f(x) = 1.2 \ln(x) + 5$ computes the wind speed in meters per second x meters above the ground, where $x \geq 1$. (*Source:* A. Miller and R. Anthes, *Meteorology.*)
(a) Find the wind speed at a height of 5 meters.
(b) Estimate the height at which the wind speed is 8 meters per second.

CHAPTER 12   Test   MyMathLab®, YouTube

1. If $f(x) = 4x^3 - 5x$ and $g(x) = x + 7$, evaluate $(g \circ f)(1)$ and $(f \circ g)(x)$.

2. Use the graph to evaluate each expression.
(a) $(f \circ g)(-1)$ (b) $(g \circ f)(1)$

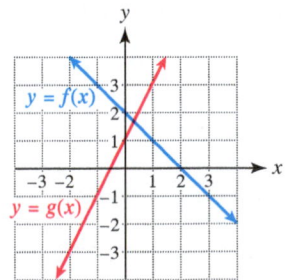

3. Explain why $f(x) = x^2 - 25$ is not a one-to-one function.

4. If $f(x) = 5 - 2x$, find $f^{-1}(x)$.

5. Use the graph of $y = f(x)$ to sketch a graph of $y = f^{-1}(x)$. Include the graph of f and the line $y = x$ in your graph.

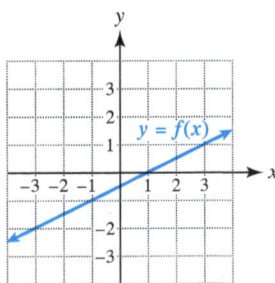

6. Use the table to write a table of values for $f^{-1}(x)$. What are the domain and range of f^{-1}?

x	1	2	3	4
$f(x)$	8	6	4	2

7. Evaluate $f(x) = 3\left(\frac{1}{4}\right)^x$ at $x = 2$.

8. Graph $f(x) = 1.5^{-x}$. State whether the graph of f illustrates exponential growth, exponential decay, or logarithmic growth.

Exercises 9 and 10: A table for a function f is given.

(a) *Determine whether f represents linear or exponential growth.*

(b) *Find a formula for $f(x)$.*

9.

x	−2	−1	0	1	2
$f(x)$	0.75	1.5	3	6	12

10.

x	−2	−1	0	1	2
$f(x)$	−4	−2.5	−1	0.5	2

11. Use the graph of $y = Ca^x$ to find C and a.

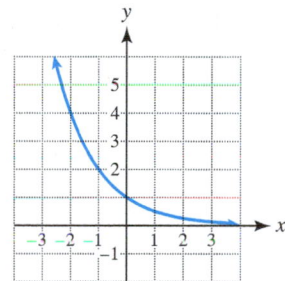

12. Find the percent change if $600 increases to $900.

13. If $1000 increases by 5% per year, what is the growth factor?

14. If a total of $750 is deposited in an account paying 7% annual interest at the end of each year, approximate the amount in the account after 5 years.

15. Let $f(x) = 1.5 \ln(x - 5)$. Approximate $f(21)$ to the nearest hundredth.

16. Evaluate $\log \sqrt{10}$ by hand.

17. Approximate $\log_2 43$ to the nearest thousandth.

18. Graph $f(x) = \log(x - 2)$. Compare this graph to the graph of $y = \log x$.

19. Write $\log \dfrac{x^3 y^2}{\sqrt{z}}$, using sums and differences of logarithms of x, y, and z.

20. Write $4 \ln x - 5 \ln y + \ln z$ as one logarithm.

21. Rewrite $\log 7^{2x}$, using the power rule.

22. Simplify $\ln e^{1-3x}$, using inverse properties.

Exercises 23–26: Solve the given equation. Approximate answers to the nearest hundredth when appropriate.

23. $2e^x = 50$

24. $3(10)^x - 7 = 143$

25. $5 \log x = 9$

26. $3 \ln 5x = 27$

27. **Modeling Data** Find values for constants a and b so that $y = a + b \log x$ models the data.

x	0.01	0.1	1	10	100
y	−1	2	5	8	11

28. **Modeling Bacteria Growth** A sample of bacteria is growing at a rate of 9% per hour and can be modeled by $f(x) = 4(1.09)^x$, where the input x represents elapsed time in hours and the output $f(x)$ is in millions of bacteria.

(a) What was the initial number of bacteria?

(b) Evaluate $f(5)$ and interpret the result.

(c) Does this function represent exponential growth or exponential decay?

(d) Estimate the elapsed time when there were 8 million bacteria.

29. **Investing** A mutual fund account containing $5000 decreases by 2% per year.

(a) Write a function A that gives the amount in the account after x years.

(b) Evaluate $A(3)$ and interpret the result.

(c) Estimate the number of years for the account to decrease to $4500.

CHAPTERS 1–12	**Cumulative Review Exercises**

1. Write the number 0.000429 in scientific notation.

2. Classify each real number as one or more of the following: natural number, whole number, integer, rational number, or irrational number.

$$-\frac{11}{7},\ -3,\ 0,\ \sqrt{6},\ \pi,\ 5.\overline{18}$$

3. Select the formula that models the data best.

x	−2	−1	0	1	2
y	−7	−5	−3	−1	1

a. $y = 3x + 1$ b. $y = x - 3$ c. $y = 2x - 3$

4. State whether the equation illustrates an identity, commutative, associative, or distributive property.

$$(5 - y) + 9 = 9 + (5 - y)$$

Exercises 5–8: Simplify the expression. Write the result with positive exponents.

5. $\left(\dfrac{1}{d^2}\right)^{-2}$ 　　　**6.** $\left(\dfrac{8a^2}{2b^3}\right)^{-3}$

7. $\dfrac{(2x^{-2}y^3)^2}{xy^{-2}}$ 　　　**8.** $\dfrac{x^{-3}y}{4x^2y^{-3}}$

9. Use the graph to express the equation of the line in slope–intercept form.

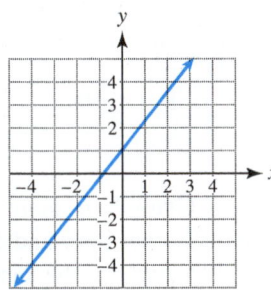

10. Find the domain of $f(x) = \dfrac{10}{x+3}$.

11. Use the table to write the formula for $f(x) = ax + b$.

x	-2	-1	0	1	2
$f(x)$	-11	-7	-3	1	5

12. Write the equation of the vertical line passing through the point $(4, 7)$.

13. Calculate the slope of the line passing through the points $(4, -1)$ and $(2, -3)$.

14. Sketch the graph of a line passing through the point $(-1, -2)$ with slope $m = 3$.

Exercises 15 and 16: Write the slope–intercept form for a line satisfying the given conditions.

15. Perpendicular to $y = -\frac{1}{7}x - 8$, passing through $(1, 1)$

16. Parallel to $y = 3x - 1$, passing through $(0, 5)$

17. Use the graph to solve the equation $y_1 = y_2$.

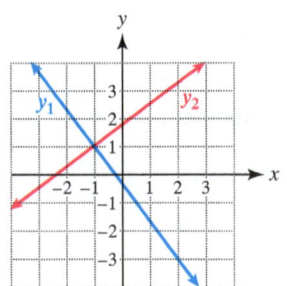

18. Use the table to solve the inequality $y < -4$, where y represents a linear function. Write the solution set as an inequality.

x	-2	-1	0	1	2
y	-24	-14	-4	6	16

Exercises 19–24: Solve the equation or inequality. Write the solutions to the inequalities in interval notation.

19. $\frac{2}{3}(x - 3) + 8 = -6$

20. $\frac{1}{3}z + 6 < \frac{1}{4}z - (5z - 6)$

21. $\left(\dfrac{t + 2}{3}\right) - 10 = \frac{1}{3}t - (5t + 8)$

22. $-10 \le -\frac{3}{5}x - 4 < -1$

23. $-2|t - 4| \ge -12$ 　　　**24.** $\left|\frac{1}{2}x - 5\right| = 3$

25. Shade the solution set in the xy-plane.

$$x + y > 3$$
$$2x - y \ge 3$$

26. Evaluate $\det A$ if $A = \begin{bmatrix} -1 & -2 \\ 3 & 4 \end{bmatrix}$.

Exercises 27–30: Solve the system of equations, if possible. Write the solution as an ordered pair or ordered triple where appropriate.

27. $4x - 3y = 1$
$5x + 2y = 7$

28. $2x - 3y = -2$
$-6x + 9y = 5$

29. $2x - y + 3z = -2$
$x + 5y - 2z = -8$
$-3x - y - 3z = 6$

30. $x + y - z = -1$
$-x - y - z = -1$
$x - 2y + z = 1$

Exercises 31–34: Factor completely.

31. $2x^3 - 4x^2 + 2x$ 　　　**32.** $4a^2 - 25b^2$

33. $8t^3 - 27$ 　　　**34.** $4a^3 - 2a^2 + 10a - 5$

Exercises 35–38: Solve the equation.

35. $6x^2 - 7x - 10 = 0$ 　　　**36.** $9x^2 = 4$

37. $x^4 - 2x^3 = 15x^2$ 　　　**38.** $5x - 10x^2 = 0$

Exercises 39 and 40: Simplify the expression.

39. $\dfrac{x^2 + 5x + 6}{x^2 - 9} \cdot \dfrac{x - 3}{x + 2}$

40. $\dfrac{x^2 - 2x - 8}{x^2 + x - 12} \div \dfrac{(x - 4)^2}{x^2 - 16}$

Exercises 41 and 42: Solve the rational equation. Check your result.

41. $\dfrac{2}{x+2} - \dfrac{1}{x-2} = \dfrac{-3}{x^2-4}$

42. $\dfrac{3y}{y^2+y-2} = \dfrac{1}{y-1} - 2$

43. Solve the equation for J.

$$P = \dfrac{J+2z}{J}$$

44. Simplify the complex fraction.

$$\dfrac{\dfrac{3}{x^2}+x}{x-\dfrac{3}{x^2}}$$

45. Suppose that y varies directly with x. If $y = 15$ when $x = 3$, find y when x is 8.

46. Divide $(3x^3 - 2x - 15)$ by $(x - 2)$.

Exercises 47–52: Simplify the expression. Assume that all variables are positive.

47. $\left(\dfrac{x^6}{y^9}\right)^{2/3}$

48. $\sqrt[3]{-x^4} \cdot \sqrt[3]{-x^5}$

49. $\sqrt{5ab} \cdot \sqrt{20ab}$

50. $2\sqrt{24} - \sqrt{54}$

51. $\sqrt[3]{a^5b^4} + 3\sqrt[3]{a^5b}$

52. $\left(5 + \sqrt{5}\right)\left(5 - \sqrt{5}\right)$

53. Rationalize the denominator.

$$\dfrac{2}{5 - \sqrt{3}}$$

54. Find the domain of f. Write your answer in interval notation.

$$f(x) = \dfrac{3}{\sqrt{x-4}}$$

Exercises 55 and 56: Solve. Check your answer.

55. $2(x + 1)^2 = 50$

56. $\sqrt{x+6} = x$

Exercises 57 and 58: Write in standard form.

57. $(-2 + 3i) - (-5 - 2i)$

58. $\dfrac{3 - i}{1 + 3i}$

59. Find the vertex on the graph of the function given by $f(x) = 3x^2 - 12x + 13$.

60. Find the maximum y-value on the parabola determined by $y = -2x^2 + 6x - 1$.

Exercises 61–64: Solve the quadratic equation by using the method of your choice.

61. $x^2 - 13x + 40 = 0$

62. $2d^2 - 5 = d$

63. $z^2 - 4z = -2$

64. $x^4 - 10x^2 + 24 = 0$

65. A graph of $y = ax^2 + bx + c$ is shown.
   (a) Solve $ax^2 + bx + c = 0$.
   (b) State whether $a > 0$ or $a < 0$.
   (c) Determine whether the discriminant is positive, negative, or zero.

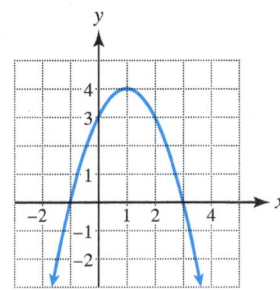

66. Solve $x^2 + 5x - 14 \geq 0$. Write your answer in interval notation.

67. For $f(x) = x^2 - 2$ and $g(x) = 2x + 1$, find each of the following.
   (a) $(f \circ g)(1)$ **(b)** $(g \circ f)(x)$

68. Find $f^{-1}(x)$ for $f(x) = \dfrac{3}{x}$.

Exercises 69 and 70: Evaluate without a calculator.

69. $\log_3 81$

70. $e^{\ln 2x}$

71. Write the expression by using sums and differences of logarithms of x and y. Assume x and y are positive.

$$\log \dfrac{\sqrt{x}}{y^2}$$

72. Write $2 \ln x + \ln 5x$ as the logarithm of a single expression.

Exercises 73 and 74: Solve the equation. Approximate answers to the nearest hundredth.

73. $6 \log x - 2 = 9$

74. $2^{3x} = 17$

APPLICATIONS

75. Population Growth The population P of a community with an annual percentage growth rate r (expressed as a decimal) after t years is given by $P = P_0(1 + r)^t$, where P_0 represents the initial population of the community. If a community with an initial population of $P_0 = 12,000$ grows to a population of $P = 14,600$ in $t = 5$ years, find the annual percentage growth rate r for this community.

76. Wing Span of a Bird The wing span L of a bird with weight W can sometimes be modeled by $L = 27.4\sqrt[3]{W}$, where L is in inches and W is in pounds. Estimate the weight of a bird with a wing span of 36 inches. (*Source:* C. Pennycuick, *Newton Rules Biology.*)

77. Investing for Retirement A college student invests $8000 in an account that pays interest annually. If the student would like this investment to be worth $1,000,000 in 45 years, what annual interest rate would the account need to pay?

78. Modeling Wind Speed Wind speeds are usually measured at heights from 5 to 10 meters above the ground. For a particular day, $f(x) = 1.4 \ln(x) + 7$ computes the wind speed in meters per second x meters above the ground, where $x \geq 1$. (*Source:* A. Miller and R. Anthes, *Meteorology.*)

(a) Find the wind speed at a height of 8 meters.

(b) Estimate the height at which the wind speed is 10 meters per second.

13 Conic Sections

Throughout history, people have been fascinated with the universe around them and compelled to understand its mysteries. Conic sections, which include parabolas, circles, ellipses, and hyperbolas, have played an important role in gaining this understanding. All celestial objects—including planets, comets, asteroids, and satellites—travel in paths described by conic sections. Today, scientists search the sky for information about the universe with enormous radio telescopes in the shape of parabolic dishes. The Hubble telescope also makes use of a parabolic mirror. As a result, our understanding of the universe has changed dramatically in recent years.

Conic sections have had a profound influence on people's understanding of their world and the cosmos. In this chapter, we introduce you to these age-old curves. (*Source: Historical Topics for the Mathematics Classroom, Thirty-first Yearbook, NCTM.*)

13.1 Parabolas and Circles

NEW VOCABULARY

☐ Conic sections
☐ Circle
☐ Radius
☐ Center
☐ Standard equation of a circle

1 Recognizing Types of Conic Sections

Conic sections are named after the different ways that a plane can intersect a cone. The three basic curves are parabolas, ellipses, and hyperbolas. A circle is a special case of an ellipse. **FIGURE 13.1** shows the three types of conic sections along with an example of the graph in the *xy*-plane associated with each.

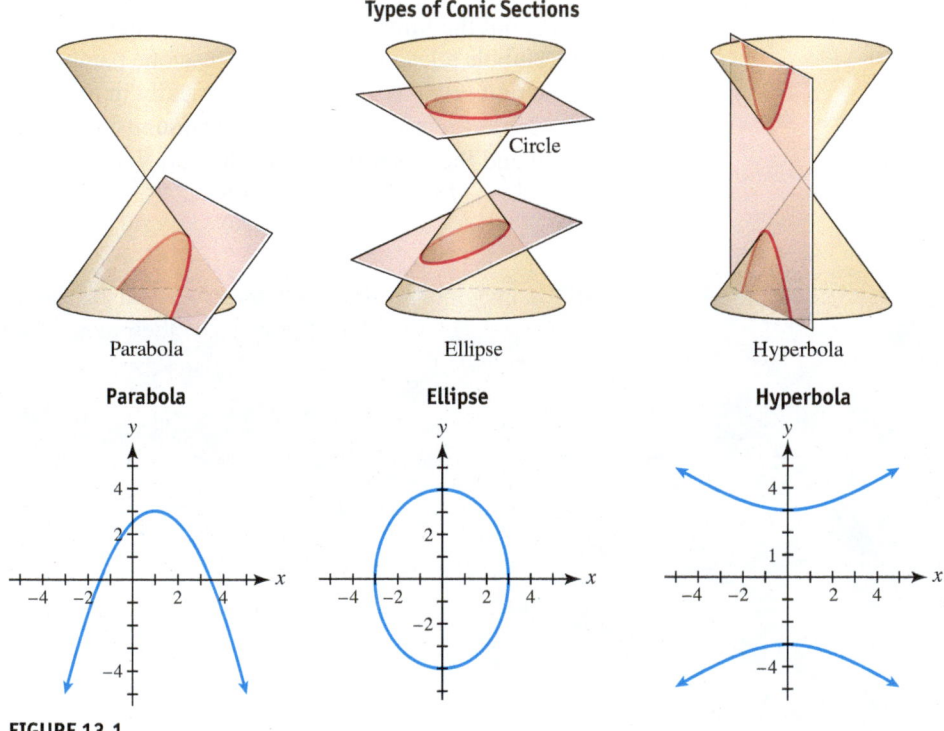

Types of Conic Sections

Parabola   Ellipse   Hyperbola

Parabola   **Ellipse**   **Hyperbola**

FIGURE 13.1

READING CHECK 1

• State the three basic types of conic sections.

2 Graphing Parabolas with Horizontal Axes of Symmetry and Writing Their Equations

Recall that the *vertex form of a parabola* with a vertical axis of symmetry is

$$y = a(x - h)^2 + k, \quad \text{Vertex form}$$

where (h, k) is the vertex. If $a > 0$, the parabola opens upward; if $a < 0$, the parabola opens downward, as shown in See the Concept 1. The preceding vertex form can also be expressed in the general form

$$y = ax^2 + bx + c. \quad \text{General form}$$

In this form, the *x*-coordinate of the vertex is $x = -\frac{b}{2a}$.

See the Concept 1   PARABOLAS WITH A VERTICAL AXIS OF SYMMETRY

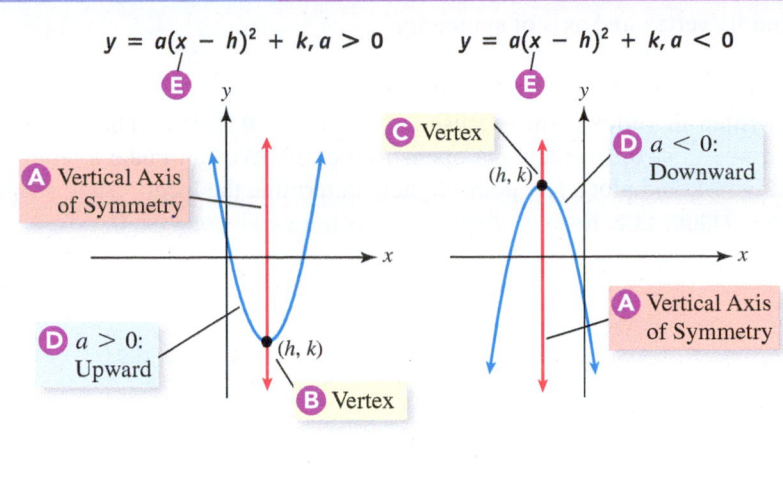

$y = a(x - h)^2 + k, a > 0$

$y = a(x - h)^2 + k, a < 0$

A The graph of this type of parabola *is* a function, with a vertical axis of symmetry.

B The vertex (h, k) is the *lowest* point on a parabola that opens *upward*.

C The vertex (h, k) is the *highest* point on a parabola that opens *downward*.

D Opens upward if $a > 0$, and downward if $a < 0$.

E If the x variable is squared, the axis of symmetry is vertical.

WATCH VIDEO IN [MML].

Interchanging the roles of x and y (and also h and k) gives equations for parabolas that open to the right or the left. In this case, their axes of symmetry are horizontal.

READING CHECK 2

• Give vertex form for a parabola with a horizontal axis of symmetry.

PARABOLAS WITH HORIZONTAL AXES OF SYMMETRY

The graph of $x = a(y - k)^2 + h$ is a parabola that opens to the right if $a > 0$ and to the left if $a < 0$. The vertex of the parabola is located at (h, k).

The graph of $x = ay^2 + by + c$ is a parabola opening to the right if $a > 0$ and to the left if $a < 0$. The y-coordinate of its vertex is given by $y = -\frac{b}{2a}$.

See the Concept 2   PARABOLAS WITH A HORIZONTAL AXIS OF SYMMETRY

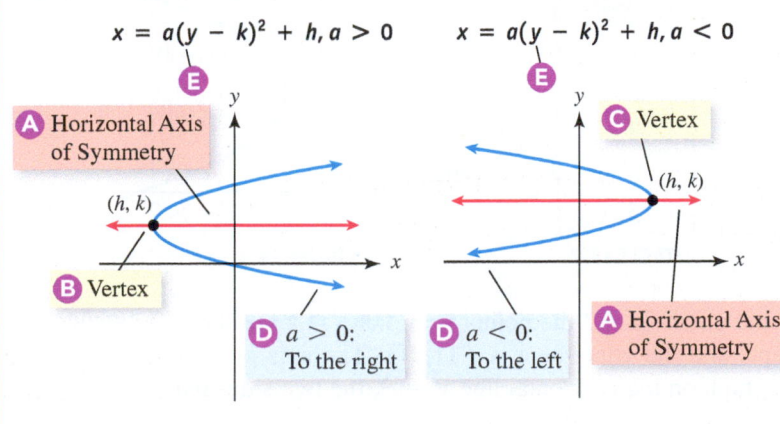

$x = a(y - k)^2 + h, a > 0$

$x = a(y - k)^2 + h, a < 0$

A The graph of this type of parabola *is not* a function. The axis of symmetry is horizontal.

B The vertex (h, k) is the *leftmost* point on a parabola that opens *to the right*.

C The vertex (h, k) is the *rightmost* point on a parabola that opens *to the left*.

D Opens to the right if $a > 0$, and to the left if $a < 0$.

E If the y-variable is squared, the axis of symmetry is horizontal.

WATCH VIDEO IN [MML].

EXAMPLE 1 **Graphing a parabola**

Graph $x = -\frac{1}{2}y^2$. Find its vertex and axis of symmetry.

Solution
The equation can be written in vertex form because $x = -\frac{1}{2}(y - \mathbf{0})^2 + \mathbf{0}$. The vertex is $(\mathbf{0}, \mathbf{0})$, and because $a = -\frac{1}{2} < 0$, the parabola opens to the left. We can make a table of values, as shown in **TABLE 13.1**, and plot a few points to help determine the location and shape of the graph, as shown in **FIGURE 13.2**. Its axis of symmetry is the x-axis, or $y = 0$.

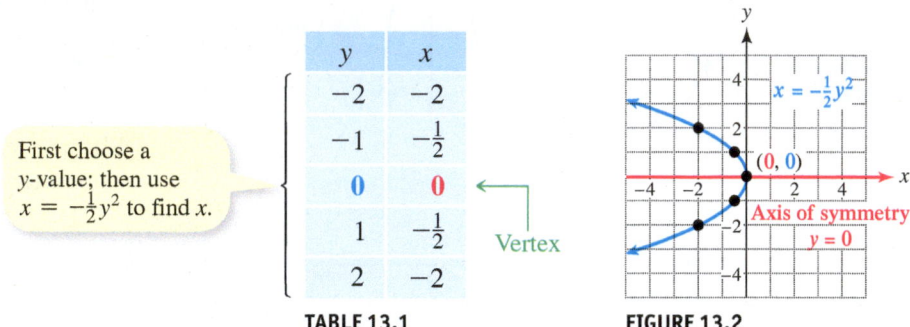

First choose a y-value; then use $x = -\frac{1}{2}y^2$ to find x.

y	x
-2	-2
-1	$-\frac{1}{2}$
$\mathbf{0}$	$\mathbf{0}$
1	$-\frac{1}{2}$
2	-2

Vertex

TABLE 13.1 **FIGURE 13.2**

Now Try Exercise 15

EXAMPLE 2 **Graphing a parabola**

Graph $x = (y - 3)^2 + 2$. Find its vertex and axis of symmetry.

Solution
Because $h = 2$ and $k = 3$ in the equation $x = a(y - k)^2 + h$, the vertex is $(\mathbf{2}, \mathbf{3})$, and because $a = 1 > 0$, the parabola opens to the right. This parabola has the same shape as $y = x^2$, except that it opens to the right rather than upward. To graph this parabola, we can make a table of values and plot a few points, as shown in **TABLE 13.2** and **FIGURE 13.3**.

First choose a y-value; then use $x = (y - 3)^2 + 2$ to find x.

y	x
1	6
2	3
3	2
4	3
5	6

Vertex

TABLE 13.2 **FIGURE 13.3**

Both the graph of the parabola and the points from **TABLE 13.2** are shown in **FIGURE 13.3**. There are no y-intercepts and the x-intercept is $(11, 0)$. The axis of symmetry is $y = 3$ because, if we fold the graph on the horizontal line $y = 3$, the two sides match.

NOTE: Sometimes, finding the x- and y-intercepts of the parabola is helpful when you are graphing. To find the x-intercept, let $y = \mathbf{0}$ in $x = (y - 3)^2 + 2$. The x-coordinate of the x-intercept is $x = (\mathbf{0} - 3)^2 + 2 = 11$. To find any y-intercepts, let $x = \mathbf{0}$ in $x = (y - 3)^2 + 2$. Here, $\mathbf{0} = (y - 3)^2 + 2$ means that $(y - 3)^2 = -2$, which has no real solutions, and thus this parabola has no y-intercepts. ∎

Now Try Exercise 23

EXAMPLE 3 **Graphing a parabola and finding its vertex**

Identify the vertex and then graph each parabola.
(a) $x = -y^2 + 1$ (b) $x = y^2 - 2y - 1$

Solution
(a) If we rewrite $x = -y^2 + 1$ as $x = -(y - 0)^2 + 1$, then $h = 1$ and $k = 0$, so the vertex is $(1, 0)$. By letting $y = 0$ in $x = -y^2 + 1$, we find that the x-coordinate of the x-intercept is $x = -0^2 + 1 = 1$. Similarly, we let $x = 0$ in $x = -y^2 + 1$ to find the y-coordinates of the y-intercepts. The equation $0 = -y^2 + 1$ has solutions -1 and 1.

The parabola opens to the left because $a = -1 < 0$. Additional points given in **TABLE 13.3** help in graphing the parabola shown in **FIGURE 13.4**.

Vertex is $(1, 0)$, so choose y-values on each side of 0.

y	x
-2	-3
-1	0
0	1
1	0
2	-3

TABLE 13.3

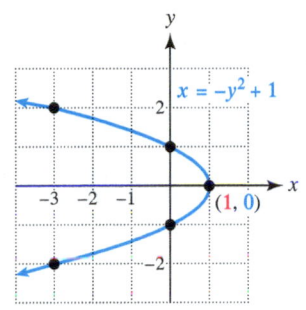

FIGURE 13.4

(b) The y-coordinate of the vertex for the graph of $x = y^2 - 2y - 1$ is given by

$$y = -\frac{b}{2a} = -\frac{-2}{2(1)} = 1.$$

To find the x-coordinate of the vertex, substitute $y = 1$ into $x = y^2 - 2y - 1$.

$$x = (1)^2 - 2(1) - 1 = -2$$

The vertex is $(-2, 1)$. The parabola opens to the right because $a = 1 > 0$. Additional points given in **TABLE 13.4** help in graphing the parabola shown in **FIGURE 13.5**. Note that the y-coordinates of the y-intercepts do not have integer values and that the quadratic formula could be used to find approximations for these ordered pairs. The x-intercept is $(-1, 0)$.

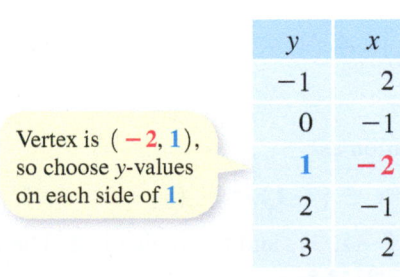

Vertex is $(-2, 1)$, so choose y-values on each side of 1.

y	x
-1	2
0	-1
1	-2
2	-1
3	2

TABLE 13.4

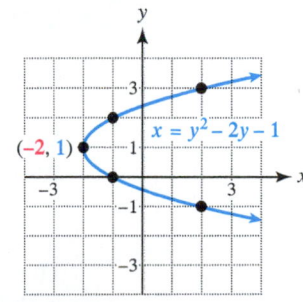

FIGURE 13.5

Now Try Exercises 17, 39

3 Graphing Circles and Writing Their Equations

GRAPHING A CIRCLE A **circle** consists of the set of points in a plane that are the same distance from a fixed point. The fixed distance is called the **radius**, and the fixed point is called the **center**. In **FIGURE 13.6**, all points lying on the blue circle are a distance of 2 units from the center $(2, 1)$. Therefore the radius of the circle equals 2.

Circle

FIGURE 13.6

Circle

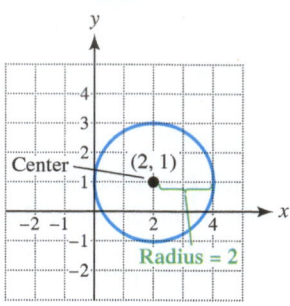

FIGURE 13.6 (repeated)

We can find the equation of the circle shown in **FIGURE 13.6** by using the distance formula. If a point (x, y) lies on the graph of a circle, its distance from the center $(2, 1)$ is 2, and

$$\sqrt{(x - 2)^2 + (y - 1)^2} = 2.$$

$$d = \sqrt{(x_2 - x_1)^2 + (y_2 - y_1)^2}$$

Squaring each side gives

$$(x - 2)^2 + (y - 1)^2 = 2^2.$$

This equation represents the standard equation for a circle with center $(2, 1)$ and radius 2.

STANDARD EQUATION OF A CIRCLE

The **standard equation of a circle** with center (h, k) and radius r is

$$(x - h)^2 + (y - k)^2 = r^2.$$

EXAMPLE 4 **Graphing a circle**

Graph $x^2 + y^2 = 9$. Find the radius and center.

Solution
The equation $x^2 + y^2 = 9$ can be written in standard form as

$$(x - 0)^2 + (y - 0)^2 = 3^2.$$

Therefore the center is $(0, 0)$ and the radius is 3. Its graph is shown in **FIGURE 13.7**.

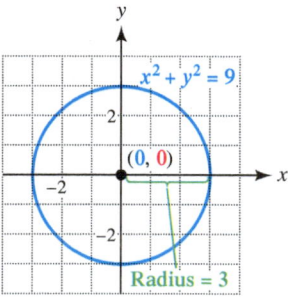

FIGURE 13.7

Now Try Exercise 65

EXAMPLE 5 **Graphing a circle**

Graph $(x + 1)^2 + (y - 3)^2 = 4$. Find the radius and center.

Solution
Write the equation as

$$(x - (-1))^2 + (y - 3)^2 = 2^2.$$

The center is $(-1, 3)$, and the radius is 2. The circle's graph is shown in **FIGURE 13.8**.

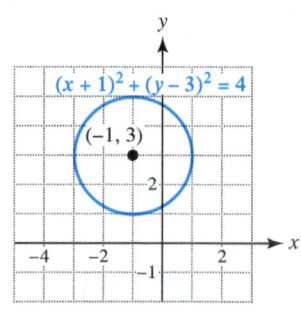

FIGURE 13.8

READING CHECK 3

- Give the center and radius of the circle whose equation is $(x - 3)^2 + (y + 1)^2 = 9$.

Now Try Exercise 69

TECHNOLOGY NOTE

Graphing Circles
The graph of a circle does not represent a function. One way to graph a circle with a graphing calculator is to solve the equation for y and obtain two equations. One equation gives the upper half of the circle, and the other equation gives the lower half.

For example, to graph $x^2 + y^2 = 4$, begin by solving for y.

$$y^2 = 4 - x^2 \qquad \text{Subtract } x^2.$$

$$y = \pm\sqrt{4 - x^2} \qquad \text{Square root property}$$

Then graph $y_1 = \sqrt{4 - x^2}$ and $y_2 = -\sqrt{4 - x^2}$. The graph of y_1 is the upper half of the circle, and the graph of y_2 is the lower half of the circle, as shown in **FIGURE 13.9**.

Upper half	Lower half	Both halves
$[-4.7, 4.7, 1]$ by $[-3.1, 3.1, 1]$	$[-4.7, 4.7, 1]$ by $[-3.1, 3.1, 1]$	$[-4.7, 4.7, 1]$ by $[-3.1, 3.1, 1]$

(a)   (b)   (c)

FIGURE 13.9

$[-4, 4, 1]$ by $[-5, 5, 1]$

FIGURE 13.10

NOTE: If a circle is not graphed in a *square viewing rectangle*, it will appear to be an oval rather than a circle. In a square viewing rectangle, a circle will appear circular. **FIGURE 13.10** shows the circle graphed in a viewing rectangle that is *not* square. ∎

COMPLETING THE SQUARE TO FIND THE CENTER AND RADIUS In the next example, we use the *method of completing the square* to find the center and radius of a circle.

EXAMPLE 6   **Finding the center of a circle**

Find the center and radius of the circle given by $x^2 + 4x + y^2 - 6y = 5$.

Solution
Begin by writing the equation as

$$(x^2 + 4x + \underline{\hspace{1em}}) + (y^2 - 6y + \underline{\hspace{1em}}) = 5.$$

To complete the square, add $\left(\frac{4}{2}\right)^2 = 4$ and $\left(\frac{-6}{2}\right)^2 = 9$ to each side of the equation.

$$(x^2 + 4x + 4) + (y^2 - 6y + 9) = 5 + 4 + 9$$

Factoring each perfect square trinomial yields

$$(x + 2)^2 + (y - 3)^2 = 18.$$

The center is $(-2, 3)$, and because $18 = \left(\sqrt{18}\right)^2$, the radius is $\sqrt{18}$, or $3\sqrt{2}$.

Now Try Exercise 71

CRITICAL THINKING 1

Does the following equation represent a circle? If so, give its center and radius.

$$x^2 + y^2 + 10y = -32$$

13.1 Putting It All Together

CONCEPT	EXPLANATION	EXAMPLE
Parabola with Horizontal Axis	Vertex form: $x = a(y - k)^2 + h$ If $a > 0$, it opens to the right. If $a < 0$, it opens to the left. The vertex is (h, k). These parabolas may also be expressed as $x = ay^2 + by + c$, where the y-coordinate of the vertex is $y = -\frac{b}{2a}$.	$x = 2(y - 1)^2 + 4$ opens to the right and its vertex is $(4, 1)$. $a > 0$, so opens right.
Standard Equation of a Circle	$(x - h)^2 + (y - k)^2 = r^2$ The radius is r and the center is (h, k).	$(x + 2)^2 + (y - 1)^2 = 16$ has center $(-2, 1)$ and radius 4.

13.1 Exercises

MyMathLab®

CONCEPTS AND VOCABULARY

1. Name the three general types of conic sections.

2. What is the difference between the parabolas given by $y = ax^2 + bx + c$ and $x = ay^2 + by + c$?

3. If a parabola has a horizontal axis of symmetry, does it represent a function?

4. Sketch a graph of a parabola with a horizontal axis of symmetry.

5. If a parabola has two y-intercepts, does it represent a function? Why or why not?

6. If $x = a(y - k)^2 + h$, what is the vertex?

7. The graph of $x = -y^2$ opens to the _____.

8. The graph of $x = 2y^2 + y - 1$ opens to the _____.

9. The graph of $(x - h)^2 + (y - k)^2 = r^2$ is a(n) _____ with center _____.

10. The graph of $x^2 + y^2 = r^2$ is a circle with center _____ and radius _____.

11. Which of the following (a–d) are the coordinates of the vertex for the parabola given by $x = 4(y + 2)^2 - 3$?
    a. $(-3, 4)$ **b.** $(3, -2)$
    c. $(-3, 2)$ **d.** $(-3, -2)$

12. Which of the following (a–d) is the equation of the axis of symmetry for the parabola given by $x = -5(y - 1)^2 + 7$?
    a. $y = 1$ **b.** $x = -5$
    c. $x = 7$ **d.** $y = 7$

13. Which of the following (a–d) is the center and radius of the circle given by $x^2 + (y - 2)^2 = 9$?
    a. $(0, -2), r = 3$
    b. $(2, 0), r = 3$
    c. $(0, 2), r = 3$
    d. $(0, 2), r = 9$

14. Which of the following (a–d) is the y-coordinate of the vertex of the parabola given by $x = y^2 - 2y + 3$?
    a. -2 **b.** 1
    c. 0   **d.** 3

GRAPHING PARABOLAS AND WRITING THEIR EQUATIONS

Exercises 15–40: Graph the parabola. Find the vertex and axis of symmetry.

15. $x = y^2$

16. $x = -y^2$

17. $x = y^2 + 1$

18. $x = y^2 - 1$

19. $y = x^2 - 1$

20. $y = 2x^2$

21. $x = 2y^2$

22. $x = \frac{1}{4}y^2$

23. $x = (y - 1)^2 + 2$

24. $x = (y - 2)^2 + 1$

25. $y = (x + 2)^2 + 1$

26. $y = (x - 4)^2 + 5$

27. $y = -2(x + 2)^2$

28. $y = (x - 1)^2 - 2$

29. $x = \frac{1}{2}(y + 1)^2 - 3$

30. $x = -2(y + 3)^2 + 1$

31. $x = -3(y - 1)^2$

32. $x = \frac{1}{4}(y + 2)^2 - 3$

33. $y = 2x^2 - x + 1$

34. $y = -x^2 + 2x + 2$

35. $x = -2y^2 + 3y + 2$

36. $x = \frac{1}{2}y^2 + y - 1$

37. $x = 3y^2 + y$

38. $x = -\frac{3}{2}y^2 - 2y + 1$

39. $x = y^2 + 2y + 1$

40. $x = y^2 - 3y - 4$

Exercises 41–44: Use the graph to determine the equation of the parabola. (Hint: Either $a = 1$ or $a = -1$.)

41.

42.

43.

44.

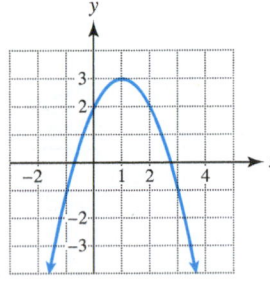

Exercises 45–48: Determine the direction that the parabola opens if it satisfies the given conditions.

45. Passing through $(2, 0)$, $(-2, 0)$, and $(0, -2)$

46. Passing through $(0, -3)$, $(0, 2)$, and $(1, 1)$

47. Vertex $(1, 2)$ passing through the point $(-1, -2)$ with a vertical axis

48. Vertex $(-1, 3)$ passing through the point $(0, 0)$ with a horizontal axis

49. What x-values are possible for the graph of the equation $x = 2y^2$?

50. What y-values are possible for the graph of the equation $x = 2y^2$?

51. **Thinking Generally** How many y-intercepts does a parabola given by

$$x = a(y - k)^2 + h$$

have if $a > 0$ and $h < 0$?

52. **Thinking Generally** Does the graph of the equation $x = ay^2 + by + c$ always have a y-intercept? Explain.

53. What is the x-intercept for the graph of the equation given by

$$x = 3y^2 - y + 1?$$

54. What are the y-intercepts for the graph of the equation given by

$$x = y^2 - 3y + 2?$$

GRAPHING CIRCLES AND WRITING THEIR EQUATIONS

Exercises 55–60: Write the standard equation of the circle with the given radius r and center C.

55. $r = 1$ $C = (0, 0)$

56. $r = 4$ $C = (2, 3)$

57. $r = 3$ $C = (-1, 5)$

58. $r = 5$ $C = (5, -3)$

59. $r = \sqrt{2}$ $C = (-4, -6)$

60. $r = \sqrt{6}$ $C = (0, 4)$

Exercises 61–64: Use the graph to find the standard equation of the circle.

61.

62.

63.

64.

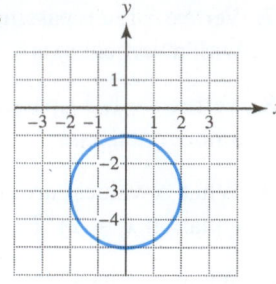

Exercises 65–74: Find the radius and center of the circle. Then graph the circle.

65. $x^2 + y^2 = 4$ **66.** $x^2 + y^2 = 1$

67. $(x - 1)^2 + (y - 3)^2 = 9$

68. $(x + 2)^2 + (y + 1)^2 = 4$

69. $(x + 5)^2 + (y - 5)^2 = 25$

70. $(x - 4)^2 + (y + 3)^2 = 16$

71. $x^2 + 6x + y^2 - 2y = -1$

72. $x^2 + y^2 + 12y + 32 = 0$

73. $x^2 + 6x + y^2 - 2y + 3 = 0$

74. $x^2 - 4x + y^2 + 4y = -3$

Exercises 75–78: Use the endpoints of the diameter to find the standard equation of the circle.

75. $(-3, 4), (5, 6)$ **76.** $(-1, 2), (3, -2)$

77. $(2, 1), (5, 5)$ **78.** $(3, -2), (1, -4)$

APPLICATIONS OF PARABOLAS

79. Radio Telescopes The Parks radio telescope has the shape of a parabolic dish, as depicted in the figure below. A cross section of this telescope can be modeled by $x = \frac{32}{11,025}y^2$, where $-105 \le y \le 105$; the units are feet. (*Source:* J. Mar, *Structure Technology for Large Radar and Telescope Systems.*)

(a) Graph the cross-sectional shape of the dish in $[-40, 40, 10]$ by $[-120, 120, 20]$.

(b) Find the depth d of the dish.

80. Train Tracks To make a curve safer for trains, parabolic curves are sometimes used instead of circular curves. See the accompanying figures. (*Source:* F. Mannering and W. Kilareski, *Principles of Highway Engineering and Traffic Analysis.*)

(a) Suppose that a curve must pass through the points $(-1, 0)$, $(0, 3)$, and $(0, -3)$, where the units are kilometers. Find an equation for the train tracks in the form

$$x = a(y - h)^2 + k.$$

(b) Find another point that lies on the train tracks.

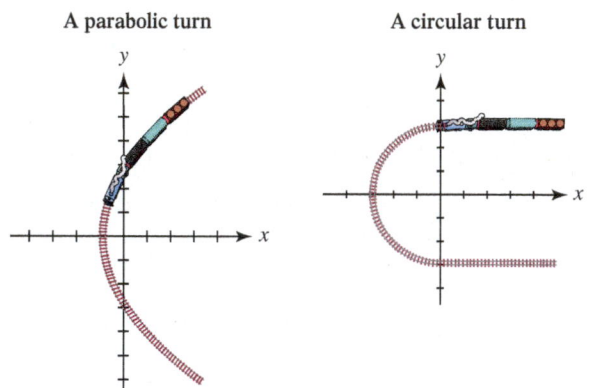

A parabolic turn   A circular turn

81. Trajectories of Comets Under certain circumstances, a comet can pass by the sun once and never return. In this situation, the comet may travel in a parabolic path, as illustrated in the figure below. Suppose that a comet's path is given by $x = -2.5y^2$, where the sun is located at $(-0.1, 0)$ and the units are astronomical units (A.U.). One astronomical unit equals 93 million miles. (*Source:* W. Thomson, *Introduction to Space Dynamics.*)

(a) Plot a point for the sun's location and then graph the path of the comet.

(b) Find the distance from the sun to the comet when the comet is located at $(-2.5, 1)$.

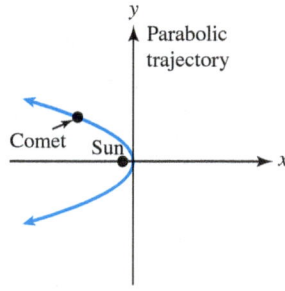

82. Speed of a Comet (Continuation of Exercise 81.) The velocity V in meters per second of a comet traveling in a parabolic trajectory around the sun is given by $V = \frac{k}{\sqrt{D}}$, where D is the comet's distance from the sun in meters and $k = 1.15 \times 10^{10}$.

(a) How does the velocity of the comet change as its distance from the sun changes?

(b) Calculate the velocity of the comet when it is closest to the sun. (*Hint:* 1 mile ≈ 1609 meters.)

WRITING ABOUT MATHEMATICS

83. Suppose that you are given the equation

$$x = a(y - k)^2 + h.$$

(a) Explain how you can determine the direction that the parabola opens.

(b) Explain how to find the axis of symmetry and the vertex.

(c) If the points $(0, 4)$ and $(0, -2)$ lie on the graph of x, what is the axis of symmetry?

(d) Generalize part (c) if $(0, y_1)$ and $(0, y_2)$ lie on the graph of x.

84. Suppose that you are given the vertex of a parabola. Can you determine the axis of symmetry? Explain.

Group Activity

Radio Telescope The U.S. Naval Research Laboratory designed a giant radio telescope weighing 3450 tons. Its parabolic dish has a diameter of 300 feet and a depth of 44 feet, as shown in the figure to the right.

(*Source:* J. Mar, *Structure Technology for Large Radio and Radar Telescope Systems.*)

(a) Determine an equation of the form $x = ay^2, a > 0$, that models a cross section of the dish.

(b) Graph your equation in an appropriate viewing rectangle.

13.2 Ellipses and Hyperbolas

Objectives

1 Graphing Ellipses and Writing Their Equations

2 Graphing Hyperbolas and Writing Their Equations

FIGURE 13.11

1 Graphing Ellipses and Writing Their Equations

One method used to sketch an ellipse is to tie the ends of a string to two nails driven into a flat board. If a pencil is placed against the string anywhere between the nails, as shown in **FIGURE 13.11**, and is used to draw a curve, the resulting curve is an ellipse. The sum of the distances d_1 and d_2 between the pencil and each of the nails is always fixed by the length of the string. The location of the nails corresponds to the *foci* of the ellipse. An **ellipse** is the set of points in a plane the sum of whose distances from two fixed points is constant. Each fixed point is called a **focus** (plural, foci) of the ellipse.

CRITICAL THINKING 1

What happens to the shape of the ellipse shown in **FIGURE 13.11** as the nails are moved farther apart? What happens to its shape as the nails are moved closer together? When would a circle be formed?

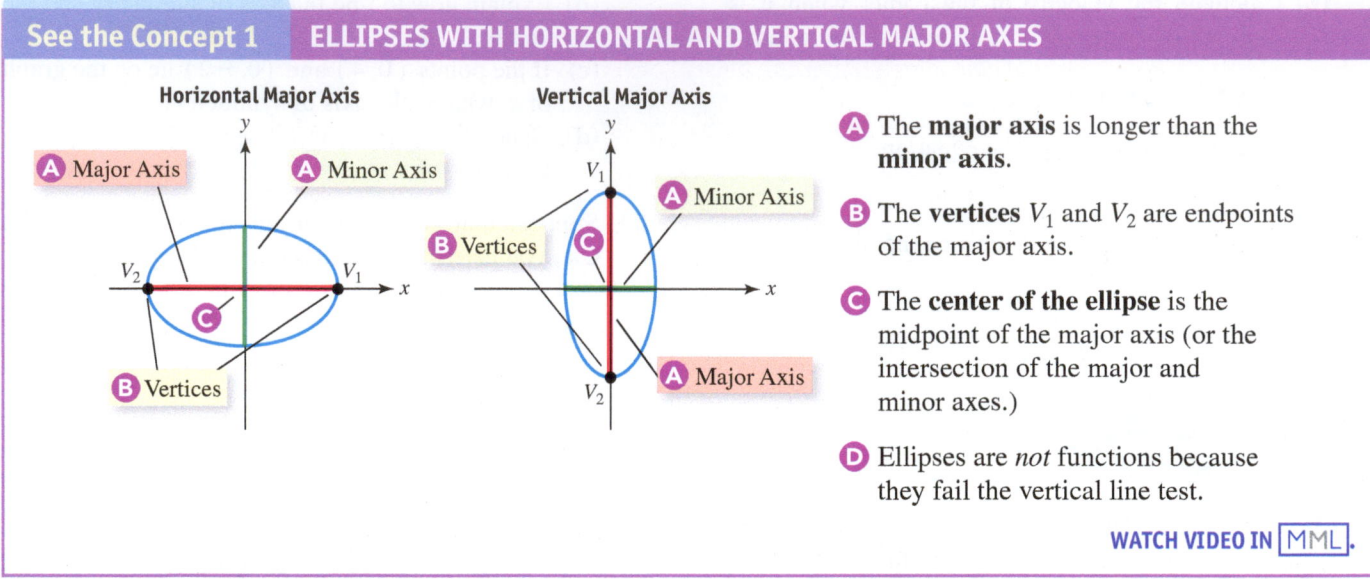

See the Concept 1   ELLIPSES WITH HORIZONTAL AND VERTICAL MAJOR AXES

Ⓐ The **major axis** is longer than the **minor axis**.

Ⓑ The **vertices** V_1 and V_2 are endpoints of the major axis.

Ⓒ The **center of the ellipse** is the midpoint of the major axis (or the intersection of the major and minor axes.)

Ⓓ Ellipses are *not* functions because they fail the vertical line test.

WATCH VIDEO IN $\boxed{\text{MML}}$.

NEW VOCABULARY

- ☐ Ellipse
- ☐ Focus of an ellipse
- ☐ Major/minor axis
- ☐ Vertices of an ellipse
- ☐ Center of an ellipse
- ☐ Hyperbola
- ☐ Focus of a hyperbola
- ☐ Branches
- ☐ Vertices of a hyperbola
- ☐ Transverse axis
- ☐ Fundamental rectangle
- ☐ Asymptotes

Some ellipses can be represented by the following equations.

STANDARD EQUATIONS FOR ELLIPSES CENTERED AT (0, 0)

The ellipse with center at the origin, *horizontal* major axis, and equation

$$\frac{x^2}{a^2} + \frac{y^2}{b^2} = 1, \qquad a > b > 0,$$

has vertices $(\pm a, 0)$ and endpoints of the minor axis $(0, \pm b)$.

The ellipse with center at the origin, *vertical* major axis, and equation

$$\frac{x^2}{b^2} + \frac{y^2}{a^2} = 1, \qquad a > b > 0,$$

has vertices $(0, \pm a)$ and endpoints of the minor axis $(\pm b, 0)$.

FIGURE 13.12(a) shows an ellipse that has a horizontal major axis; **FIGURE 13.12(b)** shows one that has a vertical major axis. The coordinates of the vertices V_1 and V_2 and endpoints of the minor axis U_1 and U_2 are labeled.

READING CHECK 1

- How can you determine from the standard equation of an ellipse if the major axis is horizontal or vertical?

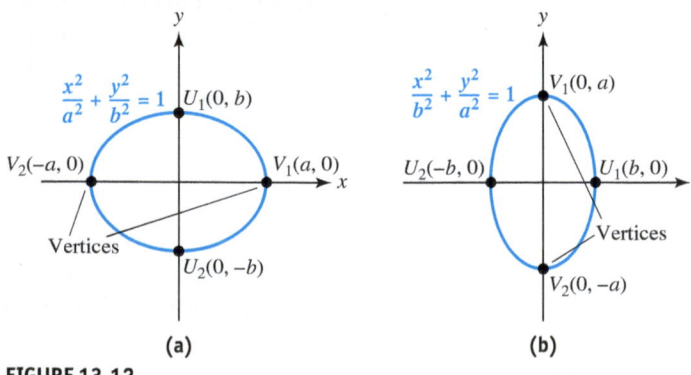

FIGURE 13.12

In the next example, we show how to sketch graphs of ellipses.

EXAMPLE 1   **Sketching ellipses**

Sketch a graph of each ellipse. Label the vertices and endpoints of the minor axes.

(a) $\dfrac{x^2}{25} + \dfrac{y^2}{4} = 1$ (b) $9x^2 + 4y^2 = 36$

Solution

(a) The equation $\dfrac{x^2}{25} + \dfrac{y^2}{4} = 1$ describes an ellipse with $a^2 = 25$ and $b^2 = 4$. (When you are deciding whether 25 or 4 represents a^2, let a^2 be the larger of the two numbers.) Thus $a = 5$ and $b = 2$, so the ellipse has a horizontal major axis with vertices $(\pm 5, 0)$ and the endpoints of the minor axis are $(0, \pm 2)$. Plot these four points and then sketch the ellipse, as shown in **FIGURE 13.13(a)**.

(b) To write $9x^2 + 4y^2 = 36$ as a standard equation, divide each side by 36.

$$9x^2 + 4y^2 = 36 \quad \text{Given equation}$$

$$\frac{9x^2}{36} + \frac{4y^2}{36} = \frac{36}{36} \quad \text{Divide each side by 36.}$$

$$\frac{x^2}{4} + \frac{y^2}{9} = 1 \quad \text{Simplify.}$$

This ellipse has a vertical major axis with $a = 3$ and $b = 2$. The vertices are $(0, \pm 3)$, and the endpoints of the minor axis are $(\pm 2, 0)$, as shown in **FIGURE 13.13(b)**.

CRITICAL THINKING 2

Suppose that $a = b$ for an ellipse centered at (0, 0). What can be said about the ellipse? Explain.

(a)

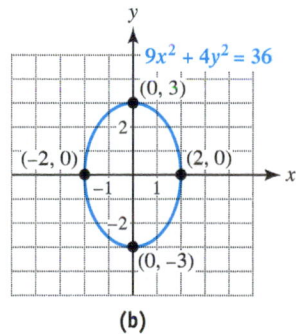

(b)

FIGURE 13.13

Now Try Exercises 15, 21

EXAMPLE 2   **Finding the standard equation of an ellipse**

Use the graph in **FIGURE 13.14** to determine the standard equation of the ellipse.

FIGURE 13.14

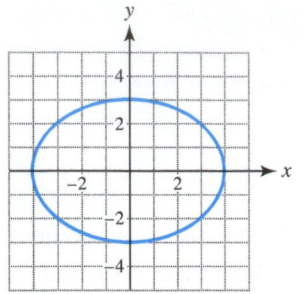

FIGURE 13.14 (repeated)

Solution

The ellipse in **FIGURE 13.14** is centered at $(0, 0)$ with a horizontal major axis. The length of the major axis is 8, so $a = 4$. The length of the minor axis is 6, so $b = 3$. Thus $a^2 = \textbf{16}$ and $b^2 = \textbf{9}$, and the standard equation of the ellipse is

$$\frac{x^2}{16} + \frac{y^2}{9} = 1.$$

Now Try Exercise 25

Astronomy
🌐 *Math in Context* Planets travel around the sun in elliptical orbits. Astronomers have measured the values of a and b for each planet. Using this information, we can find the equation of a planet's orbit, as illustrated in the next example.

EXAMPLE 3 | **Modeling the orbit of Mercury**

Mercury

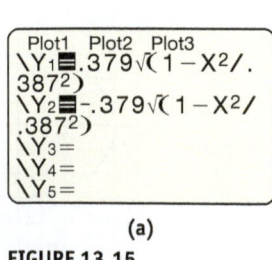

Not to Scale

The planet Mercury has one of the least circular orbits of the eight major planets, as illustrated in the margin. For Mercury, $a = 0.387$ and $b = 0.379$. The units are astronomical units (A.U.), where 1 A.U. equals 93 million miles—the distance between Earth and the sun. Graph $\frac{x^2}{a^2} + \frac{y^2}{b^2} = 1$ to model the orbit of Mercury in $[-0.6, 0.6, 0.1]$ by $[-0.4, 0.4, 0.1]$. Then plot the sun at the point $(0.08, 0)$. (*Source:* M. Zeilik, *Introductory Astronomy and Astrophysics.*)

Solution

The orbit of Mercury is given by

$$\frac{x^2}{0.387^2} + \frac{y^2}{0.379^2} = 1. \qquad \frac{x^2}{a^2} + \frac{y^2}{b^2} = 1$$

To graph an ellipse with some graphing calculators, we must solve the equation for y. Doing so results in two equations.

$$\frac{x^2}{0.387^2} + \frac{y^2}{0.379^2} = 1$$

$$\frac{y^2}{0.379^2} = 1 - \frac{x^2}{0.387^2} \qquad \text{Subtract } \frac{x^2}{0.387^2}.$$

$$\frac{y}{0.379} = \pm\sqrt{1 - \frac{x^2}{0.387^2}} \qquad \text{Square root property}$$

$$y = \pm 0.379\sqrt{1 - \frac{x^2}{0.387^2}} \qquad \text{Multiply by } 0.379.$$

The orbit of Mercury can be found by graphing these equations. See **FIGURES 13.15(a)** and **13.15(b)**. The point $(0.08, 0)$ represents the position of the sun in **FIGURE 13.15(b)**.

CRITICAL THINKING 3

Use **FIGURE 13.15** and the information in Example 3 to estimate the minimum and maximum distances that Mercury is from the sun.

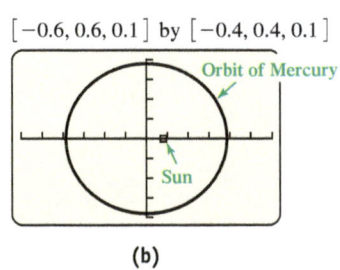

(a)

(b)

FIGURE 13.15

Now Try Exercise 49

2 Graphing Hyperbolas and Writing Their Equations

The third type of conic section is the **hyperbola**, which is the set of points in a plane the difference of whose distances from two fixed points is constant. Each fixed point is called a **focus** of the hyperbola.

See the Concept 2   **HYPERBOLAS WITH HORIZONTAL AND VERTICAL TRANSVERSE AXES**

A The **transverse axis** connects the two vertices and can be either horizontal or vertical.

B The **vertices** are the endpoints of the transverse axis.

C The **center of the hyperbola** is the midpoint of the transverse axis.

D A hyperbola has **left and right branches** if the transverse axis is horizontal and **lower and upper branches** if it is vertical.

WATCH VIDEO IN MML.

READING CHECK 2

• How can you determine from the standard equation of a hyperbola if the transverse axis is horizontal or vertical?

Many hyperbolas can be described by the following equations.

STANDARD EQUATIONS FOR HYPERBOLAS CENTERED AT (0, 0)

The hyperbola with center at the origin, *horizontal* transverse axis, and equation

$$\frac{x^2}{a^2} - \frac{y^2}{b^2} = 1$$

has vertices $(\pm a, 0)$

The hyperbola with center at the origin, *vertical* transverse axis, and equation

$$\frac{y^2}{a^2} - \frac{x^2}{b^2} = 1$$

has vertices $(0, \pm a)$.

Hyperbolas, along with the coordinates of their vertices, are shown in **FIGURE 13.16** on the next page. The two parts of the hyperbola in **FIGURE 13.16(a)** are the *left branch* and *right branch*, whereas in **FIGURE 13.16(b)**, the hyperbola has an *upper branch* and a *lower branch*. The dashed rectangle in each figure is called the **fundamental rectangle**, and its four vertices (corners) are determined by either $(\pm a, \pm b)$ or $(\pm b, \pm a)$. If its diagonals are extended, they correspond to the asymptotes of the hyperbola. The dashed lines $y = \pm \frac{b}{a}x$ and $y = \pm \frac{a}{b}x$ are **asymptotes** for the hyperbolas.

Horizontal Transverse Axis   **Vertical Transverse Axis**

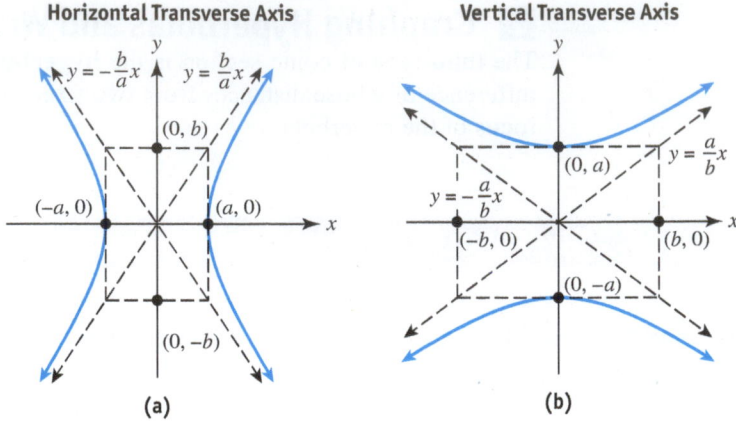

(a)   **(b)**

FIGURE 13.16 Not Functions

NOTE: A hyperbola consists of two solid (blue) curves, or branches. The dashed lines and rectangles are not part of the graph but are used as an aid for sketching the hyperbola. ∎

🌐 *Math in Context* ᴬˢᵗʳᵒⁿᵒᵐʸ One interpretation of an asymptote of a hyperbola can be based on trajectories of comets as they approach the sun. Comets travel in parabolic, elliptic, or hyperbolic trajectories. See **FIGURE 13.17**.

Elliptic Orbit   **Hyperbolic Path**   **Parabolic Path**

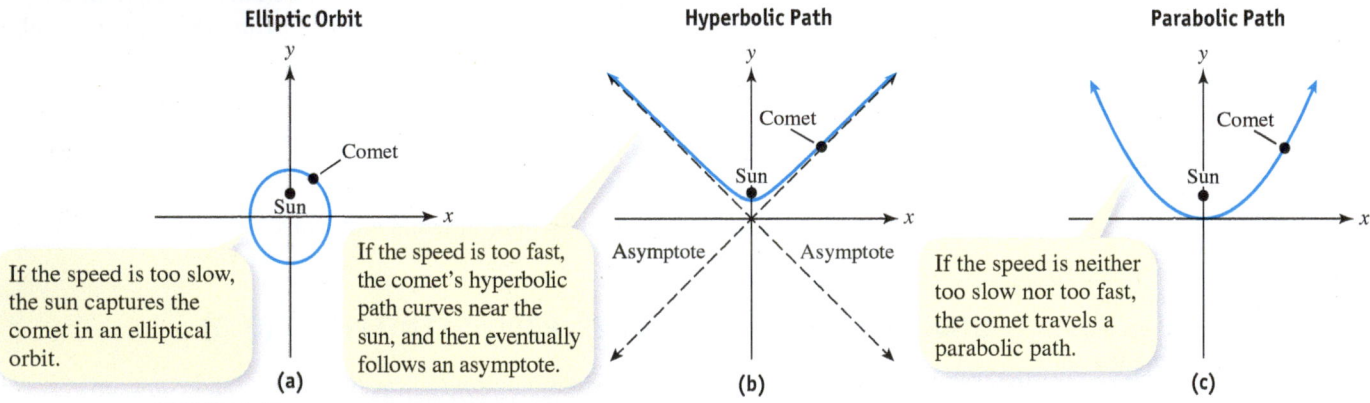

If the speed is too slow, the sun captures the comet in an elliptical orbit.

If the speed is too fast, the comet's hyperbolic path curves near the sun, and then eventually follows an asymptote.

If the speed is neither too slow nor too fast, the comet travels a parabolic path.

(a)   **(b)**   **(c)**

FIGURE 13.17

EXAMPLE 4 | **Sketching a hyperbola**

Sketch a graph of $\frac{y^2}{4} - \frac{x^2}{9} = 1$. Label the vertices and show the asymptotes.

Solution

The equation is in standard form with $a^2 = 4$ and $b^2 = 9$, so $a = 2$ and $b = 3$. It has a vertical transverse axis with vertices $(0, -2)$ and $(0, 2)$. The vertices (corners) of the fundamental rectangle are $(\pm 3, \pm 2)$, that is, $(3, 2)$, $(3, -2)$, $(-3, 2)$, and $(-3, -2)$. The asymptotes are the diagonals of this rectangle and are given by $y = \pm \frac{a}{b}x$, or $y = \pm \frac{2}{3}x$. **FIGURE 13.18** shows the hyperbola and these features. The hyperbola has upper and lower branches because the transverse axis is vertical.

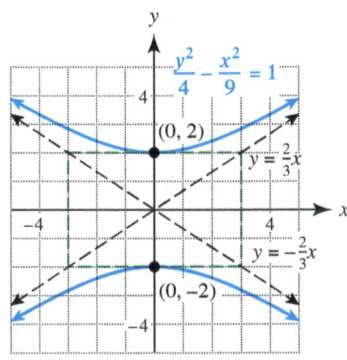

FIGURE 13.18

Now Try Exercise 29

TECHNOLOGY NOTE

Graphing a Hyperbola

The graph of a hyperbola does not represent a function. One way to graph a hyperbola with a graphing calculator is to solve the equation for y and obtain two equations. One equation gives the upper portion of the hyperbola and the other equation gives the lower portion.

For example, to graph $\frac{y^2}{4} - \frac{x^2}{8} = 1$, begin by solving for y.

$$\frac{y^2}{4} = 1 + \frac{x^2}{8} \qquad \text{Add } \tfrac{x^2}{8}.$$

$$y^2 = 4\left(1 + \frac{x^2}{8}\right) \qquad \text{Multiply by 4.}$$

$$y = \pm 2\sqrt{1 + \frac{x^2}{8}} \qquad \text{Square root property}$$

Graph $y_1 = 2\sqrt{1 + x^2/8}$ and $y_2 = -2\sqrt{1 + x^2/8}$, which give the upper and lower branches, respectively. See **FIGURE 13.19**.

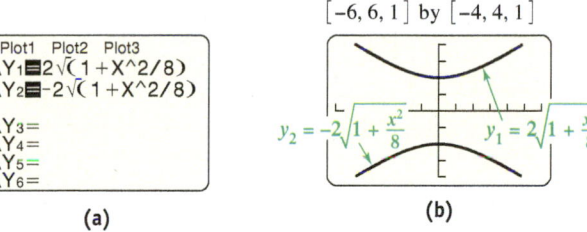

(a)   (b)

FIGURE 13.19

EXAMPLE 5   ## Determining the standard equation of a hyperbola

Use the graph in **FIGURE 13.20** to determine the standard equation of the hyperbola.

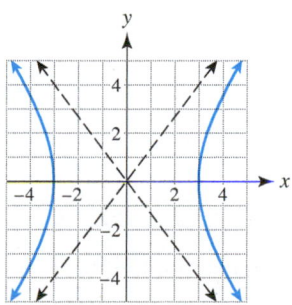

FIGURE 13.20

Solution

The hyperbola has a horizontal transverse axis, so the x^2-term must come first in the equation. The vertices of the hyperbola are $(\pm 3, 0)$, which indicates that $a = 3$ and $a^2 = 9$. The value of b can be found by noting that one of the asymptotes passes through the point $(3, 4)$. This asymptote has the equation $y = \frac{b}{a}x$ or $y = \frac{4}{3}x$, so let $b = 4$ and $b^2 = 16$. The equation of the hyperbola is

$$\frac{x^2}{9} - \frac{y^2}{16} = 1.$$

Now Try Exercise 41

13.2   Putting It All Together

CONCEPT	DESCRIPTION
Ellipses Centered at $(0, 0)$ with $a > b > 0$	**Horizontal Major Axis** Vertices: $(a, 0)$ and $(-a, 0)$ Endpoints of minor axis: $(0, b)$ and $(0, -b)$ $$\frac{x^2}{a^2} + \frac{y^2}{b^2} = 1$$ 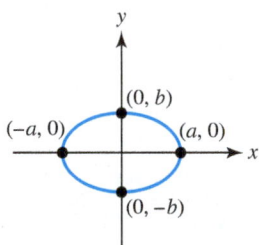 **Vertical Major Axis** Vertices: $(0, a)$ and $(0, -a)$ Endpoints of minor axis: $(-b, 0)$ and $(b, 0)$ $$\frac{x^2}{b^2} + \frac{y^2}{a^2} = 1$$ 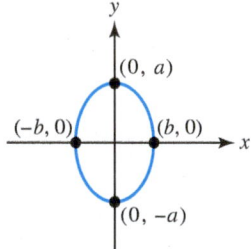
Hyperbolas Centered at $(0, 0)$ with $a > 0$ and $b > 0$	**Horizontal Transverse Axis** Vertices: $(a, 0)$ and $(-a, 0)$ Asymptotes: $y = \pm\dfrac{b}{a}x$ $$\frac{x^2}{a^2} - \frac{y^2}{b^2} = 1$$ 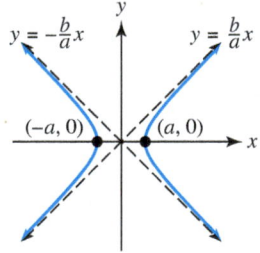 **Vertical Transverse Axis** Vertices: $(0, a)$ and $(0, -a)$ Asymptotes: $y = \pm\dfrac{a}{b}x$ $$\frac{y^2}{a^2} - \frac{x^2}{b^2} = 1$$ 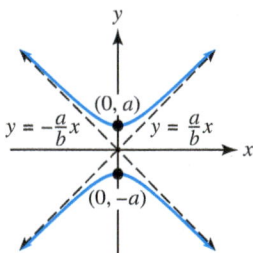

13.2 Exercises

MyMathLab®

CONCEPTS AND VOCABULARY

1. Sketch an ellipse with a horizontal major axis.

2. Sketch a hyperbola with a vertical transverse axis.

3. The ellipse whose standard equation is $\frac{x^2}{a^2} + \frac{y^2}{b^2} = 1$, $a > b > 0$, has a(n) _____ major axis.

4. The ellipse whose standard equation is $\frac{x^2}{b^2} + \frac{y^2}{a^2} = 1$, $a > b > 0$, has a(n) _____ major axis.

5. What is the maximum number of times that a line can intersect an ellipse?

6. What is the maximum number of times that a parabola can intersect an ellipse?

7. The hyperbola whose equation is $\frac{x^2}{a^2} - \frac{y^2}{b^2} = 1$ has _____ and _____ branches.

8. The hyperbola whose equation is $\frac{y^2}{a^2} - \frac{x^2}{b^2} = 1$ has _____ and _____ branches.

9. How are the asymptotes of a hyperbola related to the fundamental rectangle?

10. Could an ellipse be centered at the origin and have vertices $(4, 0)$ and $(0, -5)$?

11. Which of the following (a–d) represents the coordinates of the vertices on the ellipse $\frac{x^2}{4} + \frac{y^2}{9} = 1$?
    a. $(0, \pm 3)$ **b.** $(\pm 2, 0)$
    c. $(\pm 4, 0)$ **d.** $(0, \pm 9)$

12. Which of the following (a–d) represents the coordinates of the vertices on the hyperbola $\frac{x^2}{4} - \frac{y^2}{9} = 1$?
    a. $(0, \pm 3)$ **b.** $(\pm 2, 0)$
    c. $(\pm 4, 0)$ **d.** $(0, \pm 9)$

ELLIPSES AND THIER EQUATIONS

Exercises 13–24: Graph the ellipse. Label the vertices and endpoints of the minor axis.

13. $\frac{x^2}{9} + \frac{y^2}{25} = 1$ 14. $\frac{y^2}{9} + \frac{x^2}{25} = 1$

15. $\frac{x^2}{9} + \frac{y^2}{4} = 1$ 16. $\frac{x^2}{3} + \frac{y^2}{9} = 1$

17. $x^2 + \frac{y^2}{4} = 1$ 18. $\frac{x^2}{9} + y^2 = 1$

19. $\frac{y^2}{5} + \frac{x^2}{7} = 1$ 20. $\frac{y^2}{11} + \frac{x^2}{6} = 1$

21. $36x^2 + 4y^2 = 144$ 22. $25x^2 + 16y^2 = 400$

23. $6y^2 + 7x^2 = 42$ 24. $9x^2 + 5y^2 = 45$

Exercises 25–28: Use the graph to determine the standard equation of the ellipse.

25.

26.

27.

28.
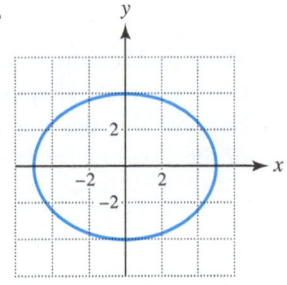

HYPERBOLAS AND THEIR EQUATIONS

Exercises 29–40: Graph the hyperbola. Show the asymptotes and vertices.

29. $\frac{x^2}{4} - \frac{y^2}{9} = 1$ 30. $\frac{y^2}{4} - \frac{x^2}{9} = 1$

31. $\frac{y^2}{25} - \frac{x^2}{16} = 1$ 32. $\frac{x^2}{25} - \frac{y^2}{16} = 1$

33. $x^2 - y^2 = 1$ 34. $y^2 - x^2 = 1$

35. $\frac{x^2}{3} - \frac{y^2}{4} = 1$ 36. $\frac{y^2}{5} - \frac{x^2}{8} = 1$

37. $9y^2 - 4x^2 = 36$ 38. $36x^2 - 25y^2 = 900$

39. $16x^2 - 4y^2 = 64$ 40. $y^2 - 9x^2 = 9$

Exercises 41–44: Use the graph to determine the standard equation of the hyperbola.

41.

42.

43.

44.
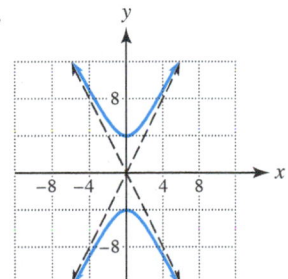

TRANSLATIONS OF ELLIPSES AND HYPERBOLAS

Exercises 45 and 46: Ellipses and hyperbolas can be translated so that they are centered at a point (h, k), rather than at the origin. These techniques are the same as those used for parabolas and circles. To translate a conic section so that it is centered at (h, k) rather than at $(0, 0)$, replace x with $(x - h)$ and replace y with $(y - k)$. For example, to center $\frac{x^2}{9} + \frac{y^2}{4} = 1$ at $(-1, 2)$, change its equation to $\frac{(x + 1)^2}{9} + \frac{(y - 2)^2}{4} = 1$. See the accompanying figures on the next page.

continued on next page

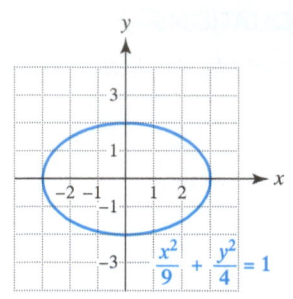

Ellipse centered at $(0, 0)$

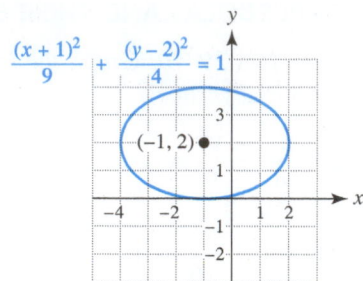

Ellipse centered at $(-1, 2)$

45. Graph each conic section. For hyperbolas, give the equations of the asymptotes.

(a) $\dfrac{(x-3)^2}{25} + \dfrac{(y-1)^2}{9} = 1$

(b) $\dfrac{(x+1)^2}{4} + \dfrac{(y+2)^2}{16} = 1$

(c) $\dfrac{(x+1)^2}{4} - \dfrac{(y-3)^2}{9} = 1$

(d) $\dfrac{(y-4)^2}{16} - \dfrac{(x+1)^2}{4} = 1$

46. Write the equation of an ellipse that has the following properties.

(a) Horizontal major axis of length 8, minor axis of length 4, and centered at $(-3, 5)$

(b) Vertical major axis of length 10, minor axis of length 6, and centered at $(2, -3)$

APPLICATIONS OF ELLIPSES AND HYPERBOLAS

47. Geometry of an Ellipse The area inside an ellipse is given by $A = \pi ab$, and its perimeter can be approximated by

$$P = 2\pi \sqrt{\dfrac{a^2 + b^2}{2}}.$$

Approximate A and P to the nearest hundredth for each ellipse.

(a) $\dfrac{x^2}{16} + \dfrac{y^2}{25} = 1$ (b) $\dfrac{x^2}{7} + \dfrac{y^2}{2} = 1$

48. Geometry of an Ellipse (Refer to the previous exercise.) If $a = b$ in the equation for an ellipse, the ellipse becomes a circle. Let $a = b$ in the formulas for the area and perimeter of an ellipse. Do the equations simplify to the area and perimeter for a circle? Explain.

49. Planet Orbit (Refer to Example 3.) Pluto's orbit is less circular than that of any of the eight planets. For Pluto (now a dwarf planet), $a = 39.44$ A.U. and $b = 38.20$. A.U.

(a) Graph the elliptic orbit of Pluto in the window $[-60, 60, 10]$ by $[-40, 40, 10]$. Plot the point $(9.81, 0)$ to show the position of the sun. Assume that the major axis is horizontal.

(b) Use the information in Exercise 47 to determine how far Pluto travels in one orbit around the sun and approximate the area inside its orbit.

50. Halley's Comet (Refer to Example 3.) The famous Halley's comet travels in an elliptical orbit with $a = 17.95$ and $b = 4.44$ and passes by Earth roughly every 76 years. The most recent pass by Earth was in February 1986. (*Source:* M. Zeilik.)

(a) Graph the orbit of Halley's comet in $[-21, 21, 5]$ by $[-14, 14, 5]$. Assume that the major axis is horizontal and that all units are in astronomical units. Plot a point at $(17.39, 0)$ to represent the position of the sun.

(b) Use the formula in Exercise 47 to estimate how many miles Halley's comet travels in one orbit around the sun.

(c) Estimate the average speed of Halley's comet in miles per hour.

51. Satellite Orbit The orbit of Explorer VII and the outline of Earth's surface are shown in the accompanying figure. This orbit is described by

$$\dfrac{x^2}{4464^2} + \dfrac{y^2}{4462^2} = 1,$$

and the surface of Earth is described by

$$\dfrac{(x-164)^2}{3960^2} + \dfrac{y^2}{3960^2} = 1.$$

Find the maximum and minimum heights of the satellite above Earth's surface if all units are miles. (*Source:* W. Thomson, *Introduction to Space Dynamics.*)

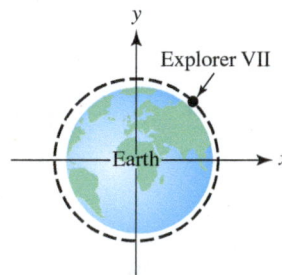

52. Weight Machines Elliptic shapes are used rather than circular shapes in some weight machines. Suppose that the ellipse shown in the figure at the top of the next column is represented by

$$\dfrac{x^2}{16} + \dfrac{y^2}{100} = 1,$$

where the units are inches. Find r_1 and r_2.

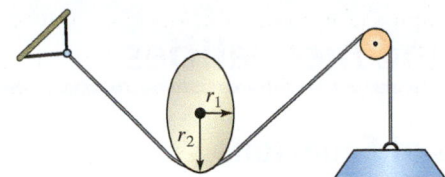

53. Arch Bridge The arch under a bridge is designed as the upper half of an ellipse, as illustrated in the accompanying figure. Its equation is modeled by

$$400x^2 + 10,000y^2 = 4,000,000$$

where the units are feet. Find the height and width of the arch.

54. Thinking Generally Suppose that the population y of a country can be modeled by the upper right branch of the hyperbola

$$\frac{x^2}{a^2} - \frac{y^2}{b^2} = 1,$$

where x represents time in years. What happens to the population after a long period of time?

55. Online Exploration Go online and find one application of ellipses. Write a short paragraph about your findings.

56. Online Exploration Go online and find one application of hyperbolas. Write a short paragraph about your findings.

WRITING ABOUT MATHEMATICS

57. Explain how the values of a and b affect the graph of $\frac{x^2}{a^2} + \frac{y^2}{b^2} = 1$. Assume that $a > b > 0$.

58. Explain how the values of a and b affect the graph of $\frac{x^2}{a^2} - \frac{y^2}{b^2} = 1$. Assume that a and b are positive.

SECTIONS 13.1 and 13.2 **Checking Basic Concepts**

1. Graph the parabola $x = (y - 2)^2 + 1$. Find the vertex and axis of symmetry.

2. Find the equation of the circle with center $(1, -2)$ and radius 2. Graph the circle.

3. Find the x- and y-intercepts on the graph of

$$\frac{x^2}{4} + \frac{y^2}{9} = 1.$$

4. Graph the following. Label any vertices and state the type of conic section that it represents.

(a) $x = y^2$

(b) $\frac{x^2}{16} + \frac{y^2}{25} = 1$

(c) $\frac{x^2}{4} - \frac{y^2}{9} = 1$

(d) $(x - 1)^2 + (y + 2)^2 = 9$

13.3 Nonlinear Systems of Equations and Inequalities

NEW VOCABULARY

☐ Nonlinear system of equations
☐ Method of substitution
☐ Nonlinear system of inequalities

1 Solving Nonlinear Systems of Equations

SOLUTIONS TO NONLINEAR SYSTEMS One way to locate the points at which the line $y = 2x$ intersects the circle $x^2 + y^2 = 5$ is to graph both equations (see **FIGURE 13.21**).

The equation describing the circle is nonlinear. Another way to locate the points of intersection is symbolically, by solving the *nonlinear system of equations*.

$$y = 2x \quad \text{Line}$$
$$x^2 + y^2 = 5 \quad \text{Circle}$$

Linear systems of equations can have no solutions, one solution, or infinitely many solutions. It is possible for a nonlinear system of equations to have *any number* of solutions. **FIGURE 13.21** shows that this nonlinear system of equations has two solutions: $(-1, -2)$ and $(1, 2)$.

Solving a System of Equations Graphically

Another solution to the system is $(1, 2)$.

One solution to the system is $(-1, -2)$.

FIGURE 13.21

METHOD OF SUBSTITUTION A symbolic technique for solving a system of equations is the **method of substitution**, which we demonstrate in the next example.

EXAMPLE 1   **Solving a nonlinear system of equations symbolically**

Solve the following system of equations symbolically. Check any solutions and compare them with **FIGURE 13.21**.

$$y = 2x$$
$$x^2 + y^2 = 5$$

Solution
Substitute $2x$ for y in the second equation $x^2 + y^2 = 5$ and solve for x.

$$x^2 + (2x)^2 = 5 \qquad \text{Let } y = 2x \text{ in the second equation.}$$
$$x^2 + 4x^2 = 5 \qquad \text{Properties of exponents}$$
$$5x^2 = 5 \qquad \text{Combine like terms.}$$
$$x^2 = 1 \qquad \text{Divide by 5.}$$
$$x = \pm 1 \qquad \text{Square root property}$$

To determine corresponding y-values, substitute $x = \pm 1$ in $y = 2x$; the solutions are $(1, 2)$ and $(-1, -2)$. To check $(1, 2)$, substitute $x = 1$ and $y = 2$ in the given equations.

$$2 \overset{?}{=} 2(1) \checkmark \quad \text{True}$$
$$(1)^2 + (2)^2 \overset{?}{=} 5 \checkmark \quad \text{True}$$

To check $(-1, -2)$, substitute $x = -1$ and $y = -2$ in the given equations.

$$-2 \stackrel{?}{=} 2(-1) \checkmark \quad \text{True}$$
$$(-1)^2 + (-2)^2 \stackrel{?}{=} 5 \checkmark \quad \text{True}$$

The solutions check and agree with **FIGURE 13.21**.

Now Try Exercise 13

GRAPHICAL AND SYMBOLIC SOLUTIONS In the next example, we solve a nonlinear system of equations graphically and symbolically. The symbolic solution gives the exact solutions.

EXAMPLE 2   **Solving a nonlinear system of equations**

Solve the nonlinear system of equations graphically and symbolically.

$$x^2 - y = 2$$
$$x^2 + y = 4$$

Solution

Graphical Solution Begin by solving each equation for y.

$$y = x^2 - 2$$
$$y = 4 - x^2$$

Graph $y_1 = x^2 - 2$ and $y_2 = 4 - x^2$. The solutions are approximately $(-1.73, 1)$ and $(1.73, 1)$, as shown in **FIGURE 13.22**. The graphs consist of two parabolas intersecting at two points.

(a)

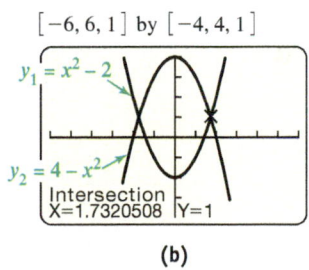
(b)

FIGURE 13.22

Symbolic Solution Solving the first equation for y gives $y = x^2 - 2$. Substitute this expression for y in the second equation and solve for x.

$$x^2 + y = 4 \qquad \text{Second equation}$$
$$x^2 + (x^2 - 2) = 4 \qquad \text{Substitute } y = x^2 - 2.$$
$$2x^2 = 6 \qquad \text{Combine like terms; add 2.}$$
$$x^2 = 3 \qquad \text{Divide by 2.}$$
$$x = \pm\sqrt{3} \qquad \text{Square root property}$$

To determine y, substitute $x = \pm\sqrt{3}$ in $y = x^2 - 2$.

$$y = (\sqrt{3})^2 - 2 = 3 - 2 = 1$$
$$y = (-\sqrt{3})^2 - 2 = 3 - 2 = 1$$

The *exact* solutions are $(\sqrt{3}, 1)$ and $(-\sqrt{3}, 1)$.

Now Try Exercise 29 (a), (b)

EXAMPLE 3   Solving a nonlinear system of equations

Solve the nonlinear system of equations symbolically and graphically.

$$x^2 - y^2 = 3$$
$$x^2 + y^2 = 5$$

Solution

Symbolic Solution Instead of using substitution on this nonlinear system of equations, we use elimination. Note that, if we add the two equations, the y-variable will be eliminated.

$x^2 - y^2 = 3$	First equation
$x^2 + y^2 = 5$	Second equation
$2x^2 \quad\;\; = 8$	Add equations.

Solving gives $x^2 = 4$, or $x = \pm 2$. To determine y, substitute **4** for x^2 in $x^2 + y^2 = 5$.

$$4 + y^2 = 5 \quad \text{or} \quad y^2 = 1$$

Because $y^2 = 1$, $y = \pm 1$. There are four solutions: $(2, 1)$, $(2, -1)$, $(-2, 1)$, and $(-2, -1)$.

Graphical Solution The graph of the first equation is a hyperbola, and the graph of the second is a circle with radius $\sqrt{5}$. The four points of intersection (solutions) are $(\pm 2, \pm 1)$, as shown in **FIGURE 13.23**.

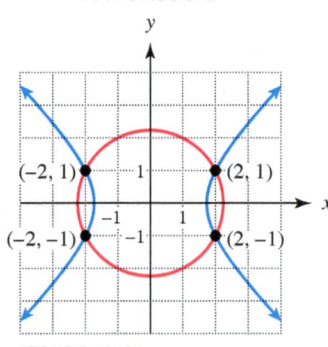

Four Solutions

FIGURE 13.23

Now Try Exercises 23, 25

🌐 **Math in Context** *Manufacturing* To describe characteristics of curved objects, we often need *nonlinear equations*. The equations of the conic sections discussed in this chapter are but a few examples of nonlinear equations. For instance, cylinders have a curved shape, as illustrated in **FIGURE 13.24**. If the radius of a cylinder is denoted r and its height h, then its volume V is given by the nonlinear equation $V = \pi r^2 h$ and its side area A is given by the nonlinear equation $A = 2\pi rh$.

If we want to manufacture a cylindrical container that holds 35 cubic inches and whose side area is 50 square inches, we need to solve the following nonlinear system of equations.

Cylindrical Container

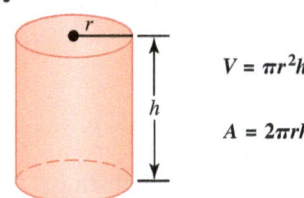

$V = \pi r^2 h$

$A = 2\pi rh$

FIGURE 13.24

> Solve to find the r and h that give a volume of 50 in³ and an area of 35 in².

$$\pi r^2 h = 35 \quad \text{The side area is 35 in}^2$$
$$2\pi rh = 50 \quad \text{The volume is 50 in}^3$$

EXAMPLE 4   Modeling the dimensions of a cylindrical container

Find the dimensions of a cylindrical container that has a volume V of 35 cubic inches and a side area A of 50 square inches by solving the following nonlinear system of equations.

$$\pi r^2 h = 35 \quad \text{The side area is 35 in}^2$$
$$2\pi rh = 50 \quad \text{The volume is 50 in}^3$$

Solution

We can find r by solving each equation for h and setting the results equal to each other. This eliminates the variable h.

$$\frac{50}{2\pi r} = \frac{35}{\pi r^2} \qquad h = \frac{50}{2\pi r} \quad \text{and} \quad h = \frac{35}{\pi r^2}$$

$$50\pi r^2 = 70\pi r \qquad \text{Eliminate fractions (cross multiply).}$$

$$50\pi r^2 - 70\pi r = 0 \qquad \text{Subtract } 70\pi r.$$

$$10\pi r(5r - 7) = 0 \qquad \text{Factor out } 10\pi r.$$

$$10\pi r = 0 \quad \text{or} \quad 5r - 7 = 0 \qquad \text{Zero-product property}$$

$$r = 0 \quad \text{or} \quad r = \tfrac{7}{5} = 1.4 \qquad \text{Solve.}$$

Because $h = \frac{50}{2\pi r}$, $r = 0$ is not possible, but we can find h by substituting **1.4** for r in the formula.

$$h = \frac{50}{2\pi\,(1.4)} \approx 5.68$$

A cylindrical container that has a volume of 35 cubic inches and a side area of 50 square inches has a radius of 1.4 inches and a height of about 5.68 inches.

Now Try Exercise 51(b)

READING CHECK 1

- How many solutions can a nonlinear system of equations have?

2 Solving Nonlinear Systems of Inequalities

RECOGNIZING THE SOLUTION SET Previously in this text, we solved systems of *linear* inequalities. A **nonlinear system of inequalities** in two variables can be solved similarly by using graphical techniques. For example, consider the nonlinear system of inequalities

$$y \geq x^2 - 2$$
$$y \leq 4 - x^2.$$

The graph of $y = x^2 - 2$ is a parabola opening upward. The solution set to $y \geq x^2 - 2$ includes all points lying on or above this parabola. See **FIGURE 13.25(a)**. Similarly, the graph of $y = 4 - x^2$ is a parabola opening downward. The solution set to $y \leq 4 - x^2$ includes all points lying on or below this parabola. See **FIGURE 13.25(b)**.

FIGURE 13.25

The solution set for this nonlinear *system* of inequalities includes all points (x, y) in *both* shaded regions. The *intersection* of the shaded regions is shown in **FIGURE 13.25(c)**.

GRAPHICAL SOLUTIONS In the next three examples, graphical solutions are found by shading the solution set in the xy-plane.

EXAMPLE 5   Solving a nonlinear system of inequalities graphically

Shade the solution set for the system of inequalities.

$$\frac{x^2}{4} + \frac{y^2}{9} < 1$$

$$y > 1$$

Solution

The solutions to $\frac{x^2}{4} + \frac{y^2}{9} < 1$ lie *inside* the ellipse $\frac{x^2}{4} + \frac{y^2}{9} = 1$. See **FIGURE 13.26(a)**. Solutions to $y > 1$ lie above the line $y = 1$, as shown in **FIGURE 13.26(b)**. The intersection of these two regions is shown in **FIGURE 13.26(c)**. Any point in this region is a solution.

Solutions Lie inside the Ellipse	Solutions Lie above the Line	Solutions Lie inside the Ellipse *and* above the Line
(a)	(b)	(c)

FIGURE 13.26

For example, the point $(0, 2)$ lies in the shaded region and is a solution to the system. Note that a dashed curve and a dashed line are used when equality is *not* included.

Now Try Exercise 39

EXAMPLE 6   Solving a nonlinear system of inequalities graphically

Shade the solution set for the following system of inequalities.

$$x^2 + y \le 4$$

$$-x + y \ge 2$$

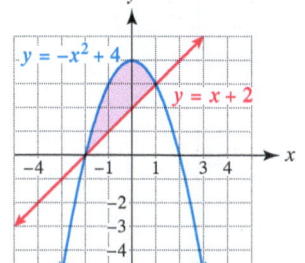

FIGURE 13.27

Solution

The solutions to $x^2 + y \le 4$ lie *on* or *below* the parabola $y = -x^2 + 4$, and the solutions to $-x + y \ge 2$ lie *on* or *above* the line $y = x + 2$. The appropriate shaded region is shown in **FIGURE 13.27**. Both the parabola and the line are solid because equality is included in both inequalities.

Now Try Exercise 37

In the next example, we use a graphing calculator to shade a region that lies above both graphs, using the "$Y_1 =$" menu. This feature allows us to shade either above or below the graph of a function.

EXAMPLE 7 **Solving a system of inequalities with a graphing calculator**

Shade the solution set for the following system of inequalities.

$$y \geq x^2 - 2$$
$$y \geq -1 - x$$

Solution

Enter $y_1 = x^2 - 2$ and $y_2 = -1 - x$, as shown in **FIGURE 13.28(a)**. Note that the option to shade above the graphs of Y_1 and Y_2 are selected to the left of Y_1 and Y_2. Then the two inequalities are graphed in **FIGURE 13.28(b)**. The solution set corresponds to the region where there is both vertical and horizontal shading.

CALCULATOR HELP

To shade the solution set to a system of inequalities, see Appendix A (pages AP-6 and AP-7).

(a)

$[-4.7, 4.7, 1]$ by $[-3.1, 3.1, 1]$

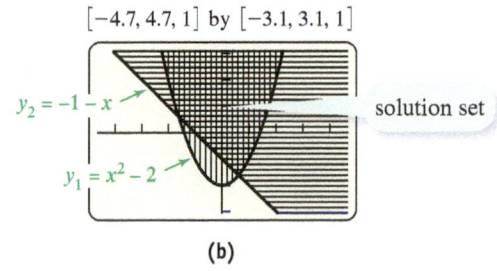

$y_2 = -1 - x$

$y_1 = x^2 - 2$

solution set

(b)

FIGURE 13.28

Now Try Exercise 49

13.3 Putting It All Together

CONCEPT	EXPLANATION
Nonlinear Systems of Equations in Two Variables	To solve the following system of equations symbolically, using *substitution*, solve the first equation for y to get $y = \mathbf{5 - x}$. $$x + y = 5 \quad \text{First equation}$$ $$x^2 - y = 1 \quad \text{Second equation}$$ Substitute $\mathbf{5 - x}$ for y in the second equation and solve the resulting quadratic equation. (*Elimination* can also be used on this system.) Then $$x^2 - (\mathbf{5 - x}) = 1 \quad \text{or} \quad x^2 + x - 6 = 0.$$ Factoring can be used to determine that the solutions are given by $$x = \mathbf{-3} \quad \text{or} \quad x = \mathbf{2}.$$ Thus $y = 5 - (\mathbf{-3}) = \mathbf{8}$ or $y = 5 - 2 = \mathbf{3}$. The solutions are $(\mathbf{-3, 8})$ and $(\mathbf{2, 3})$. Graphical support is shown in the accompanying figure. **Two Solutions** Nonlinear systems of equations can have *any number of solutions*.

continued on next page

continued from previous page

CONCEPT	EXPLANATION
Nonlinear Systems of Inequalities in Two Variables	To solve the following system of inequalities graphically, solve each inequality for y. $\quad x + y \leq 5 \quad$ or $\quad y \leq 5 - x \quad$ First inequality $\quad x^2 - y \leq 1 \quad$ or $\quad y \geq x^2 - 1 \quad$ Second inequality The solutions lie on or below the line and on or above the parabola, as shown in the figure. 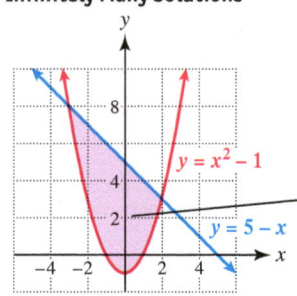 Infinitely Many Solutions The solution set to a nonlinear system of inequalities can be represented by a shaded region in the xy-plane.

13.3   Exercises

MyMathLab®

CONCEPTS AND VOCABULARY

1. How many solutions can a nonlinear system of equations have?

2. If a nonlinear system of equations has two equations, how many equations does a solution have to satisfy?

3. Determine visually the number of solutions to the following system of equations. Explain your reasoning.

$$y = x$$
$$x^2 + y^2 = 4$$

4. Describe the solution set to $x^2 + y^2 \leq 1$.

5. Does $(-2, -1)$ satisfy $5x^2 - 2y^2 > 18$?

6. Does $(3, 4)$ satisfy $x^2 - 2y \geq 4$?

7. Sketch a parabola and ellipse with four points of intersection.

8. Sketch a line and a hyperbola with two points of intersection.

SOLVING NONLINEAR SYSTEMS OF EQUATIONS

Exercises 9–12: Use the graph to estimate all solutions to the system of equations. Check each solution.

9. $x^2 + y^2 = 10$
   $\quad y = 3x$

10. $x^2 + 3y^2 = 16$
    $\quad y = -x$

11. $y^2 - x^2 = 1$
    $\quad x^2 + 3y^2 = 3$

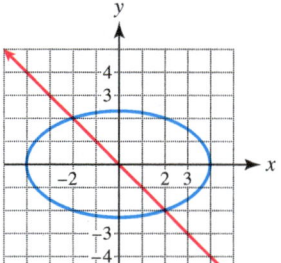

12. $\quad y = 1 - x^2$
    $\quad x^2 + y^2 = 1$

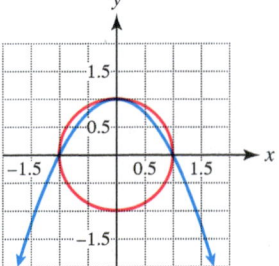

Exercises 13–24: Solve the system of equations symbolically. Check your solutions.

13.
$$y = 2x$$
$$x^2 + y^2 = 45$$

14.
$$y = x$$
$$y^2 = 3 - 2x^2$$

15.
$$x + y = 1$$
$$x^2 - y^2 = 3$$

16.
$$y - x = -1$$
$$y = 2x^2$$

17.
$$y - x^2 = 0$$
$$x^2 + y^2 = 6$$

18.
$$x^2 - y^2 = 4$$
$$x^2 + y^2 = 4$$

19.
$$3x^2 + 2y^2 = 5$$
$$x - y = -2$$

20.
$$x^2 + 2y^2 = 9$$
$$2x^2 - y^2 = -2$$

21.
$$x^2 + y^2 = 4$$
$$x^2 - 9y^2 = 9$$

22.
$$x^2 + y^2 = 15$$
$$9x^2 + 4y^2 = 36$$

23.
$$x^2 + y^2 = 10$$
$$2x^2 - y^2 = 17$$

24.
$$2x^2 + 3y^2 = 5$$
$$3x^2 - 4y^2 = -1$$

Exercises 25–28: Solve the system of equations graphically. Check your solutions.

25.
$$y = x^2 - 3$$
$$2x^2 - y = 1 - 3x$$

26.
$$x + y = 2$$
$$x - y^2 = 3$$

27.
$$y - x = -4$$
$$x - y^2 = -2$$

28.
$$xy = 1$$
$$y = x$$

Exercises 29–32: Solve the system of equations
  (a) *symbolically,*
  (b) *graphically, and*
  (c) *numerically.*

29.
$$y = -2x$$
$$x^2 + y = 3$$

30.
$$4x - y = 0$$
$$x^3 - y = 0$$

31.
$$xy = 1$$
$$x - y = 0$$

32.
$$x^2 + y^2 = 4$$
$$y - x = 2$$

SOLVING NONLINEAR SYSTEMS OF INEQUALITIES

Exercises 33–36: Shade the solution set in the xy-plane.

33. $y \geq x^2$

34. $y \leq x^2 - 1$

35. $\dfrac{x^2}{4} + \dfrac{y^2}{9} > 1$

36. $x^2 + y^2 \leq 1$

Exercises 37–44: Shade the solution set in the xy-plane. Then use the graph to select one solution.

37.
$$y > x^2 + 1$$
$$y < 3$$

38.
$$y > x^2$$
$$y < x + 2$$

39.
$$x^2 + y^2 \leq 1$$
$$y < x$$

40.
$$y > x^2 - 2$$
$$y \leq 2 - x^2$$

41.
$$x^2 + y^2 \leq 1$$
$$(x - 2)^2 + y^2 \leq 1$$

42.
$$x^2 - y \geq 2$$
$$(x + 1)^2 + y^2 \leq 4$$

43.
$$x^2 - y^2 \leq 4$$
$$x^2 + y^2 \leq 9$$

44.
$$3x + 2y < 6$$
$$x^2 + y^2 \leq 16$$

Exercises 45 and 46: Match the inequality or system of inequalities with its graph (a or b).

45. $y \leq \dfrac{1}{2}x^2$

46.
$$y \geq x^2 + 1$$
$$y \leq 5$$

a.

b.

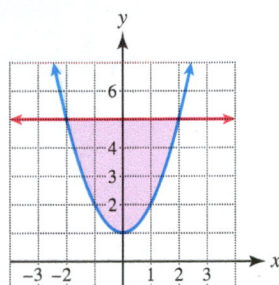

Exercises 47 and 48: Use the graph to write the inequality or system of inequalities.

47.

48.

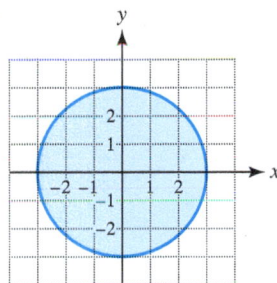

Exercises 49 and 50: Use a graphing calculator to shade the solution set to the system of inequalities.

49.
$$y \geq x^2 - 1$$
$$y \geq 2 - x$$

50.
$$y \leq 4 - x^2$$
$$y \leq x^2 - 4$$

APPLICATIONS OF NONLINEAR SYSTEMS

51. Dimensions of a Can (Refer to Example 4.) Find the dimensions of a cylindrical container with a volume of 40 cubic inches and a side area of 50 square inches **(a)** graphically and **(b)** symbolically.

52. Dimensions of a Can (Refer to Example 4.) Is it possible to design an aluminum can with volume of 60 cubic inches and side area of 60 square inches? If so, find the dimensions of the can.

53. Area and Perimeter The area of a room is 143 square feet, and its perimeter is 48 feet. Let x be the width and y be the length of the room. See the accompanying figure on the next page.
  (a) Write a nonlinear system of equations that models this situation.

continued on next page

(b) Solve the system.

54. **Dimensions of a Cone** The volume V of a cone is given by $V = \frac{1}{3}\pi r^2 h$, and the surface area S of its side is given by $S = \pi r\sqrt{r^2 + h^2}$, where h is the height and r is the radius of the base. See the accompanying figure.

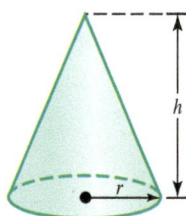

(a) Solve each equation for h.

(b) Estimate r and h graphically for a cone with volume V of 34 cubic feet and surface area S of 52 square feet.

WRITING ABOUT MATHEMATICS

55. A student *incorrectly* changes the following system of inequalities

$$\begin{aligned} x^2 - y &\geq 6 \\ 2x - y &\leq -3 \end{aligned} \quad \text{to} \quad \begin{aligned} y &\geq x^2 - 6 \\ y &\leq 2x + 3. \end{aligned}$$

The student discovers that $(1, 2)$ satisfies the second system of inequalities but not the first. Explain the student's error.

56. Explain graphically how nonlinear systems of equations can have any number of solutions. Sketch graphs of different systems with zero, one, two, and three solutions.

SECTION 13.3

Checking Basic Concepts

1. Solve the following system of equations symbolically and graphically.

$$x^2 - y = 2x$$
$$2x - y = 3$$

2. Determine visually the number of solutions to the following system of equations.

$$y = x^2 - 4$$
$$y = x$$

3. The solution set for a system of inequalities is shown in the accompanying figure to the right.
   (a) Find one ordered pair (x, y) that is a solution and one that is not.

(b) Write the system of inequalities represented by the graph.

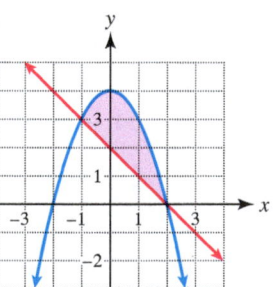

4. Shade the solution set for the following system of inequalities.

$$x^2 + y^2 \leq 4$$
$$y < 1$$

CHAPTER 13   Summary

SECTION 13.1 ■ PARABOLAS AND CIRCLES

Types of Conic Sections There are three basic types of conic sections: parabolas, ellipses, and hyperbolas.

Parabolas A parabola can have a vertical or a horizontal axis of symmetry. Two forms of an equation for a parabola with a *vertical* axis of symmetry are

$$y = ax^2 + bx + c \quad \text{and} \quad y = a(x - h)^2 + k.$$

If $a > 0$, the parabola opens upward, and if $a < 0$, it opens downward. The vertex is located at (h, k). Two forms of an equation for a parabola with a *horizontal* axis of symmetry are

$$x = ay^2 + by + c \quad \text{and} \quad x = a(y - k)^2 + h.$$

If $a > 0$, the parabola opens to the right, and if $a < 0$, it opens to the left. The vertex is located at (h, k).

Example: $x = -2(y - 4)^2 + 1$ has vertex $(1, 4)$, has axis of symmetry $y = 4$, and opens to the left.

Circles The standard equation for a circle with center (h, k) and radius r is

$$(x - h)^2 + (y - k)^2 = r^2.$$

Example: $(x + 2)^2 + (y - 3)^2 = 36$ has center $(-2, 3)$ and radius 6.

SECTION 13.2 ■ ELLIPSES AND HYPERBOLAS

Ellipses

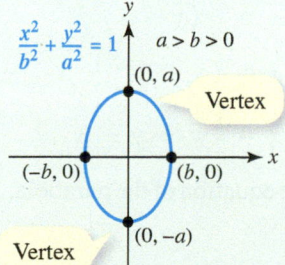

Example: $\frac{x^2}{9} + \frac{y^2}{4} = 1$ is the standard equation for an ellipse with vertices $(\pm 3, 0)$, centered at $(0, 0)$.

Hyperbolas

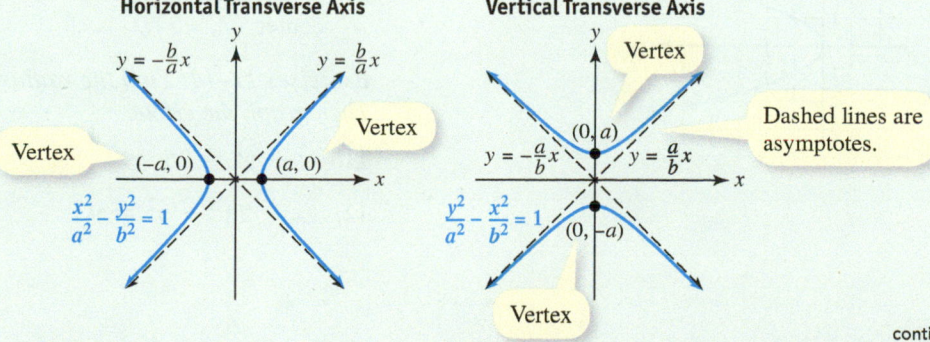

continued on next page

Example: $\frac{x^2}{9} - \frac{y^2}{4} = 1$ is the standard equation for a hyperbola with horizontal transverse axis, vertices $(\pm 3, 0)$, center $(0, 0)$, and asymptotes $y = \pm\frac{2}{3}x$.

SECTION 13.3 ■ NONLINEAR SYSTEMS OF EQUATIONS AND INEQUALITIES

Nonlinear Systems Nonlinear systems of equations can have any number of solutions. The methods of substitution and elimination can often be used to solve a nonlinear system of equations symbolically. Nonlinear systems can also be solved graphically. The solution set for a nonlinear system of two inequalities in two variables is typically a region in the xy-plane. A solution is an ordered pair (x, y) that satisfies both inequalities.

Example: Solve $y \geq x^2 - 2$

$$y \leq 4 - \tfrac{1}{2}x^2.$$

Nonlinear System of Inequalities

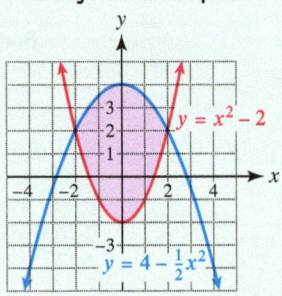

CHAPTER 13   Review Exercises

SECTION 13.1

Exercises 1–6: Graph the parabola. Find the vertex and axis of symmetry.

1. $x = 2y^2$

2. $x = -(y + 1)^2$

3. $x = -2(y - 2)^2$

4. $x = (y + 2)^2 - 1$

5. $x = -3y^2 + 1$

6. $x = \frac{1}{2}y^2 + y - 3$

7. Use the graph to determine the equation of the parabola.

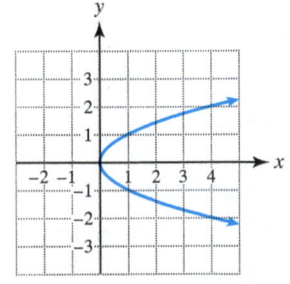

8. Use the graph to find the equation of the circle.

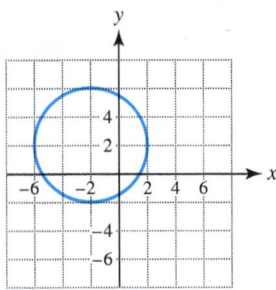

9. Write the equation of the circle with radius 1 and center $(0, 0)$.

10. Write the equation of the circle with radius 4 and center $(2, -3)$.

Exercises 11–14: Find the radius and center of the circle. Then graph the circle.

11. $x^2 + y^2 = 25$

12. $(x - 2)^2 + y^2 = 9$

13. $(x + 3)^2 + (y - 1)^2 = 5$

14. $x^2 - 2x + y^2 + 2y = 7$

SECTION 13.2

Exercises 15–18: Graph the ellipse. Label the vertices and endpoints of the minor axis.

15. $\dfrac{x^2}{4} + \dfrac{y^2}{25} = 1$ **16.** $x^2 + \dfrac{y^2}{4} = 1$

17. $25x^2 + 20y^2 = 500$ **18.** $4x^2 + 9y^2 = 36$

19. Use the graph to determine the equation of the ellipse.

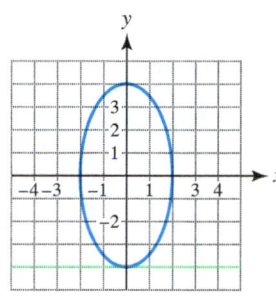

20. Use the graph to find the equation of the hyperbola.

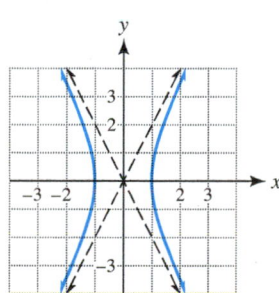

Exercises 21–24: Graph the hyperbola and include the asymptotes in your graph.

21. $\dfrac{x^2}{9} - \dfrac{y^2}{4} = 1$ **22.** $\dfrac{y^2}{25} - \dfrac{x^2}{16} = 1$

23. $y^2 - x^2 = 1$ **24.** $25x^2 - 16y^2 = 400$

SECTION 13.3

Exercises 25–28: Use the graph to estimate all solutions to the system of equations. Check each solution.

25. $x^2 + y^2 = 9$ **26.** $xy = 2$
     $x + y = 3$ $y = 2x$

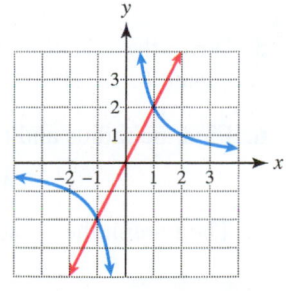

27. $x^2 - y = x$ **28.** $x^2 + y^2 = 5$
     $y = x$ $x^2 - y^2 = 3$

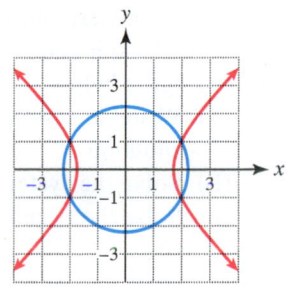

Exercises 29–32: Solve the system of equations. Check your solutions.

29. $y = x$ **30.** $x - y = 4$
     $x^2 + y^2 = 32$ $x^2 + y^2 = 16$

31. $y = x^2$ **32.** $y = x^2 + 1$
     $2x^2 + y = 3$ $2x^2 - y = 3x - 3$

Exercises 33 and 34: Solve the system graphically.

33. $2x - y = 4$ **34.** $x^2 + y = 0$
     $x^2 + y = 4$ $x^2 + y^2 = 2$

Exercises 35 and 36: Solve the system of equations
   (a) *symbolically,*
   (b) *graphically, and*
   (c) *numerically.*

35. $y = x$ **36.** $y = x^3$
     $x^2 + 2y = 8$ $x^2 - y = 0$

Exercises 37–44: Shade the solution set in the xy-plane.

37. $y \geq 2x^2$ **38.** $y < 2x - 3$

39. $y < -x^2$ **40.** $\dfrac{x^2}{9} + \dfrac{y^2}{16} \leq 1$

41. $y - x^2 \geq 1$ **42.** $x^2 + y \leq 4$
     $y \leq 2$ $3x + 2y \geq 6$

43. $y > x^2$ **44.** $\dfrac{x^2}{4} + \dfrac{y^2}{9} > 1$
     $y < 4 - x^2$ $x^2 + y^2 < 16$

Exercises 45 and 46: Use the graph to write the system of inequalities.

45.   **46.**

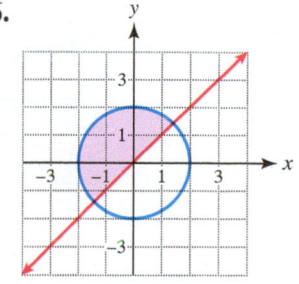

APPLICATIONS

47. Area and Perimeter The area of a desktop is 1000 square inches, and its perimeter is 130 inches. Let x be the width and y be the length of the desktop. See the figure below.
   (a) Write a system of equations that models this situation.
   (b) Solve the system graphically.
   (c) Solve the system symbolically.

48. Numbers The product of two positive numbers is 60, and their difference is 7. Let x be the smaller number and y be the larger number.
   (a) Write a system of equations whose solution gives the two numbers.
   (b) Solve the system graphically.
   (c) Solve the system symbolically.

49. Dimensions of a Container The volume of a cylindrical container is $V = \pi r^2 h$, and its surface area, *excluding* the top and bottom, is $A = 2\pi rh$. Find the dimensions of a container with $A = 100$ square feet and $V = 50$ cubic feet. Is your answer unique?

 50. Dimensions of a Container The volume of a cylindrical container is $V = \pi r^2 h$, and its surface area, *including* the top and bottom, is $A = 2\pi rh + 2\pi r^2$. Graphically find the dimensions of a container with $A = 80$ square inches and $V = 35$ cubic inches. Is your answer unique?

51. Geometry of an Ellipse The area inside an ellipse is given by $A = \pi ab$, and its perimeter P can be approximated by
$$P = 2\pi\sqrt{\frac{a^2 + b^2}{2}}.$$
   (a) Graph $\dfrac{x^2}{5} + \dfrac{y^2}{12} = 1$.
   (b) Estimate the ellipse's area and perimeter.

52. Orbit of Mars Mars has an elliptical orbit that is nearly circular, with $a = 1.524$ and $b = 1.517$, where the units are astronomical units (1 A.U. equals 93 million miles). (*Source:* M. Zeilik.)
   (a) Graph the orbit of Mars in $[-3, 3, 1]$ by $[-2, 2, 1]$. Plot the point $(0.15, 0)$ to show the position of the sun. Assume that the major axis is horizontal.
   (b) Use the information in Exercise 51 to estimate how far Mars travels in one orbit around the sun. Approximate the area inside its orbit.

CHAPTER 13   Test   MyMathLab® YouTube

1. Graph the parabola $y = -(x - 1)^2 + 2$. Find the vertex and axis of symmetry.

2. Graph the parabola $x = (y - 4)^2 - 2$. Find the vertex and axis of symmetry.

3. Use the graph to find the equation of the parabola.

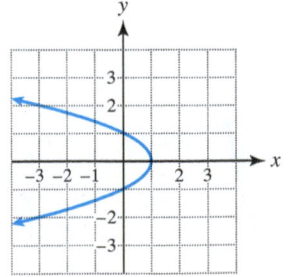

4. Use the graph to find the equation of the circle.

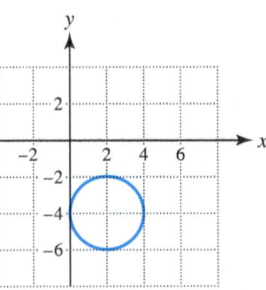

5. Write the equation of the circle with radius 10 and center $(-5, 2)$.

6. Find the radius and center of the circle given by
$$x^2 + 4x + y^2 - 6y = 3.$$
Then graph the circle.

7. Graph the ellipse $\frac{x^2}{16} + \frac{y^2}{49} = 1$. Label the vertices and endpoints of the minor axis.

8. Use the graph to determine the standard equation of the ellipse.

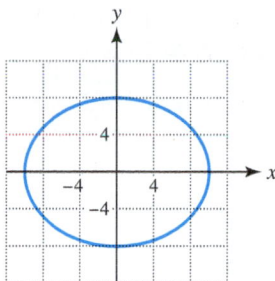

9. Graph the hyperbola $4x^2 - 9y^2 = 36$. Include the asymptotes in your graph.

10. Use the graph to estimate all solutions to the system of equations. Check each solution by substitution in the system of equations.

$$x^2 + y^2 = 16$$
$$x - y = 4$$

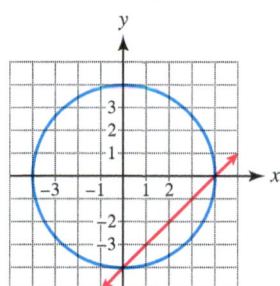

11. Solve the system of equations symbolically.

$$x - y = 3$$
$$x^2 + y^2 = 17$$

12. Solve the system of equations graphically.

$$2x^2 - y = 4$$
$$x^2 + y = 8$$

13. Shade the solution set in the xy-plane.

$$3x + y > 6$$
$$x^2 + y^2 < 25$$

14. Use the graph to write the system of inequalities.

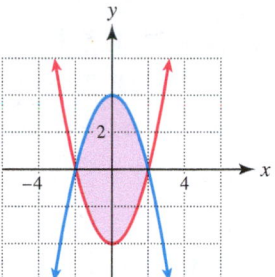

15. Area and Perimeter The area of a rectangular swimming pool is 5000 square feet, and its perimeter is 300 feet. Let x be the width and y be the length of the pool.
(a) Write a nonlinear system of equations that models this situation.
(b) Solve the system.

16. Dimensions of a Box The volume of a rectangular box with a square bottom and open top is $V = x^2y$, and its surface area is $A = x^2 + 4xy$, where x represents its width and length and y represents its height. Estimate graphically the dimensions of a box with $V = 1183$ cubic inches and $A = 702$ square inches. Is your answer unique? (*Hint:* Substitute appropriate values for V and A, and then solve each equation for y.)

17. Orbit of Uranus The planet Uranus has an elliptical orbit that is nearly circular, with $a = 19.18$ and $b = 19.16$, where the units are astronomical units (1 A.U. equals 93 million miles). (*Source:* M. Zeilik.)
(a) Graph the orbit of Uranus in $[-30, 30, 10]$ by $[-20, 20, 10]$. Plot the point $(0.9, 0)$ to show the position of the sun. Assume that the major axis is horizontal.
(b) Find the minimum distance between Uranus and the sun.

CHAPTERS 1–13   Cumulative Review Exercises

1. Evaluate $K = x^2 + y^2$ when $x = 4$ and $y = -3$.

2. Simplify $\dfrac{(a^{-2}b)^2}{a^{-1}(b^3)^{-2}}$.

3. Write 7.345×10^{-3} in standard (decimal) notation.

4. Find $f(-4)$, if $f(x) = \frac{x}{x-4}$. State the domain of f.

5. If the point $(3, 4)$ is on the graph of $y = f(x)$, then $f(\underline{}) = \underline{}$.

Exercises 6 and 7: Graph f by hand.

6. $f(x) = x^2 - x$ **7.** $f(x) = 1 - 2x$

8. Find the slope–intercept form of the line that passes through the point $(2, -2)$ and is perpendicular to the line given by $y = -\frac{2}{3}x + 1$.

9. Solve $2(1 - x) - 4x = x$.

Exercises 10–12: Solve the inequality. Write your answer in interval notation.

10. $-5 \le 1 - 2x < 3$ **11.** $x^2 - 4 \le 0$

12. $|1 - x| \ge 2$

13. Solve the system.
$$-2x + y = 1$$
$$5x - y = 2$$

14. Shade the solution set in the xy-plane.
$$x + y \le 4$$
$$x - y \ge 2$$

Exercises 15 and 16: Multiply the expression.

15. $(2x - 1)(x + 5)$ **16.** $xy(2x - 3y^2 + 1)$

Exercises 17 and 18: Factor the expression.

17. $6x^2 - 13x - 5$ **18.** $x^3 - 4x$

Exercises 19 and 20: Solve the equation.

19. $x^2 + 3x + 2 = 0$ **20.** $x^2 + 1 = -3x$

Exercises 21 and 22: Simplify the expression.

21. $\dfrac{x - 2}{x + 2} \div \dfrac{2x - 4}{3x + 6}$ **22.** $\dfrac{1}{x + 1} + \dfrac{1}{x - 1}$

Exercises 23–26: Simplify the expression. Assume all variables are positive.

23. $\sqrt{8x^2}$ **24.** $8^{2/3}$

25. $\sqrt[3]{2x} \cdot \sqrt[3]{32x^2}$ **26.** $3\sqrt{3x} + \sqrt{12x}$

27. Graph $f(x) = -\sqrt{x}$.

28. Find the distance between $(2, -3)$ and $(-2, 0)$.

29. Write $(2 + 3i)(2 - 3i)$ in standard form.

30. Solve $\sqrt{x + 2} = x$.

31. Find the vertex of the graph of $y = x^2 - 6x + 3$.

32. Write $f(x) = x^2 - 2x + 3$ in vertex form.

33. Compare the graph of $f(x) = \sqrt{x} - 4$ to the graph of $y = \sqrt{x}$.

34. Solve $x(3 - x) = 2$.

35. Solve $x^3 + x = 0$. Find all complex solutions.

36. Simplify by hand.
(a) $\log 10{,}000$ (b) $\log_2 8$
(c) $\log_3 3^x$ (d) $e^{\ln 6}$
(e) $\log 2 + \log 50$ (f) $\log_2 24 - \log_2 3$

37. If $f(x) = x^2 + 1$ and $g(x) = 2x$, find the following.
(a) $(f \circ g)(2)$ (b) $(g \circ f)(x)$

38. If $f(x) = 2 - 3x$, find $f^{-1}(x)$.

39. If a total of \$1000 is deposited in an account that pays 5% annual interest, approximate the amount in the account after 6 years.

40. Write $\log \dfrac{x^2 \sqrt{y}}{z^3}$ in terms of logarithms of x, y, and z.

41. Solve $2e^x - 1 = 17$.

42. Solve $3 + \log 4x = 5$.

43. Graph each conic section.
(a) $x = (y - 1)^2$ (b) $(x - 1)^2 + (y + 1)^2 = 4$
(c) $\dfrac{x^2}{4} + \dfrac{y^2}{25} = 1$ (d) $4x^2 - 9y^2 = 36$

44. Solve the nonlinear system of equations.
$$x^2 + y^2 = 1$$
$$x^2 + 9y^2 = 9$$

45. Shade the solution set in the xy-plane.
$$x^2 + y^2 \le 4$$
$$x^2 - y \le 2$$

APPLICATIONS

46. **Calculating Distance** The distance D in miles that a driver of a car is from home after x hours is given by $D(x) = 400 - 50x$.
(a) Evaluate $D(0)$. Interpret your answer.
(b) What is the x-intercept on the graph of D? Interpret your answer.
(c) What is the slope of the graph of D? Interpret your answer.

47. **Investment** Suppose a total of \$2000 is deposited in three accounts paying 5%, 6%, and 7% annual interest. The amount invested at 6% is \$500 more than the amount invested at 5%. The interest after 1 year is \$120. How much is invested at each rate?

48. **Area** There are 1200 feet of fence available to surround the perimeter of a rectangular garden. What dimensions for the garden give the largest area?

49. **Population** The population of a city in millions after x years is given by $P = 2e^{0.02x}$. How long will it take for the population to double?

50. **Dimensions of a Can** Find the radius of a cylindrical can with a volume V of 60 cubic inches and a side area S of 50 square inches. (*Hint:* $V = \pi r^2 h$ and $S = 2\pi rh$.)

14

Sequences and Series

In this chapter, we present sequences and series, which are essential topics because they are used to model and approximate important quantities. Complicated population growth, such as that experienced by populations of insects and other species over time, can be modeled with *sequences*. Accurate approximations for numbers such as π and e are made with *series*. For example, in about 2000 B.C., the Babylonians thought that π equaled 3.125, whereas at the same time the Chinese thought π equaled 3. By 1700, mathematicians were able to calculate π to 100 decimal places due to the discovery of series. Series are also essential to the solutions of many modern applied mathematics problems. (*Source:* P. Beckman, *A History of Pi*.)

14.1 Sequences

Objectives

1 Finding Terms of a Sequence

2 Recognizing Representations of Sequences

3 Modeling Population Growth with Sequences

NEW VOCABULARY

☐ Terms of a sequence
☐ Finite sequence
☐ Infinite sequence
☐ *nth* term
☐ General term

1 Finding Terms of a Sequence

🌐 *Math in Context* — Biology — Sequences are *ordered lists*. For example, names listed alphabetically represent a sequence. Listing populations by year is another example of a sequence. **FIGURE 14.1** shows an insect population in thousands per acre over a 6-year period. In mathematics, a sequence is a function for which valid inputs must be natural numbers. For example, we can use a function f to define this sequence by letting $f(1)$ represent the insect population after 1 year, $f(2)$ represent the insect population after 2 years, and in general let $f(n)$ represent the population after n years.

An Insect Population

$f(1) = 5000$, so after 1 year the insect population is 5000 per acre.

FIGURE 14.1

🌐 *Math in Context* — Earnings — Suppose that an individual's starting salary is $40,000 per year and that the person's salary is increased by 10% each year. This situation is modeled by the formula

$$f(n) = 40{,}000(1.10)^n.$$ Symbolic representation

We do not allow the input n to be any real number, but rather limit n to a *natural number* because the individual's salary is constant throughout a particular year. The first five *terms of the sequence* are

$$f(1), f(2), f(3), f(4), f(5).$$

They can be computed as follows.

The first five terms of the sequence are found by evaluating function f for $n = 1, 2, 3, 4, 5$.

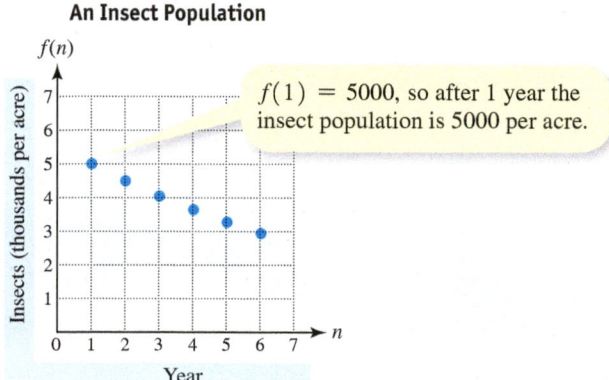

$$f(1) = 40{,}000(1.10)^1 = 44{,}000$$
$$f(2) = 40{,}000(1.10)^2 = 48{,}400$$
$$f(3) = 40{,}000(1.10)^3 = 53{,}240$$
$$f(4) = 40{,}000(1.10)^4 = 58{,}564$$
$$f(5) = 40{,}000(1.10)^5 \approx 64{,}420$$

The first five terms of a sequence

This sequence is represented *numerically* in **TABLE 14.1** and *graphically* in **FIGURE 14.2**.

Numerical and Graphical Representations of a Sequence

Table of values (numerical representation)

n	$f(n)$
1	44,000
2	48,400
3	53,240
4	58,564
5	64,420

TABLE 14.1

Plot each data point from Table 14.1 to obtain a scatterplot. (graphical representation)

FIGURE 14.2

NOTE: Graphs of sequences are scatterplots. ∎

The preceding sequence is an example of a *finite sequence* of numbers. The even natural numbers,

$$2, 4, 6, 8, 10, 12, 14, \ldots$$ Infinite sequence

are an example of an *infinite sequence* represented by $f(n) = 2n$, where n is a natural number. The three dots, or periods (called an *ellipsis*), indicate that the pattern continues indefinitely.

> ## SEQUENCES
>
> A **finite sequence** is a function whose domain is $D = \{1, 2, 3, \ldots, n\}$ for some fixed natural number n.
> An **infinite sequence** is a function whose domain is the set of natural numbers.

READING CHECK 1

• How do we denote the fourth term of a sequence?

Because sequences are functions, many of the concepts discussed in previous chapters apply to sequences. Instead of letting y represent the output, however, the convention is to write $a_n = f(n)$, where n is a natural number in the domain of the sequence. The *terms* of a sequence are

Terms of a sequence $a_1, a_2, a_3, \ldots, a_n, \ldots$ nth term

To calculate the terms of a sequence, the first term is given by $a_1 = f(1)$, the second term is given by $a_2 = f(2)$, and so on. The **nth term**, or **general term**, of a sequence is given by $a_n = f(n)$.

EXAMPLE 1   **Computing terms of a sequence**

Write the first four terms of each sequence for $n = 1, 2, 3,$ and 4.

(a) $f(n) = 2n - 1$ **(b)** $f(n) = 3(-2)^n$ **(c)** $f(n) = \dfrac{n}{n+1}$

Solution

(a) For $a_n = f(n) = 2n - 1$, we write the first four terms as follows.

$$a_1 = f(1) = 2(1) - 1 = 1$$
$$a_2 = f(2) = 2(2) - 1 = 3$$
$$a_3 = f(3) = 2(3) - 1 = 5$$
$$a_4 = f(4) = 2(4) - 1 = 7$$

The first four terms are **1, 3, 5**, and **7**.

(b) For $a_n = f(n) = 3(-2)^n$, we write the first four terms as follows.

$$a_1 = f(1) = 3(-2)^1 = -6$$
$$a_2 = f(2) = 3(-2)^2 = 12$$
$$a_3 = f(3) = 3(-2)^3 = -24$$
$$a_4 = f(4) = 3(-2)^4 = 48$$

The first four terms are -6, **12**, -24, and **48**.

(c) For $a_n = f(n) = \dfrac{n}{n+1}$, we write the first four terms as follows.

$$a_1 = f(1) = \frac{1}{1+1} = \frac{1}{2}$$

$$a_2 = f(2) = \frac{2}{2+1} = \frac{2}{3}$$

$$a_3 = f(3) = \frac{3}{3+1} = \frac{3}{4}$$

$$a_4 = f(4) = \frac{4}{4+1} = \frac{4}{5}$$

The first four terms are $\frac{1}{2}, \frac{2}{3}, \frac{3}{4}$, and $\frac{4}{5}$. Note that, although the *input* to a sequence is a natural number, the *output* need not be a natural number.

Now Try Exercises 9, 11, 13

TECHNOLOGY NOTE

Generating Sequences

Many graphing calculators can generate sequences if you change the MODE from function (Func) to sequence (Seq). In **FIGURES 14.3** and **14.4**, the sequences from Example 1 are generated. On some calculators, the sequence utility is found in the LIST OPS menus. The expression

$$\text{seq}(2n - 1, n, 1, 4)$$

represents terms 1 through 4 of the sequence $f(n) = 2n - 1$ with the variable n.

```
seq(2n-1,n,1,4)
           {1 3 5 7}
seq(3(-2)^n,n,1,
4)
     {-6 12 -24 48}
```

FIGURE 14.3

```
seq(n/(n+1),n,1,
4)▶Frac
{1/2 2/3 3/4 4/...
```

FIGURE 14.4

2 Recognizing Representations of Sequences

Because sequences are functions, they can be represented symbolically, graphically, and numerically. The next two examples illustrate such representations.

EXAMPLE 2   Using a graphical representation

Use **FIGURE 14.5** to write the terms of the sequence.

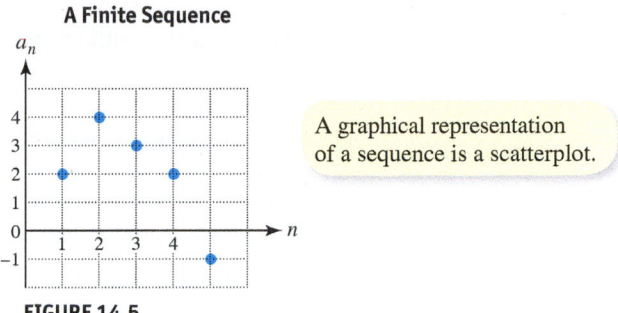

A graphical representation of a sequence is a scatterplot.

FIGURE 14.5

Solution
The points $(1, 2)$, $(2, 4)$, $(3, 3)$, $(4, 2)$, and $(5, -1)$ are shown in the graph. The terms of the sequence are $2, 4, 3, 2,$ and -1.

Now Try Exercise 33

EXAMPLE 3   Representing a sequence

In 2014, the average person in the United States used about 100 gallons of water at home each day. Give symbolic, numerical, and graphical representations for a sequence that models the total amount of water used over a 7-day period. (*Source:* U.S. Geological Survey.)

Solution
Symbolic Representation Let

$$a_n = 100n \text{ for } n = 1, 2, 3, \ldots, 7.$$

Numerical Representation **TABLE 14.2** contains the sequence.

Graphical Representation Plot the points $(1, 100)$, $(2, 200)$, $(3, 300)$, $(4, 400)$, $(5, 500)$, $(6, 600)$, and $(7, 700)$, as shown in **FIGURE 14.6**.

A numerical representation of a sequence is a table of values.

n	a_n
1	100
2	200
3	300
4	400
5	500
6	600
7	700

TABLE 14.2

Personal Water Usage

FIGURE 14.6

A graphical representation of a sequence is a scatterplot.

Now Try Exercise 43

3 Modeling Population Growth with Sequences

🌐 *Math in Context* *Entomology* A population model for a species of insect with a life span of 1 year can be described with a sequence. Suppose that each adult female insect produces, on average, r female offspring that survive to reproduce the following year. Let a_n represent the female insect population at the *beginning* of year n. Then the number of female insects is given by

$$a_n = Cr^{n-1},$$

where C is the initial population of female insects. (*Source:* D. Brown and P. Rothery, *Models in Biology.*)

EXAMPLE 4 **Modeling numbers of insects**

Suppose that the initial population of adult female insects is 500 per acre and that $r = 1.04$. Then the average number of female insects per acre at the beginning of year n is described by

$$a_n = 500(1.04)^{n-1}.$$

Represent the female insect population numerically and graphically for 7 years. Round values to the nearest whole number. Discuss the results. By what percent is the population increasing each year?

Solution

Numerical Representation **TABLE 14.3** contains *approximations* for the first 7 terms of the sequence. The insect population increases from 500 to about 633 insects per acre during this time period.

An Insect Population (per acre)

n	1	2	3	4	5	6	7
a_n	500	520	541	562	585	608	633

TABLE 14.3

Graphical Representation Plot the points $(1, 500)$, $(2, 520)$, $(3, 541)$, $(4, 562)$, $(5, 585)$, $(6, 608)$, and $(7, 633)$, as shown in **FIGURE 14.7**. These results indicate that the insect population gradually increases. Because the growth factor is 1.04, the population is increasing by 4% each year.

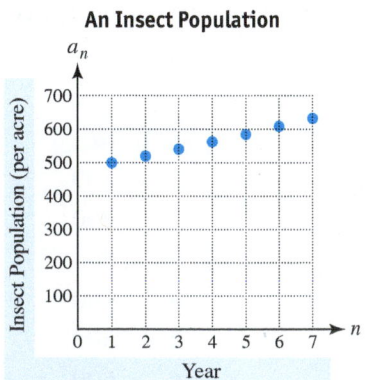

An Insect Population

FIGURE 14.7

CRITICAL THINKING 1

Explain how the value of r in Example 4 affects the population of female insects over time. Assume that $r > 0$.

Now Try Exercise 49

TECHNOLOGY NOTE

Graphs and Tables of Sequences
In sequence mode, many graphing calculators are capable of representing sequences graphically and numerically. **FIGURE 14.8(a)** shows how to enter the sequence from Example 4 to produce the table of values shown in **FIGURE 14.8(b)**.

(a)

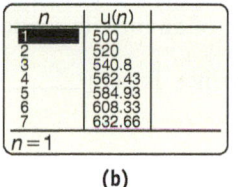

(b)

FIGURE 14.8

FIGURES 14.9(a) and **(b)** show the set-up used for graphing the sequence from Example 4. The resulting graph is shown in **FIGURE 14.9(c)**.

(a)

(b)

(c)

FIGURE 14.9

14.1 Putting It All Together

An infinite sequence is a function whose domain is the set of natural numbers. A finite sequence has the domain $D = \{1, 2, 3, \dots, n\}$ for some fixed natural number n.

REPRESENTATION	EXAMPLE
Symbolic	$a_n = n - 3$ represents a sequence. The first four terms are $-2, -1, 0,$ and 1: $$a_1 = 1 - 3 = -2, \qquad a_2 = 2 - 3 = -1,$$ $$a_3 = 3 - 3 = 0, \qquad a_4 = 4 - 3 = 1.$$
Numerical	A numerical representation for $a_n = n - 3$ with $n = 1, 2, 3,$ and 4 is shown in the table. 	n	1	2	3	4	 	a_n	-2	-1	0	1	 Finite sequence: $-2, -1, 0, 1$
Graphical	For a graphical representation of the first four terms of $a_n = n - 3$, the points $(1, -2)$, $(2, -1)$, $(3, 0)$, and $(4, 1)$ from the previous table are plotted. It is a scatterplot. Finite sequence: $-2, -1, 0, 1$

14.1   Exercises

CONCEPTS AND VOCABULARY

1. Give an example of a finite sequence.

2. Give an example of an infinite sequence.

3. An infinite sequence is a(n) _____ whose domain is the set of _____.

4. An ordered list is a(n) _____.

5. The third term in the sequence $4, -5, 6, -7, 8$ is _____.

6. The graph of a sequence is not a continuous graph but rather a(n) _____.

7. If $f(n)$ represents a sequence, the second term of the sequence is given by _____.

8. If a_n represents a sequence, the fourth term of the sequence is given by _____.

EVALUATING SEQUENCES

Exercises 9–16: Write the first four terms of the sequence for $n = 1, 2, 3,$ and 4.

9. $f(n) = n^2$

10. $f(n) = 3n + 4$

11. $f(n) = \dfrac{1}{n + 5}$

12. $f(n) = 3^n$

13. $f(n) = 5\left(\dfrac{1}{2}\right)^n$

14. $f(n) = n^2 + 2n$

15. $f(n) = 9$

16. $f(n) = (-1)^n$

Exercises 17–24: Write the first three terms of the given sequence for $n = 1, 2,$ and 3.

17. $a_n = n^3$

18. $a_n = 5 - n$

19. $a_n = \dfrac{4n}{3 + n}$

20. $a_n = 3^{-n}$

21. $a_n = 2n^2 + n - 1$

22. $a_n = n^4 - 1$

23. $a_n = -2$

24. $a_n = n^n$

Exercises 25–28: The first five terms of a sequence are shown. Write a formula $f(n)$ that represents the sequence.

25. $1, \frac{1}{2}, \frac{1}{3}, \frac{1}{4}, \frac{1}{5}, \frac{1}{6}$

26. $\frac{1}{2}, \frac{2}{3}, \frac{3}{4}, \frac{4}{5}, \frac{5}{6}$

27. $3, 5, 7, 9, 11$

28. $3, 6, 9, 12, 15$

Exercises 29 and 30: Thinking Generally Let b and c be fixed numbers (constants). Find a_1 and a_2.

29. $a_n = bn + c$

30. $a_n = \dfrac{n + b}{n - c}, c \neq 1$ and $c \neq 2$

REPRESENTING SEQUENCES

Exercises 31 and 32: Use the numerical representation to evaluate $\frac{1}{2}(a_1 + a_4)$.

31.

n	1	2	3	4	5
a_n	10	8	6	4	2

32.

n	1	2	3	4	5
a_n	-5	0	10	30	60

Exercises 33–36: Write the terms of the sequence.

33.

34.

35.

36.
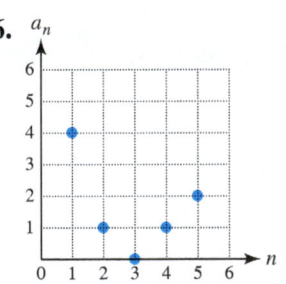

Exercises 37–42: Represent the first seven terms of the sequence numerically and graphically.

37. $a_n = n + 1$

38. $a_n = \frac{1}{2}n - \frac{1}{2}$

39. $a_n = n^2 - n$

40. $a_n = \frac{1}{2}n^2$

41. $a_n = 2^n$

42. $a_n = 2(0.5)^n$

APPLICATIONS OF SEQUENCES

43. **Solid Waste** On average, each U.S. resident generates about 30 pounds of solid waste per week. Give

symbolic, numerical, and graphical representations for a sequence that models the total amount of waste produced over a 7-week period. (*Source:* Environmental Protection Agency.)

44. **Carbon Dioxide Emitters** Because people burn fossil fuels, the United States emits about 5.8 billion metric tons of carbon dioxide per year. (A metric ton is about 2200 pounds.) Give symbolic, numerical, and graphical representations for a sequence that models the total amount of carbon dioxide emitted in billions of metric tons in the United States during a 5-year period. (*Source:* Energy Information Administration.)

45. **Geometry** The lengths of the sides of a sequence of squares are given by 1, 2, 3, and 4, as shown in the figure below. Write sequences that give the following.
    (a) the areas of the squares
    (b) the perimeters of the squares

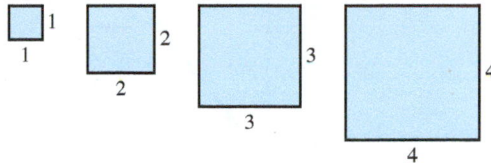

46. **Salaries** An individual's starting salary is $50,000, and the individual receives an increase of 8% per year. Give symbolic, numerical, and graphical representations for this person's salary over 5 years.

47. **Depreciation** Automobiles usually depreciate in value over time. Often, a newer automobile may be worth only 80% of its previous year's value. Suppose that a car is worth $25,000 new.
    (a) How much is it worth after 1 year? After 2 years?
    (b) Write a formula for a sequence that gives the car's value after n years.
    (c) Make a table that shows how much the car was worth each year during the first 7 years.

48. **Falling Object** The distance d that an object falls during *consecutive* seconds is shown in the table. For example, during the third second (from $n = 2$ to $n = 3$), an object falls a distance of 80 feet.

n (seconds)	1	2	3	4	5
d (feet)	16	48	80	112	144

   (a) Find values for c and b so that $d = cn + b$ models these data.
   (b) How far does an object fall during the sixth second?

49. **Modeling Insect Populations** (Refer to Example 4.) Suppose that the initial population of insects is 2048 per acre and that $r = 0.5$. Use a sequence to represent the insect population over a 7-year period
    (a) symbolically,
    (b) numerically, and
    (c) graphically.

50. **Auditorium Seating** An auditorium has 50 seats in the first row, 55 seats in the second row, 60 seats in the third row, and so on.
    (a) Make a table that shows the number of seats in the first seven rows.
    (b) Write a formula that gives the number of seats in row n.
    (c) How many seats are there in row 23?
    (d) Graph the number of seats in each row for $n = 1, 2, 3, \ldots, 10$.

WRITING ABOUT MATHEMATICS

51. Compare the graph of the function $f(x) = 2x + 1$, where x is a real number, with the graph of the sequence $f(n) = 2n + 1$, where n is a natural number.

52. Explain what a sequence is. Describe the difference between a finite and an infinite sequence.

14.2 Arithmetic and Geometric Sequences

NEW VOCABULARY

☐ Arithmetic sequence
☐ Common difference
☐ Geometric sequence
☐ Common ratio

1 Recognizing Representations of Arithmetic Sequences

DEFINING ARITHMETIC SEQUENCES If a sequence is defined by a linear function, it is an *arithmetic sequence*. For example,

$$f(n) = 2n - 3$$

represents an arithmetic sequence because $f(x) = 2x - 3$ defines a linear function.

See the Concept 1   ARITHMETIC SEQUENCE

For the arithmetic sequence $f(n) = 2n - 3$:

A If n increases by 1, **B** then $f(n)$ increases by 2.

B The difference between any two successive terms is 2, so the *common difference* is 2.

C The graph of $f(n) = 2n - 3$ consists of equally spaced points on the (blue) line $y = 2x - 3$.

D The slope of the line $y = 2x - 3$ is $m = 2$. It is equal to the *common difference* $d = 2$.

WATCH VIDEO IN $\boxed{\text{MML}}$.

READING CHECK 1

• Given a term in an arithmetic sequence, how can you find the next term and the previous term?

ARITHMETIC SEQUENCE

An **arithmetic sequence** is a linear function given by $a_n = dn + c$ whose domain is the set of natural numbers. The value of d is called the **common difference**.

NOTE: If each term after the first term is obtained by adding a fixed number to the previous term, then the sequence is an arithmetic sequence. The fixed number is the *common difference*. For example, 1, 6, 11, 16, 21, … is an arithmetic sequence because each term (after the first) is found by adding the common difference of **5** to the previous term. That is, $6 = 1 + 5$, $11 = 6 + 5$, $16 = 11 + 5$, and so on. The difference between successive terms is always 5. ∎

EXAMPLE 1   Recognizing arithmetic sequences

Determine whether f is an arithmetic sequence. If it is, identify the common difference d.
(a) $f(n) = 2 - 3n$

(b)

n	$f(n)$
1	10
2	5
3	0
4	-5
5	-10

(c) A graph of f is shown in **FIGURE 14.10**.

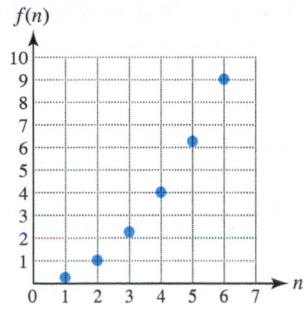

FIGURE 14.10

Solution

(a) This sequence is arithmetic because $f(x) = -3x + 2$ defines a linear function. The common difference is $d = -3$.

(b) The table reveals that each term is found by adding -5 to the previous term. This represents an arithmetic sequence with common difference $d = -5$.

(c) The sequence shown in **FIGURE 14.10** is not an arithmetic sequence because the points are not collinear. That is, the points do not lie on a line and there is no common difference.

Now Try Exercises 11, 17, 25

MAKING CONNECTIONS 1

Common Difference and Slope

The common difference d of an arithmetic sequence equals the slope of the line passing through the collinear points. For example, if $a_n = -2n + 4$, the common difference is -2, and the slope of the line passing through the points on the graph of a_n is also -2, as shown in **FIGURE 14.11**.

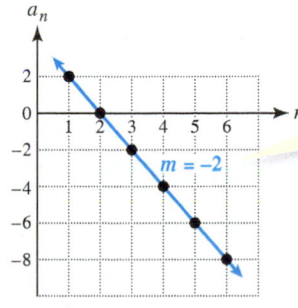

The common difference of the sequence

$$2, 0, -2, -4, -6, -8$$

is -2, which equals the slope of the blue line.

FIGURE 14.11

FINDING ARITHMETIC SEQUENCES In the next example, we write the formula for the general term of an arithmetic sequence.

EXAMPLE 2 **Finding symbolic representations**

Find the general term a_n for each arithmetic sequence.

(a) $a_1 = 3$ and $d = 4$ **(b)** $a_1 = 3$ and $a_4 = 12$

continued on next page

Solution

(a) Let $a_1 = 3$, $d = 4$, and $a_n = dn + c$. For $d = 4$, we write $a_n = 4n + c$, and to find c, we use $a_1 = 3$.

$$a_1 = 4(1) + c = 3 \quad \text{or} \quad c = -1$$

Thus $a_n = 4n - 1$.

(b) Because $a_1 = 3$ and $a_4 = 12$, the common difference d equals the slope of the line passing through the points $(1, 3)$ and $(4, 12)$, or

$$d = \frac{12 - 3}{4 - 1} = 3.$$

Therefore $a_n = 3n + c$. To find c, we use $a_1 = 3$ and obtain

$$a_1 = 3(1) + c = 3 \quad \text{or} \quad c = 0.$$

Thus $a_n = 3n$.

Now Try Exercises 29, 31

Consider the arithmetic sequence

> The common difference is 4 because the difference between successive terms is 4.

$$1, 5, 9, 13, 17, 21, 25, 29, \ldots.$$

The common difference is $d = 4$, and the first term is $a_1 = 1$. To find the second term, we add d to the first term. To find the third term, we add $2d$ to the first term, and to find the fourth term, we add $3d$ to the first term. That is,

$$a_1 = 1,$$
$$a_2 = a_1 + 1d = 1 + 1 \cdot 4 = 5,$$
$$a_3 = a_1 + 2d = 1 + 2 \cdot 4 = 9,$$
$$a_4 = a_1 + 3d = 1 + 3 \cdot 4 = 13,$$

> First four terms

and, in general, a_n is determined by

$$a_n = a_1 + (n - 1)d = 1 + (n - 1)4.$$

This result suggests the following formula.

READING CHECK 2

- Write the general term of an arithmetic sequence, given the first term and the common difference.

GENERAL TERM OF AN ARITHMETIC SEQUENCE

The nth term a_n of an arithmetic sequence is given by

$$a_n = a_1 + (n - 1)d,$$

where a_1 is the first term and d is the common difference.

EXAMPLE 3 **Finding terms of an arithmetic sequence**

If $a_1 = 5$ and $d = 3$, find a_{54}.

Solution

To find a_{54}, apply the formula $a_n = a_1 + (n - 1)d$ with $a_1 = 5$, $n = 54$, and $d = 3$.

$$a_{54} = 5 + (54 - 1)3 = 164$$

Now Try Exercise 35

2 Recognizing Representations of Geometric Sequences

DEFINING GEOMETRIC SEQUENCES The following See the Concept describes a geometric sequence.

See the Concept 2   GEOMETRIC SEQUENCE

For the geometric sequence $f(n) = 3(2)^{n-1}$:

A If n increases by 1...

n	$f(n)$
1	3
2	6
3	12
4	24
5	48

B then $f(n)$ increases by a factor of 2.

C Points lie on a curve not a line.

D The ratio of successive terms is 2.

$y = 3(2)^{x-1}$

A If n increases by 1, **B** then $f(n)$ increases by a factor of 2.

B Each new term is found by multiplying the previous term by 2, so the *common ratio* is 2.

C The graph of $f(n) = 3(2)^{n-1}$ consists of points that lie on the exponential curve $y = 3(2)^{x-1}$.

D The ratio of any two successive terms is 2, so the *common ratio* is 2.

WATCH VIDEO IN MML.

A geometric sequence with a positive common ratio is an exponential function whose domain is the set of natural numbers. Its terms reflect either *exponential growth* or *decay*.

READING CHECK 3

• Given a term in a geometric sequence, how can you find the next term and the previous term?

GEOMETRIC SEQUENCE

A **geometric sequence** is given by $a_n = a_1(r)^{n-1}$, where n is a natural number and r is neither 0 nor 1. The **common ratio** is r, and a_1 is the first term of the sequence.

NOTE: If each term after the first term is obtained by multiplying the previous term by a fixed number, the sequence is a geometric sequence. The fixed number can be either positive or negative, but it cannot be 0 or 1. ∎

EXAMPLE 4

Recognizing geometric sequences

Determine whether f is a geometric sequence. If it is, identify the common ratio.
(a) $f(n) = 2(0.9)^{n-1}$ **(c)** A graph of f is shown in **FIGURE 14.12**.
(b)

n	$f(n)$
1	8
2	4
3	2
4	1
5	$\frac{1}{2}$

FIGURE 14.12

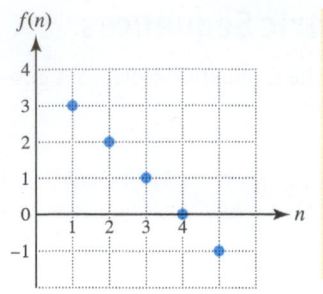

FIGURE 14.12 (repeated)

Solution

(a) This sequence is geometric because $f(x) = 2(0.9)^{x-1}$ defines an exponential function. The common ratio is $r = 0.9$.

(b) The table shows that each successive term 8, 4, 2, 1, $\frac{1}{2}$, is half the previous term. This sequence represents a geometric sequence with a common ratio of $r = \frac{1}{2}$.

(c) The sequence shown in **FIGURE 14.12** (repeated in the margin) is not a geometric sequence because the points are collinear. There is no common ratio.

Now Try Exercises 41, 45, 53

FINDING GEOMETRIC SEQUENCES In the next example, we write the formula for the general term of a geometric sequence.

EXAMPLE 5 | **Finding symbolic representations**

Find a general term a_n for each geometric sequence.
(a) $a_1 = \frac{1}{2}$ and $r = 5$
(b) $a_1 = 2$, $a_3 = 18$, and $r < 0$.

Solution

(a) Let $a_n = a_1(r)^{n-1}$. Because $a_1 = \frac{1}{2}$ and $r = 5$, we can write $a_n = \frac{1}{2}(5)^{n-1}$.

(b) $a_1 = 2$ and $a_3 = 18$, so

$$a_3 = a_1(r)^{3-1}$$
$$18 = 2(r)^2.$$

CRITICAL THINKING 1

If we are given a_1 and a_5, can we determine the common ratio of a geometric series? Explain.

This equation simplifies to

$$r^2 = 9 \text{ or } r = \pm 3.$$

It is specified that $r < 0$, so $r = -3$ and $a_n = 2(-3)^{n-1}$. Note that the common ratio r can be *negative*. If $r < 0$, successive terms alternate sign (positive or negative).

Now Try Exercises 55, 59

EXAMPLE 6 | **Finding a term of a geometric sequence**

If $a_1 = 5$ and $r = 3$, find a_{10}.

Solution

To find a_{10}, apply the formula $a_n = a_1(r)^{n-1}$ with $a_1 = 5$, $r = 3$, and $n = 10$.

$$a_{10} = 5(3)^{10-1} = 5(3)^9 = 98{,}415$$

Now Try Exercise 61

3 ▸ Solving Applications Involving Sequences

Public Health

🌐 *Math in Context* Indoor air pollution has become more hazardous as people spend 80% to 90% of their time in tightly sealed, energy-efficient buildings, which often lack proper ventilation. In the next example, we use a sequence to model proper classroom ventilation.

EXAMPLE 7 | **Modeling classroom ventilation**

Ventilation is an effective means for removing indoor air pollutants. According to the American Society of Heating, Refrigerating, and Air-Conditioning Engineers (ASHRAE), a classroom should have a ventilation rate of 900 cubic feet per hour per person.

(a) Write a sequence that gives the hourly ventilation necessary for 1, 2, 3, 4, and 5 people in a classroom. Is this sequence arithmetic, geometric, or neither?

(b) Write the general term for this sequence. Why is it reasonable to limit the domain to natural numbers?

(c) Find a_{30} and interpret the result.

Solution

(a) One person requires 900 cubic feet of air circulated per hour, two people require 1800, three people 2700, and so on. The first five terms of this sequence are

$$900, 1800, 2700, 3600, 4500.$$

> To find the next term, add 900 to the previous term.

This sequence is arithmetic, with a common difference of 900.

(b) The nth term equals $900n$, so we let $a_n = 900n$. Because we cannot have a fraction of a person, limiting the domain to the natural numbers is reasonable.

(c) The result $a_{30} = 900(\mathbf{30}) = \mathbf{27{,}000}$ indicates that a classroom with **30** people should have a ventilation rate of **27,000** cubic feet per hour.

Now Try Exercise 65

🌐 *Math in Context* (Chemistry) Chlorine is frequently added to the water to disinfect swimming pools. The chlorine concentration should remain between 1.5 and 2.5 parts per million (ppm). On a warm, sunny day, 30% of the chlorine may dissipate from the water. In the next example, we use a sequence to model the amount of chlorine in a pool at the beginning of each day. (*Source:* D. Thomas, *Swimming Pool Operator's Handbook.*)

EXAMPLE 8

Modeling chlorine in a swimming pool

A swimming pool on a warm, sunny day begins with a high chlorine content of 4 parts per million. (Assume that each day, 30% of the chlorine dissipates.)

(a) Write a sequence that models the amount of chlorine in the pool at the beginning of the first 3 days, assuming that no additional chlorine is added and that the days are warm and sunny. Is this sequence arithmetic, geometric, or neither?

(b) Write the general term for this sequence.

(c) At the beginning of what day does the chlorine first drop below 1.5 parts per million?

Solution

(a) Because 30% of the chlorine dissipates, 70% remains in the water at the beginning of the next day. If the concentration at the beginning of the first day is 4 parts per million, then at the beginning of the second day, it is

$$4 \cdot 0.70 = 2.8 \text{ parts per million,}$$

> After 1 day, 30% dissipates, or 70% remains.

and at the start of the third day, it is

$$2.8 \cdot 0.70 = 1.96 \text{ parts per million.}$$

The first three terms are 4, 2.8, 1.96. Successive terms are found by multiplying the previous term by 0.7. Thus the sequence is geometric, with common ratio 0.7.

(b) The initial amount is $a_1 = \mathbf{4}$ and the common ratio is $r = \mathbf{0.7}$, so the sequence can be represented by $a_n = \mathbf{4(0.7)}^{n-1}$.

(c) The table shown in **FIGURE 14.13** reveals that $a_4 = 4(0.7)^{4-1} \approx 1.372 < 1.5$. Thus, at the beginning of the fourth day, the chlorine level in the swimming pool drops below the recommended minimum of 1.5 parts per million.

Now Try Exercise 67

n	$u(n)$
1	4
2	2.8
3	1.96
4	1.372 ←
5	.9604
6	.67228
7	.4706

$u(n) \blacksquare 4(.7)^{\wedge}(n-1)$

FIGURE 14.13

14.2   Putting It All Together

Arithmetic sequences are linear functions, and geometric sequences with a *positive r* are exponential functions. The inputs for both are limited to the natural numbers. The graph of an arithmetic sequence consists of points that lie on a line, whereas the graph of a geometric sequence (with a positive *r*) consists of points that lie on an exponential curve.

SEQUENCE	FORMULA	EXAMPLE
Arithmetic	$a_n = dn + c$ or $a_n = a_1 + (n-1)d,$ where d is the common difference and a_1 is the first term.	If $a_n = 5n + 2$, then the common difference is $d = 5$ and the terms of the sequence are $$7, 12, 17, 22, 27, 32, 37, \ldots.$$ Each term after the first is found by adding 5 to the previous term. The general term can be written as $$a_n = 7 + 5(n-1).$$
Geometric	$a_n = a_1(r)^{n-1},$ where r is the common ratio $(r \neq 0, r \neq 1)$ and a_1 is the first term.	If $a_n = 4(-2)^{n-1}$, then the common ratio is $r = -2$ and the first term is $a_1 = 4$. The terms of the sequence are $$4, -8, 16, -32, 64, -128, 256, \ldots.$$ Each term after the first is found by multiplying the previous term by -2.

14.2   Exercises

MyMathLab®

CONCEPTS AND VOCABULARY

1. An arithmetic sequence is a(n) _____ function.

2. A geometric sequence with $r > 0$ is a(n) _____ function.

3. Give an example of an arithmetic sequence. State the common difference.

4. Give an example of a geometric sequence. State the common ratio.

5. To find successive terms in an arithmetic sequence, _____ the common difference to the _____ term.

6. To find successive terms in a geometric sequence, _____ the previous term by the _____.

7. Find the next term in the arithmetic sequence given by 3, 7, 11, 15. What is the common difference?

8. Find the next term in the geometric sequence given by 2, −4, 8, −16. What is the common ratio?

9. Write the general term a_n for a geometric sequence, using a_1 and r.

10. Write the general term a_n for an arithmetic sequence, using a_1 and d.

RECOGNIZING ARITHMETIC SEQUENCES

Exercises 11–28: (Refer to Example 1.) Determine whether f is an arithmetic sequence. Identify the common difference when possible.

11. $f(n) = 10n - 5$

12. $f(n) = -3n - 5$

13. $f(n) = 6 - n$

14. $f(n) = 6 + \frac{1}{2}n$

15. $f(n) = n^3 + 1$

16. $f(n) = 5\left(\frac{1}{3}\right)^{n-1}$

17.

n	1	2	3	4
$f(n)$	3	6	9	12

18.

n	1	2	3	4
$f(n)$	−7	−5	−3	−1

19.

n	1	2	3	4
$f(n)$	10	7	4	1

20.

n	1	2	3	4
$f(n)$	1	2	4	8

21.

n	1	2	3	4
$f(n)$	−4	0	8	12

22.

n	1	2	3	4
$f(n)$	1	2.5	4	5.5

23. $f(n)$

24. $f(n)$

25. $f(n)$

26. $f(n)$

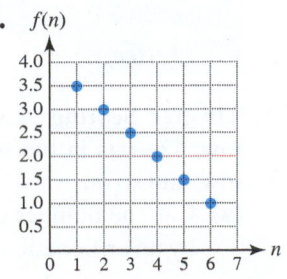

27. $f(n)$

28. $f(n)$

FINDING ARITHMETIC SEQUENCES

Exercises 29–34: (Refer to Example 2.) Find the general term a_n for the arithmetic sequence.

29. $a_1 = 7$ and $d = -2$

30. $a_1 = 5$ and $a_2 = 9$

31. $a_1 = -2$ and $a_3 = 6$

32. $a_2 = 7$ and $a_3 = 10$

33. $a_8 = 16$ and $a_{12} = 8$

34. $a_3 = 7$ and $d = -5$

Exercises 35–38: (Refer to Example 3.)

35. If $a_1 = -3$ and $d = 2$, find a_{32}.

36. If $a_1 = 2$ and $d = -3$, find a_{19}.

37. If $a_1 = -3$ and $a_2 = 0$, find a_9.

38. If $a_3 = -3$ and $d = 4$, find a_{62}.

RECOGNIZING GEOMETRIC SEQUENCES

Exercises 39–54: (Refer to Example 4.) Determine whether f is a geometric sequence. Identify the common ratio when possible.

39. $f(n) = 3^n$ **40.** $f(n) = 2(4)^n$

41. $f(n) = \frac{2}{3}(0.8)^{n-1}$ **42.** $f(n) = 7 - 3n$

43. $f(n) = 2(n-1)^2$ **44.** $f(n) = 2\left(-\frac{3}{4}\right)^{n-1}$

45.

n	1	2	3	4
$f(n)$	2	4	8	16

46.

n	1	2	3	4
$f(n)$	−6	3	−1.5	0.75

47.

n	1	2	3	4
$f(n)$	1	4	9	16

48.

n	1	2	3	4
$f(n)$	7	4	−1	−8

49.

n	1	2	3	4
$f(n)$	2	8	32	128

50.

n	1	2	3	4
$f(n)$	1	$\frac{1}{2}$	$\frac{1}{4}$	$\frac{1}{8}$

51.

52.

53.

54.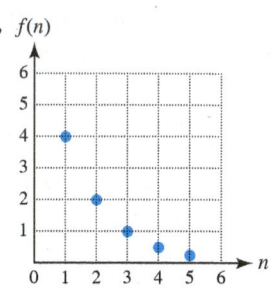

FINDING GEOMETRIC SEQUENCES

Exercises 55–60: (Refer to Example 5.) Find the general term a_n for the geometric sequence.

55. $a_1 = 1.5$ and $r = 4$ **56.** $a_1 = 3$ and $r = \frac{1}{4}$

57. $a_1 = -3$ and $a_2 = 6$ **58.** $a_1 = 2$ and $a_4 = 54$

59. $a_1 = 1$, $a_3 = 16$, and $r > 0$

60. $a_2 = 3$, $a_4 = 12$, and $r < 0$

Exercises 61–64: (Refer to Example 6.)

61. If $a_1 = 2$ and $r = 3$, find a_8.

62. If $a_1 = 4$ and $a_2 = 2$, find a_9.

63. If $a_1 = -1$ and $a_2 = 3$, find a_6.

64. If $a_3 = 5$ and $r = -3$, find a_7.

APPLICATIONS OF ARITHMETIC AND GEOMETRIC SEQUENCES

65. Room Ventilation (Refer to Example 7.) In areas that allow smoking, the ventilation rate should be 3000 cubic feet per hour per person. (*Source: ASHRAE.*)
   (a) Write a sequence that gives the hourly ventilation necessary for 1, 2, 3, 4, and 5 people in a smoking area. Is this sequence arithmetic, geometric, or neither?
   (b) Write the general term for this sequence.
   (c) Find a_{20} and interpret the result.
   (d) Give a graphical representation for eight terms of this sequence, using $n = 1, 2, 3, \ldots, 8$. Are the points collinear?

66. Salary Suppose that an employee receives a $2000 raise each year and that the sequence a_n models the employee's salary after n years. Is this sequence arithmetic, geometric, or neither? Explain.

67. Chlorine in Swimming Pools (Refer to Example 8.) Suppose that the water in a swimming pool initially has a chlorine content of 3 parts per million and that 20% of the chlorine dissipates each day.
   (a) If no additional chlorine is added, write the general term for a sequence that gives the chlorine concentration at the beginning of each day.
   (b) Give a graphical representation for this sequence, using $n = 1, 2, 3, \ldots, 8$. Are the points collinear? Is this sequence arithmetic, geometric, or neither?

68. Salary Suppose that an employee receives a 7% increase in salary each year and that the sequence a_n models the employee's salary after n years. Is this sequence arithmetic, geometric, or neither? Explain.

69. Bouncing Ball A tennis ball bounces back to 85% of the height from which it was dropped and then to 85% of the height of each successive bounce.
   (a) Write the general term for a sequence a_n that gives the maximum height of the ball on the nth bounce. Let $a_1 = 5$ feet.
   (b) Is the sequence arithmetic or geometric? Explain.
   (c) Find a_8 and interpret the result.

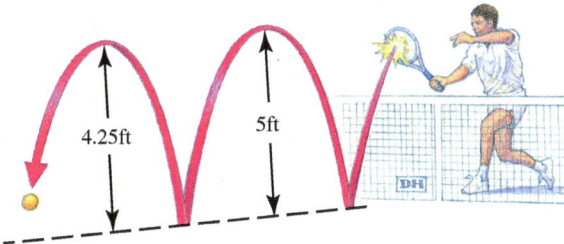

70. Falling Object The total distance D_n that an object falls in n seconds is shown in the table. Is the sequence arithmetic, geometric, or neither? Explain your reasoning.

n (seconds)	1	2	3	4	5
D_n (feet)	16	64	144	256	400

71. Theater Seating A theater has 40 seats in the first row, 42 seats in the second row, 44 seats in the third row, and so on.
   (a) Can the number of seats in each row be modeled by an arithmetic or geometric sequence? Explain.

(b) Write the general term for a sequence a_n that gives the number of seats in row n.

(c) How many seats are there in row 20?

72. Appreciation of Lake Property A certain type of lake property in northern Minnesota is increasing in value by 15% per year. Let the sequence a_n give the value of this type of lake property at the beginning of year n.

(a) Is a_n arithmetic, geometric, or neither? Explain your reasoning.

(b) Write the general term a_n for this sequence if $a_1 = \$100{,}000$.

(c) Find a_7 and interpret the result.

(d) Give a graph for a_n, where $n = 1, 2, 3, \ldots, 10$.

Exercises 73 and 74: **Recursive Sequences** *Some sequences are not defined by a formula for a_n. Instead, they are defined recursively. With a recursive formula, you must find terms a_1 through a_{n-1} before you can find a_n. For example, let*

$$a_1 = 2$$
$$a_n = a_{n-1} + 3, \quad \text{for } n \geq 2.$$

To find a_2, a_3, and a_4, we let $n = 2, 3, 4$.

$$a_2 = a_1 + 3 = 2 + 3 = 5$$
$$a_3 = a_2 + 3 = 5 + 3 = 8$$
$$a_4 = a_3 + 3 = 8 + 3 = 11$$

The first four terms of the sequence are 2, 5, 8, 11.

73. Fibonacci Sequence The Fibonacci sequence dates back to 1202 and is one of the most famous sequences in mathematics. It can be defined recursively as follows.

$$a_1 = 1, \quad a_2 = 1$$
$$a_n = a_{n-1} + a_{n-2}, \quad \text{for } n \geq 3$$

Find the first 12 terms of this sequence.

74. Insect Populations Frequently, the population of a particular insect does not continue to grow indefinitely. Instead, its population grows rapidly at first and then levels off because of competition for limited resources. In one study, the behavior of the winter moth was modeled with a sequence similar to the following, where a_n gives the population density in thousands per acre during year n.

$$a_1 = 1$$
$$a_n = 2.85a_{n-1} - 0.19a_{n-1}^2, \quad n \geq 2$$

(*Source:* G. Varley and G. Gradwell, "Population models for the winter moth.")

(a) Make a table for $n = 1, 2, 3, \ldots, 7$. Describe what happens to the population density of the winter moth.

(b) Graph the sequence for $n = 1, 2, 3, \ldots, 20$. Discuss the graph.

NOTE: Many graphing calculators are capable of generating tables and graphs for a recursive sequence. ∎

WRITING ABOUT MATHEMATICS

75. If you are given a table of values for a sequence, how can you determine if the sequence is geometric? Give an example.

76. If you are given a graph of a sequence, how can you determine if the sequence is arithmetic? Give an example.

SECTIONS 14.1 and 14.2

Checking Basic Concepts

1. Write the first four terms of the sequence defined by $a_n = \dfrac{n}{n + 4}$.

2. Represent the sequence $a_n = n + 1$ numerically and graphically for $n = 1, 2, 3, 4, 5$.

3. Use the table to determine whether the sequence is arithmetic or geometric. Write the general term for the sequence.

(a)

n	1	2	3	4	5
a_n	−2	1	4	7	10

(b)

n	1	2	3	4	5
a_n	3	−6	12	−24	48

4. Determine the general term a_n for an arithmetic sequence with $a_1 = 5$ and $d = 2$.

5. Determine the general term a_n for a geometric sequence with $a_1 = 5$ and $r = 2$.

14.3 Series

Objectives

1 Introducing Series
2 Working with Arithmetic Series
3 Working with Geometric Series
4 Using Summation Notation

NEW VOCABULARY

☐ Finite series
☐ Arithmetic series
☐ Sum of the first n terms of an arithmetic series
☐ Geometric series
☐ Sum of the first n terms of a geometric series
☐ Annuity
☐ Summation notation
☐ Index of summation
☐ Lower limit
☐ Upper limit

1 Introducing Series

Although the terms *sequence* and *series* are sometimes used interchangeably in everyday life, they represent different mathematical concepts. A sequence is an *ordered list* of numbers, whereas a series is a *summation of the terms* in a sequence.

Suppose that a person has a starting salary of $30,000 per year and receives a $2000 raise each year. Then the *sequence*

$$30{,}000, \ 32{,}000, \ 34{,}000, \ 36{,}000, \ 38{,}000$$

Sequence

A sequence is an *ordered list*.

lists these salaries over a 5-year period. The total amount earned is given by the *series*

$$30{,}000 \ + \ 32{,}000 \ + \ 34{,}000 \ + \ 36{,}000 \ + \ 38{,}000,$$

Series

A series is the *sum of the terms* of a sequence.

whose sum is $170,000.

See the Concept 1   **SEQUENCES AND SERIES**

For the first seven odd integers:

A Sequence — 1, 3, 5, 7, 9, 11, 13

A A sequence is an ordered list of terms.

B Series — $1 + 3 + 5 + 7 + 9 + 11 + 13$

B A series is the sum of the terms in a sequence.

C Sum of the series — 49

C When we evaluate the sum in **B**, we say that we are finding the "sum of series."

WATCH VIDEO IN MML.

We now define the concept of a series.

FINITE SERIES

A **finite series** is an expression of the form

$$a_1 + a_2 + a_3 + \cdots + a_n.$$

EXAMPLE 1   Twitter worldwide advertising revenues

TABLE 14.4 presents a sequence a_n that gives Twitter worldwide advertising revenues in millions of dollars n years after 2011. For example, $a_3 = 1095$ indicates that in 2014, advertising revenues were $1095 million.

Twitter Advertising

n	1	2	3	4	5
a_n	269	595	1095	1692	2444

TABLE 14.4
Source: Statista.

(a) Write a series whose sum represents the total worldwide advertising revenues for Twitter from 2012 to 2016. Find the sum of the series.

(b) Interpret $a_1 + a_2 + a_3 + a_4 + a_5$.

Solution

(a) The required series and sum are given by

$$269 + 595 + 1095 + 1692 + 2444 = 6095.$$

(b) Revenues from 2012 to 2016 were \$6095 million, or \$6.095 billion. This series represents total revenues for 5 years from 2012 to 2016.

Now Try Exercise 45

READING CHECK 1

• How do you distinguish between a sequence and a series?

2 Working with Arithmetic Series

Summing the terms of an arithmetic sequence results in an **arithmetic series**. For example, $a_n = 2n - 1$ for $n = 1, 2, 3, \ldots, 7$ defines the arithmetic sequence

$$\underbrace{1, 3, 5, 7, 9, 11, 13.}_{\textbf{Arithmetic sequence}}$$

The corresponding arithmetic *series* is

$$\underbrace{1 + 3 + 5 + 7 + 9 + 11 + 13,}_{\textbf{Arithmetic series}}$$

> An arithmetic series is the sum of the terms of an arithmetic sequence.

whose sum is 49. The following formula gives the sum of the first n terms of an arithmetic sequence. Note that the sum of the first n terms of a sequence is a finite series.

READING CHECK 2

• Use a formula to find the sum of $1 + 5 + 9 + 13 + 17 + 21$. Check your answer by adding the terms.

SUM OF THE FIRST n TERMS OF AN ARITHMETIC SEQUENCE

The **sum of the first n terms of an arithmetic sequence**, denoted S_n, is found by averaging the first and nth terms and then multiplying by n. That is,

$$S_n = a_1 + a_2 + a_3 + \cdots + a_n = n\left(\frac{a_1 + a_n}{2}\right).$$

The series $1 + 3 + 5 + 7 + 9 + 11 + 13$ consists of **7** terms, where the first term is **1** and the last term is **13**. Substituting in the formula gives

$$S_7 = 7\left(\frac{1 + 13}{2}\right) = 49,$$

which agrees with the sum obtained by adding the 7 terms.

Because $a_n = a_1 + (n - 1)d$ for an arithmetic sequence, S_n can also be written

$$S_n = n\left(\frac{a_1 + a_n}{2}\right)$$

$$= \frac{n}{2}(a_1 + a_n)$$

$$= \frac{n}{2}(a_1 + a_1 + (n - 1)d)$$

$$= \frac{n}{2}(2a_1 + (n - 1)d).$$

EXAMPLE 2 **Finding the sum of the terms of an arithmetic sequence**

Suppose that a person has a starting annual salary of $30,000 and receives a $1500 raise each year. Calculate the total amount earned after 10 years.

Solution
The sequence that gives the salary during year n is given by

$$a_n = 30{,}000 + 1500(n - 1).$$

One way to calculate the sum of the first 10 terms, denoted S_{10}, is to find a_1 and a_{10}.

$$a_1 = 30{,}000 + 1500(1 - 1) = \mathbf{30{,}000} \quad \text{— Salary after first year}$$
$$a_{10} = 30{,}000 + 1500(10 - 1) = \mathbf{43{,}500}$$
$$\text{Salary after tenth year}$$

Thus the total amount earned during this 10-year period is

$$S_{10} = 10\left(\frac{a_1 + a_{10}}{2}\right)$$

$$= 10\left(\frac{30{,}000 + 43{,}500}{2}\right)$$

$$= \$367{,}500.$$

This sum can also be found with the second formula by letting $d = \mathbf{1500}$.

$$S_n = \frac{n}{2}(2a_1 + (n - 1)d)$$

$$= \frac{10}{2}(2 \cdot 30{,}000 + (10 - 1)1500)$$

$$= 5(60{,}000 + 9 \cdot 1500)$$

$$= \$367{,}500.$$

NOTE: $\underbrace{30{,}000 + 31{,}500 + 33{,}000 + 34{,}500 + \cdots + 43{,}500}_{\text{10 terms}} = 367{,}500$ ∎

Now Try Exercise 53

EXAMPLE 3 **Finding the sum of the terms of an arithmetic sequence**

Use a formula to find the sum of the series $2 + 4 + 6 + \cdots + 100$.

Solution
The first term is $a_1 = \mathbf{2}$, and the common difference is $d = 2$. This series represents the even numbers from $\mathbf{2}$ to $\mathbf{100}$, so the number of terms is $n = \mathbf{50}$. Using the formula

$$S_n = n\left(\frac{a_1 + a_n}{2}\right), \quad \text{Sum formula}$$

we obtain

$$S_{50} = 50\left(\frac{\mathbf{2} + \mathbf{100}}{2}\right) = 2550.$$

Now Try Exercise 13

TECHNOLOGY NOTE

Sum of the First *n* Terms of a Sequence
The "seq(" utility, found on some calculators under the LIST OPS menus, generates a sequence. The "sum(" utility, found on some calculators under the LIST MATH menus, calculates the sum of the sequence inside the parentheses. To verify the result in Example 2, let $a_n = 30{,}000 + 1500(n - 1)$. The value 367,500 for S_{10} is shown in **FIGURE 14.14(a)**. The result found in Example 3 is shown in **FIGURE 14.14(b)**.

Example 2

```
sum(seq(30000+15
00(n−1),n,1,10))
              367500
```

(a)

Example 3

```
sum(seq(2n,n,1,5
0))
              2550
```

(b)

FIGURE 14.14

3 Working with Geometric Series

A **geometric series** is the sum of the terms of a geometric sequence. For example,

$$\underbrace{1, 2, 4, 8, 16, 32}_{\textbf{Geometric sequence}}$$

is a geometric sequence with $a_1 = 1$ and $r = 2$. Then

$$\underbrace{1 + 2 + 4 + 8 + 16 + 32}_{\textbf{Geometric series}}$$

> A geometric series is the sum of the terms of a geometric sequence.

is a geometric series. We can use the following formula to sum the first *n* terms of a geometric sequence.

READING CHECK 3

- Use a formula to find the sum

$$1 + 4 + 16 + 64 + 256.$$

Check your answer by adding the terms.

SUM OF THE FIRST *n* TERMS OF A GEOMETRIC SEQUENCE

If its first term is a_1 and its common ratio is r, then the **sum of the first *n* terms of a geometric sequence** is given by

$$S_n = a_1\left(\frac{1 - r^n}{1 - r}\right),$$

provided $r \neq 1$.

EXAMPLE 4 **Finding the sum of the terms of a geometric sequence**

Find each sum.
(a) $1 + 2 + 4 + 8 + 16 + 32 + 64 + 128 + 256$
(b) $\frac{1}{2} - \frac{1}{4} + \frac{1}{8} - \frac{1}{16} + \frac{1}{32}$

Solution
(a) This series is geometric, with $n = 9$, $a_1 = 1$, and $r = 2$, so

$$S_9 = 1\left(\frac{1 - 2^9}{1 - 2}\right) = 511. \qquad S_n = a_1\left(\frac{1 - r^n}{1 - r}\right)$$

(b) This series is geometric, with $n = 5$, $a_1 = \frac{1}{2} = 0.5$, and $r = -\frac{1}{2} = -0.5$, so

$$S_5 = 0.5\left(\frac{1 - (-0.5)^5}{1 - (-0.5)}\right) = \frac{11}{32} = 0.34375.$$

Now Try Exercises 19, 21

🌐 *Math in Context* ^Investment A sum of money from which regular payments are made is called an **annuity**. An annuity may be purchased with a lump sum deposit or by deposits made at various intervals. Suppose that a total of $1000 is deposited at the end of each year in an annuity account that pays an annual interest rate I expressed as a decimal. At the end of the first year, the account contains $1000. At the end of the second year, $1000 is deposited again. In addition, the first deposit of $1000 would have received interest during the second year. Therefore the value of the annuity after 2 years is

$$1000 + 1000(1 + I).$$

After 3 years, the balance is

$$1000 + 1000(1 + I) + 1000(1 + I)^2,$$

and after n years, this amount is given by

$$1000 + 1000(1 + I) + 1000(1 + I)^2 + \cdots + 1000(1 + I)^{n-1}.$$

This series is a geometric series with its first term $a_1 = 1000$ and the common ratio $r = (1 + I)$. The sum of the first n terms is given by

$$S_n = a_1\left(\frac{1 - (1 + I)^n}{1 - (1 + I)}\right) = a_1\left(\frac{(1 + I)^n - 1}{I}\right).$$

EXAMPLE 5 **Finding the future value of an annuity**

Suppose that a 20-year-old worker deposits $1000 into an annuity account at the end of each year. If the interest rate is 5%, find the future value of the annuity when the worker is 65 years old.

Solution
Let $a_1 = 1000$, $I = 0.05$, and $n = 45$. The future value of the annuity is

$$S_n = a_1\left(\frac{(1 + I)^n - 1}{I}\right)$$

$$= 1000\left(\frac{(1 + 0.05)^{45} - 1}{0.05}\right)$$

$$\approx \$159{,}700.$$

Now Try Exercise 41

4 Using Summation Notation

Summation notation is used to write series efficiently. The symbol Σ, the uppercase Greek letter *sigma*, is used to indicate a sum.

SUMMATION NOTATION

$$\sum_{k=1}^{n} a_k = a_1 + a_2 + a_3 + \cdots + a_n$$

The letter k is called the **index of summation**. The numbers 1 and n represent the subscripts of the first and last terms in the series. They are called the **lower limit** and **upper limit** of the summation, respectively.

EXAMPLE 6   **Using summation notation**

Find each sum.

(a) $\displaystyle\sum_{k=1}^{5} k^2$ (b) $\displaystyle\sum_{k=1}^{4} 5$ (c) $\displaystyle\sum_{k=3}^{6} (2k - 5)$

Solution

(a) $\displaystyle\sum_{k=1}^{5} k^2 = 1^2 + 2^2 + 3^2 + 4^2 + 5^2 = 55$

(b) $\displaystyle\sum_{k=1}^{4} 5 = 5 + 5 + 5 + 5 = 20$

(c) $\displaystyle\sum_{k=3}^{6} (2k - 5) = \underbrace{(2(3) - 5)}_{k=3} + \underbrace{(2(4) - 5)}_{k=4} + \underbrace{(2(5) - 5)}_{k=5} + \underbrace{(2(6) - 5)}_{k=6}$

$$= 1 + 3 + 5 + 7 = 16$$

Now Try Exercises 27, 29, 33

Summation notation is used frequently in statistics. The next example demonstrates how averages can be expressed in summation notation.

EXAMPLE 7   **Applying summation notation**

Express the average of the n numbers $x_1, x_2, x_3, \ldots, x_n$ in summation notation.

Solution
The average of n numbers can be written as

$$\frac{x_1 + x_2 + x_3 + \cdots + x_n}{n}.$$

This expression is equivalent to $\frac{1}{n}\left(\sum_{k=1}^{n} x_k\right)$.

Now Try Exercise 35

NOTE: $\sum_{k=1}^{n} x_k$ is equivalent to $\displaystyle\sum_{k=1}^{n} x_k$. ∎

 Public Health
🌐 *Math in Context*   Sometimes it is necessary to have air pass through more than one filter in order to clean the air properly. In the next example, we use a series to describe this situation.

EXAMPLE 8   **Modeling air filtration**

Suppose that an air filter removes 90% of the impurities entering it.
(a) Find a series that represents the amount of impurities removed by a sequence of n air filters. Express this answer in summation notation.
(b) How many air filters would be necessary to remove 99.99% of the impurities?

Impurities Passing through Air Filters

100%   10%   1%   0.1%

FIGURE 14.15

Solution

(a) The first filter removes 90% of the impurities, so 10%, or 0.1, passes through it. Of the 0.1 that passes through the first filter, 90% is removed by the second filter, while 10% of 10%, or 0.01, passes through. Then, 10% of 0.01, or 0.001, passes through the third filter. **FIGURE 14.15** depicts these results, from which we can establish a pattern. If we let 100%, or 1, represent the amount of impurities entering the first air filter, the amount *removed* by n filters equals

$$(0.9)(1) + (0.9)(0.1) + (0.9)(0.01) + (0.9)(0.001) + \cdots + (0.9)(0.1)^{n-1}.$$

In summation notation, we write this series as $\sum_{k=1}^{n} 0.9(0.1)^{k-1}$.

(b) To remove 99.99%, or 0.9999, of the impurities requires 4 air filters, because

$$\sum_{k=1}^{4} 0.9(0.1)^{k-1} = (0.9)(1) + (0.9)(0.1) + (0.9)(0.01) + (0.9)(0.001)$$

$$= 0.9 + 0.09 + 0.009 + 0.0009$$

$$= 0.9999.$$

Now Try Exercise 47

14.3   Putting It All Together

A finite sequence is an ordered list such as

$$a_1, a_2, a_3, a_4, a_5, \ldots, a_n.$$

A finite series is the summation of the terms of a sequence and can be expressed as

$$a_1 + a_2 + a_3 + a_4 + a_5 + \cdots + a_n.$$

SERIES	DESCRIPTION	EXAMPLE
Finite Arithmetic	$a_1 + a_2 + a_3 + \cdots + a_n$, where $a_n = dn + c$ or $a_n = a_1 + (n-1)d$. The sum of the first n terms is $$S_n = n\left(\frac{a_1 + a_n}{2}\right) \quad \text{or}$$ $$S_n = \frac{n}{2}(2a_1 + (n-1)d),$$ where a_1 is the first term and d is the common difference.	The series $$4 + 7 + 10 + 13 + 16 + 19 + 22$$ is obtained from the sequence $$a_n = 3n + 1 \quad \text{or} \quad a_n = 4 + 3(n-1).$$ Its sum is $$S_7 = 7\left(\frac{4 + 22}{2}\right) = 91 \quad \text{or}$$ $$S_7 = \frac{7}{2}(2 \cdot 4 + (7-1)3) = 91.$$
Finite Geometric	$a_1 + a_2 + a_3 + \cdots + a_n$, where $a_n = a_1(r)^{n-1}$ for nonzero constants a_1 and r. The sum of the first n terms is $$S_n = a_1\left(\frac{1 - r^n}{1 - r}\right),$$ where a_1 is the first term and r is the common ratio $(r \neq 1)$.	The series $$3 + 6 + 12 + 24 + 48 + 96$$ has $n = 6$, $a_1 = 3$, and $r = 2$. Its sum is $$S_6 = 3\left(\frac{1 - 2^6}{1 - 2}\right) = 189.$$

14.3 Exercises

CONCEPTS AND VOCABULARY

1. The summation of the terms of a sequence is called a(n) _____.

2. Find the sum of the series $1 + 2 + 3 + 4$.

3. The series $1 + 3 + 5 + 7 + 9$ is an example of a(n) _____ series.

4. The series $1 + 3 + 9 + 27 + 81$ is an example of a(n) _____ series.

5. If $a_1 + a_2 + a_3 + \cdots + a_n$ is an arithmetic series, its sum is $S_n =$ _____.

6. If $a_1 + a_2 + a_3 + \cdots + a_n$ is a geometric series with the common ratio $r \neq 1$, its sum is $S_n =$ _____.

7. The symbol Σ is used to indicate a(n) _____.

8. Write $\sum_{k=1}^{4} a_k$ as a sum.

9. $\sum_{n=1}^{5} a_1 + (n - 1)d$ is an example of a(n) _____ series.

10. $\sum_{n=1}^{4} a_1 r^{n-1}$ is an example of a(n) _____ series.

WORKING WITH ARITHMETIC SERIES

Exercises 11–18: Find the sum by using a formula.

11. $3 + 5 + 7 + 9 + 11 + 13$

12. $7.5 + 6 + 4.5 + 3 + 1.5 + 0 + (-1.5)$

13. $1 + 2 + 3 + 4 + \cdots + 40$

14. $1 + 3 + 5 + 7 + \cdots + 99$

15. $-7 + (-4) + (-1) + 2 + 5$

16. $89 + 84 + 79 + 74 + 69 + 64 + 59 + 54$

17. $-1 + (-4) + (-7) + (-10) + \cdots + (-22)$

18. $-2 + 3 + 8 + 13 + 18 + \cdots + 33$

WORKING WITH GEOMETRIC SERIES

Exercises 19–26: Find the sum by using a formula.

19. $3 + 9 + 27 + 81 + 243 + 729 + 2187$

20. $2 - 1 + \frac{1}{2} - \frac{1}{4} + \frac{1}{8} - \frac{1}{16} + \frac{1}{32}$

21. $1 - 2 + 4 - 8 + 16 - 32 + 64 - 128$

22. $2 + \frac{1}{2} + \frac{1}{8} + \frac{1}{32} + \frac{1}{128} + \frac{1}{512}$

23. $0.5 + 1.5 + 4.5 + 13.5 + 40.5 + 121.5$

24. $0.6 + 0.3 + 0.15 + 0.075 + 0.0375$

25. $5 + 20 + 80 + 320 + \cdots + 5120$

26. $90 + 30 + 10 + \frac{10}{3} + \frac{10}{9} + \frac{10}{27}$

USING SUMMATION NOTATION

Exercises 27–34: Write the terms of the series and find their sum.

27. $\sum_{k=1}^{4} 2k$

28. $\sum_{k=1}^{6} (k - 1)$

29. $\sum_{k=1}^{8} 4$

30. $\sum_{k=2}^{6} (5 - 2k)$

31. $\sum_{k=1}^{7} k^2$

32. $\sum_{k=1}^{4} 5(2)^{k-1}$

33. $\sum_{k=4}^{5} (k^2 - k)$

34. $\sum_{k=1}^{4} \log k$

Exercises 35–38: Write in summation notation.

35. $1^4 + 2^4 + 3^4 + 4^4 + 5^4 + 6^4$

36. $1 + \frac{1}{5^1} + \frac{1}{5^2} + \frac{1}{5^3} + \frac{1}{5^4}$

37. $1 + \frac{1}{2^2} + \frac{1}{3^2} + \frac{1}{4^2} + \frac{1}{5^2}$

38. $1 + \frac{1}{10} + \frac{1}{100} + \frac{1}{1000} + \frac{1}{10,000}$

39. Verify that $\sum_{k=1}^{n} k = \frac{n(n + 1)}{2}$ by using a formula for the sum of the first n terms of an arithmetic series.

40. Use Exercise 39 to find the sum of the series $\sum_{k=1}^{200} k$.

APPLICATIONS OF ARITHMETIC AND GEOMETRIC SEQUENCES

Exercises 41–44: Annuities (Refer to Example 5.) Find the future value of the annuity.

41. $a_1 = \$2000$ $I = 0.08$ $n = 20$

42. $a_1 = \$500$ $I = 0.15$ $n = 10$

43. $a_1 = \$10,000$ $I = 0.11$ $n = 5$

44. $a_1 = \$3000$ $I = 0.19$ $n = 45$

45. **Female CEO Compensation** The following table lists the compensation in millions of dollars for the six top-paid female CEOs of S&P 500 companies in 2015. (*Source: Statista.*)

continued on next page

Company	Xerox	Oracle	TJX
Compensation	22.2	37.7	28.7

Company	Yahoo!	Lockheed Martin	Pepsico
Compensation	42.1	33.7	22.5

(a) Write a series whose sum is the total compensation in millions for the six top-paid female CEOs.

(b) Find the sum of the series in millions of dollars.

46. Global Social Networks The following table lists the number of active users in millions of the top six global social networks in 2015. (*Source:* Statista.)

Company	Facebook Messenger	Whats App	QQ
Active Users	700	800	832

Company	Facebook	QZone	WeChat
Active Users	1490	668	549

(a) Write a series whose sum is the total number of active users in millions for the top six global social networks.

(b) Find the sum of the series in millions.

47. Air Filtration (Refer to Example 8.) Suppose that an air filter removes 80% of the impurities entering it.

(a) Find a series that represents the amount of impurities removed by a sequence of n air filters. Express the answer in summation notation.

(b) How many filters would be necessary to remove 96% of the impurities?

48. Air Filtration Suppose that an air filter removes 70% of the impurities entering it.

(a) Find a series that represents the amount of impurities removed by a sequence of n air filters. Express the answer in summation notation.

(b) How many filters would be necessary to remove 97.3% of the impurities?

49. Area A sequence of smaller squares is formed by connecting the midpoints of the sides of a larger square as shown in the figure.

(a) If the area of the largest square is 1 square unit, determine the first five terms of a sequence that describes the area of each successive square.

(b) Use a formula to sum the areas of the first 10 squares.

50. Perimeter (Refer to the previous exercise.) Use a formula to find the sum of the perimeters of the first 10 squares.

51. Stacking Logs A stack of logs is made in layers, with one log less in each layer, as shown in the accompanying figure. If the top layer has 6 logs and the bottom layer has 14 logs, what is the total number of logs in the pile? Use a formula to find this sum.

52. Stacking Logs (Refer to the previous exercise.) Suppose that a stack of logs has 15 logs in the top layer and a total of 10 layers. How many logs are in the stack?

53. Salaries (Refer to Example 2.) Suppose that an individual's starting salary is $35,000 per year and that the individual receives a $2000 raise each year. Find the total amount earned over 20 years.

54. Salaries Suppose that an individual's starting salary is $35,000 per year and that the individual receives a 10% raise each year. Find the total amount earned over 20 years.

55. Bouncing Ball A tennis ball first bounces to 75% of the height from which it was dropped and then to 75% of the height of each successive bounce. If it is dropped from a height of 10 feet, find the distance it *falls* between the fourth and fifth bounce.

56. Bouncing Ball A tennis ball first bounces to 75% of the height from which it was dropped and then to 75% of the height of each successive bounce. If it is dropped from a height of 10 feet, find the *total* distance it travels before it reaches its fifth bounce. (*Hint:* Make a sketch.)

57. Infinite Series The sum S of an infinite geometric series can be found if its common ratio r satisfies $|r| < 1$. It is given by

$$S = \frac{a_1}{1 - r}.$$

(If $|r| \geq 1$, this sum does not exist.) For example, the infinite geometric series

$$1 + \frac{1}{2} + \frac{1}{4} + \frac{1}{8} + \frac{1}{16} + \cdots$$

has $a_1 = 1$ and $r = \frac{1}{2}$. Therefore its sum S equals

$$S = \frac{1}{1 - \frac{1}{2}} = 2.$$

You might want to add terms of this series to see how increasing the number of terms results in a sum closer to 2. Find the sum of each infinite geometric series.

(a) $2 - 1 + \frac{1}{2} - \frac{1}{4} + \frac{1}{8} - \frac{1}{16} + \cdots$

(b) $1 + \frac{1}{3} + \frac{1}{9} + \frac{1}{27} + \frac{1}{81} + \cdots$

(c) $0.1 + 0.01 + 0.001 + 0.0001 + \cdots$

(d) $0.12 + 0.0012 + 0.000012 + 0.00000012 + \cdots$

58. Online Exploration Go online and find a series that can be used to calculate π. Use the first five terms of your series to estimate π.

59. Online Exploration Go online and find a series that can be used to calculate the number e. Use the first five terms of your series to estimate e.

WRITING ABOUT MATHEMATICS

60. Discuss the difference between a sequence and a series. Give an example of each.

61. Suppose that an arithmetic series has $a_1 = 1$ and a common difference of $d = 2$, whereas a geometric series has $a_1 = 1$ and a common ratio of $r = 2$. Discuss how their sums compare as the number of terms n becomes large. (*Hint:* Calculate each sum for $n = 10$, 20, and 30.)

Group Activity

Calculating π The quest for an accurate estimation for π is a fascinating story covering thousands of years. Because π is an irrational number, it cannot be represented exactly by a fraction. Its decimal expansion neither repeats nor has a pattern. The ability to compute π was essential to the development of societies because π appears in formulas used in construction, surveying, and geometry. In early historical records, π was given the value of 3. Later the Egyptians used a value of

$$\frac{256}{81} \approx 3.1605.$$

Not until the discovery of the series was an exceedingly accurate decimal approximation of π possible. In 1989, π was computed to 1,073,740,000 digits, which required 100 hours of supercomputer time. Why

would anyone want to compute π to so many decimal places? One practical reason is to test electrical circuits in new computers. If a computer has a small defect in its hardware, there is a good chance that an error will appear after it has performed trillions of arithmetic calculations during the computation of π. The series

$$\frac{\pi^4}{90} \approx \frac{1}{1^4} + \frac{1}{2^4} + \frac{1}{3^4} + \frac{1}{4^4} + \cdots + \frac{1}{n^4}$$

gives an estimate of π, where larger values of n give better approximations. (*Source:* P. Beckmann, *A History of Pi.*)

(a) Approximate π by finding the sum of the first four terms.

(b) Use a calculator to approximate π by summing the first 50 terms. Compare the result to the actual value of π.

14.4   The Binomial Theorem

NEW VOCABULARY

☐ Pascal's triangle
☐ Factorial notation
☐ Binomial coefficient
☐ Binomial theorem

1   Introducing Pascal's Triangle

We demonstrate how to expand expressions of the form $(a + b)^n$, where n is a natural number. These expressions occur in statistics, finite mathematics, computer science, and calculus. The first method that we discuss is Pascal's triangle. Expanding $(a + b)^n$ for increasing values of n gives the following results.

Expanding Binomials

$$(a + b)^0 = 1$$
$$(a + b)^1 = 1a + 1b$$
$$(a + b)^2 = 1a^2 + 2ab + 1b^2$$
$$(a + b)^3 = 1a^3 + 3a^2b + 3ab^2 + 1b^3$$
$$(a + b)^4 = 1a^4 + 4a^3b + 6a^2b^2 + 4ab^3 + 1b^4$$
$$(a + b)^5 = 1a^5 + 5a^4b + 10a^3b^2 + 10a^2b^3 + 5ab^4 + 1b^5$$

Note that $(a + b)^1$ has two terms, starting with a and ending with b; $(a + b)^2$ has three terms, starting with a^2 and ending with b^2; and, in general, $(a + b)^n$ has $n + 1$ terms, starting with a^n and ending with b^n. From left to right, the exponent on a decreases by 1 each successive term, and the exponent on b increases by 1 each successive term.

The triangle formed by the red numbers is called **Pascal's triangle**. This triangle consists of 1s along the sides, and each element inside the triangle is the sum of the two numbers above it, as shown in **FIGURE 14.16**. Pascal's triangle is usually written without variables and can be extended to include as many rows as needed.

READING CHECK 1

• Explain how to find the numbers in Pascal's triangle.

Pascal's Triangle

```
            1
          1   1
        1   2   1
      1   3   3   1
    1   4   6   4   1
  1   5  10  10   5   1
```

FIGURE 14.16

We can use this triangle to expand $(a + b)^n$, where n is a natural number. For example, the expression $(m + n)^4$ consists of five terms written as

$$(m + n)^4 = \_\_m^4 + \_\_m^3n^1 + \_\_m^2n^2 + \_\_m^1n^3 + \_\_n^4.$$

Because there are five terms, the coefficients can be found in the fifth row of Pascal's triangle, which is

$$1 \quad 4 \quad 6 \quad 4 \quad 1.$$

Thus

$$(m + n)^4 = \underline{1}\, m^4 + \underline{4}\, m^3n^1 + \underline{6}\, m^2n^2 + \underline{4}\, m^1n^3 + \underline{1}\, n^4$$
$$= m^4 + 4m^3n + 6m^2n^2 + 4mn^3 + n^4.$$

EXAMPLE 1   **Expanding a binomial**

Expand each binomial, using Pascal's triangle.

(a) $(x + 2)^5$ **(b)** $(2m - n)^3$

Solution

(a) To find the coefficients, use the sixth row (1 5 10 10 5 1) in Pascal's triangle.

$$(x + 2)^5 = \underline{1}\,x^5 + \underline{5}\,x^4 \cdot 2^1 + \underline{10}\,x^3 \cdot 2^2 + \underline{10}\,x^2 \cdot 2^3 + \underline{5}\,x^1 \cdot 2^4 + \underline{1}\,(2^5)$$
$$= x^5 + 10x^4 + 40x^3 + 80x^2 + 80x + 32$$

(b) To find the coefficients, use the fourth row (1 3 3 1) in Pascal's triangle.

$$(2m - n)^3 = \underline{1}\,(2m)^3 + \underline{3}\,(2m)^2(-n)^1 + \underline{3}\,(2m)^1(-n)^2 + \underline{1}\,(-n)^3$$
$$= 8m^3 - 12m^2n + 6mn^2 - n^3$$

Now Try Exercises 13, 15

2 ▸ Writing Factorial Notation

An alternative to Pascal's triangle is the binomial theorem, which uses **factorial notation**.

> ### *n* FACTORIAL (*n*!)
>
> For any positive integer n,
> $$n! = 1 \cdot 2 \cdot 3 \cdot \cdots \cdot n.$$
>
> We also define $0! = 1$.
>
> **NOTE:** Because multiplication is commutative, n factorial can also be defined as
> $$n! = n \cdot (n - 1) \cdot (n - 2) \cdot \cdots \cdot 2 \cdot 1. \ \blacksquare$$

Examples include the following.

$$0! = 1$$
$$1! = 1$$
$$2! = 1 \cdot 2 = 2$$
$$3! = 1 \cdot 2 \cdot 3 = 6$$
$$4! = 1 \cdot 2 \cdot 3 \cdot 4 = 24$$
$$5! = 1 \cdot 2 \cdot 3 \cdot 4 \cdot 5 = 120$$

FIGURE 14.17 supports these results. On some calculators, factorial (!) can be accessed in the MATH PRB menus.

Factorials

0!		1
1!		1
2!		2

3!		6
4!		24
5!		120

(a)   (b)

FIGURE 14.17

EXAMPLE 2 **Evaluating factorial expressions**

Simplify the expression.

(a) $\dfrac{5!}{3!2!}$ (b) $\dfrac{4!}{4!0!}$

Solution

(a) $\dfrac{5!}{3!2!} = \dfrac{1 \cdot 2 \cdot 3 \cdot 4 \cdot 5}{(1 \cdot 2 \cdot 3)(1 \cdot 2)} = \dfrac{120}{6 \cdot 2} = 10$ (b) $0! = 1$, so $\dfrac{4!}{4!0!} = \dfrac{4!}{4!(1)} = \dfrac{4!}{4!} = 1$

Now Try Exercises 25, 27

3 Finding Binomial Coefficients

The expression $_nC_r$ represents a *binomial coefficient* that can be used to calculate the numbers in Pascal's triangle.

> **BINOMIAL COEFFICIENT $_nC_r$**
>
> For n and r nonnegative integers, $n \geq r$,
>
> $$_nC_r = \dfrac{n!}{(n-r)!\,r!}$$
>
> is a **binomial coefficient**.

Values of $_nC_r$ for $r = 0, 1, 2, \ldots, n$ correspond to the $n + 1$ numbers in row $n + 1$ of Pascal's triangle.

EXAMPLE 3 **Calculating $_nC_r$**

Calculate $_3C_r$ for $r = 0, 1, 2, 3$ by hand. Check your results on a calculator. Compare these numbers with the fourth row in Pascal's triangle.

Solution

$$_3C_0 = \dfrac{3!}{(3-0)!\,0!} = \dfrac{6}{6 \cdot 1} = 1 \qquad _3C_1 = \dfrac{3!}{(3-1)!\,1!} = \dfrac{6}{2 \cdot 1} = 3$$

$$_3C_2 = \dfrac{3!}{(3-2)!\,2!} = \dfrac{6}{1 \cdot 2} = 3 \qquad _3C_3 = \dfrac{3!}{(3-3)!\,3!} = \dfrac{6}{1 \cdot 6} = 1$$

These results are supported in **FIGURE 14.18**. The fourth row of Pascal's triangle is

1　3　3　1,

which agrees with the calculated values for $_3C_r$ when $r = 0, 1, 2, 3$. On some calculators, the MATH PRB menus are used to calculate $_nC_r$.

Binomial Coefficients

3 nCr 0	
3 nCr 1	1
	3

3 nCr 2	
3 nCr 3	3
	1

(a)　　　　　　　(b)

FIGURE 14.18

Now Try Exercises 29, 33

4 Using the Binomial Theorem

EXPANDING A BINOMIAL The binomial coefficients can be used to expand expressions of the form $(a + b)^n$. To do so, we use the **binomial theorem**.

BINOMIAL THEOREM

For any positive integer n and any numbers a and b,

$$(a + b)^n = {}_nC_0 a^n + {}_nC_1 a^{n-1}b^1 + \cdots + {}_nC_{n-1} a^1 b^{n-1} + {}_nC_n b^n.$$

For example, using the results of Example 3, we write

$$\begin{aligned}
(a + b)^3 &= {}_3C_0 a^3 + {}_3C_1 a^2 b^1 + {}_3C_2 a^1 b^2 + {}_3C_3 b^3 \\
&= 1a^3 + 3a^2 b + 3ab^2 + 1b^3 \\
&= a^3 + 3a^2 b + 3ab^2 + b^3.
\end{aligned}$$

EXAMPLE 4 **Expanding a binomial**

Use the binomial theorem to expand each expression.

(a) $(x + y)^5$ (b) $(3 - 2x)^4$

Solution

(a) The coefficients are calculated as follows.

$${}_5C_0 = \frac{5!}{(5 - 0)!\,0!} = 1, \quad {}_5C_1 = \frac{5!}{(5 - 1)!\,1!} = 5, \quad {}_5C_2 = \frac{5!}{(5 - 2)!\,2!} = 10$$

$${}_5C_3 = \frac{5!}{(5 - 3)!\,3!} = 10, \quad {}_5C_4 = \frac{5!}{(5 - 4)!\,4!} = 5, \quad {}_5C_5 = \frac{5!}{(5 - 5)!\,5!} = 1$$

Using the binomial theorem, we arrive at the following result.

$$\begin{aligned}
(x + y)^5 &= {}_5C_0 x^5 + {}_5C_1 x^4 y^1 + {}_5C_2 x^3 y^2 + {}_5C_3 x^2 y^3 + {}_5C_4 x^1 y^4 + {}_5C_5 y^5 \\
&= 1x^5 + 5x^4 y + 10x^3 y^2 + 10x^2 y^3 + 5xy^4 + 1y^5 \\
&= x^5 + 5x^4 y + 10x^3 y^2 + 10x^2 y^3 + 5xy^4 + y^5
\end{aligned}$$

(b) The coefficients are calculated as follows.

$${}_4C_0 = \frac{4!}{(4 - 0)!\,0!} = 1, \quad {}_4C_1 = \frac{4!}{(4 - 1)!\,1!} = 4, \quad {}_4C_2 = \frac{4!}{(4 - 2)!\,2!} = 6,$$

$${}_4C_3 = \frac{4!}{(4 - 3)!\,3!} = 4, \quad {}_4C_4 = \frac{4!}{(4 - 4)!\,4!} = 1$$

Using the binomial theorem with $a = 3$ and $b = (-2x)$, we arrive at the following result.

$$\begin{aligned}
(3 - 2x)^4 &= {}_4C_0 (3)^4 + {}_4C_1 (3)^3 (-2x) + {}_4C_2 (3)^2 (-2x)^2 \\
&\quad + {}_4C_3 (3)(-2x)^3 + {}_4C_4 (-2x)^4 \\
&= 1(81) + 4(27)(-2x) + 6(9)(4x^2) + 4(3)(-8x^3) + 1(16x^4) \\
&= 81 - 216x + 216x^2 - 96x^3 + 16x^4
\end{aligned}$$

Now Try Exercises 41, 43

FINDING A PARTICULAR TERM OF $(a + b)^n$ The binomial theorem gives *all* of the terms of $(a + b)^n$. However, we can find a particular term by noting that the $(r + 1)$st term in the binomial expansion for $(a + b)^n$ is given by the formula

$$_nC_r\, a^{n-r} b^r,$$

for $0 \le r \le n$. The next example shows how to use this formula to find the $(r + 1)$st term of $(a + b)^n$.

EXAMPLE 5 **Finding the *k*th term in a binomial expansion**

Find the third term of $(x - y)^5$.

Solution

In this example, the $(r + 1)$st term is the *third* term in the expansion of $(x - y)^5$. That is, $r + 1 = 3$, or $r = \mathbf{2}$. Also, the exponent in the expression is $n = \mathbf{5}$. To get this binomial into the form $(a + b)^n$, we note that the first term in the binomial is $a = x$ and that the second term in the binomial is $b = -y$. Substituting the values for r, n, a, and b in the formula $_nC_r\, a^{n-r} b^r$ for the $(r + 1)$st term yields

$$_5C_2(x)^{5-2}(-y)^2 = 10x^3y^2.$$

The third term in the binomial expansion of $(x - y)^5$ is $10x^3y^2$.

Now Try Exercise 55

14.4 Putting It All Together

CONCEPT	EXPLANATION	EXAMPLE
Pascal's Triangle	$$1$$ $$1 \quad 1$$ $$1 \quad 2 \quad 1$$ $$\mathbf{1 \quad 3 \quad 3 \quad 1}$$ $$1 \quad 4 \quad 6 \quad 4 \quad 1$$ $$1 \quad 5 \quad 10 \quad 10 \quad 5 \quad 1$$	$(a + b)^3 = \mathbf{1}a^3 + \mathbf{3}a^2b + \mathbf{3}ab^2 + \mathbf{1}b^3$ (Row 4) To expand $(a + b)^n$, use row $n + 1$ in the triangle.
Factorial Notation	The expression $n!$ equals $$1 \cdot 2 \cdot 3 \cdot \,\cdots\, \cdot n.$$	$5! = 1 \cdot 2 \cdot 3 \cdot 4 \cdot 5 = 120$ $0! = 1$
Binomial Coefficient $_nC_r$	$$_nC_r = \frac{n!}{(n-r)!\,r!}$$	$$_6C_4 = \frac{6!}{(6-4)!\,4!} = \frac{6!}{2!\,4!} = \frac{720}{2 \cdot 24} = 15$$
Binomial Theorem	$(a + b)^n = {_nC_0}a^n + {_nC_1}a^{n-1}b^1 + \cdots$ $+ \; {_nC_{n-1}}a^1b^{n-1} + {_nC_n}b^n$	$(a + b)^4 = {_4C_0}a^4 + {_4C_1}a^3b + {_4C_2}a^2b^2$ $\qquad\quad + \; {_4C_3}ab^3 + {_4C_4}b^4$ $\quad = 1a^4 + 4a^3b + 6a^2b^2 + 4ab^3 + 1b^4$ $\quad = a^4 + 4a^3b + 6a^2b^2 + 4ab^3 + b^4$

14.4 Exercises

MyMathLab®

CONCEPTS AND VOCABULARY

1. How many terms result from expanding $(a + b)^4$?

2. How many terms result from expanding $(a + b)^n$?

3. To find the coefficients for the expansion of $(a + b)^3$, what row of Pascal's triangle do you use?

4. Write down the first 5 rows of Pascal's triangle.

5. $4! = $ _____

6. $1 \cdot 2 \cdot 3 \cdot 4 \cdot 5 \cdot 6 = $ _____

7. $_nC_r = $ _____

8. $(a + b)^2 = $ _____

9. Thinking Generally $\frac{n!}{(n - 1)!} = $ _____

10. Thinking Generally $_nC_n = $ _____

USING PASCAL'S TRIANGLE

Exercises 11–18: Use Pascal's triangle to expand the given expression.

11. $(x + y)^3$

12. $(x + y)^4$

13. $(2x + 1)^4$

14. $(2x - 1)^4$

15. $(a - b)^5$

16. $(3x + 2y)^3$

17. $(x^2 + 1)^3$

18. $\left(\frac{1}{2} - x^2\right)^5$

EVALUATING FACTORIAL NOTATION

Exercises 19–28: Evaluate the expression.

19. $3!$

20. $6!$

21. $7!$

22. $0!$

23. $\frac{4!}{3!}$

24. $\frac{6!}{3!}$

25. $\frac{2!}{0!}$

26. $\frac{5!}{1!}$

27. $\frac{5!}{2!3!}$

28. $\frac{6!}{4!2!}$

FINDING BINOMIAL COEFFICIENTS

Exercises 29–34: Evaluate the binomial coefficient by hand.

29. $_5C_4$

30. $_3C_1$

31. $_6C_5$

32. $_2C_2$

33. $_4C_0$

34. $_4C_3$

Exercises 35–40: Evaluate the binomial coefficient with a calculator.

35. $_{12}C_7$

36. $_{13}C_8$

37. $_9C_5$

38. $_{25}C_{14}$

39. $_{19}C_{11}$

40. $_{10}C_6$

EXPANDING USING THE BINOMIAL THEOREM

Exercises 41–52: Use the binomial theorem to expand the expression.

41. $(m + n)^3$

42. $(m + n)^5$

43. $(x - y)^4$

44. $(1 - 3x)^4$

45. $(2a + 1)^3$

46. $(x^2 - 1)^3$

47. $(x + 2)^5$

48. $(a - 3)^5$

49. $(3 + 2m)^4$

50. $(m - 3n)^3$

51. $(2x - y)^3$

52. $(2a + 3b)^4$

FINDING A PARTICULAR TERM

Exercises 53–58: The $(r + 1)$st term of the expression $(a + b)^n$, $0 \le r \le n$, is given by $_nC_r a^{n-r} b^r$. Find the specified term. Refer to Example 5.

53. The first term of $(a + b)^8$

54. The second term of $(a - b)^{10}$

55. The fourth term of $(x + y)^7$

56. The sixth term of $(a + b)^9$

57. The first term of $(2m + n)^9$

58. The eighth term of $(2a - b)^8$

WRITING ABOUT MATHEMATICS

59. Explain how to find the numbers in Pascal's triangle.

60. Compare the expansion of $(a + b)^n$ to the expansion of $(a - b)^n$. Give an example.

Checking Basic Concepts

1. Determine whether the series is arithmetic or geometric.
   (a) $\frac{1}{2} + \frac{1}{4} + \frac{1}{8} + \cdots + \frac{1}{256}$
   (b) $\frac{1}{2} + \frac{5}{2} + \frac{9}{2} + \frac{13}{2} + \frac{17}{2}$

2. Use a formula to find the sum.
   $$4 + 8 + 12 + \cdots + 48.$$

3. Use a formula to find the sum.
   $$1 - 2 + 4 - 8 + 16 - 32 + 64 - 128 + 256 - 512.$$

4. Use Pascal's triangle to expand $(x - y)^4$.

5. Use the binomial theorem to expand $(x + 2)^3$.

CHAPTER 14   Summary

SECTION 14.1 ■ SEQUENCES

Sequences An *infinite sequence* is a function whose domain is the natural numbers. A *finite sequence* is a function whose domain is $D = \{1, 2, 3, \ldots, n\}$ for some natural number n.

Example: $a_n = 2n$ is a symbolic representation of the even natural numbers. The first six terms, 2, 4, 6, 8, 10, 12, of this sequence are represented numerically and graphically in the table and figure.

Numerical Representation

n	a_n
1	2
2	4
3	6
4	8
5	10
6	12

Graphical Representation

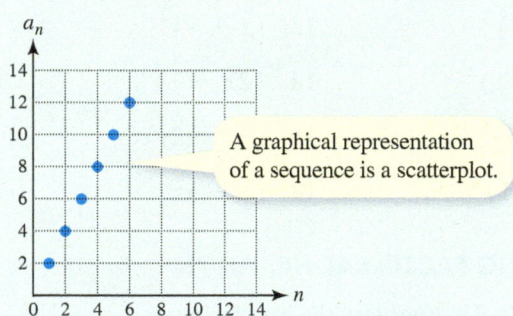

A graphical representation of a sequence is a scatterplot.

SECTION 14.2 ■ ARITHMETIC AND GEOMETRIC SEQUENCES

Two common types of sequences are arithmetic and geometric.

Arithmetic Sequence An arithmetic sequence is determined by a linear function of the form $f(n) = dn + c$ or $f(n) = a_1 + (n - 1)d$. Successive terms in an arithmetic sequence are found by adding the common difference d to the previous term.

Example: The sequence 1, 3, 5, 7, 9, 11, . . . is an arithmetic sequence with its first term $a_1 = 1$, common difference $d = 2$, and general term $a_n = 2n - 1$.

Geometric Sequence The general term for a geometric sequence is given by $f(n) = a_1 r^{n-1}$. Successive terms in a geometric sequence are found by multiplying the previous term by the common ratio r.

Example: The sequence 3, 6, 12, 24, 48, . . . is a geometric sequence with its first term $a_1 = 3$, common ratio $r = 2$, and general term $a_n = 3(2)^{n-1}$.

SECTION **14.3** ■ SERIES

Series A series results when the terms of a sequence are summed. The series associated with the sequence 2, 4, 6, 8, 10 is

$$2 + 4 + 6 + 8 + 10,$$

and its sum equals 30. An arithmetic series results when the terms of an arithmetic sequence are summed, and a geometric series results when the terms of a geometric sequence are summed. In this section, we discussed formulas for finding sums of arithmetic and geometric series. See Putting It All Together for Section 3 of this chapter.

Summation Notation Summation notation can be used to write series efficiently.

Example: $1^2 + 2^2 + 3^2 + 4^2 + 5^2 = \sum_{k=1}^{5} k^2.$

SECTION **14.4** ■ THE BINOMIAL THEOREM

Pascal's triangle may be used to find the coefficients for the expansion of $(a + b)^n$, where n is a natural number.

$$
\begin{array}{ccccccccccc}
& & & & & 1 & & & & & \\
& & & & 1 & & 1 & & & & \\
& & & 1 & & 2 & & 1 & & & \\
& & 1 & & 3 & & 3 & & 1 & & \\
& 1 & & 4 & & 6 & & 4 & & 1 & \\
1 & & 5 & & 10 & & 10 & & 5 & & 1
\end{array}
$$

Example: To expand $(x + y)^4$, use the fifth row of Pascal's triangle.

$$(x + y)^4 = \mathbf{1}x^4 + \mathbf{4}\,x^3y + \mathbf{6}\,x^2y^2 + \mathbf{4}\,xy^3 + \mathbf{1}y^4$$
$$= x^4 + 4x^3y + 6x^2y^2 + 4xy^3 + y^4$$

The binomial theorem can also be used to expand powers of binomials. See Putting It All Together for Section 4 of this chapter.

CHAPTER 14   **Review Exercises**

SECTION 14.1

Exercises 1–4: Write the first four terms of the sequence for n = 1, 2, 3, and 4.

1. $f(n) = n^3$

2. $f(n) = 5 - 2n$

3. $f(n) = \dfrac{2n}{n^2 + 1}$

4. $f(n) = (-2)^n$

Exercises 5 and 6: Write the terms of the sequence.

5.

6.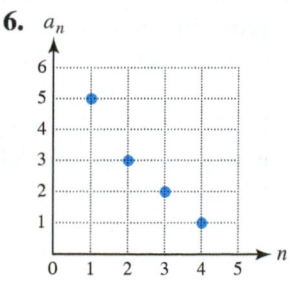

Exercises 7–10: Represent the first seven terms of the sequence numerically and graphically.

7. $a_n = 2n$

8. $a_n = n^2 - 4$

9. $a_n = 4\left(\frac{1}{2}\right)^n$

10. $a_n = \sqrt{n}$

SECTION 14.2

Exercises 11–18: Determine whether f is an arithmetic sequence. Identify the common difference when possible.

11. $f(n) = 5n - 1$

12. $f(n) = 4 - n^2$

13. $f(n) = 2^n$

14. $f(n) = 4 - \frac{1}{3}n$

15.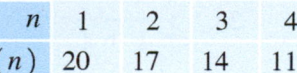

n	1	2	3	4
$f(n)$	20	17	14	11

16.

n	1	2	3	4
$f(n)$	−3	0	6	12

17. $f(n)$

18. $f(n)$

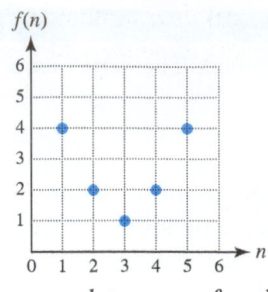

Exercises 19 and 20: Find the general term a_n for the arithmetic sequence.

19. $a_1 = -3$ and $d = 4$ **20.** $a_1 = 2$ and $a_2 = -3$

Exercises 21–28: Determine whether f is a geometric sequence. Identify the common ratio when possible.

21. $f(n) = 2(4)^n$ **22.** $f(n) = 2n^4$

23. $f(n) = 1 - 2n$ **24.** $f(n) = 5(0.7)^n$

25.

n	1	2	3	4
$f(n)$	5	4	3	1

26.

n	1	2	3	4
$f(n)$	27	−9	3	−1

27. $f(n)$

28. $f(n)$

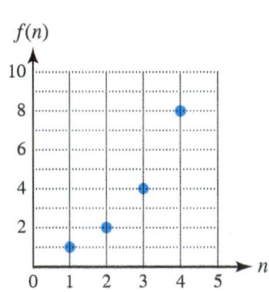

Exercises 29 and 30: Find the general term a_n for the geometric sequence.

29. $a_1 = 5$ and $r = 0.9$ **30.** $a_1 = 2$ and $a_2 = 8$

SECTION 14.3

Exercises 31–34: Find the sum, using a formula.

31. $4 + 9 + 14 + 19 + 24 + 29 + 34 + 39 + 44$

32. $4.5 + 3.0 + 1.5 + 0 - 1.5$

33. $1 - 4 + 16 - 64 + \cdots + 4096$

34. $1 + \frac{1}{2} + \frac{1}{4} + \frac{1}{8} + \frac{1}{16} + \cdots + \frac{1}{256}$

Exercises 35–38: Write the terms of the series.

35. $\displaystyle\sum_{k=1}^{5} 2k + 1$ **36.** $\displaystyle\sum_{k=1}^{4} \frac{1}{k+1}$

37. $\displaystyle\sum_{k=1}^{4} k^3$ **38.** $\displaystyle\sum_{k=2}^{7} (1 - k)$

Exercises 39–42: Write the series in summation notation.

39. $1 + 2 + 3 + \cdots + 20$

40. $1 + \frac{1}{2} + \frac{1}{3} + \cdots + \frac{1}{20}$

41. $\frac{1}{2} + \frac{2}{3} + \frac{3}{4} + \cdots + \frac{9}{10}$

42. $1^2 + 2^2 + 3^2 + 4^2 + 5^2 + 6^2 + 7^2$

SECTION 14.4

Exercises 43–46: Use Pascal's triangle to expand the given expression.

43. $(x + 4)^3$ **44.** $(2x + 1)^4$

45. $(x - y)^5$

46. $(a - 1)^6$

Exercises 47–50: Evaluate the expression.

47. $3!$ **48.** $\dfrac{5!}{3!2!}$

49. $_6C_3$ **50.** $_4C_3$

Exercises 51–54: Use the binomial theorem to expand the given expression.

51. $(m + 2)^4$ **52.** $(a + b)^5$

53. $(x - 3y)^4$ **54.** $(3x - 2)^3$

APPLICATIONS

55. Salaries An individual's starting salary is $45,000, and the individual receives a 10% raise each year. Give symbolic, numerical, and graphical representations for this person's salary over 7 years. What type of sequence is it?

56. Salaries An individual's starting salary is $45,000, and the individual receives an increase of $5000 each year. Give symbolic, numerical, and graphical representations for this person's salary over 7 years. What type of sequence is it?

57. Rain Forests Rain forests are defined as forests that grow in regions that receive more than 70 inches of rain each year. The world is losing an estimated 30 million acres of rain forests annually. Give symbolic, numerical, and graphical representations for a sequence that models the total number of acres (in millions) lost over a 7-year period. (*Source: Scientific American 2009.*)

58. Prison Escapes In 1995, there were 12,000 inmates that escaped from state institutions. Since then these escapes have declined, on average, by 10% per year. (*Source:* Bureau of Justice Statistics.)

(a) Write a sequence a_n that models the number of escapes, where $n = 1$ corresponds to 1995, $n = 2$ to 1996, and so on.

(b) Is a_n arithmetic, geometric, or neither? Explain.

(c) Find a_{19} and interpret the result.

CHAPTER 14   Test   MyMathLab® You Tube™

1. Write the first four terms of the given sequence for $n = 1, 2, 3,$ and 4.

$$f(n) = \frac{n^2}{n + 1}$$

2. Use the graph to write the terms of the sequence.

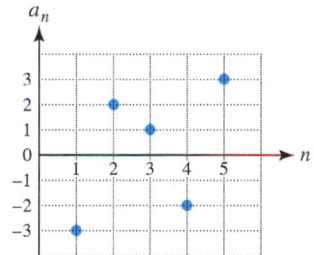

3. List the first seven terms of $a_n = n^2 - n$ in a table. Let $n = 1, 2, \ldots, 7$.

4. Expand the expression $(2x - 1)^4$.

Exercises 5 and 6: Determine whether the sequence is arithmetic or geometric. Identify either the common difference or the common ratio.

5. $f(n) = 7 - 3n$

6.

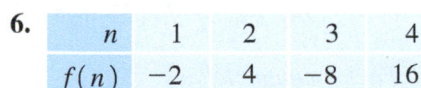

n	1	2	3	4
$f(n)$	-2	4	-8	16

7. Find the general term a_n for the arithmetic sequence if $a_1 = 2$ and $d = -3$.

8. Find the general term a_n for the geometric sequence if $a_1 = 2$ and $a_3 = 4.5$. Assume $r > 0$.

Exercises 9 and 10: Determine whether $a_n = f(n)$ is a geometric sequence. Identify the common ratio when possible.

9. $f(n) = -3(2.5)^n$

10.

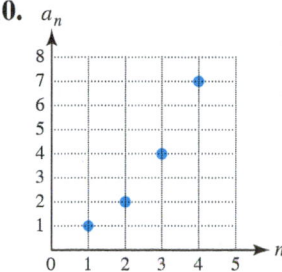

Exercises 11 and 12: Find the sum, using a formula.

11. $-1 + 2 + 5 + 8 + 11 + 14 + 17 + 20 + 23$

12. $1 - \frac{2}{3} + \frac{4}{9} - \frac{8}{27} + \frac{16}{81} - \frac{32}{243} + \frac{64}{729}$

13. Write the terms of the series $\sum_{k=2}^{7} 3k$.

14. Write the series $1^3 + 2^3 + 3^3 + \cdots + 60^3$ in summation notation.

15. Evaluate $\frac{7!}{4!\,3!}$.

16. Evaluate $_5C_3$.

17. Auditorium Seating An auditorium has 50 seats in the first row, 57 seats in the second row, 64 seats in the third row, and so on. Use a formula to find the total number of seats in the first 45 rows.

18. Mobile Addicts A mobile addict is someone who opens apps on a smartphone more than 60 times a day. In 2014, there were 176 million mobile addicts worldwide, and this number was growing by 59% per year. Give symbolic, numerical, and graphical representations for the number of mobile addicts in millions over a 5-year period, starting in 2014. What type of sequence is it? (*Source:* Business Insider.)

19. Tent Caterpillars Large numbers of tent caterpillars can defoliate trees and ruin crops. After they mature, they spin a cocoon and develop into moths that lay eggs. Suppose that an initial population of 2000 tent caterpillars doubles every 5 days.

continued on next page

(a) Write a formula for a_n that models the number of tent caterpillars after $n - 1$ five-day time periods. (*Hint:* $a_1 = 2000$, $a_2 = 4000$, and $a_3 = 8000$.)

(b) Is a_n arithmetic, geometric, or neither? Explain your reasoning.

(c) Find a_6 and interpret the result.

(d) Give a graph for a_n, where $n = 1, 2, 3, 4, 5, 6$.

CHAPTERS 1–14   Cumulative Review Exercises

1. State whether the equation illustrates an identity, commutative, associative, or distributive property.
$$29(102) = 29(100) + 29(2)$$

2. Identify the domain and range of the relation given by $S = \{(-6, 5), (-2, 1), (0, 3), (2, 0)\}$.

Exercises 3–6: Simplify the expression. Write the result using positive exponents.

3. $\dfrac{x^{-2}y^3}{(3xy^{-2})^3}$

4. $\left(\dfrac{3b}{6a^2}\right)^{-4}$

5. $\left(\dfrac{1}{z^2}\right)^{-5}$

6. $\dfrac{8x^{-3}y^2}{4x^3y^{-1}}$

7. Find the domain of $f(x) = \dfrac{-5}{x - 8}$.

8. Use the table to write the formula for $f(x) = ax + b$.

x	-2	-1	0	1	2
$f(x)$	5	3	1	-1	-3

9. Write the equation of the horizontal line that passes through the point $(2, 3)$.

10. Find the slope and the y-intercept of the graph of $f(x) = -3x + 5$.

Exercises 11 and 12: Write the slope-intercept form for a line satisfying the given conditions.

11. Perpendicular to the line $y = -\frac{2}{3}x - 4$, and passing through $(1, 4)$

12. Parallel to $y = 2x - 7$, passing through $(5, 2)$

Exercises 13–18: Solve the equation or inequality. Write the solutions to inequalities in interval notation.

13. $\frac{2}{5}(x - 4) = -12$

14. $\frac{2}{5}z + \frac{1}{4}z > 2 - (z - 1)$

15. $-3|t - 5| \le -18$

16. $\left|4 + \frac{2}{3}x\right| = 6$

17. $\dfrac{1}{4}t - (2t + 5) + 6 = \dfrac{t + 3}{4}$

18. $-3 \le \frac{2}{3}x + 5 < 11$

19. Shade the solution set in the xy-plane.
$$x - y < 4$$
$$x + 2y \ge 7$$

Exercises 20 and 21: Solve the system of equations. Write the solution as an ordered pair or ordered triple where appropriate.

20. $\begin{aligned} x - 2y &= 1 \\ -2x + 7y &= 4 \end{aligned}$

21. $\begin{aligned} x + y + z &= 5 \\ -2x - y + z &= -10 \\ x + 2y + 8z &= 1 \end{aligned}$

22. Evaluate det A.
$$A = \begin{bmatrix} 4 & -3 \\ 3 & 2 \end{bmatrix}$$

Exercises 23 and 24: Multiply the expressions.

23. $2x^3(4x^4 - 3x^3 + 5)$ 24. $(2z - 7)(3z + 4)$

Exercises 25 and 26: Factor completely.

25. $4x^2 - 9y^2$ 26. $2a^3 - a^2 + 8a - 4$

Exercises 27 and 28: Solve the equation.

27. $4x^2 - x - 3 = 0$ 28. $x^4 - 10x^3 = -24x^2$

Exercises 29 and 30: Simplify the expression.

29. $\dfrac{x^2 - 7x + 10}{x^2 - 25} \cdot \dfrac{x + 5}{x + 1}$

30. $\dfrac{x^2 + 7x + 12}{x^2 - 9} \div \dfrac{x^2 - 5x + 6}{(x - 3)^2}$

Exercises 31 and 32: Solve the rational equation. Check your result.

31. $\dfrac{2}{x + 5} = \dfrac{-3}{x^2 - 25} + \dfrac{1}{x - 5}$

32. $\dfrac{2y}{y^2 - 3y + 2} = \dfrac{1}{y - 2} + 2$

33. Solve the equation for W.

$$R = \dfrac{3C - 2W}{5}$$

34. Simplify the complex fraction.

$$\dfrac{\dfrac{1}{x^2} + \dfrac{2}{x}}{\dfrac{1}{x^2} - \dfrac{4}{x}}$$

Exercises 35 and 36: Simplify the expression. Assume that all variables are positive.

35. $\sqrt[3]{x^4 y^4} - 2\sqrt[3]{xy}$

36. $(4 + \sqrt{2})(4 - \sqrt{2})$

Exercises 37 and 38: Solve. Check your answers.

37. $8(x - 3)^2 = 200$

38. $3\sqrt{2x + 6} = 6x$

Exercises 39 and 40: Write the complex expression in standard form.

39. $(-3 + i)(-4 - 2i)$

40. $\dfrac{2 - 6i}{1 + 2i}$

41. Find the minimum y-value located on the graph of $y = 3x^2 + 8x + 5$.

42. Write the equation $y = 2x^2 + 8x + 17$ in vertex form and identify the vertex.

Exercises 43 and 44: Solve by using the method of your choice. Write any complex solutions in standard form.

43. $x^2 - 4x + 13 = 0$

44. $z^2 - 4z = 32$

45. Solve the quadratic inequality. Write your answer in interval notation.

$$x^2 + 2x - 3 < 0$$

46. For $f(x) = x^2 + 1$ and $g(x) = 3x - 2$, find each of the following.
(a) $(f \circ g)(-2)$
(b) $(g \circ f)(x)$

47. Find $f^{-1}(x)$ for the one-to-one function

$$f(x) = \dfrac{3x + 1}{2}.$$

48. Write $\ln(x^3 \sqrt{y})$ by using sums and differences of logarithms of x and y.

49. Write $2\log x - \log 4xy$ as one logarithm. Assume x and y are positive.

Exercises 50 and 51: Solve the equation. Approximate answers to the nearest hundredth.

50. $8\log x + 3 = 17$ **51.** $4^{2x} = 5$

52. Graph the parabola $x = (y - 3)^2 + 1$. Find the vertex and the axis of symmetry.

53. Find the center and the radius of the circle whose equation is $x^2 - 6x + y^2 + 2y = -6$.

Exercises 54 and 55: Graph the ellipse or hyperbola. Label the vertices and the endpoints of the minor axis on the ellipse. Show the asymptotes on the hyperbola.

54. $\dfrac{x^2}{4} + \dfrac{y^2}{9} = 1$ **55.** $\dfrac{x^2}{16} - \dfrac{y^2}{4} = 1$

Exercises 56 and 57: Use the graph to determine the equation of the ellipse or hyperbola.

56.

57.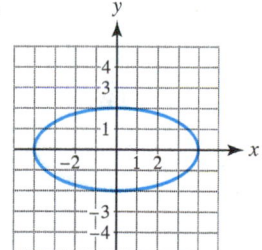

58. Solve the system of equations. Check your solutions.

$$y = x^2 + 1$$
$$x^2 + 2y = 5$$

59. Shade the solution set in the xy-plane.

$$y \geq x^2 - 2$$
$$y \leq -x$$

Exercises 60 and 61: Determine whether f is an arithmetic or a geometric sequence. If it is arithmetic, find the common difference. If it is geometric, find the common ratio.

60. $f(n) = 5 - 2n$ **61.** $f(n) = 3(0.2)^n$

62. Find the general term a_n for the arithmetic sequence where $a_1 = 2$ and $a_2 = 5$.

63. Find the general term a_n for the geometric sequence where $a_1 = 4$ and $a_2 = 12$.

Exercises 64 and 65: Find the sum using a formula.

64. $3 + 7 + 11 + 15 + 19 + \cdots + 35$

65. $1 - 2 + 4 - 8 + 16 - \cdots + 1024$

Exercises 66 and 67: Expand the binomial expression.

66. $(2x + 3)^4$ **67.** $(2a - 5b)^3$

APPLICATIONS

68. Radius of a Circle If a circle has an area of A square units, its radius r is given by $r = \sqrt{\frac{A}{\pi}}$. Approximate the radius of a circle with an area of 14 square inches to the nearest hundredth of an inch.

69. Exercise and Fluid Consumption When a person exercises, the total amount of fluid he or she will need that day increases depending on the person's weight and the duration of the exercise. To determine the number of ounces of fluid needed, divide the person's weight by 2 and then add 0.4 ounces for every minute of exercise. *(Source: Runner's World.)*
   (a) Write a function that gives the fluid requirements for a person weighing 170 pounds who exercises for x minutes a day.
   (b) If an athlete who exercises for 90 minutes requires 130 ounces of fluid, find the athlete's weight.

70. Airplane Speed An airplane travels 1080 miles into the wind in 3 hours. The return trip with the wind takes 2.7 hours. Find the average speed of the airplane and the average wind speed.

71. Working Together Suppose that one person can weed a garden in 60 minutes and a second person can weed the same garden in 90 minutes. How long would it take these two people to weed the garden if they worked together?

72. Numbers The product of two positive numbers is 96. If the larger number is subtracted from 3 times the smaller number, the result is 12. Let x be the smaller number and let y be the larger number.
   (a) Write a system of equations for this situation.
   (b) What are the two numbers?

73. Marching Band A band is marching in a triangular formation so that 1 person is in the first row, 3 people are in the second row, 5 people are in the third row, and so on. Use a formula to find the total number of musicians in the marching band if the last row contains 23 people.

APPENDIX A
Using the Graphing Calculator

1 Overview of the Appendix

This appendix provides instruction for the TI-83, TI-83 Plus, and TI-84 Plus graphing calculators that may be used in conjunction with this textbook. It includes specific keystrokes needed to work several examples from the text. Students are advised to consult the *Graphing Calculator Guidebook* provided by the manufacturer.

2 Entering Mathematical Expressions

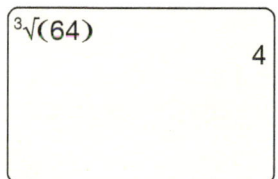

```
π
           3.141592654
√(200)
           14.14213562
10^4
                 10000
```

FIGURE A.1

EVALUATING π To evaluate π, use the following keystrokes, as shown in the first and second lines of **FIGURE A.1**. (Do *not* use 3.14 or $\frac{22}{7}$ for π.)

[2nd] [^[π]] [ENTER]

EVALUATING A SQUARE ROOT To evaluate a square root, such as $\sqrt{200}$, use the following keystrokes, as shown in the third and fourth lines of **FIGURE A.1**.

[2nd] [x²[√]] [2] [0] [0] [)] [ENTER]

EVALUATING AN EXPONENTIAL EXPRESSION To evaluate an exponential expression, such as 10^4, use the following keystrokes, as shown in the last two lines of **FIGURE A.1**.

[1] [0] [^] [4] [ENTER]

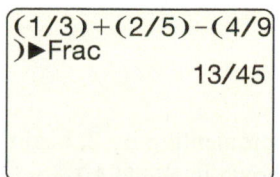

```
³√(64)
                     4
```

FIGURE A.2

EVALUATING A CUBE ROOT To evaluate a cube root, such as $\sqrt[3]{64}$, use the following keystrokes, as shown in **FIGURE A.2**.

[MATH] [4] [6] [4] [)] [ENTER]

SUMMARY: ENTERING MATHEMATICAL EXPRESSIONS

To access the *number* π, use [2nd] [^[π]].

To evaluate a *square root*, use [2nd] [x²[√]].

To evaluate an *exponential expression*, use the [^] key. To square a number, the [x²] key can also be used.

To evaluate a *cube root*, use [MATH] [4].

```
(1/3)+(2/5)-(4/9
)▶Frac
                 13/45
```

FIGURE A.3

3 Expressing Answers as Fractions

To evaluate $\frac{1}{3} + \frac{2}{5} - \frac{4}{9}$ in fraction form, use the following keystrokes, as shown in **FIGURE A.3**.

[(] [1] [÷] [3] [)] [+] [(] [2] [÷] [5] [)] [−] [(] [4] [÷] [9] [)] [MATH] [1] [ENTER]

SUMMARY: EXPRESSING ANSWERS AS FRACTIONS

Enter the arithmetic expression. To access the "Frac" feature, use the keystrokes (MATH) (1). Then press (ENTER).

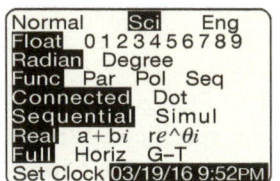

FIGURE A.4

4 Displaying Numbers in Scientific Notation

To display numbers in scientific notation, set the graphing calculator in scientific mode (Sci) by using the following keystrokes. See **FIGURE A.4**. (These keystrokes assume that the calculator is starting from normal mode.)

In scientific mode, we can display the numbers 5432 and 0.00001234 in scientific notation, as shown in **FIGURE A.5**.

SUMMARY: SETTING SCIENTIFIC MODE

If your calculator is in normal mode, it can be set in scientific mode by pressing

(MODE) (▷) (ENTER) (2nd) (MODE [QUIT]).

These keystrokes return the graphing calculator to the home screen.

FIGURE A.5

5 Entering Numbers in Scientific Notation

Numbers can be entered in scientific notation. For example, to enter 4.2×10^{-3} in scientific notation, use the following keystrokes. (Be sure to use the negation key $(-)$ rather than the subtraction key.)

This number can also be entered using the following keystrokes. See **FIGURE A.6**.

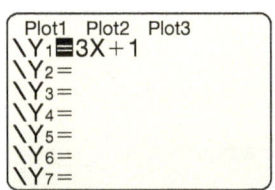

FIGURE A.6

SUMMARY: ENTERING NUMBERS IN SCIENTIFIC NOTATION

One way to enter a number in scientific notation is to use the keystrokes

(2nd) (,[EE])

to access an exponent (EE) of 10.

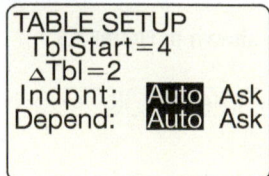

FIGURE A.7

6 Making a Table

To make a table of values for $y = 3x + 1$ starting at $x = 4$ and incrementing by 2, begin by pressing (Y =) and then entering the formula $Y_1 = 3X + 1$, as shown in **FIGURE A.7**. (See Entering a Formula on page AP-4.) To set the table parameters, press the following keys. See **FIGURE A.8**.

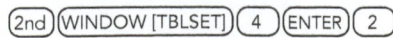

FIGURE A.8

X	Y₁	
4	13	
6	19	
8	25	
10	31	
12	37	
14	43	
16	49	

$Y_1 = 3X + 1$

FIGURE A.9

These keystrokes specify a table that starts at $x = 4$ and increments the x-values by 2. Therefore, the values of Y_1 at $x = 4, 6, 8, \ldots$ appear in the table. To create this table, press the following keys.

(2nd) (GRAPH [TABLE])

We can scroll through x- and y-values by using the arrow keys. See **FIGURE A.9**. Note that there is no first or last x-value in the table.

SUMMARY: MAKING A TABLE

1. Enter the formula for the equation using (Y =).
2. Press (2nd) (WINDOW [TBLSET]) to set the starting x-value and the increment between x-values appearing in the table.
3. Create the table by pressing (2nd) (GRAPH [TABLE]).

ZOOM MEMORY
1:ZBox
2:Zoom In
3:Zoom Out
4:ZDecimal
5:ZSquare
6:ZStandard
7↓ZTrig

FIGURE A.10

7 Setting the Viewing Rectangle (Window)

There are at least two ways to set the standard viewing rectangle of $[-10, 10, 1]$ by $[-10, 10, 1]$. The first involves pressing (ZOOM) followed by (6). See **FIGURE A.10**. The second method for setting the standard viewing rectangle is to press (WINDOW) and enter the following keystrokes. See **FIGURE A.11**.

(Be sure to use the negation key $(-)$ rather than the subtraction key.) Other viewing rectangles can be set in a similar manner by pressing (WINDOW) and entering the appropriate values. To see the viewing rectangle, press (GRAPH).

WINDOW
Xmin = -10
Xmax = 10
Xscl = 1
Ymin = -10
Ymax = 10
Yscl = 1
Xres = 1

FIGURE A.11

SUMMARY: SETTING THE VIEWING RECTANGLE

To set the standard viewing rectangle, press (ZOOM) (6). To set any viewing rectangle, press (WINDOW) and enter the necessary values. To see the viewing rectangle, press (GRAPH).

NOTE: You do not need to change "Xres" from 1. ∎

L1	L2	L3	1
1	4	------	
2	5		
3	6		
------	------		

L1 = {1, 2, 3}

FIGURE A.12

8 Making a Scatterplot or a Line Graph

To make a scatterplot with the points $(-5, -5)$, $(-2, 3)$, $(1, -7)$, and $(4, 8)$, begin by following these steps.

1. Press (STAT) followed by (1).
2. If list L1 is not empty, use the arrow keys to place the cursor on L1, as shown in **FIGURE A.12**. Then press (CLEAR) followed by (ENTER). This deletes all elements in the list. Similarly, if L2 is not empty, clear the list.
3. Input each x-value into list L1 followed by (ENTER). Input each y-value into list L2 followed by (ENTER). See **FIGURE A.13**.

L1	L2	L3	1
-5	-5	------	
-2	3		
1	-7		
4	8		
------	------		

L1(5) =

FIGURE A.13

FIGURE A.14

FIGURE A.15

$[-10, 10, 1]$ by $[-10, 10, 1]$

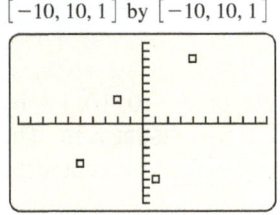

FIGURE A.16

It is essential that both lists have the same number of values—otherwise, an error message appears when a scatterplot is attempted. Before these four points can be plotted, "STAT PLOT" must be turned on. It is accessed by pressing

(2nd) (Y =) [STAT PLOT],

as shown in **FIGURE A.14**.

There are three possible "STAT PLOTS," numbered 1, 2, and 3. Any one of the three can be selected. The first plot can be selected by pressing (1). Next, place the cursor over "On" and press (ENTER) to turn "Plot1" on. There are six types of plots that can be selected. The first type is a *scatterplot* and the second type is a *line graph*, so place the cursor over the first type of plot and press (ENTER) to select a scatterplot. (To make the line graph, place the cursor over the second type of plot and press (ENTER).) The x-values are stored in list L1, so select L1 for "Xlist" by pressing (2nd)(1). Similarly, press (2nd)(2) for the "Ylist," since the y-values are stored in list L2. Finally, there are three styles of marks that can be used to show data points in the graph. We will usually use the first because it is largest and shows up the best. Make the screen appear as in **FIGURE A.15**. Before plotting the four data points, be sure to set an appropriate viewing rectangle. (See Setting the Viewing Rectangle on the previous page.) Then press (GRAPH). The data points appear as in **FIGURE A.16**.

REMARK 1: A fast way to set the viewing rectangle for any scatterplot is to select the "ZOOMSTAT" feature by pressing (ZOOM)(9). This feature automatically scales the viewing rectangle so that all data points are shown.

REMARK 2: If an equation has been entered into the (Y =) menu and selected, it will be graphed with the data. This feature is used frequently to model data.

SUMMARY: MAKING A SCATTERPLOT OR A LINE GRAPH

The following are basic steps necessary to make either a scatterplot or a line graph.

1. Use (STAT)(1) to access lists L1 and L2.
2. If list L1 is not empty, place the cursor on L1 and press (CLEAR)(ENTER). Repeat for list L2, if it is not empty.
3. Enter the x-values into list L1 and the y-values into list L2.
4. Use (2nd)(Y =) [STAT PLOT] to select appropriate parameters for the scatterplot or line graph.
5. Set an appropriate viewing rectangle. Press (GRAPH). Otherwise, press (ZOOM)(9). This feature automatically sets the viewing rectangle and plots the data.

NOTE: (ZOOM)(9) *cannot* be used to set a viewing rectangle for the graph of an equation. ∎

FIGURE A.17

9 Entering a Formula

To enter a formula, press (Y =). For example, use the following keystrokes after "$Y_1 =$" to enter $y = x^2 - 4$. See **FIGURE A.17**.

(Y =)(CLEAR)(X, T, θ, n)(^)(2)(−)(4)

Note that there is a built-in key to enter the variable X. If "$Y_1 =$" does not appear after pressing (Y =), press (MODE) and make sure the calculator is set in *function mode*, denoted "Func". See **FIGURE A.18**.

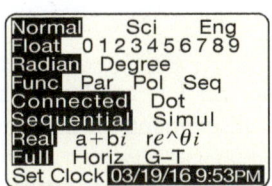

FIGURE A.18

SUMMARY: ENTERING A FORMULA

To enter a formula, press (Y =). To delete a formula, press (CLEAR).

$[-10, 10, 1]$ by $[-10, 10, 1]$

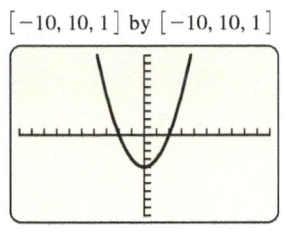

FIGURE A.19

10 Graphing an Equation

To graph an equation, such as $y = x^2 - 4$, start by pressing (Y =) and enter $Y_1 = X\char`\^2 - 4$. If there is an equation already entered, remove it by pressing (CLEAR). The equals signs in "$Y_1 =$" should be in reverse video (a dark rectangle surrounding a white equals sign), which indicates that the equation will be graphed. If the equals sign is not in reverse video, place the cursor over it and press (ENTER). Set an appropriate viewing rectangle and then press (GRAPH). The graph will appear in the specified viewing rectangle. See **FIGURES A.17** and **A.19**.

SUMMARY: GRAPHING AN EQUATION

1. Use the (Y =) menu to enter the formula.
2. Use the (WINDOW) menu to set an appropriate viewing rectangle.
3. Press (GRAPH).

11 Graphing a Vertical Line

Set an appropriate window (or viewing rectangle). Then return to the home screen by pressing

FIGURE A.20

(2nd) (MODE [QUIT]).

To graph a vertical line, such as $x = -4$, press

(2nd) (PRGM [DRAW]) (4) ((−)) (4).

See **FIGURE A.20**. Pressing (ENTER) will make the vertical line appear, as shown in **FIGURE A.21**.

$[-6, 6, 1]$ by $[-6, 6, 1]$

FIGURE A.21

SUMMARY: GRAPHING THE VERTICAL LINE $x = h$

1. Set an appropriate window by pressing (WINDOW).
2. Return to the home screen by pressing (2nd) (MODE [QUIT]).
3. Draw a vertical line by pressing (2nd) (PRGM [DRAW]) (4) (h) (ENTER).

12 Squaring a Viewing Rectangle

In a square viewing rectangle the graph of $y = x$ is a line that makes a 45° angle with the positive x-axis, a circle appears circular, and all sides of a square have the same length. An approximate square viewing rectangle can be set if the distance along the x-axis is 1.5 times the distance along the y-axis. Examples of viewing rectangles that are (approximately) square include

$$[-6, 6, 1] \text{ by } [-4, 4, 1] \quad \text{and} \quad [-9, 9, 1] \text{ by } [-6, 6, 1].$$

Square viewing rectangles can be set automatically by pressing either

(ZOOM) (4) or (ZOOM) (5).

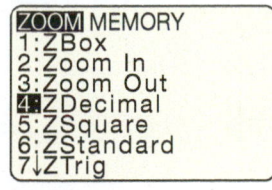

FIGURE A.22

ZOOM 4 provides a *decimal window*, which is discussed on page AP-11. See **FIGURE A.22**.

SUMMARY: SQUARING A VIEWING RECTANGLE

Either $\boxed{\text{ZOOM}}\ \boxed{4}$ or $\boxed{\text{ZOOM}}\ \boxed{5}$ may be used to produce a square viewing rectangle. An (approximately) square viewing rectangle has the form

$$[-1.5k, 1.5k, 1] \text{ by } [-k, k, 1],$$

where k is a positive number.

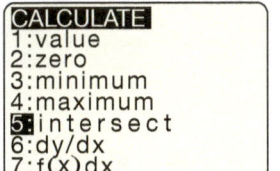

FIGURE A.23

13 Locating a Point of Intersection

To find the point of intersection for the graphs of

$$y_1 = 3(1 - x) \quad \text{and} \quad y_2 = 2,$$

start by entering Y_1 and Y_2, as shown in **FIGURE A.23**. Set the window, and graph both equations. Then press the following keys to find the intersection point.

See **FIGURE A.24**, where the "intersect" utility is being selected. The calculator prompts for the first curve, as shown in **FIGURE A.25**. Use the arrow keys to locate the cursor near the point of intersection and press $\boxed{\text{ENTER}}$. Repeat these steps for the second curve. Finally, we are prompted for a guess. For each of the three prompts, place the free-moving cursor near the point of intersection and press $\boxed{\text{ENTER}}$. The approximate coordinates of the point of intersection will be shown.

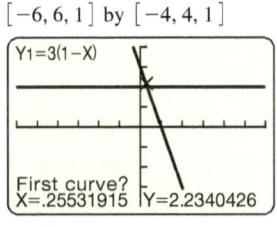

FIGURE A.24

FIGURE A.25

SUMMARY: FINDING A POINT OF INTERSECTION

1. Graph the two equations in an appropriate viewing rectangle.
2. Press $\boxed{\text{2nd}}\boxed{\text{TRACE [CALC]}}\boxed{5}$.
3. Use the arrow keys to select an approximate location for the point of intersection. Press $\boxed{\text{ENTER}}$ to make the three selections for "First curve?", "Second curve?", and "Guess?". (Note that if the cursor is near the point of intersection, you usually do not need to move the cursor for each selection. Just press $\boxed{\text{ENTER}}$ three times.)

14 Shading Inequalities

To shade the solution set for one or more linear inequalities such as $2x + y \leq 5$ and $-2x + y \geq 1$, begin by solving each inequality for y to obtain $y \leq 5 - 2x$ and $y \geq 2x + 1$. Then let $Y_1 = 5 - 2X$ and $Y_2 = 2X + 1$, as shown in **FIGURE A.26**. Position the cursor to the left of Y_1 and press $\boxed{\text{ENTER}}$ three times. The triangle that appears indicates that the calculator will shade the region below the graph of Y_1. Next locate the cursor to the left of Y_2 and press $\boxed{\text{ENTER}}$ twice. This triangle indicates that the calculator will shade the region above the graph of Y_2. After setting the viewing rectangle to $[-15, 15, 5]$ by $[-10, 10, 5]$, press $\boxed{\text{GRAPH}}$. The result is shown in **FIGURE A.27**.

FIGURE A.26

FIGURE A.27

SUMMARY: SHADING ONE OR MORE INEQUALITIES

1. Solve each inequality for y.
2. Enter each formula as Y_1 and Y_2 in the $\boxed{Y=}$ menu.
3. Locate the cursor to the left of Y_1 and press $\boxed{\text{ENTER}}$ two or three times to shade either above or below the graph of Y_1. Repeat for Y_2.
4. Set an appropriate viewing rectangle.
5. Press $\boxed{\text{GRAPH}}$.

NOTE: The "Shade" utility in the DRAW menu can also be used to shade the region *between* two graphs. ∎

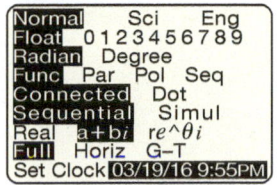

FIGURE A.28

15 Setting $a + bi$ Mode

To evaluate expressions containing square roots of negative numbers, such as $\sqrt{-25}$, set your calculator in $a + bi$ mode by using the following keystrokes.

$$\boxed{\text{MODE}}\ \boxed{\triangledown}\ \boxed{\triangledown}\ \boxed{\triangledown}\ \boxed{\triangledown}\ \boxed{\triangledown}\ \boxed{\triangledown}\ \boxed{\triangleright}\ \boxed{\text{ENTER}}\ \boxed{\text{2nd}}\ \boxed{\text{MODE [QUIT]}}$$

See **FIGURES A.28** and **A.29**.

FIGURE A.29

SUMMARY: SETTING $a + bi$ MODE

1. Press $\boxed{\text{MODE}}$.
2. Move the cursor to the seventh line, highlight $a + bi$, and press $\boxed{\text{ENTER}}$.
3. Press $\boxed{\text{2nd}}\boxed{\text{MODE [QUIT]}}$ and return to the home screen.

16 Evaluating Complex Arithmetic

Complex arithmetic can be performed much like other arithmetic expressions. This is done by entering

$$\boxed{\text{2nd}}\boxed{.\,[i]}$$

to obtain the imaginary unit i from the home screen. For example, to find the sum $(-2 + 3i) + (4 - 6i)$, perform the following keystrokes on the home screen.

$$\boxed{(}\ \boxed{(-)}\ \boxed{2}\ \boxed{+}\ \boxed{3}\ \boxed{\text{2nd}}\ \boxed{.\,[i]}\ \boxed{)}\ \boxed{+}\ \boxed{(}\ \boxed{4}\ \boxed{-}\ \boxed{6}\ \boxed{\text{2nd}}\ \boxed{.\,[i]}\ \boxed{)}\ \boxed{\text{ENTER}}$$

FIGURE A.30

The result is shown in **FIGURE A.30**. Other complex arithmetic operations are done similarly.

SUMMARY: EVALUATING COMPLEX ARITHMETIC

Enter a complex expression in the same way as you would any arithmetic expression. To obtain the complex number i, use $\boxed{\text{2nd}}\boxed{.\,[i]}$.

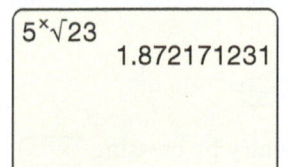

FIGURE A.31

17 Evaluating Other Mathematical Expressions

EVALUATING OTHER ROOTS To evaluate a fifth root, such as $\sqrt[5]{23}$, use the following keystrokes, as shown in **FIGURE A.31**.

$$\boxed{5}\ \boxed{\text{MATH}}\ \boxed{5}\ \boxed{2}\ \boxed{3}\ \boxed{\text{ENTER}}$$

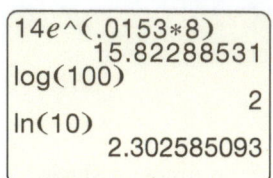

FIGURE A.32

EVALUATING THE NATURAL EXPONENTIAL FUNCTION To evaluate $14e^{0.0153(8)}$, use the following keystrokes, as shown in the first and second lines of **FIGURE A.32**.

EVALUATING THE COMMON LOGARITHMIC FUNCTION To evaluate $\log(100)$, use the following keystrokes, as shown in the third and fourth lines of **FIGURE A.32**.

(LOG)(1)(0)(0)( ))(ENTER)

EVALUATING THE NATURAL LOGARITHMIC FUNCTION To evaluate $\ln(10)$, use the following keystrokes, as shown in the last two lines of **FIGURE A.32**.

(LN)(1)(0)( ))(ENTER)

SUMMARY: OTHER MATHEMATICAL EXPRESSIONS

To evaluate a *kth root*, use (k)(MATH)(5).

To access the *natural exponential function*, use (2nd)(LN [eˣ]).

To access the *common logarithmic function*, use (LOG).

To access the *natural logarithmic function*, use (LN).

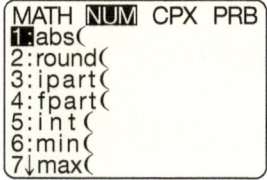

FIGURE A.33

18 Accessing the Absolute Value

To graph $y_1 = |x - 50|$, begin by entering $Y_1 = \text{abs}(X - 50)$. The absolute value (abs) is accessed by pressing

(MATH)(▷)(1).

See **FIGURE A.33**.

SUMMARY: ACCESSING THE ABSOLUTE VALUE

1. Press (MATH).
2. Position the cursor over "NUM".
3. Press (1) to select the absolute value.

---

19 Entering the Elements of a Matrix

The elements of the augmented matrix A given by

$$A = \left[\begin{array}{ccc|c} 1 & 1 & 2 & 1 \\ -1 & 0 & 1 & -2 \\ 2 & 1 & 5 & -1 \end{array} \right]$$

can be entered by using the following keystrokes on the TI-83 Plus or TI-84 Plus to define a matrix A with dimension 3×4. (*Note:* On the TI-83 the matrix menu is found by pressing (MATRX).)

(2nd)(x⁻¹ [MATRIX])(▷)(▷)(1)(3)(ENTER)(4)(ENTER)

Input the 12 elements of the matrix A, row by row. Finish each entry by pressing (ENTER). See **FIGURE A.34**. After these elements have been entered, press

(2nd)(MODE [QUIT])

FIGURE A.34

```
[A]
   [[1   1   2   1]
    [-1  0   1  -2]
    [2   1   5  -1]]
```

FIGURE A.35

to return to the home screen. To display the matrix A, press

(2nd) (x⁻¹ [MATRIX]) (1) (ENTER).

See **FIGURE A.35**.

SUMMARY: ENTERING THE ELEMENTS OF A MATRIX *A*

1. Begin by accessing the matrix A by pressing (2nd)(x⁻¹ [MATRIX])(▷)(▷)(1).
2. Enter the dimension of A by pressing (m)(ENTER)(n)(ENTER), where the dimension of the matrix is $m \times n$.
3. Input each element of the matrix, row by row. Finish each entry by pressing (ENTER). Use (2nd)(MODE [QUIT]) to return to the home screen.

NOTE: On the TI-83, replace the keystrokes (2nd)(x⁻¹ [MATRIX]) with (MATRX). ∎

---

20 Finding Reduced Row–Echelon Form

To find the reduced row–echelon form of matrix A (entered above in **FIGURE A.35**), use the following keystrokes from the home screen on the TI-83 Plus or TI-84 Plus.

(2nd)(x⁻¹ [MATRIX])(▷)(ALPHA)(APPS [B])(2nd)(x⁻¹ [MATRIX])(1)( ))(ENTER)

```
rref([A])
   [[1   0   0   1]
    [0   1   0   2]
    [0   0   1  -1]]
```

FIGURE A.36

The resulting matrix is shown in **FIGURE A.36**. On the TI-83 graphing calculator, use the following keystrokes to find the reduced row–echelon form.

(MATRX)(▷)(ALPHA)(MATRX [B])(MATRX)(1)( ))(ENTER)

SUMMARY: FINDING THE REDUCED ROW–ECHELON FORM OF A MATRIX

1. To make rref([A]) appear on the home screen, use the following keystrokes for the TI-83 Plus or TI-84 Plus graphing calculator.

(2nd)(x⁻¹ [MATRIX])(▷)(ALPHA)(APPS [B])(2nd)(x⁻¹ [MATRIX])(1)( ))

2. Press (ENTER) to calculate the reduced row–echelon form.
3. Use arrow keys to access elements that do not appear on the screen.

NOTE: On the TI-83, replace the keystrokes (2nd)(x⁻¹ [MATRIX]) with (MATRX) and (APPS [B]) with (MATRX [B]). ∎

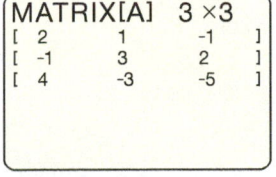

```
MATRIX[A]  3 ×3
[ 2    1    -1  ]
[ -1   3    2   ]
[ 4   -3   -5   ]
```

FIGURE A.37

21 Evaluating a Determinant

To evaluate the determinant of matrix A given by

$$A = \begin{bmatrix} 2 & 1 & -1 \\ -1 & 3 & 2 \\ 4 & -3 & -5 \end{bmatrix},$$

start by entering the 9 elements of the 3×3 matrix, as shown in **FIGURE A.37**. To compute det A, perform the following keystrokes from the home screen.

(2nd)(x⁻¹ [MATRIX])(▷)(1)(2nd)(x⁻¹ [MATRIX])(1)( ))(ENTER)

```
det([A])
                -6
```

FIGURE A.38

The results are shown in **FIGURE A.38**.

SUMMARY: EVALUATING A DETERMINANT OF A MATRIX

1. Enter the dimension and elements of the matrix A.
2. Return to the home screen by pressing (2nd)(MODE [QUIT]).
3. On the TI-83 Plus or TI-84 Plus, perform the following keystrokes.

NOTE: On the TI-83, replace the keystrokes (2nd)(x^{-1} [MATRIX]) with (MATRX). ∎

CALCULATE
```
1:value
2:zero
3:minimum
4:maximum
5:intersect
6:dy/dx
7:∫f(x)dx
```
FIGURE A.39

22 Locating an x-Intercept or Zero

To locate an x-intercept or *zero* of $f(x) = x^2 - 4$, start by entering $Y_1 = X \wedge 2 - 4$ into the (Y =) menu. Set the viewing rectangle to $[-9, 9, 1]$ by $[-6, 6, 1]$ and graph Y_1. Afterwards, press the following keys to invoke the zero finder. See **FIGURE A.39**.

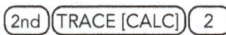

The graphing calculator prompts for a left bound. Use the arrow keys to set the cursor to the left of the x-intercept and press (ENTER). The graphing calculator then prompts for a right bound. Set the cursor to the right of the x-intercept and press (ENTER). Finally, the graphing calculator prompts for a guess. Set the cursor roughly at the x-intercept and press (ENTER). See **FIGURES A.40–A.42**. The calculator then approximates the x-intercept or zero automatically, as shown in **FIGURE A.43**. The zero of -2 can be found similarly.

$[-9, 9, 1]$ by $[-6, 6, 1]$

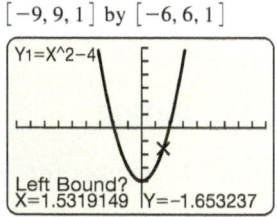

FIGURE A.40

$[-9, 9, 1]$ by $[-6, 6, 1]$

FIGURE A.41

$[-9, 9, 1]$ by $[-6, 6, 1]$

FIGURE A.42

$[-9, 9, 1]$ by $[-6, 6, 1]$

FIGURE A.43

SUMMARY: LOCATING AN x-INTERCEPT OR ZERO

1. Graph the function in an appropriate viewing rectangle.
2. Press (2nd)(TRACE [CALC])(2).
3. Select the left and right bounds, followed by a guess. Press (ENTER) after each selection. The calculator then approximates the x-intercept or zero.

FIGURE A.44

23 Setting Connected or Dot Mode

To set your graphing calculator in dot mode, press (MODE), position the cursor over "Dot," and press (ENTER). See **FIGURE A.44**. Graphs will now appear in dot mode rather than connected mode.

SUMMARY: SETTING CONNECTED OR DOT MODE

1. Press (MODE).
2. Position the cursor over "Connected" or "Dot". Press (ENTER).

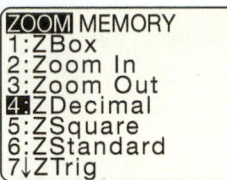

FIGURE A.45

24 Setting a Decimal Window

With a decimal window, the cursor stops on convenient x-values. In the decimal window $[-9.4, 9.4, 1]$ by $[-6.2, 6.2, 1]$ the cursor stops on x-values that are multiples of 0.2. If we reduce the viewing rectangle to $[-4.7, 4.7, 1]$ by $[-3.1, 3.1, 1]$, the cursor stops on x-values that are multiples of 0.1. To set this smaller window automatically, press (ZOOM) (4). See **FIGURE A.45**. Decimal windows are also useful when graphing rational functions with asymptotes in connected mode.

> ### SUMMARY: SETTING A DECIMAL WINDOW
>
> 1. Press (ZOOM) (4) to set the viewing rectangle $[-4.7, 4.7, 1]$ by $[-3.1, 3.1, 1]$.
> 2. A larger decimal window is $[-9.4, 9.4, 1]$ by $[-6.2, 6.2, 1]$.

25 Finding Maximum and Minimum Values

To find a minimum y-value (or vertex) on the graph of $f(x) = 1.5x^2 - 6x + 4$, start by entering $Y_1 = 1.5X\text{^}2 - 6X + 4$ from the (Y=) menu. Set the viewing rectangle and then perform the following keystrokes to find the minimum y-value.

(2nd) (TRACE [CALC]) (3)

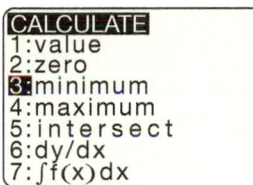

FIGURE A.46

See **FIGURE A.46**.

The calculator prompts for a left bound. Use the arrow keys to position the cursor left of the vertex and press (ENTER). Similarly, position the cursor to the right of the vertex for the right bound and press (ENTER). Finally, the graphing calculator asks for a guess between the left and right bounds. Place the cursor near the vertex and press (ENTER). See **FIGURES A.47–A.49**. The minimum value is shown in **FIGURE A.50**.

$[-4.7, 4.7, 1]$ by $[-3.1, 3.1, 1]$ $[-4.7, 4.7, 1]$ by $[-3.1, 3.1, 1]$ $[-4.7, 4.7, 1]$ by $[-3.1, 3.1, 1]$ $[-4.7, 4.7, 1]$ by $[-3.1, 3.1, 1]$

FIGURE A.47   **FIGURE A.48**   **FIGURE A.49**   **FIGURE A.50**

$[0, 50, 10]$ by $[0, 1000, 100]$

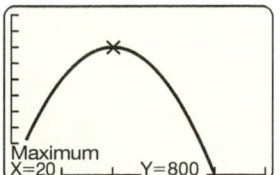

FIGURE A.51

A maximum of the function f on an interval can be found similarly, except enter

(2nd) (TRACE [CALC]) (4)

The calculator prompts for left and right bounds, followed by a guess. Press (ENTER) after the cursor has been located appropriately for each prompt. The graphing calculator will display the maximum y-value. An example is shown in **FIGURE A.51**, where $f(x) = x(80 - 2x)$.

> ### SUMMARY: FINDING MAXIMUM AND MINIMUM VALUES
>
> 1. Graph the function in an appropriate viewing rectangle.
> 2. Press (2nd) (TRACE [CALC]) (3) to find a minimum y-value.
> 3. Press (2nd) (TRACE [CALC]) (4) to find a maximum y-value.
> 4. Use the arrow keys to locate the left and right x-bounds, followed by a guess. Press (ENTER) to select each position of the cursor.

APPENDIX B
Sets

NEW VOCABULARY

- ☐ Set
- ☐ Elements
- ☐ Empty (null) set
- ☐ Universal set
- ☐ Subset
- ☐ Equal sets
- ☐ Venn diagrams
- ☐ Complement
- ☐ Union
- ☐ Intersection
- ☐ Finite/infinite set

1 Determining the Elements of a Set

A **set** is a collection of things, and the members of a set are called **elements**. A set can be described by listing its elements between braces. For example, the set W containing the *weekdays* is

Elements

$$\text{Set} \quad W = \overbrace{\{\text{Monday, Tuesday, Wednesday, Thursday, Friday}\}}.$$

This set has 5 elements. For example, Monday *is an element of* W, which is denoted

$$\text{Monday} \in W.$$

However, Sunday *is not an element of* W, which is denoted

$$\text{Sunday} \notin W.$$

If a set contains no elements, then it is called the **empty set** or **null set**. The empty set is denoted $\varnothing$, or $\{\ \}$. For example, the set Z that contains the names of U.S. states starting with the letter Z is the empty set. That is, $Z = \varnothing$ or $Z = \{\ \}$.

NOTE: Do *not* write the empty set as $\{\varnothing\}$. ■

EXAMPLE 1 **Listing the elements of a set**

Use set notation to list the elements of each set S described.
(a) The natural numbers from 1 to 12 that are odd
(b) The days of the week that start with the letter T
(c) The last names of U.S. presidents in office from 2005 to 2015

Solution
(a) The list of the natural numbers from 1 to 12 is

$$1, 2, 3, 4, 5, 6, 7, 8, 9, 10, 11, \text{ and } 12.$$

The set of odd natural numbers from this list is

$$S = \{1, 3, 5, 7, 9, 11\}.$$

(b) $S = \{\text{Tuesday, Thursday}\}$
(c) $S = \{\text{Bush, Obama}\}$

Now Try Exercises 1, 3, 5

EXAMPLE 2 **Determining the elements of sets**

Use $\in$ or $\notin$ to make each statement true.
(a) 5 _____ $\{1, 2, 3, 4, 5, 6\}$
(b) -2 _____ $\{-4, 0, 2, 4, 6\}$
(c) $\frac{1}{2}$ _____ $\{0, 0.5, 1.0, 1.5, 2.0\}$

Solution
(a) Because **5** is an element of $\{1, 2, 3, 4, \mathbf{5}, 6\}$, we write

$$5 \in \{1, 2, 3, 4, 5, 6\}.$$

(b) Because -2 is not an element of $\{-4, 0, 2, 4, 6\}$, we write

$$-2 \notin \{-4, 0, 2, 4, 6\}.$$

(c) Because $\frac{1}{2} = 0.5$, we write

$$\frac{1}{2} \in \{0, 0.5, 1.0, 1.5, 2.0\}.$$

Now Try Exercises 11, 13, 15

2 Determining the Universal Set

When discussing sets, we assume that there is a *universal set*. The **universal set** contains all elements under consideration. For example, if the universal set U is all days of the week, then the set S containing the days that start with the letter S is

$$S = \{\text{Sunday, Saturday}\}.$$

However, if the universal set U is only the weekdays, then the set S containing the days starting with the letter S is

$$\text{the } \textit{empty set, or } S = \{\ \}.$$

EXAMPLE 3 **Using different universal sets**

Determine the set O of odd integers that belong to each universal set U.
(a) $U = \{1, 2, 3, 4, 5, 6, 7, 8, 9, 10\}$
(b) $U = \{1, 6, 11, 16, 21, 26\}$
(c) $U = \{1, 2, 3, 4, \ldots\}$

Solution
(a) The odd integers in U are $1, 3, 5, 7$, and 9, so

$$O = \{1, 3, 5, 7, 9\}.$$

(b) The odd integers in $U = \{1, 6, 11, 16, 21, 26\}$ are $1, 11$, and 21. Thus

$$O = \{1, 11, 21\}.$$

(c) The three dots in $\{1, 2, 3, 4, \ldots\}$ indicate that U contains all natural numbers. Thus the set O contains all odd natural numbers, or

$$O = \{1, 3, 5, 7, 9, \ldots\}.$$

Now Try Exercises 21, 23

3 Determining Subsets

If every element in a set B is contained in a set A, then we say that B is a **subset** of A, denoted $B \subseteq A$. For example, if $A = \{1, 2, 3, 4\}$ and $B = \{2, 4\}$, then $B \subseteq A$ because every element in B belongs to A. However, A is not a subset of B, denoted $A \not\subseteq B$, because the elements 1 and 3 are in A but *not* in B. The symbol $\not\subseteq$ is read "is not a subset of."

If every element in set A is in set B and every element in set B is in set A, then A and B are **equal sets**, denoted $A = B$. Note that if $A \subseteq B$ and $B \subseteq A$, then $A = B$.

EXAMPLE 4 **Determining subsets**

Let $A = \{a, b, c, d, e\}$, $B = \{b, c, d\}$, $C = \{b, e\}$, and $D = \{e, b\}$. Determine whether each statement is true or false.
(a) $A \subseteq B$ **(b)** $B \subseteq A$ **(c)** $C = D$ **(d)** $C \not\subseteq A$ **(e)** $\varnothing \subseteq B$ **(f)** $A \subseteq A$

Solution

(a) False; the elements **a** and **e** in A are not in B, so A is *not* a subset of B.

(b) True; every element in B is in A, so B is a subset of A.

(c) True; although the elements in C and D are listed in a different order, they contain exactly the same elements, so C and D are equal.

(d) False; every element in C is in A, so C *is* a subset of A.

(e) True; the empty set, or null set, is a subset of *every* set.

(f) True; every element in A is in A, so A is a subset of itself.

   NOTE: Every set is a subset of itself. ■

Now Try Exercises 29, 31, 33, 35

4 Using Venn Diagrams and Finding Complements

Venn diagrams are often used to depict relationships among sets. A large rectangle typically represents the universal set, and subsets of the universal set are represented by regions within the universal set. In **FIGURE B.1**, the universal set U is represented by everything inside the large rectangle. The set A is represented by the red circular region within this rectangle because A is a subset of U.

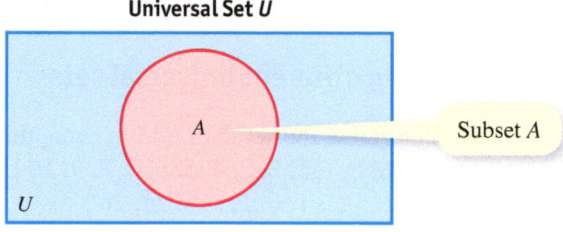

FIGURE B.1 $A \subseteq U$

   The **complement** of a set A, denoted A', is the set containing all elements in the universal set that are *not* in A. That is, if $a \notin A$, then $a \in A'$. For example, if

$$U = \{1, 2, 3, 4, 5, 6\} \quad \text{and} \quad A = \{1, 2, 3\},$$

then

$$A' = \{4, 5, 6\}$$

because the elements 4, 5, and 6 are found in U but not in A. This situation is illustrated by the Venn diagram in **FIGURE B.2**. The red region is A, and the blue region is A'. Together, the red and blue regions make up the universal set U.

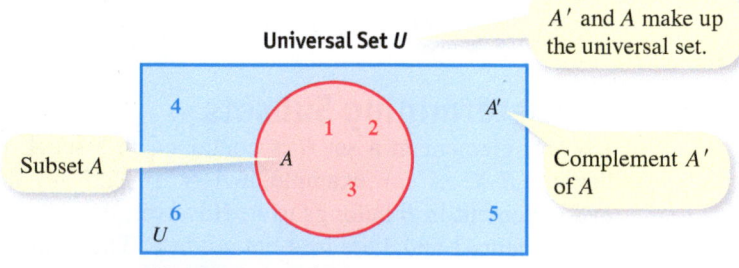

FIGURE B.2 A and A'

NOTE: Every element in U must be in either A or A', but not both. ■

EXAMPLE 5 **Determining complements**

Let the universal set be $U = \{$red, blue, yellow, green, black, white$\}$, and let two subsets of U be $A = \{$red, blue, yellow$\}$ and $B = \{$black, white$\}$. Find each of the following.
(a) A' **(b)** B' **(c)** U'

Solution
(a) The elements in U that are not in A are in $A' = \{$green, black, white$\}$.
(b) $B' = \{$red, blue, yellow, green$\}$
(c) Because the universal set U contains every element under consideration, the complement of U, or U', is empty. That is, $U' = \{\ \ \}$ or $\varnothing$.

Now Try Exercises 73, 75, 81

5 Finding Unions and Intersections

Although we do not perform arithmetic operations, such as multiplication or division, on sets, we can find the *union* or *intersection* of two or more sets. The **union** of two sets A and B, denoted $A \cup B$ and read "A union B," is the set containing any element that can be found in *either* set A *or* set B. If an element is in both A and B, then this element is listed only once in $A \cup B$. For example, if

$$A = \{1, 2, 3, 4\} \quad \text{and} \quad B = \{3, 4, 5, 6\},$$

then

$$A \cup B = \{1, 2, 3, 4, 5, 6\}. \quad \text{Elements in set } A \text{ or in set } B$$

Note that elements **3** and **4** are in both A and B, but are listed only once in $A \cup B$. A Venn diagram of this situation is shown in **FIGURE B.3**. The region that represents the union is shaded blue.

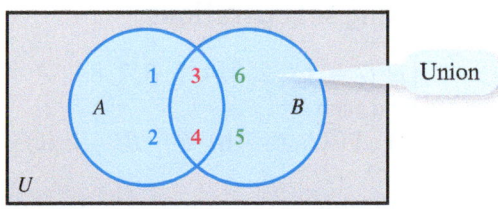

FIGURE B.3

The **intersection** of two sets A and B, denoted $A \cap B$ and read "A intersect B," is the set containing elements that can be found in *both* set A *and* set B. For example, if

$$A = \{1, 2, 3, 4\} \quad \text{and} \quad B = \{3, 4, 5, 6\},$$

then

$$A \cap B = \{3, 4\}. \quad \text{Elements in both set } A \text{ and set } B$$

Note that elements 3 and 4 belong to *both A and B*. This situation is illustrated by the Venn diagram in **FIGURE B.4** on the next page. The purple region containing elements 3 and 4 represents the intersection of A and B.

FIGURE B.4

EXAMPLE 6   **Finding unions and intersections of sets**

Let $A = \{a, b, c, x, z\}$, $B = \{a, x, y\}$, and $C = \{c, x, y, z\}$. Find each set.
(a) $A \cup B$ **(b)** $B \cap C$ **(c)** $A \cup C$ **(d)** $A \cap B \cap C$

Solution
(a) The union contains the elements belonging to either *A or B* or both. Thus

$$A \cup B = \{a, b, c, x, y, z\}.$$

(b) The intersection contains the elements belonging to both *B and C*. Thus

$$B \cap C = \{x, y\}.$$

(c) $A \cup C = \{a, b, c, x, y, z\}$
(d) For an element to be in the intersection of three sets, the element must belong to each set. The only element belonging to *A and B and C* is x. Thus $A \cap B \cap C = \{x\}$.

Now Try Exercises 63, 65, 69, 79

EXAMPLE 7   **Using set operations**

Let $U = \{1, 2, 3, 4, 5, 6, 7, 8\}$, $A = \{2, 4, 6, 8\}$, $B = \{1, 3, 5, 6\}$, and $C = \{5, 6, 7, 8\}$. Find each set.
(a) $A \cap B$ **(b)** $B \cap B'$ **(c)** $B \cup C$ **(d)** $A \cup C'$

Solution
(a) $A \cap B = \{6\}$
(b) Because $B = \{1, 3, 5, 6\}$ and $B' = \{2, 4, 7, 8\}$, $B \cap B' = \{\ \ \}$.

NOTE: The intersection of a set and its complement is always the empty set. ■

(c) $B \cup C = \{1, 3, 5, 6, 7, 8\}$
(d) Because $A = \{2, 4, 6, 8\}$ and $C' = \{1, 2, 3, 4\}$, $A \cup C' = \{1, 2, 3, 4, 6, 8\}$.

Now Try Exercises 67, 71, 77, 83

6 Recognizing Finite and Infinite Sets

Sets can either have a finite number of elements or infinitely many elements. The elements of a **finite set** can be listed explicitly, whereas the elements of an **infinite set** cannot be listed because there are infinitely many elements. For example, the set of integers *S* from -3 to 3 is a finite set with 7 elements because

$$S = \{-3, -2, -1, 0, 1, 2, 3\}. \qquad \text{Finite set}$$

In contrast, the set of natural numbers,

$$N = \{1, 2, 3, 4, \ldots\}, \qquad \text{Infinite set}$$

is an infinite set because the list continues without end.

EXAMPLE 8   **Recognizing finite and infinite sets**

Identify each set as finite or infinite.
(a) The set of integers
(b) The set of natural numbers between 1 and 5, inclusive
(c) The set of rational numbers between 1 and 5, inclusive

Solution
(a) There are infinitely many integers, so the set of integers is infinite.
(b) The set of natural numbers between 1 and 5, inclusive, is $\{1, 2, 3, 4, 5\}$. Because we can list each element of this set, it is a finite set.
(c) Because there are infinitely many rational numbers (fractions) between 1 and 5, this set is an infinite set.

Now Try Exercises 87, 89

B   Exercises

MyMathLab®

FINDING ELEMENTS OF A SET

Exercises 1–10: Use set notation to list the elements of the set described.

1. The natural numbers less than 8

2. The natural numbers greater than 4 and less than or equal to 10

3. The days of the week starting with the letter S

4. U.S. states whose names start with the letter A

5. The letters of the alphabet from A to G

6. The letters of the alphabet from D to M

7. Dogs that can fly under their own power

8. People having an income of $10 trillion in 2017

9. The even integers greater than −2 and less than 11

10. The odd integers less than 21 and greater than 12

Exercises 11–20: Insert $\in$ or $\notin$ to make the statement true.

11. 10 _____ $\{5, 10, 15, 20\}$

12. 7 _____ $\{1, 3, 5, 7, 9, 11\}$

13. 5 _____ $\{2, 4, 6, 8, 10\}$

14. −1 _____ $\{0, 1, 2, 3, 4, 5\}$

15. $\frac{1}{4}$ _____ $\{0, 0.25, 0.5, 0.75, 1\}$

16. $\frac{2}{5}$ _____ $\{0, 0.2, 0.4, 0.6, 0.8, 1\}$

17. $\frac{1}{3}$ _____ $\{0, 0.33, 0.67, 1\}$

18. $\frac{4}{2}$ _____ $\{1, 3, 4, 5, 7\}$

19. red _____ $\{$blue, green, red, yellow$\}$

20. M _____ $\{$M, T, W, R, F$\}$

Exercises 21–24: Determine the set E of even integers that belong to the given universal set U.

21. $U = \{1, 2, 3, 4, 5, 6, 7, 8, 9, 10\}$

22. $U = \{-3, -2, -1, 0, 1, 2, 3\}$

23. $U = \{1, 2, 3, 4, \ldots\}$

24. $U = \{1, 3, 5, 7, 9, 11, \ldots\}$

Exercises 25–28: Determine the set A of words starting with the letter A that belong to the given universal set U.

25. $U = \{$Apple, Orange, Pear, Apricot$\}$

26. $U = \{$Apple, Orange$\}$

27. $U = \{$Calculus, Algebra, Geometry$\}$

28. $U = \{$Calculus, Algebra, Geometry, Arithmetic$\}$

FINDING SUBSETS

Exercises 29–36: Let $A = \{1, 2, 3, 4\}$, $B = \{3, 4, 5, 6\}$, and $C = \{3, 6\}$. Determine whether the statement is true or false.

29. $A \subseteq B$

30. $B \subseteq A$

31. $C \subseteq B$

32. $\varnothing \subseteq C$

33. $C \subseteq C$

34. $A \nsubseteq B$

35. $B \nsubseteq C$

36. $C \subseteq A$

Exercises 37–44: Let $A = \{a, b, c, d, e\}$, $B = \{b, d\}$, and $C = \{a, c, e\}$. Determine whether the statement is true or false.

37. $B \subseteq A$

38. $C \subseteq A$

39. $B \nsubseteq C$

40. $C = B$

41. $C \subseteq \varnothing$

42. $C \nsubseteq B$

43. $A \subseteq A$

44. $C \nsubseteq C$

SKETCHING VENN DIAGRAMS

Exercises 45–52: Let A and B be two sets and U be the universal set. Sketch a Venn diagram and shade the region that illustrates the given set.

45. $A \cup B$

46. $A \cap B$

47. A'

48. B'

49. $A \cap U$

50. $B \cup U$

51. $(A \cap B)'$

52. $(A \cup B)'$

FINDING UNIONS, INTERSECTIONS, AND COMPLEMENTS

Exercises 53–62: Write the expression in terms of one set.

53. $\{1, 2, 3\} \cup \{3, 4\}$

54. $\{1, 2, 3\} \cup \{6\}$

55. $\{a, b, c\} \cap \{a, b, d\}$

56. $\{x, y, z\} \cap \{a, b, c\}$

57. $\{a, b, c\} \cup \{\ \}$

58. $\{a, b, c\} \cap \varnothing$

59. $\{a, b\} \cup \{c, b\} \cup \{d, a\}$

60. $\{1, 3, 5\} \cup \{4, 5\} \cup \{4, 5, 6\}$

61. $\{4, 5, 8, 9\} \cap \{3, 5, 8, 9\} \cap \{4, 5, 8\}$

62. $\{1, 2, 5\} \cap \{2, 1\} \cap \{1, 2, 3\}$

Exercises 63–86: Let

$$U = \{1, 2, 3, 4, 5, 6, 7, 8, 9, 10\},$$
$$A = \{1, 3, 5, 7, 9\}, B = \{2, 4, 6, 8, 10\},$$
$$C = \{3, 4, 5, 6, 7\}, \text{ and } D = \{4, 5, 6, 9\}.$$

Use set notation to list the elements in the given set.

63. $A \cup B$

64. $A \cap B$

65. $C \cap D$

66. $A \cup D$

67. $U \cap \varnothing$

68. $\varnothing \cup U$

69. $A \cup \varnothing$

70. $\varnothing \cap B$

71. $D \cap D'$

72. $A \cup A$

73. U'

74. A'

75. B'

76. D'

77. $A \cup D'$

78. $A' \cap B'$

79. $A \cap B \cap C$

80. $B \cup C \cup D$

81. C'

82. $D' \cup D$

83. $B \cap C$

84. $B \cap C \cap D$

85. $C \cup C'$

86. $D' \cap B \cap C$

RECOGNIZING FINITE AND INFINITE SETS

Exercises 87–92: Determine whether the given set is finite or infinite.

87. The set of whole numbers

88. The set of real numbers

89. The set of natural numbers less than 1000

90. The set of natural numbers greater than 4 and less than 20

91. The days of the week

92. The names of the students in your class

APPENDIX C
Linear Programming

NEW VOCABULARY

- ☐ Objective function
- ☐ Constraint
- ☐ Feasible solutions
- ☐ Linear programming problem
- ☐ Optimal value
- ☐ Vertex

1 Understanding Objective Functions and Constraints

🌐 *Math in Context* Business Suppose that a small business sells candy for $3 per pound and freshly ground coffee for $5 per pound. All inventory is sold by the end of the day. The revenue R collected in dollars is given by

$$R = 3x + 5y, \quad \text{Objective function}$$

where x is the pounds of candy sold and y is the pounds of coffee sold. For example, if the business sells **80** pounds of candy and **40** pounds of coffee during a day, then its revenue is

$$R = 3(80) + 5(40) = \$440.$$

The function $R = 3x + 5y$ is called an **objective function**.

Suppose also that the company cannot package more than 150 pounds of candy and coffee per day. Then the inequality

$$x + y \leq 150 \quad \text{Constraint}$$

represents a **constraint** on the objective function, which limits the company's revenue for any one day. A goal of this business might be to maximize the objective function

$$R = 3x + 5y, \quad \text{Revenue equation}$$

subject to the following constraints.

$$x + y \leq 150 \quad \text{Constraint on production}$$
$$x \geq 0, y \geq 0 \quad \text{Variables cannot be negative.}$$

Constraints

Note that the constraints $x \geq 0$ and $y \geq 0$ are included because the number of pounds of candy or coffee cannot be negative. The problem that we have described is called a *linear programming problem*. Before learning how to solve a linear programming problem, we need to discuss the region of *feasible solutions*.

2 Finding the Region of Feasible Solutions

The constraints for a linear programming problem consist of linear inequalities. These inequalities are satisfied by some points in the xy-plane but not by others. The set of solutions to these constraints is called the **feasible solutions**. For example, the *region of feasible solutions* to the constraints for the business just described is shaded in **FIGURE C.1**.

Constraints on Sales

Region of feasible solutions

FIGURE C.1

The point $(50, 50)$ lies in the shaded region and represents the business selling 50 pounds of candy and 50 pounds of coffee. In the next example, we shade the region of feasible solutions to a set of constraints.

EXAMPLE 1 **Finding the region of feasible solutions**

Shade the region of feasible solutions for the following constraints.

$$x + 2y \leq 30$$
$$2x + y \leq 30$$
$$x \geq 0, y \geq 0$$

Solution

The feasible solutions are the ordered pairs (x, y) that *satisfy all four* inequalities. They lie on or below the lines $x + 2y = 30$ and $2x + y = 30$, and on or above the line $y = 0$ and on or to the right of $x = 0$, as shown in **FIGURE C.2**. Note that the inequalities $x \geq 0$ and $y \geq 0$ restrict the region of feasible solutions to quadrant I (including the x- and y-axes.)

FIGURE C.2

Now Try Exercise 15

STUDY TIP

Be sure that you can graph inequalities (see Example 1) before you attempt to solve a linear programming problem.

3 Solving Linear Programming Problems

A **linear programming problem** consists of an *objective function* and a system of linear inequalities called *constraints*. The solution set for the system of linear inequalities is called the *region of feasible solutions*. The objective function describes a quantity that is to be optimized. The **optimal value** for a linear programming problem often results in maximum revenue or minimum cost.

When the system of constraints has only two variables, the boundary of the region of feasible solutions often consists of line segments intersecting at points called *vertices* (plural of **vertex**). To solve a linear programming problem, we use the *fundamental theorem of linear programming*.

FUNDAMENTAL THEOREM OF LINEAR PROGRAMMING

If the optimal value for a linear programming problem exists, then it occurs at a vertex of the region of feasible solutions.

The fundamental theorem of linear programming is used to solve the following linear programming problem. A justification of this theorem is given after Example 2.

EXAMPLE 2 **Maximizing an objective function**

Maximize the objective function $R = 2x + 3y$ subject to

$$x + 2y \leq 30$$
$$2x + y \leq 30$$
$$x \geq 0, y \geq 0.$$

Solution

The region of feasible solutions is shaded in **FIGURE C.2** from Example 1. Note that the vertices on the boundary of feasible solutions are $(0, 0)$, $(15, 0)$, $(10, 10)$, and $(0, 15)$. To find the maximum value of R, substitute each vertex in the formula for R. The maximum value of R is 50 when $x = 10$ and $y = 10$. See **TABLE C.1**.

Vertex	$R = 2x + 3y$
$(0, 0)$	$2(0) + 3(0) = 0$
$(15, 0)$	$2(15) + 3(0) = 30$
$(10, 10)$	$2(10) + 3(10) = 50$ ◀ Maximum R is 50.
$(0, 15)$	$2(0) + 3(15) = 45$

TABLE C.1

Now Try Exercise 33

The following steps are helpful in solving linear programming word problems.

STEPS FOR SOLVING A LINEAR PROGRAMMING WORD PROBLEM

STEP 1: Read the problem carefully. Consider making a table.

STEP 2: Write the objective function and all the constraints.

STEP 3: Sketch a graph of the region of feasible solutions. Identify all vertices.

STEP 4: Evaluate the objective function at each vertex. A maximum (or a minimum) occurs at a vertex.

NOTE: If the region is unbounded, a maximum (or minimum) may not exist. ∎

JUSTIFICATION OF THE FUNDAMENTAL THEOREM To better understand the fundamental theorem of linear programming, consider the following example. Suppose that we want to maximize profit $P = 3x + 7y$ subject to the following four constraints:

$$x \geq 1, \qquad x \leq 5, \qquad y \geq x, \qquad \text{and} \qquad y \leq 6.$$

The corresponding region of feasible solutions is shown in **FIGURE C.3** on the next page.

Each value of P determines a unique line. For example, if $P = 70$, then the equation for P becomes $3x + 7y = 70$. The resulting line, shown in **FIGURE C.4** on the next page, does not intersect the region of feasible solutions. Thus there are no values for x and y that lie in this region and result in a profit of 70. **FIGURE C.4** also shows the lines that result from letting $P = 0, 10$, and 30. If $P = 10$, then the line intersects the region of feasible solutions only at the vertex $(1, 1)$. This means that if $x = 1$ and $y = 1$, then $P = 3(1) + 7(1) = 10$. If $P = 30$, then the line $3x + 7y = 30$ intersects the region of feasibility infinitely many

CRITICAL THINKING

What is the minimum value for P subject to the given constraints?

times. However, it appears that values greater than 30 are possible for P. In **FIGURE C.5**, lines are drawn for $P = 57, 63$, and 70. Notice that there are no points of intersection for $P = 63$ or $P = 70$, but there is one vertex in the region of feasible solutions at $(5, 6)$ that gives $P = 57$. Thus the maximum value of P is 57, and this maximum occurs at a vertex of the region of feasible solutions. The fundamental theorem of linear programming generalizes this result.

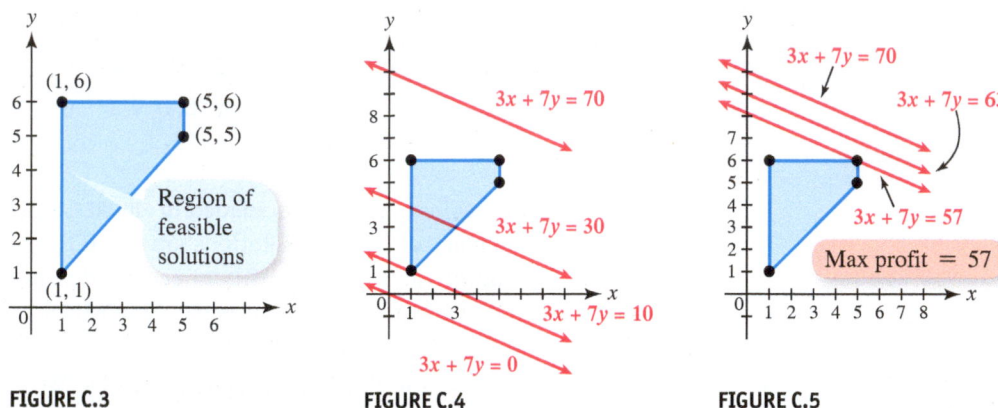

FIGURE C.3   FIGURE C.4   FIGURE C.5

EXAMPLE 3   **Minimizing the cost of vitamins**

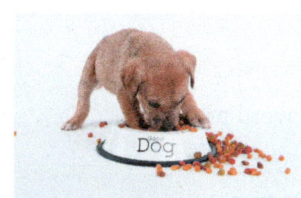

A breeder is mixing two different vitamins, Brand X and Brand Y, into pet food. Each serving of pet food should contain at least 60 units of vitamin A and 30 units of vitamin C. Brand X costs 80 cents per ounce and Brand Y costs 50 cents per ounce. Each ounce of Brand X contains 15 units of vitamin A and 10 units of vitamin C, whereas each ounce of Brand Y contains 20 units of vitamin A and 5 units of vitamin C. Determine how much of each brand of vitamin should be mixed to produce a minimum cost per serving.

Solution

STEP 1: Begin by listing the information, as illustrated in **TABLE C.2**.

Brand	Amount	Vitamin A	Vitamin C	Cost
X	x	15	10	80 cents
Y	y	20	5	50 cents
Minimum		60	30	

TABLE C.2

STEP 2: If x ounces of Brand X are purchased at 80 cents per ounce and if y ounces of Brand Y are purchased at 50 cents per ounce, then the total cost C is given by $C = 80x + 50y$. Because each ounce of Brand X contains 15 units of vitamin A and each ounce of Brand Y contains 20 units of vitamin A, the total number of units of vitamin A is $15x + 20y$. If each serving of pet food must contain at least 60 units of vitamin A, the constraint is $15x + 20y \geq 60$. Similarly, because each serving requires at least 30 units of vitamin C, $10x + 5y \geq 30$. The linear programming problem then becomes the following.

Minimize: $C = 80x + 50y$ Cost (in cents)

Subject to: $15x + 20y \geq 60$ Vitamin A constraint

$10x + 5y \geq 30$ Vitamin C constraint

$x \geq 0, y \geq 0$ Amounts cannot be negative.

STEP 3: The region containing the feasible solutions is shown in **FIGURE C.6**.

FIGURE C.6

NOTE: To determine the vertex $(2.4, 1.2)$, solve the system of equations

$$15x + 20y = 60$$
$$10x + 5y = 30$$

by using elimination. ∎

STEP 4: In **FIGURE C.6**, the vertices are $(0, 6)$, $(2.4, 1.2)$, and $(4, 0)$. Evaluate the objective function $C = 80x + 50y$ at each vertex, as shown in **TABLE C.3**.

Vertex	$C = 80x + 50y$
$(0, 6)$	$80(0) + 50(6) = 300$
$(2.4, 1.2)$	$80(2.4) + 50(1.2) = 252$ ◁ Minimum cost (cents)
$(4, 0)$	$80(4) + 50(0) = 320$

TABLE C.3

The minimum cost occurs when **2.4** ounces of Brand X and **1.2** ounces of Brand Y are mixed, at a cost of **$2.52** per serving.

Now Try Exercise 47

C   Exercises

MyMathLab®

CONCEPTS AND VOCABULARY

1. A procedure used in business to optimize quantities such as cost and profit is called _____.

2. In linear programming, the function to be optimized is called the _____ function.

3. The region in the xy-plane that satisfies the constraints is called the region of _____.

4. In linear programming, constraints typically consist of a system of _____.

5. If the optimal value for a linear programming problem exists, then it occurs at a(n) _____ of the region of feasible solutions.

6. To find the optimal value in a linear programming problem, substitute each vertex into the _____ function.

FINDING REGIONS OF FEASIBLE SOLUTIONS

Exercises 7–20: Shade the region of feasible solutions for the following constraints.

7. $x + y \leq 3$
  $x \geq 0, y \geq 0$

8. $2x + y \leq 4$
  $x \geq 0, y \geq 0$

9. $4x + 3y \leq 12$
  $x \geq 0, y \geq 0$

10. $5x + 3y \leq 15$
  $x \geq 0, y \geq 0$

11. $x \leq 5$
  $y \leq 2$
  $x \geq 0, y \geq 0$

12. $x \leq 3$
  $y \leq 4$
  $x \geq 1, y \geq 1$

13. $x + y \leq 5$
  $x + y \geq 2$
  $x \geq 0, y \geq 0$

14. $2x + y \leq 6$
  $x + y \geq 3$
  $x \geq 0, y \geq 0$

15. $3x + 2y \leq 6$
  $2x + 3y \leq 6$
  $x \geq 0, y \geq 0$

16. $5x + 3y \leq 30$
  $3x + 5y \leq 30$
  $x \geq 0, y \geq 0$

17. $x + y \leq 3$
  $x + 3y \geq 3$
  $x \geq 0, y \geq 0$

18. $3x + y \geq 6$
  $x + 2y \geq 6$
  $x \geq 0, y \geq 0$

19. $x + 2y \geq 4$
  $3x + 2y \geq 6$
  $x \geq 0, y \geq 0$

20. $4x + 3y \geq 12$
  $3x + 4y \geq 12$
  $x \geq 0, y \geq 0$

SOLVING LINEAR PROGRAMMING PROBLEMS

Exercises 21–24: Find the maximum of R on the region of feasible solutions shown in the figure.

21. $R = 4x + 5y$

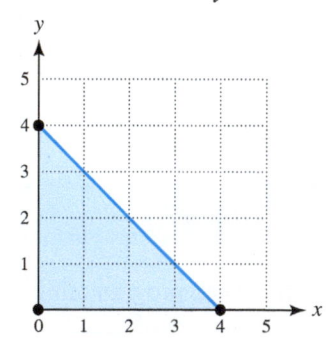

22. $R = 2x + 3y$

23. $R = x + 3y$

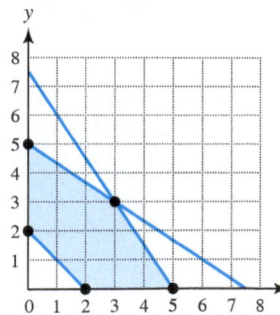

24. $R = 12x + 9y$

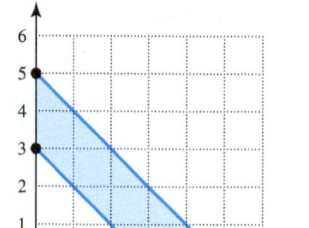

Exercises 25–28: Use the figures in Exercises 21–24 to complete Exercises 25–28, respectively, to minimize C.

25. $C = 2x + 3y$

26. $C = 3x + y$

27. $C = 5x + y$

28. $C = 2x + 7y$

Exercises 29–36: Maximize the objective function R, subject to the given constraints.

29. $R = 3x + 5y$
  $x + y \leq 150$
  $x \geq 0, y \geq 0$

30. $R = 6x + 5y$
  $x + y \leq 8$
  $x \geq 0, y \geq 0$

31. $R = 3x + 2y$
  $2x + y \leq 6$
  $x \geq 0, y \geq 0$

32. $R = x + 3y$
  $x + 2y \leq 4$
  $x \geq 0, y \geq 0$

33. $R = 12x + 9y$
  $3x + y \leq 6$
  $x + 3y \leq 6$
  $x \geq 0, y \geq 0$

34. $R = 10x + 30y$
  $x + 3y \leq 12$
  $3x + y \leq 12$
  $x \geq 0, y \geq 0$

35. $R = 4x + 5y$
  $x + y \geq 2$
  $x + 2y \leq 4$
  $x \geq 0, y \geq 0$

36. $R = 3x + 7y$
  $x + y \geq 1$
  $3x + y \leq 3$
  $x \geq 0, y \geq 0$

Exercises 37–42: Minimize the objective function C, subject to the given constraints.

37. $C = x + 2y$
  $x \leq 3, y \leq 2$
  $x \geq 0, y \geq 0$

38. $C = 3x + y$
  $x \leq 5, y \leq 3$
  $x \geq 1, y \geq 1$

39. $C = 8x + 15y$
  $x + y \geq 4$
  $x \geq 0, y \geq 0$

40. $C = x + 2y$
  $3x + 4y \geq 12$
  $x \geq 0, y \geq 0$

41. $C = 30x + 40y$
  $2x + y \leq 6$
  $x + y \geq 2$
  $x \geq 0, y \geq 0$

42. $C = 50x + 70y$
  $2x + 3y \leq 6$
  $x + y \geq 1$
  $x \geq 0, y \geq 0$

APPLICATIONS OF LINEAR PROGRAMMING

Exercises 43–50: Solve the linear programming problem.

43. Maximizing Revenue A small business sells candy for $4 per pound and coffee for $6 per pound. The business can package and sell at most a total of 100 pounds of candy and coffee per day, but at least 20 pounds of candy must be sold each day. Determine how many pounds of candy and coffee need to be sold each day to maximize revenue.

44. Maximizing Revenue An organic foods store sells almonds for $10 per pound and flax seed for $5 per pound. The business can package and sell, at most, a total of 50 pounds of almonds and flax seed per day, but it must sell at least 15 pounds of flax seed per day. Determine the maximum daily revenue.

45. Minimizing Cost It costs a business $50 to make a graphing calculator and $20 to make a scientific calculator. Each week, the company must produce at least 90 calculators. At least twice as many scientific calculators must be made as graphing calculators. Determine the minimum weekly cost.

46. Minimizing Cost It costs a business $20 to make one compact disc player and $10 to make one radio. Each week, the company must make a combined total of at least 50 compact disc players and radios. At least as many compact disc players as radios must be manufactured. Determine how many compact disc players and radios should be made to minimize weekly costs.

47. Vitamin Cost A pet owner is mixing two different brands of vitamins, Brand X and Brand Y, into pet food. Brand X costs 90 cents per ounce and Brand Y costs 60 cents per ounce. Each serving is a mixture of the two brands and should contain at least 40 units of vitamin A and 30 units of vitamin C. Each ounce of Brand X contains 20 units of vitamin A and 10 units of vitamin C, whereas each ounce of Brand Y contains 10 units of vitamin A and 10 units of vitamin C. Determine how much of each brand of vitamin should be mixed to produce a minimum cost per serving.

48. Pet Food Cost A pet owner is buying two brands of food, X and Y, for his animals. Each serving of the mixture of the two foods should contain at least 60 grams of protein and 40 grams of fat. Brand X costs 75 cents per unit and Brand Y costs 50 cents per unit. Each unit of Brand X contains 20 grams of protein and 10 grams of fat, whereas each unit of Brand Y contains 10 grams of protein and 10 grams of fat. Determine how much of each brand should be bought to obtain a minimum cost per serving.

49. Raising Animals A breeder can raise no more than 50 hamsters and mice but no more than 20 hamsters. If she sells the hamsters for $15 each and the mice for $10 each, find the maximum revenue.

50. Maximizing Profit A business manufactures two parts, X and Y. Machines A and B are needed to make each part. To make part X, machine A is needed 3 hours and machine B is needed 1 hour. To make part Y, machine A is needed 1 hour and machine B is needed 2 hours. Machine A is available 60 hours per week and machine B is available 50 hours per week. The profit from part X is $300 and the profit from part Y is $250. How many parts of each type should be made to maximize weekly profit?

APPENDIX D
Synthetic Division

NEW VOCABULARY

☐ Synthetic division

1 Performing Synthetic Division

A shortcut called **synthetic division** can be used to divide $x - k$, where k is a constant, into a polynomial. For example, to divide $x - 2$ into $3x^3 - 8x^2 + 7x - 6$, we do the following (with the equivalent long division shown at the right).

Synthetic Division	Long Division of Polynomials

$$\begin{array}{r|rrrr} 2 & 3 & -8 & 7 & -6 \\ & & 6 & -4 & 6 \\ \hline & 3 & -2 & 3 & 0 \end{array}$$

Add to find row 3.

$$\begin{array}{r} 3x^2 - 2x + 3 \\ x - 2 \overline{\smash{)}3x^3 - 8x^2 + 7x - 6} \\ \underline{3x^3 - 6x^2} \\ -2x^2 + 7x \\ \underline{-2x^2 + 4x} \\ 3x - 6 \\ \underline{3x - 6} \\ 0 \end{array}$$

Note that the blue numbers in the expression for long division correspond to the third row in synthetic division. The remainder is 0, which is the last number in the third row. The quotient is $3x^2 - 2x + 3$. Its coefficients are 3, -2, and 3 and are located in the third row. To divide $x - 2$ into $3x^3 - 8x^2 + 7x - 6$ with synthetic division, use the following steps.

STEP 1: In the top row, write 2 (the value of k) on the left and then write the coefficients of the dividend $3x^3 - 8x^2 + 7x - 6$.

STEP 2: **(a)** Copy the leading coefficient 3 of $3x^3 - 8x^2 + 7x - 6$ into the third row and multiply it by 2 (the value of k). Write the result 6 in the second row below -8. Add -8 and 6 in the second column to obtain the -2 in the third row.
(b) Repeat the process by multiplying -2 by 2 and place the result -4 below 7. Then add 7 and -4 to obtain 3.
(c) Multiply 3 by 2 and place the result 6 below the -6. Adding 6 and -6 gives 0.

STEP 3: The last number in the third row is 0, which is the remainder. The other numbers in the third row are the coefficients of the quotient, which is $3x^2 - 2x + 3$.

EXAMPLE 1 **Performing synthetic division**

Use synthetic division to divide $x^4 - 5x^3 + 9x^2 - 10x + 3$ by $x - 3$.

Solution
Because the divisor is $x - 3$, the value of k is 3.

$$\begin{array}{r|rrrrr} 3 & 1 & -5 & 9 & -10 & 3 \\ & & 3 & -6 & 9 & -3 \\ \hline & 1 & -2 & 3 & -1 & 0 \end{array}$$

The quotient is $x^3 - 2x^2 + 3x - 1$ and the remainder is **0**. This result is expressed by

$$\frac{x^4 - 5x^3 + 9x^2 - 10x + 3}{x - 3} = x^3 - 2x^2 + 3x - 1 + \frac{0}{x - 3}, \quad \text{or}$$

$$\frac{x^4 - 5x^3 + 9x^2 - 10x + 3}{x - 3} = x^3 - 2x^2 + 3x - 1.$$

Now Try Exercise 3

EXAMPLE 2 **Performing synthetic division**

Use synthetic division to divide $2x^3 - x + 5$ by $x + 1$.

Solution
Write $2x^3 - x + 5$ as $2x^3 + 0x^2 - x + 5$. The divisor $x + 1$ can be written as

$$x + 1 = x - (-1),$$

so we let $k = -1$.

$$
\begin{array}{r|rrrr}
-1 & 2 & 0 & -1 & 5 \\
   &   & -2 & 2 & -1 \\
\hline
   & 2 & -2 & 1 & 4 \\
\end{array}
$$

The remainder is **4**, and the quotient is $2x^2 - 2x + 1$. This result can also be expressed as

$$\frac{2x^3 - x + 5}{x + 1} = 2x^2 - 2x + 1 + \frac{4}{x + 1}.$$

Now Try Exercise 13

D   Exercises

MyMathLab®

PERFORMING SYNTHETIC DIVISION

Exercises 1–16: Use synthetic division to divide.

1. $\dfrac{x^2 + 3x - 1}{x - 1}$

2. $\dfrac{2x^2 + x - 1}{x - 3}$

3. $(3x^2 - 22x + 7) \div (x - 7)$

4. $(5x^2 + 29x - 6) \div (x + 6)$

5. $\dfrac{x^3 + 7x^2 + 14x + 8}{x + 4}$

6. $\dfrac{2x^3 + 3x^2 + 2x + 4}{x + 1}$

7. $\dfrac{2x^3 + x^2 - 1}{x - 2}$

8. $\dfrac{x^3 + x - 2}{x + 3}$

9. $(x^3 - 2x^2 - 2x + 4) \div (x - 4)$

10. $(2x^4 + 3x^2 - 4) \div (x + 2)$

11. $(x^3 - 3x^2 - 8x - 10) \div (x - 5)$

12. $(x^4 - 1) \div (x + 1)$

13. $(2x^4 - x) \div (x + 2)$

14. $(2x^3 + x - 1) \div (x - 3)$

15. $(b^4 - 1) \div (b - 1)$

16. $(a^2 + a) \div (a + 2.5)$

Answers to Selected Exercises

1 Introduction to Algebra

Section 1.1 (pp. 9–11)

1. natural **2.** 0 **3.** 1 **4.** composite **5.** factors
6. formula **7.** variable **8.** equals sign **9.** sum
10. product **11.** quotient **12.** difference
13. Yes **15.** No **17.** Yes **19.** No
21. Composite; $4 = 2 \cdot 2$ **23.** Neither **25.** Prime
27. Composite; $92 = 2 \cdot 2 \cdot 23$
29. Composite; $225 = 3 \cdot 3 \cdot 5 \cdot 5$
31. Prime **33.** $6 = 2 \cdot 3$ **35.** $12 = 2 \cdot 2 \cdot 3$
37. $32 = 2 \cdot 2 \cdot 2 \cdot 2 \cdot 2$ **39.** $39 = 3 \cdot 13$
41. $294 = 2 \cdot 3 \cdot 7 \cdot 7$ **43.** $300 = 2 \cdot 2 \cdot 3 \cdot 5 \cdot 5$
45. 15 **47.** 5 **49.** 4 **51.** 18 **53.** 4 **55.** 22 **57.** 12
59. 9 **61.** 5 **63.** 28 **65.** 7 **67.** 5 **69.** 0
71. 3 **73.** $x + 5$; x is the number.
75. $3s$; s is the cost of soda. **77.** $n + 5$; n is the number.
79. $p - 200$; p is the population.
81. $\frac{z}{6}$; z is the number.
83. st; s is the speed and t is the time.
85. $\frac{x + 7}{y}$; x is one number, y is the other number.
87. $5(x + 4)$; x is the number. **89.** $7A - 3$; A is the age.
91. $P = 100D$
93.

Yards (y)	1	2	3	4	5	6	7
Feet (F)	3	6	9	12	15	18	21

$F = 3y$

95. $M = 3x$; 108 mi **97.** $B = 6D$
99. $C = 12x$ **101.** 198 square feet

Section 1.2 (pp. 23–25)

1. $\frac{3}{4}; \frac{1}{4}$ **2.** 11; 21 **3.** 0 **4.** lowest **5.** True **6.** False
7. $\frac{a}{b}$ **8.** multiply **9.** $\frac{1}{a}$ **10.** $\frac{1}{5}$ **11.** $\frac{ac}{bd}$ **12.** $\frac{ad}{bc}$ **13.** $\frac{a + c}{b}$
14. $\frac{a - c}{b}$ **15.** Num: 19; Den: 21 **17.** Num: a; Den: bd
19. Num: $a + 2$; Den: $a - 5$ **21.** 4 **23.** 25 **25.** 10
27. $\frac{3}{5}$ **29.** $\frac{3}{5}$ **31.** $\frac{1}{2}$ **33.** $\frac{2}{5}$ **35.** $\frac{1}{3}$ **37.** $\frac{2}{5}$ **39.** $\frac{1}{4}$ **41.** $\frac{3}{20}$
43. 1 **45.** $\frac{3}{5}$ **47.** $\frac{12}{5}$ **49.** $\frac{3}{4}$ **51.** 1 **53.** $\frac{3a}{2b}$ **55.** $\frac{3}{16}$
57. 4 **59.** $\frac{1}{3}$ **61.** (a) $\frac{1}{5}$ (b) $\frac{1}{7}$ (c) $\frac{7}{4}$ (d) $\frac{8}{9}$
63. (a) 2 (b) 9 (c) $\frac{101}{12}$ (d) $\frac{17}{31}$ **65.** $\frac{3}{2}$ **67.** 6 **69.** 1
71. $\frac{4}{3}$ **73.** 12 **75.** $\frac{3}{10}$ **77.** $\frac{a}{2}$ **79.** 1 **81.** (a) $\frac{1}{2}$ (b) $\frac{1}{3}$
83. (a) $\frac{25}{29}$ (b) $\frac{11}{29}$ **85.** 45 **87.** 15 **89.** 24 **91.** 12
93. 24 **95.** $\frac{3}{6}, \frac{4}{6}$ **97.** $\frac{28}{36}, \frac{15}{36}$ **99.** $\frac{3}{48}, \frac{28}{48}$ **101.** $\frac{4}{12}, \frac{9}{12}, \frac{10}{12}$
103. $\frac{13}{16}$ **105.** $\frac{1}{6}$ **107.** $\frac{59}{70}$ **109.** $\frac{13}{36}$ **111.** $\frac{17}{600}$ **113.** $\frac{9}{8}$

115. 600 **117.** $4\frac{3}{4}$ ft **119.** $32\frac{5}{16}$ inches **121.** $\frac{5}{8}$ yd^2
123. $8\frac{1}{4}$ miles **125.** $\frac{1}{6250}$ **127.** $\frac{961}{1260}$

Checking Basic Concepts 1.1 & 1.2 (p. 25)

1. (a) Prime (b) Composite; $28 = 2 \cdot 2 \cdot 7$
(c) Neither (d) Composite; $180 = 2 \cdot 2 \cdot 3 \cdot 3 \cdot 5$
2. 2 **3.** 30 **4.** $x + 5$ **5.** $I = 12F$ **6.** (a) 3 (b) 8
7. (a) $\frac{5}{7}$ (b) $\frac{2}{3}$ **8.** $\frac{3}{4}$ **9.** (a) $\frac{1}{2}$ (b) $\frac{1}{4}$ (c) $\frac{2}{5}$ (d) $\frac{7}{12}$
10. $3\frac{1}{3}$ cups

Section 1.3 (pp. 31–32)

1. multiply **2.** 2 **3.** base; exponent **4.** 6^2 **5.** 8^3
6. 17; multiplication; addition **7.** 2; exponents; subtraction
8. 4; left; right **9.** False **10.** False **11.** 3^4 **13.** 2^5
15. $\left(\frac{1}{2}\right)^4$ **17.** a^5 **19.** $(x + 3)^2$ **21.** (a) 16 (b) 16
23. (a) 6 (b) 1 **25.** (a) 32 (b) 1000 **27.** (a) $\frac{4}{9}$ (b) $\frac{1}{32}$
29. 2^3 **31.** 5^2 **33.** 7^2 **35.** 10^3 **37.** $\left(\frac{1}{2}\right)^4$ **39.** $\left(\frac{2}{3}\right)^5$
41. 29 **43.** 4 **45.** 90 **47.** 3 **49.** 2 **51.** 19 **53.** 32
55. 125 **57.** 3 **59.** 4 **61.** 80 **63.** $\frac{49}{16}$ **65.** $2^3 - 8$; 0
67. $30 - 4 \cdot 3$; 18 **69.** $\frac{4^2}{2^3}$; 2 **71.** $\frac{40}{10} + 2$; 6
73. $100(2 + 3)$; 500 **75.** 26,214,400 bytes
77. (a) $k = 7$ (b) 32 **79.** (a) 8 yr (b) \$80,000

Section 1.4 (pp. 41–43)

1. $-b$ **2.** natural **3.** rational **4.** real **5.** irrational
6. True **7.** True **8.** $0.\overline{27}$ **9.** 1; 4 **10.** 4 **11.** principal
12. not equal **13.** approximately equal **14.** 0 **15.** left
16. origin **17.** (a) -9 (b) 9 **19.** (a) $-\frac{2}{3}$ (b) $\frac{2}{3}$
21. (a) -8 (b) 8 **23.** (a) $-a$ (b) a **25.** 6 **27.** $\frac{1}{2}$
29. 0.25 **31.** 0.875 **33.** 1.5 **35.** 0.05 **37.** $0.\overline{6}$ **39.** $0.\overline{7}$
41. 5 **43.** 7 **45.** 2.646 **47.** Natural, whole, integer, and
rational **49.** Natural, whole, integer, and rational
51. Whole, integer, and rational **53.** Rational
55. Rational **57.** Rational **59.** Irrational
61. Natural, integer, and rational
63. Natural, integer, and rational **65.** Rational
67.

69.

71.

$$\begin{array}{c} \text{(b) (a)} \qquad\qquad \text{(c)} \\ \hline -50\ -40\ -30\ -20\ -10\ \ 0\ \ 10\ \ 20\ \ 30\ \ 40\ \ 50 \end{array}$$

73.

$$\begin{array}{c} \text{(c)} \qquad \text{(b)} \quad \text{(a)} \\ \hline -5\ -4\ -3\ -2\ -1\ \ 0\ \ 1\ \ 2\ \ 3\ \ 4\ \ 5 \end{array}$$

75. 5.23 **77.** 8 **79.** 0 **81.** $\pi - 3$ **83.** $-b$ **85.** $<$
87. $>$ **89.** $>$ **91.** $<$ **93.** $>$ **95.** $<$ **97.** $>$ **99.** $=$
101. $-9, -2^3, -3, 0, 1$ **103.** $-2, -\frac{3}{2}, \frac{1}{3}, \sqrt{5}, \pi$
105. (a) 51% (b) 52%; Natural, rational
107. (a) Rational (b) Whole, integer, rational (c) Rational

Checking Basic Concepts 1.3 & 1.4 (p. 43)

1. (a) 5^4 (b) 7^5 **2.** (a) 8 (b) 10,000 (c) $\frac{8}{27}$
3. (a) 4^3 (b) 2^6 **4.** (a) 26 (b) 9 (c) 2 (d) $\frac{1}{2}$ (e) 4
(f) 0 **5.** $5^3 \div 3$, or $\frac{5^3}{3}$ **6.** (a) 17 (b) $-a$ **7.** (a) 0.15
(b) 0.625 **8.** (a) Natural, integer, and rational
(b) Integer and rational (c) Irrational (d) Rational
9.

$$\begin{array}{c} \text{(b)} \quad \text{(e)} \quad \text{(a)(d)} \quad \text{(c)} \\ \hline -5\ -4\ -3\ -2\ -1\ \ 0\ \ 1\ \ 2\ \ 3\ \ 4\ \ 5 \end{array}$$

10. (a) 12 (b) 0 **11.** (a) $<$ (b) $<$ (c) $<$
12. $-7, -1.6, 0, \frac{1}{3}, \sqrt{3}, 3^2$

Section 1.5 (pp. 49–50)

1. sum **2.** zero **3.** True **4.** True **5.** absolute value
6. difference **7.** addition **8.** opposite; $(-b)$ **9.** addition
10. subtraction **11.** $-25; 0$ **13.** $\sqrt{21}; 0$ **15.** $-5.63; 0$
17. 1 **19.** -7 **21.** $\frac{1}{4}$ **23.** $-\frac{9}{14}$ **25.** -1.1 **27.** 34
29. -3 **31.** 7 **33.** $-\frac{1}{14}$ **35.** $\frac{1}{2}$ **37.** 2.9 **39.** -164
41. -9 **43.** 42 **45.** -5 **47.** 50 **49.** 8.8 **51.** $\frac{1}{2}$
53. -1 **55.** 4 **57.** 2 **59.** -3 **61.** 4 **63.** 2 **65.** -1
67. 10 **69.** -150 **71.** $2 + (-5); -3$ **73.** $-5 + 7; 2$
75. $-(2^3); -8$ **77.** $-6 - 7; -13$
79. $6 + (-10) - 5; -9$ **81.** $230 **83.** 25 yards
85. 64,868 ft **87.** 49% **89.** Mount Whitney: 14,497 ft;
Death Valley: -282 ft; 14,779 ft

Section 1.6 (pp. 57–58)

1. product **2.** negative **3.** positive **4.** quotient **5.** $\frac{1}{a}$
6. reciprocal **7.** reciprocal, or multiplicative inverse
8. $-a$ **9.** $\frac{1}{b}$ **10.** positive **11.** negative **12.** 5; 8
13. -12 **15.** -18 **17.** 0 **19.** 60 **21.** $\frac{1}{4}$ **23.** -1
25. 200 **27.** -50 **29.** 120 **31.** $-\frac{3}{2}$ **33.** Negative
35. 25 **37.** -1 **39.** -16 **41.** 8 **43.** -40 **45.** -2
47. 10 **49.** -4 **51.** -32 **53.** 0 **55.** Undefined **57.** $-\frac{1}{22}$
59. $\frac{4}{15}$ **61.** $-\frac{15}{16}$ **63.** Undefined **65.** -1 **67.** $-\frac{4}{3}$ **69.** 0.5
71. 0.1875 **73.** 3.5 **75.** $5.\overline{6}$ **77.** 1.4375 **79.** 0.875
81. $\frac{1}{4}$ **83.** $\frac{4}{25}$ **85.** $\frac{5}{8}$ **87.** $\frac{11}{16}$ **89.** $2.\overline{3}; \frac{7}{3}$ **91.** $2.1; \frac{21}{10}$
93. $-1.8\overline{3}; -\frac{11}{6}$ **95.** $0.8; \frac{4}{5}$ **97.** About 380 million
99. 0.56

Checking Basic Concepts 1.5 & 1.6 (p. 58)

1. (a) 0 (b) -19 **2.** (a) $\frac{8}{9}$ (b) -3.2 **3.** (a) $-1 + 5; 4$
(b) $4 - (-3); 7$ **4.** $145\,°F$ **5.** (a) 35 (b) $\frac{4}{15}$
6. (a) -9 (b) -32 (c) 25 **7.** (a) $-\frac{15}{2}$ (b) $\frac{15}{32}$
8. $-\frac{6}{7}$ **9.** (a) -5 (b) -5 (c) -5 (d) 5
10. (a) 0.6 (b) 3.875

Section 1.7 (pp. 67–69)

1. commutative; addition **2.** commutative; multiplication
3. associative; addition **4.** associative; multiplication
5. True **6.** False **7.** distributive **8.** distributive
9. identity; addition **10.** identity; multiplication
11. $-a$ **12.** $\frac{1}{a}$ **13.** $10 + (-6)$ **15.** $6 \cdot (-5)$
17. $10 + a$ **19.** $7b$ **21.** $1 + (2 + 3)$ **23.** $(2 \cdot 3) \cdot 4$
25. $a + (5 + c)$ **27.** $x \cdot (3 \cdot 4)$
29. $a + b + c = (a + b) + c$
$\qquad\qquad\qquad = c + (a + b)$
$\qquad\qquad\qquad = c + (b + a)$
$\qquad\qquad\qquad = c + b + a$
31. 20 **33.** $ab - 8a$ **35.** $-4t + 4z$ **37.** $-5 + a$
39. $3a + 15$ **41.** $17 - a$
43. $a \cdot (b + c + d) = a \cdot ((b + c) + d)$
$\qquad\qquad\qquad\quad = a \cdot (b + c) + ad$
$\qquad\qquad\qquad\quad = ab + ac + ad$
45. $11x$ **47.** $-b$ **49.** $2a$ **51.** $-14w$
53. Commutative (multiplication)
55. Associative (addition) **57.** Distributive
59. Distributive, commutative (multiplication)
61. Distributive **63.** Associative (multiplication)
65. Distributive **67.** Identity (addition)
69. Identity (multiplication) **71.** Identity (multiplication)
73. Inverse (multiplication) **75.** Inverse (addition)
77. 30 **79.** 100 **81.** 178 **83.** 79 **85.** 90 **87.** 816
89. 1 **91.** $\frac{1}{6}$ **93.** (a) 410 (b) 9970 (c) -6300
(d) $-140,000$ **95.** (a) 19,000 (b) $-45,100$ (c) 60,000
(d) $-7,900,000$ **97.** (a) 1.256 (b) 0.96 (c) 0.0987
(d) -0.0056 (e) 120 (f) 457.8
99. Commutative (addition) **101.** 19.8 miles
103. (a) $13 \cdot (5 \cdot 2)$; multiplying by 10 is easy.
(b) Associative (multiplication)

Section 1.8 (pp. 75–76)

1. term **2.** coefficient **3.** factors; terms **4.** like
5. like **6.** distributive **7.** Yes; 91 **9.** Yes; -6 **11.** No
13. Yes; 1 **15.** No **17.** Like **19.** Like **21.** Unlike
23. Unlike **25.** Like **27.** Unlike **29.** Like **31.** $3x$
33. $14y$ **35.** $41a$ **37.** 0 **39.** Not possible
41. Not possible **43.** $3x^2$ **45.** $-y$ **47.** $3x + 2$
49. $-2z + \frac{1}{2}$ **51.** $11y$ **53.** $4z - 1$ **55.** $12y - 7z$
57. $-3x - 4$ **59.** $-6x - 1$ **61.** $-\frac{1}{3}x + \frac{2}{3}$
63. $\frac{2}{5}x + \frac{3}{5}y + \frac{1}{5}$ **65.** $0.4x^2$ **67.** $7x^2 - 7x$ **69.** $2b$
71. $7x^3 + 2y$ **73.** x **75.** 3 **77.** z **79.** $3x - 2$
81. $2z + 3$ **83.** $5x + 6x; 11x$ **85.** $x^2 + 2x^2; 3x^2$
87. $6x - 4x; 2x$ **89.** (a) $50x$ (b) 2400 cubic feet
(c) 24 minutes **91.** (a) $1570w$ (b) 65,940 square feet

Checking Basic Concepts 1.7 & 1.8 (p. 77)

1. (a) $18y$ (b) $x + 10$ **2.** $20y$ **3.** (a) $5 - x$
(b) $5x - 35$ **4.** Distributive **5.** 0 **6.** (a) 60 (b) 7
(c) 368 **7.** (a) Unlike (b) Like **8.** (a) $14z$
(b) $-3y + 3$ **9.** (a) $-3y - 3$ (b) $-14y$ (c) x (d) 35
10. $3x + 5x$; $8x$

Chapter 1 Review (pp. 82–84)

1. Prime **2.** Composite; $27 = 3 \cdot 3 \cdot 3$
3. Composite; $108 = 2 \cdot 2 \cdot 3 \cdot 3 \cdot 3$ **4.** Neither
5. 3 **6.** 6 **7.** 7 **8.** 7 **9.** $3^2 + 5$ **10.** $2^3 \div (3 + 1)$
11. $3x$, where x is the number
12. $x - 4$, where x is the number **13.** 5 **14.** 6
15. (a) $\frac{5}{8}$ (b) $\frac{3}{4}$ **16.** (a) $\frac{3}{4}$ (b) $\frac{3}{5}$ **17.** $\frac{5}{8}$ **18.** $\frac{24}{23}$ **19.** $\frac{5}{2}$
20. $\frac{3x}{2y}$ **21.** $\frac{3}{35}$ **22.** (a) $\frac{1}{8}$ (b) 1 (c) $\frac{19}{5}$ (d) $\frac{2}{3}$ **23.** 9
24. $\frac{9}{14}$ **25.** 12 **26.** $\frac{20}{27y}$ **27.** 24 **28.** 42 **29.** $\frac{1}{3}$ **30.** $\frac{1}{2}$
31. $\frac{19}{24}$ **32.** $\frac{13}{18}$ **33.** 5^6 **34.** $\left(\frac{7}{6}\right)^3$ **35.** x^5 **36.** $(a - 5)^3$
37. (a) 64 (b) 49 (c) 8 **38.** 5 **39.** 25 **40.** 7 **41.** 3
42. 25 **43.** 1 **44.** 2 **45.** 0 **46.** 1 **47.** 5 **48.** 19
49. (a) 8 (b) 3 **50.** (a) $-\frac{3}{7}$ (b) $-\frac{2}{5}$ **51.** (a) 0.8
(b) 0.15 **52.** (a) $0.\overline{5}$ (b) $0.\overline{63}$ **53.** Whole, integer, and
rational **54.** Rational **55.** Integer and rational
56. Irrational **57.** Irrational **58.** Rational
59.

60. (a) 5 (b) π (c) 0 **61.** (a) $<$ (b) $>$ (c) $<$
(d) $>$ **62.** $-3, -\frac{2}{3}, \sqrt{3}, \pi - 1, 3$ **63.** 1 **64.** -5 **65.** 1
66. -2 **67.** -44 **68.** 40 **69.** $-\frac{11}{4}$ **70.** -7 **71.** $-\frac{2}{9}$
72. $-\frac{5}{4}$ **73.** $\frac{2}{7}$ **74.** $-\frac{3}{4}$ **75.** $3 + (-5)$; -2
76. $2 - (-4)$; 6 **77.** $0.\overline{7}$ **78.** 2.2 **79.** $\frac{3}{5}$ **80.** $\frac{3}{8}$
81. $16 + 3$ **82.** $-x \cdot 14$ **83.** $(-4 + 1) + 3$
84. $x \cdot (y \cdot 5)$ **85.** $5x + 60$ **86.** $-a + 3$
87. Identity (addition) **88.** Identity (multiplication)
89. Inverse (multiplication) **90.** Inverse (addition)
91. Commutative (multiplication)
92. Associative (addition) **93.** Identity (multiplication)
94. Associative (multiplication) **95.** Distributive
96. Identity (addition) **97.** Inverse (addition)
98. Inverse (multiplication) **99.** 40 **100.** 301
101. 2475 **102.** 43.56 **103.** Yes; 55 **104.** No **105.** $-6x$
106. $15z$ **107.** $4x^2$ **108.** $3x + 1$ **109.** $9x^2 - 6$
110. 0 **111.** 5 **112.** $2x - 5$ **113.** (a) $7x$
(b) 420 square feet (c) 24 minutes
114. 16 square feet
115.

Gallons (G)	1	2	3	4	5	6
Pints (P)	8	16	24	32	40	48

$P = 8G$

116. (a) $U = 43x$ (b) 1290 million **117.** $\frac{3}{20}$
118. $200,000 **119.** $1\frac{3}{20}$ feet **120.** $15\frac{3}{8}$ miles
121. $1101 **122.** 124 °F

Chapter 1 Test (pp. 84–85)

1. (a) Prime (b) Composite; $56 = 2 \cdot 2 \cdot 2 \cdot 7$
2. $\frac{15}{7}$ **3.** $4^2 - 3$; 13 **4.** $\frac{3}{4}$ **5.** $\frac{1}{30}$ **6.** 24
7. (a) $\frac{3}{4}$ (b) $\frac{16}{45}$ (c) $\frac{2}{7}$ (d) $\frac{15}{4}$ (e) $\frac{31}{36}$ (f) $\frac{2}{13}$
8. y^4 **9.** (a) 8 (b) 71 (c) -40 (d) 9
10. 0.35 **11.** (a) Integer and rational (b) Irrational
12.

13. (a) $<$ (b) $>$ **14.** (a) -6 (b) 21 **15.** $\frac{3}{4}$
16. (a) Distributive (b) Associative (multiplication)
(c) Commutative (addition) **17.** 1734 **18.** (a) $12 - 4z$
(b) $15x - 6$ (c) $x - 19$ **19.** (a) $\frac{19}{12}x$ (b) $12\frac{2}{3}$ acres
20. $2\frac{3}{5}$ feet **21.** (a) $C = 13x$ (b) $221 **22.** $640

2 Linear Equations and Inequalities

Section 2.1 (pp. 94–96)

1. solution **2.** true **3.** false **4.** solution set **5.** solutions
6. equivalent **7.** $b + c$ **8.** subtraction **9.** bc
10. division **11.** equivalent **12.** given **13.** No **15.** Yes
17. Yes **19.** Yes **21.** No **23.** Equation; 9
25. Expression; $6x$ **27.** 22 **29.** 3 **31.** -5 **33.** 9
35. 17 **37.** $\frac{17}{10}$ **39.** 5.1 **41.** -3 **43.** $\frac{2}{3}$ **45.** a **47.** $\frac{1}{5}$
49. 6 **51.** 3 **53.** 0 **55.** 7 **57.** -6 **59.** 3 **61.** $\frac{5}{4}$ **63.** 7
65. -8.5 **67.** a

69. (a)

Hours (x)	0	1	2	3	4	5	6
Rainfall (R)	3	4	5	6	7	8	9

(b) $R = x + 3$ (c) 6 in.; yes (d) 5.25 in.
71. (a) $L = 300x$ (b) $870 = 300x$ (c) 2.9
73. 50 days **75.** (a) 50° N (b) 5.7 **77.** $25,000

Section 2.2 (pp. 106–108)

1. constant **2.** $ax + b = 0$ **3.** Exactly one
4. numerically **5.** Addition, multiplication **6.** LCD
7. None **8.** Infinitely many **9.** Yes; $a = 3, b = -7$
11. Yes; $a = \frac{1}{2}, b = 0$ **13.** No **15.** No
17. Yes; $a = 1.1, b = -0.9$ **19.** Yes; $a = 2, b = -6$
21. No
23.

x	-1	0	1	2	3	; 2
$x - 3$	-4	-3	-2	-1	0	

25.

x	0	1	2	3	4	; 2
$-3x + 7$	7	4	1	-2	-5	

27.

x	-2	-1	0	1	2	; -1
$4 - 2x$	8	6	4	2	0	

29. $\frac{3}{11}$ **31.** 23   **33.** 28   **35.** $-\frac{11}{5}$ **37.** $-\frac{7}{2}$ **39.** 3   **41.** $\frac{9}{4}$
43. $-\frac{5}{2}$ **45.** $\frac{10}{3}$ **47.** $\frac{9}{8}$ **49.** 1   **51.** 1   **53.** -0.055 **55.** 8
57. $\frac{1}{3}$ **59.** $-\frac{b}{a}$ **61.** No solutions   **63.** One solution
65. Infinitely many solutions   **67.** No solutions
69. Infinitely many solutions

71. (a)

Hours (x)	0	1	2	3	4
Distance (D)	4	12	20	28	36

(b) $D = 8x + 4$ **(c)** 28 miles; yes
(d) 2.25 hours; the bicyclist is 22 miles from home after
    2 hours and 15 minutes.
73. 2009   **75.** 2010   **77.** 2013

Checking Basic Concepts 2.1 & 2.2 (p. 108)

1. (a) No   **(b)** Yes
2.

x	3	3.5	4	4.5	5
$4x - 3$	9	11	13	15	17

; 4

3. (a) 18   **(b)** $\frac{1}{6}$ **(c)** $2.\overline{6}$ or $\frac{8}{3}$ **(d)** 3
4. (a) One solution   **(b)** Infinitely many solutions
    (c) No solutions
5. (a) $D = 300 - 75x$ **(b)** $0 = 300 - 75x$ **(c)** 4 hours

Section 2.3 (pp. 118–122)

1. Check your solution.   **2.** +   **3.** =   **4.** $n + 1$ and $n + 2$
5. $\frac{x}{100}$ **6.** 0.01   **7.** left   **8.** right   **9.** $\frac{B - A}{A} \cdot 100$
10. increase; decrease   **11.** rt **12.** distance; time
13. $2 + x = 12$; 10   **15.** $2x + 7 = 9$; 1
17. $7 - x = -2$; 9   **19.** $\frac{x}{2} = 17$; 34   **21.** 31, 32, 33
23. 34   **25.** 8   **27.** 21   **29.** 7   **31.** 210 lb   **33.** 12   **35.** 14
37. *Les Mis*: 2.6; *Phantom*: 5.6   **39.** 306   **41.** 24
43. Length: 19 inches; width: 12 inches
45. Facebook: 198; Twitter: 49   **47.** 26 hours
49. $\frac{37}{100}$; 0.37   **51.** $\frac{37}{25}$; 1.48   **53.** $\frac{69}{1000}$; 0.069   **55.** $\frac{1}{2000}$; 0.0005
57. 45%   **59.** 180%   **61.** 0.6%   **63.** 40%   **65.** 75%
67. $83.\overline{3}\%$ **69.** $24.19   **71.** $988 per credit   **73.** $900 billion
75. 150 pounds   **77.** $150   **79.** About 10.3%   **81.** $55, 0%
83. About -57% **85.** 25%; -20% **87.** $d = 8$ miles
89. $r = 20$ feet per second   **91.** $t = 5$ hours   **93.** 60 mph
95. $\frac{3}{8}$ hour   **97.** 0.6 hr at 5 mph; 0.5 hr at 8 mph
99. 54 mph and 60 mph   **101.** 60 mph   **103.** $\frac{60}{7} \approx 8.6$ mph
105. 30 ounces   **107.** $1500 at 6%; $2500 at 5%
109. 5.5 lb of the $12 brand and 2.5 lb of the $7 brand
111. 8 mph: 54 min, 9 mph: 18 min
113. Student: 56, Adult: 95   **115.** 6 gallons
117. 10 grams

Section 2.4 (pp. 131–133)

1. formula   **2.** lw **3.** $\frac{1}{2}bh$ **4.** $\frac{1}{360}$ **5.** 360   **6.** 180
7. lwh **8.** $2lw + 2wh + 2lh$ **9.** $2\pi r$ **10.** πr^2
11. $\pi r^2 h$ **12.** $\frac{1}{2}(a + b)h$ **13.** 18 ft^2 **15.** 9 in^2
17. $16\pi \approx 50.3$ cm^2 **19.** 11 mm^2 **21.** 91 in^2

23. 105 ft^2 or $\frac{35}{3}$ yd^2 **25.** $8\pi \approx 25.1$ in.   **27.** 4602 ft^2
29. 65°   **31.** 81°   **33.** 36°   **35.** 36°, 36°, 108°
37. $C = 12\pi \approx 37.7$ in.; $A = 36\pi \approx 113.1$ in^2
39. $r = 1$ in.; $A = \pi \approx 3.14$ in^2
41. $V = 2640$ in^3; $S = 1208$ in^2
43. $V = 2$ ft^3; $S = \frac{32}{3}$ ft^2 **45.** 20π in^3 **47.** 600π in^3
49. (a) $\frac{45}{32}\pi \approx 4.4$ in^3 **(b)** About 2.4 fl oz
51. $y = -3x + 2$ **53.** $y = -\frac{4}{3}x + 4$ **55.** $w = \frac{A}{l}$
57. $h = \frac{V}{\pi r^2}$ **59.** $a = \frac{2A}{h} - b$ **61.** $w = \frac{V}{lh}$
63. $b = 2s - a - c$ **65.** $b = a - c$ **67.** $a = \frac{cd}{b - d}$
69. 15 in.   **71.** 2.98   **73.** 2.24   **75.** 77 °F   **77.** -40 °F
79. -5 °C   **81.** -20 °C   **83.** 11 mpg   **85.** 2.4 mi

Checking Basic Concepts 2.3 & 2.4 (p. 133)

1. (a) $3x = 36$; 12   **(b)** $35 - x = 43$; -8
2. $-32, -31, -30$ **3.** 0.095   **4.** 125%   **5.** About 182,886
6. 6.5 hr   **7.** $5000 at 6%; $3000 at 7%   **8.** 18 gal
9. 12 in.   **10.** $A = 9\pi \approx 28.3$ ft^2; $C = 6\pi \approx 18.8$ ft
11. 30°   **12.** $l = \frac{A - \pi r^2}{\pi r}$

Section 2.5 (pp. 141–144)

1. Unit   **3.** numerator   **5.** capacity   **7.** speed   **9.** 4 ft
11. 24 in.   **13.** 1584 in.   **15.** 536 mm   **17.** 0.01459 km
19. 43.18 cm   **21.** 220.18 m   **23. (a)** 63.43 yd   **(b)** 63.22 yd
25. (a) 2.79 mi   **(b)** 2.79 mi   **27.** 8.26 m   **29.** 9.84 in.
31. 9.81 ft   **33.** 1440 in^2 **35.** 36 yd^2 **37.** 40 cm^2
39. 2500 m^2 **41.** 6 c   **43.** 41 qt   **45.** 6 pt   **47.** 80 c
49. 90 L   **51.** 1080 mL   **53.** 473.12 mL   **55.** 52.83 L
57. 19.88 gal   **59.** 4.04 c   **61.** $27\frac{1}{2}$ or 27.5 lb   **63.** 8 oz
65. 55 T   **67.** 6.5 kg   **69.** 0.55 g   **71.** 14.81 oz
73. 134.64 lb   **75.** 3827.25 mg   **77.** 659.09 g
79. 315 min   **81.** 1.5 hr   **83.** 14,760 sec   **85.** 1.15 d
87. 0.5 yr   **89.** 102 ft/sec   **91.** 0.2 m/sec   **93.** 44 ft/sec
95. 75 mi/hr   **97.** 36 ft/sec   **99.** 57 mi/hr   **101.** 11 stones
103. (a) 225 ft^2 **(b)** 25 yd^2 **(c)** $800   **105.** 16 servings
107. 0.5 L   **109.** 5.5 t   **111.** 25 servings
113. Yes (about 60.007 ft)   **115.** 3 L   **117.** 210 lb
119. 180 mi/hr   **121.** 70 mi/hr   **123.** No (48.4 km/hr)

Section 2.6 (pp. 155–158)

1. $<, \le, >, \ge$ **2.** greater than; less than   **3.** solution
4. equivalent   **5.** True   **6.** True   **7.** number line
8. $>$ **9.** $<$ **10.** $>$ **11.** Yes   **13.** No   **15.** Yes   **17.** No

19.

21.

23.

25.

27. $x < 0$ **29.** $x \le 3$ **31.** $x \ge 10$
33. $x > -2$ **35.** $x < 1$

37.

x	1	2	3	4	5
$-2x + 6$	4	2	0	-2	-4

; $x \geq 3$

39.

x	-3	-2	-1	0	1
$5 - x$	8	7	6	5	4
$x + 7$	4	5	6	7	8

; $x < -1$

41. $[6, \infty)$ **43.** $(-2, \infty)$
45. $(-\infty, 7]$

47. $x > 3$;

49. $y \geq -2$;

51. $z > 8$;

53. $t \leq -5$;

55. $x < 5$;

57. $t \leq -2$;

59. $y > -\frac{3}{20}$;

61. $z \geq -\frac{14}{3}$;

63. $x < 7$; $(-\infty, 7)$ **65.** $x \leq -\frac{4}{3}$; $\left(-\infty, -\frac{4}{3}\right]$
67. $x > -\frac{39}{2}$; $\left(-\frac{39}{2}, \infty\right)$ **69.** $x \leq \frac{3}{2}$; $\left(-\infty, \frac{3}{2}\right]$
71. $x > \frac{1}{2}$; $\left(\frac{1}{2}, \infty\right)$ **73.** $x \leq \frac{5}{4}$; $\left(-\infty, \frac{5}{4}\right]$
75. $x > -\frac{8}{3}$; $\left(-\frac{8}{3}, \infty\right)$ **77.** $x \leq -\frac{1}{3}$; $\left(-\infty, -\frac{1}{3}\right]$
79. $x < 5$; $(-\infty, 5)$ **81.** $x \leq -\frac{3}{2}$; $\left(-\infty, -\frac{3}{2}\right]$
83. $x > -\frac{9}{8}$; $\left(-\frac{9}{8}, \infty\right)$ **85.** $x \leq \frac{4}{7}$; $\left(-\infty, \frac{4}{7}\right]$
87. $x < \frac{21}{11}$; $\left(-\infty, \frac{21}{11}\right)$ **89.** $\{x | x > 1\}$
91. $\{x | x \geq -7\}$ **93.** $\{x | x < 6\}$ **95.** $x > 60$
97. $x \geq 21$ **99.** $x > 40,000$ **101.** $x \leq 70$
103. Less than 10 feet   **105.** 86 or more
107. 4.5 hours   **109.** 4 days
111. (a) $C = 1.5x + 2000$ **(b)** $R = 12x$
(c) $P = 10.5x - 2000$ **(d)** 191 or more compact discs
113. (a) After 48 minutes   **(b)** After more than 48 minutes
115. Altitudes more than 4.5 miles

Checking Basic Concepts 2.5 & 2.6 (p. 158)

1. (a) 2.3 mi   **(b)** 657 mm   **(c)** 47,520 in.   **(d)** 450 ft
2. (a) 53 yd^2 **(b)** 52,000 cm^2 **3. (a)** 42 pt   **(b)** 3400 mL
4. (a) 107.2 oz   **(b)** 0.82 g   **5. (a)** 7200 min   **(b)** 0.6 yr
6. (a) 440 ft/min   **(b)** 215.9 cm/sec

7.

8. $x \leq 1$ **9.** Yes
10.

x	-2	-1	0	1	2
$5 - 2x$	9	7	5	3	1

; $x \geq -1$

11. (a) $x > 3$ **(b)** $x \geq -35$ **(c)** $x \leq -1$
12. $x \leq 12$ **13.** More than 13 inches

Chapter 2 Review (pp. 163–166)

1. -6 **2.** 2   **3.** $\frac{9}{4}$ **4.** $-\frac{1}{2}$ **5.** 3   **6.** $-\frac{7}{3}$ **7.** -2.5 **8.** $-\frac{7}{2}$
9. Yes; $a = -4, b = 1$ **10.** No   **11.** 2   **12.** 22   **13.** $\frac{27}{5}$
14. 7   **15.** -7 **16.** $-\frac{2}{3}$ **17.** $\frac{9}{4}$ **18.** $\frac{11}{14}$ **19.** $\frac{7}{8}$ **20.** 0
21. No solutions   **22.** Infinitely many solutions
23. Infinitely many solutions   **24.** One solution
25.

x	0.5	1.0	1.5	2.0	2.5
$-2x + 3$	2	1	0	-1	-2

; 1.5

26.

x	-2	-1	0	1	2
$-(x + 1) + 3$	4	3	2	1	0

; 0

27. $6x = 72$; 12   **28.** $x + 18 = -23$; -41
29. $2x - 5 = x + 4$; 9   **30.** $x + 4 = 3x$; 2   **31.** 9
32. $-52, -51, -50$ **33.** $\frac{17}{20}$; 0.85   **34.** $\frac{7}{125}$; 0.056
35. $\frac{3}{10,000}$; 0.0003   **36.** $\frac{171}{50}$; 3.42   **37.** 89%   **38.** 0.5%
39. 230%   **40.** 100%   **41.** $d = 24$ mi   **42.** $d = 3850$ ft
43. $r = 25$ yards per second   **44.** $t = \frac{25}{3}$ hr   **45.** 7.5 m^2
46. $36\pi \approx 113.1$ ft^2 **47.** $25\pi \approx 78.5$ in^2 **48.** 40 in.
49. $18\pi \approx 56.5$ ft   **50.** 864 in^2, or 6 ft^2 **51.** 50°
52. 22.5°   **53.** $625\pi \approx 1963.5$ in^3 **54.** 1620 in^2, or 11.25 ft^2
55. 174 in^2 **56.** About 60.6 ft^2 **57.** $y = 3x - 5$
58. $y = -x + 8$ **59.** $y = \frac{z}{2x}$ **60.** $b = 3S - a - c$
61. $b = \frac{12T - 4a}{3}$ **62.** $c = \frac{ab}{d - b}$ **63.** 2.82   **64.** 3.12
65. 59°F   **66.** 45°C   **67. (a)** 1.25 m   **(b)** 76,032 in.
68. (a) 37 yd^2 **(b)** 33,454,080 ft^2
69. (a) 68 pt   **(b)** 1.2 L   **70. (a)** 3.75 T   **(b)** 0.51 g
71. (a) 12,960 min   **(b)** 3.65 days
72. (a) 88 ft/sec   **(b)** 12.7 cm/sec

73.

74.

75.

76.

77. $x < 3$ **78.** $x \geq -1$ **79.** Yes   **80.** No
81.

x	0	1	2	3	4
$5 - x$	5	4	3	2	1

$x < 2$

82.

x	1	1.5	2	2.5	3
$2x - 5$	-3	-2	-1	0	1

$x \leq 2.5$

83. $x > 3$ **84.** $x \geq -5$ **85.** $x \leq -1$ **86.** $x < \frac{23}{3}$
87. $x \leq \frac{1}{9}$ **88.** $x \geq \frac{9}{4}$ **89.** $x < 50$ **90.** $x \leq 45,000$
91. $x \geq 16$ **92.** $x < 1995$ or $x \leq 1994$

93. (a)

Time	12:00	1:00	2:00	3:00	4:00	5:00
Rainfall (R)	2	2.75	3.5	4.25	5	5.75

(b) $R = \frac{3}{4}x + 2$ **(c)** $\frac{23}{4} = 5\frac{3}{4}$ in.; yes **(d)** $\frac{77}{16} = 4\frac{13}{16}$ in.
94. \$2125 **95.** About 339.6%

96. (a)

Hours (x)	1	2	3	4	5
Distance (D)	40	30	20	10	0

(b) $D = 50 - 10x$ **(c)** 20 miles; yes
(d) 3 hours or less or from noon to 3:00 P.M.
97. $\frac{1}{6}$ hr, or 10 min **98.** 33 in. by 23 in.
99. 50 mL **100.** \$500 at 3%; \$800 at 5%
101. 25 inches or less **102.** 74 or more **103.** 6 hr
104. (a) $C = 85x + 150{,}000$ **(b)** $R = 225x$
(c) $P = 140x - 150{,}000$ **(d)** 1071 or fewer

Chapter 2 Test (pp. 166–167)

1. -6 **2.** $\frac{5}{2}$ **3.** $-\frac{16}{11}$ **4.** $\frac{25}{6}$ **5.** No solutions
6. Infinitely many solutions
7.

x	0	1	2	3	4
$6-2x$	6	4	2	0	-2

; 3

8. $x + (-7) = 6$; 13 **9.** $2x + 6 = x - 7$; -13
10. 111, 112, 113 **11.** $\frac{4}{125}$; 0.032 **12.** 34.5%
13. $r = 50$ ft/sec **14.** 7.5 in^2
15. $C = 30\pi \approx 94.2$ in.; $A = 225\pi \approx 706.9$ in^2
16. 30° **17.** $x = \frac{y - z}{3y}$ **18.** $x = \frac{20R - 4y}{5}$
19. 1620 in. **20.** 1.2 yr

21.

22. $x > 0$ **23.** $x \le 6$ **24.** $x > -\frac{1}{7}$
25. (a) $S = 2x + 5$ **(b)** 21 in. **(c)** 17.5 in. **26.** 2000 mL
27. About 54.1%

Chapters 1 and 2 Cumulative Review (pp. 167–168)

1. Composite; $3 \cdot 3 \cdot 5$ **2.** Prime **3.** $\frac{1}{2}$ **4.** $\frac{1}{9}$ **5.** $\frac{13}{24}$
6. $\frac{13}{15}$ **7.** Integer and rational **8.** Irrational **9.** 3 **10.** 10
11. 15 **12.** 9 **13.** $4x^3$ **14.** $5x + 3$ **15.** 14 **16.** -4
17. $\frac{1}{2}$ **18.** 1 **19.** Infinitely many **20.** No solutions
21. 29, 30, 31 **22.** 0.047 **23.** 17% **24.** 65 mph **25.** 20°
26. $25\pi \approx 78.5$ in^2 **27.** $x = \frac{a + 4}{3y}$ **28.** $x = 3A - y - z$
29. $x < 1$

30. $x \le 2$

31. $I = 36Y$ **32.** \$618 **33.** 300 mL **34.** 2.3%

3 Graphing Equations

Section 3.1 (pp. 175–177)

1. xy-plane **2.** origin **3.** $(0, 0)$ **4.** 4 **5.** False **6.** x; y
7. scatterplot **8.** line **9.** $(-2, -2)$, $(-2, 2)$, $(0, 0)$, $(2, 2)$
11. $(-1, 0)$, $(0, -3)$, $(0, 2)$, $(2, 0)$
13. $(1, 3)$: I; $(0, -3)$: None; $(-2, 2)$: II **15.** $(0, 6)$: None; $(8, -4)$: IV; $(-6, -6)$: III

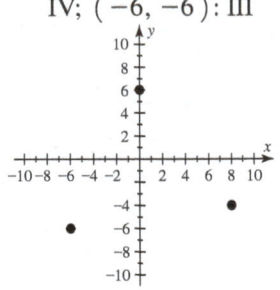

17. (a) I **(b)** III **19. (a)** None **(b)** I
21. (a) II **(b)** IV **23.** I and III
25.

27.

29.

31.

33.
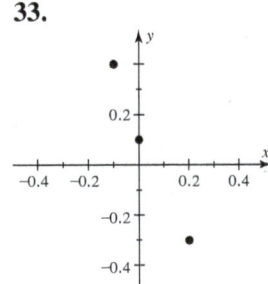

35. $(1960, 484)$, $(1980, 632)$, $(2000, 430)$, $(2010, 325)$; in 1960, there were 484 billion cigarettes consumed in the United States (answers may vary slightly).

37.

39.

41.

43. (a)

(b) The cost increased.

45. (a)

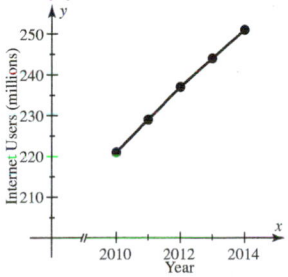

(b) The number of welfare beneficiaries increased and then decreased.

47. (a)

(b) The number of Internet users is increasing.

49. (a) The rate decreased.   **(b)** 9   **(c)** About -70%

Section 3.2 (pp. 185–187)

1. True   **2.** False   **3.** ordered pair   **4.** linear   **5.** False
6. True   **7.** graph   **8.** line   **9.** False   **10.** False   **11.** Yes
13. No   **15.** No   **17.** Yes   **19.** No

21.

x	−2	−1	0	1	2
y	−8	−4	0	4	8

23.

x	6	3	0	−3	−9
y	−2	0	2	4	8

25.

x	y
−8	−4
−4	0
0	4
4	8
8	12

27.

x	y
−6	−4
0	−2
6	0
12	2
18	4

29.

x	−3	0	3	6
y	−9	0	9	18

31.

x	8	6	4	2
y	−2	0	2	4

33.

x	y
−8	−2
−4	0
0	2
4	4

35.

x	y
$-\frac{1}{2}$	−2
$-\frac{1}{4}$	−1
0	0
$\frac{1}{4}$	1

37. They must be multiples of 5.

39.

x	−1	0	1
y	2	0	−2

Table values may vary.

41.

x	0	1	2
y	3	2	1

Table values may vary.

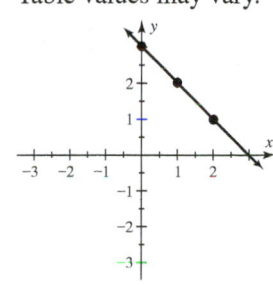

43.

x	−2	0	2
y	3	2	1

Table values may vary.

45.

47.

49.

51.

53.

55.

57.

59.

61.

63.

65.

67.

69.

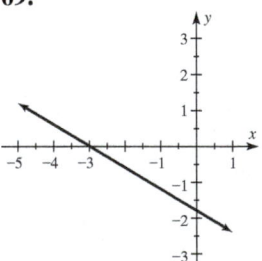

71.

x	y
−2	3
−1	0
0	−1
1	0
2	3

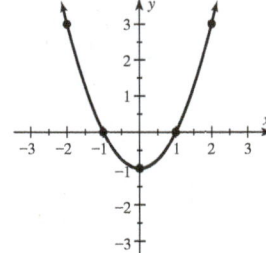

73.

x	y
−2	8
−1	2
0	0
1	2
2	8

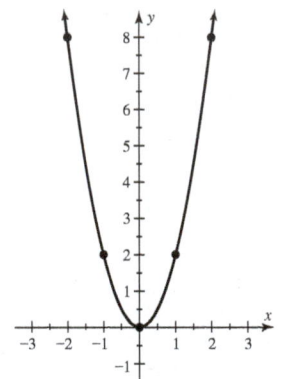

75.

x	y
−3	2
−1	0
0	−1
1	0
3	2

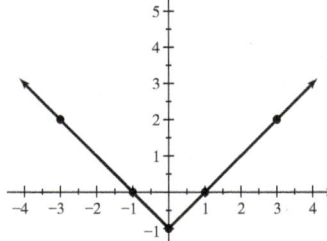

77.

x	y
0	0
1	1
4	2
9	3

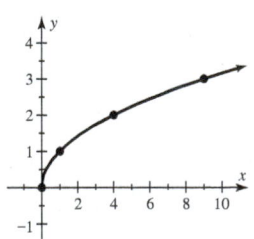

79. (a)

t	2010	2020	2030	2040	2050
P	13.2	15.0	16.7	18.5	20.3

(b) 2030

81. (a)

(b) About 37 days

83. (a) 68.3; 24.3; In 2008, Internet Explorer had 68.3% of the desktop web-browser market, and in 2014, it fell to 24.3%.

(b)

(c) 2010

Checking Basic Concepts 3.1 & 3.2 (p. 188)

1. $(-2, 2)$, II; $(-1, -2)$, III; $(1, 3)$, I; $(3, 0)$, none

2.

3.

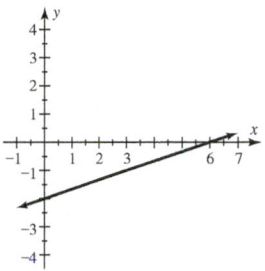

The percentage increased.

4. Yes

5.

x	-2	-1	0	1	2
y	5	3	1	-1	-3

6. (a)

(b)

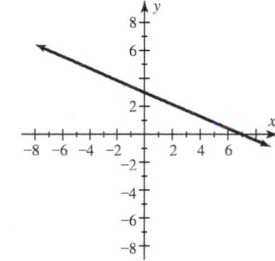

7. (a) 1; **9.** After one month, the global share was 1%; after 12 months, the global share was 9%.

(b)

(c) About 5 months

Section 3.3 (pp. 196–199)

1. x-intercept   **2.** 0   **3.** y-intercept   **4.** 0   **5.** horizontal
6. $y = b$ **7.** vertical   **8.** $x = k$ **9.** $(3, 0)$; $(0, -2)$
11. $(0, 0)$; $(0, 0)$ **13.** $(-2, 0)$, $(2, 0)$; $(0, 4)$
15. $(1, 0)$; $(0, 1)$

17.

x	-2	-1	0	1	2
y	0	1	2	3	4

$(-2, 0)$; $(0, 2)$

19.

x	-4	-2	0	2	4
y	-6	-4	-2	0	2

$(2, 0)$; $(0, -2)$

21. x-int: $(3, 0)$;
     y-int: $(0, -2)$

23. x-int: $(6, 0)$;
     y-int: $(0, -2)$

25. x-int: $(-1, 0)$;
     y-int: $(0, 6)$

27. x-int: $(7, 0)$;
     y-int: $(0, 3)$

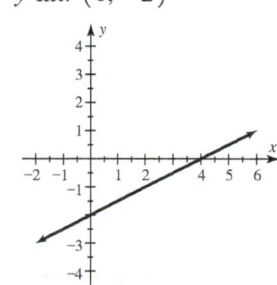

29. x-int: $(4, 0)$;
     y-int: $(0, -3)$

31. x-int: $(4, 0)$;
     y-int: $(0, -2)$

33. x-int: $(-4, 0)$;
     y-int: $(0, 3)$

35. x-int: $(3, 0)$;
     y-int: $(0, 2)$

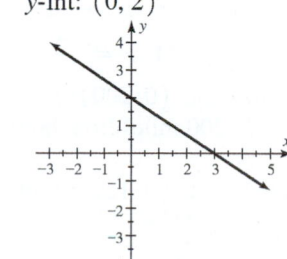

37. x-int: $(-2, 0)$; y-int: $(0, 5)$

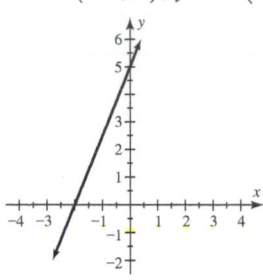

39. x-int: $\left(\frac{C}{A}, 0\right)$; y-int: $\left(0, \frac{C}{B}\right)$ **41.** $y = 1$ **43.** $x = -6$

45. (a)   **(b)**

47. (a)   **(b)**

49. (a)   **(b)**

51. $y = 4$ **53.** $x = -1$ **55.** $y = -6$ **57.** $x = 5$
59. $y = 2$; $x = 1$ **61.** $y = -45$; $x = 20$
63. $y = 5$; $x = 0$ **65.** $x = -1$ **67.** $y = -\frac{5}{6}$
69. $x = 4$ **71.** $x = -\frac{2}{3}$ **73.** $y = 0$

75. (a) y-int: $(0, 200)$; x-int: $(4, 0)$ **(b)** The driver was initially 200 miles from home; the driver arrived home after 4 hours.

77. (a) y-int: $(0, 128)$; t-int: $(4, 0)$ **(b)** The initial velocity was 128 ft/sec; the velocity after 4 seconds was 0.

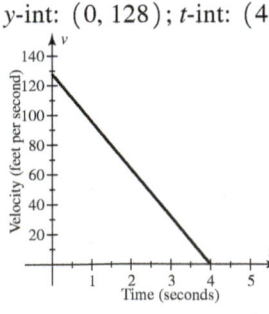

79. (a) y-int: $(0, 2000)$; x-int: $(4, 0)$ **(b)** The pool initially contained 2000 gallons; the pool was empty after 4 hours.

Section 3.4 (pp. 209–213)

1. True   **2.** True   **3.** rise; run   **4.** horizontal   **5.** vertical
6. $\frac{y_2 - y_1}{x_2 - x_1}$ **7.** positive   **8.** negative   **9.** rate   **10.** Toward
11. Positive   **13.** Zero   **15.** Negative   **17.** Undefined
19. 0; the rise always equals 0.
21. 1; the graph rises 1 unit for each unit of run.
23. 2; the graph rises 2 units for each unit of run.
25. Undefined; the run always equals 0.
27. $-\frac{3}{2}$; the graph falls 3 units for each 2 units of run.
29. -1; the graph falls 1 unit for each unit of run.

31. 2;   **33.** Undefined;

35. $-\frac{2}{3}$;   **37.** 0;

                 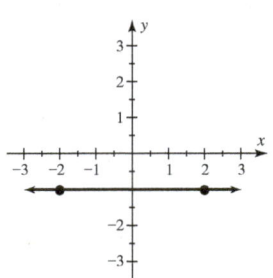

39. 1   **41.** $-\frac{6}{7}$ **43.** 0   **45.** Undefined   **47.** $\frac{13}{20}$
49. $\frac{9}{100}$ **51.** $-\frac{5}{7}$ **53.** 49

55.   **57.**

59.   **61.**

                 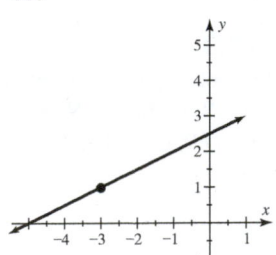

63.

x	0	1	2	3
y	-4	-2	0	2

65.

x	1	2	3	4
y	4	1	-2	-5

67.

x	-4	-2	0	2
y	0	3	6	9

69. c.   **71.** b.   **73. (a)** $m_1 = 1000$; $m_2 = -1000$
(b) $m_1 = 1000$: Water is being added to the pool at a rate of 1000 gallons per hour. $m_2 = -1000$: Water is being removed from the pool at a rate of 1000 gallons per hour. **(c)** Initially, the pool contained 2000 gallons of water. Over the first 3 hours, water was pumped into the pool at a rate of 1000 gallons per hour. For the next 2 hours, water was pumped out of the pool at a rate of 1000 gallons per hour.
75. (a) $m_1 = 50$; $m_2 = 0$; $m_3 = -50$ **(b)** $m_1 = 50$: The car is moving away from home at a rate of 50 mph. $m_2 = 0$: The car is not moving. $m_3 = -50$: The car is moving toward home at a rate of 50 mph.   **(c)** Initially, the car is at home. Over the first 2 hours, the car travels away from home at a rate of 50 mph. Then the car is parked for 1 hour. Finally, the car travels toward home at a rate of 50 mph.
77. m^3/min

79.

81.

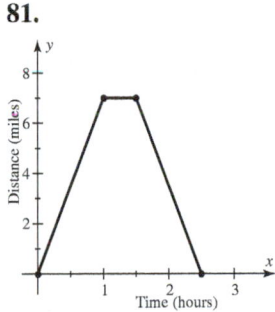

83. (a) 76,000,000   **(b)** 281,000,000   **(c)** 2,050,000/yr
85. (a) 40,800   **(b)** The celebrity gained 40,800 followers each year, on average.
87. (a) 25   **(b)** The revenue is $25 per flash drive.
89. (a)

(b) $m_1 = 20$; $m_2 = 30$; $m_3 = 40$ **(c)** $m_1 = 20$: Each mile between 0 and 5 miles is worth $20 per mile. $m_2 = 30$: Each mile between 5 and 10 miles is worth $30 per mile. $m_3 = 40$: Each mile between 10 and 15 miles is worth $40 per mile.
91. (a) 750   **(b)** Median family income increased, on average, by $750/yr over the period.
93. (a) Toward   **(b)** 10 ft   **(c)** 2 ft/min   **(d)** 5

Checking Basic Concepts 3.3 & 3.4 (pp. 213–214)

1. $(-2, 0)$; $(0, 3)$
2.

x	-2	-1	0	1	2
y	-6	-4	-2	0	2

$(1, 0)$; $(0, -2)$
3. (a) x-int: $(6, 0)$; y-int: $(0, -3)$ **(b)** y-int: $(0, 2)$

(c) x-int: $(-1, 0)$

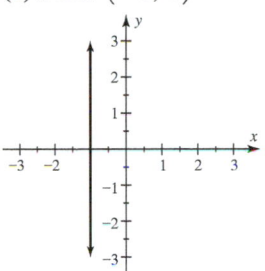

4. $y = 4$; $x = -2$ **5. (a)** $\frac{3}{4}$ **(b)** 0   **(c)** Undefined
6. $-\frac{1}{2}$
7.

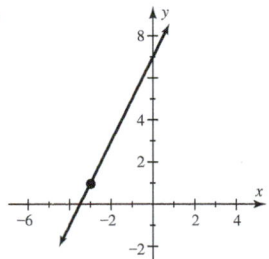

8. (a) $m_1 = 0$; $m_2 = 2$; $m_3 = -\frac{2}{3}$ **(b)** $m_1 = 0$: The depth is not changing. $m_2 = 2$: The depth increased at a rate of 2 feet per hour. $m_3 = -\frac{2}{3}$: The depth decreased at a rate of $\frac{2}{3}$ foot per hour.   **(c)** Initially, the pond had a depth of 5 feet. For the first hour, there was no change in the depth of the pond. For the next hour, the depth of the pond increased at a rate of 2 feet per hour to a depth of 7 feet. Finally, the depth of the pond decreased for 3 hours at a rate of $\frac{2}{3}$ foot per hour until it was 5 feet deep.

Section 3.5 (pp. 222–224)

1. $y = mx + b$ **2.** slope   **3.** y-coordinate of the y-intercept
4. origin   **5.** parallel   **6.** slope   **7.** perpendicular
8. reciprocals   **9.** f.   **10.** d.   **11.** a.   **12.** b.   **13.** e.
14. c.   **15.** $y = x - 1$ **17.** $y = -2x + 1$ **19.** $y = -2x$
21. $y = \frac{3}{4}x + 2$

23. $y = x + 2$

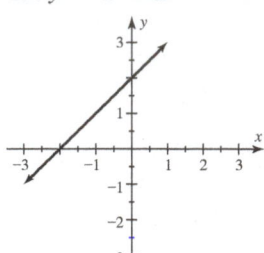

25. $y = 2x - 1$

27. $y = -\frac{1}{2}x - 2$

29. $y = \frac{1}{3}x$

31. $y = 3$

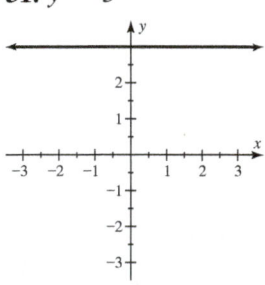

33. (a) $y = -x + 4$ **(b)** $-1; (0, 4)$
35. (a) $y = -2x + 4$ **(b)** $-2; (0, 4)$
37. (a) $y = \frac{1}{2}x + 2$ **(b)** $\frac{1}{2}; (0, 2)$
39. (a) $y = \frac{2}{3}x - 2$ **(b)** $\frac{2}{3}; (0, -2)$
41. (a) $y = \frac{1}{4}x + \frac{3}{2}$ **(b)** $\frac{1}{4}; (0, \frac{3}{2})$
43. (a) $y = -\frac{1}{3}x + \frac{2}{3}$ **(b)** $-\frac{1}{3}; (0, \frac{2}{3})$

45.

47.

49.

51.

53.

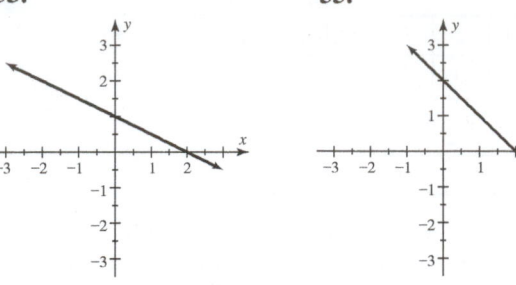

55.

57. $y = 2x + 2$ **59.** $y = x - 2$ **61.** $-\frac{1}{5}$ **63.** $\frac{5}{4}$
65. 4 **67.** $y = \frac{4}{7}x + 3$ **69.** $y = 3x$ **71.** $y = -\frac{1}{2}x + \frac{5}{2}$
73. $y = 2x$ **75.** $y = \frac{1}{3}x + \frac{1}{3}$ **77. (a)** \$25 **(b)** 25 cents
(c) $(0, 25)$; the fixed cost of renting the car is \$25 **(d)** 0.25;
the cost per mile of driving the car **79. (a)** \$7.45
(b) $C = 0.07x + 3.95$ **(c)** 67 min
81. (a) $m = 0.35; b = 164.3$ **(b)** The fixed cost of owning
the car for one month

Section 3.6 (pp. 233–236)

1. True **2.** True **3.** $y = mx + b$
4. $y - y_1 = m(x - x_1)$ or $y = m(x - x_1) + y_1$
5. 1; 3 **6.** distributive **7.** Yes; every nonvertical line has
exactly one slope and one y-intercept. **8.** No; it depends on
the point used. **9.** No **11.** Yes **13.** $y - 3 = -2(x + 2)$
15. $y - 2 = \frac{3}{4}(x - 1)$ **17.** $y + 1 = -\frac{1}{2}(x - 3)$
19. $y - 1 = 4(x + 3)$ **21.** $y + 3 = \frac{1}{2}(x + 5)$
23. $y - 30 = 1.5(x - 2010)$ **25.** $y - 4 = \frac{7}{3}(x - 2)$
27. $y = \frac{3}{5}(x - 5)$ **29.** $y - 15 = 5(x - 2003)$
31. $y = 3x - 2$ **33.** $y = \frac{1}{3}x$ **35.** $y = \frac{2}{3}x + \frac{1}{12}$
37. $y = -2x + 9$ **39.** $y = \frac{3}{5}x - 2$ **41.** $y = -16x - 19$
43. $y = -2x + 5$ **45.** $y = -x + 1$ **47.** $y = -\frac{1}{9}x + \frac{1}{3}$
49. $y = 2x - 7$ **51.** $y = 2x - 15$ **53.** $y = \frac{1}{2}x + 5$
55. $y = 2x - 3$ **57.** $y = -2x - 1$ **59.** $y = \frac{1}{2}x - \frac{5}{2}$
61. $y = 2x - 5$ **63.** $y = -\frac{1}{2}x - \frac{3}{2}$
65. $(0, -mx_1 + y_1)$. **67. (a)** $-5\,°F$ per hour
(b) $T = -5x + 45$; the temperature is decreasing at a rate of
$5\,°F$ per hour. **(c)** $(9, 0)$; at 9 A.M. the temperature was $0\,°F$.
(d)

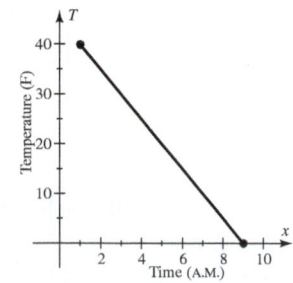

(e) At 4 A.M. the temperature was $25\,°F$.

69. (a) Entering; 300 gallons **(b)** $(0, 100)$; initially the tank contains 100 gallons. **(c)** $y = 50x + 100$; the amount of water is increasing at a rate of 50 gallons per minute. **(d)** 8 minutes **71. (a)** Toward **(b)** After 1 hour, the person is 250 miles from home. After 4 hours, the person is 100 miles from home. **(c)** 50 mph **(d)** $y = -50x + 300$; the car is traveling toward home at 50 mph. **73. (a)** In 2009, there were 700 million mobile Internet users. **(b)** $y - 700 = 200(x - 2009)$ **(c)** 1500 million; 1500 million **(d)** The number of mobile Internet users increased on average by 200 million per year. **75. (a)** $y - 1567 = -165(x - 2012)$ **(b)** 1237 **77.** Cigarette consumption decreased on average by 10.33 billion cigarettes per year. **79. (a)** $y = 0.009375x - 18.1875$ **(b)** Sea level is rising 0.009375 meter per year. **(c)** 0.75 meter

Checking Basic Concepts 3.5 & 3.6 (p. 237)

1. $y = -3x + 1$ **2.** $y = \frac{4}{5}x - 4$; $\frac{4}{5}$; $(0, -4)$
3.

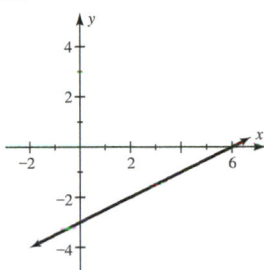

4. (a) $y = 3x - 2$ **(b)** $y = -\frac{3}{2}x$ **(c)** $y = -\frac{7}{3}x - \frac{5}{3}$ **5.** $y - 3 = -2(x + 1)$ **6.** $y = -2x + 1$ **7.** $y = 2x + 3$ **8. (a)** $y = -12x + 48$ **(b)** 12 mph **(c)** 4 P.M. **(d)** 48 miles **9. (a)** 5 inches **(b)** 2 inches per hour **(c)** $(0, 5)$; five inches of snow fell before noon **(d)** 2; the average rate of snowfall was 2 inches per hour.

Section 3.7 (pp. 242–245)

1. linear **2.** exact **3.** approximate **4.** constant **5.** constant rate of change **6.** initial amount **7.** f. **8.** d. **9.** a. **10.** c. **11.** e. **12.** b. **13.** Yes **15.** No **17.** No **19.** Exact; $y = 2x - 2$ **21.** Approximate; $y = 2x + 2$ **23.** Approximate; $y = 2$ **25.** $y = 5x + 40$ **27.** $y = -20x - 5$ **29.** $y = 8$ **31. (a)** Yes **(b)**

(c) $y = -2x + 4$

33. (a) No **(b)**

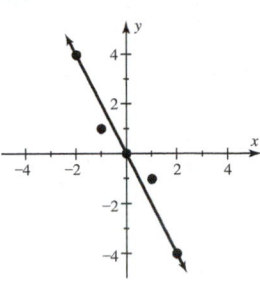

(c) $y = -2x$

35. (a) Yes **(b)**

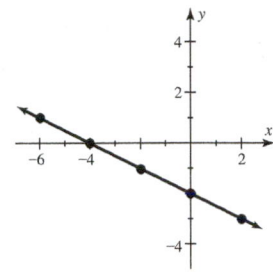

(c) $y = -\frac{1}{2}x - 2$

37. (a) No **(b)**

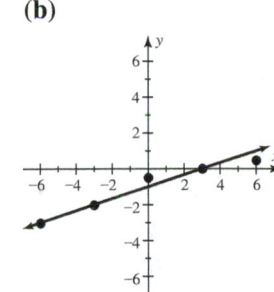

(c) $y = \frac{1}{3}x - 1$
39. $g = 5t + 200$, where g represents gallons of water and t represents time in minutes. **41.** $d = 6t + 5$, where d represents distance in miles and t represents time in hours. **43.** $p = 8t + 200$, where p represents total pay in dollars and t represents time in hours. **45.** $r = t + 5$, where r represents total number of roofs shingled and t represents time in days. **47. (a)** $-\frac{2}{45}$ acre per year **(b)** $A = -\frac{2}{45}t + 5$ **49. (a)** $y = \frac{19}{6}x - 6155.5$ **(b)** About 232,000 **51. (a)** **(b)**

(c) 20; the mileage is 20 miles per gallon. **(d)** $y = 20x$ **(e)** 140 miles

53.

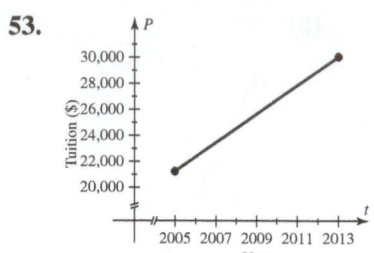

55. $y = 1106x - 2{,}196{,}295;\ (0, -2{,}196{,}295);$ no

Checking Basic Concepts 3.7 (p. 245)

1. Yes   **2.** Approximate; $y = x - 1$
3. (a) $y = 10x + 50$ **(b)** $y = -2x + 200$
4. (a)   **(b)**

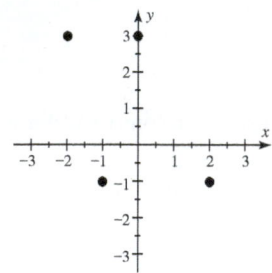

(c) $y = -\frac{1}{2}x + 1$
5. (a) $T = 0.075x$ **(b)** $5.4°\text{F}$

Chapter 3 Review (pp. 249–253)

1. $(-2, 0)$: none; $(-1, 2)$: II; $(0, 0)$: none; $(1, -2)$: IV; $(1, 3)$: I
2.

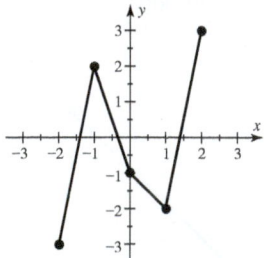

3. (a) II   **(b)** IV   **4. (a)** None   **(b)** III
5.   **6.**

7. Yes   **8.** No

9.

x	-2	-1	0	1	2
y	6	3	0	-3	-6

10.

x	4	3	2.5	2	1
y	-3	-1	0	1	3

11.

x	-2	0	2	4
y	-4	2	8	14

12.

x	-1	-3	-5	-7
y	1	2	3	4

13.   **14.**

15.   **16.**

17. $(3, 0);\ (0, -2)$ **18.** $(-2, 0)$ and $(2, 0);\ (0, -4)$

19.

x	-2	-1	0	1	2
y	4	3	2	1	0

$(2, 0);\ (0, 2)$

20.

x	-4	-2	0	2	4
y	-4	-3	-2	-1	0

$(4, 0);\ (0, -2)$

21. x-int: $(3, 0)$;
y-int: $(0, -2)$

22. x-int: $(1, 0)$;
y-int: $(0, -5)$

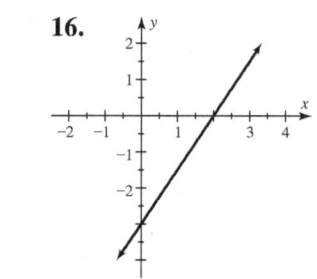

23. x-int: $(4, 0)$;
y-int: $(0, -2)$

24. x-int: $(2, 0)$;
y-int: $(0, 3)$

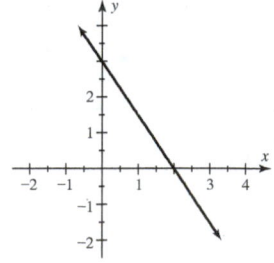

25. $y = 1$ **26.** $x = 3$

27. (a)   **(b)**

    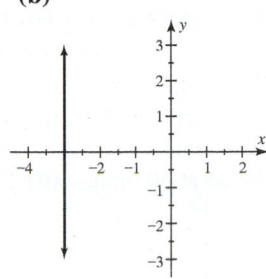

28. $x = -1; y = 1$ **29.** $y = 3; x = -2$
30. $x = 4$ **31.** $y = -5$ **32. (a)** y-int: $(0, 90)$; x-int: $(3, 0)$
(b) The driver is initially 90 miles from home; the driver
arrives home after 3 hours.   **33.** 2   **34.** $-\frac{2}{5}$
35. 0   **36.** Undefined   **37.** 3   **38.** $-\frac{1}{2}$

39. (a)   **40. (a)**

    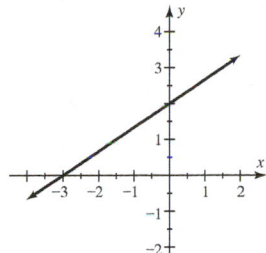

(b) -2; $(0, 0)$ **(b)** $\frac{2}{3}$; $(0, 2)$

41.   **42.**

43.

x	0	1	2	3
y	1	$\frac{3}{2}$	2	$\frac{5}{2}$

44. Positive   **45.** $y = x + 1$ **46.** $y = -2x + 2$

47. $y = 2x - 2$ **48.** $y = -\frac{3}{4}x + 3$

49. (a) $y = -x + 3$ **(b)** -1; $(0, 3)$
50. (a) $y = \frac{5}{6}x - 5$ **(b)** $\frac{5}{6}$; $(0, -5)$

51.   **52.**

53.   **54.**

    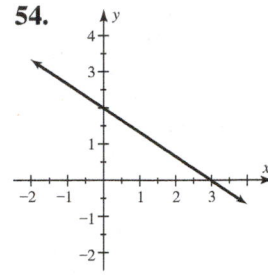

55. $y = 5x - 5$ **56.** $y = -2x$ **57.** $y = -\frac{5}{6}x + 2$

58. $y = -2x - 3$ **59.** $y = \frac{2}{3}x - 2$ **60.** $y = -\frac{1}{5}x - 2$
61. Yes   **62.** No   **63.** $y - 2 = 5(x - 1)$
64. $y + 30 = 3(x - 20)$ **65.** $y = \frac{4}{3}(x - 3)$
66. $y = 2\left(x - \frac{1}{2}\right)$ **67.** $y - 7 = 2(x - 5)$
68. $y = -\frac{2}{3}(x + 1)$ **69.** $y = 3x + 5$

70. $y = -\frac{1}{4}x + 3$ **71.** No
72. Approximate; $y = -x + 5$ **73.** $y = -2x + 40$
74. $y = 20x + 200$ **75.** $y = 50$ **76.** $y = 5x - 20$

77. (a) Yes   **(b)**

    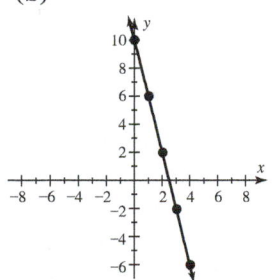

(c) $y = -4x + 10$

78. (a) No   **(b)**

    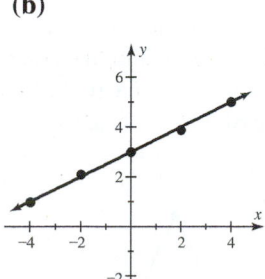

(c) $y = \frac{1}{2}x + 3$

79. (a)

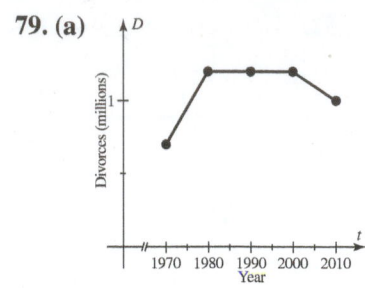

(b) The number of divorces increased significantly between 1970 and 1980, remained unchanged from 1980 to 2000, and then decreased.

80. (a) $G = 100t$
(b)

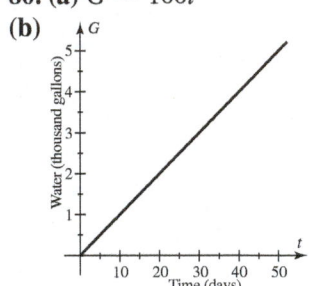

(c) 50 days

81. (a)

(b) v-int: $(0, 160)$; t-int: $(5, 0)$; the initial velocity was 160 ft/sec, and the velocity after 5 seconds was 0.

82. (a) $m_1 = 0$; $m_2 = -1500$; $m_3 = 500$ **(b)** $m_1 = 0$: The population remained unchanged. $m_2 = -1500$: The population decreased at a rate of 1500 insects per week. $m_3 = 500$: The population increased at a rate of 500 insects per week.   **(c)** For the first week, the population did not change from its initial value of 4000. Over the next two weeks, the population decreased at a rate of 1500 insects per week until it reached 1000. Finally, the population increased at a rate of 500 per week for two weeks, reaching 2000.

83.

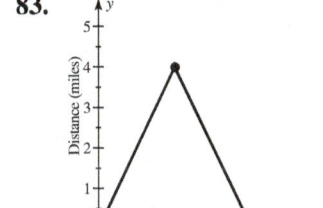

84. (a) $\frac{2}{3}$ **(b)** Revenue increased, on average, by about $0.67 billion/yr
85. (a) \$35   **(b)** 20¢   **(c)** $(0, 35)$; the fixed cost of renting the car   **(d)** 0.2; the cost for each mile driven
86. (a) Toward; the slope is negative.   **(b)** After 1 hour, the car is 200 miles from home; after 3 hours, the car is 100 miles from home.   **(c)** $y = -50x + 250$; the car is moving toward home at 50 mph.   **(d)** 150 miles; 150 miles
87. (a)

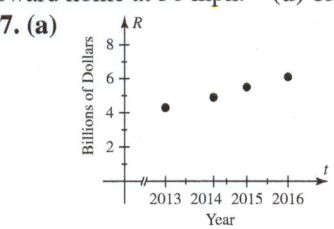

(b) $R = 0.6t - 1203.5$; Revenue increased by \$0.6 billion/yr.
(c) Yes   **(d)** \$7.3 billion
88. (a)   **(b)**

(c) 20; the mileage is 20 miles per gallon.
(d) $y = 20x$; no   **(e)** About 180 miles

Chapter 3 Test (pp. 253–254)

1. $(-2, -2)$: III; $(-2, 1)$: II; $(0, -2)$: none; $(1, 0)$: none; $(2, 3)$: I; $(3, -1)$: IV
2.

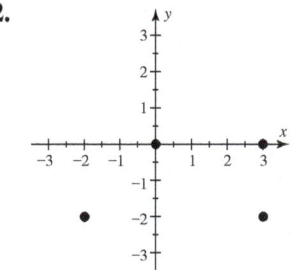

3.

x	-2	-1	0	1	2
y	-8	-6	-4	-2	0

$(2, 0)$; $(0, -4)$

4. Yes

5.

6. x-int: $(3, 0)$; y-int: $(0, -5)$

7.   **8.**

9.   **10.**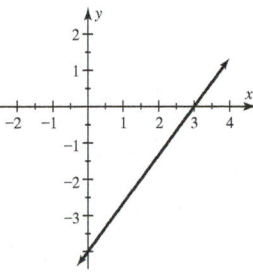

11. $y = -x + 2$ **12.** $y = 2x + \frac{1}{2}$; 2; $\left(0, \frac{1}{2}\right)$

13. $y = -5$; $x = 1$ **14.** $-\frac{2}{9}$ **15.** $y = -\frac{4}{3}x - 5$

16. $y = 3x - 11$ **17.** $y = -3x + 5$ **18.** $y = -\frac{1}{2}x$

19. $y = \frac{1}{2}x + 5$ **20.** Yes, $y = 3x - 2$

21. $y - 7 = -3(x + 2)$ **22.** Approximate; $y = x - 1$

23.

24. (a) $m_1 = 2$; $m_2 = -9$; $m_3 = 2$; $m_4 = 5$
(b) $m_1 = 2$: The population increased at a rate of 2000 fish per year. $m_2 = -9$: The population decreased at a rate of 9000 fish per year. $m_3 = 2$: The population increased at a rate of 2000 fish per year. $m_4 = 5$: The population increased at a rate of 5000 fish per year.
(c) For the first year, the population increased from an initial value of 8000 to 10,000 at a rate of 2000 fish per year. During the second year, the population dropped dramatically to 1000 at a rate of 9000 fish per year. Over the third year, the population grew to 3000 at a rate of 2000 fish per year. Finally, over the fourth year, the population grew at a rate of 5000 fish per year to reach 8000.
25. $N = 100x + 2000$

Chapters 1–3 Cumulative Review (pp. 254–255)

1. Composite; $40 = 2 \cdot 2 \cdot 2 \cdot 5$ **2.** Prime
3. $n + 10$ **4.** $n^2 - 2$ **5.** $\frac{2}{3}$ **6.** $\frac{17}{30}$ **7.** 14
8. -9 **9.** Rational   **10.** Irrational   **11.** $7x + 1$
12. $x - 4$ **13.** 2   **14.** 0   **15.** 720 ft^2 **16.** 20 ft^2

17.

x	-2	-1	0	1	2
y	10	8	6	4	2

; 1

18. $2n + 2 = n - 5$; -7 **19.** 8 hours
20. $x < 2$ **21.** $\{x \mid x > 0\}$ **22.** $\{x \mid x \le \frac{1}{2}\}$
23. $(-2, -3)$: III; $(0, 3)$: none; $(2, -2)$: IV; $(2, 1)$: I
24.

x	-2	-1	0	1	2
y	3	2.5	2	1.5	1

25.   **26.**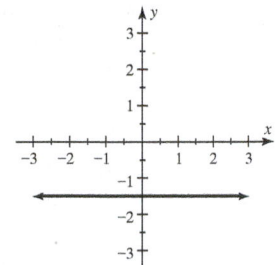

27. x-int: $(-10, 0)$; y-int: $(0, 8)$ **28.** $x = \frac{3}{2}$; $y = -2$
29. $y = 3x - 3$
30. $y = \frac{3}{5}x - 3$;

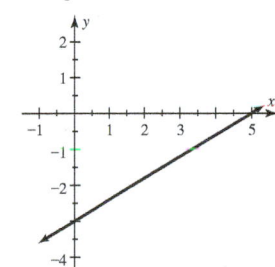

31. $y = -\frac{2}{3}x - 3$ **32.** $y = -2x + 1$
33. $y = 5x + 100$ **34.** $I = 5000x + 20{,}000$
35. $C = 8x$ **36.** $\frac{49}{100}$ **37.** \$1000 at 3%; \$2000 at 4%
38. (a) \$85   **(b)** \$25   **(c)** 30¢

4 Systems of Linear Equations in Two Variables

Section 4.1 (pp. 266–269)

1. ordered   **2.** intersection-of-graphs   **3.** no solutions; one solution; infinitely many   **4.** consistent   **5.** inconsistent
6. independent   **7.** dependent   **8.** table   **9.** the same
10. intercepts   **11.** 1   **13.** 2   **15.** -2 **17.** -1
19.

x	0	1	2	3
$y = x + 3$	3	4	5	6

; 1

21.

x	-4	-2	2	4
$y = 3 - 3x$	15	9	-3	-9

; 2

23.

x	-3	0	3	6
$2x + 3y = 6$	4	2	0	-2

; 3

25. One; consistent; independent
27. Infinitely many; consistent; dependent
29. None; inconsistent **31.** $(1, 1)$ **33.** $(2, -3)$
35. $\left(\frac{1}{5}, \frac{4}{5}\right)$ **37.** $(2, 1)$ **39.** $(3, 2)$ **41.** $(-1, 1)$
43. $(2, 4)$ **45.** $(3, 1)$
47.

x	0	1	2	3
$y = x + 2$	2	3	4	5
$y = 4 - x$	4	3	2	1

; $(1, 3)$

49. (a) $(-1, 1)$ **(b)** $(-1, 1)$ **51. (a)** $(3, 1)$ **(b)** $(3, 1)$
53. (a) $(1, 3)$ **(b)** $(1, 3)$ **55.** $(-1, -2)$ **57.** $(1, -1)$
59. $(1, 2)$ **61.** $(3, 2)$ **63.** $(3, 1)$ **65.** $(1, 2)$
67. (a) $x + y = 4, x - y = 0$ **(b)** 2, 2
69. (a) $2x + y = 7, x - y = 2$ **(b)** 3, 1
71. (a) $x = 3y, x - y = 4$ **(b)** 6, 2
73. (a) $C = 0.5x + 50$ **(b)** 60 mi **(c)** 60 mi
75. (a) $x + y = 60, x = 2y$ **(b)** $(40, 20)$; in 2013, digital downloads accounted for 40% of all music revenue and streaming subscriptions accounted for 20%.
77. 180 e-readers at $120 each
79. (a) $x - y = 4$; $2x + 2y = 28$ where x is length, y is width **(b)** $(9, 5)$; the rectangle is 9 in. $\times$ 5 in.
81. 23; yes **83.** 23; yes

Section 4.2 (pp. 276–279)

1. exact **2.** parentheses **3.** It has infinitely many solutions.
4. It has no solutions. **5.** $(3, 6)$ **7.** $(2, 1)$ **9.** $(-1, 0)$
11. $(3, 0)$ **13.** $\left(\frac{1}{2}, 0\right)$ **15.** $(0, 4)$ **17.** $(1, 2)$
19. $(6, -4)$ **21.** $(4, -3)$ **23.** $\left(1, -\frac{1}{2}\right)$ **25.** $(1, 3)$
27. $(0, -2)$ **29.** $(2, 2)$ **31.** $(-3, 5)$ **33.** No solutions
35. Infinitely many **37.** $(4, 1)$ **39.** Infinitely many
41. No solutions **43.** $\left(-\frac{14}{33}, -\frac{37}{33}\right)$ **45.** $(0, 0)$
47. No solutions **49.** No solutions
51. They are a single line.
53. (a) $L - W = 10, 2L + 2W = 72$ **(b)** $(23, 13)$
55. (a) $x = \frac{1}{2}y; x + y = 90$ **(b)** $(30, 60)$ **(c)** $(30, 60)$
57. (a) $x - y = 30, y = 0.864x$ **(b)** $(220.59, 190.59)$
59. 94 ft $\times$ 50 ft **61.** 17 and 53 **63.** 10 mph; 2 mph
65. $3.\overline{3}$ L of 20% solution, $6.\overline{6}$ L of 50% solution
67. (a) $x + y = 15, 3x + 4y = 53$ **(b)** $(7, 8)$
69. (a) $x + y = 48, 12x + 5y = 450$ **(b)** $(30, 18)$
(c) Tread climber: 360 cal, bicycle: 90 cal
71. Superior: 32,000 mi²; Michigan: 22,000 mi²

Checking Basic Concepts 4.1 & 4.2 (p. 279)

1. (a) -2 **(b)** 6 **2.** $(4, 2)$ **3.** $(2, 1)$
4. (a) No solutions **(b)** $(1, -1)$
(c) Infinitely many solutions
5. (a) $x + y = 300, 150x + 200y = 55,000$
(b) $(100, 200)$

Section 4.3 (pp. 287–290)

1. Substitution; elimination **2.** addition **3.** = **4.** =
5. It has infinitely many solutions. **6.** It has no solutions.

7. $(1, 1)$ **9.** $(-2, 1)$ **11.** No solutions **13.** Infinitely many
15. $(6, 1)$ **17.** $(-1, 4)$ **19.** $(2, 4)$ **21.** $(2, 1)$
23. $(4, -1)$ **25.** $(-4, 1)$ **27.** $\left(\frac{4}{5}, 1\right)$ **29.** $(-2, -5)$
31. $(1, 0)$ **33.** $(2, 2)$ **35.** $\left(\frac{4}{5}, \frac{9}{5}\right)$ **37.** $(3, 2)$
39. $(0, 1)$ **41.** $(2, 1)$ **43.** $(4, -3)$ **45.** $(1, 0)$
47. Infinitely many;

49. One;

51. No solutions;

53. No solutions;

55. Infinitely many;

57. healthcare: 11.7 million; ncaa: 8.7 million
59. Bicycle: 18 min; stair climber: 12 min
61. Current: 2 mph; boat: 6 mph
63. $2000 at 3%; $3000 at 5% **65.** $-43, 26$
67. (a) 20 in. $\times$ 40 in. **(b)** 20 in. $\times$ 40 in.

Section 4.4 (pp. 298–301)

1. inequality **2.** ordered pair **3.** All points below and including the line $y = k$ **4.** All points to the right of the line $x = k$. **5.** All points above and including the line $y = x$
6. test **7.** dashed **8.** solid **9.** $Ax + By = C$
10. both inequalities **11.** intersect **12.** No **13.** Yes
15. No **17.** No **19.** Yes **21.** Yes **23.** No **25.** $x > 1$
27. $y \geq 2$ **29.** $y < x$ **31.** $-x + y \leq 1$
33.

35.

37.

39.

71.

73.

41.

43.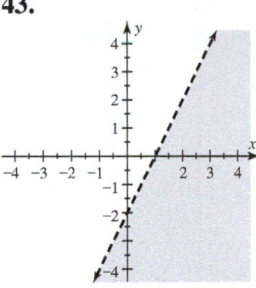

75. (a) 200 bpm; 150 bpm

(b)

45. Yes   **47.** No   **49.** Yes

51. The region containing $(1, 2)$

53. The region containing $(1, 0)$

55.

57.

(c) Possible heart rates for ages 20 to 70   **77.** 150 to 200 lb

79. (a)   **(b)**

81. Wetter and cooler regions   **83.** $7P - 5T \geq -70$

85. $7P - 5T \leq -70$
$35P - 3T \geq 140$

59.

61.

Checking Basic Concepts 4.3 & 4.4 (p. 301)

1. $(1, 1)$ **2. (a)** $(0, -1)$; one   **(b)** No solutions
(c) Infinitely many   **3.** Infinitely many

63.

65.

4. (a)   **(b)**

67.

69.

5.

6. (a) $x + y = 11.1, x - y = 5.7$ **(b)** $(8.4, 2.7)$

Chapter 4 Review (pp. 304–307)

1. 3   **2.** 2   **3. (a)** None   **(b)** Inconsistent
4. (a) One   **(b)** Consistent; independent
5. (a) Infinitely many   **(b)** Consistent; dependent
6. (a) None   **(b)** Inconsistent   **7.** $(1, 2)$
8. $(5, 2)$ **9.** $(4, 3)$ **10.** $(2, -4)$ **11.** $(2, 2)$
12. $(1, 2)$ **13.** $(2, 6)$ **14.** $(1, 1)$
15. $(4, -3)$ **16.** $(1, 2)$ **17.** $(1, 1)$
18. $(1, 2)$ **19.** $(1, 1)$ **20.** $(-2, -1)$
21. $(2, 6)$ **22.** $(2, -10)$ **23.** $(-2, 1)$
24. $(-4, -4)$ **25.** No solutions
26. No solutions   **27.** Infinitely many   **28.** $(1, 1)$
29. $(2, 1)$ **30.** $(-1, 2)$ **31.** $(11, -1)$ **32.** $(1, 0)$
33. $\left(\frac{9}{4}, \frac{7}{4}\right)$ **34.** $\left(\frac{5}{2}, 1\right)$ **35.** $(5, -7)$ **36.** $(8, 2)$
37. $(-2, 3)$ **38.** $(1, 0)$ **39.** $(1, 3)$ **40.** $(2, -1)$
41. Infinitely many   **42.** Infinitely many   **43.** No solutions
44. One   **45.** Yes   **46.** No   **47.** No   **48.** Yes   **49.** $y > 1$
50. $y \le 2x + 1$

51.

52.

53.

54.

55.
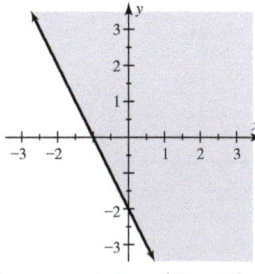

56.

57. Yes   **58.** No   **59.** The region containing $(2, -2)$
60. The region containing $(1, 3)$

61.

62.

63.

64.

65.
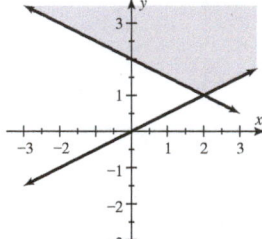

66.

67. 3100 deaths in 1912; 33,790 deaths in 2012
68. Display: $6.8 billion; native: $5.0 billion
69. (a) $C = 0.2x + 40$ **(b)** 250 mi   **(c)** 250 mi
70. 75°, 105°   **71. (a)** $2x + y = 180$, $2x - y = 40$
(b) $(55, 70)$ **(c)** $(55, 70)$ **72.** 20 ft × 24 ft
73. (a) $x + y = 10$, $80x + 120y = 920$; x is $80 rooms, y
is $120 rooms.   **(b)** $(7, 3)$ **74.** 7 lb of $2 candy; 11 lb of
$3 candy   **75.** Bicycle: 35 min; stair climber: 25 min
76. 2 mph   **77. (a)** 16 ft × 24 ft (answers may vary slightly)
(b) 16 ft × 24 ft
78.
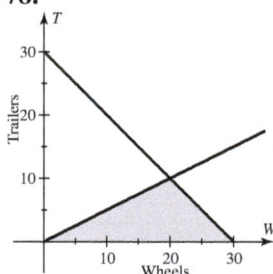

Chapter 4 Test (pp. 307–308)

1. $(1, 2)$ **2.** $(-2, -1)$ **3.** $(-1, -2)$ **4.** $(-2, 3)$
5. $(1, 3)$ **6. (a)** $(2, 1)$; one; consistent   **(b)** No solutions;
zero; inconsistent   **7. (a)** Infinitely many   **(b)** Consistent;
dependent   **8. (a)** None   **(b)** Inconsistent   **9.** $(-3, 4)$
10. No solutions   **11.** Infinitely many solutions
12. $(-1, 3)$ **13.** $y \le -\frac{1}{2}x$ **14.** Yes
15.

16.

17. 4

18.   **19.**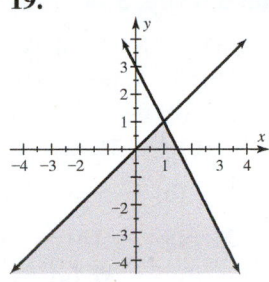

20. 3.4 million in 2012; 6.1 million in 2013
21. 7.5 L of 20% solution and 2.5 L of 60% solution
22. $\frac{2}{3}$ hr at 6 mph; $\frac{1}{3}$ hr at 9 mph

Chapters 1–4 Cumulative Review (pp. 308–309)

1. $2 \cdot 2 \cdot 2 \cdot 3 \cdot 5$ **2. (a)** 4   **(b)** 8   **3. (a)** Rational
(b) Irrational   **4. (a)** $<$ **(b)** $>$ **5. (a)** $4x^2$ **(b)** $5x - 2$
6. $\frac{2}{9}$ **7.** $\frac{1}{2}$ **8.** No solutions   **9.** 11, 12, 13, 14
10. 432 in^2 or 3 ft^2 **11.** $x = \frac{W + 7y}{3}$ **12.** $x \geq 1$
13.

14. (a)   **(b)**

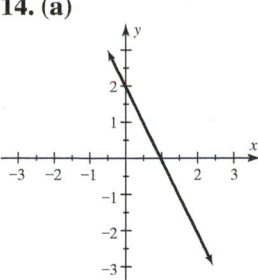

15. $(2, 0)$; $(0, -8)$
16. (a)   **(b)**

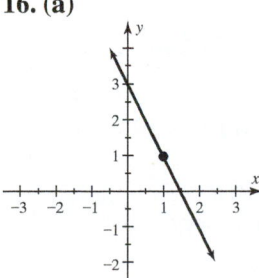

17. (a) $y = -\frac{1}{2}x - 1$ **(b)** $y = 4x - 3$
18. $y = -\frac{1}{4}x + 3$ **19.** $y = -3x + 4$ **20.** $(4, 4)$
21. $(1, -3)$ **22.** No solutions   **23.** Infinitely many solutions
24. $(-1, 2)$

25.   **26.**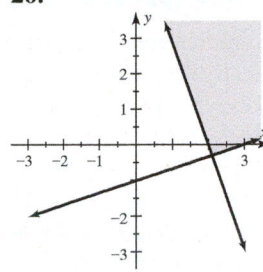

27. $F = 16.5R$ **28.** 94 °C   **29.** $245   **30.** $158.05
31. $G = -3x + 30$ **32.** $1200 at 5%; $1200 at 6%

5 Polynomials and Exponents

Section 5.1 (pp. 318–319)

1. base; exponent   **2.** 1   **3.** a^{m+n} **4.** a^{mn} **5.** $a^n b^n$ **6.** $\frac{a^n}{b^n}$
7. 64   **9.** -8 **11.** -8 **13.** 11   **15.** $\frac{1}{2}$ **17.** 1   **19.** 5   **21.** 1
23. 3^3 or 27   **25.** 4^8 or 65,536   **27.** x^9 **29.** x^6 **31.** $20x^7$
33. $-3x^3y^4$ **35.** 2^6 or 64   **37.** n^{12} **39.** x^7 **41.** $49b^2$
43. a^3b^3 **45.** 1   **47.** $-64b^6$ **49.** $x^{14}y^{21}$ **51.** $\frac{1}{27}$
53. $\frac{a^5}{b^5}$ **55.** $\frac{(x - y)^3}{27}$ **57.** $\frac{25}{(a + b)^2}$ **59.** $x^{12}y^9$ **61.** $a^{10}b^8$
63. $\frac{8x^3}{125}$ **65.** $\frac{27x^6}{125y^{12}}$ **67.** $(x + y)^4$ **69.** $(a + b)^5$ **71.** 6
73. $a^3 + 2ab^2$ **75.** $12a^5 + 6a^3b$ **77.** $x^3y^3 - x^2y^4$
79. $a = 3, b = 1, m = 1, n = 0$ (answers may vary)
81. $10x^4$ **83.** $9\pi x^4$ **85.** $8x^3$ **87.** $1157.63
89. (a) 3^2 **(b)** $72,000

Section 5.2 (pp. 328–330)

1. monomial   **2.** polynomial   **3.** degree   **4.** degree
5. binomial   **6.** trinomial   **7.** like   **8.** like   **9.** opposite
10. vertically   **11.** 2; 3   **13.** 2; -1 **15.** 2; -5 **17.** 0; 6
19. Yes; 1; 1; 1   **21.** Yes; 3; 1; 2   **23.** No   **25.** No
27. Yes; 2; 3; 6   **29.** Yes; x **31.** Yes; $-5x^3$ **33.** No
35. Yes; $2ab$ **37.** $-x + 9$ **39.** $4x^2 + 8x + 1$
41. $4y^3 + 3y^2 + 4y - 9$ **43.** $4xy + 1$ **45.** $2a^2b^3$
47. $9x^2 + x - 6$ **49.** $x^2 - 7x - 1$ **51.** $-5x^2$
53. $-3a^2 + a - 4$ **55.** $2t^2 + 3t - 4$ **57.** $4x - 2$
59. $-3x^2 + 5x + 2$ **61.** $z^3 - 6z^2 - 6z - 1$
63. $3xy + 2x^2y^2$ **65.** $-a^3b$ **67.** $-x^2 - 5x - 4$
69. $-2x^3 - 6x - 2$ **71.** 9, 9   **73.** $-93, 3$
75. (a) 200 bpm   **(b)** 100 bpm   **(c)** It decreases quickly at
first, then more slowly.   **77.** $2z^2$; 200 in^2
79. $2x^2 + x^2$ or $3x^2$; 108 ft^2 **81.** $\pi x^2 + \pi y^2$; 13π ft^2
83. (a) $m_1 \approx 0.077$; $m_2 \approx 0.083$; $m_3 \approx 0.077$. A line is
reasonable but not exact.   **(b)** For the given years, its estimates are reasonable.

Checking Basic Concepts 5.1 & 5.2 (p. 330)

1. (a) -25 **(b)** 1   **2. (a)** 10^8 **(b)** $-12x^7$ **(c)** a^6b^2
(d) $\frac{x^4}{z^{12}}$ **3. (a)** 1   **(b)** $9x^{14}$ **(c)** $10a^5 - 14a^2$
4. 3; 2; 4   **5. (a)** $2w^2h$ **(b)** 2880 in^3 **6. (a)** $3a^2 + 6$
(b) $2z^3 + 7z - 8$ **(c)** $4xy$

Section 5.3 (pp. 336–338)

1. The product rule 2. Distributive 3. term; term
4. vertically 5. x^7 7. $-12a^2$ 9. $20x^5$ 11. $4x^2y^3$
13. $-12x^3y^3$ 15. $3x + 12$ 17. $-45x - 5$ 19. $4z - z^2$
21. $-5y - 3y^2$ 23. $15x^3 - 12x$ 25. $6x^3 - 6x^2$
27. $-32t^2 - 8t - 8$ 29. $-5n^4 + n^3 - 2n^2$
31. $x^2y + xy^2$ 33. $x^4y - x^3y^2$ 35. $-a^4b + 2ab^4$
37. $x^2 + 3x$ 39. $x^2 + 4x + 4$ 41. $x^2 + 9x + 18$
43. $x^2 + 8x + 15$ 45. $x^2 - 17x + 72$
47. $6z^2 - 19z + 10$ 49. $64b^2 - 1$ 51. $10y^2 - 3y - 7$
53. $5 - 13a + 6a^2$ 55. $1 - 9x^2$ 57. $x^3 - x^2 + x - 1$
59. $4x^3 - 3x^2 + 16x - 12$ 61. $2n^3 + n^2 + 6n + 3$
63. $m^3 + 4m^2 + 4m + 1$ 65. $6x^3 - 7x^2 + 14x - 8$
67. $x^3 + 1$ 69. $4b^4 + 3b^3 + 19b^2 + 9b + 21$
71. $x^3 - x^2 - 5x + 2$ 73. $a^3 - 8$
75. $6x^4 - 2x^3 + 5x^2 - x + 1$ 77. $m \cdot n$
79. (a) $h^3 - 2h^2 - 8h$ (b) 14,175 in^3
81. (a) $50x - x^2$ (b) 625 83. $6x^2 + 12x + 6$
85. (a) $64t - 16t^2$ (b) 64; 64 (c) Yes; yes

Section 5.4 (pp. 343–345)

1. $a^2 - b^2$ 2. $a^2 + 2ab + b^2$ 3. $a^2 - 2ab + b^2$
4. $(a + b)^2$ 5. False 6. False 7. $x^2 - 9$ 9. $16x^2 - 1$
11. $1 - 4a^2$ 13. $4x^2 - 9y^2$ 15. $a^2b^2 - 25$ 17. $a^4 - b^4$
19. $x^6 - y^6$ 21. 9999 23. 391 25. 9900
27. $a^2 - 4a + 4$ 29. $4x^2 + 12x + 9$
31. $9b^2 + 30b + 25$ 33. $\frac{9}{16}a^2 - 6a + 16$
35. $1 - 2b + b^2$ 37. $25 + 10y^3 + y^6$
39. $a^4 + 2a^2b + b^2$ 41. $a^3 + 3a^2 + 3a + 1$
43. $x^3 - 6x^2 + 12x - 8$ 45. $8x^3 + 12x^2 + 6x + 1$
47. $216u^3 - 108u^2 + 18u - 1$ 49. $20x + 36$
51. $x^2 + 2x - 35$ 53. $9x^2 - 30x + 25$
55. $25x^2 + 35x + 12$ 57. $16b^2 - 25$
59. $-20x^3 + 35x^2 - 10x$ 61. $64 - 48a + 12a^2 - a^3$
63. $x^3 + 6x^2 + 9x$ 65. $x^4 - 5x^2 + 4$ 67. $a^{2n} - b^{2n}$
69. (a) $x^2 + 4x + 4$ (b) $x^2 + 4x + 4$
71. (a) $4x^2 + 12x + 9$ (b) $4x^2 + 12x + 9$
73. (a) $6x^2 + 60x + 150$ (b) $x^3 + 15x^2 + 75x + 125$
75. (a) $1 + 2x + x^2$ (b) 1.21; the money increases by
1.21 times in 2 years if the interest rate is 10%.
77. (a) $1 - 2x + x^2$ (b) 0.25; if the chance of rain on
each day is 50%, then there is a 25% chance that it will not
rain on either day.
79. (a) $32z + 256$ (b) 2176; the area of an 8-foot-wide
sidewalk around a 60×60 foot pool is 2176 ft^2.

Checking Basic Concepts 5.3 & 5.4 (p. 346)

1. (a) $-15x^3y^5$ (b) $-6x + 4x^2$ (c) $3a^3b - 6a^2b^2 + 3ab^3$
2. (a) $4x^2 + 9x - 9$ (b) $2x^4 - 2$ (c) $x^3 + y^3$
3. (a) $25x^2 - 4$ (b) $x^2 + 6x + 9$ (c) $4 - 28x + 49x^2$
(d) $t^3 + 6t^2 + 12t + 8$ 4. (a) $m^2 + 10m + 25$
(b) $m^2 + 10m + 25$

Section 5.5 (pp. 354–356)

1. $\frac{1}{a^n}$ 2. a^n 3. a^{m-n} 4. $\frac{b^m}{a^n}$ 5. $\left(\frac{b}{a}\right)^n$ 6. $1 \le b < 10$
7. (a) $\frac{1}{4}$ (b) 9 9. (a) 2 (b) $\frac{1}{1000}$ 11. (a) $\frac{1}{81}$ (b) $\frac{1}{8}$
13. (a) $\frac{1}{576}$ (b) 64 15. (a) 1 (b) 64 17. (a) 25 (b) $\frac{49}{4}$
19. (a) $\frac{1}{x}$ (b) $\frac{1}{a^4}$ 21. (a) $\frac{1}{x^2}$ (b) $\frac{1}{a^8}$ 23. (a) $\frac{y^3}{x^3}$ (b) $\frac{1}{x^3y^3}$
25. (a) $\frac{1}{16t^4}$ (b) $\frac{1}{(x+1)^7}$ 27. (a) a^8 (b) $\frac{1}{r^2t^6}$ 29. (a) $\frac{b^2}{a^4}$
(b) x^2 31. (a) a^{13} (b) $\frac{2}{z^3}$ 33. (a) $-\frac{2y^3}{3x^2}$ (b) $\frac{1}{x^3}$
35. (a) $2b$ (b) $\frac{a^3}{b^3}$ 37. (a) $\frac{x^7}{3y^8}$ (b) $\frac{4a^5}{b^6}$ 39. (a) y^5 (b) $2t^3$
41. (a) $\frac{3a^{10}}{8}$ (b) $\frac{1}{32b^9}$ 43. (a) x^2y^2 (b) a^6b^3
45. (a) $\frac{n^4}{72m^{11}}$ (b) $\frac{16x^{11}}{y^{13}}$ 47. (a) $\frac{b^2}{a^2}$ (b) $\frac{4v}{u}$
49. (a) $\frac{4}{9a^6b^6}$ (b) $\frac{16m^{14}}{25n^2}$ 51. $\frac{a^n}{a^m} = a^{n-m} = a^{-(m-n)} = \frac{1}{a^{m-n}}$
53. Thousand 55. Billion 57. Hundredth 59. -0.002
61. 45,000 63. 0.008 65. 0.000456 67. 39,000,000
69. $-500,000$ 71. 2×10^3 73. -5.67×10^2
75. 1.2×10^7 77. 4×10^{-3} 79. 8.95×10^{-4}
81. -5×10^{-2} 83. 1,500,000 85. -1.5 87. 2000
89. 0.002 91. (a) About 5.859×10^{12} mi
(b) About 2.5×10^{13} mi 93. About 9.2×10^9 mi
95. 1×10^{100} 97. (a) 1.5684×10^{13} (b) About \$49,476

Section 5.6 (pp. 362–363)

1. $\frac{a}{d} + \frac{b}{d}$ 2. $\frac{a}{d} + \frac{b}{d} - \frac{c}{d}$ 3. term 4. False 5. False
6. 9; 4; 1 7. $x + 1$; $2x^2 - 2x + 1$; 4 8. $0x^2$ 9. $2x$
11. $z^3 + z^2$ 13. $\frac{a^2}{2} - 3$ 15. $\frac{1}{3y^2} + \frac{2}{y}$ 17. $\frac{4}{x} - 7x^2$
19. $\frac{2}{y} + \frac{1}{y^2}$ 21. $3x^3 - 1 + \frac{2}{x}$ 23. $4y^2 - 1 + \frac{2}{y}$
25. $3m^2 - 2m + 4$ 27. $2x + 1 + \frac{3}{x-2}$ 29. $x + 1$
31. $x^2 + 1 + \frac{-1}{x-1}$ 33. $x^2 + 3x - 1$
35. $x^2 - x + 2 + \frac{1}{4x+1}$ 37. $x^2 + 2x + 3 + \frac{8}{x-2}$
39. $3x^2 + 3x + 3 + \frac{5}{x-1}$ 41. $x + 3 + \frac{-x-2}{x^2+1}$
43. $x + 1$ 45. $x^2 - 2x + 4$ 47. They are the same.
49. $4x$
51. $x + 2$;

Checking Basic Concepts 5.5 & 5.6 (p. 363)

1. (a) $\frac{1}{81}$ (b) $\frac{1}{2x^7}$ (c) $\frac{b^8}{16a^2}$ 2. (a) z^5 (b) $\frac{y^6}{x^3}$ (c) $\frac{x^6}{27}$
3. (a) 4.5×10^4 (b) 2.34×10^{-4} (c) 1×10^{-2}
4. (a) 47,100 (b) 0.006 5. $5a - 3$
6. $3x + 2$; R: -2 7. $x^2 + 2x + 1$; R: $x + 1$
8. (a) 9.3×10^7 (b) 500 sec (8 min 20 sec)

Chapter 5 Review (pp. 367–369)

1. 125 2. -81 3. 4 4. 11 5. -5 6. 1 7. 6^5
8. 10^{12} 9. z^9 10. y^6 11. $30x^9$ 12. a^4b^4 13. 2^{10}
14. m^{20} 15. a^3b^3 16. x^8y^{12} 17. x^7y^{11} 18. 1
19. $(r - t)^9$ 20. $(a + b)^6$ 21. $\frac{9}{(x-y)^2}$ 22. $\frac{(x+y)^3}{8}$
23. $6x^3 - 10x^2$ 24. $12x^2 + 3x^4$ 25. 7; 6 26. 5; -1
27. Yes; 1; 1; 1 28. Yes; 4; 1; 3 29. Yes; 3; 2; 2
30. No 31. $5x^2 - x + 3$ 32. $-6x^2 + 3x + 7$
33. $3x + 4$ 34. $-2x^2 - 13$ 35. $-2x^2 + 9x + 5$

36. $3x^2 - 2x - 6$ **37.** $2a^3 - a^2 + 7a$ **38.** $-2x + 12$
39. $5y^2 - 3xy$ **40.** $4xy$ **41.** $-x^5$ **42.** $-r^3t^4$
43. $-6t + 15$ **44.** $2y - 12y^2$ **45.** $18x^5 + 30x^4$
46. $-x^3 + 2x^2 - 9x$ **47.** $-a^3b + 2a^2b^2 - ab^3$
48. $a^2 + 3a - 10$ **49.** $8x^2 + 13x - 6$
50. $-2x^2 + 3x - 1$ **51.** $2y^3 + y^2 + 2y + 1$
52. $2y^4 - y^2 - 1$ **53.** $z^3 + 1$ **54.** $4z^3 - 15z^2 + 13z - 3$
55. $z^2 + z$ **56.** $2x^2 + 4x$ **57.** $z^2 - 4$ **58.** $25z^2 - 81$
59. $1 - 9y^2$ **60.** $25x^2 - 16y^2$ **61.** $r^2t^2 - 1$
62. $4m^4 - n^4$ **63.** $x^2 + 2x + 1$ **64.** $16x^2 + 24x + 9$
65. $y^2 - 6y + 9$ **66.** $4y^2 - 20y + 25$
67. $16 + 8a + a^2$ **68.** $16 - 8a + a^2$
69. $x^4 + 2x^2y^2 + y^4$ **70.** $x^2y^2 - 4xy + 4$
71. $z^3 + 15z^2 + 75z + 125$ **72.** $8z^3 - 12z^2 + 6z - 1$
73. 3599 **74.** 396 **75.** $\frac{1}{9}$ **76.** $\frac{1}{9}$ **77.** 4 **78.** $\frac{1}{1000}$
79. 36 **80.** $\frac{1}{25}$ **81.** $\frac{9}{16}$ **82.** 1 **83.** $\frac{1}{z^2}$ **84.** $\frac{1}{y^4}$ **85.** $\frac{1}{a^2}$
86. $\frac{1}{x^3}$ **87.** $\frac{1}{4t^2}$ **88.** $\frac{1}{a^3b^6}$ **89.** $\frac{1}{y^3}$ **90.** x^4 **91.** $\frac{2}{x^3}$ **92.** $\frac{2x^4}{3y^3}$
93. $\frac{a^5}{b^5}$ **94.** $4t^4$ **95.** $\frac{n^7}{72m^{12}}$ **96.** $\frac{9x^{10}}{y^{10}}$ **97.** $\frac{y^2}{x^2}$ **98.** $\frac{2v}{3u}$
99. 600 **100.** $52{,}400$ **101.** -0.0037 **102.** -0.06234
103. 1×10^4 **104.** 5.61×10^7 **105.** 5.4×10^{-5}
106. 1×10^{-3} **107.** $24{,}000{,}000$ **108.** 0.2 **109.** $\frac{5x}{3} + 1$
110. $3b^2 - 2 + \frac{1}{b^2}$ **111.** $3x + 2 + \frac{4}{x-1}$
112. $3x - 4 + \frac{6}{3x+2}$ **113.** $x^2 - 3x - 1$
114. $x^2 + \frac{-1}{2x-1}$ **115.** $x - 1 + \frac{-2x+2}{x^2+1}$
116. $x^2 + 2x + 5$ **117. (a)** 60 bpm   **(b)** 110 bpm
(c) It increases.   **118.** $6xy$; 72 ft^2 **119.** $10z^2 + 25z$; 510 in^2
120. x^4y^2 **121.** 2.64×10^9 hours
122. $\frac{4}{3}\pi x^3 + 8\pi x^2 + 16\pi x + \frac{32}{3}\pi$ **123. (a)** $96t - 16t^2$
(b) 128; after 2 seconds the ball is 128 ft high.
124. (a) $600L - L^2$ **(b)** $27{,}500$; a rectangular building
with a perimeter of 1200 ft and a side of length 50 ft has an
area of 27,500 ft^2.
125. (a) $x^2 + 10x + 25$ **(b)** $x^2 + 10x + 25$
126. (a) $16x$ **(b)** 1600

Chapter 5 Test (p. 370)

1. (a) -1 **(b)** -81 **2. (a)** -6 **(b)** $\frac{1}{64}$ **(c)** 8 **(d)** -3
3. $3; 2; 3$ **4.** $x^3 - 4x + 8$ **5.** $4x + 6$ **6.** $-3y^3 - 2y + 1$
7. $x^2 + x - 7$ **8.** $4a^3 + 2ab$ **9.** $24y^{11}$ **10.** a^5b^8 **11.** x^4
12. $\frac{a^3}{b^6}$ **13.** $a^3b - ab$ **14.** $\frac{4}{9a^4b^6}$ **15.** $\frac{2y^3}{x}$ **16.** $\frac{16}{(a+b)^4}$
17. $12x^5 - 18x^3 + 3x^2$ **18.** $2z^2 - 2z - 12$
19. $49y^4 - 9$ **20.** $9x^2 - 12x + 4$
21. $m^3 + 9m^2 + 27m + 27$ **22.** $y^3 - y + 6$ **23.** 6396
24. 0.0061 **25.** 5.41×10^3 **26.** $3x - 2 + \frac{1}{x}$
27. $x^2 - x + 1 + \frac{-1}{x+2}$ **28. (a)** $20t$ **(b)** $2t + 2000$
(c) $18t - 2000$; profit from selling t tickets
29. $12x^2$; 1200 ft^2 **30.** $3x^3 + 27x^2 + 54x$
31. (a) $88t - 16t^2$ **(b)** 120; after 3 seconds, the ball is
120 ft high.

Chapters 1–5 Cumulative Review (p. 371)

1. (a) 8 **(b)** 8 **2. (a)** 29 **(b)** $\frac{3}{2}$ **3. (a)** 7 **(b)** No solutions
4. (a) Infinitely many solutions   **(b)** -2 **5.** 68 mph
6. (a) $\frac{21}{50}$ **(b)** $\frac{19}{250}$

7.   **8.**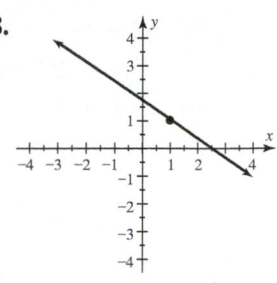

9. $y = -2x + 1$ **10.** $(2, 0)$; $(0, -3)$ **11.** $y = \frac{1}{2}x - 4$
12. $y = 3x + 1$ **13.** No solutions   **14.** $(2, -1)$
15. Infinitely many solutions   **16.** $(0, -6)$

17.   **18.**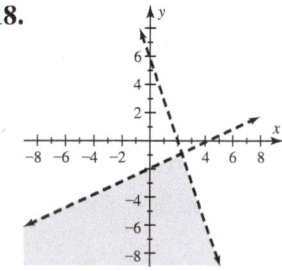

19. (a) $15x^5$ **(b)** $x^{10}y^7$ **20. (a)** $2x^2 - x + 3$
(b) $12a^3 + a - 5$ **21. (a)** $2x^2 - 11x - 21$
(b) $y^3 - 10y - 3$ **(c)** $16x^2 - 49$ **(d)** $25a^2 + 30a + 9$
22. (a) $\frac{1}{x}$ **(b)** $\frac{x^9}{8}$ **(c)** $\frac{x^4}{2y^2}$ **(d)** $\frac{x^7}{y^8}$ **23.** 2.4×10^{10}
24. 0.000000471 **25. (a)** $4x^2 - 1$ **(b)** $2x - 5 + \frac{1}{x+3}$
26. -25% **27.** 200 mL   **28. (a)** $2x^2 + 10x$
(b) $x^2 + 7x + 10$ **(c)** $2x^2 + 4x$ **(d)** $10x^2 + 42x + 20$

6 Factoring Polynomials and Solving Equations

Section 6.1 (pp. 380–382)

1. factor   **2.** a **3.** multiplying   **4.** prime; multiplication
5. greatest common factor (GCF)   **6.** grouping
7. $1, 2, x, 2x$ **8.** $2x$
9. $2(x + 2)$;

11. $z(z + 4)$;

13. $3y(y + 4)$;

15. $3x(x + 3)$ **17.** $2y^2(2y - 1)$ **19.** $2z(z^2 + 4z - 2)$
21. $3xy(2x - y)$ **23.** $6x$; $6x(1 - 3x)$
25. $4y^2$; $4y^2(2y - 3)$ **27.** $3z$; $3z(2z^2 + z + 3)$
29. x^2; $x^2(x^2 - 5x - 4)$

31. $5y^2$; $5y^2(y^3 + 2y^2 - 3y + 2)$ **33.** x; $x(y + z)$

35. ab; $ab(b - a)$ **37.** $5x^2y^3$; $5x^2y^3(y + 2x)$

39. ab; $ab(a + b + 1)$ **41.** $(x - 2)(x + 1)$

43. $(z + 4)(z + 5)$ **45.** $2x(2x^2 - 1)(x - 5)$

47. $(x^2 + 3)(x + 2)$ **49.** $(y^2 + 1)(2y + 1)$

51. $(5z^2 + 2)(3z - 1)$ **53.** $(4t^2 + 3)(t - 5)$

55. $(3r^2 - 2)(3r + 2)$ **57.** $(7x^2 - 2)(x + 3)$

59. $(y^2 - 2)(2y - 7)$ **61.** $(z^2 - 7)(z - 4)$

63. $(x^3 + 2)(2x - 3)$ **65.** $(x + y)(a + b)$

67. $3(x^2 + 1)(x + 2)$ **69.** $2y(3y^2 - 1)(y - 4)$

71. $x^2(x^2 - 3)(x + 2)$ **73.** $2x^2(x^2 - 3)(2x + 1)$

75. $xy(x + y)(x - 2)$ **77.** $2x^2y^2(x + 2)(y - x)$

79. $a\left(x^2 + \frac{b}{a}x + \frac{c}{a}\right)$ **81. (a)** $16t$ **(b)** $16t(5 - t)$

83. $2x^2 - 4x$; $2x(x - 2)$ **85.** $8y^2 - xy$; $y(8y - x)$

87. (a) 168 in^3 **(b)** $4x(x^2 - 15x + 50)$

Section 6.2 (pp. 388–390)

1. 1   **2.** $x^2 + (m + n)x + mn$; $m + n$; mn

3. mn; $m + n$ **4.** prime   **5.** $x^2 + 7x + 10$

7. $x^2 - 9x + 18$ **9.** $x^2 + x - 56$ **11.** 4, 7   **13.** 3, −10

15. −5, 10   **17.** −7, −4   **19.** $(x + 1)(x + 2)$

21. $(y + 2)(y + 2)$ **23.** Prime   **25.** $(x + 3)(x + 5)$

27. $(m + 4)(m + 9)$ **29.** $(n + 10)(n + 10)$

31. $(x - 1)(x - 5)$ **33.** $(y - 3)(y - 4)$

35. $(z - 5)(z - 8)$ **37.** $(a - 7)(a - 9)$ **39.** Prime

41. $(b - 5)(b - 25)$ **43.** $(x - 5)(x + 18)$

45. $(m - 5)(m + 9)$ **47.** Prime

49. $(n - 10)(n + 20)$ **51.** $(x - 1)(x + 23)$

53. $(a - 4)(a + 8)$ **55.** $(b + 4)(b - 5)$ **57.** Prime

59. $(x + 8)(x - 9)$ **61.** $(y + 2)(y - 17)$

63. $(z + 6)(z - 11)$ **65.** $5(x - 4)(x + 2)$

67. $-3(m + 4)(m - 1)$ **69.** $y(y - 2)(y - 5)$

71. $-x(x + 5)(x - 3)$ **73.** $3a(a + 1)(a + 6)$

75. $-2x(x^2 - 3x + 4)$ **77.** $2m^2(m - 7)(m + 2)$

79. $-3x^2(x + 1)(x - 2)$ **81.** $(x + 1)(x + 5)$

83. $(x - 1)(x - 3)$ **85.** $(6 - x)(2 + x)$

87. $(8 + x)(4 - x)$ **89.** $(x + 1)(x + k)$

91. $x + 1$;   **93.** $x + 1$, $x + 2$;

x	x^2	x
1	x	1
	x	1

x	x^2	$2x$
1	x	2
	x	2

95. $x + 1$ **97.** $(x + 2)(x + 6)$ **99.** $(x + 3)(x + 2)$

101. $(x - 5)(x - 6)$ **103.** $(x + 6)(x - 2)$

Checking Basic Concepts 6.1 & 6.2 (p. 390)

1. $4x$ **2.** $6z^2(2z - 3)$ **3. (a)** $(6y + 5)(y - 2)$

(b) $(x^2 + 5)(2x + 1)$ **(c)** $4(z^2 + 1)(z - 3)$

4. (a) $(x + 2)(x + 4)$ **(b)** $(x - 7)(x + 6)$ **(c)** Prime

(d) $4a(a + 2)(a + 3)$ **5.** $(x + 5)(x + 5)$

Section 6.3 (pp. 397–398)

1. ac; b **2.** ax^2; c **3.** +; +   **4.** +; −   **5.** −; −

6. −; +   **7.** 3; x **8.** x; $4x$ **9.** 1; 3   **10.** 1; $2x$

11. $3x$; 1   **12.** $6x$; x **13.** 5; 3   **14.** $6x$; 8

15. $(x + 3)(2x + 1)$ **17.** Prime   **19.** $(x + 1)(3x + 1)$

21. $(2x + 3)(3x + 1)$ **23.** $(x - 2)(5x - 1)$

25. $(y - 1)(2y - 5)$ **27.** Prime   **29.** $(z - 5)(7z - 2)$

31. $(t - 3)(3t + 2)$ **33.** $(3r + 2)(5r - 3)$

35. $(3m - 4)(8m + 3)$ **37.** $(5x - 1)(5x + 2)$

39. $(x + 2)(6x - 1)$ **41.** Prime

43. $(3n + 1)(7n - 1)$ **45.** $(2y + 3)(7y + 1)$

47. $(4z - 3)(7z - 1)$ **49.** $(3x - 2)(10x - 3)$

51. Prime   **53.** $(2t + 3)(9t - 2)$

55. $3(2a - 1)(2a + 3)$ **57.** $z(2z + 3)(5z + 2)$

59. $3x(4x - 3)(2x - 1)$ **61.** $2x^2(4x^2 - 3x + 1)$

63. $7x^2(2x + 1)(2x + 3)$ **65.** $(3x + 1)(x + k)$

67. $(7x + 1)(x + 2)$ **69.** $(2x - 1)(x - 2)$

71. $-(4x + 3)(2x - 1)$ **73.** $-(x + 5)(2x - 3)$

75. $-(x - 3)(5x + 1)$

77. $3x + 2$ by $2x + 1$;

$2x$	$6x^2$	$4x$
1	$3x$	2
	$3x$	2

79. $(2x + 1)(x + 3)$

Section 6.4 (pp. 404–405)

1. $(a - b)(a + b)$ **2.** $6x$; $7y$ **3.** False   **4.** False

5. $(a + b)^2$ **6.** $(a - b)^2$ **7.** $6x$ **8.** $20rt$

9. $(a + b)(a^2 - ab + b^2)$ **10.** $(a - b)(a^2 + ab + b^2)$

11. $2x$; $3y$ **12.** x; 1   **13.** −; +   **14.** +; −

15. $(x - 1)(x + 1)$ **17.** $(z - 10)(z + 10)$

19. $(2y - 1)(2y + 1)$ **21.** $(6z - 5)(6z + 5)$

23. $(3 - x)(3 + x)$ **25.** $(1 - 3y)(1 + 3y)$

27. $(2a - 3b)(2a + 3b)$ **29.** $(6m - 5n)(6m + 5n)$

31. $(9r - 7t)(9r + 7t)$ **33.** $(x + 4)^2$ **35.** Not possible

37. $(x - 3)^2$ **39.** $(3y + 1)^2$ **41.** $(2z - 1)^2$

43. Not possible   **45.** $(3x + 5)^2$ **47.** $(2a - 9)^2$

49. $(x + y)^2$ **51.** $(r - 5t)^2$ **53.** Not possible

55. $(z + 1)(z^2 - z + 1)$ **57.** $(x + 4)(x^2 - 4x + 16)$

59. $(y - 2)(y^2 + 2y + 4)$ **61.** $(n - 1)(n^2 + n + 1)$

63. $(2x + 1)(4x^2 - 2x + 1)$

65. $(m - 4n)(m^2 + 4mn + 16n^2)$

67. $(2x + 5y)(4x^2 - 10xy + 25y^2)$

69. $4(x - 2)(x + 2)$ **71.** $2(y - 7)^2$

73. $5(z + 2)(z^2 - 2z + 4)$ **75.** $xy(x - y)(x + y)$

77. $2m(m^2 - 5m + 9)$

79. $7x^2(10x - 3y)(10x + 3y)$

81. $2(2a + b)(4a^2 - 2ab + b^2)$ **83.** $4b^2(b + 3)^2$

85. $4(5r - 2t)(25r^2 + 10rt + 4t^2)$

87. $2x + 3$;

$2x$	$4x^2$	$6x$
3	$6x$	9

$2x$ 3

89. $(x - 5)(x + 5)$

Checking Basic Concepts 6.3 & 6.4 (p. 405)

1. (a) $(x - 4)(2x + 3)$ (b) $(2x + 7)(3x - 2)$

2. (a) Prime   (b) $2y(3y + 1)(y - 2)$

3. $(3x + 2)(x + 3)$ **4.** (a) $(z - 8)(z + 8)$

(b) $(3r - 2t)(3r + 2t)$ **5.** (a) $(x + 6)^2$

(b) $(3a - 2b)^2$ **6.** (a) $(m - 3)(m^2 + 3m + 9)$

(b) $(5n + 3)(25n^2 - 15n + 9)$

7. (a) $4(2x - 1)(2x + 1)$ (b) $3y(y + 2)(y^2 - 2y + 4)$

Section 6.5 (pp. 410–411)

1. Greatest common factor   **2.** GCF   **3.** Grouping
4. squares; cubes; cubes   **5.** No; a sum of squares cannot be
factored.   **6.** Yes; a sum of cubes can be factored.
7. square; grouping; FOIL   **8.** completely   **9.** $2(2x - 1)$

11. $2(y^2 - 2y + 2)$ **13.** $(z - 2)(z + 2)$

15. $(a + 2)(a^2 - 2a + 4)$ **17.** $(2b - 3)^2$

19. Not possible   **21.** $(x^2 + 5)(x - 1)$

23. $(y - 4)(y - 1)$ **25.** $(x - 3)(x + 3)(x + 4)$

27. $8(a - 2)(a^2 + 2a + 4)$

29. $2x(3x^2 + 1)(2x - 3)$

31. $2t(3t + 2)(9t^2 - 6t + 4)$

33. $2(r - 1)(r + 1)(r + 3)$ **35.** $3z^2(2z + 3)(z - 5)$

37. $2b^2(3b - 1)(2b - 1)$ **39.** $6z(1 - 2z)(1 + 2z)$

41. $3y(x - 5)^2$ **43.** $(3m - 2n)(9m^2 + 6mn + 4n^2)$

45. $3(x - 2)(x + 2)(x - 1)(x^2 + x + 1)$

47. $(5a + 3)(a - 6)$ **49.** $3r(t + 5)(t + 6)$

51. $(3b^2 + 4)(3b + 2)$ **53.** $2n(3n^2 + n - 5)$

55. $4(x - 3y)(x + 3y)$ **57.** $2a(a - 4)^2$

59. $4x(2y + 1)(4y^2 - 2y + 1)$

61. $2b(2b - 1)(2b + 1)(b + 3)$ **63.** $3x + 1$

Section 6.6 (pp. 417–419)

1. 0; 0   **2.** No; one side of the equation must be zero.
3. $2x = 0$; $x + 6 = 0$ **4.** Subtract $4x$ from each
side.   **5.** Apply the zero-product property.   **6.** solving
7. zero   **8.** 2   **9.** $ax^2 + bx + c = 0$ with $a \neq 0$
10. $x^2 - 6x + 1 = 0$ **11.** descending   **12.** 0   **13.** 0

15. $-8, 0$ **17.** 1, 2   **19.** $-\frac{5}{6}, 7$ **21.** $\frac{1}{3}, \frac{3}{7}$ **23.** 0, 5, 8

25. 0, 1   **27.** 0, 5   **29.** $-\frac{3}{2}, 0$ **31.** $-1, 1$ **33.** $-\frac{1}{2}, \frac{1}{2}$

35. $-2, -1$ **37.** 5, 7   **39.** $-2, \frac{1}{2}$ **41.** $-\frac{5}{2}, -\frac{2}{3}$

43. $-5, 5$ **45.** 0, 5   **47.** $-3, 0$ **49.** $-1, 6$

51. $-\frac{7}{2}, 2$ **53.** $-2, 1$ **55.** $-3, \frac{1}{2}$ **57.** $-\frac{1}{2}$

59. $-1, -\frac{2}{3}$ **61.** 12 ft   **63.** 4   **65.** 4

67. (a) 6 sec

(b)

Time (t)	0	1	2	3	4	5	6
Height (h)	0	80	128	144	128	80	0

; 3 sec

69. (a) 81.8 ft; 327.3 ft; when the speed doubles,
the braking distance quadruples.   (b) About 19 mph
(c) About 19 mph; yes   **71.** 40 by 50 pixels

Checking Basic Concepts 6.5 & 6.6 (p. 419)

1. (a) $9(a^2 - 2a + 3)$ (b) $7x(y^2 + 4)$

2. (a) $2z^2(3z - 2)(z - 4)$ (b) $2r^2(t - 3)(t + 3)$

3. (a) $4x(3x - 2)^2$ (b) $3(2b - 3)(4b^2 + 6b + 9)$

4. (a) $0, \frac{3}{2}$ (b) $-1, \frac{3}{5}$ **5.** $-3, 1$ **6.** After $\frac{11}{2}$ sec

Section 6.7 (pp. 423–425)

1. GCF   **2.** zero-product   **3.** factors

4. $(x^2 - y^2)^2$ **5.** $(z^2 + 1)(z^2 + 2)$

6. No; $x^4 - 1 = (x - 1)(x + 1)(x^2 + 1)$

7. Yes; $x^2 + 1$ cannot be factored further.

8. Subtract x from each side.   **9.** Factor out x^2.   **10.** One

11. $5(x - 3)(x + 2)$ **13.** $-4(y + 2)(y + 6)$

15. $-10(z + 5)(2z + 1)$ **17.** $-4(t + 3)(7t - 5)$

19. $r(r - 1)(r + 1)$ **21.** $3x(x - 2)(x + 3)$

23. $12z(2z - 1)(3z + 2)$ **25.** $x^2(x - 2)(x + 2)$

27. $t^2(t - 1)(t + 2)$ **29.** $(x^2 - 3)(x^2 - 2)$

31. $(x^2 + 3)(2x^2 + 1)$ **33.** $(y^2 + 3)^2$

35. $(x^2 - 3)(x^2 + 3)$ **37.** $(x - 3)(x + 3)(x^2 + 9)$

39. $z^3(z + 1)^2$ **41.** $(x + y)(2x - y)$

43. $(a + b)^2(a - b)^2$ **45.** $x(x + y)(x - y)$

47. $x(2x + y)^2$ **49.** (a) $x(x - 2)(x + 2)$ (b) $-2, 0, 2$

51. (a) $2y(y - 6)(y + 3)$ (b) $-3, 0, 6$

53. (a) $(x^2 + 4)(x - 1)$ (b) 1   **55.** $-8, -3$

57. $-1, 3$ **59.** $-1, 0, 4$ **61.** $-6, 0, 4$ **63.** $-6, 0, 6$

65. $-7, 0, 1$ **67.** $-3, -2, 2, 3$ **69.** $-1, 1$

71. $-3, 0, 3$ **73.** $-1, 1, 2$ **75.** 5

77. (a) $x < 7.5$ in. because the width is 15 in.
(b) $300 - 4x^2$ **(c)** 2.5 in.   **79.** 8   **81.** 18.6 trillion ft^3
83. Factoring is very difficult (answers may vary).

Checking Basic Concepts 6.7 (p. 425)

1. (a) $3(x - 4)(x + 2)$ **(b)** $-5(2y + 1)(y - 1)$
2. (a) $(z^2 - 5)(z^2 + 5)$ **(b)** $7(t - 1)(t + 1)(t^2 + 1)$
3. (a) $(x - 2)^2(x + 2)^2$ **(b)** $y(y + 10)(2y - 3)$
4. $-4, 0, 3$ **5.** 3

Chapter 6 Review (pp. 428–430)

1. $5; 5(x - 3)$ **2.** $y; y(y + 2)$ **3.** $4z^2; 4z^2(2z - 1)$
4. $3x^2; 3x^2(2x^2 + x - 4)$ **5.** $3y; 3y(3x + 5z^2)$
6. $a^2b^2; a^2b^2(b + a)$ **7.** $(x - 3)(x + 2)$
8. $y(y + 3)(x - 5)$ **9.** $(z^2 + 5)(z - 2)$
10. $(t^2 + 8)(t + 1)$ **11.** $(x^2 + 6)(x - 3)$
12. $(x - y)(a + b)$ **13.** $x^2(x^2 - 2)(x + 3)$
14. $2y(y^2 + 1)(y + 3)$ **15.** $4, 5$ **16.** $-3, 7$
17. $-9, -4$ **18.** $-25, 4$ **19.** $(x - 4)(x + 3)$
20. $(x + 4)(x + 6)$ **21.** $(x - 2)(x + 8)$
22. $(x - 7)(x + 6)$ **23.** Prime   **24.** Prime
25. $(x - 1)(x + 3)$ **26.** $(x + 10)(x + 12)$
27. $2x(x - 2)(x + 5)$ **28.** $x^2(x + 4)(x - 7)$
29. $(2 - x)(5 - x)$ **30.** $(6 - x)(4 + x)$
31. $-2(x + 5)(x - 3)$ **32.** $-x(x + 10)(x - 1)$
33. $(3x - 1)(3x + 2)$ **34.** $(x - 1)(2x + 5)$
35. $(x + 3)(3x + 5)$ **36.** $(5x - 1)(7x + 1)$
37. Prime   **38.** Prime   **39.** $(3x + 1)(8x - 5)$
40. $(x + 9)(4x - 3)$ **41.** $3x(2x + 7)(2x + 1)$
42. $2x^2(x + 3)(4x - 5)$ **43.** $(3 - 2x)(4 + x)$
44. $(1 - 2x)(1 + 5x)$ **45.** $(z - 2)(z + 2)$
46. $(3z - 8)(3z + 8)$ **47.** $(6 - y)(6 + y)$
48. $(10a - 9b)(10a + 9b)$ **49.** $(x + 7)^2$
50. $(x - 5)^2$ **51.** $(2x - 3)^2$ **52.** $(3x + 8)^2$
53. $(2t - 1)(4t^2 + 2t + 1)$
54. $(3r + 2t)(9r^2 - 6rt + 4t^2)$
55. $2x(x - 5)(x + 5)$ **56.** $3(2x + 3)(4x^2 - 6x + 9)$
57. $2x(x + 7)^2$ **58.** $2x(x - 4)(x^2 + 4x + 16)$
59. $4(3x - 2)$ **60.** $3x^2(2x + 3)$
61. $3(3y^2 - 2y + 2)$ **62.** $y(z - 3)(z + 3)$
63. $x(x - 2)(x + 2)(x + 7)$ **64.** $3x(2x + 3)^2$
65. $3a(b - 2)(b^2 + 2b + 4)$ **66.** $5x(x^2 + 4)$

67. $6x(2x - y)(2x + y)$ **68.** $y(x + 3)(x^2 - 3x + 9)$
69. $m = 0$ or $n = 0$ **70.** 0   **71.** $-9, \frac{3}{4}$ **72.** $-\frac{6}{5}, \frac{1}{4}$
73. $0, 1, 2$ **74.** $0, 7$ **75.** $-8, 8$ **76.** $-7, -2$
77. $-2, 3$ **78.** $-\frac{3}{2}, \frac{2}{5}$ **79.** $-4, 18$ **80.** $-2, \frac{5}{2}$
81. $5(x - 5)(x + 2)$ **82.** $-3(x - 3)(x + 5)$
83. $y(y - 2)(y + 2)$ **84.** $3y(y - 1)(y + 3)$
85. $2z^2(z + 2)(z + 5)$ **86.** $8z^2(z - 2)(z + 2)$
87. $(x^2 - 3)^2$ **88.** $(x - 3)(x + 3)(2x^2 + 3)$
89. $(a + 5b)^2$ **90.** $x(x + y)(x - y)$ **91.** $-\frac{1}{2}, 5$
92. $0, \frac{5}{2}, 3$ **93.** $-5, 0, 5$ **94.** $0, 3, 4$ **95.** $-2, 2$
96. $-4, 4$ **97.** -4 **98.** $-1, 1$
99. $3x + 7$;   **100.** $x + 1$ by $x + 5$;

$3x$	$9x^2$	$21x$
7	$21x$	49
	$3x$	7

x	x^2	$5x$
1	x	5
	x	5

101. $x + 1$ **102.** $(x + 1)(x + 3)$
103. $(2x + 3)(x + 6)$ **104.** 2   **105.** 8 ft by 15 ft
106. After 2 sec and 3 sec   **107. (a)** 390 ft   **(b)** 15 mph
(c) 15 mph; yes   **108. (a)** \$10,000   **(b)** \$50 or \$150
(c) \$50 or \$150; yes   **109. (a)** $2000 - 4x^2$ **(b)** 5 in.
110. $x + 2$

Chapter 6 Test (pp. 430–431)

1. $4xy; 4xy(x - 5y + 3)$ **2.** $3a^2b^2; 3a^2b^2(3a + 1)$
3. $(y + z)(a + b)$ **4.** $(x^2 - 5)(3x + 1)$
5. $(y - 2)(y + 6)$ **6.** $(2x + 5)^2$ **7.** $(z - 4)(4z - 3)$
8. $(t - 7)(2t - 3)$ **9.** Prime   **10.** Prime
11. $3x(x + 1)(2x - 1)$ **12.** $2(z - 3)(z + 3)(z^2 + 3)$
13. $4y(3y - 5)(3y + 5)$
14. $7x(x + 2)(x^2 - 2x + 4)$ **15.** $a^2(4a + 3)^2$
16. $2(b - 2)(b + 2)(b^2 + 4)$ **17.** $-4, 4$ **18.** $-4, 5$
19. $\frac{4}{3}$ **20.** $-6, 11$ **21.** $-3, 0, 3$ **22.** $-2, -1, 1, 2$
23. $3x + 5$ **24.** $(x + 2)(x + 3)$ **25. (a)** 275 ft
(b) 33 mph   **26.** 1 sec and 2 sec

Chapters 1–6 Cumulative Review (pp. 431–432)

1. $\frac{3}{7}$ **2.** $\frac{7}{10}$ **3.** 17   **4.** $-\frac{7}{2}$
5. 1;

x	-2	-1	0	1	2
$2x + 3$	-1	1	3	5	7

6. $3n - 5 = n - 7; -1$ **7.** $\frac{57}{1000}$; 0.057
8. $W = \frac{P - 2L}{2}$ **9.** $z > 2$

10.

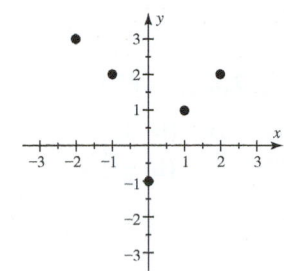

11. x-int: $\left(\frac{2}{3}, 0\right)$; y-int: $(0, -2)$

12. y-int: $(0, -2)$

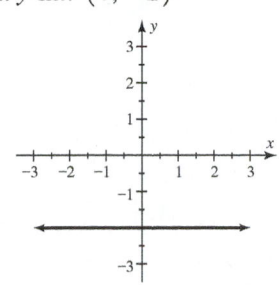

13. $y = -\frac{3}{2}x + \frac{7}{2}$ **14.** $y = \frac{4}{3}x + \frac{11}{3}$
15. $(-1, 0)$; $(0, -2)$; $y = -2x - 2$ **16.** $(1, 2)$
17. $(1, -1)$ **18.** $(-2, 5)$

19.

20.

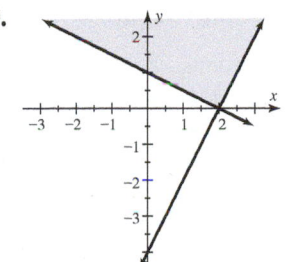

21. -16 **22.** 1 **23.** $\frac{x^{10}}{y^4}$ **24.** $-14x^5 + 21x^4$ **25.** $\frac{1}{a^2}$
26. $\frac{1}{4t^6}$ **27.** $\frac{1}{x^2y^5}$ **28.** $32x^5y^{10}$ **29.** $2x^2 + 4x$
30. $3x + \frac{-4x + 1}{x^2 + 1}$ **31.** $(x - 4)(x + 7)$
32. $(2y + 3)(3y - 4)$ **33.** $(5x - 2y)(5x + 2y)$
34. $(8x - 1)^2$ **35.** $(3t - 2)(9t^2 + 6t + 4)$
36. $-4(x - 3)(x + 2)$ **37.** $-5, 0, 5$ **38.** $-2, 1$
39. $-\frac{7}{2}, 0, \frac{7}{2}$ **40.** $-3, 3$ **41. (a)** $10x + 8x$ or $18x$
(b) 50 min **42.** $6\frac{11}{12}$ mi **43.** 25 min skiing; 35 min running
44. (a) $C = 0.25x + 20$ **(b)** 320 mi
45. (a) $2x + y = 180$; $2x - y = 20$ **(b)** $(50, 80)$, the
angles are $50°, 50°,$ and $80°$. **46.** $18xy$; 108 yd^2
47. $x + 6$ **48. (a)** 4 sec **(b)** After 1 sec and 3 sec

7 Rational Expressions

Section 7.1 (pp. 441–444)

1. $\frac{P}{Q}$; polynomials **2.** Yes; both x and $2x^2 + 1$ are polynomials. **3.** denominator **4.** a **5.** $\frac{a}{b}$ **6.** rational **7.** $-\frac{3}{7}$
9. $-\frac{4}{9}$ **11.** $-\frac{1}{4}$ **13.** Undefined **15.** Undefined **17.** -1
19. 1

21.

x	-2	-1	0	1	2
$\frac{x}{x+1}$	2	—	0	$\frac{1}{2}$	$\frac{2}{3}$

23.

x	-2	-1	0	1	2
$\frac{3x}{2x^2+1}$	$-\frac{2}{3}$	-1	0	1	$\frac{2}{3}$

25. 0 **27.** 3 **29.** $-\frac{4}{5}$ **31.** None **33.** $-5, 5$ **35.** $-3, -2$
37. $1, \frac{5}{2}$ **39.** $\frac{2}{3}$ **41.** $\frac{1}{2}$ **43.** $-\frac{2}{5}$ **45.** $-\frac{1}{3}$ **47. (a)** $\frac{1}{2}$ **(b)** $\frac{1}{2}$
49. (a) -1 **(b)** -1 **51.** $\frac{1}{2x^2}$ **53.** $\frac{4y}{3x}$ **55.** $\frac{1}{2}$ **57.** $\frac{3}{5}$
59. $\frac{x+1}{x+6}$ **61.** $\frac{5y+3}{y+2}$ **63.** -1 **65.** -1 **67.** $-\frac{1}{3}$ **69.** $-\frac{1}{2}$
71. 1 **73.** $-\frac{3x+5}{3x-5}$ **75.** $\frac{n-1}{n-5}$ **77.** $\frac{x}{6}$ **79.** $\frac{z-2}{z-3}$
81. $\frac{x+4}{3x+2}$ **83.** $\frac{1}{3x-2}$ **85.** 1 **87.** $-\frac{1}{2}$ **89.** -1
91. (a) Equation **(b)** 6 **93. (a)** Expression **(b)** $\frac{1}{x+1}$
95. (a) Expression **(b)** $x - 2$ **97. (a)** Equation **(b)** 8
99. (a)

x	-4	-3	-2	-1	0	1	2	3	4
y	-0.2	-0.25	$-0.\overline{3}$	-0.5	-1	—	1	0.5	$0.\overline{3}$

(b) 1

(c)–(e)

101. (a)

x	-4	-3	-2	-1	0	1	2	3	4
y	$1.\overline{3}$	1.5	2	—	0	0.5	$0.\overline{6}$	0.75	0.8

(b) -1 **(c)–(e)**

103. (a) $\frac{1}{2}$; when traffic arrives at a rate of 3 vehicles per
minute, the average wait is $\frac{1}{2}$ minute.

(b)

x	2	4	4.5	4.9	4.99
T	$\frac{1}{3}$	1	2	10	100

As x nears 5 vehicles per minute, a small increase in x
increases the wait dramatically.
105. (a)

x	0	12	36	72
P	0.5	4.83	6.07	6.5

(b) 50 **(c)** The population increased quickly at first, but
then leveled off.
107. (a) 6 hr **(b)** $\frac{M}{60}$ **109. (a)** $x = 5$ **(b)** As the average
rate nears 5 cars per minute, a small increase in x increases
the wait dramatically. **111.** $\frac{1}{2}$ **113. (a)** $\frac{3}{n}$ **(b)** $\frac{n-3}{n}$; $\frac{97}{100}$;
there is a 97% chance that a winning ball will not be drawn.

Section 7.2 (pp. 449–451)

1. numerators; denominators **2.** reciprocal **3.** $\frac{AC}{BD}$ **4.** $\frac{AD}{BC}$
5. $\frac{2}{5}$ **7.** $\frac{12}{7}$ **9.** $\frac{2}{3}$ **11.** $\frac{2}{11}$ **13.** 4 **15.** $\frac{8}{27}$ **17.** 10 **19.** $\frac{12}{5}$

21. 1   23. $\frac{z+1}{z+4}$ 25. $\frac{2}{3}$ 27. $\frac{x+2}{x-2}$ 29. $\frac{8(x+1)}{x^2}$

31. $\frac{x-3}{x}$ 33. $\frac{z+3}{z-7}$ 35. 1   37. $(t+1)(t+2)$

39. $\frac{x(x+4)}{x^2+4}$ 41. $\frac{z-1}{z+2}$ 43. $\frac{y}{y-1}$ 45. $\frac{x+1}{x-3}$ 47. $\frac{x-3}{x-1}$

49. $\frac{2}{2x+3}$ 51. -2 53. $\frac{z-1}{z+1}$ 55. $\frac{y+2}{2y+1}$ 57. $\frac{4(t-1)}{t^2+1}$

59. $\frac{y-3}{y-5}$ 61. $2x$ 63. $\frac{2z+1}{z+5}$ 65. $\frac{1}{t+6}$ 67. $\frac{2a+3b}{a+b}$

69. $x-y$ 71. 1   73. (a) $D=\frac{30}{x}$ (b) 300 ft; 75 ft; stopping distance on dry pavement is one-fourth as long.

75. (a) $\frac{1}{n+1}$ (b) $\frac{1}{100}$ 77. (a) $R=\frac{350(x-3)^2+50}{260(x-3)^2+150}$
(b) Six years after startup, Google's revenues were about 1.6 times Facebook's revenues.   79. (a) $54; it costs $54, on average, to make each camera when 5000 are produced.
(b) $T=50x+20,000$ (c) 270,000; it costs $270,000 to make 5000 cameras.

Checking Basic Concepts 7.1 & 7.2 (p. 451)

1. Undefined; $\frac{3}{8}$ 2. (a) $\frac{2x}{5y}$ (b) 5   (c) $\frac{x+2}{x+4}$

3. (a) $\frac{4}{9}$ (b) $\frac{2}{x-1}$ 4. (a) $\frac{5z}{6}$ (b) $x+1$

5. (a)

x	0.5	1.0	1.5	1.9
T	$\frac{2}{3}$	1	2	10

(b) As x nears 2 customers per minute, a small increase in x increases the wait dramatically.

Section 7.3 (pp. 457–459)

1. numerators; denominators   2. numerators; denominators

3. $\frac{A+B}{C}$ 4. $\frac{A-B}{C}$ 5. 1   7. $\frac{6}{5}$ 9. 1   11. $\frac{3}{7}$ 13. $\frac{1}{2}$

15. $\frac{1}{2}$ 17. $\frac{2}{3}$ 19. $\frac{3}{x}$ 21. $\frac{1}{2}$ 23. 3   25. 1   27. $\frac{4z}{4z+3}$

29. 2   31. 1   33. $\frac{1}{x-1}$ 35. $z-1$ 37. 2   39. $y-1$

41. $\frac{x}{2}$ 43. $z-2$ 45. $x-3$ 47. $\frac{7n}{2n^2-n+5}$ 49. $\frac{6}{x+3}$

51. $\frac{9}{ab}$ 53. $\frac{1}{x+y}$ 55. $\frac{10}{x-y}$ 57. 0   59. $\frac{b}{2a}$ 61. 1

63. $a-b$ 65. $\frac{1}{x+2}$ 67. $\frac{1}{3x-5}$ 69. $2x-3y$

71. It equals 7.   73. (a) $\frac{14}{n+1}$ (b) $\frac{7}{50}$; when there are 100 batteries, there are 7 chances in 50 that a defective battery is chosen.   75. $C=\frac{7x+1,500,000}{x}$

Section 7.4 (pp. 446–468)

1. Examples include 36 and 54 (answers may vary).

2. xy 3. $\frac{3}{3}$ 4. $\frac{x+1}{x+1}$ 5. 12   7. 6   9. 30   11. 72

13. $12x$ 15. $10x^2$ 17. $x(x+1)$ 19. $(2x+1)(x+3)$

21. $x(x-1)(x+1)$ 23. $(x-8)^2(x+1)$

25. $(2x-1)(2x+1)$ 27. $(x-1)(x+1)$

29. $(2x+3)(x+2)(x+3)$ 31. $3y(y+2)(y-1)$

33. $\frac{13}{10}$ 35. $\frac{2}{9}$ 37. $\frac{14}{25}$ 39. $\frac{9}{20}$ 41. $\frac{3}{9}$ 43. $\frac{15}{21}$ 45. $\frac{2x^2}{8x^3}$

47. $\frac{x-2}{x^2-4}$ 49. $\frac{x}{x^2+x}$ 51. $\frac{2x^2+2x}{x^2+2x+1}$ 53. $\frac{13}{12x}$

55. $\frac{5z-7}{z^3}$ 57. $\frac{y-x}{xy}$ 59. $\frac{a^2+b^2}{ab}$ 61. $\frac{7}{2(x+2)}$

63. $\frac{t+2}{t(t-2)}$ 65. $\frac{n^2+4n+5}{(n-1)(n+1)}$ 67. $-\frac{3}{x-3}$ 69. 0

71. $\frac{6x-4}{(x-1)^2}$ 73. $\frac{3}{2y-1}$ 75. $\frac{x-1}{x(x+2)}$ 77. $\frac{3x+5}{(x-2)(x+2)}$

79. $\frac{x+9}{x(x-3)(x+3)}$ 81. $\frac{x^2+4x-4}{x(x-2)(x+2)}$ 83. $\frac{2x+2}{(x+2)^2}$

85. $\frac{x^2+2x-1}{(x+1)(x+2)(x+3)}$ 87. $-\frac{2b}{(a+b)(a-b)}$

89. 0   91. $\frac{6}{x-2}$ 93. 0   95. $\frac{3}{x}$ 97. $\frac{x^3-x+2}{(x-1)(x+1)}$

99. $\frac{3x^2+x+4}{(x-1)(x+1)}$ 101. $\frac{1}{75}$; $\frac{1}{75}$; yes

103. $\frac{D-F}{FD}$ 105. 60 years

Checking Basic Concepts 7.3 & 7.4 (p. 468)

1. (a) 1   (b) $\frac{2-x}{3x}$ (c) z 2. (a) $15x$ (b) $4x(x+1)$

(c) $(x+1)(x-1)$ 3. (a) $\frac{6x+5}{x(x+1)}$ (b) $\frac{4}{x-3}$

(c) $-\frac{x+4}{2x+1}$ 4. $\frac{a^2+b^2}{(a-b)(a+b)}$

Section 7.5 (pp. 475–476)

1. $\frac{1}{2}\cdot\frac{4}{3}=\frac{2}{3}$ 2. $\frac{a}{b}\cdot\frac{d}{c}=\frac{ad}{bc}$ 3. fractions   4. $\dfrac{\frac{x}{2}}{\frac{1}{x-1}}$

5. $\dfrac{\frac{a}{b}}{\frac{c}{d}}$ 6. $x(x+2)$ 7. 30   9. $(x-1)(x+1)$

11. $x(2x-1)(2x+1)$ 13. $\frac{4}{5}$ 15. $\frac{10}{7}$ 17. $\frac{9}{14}$ 19. $\frac{1}{2}$

21. $\frac{3y}{x}$ 23. 3   25. $\frac{p+1}{p+2}$ 27. $\frac{5}{2}$ 29. $\frac{y}{y-3}$ 31. $\frac{4x-1}{4x+1}$

33. $\frac{x^2}{3}$ 35. 1   37. $\frac{n+m}{n-m}$ 39. $\frac{2x+y}{2x-y}$ 41. $\frac{1+b}{1-a}$

43. $\frac{q+2}{q}$ 45. $2x+3$ 47. $\frac{2x}{(x+1)(2x-1)(2x+1)}$

49. $\frac{1}{ab}$ 51. $\frac{ab}{a+b}$ 53. $\dfrac{P\left(1+\frac{r}{26}\right)^{52}-P}{\frac{r}{26}}$

55. $R=\frac{ST}{S+T}$

Section 7.6 (pp. 487–491)

1. rational   2. $\frac{2x+5}{3x}$; $\frac{2x+5}{3x}=9$ (answers may vary)

3. $ad=bc$; b; d   4. Yes   5. $12x$ 6. V; T 7. $\frac{3}{2}$ 9. $\frac{5}{2}$

11. 0   13. $\frac{1}{4}$ 15. $\frac{10}{11}$ 17. 3   19. 25   21. 5   23. $\frac{5}{16}$

25. 1   27. No solutions $\left(\text{extraneous: }-\frac{1}{2}\right)$ 29. $\frac{1}{2}$

31. -1 33. $-\frac{3}{2}, -1$ 35. 4   37. 4   39. -6 41. 6

43. -3 (extraneous: 1)   45. 1 (extraneous: -2)   47. $\frac{7}{12}$

49. $\frac{1}{6}, 5$ 51. 3   53. No solutions (extraneous: 1)   55. 3

57. No solutions (extraneous: -1)   59. No solutions

61. -2 63. Expression; 1; 1   65. Equation; 2

67. Equation; 4   69. Expression; $x+2$; 4   71. $-\frac{1}{2}; \frac{1}{2}$

73. $-1; \frac{1}{2}$ 75. (a) $-3, 1$ (b) $-3, 1$ 77. (a) -1

(b) -1 79. (a) 2   (b) 2   81. (a) $-2, 2$ (b) $-2, 2$

83. 0.300, 1.100   85. 1.084   87. $a=\frac{F}{m}$ 89. $r=\frac{V}{I}-R$

91. $b=\frac{2A}{h}$ 93. $z=\frac{15}{k-3}$ 95. $b=\frac{aT}{a-T}$

97. $x=\frac{ky}{3y+2k}$ 99. 9 cars per minute

101. $\frac{12}{7}\approx 1.7$ hours   103. $\frac{8}{3}\approx 2.7$ days

105. 20 mph; 18 mph   107. (a) 120; the braking distance is 120 feet when the slope of the road is -0.05.   (b) -0.1

109. 15 mph   111. 200 mph   113. 10 mph running; 5 mph jogging   115. (a) 1; when cars leave the ramp at a rate of 4 vehicles per minute, the wait is 1 minute.   (b) As more cars try to exit, the waiting time increases; yes
(c) $4.\overline{6}$ vehicles/min

Checking Basic Concepts 7.5 & 7.6 (p. 491)

1. (a) $\frac{5}{6}$ (b) $\frac{1}{9x^2}$ (c) $\frac{b-a}{b+a}$ (d) $\frac{r+t}{2rt}$ 2. (a) $\frac{1}{5}$

(b) -4 (c) 2   (d) $-2, \frac{1}{2}$ (e) -3 (extraneous: 1)

3. (a) $x=\frac{2(b+3y)}{a}$ (b) $m=\frac{k}{2k-1}$ 4. (a) 300; when the slope of the hill is 0.1, the braking distance is 300 feet.
(b) 0.3; the braking distance is 200 feet when the slope of the road is 0.3.

Section 7.7 (pp. 501–505)

1. A statement that two ratios are equal   **2.** $\frac{5}{6} = \frac{x}{7}$
3. It doubles.   **4.** It is halved.   **5.** constant   **6.** constant
7. Directly; if the number being fed doubles, the bill will
double.   **8.** Inversely; doubling the number of painters will
halve the time.   **9.** inverse   **10.** direct   **11.** 15   **13.** 21
15. 48   **17.** $\frac{21}{8}$ **19.** $-4, 4$ **21.** $-\frac{7}{2}, \frac{7}{2}$ **23.** $b = \frac{ad}{c}$
25. (a) $\frac{5}{8} = \frac{9}{x}$ (b) $\frac{72}{5}$ **27.** (a) $\frac{4}{8} = \frac{10}{x}$ (b) 20
29. (a) $\frac{98}{7} = \frac{x}{11}$ (b) \$154   **31.** (a) $\frac{3}{750} = \frac{7}{x}$ (b) 1750
33. (a) 2   (b) 12   **35.** (a) $\frac{3}{2}$ (b) 9   **37.** (a) $-\frac{15}{2}$
(b) -45 **39.** (a) 24   (b) 3   **41.** (a) 40   (b) 5
43. (a) 400   (b) 50   **45.** (a) $k = 0.25$
(b) $z = 8.75$ **47.** (a) $k = 11$ (b) $z = 385$
49. (a) $k = 10$ (b) $z = 350$
51. (a) Direct   (b) $y = \frac{3}{2}x$ **53.** (a) Inverse   (b) $y = \frac{36}{x}$

(c)   (c)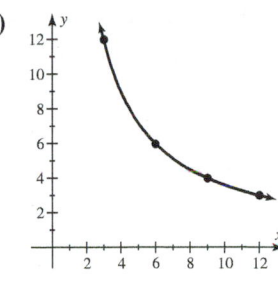

55. (a) Neither   (b) NA   (c) NA   **57.** Direct; 2
59. Neither   **61.** Inverse; 8   **63.** 47.6 minutes
65. 1.625 inches   **67.** About 2123 lb   **69.** 18.75 ft
71. (a) Direct; the ratios $\frac{R}{W}$ **73.** (a) Direct; the ratios $\frac{G}{A}$
always equal 0.012.   always equal 27.
(b) $R = 0.012W$ (b) $G = 27A$

(c) 38.4 pounds   (c) For each square-inch
increase in the cross-sec-
tional area of the hose, the
flow increases by 27 gallons
per minute.

75. (a) $F = \frac{1200}{L}$ (b) 80 pounds
77. (a) Direct   (b) $y = -19x$ (c) Negative; for each
1-mile increase in altitude, the temperature decreases by
$19°$ F.   (d) $47.5°$ F decrease   **79.** 1.8 ohms
81. $z = 0.5104x^2y^3$ **83.** About 133 pounds   **85.** About
35.2 pounds   **87.** 750   **89.** 1200 lb per square inch

Checking Basic Concepts 7.7 (p. 505)

1. (a) $\frac{18}{5}$ (b) $\frac{15}{4}$ **2.** (a) $\frac{4}{6} = \frac{8}{x}$; 12   (b) $\frac{2}{148} = \frac{5}{x}$; 370 minutes
3. 60; 6   **4.** (a) Direct; the ratios $\frac{y}{x}$ always equal $\frac{3}{2}$; $\frac{3}{2}$.
(b) Inverse; the products xy always equal 24; 24.   **5.** \$160

Chapter 7 Review (pp. 509–513)

1. $-\frac{3}{5}$ **2.** -3 **3.** Undefined   **4.** Undefined

5.

x	-2	-1	0	1	2
$\frac{3x}{x-1}$	2	$\frac{3}{2}$	0	—	6

6. $-2, 2$ **7.** $\frac{5y^3}{3x^2}$ **8.** $x - 6$ **9.** $\frac{x+3}{x+1}$ **10.** $\frac{x+4}{x+1}$
11. (a) Expression   (b) $\frac{1}{x-3}$ **12.** (a) Equation   (b) 18
13. 2   **14.** $\frac{1}{x+5}$ **15.** $\frac{1}{z+3}$ **16.** $\frac{x}{x-2}$ **17.** $\frac{5}{6}$
18. $\frac{8}{x(x+1)}$ **19.** $\frac{1}{2}$ **20.** 1   **21.** $\frac{10}{x+10}$ **22.** $\frac{1}{x-1}$ **23.** 1
24. 1   **25.** $\frac{1}{x+1}$ **26.** $\frac{2}{x-5}$ **27.** $\frac{2}{xy}$ **28.** $\frac{x}{y}$ **29.** $15x$
30. $10x^2$ **31.** $x(x-5)$ **32.** $10x^2(x-1)$ **33.** $\frac{9}{24}$
34. $\frac{16}{12x}$ **35.** $\frac{3x^2+6x}{x^2-4}$ **36.** $\frac{2x^2+4x}{2x^2+x-6}$ **37.** $\frac{19}{24}$ **38.** $\frac{7}{4x}$
39. $-\frac{1}{9x}$ **40.** $\frac{4x+3}{x(x-1)}$ **41.** $\frac{2x}{(x-1)(x+1)}$ **42.** $\frac{8-9x}{6x^2}$
43. $\frac{2x-7}{6x}$ **44.** $-\frac{1}{(x-1)(x+1)}$ **45.** $\frac{5y-x}{(x-y)(x+y)}$
46. -1 **47.** $\frac{33}{28}$ **48.** $\frac{7}{10}$ **49.** $\frac{n}{2}$ **50.** $\frac{3(p+1)}{p-1}$ **51.** $\frac{1}{3}$
52. $\frac{2-x}{2+x}$ **53.** $\frac{1}{x^2}$ **54.** $x + 3$ **55.** $\frac{20}{7}$ **56.** $\frac{8}{3}$ **57.** $\frac{1}{5}$
58. -5 **59.** 4   **60.** -2 **61.** 5   **62.** 7   **63.** $-2, 5$
64. $-3, 3$ **65.** 8   **66.** No solutions
67. No solutions (extraneous: -1)
68. $-\frac{3}{2}$ (extraneous: 0)   **69.** -3 **70.** -12
71. (a) Equation   (b) $-2, 2$ **72.** (a) Expression
(b) $x + 3$ **73.** $b = \frac{2ac}{3a-c}$ **74.** $x = \frac{y}{y-1}$ **75.** 2
76. $\frac{15}{7}$ **77.** (a) $\frac{6}{x} = \frac{13}{20}$ (b) $\frac{120}{13}$ **78.** (a) $\frac{341}{11} = \frac{x}{8}$
(b) \$248   **79.** (a) 4   (b) 20   **80.** (a) 3   (b) 15   **81.** (a) 10
(b) 2   **82.** (a) 21   (b) $\frac{21}{5}$ **83.** $k = 3$ **84.** $z = 720$
85. (a) Inverse   (b) $y = \frac{60}{x}$ **86.** (a) Direct   (b) $y = 3x$

(c)   (c)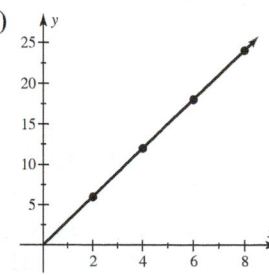

87. Direct; $\frac{1}{2}$ **88.** Inverse; 12   **89.** (a) $\frac{1}{5} = 0.2$; when the
average rate of arrival is 10 cars per minute, the average wait
is 0.2 minute, or 12 seconds.

(b)

x	5	10	13	14	14.9
T	$\frac{1}{10}$	$\frac{1}{5}$	$\frac{1}{2}$	1	10

(c) It increases dramatically.   **90.** 60 mph
91. $\frac{800}{13} \approx 61.5$ hours   **92.** 10 mph and 12 mph

93. 8 mph **94.** About 33.3 feet **95.** 200 vehicles
96. $468 **97.** 36 pounds **98.** 1.6 inches
99. 750 to 2250 seconds, or 12.5 to 37.5 minutes
100. About 771 lb **101.** 6493.8 watts **102.** 702 pounds

Chapter 7 Test (p. 513)

1. $\frac{9}{5}$ **2.** -2 **3.** $x + 5$ **4.** $x - 5$ **5.** 3 **6.** 2 **7.** $\frac{x-1}{10x}$
8. $\frac{6}{x(x+3)}$ **9.** $\frac{4x+1}{x+4}$ **10.** $\frac{t+7}{2t-3}$ **11.** $6x^2(x-1)$
12. $\frac{4x-4}{7x^2-7x}$ **13.** $-\frac{1}{y+1}$ **14.** $\frac{x^2y-x+y}{xy^2}$ **15.** $\frac{b}{15}$
16. $\frac{p}{p-2}$ **17.** $\frac{35}{2}$ **18.** 3 **19.** 1 **20.** $-1, 2$ **21.** $\frac{2}{7}$
22. -12 **23.** No solutions $\left(\text{extraneous: } \frac{1}{2}\right)$ **24.** 0 (extraneous: 5) **25.** $x = \frac{2+5y}{3y}$ **26.** $b = \frac{a}{a-1}$ **27. (a)** $\frac{7}{2}$
(b) 21 **28.** Inversely; 32 **29.** 24 hours **30.** 67.5 feet
31. $\frac{16}{5} = 3.2$; when the average arrival rate is 24 people per
hour, there are about 3 people in line, on average.

Chapters 1–7 Cumulative Review (pp. 514–515)

1. $24\pi \approx 75.4$ **2.** $2x - 2$ **3.** $\frac{2}{5}$ **4.** $\frac{3}{4}$ **5.** $2x + 2$
6. $y - 11$ **7.** -1 **8.** $x \le \frac{1}{4}$
9. x-intercept: $(3, 0)$ **10.** x-intercept: $(1, 0)$
   y-intercept: $(0, -2)$ y-intercept: none

  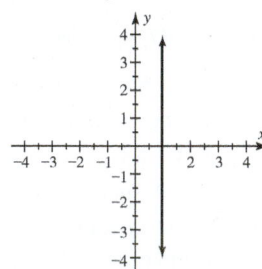

11. $y = 3x + 5$

12. $y = 2x - 1$ **13.** $y = -\frac{2}{3}x + \frac{1}{3}$ **14.** $y = \frac{2}{3}x + \frac{8}{3}$
15. $(1, -2)$ **16.** $(2, -8)$ **17.** Infinitely many solutions
18. No solutions **19.** $15z^8$ **20.** a^3b^3
21. $10y^2 - 11y - 6$ **22.** $x^4 - 2x^2y^2 + y^4$ **23.** $\frac{1}{27x^6}$
24. $\frac{2}{x^2}$ **25.** 1.23×10^{-3} **26.** $2x + 1 + \frac{4}{x-1}$
27. $(3 - x)(2 + 5x)$ **28.** $(3z - 2)(3z + 2)$
29. $(t + 8)^2$ **30.** $x(x - 4)(x + 4)$ **31.** $-7, 2$
32. $-2, 0, 2$ **33.** $\frac{1}{12x}$ **34.** $\frac{1}{2x+4}$ **35.** $\frac{x+2}{x-2}$
36. $x = \frac{z+2y}{3}$ **37.** $\frac{7}{12}$ **38.** $-1, 2$ **39.** 5.5
40. Inverse; $y = \frac{20}{x}$ **41. (a)** $21x$ **(b)** 90 min

42. 35 min running, 25 min walking
43. (a) $2x + y = 180, 2x - y = 32$ **(b)** $(53, 74)$ or 53°,
53°, 74° **44.** About 500 lb

8 Introduction to Functions

Section 8.1 (pp. 517–536)

1. function **2.** y equals f of x **3.** symbolic **4.** numerical
5. domain **6.** range **7.** one **8.** False **9.** True
10. $(3, 4)$; 3; 6 **11.** (a, b) **12.** d **13.** 1 **14.** equal
15. Yes **16.** Yes **17.** No **18.** No **19.** Yes **20.** No
21. $-6; -2$ **23.** $0; \frac{3}{2}$ **25.** $25; \frac{9}{4}$ **27.** 3; 3 **29.** 13; -22
31. $-\frac{1}{2}; \frac{2}{5}$ **33. (a)** $I(x) = 36x$ **(b)** $I(10) = 360$; there
are 360 inches in 10 yards. **35. (a)** $M(x) = \frac{x}{5280}$
(b) $M(10) = \frac{10}{5280} \approx 0.0019$; There is $\frac{10}{5280}$ mile in 10 feet.
37. (a) $A(x) = 43{,}560x$ **(b)** $A(10) = 435{,}600$; there are
435,600 square feet in 10 acres.
39. $f = \{(1, 3), (2, -4), (3, 0)\}$;
$D = \{1, 2, 3\}; R = \{-4, 0, 3\}$
41. $f = \{(a, b), (c, d), (e, a), (d, b)\}$;
$D = \{a, c, d, e\}; R = \{a, b, d\}$

43.   **45.**

47.   **49.**

51.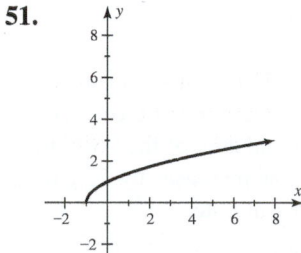

53. 3; -1 **55.** 0; 2 **57.** $-4; -3$ **59.** 5.5; 3.7
61. 26.9; in 1990, average fuel efficiency was 26.9 mpg.

63. Numerical:

x	-3	-2	-1	0	1	2	3
$y = f(x)$	2	3	4	5	6	7	8

Graphical:

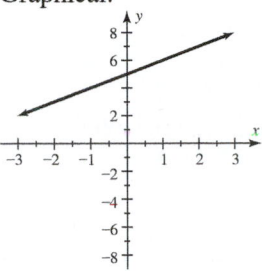

Symbolic: $y = x + 5$

65. Numerical:

x	-3	-2	-1	0	1	2	3
$y = f(x)$	-17	-12	-7	-2	3	8	13

Graphical:

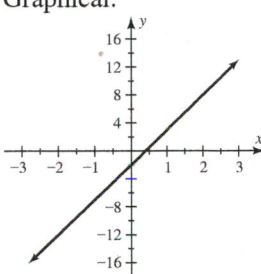

Symbolic: $y = 5x - 2$

67. Subtract $\frac{1}{2}$ from the input x to obtain the output y.

69. Divide the input x by 3 to obtain the output y.

71. Subtract 1 from the input x and then take the square root to obtain the output y.

73. $f(x) = 0.60x$

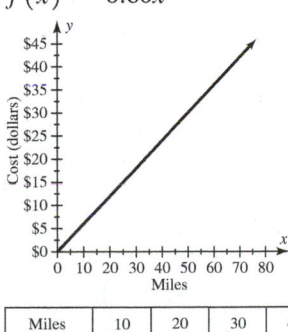

Miles	10	20	30	40	50	60	70
Cost	$6	$12	$18	$24	$30	$36	$42

75. 1400; in 2014, there were 1400 billion, or 1.4 trillion Google searches.

77. D: $-2 \le x \le 2$; R: $0 \le y \le 2$

79. D: $-2 \le x \le 4$; R: $-2 \le y \le 2$

81. D: All real numbers; R: $y \ge -1$

83. D: $-3 \le x \le 3$; R: $-3 \le y \le 2$

85. $D = \{1, 2, 3, 4\}$; $R = \{5, 6, 7\}$

87. All real numbers   **89.** All real numbers   **91.** $x \ne 5$

93. All real numbers   **95.** $x \ge 1$ **97.** All real numbers

99. $x \ne 0$ **101. (a)** 1488, 1488 whales were sighted in 2015.   **(b)** $D = \{2011, 2012, 2013, 2014, 2015\}$; $R = \{1054, 1126, 1331, 1488, 1607\}$ **(c)** Decreased from the first to the second year and then increased every year after that.   **103.** $D = \{1, 2, 3, \ldots, 20\}$; $R = \{200, 400, 600, \ldots, 4000\}$

105. Relation; $D = \{2, 3\}$; $R = \{-4, 3, 5\}$

107. Relation; Function; $D = \{-7, 3, 4\}$; $R = \{5\}$

109. Neither   **111.** Relation; Function; $D = \{-5, -2, 0, 3\}$; $R = \{-5, -2, 0, 3\}$

113. No   **115.** Yes   **117. (a)** 0.2   **(b)** Yes. Each month has one average amount of precipitation.   **(c)** 2, 3, 7, 11

119. Yes. D: All real numbers; R: All real numbers

121. No   **123.** Yes. D: $-4 \le x \le 4$; R: $0 \le y \le 4$

125. Yes. D: All real numbers; R: $y = 3$ **127.** No

129. Yes. $D = \{-6, -4, 2, 4\}$; $R = \{-4, 2\}$ **131.** Yes

133. No   **135.** The person walks away from home, then turns around and walks back a little more slowly.

137.

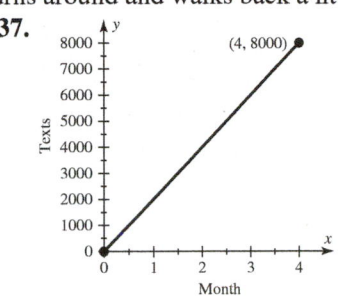

139. 10; 10
$[-10, 10, 1]$ by $[-10, 10, 1]$

141. 10; 5
$[0, 100, 10]$ by $[-50, 50, 10]$

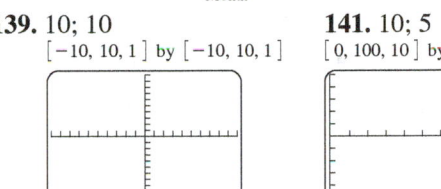

143. 16; 5
$[1980, 1995, 1]$ by $[12000, 16000, 1000]$

145.
$[-6, 6, 1]$ by $[-6, 6, 1]$

147. $[-30, 30, 5]$ by $[-50, 50, 5]$

149. $[-200, 200, 50]$ by $[-250, 250, 50]$

151. Numerical

Graphical
$[-10, 10, 1]$ by $[-10, 10, 1]$

X	Y1
-3	0
-2	1
-1	1.4142
0	1.7321
1	2
2	2.2361
3	2.4495

Y1 ■ √(X+3)

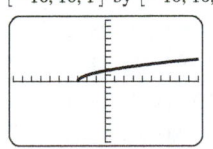

153. Numerical

Graphical (Dot Mode)
$[-10, 10, 1]$ by $[-10, 10, 1]$

X	Y1
-3	4
-2	2.3333
-1	1.5
0	1
1	.66667
2	.42857
3	.25

Y1 ■ (5−X)/(5+X)

Section 8.2 (pp. 537–553)

1. $mx + b$ **2.** b **3.** line   **4.** horizontal   **5.** 7   **6.** 0

7. True   **8.** False   **9.** Carpet costs $2 per square foot. Ten square feet of carpet costs $20.

10. The rate at which water is leaving the tank is 4 gallons per minute. After 5 minutes, the tank contains 80 gallons of water.
11. Yes; $m = \frac{1}{2}, b = -6$ **13.** No
15. Yes; $m = 0, b = -9$ **17.** Yes; $m = -9, b = 0$
19. Yes   **21.** No   **23.** Yes; $f(x) = 3x - 6$ **25.** Yes;
$f(x) = -\frac{3}{2}x + 3$ **27.** No   **29.** Yes; $f(x) = 2x$
31. Yes; $f(x) = -4$ **33.** $-16; 20$ **35.** $\frac{17}{3}; 2$
37. $-22; -22$ **39.** $-2; 0$ **41.** $-1; -4$ **43.** $1; 1$
45. $f(x) = 6x; 18$ **47.** $f(x) = \frac{x}{6} - \frac{1}{2}; 0$ **49.** d.   **51.** b.

53.

55.

57.

59.

61.

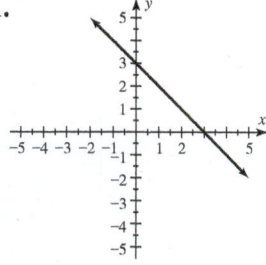

63. $f(x) = \frac{1}{16}x$ **65.** $f(t) = 65t$ **67.** $f(x) = 24$ **69.** a
71. **(a)** f multiplies the input x by -2 and then adds 1 to obtain the output y.
73. **(a)** f multiplies the input x by $\frac{1}{2}$ and then subtracts 1 to obtain the output y.

(b)

x	-2	0	2
$y = f(x)$	5	1	-3

(b)

x	-2	0	2
$y = f(x)$	-2	-1	0

(c)

(c)

75. (a)

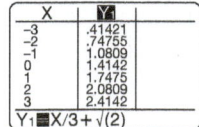

(b) $[-6, 6, 1]$ by $[-4, 4, 1]$

77. (a)

(b) $[-6, 6, 1]$ by $[-4, 4, 1]$

79. b.   **81.** c.   **83. (a)** $G(x) = \frac{X}{E}$ **(b)** $C(x) = \frac{3x}{E}$
85. -1 **87.** $(-1, 0)$ **89.** $\left(-\frac{1}{2}, \frac{1}{2}\right)$ **91.** $(20, 0)$
93. $(-8, -1)$ **95.** $\left(-1, -\frac{1}{6}\right)$ **97.** $(0.2, 0.25)$
99. $(2005, 9)$ **101.** $(2a, 2b)$
103. (a) $f(x) = -2x + 5; 1$
(b) 1   **(c)** Equal   **105. (a)** $f(x) = \frac{2}{5}x + \frac{1}{5}; 1$
(b) 1   **(c)** Equal   **107.** 79.8 years
109. 256.5 million   **111.** $50,850
113. (a) Symbolic: $f(x) = 70$
Graphical:

(b)

Hours	0	4	8	12	16	20	24
Temp. (F)	70	70	70	70	70	70	70

(c) Constant
115. $D(t) = 60t + 50$
117. (a) $K(x) = 130x$ **(b)** $A(x) = 5x$
(c) 47,450; 1825; On average, someone between ages 18 and 24 sends 47,450 texts in 1 year, while someone over 65 sends 1825 texts.   **119. (a)** 1.6 million   **(b)** 0.1 million
(c) $f(x) = 0.1x + 1.6$ **(d)** 2.2 million
121. (a) 189 million   **(b)** In 2010, there were about 145 million users.   **(c)** Users increased, on average, by 11 million per year.   **123. (a)** $V(T) = 0.5T + 137$
(b) 162 cm^3 **125.** $f(x) = 40x$; about 2.92 pounds
127. (a) 43%   **(b)** 3%   **(c)** $P(x) = 3x + 43$ **(d)** 55%

Checking Basic Concepts 8.1 & 8.2 (p. 553)

1. Symbolic: $f(x) = x^2 - 1$
   Graphical:

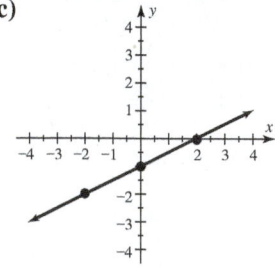

2. (a) D: $-3 \le x \le 3$; R: $-4 \le y \le 4$ **(b)** 0; 4
(c) No. The graph is not a line.   **3. (a)** Yes   **(b)** No
(c) Yes   **(d)** Yes

4. $f(-2) = 10$

5. $f(x) = \frac{1}{2}x - 1$ **6. (a)** 36.7: In 2015, the median age was about 37 years.   **(b)** 0.2: The median age is increasing by 0.2 year each year. 27.7: In 1970, the median age was 27.7 years.   **7.** $(1, -1)$

Section 8.3 (pp. 553–566)

1. $x > 1$ and $x \le 7$ (answers may vary)
2. $x \le 3$ or $x > 5$ (answers may vary)
3. No   **4.** Yes   **5.** Yes
6. Numerically, graphically, symbolically
7. Yes, no   **8.** No, yes   **9.** No, yes   **10.** Yes, no
11. No, yes   **12.** Yes, yes   **13.** $[2, 10]$ **15.** $(5, 8]$
17. $(-\infty, 4)$ **19.** $(-2, \infty)$ **21.** $[-2, 5)$
23. $(-8, 8]$ **25.** $(3, \infty)$ **27.** $(-\infty, -2] \cup [4, \infty)$
29. $(-\infty, 1) \cup [5, \infty)$ **31.** $(-3, 5]$ **33.** $(-\infty, -2)$
35. $(-\infty, 4)$ **37.** $(-\infty, 1) \cup (2, \infty)$
39. $\{x \mid -1 \le x \le 3\}$

41. $\{x \mid -2 < x < 2.5\}$

43. $\{x \mid x > 3\}$

45. $\{x \mid x \le -1 \text{ or } x \ge 2\}$

47. All real numbers

49. $[-6, 7]$ **51.** $\left(\frac{13}{4}, \infty\right)$ **53.** $(-\infty, \infty)$
55. $\left(-\infty, -\frac{11}{5}\right) \cup (-1, \infty)$ **57.** No solutions
59. $[-6, 1)$ **61.** $\left[-\frac{1}{4}, \frac{11}{8}\right)$ **63.** $[-9, 3]$
65. $\left[-4, -\frac{1}{4}\right)$ **67.** $(-1, 1]$ **69.** $[-2, 1]$ **71.** $(0, 2)$
73. $(9, 21]$ **75.** $\left[\frac{12}{5}, \frac{22}{5}\right]$ **77.** $\left(-\frac{13}{3}, 1\right)$
79. $[0, 12]$ **81.** $[-1, 2]$ **83.** $(-1, 2)$ **85.** $[-3, 1]$
87. $(-\infty, -2) \cup (0, \infty)$
89. (a) Toward, because distance is decreasing
(b) 4 hours, 2 hours   **(c)** From 2 to 4 hours
(d) During the first 2 hours
91. (a) 2   **(b)** 4   **(c)** $\{x \mid 2 \le x \le 4\}$
(d) $\{x \mid 0 \le x < 2\}$ **93.** $[1, 4]$
95. $(-\infty, -2) \cup (0, \infty)$ **97.** $[2002, 2010]$ **99.** $[1, 3]$
101. $(-\infty, \infty)$ **103.** $(-\infty, 1) \cup [3, \infty)$
105. $(c - b, d - b]$

107.

109.

111. (a) $-5 \le x \le 5$ **(b)** 2: 3; 6
(c)

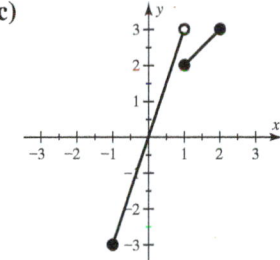

(d) $2 \le y \le 8$
113. (a) $-1 \le x \le 2$ **(b)** undefined; 0; undefined
(c)

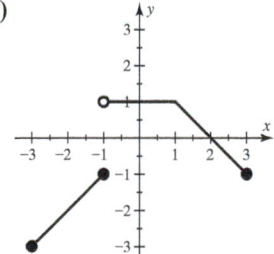

(d) $-3 \le y \le 3$
115. (a) $-3 \le x \le 3$ **(b)** -2; 1; -1
(c)

(d) $-3 \le y \le 1$
117. (a) From 2013 to 2017   **(b)** From 2013 to 2017
(c) From 2013 to 2017   **119.** From 0.5 to 1.5 miles
121. From $5.\overline{6}$ to 9 feet   **123.** From 1.5 to 2.5 miles
125. (a) $M(x) = 25x + 250$ **(b)** From 2014 to 2018
127. (a) maximum: 55 mph; minimum: 30 mph   **(b)** About 12 miles   **(c)** $f(4) = 40; f(12) = 30; f(18) = 55$

Section 8.4 (pp. 567–584)

1. domain   **2.** range   **3.** $(-\infty, \infty)$
4. $(-\infty, 5) \cup (5, \infty)$ **5.** absolute value   **6.** exponent
7. 2   **8.** 1   **9.** rational   **10.** $-\frac{1}{2}$ **11.** c.   **12.** a.
13. D: $(-\infty, \infty)$; R: $(-\infty, \infty)$
15. D: $(-\infty, \infty)$; R: $(-\infty, \infty)$
17. D: $(-\infty, \infty)$; R: $[2, \infty)$

19. $D: (-\infty, \infty); R: (-\infty, 0]$
21. $D: [-1, \infty); R: [0, \infty)$
23. $D: (-\infty, \infty); R: [0, \infty)$
25. $(-\infty, 1) \cup (1, \infty)$
27. $(-\infty, 2) \cup (2, \infty)$
29. $(-\infty, -2) \cup (-2, 2) \cup (2, \infty)$
31. $(-\infty, 0) \cup (0, 2) \cup (2, \infty)$
33. $(-\infty, 1) \cup (1, \infty)$
35. $(-\infty, -1) \cup (-1, 3) \cup (3, \infty)$
37. $D: (-\infty, \infty); R: (-\infty, \infty)$
39. $D: [-2, 2]; R: [-2, 2]$ **41.** $D: [-2, 3]; R: [-2, 2]$
43. Yes; 1; linear   **45.** Yes; 3; cubic   **47.** No
49. Yes; 2; quadratic   **51.** No   **53.** Yes; 4; fourth degree
55. 12; 0   **57.** 1; $\frac{11}{4}$ **59.** 0; 6   **61.** -14; 14   **63.** -6; 9
65. $\frac{1}{11}; -\frac{1}{7}$ **67.** $-\frac{5}{6}$; undefined   **69.** $\frac{5}{6}$; undefined
71. 1; -1 **73.** $-2; -2$ **75.** $-4; 0$ **77.** -1; undefined
79.

x	-2	-1	0	1	2
$f(x) = \frac{1}{x-1}$	$-\frac{1}{3}$	$-\frac{1}{2}$	-1	—	1

; undefined

81.

83.

85.

87.

89.

91.

93.

95.

97.
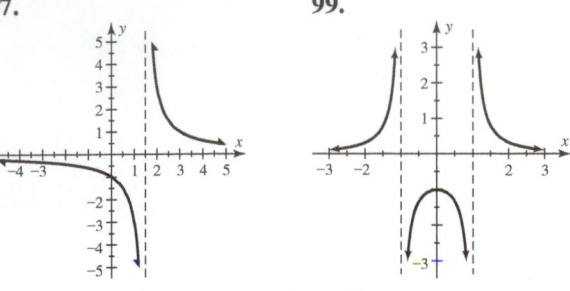

99.

101. (a) 19   **(b)** -9 **(c)** 150   **(d)** 0
103. (a) 41   **(b)** -21 **(c)** 900   **(d)** Undefined
105. (a) $2x + 3$ **(b)** -1 **(c)** $x^2 + 3x + 2$ **(d)** $\frac{x+1}{x+2}$
107. (a) $x^2 - x + 1$ **(b)** $1 - x - x^2$ **(c)** $x^2 - x^3$
(d) $\frac{1-x}{x^2}$ **109.** $a^2 - 2a$ **111.** b   **113.** d
115. $f(-2) \approx 5; f(1) \approx 0$
117. $f(-1) \approx -1; f(1) \approx 1; f(2) \approx -2$
119. $f(-3) = -63; f(1) = 3; f(4) = 10$
121. $D: [-3, 3]$

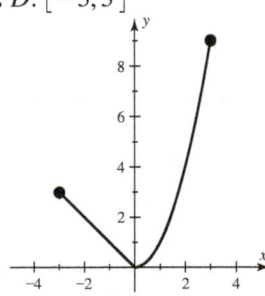

123. c.   **125.** d.   **127. (a)** No; answers may vary.
(b) Yes   **(c)** No; $0 \le t \le 10$
129. (a) 1; when cars are leaving the lot at a rate of
4 vehicles per minute, the average wait is 1 minute.
(b) As more cars try to exit, the waiting time increases;
yes.   **(c)** $4.\overline{6}$ vehicles per minute
131. (a) 75; the braking distance is 75 feet when the uphill
grade is 0.05.   **(b)** 0.15
133. (a) 123 thousand, which is close to the actual value.
(b) 478.6 thousand, which is too high; AIDS deaths did not
continue to rise as rapidly as the model predicts.
135. About 45% and 22%; after 1 day (3 days), students
remember 45% (22%) of what they have learned.
137. (a) 130; it costs $130 thousand to make 100 notebook
computers.   **(b)** The y-intercept for C is $(0, 100)$. The
company has $100 thousand in fixed costs even if it makes 0
computers. The y-intercept for R is $(0, 0)$. If the company sells
0 computers, its revenue is $0.   **(c)** $P(x) = 0.45x - 100$
(d) 223 or more

Checking Basic Concepts 8.3 & 8.4 (p. 584)

1. (a) Yes   **(b)** No
2. (a) $[-3, 1]$ **(b)** $(-\infty, -1] \cup [3, \infty)$ **(c)** $[-\frac{8}{3}, \frac{8}{3})$
3. (a) $(-\infty, \infty)$ **(b)** $(-\infty, 1) \cup (1, \infty)$ **(c)** $[0, \infty)$
4. (a) $D: [-2, 1]; R: [-3, 1]$ **(b)** 1; -3

5.

Section 8.5 (pp. 585–598)

1. $|3x + 2| = 6$ (answers may vary)
2. $|2x - 1| \leq 17$ (answers may vary)
3. Yes   **4.** Yes   **5.** Yes
6. No, it is equivalent to $-3 < x < 3$.   **7.** 2   **8.** 0
9. No, yes   **10.** Yes, no   **11.** No, yes   **12.** No, yes
13. Yes, yes   **14.** No, yes

15.

17.

19.

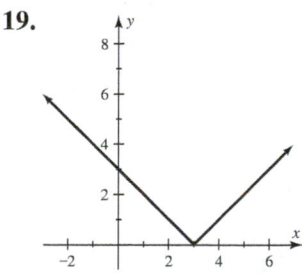

21. $-7, 7$ **23.** No solutions
25. $-\frac{9}{4}, \frac{9}{4}$ **27.** $-4, 4$ **29.** $-6, 5$ **31.** $1, 2$ **33.** $\frac{3}{4}$
35. $-8, 12$ **37.** No solutions   **39.** $-15, 18$ **41.** $-1, \frac{1}{3}$
43. $-5, \frac{3}{5}$ **45.** -6 **47.** $-3, 3$ **49.** $(-3, 3)$
51. $(-\infty, -3) \cup (3, \infty)$
53. (a) $-4, 4$ **(b)** $(-4, 4)$ **(c)** $(-\infty, -4) \cup (4, \infty)$
55. (a) $\frac{1}{2}, 2$ **(b)** $[\frac{1}{2}, 2]$ **(c)** $(-\infty, \frac{1}{2}] \cup [2, \infty)$
57. $[-3, 3]$ **59.** $(-\infty, -4) \cup (4, \infty)$
61. No solutions   **63.** $(-\infty, 0) \cup (0, \infty)$
65. $(-\infty, -\frac{7}{2}) \cup (\frac{7}{2}, \infty)$ **67.** $(-3, 5)$
69. $(-\infty, -9] \cup [-1, \infty)$ **71.** $[\frac{5}{6}, \frac{11}{6}]$ **73.** $[-10, 14]$
75. No solutions   **77.** $(-\infty, \infty)$ **79.** No solutions
81. $(-\infty, \infty)$ **83.** $(-\infty, -13] \cup [17, \infty)$
85. $[0.9, 1.1]$ **87.** $(-\infty, 9.5) \cup (10.5, \infty)$
89. (a) $-1, 3$ **(b)** $(-1, 3)$ **(c)** $(-\infty, -1) \cup (3, \infty)$
91. $0, 4$ **93. (a)** $-1, 0$ **(b)** $[-1, 0]$
(c) $(-\infty, -1] \cup [0, \infty)$ **95.** $(-\infty, -1] \cup [1, \infty)$
97. $[-2, 4]$ **99.** $(-\infty, 1) \cup (3, \infty)$ **101.** $(2, 4.\overline{6})$
103. $(-\infty, \infty)$ **105.** $\{x | -3 \leq x \leq 3\}$
107. $\{x | x < 2 \text{ or } x > 3\}$ **109.** $|x| \leq 4$
111. $|y| > 2$ **113.** $|2x + 1| \leq 0.3$ **115.** $|\pi x| \geq 7$
117. two   **119. (a)** $\{T | 19 \leq T \leq 67\}$

(b) Monthly average temperatures vary from 19 °F to 67 °F.
121. (a) $\{T | -26 \leq T \leq 46\}$ **(b)** Monthly average temperatures vary from -26°F to 46 °F.
123. (a) About 19,058 feet   **(b)** Africa and Europe
(c) South America, North America, Africa, Europe, and Antarctica   **(d)** $|E - A| \leq 5000$
125. $\{d | 2.498 \leq d \leq 2.502\}$; the diameter can vary from 2.498 to 2.502 inches.
127. $|d - 3.8| \leq 0.03$
129. Values between 19 and 21, exclusive

Checking Basic Concepts 8.5 (p. 598)

1. $-\frac{28}{3}, 12$ **2. (a)** $-\frac{2}{3}, \frac{14}{3}$ **(b)** $\left(-\frac{2}{3}, \frac{14}{3}\right)$
(c) $\left(-\infty, -\frac{2}{3}\right) \cup \left(\frac{14}{3}, \infty\right)$
3. $(0, 6)$; $(-\infty, 0] \cup [6, \infty)$
4. (a) $1, 3$ **(b)** $[1, 3]$ **(c)** $(-\infty, 1] \cup [3, \infty)$

Chapter 8 Review (pp. 602–606)

1. $-7; 0$ **2.** $-22; 2$ **3.** $-2; 1$ **4.** $5; 5$
5. (a) $P(q) = 2q$ **(b)** $P(5) = 10$; there are 10 pints in 5 quarts.   **6. (a)** $f(x) = 4x - 3$
(b) $f(5) = 17$; three less than four times 5 is 17.
7. $(3, -2)$ **8.** $4; -6$
9.

10.

11.

12.

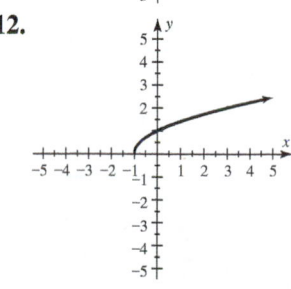

13. $1; 4$ **14.** $1; -2$ **15.** $7; -1$
16. Numerical:

x	-3	-2	-1	0	1	2	3
$y = f(x)$	-11	-8	-5	-2	1	4	7

Symbolic: $f(x) = 3x - 2$
Graphical:

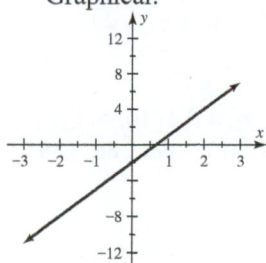

17. D: All real numbers; R: $y \leq 4$
18. D: $-4 \leq x \leq 4$; R: $-4 \leq y \leq 0$
19. Yes   **20.** No
21. $D = \{-3, -1, 2, 4\}$; $R = \{-1, 3, 4\}$; yes
22. $D = \{-1, 0, 1, 2\}$; $R = \{-2, 2, 3, 4, 5\}$; no
23. All real numbers   **24.** $x \geq 0$ **25.** $x \neq 0$
26. All real numbers   **27.** $x \leq 5$ **28.** $x \neq -2$
29. All real numbers   **30.** All real numbers
31. No   **32.** Yes   **33.** Yes; $m = -4, b = 5$
34. Yes; $m = -1, b = 7$ **35.** No
36. Yes; $m = 0, b = 6$ **37.** Yes; $f(x) = \frac{3}{2}x - 3$
38. No   **39.** 1   **40.** $f(-2) = -3$; $f(1) = 0$

41.

42.

43.

44.
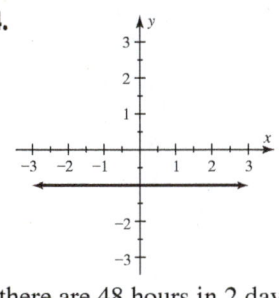

45. $H(x) = 24x$; $H(2) = 48$, there are 48 hours in 2 days.
46. (a)

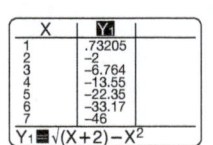

(b) Domain: $x \geq -2$
$[-10, 10, 1]$ by $[-10, 10, 1]$

47. $(0.5, -3)$ **48.** $\left(\frac{5}{12}, \frac{3}{8}\right)$
49. $[-2, 2]$

50. $(-\infty, -3]$

51. $\left(-\infty, \frac{4}{5}\right] \cup (2, \infty)$

52. $(-\infty, \infty)$

53. $[-2, 1]$ **54. (a)** -4 **(b)** 2   **(c)** $[-4, 2]$
(d) $(-\infty, 2)$ **55. (a)** 2   **(b)** $(2, \infty)$ **(c)** $(-\infty, 2)$
56. (a) 4   **(b)** 2   **(c)** $(2, 4)$
57. $\left[-3, \frac{2}{3}\right]$ **58.** $(-6, 45]$ **59.** $\left(-\infty, \frac{7}{2}\right)$
60. $[1.8, \infty)$ **61.** $(-3, 4)$ **62.** $(-\infty, 4) \cup (10, \infty)$
63. $(-5, 5)$ **64.** $[8, 28]$ **65.** $(-9, 21)$ **66.** $[88, 138]$
67. 4, 2   **68.** -10, undefined

69. $D = (-\infty, \infty)$; $R = [0, \infty)$
70. $D = (-\infty, \infty)$; $R = [0, \infty)$
71. $(-\infty, 4) \cup (4, \infty)$
72. $D = [-3, 1]$; $R = [-3, 6]$ **73.** Yes; 2; quadratic
74. Yes; 1; linear   **75.** Yes; 3; cubic
76. No   **77.** 11; 2   **78.** $-\frac{4}{5}$; undefined

79.   **80.**

81.   **82.**
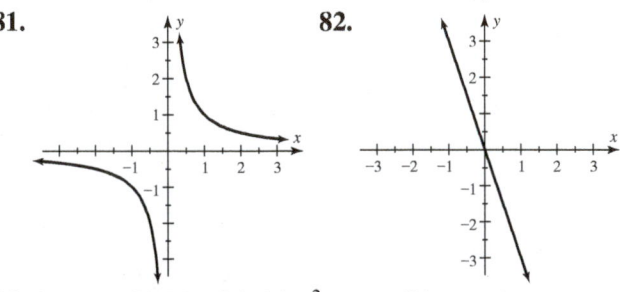

83. (a) 12   **(b)** 27   **84. (a)** $x^2 - x$ **(b)** $x + 1$
85. No, no   **86.** No, yes   **87.** No, yes   **88.** No, yes
89. (a) 0, 4   **(b)** $(0, 4)$ **(c)** $(-\infty, 0) \cup (4, \infty)$
90. (a) $-3, 1$ **(b)** $[-3, 1]$ **(c)** $(-\infty, -3] \cup [1, \infty)$
91. $-22, 22$ **92.** 1, 8   **93.** $-26, 42$ **94.** $-\frac{23}{3}, \frac{25}{3}$
95. 0, 2   **96.** $-3, \frac{9}{5}$ **97. (a)** $-8, 6$ **(b)** $[-8, 6]$
(c) $(-\infty, -8] \cup [6, \infty)$
98. (a) $-\frac{5}{2}, \frac{7}{2}$ **(b)** $\left[-\frac{5}{2}, \frac{7}{2}\right]$ **(c)** $\left(-\infty, -\frac{5}{2}\right] \cup \left[\frac{7}{2}, \infty\right)$
99. $(-\infty, -3) \cup (3, \infty)$ **100.** $(-4, 4)$ **101.** $[-3, 4]$
102. $\left(-\infty, -\frac{5}{2}\right] \cup [5, \infty)$ **103.** $[4.4, 4.6]$
104. $\left[\frac{3}{13}, \frac{7}{13}\right]$ **105.** $(-\infty, \infty)$ **106.** $\frac{3}{2}$
107. $(-\infty, -1.5] \cup [1.5, \infty)$ **108.** $[-2, 6]$
109. $|x| \leq 0.05$ **110.** $|5x - 1| > 4$
111. (a) About 25.1   **(b)** Decreased
$[1885, 1965, 10]$ by $[22, 26, 1]$

(c) -0.0492; the median age decreased by about 0.0492 year per year.
112. (a) 2.1 million   **(b)** The number of marriages each year did not change.
113. (a) $f(x) = 8x$ **(b)** 8   **(c)** The total fat increases at the rate of 8 grams per cup.
114. (a)

(b) $f(x) = -0.1675x + 350$
(c) About 15 per 1000 people (answers may vary)
115. (a) $f(2010) = 128$; in 2010, there were 128 days with an unhealthy air quality in Los Angeles.
(b) $D = \{2006, 2008, 2010, 2012, 2014\}$;
$R = \{120, 122, 128, 145, 148\}$ **(c)** It decreased.
116. (a) Linear

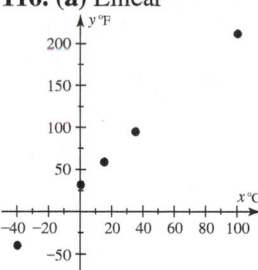

(b) $f(x) = \frac{9}{5}x + 32$; a 1 °C change equals a $\frac{9}{5}$ °F change.
(c) 68 °F   **117. (a)** 3 hours   **(b)** 2 hours and 4 hours
(c) Between 2 and 4 hours, exclusive   **(d)** 20 miles per hour
118. (a) $f(x) = -0.05x + 1.3$ **(b)** About 1.15 million
119. (a) $|A - 3.9| \le 1.7$ **(b)** $2.2 \le A \le 5.6$
120. Values between 32.2 and 37.8, exclusive

Chapter 8 Test (pp. 607–608)

1. 46; $(4, 46)$
2. $C(x) = 4x$; $C(5) = 20$, 5 pounds of candy costs $20.
3. (a)   **(b)**

(c)   **(d)**

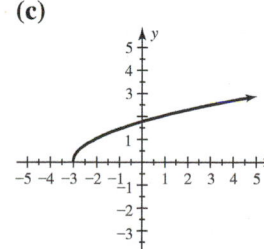

4. $0, -3$; D: $-3 \le x \le 3$; R: $-3 \le y \le 0$
5. Symbolic: $f(x) = x^2 - 5$
Numerical:

x	-3	-2	-1	0	1	2	3
$y = f(x)$	4	-1	-4	-5	-4	-1	4

Graphical:

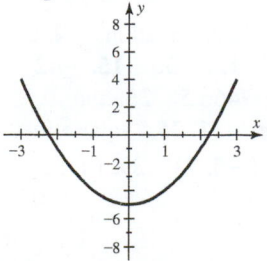

6. No, it fails the vertical line test.
7. (a) $D = \{-2, -1, 0, 5\}$ **(b)** All real numbers
(c) $x \ge -4$ **(d)** All real numbers   **(e)** $x \ne 5$
8. Yes; $f(x) = -8x + 6$
9.

10. $(-\infty, -2) \cup (1, \infty)$ **11. (a)** -5 **(b)** 5
(c) $[-5, 5]$ **(d)** $(-\infty, 5)$ **12.** $(-8, 0)$
13. $-12, 24$ **14. (a)** $[-5, 5]$ **(b)** $(-\infty, 0) \cup (0, \infty)$
15. Yes; 3; cubic   **16.** $\frac{8}{7}$; $(-\infty, 5) \cup (5, \infty)$
17. (a) 9   **(b)** $2x^3 + 2x$
18. (a) $f(x) = 0.4x + 75$ **(b)** 35 minutes
19. (a) Yes; $P(0) = 100$, $P(20) = 134$, $P(30) = 150$,
$P(50) = 180$ **(b)** Probably not because the race is over after 50 seconds; $[0, 50]$.
20. (a) $x = 25$
$[0, 25, 5]$ by $[0, 2, 0.5]$

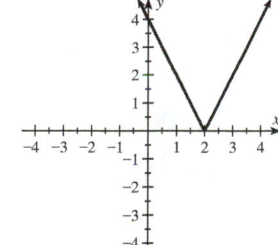

(b) 24 vehicles/min

Chapters 1–8 Cumulative Review (pp. 608–609)

1. $2^3 \cdot 3 \cdot 5$ **2.** $3n - 4 = n$; 2   **3.** $-\frac{1}{6}$ **4.** 6
5. 15   **6.** 7.5%   **7.** $a = \frac{3A + b}{2}$ **8.** $\{t \mid t > 2\}$
9.   **10.**

11. $y = -\frac{1}{2}x + 2$ **12.** $(1, 1)$
13.   **14.**

15.

16. $-x^2 + 2x - 9$ **17.** $\frac{1}{x}$ **18.** $6b$ **19.** $24t^3$
20. $\frac{1}{4x^2y^2}$ **21.** $2x^5 - 8x^4 - 10x^2$ **22.** $10x^2 - 33x - 7$

23. $y^2 - 9$ **24.** $x^2 - 8xy + 16y^2$ **25.** $25,000$
26. 2.8×10^{-2} **27.** $3x^2 - 2x + 4$
28. $3x^2 + 5x + 5 + \frac{6}{x-1}$ **29.** $5x^2y^2(2y - 3x)$
30. $(x - 1)(x + 1)(x + 3)$ **31.** $(z - 1)(2z + 3)$
32. $(4x - 5)(4x + 5)$ **33.** $(a - 2)(a^2 + 2a + 4)$
34. $(z^2 + 1)(z^2 + 6)$ **35.** $-5, 0$ **36.** 0
37. $-3, \frac{1}{2}$ **38.** $-1, 0, 1$ **39.** $x + 2$ **40.** $-\frac{5}{3}$; undefined
41. $2x - 4$ **42.** $\frac{3x + 7}{(x + 1)(x + 3)}$ **43.** -8 **44.** 10 **45.** 12
46.

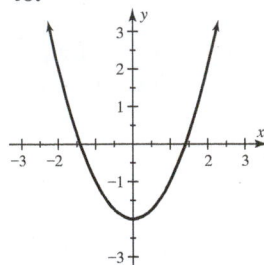

$D: (-\infty, \infty); R: [-2, \infty)$
47. $(-\infty, 5) \cup (8, \infty)$ **48.** $-1, 5$ **49.** $[-1, 5]$
50. $(-\infty, 2) \cup (6, \infty)$ **51.** $d = 72t$
52. (a) $y = \frac{1}{4}x + 2$ (b) $\frac{1}{4}$
(c) Rain is falling at a rate of $\frac{1}{4}$ inch per hour.
(d) 3 inches **53.** $2100 at 5%, $2900 at 7% **54.** 104 feet
55. (a) $20x$ (b) 50 min **56.** $1650

9 Systems of Linear Equations

Section 9.1 (pp. 620–622)

1. No; three planes cannot intersect at exactly 2 points.
2. $x + y + z = 5, 2x - 3y + z = 7, x + 2y - 4z = 2$
(answers may vary) **3.** Yes **4.** No; a solution must be an
ordered triple. **5.** Two **6.** Three **7.** no **8.** Infinitely
many **9.** $(1, 2, 3)$ **11.** $(-1, 1, 2)$ **13.** $(3, -1, 1)$
15. $\left(\frac{17}{2}, -\frac{3}{2}, 2\right)$ **17.** $(5, -2, -2)$ **19.** $(11, 2, 2)$
21. $(1, -1, 2)$ **23.** $(0, -3, 2)$ **25.** $(-1, 3, -2)$
27. $(-1, 2, 4)$ **29.** $(8, 5, 2)$ **31.** No solutions
33. $(4 - z, 2, z)$ **35.** $(3, -3, 0)$ **37.** $\left(-\frac{3}{2}, 5, -10\right)$
39. $\left(\frac{3}{2}, 1, -\frac{1}{2}\right)$ **41.** No solutions **43.** $(4 - z, 1, z)$
45. No solutions **47.** (a) $x + 2y + 4z = 10$,
$x + 4y + 6z = 15, 3y + 2z = 6$ (b) $(2, 1, 1.5)$;
a hamburger costs $2, fries $1, and a soft drink $1.50.
49. (a) $x + y + z = 180, x - z = 55, x - y - z = -10$
(b) $x = 85°, y = 65°$, and $z = 30°$ (c) These values
check. **51.** $110°, 50°, 20°$
53. (a) $a + 600b + 4c = 525, a + 400b + 2c = 365$,
$a + 900b + 5c = 805$ (b) $a = 5, b = 1, c = -20$,
$F = 5 + A - 20W$ (c) 445 fawns
55. (a) $N + P + K = 80, N + P - K = 8, 9P - K = 0$
(b) $(40, 4, 36)$; 40 pounds nitrogen, 4 pounds phosphorus,
36 pounds potassium **57.** $7500 at 4%, $8500 at 5%, and
$14,000 at 7.5%

Section 9.2 (pp. 631–634)

1. A rectangular array of numbers
2. $\begin{bmatrix} 2 & 1 & 3 \\ 0 & -4 & 2 \end{bmatrix}$ is 2×3 (answers may vary).

3. $\begin{bmatrix} 1 & 3 & | & 10 \\ 2 & -6 & | & 4 \end{bmatrix}$ is 2×3 (answers may vary).

4. 3×4 **5.** $\begin{bmatrix} 1 & 0 & | & -3 \\ 0 & 1 & | & 5 \end{bmatrix}$ (answers may vary)

6. $4, 2, -3$ **7.** 3×3 **9.** 3×2

11. $\begin{bmatrix} 1 & -3 & | & 1 \\ -1 & 3 & | & -1 \end{bmatrix}$ **13.** $\begin{bmatrix} 2 & -1 & 2 & | & -4 \\ 1 & -2 & 0 & | & 2 \\ -1 & 1 & -2 & | & -6 \end{bmatrix}$

15. $x + 2y = -6, 5x - y = 4$ **17.** $x - y + 2z = 6$,
$2x + y - 2z = 1, -x + 2y - z = 3$
19. $x = 4, y = -2, z = 7$

21. $\begin{bmatrix} 0 & 1 & 1 & 1 \\ 1 & 0 & 1 & 0 \\ 0 & 0 & 0 & 1 \\ 1 & 0 & 1 & 0 \end{bmatrix}$ **23.** $\begin{bmatrix} 0 & 1 & 0 & 0 \\ 0 & 0 & 0 & 0 \\ 1 & 1 & 0 & 1 \\ 0 & 1 & 0 & 0 \end{bmatrix}$ **25.** $(1, 3)$

27. $\left(-\frac{3}{2}, 2\right)$ **29.** $\left(\frac{7}{2}, -\frac{3}{2}\right)$ **31.** $\left(\frac{1}{2}, \frac{3}{2}\right)$ **33.** $(1, 2, 3)$
35. $(3, -3, 3)$ **37.** $(-1, 1, 0)$ **39.** $(1, 1, 1)$
41. $(-3, 2, 2)$ **43.** $(-7, 5)$ **45.** $(1, 2, 3)$
47. $(1, 0.5, -3)$ **49.** $(0.5, 0.25, -1)$
51. $(0.5, -0.2, 1.7)$ **53.** Dependent **55.** Inconsistent
57. Dependent **59.** $\left(\frac{1}{a}, \frac{1}{b}, \frac{2}{ab}\right)$ **61.** 214 pounds
63. (a) $a + 16b + 26c = 80$
     $a + 28b + 45c = 344$
     $a + 31b + 54c = 416$
(b) $a \approx -272.9, b \approx 19.8$, and $c \approx 1.4$ (c) About 216 lb
65. $\frac{1}{2}$ hour at 5 miles per hour, 1 hour at 6 miles per hour,
and $\frac{1}{2}$ hour at 8 miles per hour **67.** $500 at 3%, $1000 at
4%, and $1500 at 6% **69.** Answers may vary.

71. (a) $\begin{bmatrix} 1 & 19 & 57.5 & 32 & | & 125 \\ 1 & 26 & 65 & 42 & | & 316 \\ 1 & 30 & 72 & 48 & | & 436 \\ 1 & 30.5 & 75 & 54 & | & 514 \end{bmatrix}$

(b) $a \approx -552.272, b \approx 8.733, c \approx 2.859, d \approx 10.843$

Checking Basic Concepts 9.1 & 9.2 (pp. 634–635)

1. $(1, 3, -1)$ **2.** $(1, 2, 3)$ **3.** $(-2, 2, -1)$ **4.** $200
at 1%; $400 at 2%; $900 at 4%

Section 9.3 (pp. 640–641)

1. square **2.** number **3.** system of linear equations **4.** 0
5. -2 **7.** -53 **9.** -323 **11.** -5 **13.** -36 **15.** -42
17. -50 **19.** 0 **21.** -3555 **23.** -7466.5 **25.** abc
27. 15 square feet **29.** 52 square feet **31.** 25.5 square feet
33. $(2, -2)$ **35.** $\left(-\frac{29}{11}, -\frac{34}{11}\right)$ **37.** $(-5, 7)$ **39.** $(1, 3, 2)$
41. $(1, -1, 1)$ **43.** $(-1, 0, 4)$

Checking Basic Concepts 9.3 (p. 642)

1. (a) -1 **(b)** 17 **2.** $(-4, 6)$ **3.** 21 square units

Chapter 9 Review (pp. 643–645)

1. Yes   **2.** $(1, -1, 2)$ **3.** $(-4, 3, 2)$ **4.** $(-1, 5, -3)$
5. $(-1, 3, 2)$ **6.** $(1, 1, 1)$ **7.** No solutions
8. $(2 - z, 2z - 1, z)$

9. $\begin{bmatrix} 1 & 1 & 1 & | & -6 \\ 1 & 2 & 1 & | & -8 \\ 0 & 1 & 1 & | & -5 \end{bmatrix}$; $(-1, -2, -3)$

10. $\begin{bmatrix} 1 & 1 & 1 & | & -3 \\ -1 & 1 & 0 & | & 5 \\ 0 & 1 & 1 & | & -1 \end{bmatrix}$; $(-2, 3, -4)$

11. $\begin{bmatrix} 1 & 2 & -1 & | & 1 \\ -1 & 1 & -2 & | & 5 \\ 0 & 2 & 1 & | & 10 \end{bmatrix}$; $(-5, 4, 2)$

12. $\begin{bmatrix} 2 & 2 & -2 & | & -14 \\ -2 & -3 & 2 & | & 12 \\ 1 & 1 & -4 & | & -22 \end{bmatrix}$; $(-4, 2, 5)$

13. $(-7, 4, 2)$ **14.** $(5.4, 2.1, 9.7)$ **15.** -8 **16.** 30
17. 89   **18.** 130   **19.** 181,845   **20.** 67.688   **21.** 46 square
feet   **22.** 128 square feet   **23.** $(2, -1)$ **24.** $(-5, 3)$
25. 5489 in 1994; 4641 in 2004   **26.** \$8 tickets:
285; \$12 tickets: 195   **27. (a)** $m + 3c + 5b = 14$,
$m + 2c + 4b = 11$, $c + 3b = 5$ **(b)** Malts: \$3; cones:
\$2; bars: \$1   **28.** $100°, 65°$, and $15°$ **29.** 2 pounds of
\$1.50 candy, 4 pounds of \$2 candy, 6 pounds of \$2.50 candy
30. (a) $a + 202b + 63c = 40$, $a + 365b + 70c = 50$,
$a + 446b + 77c = 55$ **(b)** $a \approx 27.134$; $b \approx 0.061$,
$c \approx 0.009$ **(c)** About 46 inches

Chapter 9 Test (p. 645)

1. No. Three planes cannot intersect at exactly three points.
2. $(1, -2, -2)$ **3.** $(-4, 2, -5)$ **4.** No solutions
5. $(1, 2, 3)$ **6.** Infinitely many solutions: $(1, z, z)$

7. (a) $\begin{bmatrix} 2 & -4 & | & -10 \\ -3 & -2 & | & 7 \end{bmatrix}$ **(b)** $(-3, 1)$

8. (a) $\begin{bmatrix} 1 & 1 & 1 & | & 2 \\ 1 & -1 & -1 & | & 3 \\ 2 & 2 & 1 & | & 6 \end{bmatrix}$ **(b)** $\left(\frac{5}{2}, \frac{3}{2}, -2\right)$

9. 6   **10.** 114   **11.** $\left(-\frac{47}{2}, -\frac{83}{2}\right)$ **12.** 6 mph: 30 min;
7 mph: 12 min; 9 mph: 18 min   **13.** $85°, 60°$, and $35°$
14. (a) $a + 20b + 25c = 168$; $a + 24b + 40c = 270$;
$a + 30b + 50c = 405$
(b) $a = -270$, $b = 20.1$, $c = 1.44$ **(c)** About 316 lb

Chapters 1–9 Cumulative Review (pp. 646–648)

1. $2^3 \cdot 3^2 \cdot 5$ **2.** $2n + 7 = n - 2$; -9 **3.** $\frac{23}{21}$ **4.** $-\frac{1}{6}$
5. 18   **6.** $\frac{1}{10}$ **7.** $\frac{6}{5}$ **8.** 2   **9.** $\frac{31}{25}$; 1.24   **10.** 6 miles
11. $t < \frac{4}{7}$

12.

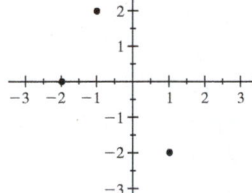

13. x-int: $(-4, 0)$; y-int: $(0, 3)$

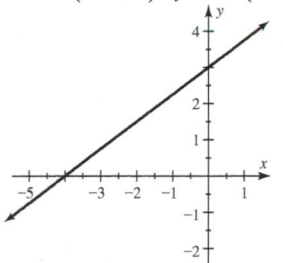

14. x-int: $(-2, 0)$; y-int: none

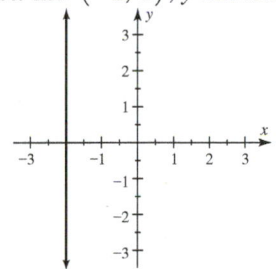

15. x-int: $(2, 0)$; y-int: $(0, 4)$; $y = -2x + 4$
16. $y = -2x + 1$ **17.** $y = \frac{3}{5}x + \frac{34}{5}$ **18.** $y = 2x + 3$
19. $P = 500x + 4000$ **20.** $(7.5, 10)$

21.   **22.**

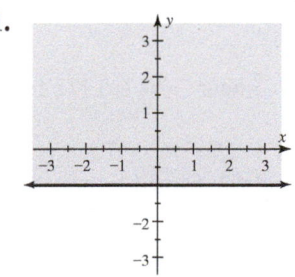

23. -76 **24.** $24t^4$ **25.** 1   **26.** $\frac{1}{4t^6}$ **27.** $2ab^3$ **28.** $\frac{a^6}{8b^6}$
29. $2a^4 - 4a^3 + 6a^2$ **30.** $5x^2 - 34x - 7$
31. $4x^2 + 12xy + 9y^2$ **32.** $a^2 - b^2$ **33.** $2x^2 - 4x + 3$
34. $x^3 - 4x^2 + 3x - 2 + \frac{1}{x - 5}$ **35.** $5ab^2(2 - 5a^2b^3)$
36. $(y - 3)(y^2 + 2)$ **37.** $(2z + 3)(3z - 1)$
38. $(2z - 3)(2z + 3)$ **39.** $(2y - 5)^2$
40. $(a - 3)(a^2 + 3a + 9)$ **41.** $(z^2 - 3)(4z^2 - 5)$
42. $ab(a + b)(2a - b)$ **43.** $-2, 1$ **44.** $-\frac{1}{3}, \frac{3}{2}$
45. $x - 4$ **46.** $\frac{x - 6}{3x - 4}$ **47.** $2(x - 1)$ or $2x - 2$ **48.** 1
49. 8   **50.** $\frac{17}{12}$ **51.** 10   **52.** 13

53. $D = (-\infty, \infty); R = [-2, \infty)$

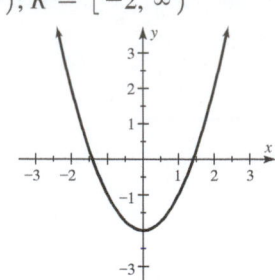

54. $2, -2$ **55.** $(3, \infty)$ **56.** $(-\infty, 0]$
57. $(-1, 5]$

58. $\left(-\infty, -\frac{3}{2}\right) \cup [3, \infty)$

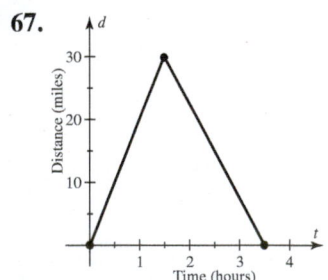

59. $[-2, 4]$ **60.** $[-16, 10]$ **61. (a)** $-3, 1$ **(b)** $[-3, 1]$
(c) $(-\infty, -3] \cup [1, \infty)$ **62.** $-6, 18$
63. $(-\infty, -6) \cup \left(\frac{8}{3}, \infty\right)$ **64.** $[4, 7]$ **65.** $(1, 1, 2)$
66. $\begin{bmatrix} 1 & 1 & -1 & | & 4 \\ -1 & -1 & -1 & | & 0 \\ 1 & -2 & 1 & | & -9 \end{bmatrix}; (-1, 3, -2)$

67.

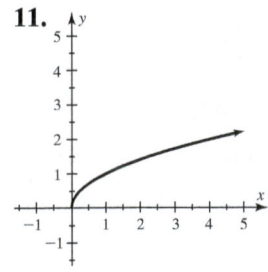

68. 3 children and 5 adults **69.** $\frac{6}{7}$ hour **70. (a)** 112 feet
(b) 5.5 seconds **71.** 61.25 feet
72. (a) $4b + 3f + 4m = 23; b + 2f + m = 7;$
$3b + f + 2m = 13$ **(b)** Burger: \$2; fries: \$1; malt: \$3

10 Radical Expressions and Functions

Section 10.1 (pp. 658–661)

1. ± 3 **2.** 3 **3.** 2 **4.** Yes **5.** b **6.** $|x|$
7. Not a real number **8.** -3 **9.** d **10.** a.
11.

12.

13. $\{x \mid x \geq 0\}$ **14.** All real numbers **15.** 3 **17.** 0.6
19. $\frac{4}{5}$ **21.** x **23.** 3 **25.** -4 **27.** $\frac{2}{3}$ **29.** $-x^3$ **31.** $3x$
33. 3 **35.** -3 **37.** Not a real number **39.** 3.32
41. 1.71 **43.** -1.48 **45.** 4 **47.** $|y|$ **49.** $|x - 5|$
51. $|x - 1|$ **53.** $|y|$ **55.** $|x^3|$ **57.** x **59.** 3, not a

real number **61.** $\sqrt{6}$, not a real number **63.** $\sqrt{20}$ or
$2\sqrt{5}, \sqrt{6}$ **65.** 1, 2 **67.** $-3, 2$ **69.** $-1, \sqrt[3]{-6}$ or $-\sqrt[3]{6}$
71. 4 **73.** 5 **75.** $[-2, \infty)$ **77.** $[2, \infty)$ **79.** $[2, \infty)$
81. $(-\infty, 1]$ **83.** $\left(-\infty, \frac{8}{5}\right]$ **85.** $(-\infty, \infty)$ **87.** $\left(-\frac{1}{2}, \infty\right)$
89. $[-2, \infty)$ **91.** $[0, \infty)$

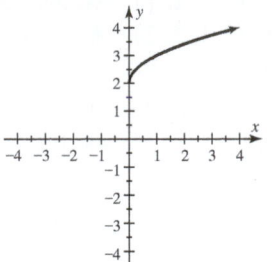

Shifted 2 units left   Shifted 2 units upward

93. $(-\infty, \infty)$

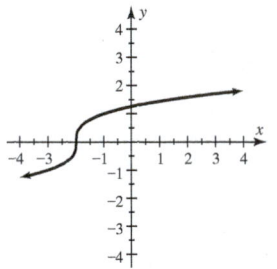

Shifted 2 units left

95.

x	$\sqrt{x} + 1$
-1	—
0	1
1	2
4	3
9	4

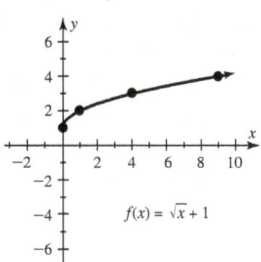

$f(x) = \sqrt{x} + 1$

97.

x	$\sqrt{3x}$
-1	—
0	0
$\frac{1}{3}$	1
$\frac{4}{3}$	2
3	3

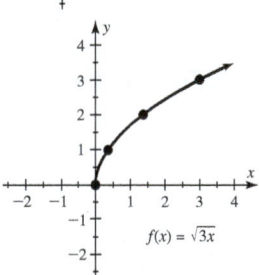

$f(x) = \sqrt{3x}$

99.

x	$2\sqrt[3]{x}$
-8	-4
-1	-2
0	0
1	2
8	4

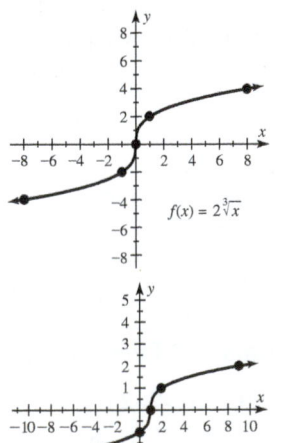

$f(x) = 2\sqrt[3]{x}$

101.

x	$\sqrt[3]{x-1}$
-7	-2
0	-1
1	0
2	1
9	2

$f(x) = \sqrt[3]{x-1}$

103. $A = 6$ **105.** 1 sec **107.** 122 mi
109. About $14 thousand; no
111. (a) $P(25,000) = \$71,246$; if \$25,000 is spent on equipment per worker, each worker will produce about \$71,246 worth of goods.

(b)

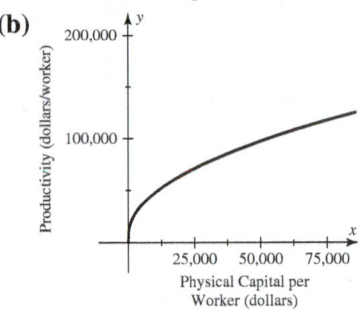

Productivity (dollars/worker)
Physical Capital per Worker (dollars)

(c) \$26,197; \$20,102; an additional \$25,000 is spent on equipment per worker, but productivity levels off. There is a point where the business starts to lose money.
113. (a) 50, about 82, about 95 (b) In 2012, about 95% of the soybeans that were planted were genetically engineered. (c) For 1990, $x = -10$ and the equation is undefined for negative values of x.

Section 10.2 (pp. 667–669)

1. 2 **2.** 2 **3.** $\frac{1}{2}$ **4.** $\frac{1}{2}$ **5.** $x^{1/2}$ **6.** $a^{4/3}$ **7.** $\sqrt[n]{a}$
8. $\left(\sqrt[n]{a}\right)^m$ or $\sqrt[n]{a^m}$ **9.** c. **10.** f. **11.** g. **12.** h. **13.** d.
14. e. **15.** a. **16.** b. **17.** 1.74 **19.** 1.71 **21.** 3.74
23. 0.55 **25.** $\sqrt[5]{7}$ **27.** $\sqrt[3]{a}$ **29.** $\sqrt[6]{x^5}$ **31.** $\sqrt{x + y}$
33. $\frac{1}{\sqrt[3]{b^2}}$ **35.** $t^{1/2}$ **37.** $(x + 1)^{1/3}$ **39.** $(x + 1)^{-1/2}$
41. $(a^2 - b^2)^{1/2}$ **43.** $x^{-7/3}$ **45.** $\sqrt{9}; 3$ **47.** $\sqrt[3]{8}; 2$
49. $\sqrt{\frac{4}{9}}; \frac{2}{3}$ **51.** $\sqrt[3]{(-8)^2}$ or $\left(\sqrt[3]{-8}\right)^2; 4$ **53.** $\sqrt[3]{8}; 2$
55. $\frac{1}{\sqrt[4]{16^3}}$ or $\frac{1}{\left(\sqrt[4]{16}\right)^3}; \frac{1}{8}$ **57.** $\frac{1}{\left(\sqrt{4}\right)^3}; \frac{1}{8}$ **59.** $\sqrt[4]{z}$ **61.** $\frac{1}{\sqrt[5]{y^2}}$
63. $\sqrt[3]{3x}$ **65.** x **67.** $2x^{2/3}$ **69.** $\frac{7}{x^{1/6}}$ **71.** x^3 **73.** xy^2
75. $y^{13/6}$ **77.** $\frac{x^4}{9}$ **79.** $\frac{y^3}{x}$ **81.** $y^{1/4}$ **83.** $\frac{1}{a^{2/3}}$ **85.** $\frac{1}{k^2}$
87. $b^{3/4}$ **89.** $p^2 + p$ **91.** $x^{5/6} - x$ **93.** $\frac{3}{x^{1/6}}$ **95.** b^3
97. (a)

x	0	20	40	60	80
$A(x)$	0	27%	37%	44%	50%

(b) The abandonment rate levels off. The longer a person watches a video, the more likely he or she will continue to watch. **99.** About 1 step/sec **101.** (a) 400 in^2
(b) $A = 100W^{2/3}$ **103.** (a) 42.4 cm; 44.7 cm
(b) First year **105.** Answers may vary.

Checking Basic Concepts 10.1 & 10.2 (p. 669)

1. (a) ± 7 (b) 7 **2.** (a) -2 (b) -3 **3.** (a) $\sqrt{x^3}$ or $\left(\sqrt{x}\right)^3$ (b) $\sqrt[3]{x^2}$ or $\left(\sqrt[3]{x}\right)^2$ (c) $\frac{1}{\sqrt[5]{x^2}}$ or $\frac{1}{\left(\sqrt[5]{x}\right)^2}$
4. $|x - 1|$ **5.** (a) 3 (b) 5 (c) $\left(\frac{12}{7}\right)^{7/12}$ or about 1.37.

Section 10.3 (pp. 676–678)

1. Yes **2.** No **3.** $\sqrt[3]{ab}$ **4.** $\frac{a}{b}$ **5.** $\frac{a}{b}$ **6.** $\frac{1}{3}$ **7.** No
8. Yes **9.** No, since $1^3 \neq 3$ **10.** Yes; $4^3 = 64$
11. 3 **13.** 10 **15.** 4 **17.** $\frac{3}{5}$ **19.** $\frac{1}{4}$ **21.** $\frac{2}{3}$ **23.** x^3

25. $\frac{\sqrt[3]{7}}{3}$ **27.** $\frac{\sqrt[4]{x}}{3}$ **29.** $\frac{x}{9}$ **31.** $\frac{x}{4}$ **33.** 3 **35.** 4 **37.** 3
39. -2 **41.** $2y$ **43.** 3 **45.** $2x^2$ **47.** $-a^2\sqrt[3]{5}$
49. $2x\sqrt[4]{y}$ **51.** $6x^3$ **53.** $4x^2y^3$ **55.** $\frac{3}{2}$ **57.** $\sqrt[3]{12ab}$
59. $5\sqrt{z}$ **61.** $\frac{1}{a}$ **63.** $\sqrt{x^2 - 16}$ **65.** $\sqrt[3]{a^3 + 1}$
67. $\sqrt{x + 1}$ **69.** 10 **71.** 2 **73.** 3 **75.** $10\sqrt{2}$ **77.** $3\sqrt[3]{3}$
79. $2\sqrt{2}$ **81.** $-2\sqrt[5]{2}$ **83.** $b^2\sqrt{b}$ **85.** $2n\sqrt{2n}$
87. $2ab^2\sqrt{3b}$ **89.** $-5xy\sqrt[3]{xy^2}$ **91.** $5\sqrt[3]{5t^2}$ **93.** $\frac{3t}{r\sqrt[4]{5r^3}}$
95. $\frac{3\sqrt[3]{x^2}}{y}$ **97.** $\frac{7a\sqrt{a}}{9}$
99. $\left(\sqrt[mn]{a^m b^m}\right)^n = (a^m b^m)^{n/mn}$
$\quad\quad = (a^m b^m)^{1/m}$
$\quad\quad = a^{m/m} b^{m/m}$
$\quad\quad = ab$
101. $\sqrt[6]{3^5}$ **103.** $2\sqrt[12]{2^5}$ **105.** $3\sqrt[12]{3^{11}}$ **107.** $x\sqrt[12]{x}$
109. $\sqrt[12]{r^{11}t^7}$ **111.** (a) $S = 5\sqrt{M}$ (b) 50 miles per hour
113. (a) 12; there is a 12% price markup 4 days prior to game day. (b) $y = 6\sqrt{x}$ (c) 12; yes **115.** (a) $\frac{9}{2}$ sec
(b) $T = \frac{\sqrt{h}}{2}$

Section 10.4 (pp. 688–690)

1. $2\sqrt{a}$ **2.** $3\sqrt[3]{b}$ **3.** like
4. Yes; $4\sqrt{15} - 3\sqrt{15} = \sqrt{15}$
5. No **6.** $\frac{\sqrt{7}}{\sqrt{7}}$ **7.** $\sqrt{t} + 5$ **8.** $\frac{5 + \sqrt{2}}{5 + \sqrt{2}}$
9. Not possible **11.** $\sqrt{7}, 2\sqrt{7}, 3\sqrt{7}$ **13.** $2\sqrt[3]{2}, -3\sqrt[3]{2}$
15. Not possible **17.** $2\sqrt[3]{xy}, xy\sqrt[3]{xy}$ **19.** $9\sqrt{3}$ **21.** $6\sqrt[3]{5}$
23. Not possible **25.** Not possible **27.** Not possible
29. $5\sqrt[3]{2}$ **31.** $8\sqrt{2}$ **33.** $6\sqrt{11}$ **35.** $2\sqrt{x} - \sqrt{y}$
37. $2\sqrt[3]{z}$ **39.** $-5\sqrt[3]{6}$ **41.** $y^2 - y$ **43.** $9\sqrt{7}$ **45.** $6\sqrt[4]{3}$
47. $7\sqrt{x}$ **49.** $8\sqrt{2k}$ **51.** $-2\sqrt{11}$ **53.** $5\sqrt[3]{2} - \sqrt{2}$
55. $-\sqrt[3]{xy}$ **57.** $3\sqrt{x + 2}$ **59.** $\sqrt{x + 2}$
61. $(x - 1)\sqrt{x + 1}$ **63.** $4x\sqrt{x}$ **65.** $\frac{\sqrt[3]{7x}}{6}$ **67.** $\frac{3\sqrt{3}}{2}$
69. $\frac{71\sqrt{2}}{10}$ **71.** $2\sqrt{2}$ **73.** $(5x - 1)\sqrt[4]{x}$
75. $(8x + 2)\sqrt{x}$ or $2\sqrt{x}(4x + 1)$
77. $(3ab - 1)\sqrt[4]{ab}$ **79.** $(n - 2)\sqrt[3]{n}$
81. $(f + g)(x) = 3\sqrt{x} + 1$; $(f - g)(x) = 7\sqrt{x} - 5$
83. $(f + g)(x) = 4\sqrt[3]{x}$; $(f - g)(x) = 2$
85. $x - \sqrt{x} - 6$ **87.** 2 **89.** 119 **91.** $x - 64$
93. $ab - c$ **95.** $x + \sqrt{x} - 56$ **97.** $\frac{\sqrt{7}}{7}$ **99.** $\frac{4\sqrt{3}}{3}$
101. $\frac{\sqrt{5}}{3}$ **103.** $\frac{\sqrt{3b}}{6}$ **105.** $\frac{t\sqrt{r}}{2r}$ **107.** $\frac{3 + \sqrt{2}}{7}$
109. $\sqrt{10} - 2\sqrt{2}$ **111.** $\frac{11 - 4\sqrt{7}}{3}$ **113.** $\sqrt{7} + \sqrt{6}$
115. $\frac{z + 3\sqrt{z}}{z - 9}$ **117.** $\frac{a + 2\sqrt{ab} + b}{a - b}$ **119.** $\sqrt{x + 1} + \sqrt{x}$
121. $\frac{3\sqrt[3]{x^2}}{x}$ **123.** $\frac{\sqrt[3]{x}}{x}$ **125.** $12\sqrt{3} \approx 20.8$ cm **127.** $2\sqrt{6}$
129. $120\sqrt{2}$ ft **131.** $\sqrt{2x}$ ft **133.** 850 ft^3/sec
135. $P = 325$; there were about 325 thousand adults incarcerated in state or federal prisons for drug law violations in the year 2000. $J = 105$; there were about 105 thousand adults incarcerated in jails for drug law violations in the year 2000.
137. $T = 90\sqrt{x} + 70$

Checking Basic Concepts 10.3 & 10.4 (p. 690)

1. (a) $\frac{1}{8}$ (b) 10 (c) $-2x\sqrt[3]{xy}$ (d) $\frac{4b^2}{5}$ **2.** $\sqrt[6]{7^5}$
3. (a) 6 (b) 3 (c) $6x^3$ **4.** (a) $7\sqrt{6} + \sqrt{7}$
(b) $5\sqrt[3]{x}$ (c) $\sqrt{x}$ **5.** (a) $(y - x)\sqrt[3]{xy}$ (b) 14
6. $\frac{\sqrt{6}}{2}$ **7.** $\frac{\sqrt{5} + 1}{2}$

Section 10.5 (pp. 695–698)

1.   **2.**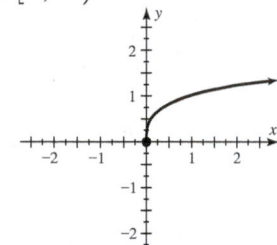

3. $\{x \mid x \geq 0\}$ **4.** All real numbers   **5.** $f(x) = x^p$, p is rational.   **6.** $f(x) = \sqrt[n]{x}$, where n is an integer greater than 1.   **7.** $\{x \mid x \geq 0\}$ **8.** All real numbers
9. $\sqrt{3} \approx 1.73$; not a real number   **11.** $\sqrt[4]{6} \approx 1.57$; not a real number   **13.** $\sqrt[5]{13} \approx 1.67$; 1   **15.** $\sqrt[6]{6} \approx 1.82$; -1
17. $f(x) = \sqrt{x}$ **19.** $f(x) = \sqrt[3]{x^2}$ **21.** $f(x) = \frac{1}{\sqrt[5]{x}}$
23. 32; 55.90   **25.** $-\frac{1}{128} \approx -0.01$; 0.04
27. 4; not possible   **29.** 4; 4
31. $[0, \infty)$ **33.** $(-\infty, \infty)$

35. $g(x)$ is greater.
$[0, 6, 1]$ by $[0, 6, 1]$

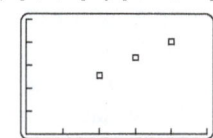

37. $f(x)$ is greater.
$[0, 6, 1]$ by $[0, 6, 1]$

39. $x^p > x^q$ **41. (a)** 6   **(b)** $\sqrt{2x}$ **(c)** $4|x|$
(d) 2   **43.** b.   **45.** c.   **47.** d.   **49.** b.
51. (a) $f(x) = (x + 2)(x - 2)$ **(b)** $f(x) = x^2 - 4$
53. (a) $f(x) = (x + \sqrt{7})(x - \sqrt{7})$
(b) $f(x) = x^2 - 7$
55. (a) $f(x) = (x + \sqrt{6})(x - \sqrt{6})$
(b) $f(x) = x^2 - 6$ **57.** 2877 in²
59. About 35%   **61.** No, for $x \geq 10$, it is less than double.
63. (a) 6 yr   **(b)** The twin in the spaceship will be 4 years younger than the twin on Earth.
65. (a) $[0, 5, 1]$ by $[0, 0.5, 0.1]$ **(b)** $k \approx 0.16$

(c) $[0, 5, 1]$ by $[0, 0.5, 0.1]$; yes   **(d)** 0.295 m²

(answers may vary)

67. (a) $k \approx 0.91$ **(b)** $[0, 1.5, 0.1]$ by $[0, 1, 0.1]$

It increases.
(c) About 0.808 m   **(d)** About 0.788; a bird weighing 0.65 kg has a wing span of about 0.788 m.

Section 10.6 (pp. 707–711)

1. Square each side.   **2.** Cube each side.   **3.** Yes
4. Check them in the given equation.   **5.** Finding an unknown side of a right triangle (answers may vary)   **6.** 5
7. $d = \sqrt{(x_2 - x_1)^2 + (y_2 - y_1)^2}$ **8.** $x^{1/2} + x^{3/4} = 2$
9. 2   **11.** x **13.** $2x + 1$ **15.** $5x^2$ **17.** 64   **19.** 81   **21.** 6
23. 48   **25.** 6   **27.** 3   **29.** 27   **31.** -2 **33.** 15   **35.** $\frac{1}{2}$
37. 0, 1   **39.** 2   **41.** -4 **43.** 9   **45.** 9   **47.** ± 7
49. ± 10 **51.** $-5, 3$ **53.** $\frac{1}{3}, \frac{11}{3}$ **55.** 4   **57.** -4
59. 1   **61.** $\frac{7}{5}$ **63.** ± 2 **65.** $\sqrt[5]{12}$ **67.** $-5, 1$ **69.** 3
71. 1.88   **73.** $-1, 0.70$ **75.** 1.79   **77.** No Solutions
79. (a–c) 16   **81. (a–c)** 11   **83.** $L = \frac{8T^2}{\pi^2}$ **85.** $A = \pi r^2$
87. Yes   **89.** Yes   **91.** Yes   **93.** No   **95.** $\sqrt{32} = 4\sqrt{2}$
97. 7   **99.** $c = 5$ **101.** $b = \sqrt{61}$ **103.** $a = 14$
105. $\sqrt{20} = 2\sqrt{5}$ **107.** $\sqrt{4000} = 20\sqrt{10}$ **109.** $\sqrt{89}$
111. 5   **113.** 4   **115.** 2, 122   **117.** $W = 8$ pounds
119. About 81 sec   **121.** About 3 miles   **123.** About 269 feet
125. 19 inches   **127.** $h \approx 16.3$ inches, $d \approx 33.3$ inches
129. (a) 121 feet   **(b)** About 336 feet   **131. (a)** About 38 miles per hour; the vehicle involved in the accident was traveling about 38 miles per hour.   **(b)** About 453 feet
133. (a) It increases by a factor of 8.   **(b)** $v = \sqrt[3]{\frac{W}{3.8}}$
(c) 20 mph   **135.** $a\sqrt{2}$ **137. (a)** $30x$
(b) $d = \sqrt{(1000 - x)^2 + 500^2}$
(c) $50\sqrt{(1000 - x)^2 + 500^2}$
(d) $30x + 50\sqrt{(1000 - x)^2 + 500^2}$
(e) $[0, 1000, 100]$ by
$[40000, 60000, 5000]$

Checking Basic Concepts 10.5 & 10.6 (p. 711)

1. (a)   **(b)**

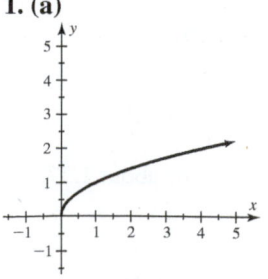

$f(-1)$ is not a real number. $f(-1) = -1$

(c)

$f(-1) = 1$

2. 3.2   **3.** $[4, \infty)$ **4. (a)** 4   **(b)** 28   **(c)** 3   **5.** 13
6. 9.6 in.   **7.** $-3, 1$

Section 10.7 (pp. 717–719)

1. $2 + 3i$ (answers may vary)   **2.** No; any real number a
can be written as $a + 0i$.   **3.** i **4.** -1 **5.** $i\sqrt{a}$
6. $10 - 7i$ **7.** $a + bi$ **8.** $1 + 2i$ **9.** 4   **10.** -5 **11.** 0
12. pure   **13.** $i\sqrt{5}$ **15.** $10i$ **17.** $12i$ **19.** $2i\sqrt{3}$
21. $3i\sqrt{2}$ **23.** 3   **25.** $-8 + 7i$ **27.** $1 - 9i$ **29.** $-10 + 7i$
31. $-13 + 13i$ **33.** $20 - 12i$ **35.** 41   **37.** 20   **39.** $1 + 5i$
41. $3 + 4i$ **43.** $-2 - 6i$ **45.** -2 **47.** $a^2 + 9b^2$
49. $-i$ **51.** i **53.** -1 **55.** 1   **57.** $3 - 4i$ **59.** $6i$
61. $5 + 4i$ **63.** -1 **65.** $1 - i$ **67.** $-\frac{6}{29} + \frac{15}{29}i$ **69.** $2 + i$
71. $-1 + 6i$ **73.** $-1 - 2i$ **75.** $-\frac{3}{2}i$ **77.** $-\frac{1}{2} + \frac{3}{2}i$
79. (a) $f(x) = (x + i)(x - i)$ **(b)** $f(x) = x^2 + 1$
81. (a) $f(x) = (x + 7i)(x - 7i)$ **(b)** $f(x) = x^2 + 49$
83. (a) $f(x) = \left(x + i\sqrt{6}\right)\left(x - i\sqrt{6}\right)$
(b) $f(x) = x^2 + 6$ **85.** $\frac{290}{13} + \frac{20}{13}i$ **87.** They are graphed
using a real axis and an imaginary axis.

Checking Basic Concepts 10.7 (p. 719)

1. (a) $8i$ **(b)** $i\sqrt{17}$ **2. (a)** $3 - 4i$ **(b)** $-2 + 3i$
(c) $5 + i$ **(d)** $\frac{3}{4} + \frac{3}{4}i$

Chapter 10 Review (pp. 722–724)

1. 2   **2.** 6   **3.** $3|x|$ **4.** $|x - 1|$ **5.** -4 **6.** -5 **7.** x^2
8. $3x$ **9.** 2   **10.** -1 **11.** x^2 **12.** $x + 1$ **13.** $\sqrt{14}$
14. $\sqrt[3]{-5}$ **15.** $\left(\sqrt{\frac{x}{y}}\right)^3$ or $\sqrt{\left(\frac{x}{y}\right)^3}$ **16.** $\frac{1}{\sqrt[3]{(xy)^2}}$ or $\frac{1}{(\sqrt[3]{xy})^2}$

17. 9   **18.** 2   **19.** 64   **20.** 27   **21.** z^2 **22.** xy^2 **23.** $\frac{x^3}{y^9}$
24. $\frac{y^2}{x}$ **25.** 8   **26.** -2 **27.** x^2 **28.** 2   **29.** $-\frac{\sqrt[3]{x}}{2}$ **30.** $\frac{1}{3}$
31. $4\sqrt{3}$ **32.** $3\sqrt{6}$ **33.** $\frac{3}{x}$ **34.** $4ab\sqrt{2a}$ **35.** $9xy$
36. $5z\sqrt[3]{z}$ **37.** $x + 1$ **38.** $\frac{2a\sqrt[4]{a}}{b}$ **39.** $2\sqrt[6]{x^5}$
40. $\sqrt[6]{r^5 t^8}$ or $t\sqrt[6]{r^5 t^2}$ **41.** $4\sqrt{3}$ **42.** $3\sqrt[3]{x}$ **43.** $-3\sqrt[3]{5}$
44. $-\sqrt[4]{y}$ **45.** $11\sqrt{3}$ **46.** $7\sqrt{2}$ **47.** $13\sqrt[3]{2}$
48. $3\sqrt{x + 1}$ **49.** $(2x - 1)\sqrt{x}$ **50.** $(b + 2a)\sqrt[3]{ab}$
51. $5 + 4\sqrt{2}$ **52.** $7 + 7\sqrt{3} - \sqrt{5} - \sqrt{15}$ **53.** 3
54. 95   **55.** $a - 2b$ **56.** $xy + \sqrt{xy} - 2$ **57.** $\frac{4\sqrt{5}}{5}$
58. $\frac{r\sqrt{t}}{2t}$ **59.** $\frac{3 - \sqrt{2}}{7}$ **60.** $\frac{5 + \sqrt{7}}{9}$
61. $\sqrt{8} + \sqrt{7}$ **62.** $\frac{a - 2\sqrt{ab} + b}{a - b}$

63.

64.

65. $f(x) = \sqrt{x}; 2$ **66.** $f(x) = \sqrt[7]{x^2}; \sqrt[7]{16}$
67.

68.

Shifted 2 units downward   Shifted 1 unit to the right

69. $[1, \infty)$ **70.** $(-\infty, 3]$ **71.** $(-\infty, \infty)$ **72.** $(-2, \infty)$
73. 2   **74.** 4   **75.** 9   **76.** 9   **77.** 8   **78.** $\frac{1}{4}$ **79.** 4.5
80. 1.62   **81.** $c = \sqrt{65}$ **82.** $b = \sqrt{39}$ **83.** $\sqrt{41}$
84. $\sqrt{52} = 2\sqrt{13}$ **85.** ± 11 **86.** ± 4 **87.** $-3, 5$
88. 4   **89.** 3   **90.** 2   **91.** ± 4 **92.** -1 **93.** 1
94. $-2, 0$ **95.** -2 **96.** $-2 + 4i$ **97.** $5 + i$ **98.** $1 + 2i$
99. $-\frac{14}{13} + \frac{5}{13}i$ **100.** $2 - 2i$ **101.** About 85 feet
102. $\sqrt{16{,}200} = 90\sqrt{2} \approx 127.3$ feet
103. About 0.79 second   **104. (a)** 5 square units
(b) $5\sqrt{5}$ cubic units   **(c)** $\sqrt{10}$ units   **(d)** $\sqrt{15}$ units
105. $r = \sqrt[210]{\frac{281}{4}} - 1 \approx 0.02$; from 1790 through 2000, the
average annual percent growth rate was about 2%.
106. (a) About 43 miles per hour   **(b)** About 34 miles per
hour; a steeper bank allows for a higher speed limit; yes
107. $\sqrt{7} \approx 2.65$ feet

Chapter 10 Test (pp. 724–725)

1. -3 **2.** $|z + 1|$ **3.** $5x^2$ **4.** $2z^2$ **5.** $2xy\sqrt[6]{y}$ **6.** 1
7. $\sqrt[5]{7^2}$ or $(\sqrt[5]{7})^2$ **8.** $\sqrt{\left(\frac{y}{x}\right)^2}$ or $\left(\sqrt[3]{\frac{y}{x}}\right)^2$ **9.** 16
10. $\frac{1}{216}$ **11.** $x^{4/3}$ **12.** $x^{7/10}$ **13.** $(-\infty, 4]$
14.

15. $8z^{3/2}$ **16.** $\frac{z}{y^{2/3}}$ **17.** 9   **18.** $\frac{y}{2}$ **19.** $4\sqrt{7} + \sqrt{5}$
20. $6\sqrt[3]{x}$ **21.** $14\sqrt{2}$ **22.** 2   **23. (a)** 27   **(b)** 7   **(c)** 3
(d) 17   **24. (a)** $\frac{2\sqrt{7}}{21}$ **(b)** $\frac{-1 + \sqrt{5}}{4}$ **25.** 2.63
26. $\sqrt{120} \approx 10.95$ **27.** $\sqrt[3]{8} = 2\sqrt{2}$ **28.** $2 - 19i$
29. $-6 + 8i$ **30.** $\frac{5}{4}$ **31.** $\frac{4}{29} + \frac{10}{29}i$ **32. (a)** About 17.25 mi
(b) About 420 ft   **33.** 1.31 pounds

Chapters 1–10 Cumulative Review (pp. 725–726)

1. 36π **2.** $D = \{-1, 0, 1\}$, $R = \{2, 4\}$ **3. (a)** $\frac{1}{a^3b^9}$
(b) x^4y^9 **(c)** r^8t^5 **4.** 4.3×10^{-4} **5.** $3; x \neq 2$
6. All real numbers

7. **8.**

9.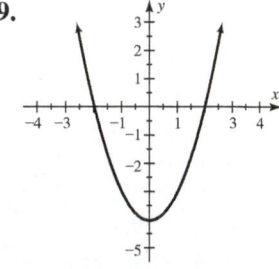

(a) D: all real numbers; R: $y \geq -4$ **(b)** 0
(c) $(-2, 0)$, $(2, 0)$ **(d)** $-2, 2$ **10.** $y = \frac{1}{2}x + \frac{5}{2}$
11. $f(x) = 1 - 3x$

12.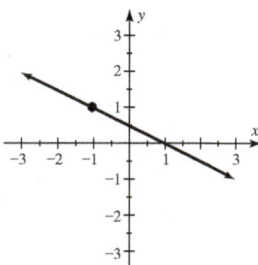

13. $\frac{6}{11}$ **14.** $(-\infty, 3]$ **15.** $[-1, 5]$ **16.** $\left[-\frac{5}{2}, 1\right]$
17. (a) $(4, 4)$ **(b)** No solutions

18.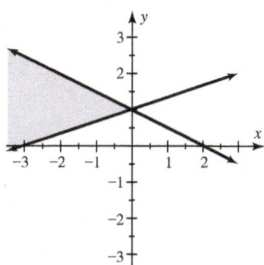

19. $(3, 2, 1)$ **20.** 4 **21.** $-4x^4 + 16x$ **22.** $x^2 - 16$
23. $5x^2 - 7x - 6$ **24.** $16x^2 + 72x + 81$
25. $(3x - 4)(3x + 4)$ **26.** $(x - 2)^2$ **27.** $3x^2(5x - 3)$
28. $(4x - 3)(3x + 1)$ **29.** $(r - 1)(r^2 + r + 1)$
30. $(x - 3)(x^2 + 5)$ **31.** $1, 2$ **32.** $-2, 0, 2$ **33.** $2x + 4$
34. $\frac{7x - 5}{x^2 - x}$ **35.** $6x$ **36.** 4 **37.** $\frac{1}{64}$ **38.** 5 **39.** $4x$
40. $x^{3/4}$ or $\sqrt[4]{x^3}$ **41.** $2x$ **42.** $6\sqrt{3x}$ **43.** $2x^2 - \sqrt{3}x - 3$
44. $(-\infty, 1]$

45.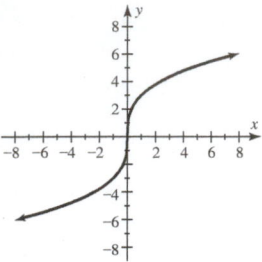

46. $\sqrt{10}$ **47.** $5 + i$ **48.** 13 **49.** 6 **50.** 28 **51.** $\frac{4}{9}, 4$
52. ± 3 **53.** 1.41 **54.** 4.06 **55.** The tank initially
contains 300 gallons of water. Water is leaving the tank
at 15 gal/min.

56.

57. $1300 at 5\%$, $700 at 4\%$ **58.** 8 in. by 8 in. by 4 in.
59. 22.4 feet **60.** $45°, 55°, 80°$

11 Quadratic Functions and Equations

Section 11.1 (pp. 737–740)

1. parabola **2.** The vertex **3.** axis of symmetry **4.** $(0, 0)$
5.

(answers may vary)

6. $-\frac{b}{2a}$ **7.** $ax^2 + bx + c$ with $a \neq 0$ **8.** vertex
9. True **10.** False **11.** False **12.** True **13.** 17
15. -8 **17.** 7 **19.** 8 **21.** $0, -4$ **23.** $-2, -2$
25. $(1, -2); x = 1;$ upward **27.** $(-2, 3); x = -2;$
downward **29.** $(2, -6)$ **31.** $(-3, 4)$
33. $(0, 3)$ **35.** $(1, 1.4)$ **37.** $(3, 9)$
39. (a) $(0, 0); x = 0$ **41. (a)** $(0, -2); x = 0$
(b) **(b)**

 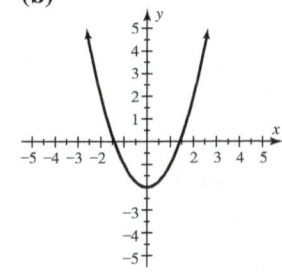

(c) $2; 4.5$ **(c)** $2; 7$

43. (a) $(0, 1)$; $x = 0$
(b)

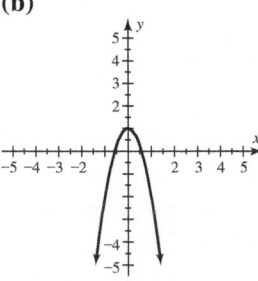

(c) -11; -26

47. (a) $(-2, 0)$; $x = -2$
(b)

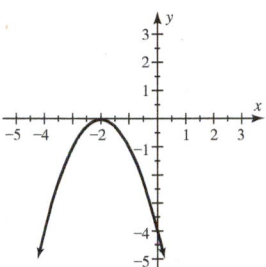

(c) 0; -25

51. (a) $(0, -3)$; $x = 0$
(b)

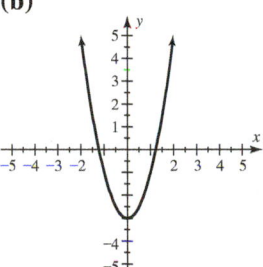

(c) 5; 15

55. (a) $(1, 1)$; $x = 1$
(b)

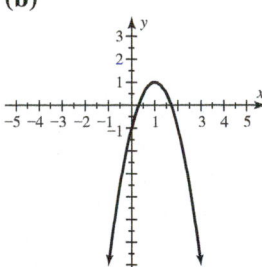

(c) -17; -7

59. Incr: $x > 2$, decr: $x < 2$

63. Incr: $x > \frac{2}{3}$, decr: $x < \frac{2}{3}$

65. Incr: $x < -\frac{3}{2}$, decr: $x > -\frac{3}{2}$

67. Incr: $x < -\frac{1}{8}$, decr: $x > -\frac{1}{8}$

69. -2 **71.** $-\frac{25}{4}$ **73.** $-\frac{7}{2}$ **75.** 6 **77.** 4
79. $-\frac{39}{8}$ **81.** $10, 10$ **83.** 3 slices **85.** d. **87.** a.
89. (a) 2 feet **(b)** 2 seconds **(c)** 66 feet
91. $\frac{66}{32} \approx 2$ seconds; about 74 feet **93. (a)** The revenue
increases at first up to 50 tickets and then it decreases.
(b) $\$2500$; 50 **(c)** $f(x) = x(100 - x)$ **(d)** $\$2500$; 50

45. (a) $(1, 0)$; $x = 1$
(b)

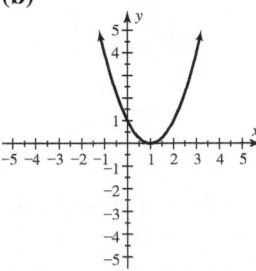

(c) 9; 4

49.
(a) $(-0.5, -2.25)$; $x = -0.5$
(b)

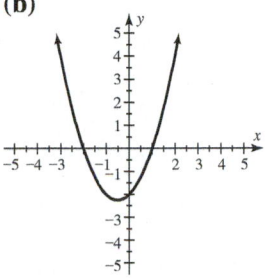

(c) 0; 10

53. (a) $(1, 1)$; $x = 1$
(b)

(c) -8; -3

57. (a) $(2, 4)$; $x = 2$
(b)

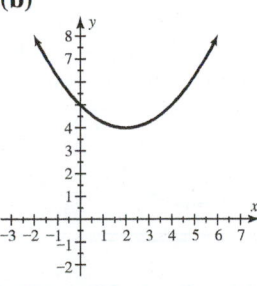

(c) 8; 4.25

61. Incr: $x < 1$, decr: $x > 1$

95. (a) $V(1) = 147.5$, $V(2) = 156$, $V(3) = 163.5$,
$V(4) = 170$; in 2011, there were 147.5 million unique Facebook visitors in one month. Other values can be interpreted similarly.   **(b)** A linear function does not model these data because these three increases are not equal (or nearly equal).
97. 300 feet by 600 feet   **99.** 42 in.   **101. (a)** 6; in 1990, emissions were 6 billion metric tons.   **(b)** 3 billion metric tons.

Section 11.2 (pp. 751–754)

1. $x^2 + 2$ **2.** $(x - 2)^2$ **3.** $(1, 2)$ **4.** $(-1, -2)$
5. $f(x) = ax^2 + bx + c$; $f(x) = a(x - h)^2 + k$
6. $y = a(x - h)^2 + k$; (h, k) **7.** downward   **8.** $-\frac{b}{2a}$
9. a.   **10.** d.
11.

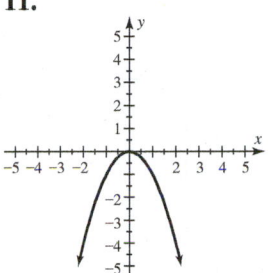

Reflected across the x-axis

13.

Narrower

15.

Wider

17.

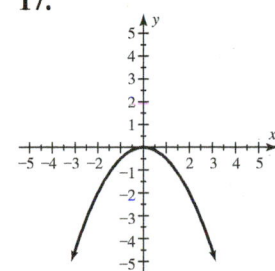

Reflected across the x-axis
and wider

19.

x	-2	-1	0	1	2
$y = x^2$	4	1	0	1	4
$y = x^2 - 3$	1	-2	-3	-2	1

Shifted 3 units downward

21.

x	-2	-1	0	1	2
$y = x^2$	4	1	0	1	4

x	1	2	3	4	5
$y = (x - 3)^2$	4	1	0	1	4

Shifted 3 units right

23. (a)

(b) $(0, -4)$
(c) Down 4 units

25. (a)

(b) $(0, 1)$
(c) Narrower and up 1 unit

27. (a)

(b) $(3, 0)$
(c) Right 3 units

31. (a)

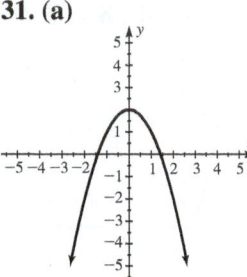

(b) $(0, 2)$
(c) Reflected across the
x-axis and up 2 units

35. (a)

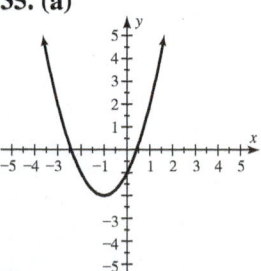

(b) $(-1, -2)$
(c) Left 1 unit and down 2 units

39. (a)

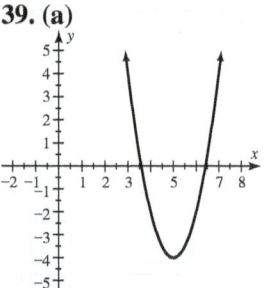

(b) $(5, -4)$
(c) Narrower, right 5 units
and down 4 units

43. Translated 1 unit right, 2
units downward, and is wider

29. (a)

(b) $(0, 0)$
(c) Reflected across the x-axis

33. (a)

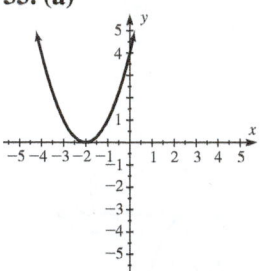

(b) $(-2, 0)$
(c) Left 2 units

37. (a)

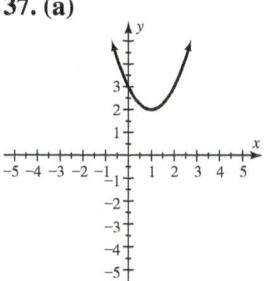

(b) $(1, 2)$
(c) Right 1 unit and up 2 units

41. (a)

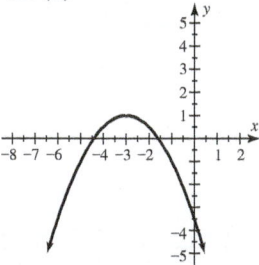

(b) $(-3, 1)$
(c) Wider, reflected across the
x-axis, left 3 units and up 1 unit

45. Translated 1 unit left,
3 units upward, opens downward, and is narrower

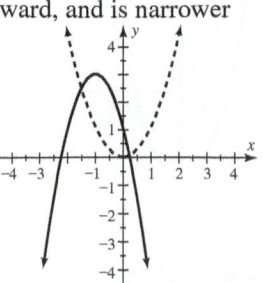

47. $[-20, 20, 2]$ by $[-20, 20, 2]$

49. $y = 3(x - 3)^2 + 4; y = 3x^2 - 18x + 31$
51. $y = -\frac{1}{2}(x - 5)^2 - 2; y = -\frac{1}{2}x^2 + 5x - \frac{29}{2}$
53. $y = (x - 1)^2 + 2$ **55.** $y = -(x - 0)^2 - 3$
57. $y = (x - 0)^2 - 3$ **59.** $y = -(x + 1)^2 + 2$
61. (a) $(1, 1)$ **(b)** $y = 4(x - 1)^2 + 1$ **63. (a)** $(-1, -2)$
(b) $y = -(x + 1)^2 - 2$ **65. (a)** $(-1, 3)$
(b) $y = -2(x + 1)^2 + 3$ **67.** 1
69. $y = (x + 1)^2 - 1; (-1, -1)$
71. $y = (x - 2)^2 - 4; (2, -4)$
73. $y = (x + 1)^2 - 4; (-1, -4)$
75. $y = (x - 2)^2 + 1; (2, 1)$
77. $y = \left(x + \frac{3}{2}\right)^2 - \frac{17}{4}; \left(-\frac{3}{2}, -\frac{17}{4}\right)$
79. $y = \left(x - \frac{7}{2}\right)^2 - \frac{45}{4}; \left(\frac{7}{2}, -\frac{45}{4}\right)$
81. $y = 3(x + 1)^2 - 4; (-1, -4)$
83. $y = 2\left(x - \frac{3}{4}\right)^2 - \frac{9}{8}; \left(\frac{3}{4}, -\frac{9}{8}\right)$
85. $y = -2(x + 2)^2 + 13; (-2, 13)$ **87.** $a = 2$
89. $a = 0.3$ **91.** $y = 2(x - 1)^2 - 3$
93. $y = 0.5(x - 1980)^2 + 6$
95. (a)   **(b)** $D(x) = \frac{1}{12}x^2$

97. (a) Decreases and then increases
(b) No, because the data decrease and then increase.
(c) Quadratic; it can model data that decrease and increase.
(d) $(1995, 600)$; it has the minimum y-value.
(e) $C(x) = 1.8(x - 1995)^2 + 600$
(f) $[1970, 2030, 10]$ by $[0, 2500, 500]$

Checking Basic Concepts 11.1 & 11.2 (p. 754)

1. (a)   **(b)**

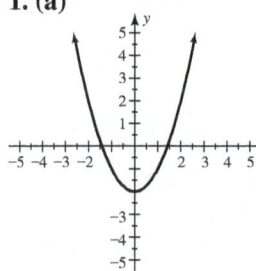

$(0, -2); x = 0$ $(1, -3); x = 1$

2. y_1 opens upward, whereas y_2 opens downward, y_1 is narrower than y_2.　**3.** 7
4. (a)　　　　　　　　**(b)**

　　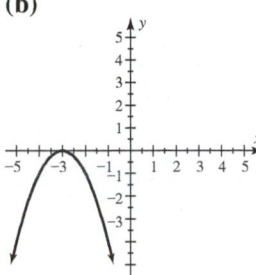

1 unit right, 2 units up　　Reflected across the
　　　　　　　　　　　x-axis, 3 units left
5. (a) $y = (x + 7)^2 - 56$　**(b)** $y = 4(x + 1)^2 - 6$

Section 11.3 (pp. 764–767)

1. $x^2 + 3x - 2 = 0$ (answers may vary); it can have 0, 1, or 2 solutions　**2.** Nonlinear　**3.** Factoring, square root property, completing the square
4.　　　　　　　　　**5.**

　　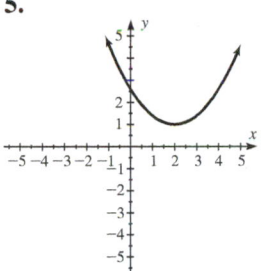

(answers may vary)　　　(answers may vary)
6. Two; the solutions are the x-coordinates of the x-intercepts.
7. ± 8; the square root property (answers may vary)
8. $\left(\frac{b}{2}\right)^2$　**9.** Yes　**11.** No　**13.** Yes　**15.** No
17. (a) $3.65, -1.65$　**(b)** $1.32, -5.32$　**19. (a)** $1.32, -0.12$
(b) $-0.28, -0.83$　**21.** $-2, 1$　**23.** No real solutions
25. $-2, 3$　**27.** -0.5　**29.** $-1, 5$　**31.** $-3, 1$
33. $-3, 3$　**35.** $-\frac{1}{2}, \frac{3}{2}$　**37.** -1　**39.** No real solutions
41. $-7, 5$　**43.** $-\frac{1}{3}, \frac{1}{2}$　**45.** $-\frac{3}{2}$　**47.** $-2, 7$　**49.** $\frac{6}{5}, \frac{3}{2}$
51. ± 12　**53.** $\pm \frac{8}{\sqrt{5}}$ or $\pm \frac{8\sqrt{5}}{5}$　**55.** $-6, 4$
57. $-7, 9$　**59.** $\frac{1 \pm \sqrt{5}}{2}$　**61.** $5 \pm \sqrt{5}$
63. $\pm 2i$　**65.** $\pm i\sqrt{3}$　**67.** $\pm 2i\sqrt{5}$　**69.** $\pm 4i$
71. $\pm 3i\sqrt{2}$　**73.** 4　**75.** $\frac{25}{4}$　**77.** $16; (x - 4)^2$
79. $\frac{81}{4}; \left(x + \frac{9}{2}\right)^2$　**81.** $-4, 6$　**83.** $-3 \pm \sqrt{11}$
85. $\frac{3 \pm \sqrt{29}}{2}$　**87.** $\frac{5 \pm \sqrt{21}}{2}$　**89.** $1 \pm \sqrt{5}$
91. $\frac{3 \pm \sqrt{41}}{4}$　**93.** $\frac{2 \pm \sqrt{11}}{2}$　**95.** $\frac{-3 \pm \sqrt{5}}{12}$
97. $-6, 2$　**99.** $-2 \pm \sqrt{2}$　**101.** $\pm \sqrt{2}$　**103.** $-5, \frac{7}{3}$
105. $4 \pm \sqrt{14}$　**107.** $\pm \sqrt{\frac{c}{a}}$　**109.** $-3, 6$　**111.** $3, 5$
113. $5, 7$　**115.** $y = \pm\sqrt{x + 1}$　**117.** $v = \pm\sqrt{\frac{2K}{m}}$
119. $r = \pm\sqrt{\frac{k}{E}}$　**121.** $f = \pm\frac{1}{2\pi\sqrt{LC}}$　**123. (a)** 30 miles
per hour　**(b)** 40 miles per hour　**125.** About 1.9 seconds; no
127. 0.5 sec, 2.5 sec　**129.** Does not happen
131. About 8 feet　**133.** 2 hours
135. (a) $x^2 + 6x - 520 = 0$　**(b)** -26 or 20; 20 feet
137. About 23 °C and 34 °C　**139.** 2014

Section 11.4 (pp. 777–781)

1. To solve quadratic equations that are written in the form $ax^2 + bx + c = 0$　**2.** Completing the square
3. $b^2 - 4ac$　**4.** One solution　**5.** Factoring, square root property, completing the square, and the quadratic formula
6. No; not when $b^2 - 4ac < 0$　**7.** $\pm\sqrt{k}$　**8.** $\pm i\sqrt{k}$
9. $-6, \frac{1}{2}$　**11.** 1　**13.** No real solutions　**15.** $-2, 8$
17. $\frac{1 \pm \sqrt{17}}{8}$　**19.** No real solutions　**21.** $\frac{1}{2}$　**23.** $\frac{3 \pm \sqrt{13}}{2}$
25. $2 \pm \sqrt{3}$　**27.** $\frac{1 \pm \sqrt{37}}{6}$　**29.** $\frac{1 \pm \sqrt{15}}{2}$　**31.** $\frac{5 \pm \sqrt{10}}{3}$
33. $(1 \pm \sqrt{2}, 0)$　**35.** $\left(-\frac{3}{2}, 0\right), (1, 0)$　**37.** None
39. None　**41.** $\left(\frac{-2 \pm \sqrt{10}}{3}, 0\right)$　**43. (a)** $a > 0$
(b) $-1, 2$　**(c)** Positive　**45. (a)** $a > 0$　**(b)** No real solutions　**(c)** Negative　**47. (a)** $a < 0$　**(b)** 2　**(c)** Zero
49. (a) 25　**(b)** 2　**51. (a)** 0　**(b)** 1　**53. (a)** $-\frac{7}{4}$　**(b)** 0
55. (a) 21　**(b)** 2
57. (a) Upward; wider　　　**59. (a)** Downward; the same
(b) $x = -1; (-1, -2)$　　**(b)** $x = 1; (1, 1)$
(c) 4, 2　　　　　　　　**(c)** 4, 2
(d) y-int: $\left(0, -\frac{3}{2}\right)$;　　**(d)** y-int: $(0, 0)$; x-int:
x-int: $(-3, 0), (1, 0)$　　$(0, 0), (2, 0)$
(e)　　　　　　　　　**(e)**

　　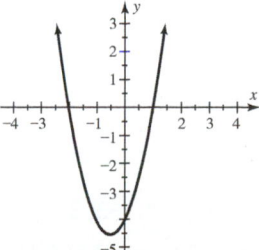

61. (a) Upward; narrower
(b) $x = -\frac{1}{2}; \left(-\frac{1}{2}, -\frac{9}{2}\right)$
(c) 36, 2
(d) y-int: $(0, -4)$; x-int: $(-2, 0), (1, 0)$
(e)

63. $\pm 3i$　**65.** $\pm 4i\sqrt{5}$　**67.** $\pm\frac{1}{2}i$　**69.** $\pm\frac{3}{4}i$　**71.** $\pm i\sqrt{6}$
73. $\pm\sqrt{3}$　**75.** $\pm i\sqrt{2}$　**77.** $\frac{1}{2} \pm i\frac{\sqrt{7}}{2}$　**79.** $-\frac{3}{4} \pm i\frac{\sqrt{23}}{4}$
81. $2 \pm \sqrt{3}$　**83.** $-\frac{1}{2} \pm i\frac{\sqrt{7}}{2}$　**85.** $-\frac{1}{5} \pm i\frac{\sqrt{19}}{5}$
87. $-\frac{3}{4} \pm i\frac{\sqrt{23}}{4}$　**89.** $-\frac{1}{2} \pm i\frac{\sqrt{15}}{2}$　**91.** $\frac{1 \pm \sqrt{3}}{2}$
93. $\frac{1}{4} \pm i\frac{\sqrt{15}}{4}$　**95.** $-1 \pm i\sqrt{3}$　**97.** $-2 \pm i$
99. $1 \pm i\sqrt{2}$　**101.** $1, 2$　**103.** $-\frac{1}{2}, 4$　**105.** $\frac{5 \pm \sqrt{17}}{2}$
107. $-\frac{1}{4} \pm \frac{3}{4}i\sqrt{7}$　**109.** $\pm\frac{1}{2}$　**111.** $\pm i\sqrt{2}$　**113.** $\frac{1}{3}$
115. 9 miles per hour　**117.** 45 miles per hour

119. (a) 6; in October 2010, Groupon's value was $6 billion.
(b) January 2011   **121.** $x \approx 8.04$, or about 1992; this
agrees with the graph.   **123.** About 1993 and 2004
$(x \approx 4.11, 15.36)$ **125.** $130 + 5\sqrt{634} \approx 256$ mph
127. (a)–(c) 11 in. by 14 in.   **129. (a)** The rate of change is
not constant.   **(b)** 75 seconds (answers may vary)
(c) 75 seconds

Checking Basic Concepts 11.3 & 11.4 (p. 781)

1. $\frac{1}{2}, 3$ **2.** $\pm\sqrt{5}$ **3.** $2 \pm \sqrt{3}$ **4.** $y = \pm\sqrt{1 - x^2}$
5. (a) $\frac{3 \pm \sqrt{17}}{4}$ **(b)** $\frac{4}{3}$ **6. (a)** 5; two real solutions
(b) -7; no real solutions   **(c)** 0; one real solution
7. (a) $\pm i\sqrt{5}$ **(b)** $-\frac{1}{2} \pm i\frac{\sqrt{11}}{2}$

Section 11.5 (pp. 788–790)

1. It has an inequality symbol rather than an equals sign.
2. No, they often have infinitely many.   **3.** No   **4.** Yes
5. $-2 < x < 4$ **6.** $x < -3$ or $x > 1$ **7.** Yes   **9.** Yes
11. No   **13.** Yes   **15.** No   **17.** No   **19. (a)** $-3, 2$
(b) $-3 < x < 2$ **(c)** $x < -3$ or $x > 2$ **21. (a)** $-2, 2$
(b) $-2 < x < 2$ **(c)** $x < -2$ or $x > 2$ **23. (a)** $-10, 5$
(b) $x < -10$ or $x > 5$ **(c)** $-10 < x < 5$
25. $[-0.128, 0.679]$ **27. (a)** $-2, 2$ **(b)** $-2 < x < 2$
(c) $x < -2$ or $x > 2$ **29. (a)** $-4, 0$ **(b)** $-4 < x < 0$
(c) $x < -4$ or $x > 0$ **31.** $[-7, -3]$
33. $(-\infty, 1) \cup (2, \infty)$ **35.** $(-\sqrt{10}, \sqrt{10})$
37. $(-\infty, 0) \cup (6, \infty)$
39. $(-\infty, 2 - \sqrt{2}] \cup [2 + \sqrt{2}, \infty)$ **41. (a)** $-2, 2$
(b) $-2 < x < 2$ **(c)** $x < -2$ or $x > 2$ **43. (a)** $\frac{-1 \pm \sqrt{5}}{2}$
(b) $\frac{-1 - \sqrt{5}}{2} < x < \frac{-1 + \sqrt{5}}{2}$
(c) $x < \frac{-1 - \sqrt{5}}{2}$ or $x > \frac{-1 + \sqrt{5}}{2}$
45. $(-3, -1)$ **47.** $(-\infty, -2.5] \cup [3, \infty)$ **49.** $[-2, 2]$
51. $(-\infty, \infty)$ **53.** $(0, 3)$ **55.** No real solutions   **57.** 2
59. $(-\infty, -1) \cup (-1, \infty)$ **61.** $[-1, 2]$ **63.** From
11 feet to 20 feet   **65. (a)** From 1131 feet to 3535 feet
(approximately)   **(b)** Before 1131 feet or after 3535 feet
(approximately)   **67. (a)** About 383; they agree (approx.).
(b) About 1969 or after   **(c)** About 1969 or after

Section 11.6 (pp. 794–795)

1. $\pm 1, \pm\sqrt{6}$ **3.** $-\sqrt[3]{2}, \sqrt[3]{\frac{5}{3}}$ **5.** $-\frac{4}{5}, -\frac{1}{3}$ **7.** $-3, 3$
9. $-\sqrt[3]{\frac{1}{3}}, \sqrt[3]{2}$ **11.** $-\frac{1}{8}, \frac{2}{5}$ **13.** 1   **15.** 1, $32^5 = 33,554,432$
17. 16, 81   **19.** 1   **21.** $-3, 6$ **23.** $-\sqrt{3}, \sqrt{3}$
25. $\pm 2, \pm 2i$ **27.** 0, $\pm i$ **29.** $\pm\sqrt{2}, \pm i$
31. $-1 \pm i$ **33.** $\pm 2i$

Checking Basic Concepts 11.5 & 11.6 (p. 795)

1. $(-\infty, -2) \cup (3, \infty)$ **2.** $\left[-1, -\frac{2}{3}\right]$ **3.** $-2, \sqrt[3]{2}$
4. $-1, 8^3 = 512$ **5.** $\pm i$

Chapter 11 Review (pp. 798–801)

1. $(-3, 4)$; $x = -3$; downward   **2.** $(1, 0)$; $x = 1$; upward
3. (a)

4. (a)

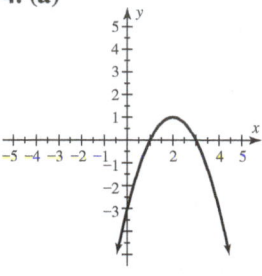

(b) $(0, -2)$; $x = 0$ **(b)** $(2, 1)$; $x = 2$
(c) -1 **(c)** 0
5. (a)

6. (a)

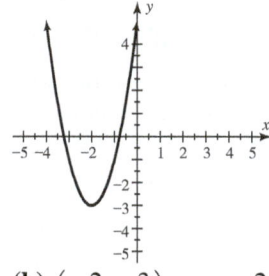

(b) $(1, 2)$; $x = 1$ **(b)** $(-2, -3)$; $x = -2$
(c) -2.5 **(c)** -1
7. $-\frac{7}{2}$ **8.** $-\frac{14}{3}$ **9.** $(2, -6)$ **10.** $(0, 5)$ **11.** $(2, 2)$
12. $(-1, 1)$
13. (a)

14. (a)

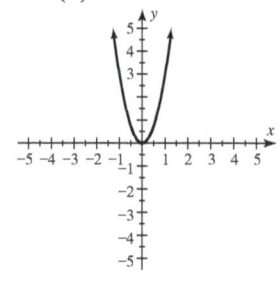

(b) Up 2 units   **(b)** Narrower
15. (a)

16. (a)

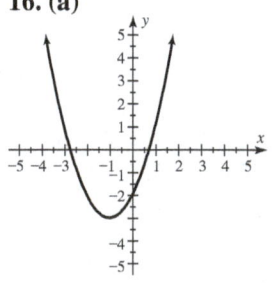

(b) Right 2 units   **(b)** Left 1 unit, down 3 units
17. (a)

18. (a)

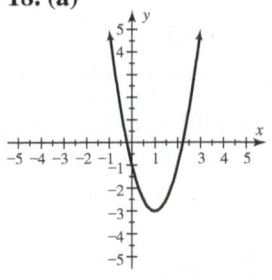

(b) Wider, left 1 unit, up   **(b)** Narrower, right 1 unit,
2 units   down 3 units

19. $y = -4(x - 2)^2 - 5$ **20.** $y = -(x + 4)^2 + 6$

21. $y = (x + 2)^2 - 11; (-2, -11)$

22. $y = \left(x - \frac{7}{2}\right)^2 - \frac{45}{4}; \left(\frac{7}{2}, -\frac{45}{4}\right)$

23. $y = 2\left(x - \frac{3}{4}\right)^2 - \frac{73}{8}; \left(\frac{3}{4}, -\frac{73}{8}\right)$

24. $y = 3(x + 1)^2 - 5; (-1, -5)$ **25.** $a = 3$

26. $a = \frac{1}{4}$ **27.** $f(x) = -5x^2 + 30x - 41; (0, -41)$

28. $f(x) = 3x^2 + 12x + 8; (0, 8)$ **29.** $-2, 3$

30. -1 **31.** No real solutions   **32.** $-4, 6$ **33.** $-10, 5$

34. $-0.5, 0.25$ **35.** $-5, 10$ **36.** $-3, 1$ **37.** $-4, 2$

38. $-1, 3$ **39.** $-5, 4$ **40.** $-8, -3$ **41.** $-\frac{2}{5}, \frac{2}{3}$

42. $\frac{4}{7}, 3$ **43.** ± 10 **44.** $\pm\frac{1}{3}$ **45.** $\pm\frac{\sqrt{6}}{2}$

46. No real solutions   **47.** $-3 \pm \sqrt{7}$ **48.** $2 \pm \sqrt{10}$

49. $1 \pm \sqrt{6}$ **50.** $\frac{-3 \pm \sqrt{11}}{2}$ **51.** $R = -r \pm \sqrt{\frac{k}{F}}$

52. $y = \pm\sqrt{\frac{12 - 2x^2}{3}}$ **53.** $3, 6$ **54.** $11, 13$ **55.** $-\frac{1}{2}, \frac{1}{3}$

56. $\frac{5 \pm \sqrt{5}}{10}$ **57.** $4 \pm \sqrt{21}$ **58.** $\frac{3 \pm \sqrt{3}}{2}$ **59.** ± 2

60. $\pm\frac{1}{2}$ **61.** $\frac{5}{2}, 3$ **62.** $\frac{3}{2}, 5$ **63.** $\frac{3 \pm \sqrt{5}}{2}$

64. $\frac{1 \pm \sqrt{5}}{4}$ **65. (a)** $a > 0$ **(b)** $-2, 3$ **(c)** Positive

66. (a) $a > 0$ **(b)** 2 **(c)** Zero   **67. (a)** $a < 0$

(b) No real solutions   **(c)** Negative   **68. (a)** $a < 0$

(b) $-4, 2$ **(c)** Positive   **69. (a)** 1 **(b)** 2 **70. (a)** 144

(b) 2 **71. (a)** -23 **(b)** 0 **72. (a)** 0 **(b)** 1

73. $-\frac{1}{2} \pm i\frac{\sqrt{19}}{2}$ **74.** $\pm 2i$ **75.** $\frac{1}{4} \pm i\frac{\sqrt{7}}{4}$ **76.** $\frac{1}{7} \pm i\frac{\sqrt{34}}{7}$

77. (a) $-2, 6$ **(b)** $-2 < x < 6$ **(c)** $x < -2$ or $x > 6$

78. (a) $-2, 0$ **(b)** $x < -2$ or $x > 0$ **(c)** $-2 < x < 0$

79. (a) $-4, 4$ **(b)** $-4 < x < 4$ **(c)** $x < -4$ or $x > 4$

80. (a) $-2, 1$ **(b)** $-2 < x < 1$ **(c)** $x < -2$ or $x > 1$

81. (a) $-1, 3$ **(b)** $-1 < x < 3$ **(c)** $x < -1$ or $x > 3$

82. (a) $-\frac{3}{2}, 5$ **(b)** $-\frac{3}{2} \leq x \leq 5$ **(c)** $x \leq -\frac{3}{2}$ or $x \geq 5$

83. $[-3, -1]$ **84.** $\left(\frac{1}{5}, 3\right)$ **85.** $\left(-\infty, \frac{1}{6}\right) \cup (2, \infty)$

86. $\left(-\infty, -\sqrt{5}\right] \cup \left[\sqrt{5}, \infty\right)$ **87.** $(-\infty, \infty)$

88. No solutions   **89.** $\pm\sqrt{5}, \pm 3$ **90.** $-\frac{1}{4}, \frac{2}{7}$

91. $1, 512$ **92.** 0 **93.** $\pm i\frac{\sqrt{2}}{2}$ **94.** $2 \pm i\sqrt{2}$

95. (a) $f(x) = x(12 - 2x)$ **(b)** 6 inches by 3 inches

96. (a) After 1 second and 1.75 seconds   **(b)** 1.375 seconds; 34.25 feet   **97. (a)** $f(x) = x(90 - 3x)$

(b) $[0, 30, 5]$ by $[0, 800, 100]$

(c) 10 or 20 rooms   **(d)** 15 rooms

98. (a) $x(x + 2) = 143$ **(b)** $x = -13$ or $x = 11$; the numbers are -13 and -11 or 11 and 13.

99. (a) $\sqrt{1728} \approx 41.6$ miles per hour   **(b)** 60 miles per hour

100. (a) $(1950, 220)$; in 1950, the per capita consumption was at a low of 220 million Btu.

(b) $[1950, 1970, 5]$ by $[200, 350, 25]$

It increased.

(c) $f(2010) = 1120$; no; the trend represented by this model did not continue after 1970.   **101.** About 11.1 inches by 11.1 inches   **102.** 50 feet   **103.** About 1.5 feet

104. About 6.0 to 9.0 inches

Chapter 11 Test (pp. 801–802)

1. $\left(1, \frac{3}{2}\right); x = 1; -3$ **2.** $-\frac{29}{4}$ **3.** $a = -\frac{1}{2}$

4. (a) Same as $y = x^2$ except shifted 1 unit right   **(b)** Same as $y = x^2$ except shifted 2 units downward

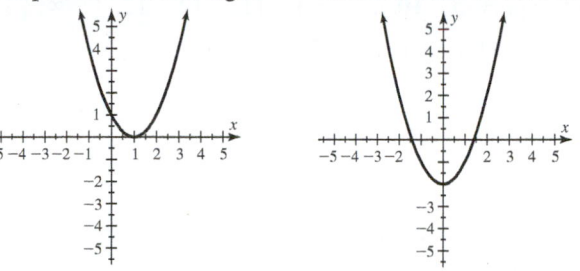

(c) Same as $y = x^2$ except it is wider, shifted right 3 units, and shifted upward 2 units

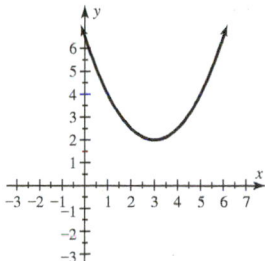

5. $y = (x - 3)^2 - 7; (3, -7); x = 3$ **6.** $-1, 2; 2$

7. $-4, \frac{1}{3}$ **8.** $-\frac{1}{2}, \frac{1}{2}$ **9.** $4 \pm \sqrt{17}$ **10.** $\frac{3 \pm \sqrt{17}}{4}$

11. $\pm\frac{4}{3}$ **12.** $m = \pm\sqrt{\frac{Fr^2}{G}}$ **13. (a)** $a < 0$ **(b)** $-3, 1$

(c) Positive   **14. (a)** -44 **(b)** No real solutions

(c) The graph of $y = -3x^2 + 4x - 5$ does not intersect the x-axis.   **15. (a)** $-1, 1$ **(b)** $-1 < x < 1$

(c) $x < -1$ or $x > 1$ **16. (a)** $-10, 20$

(b) $x < -10$ or $x > 20$ **(c)** $-10 < x < 20$

17. (a) $-\frac{1}{2}, \frac{3}{4}$ **(b)** $\left[-\frac{1}{2}, \frac{3}{4}\right]$ **(c)** $\left(-\infty, -\frac{1}{2}\right] \cup \left[\frac{3}{4}, \infty\right)$

18. $[-2, 0]$ **19.** $\sqrt[3]{2}, 1$ **20.** $-1 \pm i\frac{\sqrt{2}}{2}$

21. $-1.37, 0.69$ **22.** $\sqrt{2250} \approx 47.4$ miles per hour

23. (a) $f(x) = (x + 20)(90 - x)$ **(b)** 35

24. (a) $[0, 6, 1]$ by $[0, 150, 50]$

(b) After about 5.6 seconds

(c) 2.75 seconds; 129 feet

Chapters 1–11 Cumulative Review (pp. 802–804)

1. 1 **2.** Natural: $\sqrt[3]{8}$; whole: $0, \sqrt[3]{8}$; integer: $0, -5, \sqrt[3]{8}$; rational: $0.\overline{4}, 0, -5, \sqrt[3]{8}, -\frac{4}{3}$; irrational: $\sqrt{7}$ **3. (a)** $x^{10}y^{12}$

(b) $\frac{x}{y^8}$ **(c)** $\frac{1}{b^{10}}$ **4.** 9.29×10^6 **5.** $2; x \leq 2$ **6.** $(2, 5)$

7.

8.

9. $y = -\frac{3}{2}x + 5$ **10.** $x = -3$ **11.** -12 **12.** $\left(-\infty, \frac{7}{4}\right)$
13. $\left[\frac{1}{3}, 1\right]$ **14.** $(-1, 5]$ **15.** $(-1, 1)$
16.

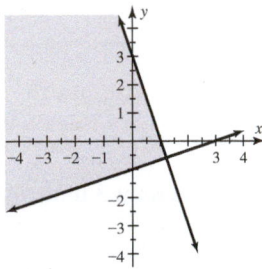

17. $(2, 2, 1)$ **18.** $6x^2 + 17x - 14$ **19.** $3x^3y + 3xy^3$
20. $x - 9$ **21.** $x(x + 1)(x - 2)$
22. $(2x + 5)(2x - 5)$ **23.** $\pm\sqrt{3}$ **24.** 1   **25.** $\frac{x + 3}{2}$
26. $-\frac{2}{x(x + 2)}$ **27.** $4x^3$ **28.** $\frac{1}{64}$ **29.** 3   **30.** $3\sqrt{2x}$
31.

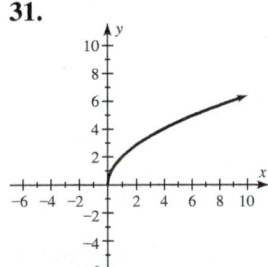

32. $\sqrt{26}$ **33.** $1 - i$ **34.** 3   **35.** 0.79
36.

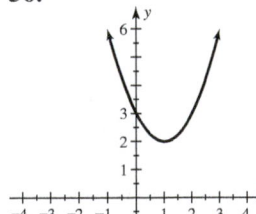

(a) $(1, 2)$ **(b)** 6   **(c)** $x = 1$ **(d)** $x > 1$
37. $f(x) = 2(x - 1)^2 - 3$ **38.** Shifted 1 unit left, 2 units
downward; narrower   **39.** $-3 \pm \sqrt{11}$ **40.** $\frac{3 \pm \sqrt{17}}{4}$
41. 1, 3   **42. (a)** $-2, 1$ **(b)** $-2 \le x \le 1$
43. $x < 1$ or $x > 2$ **44.** $\pm 4, \pm 4i$ **45.** c.   **46.** f.   **47.** g.
48. e.   **49.** d.   **50.** a.   **51.** b.   **52.** h.   **53. (a)** 300; initially,
the tank holds 300 gal.   **(b)** $(6, 0)$; after 6 minutes, the tank
is empty.   **(c)** -50; water is pumped out at 50 gal/min.
(d) $G(t) = 300 - 50t$ **54.** $2200 at 6%, $1200 at 5%,
$600 at 4%   **55.** 125 ft by 125 ft   **56.** 33 feet

12 Exponential and Logarithmic Functions

Section 12.1 (pp. 817–821)

1. $g(f(7))$ **2.** $f(g(x))$ **3.** No   **4.** inputs; outputs
5. No   **6.** one-to-one   **7.** adding 10   **8.** 7   **9.** 8; 6
10. x; y **11.** one-to-one   **12.** reflection; $y = x$
13. (a) 7   **(b)** 49   **(c)** $(g \circ f)(x) = x^2 + 3$
(d) $(f \circ g)(x) = (x + 3)^2$ **15. (a)** -65 **(b)** 126
(c) $(g \circ f)(x) = 8x^3 - 1$ **(d)** $(f \circ g)(x) = 2x^3 - 2$
17. (a) 3   **(b)** 1   **(c)** $(g \circ f)(x) = \left|\frac{1}{2}x - 2\right|$
(d) $(f \circ g)(x) = \frac{1}{2}|x - 2|$ **19. (a)** $\frac{11}{2}$ **(b)** $-\frac{1}{17}$
(c) $(g \circ f)(x) = 3 - \frac{5}{x}$ **(d)** $(f \circ g)(x) = \frac{1}{3 - 5x}$
21. (a) 77   **(b)** 122   **(c)** $(g \circ f)(x) = 16x^2 - 4x + 5$
(d) $(f \circ g)(x) = 8x^2 - 4x + 10$ **23. (a)** 1   **(b)** 2
25. (a) -1 **(b)** 1   **27. (a)** 0   **(b)** 2   **29. (a)** 2   **(b)** -3
(c) -1 **31.** $f(1) = f(-1) = 5$
33. $f(1) = f(-1) = 101$ **35.** $f(2) = f(-2) = 4$
37. Yes   **39.** No   **41.** Yes   **43.** Divide x by 7;
$f(x) = 7x; g(x) = \frac{x}{7}$ **45.** Multiply x by 2, then subtract 5;
$f(x) = \frac{x + 5}{2}; g(x) = 2x - 5$ **47.** Add 3 to x and multiply the result by 2; $f(x) = \frac{1}{2}x - 3; g(x) = 2(x + 3)$
49. Take the cube root of x and subtract 5;
$f(x) = (x + 5)^3; g(x) = \sqrt[3]{x} - 5$
51. $(f \circ f^{-1})(x) = 4\left(\frac{x}{4}\right) = x; (f^{-1} \circ f)(x) = \frac{4x}{4} = x$
53.–57. Show $(f \circ f^{-1})(x) = (f^{-1} \circ f)(x) = x$. See the
answer to Exercise 51 above.   **59.** $f^{-1}(x) = \frac{x}{12}$
61. $f^{-1}(x) = x - 8$ **63.** $f^{-1}(x) = \frac{x + 2}{5}$
65. $f^{-1}(x) = -2(x - 1)$ **67.** $f^{-1}(x) = 8 - x$
69. $f^{-1}(x) = 2x - 1$ **71.** $f^{-1}(x) = \frac{x^3}{2}$
73. $f^{-1}(x) = \sqrt[3]{x + 8}$
75.

x	0	5	10	15	20
$f^{-1}(x)$	0	1	2	3	4

Domain of f = range of f^{-1} = $\{0, 1, 2, 3, 4\}$
Range of f = domain of f^{-1} = $\{0, 5, 10, 15, 20\}$

77.

x	4	2	0	-2	-4
$f^{-1}(x)$	-5	0	5	10	15

Domain of f = range of f^{-1} = $\{-5, 0, 5, 10, 15\}$
Range of f = domain of f^{-1} = $\{-4, -2, 0, 2, 4\}$

79.

81.

83.

85.
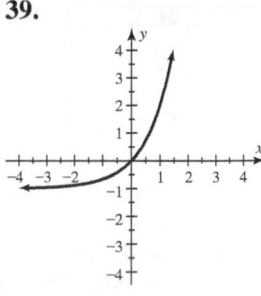

87. (a) -2 **(b)** 2 **(c)** 0 **89. (a)** -6 **(b)** 4 **(c)** -2
91. (a) 12 **(b)** -4 **(c)** $x^4 + 4x^2 + 2$ **93. (a)** $\frac{1}{2}$
(b) $-\frac{1}{x}$ **(c)** $\frac{x}{2}$ **95. (a)** 20π; after 5 seconds, the wave has a
circumference of $20\pi \approx 62.8$ feet. **(b)** $(C \circ r)(t) = 4\pi t$
97. (a) 16; in 1980, 16% of people 25 or older had
completed four or more years of college.

(b)

x	8	16	27	29
$P^{-1}(x)$	1960	1980	2000	2010

(c) 1980

99. (a) 75°; 150 **(b)** 150; one hour after midnight, there
are 150 mosquitoes per 100 square feet. **(c)** The number of
mosquitoes per 100 square feet, h hours after midnight
(d) $T(h) = -5h + 80; M(T) = 2T$
(e) $(M \circ T)(h) = -10h + 160$
101. (a) Yes, different inputs result in different outputs.
(b) $f^{-1}(x) = \frac{5}{9}(x - 32)$ converts x degrees Fahrenheit to
an equivalent temperature in degrees Celsius.
103. $f(x) = 4x; f^{-1}(x) = \frac{x}{4}$ converts x quarts to gallons.
105. (a) $C(x) = 16x$ **(b)** $T(x) = 48x$
(c) $(T \circ C)(x) = 768x$ **(d)** 2304; there are 2304 tsp in 3 gal.

Section 12.2 (pp. 834–838)

1. $f(x) = Ca^x$
2. Answers may vary.

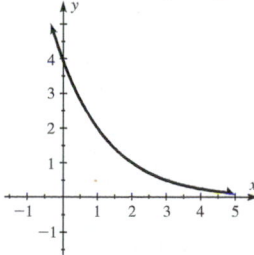

3. D: all real numbers; R: all positive real numbers
4. 32; 25 **5.** 2.718 **6.** 7.389; 9.870 (approximately)
7. factor **8.** 1.5 **9.** $\frac{B - A}{A} \times 100$ **10.** 1.35
11. $\frac{1}{9}$; 9 **13.** 5; 160 **15.** 4; $\frac{1}{8}$ **17.** 15; $\frac{5}{9}$
19. 0.17; 2.41 **21.** 5; 1.08 **23.** 1; $\frac{1}{a}$ **25.** c. **27.** d.
29. $C = 1, a = 2$ **31.** $C = 4, a = \frac{1}{4}$
33.

35.

Growth   Decay

37.

Decay

39.
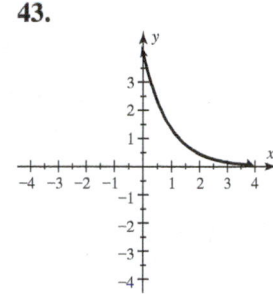

Growth

41.

43.

Growth   Decay

45. (a) 1 unit right **(b)** 1 unit down **(c)** Reflected across
the y-axis **(d)** Reflected across the x-axis
47. (a) Exponential decay **(b)** $f(x) = 64\left(\frac{1}{4}\right)^x$
49. (a) Linear growth **(b)** $f(x) = 3x + 8$
51. (a) Exponential growth **(b)** $f(x) = 4(1.25)^x$
53. (a) 100% **(b)** -50% **55. (a)** -80% **(b)** 400%
57. (a) $1200 **(b)** $2200 **(c)** 2.2
59. (a) $130 **(b)** $780 **(c)** 1.2
61. (a) $-80 **(b)** $720 **(c)** 0.9
63. $C = 9, a = 1.07, R = 7\%$
65. $C = 1.5, a = 0.45, R = -55\%$ **67.** $3551.05
69. $1,820,087.63 **71.** $792.75 **73.** Yes; this is equivalent
to having two accounts, each containing $1000 initially.
75. $788.78 **77.** $1429.24 **79.** 3.32 **81.** 0.86
83. $[-4, 4, 1]$ by $[0, 8, 1]$ **85.** $[-4, 4, 1]$ by $[0, 8, 1]$

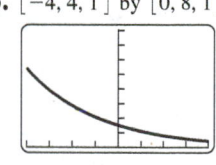

Growth   Decay

87. $f(x) = 5000(0.75)^x; f(4) \approx 1582$
89. $f(x) = 50(1.1)^x; f(4) \approx 73$ **91.** 20% is much better
93. (a) 0.6 **(b)** $B(x) = 0.07(0.6)^x$ **(c)** 0.0252; after
2 hours, blood alcohol is 0.0252 g/100 mL
95. $F(t) = 100\left(\frac{1}{2}\right)^{t/3}, 40\%$ **97. (a)** 2.724
(b) In July 2010, there were about 0.1 billion tweets per month.
(c) About 15; after 5 years, there were 15 billion tweets
per month. **99. (a)** $C = 500, a = 2$ **(b)** About 5278
thousand per milliliter **(c)** The growth is exponential and
doubles every 50 seconds.

101.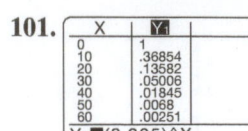

X	Y₁
0	1
10	.36854
20	.13582
30	.05006
40	.01845
50	.0068
60	.00251

$Y_1 \blacksquare (0.905)^X$

(a) 1; the probability that no vehicle will enter the intersection during a period of 0 seconds is 1 or 100%.
(b) About 30 seconds
103. (a) $f(x) = 2.7e^{0.014x}$ **(b)** 3.1 million
105. (a) $f(x) = 38e^{0.0102x}$ **(b)** 42 million
107. Answers may vary.

Checking Basic Concepts 12.1 & 12.2 (p. 839)

1. (a) 7 **(b)** $(f \circ g)(x) = 2x^2 + 9x + 6$
2.

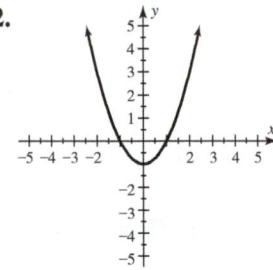

(a) No, it does not pass the horizontal line test. **(b)** No
3. $f^{-1}(x) = \frac{x + 3}{4}$ **4.** $\frac{3}{4}$
5.

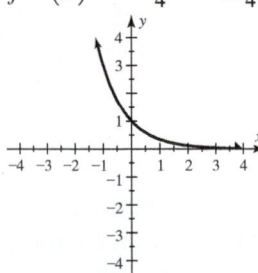

6. $C = 2; a = \frac{1}{2}$

Section 12.3 (pp. 848–851)

1. 10 **2.** e **3.** $D = \{x \mid x > 0\}$; R: all real numbers
4. D: all real numbers; $R = \{x \mid x > 0\}$ **5.** k **6.** k
7. x **8.** x **9.** $\log 5$ **10.** $\ln 5$ **11.** 0 **12.** undefined
13. (a) $10^2 = 100$ **(b)** $\log \frac{1}{10} = -1$
15. (a) $\log 1 = 0$ **(b)** $10^{-5} = \frac{1}{100,000}$
17.

x	10^{-5}	10^0	$10^{0.5}$	$10^{2.2}$
$\log x$	-5	0	0.5	2.2

19. 5 **21.** -4 **23.** 0 **25.** -2 **27.** 4.7 **29.** 4 **31.** $\frac{1}{2}$
33. Undefined **35.** -3 **37.** 2 **39.** $x^2, x \neq 0$ **41.** 5
43. $2x - 7$ **45.** 1.398 **47.** 0.161
49.

1 unit downward

51.

1 unit to the left

53.

1 unit to the right

55.

Increases faster

57.

x	$e^{0.5}$	e^{-12}	$e^{\frac{1}{2}}$	$e^{-2.4}$
$\ln x$	0.5	-12	$\frac{1}{2}$	-2.4

59. 0 **61.** $-5x$ **63.** $x^2, x \neq 0$ **65.** 1.946 **67.** -0.560
69. $[-4, 4, 1]$ by $[-4, 4, 1]$ **71.** $[-4, 4, 1]$ by $[-4, 4, 1]$

  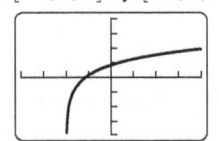

Reflected across the y-axis together with the graph of $y = \ln x$; $D = \{x \mid x \neq 0\}$

2 units to the left; $D = \{x \mid x > -2\}$

73.

x	$5^{\frac{2}{3}}$	5^{-1}	$5^{5.5}$	$5^{-0.9}$
$\log_5 x$	$\frac{2}{3}$	-1	5.5	-0.9

75. $6x$ **77.** $-\frac{3}{2}$ **79.** 8 **81.** $\frac{3}{2}$ **83.** $-\frac{2}{3}$ **85.** Undefined
87. 2 **89.** -4 **91.** -2 **93.** -2 **95.** 17
97. $(2x)^2, x \neq 0$ **99.** $0.6z, z > 0$
101.

x	1/4	1/2	1	$\sqrt{2}$	64
$\log_2 x$	-2	-1	0	1/2	6

103.

105.

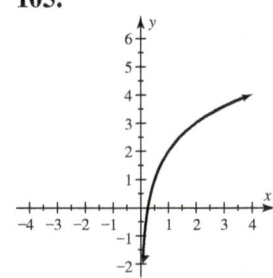

107. d. **109.** a. **111.** 120 dB; yes **113. (a)** It increases, but it doesn't double when the weight of the plane doubles.
(b) About 5.086; a 50,000-pound airplane needs a runway 5086 feet long. **115. (a)** 3.5 **(b)** 1 **(c)** It decreases by 2.5.
117. (a) 10^6; 10^8 **(b)** 100 times

Section 12.4 (pp. 857–858)

1. 4 **2.** 4 **3.** 3 **4.** $\log m + \log n$ **5.** $\log m - \log n$
6. $r \log m$ **7.** No **8.** Yes **9.** No **10.** No
11. $\log_a x = \frac{\log x}{\log a}$ or $\log_a x = \frac{\ln x}{\ln a}$ **12.** 0; 1
13. $\ln 3 + \ln 5$ **15.** $\log x + \log y$ **17.** $\log y + \log y$
19. $\log 7 - \log 3$ **21.** $\ln x - \ln y$ **23.** $\log_2 45 - \log_2 x$
25. $\log 225$ **27.** $\ln xy$ **29.** $\ln 14x^3$ **31.** $\ln xy$ **33.** $\log 5$
35. $\ln x^2$ **37.** 2 **39.** $6 \log 3$ **41.** $x \ln 2$ **43.** $\frac{1}{4} \log_2 5$
45. $\frac{1}{3} \log_4 z$ **47.** $(y - 1)\log x$ **49.** 1.2 **51.** 1.8

53. 2.1   **55.** 0.4   **57.** −1.1   **59.** log z **61.** log x^3y^2
63. 0   **65.** ln 2^x **67.** ln $x^{5/6}$ **69.** $\log_a \frac{x+1}{x-1}$
71. log x + 2 log y **73.** 4 ln x + ln y − ln z
75. $\frac{1}{3}$ log z − $\frac{1}{2}$ log y **77.** 4 log x + 3 log y
79. ln x − ln y **81.** $\frac{3}{2}\log_4 x + \frac{1}{2}\log_4 y - \log_4 z$
83. $[-6, 6, 1]$ by $[-4, 4, 1]$ $[-6, 6, 1]$ by $[-4, 4, 1]$

By the power rule, log x^3 = 3 log x
85. $[-6, 6, 1]$ by $[-4, 4, 1]$ $[-6, 6, 1]$ by $[-4, 4, 1]$

Not the same
87. 1.46   **89.** 4.64   **91.** 2.10
93. 10 log $(10^{16}x)$ = 10$(\log 10^{16} + \log x)$ = 10$(16 + \log x)$ = 160 + 10 log x

Checking Basic Concepts 12.3 & 12.4 (p. 858)

1. (a) 4   **(b)** x **(c)** −3   **(d)** $\frac{1}{2}$
2.

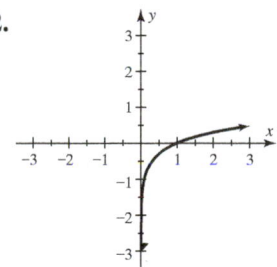

(a) $D = \{x \mid x > 0\}$; R: all real numbers   **(b)** 0
(c) Yes; for example, log $\frac{1}{10}$ = −1.   **(d)** No; negative numbers are not in the domain of log x.
3. (a) log x + log y **(b)** ln x − ln y − ln z
(c) 2 ln x **(d)** 2 log x + 3 log y − $\frac{1}{2}$ log z
4. (a) log xy **(b)** ln $\frac{2x}{y^3}$ **(c)** $\log_2 \frac{x^2y^3}{z}$

Section 12.5 (pp. 866–870)

1. Add 5 to each side.   **2.** Divide each side by 5.
3. Take the common logarithm of each side.
4. Exponentiate each side using base 10.   **5.** x
6. $x, x > 0$ **7.** 2x **8.** x + 7, x > −7
9. No; log $\frac{5}{4}$ = log 5 − log 4
10. No; log 5 − log 4 = log $\frac{5}{4}$ **11.** 1   **12.** 1
13. (a) 3x = log$_6$ 8   **(b)** 7x = ln 10
15. (a) 8x = log$_5$ 12   **(b)** x = ln 3
17. (a) 2^8 = x **(b)** e^3 = 2x
19. (a) 4^6 = 6x **(b)** e^{10} = x + 1
21. 3   **23.** 6   **25.** 6   **27.** 4   **29.** $\frac{\log 124}{0.4}$ ≈ 5.23
31. 0   **33.** ln 25 ≈ 3.22   **35.** $\frac{\ln 2}{\ln 0.4}$ ≈ −0.76
37. ln 7 ≈ 1.95   **39.** log $\frac{35}{2}$ − 2 ≈ −0.76

41. $\frac{\log 20}{2\log 3.1}$ ≈ 1.32   **43.** −1   **45.** −1, 5
47. $\frac{\ln 10}{3}$ ≈ 0.77   **49.** $\frac{2\ln 2}{1 - \ln 2}$ ≈ 4.52
51. $\frac{2\log 5}{0.5\log 4 - \log 5}$ ≈ −3.51   **53. (a)** 1   **(b)** 1
55. (a) −2   **(b)** −2   **57.** −1   **59.** 0   **61.** 5
63. $\frac{\log 1.45}{\log 0.55}$ ≈ −0.62   **65.** −1.84, 1.15   **67.** 1.31
69. 100   **71.** e^5 ≈ 148.41   **73.** 5,000,000   **75.** 16
77. $\frac{2^{2.3}}{5}$ ≈ 0.98   **79.** $10^{1.4}$ ≈ 25.12
81. $\frac{e^{11} - 1}{2}$ ≈ 29,936.57   **83.** 1   **85.** 5   **87.** 4   **89.** 3
91. $10^{1.6}$ ≈ 39.81   **93.** e − 1 ≈ 1.72   **95.** 9
97. (a) 2   **(b)** $e^{0.7}$ ≈ 2.01
99. (a) 2   **(b)** $\frac{1}{2}(10^{0.6})$ ≈ 1.99
101. (a) About 54.9; in 2018, there will be an annual shipment of 54.9 million smartwatches.   **(b)** 2020
103. 10 hours   **105.** 8 years   **107. (a)** About 126; in 2017, about 126 thousand people were waiting for liver transplants.   **(b)** Around 2021   **109.** About 20,893 pounds
111. (a) About 203; in 1975, there were about 203 thousand bluefin tuna.   **(b)** 1979   **(c)** 1979
113. a = 25, b = 3   **115. (a)** About 1.67 acres
(b) About 1.67 acres   **117. (a)** Nonlinear; they do not increase at a constant rate.   **(b)** Each year, Fitbit sales increased by a factor of 3.33.   **(c)** 2016
119. 10^{-6} watts/square centimeter   **121.** About 7 miles
123. About $42,055.97

Checking Basic Concepts 12.5 (p. 870)

1. (a) log 20 ≈ 1.30   **(b)** $\frac{\log 147}{3\log 2}$ ≈ 2.40
(c) $e^{4.1}$ ≈ 60.34   **(d)** 500   **2.** 8   **3.** 20 years

Chapter 12 Review (pp. 873–876)

1. (a) 81   **(b)** $(f \circ g)(x)$ = 50x^2 − 2   **2. (a)** −32
(b) $(f \circ g)(x)$ = $\sqrt[3]{4x^3 - 6}$ **3. (a)** 0   **(b)** 3
4. (a) 3   **(b)** −2   **(c)** −1   **5.** $f(1)$ = $f(-1)$ = 2
6. $f(0)$ = $f(2)$ = 1   **7.** No   **8.** Yes
9. $(f \circ f^{-1})(x)$ = 2$\left(\frac{x+9}{2}\right)$ − 9 = x
$(f^{-1} \circ f)(x)$ = $\frac{(2x-9)+9}{2}$ = x
10. $(f \circ f^{-1})$ = $(\sqrt[3]{x-1})^3$ + 1 = x
$(f^{-1} \circ f)(x)$ = $\sqrt[3]{(x^3 + 1) - 1}$ = x
11. $f^{-1}(x)$ = $\frac{x}{5}$ **12.** $f^{-1}(x)$ = x + 11
13. $f^{-1}(x)$ = $\frac{x-7}{2}$ **14.** $f^{-1}(x)$ = $\frac{4}{x}$
15.

x	10	8	7	3
$f^{-1}(x)$	0	1	2	3

$D = \{3, 7, 8, 10\}$; $R = \{0, 1, 2, 3\}$
16.

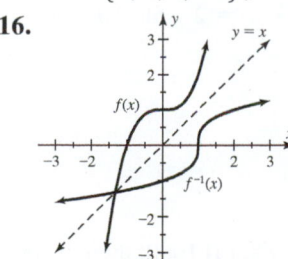

17. $\frac{1}{6}$; 36   **18.** 5; $\frac{5}{8}$ **19.** 3; $\frac{1}{81}$ **20.** 3; $\frac{1}{2}$

21.

Exponential growth

23.

Logarithmic growth

22.

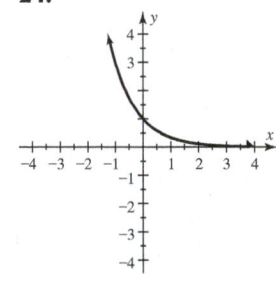

Exponential decay

24.

Exponential decay

25. (a) Exponential growth   **(b)** $f(x) = 5(2)^x$
26. (a) Linear growth   **(b)** $f(x) = 5x + 5$
27. $C = \frac{1}{2}, a = 2$ **28.** $k = 2$ **29.** -20% **30.** 1.07
31. (a) 100%   **(b)** -50% **32. (a)** $-54.\overline{54}\%$ **(b)** 120%
33. (a) $1050   **(b)** $1550   **(c)** 3.1   **34. (a)** $-$175
(b) $525   **(c)** 0.75   **35.** $f(x) = 20{,}000(0.95)^x$;
$f(2) = 18{,}050$ **36.** $f(x) = 1500(4)^x$; $f(2) = 24{,}000$
37. $2829.54   **38.** $675,340.51   **39.** 399.67   **40.** 0.71
41. 3.48   **42.** 3.89   **43.** -3 **44.** 2   **45.** -4 **46.** 2
47. 1.813   **48.** -0.163 **49.** 4.787   **50.** -0.398 **51.** 7
52. $\frac{5}{9}$ **53.** $6 - x$ **54.** $x^2, x > 0$ **55.** $\ln x + \ln y$
56. $\log x - \log y$ **57.** $2 \ln x + 3 \ln y$
58. $\frac{1}{2} \log x - 3 \log z$ **59.** $2 \log_2 x + \log_2 y - \log_2 z$
60. $\frac{1}{3} \log_3 x - \frac{1}{3} \log_3 y$ **61.** $\log 75$ **62.** $\log_4 (10x^2)$
63. $\ln \frac{x^2}{y^3}$ **64.** $\log xy$ **65.** $3 \log 6$ **66.** $2 \ln x$
67. $2x \log_2 5$ **68.** $(x + 1) \log_4 0.6$ **69.** 2   **70.** 4
71. $\ln 9 \approx 2.20$ **72.** $\frac{\log 0.2}{\log 0.85} \approx 9.90$ **73.** $e^{0.8} \approx 2.23$
74. $\frac{1}{2}e^5 \approx 74.21$ **75.** 10^{40} **76.** 100
77. $\frac{4 \log 2}{\log 3 - \log 2} \approx 6.84$ **78.** 6   **79. (a)** 3   **(b)** 3
80. (a) 4   **(b)** 4   **81. (a)** 64π; after 8 seconds, the balloon
has a surface area of $64\pi \approx 201$ in^2.
(b) $(S \circ r)(t) = 8\pi t$ **82. (a)** Yes, different inputs result in
different outputs.   **(b)** $f^{-1}(x) = \frac{x}{0.08}$ calculates the cost of
an item whose sales tax is x dollars.   **83.** 7 years
84. $a = 100, b = 50$ **85.** $C = 3, a = 2$ **86.** 10^7
87. (a) $[0, 10, 2]$ by $[0, 4, 1]$

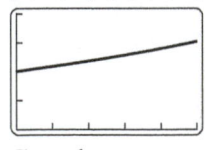

Growth

(b) About 3.3 million   **(c)** 2008   **88. (a)** 1000; there were
1000 bacteria/mL initially.   **(b)** About 495.11 minutes
89. (a) About 6.93 meters/second   **(b)** 12.18 meters

Chapter 12 Test (pp. 876–877)

1. 6; $(f \circ g)(x) = 4(x + 7)^3 - 5(x + 7)$
2. (a) 3   **(b)** 3   **3.** $f(-5) = f(5) = 0$
(answers may vary)   **4.** $f^{-1}(x) = \frac{5 - x}{2}$
5.

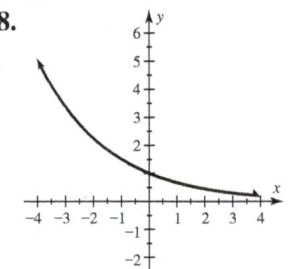

6.

x	8	6	4	2
$f^{-1}(x)$	1	2	3	4

$D = \{2, 4, 6, 8\}; R = \{1, 2, 3, 4\}$
7. $\frac{3}{16}$

8.

Exponential decay
9. (a) Exponential growth   **(b)** $f(x) = 3(2)^x$
10. (a) Linear growth   **(b)** $f(x) = 1.5x - 1$
11. $C = 1, a = \frac{1}{2}$ **12.** 50%   **13.** 1.05   **14.** $1051.91
15. 4.16   **16.** $\frac{1}{2}$ **17.** 5.426

18.

Shifted to the right 2 units
19. $3 \log x + 2 \log y - \frac{1}{2} \log z$ **20.** $\ln \frac{x^4 z}{y^5}$ **21.** $2x \log 7$
22. $1 - 3x$ **23.** $\ln 25 \approx 3.22$ **24.** $\log 50 \approx 1.70$
25. $10^{1.8} \approx 63.10$ **26.** $\frac{1}{5}e^9 \approx 1620.62$
27. $a = 5, b = 3$ **28. (a)** 4 million   **(b)** About 6.15;
after 5 hours, there were about 6.15 million bacteria.
(c) Growth   **(d)** After 8 hours
29. (a) $A(x) = 5000(0.98)^x$ **(b)** $A(3) \approx 4705.96$;
after 3 years, the account contains $4705.96.   **(c)** About 5

Chapters 1–12 Cumulative Review (pp. 877–880)

1. 4.29×10^{-4} **2.** Natural: none; whole: 0; integer: $-3, 0$;
rational: $-\frac{11}{7}, -3, 0, 5.\overline{18}$; irrational: $\sqrt{6}, \pi$ **3.** c.

4. Commutative   **5.** d^4 **6.** $\frac{b^9}{64a^6}$ **7.** $\frac{4y^8}{x^5}$ **8.** $\frac{y^4}{4x^5}$

9. $y = \frac{5}{4}x + 1$ **10.** $\{x \mid x \neq -3\}$ **11.** $f(x) = 4x - 3$

12. $x = 4$ **13.** 1

14.

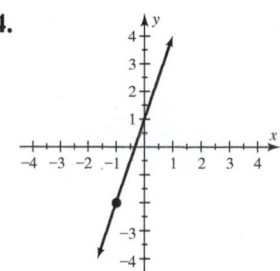

15. $y = 7x - 6$ **16.** $y = 3x + 5$ **17.** -1 **18.** $x < 0$

19. -18 **20.** $(-\infty, 0)$ **21.** $\frac{4}{15}$ **22.** $(-5, 10]$

23. $[-2, 10]$ **24.** 4, 16

25.

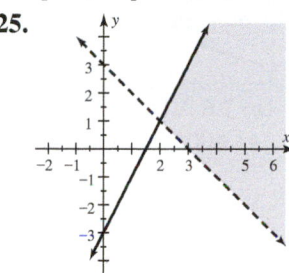

26. 2   **27.** $(1, 1)$ **28.** No solutions   **29.** $(-4, 0, 2)$

30. $(0, 0, 1)$ **31.** $2x(x - 1)^2$ **32.** $(2a - 5b)(2a + 5b)$

33. $(2t - 3)(4t^2 + 6t + 9)$ **34.** $(2a^2 + 5)(2a - 1)$

35. $-\frac{5}{6}, 2$ **36.** $-\frac{2}{3}, \frac{2}{3}$ **37.** $-3, 0, 5$ **38.** $0, \frac{1}{2}$ **39.** 1

40. $\frac{x + 2}{x - 3}$ **41.** 3   **42.** -3 **43.** $J = \frac{2z}{P - 1}$ **44.** $\frac{x^3 + 3}{x^3 - 3}$

45. 40   **46.** $3x^2 + 6x + 10 + \frac{5}{x - 2}$ **47.** $\frac{x^4}{y^6}$ **48.** x^3

49. $10ab$ **50.** $\sqrt{6}$ **51.** $(b + 3)a\sqrt[3]{a^2b}$ **52.** 20

53. $\frac{5 + \sqrt{3}}{11}$ **54.** $(4, \infty)$ **55.** $-6, 4$ **56.** 3   **57.** $3 + 5i$

58. $-i$ **59.** $(2, 1)$ **60.** $\frac{7}{2}$ **61.** 5, 8   **62.** $\frac{1 \pm \sqrt{41}}{4}$

63. $2 \pm \sqrt{2}$ **64.** $\pm 2, \pm \sqrt{6}$ **65.** **(a)** $-1, 3$ **(b)** $a < 0$

(c) Positive   **66.** $(-\infty, -7] \cup [2, \infty)$ **67.** **(a)** 7

(b) $(g \circ f)(x) = 2x^2 - 3$ **68.** $f^{-1}(x) = \frac{3}{x}$ **69.** 4

70. $2x, x > 0$ **71.** $\frac{1}{2} \log x - 2 \log y$ **72.** $\ln(5x^3)$

73. $10^{11/6} \approx 68.13$ **74.** $\frac{\log 17}{3 \log 2} \approx 1.36$ **75.** 4%

76. 2.27 pounds   **77.** About 11.3%

78. **(a)** 9.91 meters per second   **(b)** 8.52 meters

13 Conic Sections

Section 13.1 (pp. 888–891)

1. Parabola, ellipse, hyperbola   **2.** The axes of symmetry
are vertical and horizontal, respectively.   **3.** No

4.

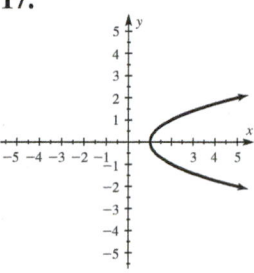

5. No; it does not pass the vertical line test.   **6.** (h, k)

7. left   **8.** right   **9.** circle; (h, k) **10.** $(0, 0)$; r

11. d.   **12.** a.   **13.** c.   **14.** b.

15.   **17.**

    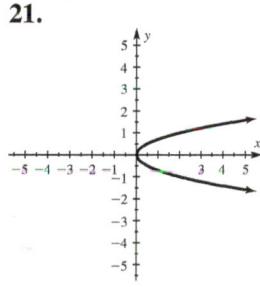

$(0, 0)$; $y = 0$ $(1, 0)$; $y = 0$

19.   **21.**

$(0, -1)$; $x = 0$ $(0, 0)$; $y = 0$

23.   **25.**

    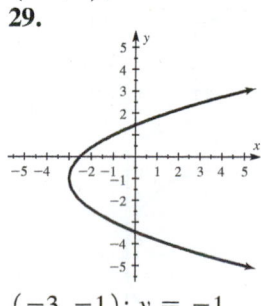

$(2, 1)$; $y = 1$ $(-2, 1)$; $x = -2$

27.   **29.**

$(-2, 0)$; $x = -2$ $(-3, -1)$; $y = -1$

31.

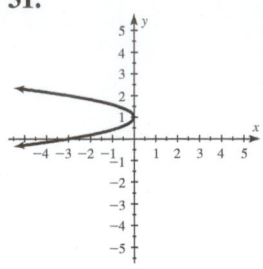

$(0, 1); y = 1$

33.

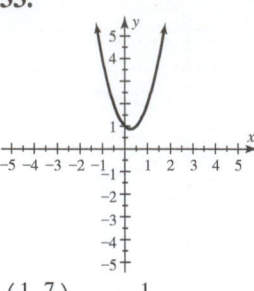

$\left(\frac{1}{4}, \frac{7}{8}\right); x = \frac{1}{4}$

35.

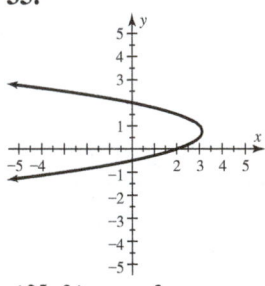

$\left(\frac{25}{8}, \frac{3}{4}\right); y = \frac{3}{4}$

37.

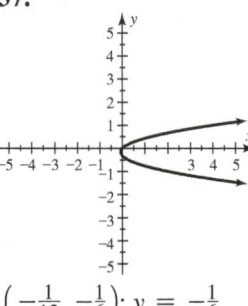

$\left(-\frac{1}{12}, -\frac{1}{6}\right); y = -\frac{1}{6}$

39.

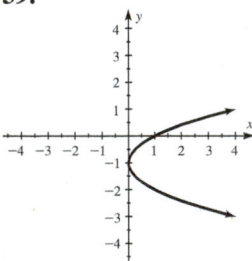

$(0, -1); y = -1$
41. $y = x^2$ **43.** $x = (y + 1)^2 - 2$ **45.** Upward
47. Downward   **49.** $x \geq 0$ **51.** Two   **53.** $(1, 0)$
55. $x^2 + y^2 = 1$ **57.** $(x + 1)^2 + (y - 5)^2 = 9$
59. $(x + 4)^2 + (y + 6)^2 = 2$ **61.** $x^2 + y^2 = 16$
63. $(x + 3)^2 + (y - 2)^2 = 1$
65. 2; $(0, 0)$

69. 5; $(-5, 5)$

67. 3; $(1, 3)$

71. 3; $(-3, 1)$

73. $\sqrt{7}; (-3, 1)$

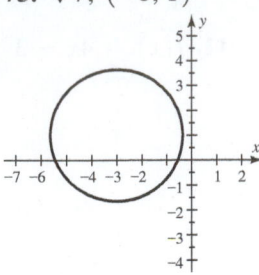

75. $(x - 1)^2 + (y - 5)^2 = 17$
77. $\left(x - \frac{7}{2}\right)^2 + (y - 3)^2 = \frac{25}{4}$
79. (a)

$[-40, 40, 10]$ by $[-120, 120, 20]$

(b) 32 ft

81. (a)

$[-1.5, 1.5, 0.5]$ by $[-1, 1, 0.5]$

(b) 2.6 A.U., or
241,800,000 miles

Section 13.2 (pp. 898–901)

1.

(answers may vary)

2.

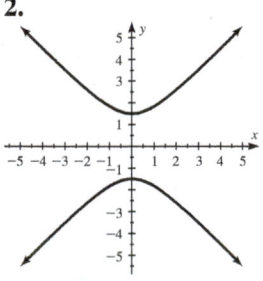

(answers may vary)

3. horizontal   **4.** vertical   **5.** 2   **6.** 4   **7.** left; right
8. lower; upper   **9.** They are the diagonals extended.
10. No   **11.** a.   **12.** b.
13.

15.

17.

19.

21.

23.

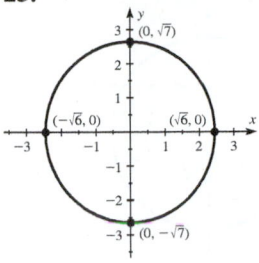

25. $\frac{x^2}{9} + \frac{y^2}{4} = 1$ **27.** $\frac{y^2}{25} + \frac{x^2}{16} = 1$

29.

31.

33.

35.

37.

39.

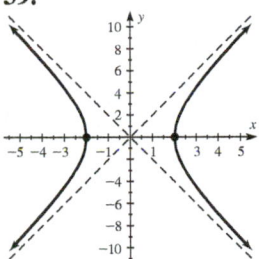

41. $x^2 - y^2 = 1$ **43.** $\frac{y^2}{4} - \frac{x^2}{9} = 1$

45. (a)

(b)

(c)

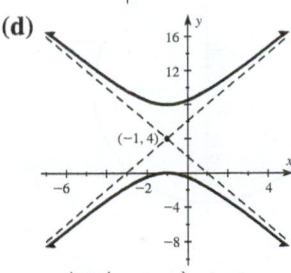

(d)

$y = \pm\frac{3}{2}(x + 1) + 3$, or
$y = \frac{3}{2}x + \frac{9}{2}, y = -\frac{3}{2}x + \frac{3}{2}$

$y = \pm 2(x + 1) + 4$, or
$y = 2x + 6, y = 2x + 2$

47. (a) $A \approx 62.83; P \approx 28.45$
(b) $A \approx 11.75; P \approx 13.33$
49. (a) $[-60, 60, 10]$ by $[-40, 40, 10]$

(b) $P \approx 243.9$ A.U., or about 2.27×10^{10} miles;
$A \approx 4733$ square A.U., or about 4.09×10^{19} square miles
51. Maximum: 668 miles; minimum: 340 miles
53. Height: 20 feet; width: 200 feet
55. Answers may vary.

Checking Basic Concepts 13.1 & 13.2 (p. 901)

1.

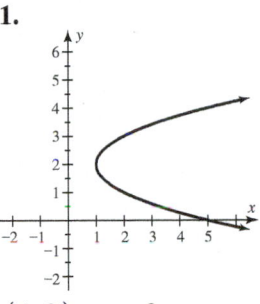

2. $(x - 1)^2 + (y + 2)^2 = 4$

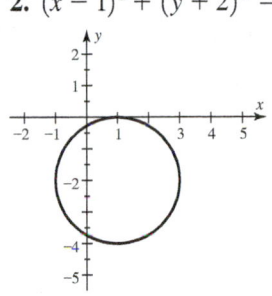

$(1, 2); y = 2$
3. x-intercepts: $(\pm 2, 0)$; y-intercepts: $(0, \pm 3)$
4. (a)

Parabola

(b)

Ellipse

(c)

Hyperbola

(d)

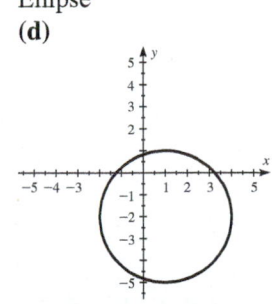

Circle (and ellipse)

Section 13.3 (pp. 908–910)

1. Any number   **2.** Two   **3.** Two; the line intersects the circle twice.   **4.** All points inside and including a circle of radius 1 centered at the origin   **5.** No   **6.** No

7.

8.

4.

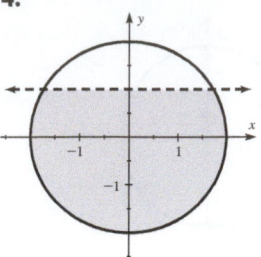

(answers may vary)   (answers may vary)

9. $(1, 3)$, $(-1, -3)$ **11.** $(0, -1)$, $(0, 1)$
13. $(3, 6)$, $(-3, -6)$ **15.** $(2, -1)$ **17.** $(-\sqrt{2}, 2)$, $(\sqrt{2}, 2)$
19. $\left(-\frac{3}{5}, \frac{7}{5}\right)$, $(-1, 1)$ **21.** No solutions   **23.** $(\pm 3, \pm 1)$
25. $(-1, -2)$, $(-2, 1)$ **27.** $(7, 3)$, $(2, -2)$
29. $(-1, 2)$, $(3, -6)$ **31.** $(-1, -1)$, $(1, 1)$
33.

35.

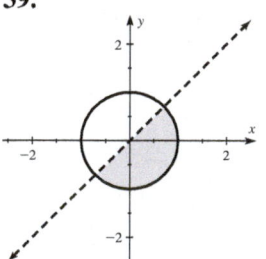

Chapter 13 Review (pp. 912–914)

1.

2.

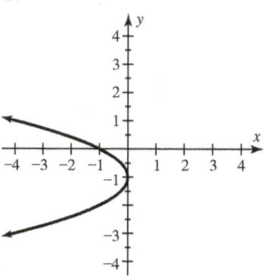

$(0, 0)$; $y = 0$ $(0, -1)$; $y = -1$

3.

4.

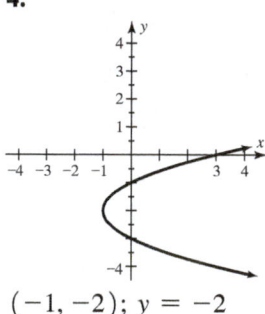

$(0, 2)$; $y = 2$ $(-1, -2)$; $y = -2$

5.

6.

37.

39.

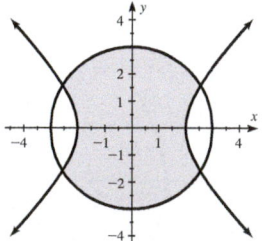

$(0, 2)$ (answers may vary) $\left(\frac{1}{2}, -\frac{1}{2}\right)$ (answers may vary)

41.

43.

$(1, 0)$ $(0, 0)$ (answers may vary)

$(1, 0)$; $y = 0$ $\left(-\frac{7}{2}, -1\right)$; $y = -1$

7. $x = y^2$ **8.** $(x + 2)^2 + (y - 2)^2 = 16$
9. $x^2 + y^2 = 1$ **10.** $(x - 2)^2 + (y + 3)^2 = 16$
11. 5; $(0, 0)$ **12.** 3; $(2, 0)$

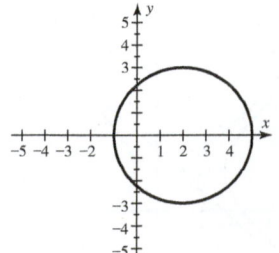

45. a.   **47.** $y \geq x^2$; $y < 4 - x$
49. $[-10, 10, 1]$ by $[-10, 10, 1]$

51. $r = 1.6$ inches; $h \approx 4.97$ inches
53. (a) $xy = 143$, $2x + 2y = 48$
(b) $x = 11$ ft, $y = 13$ ft

Checking Basic Concepts 13.3 (p. 910)

1. $(1, -1)$, $(3, 3)$ **2.** Two   **3.** (a) $(0, 3)$, $(4, 4)$
(answers may vary)   (b) $y \geq 2 - x$ and $y \leq 4 - x^2$

13. $\sqrt{5}$; $(-3, 1)$

14. 3; $(1, -1)$

15.

16.

17.

18.

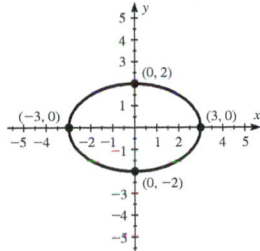

19. $\dfrac{y^2}{16} + \dfrac{x^2}{4} = 1$ **20.** $x^2 - \dfrac{y^2}{4} = 1$

21.

22.

23.

24.

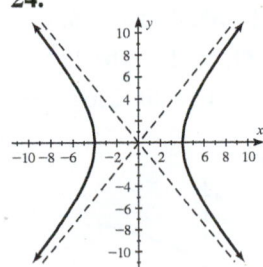

25. $(0, 3)$, $(3, 0)$ **26.** $(-1, -2)$, $(1, 2)$
27. $(0, 0)$, $(2, 2)$
28. $(-2, -1)$, $(-2, 1)$, $(2, -1)$, $(2, 1)$
29. $(-4, -4)$, $(4, 4)$ **30.** $(0, -4)$, $(4, 0)$
31. $(-1, 1)$, $(1, 1)$ **32.** $(1, 2)$, $(2, 5)$
33. $(-4, -12)$, $(2, 0)$ **34.** $(-1, -1)$, $(1, -1)$
35. $(2, 2)$, $(-4, -4)$ **36.** $(1, 1)$, $(0, 0)$

37.

38.

39.

40.

41.

42.

43.

44.

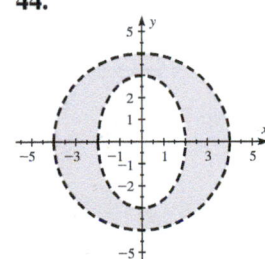

45. $y \geq x^2 - 2$, $y \leq 2 - x$ **46.** $y \geq x$, $x^2 + y^2 \leq 4$
47. **(a)** $xy = 1000$, $2x + 2y = 130$
(b) $x = 25$ inches; $y = 40$ inches
(c) $x = 25$ inches; $y = 40$ inches
48. **(a)** $xy = 60$, $y - x = 7$ **(b)** $x = 5$; $y = 12$
(c) $x = 5$; $y = 12$ **49.** $r = 1$ foot, $h \approx 15.92$ feet; yes
50. Either $r \approx 0.94$ inches, $h \approx 12.60$ inches or $r \approx 3.00$
inches, $h \approx 1.23$ inches; no
51. **(a)** $[-7.5, 7.5, 1]$ by $[-5, 5, 1]$

(b) $A \approx 24.33$, $P \approx 18.32$
52. **(a)** $[-3, 3, 1]$ by $[-2, 2, 1]$

(b) $P \approx 9.55$ A.U., or about 8.9×10^8 miles; $A \approx 7.26$
square A.U., or about 6.3×10^{16} square miles

Chapter 13 Test (pp. 914–915)

1.

2.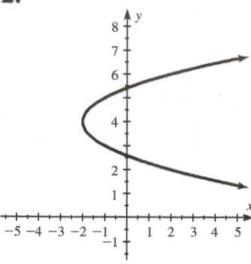

$(1, 2); x = 1$ $\qquad$ $(-2, 4); y = 4$

3. $x = -y^2 + 1$ **4.** $(x - 2)^2 + (y + 4)^2 = 4$

5. $(x + 5)^2 + (y - 2)^2 = 100$

6. $r = 4$, center $= (-2, 3)$ **7.**

 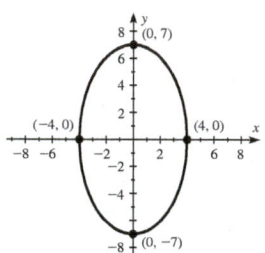

8. $\frac{x^2}{100} + \frac{y^2}{64} = 1$

9.

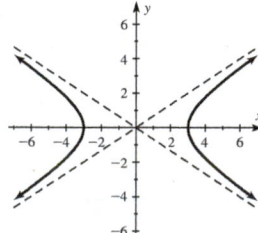

10. $(0, -4), (4, 0)$ **11.** $(-1, -4), (4, 1)$

12. $(-2, 4), (2, 4)$

13.

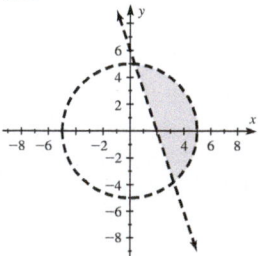

14. $y \le 4 - x^2, y \ge x^2 - 4$

15. (a) $xy = 5000, 2x + 2y = 300$

(b) 50 feet by 100 feet

16. Either $x \approx 22.08$ in., $y \approx 2.43$ in. or

$x \approx 7.29$ in., $y \approx 22.24$ in.; no

17. (a) $[-30, 30, 10]$ by $[-20, 20, 10]$

(b) 18.28 A.U., or about 1,700,040,000 miles

Chapters 1–13 Cumulative Review (pp. 915–916)

1. 25 **2.** $\frac{b^8}{a^3}$ **3.** 0.007345 **4.** $\frac{1}{2}; x \ne 4$ **5.** 3; 4

6. **7.**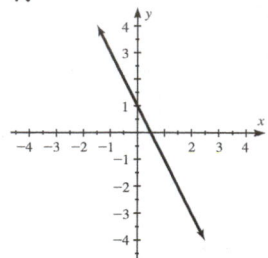

8. $y = \frac{3}{2}x - 5$ **9.** $\frac{2}{7}$ **10.** $(-1, 3]$ **11.** $[-2, 2]$

12. $(-\infty, -1] \cup [3, \infty)$ **13.** $(1, 3)$

14.

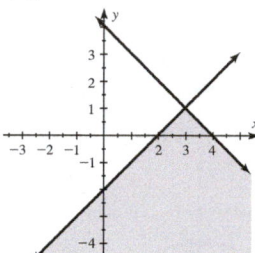

15. $2x^2 + 9x - 5$ **16.** $2x^2y - 3xy^3 + xy$

17. $(3x + 1)(2x - 5)$ **18.** $x(x + 2)(x - 2)$ **19.** $-2, -1$

20. $\frac{-3 \pm \sqrt{5}}{2}$ **21.** $\frac{3}{2}$ **22.** $\frac{2x}{x^2 - 1}$ **23.** $2x\sqrt{2}$ **24.** 4

25. $4x$ **26.** $5\sqrt{3x}$

27.

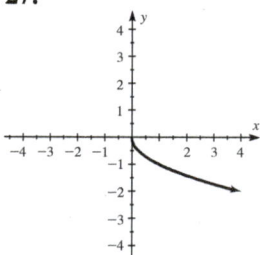

28. 5 **29.** 13 **30.** 2 **31.** $(3, -6)$

32. $f(x) = (x - 1)^2 + 2$ **33.** Shifted 4 units right

34. 1, 2 **35.** 0, $\pm i$ **36. (a)** 4 **(b)** 3 **(c)** x **(d)** 6 **(e)** 2

(f) 3 **37. (a)** 17 **(b)** $2x^2 + 2$ **38.** $\frac{2}{3} - \frac{1}{3}x$

39. \$1340.10 **40.** $2 \log x + \frac{1}{2} \log y - 3 \log z$ **41.** ln 9

42. 25

43. (a) $\qquad\qquad$ **(b)**

(c)

(d)

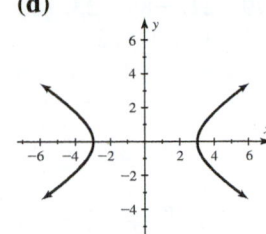

44. $(0, -1)$, $(0, 1)$

45.

46. (a) 400; initially, the driver is 400 miles from home.
(b) $(8, 0)$; after 8 hours, the driver arrives at home.
(c) -50; the driver is traveling 50 mph toward home.
47. $500 at 5%, $1000 at 6%, $500 at 7%
48. 300 ft by 300 ft **49.** $50 \ln 2 \approx 34.7$ yr
50. 2.4 in.

14 Sequences and Series

Section 14.1 (pp. 924–925)

1. 1, 2, 3, 4 (answers may vary) **2.** 1, 3, 5, 7, ... (answers may vary) **3.** function; natural numbers **4.** sequence
5. 6 **6.** scatterplot **7.** $f(2)$ **8.** a_4 **9.** 1, 4, 9, 16
11. $\frac{1}{6}, \frac{1}{7}, \frac{1}{8}, \frac{1}{9}$ **13.** $\frac{5}{2}, \frac{5}{4}, \frac{5}{8}, \frac{5}{16}$ **15.** 9, 9, 9, 9 **17.** 1, 8, 27
19. 1, $\frac{8}{5}$, 2 **21.** 2, 9, 20 **23.** $-2, -2, -2$ **25.** $f(n) = \frac{1}{n}$
27. $f(n) = 2n + 1$ **29.** $b + c$; $2b + c$
31. 7 **33.** 3, 4, 5, 3, 1 **35.** 6, 5, 4, 3, 2, 1
37.

n	1	2	3	4	5	6	7
a_n	2	3	4	5	6	7	8

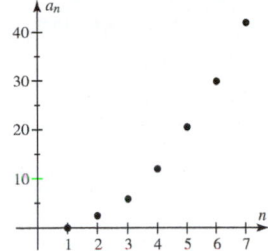

39.

n	1	2	3	4	5	6	7
a_n	0	2	6	12	20	30	42

41.

n	1	2	3	4	5	6	7
a_n	2	4	8	16	32	64	128

43. $a_n = 30n$ for $n = 1, 2, 3, \ldots, 7$

n	1	2	3	4	5	6	7
a_n	30	60	90	120	150	180	210

45. (a) 1, 4, 9, 16 **(b)** 4, 8, 12, 16
47. (a) $20,000; $16,000 **(b)** $a_n = 25,000(0.8)^n$
(c)

n	1	2	3	4	5	6	7
a_n	20,000	16,000	12,800	10,240	8192	6553.6	5242.9

49. (a) $a_n = 2048(0.5)^{n-1}$, for $n = 1, 2, 3, \ldots, 7$
(b)

n	1	2	3	4	5	6	7
a_n	2048	1024	512	256	128	64	32

(c)

Section 14.2 (pp. 932–935)

1. linear **2.** exponential **3.** $a_n = 3n + 1$; 3 (answers may vary) **4.** $a_n = 5(2)^{n-1}$; 2 (answers may vary)
5. add; previous **6.** multiply; common ratio **7.** 19; 4
8. 32; -2 **9.** $a_n = a_1(r)^{n-1}$ **10.** $a_n = a_1 + (n - 1)d$
11. Yes; 10 **13.** Yes; -1 **15.** No **17.** Yes; 3
19. Yes; -3 **21.** No **23.** Yes; 1 **25.** No **27.** Yes; 2
29. $a_n = -2n + 9$ **31.** $a_n = 4n - 6$

33. $a_n = -2n + 32$ **35.** 59   **37.** 21   **39.** Yes; 3
41. Yes; 0.8   **43.** No   **45.** Yes; 2   **47.** No   **49.** Yes; 4
51. Yes; 2   **53.** No   **55.** $a_n = 1.5(4)^{n-1}$
57. $a_n = -3(-2)^{n-1}$ **59.** $a_n = 1(4)^{n-1}$ **61.** 4374
63. 243   **65. (a)** 3000, 6000, 9000, 12,000, 15,000; arithmetic
(b) $a_n = 3000n$ **(c)** 60,000; when there are 20 people, the
ventilation rate should be 60,000 cubic feet per hour.

(d)

Yes
67. (a) $a_n = 3(0.8)^{n-1}$
     (b) No; geometric

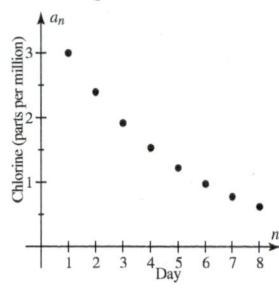

69. (a) $a_n = 5(0.85)^{n-1}$ **(b)** Geometric; the common
ratio is 0.85.   **(c)** About 1.6; on the 8th bounce, the ball
reaches a maximum height of about 1.6 ft.
71. (a) Arithmetic; the common difference is 2.
(b) $a_n = 40 + 2(n - 1)$ or $a_n = 38 + 2n$ **(c)** 78
73. 1, 1, 2, 3, 5, 8, 13, 21, 34, 55, 89, 144

Checking Basic Concepts 14.1 & 14.2 (p. 935)

1. $\frac{1}{5}, \frac{1}{3}, \frac{3}{7}, \frac{1}{2}$
2.

n	1	2	3	4	5
a_n	2	3	4	5	6

3. (a) Arithmetic; $a_n = 3n - 5$ **(b)** Geometric;
$a_n = 3(-2)^{n-1}$ **4.** $a_n = 2n + 3$ **5.** $a_n = 5(2)^{n-1}$

Section 14.3 (pp. 943–945)

1. series   **2.** 10   **3.** arithmetic   **4.** geometric
5. $n\left(\frac{a_1 + a_n}{2}\right)$ or $\frac{n}{2}(2a_1 + (n - 1)d)$ **6.** $a_1\left(\frac{1 - r^n}{1 - r}\right)$
7. sum   **8.** $a_1 + a_2 + a_3 + a_4$ **9.** arithmetic
10. geometric   **11.** 48   **13.** 820   **15.** -5 **17.** -92

19. 3279   **21.** -85 **23.** 182   **25.** 6825
27. $2 + 4 + 6 + 8$; 20
29. $4 + 4 + 4 + 4 + 4 + 4 + 4 + 4$; 32
31. $1 + 4 + 9 + 16 + 25 + 36 + 49$; 140
33. $12 + 20$; 32   **35.** $\sum_{k=1}^{6} k^4$ **37.** $\sum_{k=1}^{5} \frac{1}{k^2}$
39. $\sum_{k=1}^{n} k = n\left(\frac{a_1 + a_n}{2}\right) = n\left(\frac{1 + n}{2}\right) = \frac{n(n + 1)}{2}$
41. \$91,523.93   **43.** \$62,278.01
45. (a) $22.2 + 37.7 + 28.7 + 42.1 + 33.7 + 22.5$
(b) 186.9   **47. (a)** $\sum_{k=1}^{n} 0.8(0.2)^{k-1}$ **(b)** 2
49. (a) $1, \frac{1}{2}, \frac{1}{4}, \frac{1}{8}, \frac{1}{16}$ **(b)** $\frac{1023}{512}$ **51.** 90 logs   **53.** \$1,080,000
55. About 3.16 feet   **57. (a)** $\frac{4}{3}$ **(b)** $\frac{3}{2}$ **(c)** $0.\overline{1} = \frac{1}{9}$
(d) $0.\overline{12} = \frac{4}{33}$ **59.** Answers may vary.

Section 14.4 (p. 951)

1. 5   **2.** $n + 1$ **3.** 4
4.
```
            1
          1   1
        1   2   1
      1   3   3   1
    1   4   6   4   1
```
5. 24   **6.** $6! = 720$ **7.** $\frac{n!}{(n - r)! \, r!}$ **8.** $a^2 + 2ab + b^2$
9. n **10.** 1   **11.** $x^3 + 3x^2y + 3xy^2 + y^3$
13. $16x^4 + 32x^3 + 24x^2 + 8x + 1$
15. $a^5 - 5a^4b + 10a^3b^2 - 10a^2b^3 + 5ab^4 - b^5$
17. $x^6 + 3x^4 + 3x^2 + 1$ **19.** 6   **21.** 5040   **23.** 4
25. 2   **27.** 10   **29.** 5   **31.** 6   **33.** 1   **35.** 792   **37.** 126
39. 75,582   **41.** $m^3 + 3m^2n + 3mn^2 + n^3$
43. $x^4 - 4x^3y + 6x^2y^2 - 4xy^3 + y^4$
45. $8a^3 + 12a^2 + 6a + 1$
47. $x^5 + 10x^4 + 40x^3 + 80x^2 + 80x + 32$
49. $81 + 216m + 216m^2 + 96m^3 + 16m^4$
51. $8x^3 - 12x^2y + 6xy^2 - y^3$ **53.** a^8 **55.** $35x^4y^3$
57. $512m^9$

Checking Basic Concepts 14.3 & 14.4 (p. 952)

1. (a) Geometric   **(b)** Arithmetic   **2.** 312
3. -341 **4.** $x^4 - 4x^3y + 6x^2y^2 - 4xy^3 + y^4$
5. $x^3 + 6x^2 + 12x + 8$

Chapter 14 Review (pp. 953–955)

1. 1, 8, 27, 64   **2.** 3, 1, -1, -3 **3.** $1, \frac{4}{5}, \frac{3}{5}, \frac{8}{17}$
4. $-2, 4, -8, 16$ **5.** $-2, 0, 4, 2$ **6.** 5, 3, 2, 1
7.

n	1	2	3	4	5	6	7
a_n	2	4	6	8	10	12	14

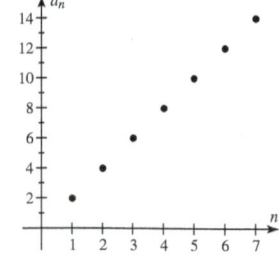

8.

n	1	2	3	4	5	6	7
a_n	−3	0	5	12	21	32	45

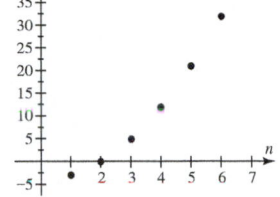

9.

n	1	2	3	4	5	6	7
a_n	2	1	0.5	0.25	0.125	0.0625	0.0313

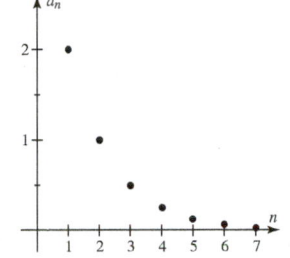

10.

n	1	2	3	4	5	6	7
a_n	1	1.4142	1.7321	2	2.2361	2.4495	2.6458

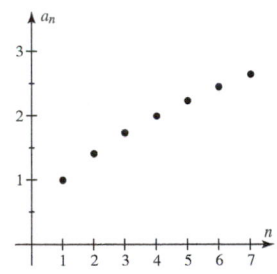

11. Yes; 5   **12.** No   **13.** No   **14.** Yes; $-\frac{1}{3}$ **15.** Yes; −3
16. No   **17.** Yes; −1   **18.** No   **19.** $a_n = 4n - 7$
20. $a_n = -5n + 7$ **21.** Yes; 4   **22.** No   **23.** No
24. Yes; 0.7   **25.** No   **26.** Yes; $-\frac{1}{3}$ **27.** No   **28.** Yes; 2
29. $a_n = 5(0.9)^{n-1}$ **30.** $a_n = 2(4)^{n-1}$ **31.** 216   **32.** 7.5
33. 3277   **34.** $\frac{511}{256}$ **35.** $3 + 5 + 7 + 9 + 11$
36. $\frac{1}{2} + \frac{1}{3} + \frac{1}{4} + \frac{1}{5}$ **37.** $1 + 8 + 27 + 64$
38. $-1 + (-2) + (-3) + (-4) + (-5) + (-6)$
39. $\sum_{k=1}^{20} k$ **40.** $\sum_{k=1}^{20} \frac{1}{k}$ **41.** $\sum_{k=1}^{9} \frac{k}{k+1}$ **42.** $\sum_{k=1}^{7} k^2$
43. $x^3 + 12x^2 + 48x + 64$
44. $16x^4 + 32x^3 + 24x^2 + 8x + 1$
45. $x^5 - 5x^4y + 10x^3y^2 - 10x^2y^3 + 5xy^4 - y^5$
46. $a^6 - 6a^5 + 15a^4 - 20a^3 + 15a^2 - 6a + 1$
47. 6   **48.** 10   **49.** 20   **50.** 4
51. $m^4 + 8m^3 + 24m^2 + 32m + 16$
52. $a^5 + 5a^4b + 10a^3b^2 + 10a^2b^3 + 5ab^4 + b^5$
53. $x^4 - 12x^3y + 54x^2y^2 - 108xy^3 + 81y^4$
54. $27x^3 - 54x^2 + 36x - 8$

55. $a_n = 45,000(1.1)^{n-1}$ for $n = 1, 2, 3, \ldots, 7$; geometric

n	1	2	3	4	5	6	7
a_n	45,000	49,500	54,450	59,895	65,885	72,473	79,720

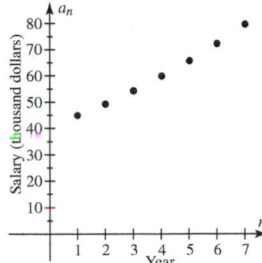

56. $a_n = 45,000 + 5000(n - 1)$ for $n = 1, 2, 3, \ldots, 7$; arithmetic

n	1	2	3	4	5	6	7
a_n	45,000	50,000	55,000	60,000	65,000	70,000	75,000

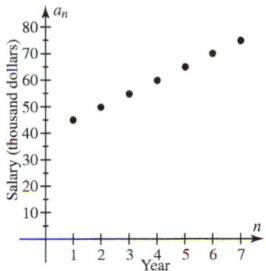

57. $a_n = 30n$ for $n = 1, 2, 3, \ldots, 7$

n	1	2	3	4	5	6	7
a_n	30	60	90	120	150	180	210

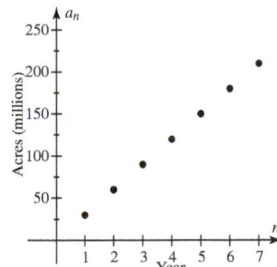

58. **(a)** $a_n = 12,000(0.9)^{n-1}$ **(b)** Geometric; the common ratio is 0.9.   **(c)** About 1800; in 2013, there were about 1800 escapes.

Chapter 14 Test (pp. 955–956)

1. $\frac{1}{2}, \frac{4}{3}, \frac{9}{4}, \frac{16}{5}$ **2.** −3, 2, 1, −2, 3
3.

n	1	2	3	4	5	6	7
a_n	0	2	6	12	20	30	42

4. $16x^4 - 32x^3 + 24x^2 - 8x + 1$
5. Arithmetic; −3   **6.** Geometric; −2
7. $a_n = 2 - 3(n - 1)$ or $a_n = 5 - 3n$
8. $a_n = 2(1.5)^{n-1}$ **9.** Yes; 2.5   **10.** No   **11.** 99
12. $\frac{463}{729}$ **13.** $6 + 9 + 12 + 15 + 18 + 21$
14. $\sum_{k=1}^{60} k^3$ **15.** 35   **16.** 10   **17.** 9180

18. $a_n = 176(1.59)^{n-1}$ for $n = 1, 2, 3, 4, 5$; geometric

n	1	2	3	4	5
a_n	176	279.84	444.95	707.46	1124.87

19. (a) $a_n = 2000(2)^{n-1}$ **(b)** Geometric; the common ratio is 2.   **(c)** 64,000; after 25 days, there are 64,000 caterpillars.

(d)

Chapters 1–14 Cumulative Review (pp. 956–958)

1. Distributive   **2.** $D = \{-6, -2, 0, 2\}$; $R = \{0, 1, 3, 5\}$

3. $\frac{y^9}{27x^5}$ **4.** $\frac{16a^8}{b^4}$ **5.** z^{10} **6.** $\frac{2y^3}{x^6}$ **7.** $D = \{x \mid x \neq 8\}$

8. $f(x) = -2x + 1$ **9.** $y = 3$ **10.** $-3; 5$

11. $y = \frac{3}{2}x + \frac{5}{2}$ **12.** $y = 2x - 8$ **13.** -26

14. $\left(\frac{20}{11}, \infty\right)$ **15.** $(-\infty, -1] \cup [11, \infty)$ **16.** $-15, 3$

17. $\frac{1}{8}$ **18.** $[-12, 9)$

19.

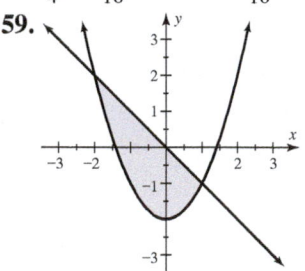

20. $(5, 2)$ **21.** $(3, 3, -1)$ **22.** 17

23. $8x^7 - 6x^6 + 10x^3$ **24.** $6z^2 - 13z - 28$

25. $(2x - 3y)(2x + 3y)$ **26.** $(a^2 + 4)(2a - 1)$

27. $-\frac{3}{4}, 1$ **28.** $0, 4, 6$ **29.** $\frac{x - 2}{x + 1}$ **30.** $\frac{x + 4}{x - 2}$ **31.** 12

32. $\frac{1}{2}, 3$ **33.** $W = \frac{3C - 5R}{2}$ **34.** $\frac{1 + 2x}{1 - 4x}$

35. $(xy - 2)\sqrt[3]{xy}$ **36.** 14   **37.** $-2, 8$ **38.** $\frac{3}{2}$

39. $14 + 2i$ **40.** $-2 - 2i$ **41.** $-\frac{1}{3}$

42. $y = 2(x + 2)^2 + 9$; $(-2, 9)$ **43.** $2 \pm 3i$

44. $-4, 8$ **45.** $(-3, 1)$ **46. (a)** 65

(b) $(g \circ f)(x) = 3x^2 + 1$ **47.** $f^{-1}(x) = \frac{2x - 1}{3}$

48. $3 \ln x + \frac{1}{2} \ln y$ **49.** $\log \frac{x}{4y}$ **50.** 56.23   **51.** 0.58

52.

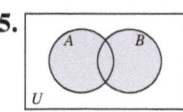

$(1, 3)$; $y = 3$

53. $(3, -1)$; 2

54.   **55.**

56. $\frac{y^2}{4} - \frac{x^2}{16} = 1$ **57.** $\frac{x^2}{16} + \frac{y^2}{4} = 1$ **58.** $(1, 2), (-1, 2)$

59.

60. Arithmetic; -2 **61.** Geometric; 0.2

62. $a_n = 3n - 1$ **63.** $a_n = 4(3)^{n-1}$ **64.** 171

65. 683   **66.** $16x^4 + 96x^3 + 216x^2 + 216x + 81$

67. $8a^3 - 60a^2b + 150ab^2 - 125b^3$

68. $\sqrt{\frac{14}{\pi}} \approx 2.11$ inches

69. (a) $f(x) = 0.4x + 85$ **(b)** 188 pounds

70. Airplane: 380 mph; wind: 20 mph

71. 36 minutes   **72. (a)** $xy = 96$, $3x - y = 12$

(b) 8 and 12   **73.** 144

B Sets

(pp. AP-17–AP-18)

1. $\{1, 2, 3, 4, 5, 6, 7\}$ **3.** $\{$Sunday, Saturday$\}$

5. $\{$A, B, C, D, E, F, G$\}$ **7.** $\varnothing$ **9.** $\{0, 2, 4, 6, 8, 10\}$

11. $\in$ **13.** $\notin$ **15.** $\in$ **17.** $\notin$ **19.** $\in$

21. $E = \{2, 4, 6, 8, 10\}$ **23.** $E = \{2, 4, 6, 8, 10, \ldots\}$

25. $A = \{$Apple, Apricot$\}$ **27.** $A = \{$Algebra$\}$

29. False   **31.** True   **33.** True   **35.** True   **37.** True

39. True   **41.** False   **43.** True

45.   **47.**

49. **51.**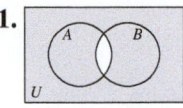

53. $\{1, 2, 3, 4\}$ **55.** $\{a, b\}$ **57.** $\{a, b, c\}$
59. $\{a, b, c, d\}$ **61.** $\{5, 8\}$
63. $\{1, 2, 3, 4, 5, 6, 7, 8, 9, 10\}$ **65.** $\{4, 5, 6\}$
67. $\varnothing$ **69.** $\{1, 3, 5, 7, 9\}$ **71.** $\varnothing$ **73.** $\varnothing$
75. $\{1, 3, 5, 7, 9\}$
77. $\{1, 2, 3, 5, 7, 8, 9, 10\}$ **79.** $\varnothing$ **81.** $\{1, 2, 8, 9, 10\}$
83. $\{4, 6\}$ **85.** $\{1, 2, 3, 4, 5, 6, 7, 8, 9, 10\}$ **87.** Infinite
89. Finite **91.** Finite

C Linear Programming

(pp. AP-23–AP-25)

1. linear programming
3. feasible solutions **5.** vertex

7.   **9.**

11.   **13.**

15.   **17.**

19.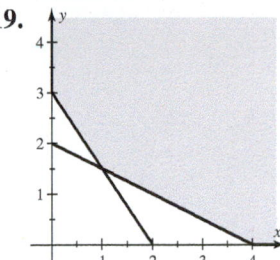

21. $R = 31$ **23.** $R = 15$ **25.** $C = 5$ **27.** $C = 2$
29. $R = 750$ **31.** $R = 12$ **33.** $R = 31.5$ **35.** $R = 16$
37. $C = 0$ **39.** $C = 32$ **41.** $C = 60$
43. 20 pounds of candy, 80 pounds of coffee **45.** $1800
47. 1 ounce Brand **X**, 2 ounces Brand **Y** **49.** $600

D Synthetic Division

(p. AP-27)

1. $x + 4 + \frac{3}{x - 1}$ **3.** $3x - 1$ **5.** $x^2 + 3x + 2$
7. $2x^2 + 5x + 10 + \frac{19}{x - 2}$ **9.** $x^2 + 2x + 6 + \frac{28}{x - 4}$
11. $x^2 + 2x + 2$ **13.** $2x^3 - 4x^2 + 8x - 17 + \frac{34}{x + 2}$
15. $b^3 + b^2 + b + 1$

Glossary

absolute value A nonnegative number, written $|a|$, that is equal to the distance of a from the origin on the number line.

absolute value equation An equation that contains an absolute value.

absolute value function The function defined by $f(x) = |x|$.

absolute value inequality An inequality that contains an absolute value.

addends In an addition problem, the two numbers that are added.

addition property of equality If a, b, and c are real numbers, then $a = b$ is equivalent to $a + c = b + c$.

additive identity The number 0.

additive inverse (opposite) The additive inverse, or opposite, of a number a is $-a$.

adjacency matrix A matrix that may be used to represent a social network or a map showing distances between cities.

algebraic expression An expression consisting of numbers, variables, operation symbols, such as $+$, $-$, $\times$, and $\div$, and grouping symbols, such as parentheses.

annuity A sum of money from which regular payments are made.

approximately equal The symbol $\approx$ indicates that two quantities are nearly equal.

arithmetic sequence A linear function given by $a_n = dn + c$ whose domain is the set of natural numbers.

arithmetic series The sum of the terms of an arithmetic sequence.

associative property for addition For any real numbers a, b, and c, $(a + b) + c = a + (b + c)$.

associative property for multiplication For any real numbers a, b, and c, $(a \cdot b) \cdot c = a \cdot (b \cdot c)$.

asymptotes of a hyperbola The two lines determined by the diagonals of the hyperbola's fundamental rectangle.

augmented matrix A matrix used to represent a system of linear equations; a vertical line is positioned in the matrix where the equals signs occur in the system of equations.

average The result of adding up the numbers of a set and then dividing the sum by the number of elements in the set.

axis of symmetry of a parabola The line passing through the vertex of the parabola that divides the parabola into two symmetric parts.

base The value of b in the expression b^n.

base e If the base of an exponential expression is e (approximately 2.72), then we say this expression has base e.

basic complex fraction A complex fraction where both the numerator and denominator are single fractions.

basic principle of fractions When simplifying fractions, the principle that states $\frac{a \cdot c}{b \cdot c} = \frac{a}{b}$.

basic rational equation A rational equation that has a single rational expression on each side of the equals sign.

binary operation An operation that requires two numbers to calculate an answer.

binomial A polynomial with two terms.

binomial coefficient The expression $_nC_r = n!/((n - r)!r!)$, where n and r are nonnegative integers, $n \geq r$, that can be used to calculate the numbers in Pascal's triangle.

binomial theorem A theorem that provides a formula to expand expressions of the form $(a + b)^n$.

braces { } Used to enclose the elements of a set.

branches A hyperbola has two branches, a left branch and a right branch, or an upper branch and a lower branch.

byte A unit of computer memory, capable of storing one letter of the alphabet.

capacity A measure of the amount of a substance that a container can hold.

center of a circle The point that is a fixed distance from all the points on a circle.

center of an ellipse The midpoint of the major axis.

change of base formula A formula used to evaluate a logarithm of one base by using a logarithm of a different base and given by $\log_a x = \frac{\log x}{\log a}$ or $\log_a x = \frac{\ln x}{\ln a}$.

circle The set of points in a plane that are the same distance from a fixed point.

circumference The perimeter of a circle.

coefficient The numeric constant of a term.

coefficient of a monomial The number in a monomial.

common difference The value of d in an arithmetic sequence, $a_n = dn + c$.

common logarithmic function The function given by $f(x) = \log x$.

common logarithm of a positive number x Denoted $\log x$, it may be calculated as follows: If x is expressed as $x = 10^k$, then $\log x = k$, where k is a real number. That is, $\log 10^k = k$.

common ratio The value of r in a geometric sequence, $a_n = a_1(r)^{n-1}$.

commutative property for addition For any real numbers a and b, $a + b = b + a$.

commutative property for multiplication For any real numbers a and b, $a \cdot b = b \cdot a$.

complement The complement of set A is the set containing all elements in the universal set that are not in A, denoted A'.

completing the square method An important technique in mathematics that involves adding a constant to a binomial so that a perfect square trinomial results.

complex conjugate The complex conjugate of $a + bi$ is $a - bi$.

complex fraction A rational expression that contains fractions in its numerator, denominator, or both.

complex number A complex number can be written in standard form as $a + bi$, where a and b are real numbers and i is the imaginary unit.

composite function If f and g are functions, then g of f, or the composition of g and f, is defined by $(g \circ f)(x) = g(f(x))$ and is read "g of f of x."

composite number A natural number greater than 1 that is not a prime number.

composition Replacing a variable with an algebraic expression in function notation is called composition of functions; when functions are applied in sequence.

compound inequality Two inequalities joined by the word *and* or the word *or*.

compound interest A type of interest paid at the end of each year using the formula $A = P(1 + r)^t$, where P is the amount deposited, r is the annual interest rate in decimal form, and t is the number of years for the account to contain A dollars.

conic section The curve formed by the intersection of a plane and a cone (or cones).

conjugate The conjugate of $a + b$ is $a - b$.

consistent system with dependent equations A system of linear equations with infinitely many solutions.

consistent system with independent equations A system of linear equations with exactly one solution.

constant function A linear function with $m = 0$ that can be written as $f(x) = b$.

constant of proportionality (constant of variation) In the equation $y = kx$, the nonzero number k.

constraint In linear programming, an inequality which limits the objective function.

continuous growth Growth in a quantity that is directly proportional to the amount present.

contradiction An equation that is always false regardless of the values of any variables.

Cramer's rule A method that uses determinants to solve linear systems of equations.

cube root The number b is a cube root of a if $b^3 = a$.

cube root function The function defined by $f(x) = \sqrt[3]{x}$.

cubic function A function f of degree 3 represented by $f(x) = ax^3 + bx^2 + cx + d$, where $a, b, c,$ and d are constants and $a \neq 0$.

decay factor The value of a in an exponential function, $f(x) = Ca^x$, when $0 < a < 1$.

degree A degree (°) is $\frac{1}{360}$ of a revolution.

degree of a monomial The sum of the exponents of the variables.

degree of a polynomial The degree of the term (or monomial) with highest degree.

dependent equations Equations in a linear system that has infinitely many solutions.

dependent variable The variable that represents the output of a function.

determinant A real number associated with a square matrix.

diagrammatic representation A function represented by a diagram.

difference The answer to a subtraction problem.

difference of two cubes Expression in the form $a^3 - b^3$, which can be factored as $(a - b)(a^2 + ab + b^2)$.

difference of two squares Expression in the form $a^2 - b^2$, which can be factored as $(a - b)(a + b)$.

dimension of a matrix The size expressed in number of rows and columns. For example, if a matrix has m rows and n columns, its dimension is $m \times n$ (m by n).

directly proportional (varies directly) A quantity y is directly proportional to x if there is a nonzero number k such that $y = kx$.

discriminant The expression $b^2 - 4ac$ in the quadratic formula.

distance The distance d between the points (x_1, y_1) and (x_2, y_2) in the xy-plane is $d = \sqrt{(x_2 - x_1)^2 + (y_2 - y_1)^2}$.

distributive properties For any real numbers $a, b,$ and c, $a(b + c) = ab + ac$ and $a(b - c) = ab - ac$.

dividend In a division problem, the number being divided.

divisor In a division problem, the number being divided *into* the dividend.

domain The set of all x-values of the ordered pairs in a function.

element of a matrix Each number in a matrix.

elements of a set The members of a set.

elimination method A symbolic method used to solve a system of equations that is based on the property that if "equals are added to equals the results are equal."

ellipse The set of points in a plane the sum of whose distances from two fixed points is constant.

empty set (null set) A set that contains no elements.

equal sets If every element in a set A is in set B and every element in set B is in set A, then A and B are equal sets, denoted $A = B$.

equation A mathematical statement that two algebraic expressions are equal.

equation of a line Point-slope form and slope-intercept form are examples of an equation of a line.

equivalent equations Equations that have the same solution set.

equivalent fractions Fractions that name the same number.

even root The nth root, $\sqrt[n]{a}$, where n is even.

expansion of a determinant by minors A method of finding a 3×3 determinant by using determinants of 2×2 matrices.

exponent The value of n in the expression b^n.

exponential decay When $0 < a < 1$, the graph of $f(x) = Ca^x$ models exponential decay.

exponential equation An equation that has a variable as an exponent.

exponential expression An expression that has an exponent.

exponential function with base a and coefficient C A function represented by $f(x) = Ca^x$, where $a > 0, C > 0,$ and $a \neq 1$.

exponential growth When $a > 1$, the graph of $f(x) = Ca^x$ models exponential growth.

extraneous solution A solution that does not satisfy the given equation.

factorial notation $n! = 1 \cdot 2 \cdot 3 \cdot \cdots \cdot n$ for any positive integer.

factoring a polynomial The process of writing a polynomial as a product of lower degree polynomials.

factoring by grouping A technique that uses the associative and distributive properties by grouping four terms of a polynomial in such a way that the polynomial can be factored even though its greatest common factor is 1.

factors In a multiplication problem, the two numbers multiplied.

feasible solutions In linear programming, the set of solutions that satisfy the constraints.

finite sequence A function with domain $D = \{1, 2, 3, \ldots, n\}$ for some fixed natural number n.

finite series A series that contains a finite number of terms, and that can be expressed in the form $a_1 + a_2 + a_3 + \cdots + a_n$ for some n.

finite set A set whose elements can be listed explicitly.

focus (plural: **foci**) A fixed point used to determine the points that form a parabola, an ellipse, or a hyperbola.

FOIL A method for multiplying two binomials $(a + b)$ and $(c + d)$. Multiply **F**irst terms ac, **O**utside terms ad, **I**nside terms bc, and **L**ast terms bd; then combine like terms.

formula A special type of equation used to calculate one quantity from given values of other quantities.

function A set of ordered pairs (x, y), where each x-value corresponds to exactly one y-value.

function notation The notation $y = f(x)$, where the input x produces output y.

fundamental rectangle The rectangle of a hyperbola whose four vertices are determined by either $(\pm a, \pm b)$ or $(\pm b, \pm a)$, where $\frac{x^2}{a^2} - \frac{y^2}{b^2} = 1$ or $\frac{y^2}{a^2} - \frac{x^2}{b^2} = 1$.

Gauss–Jordan elimination A method used to solve a linear system in which matrix row transformations are applied to an augmented matrix.

general term (nth term) of a sequence a_n, where n is a natural number in the domain of a sequence $a_n = f(n)$.

geometric sequence An exponential function given by $a_n = a_1(r)^{n-1}$, where n is a natural number and $r \neq 0$ or 1.

geometric series The sum of the terms of a geometric sequence.

graphical representation A graph of a function.

graphical solution A solution to an equation obtained by graphing.

greater than If a real number b is located to the right of a real number a on the number line, we say that b is greater than a, and write $b > a$.

greater than or equal to If a real number a is greater than or equal to b, denoted $a \geq b$, then either $a > b$ or $a = b$ is true.

greatest common factor (GCF) of a polynomial The term with the highest degree and greatest coefficient that is a factor of all terms in the polynomial.

growth factor The value of a in the exponential function, $f(x) = Ca^x$, when $a > 1$.

half-life The time it takes for an amount to decay to half its original amount.

horizontal line test If every horizontal line intersects the graph of a function f at most once, then f is a one-to-one function.

hyperbola The set of points in a plane the difference of whose distances from two fixed points is constant.

identity An equation that is always true regardless of the values of any variables.

identity property of 1 If any number a is multiplied by 1, the result is a, that is, $a \cdot 1 = 1 \cdot a = a$.

identity property of 0 If 0 is added to any real number a, the result is a, that is, $a + 0 = 0 + a = a$.

imaginary number A complex number $a + bi$ with $b \neq 0$.

imaginary part The value of b in the complex number $a + bi$.

imaginary unit A number denoted i whose properties are $i = \sqrt{-1}$ and $i^2 = -1$.

improper fraction A fraction whose numerator is greater than or equal to its denominator in absolute value.

inch, foot, yard, mile The U.S. units of measurement for length.

inconsistent system A system of linear equations that has no solution.

independent equations Equations in a linear system that have different graphs.

independent variable The variable that represents the input of a function.

index The value of n in the expression $\sqrt[n]{a}$.

index of summation The variable k in the expression $\sum_{k=1}^{n}$.

inequality When the equals sign in an equation is replaced with any one of the symbols $<, \leq, >$, or $\geq$, an inequality results.

infinite sequence A function whose domain is the set of natural numbers.

infinite set A set with infinitely many elements.

infinity Values that increase without bound.

input An element of the domain of a function.

integers A set of numbers including natural numbers, their opposites, and 0, or $\ldots, -3, -2, -1, 0, 1, 2, 3, \ldots$.

intercept form A linear equation in the form $\frac{x}{a} + \frac{y}{b} = 1$.

intersection The set containing elements that belong to *both A and B*, denoted $A \cap B$ and read "*A* intersect *B*."

intersection-of-graphs method A graphical technique for solving two equations.

interval notation A notation used to represent inequalities that eliminates the need to draw an entire number line.

inverse function If f is a one-to-one function, then f^{-1} is the inverse function of f, if $(f^{-1} \circ f)(x) = f^{-1}(f(x)) = x$ for every x in the domain of f, and $(f \circ f^{-1})(x) = f(f^{-1}(x)) = x$ for every x in the domain of f^{-1}.

inversely proportional (varies inversely) A quantity y is inversely proportional to x if there is a nonzero number k such that $y = \frac{k}{x}$.

irrational numbers Real numbers that cannot be expressed as fractions, such as π or $\sqrt{2}$.

Isosceles triangle A triangle with at least two sides of the same length.

joint variation A quantity z varies jointly with x and y if there is a nonzero number k such that $z = kxy$.

leading coefficient In a polynomial of one variable, the coefficient of the monomial with highest degree.

least common denominator (LCD) The LCD of two or more fractions is the smallest number that is divisible by every denominator.

least common multiple (LCM) The smallest number that two or more numbers will divide into evenly.

less than If a real number a is located to the left of a real number b on the number line, we say that a is less than b and write $a < b$.

less than or equal to If a real number a is less than or equal to b, denoted $a \leq b$, then either $a < b$ or $a = b$ is true.

like radicals Radicals that have the same index and the same radicand.

like terms Two terms, or monomials, that contain the same variables raised to the same powers.

linear equation An equation that can be written in the form $ax + b = 0$, where $a \neq 0$.

linear equation in two variables An equation that can be written in the form $Ax + By = C$, where A, B, and C are fixed numbers and A and B are not both equal to 0.

linear function A function f represented by $f(x) = mx + b$, where m and b are constants.

linear inequality A linear inequality results whenever the equals sign in a linear equation is replaced with any one of the symbols $<, \leq, >,$ or $\geq$.

linear inequality in two variables When the equals sign in a linear equation in two variables is replaced with $<, \leq, >,$ or $\geq$, a linear inequality in two variables results.

linear polynomial A polynomial of degree 1 that can be written as $ax + b$, where $a \neq 0$.

linear programming problem A problem consisting of an objective function and a system of linear inequalities called constraints.

linear system in three variables A system of three equations in which each equation can be written in the form $ax + by + cz = d$; an ordered triple (x, y, z) is a solution to the system of equations if the values for x, y, and z make *all three* equations true.

line graph The resulting graph when consecutive data points in a scatterplot are connected with straight line segments.

logarithm with base a of a positive number x Denoted $\log_a x$, it may be calculated as follows: If x can be expressed as $x = a^k$, then $\log_a x = k$, where $a > 0, a \neq 1$, and k is a real number. That is, $\log_a a^k = k$.

logarithmic function with base a The function represented by $f(x) = \log_a x$.

logistic function A function used to model growth of a population.

lower limit In summation notation, the number representing the subscript of the first term of the series.

lowest terms A fraction is in lowest terms if its numerator and denominator have no factors in common.

main diagonal In an augmented matrix, the diagonal set of numbers from the upper left of the matrix to the lower right.

major axis The longer axis of an ellipse, which connects the vertices.

mass A measure of the amount of matter in an object.

matrix A rectangular array of numbers.

matrix row transformations Operations performed on rows of an augmented matrix that result in an equivalent system of linear equations.

method of substitution A symbolic method for solving a system of equations in which one equation is solved for one of the variables and then the result is substituted into the other equation.

milligram, gram, kilogram The metric units of measurement for mass.

milliliter, liter The metric units of measurement for capacity.

millimeter, centimeter, meter, kilometer The metric units of measurement for length.

minor axis The shorter axis of an ellipse.

minors The 2×2 matrices that are used to find a determinant of a 3×3 matrix.

monomial A number, a variable, or a product of numbers and variables raised to natural number powers.

multiplication property of equality If a, b, and c are real numbers with $c \neq 0$, then $a = b$ is equivalent to $ac = bc$.

multiplicative identity The number 1.

multiplicative inverse (reciprocal) The multiplicative inverse of a nonzero number a is $1/a$.

name of function In the function given by $f(x)$, we call the function f.

natural exponential function The function represented by $f(x) = e^x$, where $e \approx 2.71828$.

natural logarithm The base-e logarithm, denoted either $\log_e x$ or $\ln x$.

natural numbers The set of (counting) numbers expressed as $1, 2, 3, 4, 5, 6, \ldots$.

negative infinity Values that decrease without bound.

negative reciprocals Slopes of two lines satisfy $m_1 = -\frac{1}{m_2}$, or $m_1 \cdot m_2 = -1$. These lines are perpendicular.

negative sign The symbol $(-)$ used to form the opposite of a number.

negative slope On a graph, the slope of a line that falls from left to right.

negative square root of a Denoted $-\sqrt{a}$.

nonlinear data If data points do not lie on a (straight) line, the data are nonlinear.

nonlinear function A function that is *not* a linear function; its graph is not a line.

nonlinear system of equations Two or more equations at least one of which is nonlinear.

nonlinear system of inequalities Two or more inequalities at least one of which is nonlinear.

nonreal complex number A complex number $a + bi$ with $b \neq 0$; sometimes called an imaginary number.

nth root The number b is an nth root of a if $b^n = a$, where n is a positive integer.

nth term (general term) of a sequence See general term (nth term) of a sequence.

null set (empty set) A set that contains no elements.

numerical representation A table of values for a function.

numerical solution A solution often obtained by using a table of values.

objective function The given function to be optimized in a linear programming problem.

odd root The nth root, $\sqrt[n]{a}$, where n is odd.

one-to-one function A function f in which for any c and d in the domain of f, $c \neq d$ implies that $f(c) \neq f(d)$. That is, different inputs always result in different outputs.

opposite (additive inverse) The opposite, or additive inverse, of a number a is $-a$.

opposite of a polynomial The polynomial obtained by negating each term in a given polynomial.

optimal value In linear programming, the value that maximizes or minimizes the objective function.

ordered pair A pair of numbers written in parentheses (x, y), in which the order of the numbers is important.

ordered triple Can be expressed as (x, y, z), where x, y, and z are numbers and represent a solution to a linear system in three variables.

origin On the number line, the point associated with the real number 0; in the xy-plane, the point where the axes intersect, $(0, 0)$.

ounce, cup, pint, quart, gallon The U.S. units of measurement for capacity.

ounce, pound, ton The U.S. units of measurement for weight.

output An element of the range of a function.

parabola The $\cup$-shaped graph of a quadratic function that opens either upward or downward.

parallel lines Two or more lines in the same plane that never intersect; they have the same slope.

Pascal's triangle A triangle made up of numbers in which there are 1s along the sides and each element inside the triangle is the sum of the two numbers above it.

percent change If a quantity changes from x to y, then the percent change is $[(y - x)/x] \times 100$.

perfect cube An integer with an integer cube root.

perfect nth power The value of a if there exists an integer b such that $b^n = a$.

perfect square An integer with an integer square root.

perfect square trinomial A trinomial that can be factored as the square of a binomial, for example, $a^2 + 2ab + b^2 = (a + b)^2$.

perpendicular lines Two lines in a plane that intersect to form a right $(90°)$ angle.

piecewise-defined function A function that is defined by using different formulas on different intervals of its domain.

piecewise-defined linear function A piecewise-defined function in which every piece is a linear function.

point–slope form The line with slope m passing through the point (x_1, y_1), given by the equation $y - y_1 = m(x - x_1)$ or, equivalently, $y = m(x - x_1) + y_1$.

polynomial A monomial or the sum of two or more monomials.

polynomial functions of one variable Functions that are defined by a polynomial in one variable.

polynomials in one variable Polynomials that contain one variable.

positive slope On a graph, the slope of a line that rises from left to right.

power function A function that can be represented by $f(x) = x^p$, where p is a rational number.

prime factorization A number written as a product of prime numbers.

prime number A natural number greater than 1 that has *only* itself and 1 as natural number factors.

prime polynomial A polynomial with integer coefficients that cannot be factored by using integer coefficients.

principal nth root of a Denoted $\sqrt[n]{a}$.

principal square root The square root of a that is nonnegative, denoted $\sqrt{a}$.

probability A real number between 0 and 1, inclusive. A probability of 0 indicates that an event is impossible, whereas a probability of 1 indicates that an event is certain.

product The answer to a multiplication problem.

proportion A statement that two ratios are equal.

pure imaginary number A complex number $a + bi$ with $a = 0$ and $b \neq 0$.

Pythagorean theorem If a right triangle has legs a and b with hypotenuse c, then $a^2 + b^2 = c^2$.

quadrants The four regions determined by the xy-plane.

quadratic equation An equation that can be written in the form $ax^2 + bx + c = 0$, where a, b, and c are constants, with $a \neq 0$.

quadratic formula The solutions to the quadratic equation, $ax^2 + bx + c = 0, a \neq 0$, are given by $x = \frac{-b \pm \sqrt{b^2 - 4ac}}{2a}$.

quadratic function A function f represented by the equation $f(x) = ax^2 + bx + c$, where a, b, and c are constants, with $a \neq 0$.

quadratic inequality If the equals sign in a quadratic equation is replaced with $>$, $\geq$, $<$, or $\leq$, a quadratic inequality results.

quadratic polynomial A polynomial of degree 2 that can be written as $ax^2 + bx + c$, with $a \neq 0$.

quotient The answer to a division problem.

radical expression An expression that contains a radical sign.

radical sign The symbol $\sqrt{}$ or $\sqrt[n]{}$ for some positive integer n.

radicand The expression under the radical sign.

radius The fixed distance between the center and any point on a circle.

range The set of all y-values of the ordered pairs in a function.

rate of change Slope can be interpreted as a rate of change. It indicates how fast the graph of a line is changing.

ratio A comparison of two quantities, expressed as a quotient.

rational equation An equation that contains one or more rational expressions.

rational expression A polynomial divided by a nonzero polynomial.

rational function A function defined by $f(x) = p(x)/q(x)$, where $p(x)$ and $q(x)$ are polynomials and the domain of f includes all x-values such that $q(x) \neq 0$.

rational number Any number that can be expressed as the ratio of two integers p/q, where $q \neq 0$; a fraction.

rationalizing the denominator The process of removing radicals from a denominator so that the denominator contains only rational numbers.

real numbers All rational and irrational numbers; any number that can be represented by decimal numbers.

real part The value of a in the complex number $a + bi$.

reciprocal (multiplicative inverse) The reciprocal of a nonzero number a is $1/a$.

rectangular coordinate system (xy-plane) The xy-plane is used to plot points and graph data.

reduced row–echelon form A matrix form for representing a system of linear equations in which there are 1s on the main diagonal with 0s above and below each 1.

reflection If the point (x, y) is on the graph of a function, then $(x, -y)$ is on the graph of its reflection across the x-axis.

relation A set of ordered pairs.

rise The change in y between two points on a line, that is, $y_2 - y_1$.

root function In the power function $f(x) = x^p$, if $p = \frac{1}{n}$, where $n \geq 2$ is an integer, then f is also a root function, which is given by $f(x) = \sqrt[n]{x}$.

run The change in x between two points on a line, that is, $x_2 - x_1$.

scatterplot A graph of distinct points plotted in the xy-plane.

scientific notation A real number a written as $b \times 10^n$, where $1 \leq |b| < 10$ and n is an integer.

second, minute, hour, day, week, month, year Units of time used in both the U.S. and metric systems of measurement.

set A collection of things.

set-builder notation Notation to describe a set of numbers without having to list all of the elements. For example, $\{x \mid x > 5\}$ is read as "the set of all real numbers x such that x is greater than 5."

slope The ratio of the change in y (rise) to the change in x (run) along a line. The slope m of a line passing through the points (x_1, y_1) and (x_2, y_2) is $m = (y_2 - y_1)/(x_2 - x_1)$, where $x_1 \neq x_2$.

slope–intercept form The line with slope m and y-intercept $(0, b)$ is given by $y = mx + b$.

solution Each value of the variable that makes the equation true.

solution set The set of all solutions to an equation.

solution to a system In a system of two equations in two variables, an ordered pair, (x, y), that makes *both* equations true.

speed A rate that gives a distance traveled in an amount of time.

square matrix A matrix in which the number of rows and the number of columns are equal.

square root The number b is a square root of a number a if $b^2 = a$.

square root function The function given by $f(x) = \sqrt{x}$, where $x \geq 0$.

square root property If k is a nonnegative number, then the solutions to the equation $x^2 = k$ are given by $x = \pm\sqrt{k}$. If $k < 0$, then this equation has no real solutions.

standard equation of a circle The standard equation of a circle with center (h, k) and radius r is $(x - h)^2 + (y - k)^2 = r^2$.

standard form (of a linear equation in two variables) The form $Ax + By = C$, where A, B, and C are constants, with A and B not both 0.

standard form of a complex number $a + bi$, where a and b are real numbers.

standard form of a quadratic equation The equation given by $ax^2 + bx + c = 0$, where $a \neq 0$.

standard viewing rectangle of a graphing calculator Settings Xmin = −10, Xmax = 10, and Xscl = 1, with Ymin = −10, Ymax = 10, and Yscl = 1. This window is denoted $[-10, 10, 1]$ by $[-10, 10, 1]$

subscript The symbol x_1 has a subscript of 1 and is read "x sub one" or "x one".

subset If every element in a set B is contained in a set A, then we say that B is a subset of A, denoted $B \subseteq A$.

sum The answer to an addition problem.

sum of the first n terms of an arithmetic sequence Denoted S_n, is found by averaging the first and nth terms and then multiplying by n.

sum of the first n terms of a geometric sequence Given by $S_n = a_1(1 - r^n)/(1 - r)$, if its first term is a_1 and its common ratio is r, provided $r \neq 1$.

sum of two cubes Expression in the form $a^3 + b^3$, which can be factored as $(a + b)(a^2 - ab + b^2)$.

summation notation Notation in which the uppercase Greek letter sigma represents the sum, for example,

$$\sum_{k=1}^{n} a_k = a_1 + a_2 + a_3 + \cdots + a_n.$$

symbolic representation Representing a function with a formula; for example, $f(x) = x^2 - 2x$.

symbolic solution A solution to an equation obtained by using properties of equations; the resulting solution set is exact.

synthetic division A shortcut that can be used to divide $x - k$, where k is a number, into a polynomial.

system of linear equations in two variables A system of equations in which each equation can be written as $Ax + By = C$.

system of linear inequalities in two variables Two or more linear inequalities to be solved at the same time, the solution to which must satisfy each inequality.

table of values An organized way to display the inputs and outputs of a function; a numerical representation.

term A number, a variable, or a product of numbers and variables raised to powers.

terms of a sequence $a_1, a_2, a_3, \ldots a_n, \ldots$ where the first term is $a_1 = f(1)$, the second term is $a_2 = f(2)$, and so on.

test point When graphing the solution set to an inequality, a point chosen to determine which region of the xy-plane to include in the solution set.

test value A real number chosen to determine the solution set to an inequality.

three-part inequality A compound inequality written in the form $a < x < b$, where $\leq$ may replace $<$.

translation The shifting of a graph upward, downward, to the right, or to the left in such a way that the shape of the graph stays the same.

transverse axis In a hyperbola, the line segment that connects the vertices.

trinomial A polynomial with three terms.

unary operation An operation that requires only one number.

undefined slope The slope of a line that is vertical.

union Denoted $A \cup B$ and read "A union B," it is the set containing any element that can be found in *either A or B*.

unit fraction A fraction that is equivalent to 1.

universal set A set that contains all elements under consideration.

upper limit In summation notation, the number representing the subscript of the last term of the series.

variable A symbol, such as x, y, or z, used to represent any unknown quantity.

varies directly A quantity y varies directly with x if there is a nonzero number k such that $y = kx$.

varies inversely A quantity y varies inversely with x if there is a nonzero number k such that $y = \frac{k}{x}$.

varies jointly A quantity z varies jointly with x and y if there is a nonzero number k such that $z = kxy$.

Venn diagrams Diagrams used to depict relationships between sets.

verbal representation A description, in words, of what a function computes.

vertex The lowest point on the graph of a parabola that opens upward or the highest point on the graph of a parabola that opens downward.

vertex form of a parabola The vertex form of a parabola with vertex (h, k) is $y = a(x - h)^2 + k$, where $a \neq 0$ is a constant.

vertical asymptote A vertical asymptote typically occurs in the graph of a rational function when the denominator of a rational expression equals 0 but the numerator does not equal 0; it can be represented by a vertical line in the graph of a rational function.

vertical line test If every vertical line intersects a graph at no more than one point, then the graph represents a function.

vertices of an ellipse The endpoints of the major axis.

vertices of a hyperbola The endpoints of the transverse axis.

viewing rectangle (window) On a graphing calculator, the window that determines the x- and y-values shown in the graph.

volume A measure of the amount of a substance.

whole numbers The set of numbers 0, 1, 2, 3, 4, 5, ...

weight A measure of the force on an object due to gravity.

x-axis The horizontal axis in the xy-plane.

x-coordinate The first value in an ordered pair.

x-intercept A point where a graph intersects the x-axis.

Xmax Regarding the viewing rectangle of a graphing calculator, Xmax is the maximum x-value along the x-axis.

Xmin Regarding the viewing rectangle of a graphing calculator, Xmin is the minimum x-value along the x-axis.

Xscl Regarding the viewing rectangle of a graphing calculator, the distance between consecutive tick marks on the x-axis.

xy-plane (rectangular coordinate system) The system used to plot points and graph data.

y-axis The vertical axis in the xy-plane.

y-coordinate The second value in an ordered pair.

y-intercept A point where a graph intersects the y-axis.

Ymax Regarding the viewing rectangle of a graphing calculator, Ymax is the maximum y-value along the y-axis.

Ymin Regarding the viewing rectangle of a graphing calculator, Ymin is the minimum y-value along the y-axis.

Yscl Regarding the viewing rectangle of a graphing calculator, the distance between consecutive tick marks on the y-axis.

zero of a polynomial Any value that results in 0 when it is substituted for the variable in a polynomial; for example, the zeros of $x^2 - 4$ are 2 and -2.

zero-product property If the product of two numbers is 0, then at least one of the numbers must be 0, that is, $ab = 0$ implies $a = 0$ or $b = 0$ (or both).

zero slope The slope of a line that is horizontal.

Photo Credits

Cover, Sergojpg/Fotolia; Oleg_P/Shutterstock; Jethuynh.com/Moment Open/Getty Images *First endsheet:* **upper right,** Monkey Business Images/Shutterstock **upper left,** Wendy Rockswold **middle right,** dotshock. Shutterstock **lower left,** Wendy Rockswold **lower right,** Wendy Rockswold *Second endsheet:* **upper left,** Tommaso Lizzul/Fotolia **upper right,** IQoncept/Shutterstock **middle left,** Jules Selmes/Pearson Education, Inc. **lower right,** Wendy Rockswold **lower left,** Little_Desire/Shutterstock **iii,** Wendy Rockswold **iii,** Terry Krieger

Chapter 1   1, Prod-akszyn/Shutterstock **24,** Monkey Business Images/Shutterstock **37,** Wendy Rockswold **59,** Wendy Rockswold **69,** Johns Hopkins University Applied Physics Laboratory/Southwest Research Institute/Goddard Space Flight Center/NASA

Chapter 2   86, Wendy Rockswold **100,** Wendy Rockswold **119,** Peter Zaharov/Shutterstock **119,** Levent Konuk/Shutterstock **122,** VikOl/Shutterstock **139,** Wendy Rockswold **148,** Wendy Rockswold **153,** Wendy Rockswold

Chapter 3   169, Wendy Rockswold **171,** Lculig/123RF **225,** Wendy Rockswold **244,** Graeme Shannon/Shutterstock

Chapter 4   256, Sergey Nivens/Fotolia **262,** Little_Desire/Shutterstock **263,** Jules Selmes/Pearson Education, Inc. **269,** Nikola VolrÃ¡bovÃ¡/123RF **273,** Gary Arbach/123RF **285,** Wendy Rockswold **307,** Wendy Rockswold

Chapter 5   310, lzf/Shutterstock **320,** T-Design/Shutterstock **341,** Areipa.lt/Shutterstock **351,** Fusebulb/Shutterstock **356,** Viktar Malyshchyts/Shutterstock **357,** Wendy Rockswold

Chapter 6   372, Minicel73/Fotolia **415,** Tony Bowler/Shutterstock

Chapter 7   433, Bart Everett/Shutterstock **438,** Wendy Rockswold **439,** Wendy Rockswold **444,** Jfergusonphotos/Fotolia **447,** Wendy Rockswold **456,** Wendy Rockswold **492,** Wendy Rockswold **504,** Wendy Rockswold **505,** Wendy Rockswold **505,** Wendy Rockswold **512,** Wendy Rockswold

Chapter 8   516, Wendy Rockswold **528,** Mikhail P./Shutterstock **534,** Wendy Rockswold **541,** Dmitry Nikolaev/Shutterstock **542,** Wendy Rockswold **543,** Szaffy/E+/Getty Images **560,** Minerva Studio/Shutterstock **570,** dotshock/Shutterstock **575,** Chanpipat/Shutterstock **576,** Wendy Rockswold

Chapter 9   610, Betty Shelton/123RF **613,** Sascha Burkard/Fotolia **622,** Pictureguy66/123RF **624,** Goodluz/Shutterstock **638,** Celeborn/Shutterstock

Chapter 10   649, Wendy Rockswold **651,** Wendy Rockswold **663,** Georgejmclittle/Fotolia **665,** Martin Harvey/Digital Vision/Gettylmages **669,** Wendy Rockswold **672,** Janenet/Fotolia **682,** Baloncici/Shutterstock **690,** Wendy Rockswold **702,** Wendy Rockswold **706,** Wendy Rockswold **709,** Di Studio/Fotolia **709,** Wendy Rockswold **719,** Wendy Rockswold

Chapter 11   727, Toria/Shutterstock **733,** Tommaso Lizzul/Fotolia **740,** Levent Konuk/Shutterstock

Chapter 12   805, IQoncept/Shutterstock **820,** Jim Barber/Shutterstock **831,** Wendy Rockswold **838,** Holbox/Shutterstock686 **862,** ChrisHill/Shutterstock **865,** Froemic/Fotolia

Chapter 13   881, Cardens Design/Shutterstock **896,** MarcelClemens/Shutterstock

Chapter 14   917, Wendy Rockswold

Appendix   AP-22, Africa Studio/Fotolia

Index

Formulas from Geometry

Rectangle
$A = lw$
$P = 2l + 2w$

Triangle
$A = \frac{1}{2}bh$
$P = a + b + c$

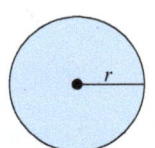

Circle
$A = \pi r^2$
$C = 2\pi r$

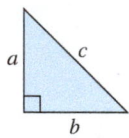

Pythagorean Theorem
$a^2 + b^2 = c^2$

Trapezoid
$A = \frac{1}{2}(a + b)h$

**Sum of the Angles
in a Triangle**
$A + B + C = 180°$

Rectangular Prism
$V = lwh$
$S = 2lw + 2lh + 2wh$

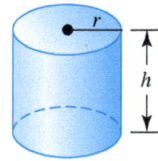

Circular Cylinder
$V = \pi r^2 h$
$S = 2\pi rh + 2\pi r^2$

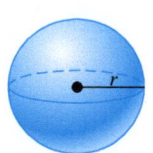

Sphere
$V = \frac{4}{3}\pi r^3$
$S = 4\pi r^2$

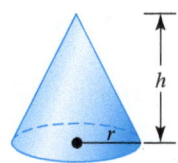

Cone
$V = \frac{1}{3}\pi r^2 h$
$S = \pi r\sqrt{r^2 + h^2} + \pi r^2$

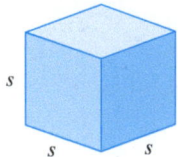

Cube
$V = s^3$
$S = 6s^2$

Square-Based Prism
$V = \frac{1}{3}s^2 h$
$S = s^2 + 2sl$

Formulas from Algebra

Basic Properties
$a + b = b + a$

$ab = ba$

$(a + b) + c = a + (b + c)$

$(ab)c = a(bc)$

$a(b + c) = ab + ac$

$\dfrac{a}{c} + \dfrac{b}{c} = \dfrac{a + b}{c}$

$\dfrac{a}{b} \times \dfrac{c}{d} = \dfrac{ac}{bd}$

$\dfrac{a}{b} \div \dfrac{c}{d} = \dfrac{ad}{bc}$

Exponents
$a^m a^n = a^{m+n}$

$(a^m)^n = a^{mn}$

$(ab)^n = a^n b^n$

$\left(\dfrac{a}{b}\right)^n = \dfrac{a^n}{b^n}$

$\dfrac{a^m}{a^n} = a^{m-n}$

$a^{-n} = \dfrac{1}{a^n}$

$\dfrac{a^{-m}}{a^{-n}} = \dfrac{a^n}{a^m}$

$\left(\dfrac{a}{b}\right)^{-n} = \left(\dfrac{b}{a}\right)^n$

Logarithms
$\log_a x = k$ means $x = a^k$

$\log_a 1 = 0$ and $\log_a a = 1$

$\log_a m + \log_a n = \log_a mn$

$\log_a m - \log_a n = \log_a \dfrac{m}{n}$

$\log_a m^r = r \log_a m$

$\log_a a^x = x$

$a^{\log_a x} = x \quad (x > 0)$

$\log_a x = \dfrac{\log_b x}{\log_b a}$ (change of base)

Radicals
$a^{1/n} = \sqrt[n]{a}$

$a^{m/n} = \sqrt[n]{a^m}$

$a^{m/n} = \left(\sqrt[n]{a}\right)^m$

$\sqrt[n]{ab} = \sqrt[n]{a} \cdot \sqrt[n]{b}$

$\sqrt[n]{\dfrac{a}{b}} = \dfrac{\sqrt[n]{a}}{\sqrt[n]{b}}$

Special Factoring
$a^2 - b^2 = (a - b)(a + b)$

$a^2 + 2ab + b^2 = (a + b)^2$

$a^2 - 2ab + b^2 = (a - b)^2$

$a^3 + b^3 = (a + b)(a^2 - ab + b^2)$

$a^3 - b^3 = (a - b)(a^2 + ab + b^2)$

Constants
$\pi \approx 3.141593$

$e \approx 2.718282$

$i = \sqrt{-1}$

$i^2 = -1$

Distance Formula
$d = \sqrt{(x_2 - x_1)^2 + (y_2 - y_1)^2}$

Midpoint Formula
$M = \left(\dfrac{x_1 + x_2}{2}, \dfrac{y_1 + y_2}{2}\right)$

Interval Notation
$x < 4$: $\qquad (-\infty, 4)$

$1 < x \le 6$: $\quad (1, 6]$

$x \ge 5$: $\qquad [5, \infty)$

Slope of a Line
$m = \dfrac{\text{rise}}{\text{run}} = \dfrac{y_2 - y_1}{x_2 - x_1} \quad (x_1 \neq x_2)$

Point–Slope Form
$y - y_1 = m(x - x_1)$, or

$y = m(x - x_1) + y_1$

Distance, rate, and time
$d = rt$

Vertical Line
$x = k$

Horizontal Line
$y = b$

Slope of Parallel Lines
$m_1 = m_2$

Slope of Perpendicular Lines
$m_1 \cdot m_2 = -1$, or

$m_1 = -\dfrac{1}{m_2}$

Quadratic Formula
$x = \dfrac{-b \pm \sqrt{b^2 - 4ac}}{2a}$

Vertex Formula
$x = -\dfrac{b}{2a}$

Vertex (Standard) Form
$y = a(x - h)^2 + k$

Slope–Intercept Form
$y = mx + b$

Standard Equation of a Circle
$(x - h)^2 + (y - k)^2 = r^2$